W9-CPV-737

Harper's
Biochemistry

Twenty-second Edition

a LANGE medical book

Harper's Biochemistry
Twenty-second Edition

Robert K. Murray, MD, PhD
Professor of Biochemistry
University of Toronto

Daryl K. Granner, MD
Professor and Chairman
Department of Molecular Physiology and Biophysics
Professor of Medicine
Vanderbilt University
Nashville, Tennessee

Peter A. Mayes, PhD, DSc
Reader in Biochemistry
Royal Veterinary College
University of London

Victor W. Rodwell, PhD
Professor of Biochemistry
Purdue University
West Lafayette, Indiana

APPLETON & LANGE
Norwalk, Connecticut/San Mateo, California

0-8385-3640-9

Notice: Our knowledge in clinical sciences is constantly changing. As new
information becomes available, changes in treatment and in the use of drugs
become necessary. The authors and the publisher of this volume have taken
care to make certain that the doses of drugs and schedules of treatment are
correct and compatible with the standards generally accepted at the time of
publication. The reader is advised to consult carefully the instruction
and information material included in the package insert of each drug or
therapeutic agent before administration. This advice is especially
important when using new or infrequently used drugs.

Copyright © 1990 by Appleton & Lange
A Publishing Division of Prentice Hall
Copyright © 1988 by Appleton & Lange
Previous editions © Lange Medical Publications

All rights reserved. This book, or any parts thereof, may not be used or
reproduced in any manner without written permission. For information,
address Appleton & Lange, 25 Van Zant Street, East Norwalk, Connecticut 06855.

90 91 92 93 94 / 10 9 8 7 6 5 4 3 2 1

Prentice Hall International (UK) Limited, *London*
Prentice Hall of Australia Pty. Limited, *Sydney*
Prentice Hall Canada, Inc., *Toronto*
Prentice Hall Hispanoamericana, S.A., *Mexico*
Prentice Hall of India Private Limited, *New Delhi*
Prentice Hall of Japan, Inc., *Tokyo*
Simon & Schuster Asia Pte. Ltd., *Singapore*
Editora Prentice Hall do Brasil Ltda., *Rio de Janeiro*
Prentice Hall, *Englewood Cliffs, New Jersey*

ISBN: 0-8385-3640-9
ISSN: 0734-9866

Production Editor: Christine Langan
Designer: Steven Byrum

PRINTED IN THE UNITED STATES OF AMERICA

Table of Contents

Section I. Structures & Functions of Proteins & Enzymes

Section II. Bioenergetics & the Metabolism of Carbohydrates & Lipids

Section III. Metabolism of Proteins & Amino Acids _____

Section VI. Special Topics _____

Preface

Harper's Biochemistry provides concise yet comprehensive coverage of the principles of biochemistry and molecular biology. It also offers numerous examples of why a knowledge of biochemistry is imperative for understanding how health is maintained, and for understanding the causes and rational treatments of many diseases, these latter being major concerns of physicians and other health care workers.

Changes in the Twenty-second Edition

The two central goals guiding the preparation of this new edition have been to reflect the latest advances in biochemistry that are important to medicine and to provide medical and other students of the health sciences with a book that is both user-friendly and interesting. The following features reflect these objectives:

- All chapters have been revised to reflect important advances in biochemical knowledge.
- Informative declarative statements have been introduced at the beginning of most paragraphs to highlight the major principles or points made in the text.
- Details concerning relatively minor pathways of metabolism have been greatly reduced.
- Many figures have been modified to improve their clarity.
- New chapters on Water and pH, Water-Soluble Vitamins, Fat-Soluble Vitamins, Oxidation of Fatty Acids and Ketogenesis, Nutrition and Metabolism of Xenobiotics have been included.
- The final chapter, also new, is entitled Biochemistry and Disease. It discusses briefly the causes and mechanisms of disease from a biochemical perspective and illustrates the application of biochemistry to medicine by the use of eight case histories, selected to cover the major categories of disease. It, along with the Appendix on Laboratory Tests, should help to bridge the gap between biochemistry classes and clinics.

Organization of This Book

This text is divided into three introductory chapters followed by six main sections.

Section I is devoted to proteins and enzymes, the work-horses of the body. Because most reactions in the human cell are catalyzed by enzymes, it is vital to understand the properties of enzymes before moving on to other topics.

Section II sets out how various cellular reactions either utilize or produce energy, and traces the pathways by which carbohydrates and lipids are synthesized and degraded. The functions of the various members of these two classes of molecules are also described.

Section III deals with the amino acids and their fates, and shows how the metabolism of amino acids also uses and yields energy.

Section IV describes the structure and functions of the nucleic acids, covering the representation DNA → RNA → proteins. This section also describes the principles of recombinant DNA technology, a topic with tremendous implications for biomedical science.

Section V discusses hormones and the key roles that they play in intercellular communication and metabolic regulation. In order to affect cells, hormones must interact with the plasma membranes of cells; thus, membrane structure and function are addressed initially in this section.

Section VI consists of seven special topics, including the new chapters on metabolism of xenobiotics and on biochemistry and disease.

The **Appendix** briefly discusses the interpretation of biochemical laboratory tests and also presents their major diagnostic applications.

Acknowledgments

The authors extend their appreciation to Mr. Alexander Kugushev; many of the changes in this edition would not have occurred without his constant interest, advice, prodding, and encouragement.

The authors wish to extend their gratitude to professional colleagues and friends throughout the world who have conveyed suggestions for improvement and corrections to us. We wish to encourage their continued interest and effort. Comments from students are especially welcome.

The authors are most gratified by the broad base of acceptance and support this book has received all over the world. Several editions of the English language version have been reprinted in Japan, Lebanon, Taiwan, the Philippines, and Korea. In addition, there are now translations in Italian, Spanish, French, Portugese, Japanese, Polish, German, Indonesian, Serbo-Croatian, and Greek.

—RKM
—DKG
—PAM
—VWR

May, 1990

Biochemistry & Medicine

1

Robert K. Murray, MD, PhD

INTRODUCTION

Biochemistry is the science concerned with the various molecules and their chemical reactions that occur in living cells and organisms. Anything more than an extremely superficial comprehension of life—in all its diverse manifestations—demands a knowledge of biochemistry. In addition, medical students who acquire a sound knowledge of biochemistry will be in a position to confront, in practice and research, the 2 central concerns of the health sciences: (1) the understanding and **maintenance of health** and (2) the understanding and effective **treatment of disease.**

Biochemistry is the Chemistry of Life

Biochemistry can be defined more formally as **the science concerned with the chemical basis of life** (Gk *bios* "life").

The cell is the structural unit of living systems. Consideration of this concept leads to a functional definition of biochemistry as **the science concerned with the chemical constituents of living cells and with the reactions and processes they undergo.** By this definition, biochemistry encompasses wide areas of **cell biology** and all of **molecular biology.**

The Aim of Biochemistry Is To Describe and Explain, in Molecular Terms, All Chemical Processes of Living Cells

The major objective of biochemistry is **the complete understanding at the molecular level of all of the chemical processes associated with living cells.**

To achieve this objective, biochemists have sought to **isolate the numerous molecules** found in cells, **determine their structures,** and **analyze how they function.** To give one example, the efforts of many biochemists to understand the molecular basis of **contractility**—a process associated primarily, but not exclusively, with muscle cells—have entailed purification of many molecules, both simple and complex, followed by detailed structure-function studies. Through these efforts, some of the features of the molecular basis of muscle contraction have been revealed.

A further objective of biochemistry is to attempt to **understand how life began.** Knowledge of this fascinating subject is still embryonic.

The scope of biochemistry is as wide as life itself. Wherever there is life, chemical processes are occurring. Biochemists study the chemical processes that occur in microorganisms, plants, insects, fish, birds, lower and higher mammals, and human beings. Students in the biomedical sciences will be particularly interested in the biochemistry of the 2 latter groups. However, an appreciation of the biochemistry of less complex forms of life is often of direct relevance to human biochemistry. For instance, contemporary theories on the **regulation of the activities of genes and of enzymes** in humans emanate from pioneering studies on bread molds and on bacteria. The field of **recombinant DNA** emerged from studies on bacteria and their viruses. Their **rapid multiplication times** and the ease of extracting their genetic material make them suitable for **genetic analyses** and **manipulations.** Knowledge gained from the study of viral genes responsible for certain types of cancer in animals (**viral oncogenes**) has provided profound insights into how human cells become cancerous.

A Knowledge of Biochemistry Is Essential to All Life Sciences, Including Medicine

The biochemistry of the nucleic acids lies at the heart of **genetics;** in turn, the use of genetic approaches has been critical for elucidating many areas of biochemistry. **Physiology,** the study of body function, overlaps with biochemistry almost completely. **Immunology** employs numerous biochemical techniques, and many immunologic approaches have found wide use by biochemists. **Pharmacology** and **pharmacy** rest on a sound knowledge of biochemistry and physiology; in particular, most drugs are metabolized by enzyme-catalyzed reactions, and the complex interactions among drugs are best understood biochemically. Poisons act on biochemical reactions or processes; this is the subject matter of **toxicology.** Biochemical approaches are being used increasingly to study basic aspects of **pathology** (the study of disease), such as inflammation, cell injury, and cancer. Many workers in **microbiology, zoology,** and **botany** employ biochemical approaches almost exclusively.

These relationships are not surprising, because life as we know it depends on biochemical reactions and processes. In fact, the old barriers among the life sciences are breaking down, and biochemistry is increasingly becoming their common language.

A Reciprocal Relationship Between Biochemistry and Medicine Has Stimulated Mutual Advances

As stated at the beginning of this chapter, the 2 major concerns for workers in the health sciences—and particularly physicians—are (1) the understanding and maintenance of **health** and (2) the understanding and effective treatment of **diseases.** Biochemistry impacts enormously on both of these fundamental concerns of medicine. In fact, the interrelationship of biochemistry and medicine is a wide, 2-way street. Biochemical studies have illuminated many aspects of health and disease, and, conversely, the study of various aspects of health and disease has opened up new areas of biochemistry.

This relationship between medicine and biochemistry has important philosophical implications for the former. As long as medical treatment is firmly grounded in a knowledge of biochemistry and other relevant basic sciences (eg, physiology, microbiology, nutrition), the practice of medicine will have a rational basis that can be adapted to accommodate new knowledge. This contrasts with unorthodox health cults, which are often founded on little more than myth and wishful thinking and lack any intellectual basis.

NORMAL BIOCHEMICAL PROCESSES ARE THE BASIS OF HEALTH

The World Health Organization (WHO) defines health as a state of "complete physical, mental and social well-being and not merely the absence of disease and infirmity." From a strictly biochemical viewpoint, **health** may be considered **that situation in which all of the many thousands of intra- and extracellular reactions that occur in the body are proceeding at rates commensurate with its maximal survival in the physiologic state.**

Biochemical Research Impacts on Nutrition and Preventive Medicine

One major prerequisite for the maintenance of health is that there be **optimal dietary intake of a number of chemicals;** the chief of these are **vitamins,** certain **amino acids,** certain **fatty acids,** various **minerals,** and **water.** Because much of the subject matter of both biochemistry and **nutrition** is concerned with the study of various aspects of these chemicals, there is a very close relationship between these 2 sciences. Moreover, as attempts are made to curb the rising costs of medical care, more emphasis is likely to be placed on systematic attempts to maintain health and forestall disease, ie, on **preventive medicine.**

Thus, nutritional approaches to, for example, the prevention of atherosclerosis and cancer are likely to receive increasing emphasis. Understanding nutrition depends to a great extent on a knowledge of biochemistry.

All Disease Has a Biochemical Basis

All diseases are manifestations of abnormalities of molecules, chemical reactions, or processes. The major factors responsible for causing diseases in animals and humans are listed in Table 1–1. All of them affect one or more critical chemical reactions or molecules in the body.

Biochemical Studies Contribute to Diagnosis, Prognosis, and Treatment

There is a wealth of documentation of the uses of biochemistry in prevention, diagnosis, and treatment of disease; many examples will be cited throughout this text. However, at this juncture only 7 brief examples are presented in order to illustrate the breadth of the subject and to stimulate the reader's interest.

(1) Humans must ingest a number of complex organic molecules called **vitamins** in order to maintain health. Vitamins are generally converted in the body to more complex molecules (coenzymes) that play key roles in many cellular reactions. If a particular vitamin is deficient in the diet, a deficiency disease may result, such as scurvy or rickets (results of lack of intake of vitamins C and D, respectively). The elucidation of the key roles played by the vitamins or their biologically active derivatives in animal and human cells has been a major concern of biochemists and nutritionists since the turn of the century. Once a disease was established as resulting from a vitamin deficiency, it became rational to treat it by administration of the appropriate vitamin.

Table 1–1. The major causes of diseases. All of the causes listed act by influencing various biochemical mechanisms in the cell or in the body.[*]

1. **Physical agents:** Mechanical trauma, extremes of temperature, sudden changes in atmospheric pressure, radiation, electric shock.
2. **Chemical agents and drugs:** Certain toxic compounds, therapeutic drugs, etc.
3. **Biologic agents:** Viruses, rickettsiae, bacteria, fungi, higher forms of parasites.
4. **Oxygen lack:** Loss of blood supply, depletion of the oxygen-carrying capacity of the blood, poisoning of the oxidative enzymes.
5. **Genetic:** Congenital, molecular.
6. **Immunologic reactions:** Anaphylaxis, autoimmune disease.
7. **Nutritional imbalances:** Nutritional deficiencies, nutritional excesses.
8. **Endocrine imbalances:** Hormonal deficiencies, hormonal excesses.

[*]Adapted, with permission, from Robbins SL, Cotram RS, Kumar V: *The Pathologic Basis of Disease,* 3rd ed. Saunders, 1984.

(2) The fact that many plants in Africa are **deficient in one or more essential amino acids** (ie, amino acids that must be supplied in the diet in order to maintain health) helps explain the debilitating malnutrition (kwashiorkor) suffered by those who depend on such plants as major dietary sources of protein. Treatment of deficiencies of essential amino acids is rational but, unfortunately, not always feasible. It consists of providing a well-balanced diet containing adequate amounts of all of the essential amino acids.

(3) Greenland Inuit consume large quantities of fish oils rich in certain **polyunsaturated fatty acids (PUFA)** and are known to have low plasma levels of cholesterol and a low incidence of **atherosclerosis.** These observations have stimulated great interest in the use of PUFA to reduce plasma levels of cholesterol.

The vitamin deficiency diseases and the essential amino acid deficiencies are examples of nutritional imbalances (see Table 1–1). Atherosclerosis may be considered as a nutritional imbalance, but other factors (eg, genetic) are also involved.

(4) The condition known as **phenylketonuria (PKU),** if untreated, may lead to severe mental retardation in infancy. The biochemical basis of PKU has been known for more than 30 years; the disorder is genetically determined (see Table 1–1) and results from low or absent activity of the enzyme that converts the amino acid phenylalanine to the amino acid tyrosine. As a consequence of the enzyme deficiency, phenylalanine and some its metabolites, such as phenylketones, accumulate in the tissues and apparently damage the developing central nervous system. When the nature of the biochemical lesion in PKU was revealed, it became rational to treat the disease by placing affected infants on a diet low in phenylalanine. Once biochemical screening tests for diagnosing PKU at birth became available, effective treatment could be started immediately.

(5) **Cystic fibrosis** is a common genetic disease (see Table 1–1) of the exocrine glands and of the eccrine sweat glands. It is characterized by abnormally viscous secretions that plug up the secretory ducts of the pancreas and the bronchioles. In addition, patients with cystic fibrosis exhibit elevated amounts of chloride in their sweat. Victims often die at an early age from lung infections. Very recently (1989), the gene implicated in the disease has been isolated and sequenced. The normal gene codes for a transmembrane protein, 1480 amino acids in length, which appears to be either a chloride channel or possibly a regulator of a chloride channel. The abnormality in approximately 70% of patients with cystic fibrosis appears to be a deletion of 3 bases in DNA, which results in the transmembrane protein lacking amino acid number 508, a phenylalanine residue. How this deletion impairs the function of the transmembrane protein and results in the excessively thick mucus has yet to be worked out. This important work should facilitate the detection of carriers of the cystic fibrosis gene and it is hoped will lead to more rational treatment of the disease than exists at present. For instance, it may be possible to design drugs that can correct the abnormality in the transmembrane protein; likewise, it may be possible to introduce the normal gene into lung cells by gene therapy.

(6) Analysis of the mechanism of action of the bacterial toxin that causes **cholera** has provided important insights into how the clinical manifestations of this disease (copious diarrhea and loss of salt and water) are brought about.

(7) The finding that the mosquitoes which transmit the parasites (plasmodia) causing **malaria** can develop biochemically based resistance to the action of insecticides has important consequences for attempts to eradicate this disease. This example and the previous one are instances of diseases caused by biologic agents (see Table 1–1).

Many Biochemical Studies Illuminate Disease Mechanisms, and Diseases Inspire Research in Specific Areas of Biochemistry

The initial observations made by the English physician Archibald Garrod on a small group of **inborn errors of metabolism** in the early 1900s stimulated the investigation of the biochemical pathways affected in these conditions. Efforts to understand the basis of the genetic disease known as familial hypercholesterolemia, which results in severe atherosclerosis at an early age, have led to dramatic progress in knowledge of cell receptors and of mechanisms of uptake of cholesterol into cells. The ongoing studies of **oncogenes** in cancer cells have directed attention to the molecular mechanisms involved in the control of normal cell growth. These and many other possible examples illustrate how observations originally made in the clinic can open up whole areas of cell function for basic biochemical research.

THIS TEXT WILL HELP RELATE BIOCHEMICAL KNOWLEDGE TO CLINICAL PROBLEMS

The final chapter pulls together many of the concepts regarding the relationships of biochemistry and disease that are disseminated throughout the text. In addition, the biochemical mechanisms operating in specific diseases brought about by each of the eight causes listed in Table 1-1 are considered as examples. The appendix discusses basic considerations used in the interpretation of the results of the biochemical laboratory tests routinely ordered in clinical practice. The overall purpose of these efforts is to assist and encourage you to translate your knowledge of biochemistry into effective clinical use.

The uses of biochemical investigations in relation to diseases may be summarized under 5 categories. Such investigations can

(1) reveal their **causes,**

(2) suggest rational and effective **treatments,**

(3) make available screening tests for **early diagnosis,**

(4) assist in **monitoring progress,**
(5) help in **assessing response to therapy.**

An appendix to this text describes the most important **routine** biochemical diagnostic tests used to investigate various diseases (ie, those used under categories 3, 4, and 5 above). It will serve as a useful reference when the biochemical diagnosis of diseases (eg, myocardial infarction, acute pancreatitis) is discussed.

Biomolecules & Biochemical Methods

2

Robert K. Murray, MD, PhD

INTRODUCTION

This chapter has 5 objectives. The first is to indicate **the composition of the body and the major classes of molecules** found in it. These molecules make up the principal subjects of this text.

The cell is the major structural and functional unit of biology. Most chemical reactions occurring in the body take place within cells. Thus, the second objective is to give a concise account of the **components of cells** and how they may be isolated; the details of how these components function form much of the fabric of this text.

The third objective concerns the fact that biochemistry is an experimental science. It is important to have an understanding and appreciation of the **experimental approach and methods** used in biochemistry, lest its study become an exercise in rote learning. Moreover, biochemistry is not an immutable corpus of knowledge, but a constantly evolving field. Advances, like those in other biomedical areas, depend on the experimental approach and technologic innovation.

The fourth objective is to summarize briefly the **principal achievements** that have been made in biochemistry. The bird's eye view of the science that will be presented will assist in imparting to the reader a sense of the overall direction of the remainder of the text.

The fifth objective is to indicate **how little is known about certain areas,** eg, about development, differentiation, brain function, cancer, and many other human diseases. Perhaps this will serve as a stimulus to some readers to contribute to research in these areas.

THE HUMAN BODY IS COMPOSED OF A FEW ELEMENTS THAT COMBINE TO FORM A GREAT VARIETY OF MOLECULES

C, H, O, and N Are the Major Elements

The elementary composition of the human body has been determined, and the major findings are listed in Table 2–1. **Carbon, oxygen, hydrogen,** and **nitrogen** are the major constituents of most biomolecules. **Phosphate** is a component of the nucleic acids and other molecules and is also widely distributed in its ionized form in the human body. **Calcium** plays a key

Table 2–1. Approximate elementary composition of the human body (dry weight basis).*

Element	Percent	Element	Percent
Carbon	50	Potassium	1
Oxygen	20	Sulfur	0.8
Hydrogen	10	Sodium	0.4
Nitrogen	8.5	Chlorine	0.4
Calcium	4	Magnesium	0.1
Phosphorus	2.5	Iron	0.01
		Manganese	0.001
		Iodine	0.00005

*Reproduced, with permission, from West ES, Todd WR: *Textbook of Biochemistry,* 3rd ed. Macmillan, 1961.

role in innumerable biologic processes and is the focus of much current research. The elements listed in column 3 of the table fulfill diverse roles. Most are encountered on an almost daily basis in medical practice in dealing with patients with electrolyte imbalances (K^+, Na^+, Cl^-, and Mg^{2+}), iron-deficiency anemia (Fe^{2+}), and thyroid diseases (I^-).

The Five Major Biopolymers Are DNA, RNA, Proteins, Polysaccharides, and Complex Lipids

As shown in Table 2–2, the major **complex biomolecules** found in the cells and tissues of higher animals (including humans) are **DNA, RNA, proteins, polysaccharides,** and **lipids.** These complex molecules are constructed from **simple biomolecules,** which are also listed. The building blocks of DNA and RNA (collectively known as the nucleic acids) are **deoxynucleotides** and **ribonucleotides,** respectively. The building blocks of proteins are **amino acids.** Polysaccharides are built up from simple carbohydrates; in the case of **glycogen** (the principal polysaccharide found in human tissues), the carbohydrate is **glucose. Fatty acids** may be considered to be the building blocks of lipids, although lipids are not polymers of fatty acids. DNA, RNA, proteins, and polysaccharides are referred to as **biopolymers** because they are composed of repeating units of their building blocks (the monomers). These complex molecules essentially make up the "stuff of life"; most of this text will be largely concerned with descriptions of various biochemical features of these biopolymers and mono-

Table 2–2. The major complex organic biomolecules of cells and tissues. The nucleic acids, proteins, and polysaccharides are biopolymers, constructed from the building blocks shown. The lipids are not generally biopolymers, and not all lipids have fatty acids as building blocks.

Biomolecule	Building Block	Major Functions
DNA	Deoxynucleotide	Genetic material
RNA	Ribonucleotide	Template for protein synthesis
Protein	Amino acids	Numerous; usually they are the molecules of the cell that carry out work (eg, enzymes, contractile elements)
Polysaccharide (glycogen)	Glucose	Short-term storage of energy as glucose
Lipids	Fatty acids	Numerous, eg, membrane components and long-term storage of energy as triacylglycerols

mers. The same complex molecules are also generally found in lower organisms, although the building blocks in certain cases may differ from those shown in Table 2–2. For instance, bacteria do not contain glycogen or triacylglycerols, but they do contain other polysaccharides and lipids.

Protein, Fat, Carbohydrate, Water, and Minerals are the Chief Components of the Human Body

The elementary composition of the human body is given above. Its chemical composition is shown in Table 2–3; **protein, fat, carbohydrate, water,** and **minerals** are the chief components. Water constitutes the major component, although its amount varies widely among different tissues. Its polar nature and ability to form hydrogen bonds render water ideally suited for its function as the solvent of the body. A detailed account of the properties of water is presented in Chapter 3.

Table 2–3. Normal chemical composition for a man weighing 65 kg.*

	Kg	Percent
Protein	11	17.0
Fat	9	13.8
Carbohydrate	1	1.5
Water†	40	61.6
Minerals	4	6.1

*Reproduced, with permission, from Davidson SD, Passmore R, Brock JF: *Human Nutrition and Dietetics,* 5th ed. Churchill Livingstone, 1973.
†The value for water can vary widely among different tissues, being as low as 22.5% for marrow-free bone. The percentage comprised by water also tends to diminish as body fat increases.

THE CELL IS THE BASIC UNIT OF BIOLOGY

The cell was established as the fundamental unit of biologic activity by Schleiden and Schwann and other pioneers, such as Virchow, in the 19th century. However, in the years immediately after World War II, 3 developments helped usher in a period of unparalleled activity in biochemistry and cell biology. These were (1) the increasing availability of the **electron microscope;** (2) the introduction of methods permitting **disruption of cells** under relatively mild conditions that preserved function; and (3) the increasing availability of the high-speed, refrigerated **ultracentrifuge,** capable of generating centrifugal forces sufficient to separate the constituents of disrupted cells from one another without overheating them. Use of the electron microscope revealed many previously unknown or poorly observable cellular components, while disruption and ultracentrifugation permitted their isolation and analysis in vitro.

A Rat Hepatocyte Illustrates Features Common to Many Eukaryotic Cells

A diagram of the structure of a liver cell (hepatocyte) of a rat is shown in Fig 2–1; this cell is probably the most studied of all cells from a biochemical viewpoint, partly because of its availability in relatively large amounts, suitability for fractionation studies, and diversity of functions. The hepatocyte contains the major **organelles** found in eukaryotic cells (Table 2–4); these include the nucleus, mitochondria, endoplasmic reticulum, free ribosomes, Golgi apparatus, lysosomes, peroxisomes, plasma membrane, and certain cytoskeletal elements.

Physical Techniques Are Used To Disrupt Cells and To Isolate Intracellular Molecules and Subcellular Organelles

In order to study the function of any organelle in depth, it is first necessary to isolate it in relatively pure form, free of significant contamination by other organelles. The usual process by which this is achieved is called **subcellular fractionation** and generally entails 3 procedures: extraction, homogenization, and centrifugation. Much of the pioneering work in this area was done using rat liver.

Extraction: As a first step toward isolating a specific organelle (or molecule), it is necessary to extract it from the cells in which it is located. **Most organelles and many biomolecules are labile and subject to loss of biologic activities: they must be extracted using mild conditions** (ie, employment of aqueous solutions and avoidance of extremes of pH and osmotic pressure and of high temperatures). In fact, most procedures for isolating organelles are performed at about 0–4 °C (eg, in a cold room or using material kept on ice). Significant losses of activity can occur at room temperature, partly owing to the action of various di-

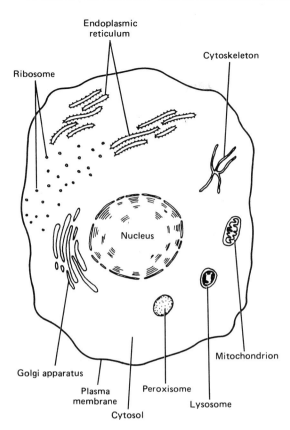

Figure 2–1. Schematic representation of a rat liver cell with its major organelles.

gestive enzymes (proteases, nucleases, etc) liberated when cells are disrupted. A common solution for extraction of organelles consists of sucrose, 0.25 mol/L (isosmotic), adjusted to pH 7.4 by TRIS (tris[hydroxymethyl]aminomethane) hydrochloric acid buffer, 0.05 mol/L, containing K^+ and Mg^{2+} ions at near physiologic concentrations; this solution is conveniently called STKM. Not all solvents used for extraction are as mild as STKM; eg, **organic solvents** are used for the extraction of lipids and of nucleic acids.

Homogenization: To extract an organelle (or biomolecule) from cells, it is first necessary to disrupt the cells under mild conditions. Organs (eg, liver, kidney, brain) and their contained cells may be conveniently disrupted by the process of homogenization, in which a manually operated or motor-driven pestle is rotated within a glass tube of suitable dimensions containing minced fragments of the organ under study and a suitable homogenizing medium, such as STKM. The controlled rotation of the pestle exerts mechanical shearing forces on cells and disrupts them, liberating their constituents into the sucrose. The resulting suspension, containing many intact organelles, is known as a homogenate.

Centrifugation: Subfractionation of the contents of a homogenate by differential centrifugation has been a technique of central importance in biochemistry. The classic method uses a series of 3 different centrifugation steps at successively greater speeds (Fig 2–2), each yielding a pellet and a supernatant. The supernatant from each step is subjected to centrifugation in the next step. This procedure provides 3 pellets, named the nuclear, mitochondrial, and microsomal fractions. None of these fractions are composed of absolutely pure organelles. However, it has been well established by the use of the electron microscope and by measurements of suitable "marker" enzymes and chemical components (eg, DNA and RNA) that the major constituents of each of these 3 fractions are nuclei, mitochondria, and microsomes, respectively. A "marker" enzyme or chemical is one that is almost exclusively confined to one particular organelle, eg, acid phosphatase to lysosomes and DNA to the nucleus (Table 2–4). The marker can thus serve to indicate the presence or absence in any particular fraction of the organelle in which it is contained. The **microsomal fraction (microsomes)** contains mostly a mixture of smooth endoplasmic reticulum, rough endoplasmic reticulum (ie, endoplasmic reticulum with attached ribosomes), and free ribosomes. The contents of the final supernatant correspond approximately to those of the **cell sap (cytosol)**. Modifications of this basic approach, using different homogenization media or different protocols or methods of centrifugation (eg, the use of gradients—either continuous or discontinuous—of sucrose), have permitted the isolation in more or less pure form of all of the organelles illustrated in Fig 2–1 and listed in Table 2–4. The scheme described above is applicable in general terms to most organs and cells; however, cell fractionations of this type must be assessed by the use of measurements of marker enzymes and chemicals and by the electron microscope until the overall procedure can be considered to be standardized.

The importance of subcellular fractionation studies in the development of biochemistry and cell biology cannot be overemphasized. It has been one of the major components of the experimental approach (see below), and—largely because of its application—the functions of the organelles indicated in Table 2–4 have been elucidated. The information on function summarized in this table represents one of the major achievements of biochemical research (see below).

The Experimental Approach Has Three Components

There are 3 major components to the experimental approach used in biochemistry: (1) **isolation of biomolecules** and organelles (see **Centrifugation,** above) found in cells; (2) **determination of the structures of biomolecules;** and (3) **analyses, using various preparations, of the function and metabolism** (ie, synthesis and degradation) **of biomolecules.**

Table 2–4. Major intracellular organelles and their functions. Only the major functions associated with each organelle are listed. In a number of instances, many other pathways, processes, or reactions may occur in the organelle.

Organelle or Fraction*	Marker	Major Functions
Nucleus	DNA	Site of chromosomes Site of DNA-directed RNA synthesis (**transcription**)
Mitochondrion	Glutamic dehydrogenase	Citric acid cycle, oxidative phosphorylation
Ribosome*	High content of RNA	Site of protein synthesis (**translation** of mRNA into protein)
Endoplasmic reticulum	Glucose-6-phosphatase	Membrane-bound ribosomes are a major site of protein synthesis Synthesis of various lipids Oxidation of many xenobiotics (cytochrome P-450)
Lysosome	Acid phosphatase	Site of many hydrolases (enzymes catalyzing degradative reactions)
Plasma membrane	Na^+/K^+-ATPase 5′-Nucleotidase	Transport of molecules in and out of cells Intercellular adhesion and communication
Golgi apparatus	Galactosyl transferase	Intracellular sorting of proteins Glycosylation reactions Sulfation reactions
Peroxisome	Catalase Uric acid oxidase	Degradation of certain fatty acids and amino acids Production and degradation of hydrogen peroxide
Cytoskeleton*	No specific enzyme markers†	Microfilaments, microtubules, intermediate filaments
Cytosol*	Lactate dehydrogenase	Enzymes of glycolysis, fatty acid synthesis

*An organelle can be defined as a subcellular entity that is membrane-limited and is isolated by centrifugation at high speeds. According to this definition, ribosomes, the cytoskeleton, and the cytosol are not organelles. However, they are considered in this table along with the organelles because they also are usually isolated by centrifugation. They can be considered subcellular entities or fractions. An organelle as isolated by one cycle of differential centrifugation is rarely pure; to obtain a pure fraction usually requires at least a number of cycles.
†The cytoskeletal fractions can be recognized by electron microscopy or by analysis by electrophoresis of characteristic proteins that they contain.

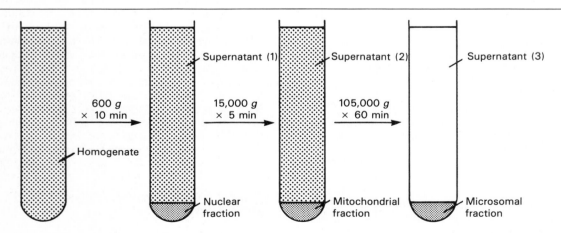

Figure 2–2. Scheme of separation of subcellular fractions by differential centrifugation. The homogenized tissue (eg, liver) is first subjected to low-speed centrifugation, which yields the nuclear fraction (containing both nuclei and unruptured cells) and supernatant (*1*). The latter is decanted and subjected to centrifugation at an intermediate speed, yielding the mitochondrial fraction (containing mitochondria, lysosomes, and peroxisomes) and supernatant (*2*). The latter is next decanted and subjected to high-speed centrifugation, yielding the microsomal fraction (containing a mixture of free ribosomes and smooth and rough endoplasmic reticulum) and the final clear solution, supernatant (*3*). The latter corresponds approximately to the cytosol or cell sap. By various modifications of this basic approach, it is generally possible to isolate each cell organelle in relatively pure form.

It Requires Isolation of Biomolecules

As in the case of organelles, elucidation of the function of any biomolecule requires that it first be isolated in pure form. Table 2–5 lists the major methods that are used to separate and purify biomolecules. No details of these methods will be given here; certain of them are described briefly at various places in this text. A combination of the successive use of several of them is almost always needed to purify a biomolecule to **homogeneity** (freedom from contamination by any other biomolecule).

It is important to appreciate that advances in biochemistry are dependent upon the development of new methods of analysis, purification, and structural determination. For instance, the field of lipid biochemistry was turned on its heels by the introduction of **gas-liquid** and **thin-layer chromatography.** The analysis of membrane and many other proteins was extremely difficult until the introduction of **sodium dodecyl sulfate-polyacrylamide gel electrophoresis** (SDS-PAGE); the introduction of the detergent SDS permitted the "solubilization" for electrophoresis of many proteins that were previously rather insoluble. The development of methods for **sequencing and cloning DNA** has had a revolutionary impact on the study of the nucleic acids and of biology in general.

It Requires Determination of the Structure of Biomolecules

Once a biomolecule has been purified, it is then necessary to determine its structure. This should allow detailed correlations to be made between structure and function. The major methods used to analyze the structures of biomolecules are listed in Table 2–6. They will be familiar to the reader who has some knowledge of organic chemistry. The known specificity of certain enzymes makes them very powerful tools in the elu-

Table 2–5. Major methods used to separate and purify biomolecules. Most of these methods are suitable for analyzing the components present in cell extracts and other biochemical materials. The sequential use of a number of these techniques will generally permit purification of most biomolecules. The reader is referred to texts on methods of biochemical research for details regarding each.

Salt fractionation (eg, precipitation with ammonium sulfate)
Chromatography
  Paper
  Ion exchange (anion and cation exchange)
  Affinity
  Thin-layer
  Gas-liquid
  High-pressure liquid
Gel filtration
Electrophoresis
  Paper
  High-voltage
  Agarose
  Cellulose acetate
  Starch gel
  Polyacrylamide
  SDS-polyacrylamide
Ultracentrifugation

Table 2–6. Principal methods used for determining the structures of biomolecules.

Elemental analysis
Ultraviolet, visible, infrared, and NMR spectroscopy
Use of acid or alkaline hydrolysis to degrade the biomolecule under study into its basic constituents
Use of a battery of enzymes of known specificity to degrade the biomolecule under study (eg, proteases, nucleases, glycosidases)
Mass spectrometry
Specific sequencing methods (eg, for proteins and nucleic acids)
x-Ray crystallography

cidation of the structural features of certain biomolecules. Improvements in resolution afforded by theoretic and technologic advances are increasingly making **mass spectrometry** and **nuclear magnetic resonance (NMR) spectroscopy** the routine methods of choice for structural determination. For instance, the structures of the extremely complex carbohydrate chains found in certain biomolecules, such as glycoproteins, can now often be elucidated by high-resolution NMR spectrometry. The most detailed information about the structure of biomolecules is provided by **x-ray diffraction and crystallography.** Their use was critical in revealing the detailed structures of various proteins and enzymes and the double helical nature of DNA.

It Requires Analysis, Using Various Preparations, of the Function and Metabolism of Biomolecules

Initial biochemical research on humans and animals was performed at the level of the whole animal. Examples were studies of respiration and of the fate of ingested compounds. It soon became apparent that the whole animal was too complex to permit definitive answers to be given to many questions. Accordingly, simpler in vitro preparations were developed that removed many of the complications experienced at the level of the whole animal. Table 2–7 summarizes the various types of preparations that are now available to study biochemical processes; most of the facts presented in this text have been obtained by their use. The items in the list are arranged in decreasing order of complexity. Just as the use of whole animals has its drawbacks, the other preparations also have limitations. Erroneous results (artifacts) can be derived from the use of in vitro approaches; eg, homogenization of cells can liberate enzymes that may partially digest cellular molecules.

STRATEGIES FOR STUDYING BIOCHEMICAL REACTIONS ARE COMPLEX & MULTILEVEL

Much of this text will be concerned with complex biochemical processes (eg, protein synthesis and muscle contraction), including metabolic pathways. A

Table 2–7. Hierarchy of preparations used to study biochemical processes.

Method	Comments
Studies at the whole animal level	These can include (1) removal of an organ (eg, hepatectomy). (2) alterations of diet (eg, fasting-feeding). (3) administration of a drug (eg, phenobarbital). (4) administration of a toxin (eg, carbon tetrachloride). (5) use of an animal with a specific disease (eg, diabetes mellitus). (6) use of sophisticated techniques such as NMR spectroscopy and positron emission tomography. Studies at this level are often physiologic but can be difficult to interpret because of the interplay among organs mediated by the circulation and the nervous system.
Isolated perfused organ	Liver, heart, and kidney are particularly suitable. This method allows study of an organ removed from the influence of other organs or the nervous system. The use of perfusion generally ensures that organ function is maintained for at least several hours.
Tissue slice	Liver slices have been especially used. Removes the organ slice from other influences, but the preparations tend to deteriorate within a few hours, partly because of inadequate supply of nutrients.
Use of whole cells	(1) Particularly applicable to blood cells, which can be purified relatively easily. (2) Use of cells in tissue culture is indispensable in many areas of biology.
Homogenate	(1) Ensures a cell-free preparation. (2) Specific compounds can be added or removed (eg, by dialysis) and their effects studied. (3) Can be subfractionated by centrifugation to yield individual cell organelles.
Isolated cell organelles	Extensively used to study the function of mitochondria, the endoplasmic reticulum, ribosomes, etc.
Subfractionation of organelles	Extensively used, eg, in studies of mitochondrial function.
Isolation and characterization of metabolites and enzymes	A vital part of the analysis of any chemical reaction or pathway.
Cloning of genes for enzymes and proteins	Isolation of the cloned gene is vital for studying the details of its structure and regulation; it can also reveal the amino acid sequence of the enzyme or protein for which it codes.

metabolic pathway is a series of reactions responsible for the **synthesis** of a more complex compound from one or more simple compounds, or for the **degradation** of a compound to its end product. The existence of a complex biochemical process or of certain metabolic pathways can be inferred from observations made at the level of the whole animal. For instance, direct observations on our fellow humans indicate that muscles contract. We know that glucose serves as a source of energy for humans and other animals and can thus infer that it must be degraded (metabolized) in the body to yield energy. However, to understand fully how glucose is metabolized by human cells—and knowledge of this is still far from complete—requires analyses at a variety of levels. Fig 2–3 shows the various types of observations and analyses that are required in order to comprehend biochemical processes, such as the initial breakdown of glucose to yield energy (a process known as **glycolysis**). The scheme shown applies at a general level to all of the major biochemical processes discussed in this text and thus represents an overall strategy for elucidating biochemical processes; it should be borne in mind when each of the major biochemical processes described in this text (eg, glycolysis, fatty acid oxidation) is under consideration, although not every point listed will always be relevant.

A number of important points concerning Table 2–7 and Fig 2–3 merit discussion. (1) Despite the possibility of artifacts, it is absolutely necessary to **isolate** and **identify** each of the components of a biochemical process in pure form in order to understand the process at a molecular level. Numerous examples of this will be encountered subsequently. (2) It is also important to be able to **reconstitute** the process under study in vitro by the systematic reassembly of its individual components. If the process does not function when its parts are reassembled, one explanation is that some critical component has escaped identification and has not been added back. (3) Recent technologic advances (eg, in NMR spectroscopy and positron emission tomography ([PET] scanning) have permitted detection of certain biomolecules at the whole-organ level and monitoring of changes of their amounts with time. Such developments indicate that it is becoming possible to make sophisticated analyses of many biochemical processes at the in vivo level. (4) When the results obtained using many different levels of approach are consistent, then one is justified in concluding that real progress has been made in understanding the biochemical process under study. If major inconsistencies are obtained using the various approaches, then their causes must be investigated until rational explanations are ob-

Inference of existence of biochemical process or metabolic pathway from observations made at the whole-animal level
↓
Analyses of its control mechanisms in vivo
↓
Analyses of effects on it of specific diseases (eg, inborn errors of metabolism, cancer, etc)
↓
Its localization to one or more organs
↓
Its localization to one or more cellular organelles or subcellular fractions
↓
Delineation of the number of reactions involved in it
↓
Purification of its individual substrates, products, enzymes, and cofactors or other components
↓
Analyses of its control mechanisms in vitro
↓
Establishment of reaction mechanisms involved in it
↓
Its reconstitution
↓
Studies of it at the gene level by the methods of recombinant DNA

Figure 2–3. Scheme of the general strategy used to analyze a biochemical process or metabolic pathway. The approaches used need not necessarily be performed in the precise sequence listed here. However, it is by their employment that the details of a biochemical process or pathway are usually elucidated. The scheme thus applies at a general level to all of the major metabolic pathways discussed in subsequent chapters of this text.

tained. (5) The preparations and levels of analysis outlined can be used to study biochemical alterations in animals with **altered metabolic states** (eg, fasting, feeding) or **specific diseases** (eg, diabetes mellitus, cancer). (6) Most of the methods and approaches indicated can be applied to studies of normal or diseased human cells or tissues. However, care must be taken to obtain such material in a fresh condition, and particular attention must be paid to those ethical considerations that apply to human experimentation.

Radioactive and Heavy Isotopes Have Made Major Contributions to Elucidation of Biochemical Processes

The introduction of the use of isotopes into biochemistry in the 1930s had a dramatic impact; consequently, their use deserves special mention. Prior to their employment, it was very difficult to "tag" biomolecules so that their metabolic fates could be monitored conveniently. Pioneering studies, particularly by Schoenheimer and colleagues, applied the use of certain **stable isotopes** (eg, ^{2}D, ^{15}N) combined with their detection by mass spectrometry to many biochemical problems. For instance, certain amino acids, sugars, and fatty acids could be synthesized containing a suitable stable isotope and then administered to an animal or added to an in vitro preparation to follow their metabolic fates (eg, half-lives, conversion to other biomolecules). Compounds labeled with suitable stable isotopes were used to investigate many aspects of the metabolism of proteins, carbohydrates, and lipids. From such studies, it has become apparent that metabolism is a very active process, with most of the compounds in the cell being continuously synthesized and degraded, although at widely differing rates. These findings were epitomized by Schoenheimer as "the dynamic nature of metabolism."

The subsequent introduction of **radioactive isotopes** and of instruments enabling their measurement was also extremely important. The principal stable and radioactive isotopes used in biologic systems are listed in Table 2–8. The use of isotopes, both stable and radioactive, is critical to the development of every area of biochemistry. Investigations of complex and simple biomolecules rely heavily on their use, either in vivo or in vitro. The tremendous progress made recently in sequencing nucleic acids and in measuring extremely small amounts of compounds found in biologic systems by radioimmunoassay has also depended on their employment.

NUMEROUS MAJOR ACHIEVEMENTS CHARACTERIZE THE CONTRIBUTIONS OF BIOCHEMISTRY TO CELL SCIENCE AND MEDICINE

Following is a summary of the principal achievements made in the field of biochemistry, particularly in relation to human biochemistry.
- The overall **chemical composition** of cells, tissues, and the body has been determined, and the

Table 2–8. Principal isotopes used in biochemical research.

Stable Isotopes	Radioactive Isotopes
^{2}D	^{3}H
^{15}N	^{14}C
^{18}O	^{32}P
	^{35}S
	^{35}Ca
	^{125}I
	^{131}I

principal compounds present have been isolated and their structures established.

- The **functions** of many of the simple biomolecules are understood, at least at a general level, and are described later in this text. The functions of the major complex biomolecules have also been established. Of central interest is that DNA is known to be the genetic material and to transmit its information to one type of RNA (messenger RNA, or mRNA), which in turn dictates the linear sequence of amino acids in proteins. The flow of information from DNA may be conveniently represented as DNA → RNA → protein.
- The **principal organelles** of animal cells have been isolated and their major functions established.
- Almost all of the reactions occurring in cells have been found to be catalyzed by **enzymes;** many enzymes have been purified and studied, and the broad features of their mechanisms of action have been revealed.
- The **metabolic pathways** involved in the synthesis and degradation of the major simple and complex biomolecules have been delineated. In general, the pathway of synthesis of a compound has been found to be distinct from its pathway of degradation.
- A number of aspects of the **regulation of metabolism** have been clarified.
- The broad features of how cells **conserve and utilize energy** have been recognized.
- Many aspects of the structure and function of the various **membranes** found in cells are understood; proteins and lipids are their major components.
- Considerable information is available at a general level on how the major **hormones** act.
- Biochemical bases for a considerable number of **diseases** have been discovered.

MUCH REMAINS TO BE LEARNED

While it is important to know that much biochemical information has been accumulated, it is just as important to appreciate how little is known in many areas. Probably the 2 major problems to be solved concern establishing the biochemical bases of **development and differentiation** and of **brain function.** Although the chemical nature of the genetic material is now well established, almost nothing is known about the mechanisms that turn genes on and off during development. Understanding **gene regulation** is also a key area in learning how cells differentiate and how they become cancerous. Knowledge of **cell division and growth—** both normal and malignant—and of their regulation is very primitive. Virtually nothing is known concerning the biochemical bases of **complex neural phenomena** such as consciousness and memory. Only very limited information is available concerning mechanisms of **cell secretion.** Despite some progress, the molecular bases of **most major genetic diseases** are unknown, but approaches provided by recombinant DNA technology suggest that remarkable progress will be made in this area during the next few years. The human genome may be sequenced by the turn of the next century or earlier; the information made available by this massive endeavor will have a tremendous impact on human biology and medicine.

REFERENCES

Freifelder D: *Physical Biochemistry: Applications to Biochemistry and Molecular Biology.* Freeman, 1982.

Fruton JS: *Molecules and Life; Historical Essays on the Interplay of Chemistry and Biology.* Wiley-Interscience, 1972.

Weinberg RA: The molecules of life. *Sci Am* (Oct) 1985;**253:** 48. [Entire issue is of interest.]

Water & pH

Victor W. Rodwell, PhD

INTRODUCTION

Biochemistry is concerned, for the most part, with the properties and reactions of organic compounds. However, in living cells most biochemicals exist and most reactions occur in an aqueous environment. Water is an active participant in many biochemical reactions and is an important determinant of the properties of macromolecules such as proteins. Water dissociates to form OH^- and H^+. The term **pH** is used to denote the concentration of hydrogen ions in cells and body fluids and is a key concept in biochemistry and medicine. The functional groups (amino and carboxyl groups, etc) of biomolecules dissociate at specific pH values, and many of the biologic and physical properties of these molecules depend upon this dissociation.

BIOMEDICAL IMPORTANCE

A fundamental biologic principle is that the **constancy of the internal environment** of the body must be kept within relatively narrow limits if health is to be maintained. This applies to the overall distribution of **water** in the body and also to maintenance of **pH** and of the concentrations of **various electrolytes** (eg, Na^+, K^+, Ca^{2+}, Mg^{2+}, and phosphate) in the body. **Total body water** in men varies from 55 to 65% of body weight, the lower figure applying to obese individuals; values for women average about 10% less. Two-thirds of total body water is **intracellular fluid** (ICF); the remainder is **extracellular fluid** (ECF), with approximately 25% of the ECF being in the plasma.

Regulation of water balance is complex but depends principally upon hypothalamic mechanisms for controlling thirst, antidiuretic hormone (ADH), and the activity of the kidneys. States of **depletion of water** and of **excess of body water** are quite common. In many cases, these are accompanied by depletion or excess of **sodium.** Causes of depletion of water are (1) decreased intake (eg, in coma) and (2) increased loss (eg, renal loss in diabetes mellitus, skin loss in severe sweating, and gastrointestinal loss in severe diarrhea in infants and in cholera). Causes of excess of body water are (1) increased intake (eg, excessive administration of intravenous fluids) and (2) decreased excretion (eg, in severe renal failure).

The **pH** of the extracellular fluid is maintained between 7.35 and 7.45 in health. The **bicarbonate** (HCO_3^-/H_2CO_3) buffer system is of particular importance in this regard. When the pH is less than 7.35, a state of **acidosis** exists. Conversely, when the pH is above 7.45, a state of **alkalosis** exists. The existence of a **disturbance of acid-base balance** can usually be recognized if measurements of the pH of arterial blood, of the PCO_2, and of the CO_2 content of venous blood (or calculation of HCO_3^-) are performed. There are numerous medical causes of acidosis (diabetic ketoacidosis, lactic acidosis, etc) and of alkalosis (vomiting of acidic gastric contents, treatment with certain diuretics, etc). Accurate diagnosis and appropriate treatment of both conditions of water imbalance and of acid-base disturbances rely in part upon an understanding of the basic concepts that are dealt with in this chapter. It is therefore appropriate to consider those properties of water that enable it to play such a key role in biochemistry.

Water Is a Slightly Skewed Tetrahedral Molecule

The water molecule is an irregular tetrahedron with oxygen at its center (Fig 3–1). The 2 bonds with hydrogen are directed toward 2 corners of the tetrahedron, while the unshared electrons on the 2 sp^3-hybridized orbitals occupy the 2 remaining corners. The angle between the 2 hydrogen atoms (105 degrees) is slightly less than the tetrahedral angle (109.5 degrees), forming a slightly skewed tetrahedron.

An Unequal Distribution of Charge on the Water Molecule Forms a Dipole

Because of its skewed tetrahedral structure electrical charge is not uniformly distributed about the water molecule. The side of the oxygen opposite to the 2 hydrogens is relatively rich in electrons, while on the other side the relatively unshielded hydrogen nuclei form a region of local positive charge. The term **dipole** denotes molecules such as water that have electrical charge (electrons) unequally distributed about their structure.

Like Water, Ammonia Is a Tetrahedral Dipolar Molecule

In ammonia, the bond angles between the hydrogens (107 degrees) approach the tetrahedral angle even more closely than in water (Fig 3–2). Many bio-

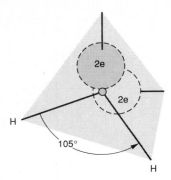

Figure 3–1. Tetrahedral structure of water.

chemicals are dipoles. Examples include alcohols, phospholipids, amino acids, and nucleic acids.

SKEWED STRUCTURE AND DIPOLARITY ALLOW H_2O A FLEXIBLE ROLE IN REACTIONS

Liquid Water Is Stabilized by Intermolecular Hydrogen Bonds That Confer a Loose Macromolecular Structure Suggestive of the More Ordered Structure of Ice

Again because of its dipolar character, water molecules can assume ordered arrangements (recall a snowflake). Molecular ordering of water molecules is not restricted to ice, however. Liquid water exhibits macromolecular structure that parallels the geometric disposition of water molecules in ice. The ability of water molecules to associate with one another in both solid and liquid states arises from the dipolar character of water. It remains a liquid rather than a solid because of the transient nature of these macromolecular complexes (the half-life for association-dissociation of water molecules is about 1 microsecond). In the solid state, each water molecule is associated with 4 other water molecules. In the liquid state, the number is somewhat

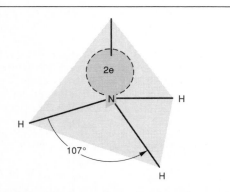

Figure 3–2. Tetrahedral structure of ammonia.

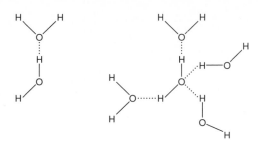

Figure 3–3. *Left:* Association of 2 dipolar water molecules. The dotted line represents a hydrogen bond. *Right:* Association of a central water molecule with 4 other water molecules by hydrogen bonding. This structure is typical of ice and, to a lesser extent, of liquid water.

less (about 3.5). With the exception of the transient nature of intermolecular interactions in liquid water, it thus resembles ice in its macromolecular structure more closely than may at first be imagined.

The dipolar character of water molecules favors their mutual association in ordered arrays with a precise geometry dictated by the internal geometry of the water molecule (Fig 3–3).

The electrostatic interaction between the hydrogen nucleus of one water molecule and the unshared electron pair of another is termed a hydrogen bond. Compared to covalent bonds, hydrogen bonds are quite weak. To break a hydrogen bond in liquid water requires about 4.5 kcal of energy per mole—about 4% of the energy required to rupture the O–H bond in water (110 kcal/mol).

Although Individually Weak, Hydrogen Bonds Play Major Roles in Biochemistry, Including Stabilizing Proteins and Nucleic Acids

Hydrogen bonds play significant roles in biochemistry because they can be formed in large numbers. Multiple hydrogen bonds confer significant structure not only upon water but also upon other dipolar molecules as diverse as alcohols, DNA, and proteins. Fig 3–4 illustrates hydrogen bonds formed between representative biochemicals.

Hydrogen bonds are not restricted to water molecules. The hydrogens of nitrogen atoms can also participate in hydrogen bonding. This topic will be considered again in connection with the 3-dimensional structure of proteins and with base pairing in DNA.

Water Molecules Exhibit a Slight, but Physiologically Important Tendency to Dissociate, Forming Small Quantities of OH^- and H^+ Ions

Water molecules have a limited tendency to dissociate (ionize) into H^+ and OH^- ions:

$$H_2O \rightleftarrows H^+ + OH^-$$

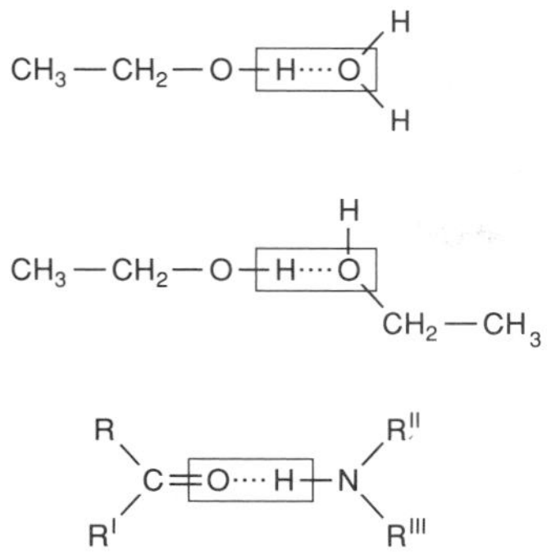

Figure 3–4. Formation of hydrogen bonds between an alcohol and water, between 2 molecules of ethanol, and between the peptide carbonyl oxygen and the hydrogen on the peptide nitrogen of an adjacent peptide.

Since ions continuously recombine to form water molecules, and vice versa, it cannot be stated whether an individual hydrogen or oxygen is present as an ion or as part of a water molecule. At one instant it is an ion; an instant later, part of a molecule. Fortunately, individual ions or molecules need not be considered. Since 1 g of water contains 3.46×10^{22} molecules, the ionization of water can be described statistically. It is sufficient to know the **probability** that a hydrogen will be present as an ion or as part of a water molecule.

To state that the probability that a hydrogen exists as an ion is 0.01 means that a hydrogen atom has one chance in 100 of being an ion and 99 chances out of 100 of being in a water molecule. The actual probability of a hydrogen atom in pure water existing as a hydrogen ion is approximately 0.0000000018, or 1.8×10^{-9}. Consequently, the probability of its being part of a molecule is almost unity. Stated another way, for every hydrogen ion and hydroxyl ion in pure water, there are 1.8 billion or 1.8×10^{9} water molecules. Hydrogen and hydroxyl ions nevertheless contribute significantly to the properties of water.

The tendency of water to dissociate is expressed as follows:

$$K = \frac{[H^+]\,[OH^-]}{[H_2O]}$$

where the bracketed terms represent the molar concentrations of hydrogen ions, hydroxyl ions, and undissociated water molecules,* and K is termed the **dissociation constant.** To calculate the dissociation constant for water, recall that 1 mol of water weighs 18 g. One liter (L) (1000 g) of water therefore contains $1000 \div 18 = 55.56$ mol. Pure water is thus 55.56 molar. Since the probability that a hydrogen in pure water will exist as an H^+ ion is 1.8×10^{-9}, the molar concentration of H^+ ions (or of OH^- ions) in pure

water is calculated by multiplying the probability, 1.8×10^{-9}, by the molar concentration of water, 55.56 mol/L. This result is 1.0×10^{-7} mol/L.

We can now calculate K for water:

$$K = \frac{[H^+]\,[OH^-]}{[H_2O]} = \frac{[10^{-7}]\,[10^{-7}]}{[55.56]}$$
$$= 0.018 \times 10^{-14} = 1.8 \times 10^{-16}\ mol/L$$

The high concentration of molecular water (55.56 mol/L) is not significantly affected by dissociation. It is therefore convenient to consider it as essentially constant. This constant may then be incorporated into the dissociation constant, K, to provide a new constant, K_w, termed the **ion product** for water. The relationship between K_w and K is shown below:

$$K = \frac{[H^+]\,[OH^-]}{[H_2O]} = 1.8 \times 10^{-16}\ mol/L$$
$$K_w = (K)\,[H_2O] = [H^+]\,[OH^-]$$
$$= (1.8 \times 10^{-16}\ mol/L)\,(55.56\ mol/L)$$
$$= 1.00 \times 10^{-14}\ (mol/L)^2$$

Note that the dimensions of K are moles per liter and of K_w moles^2 per liter^2. As its name suggests, the ion product, K_w, is numerically equal to the product of the molar concentrations of H^+ and OH^-:

$$K_w = [H^+]\,[OH^-]$$

At 25 °C, $K_w = (10^{-7})^2 = 10^{-14}$ (mol/L)^2. At temperatures below 25 °C, K_w is less than 10^{-14}, and at temperatures above 25 °C, greater than 10^{-14}. For example, at the temperature of the human body (37 °C), the concentration of H^+ in pure water is slightly more than 10^{-7} mol/L. Within the stated limitations of the effect of temperature, **$K_w = 10^{-14}$ (mol/L)^2 for all aqueous solutions**—even those that contain acids or bases. We shall use this constant in the calculation of pH values for acidic and basic solutions.

THE pH IS THE NEGATIVE LOG OF THE HYDROGEN ION CONCENTRATION, OR MORE ACCURATELY, OF THE HYDROGEN ION ACTIVITY

The term **pH** was introduced in 1909 by Sorensen, who defined pH as **the negative log of the hydrogen ion concentration:**

$$pH = -\log\,[H^+]$$

This definition—while not rigorous*—is adequate for most biochemical purposes. To calculate the pH of a solution:

(1) Calculate hydrogen ion concentration, $[H^+]$.

---

*Strictly speaking, the bracketed terms represent molar activity rather than molar concentration.

*pH $= -\log$ (H^+ activity).

(2) Calculate the base 10 logarithm of $[H^+]$.

(3) pH is the negative of the value found in step 2. For example, for pure water at 25 °C:

$$\textbf{pH} = -\textbf{log } [\textbf{H}^+] = -\textbf{log } 10^{-7} = -(-7) = 7.0$$

Low pH values correspond to high concentrations of H^+, and high pH values to low concentrations of H^+.

Acids are **proton donors,** and bases are **proton acceptors.** A distinction is made, however, between strong acids (eg, HCl, H_2SO_4), which completely dissociate even in strongly acidic solutions (low pH), and **weak acids,** which dissociate only partially in acidic solutions. A similar distinction is made between **strong bases** (eg, KOH, NaOH) and **weak bases** (eg, $Ca[OH]_2$). Only strong bases are dissociated at high pH. Many biochemicals are **weak acids.** Exceptions include phosphorylated intermediates, which possess the strongly acidic primary phosphoric acid group.

The following examples illustrate how to calculate the pH of acidic and basic solutions.

Example: What is the pH of a solution whose hydrogen ion concentration is 3.2×10^{-4} mol/L?

$$\begin{aligned}
\textbf{pH} &= -\textbf{log } [\textbf{H}^+] \\
&= -\textbf{log } (3.2 \times 10^{-4}) \\
&= -\textbf{log } (3.2) - \textbf{log } (10^{-4}) \\
&= -0.5 + 4.0 \\
&= 3.5
\end{aligned}$$

Example: What is the pH of a solution whose hydroxide ion concentration is 4.0×10^{-4} mol/L?

To approach this problem, we define a quantity **pOH** that is equal to $-\log [OH^-]$ and that may be derived from the definition of K_w:

$$\textbf{K}_\textbf{w} = [\textbf{H}^+] [\textbf{OH}^-] = 10^{-14}$$

therefore: $\textbf{log } [\textbf{H}^+] + \textbf{log } [\textbf{OH}^-] = \textbf{log } 10^{-14}$

or: $\quad\quad \textbf{pH} + \textbf{pOH} = 14$

To solve the problem by this approach:

$$\begin{aligned}
[\textbf{OH}^-] &= 4.0 \times 10^{-4} \\
\textbf{pOH} &= -\textbf{log } [\textbf{OH}^-] \\
&= -\textbf{log } (4.0 \times 10^{-4}) \\
&= -\textbf{log } (4.0) - \textbf{log } (10^{-4}) \\
&= -0.60 + 4.0 \\
&= 3.4
\end{aligned}$$

Now: $\quad \textbf{pH} = 14 - \textbf{pOH} = 14 - 3.40$
$$= 10.6$$

Example: What is the pH of (a) 2.0×10^{-2} mol/L KOH, (b) 2.0×10^{-6} mol/L KOH? The OH^- arises from 2 sources: KOH and water. Since pH is determined by the **total** $[H^+]$ (and pOH by the **total** $[OH^-]$), both sources must be considered. In the first case, the contribution of water to the total $[OH^-]$ is negligible. The same cannot be said for the second case:

	Concentration (mol/L)	
	(a)	**(b)**
Molarity of KOH	2.0×10^{-2}	2.0×10^{-6}
$[OH^-]$ from KOH	2.0×10^{-2}	2.0×10^{-6}
$[OH^-]$ from water	1.0×10^{-7}	1.0×10^{-7}
Total $[OH^-]$	2.00001×10^{-2}	2.1×10^{-6}

Once a decision has been reached about the significance of the contribution by water, pH may be calculated as above.

In the above examples, it was assumed that the strong base KOH was completely dissociated in solution and that the molar concentration of OH^- ions was thus equal to the molar concentration of KOH. This assumption is valid for relatively dilute solutions of **strong** bases or acids but **not for solutions of weak bases or acids.** Since these weak electrolytes dissociate only slightly in solution, we must calculate the concentration of H^+ (or $[OH^-]$) produced by a given molarity of the acid (or base) using the **dissociation constant** before calculating total $[H^+]$ (or total $[OH^-]$), and subsequently calculating the pH.

The Presence on Biomolecules of Functional Groups that Behave as Weak Acids Has Profound Physiologic Significance

Many biochemicals possess functional groups that are weak acids or bases. One or more of these functional groups—carboxyl groups, amino groups, or the secondary phosphate dissociation of phosphate esters—are present in all proteins and nucleic acids, most coenzymes, and most intermediary metabolites. The dissociation behavior (protonic equilibria) of weakly acidic and weakly basic functional groups is therefore fundamental to understanding the influence of intracellular pH on the structure and biochemical activity of these compounds. Their separation and identification in research and clinical laboratories is also facilitated by knowledge of the dissociation behavior of their functional groups.

We term the protonated form of an acid (eg, HA or RNH_3^+) the **acid** and the unprotonated form (eg, A^- or RNH_2) its **conjugate base** (Table 3–1). Similarly,

Table 3–1. Examples of weak acids and their conjugate bases.

Acid	Conjugate Base
CH_3COOH $CH_3NH_3^+$	CH_3COO^- CH_3NH_2
OH	O^-
H$^+$N NH	N NH

we may refer to a **base** (eg, A^- or RNH_2) and its **conjugate acid** (eg, HA or $RNH_3{}^+$) (Latin *coniungere* "to join together").

The relative strengths of weak acids and of weak bases are expressed quantitatively as their **dissociation constants,** which express their tendency to ionize. Shown below are the expressions for the dissociation constant (K) for 2 representative weak acids, $R-COOH$ and $R-NH_3{}^+$.

$$R\text{-}COOH \rightleftharpoons R\text{-}COO^- + H^+$$

$$K = \frac{[R\text{-}COO^-]\,[H^+]}{[R\text{-}COOH]}$$

$$R\text{-}NH_3{}^+ \rightleftharpoons R\text{-}NH_2 + H^+$$

$$K = \frac{[R\text{-}NH_2]\,[H^+]}{[R\text{-}NH_3{}^+]}$$

Since the numerical values of K for weak acids are negative exponential numbers, it is convenient to express K as pK, where

$$pK = -\log K$$

Note that pK is related to K as pH is to H^+ concentration. Table 3–2 lists illustrative K and pK values for a monocarboxylic, a dicarboxylic, and a tricarboxylic acid. Observe that the **stronger acid groups have lower pK values.**

From the above equations that relate K to $[H^+]$ and to the concentrations of undissociated acid and its conjugate base, note that when

$$[R\text{-}COO^-] = [R\text{-}COOH]$$

or when

$$[R\text{-}NH_2] = [R\text{-}NH_3{}^+]$$

then

$$K = [H^+]$$

In words, **when the associated (protonated) and dissociated (conjugate base) species are present in equal concentrations, the prevailing hydrogen ion concentration $[H^+]$ is numerically equal to the dissociation constant, K.** If the logarithms of both sides of the above equation are taken and both sides are

multiplied by -1, the expressions would be as follows:

$$K = [H^+]$$

$$-\log K = -\log [H^+]$$

Now, $-\log K$ is defined as pK, and $-\log [H^+]$ is the definition of pH. Consequently, the equation may be rewritten as

$$pK = pH$$

ie, **the pK of an acid group is that pH at which the protonated and unprotonated species are present at equal concentrations.** The pK for an acid may be determined experimentally by adding 0.5 equivalent of alkali per equivalent of acid. The resulting pH will be equal to the pK of the acid.

The Behavior of Weak Acids and of Buffers, Which Are Solutions of Weak Acids and Their Salts, Is Expressed by the Henderson-Hasselbalch Equation

The pH of a solution containing a weak acid is related to its acid dissociation constant, as shown above for the weak acid water. The relationship can be stated in the convenient form of the **Henderson-Hasselbalch** equation, derived below.

A weak acid, HA, ionizes as follows:

$$HA \rightleftharpoons H^+ + A^-$$

The equilibrium constant for this dissociation is written:

$$K = \frac{[H^+]\,[A^-]}{[HA]}$$

Cross multiply

$$[H^+]\,[A^-] = K\,[HA]$$

Divide both sides by $[A^-]$

$$[H^+] = K\,\frac{[HA]}{[A^-]}$$

Take the log of both sides

$$\log [H^+] = \log \left(K\,\frac{[HA]}{[A^-]} \right)$$

$$= \log K + \log \frac{[HA]}{[A^-]}$$

Multiply through by -1

$$-\log [H^+] = -\log K - \log \frac{[HA]}{[A^-]}$$

Table 3–2. Dissociation constants and pK values for representative carboxylic acids.

Acid	K		pK
Acetic		1.76×10^{-5}	4.75
Glutaric	(1st)	4.58×10^{-5}	4.34
	(2nd)	3.89×10^{-6}	5.41
Citric	(1st)	8.40×10^{-4}	3.08
	(2nd)	1.80×10^{-5}	4.74
	(3rd)	4.00×10^{-6}	5.40

Substitute pH and pK for $-\log[H^+]$ and $-\log K$, respectively; then

$$pH = pK - \log \frac{[HA]}{[A^-]}$$

Then, to remove the minus sign, invert the last term.

$$pH = pK + \log \frac{[A^-]}{[HA]}$$

The Henderson-Hasselbalch equation has proved to be an expression of great predictive value in protonic equilibria. For example,

(1) When an acid is exactly half neutralized, [A⁻] = [HA]. Under these conditions,

$$pH = pK + \log \frac{[A^-]}{[HA]} = pK + \log \frac{1}{1} = pK + 0$$

Therefore, at half neutralization, pH = pK.

(2) When the ratio [A⁻]/[HA] = 100 to 1,

$$pH = pK + \log \frac{[A^-]}{[HA]}$$

$$pH = pK + \log 100/1 = pK + 2$$

(3) When the ratio [A⁻]/[HA] = 1 to 10,

$$pH = pK + \log \frac{1}{10} + pK + (-1)$$

If the equation is evaluated at several ratios of [A⁻]/[HA] between the limits 10^3 and 10^{-3}, and the calculated pH values plotted, the result obtained describes the titration curve for a weak acid (Fig 3–5).

Solutions of Weak Acids and Their Salts Buffer the pH When Protons Are Added or Removed

Solutions of weak acids and their conjugate bases (or of weak bases and their conjugate acids) exhibit the phenomenon of **buffering—the tendency of a solution to resist more effectively a change in pH following addition of a strong acid or base than does an equal volume of water.** The phenomenon of buffering is best illustrated by titrating a weak acid or base using a pH meter. Alternatively, we may calculate the pH shift that accompanies addition of acid or base to a buffered solution. In the example, the buffered solution (a mixture of a weak acid, pK = 5.0, and its conjugate base) is present initially at one of 4 pH values. We will calculate the pH shift that results when 0.1 meq of KOH is added to 1 meq of each of these solutions:

Initial pH	5.00	5.37	5.60	5.86
[A⁻]$_{initial}$	0.50	0.70	0.80	0.88
[HA]$_{initial}$	0.50	0.30	0.20	0.12
([A⁻]/[HA])$_{initial}$	1.00	2.33	4.00	7.33
Addition of 0.1 meq of KOH produces				
[A⁻]$_{final}$	0.60	0.80	0.90	0.98
[HA]$_{final}$	0.40	0.20	0.10	0.02
([A⁻]/[HA])$_{final}$	1.50	4.00	9.00	49.0
log ([A⁻]/[HA])$_{final}$	0.176	0.602	0.95	1.69
Final pH	5.18	5.60	5.95	6.69
ΔpH	0.18	0.60	0.95	1.69

Observe that the pH change per milliequivalent of OH⁻ added varies greatly depending on the pH. At pH values close to pK, the solution resists changes in pH most effectively, and it is said to exert a **buffering effect. Solutions of weak acids and their conjugate bases buffer most effectively in the pH range pK ± 2.0 pH units.** This means that to buffer a solution at pH X, a weak acid or base whose pK is no more than 2.0 pH units removed from pH X should be used.

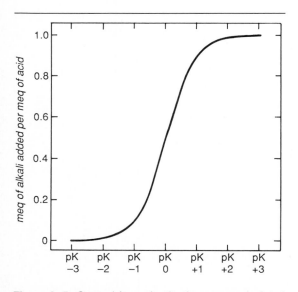

Figure 3–5. General form of a titration curve calculated from the Henderson-Hasselbalch equation.

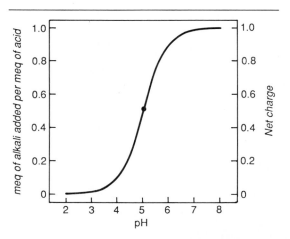

Figure 3–6. Titration curve for an acid of the type HA. The dot (•) indicates the pK, 5.0.

Shown in Fig 3–6 is the net charge on one molecule of the acid as a function of pH. A fractional charge of −0.5 means not that an individual molecule bears a fractional charge but that the statistical probability that a given molecule has a unit negative charge is 0.5. Consideration of the net charge on macromolecules as a function of pH provides the basis for many separatory techniques, including the electrophoretic separation of amino acids, plasma proteins, and abnormal hemoglobins.

In living cells, phosphate and bicarbonate, in addition to proteins, constitute the major buffers.

REFERENCES

Segel IM: *Biochemical Calculations*. Wiley, 1968.

Amino Acids

4

Victor W. Rodwell, PhD

INTRODUCTION

Living cells produce **macromolecules (proteins, nucleic acids, polysaccharides)** that serve as structural components, catalysts, hormones, receptors, or repositories of genetic information. These macromolecules are **biopolymers** constructed of **monomer units,** or building blocks. For nucleic acids, the monomer units are **nucleotides;** for complex polysaccharides, **sugar derivatives;** and for proteins, **L-α-amino acids.**

While proteins may also contain substances in addition to amino acids (eg, heme, carbohydrate, lipid), their 3-dimensional structure and therefore their biologic properties are determined largely by the **kinds of amino acids present,** the **order in which they are linked together** in a polypeptide chain, and thereby the **spatial relationship of one amino acid to another.**

Amino acids play additional roles in cells. Some compounds of biologic importance that arise from amino acids are listed in Tables 4–4 and 4–5.

BIOMEDICAL IMPORTANCE

Certain amino acids appear to be involved in the **transmission of impulses in the nervous system;** glycine and glutamic acid are examples. **Essential amino acids** must be supplied in the diet, since our bodies cannot synthesize them in amounts adequate to support growth (infants) or to maintain health (adults). The metabolism of amino acids gives rise to many compounds of biomedical importance. For instance, decarboxylation of certain amino acids produces the corresponding **amines,** some of which (eg, histamine and γ-aminobutyrate [GABA]) have important biologic functions. A number of diseases are due to abnormalities of the **transport** of amino acids into cells. Many of these conditions are characterized by the presence of greatly increased amounts of one or more amino acids in the urine, and for this reason they are often referred to as the **aminoacidurias.**

ALL AMINO ACIDS HAVE AT LEAST TWO FUNCTIONAL GROUPS

Amino acids contain both amino and carboxylic acid functional groups. In an α-amino acid, both are attached to the same (α) carbon atom (Fig 4–1).

Although about 300 amino acids occur in nature, only 20 of these occur in proteins (Table 4–3). Complete hydrolysis* of proteins produces 20 L-α-amino acids (Table 4–2). The same 20 amino acids are present in proteins from all forms of life—plant, animal, or microbial. The reason for this becomes apparent when the universality of the genetic code is discussed (see Chapter 30). However, some proteins contain amino acid derivatives formed after incorporation of the amino acid into the protein molecule (see Table 6–4).

With the exception of glycine, for which R = a hydrogen atom (Fig 4–1), all 4 groups linked to the α-carbon atom of amino acids are different. This tetrahedral orientation of 4 different groups about the α-carbon atom confers optical activity (the ability to rotate the plane of plane-polarized light) on amino acids. Although some amino acids found in proteins are dextrorotatory and some levorotatory at pH 7.0, all have the **absolute configurations** of L-glyceraldehyde and hence are **L-α-amino acids.**

The General Chemical Reactions of Amino Acids Can Be Predicted from a Knowledge of Organic Functional Group Chemistry

The carboxyl and amino groups of amino acids exhibit all the expected reactions of these functions, eg, salt formation, esterification, and acylation.

---

*Hydrolysis = rupture of a covalent bond with addition of the atoms of water.

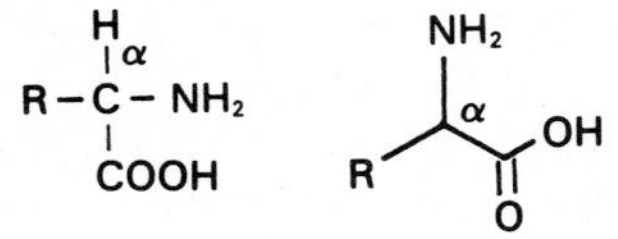

Figure 4–1. Two representations of an α-amino acid.

PROTONIC EQUILIBRIA OF AMINO ACIDS

Depending on the pH of the Surrounding Medium, an Amino Acid May Have Positive, Negative, or Zero Net Charge

Amino acids bear at least 2 ionizable weak acid groups, a $-COOH$ and an $-NH_3^+$. In solution, 2 forms of these groups, one charged and one uncharged, exist in protonic equilibrium:

$$R-COOH \rightleftarrows R-COO^- + H^+$$
$$R-NH_3^+ \rightleftarrows R-NH_2 + H^+$$

$R-COOH$ and $R-NH_3^+$ are the protonated, or acidic, partners in these equilibria. $R-COO^-$ and $R-NH_2$ are the **conjugate bases** (proton acceptors) of the corresponding acids. Although both $R-COOH$ and $R-NH_3^+$ are weak acids, $R-COOH$ is a far stronger acid than is $R-NH_3^+$. At the pH of blood plasma or the intracellular space (7.4 and 7.1, respectively), carboxyl groups exist almost entirely as **carboxylate ions,** $R-COO^-$. At these pH values, most amino groups are predominantly in the associated (protonated) form, $R-NH_3^+$. The prevalent ionic species of amino acids in

Table 4–2. Classification of the L-α-amino acids present in proteins on the basis of the relative polarities of their R groups. A nonpolar group has little or no charge difference from one region to another, whereas a polar group has a relatively large charge difference in different regions.

Nonpolar	Polar
Alanine	Arginine
Isoleucine	Asparagine
Leucine	Aspartic acid
Methionine	Cysteine
Phenylalanine	Glutamic acid
Proline	Glutamine
Tryptophan	Glycine
Valine	Histidine
	Lysine
	Serine
	Threonine
	Tyrosine

blood and most tissues should be represented as shown in Fig 4–2A. Structure B (Fig 4–2) cannot exist at *any* pH. At a pH sufficiently low to protonate the carboxyl group, the more weakly acidic amino group would also be protonated. The approximate pK_a values for α-carboxyl and α-amino groups of an α-amino acid are 2 and 10, respectively (Table 4–1). At a pH 2 units below its pK_a, an acid is approximately 99% protonated. If the pH is gradually raised, the proton from the carboxylic acid will be lost long before that from the $R-NH_3^+$. At any pH sufficiently high for the uncharged conjugate base of the amino group to predominate, a carboxyl group is present as the carboxylate ion ($R-COO^-$). The B representation, however, is used for many equations not involving protonic equilibria.

Table 4–1. Weak acid groups of the amino acids present in proteins.

	Conjugate Acid	Conjugate Base	Approximate pK_a
α-Carboxyl	$R-COOH$	$R-COO^-$	2.1 ± 0.5
Non-α-carboxyl (aspartate, glutamate)	$R-COOH$	$R-COO^-$	4.0 ± 0.3
Imidazolium (histidine)	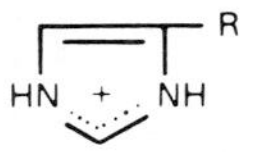	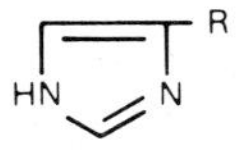	6.0
α-Amino	$R-NH_3^+$	$R-NH_2$	9.8 ± 1.0
ε-Amino (lysine)	$R-NH_3^+$	$R-NH_2$	10.5
Phenolic OH (tyrosine)	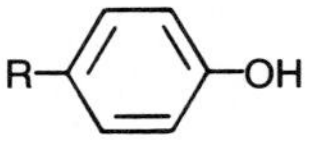	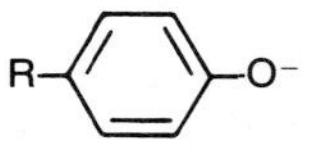	10.1
Guanidinium (arginine)	$\overset{H}{\underset{}{}} \, \overset{NH_2}{\underset{}{}}$ $R-N-C-NH_2^+$	$\overset{H}{\underset{}{}} \, \overset{NH}{\underset{}{}}$ $R-N-C-NH_2$	12.5
Sulfhydryl (cysteine)	$R-SH$	$R-S^-$	8.3

Figure 4–2. Ionically correct structure for an amino acid at or near physiologic pH (*A*). The uncharged structure (*B*) cannot exist at any pH but may be used as a convenience when discussing the chemistry of amino acids.

The Relative Strengths of Weak Acids Are Expressed in Terms of pK$_a$'s

Relative acid strengths of weak acids are expressed by their acid dissociation constant, K$_a$, or by their pK$_a$, the negative log of the dissociation constant:

$$pK_a = -\log K_a^*$$

Table 4–1 lists pK values for the functional groups present on the 20 amino acids of proteins.

The **net charge** (the algebraic sum of all the positively and negatively charged groups present) of an amino acid **depends upon the pH, or proton concentration, of the surrounding solution.** The ability to alter the charge on amino acids or their derivatives by manipulating the pH facilitates the physical separation of amino acids, peptides, and proteins.

The pH at Which an Amino Acid Bears No Net Charge and Hence Does Not Move in a Direct Current Electrical Field Is Its Isoelectric pH (pI)

For an aliphatic amino acid such as alanine, the isoelectric species is the form shown in Fig 4–3.

The isoelectric pH is the pH midway between pK values on either side of the isoelectric species. For an amino acid with only 2 dissociating groups, there can be no possible ambiguity, as is shown below in the calculation of the pI for alanine. Since pK$_1$ (R–COOH) = 2.35 and pK$_2$ (R–NH$_3$$^+$) = 9.69, the isoelectric pH (pI) of alanine is:

$$pI = \frac{pK_1 + pK_2}{2} = \frac{2.35 + 9.69}{2} = 6.02$$

Calculation of pI for a compound with more than 2 dissociable groups carries more possibility for error. For example, from consideration of Fig 4–4, what would be the isoelectric pH (pI) for aspartic acid?

First, write out all possible ionic structures for a compound in the order in which they occur proceeding from strongly acidic to basic solution (eg, as for aspartic acid in Fig 4–4). Next, identify the isoionic, zwitterionic, or neutral representation (as in Fig 4–4B). The pI is the pH at the midpoint between the pK values on either side of the isoionic species. In this example,

$$pI = \frac{2.09 + 3.86}{2} = 2.98$$

This approach works equally well for amino acids with additional dissociating groups, eg, lysine or histidine. After writing the formulas for all possible charged species of the basic amino acids lysine and arginine, observe that

$$pI = \frac{pK_2 + pK_3}{2}$$

For lysine, pI is 9.7; for arginine, pI is 10.8. The student should determine the pI for histidine.

Determining pK values on either side of the zwitterion by inspection of charged structures is not limited to amino acids. It may be applied to calculating the charge on a molecule with any number of dissociating groups. The ability to perform calculations of this type is of value in the clinical laboratory to predict the mobility of compounds in electrical fields and to select appropriate buffers for separations. For example, a buffer at pH 7.0 will separate 2 molecules with pI's of 6 and 8, respectively, because the molecule with pI = 6 will have a larger net negative charge at pH 7.0 than the molecule with pI = 8. Similar considerations apply to understanding separations on ionic supports such as positively or negatively charged polymers (eg, DEAE cellulose or Dowex 1 resin).

The Solubility & Melting Points of Amino Acids Reflect Their Ionic Character

Since multiple charged groups are present on amino acids, they are readily solvated by, and hence soluble in, polar solvents such as water and ethanol, but insoluble in nonpolar solvents such as benzene, hexane, and ether. Their high melting points (>200 °C) reflect the high energy needed to disrupt the ionic forces that stabilize the crystal lattice.

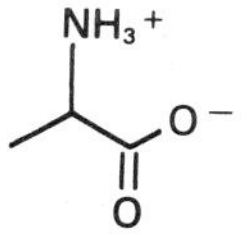

Figure 4–3. Isoelectric or "zwitterionic" structure of alanine. Although charged, the zwitterion bears no *net* charge and hence does not migrate in a direct current electric field.

*For convenience, the a subscript in K$_a$ and pK$_a$ will be implied but dropped hereafter from the notation.

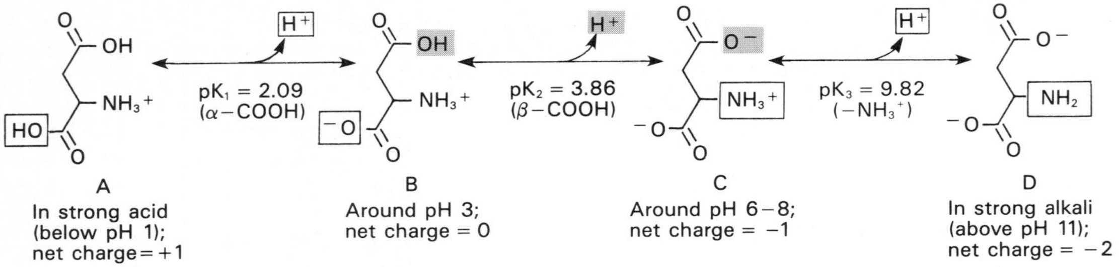

Figure 4–4. Protonic equilibria of aspartic acid.

THE AMINO ACIDS IN PROTEINS MAY BE CLASSIFIED ON THE BASIS OF THE RELATIVE POLARITIES OF THEIR R GROUPS

The amino acids in proteins may be divided into 2 broad groups on the basis of whether the R groups attached to the α-carbon atoms are polar or nonpolar. The single-letter abbreviations (Table 4–3) are used to represent extremely long sequences of amino acids (eg, for listing the complete sequence of amino acids in a protein).

Amino acids that occur in free or combined states (but not in proteins) fulfill important roles in metabolic processes (Tables 4–4 and 4–5). For example, ornithine, citrulline, and argininosuccinate participate in the formation of urea. Over 20 D-amino acids occur naturally. These include the D-alanine and D-glutamate of certain bacterial cell walls and a variety of D-amino acids in antibiotics.

THE PROPERTIES OF INDIVIDUAL AMINO ACIDS ARE DICTATED BY THE NATURE OF THEIR α-R GROUPS

Glycine, the smallest of the amino acids, can fit into regions of the 3-dimensional structure of proteins inaccessible to other amino acids and occurs in regions where peptides bend sharply.

The aliphatic R groups of alanine, valine, leucine, and isoleucine and the aromatic R groups of phenylalanine, tyrosine, and tryptophan are hydrophobic, a property that has important consequences for the ordering of water molecules in proteins in their immediate neighborhood. These amino acids typically occur primarily in the interior of cytosolic proteins.

The charged R groups of basic and acidic amino acids perform key roles in stabilizing specific protein conformations via formation of salt bonds. For example, rupture and reformation of salt bonds accompany oxygenation and deoxygenation of hemoglobin (see Chapter 7). In addition, amino acids with positively or negatively charged R groups function in the "charge relay" systems that transmit charges across considerable distances during enzymatic catalysis. Finally, his-

tidine occupies a unique and important place in enzymatic catalysis, since the pK of its imidazole proton permits it, at pH 7.0, to function as either a base or an acid catalyst.

The primary alcohol group of serine and the primary thioalcohol (−SH) group of cysteine are excellent nucleophiles and may function as such during enzymatic catalysis. While the secondary alcohol group of threonine also is a good nucleophile, it is not known to perform this role in catalysis. In addition to its catalytic role, the −OH of serine and of tyrosine function in regulation of the activity of certain enzymes whose catalytic activity depends upon the phosphorylation state of specific seryl or tyrosyl residues.

Amino acids do not absorb visible light (ie, they are colorless) and, with the exceptions of the aromatic amino acids tryptophan, tyrosine, phenylalanine, and histidine, do not absorb ultraviolet light of a wavelength above 240 nm. As shown in Fig 4–5, above 240 nm, **most of the ultraviolet absorption of proteins is due to their tryptophan content.**

REACTION WITH NINHYDRIN OR FLUORESCAMINE PROVIDES A MEANS BY WHICH TO VISUALIZE AMINO ACIDS

Ninhydrin (Fig 4–6) oxidatively decarboxylates α-amino acids to CO_2, NH_3, and an aldehyde with one less carbon atom than the parent amino acid. The reduced ninhydrin then reacts with the liberated ammonia, forming a blue complex that maximally absorbs light of wavelength 570 nm. This blue color forms the basis of a **quantitative test for α-amino acids** that can detect as little as 1 μg of amino acid. Amines other than α-amino acids also react with ninhydrin, forming a blue color, but without evolving CO_2. The evolution of CO_2 thus indicates an α-amino acid. NH_3 and peptides also react but more slowly than α-amino acids. Proline and 4-hydroxyproline produce a yellow color with ninhydrin.

Fluorescamine (Fig 4–7), an even more sensitive reagent, can detect nanogram quantities of an amino acid. Like ninhydrin, fluorescamine forms a complex with amines other than amino acids.

Table 4-3. L-α-Amino acids present in proteins.*

Name	Symbol	Structural Formula
With Aliphatic Side Chains		
Gycine	Gly [G]	$H-CH-COO^-$ $_+NH_3$
Alanine	Ala [A]	$CH_3-CH-COO^-$ $_+NH_3$
Valine	Val [V]	$\begin{array}{c}H_3C\\ \quad \rangle CH-CH-COO^- \\ H_3C \quad\quad _+NH_3\end{array}$
Leucine	Leu [L]	$\begin{array}{c}H_3C\\ \quad \rangle CH-CH_2-CH-COO^- \\ H_3C \quad\quad\quad\quad _+NH_3\end{array}$
Isoleucine	Ile [1]	$\begin{array}{c}CH_3\\ CH_2\\ CH-CH-COO^- \\ CH_3 \quad _+NH_3\end{array}$
With Side Chains Containing Hydroxylic (OH) Groups		
Serine	Ser [S]	$CH_2-CH-COO^-$ $OH \quad _+NH_3$
Threonine	Thr [T]	$CH_3-CH-CH-COO^-$ $OH \quad _+NH_3$
Tyrosine	Tyr [Y]	See below.
With Side Chains Containing Sulfur Atoms		
Cysteine†	Cys [C]	$CH_2-CH-COO^-$ $SH \quad _+NH_3$
Methionine	Met [M]	$CH_2-CH_2-CH-COO^-$ $S-CH_3 \quad _+NH_3$
With Side Chains Containing Acidic Groups or Their Amides		
Aspartic acid	Asp [D]	$^-OOC-CH_2-CH-COO^-$ $_+NH_3$
Asparagine	Asn [N]	$H_2N-C-CH_2-CH-COO^-$ $\quad\quad \overset{\|}{O} \quad\quad _+NH_3$
Glutamic acid	Glu [E]	$^-OOC-CH_2-CH_2-CH-COO^-$ $_+NH_3$
Glutamine	Gln [Q]	$H_2N-C-CH_2-CH_2-CH-COO^-$ $\quad\quad \overset{\|}{O} \quad\quad\quad\quad _+NH_3$

Table 4-3 (cont'd). L-α-Amino acids present in proteins.*

Name	Symbol	Structural Formula
With Side Chains Containing Basic Groups		
Arginine	Arg [R]	$H-N-CH_2-CH_2-CH_2-CH-COO^-$; $C=NH_2$; $\overset{+}{N}H_2$; with $\overset{+}{N}H_3$
Lysine	Lys [K]	$CH_2-CH_2-CH_2-CH_2-CH-COO^-$; $\overset{+}{N}H_3$; $\overset{+}{N}H_3$
Histidine	His [H]	imidazole ring $HN\quad \overset{+}{N}H$ $-CH_2-CH-COO^-$; $\overset{+}{N}H_3$
Containing Aromatic Rings		
Histidine	His [H]	See above.
Phenylalanine	Phe [F]	(benzene ring)$-CH_2-CH-COO^-$; $\overset{+}{N}H_3$
Tyrosine	Tyr [Y]	$HO-$(benzene ring)$-CH_2-CH-COO^-$; $\overset{+}{N}H_3$
Tryptophan	Trp [W]	(indole ring, N–H)$-CH_2-CH-COO^-$; $\overset{+}{N}H_3$
Imino Acids		
Proline	Pro [P]	(pyrrolidine ring) $\overset{+}{N}H_2$ $-COO^-$

*Except for hydroxylysine (Hyl) and hydroxyproline (Hyp), which are incorporated into polypeptide linkages as lysine and proline and subsequently hydroxylated, specific transfer RNA molecules exist for all the amino acids listed in Table 4–3. Their incorporation into proteins is thus under direct genetic control.

†Cystine consists of 2 cysteine residues linked by a disulfide bond:

$$\overset{NH_3^+}{\underset{}{|}}$$
$$^-OCC-CH-CH_2-S-S-CH_2-CH-COO^-$$
$$\underset{NH_3^+}{|}$$

A VARIETY OF SEPARATORY TECHNIQUES ARE APPLICABLE TO AMINO ACIDS

Chromatography

In all chromatographic separations, molecules are **partitioned between a stationary and a mobile phase** (Table 4–6). **Separation depends on the relative tendencies of molecules in a mixture to associate more strongly with one or the other phase.**

While these separatory techniques are discussed principally with respect to amino acids, their use is by no means restricted to these molecules.

Paper Chromatography

While largely supplanted by more sophisticated techniques, paper chromatography still finds application in amino acid separations. Samples are applied at a marked point about 5 cm from the end of a filter paper strip. The strip is then suspended in a sealed vessel that contains the chromatographic solvent (Fig 4–8).

Solvents for amino acid separations are mixtures of

Table 4–4. Examples of α-amino acids not present in proteins but essential in mammalian metabolism.

Common and Systematic Names	Formula at Neutral pH	Significance
Homocysteine (2-amino-4-mercaptobutanoic acid)	CH_2—CH_2—CH—COO$^-$ $\mid$ $\mid$ SH $_+NH_3^+$	Intermediate in methionine biosynthesis
Cysteine sulfinic acid (2-amino-3-sulfinopropanoic acid)	CH_2—CH—COO$^-$ $\mid$ $\mid$ SO_2 $_+NH_3$	Intermediate in cysteine catabolism.
Homoserine (2-amino-4-hydroxybutanoic acid)	CH_2—CH_2—CH—COO$^-$ $\mid$ $\mid$ OH $_+NH_3$	Intermediate in threonine, aspartate, and methionine metabolism.
Ornithine (2,5-bisaminopentanoic acid)	CH_2—CH_2—CH_2—CH—COO$^-$ $\mid$ $\mid$ $_+NH_3$ $_+NH_3$	Intermediate in threonine, aspartate, and methionine metabolism.
Citruline (2-amino-5-ureidopentanoic acid)	CH_2—CH_2—CH_2—CH—COO$^-$ $\mid$ $\mid$ NH $_+NH_3$ $\mid$ C=O $\mid$ NH_2	Intermediate in the biosynthesis of urea.
Argininosuccinic acid	$_+$ CH_2—CH_2—CH_2—CH—COO$^-$ NH $\mid$ $\parallel$ $_+NH_3$ HN—C—NH— $^-$OOC—CH_2—C—COO$^-$	Intermediate in the biosynthesis of urea.
Dopa (3,4-dihydroxyphenylalanine)	HO—⟨ring⟩—CH_2—CH—COO$^-$ (ring with HO) $\mid$ $_+NH_3$	Precursor of melanin.
3-Monoiodotyrosine	HO—⟨ring, I⟩—CH_2—CH—COO$^-$ $\mid$ $_+NH_3$	Precursor of thyroid hormones.
3,5-Diiodotyrosine	HO—⟨ring, I, I⟩—CH_2—CH—COO$^-$ $\mid$ $_+NH_3$	Precursor of thyroid hormones
3,5,3′-Triiodothyronine (T$_3$)	HO—⟨ring, I⟩—O—⟨ring, I, I⟩—CH_2—CH—COO$^-$ $\mid$ $_+NH_3$	Precursor of thyroid hormones.
Thyroxine (3,5,3′,5′-tetraiodothyronine; T$_4$)	HO—⟨ring, I, I⟩—O—⟨ring, I, I⟩—CH_2—CH—COO$^-$ $\mid$ $_+NH_3$	Precursor of thyroid hormones.

water, alcohols, and acids or bases. The more polar components of the solvent associate with the cellulose and form the stationary phase. The less polar components constitute the mobile phase. This is **normal partition chromatography.** For **reversed-phase partition chromatography,** the polarities of the mobile and stationary phases are reversed (eg, by first dipping the paper in a solution of a silicone). Reversed-phase partition chromatography is used to separate nonpolar peptides or lipids, not polar compounds such as amino

Table 4–5. Examples of amino acids with non-α-amino groups important in mammalian metabolism.

Common and Systematic Names	Formula at Neutral pH	Significance
β-Alanine (3-aminopropanoate)	$CH_2—CH_2—COO^-$ $\|$ $^+NH_3$	Part of coenzyme A and of the vitamin pantetheine.
Taurine (2-aminoethylsulfonate)	$CH_2—CH_2—SO_3^-$ $\|$ $^+NH_3$	Occurs in bile combined with bile acids.
γ-Aminobutyrate (GABA) (4-aminobutanoate)	$CH_2—CH_2—CH_2—COO^-$ $\|$ $^+NH_3$	Neurotransmitter formed from glutamate in brain tissue.
β-Aminoisobutyrate (2-methyl-3-aminopropanoate)	$H_3N^+—CH_2—CH—COO^-$ $\|$ CH_3	End product of pyrimidine catabolism in urine of some persons.

acids. The solvent may migrate up or down the paper (ascending or descending chromatography). When it has migrated almost to the end, the strip is dried and treated to allow visualization of the molecules of interest (eg, for amino acids, with 0.5% ninhydrin in acetone followed by heating at 90–110 °C for a few minutes). Amino acids with large nonpolar side chains (Leu, Ile, Phe, Trp, Val, Met, Try) migrate farther than those with shorter nonpolar side chains (Pro, Ala, Gly) or with polar side chains (Thr, Glu, Ser, Arg, Asp, His, Lys, Cys) (See Fig 4–9.) This reflects the greater relative solubility of polar molecules in the hydrophilic stationary phase and of nonpolar molecules in organic solvents. For a nonpolar series (Gly, Ala, Val, Leu), increasing length of the nonpolar side chain, which increases nonpolar character, results in increased mobility.

The ratio of the distance traveled by an amino acid to that traveled by the solvent front, both measured from the point of application of the amino acid mixture, is called the **R_f value** (mobility relative to the solvent front). R_f values for a given amino acid vary with experimental conditions, eg, the solvent used. Although it is possible to identify an amino acid tentatively by its R_f value alone, it is preferable to chromatograph known amino acid standards simultaneously with the unknown mixture. Then mobility may be expressed relative to that of a standard (eg, as R_{ala} rather than as R_f). Mobilities expressed relative to a standard vary less than R_f values from experiment to experiment.

Quantitation of amino acids may be accomplished by cutting out each spot, eluting with a suitable solvent, and performing a quantitative colorimetric (ninhydrin) analysis. Alternatively, the paper may be sprayed with ninhydrin and the color densities of the spots measured with a recording transmittance or reflectance photometer.

For **2-dimensional paper chromatography,** sample is applied to one corner of a square sheet of paper or other suitable medium and chromatographed in one solvent mixture. The sheet is then removed, dried, turned through 90 degrees, and chromatographed in a second solvent (Fig 4–10).

Thin-Layer Chromatography

There are 2 distinct classes of thin-layer chromatography (TLC). Partition TLC (PTLC) closely resembles partition chromatography on paper. For PTLC on powdered cellulose or other relatively inert supports, the solvent systems and detection reagents used for paper chromatography are fully applicable. Reversed-phase PTLC also is possible.

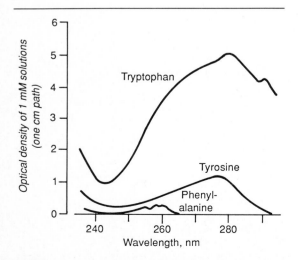

Figure 4–5. The ultraviolet absorption spectra of tryptophan, tyrosine, and phenylalanine.

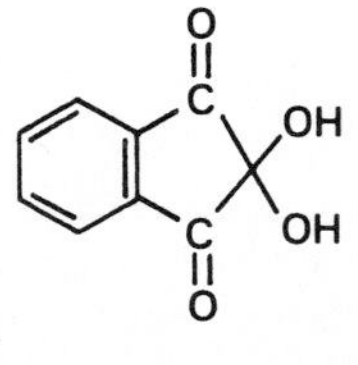

Figure 4–6. Ninhydrin.

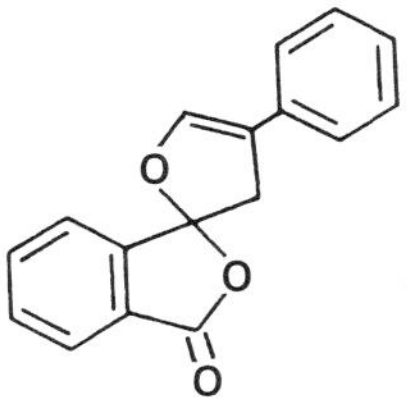

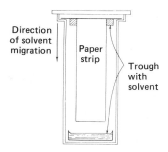

Figure 4–7. Fluorescamine.

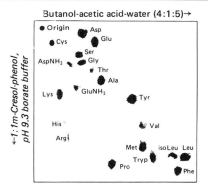

Figure 4–8. Descending paper chromatography.

Adsorption TLC (ATLC) bears no similarity to paper chromatography and is based on entirely different principles. ATLC depends on the ability of the solvent (which need not be binary or more complex) to elute sample components from adsorption sites on an activated sorbent such as heated silica gel. ATLC is applicable to nonpolar materials, such as lipids, and hence not to amino acids or most peptides.

Ion Exchange Chromatography

Analysis of amino acid residues after hydrolysis of a polypeptide generally involves **automated ion exchange chromatography.** Complete separation, identification, and quantitation require less than 3 hours. The procedure of Moore and Stein uses a short and long column containing the Na^+ form of a sulfonated polystyrene resin. When acid hydrolysate at pH 2 is applied to the columns, the amino acids bind via cation exchange with Na^+. The columns are then eluted with sodium citrate under preprogrammed conditions of pH and temperature. The short column requires a single elution buffer; the long column, two. Eluted material is reacted with ninhydrin reagent, and color densities are monitored in a flow-through colorimeter. Data are displayed on a cathode ray tube with computer-linked integration of peak areas (Fig 4–11).

High-Voltage Electrophoresis (HVE)

Separations of amino acids, polypeptides, and other ampholytes (molecules whose net charge depends on the pH of the surrounding medium) in a direct current field have many applications in biochemistry. For amino acids, paper sheets or thin layers of powdered cellulose are most frequently used as supports. For large polypeptides or proteins, a crosslinked polyacryl-

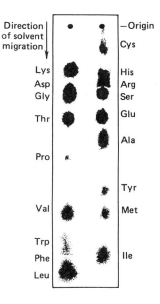

Figure 4–9. Identification of amino acids present in proteins. After descending paper chromatography in butanol-acetic acid, spots were visualized with ninydrin.

Figure 4–10. Two-dimensional chromatogram of protein amino acids. (Redrawn, slightly modified, from Levy AI, Chung D: Two-dimensional chromatography of amino acids on buffered papers. *Anal Chem* 1953;**25**:396. Copyright © 1953 by American Chemical Society; reproduced with permission.)

Table 4–6. Phase relationships for chromatography.

Form of Chromatography	Stationary Phase	Mobile Phase
Partition chromatography on solid subborts; gel filtration.	Liquid	Liquid
Ion exchange chromatography.	Solid	Liquid
Partition chromatography between thin layer of liquid on support and mobile gas.	Liquid	Gas

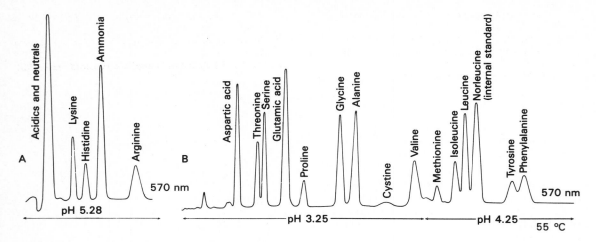

Figure 4–11. Automated analysis of an acid hydrolysate of a protein on Moore-Stein Dowex 50 columns (at 55 °C). *A:* Short (5.0 × 0.9 cm) column used to resolve basic amino acids at pH 5.28. Time required = 60 minutes. *B:* Longer (55 × 0.9 cm) column used to resolve neutral and acidic amino acids by elution first with pH 3.25 and then with pH 4.25 buffer. An internal standard of norleucine is included for reference. Basic amino acids remain bound to the column. Time required = 180 minutes. Emerging samples are automatically reacted with ninhydrin and the optical density of samples recorded at 570 nm and 440 nm. The latter wavelength is used solely to detect proline and hydroxyproline. Ordinate = optical density plotted on a log scale. Abscissa = time in minutes. (Courtesy of Professor ET Mertz, Purdue University.)

amide gel is used. For nucleotide oligomers, both agarose and polyacrylamide supports are used.

Separations in a 2000- to 5000-volt direct current field for 0.5–2 hours depend upon the net charge on the ampholyte and its molecular weight. For molecules with identical charge, those of lower molecular weight migrate farther. Net charge, however, is the more important factor in determining separation. Applications include amino acids, low molecular weight polypeptides, certain proteins, nucleotides, and phosphosugars. Samples are applied to the support, which is then moistened with buffer of an appropriate pH and

connected to buffer reservoirs by paper wicks. The paper may be covered by a glass plate or immersed in a hydrocarbon coolant. When current is applied, molecules with a net negative charge at the selected pH migrate toward the anode and those with a net positive charge toward the cathode. For visualization, the dried **electropherogram** is treated with ninhydrin (amino acids, peptides) or with ethidium bromide and viewed under ultraviolet light (nucleotide oligomers), etc. The choice of pH is dictated by the pK values of the dissociating groups on the molecules in the mixture.

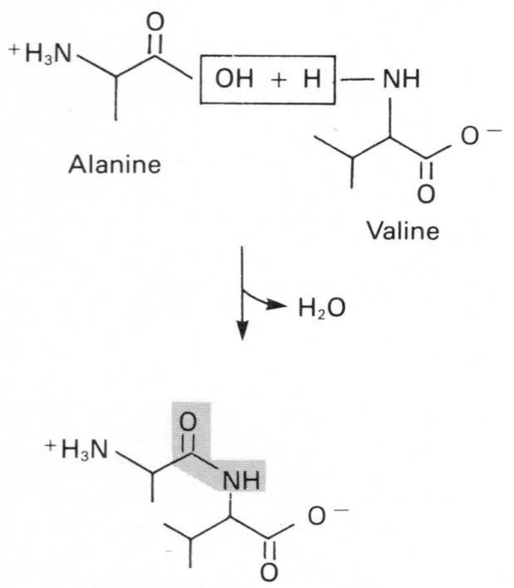

THE MOST IMPORTANT REACTION OF AMINO ACIDS IS THE FORMATION OF THE PEPTIDE BOND

In principle, peptide bond formation involves removal of 1 mol of water between the α-amino group of one amino acid and the α-carboxyl group of a second amino acid (Fig 4–12). This reaction does not, however, proceed as written, since the equilibrium constant strongly favors peptide bond hydrolysis. To synthesize peptide bonds between 2 amino acids, the carboxyl group must first be **activated.** Chemically, this may involve prior conversion to an acid chloride. **Biologically, activation involves initial condensation with ATP** (see Chapter 30.)

Figure 4–12. Amino acids united by a peptide bond (shaded).

REFERENCES

Barrett GC: *Chemistry and Biochemistry of the Amino Acids.* Chapman & Hall, 1985.

Davies JS: *Amino Acids and Peptides*. Chapman & Hall, 1985.

Gehrke CW, Kuo KCT, Zumwalt RW, Kenneth CT: *Amino Acid Analysis by Gas Chromatography*. 3 vols. CRC Press, 1987.

Hancock WS: *Handbook of HPLC for the Separation of Amino Acids, Peptides, and Proteins*. 2 vols. CRC Press, 1984.

Hugli TE: *Techniques in Protein Chemistry*. Academic Press, 1989.

Rattenbury JM: *Amino Acid Analysis*. Halstead Press, 1981.

INTRODUCTION

When the amino and carboxyl groups of amino acids combine to form peptide bonds, the constituent amino acids are termed amino acid residues. **A peptide consists of 2 or more amino acid residues linked by peptide bonds.** Peptides of more than 10 amino acid residues are termed **polypeptides.**

BIOMEDICAL IMPORTANCE

Peptides are of immense biomedical interest, particularly in **endocrinology.** Many major hormones are peptides and may be given to patients to correct corresponding deficiency states (eg, administration of **insulin** to patients with diabetes mellitus). Certain **antibiotics** are peptides (eg, valinomycin and gramicidin A), as are a few antitumor agents (eg, bleomycin). Rapid chemical synthesis and recombinant DNA technology have facilitated the manufacture of substantial amounts of peptide hormones, many of which are present in the body in relatively minute concentrations and thus difficult to isolate in quantities sufficient for therapy. The same technology allows the synthesis of other peptides, also available from natural sources in only small amounts (eg, certain viral peptides and proteins), for use in **vaccines.**

PEPTIDES ARE FORMED FROM L-α-AMINO ACIDS LINKED BY PEPTIDE BONDS

Figure 5–1 shows a tripeptide made up of alanine, cysteine, and valine. Note that **a tripeptide is one with 3 residues, not 3 peptide bonds.** By convention, peptide structures are written with the **N-terminal residue** (the residue with a free α-amino group) **at the left** and with the **C-terminal residue** (the residue with a free α-carboxyl group) **at the right.** This peptide has a **single** free α-amino group and a **single** free α-carboxyl group. However, in some peptides, the terminal amino or carboxyl groups may be derivatized (eg, an

N-formyl amine or an amide of the carboxyl group) and thus not free.

Peptide Structures May Be Drawn by a Simple Method

For a simple way to write peptide structures, first draw its "backbone" of linked α-NH_2, α-COOH, and α-carbon atoms. Next, insert the appropriate side chains on the α-carbon atoms (see below).

(1) Write a zig-zag of arbitrary length, and insert the N-terminal amino group:

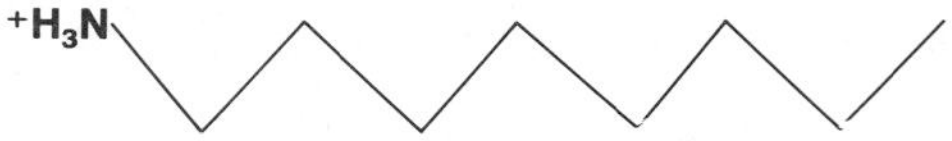

(2) Insert the α-carbon, α-carboxyl, and α-amino groups:

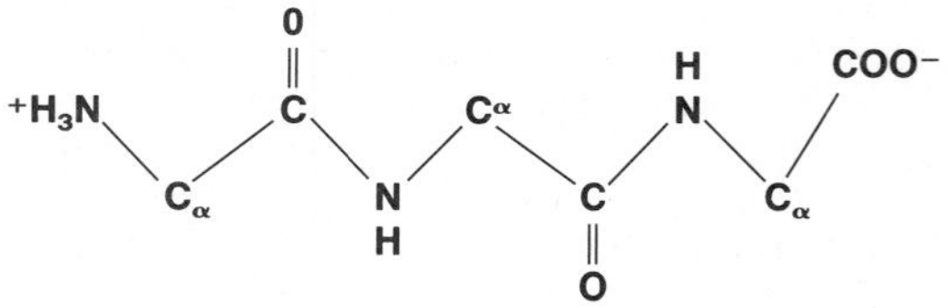

(3) Add the appropriate R groups (shaded) and α-hydrogens to the α-carbon atoms:

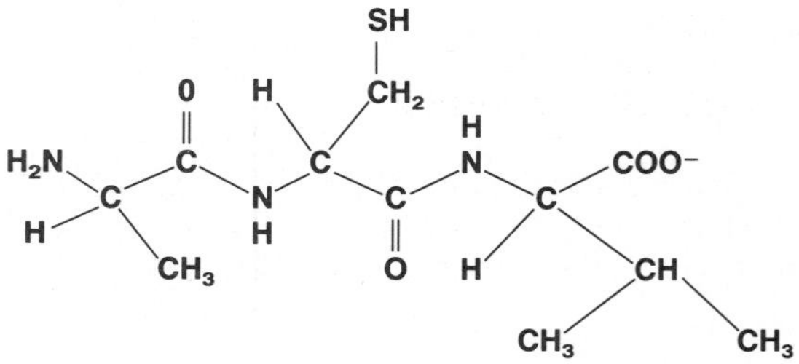

The Sequence of Amino Acids Determines the Primary Structure of a Peptide

When the **number, structure,** and **order** of all of the amino acid residues in a polypeptide are known, its primary structure has been determined. Peptides are named as derivatives of their C-terminal amino acid.

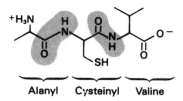

Figure 5–1. Structural formula for a tripeptide. Peptide bonds shaded for emphasis.

Three- and One-Letter Abbreviations Are Used to Name the Amino Acids in Peptides

Since polypeptides (proteins) may contain 100 or more residues, either the 3-letter or one-letter abbreviation for the amino acids shown in column 2 of Table 4–3 are used to represent primary structures (Fig 5–2).

Three-letter abbreviations for amino acid residues linked by straight lines represent a primary structure that is known and unambiguous. These lines are omitted for single-letter abbreviations. Where there is uncertainty about the precise **order** of a portion of a polypeptide, the questionable residues are enclosed in brackets and separated by commas (Fig 5–3).

A Change in the Primary Structure of a Peptide Can Alter its Biologic Activity

Substitution of a single amino acid for another in a linear sequence of possibly 100 or more amino acids may reduce or abolish biologic activity, with potentially serious consequences (eg, sickle cell disease). Many inherited metabolic errors involve a single change of this type. Powerful new methods to determine protein and DNA structure have greatly increased our understanding of the biochemical basis for many inherited metabolic diseases.

PEPTIDES ARE POLYELECTROLYTES WHOSE NET CHARGE DEPENDS ON THE pH OF THE SURROUNDING MEDIUM

The peptide (amide) bond is uncharged at any pH of physiologic interest. Formation of peptides from amino acids at pH 7.4 is therefore accompanied by a net

Glu-Ala-Lys-Gly-Tyr-Ala

E A K G Y A

Figure 5–2. Use of 3-letter and one-letter abbreviations for amino acid residues to represent the primary structure of a hexapeptide with glutamate (Glu, E) at the N terminus and alanine (Ala, A) at the C terminus.

Glu-Lys-(Ala,Gly,Tyr)-His-Ala

Figure 5–3. A heptapeptide containing a region of uncertain primary structure (in parentheses).

loss of one positive and one negative charge per peptide bond formed. Peptides are, however, charged molecules at physiologic pH owing to their C- and N-terminal groups and to functional groups present in polar amino acid residues attached to the α-carbon atoms (see Table 4–1).

The Number of Possible Conformations of a Peptide Is Constrained by Noncovalent Forces

A large number of conformations (spatial arrangements) are possible for a polypeptide. However, in solution a narrow range of conformations tends to predominate. These favored conformations reflect factors such as steric hindrance, coulombic interactions, hydrogen bonding, and hydrophobic interactions (see Chapter 6). As is the case for proteins, specific conformations are required for physiologic activity of polypeptides.

MANY SMALL PEPTIDES HAVE PHYSIOLOGIC ACTIVITY

Animal, plant, and bacterial cells contain a variety of low molecular weight polypeptides (3–100 amino acid residues) having profound physiologic activity. Some, including most mammalian polypeptide hormones, contain only peptide bonds formed between α-amino and α-carboxyl groups of the 20 L-α-amino acids of proteins. However, additional amino acids or derivatives of the protein amino acids may also be present in polypeptides.

The short polypeptide **bradykinin** is a smooth muscle hypotensive agent liberated from specific plasma proteins by proteolysis.

Arg-Pro-Pro-Gly-Phe-Ser-Pro-Phe-Arg
Bradykinin

Glutathione (Fig 5–4), an atypical tripeptide in which the N-terminal glutamate is linked to cysteine via a non-α-peptidyl bond, is required by several enzymes. Glutathione and the enzyme glutathione reductase participate in the formation of the correct disulfide bonds of many proteins and polypeptide hormones (see Chapter 6).

Polypeptide antibiotics elaborated by fungi contain both D- and L-amino acids and amino acids not present in proteins. Examples include tyrocidine and gramicidin S, cyclic polypeptides that contain D-phenylalanine, and the nonprotein amino acid ornithine. These polypeptides are not synthesized on ribosomes.

Thyrotropin-releasing hormone (TRH) (Fig 5–5)

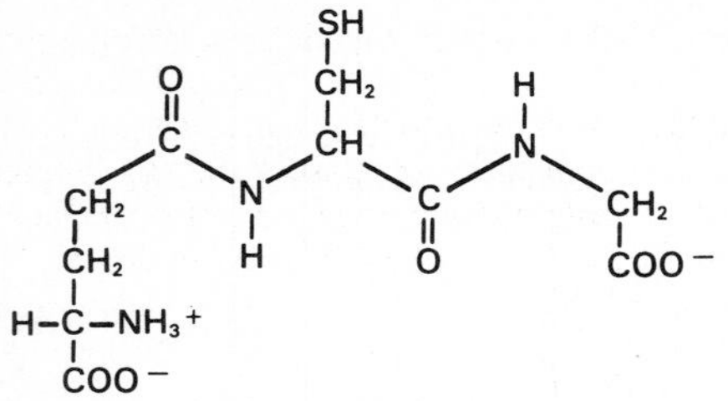

Figure 5–4. Glutathione (γ-glutamyl-cysteinyl-glycine).

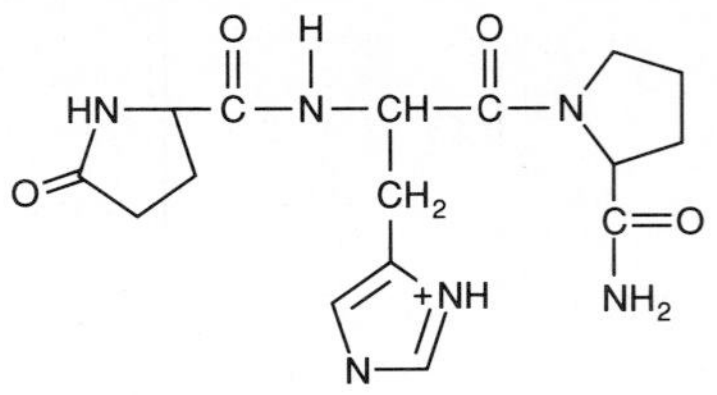

Figure 5–5. Pyro-glutamyl-histidyl-prolinamide (TRH)

illustrates yet another variant. The N-terminal glutamate is cyclized to pyroglutamic acid, and the C-terminal prolyl carboxyl is amidated.

A mammalian polypeptide may contain more than one physiologically potent polypeptide. Within the primary structure of **β-lipotropin**—a hypophyseal hormone that stimulates the release of fatty acids from adipose tissue—are sequences of amino acids that are common to several other polypeptide hormones with diverse physiologic activities (Fig 5–6). The large polypeptide is a precursor of the smaller polypeptides.

COMPLEX MIXTURES OF PEPTIDES MAY BE SEPARATED BY ELECTROPHORESIS OR BY CHROMATOGRAPHY

Chromatography & High-Voltage Electrophoresis (HVE)

These techniques (see Chapter 4), which separate on the basis of charge, are applicable to low molecular weight polypeptides as well as to amino acids. The pK value for the C-terminal carboxyl group of a polypeptide is higher than that of the α-carboxyl group in the corresponding amino acid (ie, the peptide COOH is a weaker acid). Conversely, the N-terminal amino group is a stronger acid (has a lower pK) than the amino acid from which it has been derived (Table 5–1).

Gel Filtration

Automated sequencing utilizes small numbers of large (30- to 100-residue) peptides. However, many denatured, high molecular weight polypeptides may be insoluble owing to exposure during denaturation of previously buried hydrophobic residues. While insolubility can be overcome by urea, alcohols, organic acids, or bases, these restrict the subsequent use of ion exchange techniques for peptide purification. Gel filtration of large hydrophobic peptides, however, may be performed in 1–4 molar formic or acetic acid.

Reversed-Phase High Performance Liquid Chromatography

A powerful technique for purification of high molecular weight, nonpolar peptides is high performance

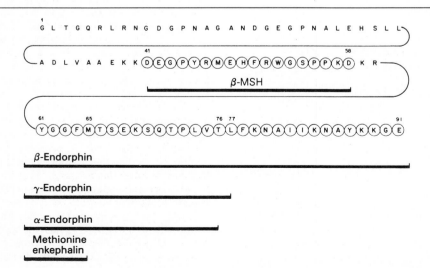

Figure 5–6. Primary structure of β-lipotropin. Residues 41–58 are melanocyte-stimulating hormone (β-MSH). Residues 61–91 contain the primary structures of the indicated endorphins.

Table 5–1. pK values for glycine and glycine peptides.

	pK (COOH)	pK (NH$_3^+$)
Gly	2.34	9.60
Gly-Gly	3.12	8.17
Gly-Gly-Gly	3.26	7.91

liquid chromatography on nonpolar materials with elution by polar solvents (reversed-phase HPLC). Fig 5–7 illustrates resolution of cyanogen bromide fragments of human fetal globin by this technique. Gel filtration and reversed-phase HPLC are used in conjunction to purify complex mixtures of peptides that result from partial digestion of proteins.

High-Voltage Electrophoresis (HVE) on Molecular Sieves

Molecular sieving may be used in conjunction with charge separation to facilitate separation. While starch and agarose are used, most commonly the support is a cross linked polymer of acrylamide (CH$_2$ = CH·CONH$_2$). For **polyacrylamide gel electrophoresis (PAGE),** protein solutions are applied to buffered tubes or slabs of polyacrylamide cross linked 2–10% by inclusion of methylene bisacrylamide (bis) (CH$_2$ = CONH)$_2$—CH$_2$ or similar cross linking reagents. Direct current is then applied. Visualization is by staining with Coomassie blue dye or Ag$^+$ (polypeptides), ethidium bromide (polynucleotides), etc. A popular variant is PAGE under denaturing conditions. Proteins are boiled and subsequently electrophoresed in the presence of the denaturing agents, urea or sodium dodecyl sulfate (SDS). This negatively charged molecule (CH$_3$—(CH$_2$)$_{11}$—SO$_3^{2-}$) coats proteins in the approximate proportion of one SDS per 2 peptide bonds. This "swamps" the native charge of the protein, making it strongly negative. Subsequent separations thus are based strictly on molecular size. SDS-PAGE is widely used to establish subunit molecular weights of proteins by comparison of mobilities with those of standards of known molecular weight.

THE FIRST STEP IN DETERMINING THE STRUCTURE OF A PEPTIDE IS TO OBTAIN ITS AMINO ACID COMPOSITION

The peptide bonds linking the amino acids are first broken by hydrolysis. Since peptide bonds are stable at neutral pH, acid or base is employed. Enzymatic catalysis is relatively unsuitable for complete hydrolysis. No procedure completely hydrolyzes proteins without partial loss of certain amino acid residues. The method of choice is **hydrolysis** in 6N HCl at 110 °C in a sealed, evacuated tube. Under these conditions, all of the tryptophan and cysteine and most of the cystine are destroyed. If metals are present, methionine and tyrosine are partially lost. Glutamine and asparagine are quantitatively deamidated to glutamate and aspartate. Recovery of serine and threonine is incomplete and decreases with increasing time of hydrolysis. Finally, certain bonds between neutral residues (Val-Val, Ile-Ile, Val-Ile, Ile-Val) are only 50% hydrolyzed after 20 hours. Typically, replicate samples are hydrolyzed for 24, 48, 72, and 96 hours. Serine and threonine data are then plotted on semilog paper and extrapolated back to zero time of hydrolysis. Valine and isoleucine are estimated from 96-hour data. Dicarboxylic acids and their amides are determined together and are reported collectively as "Glx" or "Asx." Cysteine and cystine are converted to an acid-stable derivative (eg, cysteic acid) prior to hydrolysis. **Base-catalyzed hydrolysis,** which destroys serine, threonine, arginine, and cysteine and racemizes all amino acids, is employed to analyze for tryptophan. Following hydrolysis, amino acid composition may be determined by **automated ion exchange chromatography** (see Fig 4–11) **or by HPLC.**

The First Protein Was Sequenced by Fred Sanger Using the Reagent That Bears His Name

Over 3 decades have elapsed since Sanger determined the complete primary structure of the polypeptide hormone insulin. Sanger's approach was first to separate the 2 polypeptide chains, A and B, of insulin and then to convert them by specific enzymatic cleavage into smaller peptides that contained regions of overlapping sequence. Using 1-fluoro-2,4-dinitrobenzene (Fig 5–8), he then removed and identified, one at a time, the N-terminal amino acid residues of these peptides. By comparing the sequences of overlapping peptides, he was able to deduce an unambiguous primary structure for both the A and the B

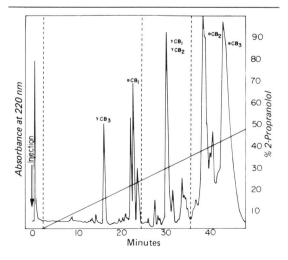

Figure 5–7. Reversed-phase HPLC elution profile for cyanogen bromide (CB) fragments of human fetal globin. Designations are for arbitrarily numbered fragments from the α and γ chains. (Courtesy of JD Pearson et al, Department of Biochemistry, Purdue University.)

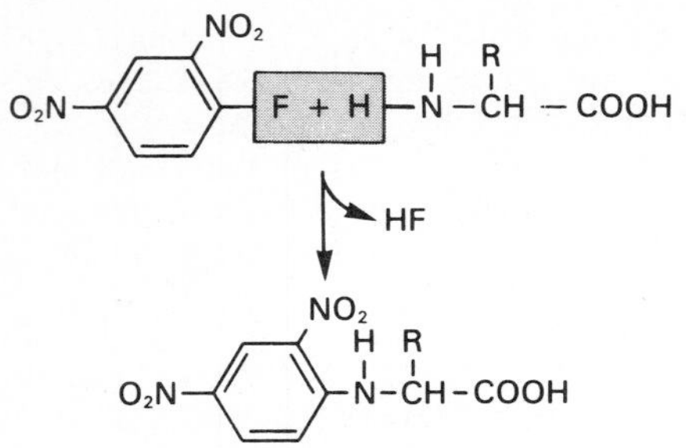

Figure 5–8. Reaction of an amino acid with 1-fluoro-2,4-dinitrobenzene (Sanger's reagent). The reagent is named for the Nobel laureate (1958) biochemist Frederick Sanger, who used it to determine the primary structure of insulin. This quantitatively arylates all free amino groups, producing intensely yellow 2,4-dinitrophenyl amino acids. These derivatives are readily quantitated by spectrophotometry. In addition to the N-terminal residue, the ε-amino groups of lysine, the imidazole of histidine, the OH of tyrosine, and the SH of cysteine also react with fluorodinitrobenzene. Since the dinitrophenyl group is resistant to removal by acid hydrolysis, it was used to determine the N-terminal amino acid of polypeptides.

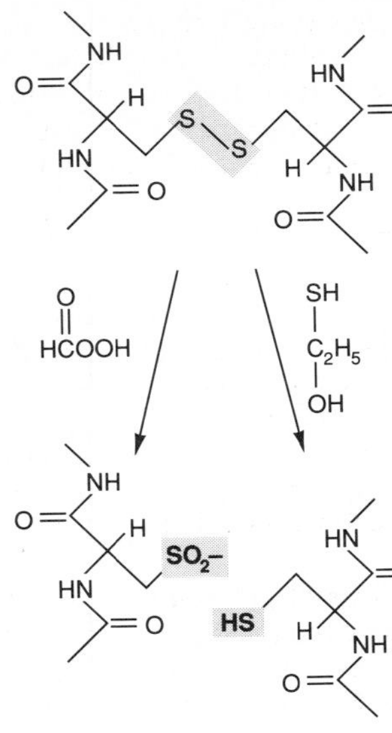

Figure 5–9. Oxidative cleavage of adjacent polypeptide chains linked by disulfide bonds (shaded) by performic acid (left) or reductive cleavage by β-mercaptoethanol (right) forms 2 peptides that incorporate cysteic acid residues or cysteinyl residues, respectively.

chains. While the overall strategy of Sanger's approach remains valid, 2 techniques have revolutionized determination of the primary structures of polypeptides (proteins). The first was the introduction, in 1967, of a procedure for the **automated sequential removal and identification of N-terminal amino acid residues as their phenylthiohydantoin derivatives.** The second was the independent introduction by Sanger and by Maxam and Gilbert of techniques for **rapid sequencing of the DNA of the gene that codes for the protein in question.** At present, optimal strategy is to utilize both approaches simultaneously. The automated Edman technique, while rapid compared to the methods used by Sanger, is slow relative to DNA sequencing methods, and difficulties may be encountered. On the other hand, DNA sequencing techniques do not infallibly yield unambiguous primary structures for the protein of interest. A major complication in sequencing eukaryotic genes is the presence of introns within the gene that are not expressed in the mature protein. A major advantage of DNA sequencing, however, is the relative ease of detection and sequencing of regions of precursor molecules that may have eluded detection by the automated Edman technique owing to maturation of the protein prior to its isolation. DNA sequencing and the automated Edman technique therefore are complementary techniques that have revolutionized and vastly expanded our knowledge of the primary structures of proteins.

The Primary Structures of Polypeptides Are Determined by the Automated Edman Technique

Since many proteins consist of more than one polypeptide chain associated by noncovalent forces or disulfide bridges, the first step may be to dissociate and

separate individual polypeptide chains. Denaturing agents (urea, guanidine hydrochloride) disrupt hydrogen bonds and dissociate noncovalently associated polypeptides. Oxidizing and reducing agents disrupt disulfide bridges (Fig 5–9). Polypeptides are then separated by chromatography.

Large Polypeptides Must First Be Cleaved Into Peptides of a Size Suitable for Automated Sequencing

Automated sequencing instruments (sequenators) operate most efficiently on polypeptides 20–60 residues long. This has therefore influenced the techniques used to cleave polypeptides and to purify the resulting fragments. Emphasis has shifted to production of small numbers of large (30- to 100-residue) fragments. Highly specific and complete cleavage at a restricted number of sites is therefore desired. Cleavage with cyanogen bromide (CNBr), trypsin, or o-iodosobenzene meets these requirements.

CNBr: Cysteine residues are first modified with iodoacetic acid. CNBr then cleaves Met-peptide bonds specifically and, in most instances, quantitatively, on the COOH side of Met. Since Met is comparatively rare in polypeptides, this usually generates peptide fragments of the desired size range.

Trypsin: Trypsin cleaves on the COOH side of Lys and Arg residues. If Lys residues are first derivatized with citraconic anhydride (a reversible reaction) to change the charge on Lys residues from positive to negative, trypsin now cleaves only after Arg. Derivatization of Arg residues is less useful because of the relative abundance of Lys residues. It is, however, useful for subsequent cleavage of CNBr fragments.

o-Iodosobenzene: o-Iodosobenzene cleaves specifically and quantitatively at the comparatively rare Trp-X residues. It requires no prior protection of other residues.

Hydroxylamine: Hydroxylamine cleaves comparatively rare Asn-Gly bonds, although generally not in quantitative yield.

Protease V8: *Staphylococcus aureus* protease V8 cleaves Glu-X peptide residues with a preference for situations where X is hydrophobic. Glu-Lys resists cleavage. This reaction is useful for subsequent degradation of CNBr fragments.

Mild Acid Hydrolysis: This cleaves the rare Asp-Pro bond.

Two or 3 digests of the original polypeptide, normally at Met, Trp, Arg, and Asn-Gly, combined with appropriate subdigests of the resulting fragments, usually permit determination of the entire primary structure of the polypeptide. Barring unusual difficulties in purifying fragments, this can be accomplished with a few micromoles of polypeptide.

The Complex Mixture of Peptides Produced by Polypeptide Cleavage Must Be Separated Into Homogeneous Peptides, Prior to Sequencing

Fragment purification is achieved chiefly by gel filtration in acetic or formic acid by reversed phase HPLC (Fig 5–7), or by ion exchange chromatography on phosphocellulose or sulfophenyl Sephadex in solutions of phosphoric acid.

Sequencing Employs the Edman Reagent & the Edman Reaction

Automated sequencing employs phenylisothiocyanate (Edman reagent) and a sequence of reactions that result in the removal of the N-terminal residue as its phenylthiohydantoin derivative (the Edman reaction, Fig 5–10). The heart of the instrument is a spinning cup reaction chamber in which reactions occur in a thin film of solution on the cup wall. This facilitates extractions and subsequent removal of solvents. Fully automated apparatus can sequence up to 30–40 residues (or, in exceptional cases, up to 60 or even 80 residues) in one continuous operation. The apparatus is programmed to perform sequential Edman degradations on the N-terminal residue of a polypeptide. After the initial N-terminal amino acid has been removed, separated, and identified, an Edman derivative (Fig 5–10) of the next one in the sequence is formed, etc. The phenylthiohydantoin derivatives are separated by HPLC and identified by their order and position of elution.

Deduction of the Complete Primary Structure Requires Sequencing Overlapping Peptides

To have sequenced all the CNBr peptides from a protein does not, by itself, provide its complete pri-

Phenylisothiocyanate (Edman reagent) and a peptide

A phenylthiohydantoic acid

H^+, nitro-methane → H_2O

A phenylthiohydantoin and a peptide shorter by one residue

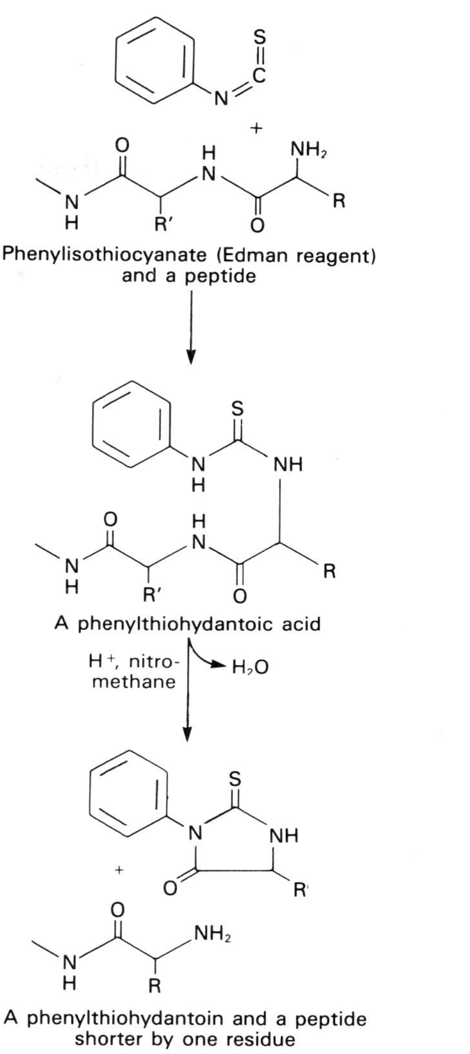

Figure 5–10. Phenylisothiocyanate converts an amino acid (or of the N-terminal residue of a polypeptide) to a phenylthiohydantoin. Phenylisothiocyanate reacts with the amino groups of amino acids and peptides, yielding phenylthiohydantoic acids. On treatment with acid in nonhydroxylic solvents, these cyclize to phenylthiohydantoins. The principal use of this reaction, which identifies the N-terminal residues of a peptide, is in automated sequencing of polypeptides.

mary structure since the order in which these peptides occur in the protein is not known. To deduce an unambiguous primary structure, **additional peptides whose N- and C-termini overlap those of the CNBr peptides** must also be prepared and sequenced. These additional peptides are produced by techniques (eg, digestion with chymotrypsin) that cleave the protein at places other than Met residues. Deduction of an unambiguous primary structure by comparison of peptide sequences then involves a process analogous to the assembly of a jigsaw puzzle (Fig 5–11).

To locate disulfide bonds, peptides from untreated and from reduced or oxidized protein are separated by 2-dimensional chromatography or by electrophoresis and chromatography (fingerprinting). Vizualization with ninhydrin reveals **2 fewer peptides** in the digest from untreated protein and **one new peptide** in the digest from treated protein. With knowledge of the primary structure of these peptides, the positions of disulfide bonds can then be inferred.

PEPTIDES ARE SYNTHESIZED BY AUTOMATED TECHNIQUES

Fig 5–12, which illustrates synthesis of a representative dipeptide A-B by the Merrifield solid-phase technique, summarizes all the reactions required to synthesize a peptide of any desired length. These steps in the procedure are:

1. Block the N-termini of amino acid A (open symbol) and amino acid B (shaded symbol) with the *t*-butyloxycarbonyl [*t*-BOC] group (▪):

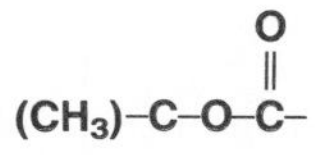

$$\text{(CH}_3)\text{–C–O–}\overset{\overset{\displaystyle O}{\|}}{\text{C}}\text{–}$$

forming *t*-BOC-A and *t*-BOC-B.

2. Activate the carboxyl group of *t*-BOC-B with dicyclohexyl carbodiimide (DCC) (▶):

$$C_6H_6\text{–N}=\text{C}=\text{N–}C_6H_6$$

3. React the carboxyl group of amino acid A (which will become the C-terminal residue of the peptide) with an activated, insoluble polystyrene resin (◎).

4. Remove the blocking group from *t*-BOC-A with room temperature trifluoroacetic acid (TFA, F_3C-COOH).

[**Note:** In practice, steps 3 and 4 may be omitted since resins with any given *t*-BOC-amino acid connected via an ester bond to a phenylacetamidomethyl (PAM) "linker" molecule attached to the polystyrene resin are commercially available.]

5. Condense the activated carboxyl group of *t*-BOC-B with the free amino group of immobilized A.

6. Remove the *t*-BOC blocking group with TFA (see step 4).

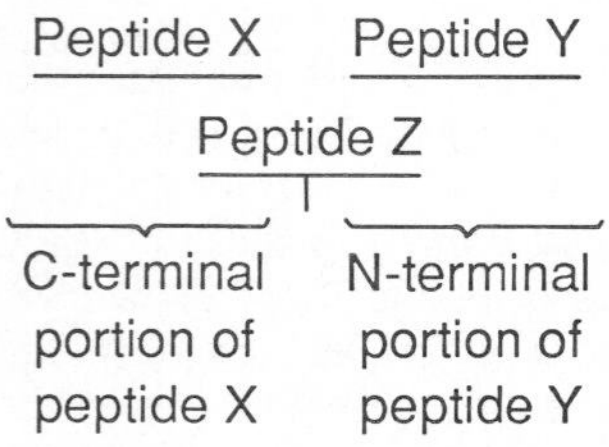

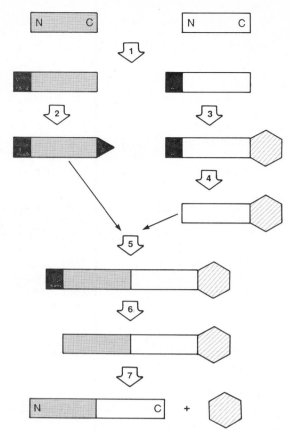

Figure 5–12. Symbolic representation of the synthesis of a generic dipeptide by the solid-phase synthesis technique pioneered by Merrifield. See accompanying text for explanations of symbols.

7. Liberate the dipeptide A-B from the resin particle by treating at −2 °C with HF in dichloromethane.

The initial achievements of the Merrifield technique were the synthesis of the A chain (21 residues) and B chain (30 residues) of insulin in 11 days and of the enzyme pancreatic ribonuclease in 18% overall yield. Subsequent improvements have reduced the time for synthesis of a peptide bond to about 1 hour and have increased yields significantly. This has initiated new prospects, not only for confirming *de novo* synthesis of the primary structures of proteins, but for immunology, for producing vaccines and polypeptide hormones, and conceivably also for treating selected inborn errors of metabolism.

Peptide X	Peptide Y
Peptide Z	
C-terminal portion of peptide X	N-terminal portion of peptide Y

Figure 5–11. The overlapping peptide Z is used to deduce that peptides X and Y are present in the original protein in the order X → Y, not Y → X.

REFERENCES

Allen G: Sequencing of proteins and Peptides. North Holland, 1981.

Bhwon AS: Protein/Peptide Sequence Analysis: Current Methodologies. CRC Press, 1988.

Cantor CR, Schimmel PR: Biophysical Chemistry, Part I: The Conformation of Macromolecules. Freeman, 1980.

Dayhoff M (editor): Atlas of Protein Sequence and Structure.

Vol 5. National Biomedical Research Foundation, Washington, DC.

Doolittle RF: Of urfs and orfs: a Primer on How to Analyze Derived Amino Acid Sequences. University Science Books, 1987.

Hewick RM, Hunkapiller MW, Hood LE, Dryer WJ: A gas-liquid solid phase peptide and protein seqenator. *J Biol Chem* 1981;**256:**7990.

Hunkapiller MW, Hood LE: Protein sequence analysis: automated microsequencing. *Science* 1983;**219:**650.

Hunkapiller MW, Strickler JE, Wilson KJ: *Science* 1984; **226:**304.

Lipman DJ, Pearson WR: Similar amino acid sequences: chance or common ancestry? *J Mol Biol* 1981;**16:**9.

Mahoney WC, Smith PK, Hermodson MA: Fragmentation of proteins with *o*-iodosobenzoic acid: Chemical mechanism and identification of *o*-iodosobenzoic acid as a reactive contaminant that modifies tyrosyl residues. *Biochemistry* 1981;**20:**443; *Methods Enzymol.* Vols. 47,48,49,91 (published continuously).

Merrifield RB: Solid phase synthesis. *Science* 1986;**232:** 341.

Pearson JD et al: Reversed-phase supports for the resolution of large denatured protein fragments. *J Chromatogr* 1981; **207:**325.

Regnier FE, Gooding KM: High performance liquid chromatography of proteins. *Anal Biochem* 1980;**103:**1.

6

Proteins: Structure & Properties

Victor W. Rodwell, PhD

INTRODUCTION

Proteins are high-molecular weight polypeptides. The dividing line between large polypeptides and small proteins is customarily drawn between MW 8000 and 10,000. **Simple proteins** contain only amino acids. **Complex proteins** contain additional, non-amino acid materials, such as heme, vitamin derivatives, lipid, or carbohydrate. This chapter considers simple proteins. Specific complex proteins such as heme proteins, glycoproteins, and lipoproteins are considered elsewhere, as are the properties of simple proteins of highly individual structure such as collagen and contractile proteins.

BIOMEDICAL IMPORTANCE

Proteins play a central role in cell function and cell structure. Analyses of certain proteins and enzymes of the blood are widely used for diagnostic purposes. For example, electrophoretic analysis of the plasma albumin:globulin ratio is an integral part of the diagnostic workup for liver disease. Analyses of plasma lipoproteins and of plasma immunoglobulins by electrophoresis and other methods are commonly employed to diagnose specific types of hyperlipoproteinemias and specific types of immune disorders, respectively. Normal human urine is usually free of protein, so detection of significant proteinuria is generally an important indicator of renal diseases, such as the various forms of nephritis.

PROTEINS ARE CLASSIFIED IN WAYS THAT ILLUSTRATE DIFFERENT FEATURES OF THEIR STRUCTURAL FUNCTIONS

Since no universally satisfactory system of protein classification exists, several mutually contradictory protein classification systems persist in current use. All are of comparatively limited value in assisting our understanding of many key properties of proteins. The persistence of these systems and terms—particularly in the clinical laboratory—dictates, however, brief consideration. Discussed below are salient features of protein classification systems based on solubility, shape, function, physical properties, and 3-dimensional structure.

Classification by Solubility

A classification system based on solubility developed in 1907–1908 is still in limited use today, particularly in clinical biochemistry (Table 6–1). The lines of demarcation between the classes are not stringent. For example, a clear distinction between albumins and globulins cannot be made solely on the basis of their solubilities in water or salt solutions.

Classification by Overall Shape

Two broad classes of proteins may be distinguished by their **axial ratios** (ratios of length to breadth). **Globular proteins,** which have axial ratios less than 10 and generally not over 3–4, have compactly folded and coiled polypeptide chains. Examples include insulin, plasma albumins and globulins, and many enzymes. **Fibrous proteins,** which have axial ratios greater than 10, have groups of polypeptide chains coiled in a spiral or helix and crosslinked covalently or by hydrogen bonds. Examples include keratin, myosin, collagen, and fibrin.

Classification by Function

Proteins may be classified according to their biologic functions—for example, as structural, catalytic, or transport proteins (Table 6–2). Catalytic proteins (enzymes), which comprise the majority of protein types, are themselves classified by the type of reaction they catalyze.

Classification by Physical Properties

For certain proteins of medical interest, specialized systems of classification distinguish between closely related proteins. For example, 2 systems of nomenclature for plasma lipoproteins are in wide use, and a third is under consideration. One distinguishes lipoproteins as "origin," α_1-, α_2-, β-, and γ-lipoproteins, on the basis of their electrophoretic mobility at pH 8.6. Lipoproteins are also classified on the basis of their sedimentation behavior as chylomicrons, VLDL, LDL, HDL, and VHDL. In addition, 6 broad classes of plasma lipoproteins can be differentiated based on whether apoprotein A, B, C, D, E, or F is present. The apoproteins may be differentiated by immunologic criteria.

Table 6–1. Classification of proteins based on their solubilities.

Albumins	Soluble in water and salt solutions. No distinctive amino acids.
Globulins	Sparingly soluble in water but soluble in salt solutions. No distinctive amino acids.
Prolamines	Soluble in 70–80% ethanol but insoluble in water and absolute ethanol. Arginine-rich.
Histones	Soluble in salt solutions.
Scleroproteins	Insoluble in water or salt solutions. Rich in Gly, Ala, Pro.

Classification by 3-Dimensional Structure

Proteins may be distinguished by whether or not they possess quaternary structure (see below). In addition, similarities in structure, revealed primarily by x-ray crystallography, provide a potentially valuable basis for protein classification. For instance, proteins that bind nucleotides share a "nucleotide-binding domain" of tertiary structure, and these proteins may be evolutionarily related. As more protein crystal structures are solved, additional common domain structures become recognizable.

THE PRIMARY STRUCTURE OF PROTEINS DERIVES FROM COVALENT LINKAGE OF L-α-AMINO ACIDS BY α-PEPTIDE BONDS

While the central role of peptide bonds in proteins was deduced long ago by multiple lines of evidence, the most convincing proof that only peptide bonds were needed was the total chemical synthesis of insulin and ribonuclease solely by linking amino acids via peptide bonds.

The Bond That Connects the Carbonyl Carbon and Nitrogen Atoms of a Peptide Bond Has Partial Double-Bond Character

While peptides are written with a single bond connecting α-carboxyl and α-nitrogen atoms, the carbon-

Table 6–2. Principal functions of the proteins.

Function	Protein (Examples)
Catalytic role	Enzymes
Contraction	Actin, myosin
Gene regulation	Histones, repressor proteins
Hormonal role	Insulin
Protection	Fibrin, immunoglobulins, interferon
Regulatory role	Calmodulin
Structural role	Collagen, elastin, keratins
Transport	Albumin (of bilirubin, fatty acids, etc), hemoglobin (oxygen), lipoproteins (various lipids), transferrin (iron)

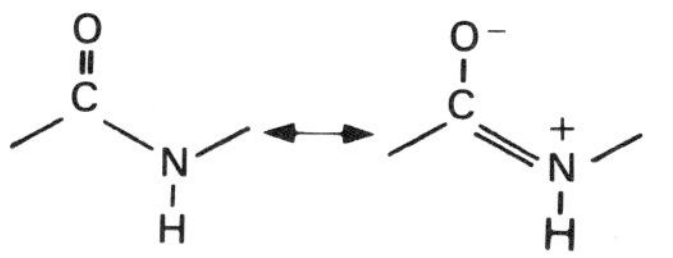

Figure 6–1. Resonance stabilization of the peptide bond confers partial double-bond character, and hence rigidity, on the C—N bond.

nitrogen bond in fact has partial double-bond character (Fig 6–1). There is no freedom of rotation about the bond that connects the C and N atoms, and all 4 of the atoms shown in Fig 6–1 lie in the same plane (ie, are coplanar). There is, by contrast, ample freedom of rotation about the remaining bonds of the polypeptide backbone. These concepts are summarized in Fig 6–2, where the bonds having free rotation are circled by arrows and the coplanar atoms are in shaded boxes. This semirigidity has important consequences for orders of protein structure above the primary level.

In addition to the peptide bonds that form the "backbone" of a polypeptide, additional covalent and noncovalent bonds contribute stability to polypeptides.

Disulfide Bonds Formed From Cysteine Residues Form Covalent Bonds Within and Between Polypeptides of Certain Proteins

The disulfide bond formed between 2 cysteine residues links 2 portions of polypeptide chains through a cystine residue. The cystine bond is resistant to condi-

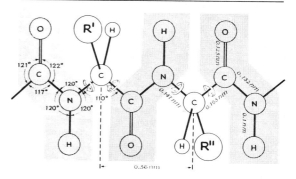

Figure 6–2. Dimensions of a fully extended polypeptide chain. The 4 atoms in shaded boxes, which are coplanar, comprise the polypeptide bond. The unshaded atoms are the α-carbon atom, the α-hydrogen atom, and the α-R group of the particular amino acid. Free rotation can occur about the bonds connecting the α-carbon with the α-nitrogen and α-carbonyl functions (white arrows). The extended polypeptide chain is thus a semirigid structure with two-thirds of the atoms of the backbone held in a fixed planar relationship one to another. The distance between adjacent α-carbon atoms is 0.36 nm. The interatomic distances and bond angles, which are not equivalent, are also shown. (Redrawn and reproduced, with permission, from Pauling L. Corey LP, Branson HR: *Proc Natl Acad Sci USA* 1951;**37**:205.)

tions usual for protein denaturation. Performic acid (oxidizes S–S bonds) or β-mercaptoethanol (reduces the S–S bonds, generating 2 cysteine residues) separates polypeptide chains linked by disulfide bonds without affecting primary structure (see Fig 5–9).

NON-COVALENT BONDS ALSO STABILIZE PROTEINS

Three major types of non-covalent bonds make major contributions to stabilization of protein structures.

Multiple Hydrogen Bonds Stabilize the Overall Structure of Proteins

Hydrogen bonds formed between bonding residues in the side chains of peptide-linked amino acids, between the hydrogen and oxygen atoms of the peptide bonds themselves, and between polar residues on the surfaces of proteins and water all play important roles in the maintenance of protein structure above the primary order (see α-helix and β-pleated sheet, below).

Hydrophobic Interactions Contribute Stability to the Interiors of Proteins

The nonpolar side chains of neutral amino acids tend to associate in proteins. Since the relationship is not stoichiometric, no true bond exists. Nonetheless, their large number dictates that these interactions play a significant role in maintaining protein structure.

Electrostatic Bonds Link Surface Residues in Proteins

Salt, or electrostatic, bonds are formed between oppositely charged groups in the side chains of amino acids or between N-terminal and C-terminal residues and other oppositely charged groups. For example, the ε-amino group of lysine bears a net charge of +1 at physiologic pH and the non-α-carboxyl of aspartate and glutamate a net charge of −1. These may therefore interact electrostatically to stabilize proteins.

Protein Denaturants Rupture Noncovalent Bonds in Proteins With Loss of Biologic Activity

Denaturation of proteins by reagents such as urea, sodium dodecyl sulfate (SDS), mild H^+, or OH^- ruptures hydrogen, hydrophobic, and electrostatic bonds, but not peptide or disulfide bonds.

THE α-HELIX IS ONE STRUCTURAL THEME FOR MAINTAINING ORDERED CONFORMATIONS IN PROTEINS

The discovery that polypeptide chains have highly ordered conformations maintained by hydrogen bonds was a major conceptual advance. Although the existence of these highly ordered structures was subsequently confirmed by x-ray crystallography, their existence was first proposed from theoretical considerations.

X-Ray data indicated that hair and wool α-keratins had repeating units spaced 0.5–0.55 nm along their longitudinal axis. However, no dimension of the extended polypeptide chain measures 0.5–0.55 nm (Fig 6–2). This apparent anomaly was resolved by Pauling and Corey, who proposed that the polypeptide chain of α-keratin was arranged as an α-helix (Fig 6–3). In this structure, the R groups protrude outward from the center of the helix (Fig 6–4). There are 3.6 amino acid residues per turn of the helix, and the distance traveled per turn is 0.54 nm. The spacing per amino acid residue is 0.15 nm. Features of the α-helix (Fig 6–5) are:

(1) The α-helix is stabilized by interresidue hydrogen bonds formed between the H atom of a peptide N and the carbonyl O of the residue fourth in line behind in the primary structure.

(2) Each peptide bond participates in the H bonding. This confers maximum stability.

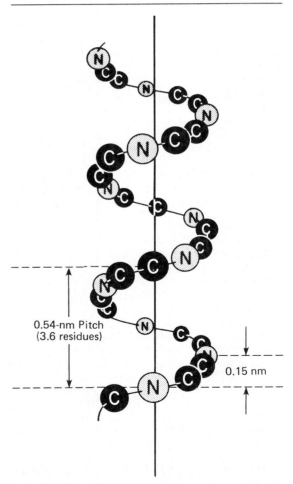

Figure 6–3. Representation of the helical orientation of the main chain of a peptide about the axis of an α-helix.

0.54-nm Pitch (3.6 residues)

0.15 nm

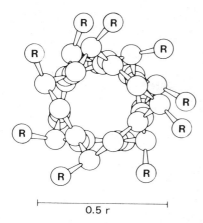

0.5 r

Figure 6–4. View down the axis of an α-helix. The side chains (R) are on the outside of the helix. The van der Waals radii of the atoms are larger than shown here; hence, there is almost no free space inside the helix. (Slightly modified and reproduced, with permission, from Stryer L: *Biochemistry,* 2nd ed. Freeman, 1981. Copyright © 1981 by W.H. Freeman and Co.)

(3) All of the main chain peptide N and carbonyl O residues are hydrogen bonded, thus greatly reducing the hydrophilic (increasing the hydrophobic) nature of the α-helical region.

(4) An α-helix forms spontaneously, since it is the lowest energy, most stable conformation for a polypeptide chain.

(5) When the residues are L-amino acids, the right-handed helix that occurs in proteins is significantly more stable than the left-handed helix.

(6) Amino acids that disrupt the α-helix include proline (the N-atom is part of a rigid ring and no rotation of the N–C bond can occur) and amino acids with charged or bulky R groups that electrostatically or physically interfere with helix formation (Table 6–3).

PARALLEL AND ANTIPARALLEL PLEATED SHEETS CONSTITUTE A SECOND TYPE OF RECOGNIZABLY ORDERED STRUCTURE

Pauling and Corey also proposed a second ordered structure, the β-pleated sheet (β because it was their second structure, the α-helix being the first). Whereas in the α-helix the polypeptide chain is condensed, in the β-pleated sheet it is almost fully extended (Fig 6–6). When the adjacent polypeptide chains run in opposite directions (N to C terminus), the structure is termed an **antiparallel** β-pleated sheet (Fig 6–6). When the chains run in the same direction, it is termed **parallel** (not shown).

Both parallel and antiparallel regions of β-pleated structure are present in many proteins. From 2 to 5 adjacent strands of polypeptide may combine to form these structures. Fig 6–7 illustrates a region of

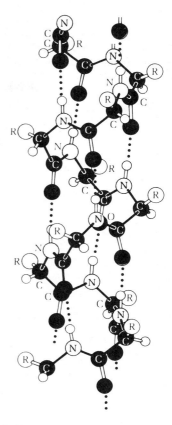

Figure 6–5. The hydrogen bonds (dots) formed between H and O atoms stabilize a polypeptide in an α-helical conformation. (Reprinted with permission, from Haggis GH et al. *Introduction to Molecular Biology.* Wiley, 1964.)

ribonuclease in which 3 sections of polypeptide chain form a β-pleated sheet structure.

While the α-helix is stabilized by hydrogen bonding between peptide bonds 4 residues apart in a primary structural sense, the β-pleated sheet is stabilized by hydrogen bonds between peptides far removed from one another in a primary structural sense (Fig 6–7).

Table 6–3. Effect of various amino acid residues on α-helix formation.

Promote α-Helix	Destabilize α-Helix	Terminate α-Helix
Ala	Arg	Pro
Asn	Asp	Hyp
Cys	Glu	
Gln	Gly	
His	Lys	
Leu	Ile	
Met	Ser	
Phe	Thr	
Trp		
Tyr		
Val		

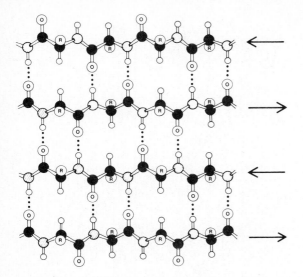

Figure 6–6. In an antiparallel β-pleated sheet, adjacent strands run in opposite directions. Hydrogen bonds between NH and CO groups of adjacent strands stabilize the structure. The side chains (R) are above and below the plane of the sheet. ● Carbon atoms; Ⓝ nitrogen atoms; ○ hydrogen atoms. (Modified and reproduced, with permission, from Stryer L: *Biochemistry,* 2nd ed. Freeman, 1981. Copyright © 1981 by W.H. Freeman and Co.)

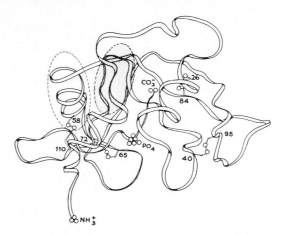

Figure 6–7. The main chain of bovine pancreatic ribonuclease consists of a single chain of 124 residues cross linked at 4 places by disulfide bridges. A region of α-helix is indicated by the dotted oval, and a region of pleated sheet is shaded. Other portions of the molecule are predominantly random coil. The active site is indicated by the phosphate ion (PO_4^{3-}). (Adapted from Kartha G. Bello J, Harker D: Tertiary structure of ribonuclease. *Nature* 1967; **213**:862.) The protein has been chemically synthesized in its entirety.

β-BENDS PERMIT POLYPEPTIDES TO FORM GLOBULAR MASSES

Since polypeptides typically form compact, globular masses, there must be ways to change the direction of a polypeptide chain. One way is to form a tight loop, a β-bend, in which the carbonyl oxygen of one residue hydrogen bonds to the amide hydrogen of the residue 3 residues ahead. Proline and glycine occur frequently in β-bends. Part of the geometry of the β-bend is preformed in proline, which is conformationally more restricted than other amino acids. By contrast, its small R group permits glycine to act as a flexible "hinge" between regions of polypeptide whose R groups would otherwise tend to favor a more extended conformation.

REGIONS OF PROTEINS NOT ORGANIZED AS HELICES, PLEATED SHEETS, OR β-BENDS ARE SAID TO BE IN RANDOM COIL CONFORMATION

Considerable portions of a protein may be present in a random-coil conformation (Fig 6–7). The term "random" is unfortunate, since it may imply less biologic significance than more highly repeating regions. For biologic function, regions of random coil are of equal importance to those of α-helix or β-pleated sheet.

ASPECTS OF PROTEIN STRUCTURE ARE CONSIDERED IN TERMS OF FOUR DIVISIONS, OR ORDERS

Primary Structure
Primary structure, as for peptides, refers to the order of the amino acids in the polypeptide chain or chains and the location of disulfide bonds, if these are present.

Secondary Structure
Secondary structures, the steric relationship of amino acids close together in a primary structural sense, may be regular (eg, α-helix, β-pleated sheet), or it may exhibit few regularities (eg, random coil).

Tertiary Structure
The overall arrangement and interrelationship of the various regions, or domains, and individual amino acid residues of a single polypeptide chain is the tertiary structure of the protein. While the division between secondary and tertiary structure is not clear-cut, tertiary structure considers the spatial relationship of amino acid residues that are, in general, far apart in a primary structural sense.

Quaternary Structure
Proteins possess quaternary structure if they consist of **2 or more polypeptide chains united by forces other than covalent bonds** (ie, not peptide or disulfide bonds). The forces that stabilize these aggregates are hydrogen bonds and electrostatic (salt) bonds formed between residues on the surfaces of the poly-

peptide chains. Such proteins are termed **oligomers,** and the individual polypeptide chains of which they are composed are variously termed **protomers, monomers,** or **subunits.**

Many oligomeric proteins contain 2 or 4 protomers and are termed dimers or tetramers, respectively. Oligomers containing more than 4 protomers are also common, particularly among regulated enzymes (eg, aspartate transcarbamoylase). Oligomeric proteins play special roles in intracellular regulation, because the protomers can assume different spatial orientations relative to one another with resulting changes in the properties of the oligomer.

Certain Proteins Form Macromolecular Complexes

Aggregation of different functional proteins—each of which alone has all 4 orders of structure—into multifunctional macromolecular complexes is encountered in electron transport, in fatty acid biosynthesis, and in pyruvate metabolism.

THE SECONDARY AND TERTIARY STRUCTURE OF A PROTEIN IS DETERMINED BY THE PRIMARY STRUCTURE OF THE POLYPEPTIDE CHAIN

Once the chain has been formed, the R groups direct the specific regional folding (secondary structure) and specific aggregation of the regions (tertiary structure). This is shown by a classic experiment. Treatment of ribonuclease with a mild reducing agent (β-mercaptoethanol) and a denaturing agent (urea or guanidine; see below) inactivates it as it assumes a random coil conformation. Slow removal of the denaturing agent and gentle reoxidation to re-form the S–S bonds lead to almost complete restoration of enzymatic activity. **It is therefore not necessary to postulate independent genetic control of orders of protein structure above the primary level, since the primary structure specifies the secondary, tertiary, and (when present) quaternary structure (ie, conformation) of a protein.** The native conformation of a protein such as ribonuclease appears to be that which is thermodynamically most stable for a given environment, eg, a hydrophilic versus hydrophobic one.

The structure of a protein may be modified during posttranslational processing, such as the conversion of a preproenzyme to the catalytically active form or removal of the "leader peptide" that directs exported proteins through membranes.

The Comparatively Weak Forces Responsible for Maintaining Secondary, Tertiary, and Quaternary Structure of Proteins Are Readily Disrupted With Resulting Loss of Biologic Activity

Disruption of native structure is termed denaturation. Physically, denaturation may be viewed as ran-

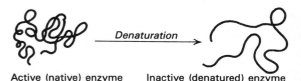

Figure 6–8. Representation of denaturation of a protomer.

domizing the conformation of a polypeptide chain without affecting its primary structure. For a protomer, the process may be represented as shown in Fig 6–8. For an oligomeric protein, denaturation may involve dissociation of the protomers with or without accompanying changes in protomer conformation.

The biologic activity of most proteins is destroyed by strong acids or bases, heat, ionic detergents (amphipaths), chaotropic agents (urea, guanidine), heavy metals (Ag, Pb, Hg), or organic solvents. Denatured proteins generally are less soluble and they precipitate. This is used to advantage in the clinical laboratory. Blood or serum to be analyzed for small molecules (eg, glucose, uric acid, drugs) generally is first treated with trichloroacetic, phosphotungstic, or phosphomolybdic acid to precipitate proteins. These are removed by centrifugation, and the protein-free supernatant liquid is then analyzed.

The heat, acid, and protease lability of most enzymes provides a preliminary test to determine whether a reaction is enzyme catalyzed. If a cell extract having catalytic activity loses this activity when boiled, acidified and reneutralized, or treated with a protease, the catalyst probably was an enzyme.

Frequently, the denaturation of an enzyme is influenced by the presence of its substrate. If a conformational change occurs when substrate is bound, the new conformation may be either more stable or less stable than before.

THE PRIMARY STRUCTURES OF PROTEINS ARE DETERMINED BY THE METHODS USED TO SEQUENCE PEPTIDES

Complex proteins are first treated to remove prosthetic groups (eg, heme), and disulfide bonds are oxidized to yield linear polypeptides (see Fig 5–9). The methods used to sequence these polypeptides are discussed in Chapter 5. While most proteins contain only the amino acids listed in Table 4–3, derivatives of these amino acids also occur in certain proteins (Tables 6–4 and 6–5). While discussion of the methods used to identify these amino acid derivatives lies beyond the scope of this chapter, their presence can complicate the determination of primary structure.

The Secondary & Tertiary Structure of Proteins Is Revealed by x-Ray Crystallography

Techniques formerly used to infer the presence of helical structures in proteins (eg, optical rotatory dis-

Table 6–4. Modified α-COOH and α-NH$_2$ groups of proteins.*

α-COOH	α-NH$_2$		
Amide	N-Formyl	N-Acetyl	N-Methyl
Asp Glu Gly His Met Phe Pro Tyr Val	Gly Met	Ala Asp Gly Met Ser Thr Val	Ala Asp Gly Met Ser Thr

*Modified and reproduced, with permission, from Uy R, Wold F: Posttranslational covalent modification of proteins. *Science* 1977;**198**:890. Copyright © 1977 by the American Association for the Advancement of Science.

Table 6–5. Modified non-α functional groups of proteins.*

−OH	Non-α-N		
PO$_3$H$_2$	N-Methyl	N-Dimethyl	N-Trimethyl
Ser Thr Tyr	Arg His Lys	Arg Lys	Lys

*Modified and reproduced, with permission, from Uy R, Wold F: Posttranslational covalent modification of proteins. *Science* 1977;**198**:890. Copyright © 1977 by the American Association for the Advancement of Science.

persion, tritium exchange of labile protons) have been largely supplanted by x-ray crystallography. x-Rays are directed at a crystal of protein and generally also at one of its derivatives that contains an added heavy metal ion. The rays are scattered in a pattern that depends upon the electron densities in different parts of the protein. Images, collected on a photographic plate, are translated into electron density maps which, when superimposed one on another, permit the crystallographer to construct a faithful model of the protein in question. Although time consuming, expensive, and requiring highly specialized training, x-ray crystallography reveals detailed, precise views of the orientations of all the amino acids in many proteins. Its contributions to our present-day concepts of protein structure can hardly be overemphasized.

An emerging technique, magnetic resonance imaging (MRI) has recently been applied to determining the 3-dimensional structure of a small protein. MRI appears to hold promise for extension to larger proteins.

Physical Methods are Used to Study the Quaternary Structure of Proteins

Determining the quaternary structure of oligomeric proteins encompasses determining the number and kind of protomers present, their mutual orientation, and the interactions that unite them. Providing that oligomers do not undergo denaturation during the procedure used to determine molecular weight, many methods can yield molecular weight data. These same techniques may be used to determine protomer molecular weight if the oligomer is first denatured.

Ultracentrifugation, Gel Filtration, & Gel Electrophoresis Reveal the Molecular Weights of Oligomeric Proteins

Developed by Svedberg, ultracentrifugation measures sedimentation rate in an ultracentrifugal field of around $10^5 \times g$. It has tended in recent years to be replaced by less complex techniques.

In a related technique, sucrose density gradient centrifugation, protein standards and unknowns are

Table 6–6. Quaternary structures of selected enzymes.*

Enzyme (Oligomer)	Number of Protomers	Molecular Weight of Protomer
Chicken heart aspartate transaminase	2	50,000
Pigeon liver fatty acid synthase	2	230,000
Rabbit liver fructose diphosphatase	2†	29,000
	2†	37,000
Rat liver ornithine transaminase	4	33,000
Pig heart propionyl-CoA carboxylase	4	175,000
Beef heart, liver, or muscle LDH	4†	35,000
Beef heart mitochondrial ATPase	10	26,000
Escherichia coli glutamine synthase	12	48,500
Chicken liver acetyl-CoA carboxylase	2†	4,100,000
	10†	409,000

*Adapted from Klotz IM, Langerman, NR, Darnall DW: Quaternary structure of enzymes. *Annu Rev Biochem* 1970;**39**:25.
†Nonidentical subunits.

layered over a 5–20% sucrose gradient in a plastic tube. The gradient may be prepared by successive freezing and thawing of 20% sucrose. Following centrifugation overnight at around $10^5 \times g$, a small hole is punched in the bottom of the tube, the contents are collected in a set of small tubes, the relative positions of the proteins in the gradient are determined, and their mobilities are computed.

The Molecular Weight of a Globular Protein May Be Inferred From Its Mobility on a Molecular Sieve

Columns of Sephadex or similar matrices that have "pores" of known size range are calibrated using proteins of known molecular weight. The molecular weight of an unknown protein is then calculated from its elution position relative to these standards. Large errors may result if the protein is highly asymmetric or interacts strongly with the materials from which the molecular sieve has been manufactured.

Polyacrylamide Gel Electrophoresis (PAGE) Provides Yet Another Way To Determine Molecular Weight

Protein standards are separated by electrophoresis in 5–15% cross linked gels of varying porosity. Gels are stained for protein, generally with Coomassie blue stain or silver, and the molecular weight is estimated relative to the mobility of the standards. The most common application of this technique, SDS-PAGE, determines protomer molecular weight by first denaturing the oligomer (eg, by boiling in a detergent in the presence of β-mercaptoethanol) and separating on gels that contain the ionic detergent sodium dodecyl sulfate (SDS).

PROTEIN COMPLEXES MAY BE VISUALIZED BY ELECTRON PHOTOMICROGRAPHY

Magnifications as high as 100,000 diameters obtained with the electron microscope permit visualization of proteins of high molecular weight such as virus particles, enzyme complexes, and oligomeric proteins.

Table 6–6 lists examples of the numbers and molecular weights of protomers contributing to the quaternary structures of selected enzymes.

REFERENCES

Advances in Protein Chemistry. Academic Press, 1944 to date. [Annual publication.]

Celis JE, Bravo R: Two-dimensional Gel Electrophoresis of Proteins. Methods and Applications. Academic Press, 1984.

Chothia C. Principles that determine the structure of proteins. *Annu Rev Biochem* 1984;**53**:537.

Dunn MJ: Gel Electrophoresis of Proteins. Wright, 1986.

Franks F: Characterization of Proteins. Wright, 1986.

Kline AD, Braun W, Wuthrich K. Studies by ^{1}H NMR and distance geometry of the solution conformation of the α-amylase inhibitor tendamistat. *J Mol Biol* 1986;**189**:377.

Lennarz WJ (editor): The Biochemistry of Glycoproteins and Proteoglycans. Plenum Press, 1980.

L'Italien JJ: Proteins: Structure and Function. Plenum Press, 1987.

Rose CD, Geselowitz AR, Lesser GJ, Lee RM, Zehfus MH. Hydrophobicity of amino acid residues in globular proteins. *Science* 1985;**229**:834.

Schlessinger DH: Macromolecular Sequencing and Analysis: Selected Methods and Applications. AR Liss, 1988.

Schott H: Affinity Chromatography. Template Chromatography of Nucleic Acids and Proteins. Marcel Dekker, 1984.

Wittmann-Liebold B, Solnikow J, Erdmann VA: Advanced Methods in Protein Microsequence Analysis. Springer Verlag, 1986.

7 Proteins: Myoglobin & Hemoglobin

Victor W. Rodwell, PhD

INTRODUCTION

As an example of protein structure-function relationships, this chapter discusses myoglobin and hemoglobin, proteins which are of significance both in their own right and for the insights they provide into the ways in which the structures of proteins conform to, or dictate, their biologic functions.

BIOMEDICAL IMPORTANCE

Heme proteins function in oxygen binding, oxygen transport, electron transport, and photosynthesis. Detailed study of hemoglobin and myoglobin illustrates structural themes common to many proteins. In a sense, their greatest medical significance is that this knowledge eloquently illustrates protein structure-function relationships. In addition, it provides insight into the molecular basis of genetic diseases such as **sickle cell disease** (a result of altered surface properties of the hemoglobin β-subunit), and the **thalassemias** (chronic, familial hemolytic diseases characterized by defective synthesis of hemoglobin). **Cyanide** and **carbon monoxide** kill because they disrupt the physiologic function of the heme proteins cytochrome oxidase and hemoglobin, respectively. Finally, stabilization of the quaternary structure of deoxyhemoglobin by 2,3-bisphosphoglycerate (BPG) is central to understanding the mechanisms of **high-altitude sickness** and of **adaptation to high altitudes.**

THE HEME PROSTHETIC GROUP & FERROUS IRON ATOM OF MYOGLOBIN & HEMOGLOBIN CONFER THE ABILITY TO STORE & TO TRANSPORT OXYGEN

Myoglobin and hemoglobin provide clear insights into the relationships between protein structure and physiologic function. They possess as their prosthetic group **heme, a cyclic tetrapyrrole** that accounts for their red color. Tetrapyrroles consist of 4 molecules of pyrrole (Fig 7–1) linked in a planar ring by four α-methylene bridges. The β substituents determine whether a tetrapyrrole is heme or a related compound. In heme, these are methyl (M), vinyl (V), and propionate (Pr) groups arranged in the order M,V,M,V,M,Pr,Pr,M (Fig 7–2). One atom of ferrous iron (Fe^{2+}) is at the center of this planar ring. Other proteins with tetrapyrrole prosthetic groups (and their associated metal ions) include cytochromes (Fe^{2+} and Fe^{3+}), the enzymes catalase and tryptophan pyrrolase, and chlorophyll (Mg^{2+}). In cytochromes, oxidation and reduction of the iron atom are essential to their biologic function. By contrast, **oxidation of the Fe^{2+} of myoglobin or hemoglobin destroys biologic activity.**

Oxymyoglobin of Red Muscle Provides a Reserve of Stored Oxygen

Myoglobin of red muscle tissue stores oxygen. Under conditions of oxygen deprivation (eg, severe exercise), this oxygen is released for use by muscle mitochondria for oxygen-dependent synthesis of ATP.

With 2 Exceptions, Polar Residues Are on the Surface & Nonpolar Residues in the Interior of Myoglobin

Myoglobin, a single polypeptide chain of MW 17,000, is unexceptional with respect to its 153 aminoacyl residues. Clear differences, however, are apparent in their spatial distribution. The surface is polar and the interior nonpolar, a pattern characteristic of globular proteins. Residues with both polar and nonpolar regions (eg, Thr, Trp, and Tyr) orient their nonpolar regions inward. Apart from 2 histidyl residues that function in oxygen binding, the interior of myoglobin contains only nonpolar residues (eg, Leu, Val, Phe, and Met).

Myoglobin, the First Protein Whose Structure Was Solved by x-Ray Crystallography, Is Rich in α-Helix

Myoglobin is a compact, roughly spherical molecule measuring 4.5 × 3.5 × 2.5 nm (Fig 7–3). Its conformation is atypical, however. Approximately 75% of the residues are present in 8 right-handed α-helices from 7 to 20 residues in length. Starting at the N terminus, these are termed helices A through H. Interhelical regions are identified by the letters of the 2 helical regions they connect. Individual residues are designated by a letter for the helix in which they reside and a number that indicates their distance from the N terminus of that helix. For example, "His F8" refers to the eighth residue in the F helix and identifies it as a histidyl residue. Residues far distant in a primary structural sense (eg, in different helices) may neverthe-

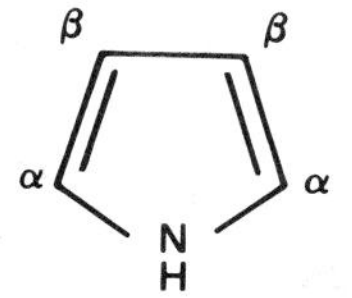

Figure 7–1. Pyrrole. The α carbons are linked by methylene bridges in a tetrapyrrole. The β carbons bear the substituents of a specific tetrapyrrole such as heme.

less be spatially close together, eg, the F8 (proximal) and E7 (distal) histidyl residues (Fig 7–3).

The secondary-tertiary structure of myoglobin in solution closely resembles that of crystalline myoglobin. They exhibit virtually identical absorption spectra; crystalline myoglobin binds oxygen; and the amount of α-helix in solution (estimated by optical rotatory dispersion and circular dichroism) closely approximates that revealed by x-ray analysis.

In the Presence of Heme, the Information Implicit in the Primary Structure of Apomyoglobin Insures Correct Protein Folding

When apomyoglobin (myoglobin with the heme removed) is prepared by lowering the pH to 3.5, its α-helical content decreases dramatically. Subsequent addition of urea removes all α-helical content. If the urea is then removed by dialysis and heme is added, full α-helical content is restored, and addition of Fe^{2+} restores full biologic (oxygen binding) activity. The primary structural information implicit in apomyoglo-

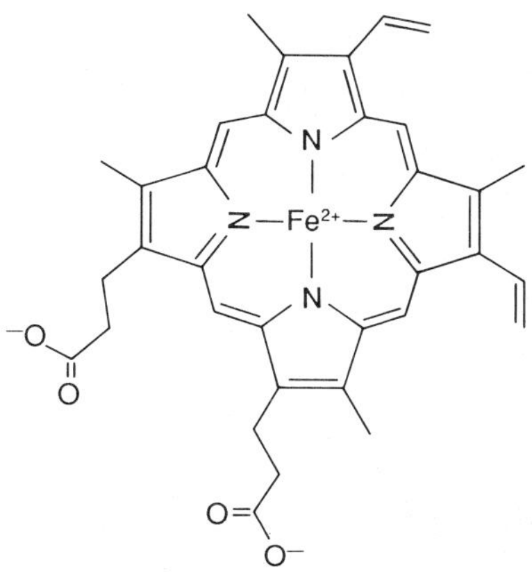

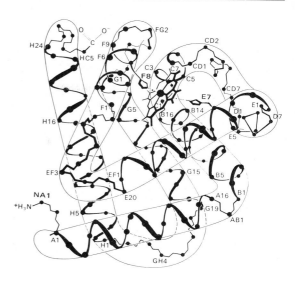

Figure 7–3. A model of myoglobin at low resolution. Only the α-carbon atoms are shown. (Based on Dickerson RE in: *The Proteins,* 2nd ed. Vol. 2. Neurath H [editor]. Academic Press, 1964. Reproduced with permission.)

bin therefore can, in the presence of heme, specify folding of the protein to its native, biologically active conformation. This important concept extends to other proteins: **the primary structure of a protein dictates its secondary-tertiary conformation.**

Histidyl Residues F8 & F9 Perform Unique Roles in Oxygen Binding by Myoglobin

The heme of myoglobin, which resides in a crevice between helices E and F (Fig 7–3), is oriented with its polar propionate substituents on the surface and the remainder in the interior surrounded by nonpolar residues, except for His F8 and His E7. The fifth coordination position of the iron atom is linked to a ring nitrogen of the **proximal histidine,** His F8 (Fig 7–4). While not linked to the sixth coordination position of the iron, the **distal histidine** (His E7) lies on the side of the heme ring across from His F8 (Fig 7–4).

The Iron Is Located Slightly Outside the Plane of the Heme Ring & Moves Toward This Plane When Oxygen Is Bound

In unoxygenated myoglobin, the heme iron resides about 0.03 nm (0.3 Å) outside the plane of the ring in the direction of His F8. In oxygenated myoglobin, an oxygen atom occupies the sixth coordination position of the iron atom, which then lies only about 0.01 nm (0.1 Å) outside the plane of the heme. Oxygenation of myoglobin is therefore accompanied by movement of the iron atom, and consequently movement of His F8 and residues covalently linked to His F8, toward the plane of the ring. This motion brings about a new conformation of portions of the protein.

Figure 7–2. Heme. The pyrrole rings and methylene bridge carbons are coplanar, and the iron atom (Fe^{2+}) resides in almost the same plane. The fifth and sixth coordination positions of Fe^{2+} are directed perpendicular to, and directly above and below, the plane of the heme ring. Observe the nature of the substituent groups on the β carbons of the pyrrole rings, the central iron atom, and the location of the polar side of the heme ring (at about 7 o'clock) that faces the surface of the myoglobin molecule.

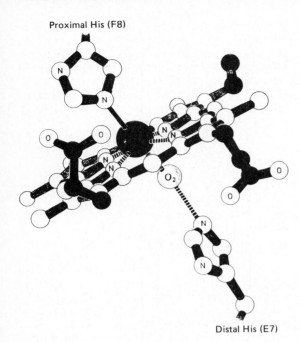

Proximal His (F8)

Distal His (E7)

Figure 7–4. Addition of oxygen to heme iron in oxygenation. Shown also are the imidazole side chains of the 2 important histidine residues of globin that attach to the heme iron. (Reproduced, with permission, from Harper HA et al: *Physiologische Chemie*. Springer-Verlag, 1975.)

Apomyoglobin Provides a Hindered Environment for Heme Iron, Which Greatly Decreases Its Affinity for Carbon Monoxide

When oxygen bonds to myoglobin, the bond between one oxygen atom and the Fe^{2+} is perpendicular to the plane of the heme ring. The second oxygen atom bonds at an angle of 121 degrees to the plane of the heme and oriented away from the distal histidine (Fig 7–5).

Carbon monoxide (CO) bonds to isolated heme about 25,000 times more strongly than does oxygen. Since the atmosphere contains traces of CO and normal catabolism of heme itself forms small quantities of CO, why then does not CO (rather than O_2) occupy the sixth coordination position of the heme iron of myoglobin? The answer lies in the **hindered environment** of heme in myoglobin. The preferred orientation for CO bonded to heme iron is with all 3 atoms (Fe, C, O) perpendicular to the heme ring (Fig 7–5). While this orientation is possible for isolated heme, in myoglobin **the distal histidine sterically hinders bonding of CO** at this angle (Fig 7–6). This forces CO to bond in a less favored configuration and reduces the strength of the heme-CO bond by over 2 orders of magnitude to about 200 times that of the heme-O_2 bond. A small portion (about 1%) of myoglobin nevertheless normally is present in the form of myoglobin-CO.

Figure 7–5. Preferred angles for bonding of oxygen and of carbon monoxide to the iron atom of heme (solid line).

THE OXYGEN DISSOCIATION CURVES FOR MYOGLOBIN & HEMOGLOBIN ILLUSTRATE THEIR SUITABILITY FOR THEIR UNIQUE PHYSIOLOGIC ROLES

Why is myoglobin unsuitable as an oxygen transport protein but effective as an oxygen storage protein? The quantity of oxygen bound to myoglobin (expressed as "percent saturation") depends upon the oxygen concentration (expressed as Po_2, or partial pressure of oxygen) in the immediate environment of the heme iron. The relationship between Po_2 and the quantity of oxygen bound may be expressed graphically as an oxygen saturation (oxygen dissociation) curve. **For myoglobin, the shape of the oxygen adsorption isotherm is hyperbolic** (Fig 7–7). Since Po_2 in the lung capillary bed is 100 mm Hg, myoglobin could effectively load oxygen in the lungs. However, the Po_2 of venous blood is 40 mm Hg and that of active muscle about 20 mm Hg. Since myoglobin cannot deliver a large fraction of its bound oxygen even at 20 mm Hg, it cannot serve as an effective vehicle for delivery of oxygen from lungs to peripheral tissues. However, the oxygen deprivation that accompanies severe physical exercise can lower the Po_2 of muscle tissue to as little as 5 mm Hg. At 5 mm, myoglobin readily releases its bound

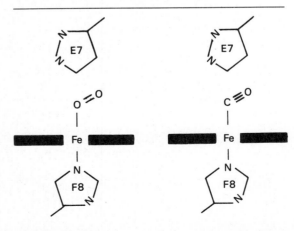

Figure 7–6. Angles for bonding of oxygen and carbon monoxide to the heme iron of myoglobin. The distal E7 histidine hinders bonding of CO at the preferred (180-degree) angle to the plane of the heme ring.

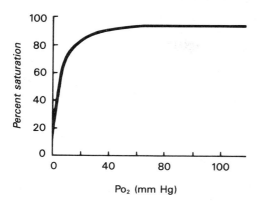

Figure 7–7. Oxygen saturation curve for myoglobin. Observe the relationship between percentage saturation and the partial pressures representative of lungs (100 mm Hg), tissues (20 mm Hg), and actively working muscle (5 mm Hg).

oxygen for oxidative synthesis of ATP by muscle mitochondria.

HEMOGLOBIN TRANSPORTS OXYGEN, CARBON DIOXIDE, & PROTONS BETWEEN LUNGS & TISSUES

Hemoglobins of vertebrate erythrocytes perform 2 major biologic functions: (1) transport of O_2 from the respiratory organ to peripheral tissues, and (2) transport of CO_2 and protons from peripheral tissues to the respiratory organ for subsequent excretion. While the comparative biochemistry of vertebrate hemoglobins provides fascinating insights, we shall here be concerned solely with human hemoglobins.

THE ALLOSTERIC PROPERTIES OF HEMOGLOBINS ARE CONSEQUENCES OF THEIR QUATERNARY STRUCTURES

The properties of individual hemoglobins are inextricable consequences of their quaternary (as well as of their secondary-tertiary) structure. The quaternary structure of hemoglobin confers striking additional properties (absent from myoglobin) that adapt it to its unique biologic roles and permit precise regulation of its properties. The **allosteric** (Gk *allos* "other," *steros* "space") properties of hemoglobin provide, in addition, a model for understanding other allosteric proteins.

Unlike Myoglobin, Hemoglobin Is a Tetrameric Protein

Unlike myoglobin, which lacks quaternary structure, hemoglobins are tetrameric proteins comprised of 2 each of 2 different polypeptides or monomer units

(termed α, β, γ, δ, S, etc). While similar in overall length, the α (141 residue) and β (146 residue) polypeptides of hemoglobin A (HbA) are encoded by different genes and have different primary structures. By contrast, the primary structures of β-, γ-, and δ- chains of human hemoglobins have highly conserved primary structures. The tetrameric structures of common hemoglobins are: HbA (normal adult hemoglobin) = $\alpha_2\beta_2$, HbF (fetal hemoglobin) = $\alpha_2\gamma_2$, HbS (sickle cell hemoglobin) = $\alpha_2 S_2$, and HbA_2 (a minor adult hemoglobin) = $\alpha_2\delta_2$.

Myoglobin & the β Subunits of Hemoglobin Share Almost Identical Secondary-Tertiary Structures

Despite differences in the type and number of amino acids present in myoglobin and the β polypeptide of HbA, they exhibit almost identical secondary-tertiary structures. This striking similarity, which extends to the location of the heme and the 8 helical regions, results in part from the substitution of amino acids of similar properties at equivalent points in the primary structures of myoglobin and of the β subunit of HbA. The β polypeptide also closely resembles myoglobin despite the presence of 7 rather than 8 helical regions. As for myoglobin, hydrophobic residues are internal and (again with the exception of 2 His residues per subunit) hydrophilic residues are surface features of both the α and β subunits of HbA.

Oxygenation of Hemoglobin Is Accompanied by Movement of the Iron Toward the Heme Ring & by Conformational Changes Transmitted Via the Proximal Histidine to the Adjoining Apoprotein

Hemoglobins bind 4 oxygen atoms per tetramer (one per subunit heme), and their oxygen saturation curves are sigmoidal (Fig 7–8). The facility with which O_2 binds to hemoglobin depends on whether other O_2 atoms are present on the same tetramer. If O_2 is already present, binding of subsequent O_2 atoms occurs more readily. Hemoglobin thus exhibits **cooperative binding kinetics,** a property that permits it to bind a maximal quantity of O_2 at the respiratory organ and to deliver a maximal quantity of O_2 at the prevailing Po_2 of the peripheral tissues. Compare, for example, these quantities at the Po_2 of human lungs (100 mm Hg) and tissues (20 mm Hg) with myoglobin (Fig 7–8).

We Compare the Relative Affinities of Different Hemoglobins by Their P_{50}, the Partial Pressure That Half-Saturates Them With Oxygen

Depending on the organism, P_{50} can vary widely but in all instances is above the peripheral tissue Po_2 in the

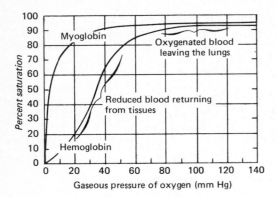

Figure 7–8. Oxygen-binding curves of hemoglobin and myoglobin. Arterial oxygen tension is about 100 mm Hg; mixed venous oxygen tension is about 40 mm Hg; capillary (active muscle) oxygen tension is about 20 mm Hg; and the minimum oxygen tension required for cytochromes is about 5 mm Hg. Association of chains into a tetrameric structure (hemoglobin) results in much greater oxygen delivery than would be possible with single chains. (Modified, with permission, from Stanbury JB, Wyngaarden JB, Fredrickson DS [editors]: *The Metabolic Basis of Inherited Disease,* 4th ed. McGraw-Hill, 1978.)

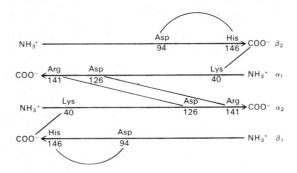

Figure 7–9. Salt links between and within subunits in deoxyhemoglobin. These noncovalent, electrostatic interactions are disrupted on oxygenation. (Slightly modified and reproduced, with permission, from Stryer L: *Biochemistry,* 2nd ed. Freeman, 1981.)

organism in question. Human fetal hemoglobin (HbF) provides an illustrative example. For HbA, $P_{50} = 26$ mm Hg; for HbF, $P_{50} = 20$ mm Hg. This difference permits HbF to extract oxygen from the HbA of placental blood. Postpartum, however, HbF is unsuitable, since its high affinity for O_2 dictates that it can deliver less O_2 to the tissues.

The human fetus initially synthesizes not α and β chains, but ζ and ε chains. By the end of the first trimester, α have replaced ζ subunits, and γ have replaced ε peptides. Hemoglobin F, the hemoglobin of late fetal life, thus has the composition $\alpha_2 \gamma_2$. Beta subunits, whose synthesis starts in the third trimester, do not completely replace γ until some weeks postpartum.

Gross Conformation Changes Accompany Oxygenation of Hemoglobin

Binding of O_2 is accompanied by the rupture of salt bonds between the carboxyl termini of all 4 subunits (Fig 7–9). Subsequent O_2 binding is facilitated since it involves rupture of fewer salt bonds. These changes also profoundly alter hemoglobin's secondary-tertiary and quaternary structures. One pair of α/β subunits rotates with respect to the other α/β pair, compacting the tetramer and increasing the affinity of the hemes for O_2 (Figs 7–10 and 7–11).

The quaternary structure of partially oxygenated hemoglobin is described as the T (taut) state and that of oxygenated hemoglobin (HbO₂) as the R (relaxed) state (Fig 7–12). R and T are also used to describe the quaternary structures of allosteric enzymes, where the T state has the lower substrate affinity.

Oxygenation of Hemoglobin Is Accompanied by Conformational Changes in the Neighborhood of the Heme Group

On oxygenation, the iron atoms of deoxyhemoglobin (which lie about 0.06 nm beyond the plane of the heme ring) move into the plane of the heme ring (Fig 7–13). This motion is transmitted to the proximal (F8) histidine, which moves toward the plane of the ring, and to residues attached to His F8.

Following Release of Oxygen At the Tissues, Hemoglobin Transports Carbon Dioxide & Protons to the Lungs

In addition to transporting oxygen from the lungs to peripheral tissues, hemoglobin facilitates the **transport of CO_2** from tissues to the lungs for exhalation. Hemoglobin can bind CO_2 directly when oxygen is released, and about 15% of the CO_2 carried in blood is

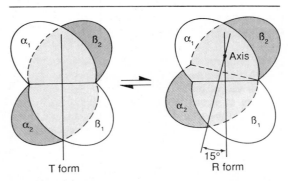

Figure 7–10. During the transition of the T form to the R form of hemoglobin, one pair of rigid subunits (α_2/β_2) rotates through 15 degrees relative to the other rigid pair (α_1/β_1). The axis of rotation is eccentric, and the α_2/β_2 pair also shifts toward the axis somewhat. In the diagram, the α_1/β_1 pair is unshaded and held fixed, while the shaded α_2/β_2 pair rotates and shifts.

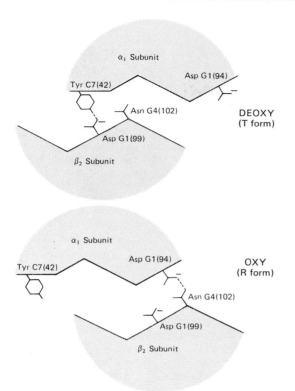

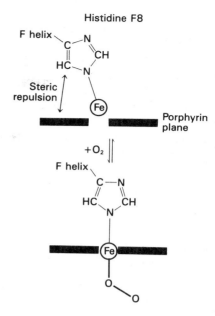

Figure 7–13. The iron atom moves into the plane of the heme on oxygenation. Histidine F8 and its associated residues are pulled along with the iron atom. (Slightly modified and reproduced, with permission, from Stryer L: *Biochemistry*, 2nd ed. Freeman, 1981.)

Figure 7–11. Changes at the α_1/β_2 contact on oxygenation. The contact "clicks" from one dovetailing area to another, involving a switch from one hydrogen bond to a second. The other bonds are nonpolar. (Reproduced, with permission, from Perutz MF: Molecular pathology of human hemoglobin: Stereochemical interpretation of abnormal oxygen affinities. *Nature* 1971;**232:408.**)

carried directly on the hemoglobin molecule. However, as CO_2 is absorbed in blood, the carbonic anhydrase in erythrocytes catalyzes the formation of carbonic acid (Fig 7–14). Carbonic acid rapidly dissociates into bicarbonate and a proton; the equilibrium is toward the dissociation. To avoid the extreme danger of increasing the acidity of blood, there must exist a

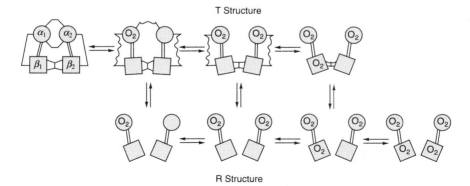

Figure 7–12. Transition from the T structure to the R structure increases in likelihood as each of the 4 heme groups is oxygenated. In this model, salt bridges (thin lines) linking the subunits in the T structure break progressively as oxygen is added, and even those salt bridges that have not yet ruptured are progressively weakened (wavy lines). The transition from T to R does not take place after a fixed number of oxygen molecules have been bound, but it becomes more probable with each successive oxygen bound. The transition between the 2 structures is influenced by several factors, including protons, carbon dioxide, chloride, and BPG. The higher their concentration, the more oxygen must be bound to trigger the transition. Fully oxygenated molecules in the T structure and fully deoxygenated molecules in the R structure are not shown, because they are too unstable to exist in significant numbers. (Modified and redrawn, with permission, from Perutz MF: Hemoglobin structure and respiratory transport. *Sci Am* [Dec] 1978;**239:**92.)

$$CO_2 + H_2O \rightleftharpoons H_2CO_3 \rightleftharpoons HCO_3^- + H^+$$

CARBONIC ANHYDRASE | Carbonic (Spontaneous) acid

Figure 7-14. The formation of carbonic acid, catalyzed by erythrocyte carbonic anhydrase, and the dissociation of carbonic acid to bicarbonate ion and a proton.

buffering system to absorb this excess proton. **Hemoglobin binds 2 protons for every 4 oxygen molecules lost** and so contributes significantly to the buffering capacity of blood (Fig 7–15). In the lungs, the process is reversed—ie, **as oxygen binds to the deoxygenated hemoglobin, protons are released** and combine with bicarbonate, forming carbonic acid. With the aid of carbonic anhydrase, the carbonic acid forms CO_2, which is exhaled. Thus, the **binding of oxygen forces the exhalation of CO_2.** This reversible phenomenon is called the **Bohr effect.** The Bohr effect is a property of tetrameric hemoglobin and is dependent upon its heme-heme interaction or cooperative effects. Myoglobin does not exhibit any Bohr effect.

The Protons Responsible for the Bohr Effect Are Generated by the Rupture of Salt Bonds During Binding of Oxygen to the T Structure of Deoxyhemoglobin

The protons responsible for the Bohr effect are gen-

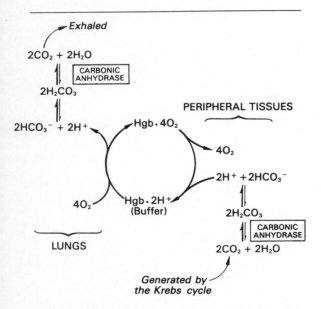

Figure 7-15. The Bohr effect. The carbon dioxide generated in peripheral tissues combines with water to form carbonic acid which dissociates into protons and bicarbonate ions. The deoxygenated hemoglobin acts as a buffer by binding protons and delivering them to the lungs. In the lungs, the binding of oxygen by hemoglobin releases protons from the hemoglobin. The protons combine with bicarbonate ion, generating carbonic acid, which with the aid of carbonic anhydrase, becomes carbon dioxide. The carbon dioxide is exhaled from the lungs.

erated by the breaking of salt bridges during the binding of oxygen to the T structure. The protons are released from the N atoms of β chain His residues HC3 (146). These released protons drive bicarbonate toward carbonic acid, which is then released as CO_2 in alveolar blood (Fig 7–15). Conversely, upon the release of oxygen, the T structure and its salt bridges are re-formed, requiring protons to bind to the β chain HC3 residues. Thus, the presence of protons from peripheral tissues favors the formation of salt bridges by protonating the terminal His residue of the β subunits. Re-formation of the salt bridges facilitates the release of oxygen from oxygenated (R form) hemoglobin. Overall, **an increase in protons causes oxygen release,** while **an increase in oxygen causes proton release.** The former can be represented in an oxygen dissociation curve by a rightward shift in the dissociation curve upon increasing hydrogen ions (protons).

2,3-Bisphosphoglycerate (BPG) in the Central Cavity of Deoxyhemoglobin Forms Salt Bonds That Stabilize the T Structure of Hemoglobin

In peripheral tissues, an oxygen shortage causes an increased accumulation of 2,3-bisphosphoglycerate (BPG) (Fig 7–16). This compound is formed from the glycolytic intermediate 1,3-bisphosphoglycerate. One molecule of BPG is bound per hemoglobin tetramer in a central cavity formed by all 4 subunits. The central cavity is of sufficient size for BPG only when the space between the H helices of the beta chains is wide enough, ie, when the hemoglobin molecule is in the **T-form.** BPG is bound by salt bridges between its oxygen atoms and both β chains via their N-terminal amino groups (Val NA1), Lys EF6, and His H21 residues (Fig 7–17). Thus, **BPG stabilizes the T or deoxygenated form of hemoglobin by cross linking the β chains** and contributing additional salt bridges that must be broken for the T form to click into the R form.

BPG binds more weakly to fetal hemoglobin than to adult hemoglobin because the H21 residue of the γ chain of fetal hemoglobin is Ser rather than His and cannot form a salt bridge with BPG in the central cavity. Therefore, **BPG has a less profound effect on the**

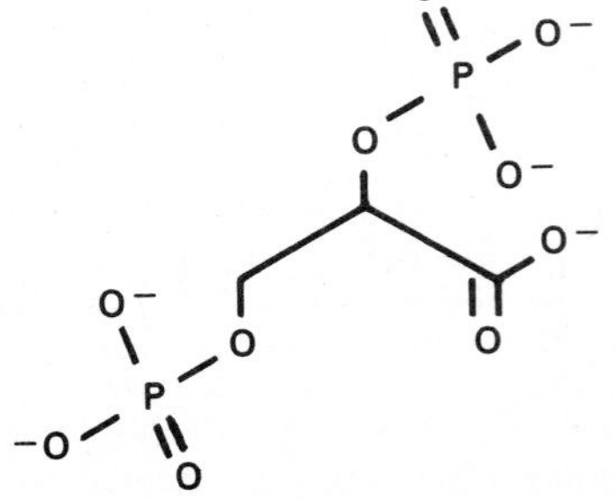

Figure 7-16. Structure of 2,3-bisphosphoglycerate (BPG).

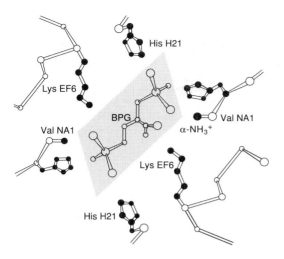

Figure 7–17. Mode of binding of 2,3-bisphosphoglycerate to human deoxyhemoglobin. BPG interacts with 3 positively charged groups on each β chain. (Based on Arnone A: X-ray diffraction study of binding of 2,3-diphosphoglycerate to human deoxyhemoglobin. *Nature* 1972;**237**:146. Reproduced with permission.)

stabilization of the T form of fetal hemoglobin and is responsible for fetal hemoglobin appearing to have a higher affinity for oxygen than does adult hemoglobin.

The trigger for the R to T transition of hemoglobin is the **movement of the iron in and out of the plane of the porphyrin ring.** Both steric and electrostatic factors mediate this trigger with a free energy of about 3000 calories per mole. Thus, a minimal change in the position of Fe^{2+} relative to the porphyrin ring induces significant switching of the conformations of hemoglobin and crucially affects its biologic function in response to an environmental signal.

SEVERAL HUNDRED MUTANT HUMAN HEMOGLOBINS, MOST OF THEM BENIGN, HAVE BEEN IDENTIFIED

Mutations in the genes that code for the α or β chains potentially can affect the biologic function of hemoglobin. Of the several hundred known mutant human hemoglobins (most of them benign) several in which biologic function is altered are described below. When biologic function is altered owing to a mutation in hemoglobin, the condition is known as a **hemoglobinopathy.**

In Hemoglobin M Variants, Tyrosyl Residues Replace the Proximal or Distal His Residues of α or β Subunits

Heme iron is stabilized in the Fe^{3+} state, since it forms a tight ionic complex with the phenolate anion of Tyr. **Methemoglobinemia** results, since ferri heme cannot bind O_2. In α-chain hemoglobin M variants, the R–T equilibrium favors the T form. Oxygen affinity is reduced, and a Bohr effect is absent. β-Chain hemoglobin M variants exhibit R–T switching, and a Bohr effect is therefore present.

Mutations (eg, hemoglobin Chesapeake) that favor the R form exhibit **increased oxygen affinity.** They therefore fail to deliver adequate oxygen to peripheral tissues. The resulting tissue hypoxia leads to **polycythemia** (increased concentration of erythrocytes).

In Hemoglobin S, a Valyl Residue Replaces the Normal Glutamyl Residue at Position 6 of the β Chain

Hemoglobin S is a substitution of Glu A2(6)β ie, residue 6 of the β chain by Val. The A2 residue, Glu or Val, is on the surface of the hemoglobin, exposed to

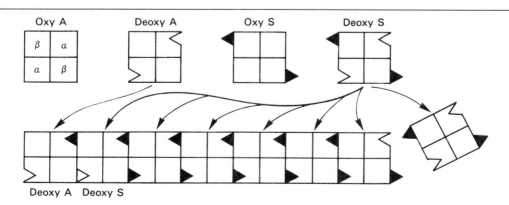

Figure 7–18. Representation of the sticky patch (▲) on hemoglobin S and its "receptor" (△) on deoxyhemoglobin A and deoxyhemoglobin S. The complementary surfaces allow deoxyhemoglobin S to polymerize into a fibrous structure, but the presence of deoxyhemoglobin A will terminate the polymerization by failing to provide sticky patches. (Modified and reproduced, with permission, from Stryer L: *Biochemistry*, 2nd ed. Freeman, 1981.)

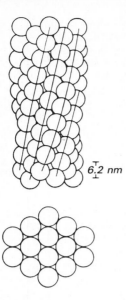

Figure 7–19. Proposed helical structure of a fiber of aggregated deoxyhemoglobin S. (Reproduced, with permission, from Maugh T II: A new understanding of sickle cell emerges. *Science* 1981;**211**:265. Copyright © 1981 by the American Association for the Advancement of Science.)

water. The substitution in hemoglobin S replaces the polar glutamate residue with a nonpolar one and generates a **"sticky patch"** on the surface of the β chain. The sticky patch is present on oxygenated and deoxygenated hemoglobin S, but not on hemoglobin A. On the surface of deoxygenated hemoglobin, there exists a **complement to the sticky patch,** but in oxygenated hemoglobin this complementary site is masked (Fig 7–18).

Deoxyhemoglobin S Can Form a Fibrous Structure That Physically Distorts the Erythrocyte

When hemoglobin S is deoxygenated, the sticky patch of hemoglobin S can bind to the complementary patch on another deoxygenated hemoglobin molecule. This binding causes a **polymerization of deoxyhemoglobin S, forming long fibrous precipitates** that mechanically distort (sickle) the red cell, causing lysis and multiple secondary clinical effects. Thus, if hemoglobin S can be maintained in an oxygenated state, or if the concentration of deoxygenated hemoglobin S can be minimized, formation of these polymers of deoxygenated hemoglobin S will not occur and "sickling" can be prevented. Clearly, it is the **T form of hemoglobin S that is subject to polymerization.** It is interesting, but of no therapeutic use, that the ferric ion in methemoglobin A remains in the plane of the porphyrin ring and thus stabilizes the R form of hemoglobin. The same occurs in sickle hemoglobin: ie, hemoglobin S in the ferric state (methemoglobin S) will not polymerize into fibers, since it is stabilized in the R form.

Although deoxyhemoglobin A contains the receptor sites for the sticky patch present on oxygenated or deoxygenated hemoglobin S (Fig 7–18), the binding of sticky hemoglobin S to deoxyhemoglobin A cannot extend the polymer, since the latter does not itself have a sticky patch to promote binding to still another hemoglobin molecule. Therefore, the **binding of deoxyhemoglobin A to either the R or the T form of hemoglobin S will terminate the polymerization.**

The polymerization of deoxyhemoglobin S forms a helical fibrous structure, each hemoglobin molecule making contact with 4 neighbors in a tubular helix (Fig 7–19). These tubular fibers distort the erythrocyte so that they take on the shape of a sickle (Fig 7–20) and

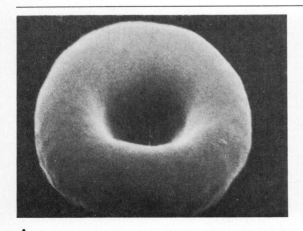

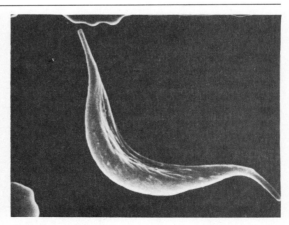

A **B**

Figure 7–20. Scanning electron micrograph of normal (*A*) and sickle (*B*) red blood cells. The change of the β-globin molecule that causes this structural alteration results from a single base mutation in DNA, T to A, which results in the substitution of valine for glutamate in the β-globin molecule.

are vulnerable to lysis as they penetrate the interstices of the splenic sinusoids.

Thalassemias Are Anemias Characterized by Reduced Synthesis of the α or β Chains of Hemoglobin

In the thalassemias, the synthesis of either the α (α-thalassemias) or β (β-thalassemias) chains of hemoglobin is reduced. This results in anemia, which may be severe. Considerable progress has been made in recent years in elucidating the molecular mechanisms responsible for the production of the thalassemias.

REFERENCES

Bunn HF, Forget BG: *Hemoglobin: Molecular, Genetic, and Clinical Aspects.* Saunders, 1986.

Dickerson RE, Geis I: *Hemoglobin.* Benjamin/Cummings, 1983.

Embury SH: The clinical pathology of sickle-cell disease. *Annu Rev Med* 1986;**37**:361.

Embury SH, Scharf SJ, Saiki RK, Gholson MA, Golbus M, Arnheim N, Erlich HA: Rapid prenatal diagnosis of sickle cell anemia by a new method of DNA analysis. *New Engl J Med* 1987;**316**:656.

Fermi G, Perutz MF, Shaanan B, Fourme R: The crystal structure of human deoxyhemoglobin at 1.74 angstrom resolution. *J Mol Biol* 1984;**175**:159.

Friedman JM: Structure, dynamics, and reactivity in hemoglobin. *Science* 1985;**228**:1273.

Saiki RK, Scharf S, Faloona F, Mullis KB, Horn GT, Erlich HA, Arnheim N: Enzymatic amplification of beta-globin genomic sequences and restriction site analysis for diagnosis of sickle-cell anemia. *Science,* 1985;**230**:1350.

Schacter L, Warth JA, Gordon EM, Prasad A, Klein BL: Altered amount and activity of superoxide dismutase in sickle cell anemia. *FASEB J* 1988;**2**:237.

Weatherall DJ, Clegg JB, Higgs DR, Wood WG: The hemoglobinopathies. In *The Metabolic Basis of Inherited Disease,* 6th Ed. Scriver CR, Beaudet AL, Sly WS, Valle D (editors). McGraw-Hill, p. 2281, 1989.

8

Enzymes: General Properties

Victor W. Rodwell, PhD

INTRODUCTION

Catalysts accelerate chemical reactions, undergo physical change during a reaction, but revert to their original state when the reaction is complete. While a few instances of catalysis by RNA molecules are known, almost all **enzymes** are **protein catalysts**. Most biochemical reactions would occur extremely slowly were it not for catalysis by enzymes. In contrast to nonprotein catalysts (H^+, OH^-, metal ions), each enzyme catalyzes a small number of reactions, frequently only one. Enzymes are thus **reaction-specific** catalysts. Essentially all biochemical reactions are enzyme-catalyzed.

BIOMEDICAL IMPORTANCE

Enzymes make life on Earth possible and hence impact many fields of the biomedical sciences. Certain diseases (the inborn errors of metabolism) are due to genetically determined abnormalities in the synthesis of enzymes. When cells are injured (eg, by impairment of blood supply or by inflammation), certain enzymes leak into the plasma. Measurement of the activity of such enzymes in plasma has become an integral part of the diagnosis of important medical disorders (eg, myocardial infarction). Diagnostic enzymology is the area of medicine involving the use of enzymes to assist in diagnosis and management. Enzymes can also be used in therapy.

THE INTERNATIONAL UNION OF BIOCHEMISTRY (IUB) SYSTEM CLASSIFIES ENZYMES BY REACTION TYPE AND REACTION MECHANISM

Enzymes initially were named by adding the suffix -ase to the name of the substrate on which they acted. Enzymes that hydrolyzed starch (amylon) were termed amylases; those that hydrolyzed fat (lipos), lipases; and those that hydrolyzed proteins, proteinases. Somewhat later, enzymes that catalyzed similar reactions were given names that indicated the type of chemical reaction catalyzed. These were termed dehydrogenases, oxidases, decarboxylases, acylases, etc.

The **International Union of Biochemistry (IUB) Nomenclature System,** while complex, is unambiguous. Its underlying principle, that of naming and classifying enzymes on the basis of **chemical reaction type and reaction mechanism,** can integrate information from divergent areas of metabolism. Major features of the IUB system are:

(1) Reactions and the enzymes that catalyze them form 6 classes, each with from 4 to 13 subclasses.

(2) The enzyme name has 2 parts. The first names the substrate or substrates. The second, ending in -ase, indicates the type of reaction catalyzed.

(3) Additional information, if needed to clarify the reaction, may follow in parentheses; eg, the enzyme catalyzing L-malate + NAD^+ = pyruvate + CO_2 + NADH + H^+ is designated 1.1.1.37 L-malate:NAD^+ oxidoreductase (decarboxylating).

(4) Each enzyme has a code number (EC) that characterizes the reaction type as to class (first digit), subclass (second digit), and subsubclass (third digit). The fourth digit is for the specific enzyme. Thus, EC 2.7.1.1 denotes class 2 (a transferase), subclass 7 (transfer of phosphate), subsubclass 1 (an alcohol functions as the phosphate acceptor). The final digit denotes hexokinase, or ATP:D-hexose 6-phosphotransferase, an enzyme catalyzing phosphate transfer from ATP to the hydroxyl group on carbon 6 of glucose.

FOR CATALYTIC ACTIVITY, MANY ENZYMES REQUIRE A COENZYME

Many enzymes require a specific, heat-stable, low molecular weight organic molecule, a coenzyme. The **holoenzyme** (complete catalytic entity) consists of the **apoenzyme** (protein part) plus the bound **coenzyme**. A coenzyme may bind covalently or noncovalently to the apoenzyme. The term "prosthetic group" denotes a covalently bonded coenzyme. Reactions that require coenzymes include oxidoreductions, group transfer and isomerization reactions, and reactions that form covalent bonds (IUB classes 1, 2, 5, and 6). Lytic reactions, including the hydrolytic reactions catalyzed by digestive enzymes, do not require coenzymes.

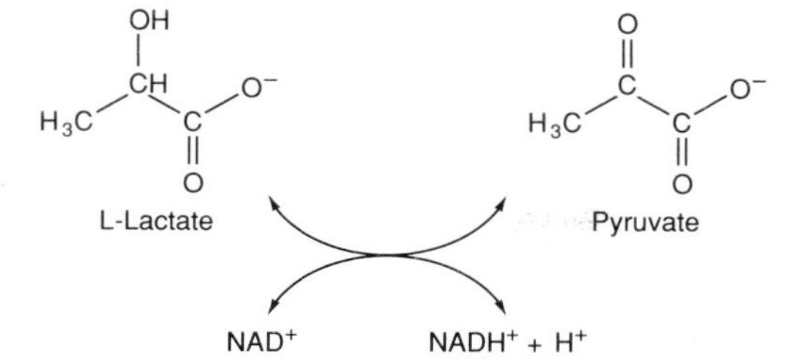

Figure 8–1. NAD$^+$ acting as cosubstrate in an oxidoreduction reaction.

Coenzymes May Be Regarded as Second Substrates

The coenzyme may be regarded as a second substrate or cosubstrate for 2 reasons. First, the chemical changes in the coenzyme exactly counterbalance those taking place in the substrate. For example, in oxidoreduction reactions, when one molecule of substrate is oxidized, one molecule of coenzyme is reduced (Fig. 8–1). Similarly, in transamination reactions pyridoxal phosphate acts as a second substrate in 2 concerted reactions and as carrier for transfer of an amino group between different α-keto acids.

A second reason to accord equal emphasis to the coenzyme is that this aspect of the reaction may be of greater fundamental physiologic significance. For example, the importance of the ability of muscle working anaerobically to convert pyruvate to lactate does not reside in pyruvate or lactate. The reaction serves merely to oxidize NADH to NAD$^+$. Without NAD$^+$, glycolysis cannot continue and anaerobic ATP synthesis (and hence work) ceases. Under anaerobic conditions, reduction of pyruvate to lactate reoxidizes NADH and permits synthesis of ATP. Other reactions can serve this function equally well. For example, in bacteria or yeast growing anaerobically, metabolites derived from pyruvate serve as oxidants for NADH and are themselves reduced (Table 8–1).

Coenzymes Function as Group-Transfer Reagents

Biochemical group transfer reactions of the type:

$$D-G + A = A-G + D$$

in which a functional group, G, is transferred from a donor molecule, D–G, to an acceptor molecule, A,

usually involve a coenzyme either as the ultimate acceptor (eg, dehydrogenation reactions) or as an intermediate group carrier (eg, transamination reactions). The display illustrates the latter concept.

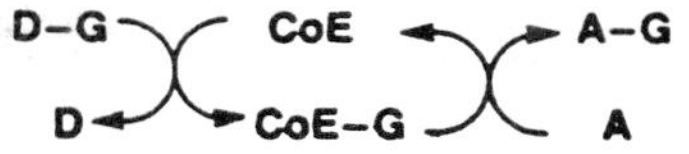

While this suggests formation of a single CoE–G complex during the course of the overall reaction, several intermediate CoE–G complexes may be involved in a particular reaction (eg, transamination).

When the group transferred is hydrogen, it is customary to represent only the left "half reaction":

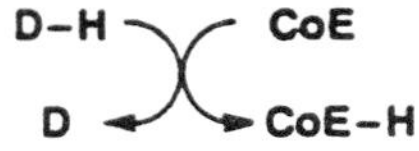

That this actually represents only a special case of general group transfer can best be appreciated in terms of the reactions that occur in intact cells (Table 8–1). These can be represented as follows:

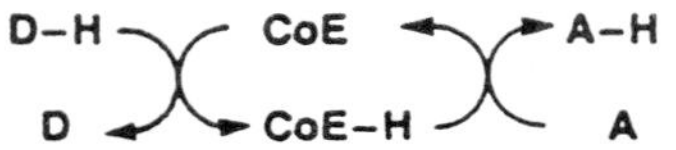

Coenzymes Can Be Classified According to the Group Whose Transfer They Facilitate

Based on the above concept, we might classify coenzymes as follows:

For transfer of groups other than H
   Sugar phosphates
   CoASH
   Thiamin pyrophosphate
   Pyridoxal phosphate
   Folate coenzymes
   Biotin
   Cobamide (B$_{12}$) coenzymes
   Lipoic acid
For transfer of H
   NAD$^+$, NADP$^+$
   FMN, FAD
   Lipoic acid
   Coenzyme Q

Many Coenzymes Are Derivatives of B Vitamins & of Adenosine Monophosphate

B vitamins form part of the structure of many coenzymes. The B vitamins **nicotinamide, thiamin, riboflavin,** and **pantothenic acid** are essential constituents of coenzymes for biologic oxidations and reductions, and **folic acid** and **cobamide** coenzymes function in one-carbon metabolism. Many coenzymes contain adenine, ribose, and phosphate and are derivatives of adenosine monophosphate (AMP). Examples include NAD$^+$ and NADP$^+$ (Fig. 8–2).

Table 8–1. Mechanisms for anaerobic regeneration of NAD$^+$.

Oxidant	Reduced Product	Life Form
Pyruvate	Lactate	Muscle, lactic bacteria
Acetaldehyde	Ethanol	Yeast
Dihydroxyacetone phosphate	α-Glycerophosphate	*Escherichia coli*
Fructose	Mannitol	Heterolactic bacteria

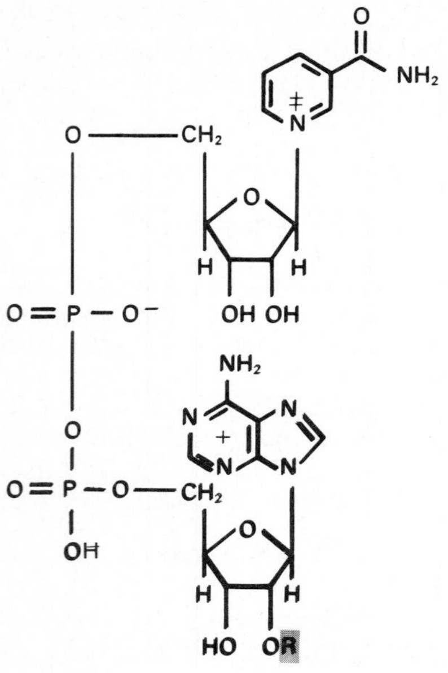

Figure 8–2. NAD(P)$^+$. In NAD$^+$, R = H; in NADP$^+$, R = $-OPO_3^{2-}$.

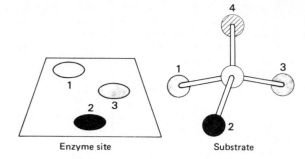

Figure 8–3. Representation of 3-point attachment of a substrate to a planar active site of an enzyme.

Since Substrates Contact Enzymes at Three Points, Enzymes Act as Stereospecific Catalysts

Most substrates form at least 3 bonds with enzymes. This **"3-point attachment"** can confer asymmetry on an otherwise symmetric molecule. Fig. 8–3 shows a substrate molecule, represented as a carbon atom having 3 different groups, about to attach at 3 points to an enzyme site. If the site can be approached only from one side and only complementary atoms and sites can interact (valid assumptions for actual enzymes), the molecule can bind in only one way. The reaction may be confined to the atoms bound at sites 1 and 2, even though atoms 1 and 3 are identical. By mentally turning the substrate molecule in space, note that it can attach at 3 points to one side of the planar site with only one orientation. Consequently, atoms 1 and 3, although identical, become distinct when the substrate is attached to the enzyme. A chemical change thus can involve atom 1 but not atom 3, or vice versa. This can explain why the enzyme-catalyzed reduction of optically inactive pyruvate forms L- and not D, L-lactate.

MOST ENZYMES CATALYZE EITHER A SPECIFIC REACTION OR, IN SOME INSTANCES, A SPECIFIC REACTION TYPE

The ability of an enzyme to catalyze one specific reaction and essentially no others is perhaps its most significant property. Rates of metabolic processes may thus be regulated by changes in the catalyt-

ic efficiency of specific enzymes. However, most enzymes catalyze the same type of reaction (phosphate transfer, oxidation-reduction, etc) with a small number of structurally related substrates. Reactions with alternative substrates tend to take place if these are present in high concentration. Whether all of the possible reactions will occur in living organisms depends on the relative concentration of alternative substrates in the cell and the relative affinity of the enzyme for those substrates.

Most Enzymes Exhibit Absolute Optical Specificity for at Least Portions of Their Substrates

Except for epimerases (racemases), which catalyze interconversion of optical isomers, **enzymes generally exhibit absolute optical specificity for at least a portion of a substrate molecule.** Thus, enzymes of the glycolytic and direct oxidative pathways catalyze the interconversion of D- but not L-phosphosugars. With few exceptions (eg, kidney D-amino acid oxidase), most mammalian enzymes act on the L-isomers of amino acids.

Optical specificity may extend to a portion of the substrate molecule or to its entirety. Glycosidases illustrate both extremes. These catalyze hydrolysis of glycosidic bonds between sugars and alcohols, are highly specific for the sugar portion and for the linkage (α or β), but are relatively nonspecific for the aglycone (alcohol portion).

Enzymes Exhibit A High Order of Specificity for the Type of Reaction They Catalyze

Lytic enzymes act on specific chemical groupings, eg, glycosidases on glycosides, pepsin and trypsin on peptide bonds, and esterases on esters. A large number of different peptide substrates may be attacked, lessening the number of digestive enzymes otherwise required. Many proteases catalyze hydrolysis of esters also. While this is of limited physiologic importance, the use of esters as synthetic substrates has contributed significantly to the study of the mechanism of action of proteases.

Certain lytic enzymes exhibit higher specificity.

Chymotrypsin hydrolyzes peptide bonds in which the carboxyl group is contributed by the aromatic amino acids phenylalanine, tyrosine, or tryptophan. Carboxypeptidases and aminopeptidases remove amino acids one at a time from the carboxy- or amino-terminal end of polypeptide chains, respectively.

Although a few oxidoreductases utilize either NAD^+ or $NADP^+$ as electron acceptor, most use exclusively one or the other. In general, **oxidoreductases functional in biosynthetic processes in mammalian systems (eg, fatty acid or sterol synthesis) use NADPH as reductant, while those functional in degradative processes (eg, glycolysis, fatty acid oxidation) use NAD^+ as oxidant.**

AN ENZYME'S CATALYTIC ACTIVITY PROVIDES A SELECTIVE & SENSITIVE PROBE FOR ITS DETECTION

The small quantity of enzymes in cells complicates measuring the amount of an enzyme in tissue extracts or fluids. Fortunately, the catalytic activity of an enzyme provides a sensitive and specific probe for its own measurement.

To measure the amount of an enzyme in a sample of tissue extract or other biologic fluid, the rate of the reaction catalyzed by the enzyme in the sample is measured. Under appropriate conditions, **the measured rate of the reaction is proportionate to the quantity of enzyme present.** Since it is difficult to determine the number of molecules or mass of enzyme present, results are expressed in **enzyme units.** Relative amounts of enzyme in different extracts may then be compared. Enzyme units are best expressed in micromoles (μmol; 10^{-6} mol), nanomoles (nmol; 10^{-9} mol), or picomoles (pmol; 10^{-12} mol) of substrate reacting or product produced per minute.

NAD+-Dependent Dehydrogenases May Be Assayed by Measuring the Change in Absorbancy at 340 nm That Accompanies the Interconversion of NAD+ & NADH

In reactions involving NAD^+ or $NADP^+$ (dehydrogenases), advantage is taken of the property of NADH or NADPH (but not NAD^+ or $NADP^+$) to absorb light of wavelength 340 nm (Fig 8–4). When NADH is oxidized to NAD^+ (or vice versa), the optical density (OD) at 340 nm changes. Under specified conditions, the rate of change in OD depends directly on the enzyme activity (Fig 8–5).

A calibration curve (Fig 8–6) is prepared by plotting the slopes of the lines (velocities) in Fig 8–5 versus the quantity of enzyme added. The quantity of enzyme present in an unknown solution may then be calculated from the observed rate of change in OD at 340 nm.

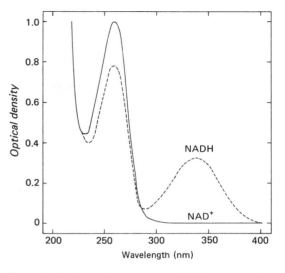

Figure 8–4. Absorption spectra of NAD^+ and NADH. Densities are for a 44 mg/L solution in a cell of 1-cm light path. $NADP^+$ and NADPH have spectra analogous to those of NAD^+ and NADH, respectively.

The Quantitative Analysis of Many Enzymes Can Be Simplified by "Coupling" the Reaction to One Catalyzed by an NAD+-Dependent Dehydrogenase

In the above example, the rate of formation of a product (NADH) was measured to determine enzyme activity. Enzymes other than dehydrogenases may also

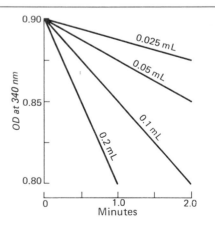

Figure 8–5. Assay of an NADH- or NADPH-dependent dehydrogenase. The rate of change in OD at 340 nm due to conversion of reduced to oxidized coenzyme is observed. Oxidized substrate (S), reduced coenzyme (NADH), and buffer are added to a cuvette. Light of 340-nm wavelength is then passed through it. Initially, the OD is high, since NADH (or NADPH) absorbs at 340 nm. On addition of 0.025–0.2 mL of a standard enzyme solution, the OD decreases.

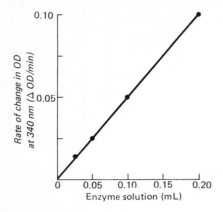

Figure 8–6. Calibration curve for enzymatic analysis. The slopes of the lines in Fig. 8–5 are plotted versus the quantity of enzyme.

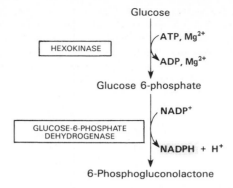

Figure 8–7. Coupled assay for hexokinase activity. The reaction is coupled to that catalyzed by glucose-6-phosphate dehydrogenase. Glucose-6-phosphate dehydrogenase, glucose, ATP, Mg^{2+}, and $NADP^+$ all are added in excess. The quantity of hexokinase present then determines the rate of the overall coupled reaction and therefore the rate of formation of NADPH, which can be measured at 340 nm.

be assayed by measuring the rate of appearance of a product (or, less commonly, the rate of disappearance of a substrate). The physicochemical properties of the product or substrate determine the specific method for quantitation. It often is convenient to "couple" the product of a reaction to a dehydrogenase for which this product is a substrate (Fig 8–7).

PURE ENZYMES ARE ESSENTIAL FOR UNDERSTANDING THEIR STRUCTURE, FUNCTION, REACTION MECHANISM, AND REGULATION

Knowledge about the reactions and chemical intermediates in metabolic pathways and of the regulatory mechanisms that operate at the level of catalysis derives to a great extent from studies of purified enzymes. Reliable information concerning the kinetics, cofactors, active sites, structure, and mechanism of action also requires highly purified enzymes.

An Effective Protocol for Purifying an Enzyme Increases the Enzyme's Specific Activity with Good Recovery at Each Step

The progress of a typical enzyme purification for a liver enzyme with good recovery and 490-fold overall

purification is shown in Table 8–2. Note how specific activity and recovery of initial activity are calculated. The aim is to achieve the maximum specific activity (enzyme units per milligram of protein) with the best possible recovery of initial activity.

The objective of enzyme purification is to isolate a specific enzyme from a crude cell extract containing many other components. Small molecules may be removed by dialysis or gel filtration, nucleic acids by precipitation with the antibiotic streptomycin, etc. The problem is to separate the desired enzyme from hundreds of chemically and physically similar proteins.

Classic Procedures for Enzyme Purification Include Selective Precipitation, Ion-Exchange Chromatography, & Size Exclusion Chromatography

Useful classic purification procedures include precipitation with varying salt concentrations (generally ammonium or sodium sulfate) or solvents (acetone or ethanol), differential heat or pH denaturation, differential centrifugation, gel filtration, and electrophoresis. Selective adsorption and elution of proteins from the cellulose anion exchanger diethylaminoethyl (DEAE)

Table 8–2. Summary of a typical enzyme purification scheme.

Enzyme Fraction	Total Activity (pU)	Total Protein (mg)	Specific Activity (pU/mg)	Overall Recovery (%)
Crude liver homogenate	100,000	10,000	10	(100)
100,000 × g supernatant liquid	98,000	8,000	12.2	98
40–50% $(NH_4)_2SO_4$ precipitate	90,000	1,500	60	90
20–35% acetone precipitate	60,000	250	240	60
DEAE column fractions 80–110	58,000	29	2,000	58
43–48% $(NH_4)_2SO_4$ precipitate	52,000	20	2,600	52
First crystals	50,000	12	4,160	50
Recrystallization	49,000	10	4,900	49

cellulose and the cation exchanger carboxymethylcellulose (CMC) have also been extremely successful for extensive and rapid purification. Separation of proteins on molecular sieves such as Sephadex that segregate proteins on the basis of their size is also widely used. These methods are relatively unselective, however, since they do not, except in combination, resolve a single protein from all others. This is more readily achieved by affinity chromatography.

Affinity Chromatographic Techniques Employ Immobilized Ligands that Interact With Specific Regions of an Enzyme

The salient feature of affinity chromatography is its ability to remove one particular protein, or at most, a small number of particular proteins, selectively from a complex protein mixture. The technique employs an immobilized ligand that interacts specifically with the enzyme whose purification is desired. When the protein mixture is exposed to this immobilized ligand, the only proteins that bind are those which interact strongly with the ligand. Unwanted protein flows though the column and is discarded. The desired protein is then eluted from the immobilized ligand, generally with a high concentration of salt or of the soluble form of the ligand. Purifications achieved by affinity chromatographic techniques are impressive, often surpassing that possible by successive application of numerous classic techniques.

Favored ligands are substrate and coenzyme derivatives covalently attached to a support such as Sephadex. Attachment may be direct or via a linker molecule 3–8 carbon atoms in length. Difficulties may arise if the mode of ligand attachment precludes its ability to interact with the enzyme—a difficulty that linker molecules may help to circumvent. The introduction of a hydrophobic "linker" however, may complicate the separation by introducing an element of hydrophobic ligand chromatography (see below). Examples of successful affinity chromatography include purification of many different dehydrogenases on NAD^+ affinity supports. While many dehydrogenases may be bound and may be eluted together when the column is treated with soluble NAD^+, subsequent application of substrate (rather than coenzyme) affinity supports or elution with an "abortive ternary mixture" of one product and one substrate have proved successful in many instances.

Dye-Ligand & Hydrophobic Chromatography Exploit Principles Analogous to Those of Affinity Chromatography

Dye-ligand chromatography on supports such as blue, green, or red Sepharose and hydrophobic ligand chromatography on support such as octyl- or phenyl-Sepharose are techniques closely related to affinity chromatography. The former employs as the immobilized ligand an organic dye that serves as an analogue of substrate, coenzyme, or allosteric effector. Elution generally is achieved using salt gradients.

In hydrophobic ligand chromatography, an alkyl or aryl hydrocarbon is attached to a support such as Sephadex. Retention of proteins on these supports involves hydrophobic interactions between the alkyl chain and hydrophobic regions on the protein. Proteins are applied in solutions that contain a high concentration of a salt (eg, $[NH_4]_2SO_4$) and are eluted with *decreasing* gradients of the same salt.

Polyacrylamide Gel Electrophoresis Detects Contaminants in Enzyme Preparations

Protein homogeneity is best assessed by polyacrylamide gel electrophoresis (PAGE) under several conditions. One-dimensional PAGE of the native protein will, if sufficient sample is applied, reveal major and minor protein contaminants. In 2-dimensional (O'Farrell) PAGE, the first dimension separates denatured proteins on the basis of their pI values by equilibrating them in an electrical field that contains urea and a pH gradient maintained by polymerized ampholytes. The second dimension then separates proteins, after treatment with SDS, on the basis of the molecular sizes of their protomer units (if present).

INTRACELLULAR DISTRIBUTION OF AN ENZYME IS INFERRED FROM HISTOCHEMICAL TECHNIQUES & BY ISOLATION & ASSAY OF SUBCELLULAR ORGANELLES

The spatial arrangement and compartmentalization of enzymes, substrates, and cofactors within the cell are of cardinal significance. In liver cells, for example, the enzymes of glycolysis are located in the cytoplasm, whereas enzymes of the citric acid cycle are in the mitochondria. The distribution of enzymes among subcellular organelles may be studied following fractionation of cell homogenates by high-speed centrifugation. The enzyme content of each fraction is then examined.

Localization of a particular enzyme in a tissue or cell in a relatively unaltered state is frequently accomplished by histochemical procedures ("histoenzymology"). Thin (2- to 10 μm) frozen sections of tissue are treated with a substrate for a particular enzyme. Where the enzyme is present, the product of the enzyme-catalyzed reaction is formed. If the product is colored and insoluble, it remains at the site of formation and localizes the enzyme. Histoenzymology provides a graphic and relatively physiologic picture of patterns of enzyme distribution.

ISOZYMES ARE PHYSICALLY DISTINCT FORMS OF THE SAME CATALYTIC ACTIVITY

While terms such as "malate dehydrogenase" or "glucose-6-phosphatase" appear to describe a single catalytic entity, they encompass all proteins that catalyze the oxidation of malate to oxaloacetate or the hydrolysis of glucose 6-phosphate to glucose and P_i, respectively. When techniques for purification of enzymes were applied, for example, to malate dehydrogenase from different sources (eg, rat liver and *Escherichia coli*), it became apparent that while rat liver and *E coli* malate dehydrogenase both catalyze the same reaction, their physical and chemical properties exhibited many significant differences. Physically distinct forms of the same catalytic activity may also be present in different tissues of the same organism, in different cell types in subcellular compartments, or within a procaryote such as *E coli*. This discovery followed from the application of electrophoretic separation procedures to separation of electrophoretically distinct forms of a particular enzymatic activity.

While the term "isozyme" (isoenzyme) encompasses all the above examples of physically distinct forms of a given catalytic activity, in practice, and particularly in clinical medicine, "isozyme" has a more restricted meaning, namely, the physically distinct and separable forms of a given enzyme present in different cell types or subcellular compartments of a human being. Isozymes are common in sera and tissues of all vertebrates, insects, plants, and unicellular organisms. Both the kind and the number of enzymes involved are equally diverse. Isozymes of numerous dehydrogenases, oxidases, transaminases, phosphatases, transphosphorylases, and proteolytic enzymes are known. Different tissues may contain different isozymes, and these isozymes may differ in their affinity for substrates.

The Ability to Separate & Identify Isozymes Is of Diagnostic Value

Medical interest in isozymes was stimulated by the discovery that **human sera contained several lactate dehydrogenase isozymes and that their relative proportions changed significantly in certain pathologic conditions.** Subsequently, many additional examples of changes in isozyme proportions as a result of disease have been described.

Serum lactate dehydrogenase isozymes may be visualized by subjecting a serum sample to electrophoresis, usually at pH 8.6, on a starch, agar, or polyacrylamide gel support. The isozymes have different charges at this pH and migrate to 5 distinct regions of the electropherogram. Isozymes are then localized by means of their ability to catalyze reduction of a colorless dye to an insoluble, colored form.

A typical dehydrogenase isoenzyme assay reagent contains the following:

(1) Reduced substrate (eg, lactate).
(2) Coenzyme (NAD^+).
(3) Oxidized dye (eg, nitroblue tetrazolium salt [NBT]).
(4) An intermediate electron carrier to transport electrons between NADH and the dye (eg, phenazine methosulfate [PMS]).
(5) Buffer; activating ions if required.

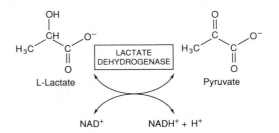

Figure 8–8. The L-lactate dehydrogenase reaction.

Lactate dehydrogenase catalyzes transfer of 2 electrons and one H^+ from lactate to NAD^+ (Fig. 8–8). The reaction proceeds at a measurable rate only in the presence of lactate dehydrogenase. When the assay mixture is spread on the electropherogram and incubated at 37 °C, concerted electron transfer reactions take place only in those regions where lactate dehydrogenase is present (Fig. 8–9). The relative intensities of the colored bands may then be quantitated by a scanning photometer (Fig. 8–10). The most negative isoenzyme is termed I_1.

Isozymes Are Products of the Expression of Closely Related Genes

Oligomeric enzymes with dissimilar protomers can exist in several forms. Frequently, one tissue produces one protomer predominantly and another tissue a different protomer. If these can combine in various ways to construct an active enzyme (eg, a tetramer), isozymes of that enzymatic activity are formed.

Lactate dehydrogenase isozymes differ at the level of quaternary structure. The oligomeric lactate dehydrogenase molecule (MW 130,000) consists of 4 protomers of 2 types, H and M (MW about 34,000). Only the tetrameric molecule possesses catalytic activity. If

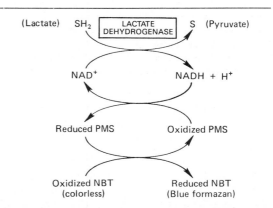

Figure 8–9. Coupled reactions in detection of lactate dehydrogenase activity on an electropherogram.

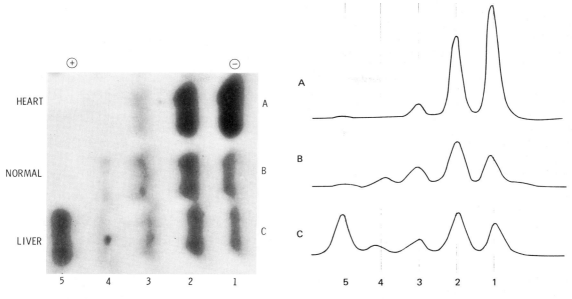

Figure 8–10. Normal and pathologic patterns of lactate dehydrogenase (LDH) isozymes in human serum. LDH isozymes of serum were separated on cellulose acetate at pH 8.6 and stained for enzyme. The photometer scan shows the relative proportion of the isozymes. Pattern A is serum from a patient with a myocardial infarct, B is normal serum, and C is serum from a patient with liver disease. (Courtesy of Dr Melvin Black and Mr Hugh Miller, St Luke's Hospital, San Francisco.)

order is unimportant, these protomers might be combined in the following 5 ways:

> HHHH
> HHHM
> HHMM
> HMMM
> MMMM

Market used conditions known to disrupt and reform quaternary structure to clarify the relationships between the lactate dehydrogenase isozymes. Splitting and reconstitution of lactate dehydrogenase-I_1 or lactate dehydrogenase-I_5 produced no new isozymes. These therefore consist of a single type of protomer. When a mixture of lactate dehydrogenase-I_1 and lactate dehydrogenase-I_5 was subjected to the same treatment, lactate dehydrogenase-I_2, -I_3, and -I_4 were generated. The proportions of the isozymes found are those which would result if the relationship were:

Lactate Dehydrogenase Isozyme	Subunits
I_1	HHHH
I_2	HHHM
I_3	HHMM
I_4	HMMM
I_5	MMMM

Syntheses of H and M subunits are controlled by distinct genetic loci that are differentially expressed in different tissues, eg, the heart and skeletal muscle.

THE QUANTITATIVE ANALYSIS OF CERTAIN PLASMA ENZYMES IS OF DIAGNOSTIC SIGNIFICANCE

Certain enzymes, proenzymes, and their substrates are present at all times in the circulation of normal individuals and perform a physiologic function in blood. Examples of **functional plasma enzymes** include lipoprotein lipase, pseudocholinesterase, and the proenzymes of blood coagulation and of blood clot dissolution. They generally are synthesized in the liver but are present in blood in equivalent or higher concentrations than in tissues.

At the name implies, **nonfunctional plasma enzymes** perform no known physiologic function in blood. Their substrates frequently are absent from plasma, and the enzymes themselves are present in the blood of normal individuals at levels up to a millionfold lower than in tissues. Their presence in plasma at levels elevated above normal values suggests an increased rate of tissue destruction. Measurement of these nonfunctional plasma enzyme levels can thus provide the physician with valuable diagnostic and prognostic clinical evidence.

Nonfunctional plasma enzymes include those in exocrine secretions and true intracellular enzymes. Exocrine enzymes—pancreatic amylase, lipase, bile alkaline phosphatase, and prostatic acid phosphatase—diffuse into the plasma. True intracellular enzymes normally are absent from the circulation.

Low Levels of Nonfunctional Plasma Enzymes Result from Normal Destruction of Cells

Low levels of nonfunctional enzymes found ordinarily in plasma apparently arise from the routine, normal destruction of erythrocytes, leukocytes, and other cells. With accelerated cell death, soluble enzymes enter the circulation. Although elevated plasma enzyme levels are generally interpreted as evidence of cellular necrosis, vigorous exercise also releases significant quantities of muscle enzymes.

Levels of Nonfunctional Plasma Enzymes Have Long Been Used as a Diagnostic Aid

Practicing physicians have long made use of quantitation of the levels of certain nonfunctional plasma enzymes. This valuable diagnostic and prognostic information is in most instances obtained on fully automated equipment. Table 8–3 contains a list of the principal enzymes employed in the field of diagnostic enzymology. Further details on the uses of these enzymes are given in the Appendix, including discussions of the important concepts of the sensitivity and specificity of diagnostic tests.

RESTRICTION ENDONUCLEASES PROVIDE A NEW APPROACH TO DIAGNOSIS OF GENETIC DISEASES

The diagnosis of genetic diseases received tremendous impetus from recombinent DNA technology. While all molecular diseases have long been known to be a consequence of altered DNA, techniques for direct examination of DNA sequences have only recently become available. Development of hybridization probes for DNA fragments has led to techniques of sensitivity sufficient for prenatal screening for hereditary disorders by restriction enzyme mapping of DNA derived from fetal cells in the amniotic fluid.

In principle, DNA probes can be constructed for the diagnosis of most genetic diseases. For example, for prenatal detection of thalassemias (characterized by defects in the synthesis of hemoglobin subunits; see Chapter 7), a probe constructed against a portion of the gene for a normal hemoglobin subunit can detect a shortened or absent restriction fragment arising from a deletion in that gene, as occurs in some α-thalassemias and certain rare types of β- and β,δ-thalassemias. Alternatively, a synthetic cDNA probe has been constructed that hybridizes to a β-globin sequence that contains a nonsense mutation present in certain β-thalassemias, but not to the normal β-globin gene. Absence of the plasma protease inhibitor α_1-antitrypsin is associated with emphysema and with infantile liver cirrhosis. The presence of inactive α_1-antitrypsin has been detected by a probe constructed against the inactive allele that contains a point mutation in the α_1-antitrypsin gene.

Hybridization probes may also be used to detect genetic alterations that lead to the loss of a restriction endonuclease site (see Chapter 38). For example, the point mutation of the Glu codon (GAG) to the Val codon (GTG) characteristic of sickle cell disease can be detected in the β-globin gene from cells in as little as 10 mL of amniotic fluid using the restriction endonuclease Mst II or Sau I.

PROBED RESTRICTION DIGESTS CAN REVEAL HOMOLOGOUS GENES WITH DIFFERENT BASE SEQUENCES

DNA probes may also be used to detect DNA sequences tightly linked to, but not actually within, the gene of interest. The analyses may be extended to detection of chromosome-specific variations (differences in sequence between homologous chromosomes). Digestion of the DNA with a restriction endonuclease generates different restriction maps (patterns of DNA fragments) from homologous genes that contain different base sequences. This phenomenon is termed "**restriction fragment length polymorphism**" (RFLP). For a genetic disease linked to a restriction length polymorphism, a human carrier of the disease will bear one chromosome with a normal gene and one with the defective gene. When the restriction fragments are resolved and probed, 2 hydridizing bands are detected (as distinct from a single band where both genes are identical). Offspring who inherit the disease-bearing chromosome exhibit only a single hybridizing band that differs from the band produced by normal chromosomes. The phenomenon of associated RFLP has been

Table 8–3. Principal serum enzymes used in clinical diagnosis. Many of the enzymes are not specific for the disease listed; further details of the range of conditions in which the activities of these enzymes are affected are given in the Appendix.

Serum Enzyme	Major Diagnostic Use
Aminotransferases Aspartate aminotransferase (AST, or SGOT)	Myocardial infarction
Alanine aminotransferase (ALT, or SGPT)	Viral hepatitis
Amylase	Acute pancreatitis
Ceruloplasmin	Hepatolenticular degeneration (Wilson's disease)
Creatine phosphokinase	Muscle disorders and myocardial infarction
γ-Glutamyl transpeptidase	Various liver diseases
Lactate dehydrogenase (isozymes)	Myocardial infarction
Lipase	Acute pancreatitis
Phosphatase, acid	Metastatic carcinoma of the prostate
Phosphatase, alkaline (isozymes)	Various bone disorders, obstructive liver diseases

applied to the analysis of sickle cell trait (based on an associated Hpa I RFLP) and of β-thalassemia (linked to a Hind III and a Bam HI RFLP).

A screen based on RFLPs has been developed for the detection of infant phenylketonuria (see Chapter 32). Note, however, that since the RFLP does not cause the disease but is merely located near the defective gene, this general approach is not infallible. RFLPs are points of departure for identification of the gene responsible for linked diseases. This approach has already been applied to a screen for the mutational events involved in formation of retinoblastoma tumors and Huntington's disease.

Further examples of the uses of restriction enzymes for diagnosis are given in Chapter 36.

REFERENCES

Methods in Enzymology. Over 130 volumes, 1955–present.
Advances in Enzymology. Issued annually.
Boyer PD, Lardy H, Myrbach K (editor): *The Enzymes,* 3rd ed. 7 vols. Academic Press, 1970–1973.
Bergmeyer HH, Bergmeyer J, Grass M: *Methods of Enzymic Analysis, Vol. XI, Antigens and Antibodies.* VCH Publishers, 1986.
Fersht A: *Enzyme Structure and Mechanism.* 2nd ed. Freeman, 1985.
Freifelder D: *Physical Biochemistry: Applications to Biochemistry and Molecular Biology.* Freeman, 1982.
Kaiser T, Lawrence DS, Rokita SZ: The chemical modification of enzyme specificity. *Ann Rev Biochem* 1985;**54**:597.
Naqui A: Where are the asymptotes of Michaelis-Menten? *Trends Biochem Sci* 1986;**11**:64.
Scopes R. *Protein Purification: Principles and Practice.* Springer-Verlag, 1982.
Tijssen P: *Practice and Theory of Enzyme Immunoassays.* Elsevier, 1985.

9 Enzymes: Kinetics

Victor W. Rodwell, PhD

INTRODUCTION

In this chapter we consider the chemical nature of the catalysis carried out by enzymes and the nature of the enzyme-substrate interaction responsible for the reaction specificity of these biologic catalysts.

BIOMEDICAL IMPORTANCE

Of the major factors (enzyme and substrate concentration, temperature, pH, and inhibitors) that affect assays of enzyme activity, the last 3 are of particular clinical interest. Because enzyme activities increase as temperature increases, the rates of metabolic processes increase significantly during fevers. If fevers continued indefinitely, the results would be fatal, in part because of metabolic overdemand. On the other hand, lowering body temperature (hypothermia)—and consequently the activities of most enzymes—proves useful when it is important to reduce overall metabolic demand (eg, during open-heart surgery or transportation of organs for transplantation surgery).

A cardinal biologic principle is that of homeostasis—maintenance of the internal milieu of the body very close to its normal conditions. Relatively small changes in pH can affect the activities of many enzymes or proteins. Most drugs act by affecting enzyme-catalyzed reactions. Many drugs resemble natural substrates and so act as competitive inhibitors of enzyme activity. Understanding of much of pharmacology and toxicology depends on a thorough knowledge of the basics of enzyme inhibition.

THE FORMATION AND DECAY OF TRANSITION STATES IN THE OVERALL REACTION ARE ASSOCIATED WITH DISCRETE CHANGES IN FREE ENERGY

In a displacement reaction an entering group Y displaces a leaving group X:

$$Y + R-X \rightleftharpoons Y-R + X$$

This overall reaction proceeds via 2 half reactions:

(1) formation of a **transition state** in which Y and X both are attached to R, and (2) decay of the transition state to form products:

$$Y + R-X \rightleftharpoons \underbrace{Y \cdots R \cdots X}_{\substack{\text{Transition} \\ \text{state}}} \rightleftharpoons Y-R + X$$

As with all chemical reactions, changes in free energy are associated with each half reaction. We define ΔG_F as the change in free energy associated with **formation** of the transition state and ΔG_D as the change in free energy associated with **decay** of the transition state to form products:

$$\Delta G_F = \Delta G_F^0 + RT \ln \frac{[Y \cdots R \cdots X]}{[Y][R-X]}$$

$$\Delta G_D = \Delta G_D^0 + RT \ln \frac{[Y-R][Y]}{[Y \cdots R \cdots X]}$$

where the change in free energy for the **overall** reaction, ΔG, is the sum of the changes in free energy for both half reactions:

$$\Delta G = \Delta G_F + \Delta G_D$$

As for any equation with 2 terms, it is not possible to infer the sign or magnitude of either ΔG_F or ΔG_D by inspecting the algebraic sign and magnitude of ΔG. Stated another way, we are unable simply from consideration of the change in free energy for the overall reaction, ΔG, to infer anything whatever concerning the free energy changes associated with formation and decay of transition states. Since catalysis is intimately associated with ΔG_F and ΔG_D, it follows that the thermodynamics of the overall reaction (ΔG) can tell us nothing of the path a reaction follows (ie, its mechanism). This is the task of kinetics.

Reaction Profiles Illustrate the Free Energy Changes Associated with the Formation & Decay of Transition States

The reaction profiles in Figs 9–1 and 9–2 illustrate the relationship between ΔG, ΔG_F, and ΔG_D. Note

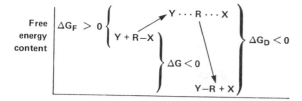

Figure 9–1. Reaction profile for a displacement reaction with a negative overall change in free energy, ie, $\Delta G < 0$.

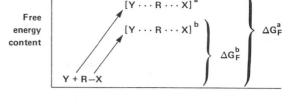

Figure 9–3. Reaction profile for formation of 2 different transition states, $[Y \cdots R \cdots X]^a$ and $[Y \cdots R \cdots X]^b$, and their associated free energies of formation, ΔG_F^a and ΔG_F^b.

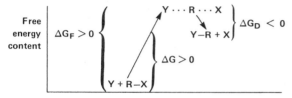

Figure 9–2. Reaction profile for a displacement reaction with a positive overall change in free energy, ie, $\Delta G > 0$.

that whereas in Fig 9–1 ΔG is negative ($\Delta G < 0$) and in Fig 9–2 ΔG is positive ($\Delta G > 0$), in both instances, ΔG_F is positive ($\Delta G_F > 0$) and ΔG_D is negative ($\Delta G_D < 0$). Therefore, as stated above, **from the sign and magnitude of ΔG we cannot infer the sign or magnitude of either ΔG_F or ΔG_D.**

Enzyme-Catalyzed Reactions Involve Transition States at Lower Energy Levels Than in the Uncatalyzed Reaction

Fig 9–3 depicts the reaction profiles for 2 different transition states in the same overall reaction. In both instances, the magnitude of the free energy of formation of the transition state represents the energy barrier for the overall reaction. The energy barrier for the reaction that proceeds via the transition state $[Y \cdots R \cdots X]^b$ is thus lower than the energy barrier for the reaction that proceeds via the transition state $[Y \cdots R \cdots X]^a$. **Catalysts alter the free energy content of the transition state.** In the above example, $[Y \cdots R \cdots X]^a$ represents the transition state for the **noncatalyzed** reaction, while $[Y \cdots R \cdots X]^b$ represents the transition state for the **catalyzed** reaction. All catalysts, including enzymes, lower the free energy of formation, ΔG_F, of the transition state. Note further that since catalysis has no effect on ΔG, **the change in free energy for the overall reaction is independent of catalysis.** Since the equilibrium constant for a chemical reaction is a function of the standard free energy change for a reaction:

$$\Delta G^0 = -RT \ln K_{eq}$$

it follows that **enzymes and other catalysts have no effect on the equilibrium constant for a reaction.**

TO REACT, MOLECULES MUST APPROACH WITHIN BOND-FORMING DISTANCE WITH SUFFICIENT ENERGY TO OVERCOME THE ENERGY BARRIERS BETWEEN TRANSITION STATES

The **kinetic,** or **collision, theory** for chemical reactions incorporates 2 key concepts:

(1) In order to react, molecules must collide (ie, be within bond-forming distance of one another).

(2) For a collision to result in a reaction, the reacting molecules must possess sufficient energy to overcome the energy barrier for the reaction.

Factors That Increase Collision Frequency & Kinetic Energy Increase Reaction Rates

If molecules have sufficient energy to react, anything that increases the frequency of collision between molecules will increase their rate of reaction. Conversely, factors that decrease either collision frequency or kinetic energy will decrease the rate of reaction. If some molecules in the population have insufficient energy to react, increasing the temperature, which increases kinetic energy, will increase the rate of the reaction. These concepts are illustrated in Fig 9–4. In A none, in B portion, and in C all of the molecules have sufficient kinetic energy to overcome the energy barrier for reaction.

In the absence of enzymatic catalysis, many chemical reactions proceed exceedingly slowly at the tem-

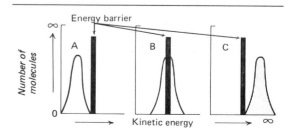

Figure 9–4. The energy barrier for chemical reactions.

perature of living cells. However, even at this temperature molecules are in motion and undergo collisions. **They fail to react rapidly, because most possess insufficient kinetic energy to overcome the energy barrier for reaction.** At a considerably higher temperature (and higher kinetic energy), the reaction will occur more rapidly. That the reaction takes place at all shows that it is spontaneous ($\Delta G < 0$). At the lower temperature, it is spontaneous but slow; at the higher temperature, it is spontaneous and fast. **Enzymes make spontaneous reactions proceed rapidly under the conditions prevailing in living cells.**

Most Chemical Reactions of Biologic Interest Involve Formation or Rupture of Covalent Bonds

For the group transfer reaction:

$$D–G + A \rightleftharpoons A–G + D$$

a group, G, is transferred from a donor, D–G, to an acceptor, A. The overall reaction involves both rupture of the D–G bond and formation of a new A–G bond.

Enzyme-catalyzed group transfer reactions may be represented as follows:

This emphasizes 3 important features of enzyme-catalyzed group transfer reactions:

(1) Each half reaction involves both the rupture and formation of a covalent bond.

(2) The enzyme is a reactant coequal with D–G and A.

(3) Whereas in the overall reaction the enzyme acts catalytically (ie, is required only in trace quantities and may be recovered unchanged when the reaction is complete), for each of the half reactions, the enzyme is a stoichiometric reactant (ie, it is required in a 1:1 molar ratio with the other reactants).

Many additional biochemical reactions may be considered as special cases of group transfer in which D, A, or both may be absent. For example, isomerization reactions (eg, the interconversion of glucose 6-phosphate and glucose 1-phosphate) might be represented as reactions in which both D and A are absent:

SEVERAL PARTIAL REACTIONS AND ENZYME-SUBSTRATE COMPLEXES PARTICIPATE IN AN ENZYME-CATALYZED REACTION

The above representations fail to emphasize yet another key feature of enzyme-catalyzed reactions—**participation in the overall reaction of 2 or more intermediate forms of EnzS complex and the consequent participation of a set of several sequential half reactions.** A representation of a group transfer reaction that emphasizes these features might be

in which EnzG, EnzG*, and EnzG** represent successive EnzS complexes in the overall reaction.

Binding of Reactants to an Enzyme Raises Their Local Concentration & Brings Them Within Bond-Forming Distance

For any of the above reactions to occur, all of the reactants must come within bond-forming (or bond-breaking) distance of one another. For homogeneous solution chemistry in the absence of catalysts, the concentrations of the reacting molecules are constant throughout the solution. This condition no longer obtains following introduction of a catalyst. A catalyst must have surface domains that bind the reacting molecules. While this binding is a reversible process, the overall equilibrium constant for binding strongly favors the bound rather than the free forms of the reacting molecules. Qualitatively, we might represent this as follows:

Reactant + Catalyst $\rightleftharpoons$ Reactant-catalyst complex

Quantitatively, we might express the tightness of association between a reactant, R, and a catalyst, C, in terms of the dissociation constant, K_d, for the R–C complex, or the equilibrium constant for the reaction:

$$R–C \rightleftharpoons R + C$$

$$K_d = \frac{[R]\,[C]}{[R–C]}$$

A low value for K_d thus represents a tight R–C complex.

One important consequence is that when a **reactant binds to a catalyst this raises the concentration of the reactant in a localized area of the solution** well above that of its concentration in free solution. Thus, we are no longer dealing with homogeneous but with heterogeneous solution chemistry.

If the catalyst for a bimolecular (2-reactant) reaction binds both reactants, the local concentration of each reactant is increased by a factor that depends on its individual affinity (K_d value) for the catalyst. Since the rate of the overall bimolecular reaction

$$A + B \rightarrow A\text{-}B$$

is, as we shall see below, proportionate to the concentrations of both A and B, **binding of both A and B by the catalyst can result in an enormous** (several thousand-fold) **increase in overall reaction rate.**

The Region of an Enzyme Concerned with Substrate Binding & Catalysis Is Termed the "Active Site"

A key property of enzymes is their ability to bind one or (more frequently) both reactants in a bimolecular reaction with an accompanying increase in local reactant concentration, and hence in local reaction rate. Enzymes are both extremely efficient and highly selective catalysts. To understand these distinctive properties of enzymes, we must introduce the concept of the "active" or "catalytic" site.*

The large size of proteins relative to substrates led to the concept that a restricted region of the enzyme, the "active site," was concerned with catalysis. Initially, it was puzzling why enzymes were so large when only a portion of their structure appeared to be required for substrate binding and catalysis. Today, we recognize that a far greater portion of the protein interacts with the substrate than was formerly supposed. When the need for allosteric sites of equal size also arises, the size of enzymes should no longer be surprising.

Many Properties of Enzymes Can Be Understood in Terms of the Rigid "Lock-and-Key" Model of an Active Site

The model of a catalytic site proposed by Emil Fischer visualized interaction between substrate and enzyme in terms of a "lock and key" analogy. This **lock and key,** or **rigid template, model** (Fig. 9–5), is

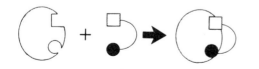

Figure 9–5. Representation of formation of an EnzS complex according to the Fischer template hypothesis.

---

*While many texts equate the active and catalytic sites of enzymes, there are on enzymes other "active" sites concerned with regulation of enzyme activity rather than with the intermediary enzymology of the catalytic process per se. We therefore use the term "catalytic site" to avoid ambiguity.

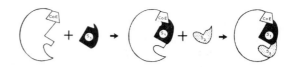

Figure 9–6. Representation of sequential adsorption of a coenzyme (CoE) and of 2 substrates (S_1 and S_2) to an enzyme in terms of the template hypothesis. The coenzyme is assumed to bear a group essential for binding the first substrate (S_1), which in turn facilitates binding of S_2.

---

still useful for understanding certain properties of enzymes—for example, the ordered binding of 2 or more substrates (Fig 9–6) or the kinetics of a simple substrate saturation curve.

In the "Induced Fit" Model of a Catalytic Site, the Substrate Induces a Conformational Change in the Enzyme That "Creates" the Catalytic Site

An unfortunate feature of the Fischer model is the implied rigidity of the catalytic site. A more general model is the **"induced fit" model** of Koshland. This model has considerable experimental support. In the Fischer model, the catalytic site is presumed to be preshaped to fit the substrate. In the induced fit model, the substrate induces a conformational change in the enzyme. This aligns amino acid residues or other groups on the enzyme in the correct spatial orientation for substrate binding, catalysis, or both.

In the example (Fig 9–7), hydrophobic groups (hatched) and charged groups (stippled) both are involved in substrate binding. A phosphoserine (— P) and the —SH of a cysteine residue are involved in catalysis. Other residues involved in neither process are represented by Lys and Met residues. In the absence of substrate, the catalytic and the substrate-binding groups are several bond distances apart. Approach of the substrate induces a conformational change in the enzyme protein, aligning the groups correctly for substrate binding and for catalysis. At the same time, the spatial orientations of other regions are also altered—the Lys and Met are now closer together (Fig 9–7).

Substrate analogs may cause some, but not all, of the correct conformational changes (Fig 9–8). On attach-

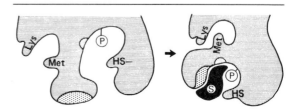

Figure 9–7. Two-dimensional representation of an induced fit by a conformational change in the protein structure. Note the relative positions of key residues before and after the substrate is bound.

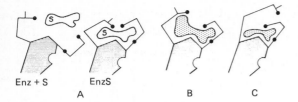

Figure 9–8. Representation of conformational changes in an enzyme protein when binding substrate (*A*) or inactive substrate analogs (*B, C*). (After Koshland.)

ment of the true substrate (A), all groups (shown as closed circles) are brought into correct alignment. Attachment of a substant analog that is too "bulky" (Fig 9–8B) or too "slim" (Fig 9–8C) induces incorrect alignment. One final feature is the site shown as a small notch on the right. One may visualize a regulatory molecule attaching at this point and "holding down" one of the polypeptide arms bearing a catalytic group. Substrate binding, but not catalysis, might then occur. As illustrated by the induced fit model, the residues that comprise the catalytic site may be distant one from another in the primary structure but spatially close in the 3-dimensional (tertiary) structure.

The Catalytic Sites of Lysozyme, Ribonuclease, & Many Other Enzymes Are in Clefts in the Enzyme Surface

Lysozyme, present in tears, nasal mucus, sputum, tissues, gastric secretions, milk, and egg white, catalyzes the hydrolysis of β-1,4 linkages of N-acetylneuraminic acid in proteoglycans and glycosaminoglycans. It performs the function, in tears and nasal mucus, of destroying the cell walls of many airborne gram-positive bacteria. Lysozyme consists of a single polypeptide chain of 129 residues. Since there is no coenzyme or metal ion, catalysis, specificity, and 3-

dimensional structure are determined solely by these amino acid residues. There are small regions of pleated sheet, little α-helix, and large regions of random coil. The molecule bears a deep central cleft that harbors a catalytic site with 6 subsites (Fig 9–9) that bind various substrates or inhibitors. The residues responsible for bond cleavage lie between sites D and E close to the carboxyl groups of Asp 52 and Glu 35. Glu 35 apparently protonates the acetal bond of the substrate, while the negatively charged Asp 52 stabilizes the resulting carbonium ion from the back side.

The catalytic site of ribonuclease lies within a cleft similar to that of lysozyme, across which lie 2 residues, His 12 and His 119. These previously were implicated by chemical evidence as being at the catalytic site. Both residues are near the binding site for uridylic acid (Fig 9–10).

A Common Primary Structural Motif Is Apparent in the Catalytic Sites of Several Hydrolytic Enzymes

The primary structures of catalytic sites of hydrolytic enzymes exhibit many similarities (Table 9–1). This implies that the number of bond-breaking mechanisms operating in biologic systems is relatively small. In view of these similarities, it is perhaps not surprising that the amino acid sequences near the catalytic sites of the same enzyme from different species bear even greater similarity.

OVER A LIMITED RANGE OF TEMPERATURES, THE VELOCITY OF ENZYME-CATALYZED REACTIONS INCREASES AS TEMPERATURE RISES

Fig 9–11 illustrates the effect of temperature on a typical enzyme-catalyzed reaction. The reaction rate initially increases as temperature rises, due to in-

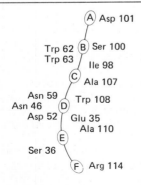

Figure 9–9. Schematic representation of the catalytic site in the cleft region of lysozyme. A to F represent the glycosyl moieties of a hexasaccharide. Some residues in the cleft region are shown with their numbers in the lysozyme sequence. (Adapted from Koshland.)

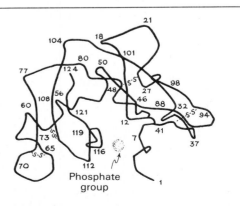

Figure 9–10. Structure of ribonuclease as determined by x-ray diffraction. Numbers refer to specific residues. See also Fig 6–7.

Table 9–1. Amino acid sequences in the neighborhood of the catalytic sites of several bovine proteases. Regions shown are those on either side of the catalytic site seryl (S) and histidyl (H) residues.*

Enzyme	Sequence Around Serine Ⓢ															Sequence Around Histidine Ⓗ								
Trypsin	D	S	C	Q	D	G	Ⓢ	G	G	P	V	V	C	S	G	K	V	V	S	A	A	Ⓗ	C	Y	K	S	G
Chymotrypsin A	S	S	C	M	G	D	Ⓢ	G	G	P	L	V	C	K	K	N	V	V	T	A	A	Ⓗ	G	G	V	T	T
Chymotrypsin B	S	S	C	M	G	D	Ⓢ	G	G	P	L	V	C	Q	K	N	V	V	T	A	A	Ⓗ	C	G	V	T	T
Thrombin	D	A	C	E	G	D	Ⓢ	G	G	P	F	V	M	K	S	P	V	L	T	A	A	Ⓗ	C	L	L	Y	P

*Reproduced, with permission, from Dayhoff MO [editor]: *Atlas of Protein Sequence and Structure.* Vol 5. National Biomedical Research Foundation, 1972.

creased kinetic energy of the reacting molecules. Eventually, however, the kinetic energy of the enzyme exceeds the energy barrier for breaking the weak hydrogen and hydrophobic bonds that maintain its secondary-tertiary structure. At this temperature denaturation, with an accompanying precipitate loss of catalytic activity, predominates. Enzymes therefore exhibit an optimal temperature. This is not, however, a fundamental parameter, since the optimal temperature depends upon the duration of the assay used to determine it; ie, the longer an enzyme is maintained at a temperature at which its structure is marginally stable, the more likely it is to be denatured.

The factor by which the rate of a biologic process increases for a 10 °C temperature rise is the Q_{10} or **temperature coefficient.** The rate of many biologic processes, for example, the rate of contraction of an excised heart, approximately doubles with a 10 °C rise in temperature ($Q_{10} = 2$).

For most enzymes, optimal temperatures are at or above those of the cells in which the enzymes occur. For example, enzymes from microorganisms adapted to growth in natural hot springs may exhibit optimal temperatures close to the boiling point of water.

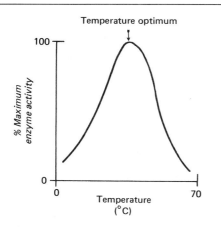

Figure 9–11. Effect of temperature on the velocity of a hypothetical enzyme-catalyzed reaction.

pH AFFECTS ENZYME ACTIVITY BY CHANGING THE CHARGE ON IONIZABLE GROUPS OF THE ENZYME, AND OFTEN OF THE SUBSTRATE ALSO

When enzyme activity is measured at several pH values, optimal activity typically is observed between pH values of 5.0 and 9.0. However, a few enzymes, for example, pepsin, are active at pH values well outside this range.

The shape of pH-activity curves is determined by:

(1) Enzyme denaturation at high or low pH.

(2) Alterations in the charged state of the enzyme and/or substrates. For the enzyme, pH can affect activity by changing the structure or by changing the charge on a residue functional in substrate binding or catalysis. To illustrate, consider a negatively charged enzyme (Enz⁻) reacting with a positively charged substrate (SH⁺):

$$Enz^- + SH^+ \rightarrow EnzSH$$

At low pH, Enz⁻ protonates and loses its negative charge:

$$Enz^- + H^+ \rightarrow EnzH$$

At high pH, SH⁺ ionizes and loses its positive charge:

$$SH^+ \rightarrow S + H^+$$

Since, by definition, the only forms that will interact are SH⁺ and Enz⁻, extreme pH values will lower the effective concentration of Enz⁻ and SH⁺, thus lowering the reaction velocity (Fig 9–12). Only in the cross-hatched area are both Enz and S in the appropriate ionic state, and the maximal concentrations of Enz and S are correctly charged at X.

Enzymes may also undergo changes in conformation when the pH is varied. A charged group distal to the region where the substrate is bound may be necessary to maintain an active tertiary or quarternary structure. As the charge on this group is changed, the protein may unravel, become more compact, or dissociate into protomers—all with resulting loss of activity. Depending upon the severity of these changes, activity

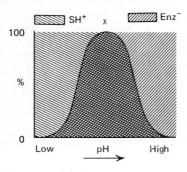

Figure 9–12. Effect of pH on enzyme activity.

may or may not be restored when the enzyme is returned to its optimal pH.

IF REACTANTS ARE PRESENT IN APPROXIMATELY EQUIMOLAR PROPORTIONS, REACTION RATES ARE PROPORTIONAL TO THE CONCENTRATIONS OF ALL REACTANTS

At high reactant concentrations, both the number of molecules with sufficient energy to react and their frequency of collision are high. This is true whether all or only a fraction of the molecules have sufficient energy to react. Consider reactions involving 2 different molecules, A and B:

$$A + B \rightarrow AB$$

Doubling the concentration either of A or of B will double the reaction rate. Doubling the concentration of both A and B will increase the probability of collision 4-fold. The reaction rate therefore increases 4-fold. **The reaction rate is proportionate to the concentrations of the reacting molecules.** Square brackets ([]) are used to denote molar concentrations;* $\propto$ means "proportionate to." The rate expression is:

$$Rate \propto [reacting\ molecules]$$

or

$$Rate \propto [A]\ [B]$$

For the situation represented by

$$A + 2B \rightarrow AB_2$$

the rate expression is:

$$Rate \propto [A]\ [B]\ [B]$$

---

*Strictly speaking, molar activities rather than concentrations should be used.

or

$$Rate \propto [A]\ [B]^2$$

For the general case when n molecules of A react with m molecules of B

$$nA + mB \rightarrow A_nB_m$$

the rate expression is:

$$Rate \propto [A]^n[B]^m$$

THE EQUILIBRIUM CONSTANT IS THE RATIO OF THE RATE CONSTANT FOR THE FORWARD REACTION DIVIDED BY THE RATE CONSTANT FOR THE REVERSE REACTION

Since all chemical reactions are reversible, for the reverse reaction where A_nB_m forms n molecules of A and m molecules of B

$$A_nB_m \rightarrow nA + mB$$

the appropriate rate expression is:

$$Rate \propto [A_nB_m]$$

Reversibility is represented by double arrows,

$$nA + mB \rightleftharpoons A_nB_m$$

This expression reads: "n molecules of A and m molecules of B are in equilibrium with A_nB_m." The "proportionate to" symbol ($\propto$) may be replaced with an equality sign by inserting a proportionality constant, k, characteristic of the reaction under study. For the general case

$$nA + mB \rightleftharpoons A_nB_m$$

expressions for the rates of the forward reaction (Rate$_1$) and back reaction (Rate$_{-1}$) are

$$Rate_1 = k_1[A]^n[B]^m$$

and

$$Rate_{-1} = k_{-1}[A_nB_m]$$

When the rates of the forward and back reactions are equal, the system is said to be **at equilibrium,** ie,

$$Rate_1 = Rate_{-1}$$

Then

$$k_1[A]^n[B]^m = k_{-1}[A_nB_m]$$

and

$$\frac{k_1}{k_{-1}} = \frac{[A_nB_m]}{[A]^n[B]^m} = K_{eq}$$

The ratio of k_1 to k_{-1} is termed the **equilibrium constant, K_{eq}.** The following important properties of a system at equilibrium should be kept in mind.

(1) The equilibrium constant is the ratio of the reaction rate **constants** k_1/k_{-1}.

(2) At equilibrium, the reaction **rates** (not the reaction rate constants) of the forward and back reactions are equal.

(3) **Equilibrium is a dynamic state.** Although no **net** change in concentration of reactant or product molecules occurs at equilibrium, A and B are continually being converted to A_nB_m and vice versa.

(4) **The equilibrium constant may be given a numerical value if the concentrations of A, B, and A_nB_m at equilibrium are known.**

The Standard Change in Free Energy (ΔG^0) for a Reaction Can Be Calculated from the Equilibrium Constant

The equilibrium constant is related to ΔG^0 as follows:

$$\Delta G^0 = -RT \text{ in } K_{eq}$$

R is the gas constant and T the absolute temperature. Since these are known, **knowledge of the numerical value of K_{eq} permits one to calculate a value for ΔG^0.** If the equilibrium constant is greater than 1, the reaction is spontaneous; ie, the reaction as written (from left to right) is favored. If it is less than 1, the opposite is true; ie, the reaction is more likely to proceed from right to left. Note, however, that although **the equilibrium constant for a reaction indicates the direction in which a reaction is spontaneous,** it does not indicate whether it will take place rapidly. That is, **it does not tell us anything about the magnitude of the energy barrier** for the reaction (ie, ΔG_F; see above). This follows because K_{eq} determines ΔG^0, previously shown to concern only initial and final states. **Reaction rates depend on the magnitude of the energy barrier, not on the magnitude of ΔG^0.**

Most factors affecting the velocity of enzyme-catalyzed reactions do so by **changing local reactant concentration.**

Enzymes Do Not Affect the Equilibrium Constants of the Reactions They Catalyze

The enzyme is a reactant that combines with substrate to form an **enzyme-substrate complex, EnzS,** which decomposes to form a product, P, and free enzyme. In its simplest form, this may be represented

$$Enz + S \underset{k_{-1}}{\overset{k_1}{\rightleftharpoons}} Enz + P$$

While the rate expressions for the forward, back, and *overall* reactions include the term [Enz],

$$Enz + S \underset{k_{-1}}{\overset{k_1}{\rightleftharpoons}} Enz + P$$

$$Rate_1 = k_1[Enz][S]$$

$$Rate_{-1} = k_{-1}[Enz][P]$$

in the expression for the *overall* equilibrium constant, [Enz] cancels out.

$$K_{eq} = \frac{k_1}{k_{-1}} = \frac{[Enz][P]}{[Enz][S]} = \frac{[P]}{[S]}$$

The enzyme concentration thus has no effect on the equilibrium constant. Stated another way, since enzymes affect rates, not rate constants, they cannot affect K_{eq}, which is a ratio of rate constants. **The K_{eq} of a reaction is the same regardless of whether equilibrium is approached with or without enzymatic catalysis** (recall ΔG^0). Enzymes change the reaction path but do not affect the initial and final equilibrium concentrations of the reactants and products, the factors that determine K_{eq} and ΔG^0.

THE INITIAL RATE OF AN ENZYME-CATALYZED REACTION IS DIRECTLY PROPORTIONAL TO THE CONCENTRATION OF ENZYME

The initial rate of a reaction is the rate measured before sufficient product has been formed to permit the reverse reaction to occur. The initial rate of an enzyme-catalyzed reaction is always proportional to the concentration of enzyme. Note, however, that this statement holds only for initial rates.

IN MANY SITUATIONS OF PHYSIOLOGIC INTEREST, THE SUBSTRATE CONCENTRATION AFFECTS THE RATE OF AN ENZYME-CATALYZED REACTION

In the following discussion, enzyme reactions are treated as if they had a single substrate and a single product. While this is the case for some enzyme-catalyzed reactions, most enzyme-catalyzed reactions have 2 or more substrates and products. This consideration does not, however, invalidate the discussion.

If the concentration of a substrate [S] is increased while all other conditions are kept constant, the **measured initial velocity, v_i** (the velocity measured when very little substrate has reacted), increases to a maximum value, V_{max}, and no further (Fig 9–13).

The velocity increases as the substrate concentration is increased up to a point where the enzyme is said to be "saturated" with substrate. The measured initial velocity reaches a maximal value and is unaffected by further increases in substrate concentration, because substrate is present in large molar excess over the en-

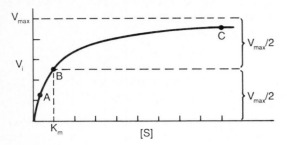

Figure 9–13. Effect of substrate concentration on the velocity of an enzyme-catalyzed reaction.

zyme. For example, if an enzyme with a molecular weight of 100,000 acts on a substrate with a molecular weight of 100 and both are present at a concentration of 1 mg/mL, there are 1000 mol of substrate for every mole of enzyme. More realistic figures might be

$$[Enz] = 0.1 \ \mu g/mL = 10^{-9} \ molar$$
$$[S] = 0.1 \ mg/mL = 10^{-3} \ molar$$

giving a 10^6 molar excess of substrate over enzyme. Even if [S] is decreased 100-fold, substrate is still present in 10,000-fold molar excess over enzyme.

The situations at points A, B, and C in Fig 9–13 are illustrated in Fig 9–14. At points A and B only a portion of the enzyme present is combined with substrate, even though there are many more molecules of substrate than of enzyme. This is because the equilibrium constant for the reaction Enz + S ⇌ EnzS (formation of the EnzS complex) is not infinitely large. **At point A or B, increasing or decreasing [S] will therefore increase or decrease the amount of Enz associated with S as EnzS, and v_i will thus depend on [S].**

At C, essentially all the enzyme is combined with substrate, so that a further increase in [S], although it increases the frequency of collision between Enz and S, cannot result in increased rates of reaction, since no free enzyme is available to react.

Case B depicts a situation of major theoretical interest where exactly half the enzyme molecules are "saturated with" substrate. The velocity is accordingly **half the maximal velocity** ($V_{max}/2$) attainable at that particular enzyme concentration.

The Michaelis-Menten Equation Models the Effects of Substrate Concentration on the Rates of Many Enzyme-Catalyzed Reactions

The substrate concentration that produces half-maximal velocity, termed the K_m value or Michaelis constant, may be determined experimentally by graphing v_i as a function of [S] (Fig 9–13). **K_m has the dimensions of molar concentration.**

When [S] is approximately equal to the K_m, v_i is very responsive to changes in [S], and the enzyme is working at precisely half-maximal velocity. In fact, many enzymes possess K_m values which approximate the physiologic concentration of their substrates.

The Michaelis-Menten expression

$$v_i = \frac{V_{max}[S]}{K_m + [S]}$$

describes the behavior of many enzymes as substrate concentration is varied. The dependence of the initial velocity of an enzyme-catalyzed reaction on [S] and on K_m may be illustrated by evaluating the Michaelis-Menten equation as follows:

(1) When [S] is very much less than K_m (point A in Figs 9–13 and 9–14). Adding [S] to K_m in the denominator now changes its value very little, so that the [S] term can be dropped from the denominator. Since V_{max} and K_m are both constants, we can replace their ratio by a new constant, K:

$$v_i = \frac{V_{max}[S]}{K_m + [S]} \ ; \ v_i \sim \frac{V_{max}[S]}{K_m} \sim \frac{V_{max}}{K_m}[S] \sim K[S]$$

[$\sim$ means "approximately equal to."]

In other words, **when the substrate concentration**

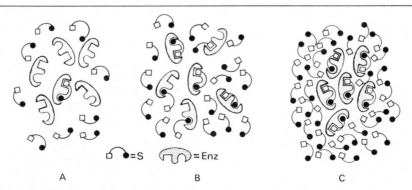

Figure 9–14. Representation of an enzyme at low (A), at high (C), and at the K_m concentration of substrate (B). Points A, B, and C correspond to those of Fig 9–13.

is considerably below that required to produce half-maximal velocity (the K_m value), the initial velocity, v_i, depends upon the substrate concentration, [S].

(2) **When [S] is very much greater than K_m** (point C in Figs 9–13 and 9–14). Now adding K_m to [S] in the denominator changes the value of the denominator very little, so that the term K_m can be dropped from the denominator:

$$v_i = \frac{V_{max}[S]}{K_m + [S]} \; ; \; v_i \sim \frac{V_{max}[S]}{[S]} \sim V_{max}$$

This states that **when the substrate concentration [S] far exceeds the K_m value, the initial velocity, v_i, is maximal, V_{max}.**

(3) **When [S] $= K_m$** (point B in Figs 9–13 and 9–14).

$$v_i = \frac{V_{max}[S]}{K_m + [S]} \; ; \; v_i = \frac{V_{max}[S]}{[S] + [S]} = \frac{V_{max}[S]}{2\,[S]} = \frac{V_{max}}{2}$$

This states that **when the substrate concentration is equal to the K_m value, the initial velocity, v_i, is half-maximal.** It also tells how **to evaluate K_m**, namely, to **determine experimentally the substrate concentration at which the initial velocity is half-maximal.**

The Michaelis-Menten Equation Is Rearranged into the Form of a Straight Line To Determine K_m & V_{max}

Since many enzymes give saturation curves that do not readily permit evaluation of V_{max} (and hence of K_m) when v_i is plotted versus [S], it is convenient to rearrange the Michaelis-Menten expression to simplify evaluation of K_m and V_{max}. The Michaelis-Menten equation may be inverted and factored as follows:

$$v_i = \frac{V_{max}[S]}{K_m + [S]}$$

Invert:
$$\frac{1}{v_i} = \frac{K_m + [S]}{V_{max}[S]}$$

Factor:
$$\frac{1}{v_i} = \frac{K_m}{V_{max}} \cdot \frac{1}{[S]} + \frac{[S]}{V_{max}[S]}$$

Simplify:
$$\frac{1}{v_i} = \frac{K_m}{V_{max}} \cdot \frac{1}{[S]} + \frac{1}{V_{max}}$$

This is the equation for a **straight line**

$$y = a \cdot x + b$$

where
$$y = \frac{1}{v_i} \text{ and } x = \frac{1}{[S]}$$

If y, or $1/v_i$, is plotted as a function of x, or $1/[S]$, the y intercept, b, is $1/V_{max}$, and the slope, a, is

K_m/V_{max}. The negative x intercept may be evaluated by setting y = 0. Then

$$x = -\frac{b}{a} = -\frac{1}{K_m}$$

Such a plot is called a double-reciprocal plot; ie, the reciprocal of v_i ($1/v_i$) is plotted versus the reciprocal of [S]$(1/[S])$.

K_m may be estimated from the **double-reciprocal or Lineweaver-Burk plot** (Fig 9–15) using either the slope and y intercept or the negative x intercept. Since [S] is expressed in molarity, **the dimensions of K_m are molarity or moles per liter.** Velocity, v_i, may be expressed in any units, since **K_m is independent of [Enz].** The double-reciprocal treatment requires relatively few points to define K_m and is the method most often used to determine K_m.

Experimentally, use of the Lineweaver-Burk approach to evaluate K_m can give rise to unwarranted emphasis on data gathered at low substrate concentrations. This is the case if the substrate concentrations selected for study differ by a constant increment. This shortcoming may be circumvented by selecting substrate concentrations the reciprocals of which differ by constant increments.

An alternative approach to the experimental evaluation of K_m and of V_{max} is that of Eadie and Hofstee. The Michaelis-Menten equation may be rearranged:

$$\frac{v_i}{[S]} = -v_i \cdot \frac{1}{K_m} + \frac{V_{max}}{K_m}$$

To evaluate K_m and V_{max}, plot $v_i/[S]$ (y axis) versus v_i (x axis). The y intercept is then V_{max}/K_m, and the x intercept is V_{max}. The slope is $-1/K_m$.

While both the Lineweaver-Burk and Eadie-Hofstee approaches are useful in selected instances, rigorous determination of K_m and of V_{max} requires statistical treatment.

Apart from their usefulness in interpretation of the mechanisms of enzyme-catalyzed reactions, K_m values are of considerable practical value. At a substrate concentration of 100 times K_m, an enzyme will act at essentially maximum rate, and therefore the **maximal velocity (V_{max}) will reflect the amount of active enzyme present.** This situation is generally desirable in

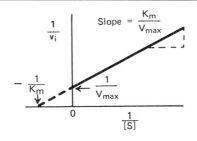

Figure 9–15. Double reciprocal or Lineweaver-Burk plot of 1/v, versus 1/[S] used to evaluate K_m and V_{max}.

the assay of enzymes. The **K_m value tells how much substrate to use in order to measure V_{max}.** Double reciprocal treatments also find extensive application in the evaluation of enzyme inhibitors.

In Specific Instances, K_m Can Approximate a Binding Constant

The **affinity** of an enzyme for its substrate is equal to the **inverse of the dissociation constant, K_d, for the enzyme-substrate complex, EnzS.**

$$Enz + S \underset{k_{-1}}{\overset{k_1}{\rightleftharpoons}} EnzS$$

$$K_d = \frac{k_{-1}}{k_1}$$

Stated another way, the smaller the tendency of the substrate and enzyme to dissociate, the greater is the affinity of the enzyme for the substrate.

The K_m value of an enzyme for its substrate may also serve as a measure of its K_d. However, for this to be true, an assumption included in the derivation of the Michaelis-Menten expression must be valid. The derivation assumed that the first step of the enzyme-catalyzed reaction

$$Enz + S \underset{k_{-1}}{\overset{k_1}{\rightleftharpoons}} EnzS$$

is fast and always at equilibrium. In other words, the rate of dissociation of EnzS to Enz + S must be much faster than its dissociation to enzyme + product:

$$EnzS \underset{k_{-2}}{\overset{k_2}{\rightleftharpoons}} Enz + P$$

In the Michaelis-Menten expression, the [S] that gives $v_i = V_{max}/2$ is

$$[S] = \frac{k_2 + k_{-1}}{k_1} = K_m$$

But when $\qquad k_{-1} \gg k_2$

then $\qquad k_2 + k_{-1} \sim k_{-1}$

and $\qquad [S] = \frac{k_{-1}}{k_1} \sim K_d$

Under these conditions, $1/K_m = 1/K_d = $ **affinity.** If $k_2 + k_{-1} \not\propto k_{-1}$, then $1/K_m$ underestimates the affinity, $1/K_d$.

THE HILL EQUATION MODELS THE EFFECT OF SUBSTRATE CONCENTRATION ON ALLOSTERIC ENZYMES

Certain enzymes and other ligand-binding proteins such as hemoglobin do not exhibit classic Michaelis-

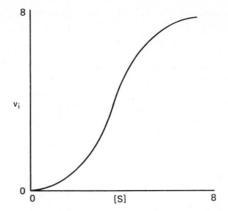

Figure 9–16. Sigmoid saturation kinetics.

Menten saturation kinetics. When [S] is plotted versus v_i, the saturation curve is sigmoid (Fig 9–16). This generally indicates cooperative binding of substrate to multiple sites. Binding at one site affects binding at the others, as described in Chapter 7 for hemoglobin.

For sigmoid substrate saturation kinetics, the methods of graphic evaluation of the substrate concentration that produces half-maximal velocity discussed above are invalid (straight lines are not produced). To evaluate sigmoid saturation kinetics, we employ a graphic representation of the Hill equation, an equation originally derived to describe the cooperative binding of O_2 to hemoglobin. Written in the form of a straight line, the Hill equation is

$$\log \frac{v_i}{V_{max} - v_i} = n\log [S] - \log k'$$

where k' is a complex constant. The equation states that, when [S] is low compared to k', the reaction velocity increases as the nth power of [S]. Fig 9–17

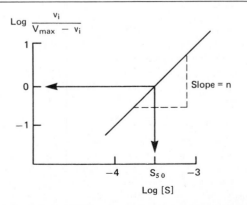

Figure 9–17. Graphic evaluation of the Hill equation to determine the substrate concentration that produces half-maximal velocity. Employed when substrate saturation kinetics are sigmoid.

illustrates a Hill plot of kinetic data for an enzyme with cooperative binding kinetics. A plot of $v_i/V_{max} - v_i$ versus log [S] yields a straight line with slope = n, where n is an empirical parameter whose value depends on the number of substrate-binding sites and the number and type of interactions between these binding sites. When n = 1, the binding sites act independently of one another. If n > 1, the sites are cooperative; and the greater the value of n, the stronger is the cooperativity and thus the more "sigmoid" are the saturation kinetics. If n < 1, the sites are said to exhibit negative cooperativity.

At half-maximal velocity ($v_i = V_{max}/2$), $v_i/(V_{max} - v_i) = 1$, and hence log $v_i/(V_{max} - v_i) = 0$. Thus, to determine S_{50} (the concentration of substrate that produces half-maximal velocity), drop a perpendicular line to the x axis from the point where log $v_i/(V_{max} - v_i) = 0$.

THE DISTINCTION BETWEEN COMPETITIVE AND NONCOMPETITIVE INHIBITORS OF ENZYMES RESTS ON WHETHER INCREASING THE SUBSTRATE CONCENTRATION ABOLISHES INHIBITION

While we distinguish inhibitors of enzyme activity on the basis of whether the inhibition is or is not relieved by increasing the substrate concentration, many inhibitors do not exhibit the idealized properties of pure competitive or noncompetitive inhibition discussed below. An alternative way to classify inhibitors is by their site of action. Some bind to the enzyme at the same site as does the substrate (the catalytic site); others bind at some site (an allosteric site) away from the catalytic site.

Competitive Inhibitors Typically Bear a Close Structural Similarity to a Substrate

Classic competitive inhibition occurs at the substrate-binding (catalytic) site. The chemical structure of a substrate analog inhibitor (I) generally resembles that of the substrate (S). It therefore combines reversibly with the enzyme, forming an enzyme-inhibitor (EnzI) complex rather than an EnzS complex. When both the substrate and this type of inhibitor are present, they compete for the same binding sites on the enzyme surface. A much studied case of competitive inhibition is that of malonate (I) with succinate (S) for succinate dehydrogenase.

Succinate dehydrogenase catalyzes formation of fumarate by removal of one hydrogen atom from each α-carbon atom of succinate (Fig 9–18).

Malonate ($^-OOC–CH_2–COO^-$) can combine with the dehydrogenase, forming an EnzI complex. This cannot be dehydrogenated, since there is no way to remove even one H atom from the single α-carbon atom of malonate without forming a pentavalent car-

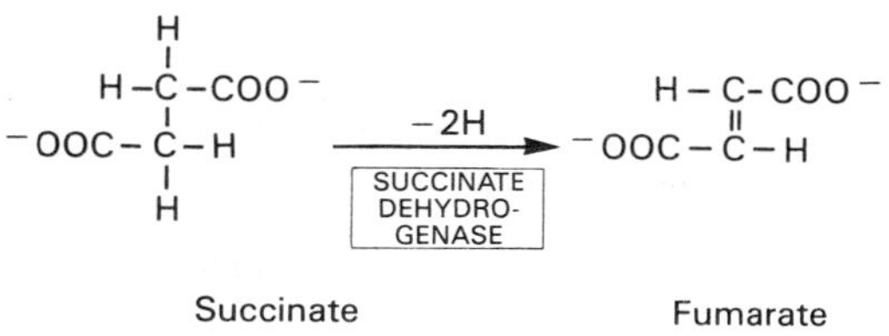

Figure 9–18. The succinate dehydrogenase reaction.

bon atom. The only reaction the EnzI complex can undergo is decomposition back to free enzyme plus inhibitor. For the reversible reaction,

$$EnzI \underset{k_{-1}}{\overset{k_1}{\rightleftharpoons}} Enz + I$$

the equilibrium constant, K_i, is

$$K_i = \frac{[Enz][I]}{[EnzI]} = \frac{k_1}{k_{-1}}$$

The action of competitive inhibitors may be understood in terms of the following reactions:

$$Enz \overset{\pm I}{\underset{\pm S}{\rightleftharpoons}} \begin{array}{l} EnzI \text{ (inactive)} \rightarrow\times\rightarrow Enz + P \\ EnzS \text{ (active)} \rightarrow Enz + P \end{array}$$

The rate of product formation, which is what generally is measured, depends solely on the concentration of EnzS. Suppose I binds very tightly to the enzyme (K_i = a small number). There now is little free enzyme (Enz) available to combine with S to form EnzS and eventually Enz + P. The reaction rate (formation of P) will therefore be slow. For analogous reasons, an equal concentration of a less tightly bound inhibitor (K_i = a larger number) will not decrease the rate of the catalyzed reaction so markedly. Suppose that, at a fixed concentration of I, more S is added. This increases the probability that Enz will combine with S rather than with I. The ratio of EnzS to EnzI and the reaction rate also rise. At a sufficiently high concentration of S, the concentration of EnzI should be vanishingly small. If so, the rate of the catalyzed reaction will be the same as in the absence of I (Fig 9–19).

The Effectiveness of Different Competitive Inhibitors Is Evaluated by a Double Reciprocal Plot of $1/v_i$ Versus $1/[S]$

Fig 9–19 represents a typical case of competitive inhibition shown graphically in the form of a Lineweaver-Burk plot. The reaction velocity (v_i) at a fixed concentration of inhibitor was measured at various concentrations of S. The lines drawn through the experimental points coincide at the y axis. Since the y intercept is $1/V_{max}$, this states that **at an infinitely high concentration of S ($1/[S] = 0$), v_i is the same as in the absence of inhibitor.** However, the intercept on

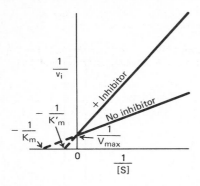

Figure 9–19. Lineweaver-Burk plot of classic competitive inhibition. Note the complete relief of inhibition at high [S] (low 1/[S]).

the x axis (which is related to K_m) varies with inhibitor concentration and becomes a larger number ($-1/K'_m$ is smaller than $-1/K_m$) in the presence of the inhibitor. Thus, **a competitive inhibitor raises the apparent K_m (K'_m) for the substrate.** Since K_m is the substrate concentration at which the concentration of free enzyme is equal to the concentration of enzyme present as EnzS, substantial free enzyme is available to combine with inhibitor. For simple competitive inhibition, the intercept on the x-axis is

$$x = \frac{1}{K_m \left(1 + \frac{[I]}{K_i} \right)}$$

K_m may be evaluated in the absence of I, and K_i evaluated using the above equation. If the number of moles of I added is much greater than the number of moles of enzyme present, [I] may generally be taken as the added (known) concentration of inhibitor. The K_i values for a series of substrate analog (competitive) inhibitors indicate which are most effective. **At a low concentration, those with the lowest K_i values will cause the greatest degree of inhibition.**

Many clinically efficacious drugs act as competitive inhibitors of important enzyme activities in microbial and animal cells.

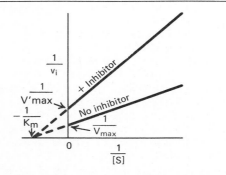

Figure 9–20. Lineweaver-Burk plot for reversible noncompetitive inhibition.

Reversible Noncompetitive Inhibitors Lower V_{max} but Do Not Affect K_m

As the name implies, in this case no competition occurs between S and I. The inhibitor usually bears little or no structural resemblance to S and may be assumed to bind to a different domain on the enzyme. **Reversible noncompetitive inhibitors lower the maximum velocity attainable with a given amount of enzyme (lower V_{max}) but usually do not affect K_m.** Since I and S may combine at different sites, formation of both EnzI and EnzIS complexes is possible. Since EnzIS may break down to form product at a slower rate than does EnzS, the reaction may be slowed but not halted. The following competing reactions may occur:

$$
\begin{array}{ccc}
 & \xrightarrow{+I} \text{EnzI} \xleftarrow{+S} & \\
\text{Enz} & & \text{EnzIS} \rightarrow \text{Enz} + \text{P} \\
 & \xrightarrow{+S} \text{EnzS} \xleftarrow{+I} & \\
 & \downarrow & \\
 & \text{Enz} + \text{P} &
\end{array}
$$

If S has equal affinity both for Enz and for EnzI (I does not affect the affinity of Enz for S), the results shown in Fig 9–20 are obtained when $1/v_i$, is plotted against 1/[S] in the presence and absence of inhibitor. (It is assumed that there has been no significant alteration of the conformation of the active site when I is bound.)

Many Enzyme "Poisons" Act as Irreversible, Noncompetitive Inhibitors of Enzyme Activity

A variety of enzyme "poisons," eg, iodoacetamide, heavy metal ions (Ag^+, Hg^{2+}), oxidizing agents, etc, reduce enzyme activity. Since these inhibitors bear no structural resemblance to the substrate, an increase in substrate concentration generally does not relieve this inhibition. The presence of one or more substrates or products, however, may protect the enzyme against inactivation. Kinetic analysis of the type discussed above may not distinguish between enzyme poisons and true reversible noncompetitive inhibitors. Reversible noncompetitive inhibition is, in any case, rare. Unfortunately this is not always appreciated, since both reversible and irreversible noncompetitive inhibition exhibit similar kinetics.

THE ACTIVITIES OF KEY ENZYMES CAN BE REGULATED BY POSITIVE OR NEGATIVE MODULATOR MOLECULES

Small molecule **modulators** that decrease catalytic activity are termed **negative modulators;** those which increase activity are called **positive modulators.** These are discussed further in subsequent chapters.

REFERENCES

Bender ML, Bergeron RJ, Komiyama M: *The Bioorganic Chemistry of Enzymatic Catalysis.* Wiley-Interscience, 1984.

Cabral F, Barlow SB: Mechanisms by which mammalian cells acquire resistance to drugs that affect microtubule assembly. *FASEB J* 1988;**3:**1593.

Czeh TR: RNA as an enzyme. *Sci Amer* 1986;**255:**64.

Dugas H, Penny C. *Bioorganic Chemistry: A Chemical Approach to Enzyme Action.* Springer-Verlag, 1981.

Fersht AR, Leatherbarrow RJ, Wells TNC: Binding energy and catalysis: a lesson from protein engineering of the tyrosyl-tRNA synthetase. *Trends Biochem Sci* 1986;**11:** 321.

Freeman RB, Hawkins HC (editors): *The Enzymology of Post-translational Modification of Proteins.* Academic Press, 1985.

Segel IH: *Enzyme Kinetics.* Wiley, 1975.

Walsh CT: Suicide substrates, mechanism-based enzyme inactivators: recent developments. *Annu Rev Biochem* 1984;**53:**493.

See also references in Ch. 7.

10 Enzymes: Mechanisms of Action

Victor W. Rodwell, PhD

INTRODUCTION

This chapter uses a detailed description of catalysis by the proteolytic enzyme chymotrypsin to illustrate principles of catalysis that apply to enzymes in general.

BIOMEDICAL IMPORTANCE

The molecular events that accompany conversion of substrates to products constitute the subject matter of the mechanism of enzyme action. These studies lead to a rational approach to therapy and drug design—areas of great potential for development in the immediate future. High-resolution structural information obtained by x-ray crystallography, combined with mechanistic information, now facilitates the design of drugs to inhibit specific enzymes such as HMG-CoA reductase, the pacemaker enzyme of cholesterol biosynthesis. Mechanistic studies also suggest ways in which recombinant DNA technology and site-directed mutagenesis may be used to modify enzyme specificity and/or catalytic efficiency. These techniques ultimately will facilitate the design and introduction into human subjects of enzymes with specific desired properties.

THE "INTERMEDIARY ENZYMOLOGY" OF CHYMOTRYPSIN ILLUSTRATES GENERAL FEATURES OF ENZYMIC CATALYSIS

Chymotrypsin catalyzes hydrolysis of peptide bonds in which the carboxyl group is contributed by an aromatic amino acid (Phe, Tyr, or Trp) or by one with a bulky nonpolar R group (Met).

Like many other proteases, chymotrypsin also catalyzes the hydrolysis of certain esters. While the ability of chymotrypsin to catalyze ester hydrolysis is of no physiologic significance, it facilitates study of the catalytic mechanism. The molecular events in catalysis are termed "intermediary enzymology."

The synthetic substrate *p*-nitrophenylacetate (Fig 10–1) facilitates colorimetric analysis of chymotrypsin activity because hydrolysis of *p*-nitrophenylacetate

releases *p*-nitrophenol. In alkali, this converts to the yellow *p*-nitrophenylate anion.

STOP-FLOW KINETICS REVEALS THE "INTERMEDIARY ENZYMOLOGY" OF CHYMOTRYPSIN

The kinetics of chymotrypsin hydrolysis of *p*-nitrophenylacetate can be studied in a "stop-flow" apparatus. These stop-flow experiments use substrate quantities of enzyme (roughly equimolar quantities of enzyme and of substrate) and measure events that occur in the first few milliseconds after enzyme and substrate are mixed. The stop-flow apparatus has 2 syringes: one for chymotrypsin and the other for *p*-nitrophenylacetate. Instantaneous mixing of enzyme and substrate is achieved by a mechanical device that rapidly and simultaneously expels the contents of both syringes into a single narrow tube that passes through a spectrophotometer. The optical density as a function of time after mixing is displayed on an oscilloscopic screen.

Release of the *p*-Nitrophenylate Anion is Biphasic

Release of *p*-nitrophenylate anion takes place in 2 distinct phases (Fig 10–2): (1) a "burst" phase, characterized by rapid liberation of *p*-nitrophenylate anion; and (2) a subsequent, slower release of additional *p*-nitrophenylate anion.

The biphasic character of the release of *p*-nitrophenylate anion is comprehensible in terms of the successive steps in catalysis shown in Fig 10–3.

The Slow Step in Catalysis by Chymotrypsin is Hydrolysis of The Chymotrypsin-Acetate (CT-Ac) Complex

Once all of the available chymotrypsin has been converted to CT-Ac, no further release of *p*-nitrophenylate anion can occur until more free chymotrypsin is liberated by the slow, hydrolytic removal of acetate anion from the CT-Ac complex (Fig 10–3). The "burst" phase of *p*-nitrophenylate anion release (Fig 10–2) corresponds to the conversion of all of the available free chymotrypsin to the CT-Ac complex with simultaneous release of *p*-nitrophenylate anion. The

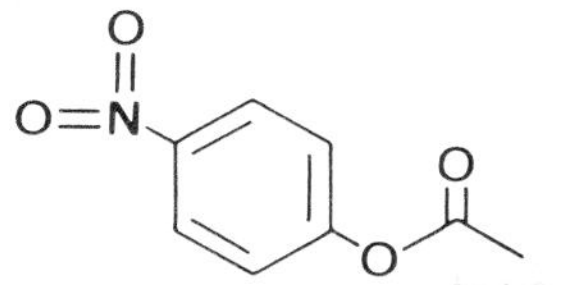

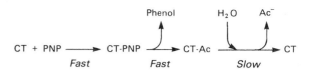

Figure 10–1. *p*-Nitrophenylacetate.

Figure 10–3. Intermediate steps in catalysis of the hydrolysis of *p*-nitrophenylacetate by chymotrypsin. CT, chymotrypsin; PNP, *p*-nitrophenylacetate; CT-PNP, chymotrypsin-*p*-nitrophenylacetate complex; CT-Ac, chymotrypsin-acetate complex phenol, *p*-nitrophenylate anion; Ac⁻, acetate anion. Formation of the CT-PNP and CT-Ac complexes is fast relative to the hydrolysis of the CT-Ac complex.

subsequent release of *p*-nitrophenylate anion (PNP) that follows the burst phase results from the slow liberation of free chymotrypsin by hydrolysis of the CT-Ac complex. This free chymotrypsin then is available for further formation of the CT-PNP and CT-Ac complexes with attendant release of PNP. Indeed, the magnitude of the "burst" phase (ie, moles of *p*-nitrophenylate anion released) is directly proportionate to the number of moles of chymotrypsin present initially.

Serine 195 Plays a Key Role in Catalysis by Chymotrypsin

The acyl group of the acyl-CT intermediate is linked to a highly reactive seryl residue—serine 195—of chymotrypsin. The essential nature and high reactivity of Ser 195 are evidenced by its ability (but not by the ability of the remaining 27 seryl residues of chymotrypsin) to react with diisopropylphosphofluoridate (DIPF) (Fig 10–4). Analogous reactions occur with other serine proteases.

Derivatization of Ser 195 inactivates chymotrypsin. Many other proteases are inactivated by DIPF by an analogous mechanism. These are termed "serine proteases."

A Charge Relay Network Functions as a Proton Shuttle During Catalysis by Chymotrypsin

The charge relay network of chymotrypsin involves 3 aminoacyl residues that are far apart in a primary structural sense but within bond-forming distance of one another in a tertiary structural sense. These residues are Asp 102, His 57, and Ser 195. While most of the charged residues of chymotrypsin are at the surface

of the molecule, those of the charge relay network are "buried" in the otherwise nonpolar interior of the molecule. The three residues are aligned in the order: Asp 102—His 57—Ser 195.

Recall that Ser 195 is the residue that is acylated during catalysis by chymotrypsin. The approach of the acetate anion (derived from *p*-nitrophenylacetate) to the oxygen atom on the R group of Ser 195 triggers sequential proton shifts that "shuttle" protons from Ser 195 through His 57 to Asp 102 (Fig 10–5).

During deacylation of the acyl–Ser 195 intermediate, protons shuttle in the reverse direction. An analogous series of proton shifts is believed to accompany hydrolysis of a physiologic chymotrypsin substrate such as a peptide.

MANY PROTEASES ARE SECRETED AS CATALYTICALLY INACTIVE PROENZYMES OR ZYMOGENS

Certain proteins are manufactured and secreted in the from of inactive precursor proteins known as **proproteins.** When the proteins are enzymes, the proproteins are termed **proenzymes** or **zymogens.** Conversion of a proprotein to the mature protein involves selective proteolysis. This converts the proprotein by one or more successive proteolytic "clips" to a form in which the characteristic activity of the mature protein (its enzymatic activity) is expressed. Examples of proteins manufactured as proproteins include the hormone insulin (proprotein = proinsulin), the digestive enzymes pepsin, trypsin, and chymotrypsin (proproteins = pepsinogen, trypsinogen, and chymotrypsinogen, respectively), several factors of the blood clotting and of the blood clot dissolution cascades (see Chapter 58), and the connective tissue protein collagen (proprotein = procollagen).

The conversion of prochymotrypsin (pro-CT), a 245-aminoacyl residue polypeptide, to the active enzyme α-chymotrypsin involves 3 proteolytic clips and the formation of an active intermediate known as π-chymotrypsin (π-CT)(Fig 10–6).

In α-chymotrypsin, the A, B, and C chains (Fig 10–6) remain associated owing to the presence in α-CT of 2 interchain disulfide bonds (Fig 10–7).

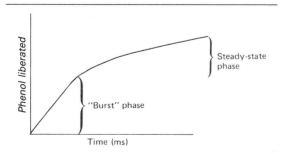

Figure 10–2. Kinetics of release of *p*-nitrophenylate anion when chymotrypsin hydrolyzes *p*-nitrophenylacetate in a stop-flow apparatus. "Phenol liberated" was calculated from optical density.

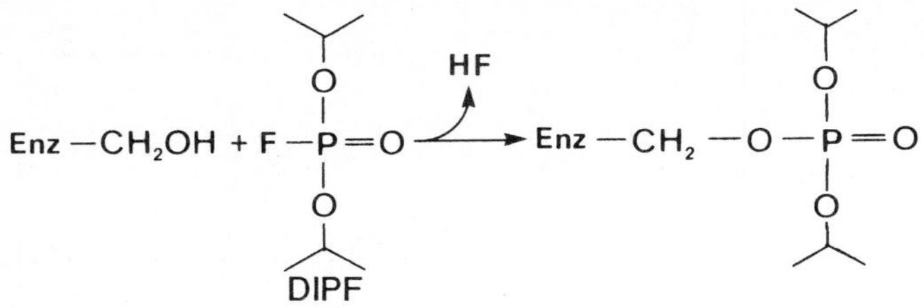

Figure 10–4. Reaction of the primary hydroxyl of Ser 195 of chymotrypsin with diisopropylphosphofluoridate (DIPF).

Proenzymes Facilitate Rapid Mobilization of an Activity in Response to Physiologic Demand

Why are certain proteins secreted in an inactive form? Certain proteins are needed at essentially all times. Others (for example, the enzymes of blood clot formation and dissolution) are needed only intermittently. Furthermore, when these intermittently needed enzymes are required, they frequently are needed rapidly. Certain physiologic processes such as digestion are intermittent but fairly regular and predictable (although this may not have been the case for primitive humans). Others (for example, blood clot formation, clot dissolution, and tissue repair) need only to be brought "on line" in response to pressing physiologic or pathophysiologic need. It may readily be appreciated that the processes of blood clot formation and dissolution must be temporally coordinated to achieve homeostasis. In addition, the synthesis of proteases as catalytically inactive precursor proteins serves to protect the tissue of origin (eg, the pancreas) from autodigestion. (Autodigestion can occur in pancreatitis.)

De novo synthesis of the required proteins might not be sufficiently rapid to respond to a pressing pathophysiologic demand such as the loss of blood. Moreover, an adequate and complete pool of the precursor amino acids must be available. Furthermore, the secretion process may be slow relative to the physiologic demand.

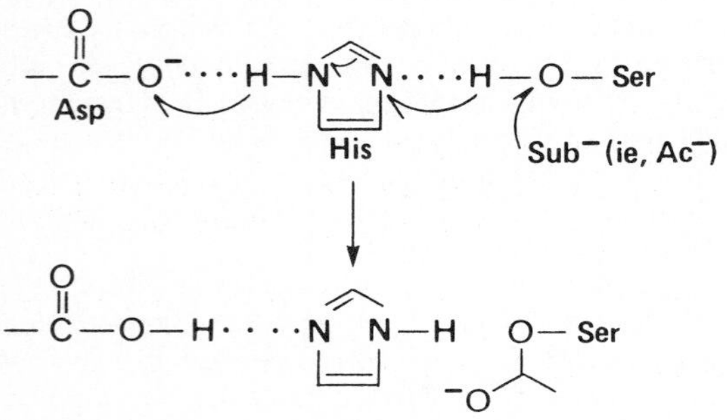

Activation of Prochymotrypsin Requires Selective Proteolysis

The example of conversion of a proprotein to its mature, physiologically active form discussed below illustrates the following general principles of proprotein to protein conversions:

(1) The process involves selective proteolysis, which in some instances requires only a single proteolytic clip.

(2) The polypeptide products may separate or may remain associated in the mature protein.

(3) The process may (or may not) be attended by a significant change in molecular weight.

(4) A major consequence of selective proteolysis is the attainment of a new conformation.

(5) If the proprotein is an enzyme, the above conformational change generates the catalytic site of the enzyme. Indeed, selective proteolysis of a proenzyme may be viewed as a process that triggers essential conformational changes that "create" the catalytic site.

Note that while His 57 and Asp 102 reside on the B peptide of α-chymotrypsin, Ser 195 resides on the C peptide (Fig 10–6). The selective proteolysis of prochymotrypsin (chymotrypsinogen) thus facilitates approximation of the 3 residues of the charge relay network. This illustrates how selective proteolysis can give rise to the catalytic site. Note also that contact and catalytic residues can be located on different peptide chains but still be within bond-forming distance of bound substrate.

Figure 10–5. Operation of the proton shuttle of chymotrypsin during acylation of Ser 195 by the substrate (Sub⁻).

SPECIFIC ENZYMES BIND SUBSTRATES IN EITHER A RANDOM OR AN ORDERED MANNER

Most enzymes catalyze a reaction between 2 or more substrates, yielding one or more products. For some enzymes, all substrates must be present simultaneously for the reaction to occur. For others, the enzyme first alters one substrate and then catalyzes its reaction with a second substrate. The order in which an enzyme binds its substrates may be **random** or **ordered** (Fig 10–8).

Many reactions that require coenzymes proceed by "ping-pong" mechanisms (so termed because the en-

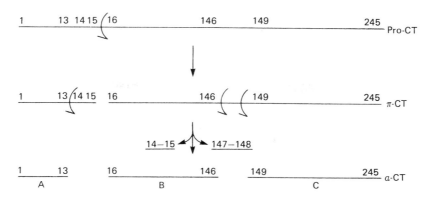

Figure 10–6. Conversion of prochymotrypsin (pro-CT) to π-chymotrypsin (π-CT) and subsequently to the mature catalytically active enzyme α-chymotrypsin (α-CT).

zyme alternates between forms E and E′) (Figs 10–9 and 10–10).

While a coenzyme frequently may be regarded as a second substrate, certain coenzymes (eg, pyridoxal phosphate) are covalently bound to the enzyme or bound noncovalently so tightly that dissociation rarely occurs (eg, thiamin pyrophosphate). In these cases, we regard the enzyme-coenzyme complex as the enzyme.

RESIDUES AT THE CATALYTIC SITE CAN ACT AS ACID-BASE CATALYSTS

Once substrate has bound at the catalytic site, the charged (or chargeable) functional groups of the side chains of nearby aminoacyl residues may participate in catalysis by functioning as acidic or basic catalysts.

There are 2 broad categories of acid-base catalysis by enzymes: **general** and **specific** acid (or base) catalysis. Reactions whose rates vary in response to changes in H^+ or H_3O^+ concentration but are independent of the concentrations of other acids or bases present in the solution are said to be subject to **specific acid** or **specific base catalysis.** Reactions whose rates are responsive to all the acids (proton donors) or bases (proton acceptors) present in solution are said to be subject to **general acid** or **general base catalysis.**

Measuring Reaction Rate at Varied pH & Buffer Concentrations Distinguishes Between General & Specific Acid-Base Catalysis

If the rate of the reaction changes as a function of pH at constant buffer concentration, the reaction is said to

be **specific base catalyzed** (if the pH is above 7) or **specific acid catalyzed** (if the pH is below 7). If the reaction rate at constant pH increases as the buffer concentration increases, the reaction is said to be subject to **general base catalysis** (if the pH is above 7) or **general acid catalysis** (if the pH is below 7).

As an example of specific acid catalysis, consider the conversion of a substrate (S) to a product (P). This occurs in 2 steps—a rapid, reversible proton transfer step,

$$S + H_3O^+ \rightleftharpoons SH^+ + H_2O$$

followed by a slower, and therefore rate-determining, step of rearrangement of the protonated substrate to product

$$SH^+ + H_2O \rightarrow P + H_3O^+$$

Increasing the concentration of hydronium ion $[H_3O^+]$ increases the reaction rate by elevating the concentration of SH^+, the conjugate acid of the substrate, which is the substrate for the rate-determining step in the overall reaction. Stated mathematically,

$$Rate = \frac{d[P]}{dt} = k[SH^+]$$

where P = the product, t = time, k = the specific rate constant, and $[SH^+]$ = the concentration of the conjugate acid of the substrate.

Since the concentration of SH^+ depends upon both the concentration of S and the concentration of H_3O^+,

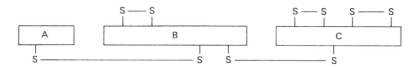

Figure 10–7. Intra- and interchain polypeptide bonds of α-chymotrypsin (α-CT).

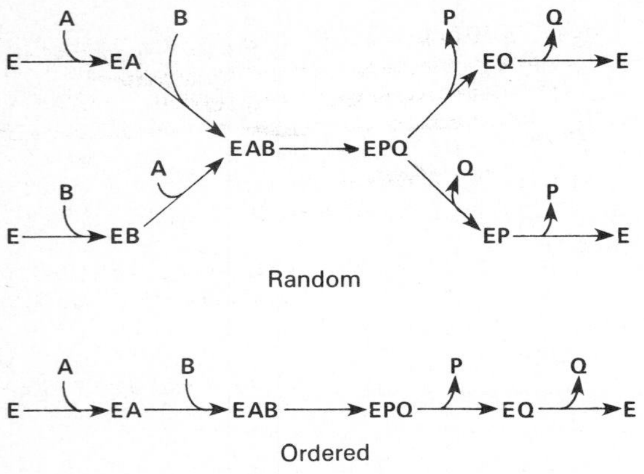

Figure 10–8. Random and ordered addition of substrates A and B and dissociation of products P and Q from an enzyme E.

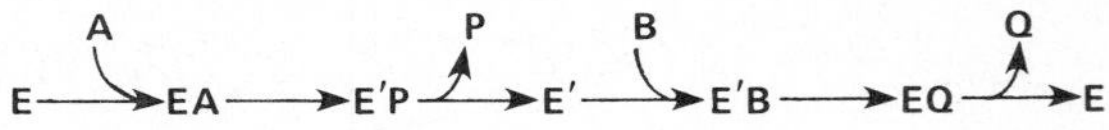

Figure 10–9. Generalized "ping-pong" mechanism for enzymatic catalysis.

the general rate expression for specific acid-catalyzed reactions is

$$\frac{d[P]}{dt} = k'[S][H_3O^+]$$

Note that it is a requirement of specific acid catalysis that the rate expression contain *only* terms for S and for H_3O^+.

Next consider that, in addition to the specific acid catalysis described above, there is also catalysis by imidazolium ion of an imidazole buffer. Since imidazole is a weak acid (pK_a about 7), it is a poor proton donor; hence the reaction

$$S + Imidazole \cdot H^+ \rightarrow SH^+ + Imidazole$$

is slow and is rate-determining for the overall reaction. Note that the fast and slow steps are reversed when the mechanism changes from specific to general acid catalysis. The rate expressions for general acid catalysis frequently are complex and for this reason are not discussed here.

METAL IONS MAY FACILITATE SUBSTRATE BINDING & CATALYSIS

Over 25% of all enzymes contain tightly bound metal ions or require them for activity. The functions of these metal ions are studied by x-ray crystallography,

magnetic resonance imaging (MRI), and electron spin resonance (ESR). Coupled with knowledge of the formation and decay of metal complexes and of reactions within the coordination spheres of metal ions, this provides insight into the roles of metal ions in enzymatic catalysis. These roles are considered below.

Metalloenzymes contain a definite quantity of functional metal ion that is retained throughout purification. **Metal-activated enzymes** bind metals less tightly but require added metals. The distinction between metalloenzymes and metal-activated enzymes thus rests on the affinity of a particular enzyme for its metal ion. The mechanisms whereby metal ions perform their functions appear to be similar in metalloenzymes and metal-activated enzymes.

Ternary Complexes With Metals Function in Catalysis

For ternary (3-component) complexes of the catalytic site (Enz), a metal ion (M), and substrate (S) that exhibit 1:1:1 stoichiometry, 4 schemes are possible:

Enz–S–M
Substrate-bridge complex

M–Enz–S
Enzyme-bridge complex

Enz–M–S
Simple metal-bridge complex

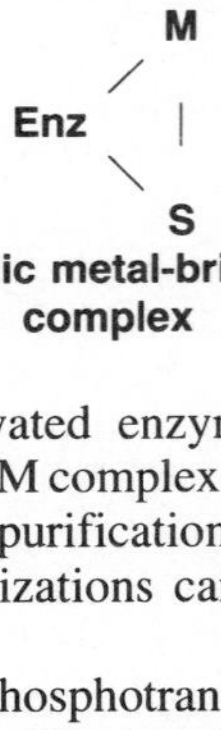

$$Enz \begin{matrix} M \\ | \\ S \end{matrix}$$

Cyclic metal-bridge complex

All 4 are possible for metal-activated enzymes. Metalloenzymes cannot form the EnzSM complex, because they retain the metal throughout purification (ie, are already as EnzM). Three generalizations can be stated:

(1) Most but not all kinases (ATP:phosphotransferases) form substrate-bridge complexes of the type Enz-nucleotide-M.

(2) Phosphotransferases using pyruvate or phosphoenolpyruvate as substrate, enzymes catalyzing other reactions of phosphoenolpyruvate, and carboxylases form metal-bridge complexes.

(3) A given enzyme may form one type of bridge

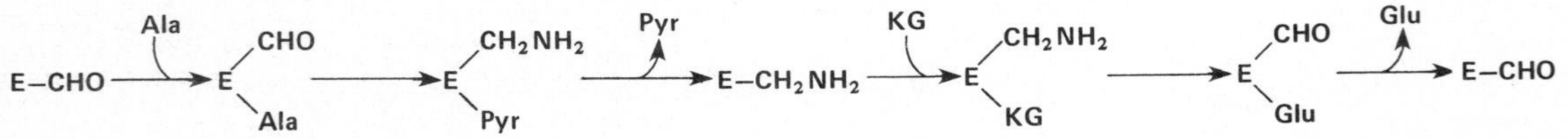

Figure 10–10. "Ping-pong" mechanism for transamination. E—CHO and E—CH₂NH₂ represent the enzyme-pyridoxal phosphate and enzyme-pyridoxamine phosphate complexes, respectively. (Ala, alanine; Pyr, pyruvate; KG, α-ketoglutarate; Glu, glutamate.)

complex with one substrate and a different type with another.

Enzyme-Bridge Complexes (MEnzS): The metals in enzyme-bridge complexes are presumed to perform structural roles maintaining an active conformation (eg, glutamine synthase) or to form a metal bridge to a substrate (eg, pyruvate kinase). In addition to its structural role, the metal ion in pyruvate kinase appears to hold one substrate (ATP) in place and to activate it.

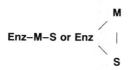

Substrate-Bridge Complexes (EnzSM): The formation of ternary substrate-bridge complexes of nucleoside triphosphates with enzyme, metal, and substrate appears attributable to displacement of H_2O from the coordination sphere of the metal by ATP:

$$ATP^{4-} + M(H_2O)_6{}^{2+} \rightleftharpoons ATP{-}M(H_2O)_3{}^{2-} + 3H_2O$$

Substrate then binds, forming the ternary complex:

$$ATP{-}M(H_2O)_3{}^{2-} + Enz \rightleftharpoons Enz{-}ATP{-}M(H_2O)_3{}^{2-}$$

In phosphotransferase reactions, metal ions are thought to activate the phosphorus atoms and form a rigid, polyphosphate-adenine complex of appropriate conformation in the active, quaternary complex.

Metal-Bridge Complexes:

$$Enz{-}M{-}S \text{ or } Enz \overset{\displaystyle M}{\underset{\displaystyle S}{\diagup \mid \diagdown}}$$

Crystallographic and sequencing data have established that a His residue is concerned with metal binding at the active site of many proteins (eg, carboxypeptidase A, cytochrome c, rubredoxin, metmyoglobin, and methemoglobin; see Chapter 7). For binary (2-component) EnzM complexes, the rate-limiting step is in many cases the departure of water from the coordination sphere of the metal ion. For many peptidases, activation by metal ions is a slow process requiring many hours. The slow reaction probably is conformational rearrangement of the binary EnzM complex to an active conformation, eg,

Metal binding:

$$Enz + M(H_2O)_6 \xrightarrow{\text{Rapid}} Enz{-}M(H_2O)_{6-n} + nH_2O$$

Rearrangement to active conformation (Enz*):

$$Enz{-}M(H_2O)_{6-n} \xrightarrow{\text{Slow}} Enz^*{-}M(H_2O)_{6-n}$$

For metalloenzymes, however, the ternary metal-bridge complex must be formed by combination of the substrate (S) with the binary EnzM complex:

$$Enz{-}M + S \rightleftharpoons Enz{-}M{-}S \text{ or } Enz \overset{\displaystyle M}{\underset{\displaystyle S}{\diagup \mid \diagdown}}$$

Metal Ions Perform Multiple Functions in Catalysis

Metal ions may participate in each of the 4 mechanisms by which enzymes are known to accelerate the rates of chemical reactions: (1) general acid-base catalysis, (2) covalent catalysis, (3) approximation of reactants, and (4) induction of strain in the enzyme or substrate. Other than iron and manganese, which function in heme proteins, the metal ions most commonly concerned in enzymatic catalysis are Mg^{2+}, Mn^{2+}, and Ca^{2+}, although other metal ions (eg, K^+) are important for the activity of certain enzymes.

Metal ions, like protons, are Lewis acids (electrophiles) and can share an electron pair, forming a sigma bond. Metal ions may also be considered "super acids," since they exist in neutral solution, frequently have a positive charge of >1, and may form pi bonds. In addition (and unlike protons), metals can serve as 3-dimensional templates for orientation of basic groups on the enzyme or substrate.

Metal ions can also accept electrons via sigma or pi bonds to activate electrophiles or nucleophiles (general acid-base catalysis). By donating electrons, metals can activate nucleophiles or act as nucleophiles themselves. The coordination sphere of a metal may bring together enzyme and substrate (approximation) or form chelate-producing distortion in either the enzyme or substrate (strain). A metal ion may also "mask" a

Table 10–1. Selected examples of the roles of metal ions in the mechanism of action of enzymes.[*]

Enzyme	Role of Metal Ion
Histidine deaminase	Masking a nucleophile
Kinases, lyases, pyruvate decarboxylase	Activation of an electrophile
Carbonic anhydrase	Activation of a nucleophile
Cobamide enzymes	Metal acts as a nucleophile
Pyruvate carboxylase, carboxypeptidase, alcohol dehydrogenase	π-Electron withdrawal
Nonheme iron proteins	π-Electron donation
Pyruvate kinase, pyruvate carboxylase, adenylate kinase	Metal ion gathers and orients ligands
Phosphotransferase, D-xylose isomerase, hemoproteins	Strain effects

[*]Adapted from Mildvan AS: Metals in enzyme catalysis. Vol 2. Page 456 in: *The Enzymes.* Boyer PD, Lardy H, Myrbäck K (editors). Academic Press, 1970.

nucleophile and thus prevent an otherwise likely side reaction. Finally, stereochemical control of the course of an enzyme-catalyzed reaction may be achieved by the ability of the metal coordination sphere to act as a 3-dimensional template to hold reactive groups in a specific steric orientation (Table 10–1).

REFERENCES

Fersht A: *Enzyme Structure and Mechanism,* 2nd ed. Freeman, 1985.

Freeman RB, Hawkins HC (editors): *The Enzymology of Post-translational Modification of Proteins.* Academic Press, 1985.

Purich DL (editor): Enzyme kinetics and mechanisms. Parts A and B in: *Methods in Enzymology.* Vol 63, 1979; Vol 64, 1980. Academic Press.

See also references in Chapters 7 and 8.

Enzymes: Regulation of Activities

11

Victor W. Rodwell, PhD

INTRODUCTION

In this chapter, mechanisms by which metabolic processes are regulated via enzymes are illustrated by selected examples. The intent is to characterize overall patterns of regulation. Throughout this book, reference is made to many other specific examples to illustrate these diverse features of metabolic regulation.

BIOMEDICAL IMPORTANCE

The mechanisms by which cells and intact organisms regulate and coordinate overall metabolism are of concern to workers in areas of the biomedical sciences as diverse as cancer, heart disease, aging, microbial physiology, differentiation, metamorphosis, hormone action, and drug action. In all of these areas, important examples of normal or abnormal regulation of enzymes are to be found. For example, **many cancer cells exhibit abnormalities in the regulation of their enzyme complement** (lack of induction or repression). This illustrates the well-established conclusion that alterations of gene control are fundamental events in cancer cells. Again, certain oncogenic viruses contain a gene that codes for a tyrosine-protein kinase. When this kinase is expressed in host cells, it can phosphorylate many proteins and enzymes that are normally not phosphorylated and thus lead to dramatic changes in cell phenotype. A change of this nature appears to lie at the heart of certain types of viral oncogenic transformation. Drug action provides another important example involving enzyme regulation. **Enzyme induction is one important biochemical cause of a drug interaction,** the situation in which the administration of one drug results in a significant change in the metabolism of another (see Chapter 60).

REGULATION OF METABOLISM ACHIEVES HOMEOSTASIS

The concept of homeostatic regulation of the internal milieu advanced by Claude Bernard in the late 19th century stressed the ability of animals to maintain the constancy of their intracellular environments despite changes in the external environment. While this concept predates modern enzymology, it in fact implies that enzyme-catalyzed reactions proceed at rates responsive to changes in the internal and external environment. A cell or organism might be defined as **diseased** when it responds inadequately or incorrectly to an internal or external stress. Knowledge of factors affecting the rates of enzyme-catalyzed reactions is essential both to understand the mechanism of homeostasis in normal cells and to comprehend the molecular basis of disease.

Metabolite Flow Tends To Be Unidirectional

All chemical reactions, including enzyme-catalyzed reactions, are to some extent reversible.* Within living cells, however, reversibility may not obtain, because reaction products are promptly removed by additional enzyme-catalyzed reactions. Metabolite flow in living cells is analogous to the flow of water in a pipe. Although the pipe can transfer water in either direction, in practice the flow is unidirectional. Metabolite flow in living cells also is largely unidirectional. True equilibrium, far from being characteristic of life, is approached only when cells die. The living cell is a dynamic steady-state system maintained by a unidirectional flow of metabolites (Fig 11–1). In mature cells the mean concentrations of metabolites remain relatively constant over considerable periods of time.† The flexibility of this steady-state system is well illustrated by the delicate shifts and balances by which organisms maintain the constancy of the internal environment despite wide variations in food, water, and mineral intake, work output, or external temperature.

Reaction Rates Must Respond to Changing Physiologic Need

For life to proceed in orderly fashion, metabolite flow through anabolic and catabolic pathways must be regulated. All requisite chemical events must proceed

---

*A readily reversible reaction has a small numerical value of ΔG. One with a large negative value for ΔG might be termed "effectively irreversible" in most biochemical situations.
†Short-term oscillations of metabolite concentrations and of enzyme levels do occur, however, and are of profound physiologic importance.

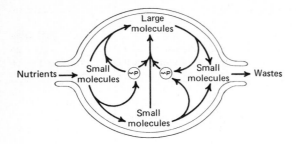

Figure 11–1. An idealized cell in steady state.

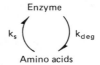

Figure 11–2. Enzyme quantity is determined by the net balance between enzyme synthesis and enzyme degradation.

at rates consistent with the requirements of the intact organism in relation to its environment. ATP production, synthesis of macromolecular precursors, transport, secretion, and tubular reabsorption all must respond to subtle changes in the environment of the cell, organ, or intact animal. These processes must be coordinated and must respond to short-term changes in the external environment (eg, addition or removal of a nutrient) as well as to periodic intracellular events (eg, DNA replication). Until recently, the molecular details of regulation were best understood in bacteria, which lack the complexities of hormonal or neural control and in which genetic studies can readily be conducted to analyze molecular events. Our understanding of molecular regulation in animal cells, however, is presently in a state of rapid expansion. While it is clear that metabolic regulation in mammals differs significantly from superficially similar phenomena in bacteria, regulation of metabolic processes in bacteria will be discussed because it provides a conceptual framework for considering regulation in humans.

THREE GENERAL MECHANISMS REGULATE ENZYME ACTIVITY

Net flow of carbon through any enzyme-catalyzed reaction may be influenced (1) by changing the absolute quantity of enzyme present, (2) by altering the pool size of reactants other than enzyme, and (3) by altering the catalytic efficiency of the enzyme. All 3 options are exploited in most forms of life.

Rates of Synthesis & Degradation Determine Enzyme Quantity

The absolute quantity of an enzyme present is determined by its rate of synthesis (k_s) and rate of degradation (k_{deg}) (Fig 11–2). The quantity of an enzyme in a cell may be raised either by an increase in its rate of synthesis (increase in k_s), by a decrease in its rate of degradation (decrease in k_{deg}), or by both. Similarly, a lower quantity of enzyme can result from a decrease in k_s, an increase in k_{deg}, or both. Examples of changes in both k_s and k_{deg} occur in human subjects. In all forms of life, enzyme (protein) synthesis from amino acids and enzyme (protein) degradation to amino acids are distinct processes catalyzed by entirely different sets of

enzymes. Independent regulation of enzyme synthesis and enzyme degradation is thus readily achieved.

Enzyme Synthesis May Be Triggered in Response to an Inducer

Cells can synthesize specific enzymes in response to the presence of specific low molecular weight inducers. For example, *Escherichia coli* grown on glucose will not ferment lactose owing to the absence of the enzyme β-galactosidase, which hydrolyzes lactose to galactose and glucose. If lactose or certain other β-galactosides are added to the growth medium, synthesis of the β-galactosidase is induced and the culture can ferment lactose.

Although many inducers are substrates for the enzymes they induce, compounds structurally similar to the substrate may be inducers but not substrates. These are termed **gratuitous inducers.** Conversely, a compound may be a substrate, but not an inducer. Frequently, a compound induces several enzymes of a catabolic pathway (eg, β-galactoside permease and β-galactosidase are both induced by lactose).

Enzymes whose concentration in a cell is independent of added inducer are termed **constitutive.** A particular enzyme may be constitutive in one strain, inducible in another, and absent in a third. Cells capable of being induced for a particular enzyme or other protein usually contain a small measurable **basal level** of that protein even when grown in the absence of added inducer. The genetic heritage of the cell determines both the nature and magnitude of the response to an inducer. "Constitutive" and "inducible" are therefore relative terms, like "hot" and "cold," that represent the extremes of a spectrum of responses.

Enzyme induction also occurs in eukaryotes. Examples of inducible enzymes in animals include tryptophan pyrrolase, threonine dehydrase, tyrosine-α-ketoglutarate transaminase, invertase, enzymes of the urea cycle, HMG-CoA reductase, and cytochrome P-450.

Enzyme Synthesis May Be Repressed by the Presence of an End Product

The presence in the reaction medium of the metabolite being biosynthesized may curtail new synthesis of that metabolite via **repression.** A small molecule such as a purine or amino acid, acting as a **corepressor,** can ultimately block synthesis of the enzymes involved in its own biosynthesis. For example, in *Salmonella ty-phimurium,* addition of histidine (His) represses synthesis of all the enzymes of His biosynthesis, and addition of leucine (Leu) represses synthesis of the first 3 enzymes unique to Leu biosynthesis. Following removal or exhaustion of an essential biosynthetic intermediate from the medium, enzyme biosynthesis again occurs. This constitutes **derepression.**

The above examples illustrate **product feedback repression** characteristic of biosynthetic pathways in bacteria. **Catabolite repression,** a related phenomenon, refers to the ability of an intermediate in a sequence of **catabolic** enzyme-catalyzed reactions to repress synthesis of catabolic enzymes. This effect was first noted in cultures of *E coli* growing on a carbon source (X) other than glucose. Addition of glucose repressed synthesis of the enzymes concerned with catabolism of X. This phenomenon was initially termed the "glucose effect." Since oxidizable nutrients other than glucose produce similar effects, the term **catabolite repression** was adopted. Catabolite repression is mediated by cAMP. The molecular mechanisms of induction, repression, and derepression are discussed in Chapter 41.

PROTEIN TURNOVER IS CHARACTERISTIC OF ALL FORMS OF LIFE

The combined processes of enzyme synthesis and degradation constitute **enzyme turnover.** Turnover of protein was recognized as a characteristic property of all mammalian cells long before it was shown also to occur in bacteria. The existence of protein (enzyme) turnover in humans was deduced from dietary experiments well over a century ago. It was Schoenheimer's classic work, however, just prior to and during World War II, that conclusively established that turnover of cellular protein occurred throughout life. By measuring the rates of incorporation of ^{15}N-labeled amino acids into protein and the rates of loss of ^{15}N from protein, Schoenheimer deduced that body proteins are in a state of "dynamic equilibrium," a concept since extended to other body constituents, including lipids and nucleic acids.

Rates of Degradation of Specific Enzymes Are Subject to Regulation

The degradation of mammalian proteins by ATP and ubiquitin-dependent—and by ATP-independent—pathways is discussed in some detail at the beginning of Chapter 31. The susceptibility of an enzyme to proteolytic degradation depends upon its conformation. The presence or absence of substrates, coenzymes, or metal ions, which can alter protein conformation, alters proteolytic susceptibility. The concentrations of substrates, coenzymes, and possibly ions in cells may thus determine the rates at which specific enzymes are degraded. Arginase and tryptophan oxygenase (tryp-

tophan pyrrolase) illustrate these concepts. Regulation of liver arginase levels can involve a change either in k_s or in k_{deg}. After a protein-rich diet is ingested, liver arginase levels rise owing to an increased rate of arginase synthesis. Liver arginase levels also rise in starved animals. Here, however, it is arginase degradation that is decreased, while k_s remains unchanged. In a second example, injection of glucocorticoids and ingestion of Trp both elevate levels of tryptophan oxygenase in mammals. The hormone raises the rate of oxygenase synthesis (raises k_s). Trp, however, has no effect on k_s but lowers k_{deg} by stabilizing the oxygenase toward proteolytic digestion. Contrast these 2 examples with enzyme induction in bacteria. For arginase, the increased intake of nitrogen on a high-protein diet may elevate liver arginase levels (see Chapter 31). The increased rate of arginase synthesis thus superficially resembles that of substrate induction in bacteria. For tryptophan pyrrolase, however, even though Trp may act as an inducer in bacteria (affects k_s), its effect in mammals is solely on the enzyme degradative process (lowers k_{deg}).

Enzyme levels in mammalian tissues may be altered by a wide range of physiologic, hormonal, or dietary manipulations. Examples are known for a variety of tissues and metabolic pathways (Table 11–1), but knowledge of the molecular details that account for these changes is fragmentary.

Glucocorticoids increase the concentration of tyrosine transaminase by stimulating k_s. This was the first clear case of a hormone regulating the synthesis of a mammalian enzyme. Insulin and glucagon—despite their mutually antagonistic physiologic effects—both independently increase k_s 4- to 5-fold. The effect of glucagon probably is mediated via cAMP, which mimics the effect of the hormone in organ cultures of rat liver.

Regulatory Advantages Result From Synthesis of Certain Enzymes as Inactive Precursors

Enzyme activity can be regulated by converting an inactive proenzyme to a catalytically active form. To become catalytically active, the proenzyme must undergo limited proteolysis, a process accompanied by conformational changes that either reveal or "create" the catalytic site (see Chapter 9). Synthesis as a catalytically inactive proenzyme is characteristic of digestive enzymes and enzymes of blood coagulation and of blood clot dissolution (see Chapter 58).

MULTIPLE OPTIONS ARE AVAILABLE FOR REGULATING THE CATALYTIC ACTIVITY OF PREFORMED ENZYMES

If a physiologic manipulation alters the level of enzyme activity, we must enquire whether the quantity of enzyme has changed or whether the enzyme is a more

Table 11–1. Selected examples of rat liver enzymes that adapt to an environmental stimulus by changes in activity.*

Enyzme	t$_{1/2}$ (hours)	Stimulus	Fold Change
Amino acid metabolism			
Arginase	100–120	Starvation or glucocorticoids.	+2
		Change from high- to low-protein diet.	−2
Serine dehydratase	20	Glucagon or dietary amino acids.	+100
Histidase	60	Change from low- to high-protein diet.	+20
Carbohydrate metabolism			
Glucose-6-phosphate dehydrogenase	15	Thyroid hormone or change from fasting state to ingestion of a high-carbohydrate diet.	+10
α-Glycerophosphate dehydrogenase	100	Thyroid hormone.	+10
Fructose 1,6-phosphatase		Glucose.	+10
Lipid metabolism			
Citrate cleavage enzyme		Change from starvation to ingestion of a high-carboydrate, low-fat diet.	
Fatty acid synthase		Starvation.	+30
		Change from starvation to ingestion of a fat-free diet.	−10
			+30
HMG-CoA reductase	2–3	Fasting or 5% cholesterol diet.	−10
		Twenty-four-hour diurnal variation.	±5
		Insulin or thyroid hormone.	+2 to 10
Purine or pyrimidine metabolism			
Xanthine oxidase		Change to high-protein diet.	−10
Aspartate transcarbamoylase	60	One percent orotic acid diet.	+2
Dihydroorotase	12	One percent orotic acid diet.	+3

*Data, with the exception of those for HMG-CoA reductase, from Schimke RT, Doyle D: Control of enzyme levels in animal tissues. *Annu Rev Biochem* 1970;**39**:929.

efficient or less efficient catalyst. **We shall refer to all changes in enzyme activity that occur without change in the quantity of enzyme present as "effects on catalytic efficiency."**

ENZYME COMPARTMENTATION FACILITATES REGULATION OF METABOLISM

The importance of compartmentation of metabolic processes in eukaryotic cells, including those of mammals, cannot be overemphasized. Localization of specific metabolic processes in the **cytosol** or in **cellular organelles** facilitates regulation of these processes independent of processes proceeding elsewhere. The extensive compartmentation of metabolic processes characteristic of higher forms of life thus confers the potential for finely tuned regulation of metabolism. At the same time, it poses problems with respect to translocation of metabolites across compartmental barriers. This is achieved via "**shuttle mechanisms**" that convert the metabolite to a form permeable to the compartmental barrier. This is followed by transport and conversion back to the original form on the other side of the barrier. Consequently, these interconversions require, for example, cytosolic and mitochondrial forms

of the same catalytic activity. Since these 2 forms of the enzyme are physically separated, their independent regulation is facilitated. The role of shuttle mechanisms in achieving equilibrium of metabolic pools of reducing equivalents, of citric acid cycle intermediates, and of other amphibolic intermediates is discussed in Chapter 18.

ENZYMES WHICH CATALYZE A SERIES OF METABOLIC REACTIONS CAN FORM A MACROMOLECULAR COMPLEX

Organization of a set of enzymes that catalyze a protracted sequence of metabolic reactions into a macromolecular complex coordinates the enzymes and channels intermediates along a metabolic path. Appropriate alignment of the enzymes can facilitate transfer of product between enzymes without prior equilibration with metabolic pools. This permits a finer level of metabolic control than is possible with the isolated components of the complex. In addition, conformational changes in one component of the complex may be transmitted by protein-protein interactions to other enzymes of the complex. Amplification of regulatory effects thus is possible.

LOCAL CONCENTRATIONS OF SUBSTRATES, COENZYMES, & CATIONS CAN REGULATE ENZYMES

The mean intracellular concentration of a substrate, coenzyme, or metal ion may have little meaning for the in vivo behavior of an enzyme. Information on the concentrations of essential metabolites **in the immediate neighborhood of the enzyme in question** is needed. However, even measuring metabolite concentrations in different cellular compartments does not account for local discontinuities in metabolite concentrations within compartments brought about by factors such as proximity to the site of entry or production of a metabolite. Finally, little consideration generally is given to the discrepancy between total and free metabolite concentrations. For example, while the total concentration of 2,3-bisphosphoglycerate in erythrocytes is extremely high, the concentration of free bisphosphoglycerate is comparable to that of other tissues. Similar considerations apply to other metabolites in the presence of proteins that bind them effectively and reduce their concentrations in the free state.

An assumption of the Michaelis-Menten kinetic approach was that the concentration of total substrate was essentially equal to the concentration of free substrate. As noted above, this assumption may well be invalid in vivo, where concentrations of free substrates may approach those of enzyme concentrations.

Metal ions, which perform catalytic and structural roles in over one-fourth of all known enzymes (see Chapter 10), may also fulfill regulatory roles, particularly for reactions where ATP is a substrate. Where the ATP-metal ion complex is the substrate, maximal activity typically is observed at molar ratio of ATP to metal of about unity. Excess metal or excess ATP is inhibitory. Since nucleoside di- and triphosphates form stable complexes with divalent cations, intracellular concentrations of the nucleotides can influence intracellular concentrations of free metal ions and hence the activity of certain enzymes.

CERTAIN ENZYMES ARE REGULATED BY ALLOSTERIC EFFECTORS

The catalytic activity of certain **regulatory enzymes** is modulated by low molecular weight **allosteric effectors** that generally have little or no structural similarity to the substrates or coenzymes for the regulated enzyme. **Feedback inhibition** refers to the inhibition of the activity of an enzyme in a biosynthetic pathway by an end product of that pathway. For biosynthesis of D from A, catalyzed by enzymes Enz_1 through Enz_3,

$$\begin{array}{cccc} Enz_1 & Enz_2 & Enz_3 \\ A \rightarrow & B \rightarrow & C \rightarrow & D \end{array}$$

a high concentration of D typically inhibits conversion of A to B. This involves not simple "backing up" of

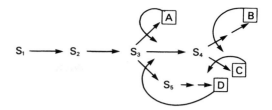

Figure 11–3. Sites of feedback inhibition in a branched biosynthetic pathway. S_1–S_5 are intermediates in the biosynthesis of end products A–D. Straight arrows represent enzymes catalyzing the indicated conversions. Curved arrows represent feedback loops and indicate probable sites of feedback inhibition by specific end products.

intermediates but the ability of D to bind to and inhibit Enz_1. D thus acts as a **negative allosteric effector** or **feedback inhibitor** of Enz_1. **Feedback inhibition** of Enz_1 by D therefore regulates the synthesis of D. Typically, D binds to the sensitive enzyme at an **allosteric site** remote from the catalytic site.

The kinetics of feedback inhibition may be competitive, noncompetitive, partially competitive, uncoupled, or mixed. Feedback inhibition is commonest in biosynthetic pathways. **Frequently the feedback inhibitor is the last small molecule before a macromolecule** (eg, amino acids before proteins, nucleotides before nucleic acids). **Feedback regulation generally occurs at the earliest functionally irreversible* step unique to a particular biosynthetic sequence;** a much-studied example is inhibition by CTP of aspartate transcarbamoylase (see below, and Chapter 36).

Frequently, a biosynthetic pathway is branched, with the initial portion serving for synthesis of 2 or more essential metabolites. Fig 11–3 shows probable sites of simple feedback inhibition in a branched biosynthetic pathway (eg, for amino acids, purines or pyrimidines). S_1, S_2, and S_3 are precursors of all 4 end products (A, B, C, and D), S_4 is a precursor of B and C, and S_5 a precursor solely of D. The sequences:

$$\begin{array}{l} S_3 \rightarrow A \\ S_4 \rightarrow B \\ S_4 \rightarrow C \\ S_3 \rightarrow S_5 \rightarrow D \end{array}$$

thus constitute linear reaction sequences that might be expected to be feedback-inhibited by their end products. Again, nucleotide biosynthesis (see Chapter 36) provides specific examples.

Multiple Feedback Loops Regulate Multiple-Product Pathways

Additional fine control is provided by multiple feedback loops (Fig 11–4). For example, if B is present in excess, the requirement for S_2 decreases. The ability

---

*One strongly favored (in thermodynamic terms) in a single direction, ie, one with a large negative ΔG.

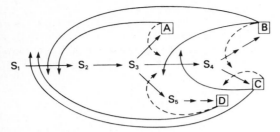

Figure 11-4. Multiple feedback inhibition in a branched biosynthetic pathway. Superimposed on simple feedback loops (dashed, curved arrows) are multiple feedback loops (solid, curved arrows) that regulate enzymes common to biosynthesis of several end products.

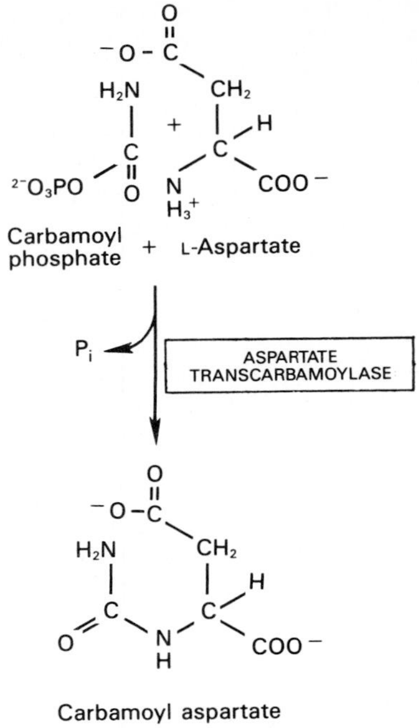

of B to decrease production of S_2 thus confers a biologic advantage. However, if excess B inhibits not only the portion of the pathway unique to its own synthesis but also portions common to that for synthesis of A, C, or D, excess B should curtail synthesis of all 4 end products. Clearly, this is undersirable. Mechanisms have, however, evolved to circumvent this difficulty.

In **cumulative feedback inhibition,** the inhibitory effect of 2 or more end products on a single regulatory enzyme is strictly additive.

In **concerted** or **multivalent feedback inhibition,** complete inhibition occurs only when 2 or more end products both are present in excess.

In **cooperative feedback inhibition,** a single end product present in excess inhibits the regulatory enzyme, but the inhibition when 2 or more end products are present far exceeds the additive effects of cumulative feedback inhibition.

The Most Intensively Studied Allosteric Enzyme Is Aspartate Transcarbamoylase

Aspartate transcarbamoylase catalyzes the first reaction unique to pyrimidine biosynthesis (Fig 11-5). Aspartate transcarbamoylase (ATCase) is **feedback-inhibited by cytidine triphosphate (CTP).** Following treatment with mercurials, ATCase loses its sensitivity to inhibition by CTP but retains its full activity for carbamoyl aspartate synthesis. This suggests that CTP is bound at a different (allosteric) site from either substrate. ATCase consists of 2 catalytic and 3 or 4 regulatory protomers. Each catalytic protomer contains 4 aspartate (substrate) sites and each regulatory protomer at least 2 CTP (regulatory) sites. Each type of protomer is subject to independent genetic control, as shown by the production of mutants lacking normal feedback control by CTP and, from these, of revertants with essentially normal regulatory properties.

Allosteric & Catalytic Sites Are Spatially Distinct

About 1963, Monod noted the lack of structural similarity between a feedback inhibitor and the sub-

Figure 11-5. The aspartate transcarbamoylase (ATCase) reaction.

strate for the enzyme whose activity it regulated. Since the effectors are not isosteric with a substrate but **allosteric** ("occupy another space"), he proposed that enzymes whose activity is regulated by **allosteric effectors** (eg, feedback inhibitors) bind the effector at an **allosteric site** that is physically distinct from the catalytic site. **Allosteric enzymes** thus are enzymes whose activity at the catalytic site may be modulated by the presence of allosteric effectors at an allosteric site. Lines of evidence that support the existence of physically distinct allosteric sites on regulated enzymes include the following:

(1) Regulated enzymes modified by chemical or physical techniques frequently become insensitive to their allosteric effectors without alteration of their catalytic activity. Selective denaturation of allosteric sites has been achieved by treatment with mercurials, urea, x-rays, proteolytic enzymes, extremes of ionic strength or pH, aging at 0-5 °C, by freezing, or by heating.

(2) Allosteric effectors frequently protect the **catalytic** site from denaturation under conditions where the substrates themselves do not protect. Since it seems unlikely that an effector bound at the catalytic site would protect when substrates do not, this suggests a second, allosteric site elsewhere on the enzyme molecule.

(3) In certain bacterial and mammalian cell mutants, the regulated enzymes have altered regulatory properties but identical catalytic properties to those of the wild-type from which the mutant derived. The structures of the allosteric and catalytic sites thus are genetically distinct.

(4) Binding studies of substrates and of allosteric effectors to regulated enzymes show that each may bind independently of the other.

(5) In certain cases (eg, ATCase), the allosteric site is present on a different protomer from the catalytic site.

Allosteric Enzymes Typically Exhibit Sigmoidal Substrate Saturation Kinetics

Fig 11–6 illustrates the rate of a reaction catalyzed by a typical allosteric enzyme measured at several concentrations of substrate in the presence and absence of an allosteric inhibitor. In the absence of the allosteric inhibitor, hyperbolic saturation kinetics are observed. In its presence, the substrate saturation curve is distorted from a hyperbola into a sigmoid, which at high substrate concentrations may merge with the hyperbola. Note the analogy to the relationship between the O_2 saturation curves for myoglobin and hemoglobin (see Chapter 7).

On kinetic analysis, feedback inhibition may appear to be competitive, noncompetitive, partially competitive, or of other types. If, at high concentrations of S, comparable activity is observed in the presence or absence of the allosteric inhibitor, the kinetics superficially resemble those of competitive inhibition. However, since the substrate saturation curve is sigmoid rather than hyperbolic, it is not possible to obtain meaningful results by graphing data for allosteric inhibition by the double reciprocal technique. This method of analysis was developed for substrate competitive inhibition **at the catalytic site.** Since allosteric inhibitors act at a different (allosteric) site, that kinetic model is invalid.

The sigmoid character of the v versus S curve in the presence of an allosteric inhibitor reflects the phenomenon of **cooperativity.** At low concentrations of S, the activity in the presence of the inhibitor is low relative to that in its absence. However, as S is increased, the extent of inhibition becomes relatively less severe. The kinetics are consistent with the presence of 2 or more interacting substrate-binding sites, where the presence of a substrate molecule at one catalytic site facilitates binding of a second substrate molecule at a second bite. Cooperativity of substrate binding has been described in Chapter 7 for hemoglobin. The sigmoid O_2 saturation curve results from cooperative interactions between four O_2 binding sites located on different protomers.

Allosteric Effects May Be Principally on K_m or on V_{max}

Reference to the kinetics of allosteric inhibition as "competitive" or "noncompetitive" with substrate carries mechanistic implications that are misleading. We refer instead to 2 classes of regulated enzymes, K-series and V-series enzymes. For K-series allosteric enzymes, the substrate saturation kinetics are competitive in the sense that K_m is raised (decreased affinity for substrate) without effect on V_{max}. For V-series allosteric enzymes, the allosteric inhibitor lowers V_{max} (lowered catalytic efficiency) without affecting the apparent K_m. Alterations in K_m or V_{max} probably result from conformational changes at the catalytic site induced by binding of the allosteric effector at the allosteric site. For a K-series allosteric enzyme, this conformational change may weaken the bonds between substrate and substrate-binding residues. For a V-series allosteric enzyme, the primary effect may be to alter the orientation of catalytic residues so as to lower V_{max}. Intermediate effects on K_m and V_{max}, however, may be observed consequent to these conformational changes.

Cooperative Binding of a Substrate Confers a Significant Physiologic Advantage

The consequences of cooperative substrate-binding kinetics are analogous to those resulting from the cooperative binding of O_2 to hemoglobin. At low substrate concentrations, the allosteric effector is an effective inhibitor. It thus regulates most effectively at the time of greatest need, ie, when intracellular concentrations of substrates are low. As more substrate becomes available, stringent regulation is less necessary. As substrate concentration rises, the degree of inhibition therefore lessens, and more product is formed. As with hemoglobin, the sigmoid substrate saturation curve in the presence of inhibitor also means that relatively small changes in substrate concentration result in large changes in activity. Sensitive control of catalytic activity thus is achieved by small changes in substrate concentration. Finally, by analogy with the differing O_2 saturation curves of hemoglobins from different

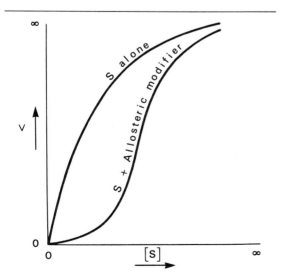

Figure 11–6. Sigmoid saturation curve for substrate in the presence of an allosteric inhibitor.

species, regulatory enzymes from different sources may have sigmoid saturation curves shifted to the left or right to accommodate to the range of prevailing in vivo concentrations of substrate.

FEEDBACK REGULATION IS NOT SYNONYMOUS WITH FEEDBACK INHIBITION

In both mammalian and bacterial cells, end products "feed back" and control their own synthesis. In many instances, this involves feedback inhibition of an early biosynthetic enzyme. We must, however, distinguish between **feedback regulation,** a phenomenologic term devoid of mechanistic implications, and **feedback inhibition,** a mechanism for regulation of many bacterial and mammalian enzymes. For example, dietary cholesterol restricts the synthesis of cholesterol from acetate in mammalian tissues. This feedback regulation, however, does not appear to involve feedback inhibition of an early enzyme of cholesterol biosynthesis. An early enzyme (HMG-CoA reductase) is affected, but the mechanism involves curtailment by cholesterol or a cholesterol metabolite of the expression of the genes that code for the formation of HMG-CoA reductase. Cholesterol added directly to HMG-CoA reductase has no effect on its catalytic activity.

REVERSIBLE, COVALENT MODIFICATION REGULATES KEY MAMMALIAN ENZYMES

Reversible modulation of the catalytic activity of enzymes can occur by covalent attachment of a phosphate group (predominates in mammals) or a nucleotide (predominates in bacteria). Enzymes that undergo covalent modification with attendant modulation of their activity are termed "interconvertible enzymes" (Fig 11–7).

Interconvertible enzymes exist in 2 activity states, one of high and the other of low catalytic efficiency. Depending on the enzyme concerned, the phospho- or the dephosphoenzyme may be the more active catalyst (Table 11–2).

Table 11–2. Examples of mammalian enzymes whose catalytic activity is altered by covalent phosphorylation-dephosphorylation. E, dephosphoenzyme; EP, phosphoenzyme.

Enzyme	Activity State	
	Low	High
Acetyl-CoA carboxylase	EP	E
Glycogen synthase	EP	E
Pyruvate dehydrogenase	EP	E
HMG-CoA reductase	EP	E
Glycogen phosphorylase	E	EP
Citrate lyase	E	EP
Phosphorylase b kinase	E	EP
HMG-CoA reductase kinase	E	EP

Enzymes May Have Multiple Phosphorylation Sites

A specific Ser residue is phosphorylated, forming an O-phosphoseryl residue, or, in rarer cases, a tyrosyl residue is phosphorylated to form an O-phosphotyrosyl residue. While an interconvertible enzyme may contain many Ser or Tyr residues, phosphorylation is highly selective and occurs at only a small number of sites. These sites probably do not form part of the catalytic site, at least in a primary structural sense, and thus constitute another example of an allosteric site.

Protein Kinases & Phosphatases Are Converter Proteins

Phosphorylation and dephosphorylation are catalyzed by protein kinases and protein phosphatases (converter proteins), respectively (Fig 11–8). In specific instances, the converter proteins themselves may be interconvertible enzymes (Table 11–2). Thus, there are protein kinase kinases and protein kinase phosphatases that catalyze the interconversion of these converter proteins. Evidence that protein phosphatases are also interconvertible proteins is less convincing, although their activity is regulated. The activity of both protein kinases and protein phosphatases is under hormonal and neural control, although the precise details by which these agents act are in most instances far from clear.

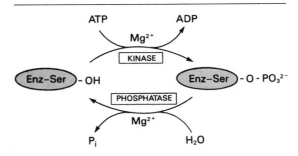

Figure 11–8. Covalent modification of a regulated enzyme by phosphorylation-dephosphorylation of a Ser residue.

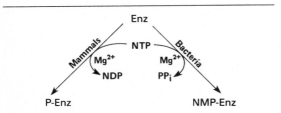

Figure 11–7. Regulation of enzyme activity by covalent modification. *Left:* phosphorylation. *Right:* nucleotidylation. For both processes the nucleoside triphosphate (NTP) generally is ATP.

Regulation Via Phosphorylation-Dephosphorylation Consumes ATP

The reactions shown in Fig 11–8 resemble those for interconversion of glucose and glucose 6-phosphate or of fructose 6-phosphate and fructose 1,6-bisphosphate (see Chapter 19). The net result of phosphorylating and then dephosphorylating 1 mol of substrate (enzyme or sugar) is the hydrolysis of 1 mol of ATP.

The activities of the kinases (catalyzing reactions 1 and 3) and of the phosphatases (catalyzing reactions 2 and 4) may themselves be regulated, for if not, they would act together to catalyze uncontrolled hydrolysis of ATP.

1.	Glucose + ATP → ADP + Glucose 6-P
2.	H_2O + Glucose 6-P → P_i + Glucose

Net:	H_2O + ATP → ADP + P_i

3.	Enz–Ser–OH + ATP → ADP + Enz–Ser–O–P
4.	H_2O + Enz–Ser–O–P → P_i + Enz–Ser–OH

Net:	H_2O + ATP → ADP + P_i

Covalent Modification, Like Feedback Inhibition, Regulates Metabolite Flow

Regulation of enzyme activity by phosphorylation-dephosphorylation has analogies to regulation by feed-back inhibition. Both provide for short-term regulation of metabolite flow in response to specific physiologic signals; both act without altering gene expression. Both act on early enzymes of a protracted (often biosynthetic) metabolic sequence; and both act at allosteric rather than catalytic sites. Feedback inhibition, however, involves a single protein and lacks hormonal and neural features. By contrast, regulation of mammalian enzymes by phosphorylation-dephosphorylation involves several proteins and ATP or another nucleoside triphosphate and is under direct neural and hormonal control.

REFERENCES

Crabtree B, Newsholme EA: A systematic approach to describing and analyzing metabolic control systems. *Trends Biochem Sci* 1987;**12**:4.

Kacser H, Porteus JW: Control of metabolism: What have we to measure? *Trends Biochem Sci* 1987;**12**:5.

Nestler EJ, Greengard P: Protein phosphorylation in the brain. *Nature* 1983;**305**:583.

Soderling TR: Role of hormones and protein phosphorylation in metabolic regulation. *Fed Proc* 1982;**41**:2615.

Scriver CR et al (editors): *The Metabolic Basis of Inherited Disease*, 6th ed. McGraw-Hill, 1989.

Weber G (editor): *Advances in Enzyme Regulation*. Pergamon Press, 1963–1990.

Section II.
Bioenergetics & the Metabolism of Carbohydrates & Lipids

Bioenergetics

12

Peter A. Mayes, PhD, DSc

INTRODUCTION

Bioenergetics, or biochemical thermodynamics, is the study of the energy changes accompanying biochemical reactions. It provides the underlying principles to explain why some reactions may occur while others do not. Nonbiologic systems may utilize heat energy to perform work, but **biologic systems are essentially isothermic and use chemical energy to power the living processes.**

BIOMEDICAL IMPORTANCE

Suitable fuel is required to provide the energy that enables the animal to carry out its normal processes. How the organism obtains this energy from its food is basic to the understanding of **normal nutrition** and **metabolism.** Death from **starvation** occurs when available energy reserves are depleted, and certain forms of malnutrition are associated with energy imbalance (**marasmus**). The rate of energy release, measured by the metabolic rate, is controlled by the **thyroid hormones, whose malfunction is a cause of disease.** Storage of surplus energy results in **obesity,** one of the most common diseases of occidental society.

FREE ENERGY IS THE USEFUL ENERGY IN A SYSTEM

Change in **free energy (ΔG)** is that portion of the total energy change in a system that is available for doing work; ie, it is the useful energy, also known in chemical systems as the **chemical potential.**

Biologic Systems Conform to the General Laws of Thermodynamics

The first law of thermodynamics states that **the total energy of a system, including its surroundings, remains constant.** This is also the law of conservation of energy. It implies that within the total system, energy is neither lost nor gained during any change. However, within that total system, energy may be transferred from one part to another or may be transformed into another form of energy. For example, chemical energy may be transformed into heat, electrical energy, radiant energy, or mechanical energy in living systems.

The second law of thermodynamics states that **the total entropy of a system must increase if a process is to occur spontaneously. Entropy** represents the extent of disorder or randomness of the system and becomes maximum in a system as it approaches true equilibrium. Under conditions of constant temperature and pressure, the relationship between the free energy change (ΔG) of a reacting system and the change in entropy (ΔS) is given by the following equation which combines the 2 laws of thermodynamics:

$$\Delta G = \Delta H - T\Delta S$$

where ΔH is the change in **enthalpy** (heat) and T is the absolute temperature.

Under the conditions of biochemical reactions, because ΔH is approximately equal to ΔE, the total change in internal energy of the reaction, the above relationship may be expressed in the following way:

$$\Delta G = \Delta E - T\Delta S$$

If ΔG is negative in sign, the reaction proceeds spontaneously with loss of free energy; ie, it is **exergonic.** If, in addition, ΔG is of great magnitude, the reaction goes virtually to completion and is essentially irreversible. On the other hand, **if ΔG is positive, the reaction proceeds only if free energy can be gained;** ie, it is **endergonic.** If, in addition, the magnitude of ΔG is great, the system is stable with little or no tendency for a reaction to occur. **If ΔG is zero, the system is at equilibrium and no net change takes place.**

When the reactants are present in concentrations of 1.0 mol/L, **ΔG⁰ is the standard free energy change.** For biochemical reactions, a standard state is defined as having a pH of 7.0. **The standard free energy change at this standard state is denoted by ΔG⁰′.**

The standard free energy change can be calculated from the equilibrium constant K'_{eq}

$$\Delta G^{0\prime} = -2.303 \; RT \; \log K'_{eq}$$

where R is the gas constant and T is the absolute temperature (see p 69). It is important to note that **ΔG may be larger or smaller than ΔG⁰′ depending on the concentrations of the various reactants.**

In a biochemical reaction system, it must be appreciated that an enzyme only speeds up the attainment of equilibrium; **it never alters the final concentrations of the reactants at equilibrium.**

ENDERGONIC PROCESSES PROCEED BY COUPLING TO EXERGONIC PROCESSES

The vital processes—eg, synthetic reactions, muscular contraction, nerve impulse conduction, and active transport—obtain energy by chemical linkage, or **coupling,** to oxidative reactions. In its simplest form, this type of coupling may be represented as shown in Fig 12–1.

The conversion of metabolite A to metabolite B occurs with release of free energy. It is coupled to another reaction, in which free energy is required to convert metabolite C to metabolite D. As some of the energy liberated in the degradative reaction is transferred to the synthetic reaction in a form other than heat, the normal chemical terms exothermic and endothermic cannot be applied to these reactions. Rather, the terms **exergonic and endergonic are used to indicate that a process is accompanied by loss or gain, respectively, of free energy, regardless of the form of energy involved.** In practice, an endergonic process cannot exist independently but must be a component of a coupled exergonic/endergonic system where the **overall net change is exergonic.** The exergonic reactions are termed **catabolism** (the breakdown or oxidation of fuel molecules), whereas the synthetic reactions that build up substances are termed **anabolism.** The total of all of the catabolic and anabolic processes is **metabolism.**

If the reaction shown in Fig 12–1 is to go from left to right, then the overall process must be accompanied by loss of free energy as heat. One possible mechanism of coupling could be envisaged if a common obligatory intermediate (I) took part in both reactions, ie,

$$A + C \rightarrow I \rightarrow B + D$$

Some exergonic and endergonic reactions in biologic systems are coupled in this way. It should be appreciated that this type of system has a built-in mechanism for biologic control of the rate at which oxidative processes are allowed to occur, since the existence of a common obligatory intermediate allows the rate of utilization of the product of the synthetic path (D) to determine by mass action the rate at which A is oxidized. Indeed, these relationships supply a basis for the concept of **respiratory control,** the process that prevents an organism from burning out of control. An extension of the coupling concept is provided by dehydrogenation reactions, which are coupled to hydrogenations by an intermediate carrier (Fig 12–2).

An alternative method of coupling an exergonic to an endergonic process is to synthesize a compound of high-energy potential in the exergonic reaction and to incorporate this new compound into the endergonic reaction, thus effecting a transference of free energy from the exergonic to the endergonic pathway (Fig 12–3).

In Fig 12–3, ~Ⓔ is a compound of high potential energy and Ⓔ is the corresponding compound of low potential energy. The biologic advantage of this mechanism is that Ⓔ, unlike I in the previous system, need

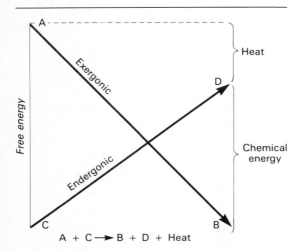

Figure 12–1. Coupling of an exergonic to an endergonic reaction.

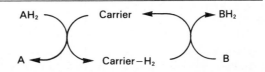

Figure 12–2. Coupling of dehydrogenation and hydrogenation reactions by an intermediate carrier.

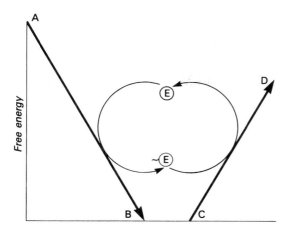

Figure 12–3. Transfer of free energy from an exergonic to an endergonic reaction via a high-energy intermediate compound.

not be structurally related to A, B, C, or D. This would allow Ⓔ to serve as a transducer of energy from a wide range of exergonic reactions to an equally wide range of endergonic reactions or processes, as shown in Fig 12–4.

In the living cell, the principal high-energy intermediate or carrier compound (designated ~Ⓔ) is **adenosine triphosphate (ATP).**

HIGH-ENERGY PHOSPHATES PLAY A CENTRAL ROLE IN ENERGY CAPTURE AND TRANSFER

In order to maintain living processes, all organisms must obtain supplies of free energy from their environment. **Autotrophic** organisms couple their metabolism to some simple exergonic process in their sur-

[figure at top right]

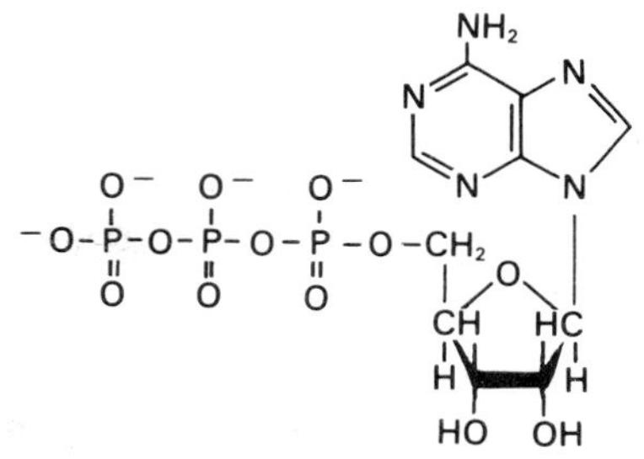

Figure 12–5. Adenosine triphosphate (ATP).

roundings; eg, green plants utilize the energy of sunlight, and some autotrophic bacteria utilize the reaction $Fe^{2+} \rightarrow Fe^{3+}$. On the other hand, **heterotrophic** organisms obtain free energy by coupling their metabolism to the breakdown of complex organic molecules in their environment. In all of these processes, ATP plays a central role in the transference of free energy from the exergonic to the endergonic processes (Figs 12–3 and 12–4). As can be seen from Fig 12–5, ATP is a specialized nucleotide containing adenine, ribose, and 3 phosphate groups. In its reactions in the cell, it functions as the Mg^{2+} complex (Fig 12–6).

The importance of phosphates in intermediary metabolism became evident with the discovery of the chemical details of glycolysis and of the role of ATP, adenosine diphosphate (ADP), and inorganic phosphate (P_i) in this process (see p 168). ATP was considered a means of transferring phosphate radicals in the process of phosphorylation. The role of ATP in biochemical energetics was indicated in experiments demonstrating that ATP and creatine phosphate were broken down during muscular contraction and that their resynthesis depended on supplying energy from oxidative processes in the muscle. It was not until Lipmann introduced the concept of "high-energy phosphates" and the "high-energy phosphate bond" that the role of these compounds in bioenergetics was clearly appreciated.

The Intermediate Value for the Free Energy of Hydrolysis of ATP Compared to Other Organophosphates Has Important Bioenergetic Significance

The standard free energy of hydrolysis of a number of biochemically important phosphates is shown in

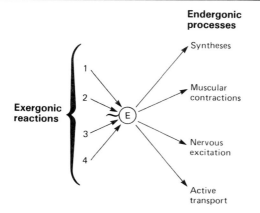

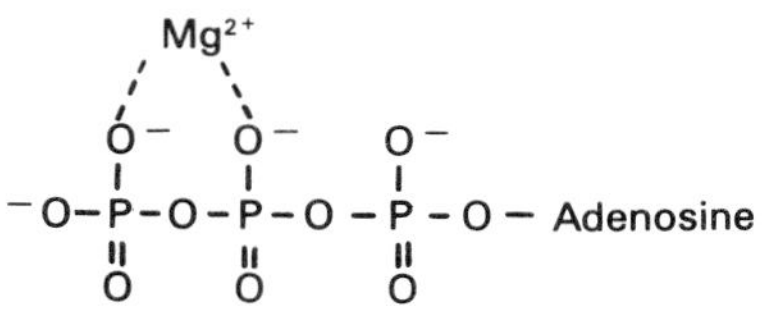

Figure 12–4. Transduction of energy through a common high-energy compound to energy-requiring (endergonic) biologic processes.

[figure bottom right: Mg²⁺ complex structure]

Figure 12–6. The magnesium complex of ATP. (Mg-ADP is similar.)

Table 12–1. Standard free energy of hydrolysis of some organophosphates of biochemical importance.*†

Compound	ΔG⁰′ kJ/mol	ΔG⁰′ kcal/mol
Phosphoenolpyruvate	−61.9	−14.8
Carbamoyl phosphate	−51.4	−12.3
1,3-Bisphosphoglycerate (to 3-phosphoglycerate)	−49.3	−11.8
Creatine phosphate	−43.1	−10.3
ATP → ADP + P$_i$	−30.5	−7.3
ADP → AMP + P$_i$	−27.6	−6.6
Pyrophosphate	−27.6	−6.6
Glucose 1-phosphate	−20.9	−5.0
Fructose 6-phosphate	−15.9	−3.8
AMP	−14.2	−3.4
Glucose 6-phosphate	−13.8	−3.3
Glycerol 3-phosphate	−9.2	−2.2

*P$_i$, inorganic orthophosphate.
†Values for ATP and most others taken from Krebs and Kornberg (1957).

Table 12–1. An estimate of the comparative tendency of each of the phosphate groups to transfer to a suitable acceptor may be obtained from the ΔG⁰′ of hydrolysis (measured at 37 °C). It may be seen from the table that the value for the hydrolysis of the terminal phosphate of ATP of −30.5 kJ/mol divides the list into 2 groups. One group of **low-energy phosphates,** exemplified by the ester phosphates found in the intermediates of glycolysis, has ΔG⁰′ values smaller than that of ATP, while in the other group, designated **high-energy phosphates,** the value is higher than that of ATP. The components of this latter group, including ATP and ADP, are usually anhydrides (eg, the 1-phosphate of 1,3-bisphosphoglycerate), enolphosphates (eg, phosphoenolpyruvate), and phosphoguanidines (eg, creatine phosphate, arginine phosphate). **The intermediate position of ATP allows it to play an important role in energy transfer.** Other biologically important compounds that are classed as "high-energy compounds" are thiol esters involving coenzyme A (eg, acetyl-CoA), acyl carrier protein, amino acid esters involved in protein synthesis, S-adenosylmethionine (active methionine), UDPGlc (uridine diphosphate glucose), and PRPP (5-phosphoribosyl-1-pyrophosphate).

High-Energy Phosphates Are Designated by the Symbol ~℗

To indicate the presence of the high-energy phosphate group, Lipmann introduced the symbol ~℗, indicating **high-energy phosphate bond.** The symbol indicates that the group attached to the bond, on transfer to an appropriate acceptor, results in transfer of the larger quantity of free energy. For this reason, the term **group transfer potential** is preferred by some to "high-energy bond." Thus, ATP contains 2 high-energy phosphate groups and ADP contains one, whereas the phosphate in AMP (adenosine monophosphate) is of the low-energy type, since it is a normal ester link (Fig 12–7).

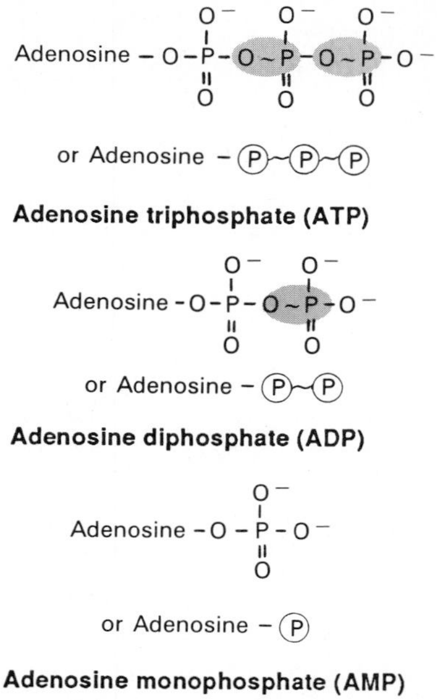

Adenosine triphosphate (ATP)

Adenosine diphosphate (ADP)

Adenosine monophosphate (AMP)

Figure 12–7. Structure of ATP, ADP, and AMP showing the position and the number of high-energy phosphates (~).

HIGH-ENERGY PHOSPHATES ACT AS THE "ENERGY CURRENCY" OF THE CELL

As a result of its position midway down the list of standard free energies of hydrolysis (Table 12–1), ATP is able to act as a donor of high-energy phosphate to those compounds below it in the table. Likewise, provided the necessary enzymatic machinery is available, ADP can accept high-energy phosphate to form ATP from those compounds above ATP in the table. In effect, an **ATP/ADP cycle** connects those processes which generate ~℗ to those processes that utilize ~℗ (Fig 12–8). Thus, **ATP is continuously consumed and regenerated.**

There are 3 major sources of ~℗ taking part in **energy conservation** or **energy capture:** (1) **Oxidative phosphorylation.** This is the greatest quantitative source of ~℗ in aerobic organisms. The free energy to drive this process comes from respiratory chain oxidation within mitochondria (see p 118). (2) **Glycolysis.** A net formation of 2 ~℗ results from the formation of lactate from one molecule of glucose generated in 2 reactions catalyzed by phosphoglycerate kinase and pyruvate kinase, respectively (see Fig 19–2). (3) **The citric acid cycle.** One ~℗ is generated directly in the cycle at the succinyl thiokinase step (see Fig 18–3).

Another group of compounds, **phosphagens,** act as storage forms of high-energy phosphate. These include creatine phosphate, occurring in vertebrate muscle and brain, and arginine phosphate, occurring in invertebrate muscle. Under physiologic conditions,

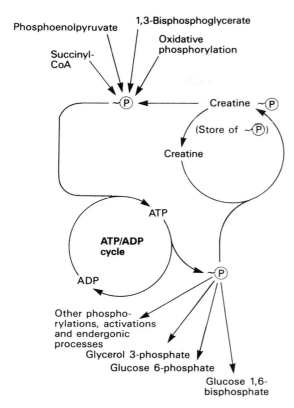

Figure 12–8. Role of ATP/ADP cycle in transfer of high-energy phosphate. Note that ~ⓅP **does not exist in a free state but is transferred in the reactions shown.**

phosphagens permit ATP concentrations to be maintained in muscle when ATP is rapidly being utilized as a source of energy for muscular contraction. On the other hand, when ATP is plentiful, their concentration can build up to act as a store of high-energy phosphate (Fig 12–9). In muscle, a **creatine phosphate shuttle** has been described that transports high-energy phosphate from mitochondria to the sarcolemma and that acts as a high-energy phosphate buffer. In the myocardium, this buffer may be of significance in affording immediate protection against the effects of infarction.

When ATP acts as a phosphate donor to form those compounds of lower free energy of hydrolysis (Table

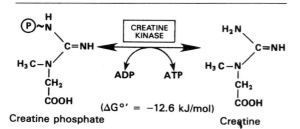

Figure 12–9. Transfer of high-energy phosphate between ATP and creatine.

12–1), the phosphate group is invariably converted to one of low energy, eg,

$$\text{Glyerol} + \text{Adenosine} - Ⓟ \sim Ⓟ \sim Ⓟ \xrightarrow{\text{GLYCEROL KINASE}}$$

$$\text{Glycerol} - Ⓟ + \text{Adenosine} - Ⓟ \sim Ⓟ$$

ATP Allows the Coupling of Thermodynamically Unfavorable Reactions to Favorable Ones

The energetics of coupled reactions are depicted in more detail in Fig 12–1 or 12–3. Such a reaction is the first in the glycolytic pathway (see Fig 19–2), the phosphorylation of glucose to glucose 6-phosphate, which is highly endergonic and cannot proceed as such under physiologic conditions.

(1) Glucose + P_i → Glucose 6-phosphate + H_2O
$$(\Delta G^{0\prime} = +13.8 \text{ kJ/mol})$$

To take place, the reaction must be coupled with another reaction that is more exergonic than the phosphorylation of glucose is endergonic. Such a reaction is the hydrolysis of the terminal phosphate of ATP.

(2) ATP → ADP + P_i ($\Delta G^{0\prime} = -30.5$ kJ/mol)

When (1) and (2) are coupled in a reaction catalyzed by hexokinase, phosphorylation of glucose readily proceeds in a highly exergonic reaction that under physiologic conditions is far from equilibrium and thus irreversible for practical purposes.

$$\text{Glucose} + \text{ATP} \xrightarrow{\text{HEXOKINASE}}$$

$$\text{Glucose 6-phosphate} + \text{ADP}$$
$$(\Delta G^{0\prime} = -16.7 \text{ kJ/mol})$$

Many "activation" reactions follow this pattern.

Adenylate Kinase Allows Interconversion of Adenine Nucleotides

The enzyme adenylate kinase (myokinase) is present in most cells. It catalyzes the interconversion of ATP and AMP on the one hand and ADP on the other:

$$\text{ATP} + \text{AMP} \xleftrightarrow{\text{ADENYLATE KINASE}} \text{2ADP}$$

This reaction has 3 functions:

(1) It allows high-energy phosphate in ADP to be used in the synthesis of ATP.

(2) It allows AMP, formed as a consequence of several activating reactions involving ATP, to be recovered by rephosphorylation to ADP.

(3) It allows AMP to increase in concentration when ATP becomes depleted and act as a metabolic (allosteric) signal to increase the rate of catabolic reac-

tions, which in turn leads to the generation of more ATP (see p 195).

When ATP Reacts to Form AMP, Inorganic Pyrophosphate (PP$_i$) is Formed

This occurs, for example, in the activation of long-chain fatty acids:

$$\text{ATP} + \text{CoA} \cdot \text{SH} + \text{R} \cdot \text{COOH} \xrightarrow{\boxed{\begin{array}{c}\text{ACYL–CoA}\\\text{SYNTHETASE}\end{array}}}$$
$$\text{AMP} + \text{PP}_i + \text{R} \cdot \text{CO–SCoA}$$

This reaction is accompanied by loss of free energy as heat, which ensures that the activation reaction will go to the right; this is further aided by the hydrolytic splitting of PP$_i$, catalyzed by **inorganic pyrophosphatase,** a reaction that itself has a large $\Delta G^{0\prime}$ of -27.6 kJ/mol. Note that activations via the pyrophosphate pathway result in the loss of 2 $\sim$℗ rather than one $\sim$℗ as occurs when ADP and P$_i$ are formed.

$$\text{PP}_i + \text{H}_2\text{O} \xrightarrow{\boxed{\begin{array}{c}\text{INORGANIC}\\\text{PYROPHOSPHATASE}\end{array}}} 2\ \text{P}_i$$

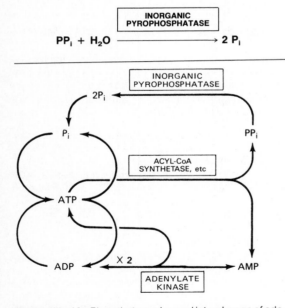

Figure 12–10. Phosphate cycles and interchange of adenine nucleotides.

A combination of the above reactions makes it possible for phosphate to be recycled and the adenine nucleotides to interchange (Fig 12–10).

Other Nucleoside Triphosphates Take Part in the Transfer of High-Energy Phosphate

By means of the enzyme nucleoside diphosphate kinase, nucleoside triphosphates similar to ATP but containing an alternative base to adenine can be synthesized from their diphosphates, eg,

$$\boxed{\begin{array}{c}\text{NUCLEOSIDE}\\\text{DIPHOSPHATE}\\\text{KINASE}\end{array}}$$

$$\text{ATP} + \text{UDP} \longleftrightarrow \text{ADP} + \text{UTP}$$
$$\text{(uridine triphosphate)}$$

$$\text{ATP} + \text{GDP} \longleftrightarrow \text{ADP} + \text{GTP}$$
$$\text{(guanosine triphosphate)}$$

$$\text{ATP} + \text{CDP} \longleftrightarrow \text{ADP} + \text{CTP}$$
$$\text{(cytidine triphosphate)}$$

All of these triphosphates take part in phosphorylations in the cell. Similarly, nucleoside monophosphate kinases, specific for each purine or pyrimidine nucleoside, catalyze the formation of nucleoside diphosphates from the corresponding monophosphates

$$\boxed{\begin{array}{c}\text{SPECIFIC NUCLEOSIDE}\\\text{MONOPHOSPHATE KINASE}\end{array}}$$
$$\text{ATP} + \text{Nucleoside} - ℗ \longleftrightarrow$$
$$\text{ADP} + \text{Nucleoside} - ℗ \sim ℗$$

Thus, adenylate kinase is a specialized monophosphate kinase.

REFERENCES

Ernster L (editor): *Bioenergetics.* Elsevier, 1984.

Harold FM: *The Vital Force: A Study of Bioenergetics.* Freeman, 1986.

Klotz IM: *Introduction to Biomolecular Energetics.* Academic Press, 1986.

Krebs HA, Kornberg HL: *Energy Transformations in Living Matter.* Springer, 1957.

Lehninger AL: *Bioenergetics: The Molecular Basis of Biological Energy Transformations,* 2nd ed. Benjamin, 1971.

Biologic Oxidation

13

Peter A. Mayes, PhD, DSc

INTRODUCTION

Chemically, **oxidation is defined as the removal of electrons** and, **reduction is the gain of electrons,** as illustrated by the oxidation of ferrous to ferric ion.

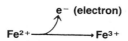

It follows that **oxidation is always accompanied by reduction of an electron acceptor.** This principle of oxidation-reduction applies equally to biochemical systems and is an important concept underlying understanding of the nature of biologic oxidation. It will be appreciated that many biologic oxidations can take place without the participation of molecular oxygen, eg, dehydrogenations.

BIOMEDICAL IMPORTANCE

Although certain bacteria (anaerobes) survive in the absence of oxygen, the life of higher animals is absolutely dependent upon a supply of oxygen. The principal use of oxygen is in **respiration,** which may be defined as the process by which cells derive energy in the form of ATP from the controlled reaction of hydrogen with oxygen to form water. In addition, molecular oxygen is incorporated into a variety of substrates by enzymes designated as **oxygenases**; many drugs, pollutants, and chemical carcinogens (xenobiotics) are metabolized by enzymes of this class, known as the **cytochrome P-450 system. Administration of oxygen** can be lifesaving in the treatment of patients with respiratory or circulatory failure and, occasionally, administration of oxygen at high pressure (hyperbaric oxygen therapy) has proved of value, although this can result in oxygen toxicity.

IN REDOX SYSTEMS FREE ENERGY CHANGES CAN BE EXPRESSED IN TERMS OF REDOX POTENTIAL

In reactions involving oxidation and reduction, the free energy exchange is proportionate to the tendency of reactants to donate or accept electrons. Thus, in addition to expressing free energy change in terms of $\Delta G^{0\prime}$ (see Chapter 12), it is possible, in an analogous manner, to express it numerically as an **oxidation-reduction** or **redox potential** $(E_o\prime)$. It is usual to compare the redox potential of a system (E_o) against the potential of the hydrogen electrode, which at pH 0 is designated as 0.0 volts. However, for biologic systems it is normal to express the redox potential $(E_o\prime)$ at pH 7.0, at which pH the electrode potential of the hydrogen electrode is -0.42 volts. The redox potentials of some redox systems of special interest in mammalian biochemistry are shown in Table 13–1. The list of redox potentials shown in the table allows prediction of the direction of flow of electrons from one redox couple to another.

ENZYMES INVOLVED IN OXIDATION & REDUCTION ARE DESIGNATED OXIDOREDUCTASES

In the following account, oxidoreductases are classified into 4 groups: oxidases, dehydrogenases, hydroperoxidases, and oxygenases.

OXIDASES USE OXYGEN AS A HYDROGEN ACCEPTOR

Oxidases catalyze the removal of hydrogen from a substrate using oxygen as a hydrogen acceptor.* They form water or hydrogen peroxide as a reaction product (Fig 13–1).

Some Oxidases Contain Copper

Cytochrome oxidase is a hemoprotein widely distributed in many plant and animal tissues. It is the terminal component of the chain of respiratory carriers found in mitochondria and is therefore responsible for the reaction whereby electrons resulting from the oxidation of substrate molecules by dehydrogenases are transferred to their final acceptor, oxygen. The enzyme is poisoned by carbon monoxide, cyanide, and hydrogen sulfide. It has also been termed cytochrome a_3. It was formerly assumed that cytochrome a and

---

*Sometimes the term "oxidase" is used collectively to denote all enzymes that catalyze reactions involving molecular oxygen.

Table 13-1. Some redox potentials of special interest in mammalian oxidation systems.

System	E_o. volts
Succinate/α-ketoglutarate	-0.67
H^+/H_2	-0.42
$NAD^+/NADH$	-0.32
Lipoate; ox/red	-0.29
Acetoacetate/β-hydroxybutyrate	-0.27
Pyruvate/lactate	-0.19
Oxaloacetate/malate	-0.17
Flavoprotein-old yellow enzyme; ox/red	-0.12
Fumarate/succinate	$+0.03$
Cytochrome b; Fe^{3+}/Fe^{2+}	$+0.08$
Ubiquinone; ox/red	$+0.10$
Cytochrome c; Fe^{3+}/Fe^{2+}	$+0.22$
Cytochrome a; Fe^{3+}/Fe^{2+}	$+0.29$
Oxygen/water	$+0.82$

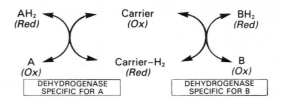

Figure 13-2. Oxidation of a metabolite catalyzed by dehydrogenases, not involving a respiratory chain.

cytochrome a_3 were separate compounds, since each has a distinct spectrum and different properties with respect to the effects of carbon monoxide and cyanide. More recent studies show that the 2 cytochromes are combined with the same protein, and the complex is known as **cytochrome aa₃.** It contains 2 molecules of heme, each having one Fe atom that oscillates between Fe^{3+} and Fe^{3+} during oxidation and reduction. Also, 2 atoms of Cu are present, each associated with a heme unit.

Phenolase (tyrosinase, polyphenol oxidase, catechol oxidase) is a copper-containing enzyme of broad specificity. It is able to convert monophenols or *o*-diphenols to *o*-quinones. A number of other enzymes also contain copper.

Other Oxidases are Flavoproteins

Flavoprotein enzymes contain **flavin mononucleotide (FMN) or flavin adenine dinucleotide (FAD) as prosthetic groups. FMN and FAD are formed in the body from the vitamin riboflavin** (see Chapter 53).

FMN and FAD are usually tightly—but not covalently—bound to their respective apoenzyme protein. Many flavoprotein enzymes contain one or more metals as essential cofactors and are known as **metalloflavoproteins.**

Enzymes belonging to this group of oxidases include **L-amino acid oxidase,** an FMN-linked enzyme found in kidney with general specificity for the oxidative deamination of the naturally occuring L-amino acids. **Xanthine oxidase** has a wide distribution, oc-

curring in milk, small intestine, kidney, and liver. It contains molybdenum and plays an important role in the conversion of purine bases to uric acid (see p. 342). It is of particular significance in the liver and kidney of birds, which excrete uric acid as the main nitrogenous end product, not only of purine metabolism but also of protein and amino acid catablism.

Aldehyde dehydrogenase is an FAD-linked enzyme present in mammalian livers. It is a metalloflavoprotein containing molybdenum and nonheme iron and acts upon aldehydes and N-heterocyclic substrates.

Of interest because of its use in estimating glucose is **glucose oxidase,** an FAD-specific enzyme prepared from fungi.

The mechanisms of oxidation and reduction of these enzymes are complex. However, evidence points to reduction of the isoalloxazine ring taking place in 2 steps via a semiquinone (free radical) intermediate (Fig 13-3).

DEHYDROGENASES CANNOT USE OXYGEN AS A HYDROGEN ACCEPTOR

There are a large number of enzymes in this class. They perform 2 main functions:

(a) Transfer of hydrogen from one substrate to another in a coupled oxidation-reduction reaction (Fig 13-2). These dehydrogenases are specific for their substrates but often utilize the same coenzyme or hydrogen carrier as other dehydrogenases. As the reactions are reversible, these properties enable reducing equivalents to be freely transferred within the cell. This type of reaction, which enables a substrate to be oxidized at the expense of another, is particularly useful in enabling oxidative processes to occur in the **absence of oxygen.**

(b) As components in a **respiratory chain** of electron transport from substrate to oxygen (Fig 13-4).

Some Dehydrogenases Are Dependent on Nicotinamide Coenzymes

A large number of dehydrogenases fall into this category. They are specific for either **nicotinamide adenine dinucleotide** (NAD⁺) or **nicotinamide adenine**

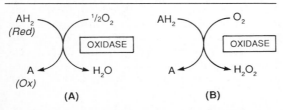

(A)   **(B)**

Figure 13-1. Oxidation of a metabolite catalyzed by an oxidase (a) forming H_2O, (b) forming H_2O_2.

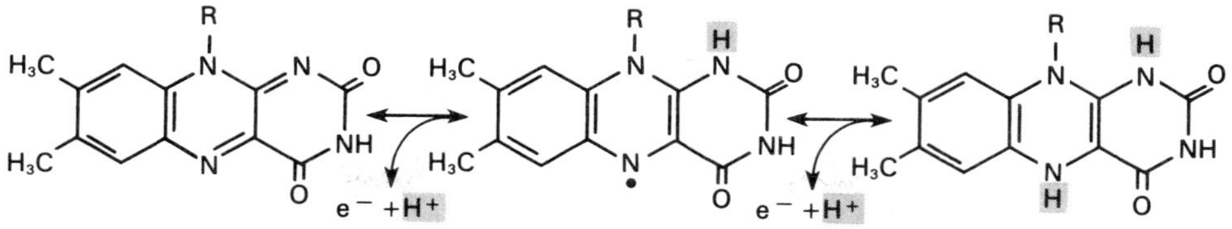

Figure 13–3. Reduction of isoalloxazine ring in flavin nucleotides.

dinucleotide phosphate ($NADP^+$) as coenzymes. However, some dehydrogenases can use either NAD^+ or $NADP^+$. **NAD^+ and $NADP^+$ are formed in the body from the vitamin niacin** (see Fig 53–4). The coenzymes are reduced by the specific substrate of the dehydrogenase and reoxidized by a suitable electron acceptor (Fig 13–5). They may freely and reversibly dissociate from their respective apoenzymes.

Generally, **NAD-linked dehydrogenases catalyze oxidoreduction reactions in the oxidative pathways of metabolism,** particularly in glycolysis, in the citric acid cycle, and in the respiratory chain of mitochondria. NADP-linked dehydrogenases are found characteristically in **reductive syntheses,** as in the extramitochondrial pathway of fatty acid synthesis and steroid synthesis. They are also to be found as coenzymes to the dehydrogenases of the pentose phosphate pathway. Some nicotinamide coenzyme-dependent dehydrogenases have been found to contain zinc, notably alcohol dehydrogenase from liver and glyceraldehyde-3-phosphate dehydrogenase from skeletal muscle. The zinc ions are not considered to take part in the oxidation and reduction.

Other Dehydrogenases Are Dependent on Riboflavin

The flavin groups associated with these dehydrogenases are similar to FMN and FAD, occurring in oxidases. They are generally more tightly bound to their apoenzymes than are the nicotinamide coenzymes. **Most of the riboflavin-linked dehydrogenases are concerned with electron transport in (or to) the respiratory chain. NADH dehydrogenase** is a member of the respiratory chain acting as a carrier of electrons between NADH and the more electropositive components (see Fig 14–2). Other dehydrogenases such as **succinate dehydrogenase, acyl-CoA dehydrogenase,** and **mitochondrial glycerol-3-phosphate dehydrogenase** transfer reducing equivalents directly from the substrate to the respiratory chain (see Fig 14–3). Another role of the flavin-dependent dehydrogenases is in the dehydrogenation (by dihydrolipoyl dehydrogenase) of reduced lipoate, an intermediate in the oxidative decarboxylation of pyruvate and α-ketoglutarate (see Fig 14–3). In this particular instance, owing to the low redox potential, the flavoprotein (FAD) acts as a hydrogen carrier from reduced lipoate to NAD (see Fig 19–5). The **electron-transferring flavoprotein** is an intermediary carrier of electrons between acyl-CoA dehydrogenase and the respiratory chain (see Fig 14–3).

Cytochromes May also Be Regarded as Dehydrogenases

Except for cytochrome oxidase (previously described), the cytochromes are also classified as dehydrogenases. Their identification and study are facilitated by the presence in the reduced state of characteristic absorption bands that disappear on oxidation. In the respiratory chain, they are involved as **carriers of electrons from flavoproteins on the one hand to cytochrome oxidase on the other** (see Fig 14–3). The cytochromes are iron-containing hemoproteins in which the iron atom oscillates between Fe^{3+} and Fe^{2+} during oxidation and reduction. Several identifiable cytochromes occur in the respiratory chain, ie, cytochromes b, c_1, c, a, and a_3 (cytochrome oxidase). Of these, only cytochrome c is soluble. Besides the respiratory chain, cytochromes are found in other locations, eg, the endoplasmic reticulum (cytochromes P-450 and b_5), plant cells, bacteria, and yeasts.

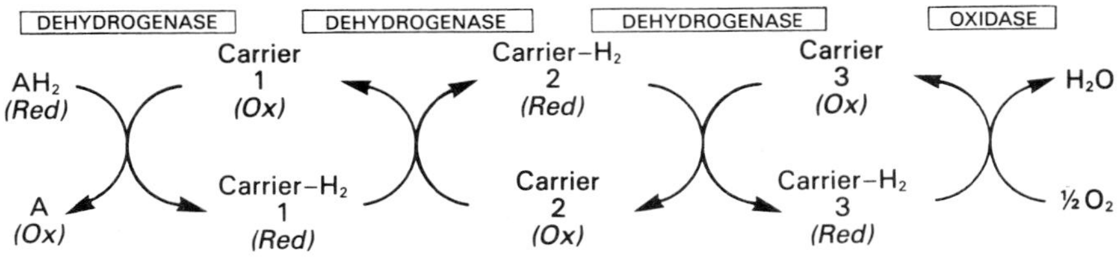

Figure 13–4. Oxidation of a metabolite by dehydrogenases and finally by an oxidase in a respiratory chain.

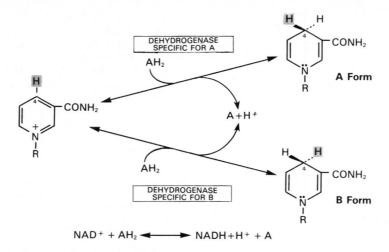

$$NAD^+ + AH_2 \longleftrightarrow NADH + H^+ + A$$

Figure 13–5. Mechanism of oxidation and reduction of nicotinamide coenzymes. There is stereospecificity about position 4 of nicotinamide when it is reduced by a substrate AH_2. One of the hydrogen atoms is removed from the substrate as a hydrogen nucleus with 2 electrons (hydride ion, H^-) and is transferred to the 4 position, where it may be attached in either the A or the B position according to the specificity determined by the particular dehydrogenase catalyzing the reaction. The remaining hydrogen of the hydrogen pair removed from the substrate remains free as a hydrogen ion.

HYDROPEROXIDASES USE HYDROGEN PEROXIDE OR AN ORGANIC PEROXIDE AS SUBSTRATE

Two types of enzymes fall into this category: **peroxidases** and **catalase.** These 2 types are found both in animals and in plants.

Hydroperoxidases protect the body against harmful peroxides. Accumulation of peroxides can lead to generation of free radicals, which in turn can disrupt membranes etc., and possibly cause cancer and atherosclerosis (see Chapter 16).

Peroxidases Reduce Peroxides Using Several Substances as Electron Acceptor

Although originally considered to be plant enzymes, peroxidases are found in milk and in leukocytes, platelets, and other tissues involved in eicosanoid metabolism (see p 231). The prosthetic group is protoheme, which, unlike the situation in most hemoproteins, is only loosely bound to the apoprotein. In the reaction catalyzed by peroxidase, hydrogen peroxide is reduced at the expense of several substances that will act as electron acceptors, such as ascorbate, quinones, and cytochrome c. The reaction catalyzed by peroxidase is complex, but the overall reaction is as follows:

$$\boxed{\text{PEROXIDASE}}$$
$$H_2O_2 + AH_2 \longrightarrow 2H_2O + A$$

In erythrocytes and other tissues, the enzyme **glutathione peroxidase,** containing **selenium** as a prosthetic group, catalyzes the destruction of H_2O_2 and lipid hydroperoxides by reduced glutathione, protecting membrane lipids and hemoglobin against oxidation by peroxides (see p 185).

Catalase Uses Hydrogen Peroxide as Electron Donor & Electron Acceptor

Catalase is a hemoprotein containing 4 heme groups. In addition to possessing peroxidase activity, it is able to use one molecule of H_2O_2 as a substrate electron donor and another molecule of H_2O_2 as oxidant or electron acceptor. Under most conditions in vivo, the peroxidase activity of catalase seems to be favored.

$$\boxed{\text{CATALASE}}$$
$$2H_2O_2 \longrightarrow 2H_2O + O_2$$

Catalase is found in blood, bone marrow, mucous membranes, kidney, and liver. Its function is assumed to be the **destruction of hydrogen peroxide** formed by the action of oxidases. Microbodies or **peroxisomes** are found in many tissues, including liver. They are rich in oxidases and in catalase, which suggests that there may be a biologic advantage in grouping the enzymes that produce H_2O_2 with the enzyme that destroys it (Fig 13–6). In addition to the peroxisomal enzymes, mitochondrial and microsomal electron transport systems as well as xanthine oxidase must be considered as sources of H_2O_2.

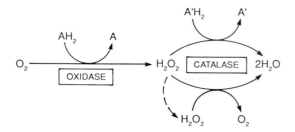

Figure 13–6. Role of catalase in the destruction of hydrogen peroxide.

OXYGENASES CATALYZE THE DIRECT TRANSFER & INCORPORATION OF OXYGEN INTO A SUBSTRATE MOLECULE

Oxygenases are concerned with the synthesis or degradation of many different types of metabolites rather than taking part in reactions that have as their purpose the provision of energy to the cell. Enzymes in this group catalyze the incorporation of oxygen into a substrate molecule. This takes place in 2 steps: (1) oxygen binding to the enzyme at the active site, and (2) the reaction in which the bound oxygen is reduced or transferred to the substrate. Oxygenases may be divided into 2 subgroups:

Dioxygenases (Oxygen Transferases, True Oxygenases) Incorporate Both Oxygen Atoms Into the Substrate

The basic reaction is

$$A + O_2 \rightarrow AO_2$$

Examples of this type include enzymes that contain iron such as **homogentisate dioxygenase** (oxidase, see p 294) and **3-hydroxyanthranilate dioxygenase** (oxidase, see p 299) from the supernatant fraction of the liver, and enzymes utilizing heme such as **L-tryptophan dioxygenase** (tryptophan pyrrolase, see p 299) from the liver.

Monooxygenases (Mixed-Function Oxidases, Hydroxylases) Incorporate Only One Atom of Oxygen Into the Substrate

The other oxygen atom is reduced to water, an additional electron donor or cosubstrate being necessary for this purpose

$$A-H + O_2 + ZH_2 \rightarrow A-OH + H_2O + Z$$

Microsomal Cytochrome P-450 Monooxygenase Systems Are Important for the Hydroxylation of Many Drugs

These monooxygenases are found in the microsomes of the liver together with **cytochrome P-450** and **cytochrome b_5.** Both NADH and NADPH donate reducing equivalents for the reduction of these cytochromes (Fig 13–7), which in turn are oxidized by substrates in a series of enzymatic reactions collectively known as the **hydroxylase cycle** (Fig 13–8).

$$\boxed{\text{Hydroxylase}}$$

DRUG–H + O_2 + 2Fe^{2+} + 2H$^+$ $\longrightarrow$
(P-450)
DRUG–OH + H_2O + 2Fe^{3+}
(P-450)

Among the drugs metabolized by this system are benzpyrene, aminopyrine, aniline, morphine, and benzphetamine. Many drugs such as phenobarbital have the ability to induce the formation of microsomal enzymes and of cytochrome P-450.

Mitochondrial Cytochrome P-450 Monooxygenase Systems Catalyze Steroidal Hydroxylations

These systems are found in steroidogenic tissues such as adrenal cortex, testis, ovary, and placenta and are concerned with the biosynthesis of steroid hormones from cholesterol (hydroxylation at C_{22} and C_{20} in side-chain cleavage and at the 11β and 18 positions). Renal systems catalyze 1α- and 24-hydroxylations of 25-hydroxycholecalciferol, and the liver catalyzes 26-

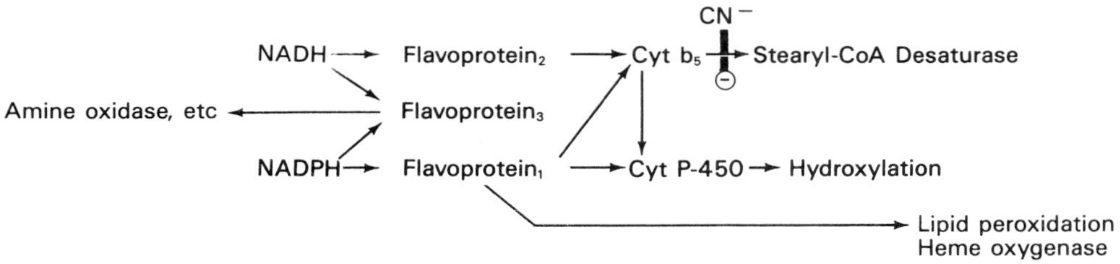

Figure 13–7. Electron transport chain in microsomes. Cyanide (CN$^-$) inhibits the indicated step.

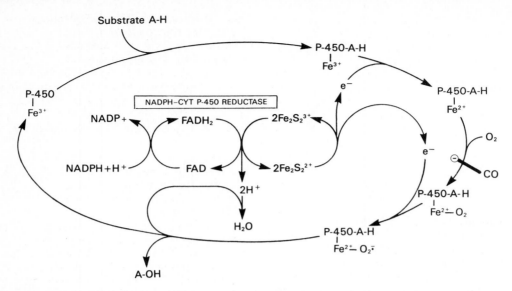

Figure 13–8. Cytochrome P-450 hydroxylase cycle in microsomes. The system shown is typical of steroid hydroxylases of the adrenal cortex. Liver microsomal cytochrome P-450 hydroxylase does not require the iron-sulfur protein Fe_2S_2. Carbon monoxide (CO) inhibits the indicated step.

hydroxylation in bile acid biosynthesis. In the adrenal cortex, mitochondrial cytochrome P-450 is 6 times more abundant than cytochromes of the respiratory chain. The monooxygenase system consists of 3 components situated on the inside of the inner mitochondrial membrane: an NADP-specific FAD containing flavoprotein, and Fe_2S_2 protein (adrenodoxin), and cytochrome P-450 (Fig 13–9).

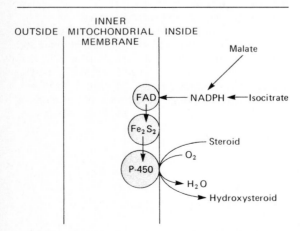

Figure 13–9. Mitochondrial cytochrome P-450 monooxygenase system. Fe_2S_2, iron-sulfur protein (adrenodoxin). Note that because NADP(H) cannot penetrate the mitochrondrial membrane, sources of reducing equivalents are confined to substrates such as malate and isocitrate for which there are intramitochondrial NADP-specific dehydrogenases.

THE SUPEROXIDE FREE RADICAL MAY ACCOUNT FOR OXYGEN TOXICITY

Oxygen is a potentially toxic substance, the toxicity of which has hitherto been attributed to the formation of H_2O_2. Recently, however, the ease with which oxygen can be reduced in tissues to the superoxide anion free radical ($O_2^- \cdot$) and the occurrence of **superoxide dismutase** in aerobic organisms (although not in obligate anaerobes) have suggested that the toxicity of oxygen is due to its conversion to superoxide. However, no direct evidence of superoxide toxicity has yet been obtained.

Superoxide is formed when reduced flavins, present, for example, in xanthine oxidase, are reoxidized univalently by molecular oxygen. It is also formed during univalent oxidations with molecular oxygen in the respiratory chain:

$$\text{EnzH}_2 + \text{O}_2 \rightarrow \text{EnzH} + \text{O}_2^- \cdot + \text{H}^+$$

Superoxide can reduce oxidized cytochrome c:

$$\text{O}_2^- \cdot + \text{Cyt c (Fe}^{3+}) \rightarrow \text{O}_2 + \text{Cyt c (Fe}^{2+})$$

or be removed by the presence of the specific enzyme superoxide dismutase

$$\text{O}_2^- \cdot + \text{O}_2^- \cdot + 2\text{H}^+ \xrightarrow{\boxed{\text{SUPEROXIDE DISMUTASE}}} \text{H}_2\text{O}_2 + \text{O}_2$$

In this reaction, superoxide acts as both oxidant and reductant. The chemical effects of superoxide in the tissues are amplified by free-radical chain reactions. It has been proposed that $O_2^- \cdot$ bound to cytochrome P-450 is an intermediate in the activation of oxygen in hydroxylation reactions (Fig 13–8).

The function of superoxide dismutase seems to be that of protecting aerobic organisms against the potential deleterious effects of superoxide. The enzyme occurs in several different compartments of the cell. The cytosolic enzyme is composed of 2 similar subunits, each one containing one equivalent of Cu^{2+} and Zn^{2+}, whereas the mitochondrial enzyme contains Mn^{2+}, similar to the enzyme found in bacteria. This finding supports the hypothesis that mitochondria have evolved from a prokaryote that entered into symbiosis with a protoeukaryote. The dismutase is present in all major aerobic tissues. Although exposure of animals to an atmosphere of 100% oxygen causes an adaptive increase of the enzyme, particularly in the lungs, prolonged exposure leads to lung damage and death. Antioxidants, eg, α-tocopherol (vitamin E), also act as scavengers of free radicals such as $O_2^- \cdot$ and reduce the toxicity of oxygen (see Chapter 54).

4 MAJOR POINTS SUMMARIZE BIOLOGIC OXIDATION

1. In biologic systems, as in chemical systems, oxidation (loss of electrons) is always accompanied by reduction of an electron acceptor.
2. Oxidoreductases are classified into 4 groups: oxidases, dehydrogenases, hydroperoxidases, and oxygenases.
3. Oxidases and dehydrogenases have a variety of roles in metabolism, but both classes of enzymes play major roles in respiration.
4. Hydroperoxidases protect the body against damage by free radicals and oxygenases mediate the hydroxylation of drugs.

REFERENCES

Bonnett R: Oxygen activation and tetrapyrroles. *Essays Biochem* 1981;**17**:1.
Ernster L (editor): *Bioenergetics.* Elsevier, 1984.
Fleisher S, Packer L (editors): Biological oxidations, microsomal, cytochrome P-450, and other hemoprotein systems. In: *Methods in Enzymology.* Vol 52. Biomembranes, part C. Academic Press, 1978.
Friedovich I: Superoxide dismutases. *Annu Rev Biochem* 1975;**44**:147.
Nicholls DG: *Cytochromes and Cell Respiration.* Carolina Biological Supply Company, N. Carolina, 1984.
Salemme FR: Structure and function of cytochromes c. *Annu Rev Biochem* 1977;**46**:299.
Tolbert NE: Metabolic pathways in peroxisomes and glyoxysomes. *Annu Rev Biochem* 1981;**50**:133.
Tyler DD, Sutton CM: Respiratory enzyme systems in mitochondrial membranes. Page 33 in: *Membrane Structure and Function.* Vol 5. Bittar EE (editor). Wiley, 1984.
White RE, Coon MJ: Oxygen activation by cytochrome P-450. *Annu Rev Biochem* 1980;**49**:315.

14

Oxidative Phosphorylation & Mitochondrial Transport Systems

Peter A. Mayes, PhD, DSc

INTRODUCTION

The mitochondrion has appropriately been termed the "powerhouse" of the cell, since it is within this organelle that most of the capture of energy derived from respiratory oxidation takes place. The system in mitochondria that couples respiration to the generation of the high-energy intermediate, ATP, is termed **oxidative phosphorylation.**

BIOMEDICAL IMPORTANCE

Oxidative phosphorylation enables aerobic organisms to capture a far greater proportion of the available free energy of respiratory substrates compared with anaerobic organisms. The chemiosmotic theory offers an insight into how this is accomplished. A number of drugs (eg, **amobarbital**) and poisons (eg, **cyanide, carbon monoxide**) inhibit oxidative phosphorylation, usually with fatal consequences. A number of inherited defects of mitochondria involving components of the respiratory chain and oxidative phosphorylation have been reported. Patients present with **myopathy, encephalopathy,** and often have **lactacidosis.**

THE RESPIRATORY CHAIN COLLECTS AND OXIDIZES REDUCING EQUIVALENTS

All of the useful energy liberated during the oxidation of fatty acids and amino acids and nearly all of that released from the oxidation of carbohydrate is made available within the mitochondria as reducing equivalents ($-H$ or electrons). The mitochondria contain the series of catalysts known as the respiratory chain that collect and transport reducing equivalents and direct them to their final reaction with oxygen to form water. Also present is the machinery for trapping the liberated free energy as high-energy phosphate. Mitochondria also contain the enzyme systems responsible for producing most of the reducing equivalents in the first place, ie, the enzymes of β-oxidation and of the citric acid cycle. The latter is the final common metabolic pathway for the oxidation of all the major foodstuffs. These relationships are shown in Fig 14–1.

THE COMPONENTS OF THE RESPIRATORY CHAIN ARE ARRANGED IN ORDER OF INCREASING REDOX POTENTIAL

The major components of the respiratory chain are shown in Fig 14–2. **Hydrogen or electrons flow through the chain in steps from the more electronegative components to the more electropositive oxygen** through a redox span of 1.1 volts from NAD^+/NADH to $O_2/2H_2O$ (see Table 13–1).

The main respiratory chain in mitochondria proceeds from the NAD-linked dehydrogenase systems, through flavoproteins and cytochromes, to molecular oxygen. Not all substrates are linked to the respiratory chain through NAD-specific dehydrogenases; some, because their redox potentials are more positive (eg, fumarate/succinate; see Table 13–1), are linked directly to flavoprotein dehydrogenases, which in turn are linked to the cytochromes of the respiratory chain (Fig 14–3).

In recent years, it has become clear that an additional carrier is present in the respiratory chain linking the flavoproteins to cytochrome b, the member of the cytochrome chain of lowest redox potential. This substance, which has been named **ubiquinone** or **Q (coenzyme Q;** see Fig 14–4), exists in mitochondria in the oxidized quinone form under aerobic conditions and in the reduced quinol form under anaerobic conditions. Q is a constituent of the mitochondrial lipds; the other lipids are predominantly phospholipids that constitute part of the mitochondrial membrane. The structure of Q is very similar to vitamin K and vitamin E. It is also similar to plastoquinone, found in chloroplasts. All of these substances are characterized by the possession of a polyisoprenoid side chain. In mitochondria, there is a large stoichiometric excess of Q compared with other members of the respiratory chain; this suggests that Q is a mobile component of the respiratory chain and that it collects reducing equivalents from the more fixed flavoprotein complexes and passes them on to the cytochromes.

An additional component found in respiratory chain preparations is the **iron-sulfur protein (FeS; nonheme iron).** It is associated with the flavoproteins

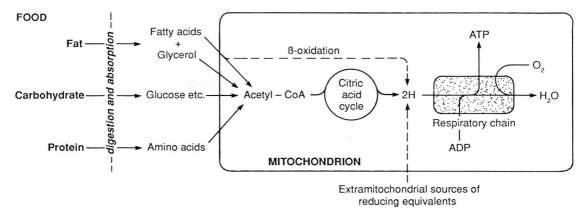

Figure 14–1. Role of the respiratory chain of mitochondria in the conversion of food energy to ATP. Oxidation of the major foodstuffs leads to the generation of reducing equivalents (2H) that are collected by the respiratory chain for oxidation and coupled generation of ATP.

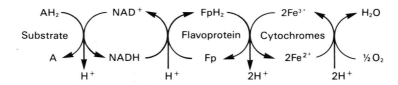

Figure 14–2. Transport of reducing equivalents through the respiratory chain.

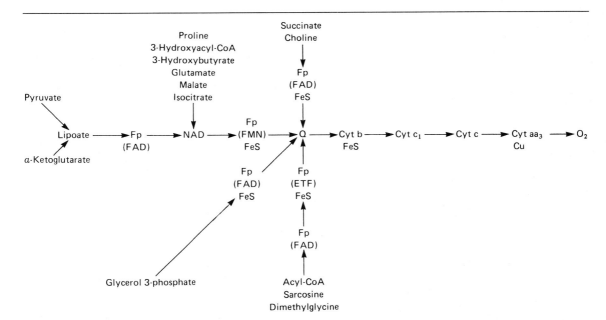

Figure 14–3. Components of the respiratory chain in mitochondria. FeS occurs in the sequences on the O_2 side of Fp or Cyt b. Cyt, cytochrome: ETF, electron-transferring flavoprotein; FeS, iron-sulfur protein, Fp, flavoprotein; Q, ubiquinone.

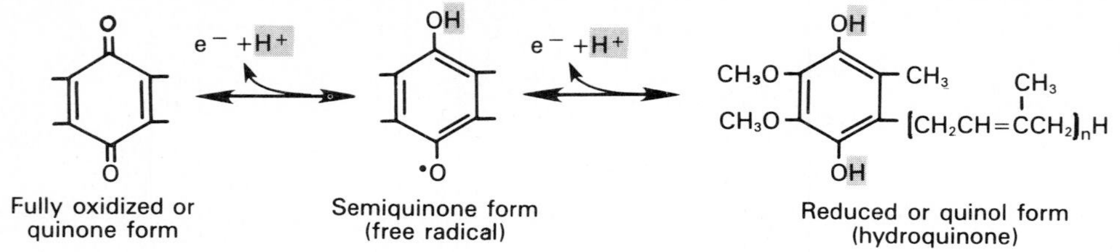

Figure 14-4. Structure of ubiquinone (Q). n = Number of isoprenoid units, which varies from 6 to 10, ie, Q_{6-10}.

(metalloflavoproteins) and with cytochrome b. The sulfur and iron are thought to take part in the oxidoreduction mechanism, which involves only a single e^- change (Fig 14–5).

A current view of the sequence of the principal components of the respiratory chain is shown in Fig 14–3. At the electronegative end of the chain, dehydrogenase enzymes catalyze the transfer of electrons from substrates to NAD of the chain. Several differences exist in the manner in which this is carried out. The α-keto acids pyruvate and ketoglutarate have complex dehydrogenase systems involving lipoate and FAD prior to the passage of electrons to NAD of the respiratory chain. Electron transfers from other dehydrogenases such as $L(+)$-3-hydroxyacyl-CoA, $D(-)$-3-hydroxybutyrate, proline, glutamate, malate, and isocitrate dehydrogenases appear to couple directly with NAD of the respiratory chain.

The reduced NADH of the respiratory chain is in turn oxidized by a metalloflavoprotein enzyme— **NADH dehydrogenase.** This enzyme contains FeS and FMN, is tightly bound to the respiratory chain, and passes reducing equivalents onto Q. Q is also the collecting point in the respiratory chain for reducing equivalents derived from other substrates that are linked directly to the respiratory chain through flavoprotein dehydrogenases. These substrates include

succinate, choline, glycerol 3-phosphate, sarcosine, dimethylglycine, and acyl-CoA (Fig 14–3). The flavin moiety of all these dehydrogenases appears to be FAD.

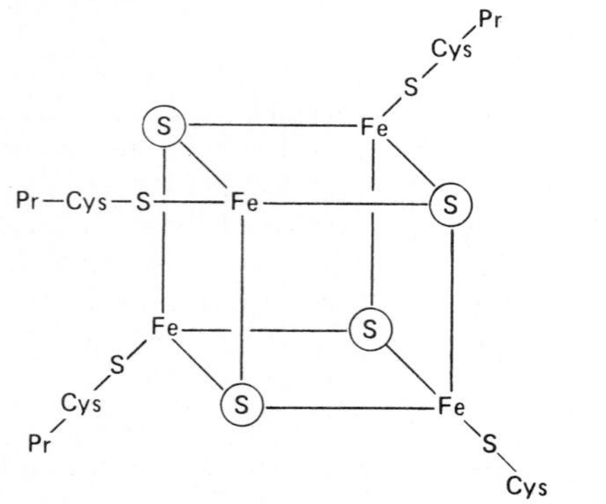

Electrons flow from Q, through the series of cytochromes shown in Fig 14–3, to molecular oxygen. The cytochromes are arranged in order of increasing redox potential. The terminal cytochrome aa_3 (cytochrome oxidase) is responsible for the final combination of reducing equivalents with molecular oxygen. It has been noted that this enzyme system contains copper, a component of several oxidase enzymes. Cytochrome oxidase has a very high affinity for oxygen, which allows the respiratory chain to function at the maximum rate until the tissue has become virtually depleted of O_2. Since this is an irreversible reaction (the only one in the chain), it gives direction to the movement of reducing equivalents in the respiratory chain and to the production of ATP, to which it is coupled.

The structural organization of the respiratory chain has been the subject of considerable study. Of significance is the finding of nearly constant molar proportions between the components. Functionally and structurally, the components of the respiratory chain are present in the inner mitochondrial membrane as 4 **protein-lipid respiratory chain complexes.** These findings have suggested that these components have a definite spatial orientation in the membranes. Cytochrome c is the only soluble cytochrome and, together with Q, seems to be a more mobile component of the respiratory chain connecting the fixed complexes (Fig 14–6).

THE RESPIRATORY CHAIN IS RESPONSIBLE FOR MOST OF THE ENERGY CAPTURED IN METABOLISM

ADP is a molecule that captures, in the form of **high-energy phosphate,** some of the free energy released by catabolic processes. The resulting **ATP** passes on this free energy to drive those processes requiring energy. Thus, ATP has been called the **energy "currency"** of the cell (see Fig 12–8).

There is a net capture of 2 high-energy phosphate groups directly in the glycolytic reactions, equivalent to approximately 61 kJ/mol of glucose (see Table 19–1). Since 1 mol of glucose yields approximately 2780 kJ on complete combustion, the energy captured by

Figure 14-5. Iron-sulfer-protein complex (Fe_4S_4). Ⓢ, acid-labile sulfur; Pr, apoprotein; Cys, cysteine residue. Some iron-sulfur proteins contain 2 iron atoms and 2 sulfur atoms (Fe_2S_2).

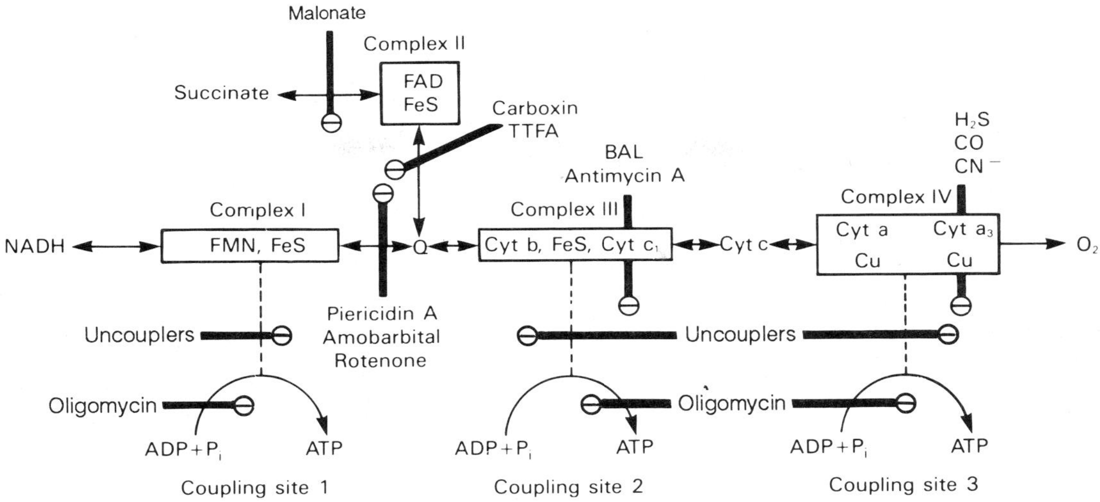

Figure 14–6. Proposed sites of inhibition ($\ominus$) of the respiratory chain by specific drugs, chemicals, and antibiotics. The sites that appear to support phosphorylation are indicated. BAL, dimercaprol TTFA is an Fe-chelating agent. Complex I, NADH:ubiquinone oxidoreductase; complex II, succinate:ubiquinone oxidoreductase; complex III, ubiquinol; ferricytochrome c oxidoreductase; complex IV, ferrocytochrome c:oxygen oxidoreductase. Other abbreviations as in Fig 14–3.

phosphorylation in glycolysis is small. The reactions of the citric acid cycle, the final pathway for the complete oxidation of glucose, include one phosphorylation step, the conversion of succinyl-CoA to succinate, which allows the capture of only 2 more high-energy phosphates per mole of glucose. All of the phosphorylations described occur **at the substrate level.** Examination of intact respiring mitochondria reveals that when substrates are oxidized via an NAD-linked dehydrogenase and the respiratory chain, 3 mol of inorganic phosphate are incorporated into 3 mol of ADP to form 3 mol of ATP per ½ mol of O_2 consumed: ie, the P:O ratio = 3 (Fig 14–6). On the other hand, when a substrate is oxidized via a flavoprotein-linked dehydrogenase, only 2 mol of ATP are formed; ie, P:O = 2. These reactions are known as **oxidative phosphorylation at the respiratory chain level.** Dehydrogenations in the pathway of catabolism of glucose in both glycolysis and the citric acid cycle, plus phosphorylations at the substrate level, can now account for nearly 42% of the free energy resulting from the combustion of glucose, captured in the form of high-energy phosphate. It is evident that the **respiratory chain is responsible for a large proportion of total ATP formation.**

RESPIRATORY CONTROL ENSURES A CONSTANT SUPPLY OF ATP

The rate of respiration of mitochondria can be controlled by the concentration of ADP. This is because **oxidation and phosphorylation are tightly coupled;** ie, oxidation cannot proceed via the respiratory chain without concomitant phosphorylation of ADP. Chance

and Williams have defined 5 conditions that can control the rate of respiration in mitochondria (Table 14–1).

Generally, most cells in the resting state are in state 4, and respiration is controlled by the availability of ADP. When work is performed, ATP is converted to ADP, allowing more respiration to occur, which in turn replenishes the store of ATP (Fig 14–7). It would appear that under certain conditions the concentration of inorganic phosphate could also affect the rate of functioning of the respiratory chain. As respiration increases (as in exercise), the cell approaches state 3 or state 5 when either the capacity of the respiratory chain becomes saturated or the PO_2 decreases below the K_m for cytochrome a_3. There is also the possibility that the ADP/ATP transporter (see p 125), which facilitates entry of cytosolic ADP into the mitochondrion, becomes rate limiting.

Thus, the manner in which biologic oxidative processes allow the free energy resulting from the oxidation of foodstuffs to become available and to be captured is stepwise, efficient (40–45%), and controlled—rather than explosive, inefficient, and uncontrolled. The remaining free energy that is not captured as high-energy phosphate is liberated as **heat.**

Table 14–1. States of respiratory control.

	Conditions Limiting the Rate of Respiration
State 1	Availability of ADP and substrate
State 2	Availability of substrate only
State 3	The capacity of the respiratory chain itself, when all substrates and components are present in saturating amounts
State 4	Availability of ADP only
State 5	Availability of oxygen only

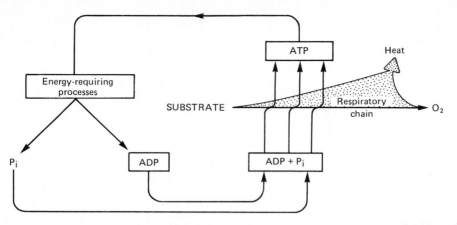

Figure 14–7. The role of ADP in respiratory control.

This need not be considered as "wasted," since it ensures the respiratory system as a whole is sufficiently exergonic to be removed from equilibrium, allowing continuous unidirectional flow and constant provision of ATP. In the warm-blooded animal it contributes to maintenance of body temperature.

MANY WELL-KNOWN POISONS ARE INHIBITORS OF THE RESPIRATORY CHAIN

Much information about the respiratory chain has been obtained by the use of inhibitors, and their proposed loci of action are shown in Fig 14–6. For descriptive purposes, they may be divided into inhibitors of the respiratory chain proper, inhibitors of oxidative phosphorylation, and uncouplers of oxidative phosphorylation.

Inhibitors that arrest respiration by blocking the respiratory chain act at 3 loci. The first is inhibited by **barbiturates** such as **amobarbital,** by the antibiotic **piericidin A,** and by the fish poison **rotenone.** These inhibitors prevent the oxidation of substrates that communicate directly with the respiratory chain via an NAD-linked dehydrogenase, eg, 3-hydroxybutyrate.

Dimercaprol and **antimycin A** inhibit the respiratory chain between cytochrome b and cytochrome c. The classic poisons H_2S, **carbon monoxide,** and **cyanide** inhibit cytochrome oxidase. **Carboxin** and **TTFA** specifically inhibit transfer of reducing equivalents from succinate dehydrogenase to Q, whereas **malonate** is a competitive inhibitor of succinate dehydrogenase.

The antibiotic **oligomycin** completely blocks oxidation and phosphorylation in intact mitochondria. However, in the additional presence of the uncoupler **dinitrophenol,** oxidation proceeds without phosphorylation, indicating that oligomycin does not act directly on the respiratory chain but subsequently on a step in phosphorylation (see Fig 14–8).

Atractyloside inhibits oxidative phosphorylation that is dependent on the transport of adenine nucleotides across the inner mitochondrial membrane. It is considered to inhibit the transporter of ADP into the mitochondrion and of ATP out of the mitochondrion (see Fig 14–16).

The action of **uncouplers** is to dissociate oxidation in the respiratory chain from phosphorylation. This results in respiration becoming uncontrolled, since the concentration of ADP or P_i no longer limits the rate of respiration. The uncoupler that has been used most frequently is 2,4-dinitrophenol, but other compounds act in a similar manner, including dinitrocresol, pen-

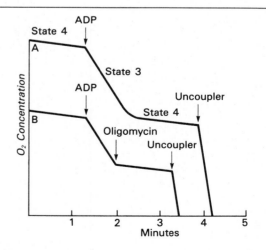

Figure 14–8. Respiratory control in mitochondria. Experiment A shows the basic state of respiration in state 4 that is accelerated upon addition of ADP. When the exogenous ADP has been phosphorylated to ATP, respiration reverts to state 4. The addition of uncoupler, eg, dinitrophenol, releases respiration from phosphorylation. In experiment B, addition of oligomycin blocks phosphorylation of added ADP and therefore of respiration as well. Addition of uncoupler again releases respiration from phosphorylation.

tachlorophenol, and CCCP (*m*-chlorocarbonyl cyanide phenylhydrazone). The latter is about 100 times as active as dinitrophenol.

THE CHEMIOSMOTIC THEORY EXPLAINS THE MECHANISM OF OXIDATIVE PHOSPHORYLATION

Two principal hypotheses have been advanced to account for the coupling of oxidation and phosphorylation. The **chemical hypothesis** postulated direct chemical coupling at all stages of the process, as in the reactions that generate ATP in glycolysis. However, energy-rich intermediates linking oxidation with phosphorylation were never isolated and the hypothesis has become discredited.

Other hypotheses have been advanced in which it is envisaged that energy from oxidation is conserved in **conformational changes** of molecules, which in turn lead to the generation of high-energy phosphate bonds.

The **chemiosmotic theory** postulates that oxidation of components in the respiratory chain generates hydrogen ions, which are ejected to the outside of a coupling membrane in the mitochondrion. The electrochemical potential difference resulting from the

asymmetric distribution of the hydrogen ions (protons, H^+) is used to drive the mechanism responsible for the formation of ATP (Fig 14–9).

The Respiratory Chain Is a Proton Pump

According to Mitchell, the primary event in oxidative phosphorylation is the translocation of protons (H^+) to the exterior of a coupling membrane (ie, the mitochondrial inner membrane), driven by oxidation in the respiratory chain. Each of the respiratory chain complexes I, III, and IV (see Fig 14–6) acts as a **proton pump.** It is also postulated that the membrane is impermeable to ions in general but particularly to protons, which accumulate outside the membrane, creating an **electrochemical potential difference across the membrane** ($\Delta\mu_{H^+}$). This consists of a chemical potential (difference in pH) and an electrical potential. The electrochemical potential difference is used to drive a **membrane-located ATP synthase** which in the presence of P_i + ADP forms ATP (Fig 14–9). Thus, there is no high-energy intermediate that is common to both oxidation and phosphorylation as in the chemical hypothesis.

It was originally proposed that the respiratory chain was folded into 3 oxidation/reduction (o/r) loops in the

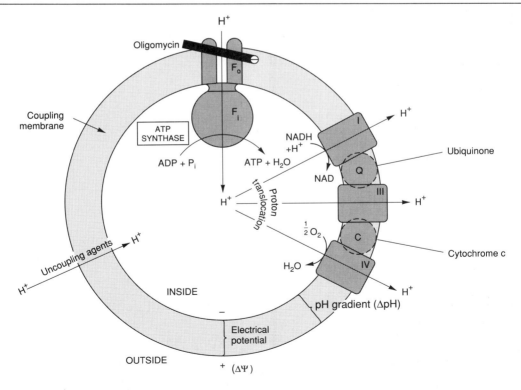

Figure 14–9. Principles of the chemiosmotic theory of oxidative phosphorylation. F_1, F_0, protein subunits responsible for phosphorylation. The main proton circuit is created by the coupling of oxidation to proton translocation from the inside to the outside of the membrane, driven by the respiratory chain complexes I, III, and IV, each of which acts as a *proton pump.* Uncoupling agents such as dinitrophenol allow leakage of H^+ across the membrane, thus collapsing the electrochemical proton gradient. Oligomycin specifically blocks conduction of H^+ through F_0.

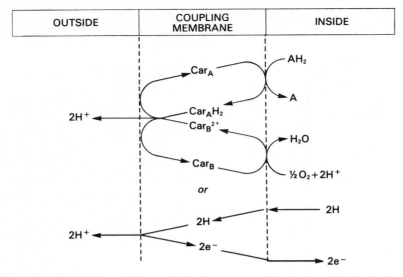

Figure 14–10. Proton-translocating oxidation/reduction (o/r) loop (chemiosmotic theory).

membrane. An idealized single loop consisting of a hydrogen carrier and an electron carrier is shown in Fig 14–10. However, this original formulation has been difficult to reconcile in every respect with the known components of the respiratory chain complexes, which span the membrane; eg, complex IV does not contain a hydrogen carrier. Furthermore, the precise number of protons pumped by each complex for each electron transported is not known with certainty. A possible

mechanism to account for proton pumping by complex III, the Q cycle, is shown in Fig 14–11.

Scattered over the surface of the inner membrane are the phosphorylating subunits responsible for the production of ATP (Fig 14–12). These consist of several proteins, collectively known as an F_i subunit, which project into the matrix and which contain the ATP synthase (Fig 14–9). These subunits are attached, possibly by a stalk, to a membrane protein subunit known as

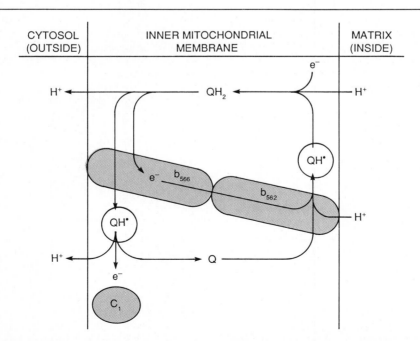

Figure 14–11. Proton-motive "Q" cycle. QH· is anchored on each side of the membrane by attachment to a Q-binding protein, whereas QH_2 and Q are mobile. Cytochromes are shown respectively as b, c_1.

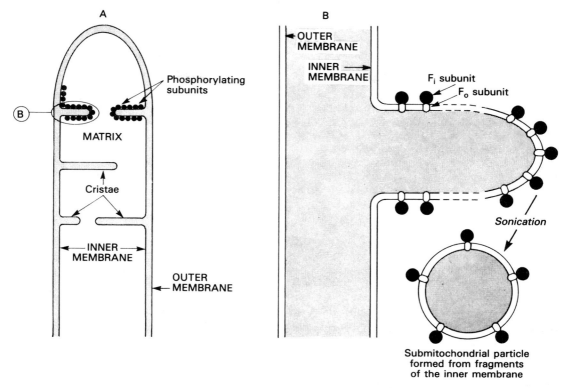

Figure 14–12. Structure of the mitochondrial membranes. Submitochondrial particles are "inside out" and allow study of an enclosed membrane system where the phosphorylating subunits are on the outside and where the proton gradient is reversed.

F_o, which probably extends through the membrane (Fig 14–9). Protons pass through the F_o–F_i complex, leading to the formation of ATP from ADP and P_i. It is of interest that similar phosphorylating units are found inside the plasma membrane of bacteria but outside the thylakoid membrane of chloroplasts. It is significant that the proton gradient is from outside to inside in mitochondria and bacteria but in the reverse direction in chloroplasts.

The mechanism of coupling of proton translocation to the anisotropic (vectorial) ATP synthase system is conjectural. One model suggested by Mitchell is shown in Fig 14–13. A proton pair attacks one oxygen of P_i to form H_2O and an active form of P_i, which immediately combines with ADP to form ATP. Other studies have suggested that ATP synthesis is not the main energy-requiring step—rather it is the release of ATP from the active site. This may involve conformational changes in the F_i subunit.

Experimental Findings Support the Chemiosmotic Theory

(1) Addition of protons (acid) to the external medium of mitochondria leads to the generation of ATP.

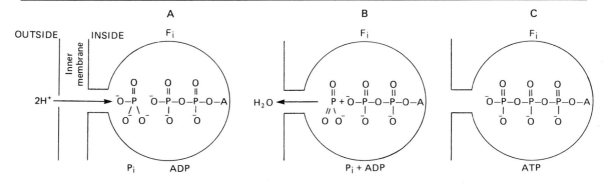

Figure 14–13. Proton-translocating ATP synthase (Mitchell).

(2) Oxidative phosphorylation does not occur in soluble systems where there is no possibility of a vectorial ATP synthase. A closed membrane must be present in order to obtain oxidative phosphorylation (Fig 14–9).

(3) The respiratory chain contains components organized in a sided manner (transverse asymmetry) as required by the chemiosmotic theory.

The Chemiosmotic Theory Can Account for the Phenomenon of Respiratory Control

The electrochemical potential difference across the membrane, once established as a result of proton translocation, inhibits further transport of reducing equivalents through the respiratory chain unless discharged by back-translocation of protons across the membrane through the vectorial ATP synthase. This in turn depends on availability of ADP and P_i.

It Explains the Action of Uncouplers

These compounds (eg, dinitrophenol) are amphipathic (see p 148) and increase the permeability of mitochondria to protons (Fig 14–9), thus reducing the electrochemical potential and short-circuiting the ATP synthase. Thus, oxidation can proceed without phosphorylation.

It Explains the Existence of Mitochondrial Exchange Transporter Systems

These are a consequence of the coupling membrane, which must be impermeable to protons and other ions in order to maintain the electrochemical gradient (see below).

THE RELATIVE IMPERMEABILITY OF THE INNER MITOCHONDRIAL MEMBRANE NECESSITATES THE PRESENCE OF EXCHANGE TRANSPORTERS

Exchange diffusion systems are present in the membrane for exchange of anions against OH^- ions and cations against H^+ ions. Such systems are necessary for uptake and output of ionized metabolites while preserving electrical and osmotic neutrality.

The Locations of Specific Enzymes Act as Markers of Compartments Separated by the Mitochondrial Membranes

Mitochondria have an outer membrane that is permeable to most metabolites, an inner membrane which is selectively permeable and which is thrown into folds or cristae, and a matrix within the inner membrane (Fig 14–12). The outer membrane may be removed by treatment with digitonin and is characterized by the presence of monoamine oxidase, acyl-CoA synthetase, glycerophosphate acyltransferase, monoacyl glycerophosphate acyltransferase, and phospholipase A_2. Adenylate kinase and creatine kinase are found in the intermembrane space. The phospholipid cardiolipin is concentrated in the inner membrane.

The soluble enzymes of the citric acid cycle and the enzymes of β-oxidation of fatty acids are found in the matrix, necessitating mechanisms for transporting metabolites and nucleotides across the inner membrane. Succinate dehydrogenase is found on the inner surface of the inner mitochondrial membrane, where it transports reducing equivalents into the respiratory chain at ubiquinone, bypassing respiratory chain complex I. Hydroxybutyrate dehydrogenase is also bound to the matrix side of the inner mitochondrial membrane. Glycerol-3-phosphate dehydrogenase is found on the outer surface of the inner membrane, where it is suitably located to participate in the glycerophosphate shuttle (Fig 14–14).

Oxidation of Extramitochondrial NADH is Mediated by Substrate Shuttles

NADH cannot penetrate the mitochondrial membrane, but it is produced continuously in the cytosol by 3-phosphoglyceraldehyde dehydrogenase, an enzyme in the glycolysis sequence (see Fig 19–2). However, under aerobic conditions, extramitochondrial NADH does not accumulate and is presumed to be oxidized by the respiratory chain in mitochondria. Several possible mechanisms have been considered to permit this process. These involve transfer of reducing equivalents

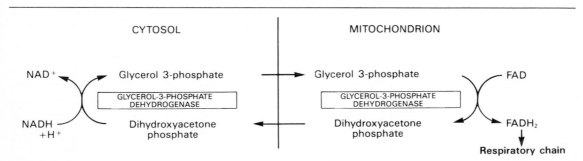

Figure 14–14. Glycerophosphate shuttle for transfer of reducing equivalents from the cytosol into the mitochondrion.

through the mitochondrial membrane via substrate pairs, linked by suitable dehydrogenases. It is necessary that the specific dehydrogenase be present on both sides of the mitochondrial membrane. The mechanism of transfer using the **glycerophosphate shuttle** is shown in Fig 14–14. It is to be noted that since the mitochondrial enzyme is linked to the respiratory chain via a flavoprotein rather than NAD, only 2 rather than 3 mol of ATP are formed per atom of oxygen consumed. In some species, the activity of the FAD-linked enzyme decreases after thyroidectomy and increases after administration of thyroxine. Although this shuttle is present in insect flight muscle and in white muscle and might be important in liver, in other tissues (eg, heart muscle) the mitochondrial glycerol-3-phosphate dehydrogenase is deficient. It is therefore believed that a transport system involving malate and cytosolic and mitochondrial malate dehydrogenase is of more universal utility. The **malate "shuttle"** system is shown in Fig 14–15. The complexity of this system is due to the impermeability of the mitochondrial membrane to oxaloacetate, which must react with glutamate and transaminate to aspartate and α-ketoglutarate before transport through the mitochondrial membrane and reconstitution to oxaloacetate in the cytosol.

Ion Transport in Mitochondria is Energy Linked

Actively respiring mitochondria in which oxidative phosphorylation is taking place maintain or accumulate cations such as K^+, Na^+, Ca^{2+}, and Mg^{2+}, and P_i. Uncoupling with dinitrophenol leads to loss of ions from the mitochondria, but the ion uptake is not inhibited by oligomycin, suggesting that the energy need not be supplied by phosphorylation of ADP. It is envisaged that a primary proton pump drives cation exchange.

Transporter Systems Preserve Electrical and Osmotic Neutrality on Either Side of the Mitochondrial Membrane (See Fig 14–16)

The inner bilipoid mitochondrial membrane is freely permeable to uncharged small molecules, such as oxygen, water, CO_2, and NH_3, and to monocarboxylic acids, such as 3-hydroxybutyric, acetoacetic, and acetic. Long-chain fatty acids are transported into mitochondria via the carnitine system (see Fig 24–1), and there is also a special carrier for pyruvate involving a symport that utilizes the H^+ gradient from outside to inside the mitochondrion. However, dicarboxylate and tricarboxylate anions and amino acids require specific transporter or carrier systems to facilitate their transport across the membrane. It appears that monocarboxylate anions penetrate more readily, because of the lesser degree of dissociation of these acids. It is the undissociated and more lipid-soluble acid that is thought to be the molecular species that penetrates the lipid membrane.

The transport of di- and tricarboxylate anions is closely linked to that of inorganic phosphate, which penetrates readily as the $H_2PO_4^-$ ion in exchange for OH^-. The net uptake of malate by the dicarboxylate transporter requires inorganic phosphate for exchange in the opposite direction. The net uptake of citrate, isocitrate, or *cis*-aconitate by the tricarboxylate transporter requires malate in exchange. α-Ketoglutarate transport also requires an exchange with malate. Thus, by the use of exchange mechanisms, osmotic balance is maintained. It will be appreciated that citrate transport across the mitochondrial membrane depends not only on malate transport but on the transport of inorganic phosphate as well. The adenine nucleotide transporter allows the exchange of ATP and ADP but not AMP. It is vital in allowing ATP exit from mito-

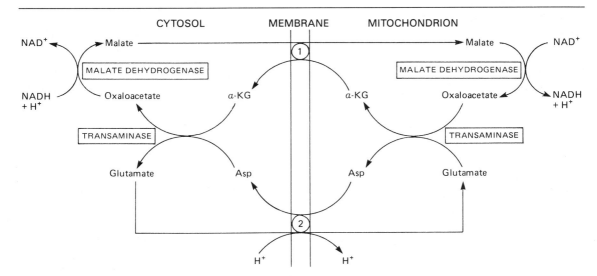

Figure 14–15. Malate shuffle for transfer of reducing equivalents from the cytosol into the mitochondrion. 1, Ketoglutarate transporter; 2, glutamate-aspartate transporter (note the proton symport).

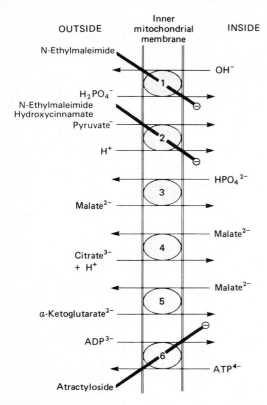

Figure 14–16. Transporter systems in the mitochondrial membrane. 1, Phosphate transporter; 2, pyruvate symport; 3, dicarboxylate transporter; 4, tricarboxylate transporter, 5, α-ketoglutarate transporter; 6, adenine nucleotide transporter. N-Ethylmaleimide, hydroxycinnamate, and atractyloside inhibit (⊖) the indicated systems. Also present (but not shown) are transporter systems for glutamate/aspartate (Fig 14–15), glutamine, ornithine, and carnitine (Fig 24–1).

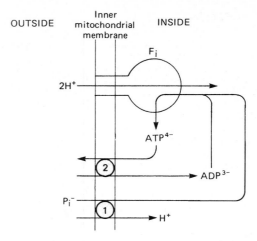

Figure 14–17. Combination of phosphate transporter (1) with the adenine nucleotide transporter (2) in ATP synthesis. The H^+/P_i, symport shown is equivalent to the P_i/OH^- antiport shown in Fig 14–16. Three protons are taken into the mitochondrion for each ATP exported. However, only 2 protons are taken in when ATP is used inside the mitochondrion.

chondria to the sites of extramitochondrial utilization and in allowing the return of ADP for ATP production within the mitochondrion (Fig 14–17). Na^+ can be exchanged for H^+, driven by the proton gradient. It is believed that active uptake of Ca^{2+} by mitochondria occurs with a net charge transfer of 1 (Ca^+ uniport), possibly through a Ca^{2+}/H^+ antiport. Calcium release from mitochondria is facilitated by exchange with Na^+.

Ionophores Permit Specific Cations To Penetrate Membranes

Ionophores are so termed because of their ability to complex specific cations and facilitate their transport through biologic membranes. This property of ionophoresis is due to their lipophilic character, which allows penetration of lipoid membranes such as the mitochondrial membrane. An example is the antibiotic **valinomycin,** which allows penetration of K^+ through the mitochondrial membrane and then dis-

charges the membrane potential between the inside and the outside of the mitochondrion. **Nigericin** also acts as an ionophore for K^+ but in exchange for H^+. It therefore abolishes the pH gradient across the membrane. In the presence of both valinomycin and nigericin, both the membrane potential and the pH gradient are eliminated, and phosphorylation is therefore completely inhibited. The classic uncouplers such as dinitrophenol are, in fact, proton ionophores.

A Proton-Translocating Transhydrogenase Is a Source of Intramitochondrial NADPH

This energy-linked transhydrogenase, a protein in the inner mitochondrial membrane, couples the passage of protons down the electrochemical gradient from outside to inside the mitochondrion with the transfer of H from intramitochondrial NADH to form NADPH. It appears to function as an energy-linked redox buffer and as a source of NADPH for intramitochondrial enzymes, such as glutamate dehydrogenase and hydroxylases, involved in steroid synthesis.

Dysfunction of the Respiratory Chain is a Cause of Disease

The condition of **fatal infantile mitochondrial myopathy and renal dysfunction** involves severe diminution or absence of most oxidoreductases of the respiratory chain. **MELAS** (mitochondrial myopathy, encephalopathy, lactacidosis, and stroke) is an inherited condition due to NADH: ubiquinone oxidoreductase (complex I) or cytochrome oxidase deficiency. Diseases involving deficiencies of most mitochondrial enzymes have been described (Scholte).

REFERENCES

Boyer PD: The unusual enzymology of ATP synthase. *Biochemistry* 1987;**26**:8503.

Cross RL: The mechanism and regulation of ATP synthesis by F_i-ATPases. *Annu Rev Biochem* 1981;**50**:681.

Hatefi Y: The mitochondrial electron transport and oxidative phosphorylation system. *Annu Rev Biochem* 1985;**54**:1015.

Hinkle PC, McCarty RE: How cells make ATP. *Sci Am* (March) 1978;**238**:104.

Mitchell P: Keilin's respiratory chain concept and its chemiosmotic consequences. *Science* 1979;**206**:1148.

Nicholls DG: *Bioenergetics: An Introduction to the Chemiosmotic Theory.* Academic Press, 1982.

Prince RC: The proton pump of cytochrome oxidase. *TIBS* 1988;**13**:159.

Scholte HR, et al: Defects in oxidative phosphorylation. Biochemical investigations in skeletal muscle and expression of the lesion in other cells. *J Inher Metab Dis* 1987;**10**, Suppl 1:81.

Tyler DD: The mitochondrial ATP synthase. Page 117 in: *Membrane Structure and Function.* Vol 5. Bittar EE (editor). Wiley, 1984.

Tyler DD, Sutton CM: Mitochondrial transporting systems. Page 181 in: *Membrane Structure and Function.* Vol 5. Bittar EE (editor). Wiley, 1984.

Carbohydrates of Physiologic Significance

Peter A. Mayes, PhD, DSc

INTRODUCTION

Carbohydrates are widely distributed in plants and animals, where they fulfill both structural and metabolic roles. In plants, glucose is synthesized from carbon dioxide and water by photosynthesis and stored as starch or is converted to the cellulose of the plant framework. Animals can synthesize some carbohydrate from fat and protein, but the bulk of animal carbohydrate is derived ultimately from plants.

BIOMEDICAL IMPORTANCE

Knowledge of the structure and properties of the physiologically significant carbohydrates is essential to understanding their role in the economy of the mammalian organism. The sugar **glucose** is the most important carbohydrate. It is as glucose that the bulk of dietary carbohydrate is absorbed into the bloodstream or into which it is converted in the liver, and it is from glucose that all other carbohydrates in the body can be formed. Glucose is a major fuel of the tissues of mammals (except ruminants) and a universal fuel of the fetus. It is converted to other carbohydrates having highly specific functions, eg, **glycogen** for storage; **ribose** in nucleic acids; **galactose** in lactose of milk, in certain complex lipids, and in combination with protein in glycoprotein and proteoglycans. Diseases associated with carbohydrates include **diabetes mellitus, galactosemia, glycogen storage diseases,** and **milk intolerance.**

CARBOHYDRATES ARE ALDEHYDE OR KETONE DERIVATIVES OF POLYHYDRIC ALCOHOLS

They are classified as follows:

Monosaccharides Are Those Carbohydrates That Cannot Be Hydrolyzed Into a Simpler Form: They may be subdivided into **trioses, tetroses, pentoses, hexoses, heptoses,** or **octoses,** depending upon the number of carbon atoms they possess; and as **aldoses** or **ketoses,** depending upon whether the aldehyde or ketone group is present. Examples are:

		Aldoses	**Ketoses**
Trioses	$(C_3H_6O_3)$	Glycerose	Dihydroxyacetone
Tetroses	$(C_4H_8O_4)$	Erythrose	Erythrulose
Pentoses	$(C_5H_{10}O_5)$	Ribose	Ribulose
Hexoses	$(C_6H_{12}O_6)$	Glucose	Fructose

Disaccharides Yield 2 Molecules of the Same or of Different Monosaccharide(s) When Hydrolyzed: Examples are sucrose, lactose, and maltose.

Oligosaccharides Yield 3–6 Monosaccharide Units on Hydrolysis: Maltotriose* is an example.

Polysaccharides Yield More Than 6 Molecules of Monosaccharides on Hydrolysis: Examples of polysaccharides, which may be linear or branched, are the starches and dextrins. These are sometimes designated as hexosans or pentosans, depending upon the identity of the monosaccharides they yield on hydrolysis.

GLUCOSE IS THE DOMINANT MONOSACCHARIDE

The Structure of Glucose Can Be Represented in 3 Ways

The straight-chain structural formula (aldohexose, Fig 15–1A) can account for some of the properties of glucose, but a cyclic structure is favored on thermodynamic grounds and accounts for the remainder of its chemical properties. For most purposes, the structural formula may be represented as a simple ring in perspective as proposed by Haworth (Fig 15–1B). X-Ray diffraction analysis shows that the 6-membered ring containing one oxygen atom is actually in the form of a chair (Fig 15–1C).

Sugars Exhibit Various Forms of Isomerism

Compounds that have the same structural formula but differ in spatial configuration are known as

---

*Note that this is not a true triose but a trisaccharide containing 3 α-glucose residues.

A

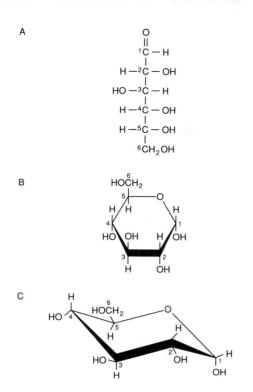

B

C

Figure 15–1. α-D-glucose. A, straight chain form; B, Haworth projection; C, chair form.

L-Glycerose
(L-glyceraldehyde)

D-Glycerose
(D-glyceraldehyde)

L-Glucose

D-Glucose

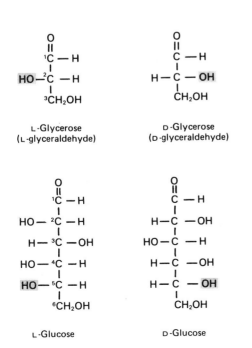

Figure 15–2. D- and L-isomerism of glycerose and glucose.

Pyran

Furan

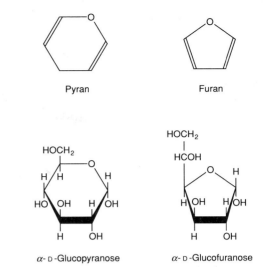

α-D-Glucopyranose

α-D-Glucofuranose

Figure 15–3. Pyranose and furanose forms of glucose.

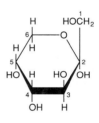

α-D-Fructofuranose

Figure 15–4. Pyranose and furanose forms of fructose.

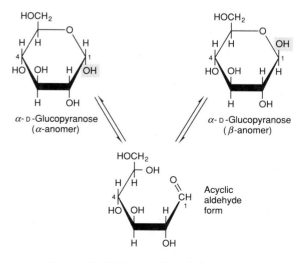

α-D-Glucopyranose
(α-anomer)

α-D-Glucopyranose
(β-anomer)

Acyclic
aldehyde
form

Figure 15–5. Mutarotation of glucose.

HOCH2
HO
H
OH
α-D-Galactose
α-D-Glucose
α-D-Mannose

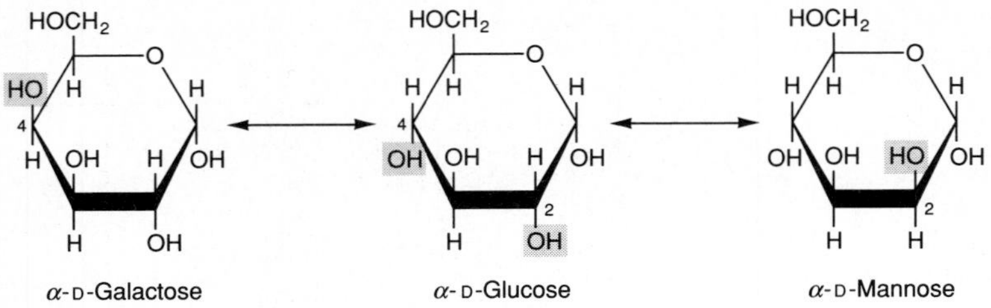

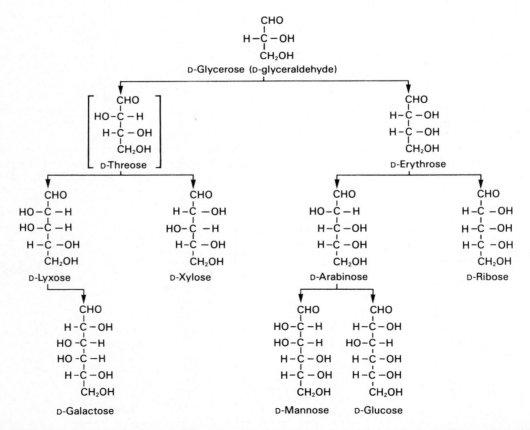

Figure 15–6. Epimerization of glucose.

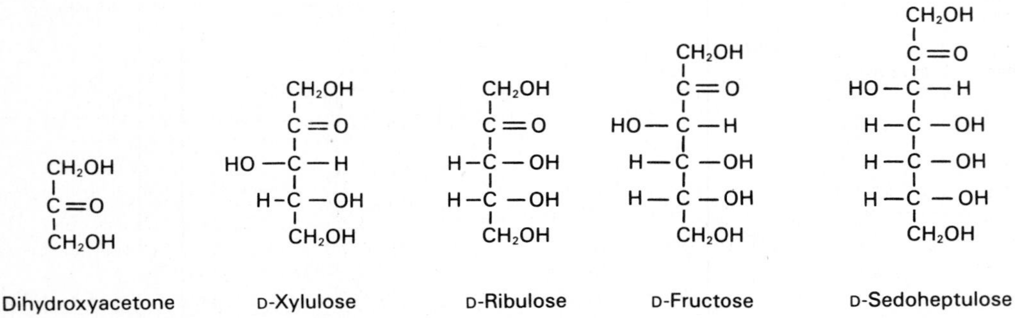

Dihydroxyacetone
D-Xylulose
D-Ribulose
D-Fructose
D-Sedoheptulose

Figure 15–7. Examples of ketoses of physiologic significance.

D-Glycerose (D-glyceraldehyde)
D-Threose
D-Erythrose
D-Lyxose
D-Xylose
D-Arabinose
D-Ribose
D-Galactose
D-Mannose
D-Glucose

Figure 15–8. The structural relations of the aldoses, D series. D-Threose is not of physiologic significance. The series is built up by the theoretical addition of a CH_2O unit to the $-CHO$ group of the sugar.

Table 15–1. Pentoses of physiologic importance.

Sugar	Where Found	Biochemical Importance	Clinical Significance
D-Ribose	Nucleic acids.	Structural elements of nucleic acids and coenzymes, eg, ATP, NAD, NADP, flavoproteins. Intermediate in pentose phosphate pathway.	
D-Ribulose	Formed in metabolic processes.	Intermediate in pentose phosphate pathway.	
D-Arabinose	Gum arabic. Plum and cherry gums.	Constituent of glycoproteins.	
D-Xylose	Wood gums, proteoglycans, glycosaminoglycans.	Constituent of glycoproteins.	
D-Lyxose	Heart muscle.	A constituent of a lyxoflavin isolated from human heart muscle.	
L-Xylulose	Intermediate in uronic acid pathway.		Found in urine in essential pentosuria.

stereoisomers. The presence of asymmetric carbon atoms (carbon atoms attached to 4 different atoms or groups) allows the formation of isomers. The number of possible isomers of a compound depends on the number of asymmetric carbon atoms (n) and is equal to 2^n. Glucose, with 4 asymmetric carbon atoms, therefore has 16 isomers. The more important types of isomerism found with glucose are as follows:

(1) D and L Isomerism: The designation of an isomer as the D form or of its mirror image as the L form is determined by its spatial relationship to the parent compound of the carbohydrate family, the 3-carbon sugar glycerose (glyceraldehyde). The L and D forms of this sugar are shown in Fig 15–2 together with the corresponding isomers of glucose. The orientation of the –H and –OH groups around the carbon atom **adjacent** to the terminal primary alcohol carbon (eg, carbon atom 5 in glucose) determines whether the sugar belongs to the D or L series. When the –OH group on this carbon is on the right (as seen in Fig 15–2), the sugar is a member of the D series; when it is on the left, it is a member of the L series. Most of the monosaccharides occurring in mammals are of the D configuration, and enzymes responsible for their metabolism are specific for this configuration.

The presence of asymmetric carbon atoms also confers **optical activity** on the compound. When a beam of plane-polarized light is passed through a solution of an **optical isomer,** it will be rotated either to the right, dextrorotatory (+), or to the left, levorotatory (−). A compound may be designated D(−), D(+), L(−), or L (+), indicating structural relationship to D or L glycerose but not necessarily exhibiting the same optical rotation. For example, the naturally occurring form of fructose is the D(−) isomer.

When equal amounts of D and L isomers are present, the resulting mixture has no optical activity, since the activities of each isomer cancel one another. Such a mixture is said to be a **racemic**—or DL—mixture. Synthetically produced compounds are necessarily racemic because the opportunities for the formation of each optical isomer are identical.

(2) Pyranose and Furanose Ring Structures: This terminology is based on the fact that the stable ring structures of monosaccharides are similar to the ring structures of either pyran or furan (Fig 15–3).

Table 15–2. Hexoses of physiologic importance.

Sugar	Source	Importance	Clinical Significance
D-Glucose	Fruit juices. Hydrolysis of starch, cane sugar, maltose, and lactose.	The "sugar" of the body. The sugar carried by the blood, and the principal one used by the tissues.	Present in the urine (glycosuria) in diabetes mellitus owing to raised blood glucose (hyperglycemia).
D-Fructose	Fruit juices. Honey. Hydrolysis of cane sugar and of inulin (from the Jerusalem artichoke).	Can be changed to glucose in the liver and intestine and so used in the body.	Hereditary fructose intolerance leads to fructose accumulation and hypoglycemia.
D-Galactose	Hydrolysis of lactose.	Can be changed to glucose in the liver and metabolized. Synthesized in the mammary gland to make the lactose of milk. A constituent of glycolipids and glycoproteins.	Failure to metabolize leads to galactosemia and cataract.
D-Mannose	Hydrolysis of plant mannans and gums.	A constituent of many glycoproteins.	

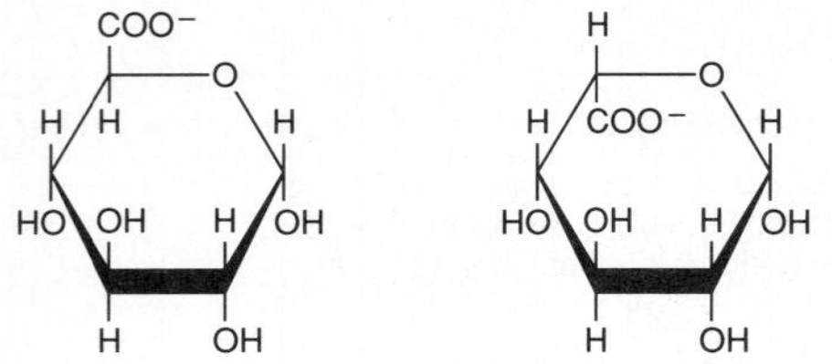

Figure 15–9. α-D-Glucuronate (*left*) and β-L-iduronate (*right*).

Ketoses may also show ring formation (eg, D-fructofuranose or D-fructopyranose) (Fig 15–4). In the case of glucose in solution, more than 99% is in the pyranose form; thus, less than 1% is in the furanose form.

(3) α and β Anomers: The ring structure of an aldose is a hemiacetal, since it is formed by combination of an aldehyde and an alcohol group (Fig 15–5). Similarly, the ring structure of a ketose is a hemiketal. Crystalline glucose is α-D-glucopyranose. The cyclic structure is retained in solution, but isomerism takes place about position 1, the carbonyl or **anomeric carbon atom,** to give a mixture of α-glycopyranose (36%) and β-glucopyranose (63%), the remaining 1% represented mainly by α and β anomers of glucofuranose. This equilibration is accompanied by optical rotation (**mutarotation**) as the hemiacetal ring opens and re-forms with change of position of the –H and –OH groups on carbon 1. The change probably takes place via a hydrated straight-chain acyclic molecule, although polarography has indicated that glucose exists only to the extent of 0.0025% in the acyclic form. The optical rotation of glucose in solution is dextrorotatory; hence, the alternative name of **dextrose,** often used in clinical practice.

(4) Epimers: Isomers differing as a result of variations in configuration of the –OH and –H on carbon atoms 2, 3, and 4 of glucose are known as epimers. Biologically, the most important epimers of glucose

are mannose and galactose, formed by epimerization at carbons 2 and 4, respectively (Fig 15–6).

(5) Aldose-Ketose Isomerism: Fructose has the same molecular formula as glucose but differs in its structural formula, since there is a potential keto group in position 2 of fructose (Fig 15–7), whereas there is a potential aldehyde group in position 1 of glucose (Fig 15–8).

Many Monosaccharides Are Physiologically Important

Triose derivatives are formed in the course of the metabolic breakdown of glucose by the glycolysis pathway. Derivatives of trioses, tetroses, and pentoses and of a 7-carbon sugar (sedoheptulose) are formed in the breakdown of glucose via the pentose phosphate pathway. Pentose sugars are important constituents of nucleotides, nucleic acids, and many coenzymes (Table 15–1). Of the hexoses, glucose, galactose, fructose, and mannose are physiologically the most important (Table 15–2).

The structures of the aldo sugars of biochemical significance are shown in Fig 15–8. Five keto sugars which are important in metabolism are shown in Fig 15–7.

Of additional significance are carboxylic acid derivatives of glucose such as D-glucuronate (important in glucuronide formation and present in glycosaminoglycans) and its metabolic derivatives, L-iduronate (present in glycosaminoglycans) (Fig 15–9) and L-gulonate (a member of the uronic acid pathway; see Fig 22–1).

Sugars Form Glycosides With Other Compounds and With Each Other

Glycosides are compounds formed from a condensation between the hydroxyl group of the anomeric carbon of a monosaccharide, or monosaccharide residue, and a second compound that may, or may not (in the case of an **aglycone**), be another monosaccharide.

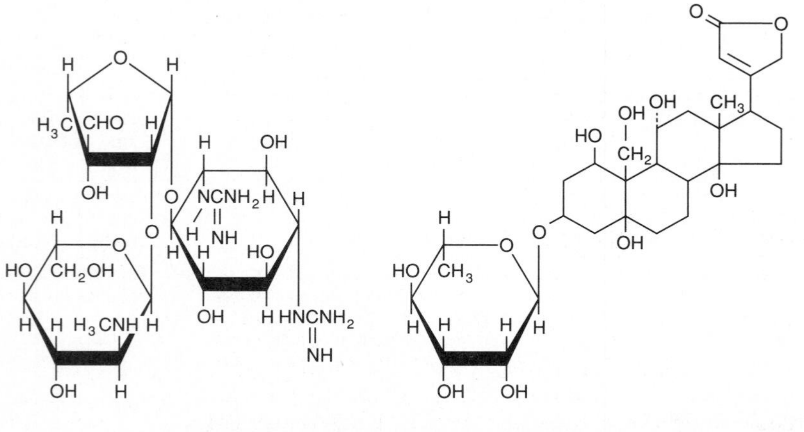

Figure 15–10. Streptomycin (*left*) and ouabain (*right*).

Figure 15–11. 2-Deoxy-D-ribofuranose (β form).

Figure 15–12. Glucosamine (2-amino-D-glucopyranose) (α form). Galactosamine is 2-amino-D-galactopyranose. Both glucosamine and galactosamine occur as N-acetyl derivatives in more complex carbohydrates, eg, glycoproteins.

Maltose

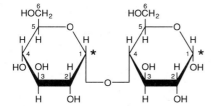

O-α-D-Glucopyranosyl-(1→4)-α-D-glucopyranose

Trehalose

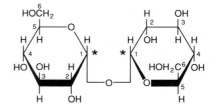

O-α-D-Glucopyranosyl-(1→1)-α-D-glucopyranoside

Sucrose

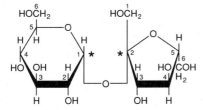

O-α-D-Glucopyranosyl-(1→2)-β-D-fructofuranoside

Cellobiose

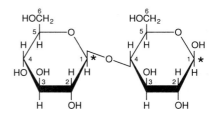

O-β-D-Glucopyranosyl-(1→4)-β-D-glucopyranose

Lactose

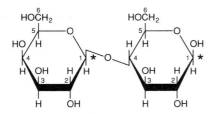

O-β-D-Galactopyranosyl-(1→4)-β-D-glucopyranose

Figure 15–13. Structures of representative disaccharides. The -α and -β refer to the configuration at the anomeric carbon atom (*). When the anomeric carbon of the second residue takes part in the formation of the glycosidic bond, the residue becomes a glycoside known as a furanoside or pyranoside. As the sugar no longer has an anomeric carbon with a free potential aldehyde or ketone group, it no longer exhibits reducing properties, as do most other sugars.

Table 15–3. Disaccharides.

Sugar	Source	Clinicial Significance
Maltose	Digestion by amylase or hydrolysis of starch. Germinating cereals and malt.	
Lactose	Milk. May occur in urine during pregnancy.	In lactase deficiency, malabsorption leads to diarrhea and flatulence.
Sucrose	Cane and beet sugar. Sorghum. Pineapple. Carrot roots.	In sucrase deficiency, malabsorption leads to diarrhea and flatulence.
Trehalose	Fungi and yeasts. The major sugar of insect hemolymph.	

If the second group is a hydroxyl, the *O*-**glycosidic bond** is an **acetal** link because it results from a reaction between a hemiacetal group (formed from an aldehyde and an –OH group) and another –OH group. If the hemiacetal portion is glucose, the resulting compound is a **glucoside;** if galactose, a **galactoside,** etc. If the second group is an amine, an *N*-glycosidic bond is formed, eg, between adenine and ribose in nucleotides such as ATP (see Fig 12–5).

Glycosides are found in many drugs and spices and in the constituents of animal tissues. The aglycone may be methanol, glycerol, a sterol, a phenol, or a base such as adenine. The glycosides that are important in medicine because of their action on the heart (**cardiac glycosides**) all contain steroids as the aglycone component. These include derivatives of digitalis and strophanthus such as **ouabain,** an inhibitor of the NA+/K+-ATPase of cell membranes. Other glycosides include antibiotics such as **streptomycin** (Fig 15–10).

Deoxy Sugars Lack an Oxygen

Deoxy sugars are those in which a hydroxyl group attached to the ring structure has been replaced by a hydrogen atom. They are obtained on hydrolysis of certain substances that are important in biologic processes. An example is the **deoxyribose** (Fig 15–11) occurring in nucleic acids (DNA).

Also found as a carbohydrate of glycoproteins is L-fucose (see Fig 15–17), and of importance as an inhibitor of glucose metabolism is 2-deoxyglucose.

Amino Sugars (Hexosamines) Are Components of Glycoproteins Gangliosides and Glycosamino-glycans

Examples of amino sugars are D-glucosamine (Fig 15–12), D-galactosamine, and D-mannosamine, all of which have been identified in nature. Glucosamine is a constituent of hyaluronic acid. Galactosamine (chondrosamine) is a constituent of chondroitin (see Chapter 54).

Several **antibiotics** (erythromycin, carbomycin) contain amino sugars. The amino sugars are believed to be related to the antibiotic activity of these drugs.

THE MOST IMPORTANT DISACCHARIDES ARE MALTOSE, SUCROSE, AND LACTOSE

The disaccharides are sugars composed of 2 monosaccharide residues united by a glycosidic linkage (Fig 15–13). Their chemical name reflects their component

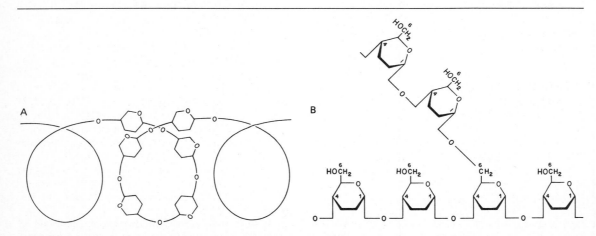

Figure 15–14. Structure of starch. *A:* Amylose, showing helical coil structure. *B:* Amylopectin, showing 1 → 6 branch point.

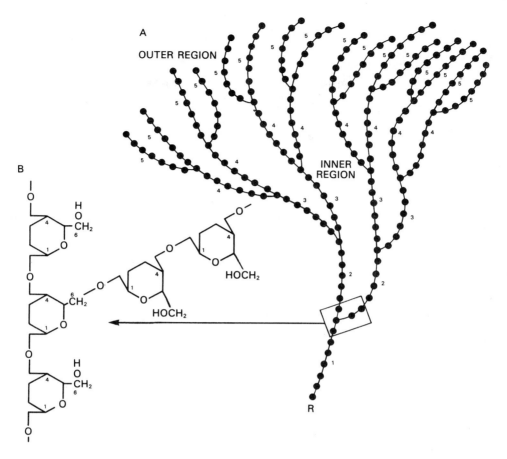

Figure 15–15. The glycogen molecule. *A:* General structure. *B:* Enlargement of structure at a branch point. The numbers in *A* refer to equivalent stages in the growth of the macromolecule. R, primary glucose residue containing the only free reducing group on C_1. The branching is more variable than shown, the ratio of $1 \rightarrow 4$ to $1 \rightarrow 6$ bonds being from 10 to 18.

monosaccharides. The physiologically important disaccharides are maltose, sucrose, lactose, and trehalose (Table 15–3).

Hydrolysis of sucrose yields a crude mixture called "invert sugar" because the strongly levorotatory fructose thus produced changes (inverts) the previous dextrorotatory action of the sucrose.

POLYSACCHARIDES HAVE STORAGE AND STRUCTURAL FUNCTIONS

Polysaccharides include the following physiologically important carbohydrates.

Starch is formed of an α-glucosidic chain. Such a compound, yielding only glucose on hydrolysis, is a homopolymer called a **glucosan** or **glucan.** It is the most important food source of carbohydrate and is found in cereals, potatoes, legumes, and other vegetables. The 2 chief constituents are **amylose** (15–20%), which has a nonbranching helical structure (Fig 15–14), and **amylopectin** (80–85%), which consists of

branched chains composed of 24–30 glucose residues united by $1 \rightarrow 4$ linkages in the chains and by $1 \rightarrow 6$ linkages at the branch points.

Glycogen (Fig 15–15) is the storage polysaccharide of the animal body. It is often called animal starch. It is a more highly branched structure than amylopectin and has chains of 10-18-α-D-glucopyranose residues (in α[$1 \rightarrow 4$]-glucosidic linkage) with branching by means of α($1 \rightarrow 6$)-glucosidic bonds.

Inulin is a starch found in tubers and roots of dahlias, artichokes, and dandelions. It is hydrolyzable to fructose, and hence it is a fructosan. This starch, unlike potato starch, is easily soluble in warm water and has been used in physiologic investigation for determination of the rate of glomerular filtration.

Dextrins are substances formed in the course of the hydrolytic breakdown of starch. Limit dextrins are the first formed products as hydrolysis reaches a certain degree of branching.

Cellulose is the chief constituent of the framework of plants. It is not soluble in ordinary solvents and consists of β-D-glucopyranose units linked by β($1 \rightarrow 4$) bonds to form long, straight chains strengthened by cross-linked hydrogen bonds. Cellulose cannot be digested by many mammals, including humans, because of the absence of a hydrolase that attacks the β linkage. Thus, it is an important source of "bulk" in the diet. In the gut of

Chitin

N-Acetylglucosamine   N-Acetylglucosamine

Hyaluronic acid

β-Glucuronic acid   N-Acetylglucosamine

Chondroitin 4-Sulfate
(Note: There is also a 6-sulfate)

β-Glucuronic acid   N-Acetylgalactosamine sulfate

Heparin

Sulfated glucosamine   Sulfated iduronic acid

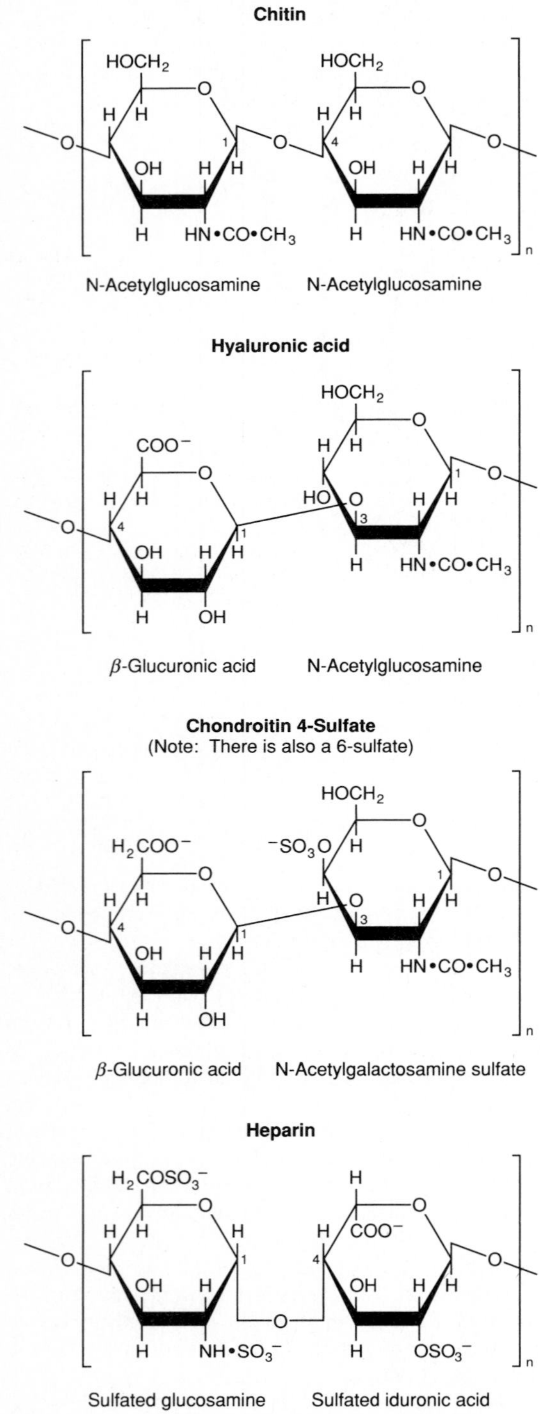

Figure 15–16. Structure of some complex polysaccharides.

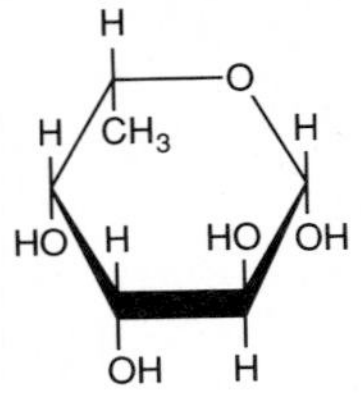

Figure 15–17. β-L-Fucose (6-deoxy-β-L-galactose).

ruminants and other herbivores, there are microorganisms that can attack the β linkage, making cellulose available as a major calorigenic source.

Chitin is an important structural polysaccharide of invertebrates. It is found, for example, in the exoskeletons of crustaceans and insects. Structurally, chitin consists of N-acetyl-D-glucosamine units joined by $\beta(1 \rightarrow 4)$-glycosidic linkages (Fig 15–16).

Glycosaminoglycans (mucopolysaccharides) consist of chains of complex carbohydrates characterized by their content of **amino sugars** and **uronic acids.** When these chains are attached to a protein molecule, the compound is known as a **proteoglycan.** As the ground or packing substance, they are associated with the structural elements of the tissues such as bone, elastin, and collagen. Their property of holding large quantities of water and occupying space, thus cushioning or lubricating other structures, is assisted by the large number of –OH groups and negative charges on the molecules, which, by repulsion, keep the carbohydrate chains apart. Examples are **hyaluronic acid, chondroitin sulfate,** and **heparin** (Fig 15–16), discussed in detail in Chapter 57.

Glycoproteins (mucoproteins) occur in many different situations in fluids and tissues, including the cell membranes (see Chapters 43 and 57). They are proteins containing carbohydrates in varying amounts attached as short or long (up to 15 units) branched or unbranched chains. Such chains are usually called oligosaccharide chains. Constituent carbohydrates include

Hexoses
Mannose (Man)   Galactose (Gal)

Acetyl hexosamines
N-Acetylglucosamine   N-Acetylgalactosamine
(GlcNAc)   (GalNAc)

Pentoses
Arabinose (Ara)   Xylose (Xyl)

Methyl pentose
L-Fucose (Fuc; see Fig 15–17)

Sialic acids
N-Acyl derivatives of neuraminic acid, eg, N-acetylneuraminic acid (NeuAc; see Fig 15–18), the predominant sialic acid

Glucose is not found in mature glycoproteins apart from collagen, and, in contrast to the glycosaminoglycans and proteoglycans, uronic acids are absent.

The **sialic acids** are N- or O-acyl derivatives of neu-

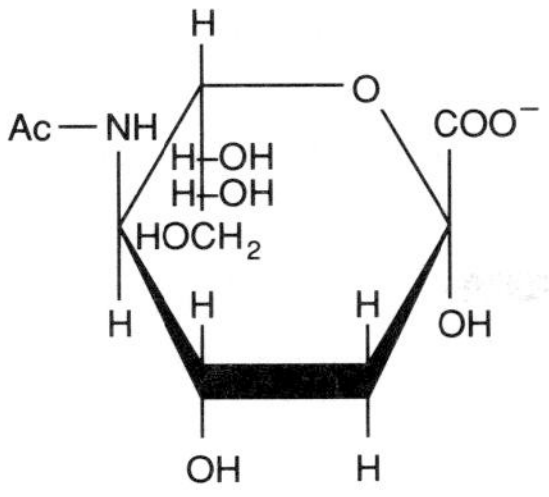

Figure 15–18. Structure of N-acetylneuraminic acid, a sialic acid (Ac = CH_3–CO–).

raminic acid (Fig 15–18). **Neuraminic acid** is a 9-carbon sugar derived from mannosamine (an epimer of glucosamine) and pyruvate. Sialic acids are constituents of both **glycoproteins** and **gangliosides.**

CARBOHYDRATES OCCUR IN CELL MEMBRANES

The lipid structure of the cell membrane is described in Chapters 16 and 43. However, analysis of mammalian cell membrane components indicates that approximately 5% are carbohydrates, present in glycoproteins and glycolipids. Their presence on the outer surface of the plasma membrane (the **glycocalyx**) has

been shown with the use of plant **lectins,** protein agglutinins that bind specifically with certain glycosyl residues. For example, **concanavalin A** has a specificity toward α-glucosyl and α-mannosyl residues.

Glycophorin is a major integral membrane glycoprotein of human erythrocytes. It has 130 amino acid residues and spans the lipid membrane, having free polypeptide portions outside both the external and internal (cytoplasmic) surfaces. Carbohydrate chains are only attached to the N-terminal portion outside the external surface (see Chapter 43).

REFERENCES

Advances in Carbohydrate Chemistry. Academic Press, 1945–current.

Collins PM (editor): *Carbohydrates.* Chapman & Hall, 1987.

Ferrier RJ, Collins PM: *Monosaccharide Chemistry.* Penguin Books, 1972.

Hughes RC: The complex carbohydrates of mammalian cell surfaces and their biological roles. *Essays Biochem* 1975; **11:**1.

Lindahl U, Höök M: Glycosaminoglycans and their binding to biological macromolecules. *Annu Rev Biochem* 1978; **47:**385.

Pigman WW, Horton D (editors): *The Carbohydrates.* Vols 1A and 1B. Academic Press, 1972.

Rees DA: *Polysaccharide Shapes.* Wiley, 1977.

Sharon N: Carbohydrates. *Sci Am* 1980;**245:**90.

16 | Lipids of Physiologic Significance

Peter A. Mayes, PhD, DSc

INTRODUCTION

The lipids are a heterogeneous group of compounds related, either actually or potentially, to the fatty acids. They have the common property of being (1) relatively **insoluble in water** and (2) **soluble in nonpolar solvents** such as ether, chloroform, and benzene. Thus, the lipids include fats, oils, waxes, and related compounds.

Lipids are **important dietary constituents** not only because of their high energy value but also because of the fat-soluble vitamins and the essential fatty acids contained in the fat of natural foods.

BIOMEDICAL IMPORTANCE

In the body, fat serves as an efficient source of energy—both directly, and potentially when stored in **adipose tissue.** It serves as a **thermal insulator** in the subcutaneous tissues and around certain organs, and nonpolar lipids act as **electrical insulators** allowing rapid propagation of depolarization waves along myelinated nerves. The fat content of **nerve tissue** is particularly high. Combinations of fat and protein (lipoproteins) are important cellular constituents, occurring both in the cell **membrane** and in the mitochondria within the cytoplasm, and serving also as the means of **transporting lipids** in the blood. A knowledge of lipid biochemistry is important in understanding many current biomedical areas of interest, eg, **obesity, atherosclerosis,** and the role of various **polyunsaturated fatty acids** in nutrition and health.

LIPIDS ARE CLASSIFIED AS SIMPLE OR COMPLEX

The following classification of lipids is modified from Bloor:

A. Simple Lipids: Esters of fatty acids with various alcohols.

1. Fats—Esters of fatty acids with glycerol. A fat in the liquid state is known as an oil.

2. Waxes—Esters of fatty acids with higher molecular weight monohydric alcohols.

B. Complex Lipids: Esters of fatty acids containing groups in addition to an alcohol and a fatty acid.

1. Phospholipids—Lipids containing, in addition to fatty acids and an alcohol, a phosphoric acid residue. They frequently have nitrogen-containing bases and other substituents.

(a) Glycerophospholipids—The alcohol is glycerol.

(b) Sphingophospholipids—The alcohol is sphingosine.

2. Glycolipids (glycosphingolipids)—Lipids containing a fatty acid, sphingosine, and carbohydrate.

3. Other complex lipids—Lipids such as sulfolipids and aminolipids. Lipoproteins may also be placed in this category.

C. Precursor and Derived Lipids: These include fatty acids, glycerol, steroids, alcohols in addition to glycerol and sterols, fatty aldehydes, and ketone bodies (see Chapter 24), hydrocarbons, lipid-soluble vitamins, and hormones.

Because they are uncharged, acylglycerols (glycerides), cholesterol, and cholesteryl esters are termed **neutral lipids.**

FATTY ACIDS ARE ALIPHATIC CARBOXYLIC ACIDS

Fatty acids occur mainly as esters in natural fats and oils but do occur in the unesterified form as **free fatty acids,** a transport form found in the plasma. Fatty acids that occur in natural fats are usually straight-chain derivatives and contain an **even number** of carbon atoms, because they are synthesized from 2-carbon units. The chain may be **saturated** (containing no double bonds) or **unsaturated** (containing one or more double bonds).

Fatty Acids Are Named After Corresponding Hydrocarbons

The most frequently used systematic nomenclature is based on naming the fatty acid after the hydrocarbon with the same number of carbon atoms, **-oic** being substituted for the final **e** in the name of the hydrocarbon (Genevan system). Thus, saturated acids end in **-anoic,** eg, octanoic acid, and unsaturated acids with double bonds end in **-enoic,** eg, octadecenoic acid (oleic acid).

Carbon atoms are numbered from the carboxyl carbon (carbon No. 1). The carbon atom adjacent to the

18:1;9 *or* Δ^9 18:1

$$\overset{18}{C}H_3(CH_2)_7\overset{10}{C}H = \overset{9}{C}H(CH_2)_7\overset{1}{C}OOH$$

or

$\omega 9, C18:1$ *or* n−9, 18:1

$$\underset{n \;\; 17}{\overset{\omega \;\; 2}{C}H_3}\underset{}{CH_2}\overset{3}{CH_2}\overset{4}{CH_2}\overset{5}{CH_2}\overset{6}{CH_2}\overset{7}{CH_2}\overset{8}{CH_2}\overset{9}{CH} = \underset{10}{\overset{10}{C}H}(CH_2)_7\underset{1}{\overset{18}{C}OOH}$$

Figure 16–1. Oleic acid, n − 9 (n minus 9).

carboxyl carbon (No. 2) is also known as the α-carbon. Carbon atom No. 3 is the β-carbon, and the end methyl carbon is known as the ω-carbon or n-carbon atom.

Various conventions are in use for indicating the number and position of the double bonds; eg, Δ^9 indicates a double bond between carbon atoms 9 and 10 of the fatty acid; ω9 indicates a double bond on the ninth carbon counting from the ω-carbon atom. Widely used conventions to indicate the number of carbon atoms, the number of double bonds, and the positions of the double bonds are shown in Fig 16–1.

In animals, additional double bonds are introduced only **between the existing double bond (eg, ω9, ω6, or ω3) and the carboxyl carbon,** leading to 3 series of fatty acids known as the ω9, ω6, and ω3 families, respectively.

Saturated Fatty Acids Contain No Double Bonds

Saturated fatty acids may be envisaged as based on acetic acid as the first member of the series. Examples are shown in Table 16–1.

Other higher members of the series are known to occur, particularly in waxes. A few branched-chain fatty acids have also been isolated from both plant and animal sources.

Unsaturated Fatty Acids Contain One or More Double Bonds (See Table 16–2.)

These may be further subdivided according to degree of unsaturation.

A. Monounsaturated (Monoethenoid, Monoenoic) acids, containing one double bond.

B. Polyunsaturated (Polyethenoid, Polyenoic Acids, containing two or more double bonds.

C. Eicosanoids: These compounds, derived from eicosa- (20-C) polyenoic fatty acids, comprise the **prostanoids** and **leukotrienes (LT).** Prostanoids include **prostaglandins (PG), prostacyclins (PGI),** and **thromboxanes (TX).** The term "prostaglandins" is often used loosely to include all prostanoids.

Prostaglandins were originally discovered in seminal plasma but are now known to exist in virtually every mammalian tissue, acting as local hormones; they have important physiologic and pharmacologic activities. They are synthesized in vivo by cyclization of the center of the carbon chain of 20-C (eicosanoic) polyunsaturated fatty acids (eg, arachidonic acid) to form a cyclopentane ring (Fig 16–2). A related series of compounds, the **thromboxanes,** discovered in platelets, have the cyclopentane ring interrupted with an oxygen atom (oxane ring) (Fig 16–3). Three different eicosanoic fatty acids give rise to 3 groups of

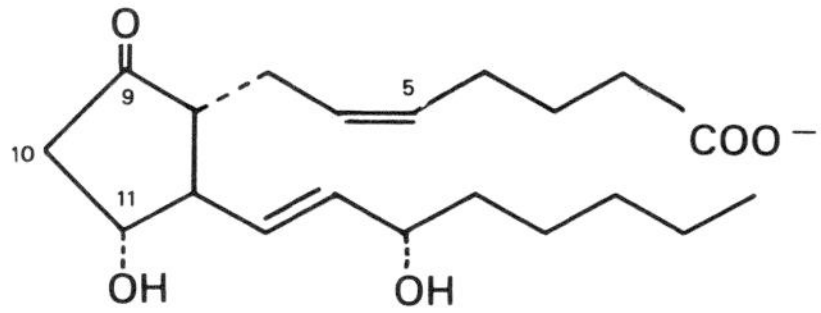

Table 16–1. Saturated fatty acids.

Common Name	Number of C Atoms	
Formic*	1	Takes part in the metabolism of "C_1" units (formate)
Acetic	2	Major end product of carbohydrate fermentation by rumen organisms†
Propionic	3	An end product of carbohydrate fermentation by rumen organisms†
Butyric	4	In certain fats in small amounts (especially butter). An end product of carbohydrate fermentation by rumen organisms.†
Valeric	5	
Caproic	6	
Caprylic (octanoic)	8	In small amounts in many fats (including butter), especially those of plant origin
Capric (decanoic)	10	
Lauric	12	Spermaceti, cinnamon, palm kernel, coconut oils, laurels
Myristic	14	Nutmeg, palm kernel, coconut oils, myrtles
Palmitic	16	Common in all animal and plant fats
Stearic	18	
Arachidic	20	Peanut (arachis) oil
Behenic	22	Seeds
Lignoceric	24	Cerebrosides, peanut oil

*Strictly, not an alkyl derivative.
†Also in the colon of humans.

Figure 16–2. Prostaglandin E$_2$ (PGE$_2$).

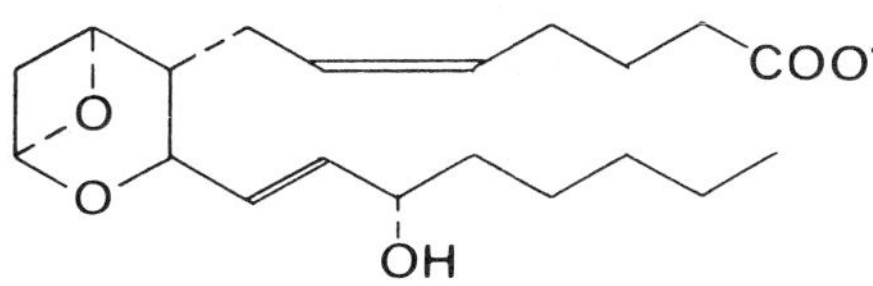

Figure 16–3. Thromboxane A$_2$ (TXA$_2$).

Table 16–2. Unsaturated fatty acids of physiologic and nutritional significance.

Number of C Atoms and Number and Position of Double Bonds	Series	Common Name	Sytematic Name	Occurrence
Monoenoic acids (one double bond)				
16:1;9	ω7	Palmitoleic	cis-9-Hexadecenoic	In nearly all fats.
18:1;9	ω9	Oleic	cis-9-Octadecenoic	Possibly the most common fatty acid in natural fats.
18:1;9	ω9	Elaidic	trans-9-Octadecenoic	Hydrogenated and ruminant fats.
22:1;13	ω9	Erucic	cis-13-Docosenoic	Rape and mustard seed oils.
24:1;15	ω9	Nervonic	cis-15-Tetracosenoic	In cerebrosides.
Dienoic acids (2 double bonds)				
18:2;9,12	ω6	Linoleic	all-cis-9,12-Octadecadienoic	Corn, peanut, cottonseed, soybean, and many plant oils.
Trienoic acids (3 double bonds)				
18:3;6,9,12	ω6	γ-Linolenic	all-cis-6,9,12-Octadecatrienoic	Some plants, eg, oil of evening primrose; minor fatty acid in animals.
18:3;9,12,15	ω3	α-Linolenic	all-cis-9,12,15-Octadecatrienoic	Frequently found with linoleic acid but particularly in linseed oil
Tetraenoic acids (4 double bonds)				
20:4;5,8,11,14	ω6	Arachidonic	all-cis-5,8,11,14-Eicosatetraenoic	Found with linoleic acid particularly in peanut oil; important component of phospholipids in animals.
Pentaenoic acids (5 double bonds)				
20:5;5,8,11,14,17	ω3	Timnodonic	all-cis-5,8,11,14,17-Eicosapentaenoic	Important component of fish oils, eg, cod liver oil.
22:5;7,10,13,16,19	ω3	Clupanodonic	all-cis-7,10,13,16,19-Docosapentaenoic	Fish oils, phospholipids in brain.
Hexaenoic acids (6 double bonds)				
22:6;4,7,10,13,16,19	ω3	Cervonic	all-cis-4,7,10,13,16,19-Docosahexaenoic	Fish oils, phospholipids in brain.

eicosanoids characterized by the number of double bonds in the side chains, eg, PG_1, PG_2, PG_3. Variations in the substituent groups attached to the rings give rise to different types in each series of prostaglandins and thromboxanes, labeled A, B, etc. For example, the "E" type of prostaglandin (as in PGE_2) has a keto group in position 9, whereas the "F" type has a hydroxyl group in this position. The **leukotrienes** are a third group of eicosanoid derivatives formed via the

lipoxygenase pathway rather than cyclization of the fatty acid chain (Fig 16–4). First described in leukocytes, they are characterized by the presence of 3 conjugated double bonds.

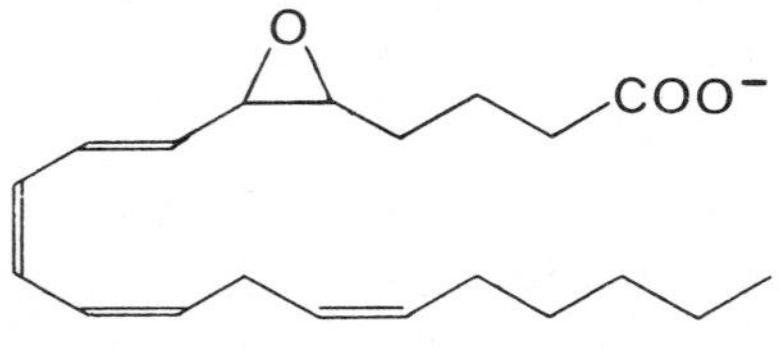

Most Naturally Occurring Unsaturated Fatty Acids Have cis Double Bonds

The carbon chains of saturated fatty acids form a zigzag pattern when extended, as at low temperatures. At higher temperatures, some bonds rotate, causing chain shortening, which explains why biomembranes become thinner with increase in temperature. A type of **geometric isomerism** occurs in unsaturated fatty acids, depending on the orientation of atoms or groups around the axis of double bonds. If the acyl chains are on the same side of the bond, it is cis-, as in oleic acid; if on opposite sides, trans-, as in elaidic acid, the unnatural isomer of oleic acid (Fig 16–5). Naturally oc-

Figure 16–4. Leukotriene A_4 (LTA_4).

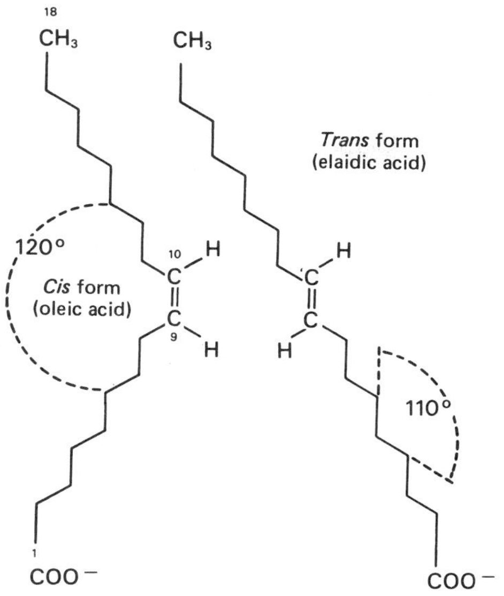

Figure 16–5. Geometric isomerism of Δ^9, 18 : 1 fatty acids (oleic and elaidic acids).

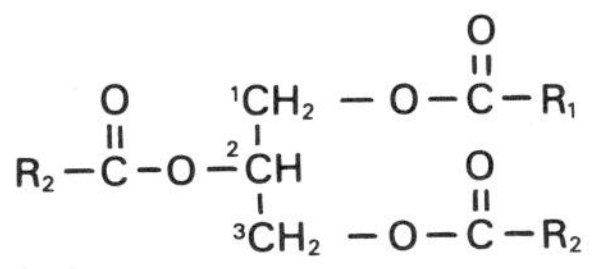

Figure 16–6. Triacylglycerol.

which must be fluid at all environmental temperatures, are more unsaturated than storage lipids. Lipids in tissues that are subject to cooling, eg, in hibernators or in the extremities of animals, are more unsaturated.

Certain Alcohols & Aldehydes Are Associated With Natural Lipids

Alcohols: Alcohols associated with lipids include glycerol, cholesterol, and higher alcohols (eg, cetyl alcohol, $C_{16}H_{33}OH$), usually found in the waxes, and the polyisoprenoid alcohol dolichol (see Fig 16–27).

Fatty Aldehydes: The fatty acids may be reduced to fatty aldehydes. These compounds are found either combined or free in natural lipids.

TRIACYLGLYCEROLS (TRIGLYCERIDES)* ARE THE MAIN STORAGE FORMS OF FATTY ACIDS

The triacylglycerols, are esters of the alcohol-glycerol and fatty acids. In naturally occurring fats, the proportion of triacylglycerol molecules containing the same fatty acid residue in all 3 ester positions is very small. They are nearly all **mixed acylglycerols.**

In Fig 16–6, if all 3 fatty acids represented by R were stearic acid, the fat would be known as tristearin, since it consists of 3 stearic acid residues esterified with glycerol. Examples of mixed acylglycerols are shown in Figs 16–7 and 16–8.

---

*According to the current standardized terminology of the International Union of Pure and Applied Chemistry (IUPAC) and the International Union of Biochemistry (IUB), the monoglycerides, diglycerides, and triglycerides are to be designated monoacylglycerols, diacylglycerols, and triacylglycerols, respectively.

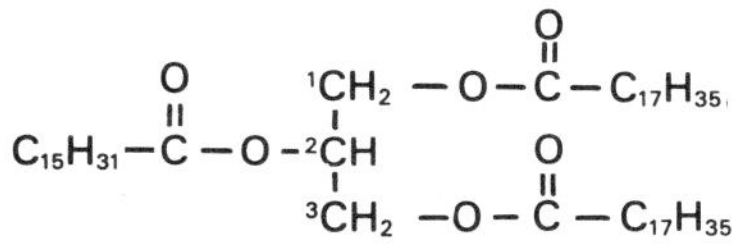

---

curring unsaturated long-chain fatty acids are nearly all of the *cis* configuration, the molecules being "bent" 120 degrees at the double bond. Thus, oleic acid has an L-shape, whereas elaidic acid remains "straight" at its *trans* double bond. Increase in the number of *cis* double bonds in a fatty acid leads to a variety of possible spatial configurations of the molecule, eg, arachidonic acid, with 4 double *cis* bonds, may have "kinks" or a "U" shape. This may have profound significance on molecular packing in membranes and on the positions occupied by fatty acids in more complex molecules such as phospholipids. The presence of *trans* double bonds will alter these spatial relationships. *Trans* fatty acids are present in certain foods. Most arise as a by-product during the saturation of fatty acids in the process of hydrogenation, or "hardening," of natural oils in the manufacture of margarine. An additional small contribution comes from the ingestion of ruminant fat that contains *trans* fatty acids arising from the action of microorganisms in the rumen.

Both Physical & Physiological Properties of Fatty Acids Are Determined by Their Chain Length & Degree of Unsaturation

Thus, the melting points of even-numbered-carbon fatty acids increase with chain length and decrease according to unsaturation. A triacylglycerol containing all saturated fatty acids of 12 C or more is solid at body temperature, whereas if all 3 fatty acid residues are 18 : 2, it is liquid to below 0 °C. In practice, natural acylglycerols contain a mixture of fatty acids tailored to suit their functional roles. The membrane lipids,

Figure 16–7. 1,3-Distearopalmitin.

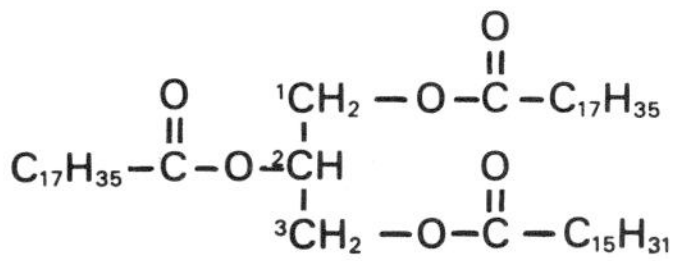

Figure 16–8. 1,2-Distearopalmitin.

Carbon Atoms 1 & 3 of Glycerol Are Not Identical

When it is required to number the carbon atoms of glycerol unambiguously, the -sn- (stereochemical numbering) system is used, eg, 1,2-distearyl-3-palmityl-sn-glycerol (shown as a projection formula also in Fig 16–9). It is important to realize that carbons 1 and 3 of glycerol are not identical when viewed in 3 dimensions. Enzymes readily distinguish between them and are nearly always specific for one or the other carbon; eg, glycerol is always phosphorylated on sn-3 by glycerokinase to give glycerol 3-phosphate and not glycerol 1-phosphate.

Partial acylglycerols consisting of mono- and diacylglycerols wherein a single fatty acid or 2 fatty acids are esterified with glycerol are also found in the tissues. These are of particular significance in the synthesis and hydrolysis of triacylglycerols.

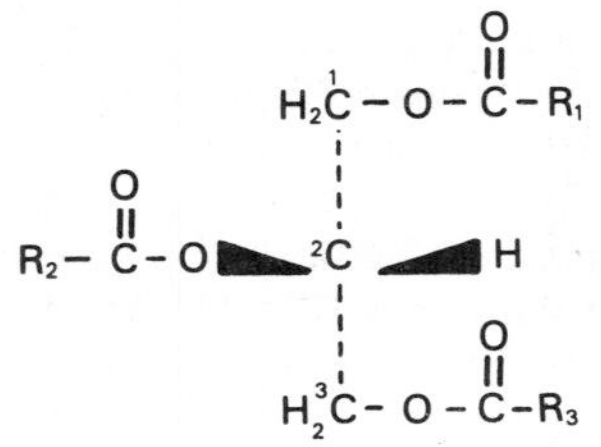

Figure 16–9. Triacyl-sn-glycerol.

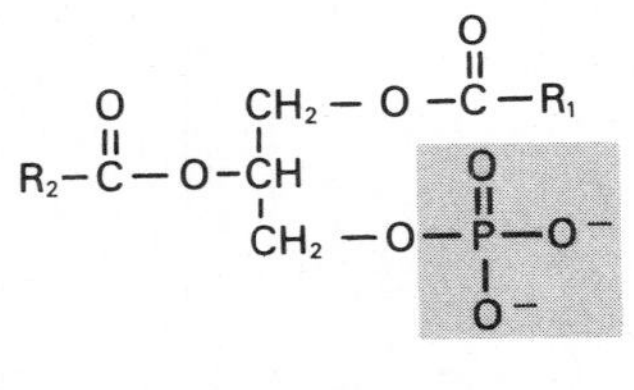

Figure 16–10. Phosphatidic acid.

PHOSPHOLIPIDS ARE THE MAIN LIPID CONSTITUENTS OF MEMBRANES

The phospholipids include the following: (1) phosphatidic acid and phosphatidylglycerol, (2) phosphatidylcholine, (3) phosphatidylethanolamine, (4) phosphatidylinositol, (5) phosphatidylserine, (6) lysophospholipids, (7) plasmalogens, and (8) sphingomyelins. All of these are **phosphoglycerols** apart from the sphingomyelins, which do not contain glycerol. They may be regarded as derivatives of **phosphatidic acid** (Fig 16–10), in which the phosphate is esterified with the –OH of a suitable alcohol.

Phosphatidic acid is important as an intermediate in the synthesis of triacylglycerols as well as phosphoglycerols but is not found in any great quantity in tissues.

Cardiolipin is a Major Lipid of Mitochondrial Membranes

Phospatidic acid is a precursor of **phophatidylglycerol** which, in turn, gives rise to **cardiolipin** (Fig 16–11) in mitochondria.

Phosphatidylcholines (Lecithins) Occur in Cell Membranes

These are phosphoglycerols containing choline (Fig 16–12). They are the most abundant phospholipids of the cell membrane and represent a large proportion of the body's store of choline. Choline is important in nervous transmission and as a store of labile methyl groups. **Dipalmitoyl lecithin** is a very effective surface-active agent, preventing adherence, due to surface tension, of the inner surfaces of the lungs. Its absence from the lungs of premature infants causes **respiratory distress syndrome.** However, most phospholipids have a saturated acyl radical in the C_1 position but an unsaturated radical in the C_2 position of glycerol.

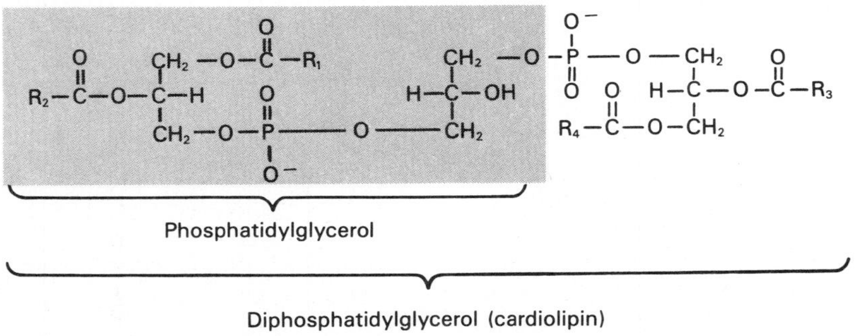

Phosphatidylglycerol

Diphosphatidylglycerol (cardiolipin)

Figure 16–11. Cardiolipin (diphosphatidylglycerol).

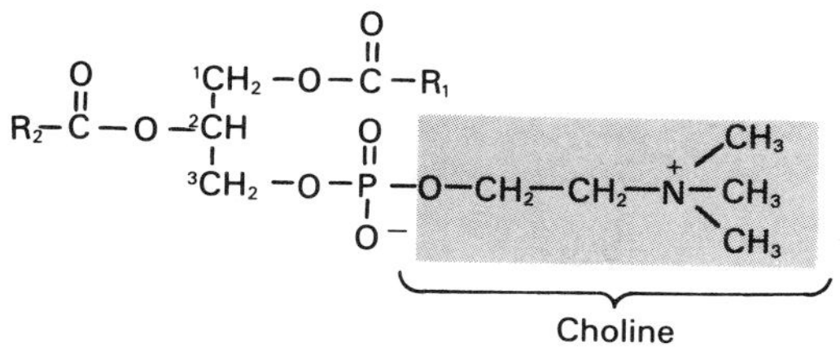

Figure 16–12. 3-Phosphatidylcholine.

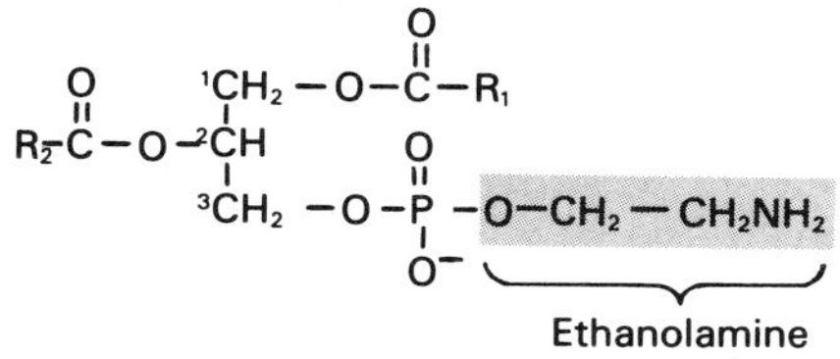

Figure 16–13. 3-Phosphatidylethanolamine.

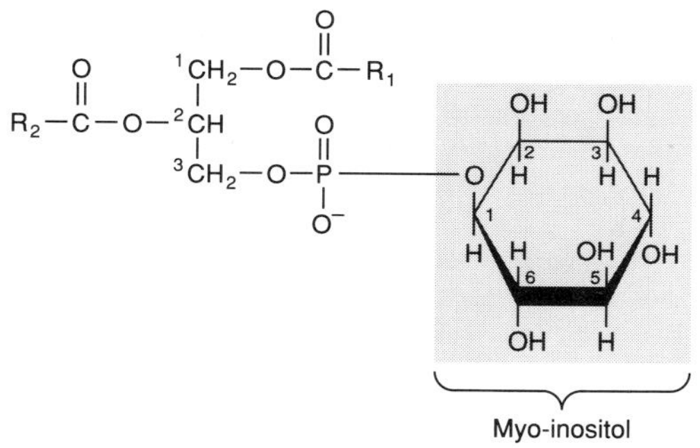

Figure 16–14. 3-Phosphatidylinositol.

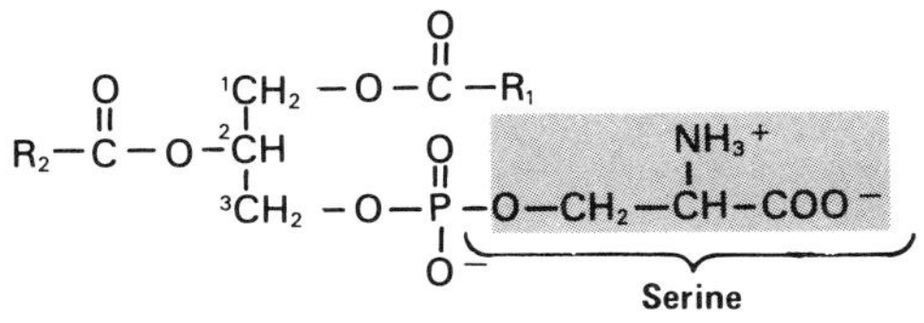

Figure 16–15. 3-Phosphatidylserine.

Phosphatidylethanolamine (Cephalin)

This differs from phosphatidylcholine only in that ethanolamine replaces choline (Fig 16–13).

Phosphatidylinositol is a Precursor of Second Messengers

The inositol is present as the stereoisomer, myo-inositol (Fig 16–14). **Phosphatidylinositol 4,5-bisphosphate** is an important constituent of cell membrane phospholipids; upon stimulation by a suitable hormone agonist, it is cleaved into **diacylglycerol** and **isositol trisphosphate,** both of which act as internal signals or second messengers (see p 475).

Phosphatidylserine

Phosphatidylserine, which contains the amino acid serine rather than ethanolamine, is found in most tissues (Fig 16–15). Phospholipids containing threonine have also been isolated.

Lysophospholipids are Intermediates in the Metabolism of Phosphoglycerols

These are phosphoacylglycerols containing only one acyl radical, eg, **lysolecithin,** important in the metabolism and interconversion of phospholipids (Fig 16–16).

Plasmalogens Occur in Brain & Muscle

These compounds constitute as much as 10% of the phospholipids of brain and muscle. Structurally, the plasmalogens resemble phosphatidylethanolamine but possess an ether link on the C_1 carbon instead of the normal ester link found in most acylglycerols. Typically, the alkyl radical is an unsaturated alcohol (Fig 16–17). In some instances, choline, serine, or inositol may be substituted for ethanolamine.

Sphingomyelins Also Occur in the Nervous System

Sphingomyelins are found in large quantities in brain and nerve tissue. On hydrolysis, the sphingomyelins yield a fatty acid, phosphoric acid, choline, and a complex amino alcohol, **sphingosine** (Fig 16–18). **No glycerol is present.** The combination sphingosine plus fatty acid is known as **ceramide,** a structure also found in the glycolipids (see below).

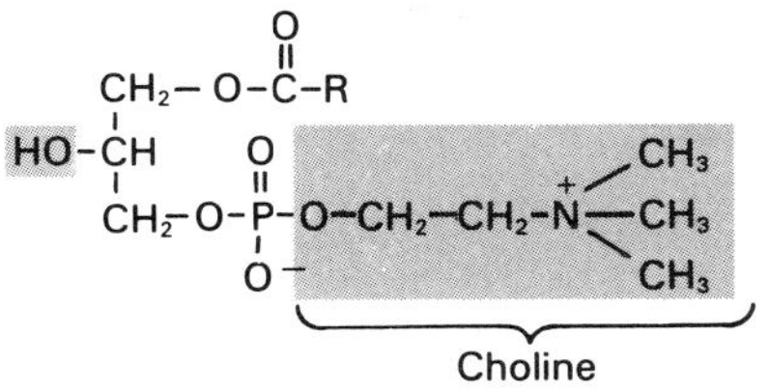

Figure 16–16. Lysolecithin.

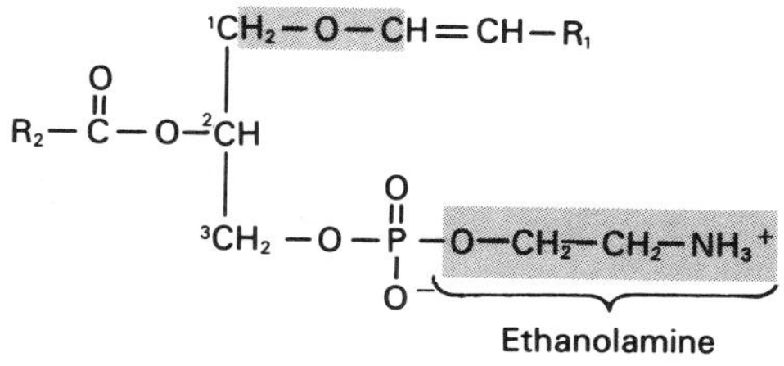

Figure 16–17. Plasmalogen (phosphatidal ethanolamine).

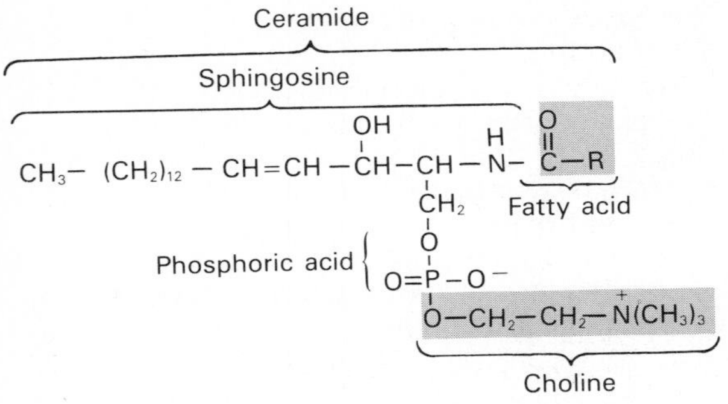

Ceramide

Sphingosine

Figure 16–18. A sphingomyelin.

GLYCOLIPIDS (GLYCOSPHINGOLIPIDS) ARE IMPORTANT IN NERVE TISSUES AND IN THE CELL MEMBRANE

Glycolipids are widely distributed in every tissue of the body, particularly in nervous tissue such as brain. They occur particularly in the outer leaflet of the plasma membrane where they contribute to **cell surface carbohydrates.**

The major glycolipids found in animal tissues are glycosphingolipids. They contain ceramide and one or more sugars. The two simplest are **galactosylceramide** and **glucosylceramide.** Galactosylceramide is a major glycosphingolipid of brain and other nervous tissue, but it is found in relatively low amounts elsewhere. It contains a number of characteristic C_{24} fatty acids. Galactosylceramide (Fig 16–19) can be converted to sulfogalactosylceramide (classic **sulfatide**), which is present in high amounts in myelin. Glucosylceramide is the predominant simple glycosphingolipid of extraneural tissues, but it also occurs in brain in small amounts. The more complex glycosphingolipids are **gangliosides** derived from glucosylceramide. A ganglioside is a glycosphingolipid that contains in addition one or more molecules of **a sialic acid.** Neuraminic acid (NeuAc; see Chapter 15) is the principal sialic acid found in human tissues. Gangliosides are also present in nervous tissues in high concentration. They appear to have receptor and other functions. The simplest ganglioside found in tissues is G_{M3}, which contains ceramide, one molecule of glucose, one molecule of galactose, and one molecule of NeuAc. In the shorthand nomenclature used, G represents ganglioside; M is a monosialo-containing species; and the subscript 3 is an arbitrary number assigned on the basis of chromatographic migration. The structure of a more complex ganglioside derived from G_{M3}, named G_{M1}, is shown in Fig 16–20. G_{M1} is a compound of considerable biologic interest as it is known to be the receptor in human intestine for cholera toxin. Other gangliosides can contain anywhere from one to 5 molecules of sialic acid, giving rise to di-, trisialogangliosides, etc.

STEROIDS PLAY MANY PHYSIOLOGICALLY IMPORTANT ROLES

Cholesterol is probably the best known steroid because of its association with **atherosclerosis.** However, biochemically it is also of significance because it is the precursor of a large number of equally important steroids which include the bile acids, adrenocortical hormones, sex hormones, D vitamins, cardiac glycosides, sitosterols of the plant kingdom, and some alkaloids.

All of the steroids have a similar cyclic nucleus resembling phenanthrene (rings A, B, and C) to which a cyclopentane ring (D) is attached. The carbon positions on the steroid nucleus are numbered as shown in Fig 16–21.

It is important to realize that in structural formulas of steroids, a simple hexagonal ring denotes a completely saturated 6-carbon ring with all valences satisfied by hydrogen bonds unless shown otherwise; ie, it is not a benzene ring. All double bonds are shown as such. Methyl side chains are shown as single bonds unattached at the farther (methyl) end. These occur typically at positions 10 and 13 (constituting C atoms 19 and 18). A side chain at position 17 is usual (as in

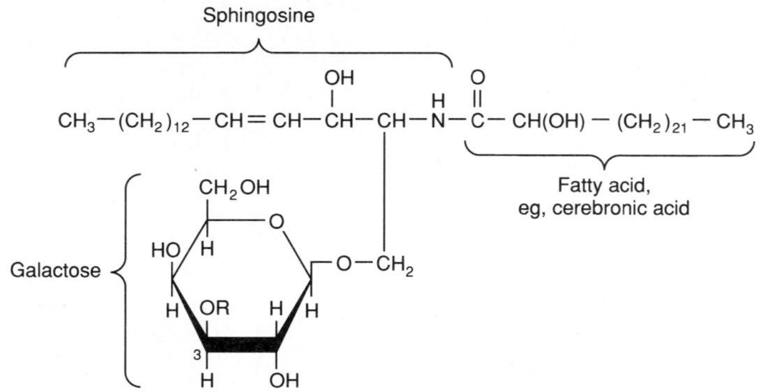

Figure 16–19. Structure of galactosylceramide (galactocerebroside, R = H), and sulfogalactosylceramide (a sulfatide, R = SO_4^{2-}).

Ceramide–Glucose–Galactose–N-Acetylgalactosamine–Galactose
(Acyl-
sphingo- |
sine)   NeuAc

or

Cer–Glc–Gal–GalNAc–Gal
                    |
                  NeuAc

Figure 16–20. G_{M1} ganglioside, a monosialoganglioside.

cholesterol). If the compound has one or more hydroxyl groups and no carbonyl or carboxyl groups, it is a **sterol,** and the name terminates in -ol.

Because of Asymmetry in the Steroid Molecule Many Stereoisomers are Possible

Each of the 6-carbon rings of the steroid nucleus is capable of existing in the 3-dimensional conformation either of a "chair" or a "boat" (Fig 16–22).

In naturally occurring steroids, virtually all the rings are in the "chair" form, which is the more stable conformation. With respect to each other, the rings can be either *cis* or *trans* (Fig 16–23).

The junction between the A and B rings can be *cis* or *trans* in naturally occurring steroids. That between B and C is *trans* and the C/D junction is *trans* except in cardiac glycosides and toad poisons. Bonds attaching substituent groups above the plane of the rings are shown with bold solid lines (β), whereas those bonds attaching groups below are indicated with broken lines (α). The A ring of a 5α steroid is always *trans* to the B ring, whereas it is *cis* in a 5β steroid. The methyl groups attached to C_{10} and C_{13} are invariably in the β configuration.

Cholesterol is a Significant Constituent of Many Tissues

Cholesterol is widely distributed in all cells of the body, but particularly in nervous tissue. It is a major constituent of the plasma membrane and of plasma lipoproteins. It is often found in combination with fatty acids as cholesteryl ester. It occurs in animal fats but not in plant fats. Cholesterol is designated as 3-hydroxy-5,6-cholestene (Fig 16–24).

Ergosterol Is a Precursor of Vitamin D

Ergosterol occurs in plants and yeast and is important as a precursor of vitamin D (Fig 16–25). When irradiated with ultraviolet light, it acquires antirachitic properties consequent to the opening of ring B (see Fig 54–6).

Coprosterol Is Found in Feces

Coprosterol (coprostanol) occurs in feces as a result of the reduction of the double bond of cholesterol between C_5 and C_6 by bacteria in the intestine.

Polyprenoids Share the Same Parent Compound as Cholesterol

Although not steroids, these compounds are related because they are synthesized, like cholesterol (see Fig 28–3), from 5-carbon isoprene units (Fig 16–26). They include **ubiquinone** (see p 116), a member of the respiratory chain in mitochondria, and the long-chain alcohol **dolichol** (Fig 16–27), which takes part in glycoprotein synthesis by transferring carbohydrate residues to asparagine residues of the polypeptide (see Chapter 57). Plant-derived isoprenoid compounds include rubber, camphor, the fat-soluble vitamins A, D, E, and K, and β-carotene (provitamin A).

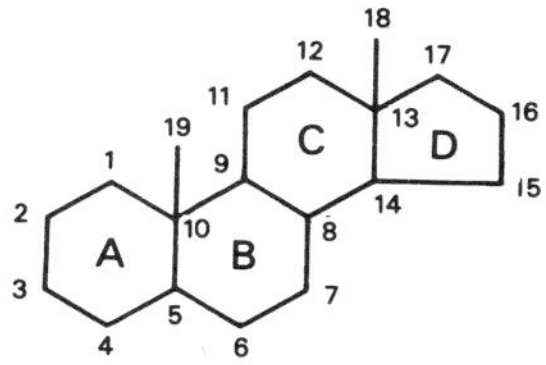

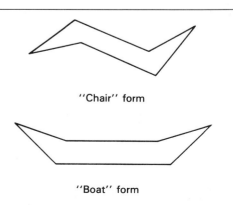

"Chair" form

"Boat" form

Figure 16–22. Conformations of stereoisomers.

Figure 16–21. The steroid nucleus.

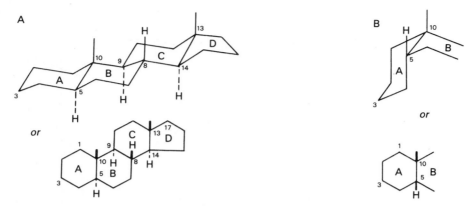

Figure 16–23. Generalized steroid nucleus, showing (A) an all-*trans* configuration between adjacent rings and (B) a *cis* configuration between rings A and B.

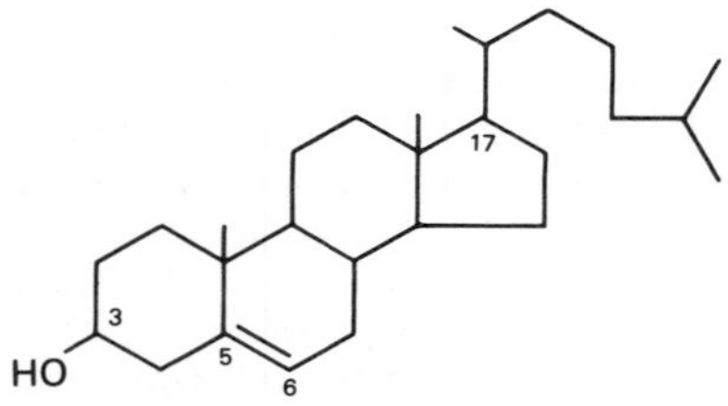

Figure 16–24. Cholesterol.

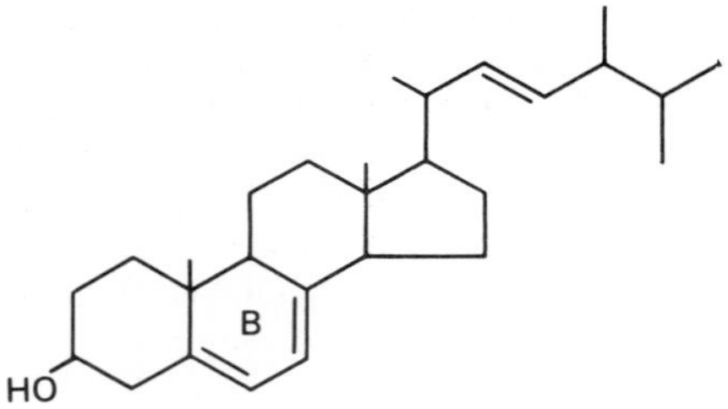

Figure 16–25. Ergosterol.

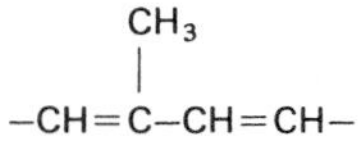

$$CH_3$$
$$-CH=C-CH=CH-$$

Figure 16–26. Isoprene unit.

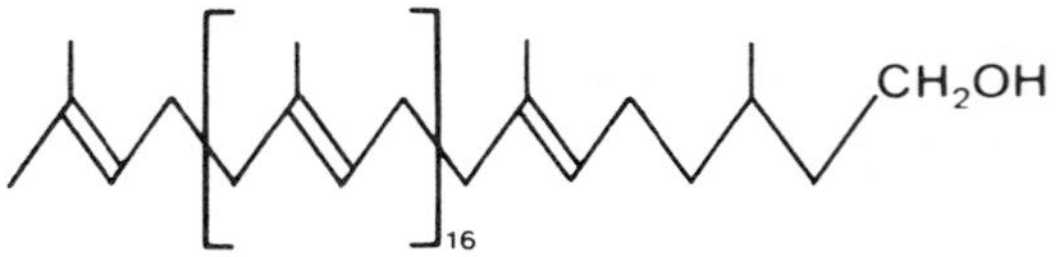

$$CH_2OH$$

Figure 16–27. Dolichol—a C_{95} alcohol.

(1) Initiation: Production of $R^{\cdot}$ from a precursor.

ROOH + metal$^{(n)+}$ → ROO$^{\cdot}$ + metal$^{(n-1)+}$ + H+
X$^{\cdot}$ + RH → R$^{\cdot}$ + XH

(2) Propagation:

R$^{\cdot}$ + O$_2$ → ROO$^{\cdot}$
ROO$^{\cdot}$ + RH → ROOH + R$^{\cdot}$, etc.

(3) Termination:

ROO$^{\cdot}$ + ROO$^{\cdot}$ → ROOR + O$_2$
ROO$^{\cdot}$ + R$^{\cdot}$ → ROOR
R$^{\cdot}$ + R$^{\cdot}$ → RR

LIPID PEROXIDATION IS A SOURCE OF FREE RADICALS IN VIVO

Peroxidation (**autooxidation**) of lipids exposed to oxygen is responsible not only for deterioration of foods (**rancidity**) but also for damage to tissues in vivo, where it may be a cause of cancer, inflammatory diseases, atherosclerosis, aging, etc. The deleterious effects are initiated by free radicals (ROO$^{\cdot}$, RO$^{\cdot}$, OH$^{\cdot}$) produced during peroxide formation from fatty acids containing methylene-interrupted double bonds, ie, those found in the naturally occurring polyunsaturated fatty acids (Fig 16–28). Lipid peroxidation is a chain reaction providing a continuous supply of free radicals that initiate further peroxidation. The whole process can be depicted as follows:

Since the molecular precursor for the initiation process is generally the hydroperoxide product ROOH, lipid peroxidation is a branching chain reaction with potentially devastating effects. To control and reduce lipid peroxidation both humans and nature invoke the use of **antioxidants.** Propyl gallate, butylated hydroxyanisole (BHA), and butylated hydroxytoluene (BHT) are antioxidants used as food additives. Naturally occurring antioxidants include vitamin E (tocopherol), which is lipid soluble, and urate and vitamin C, which are water soluble. β-Carotene is an antioxidant at low P_{O_2}. Antioxidants fall into 2 classes: (1) preventive antioxidants, which reduce the rate of chain

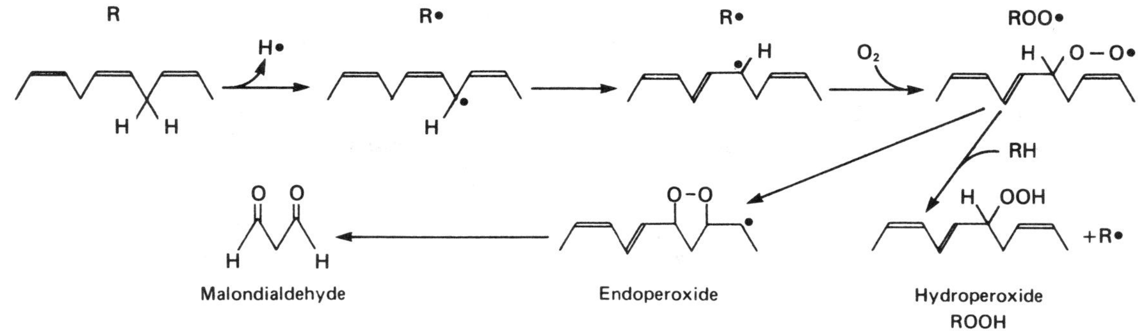

Figure 16–28. Lipid peroxidation. The reaction is initiated by light or by metal ions. Malondialdehyde is only formed by fatty acids with 3 or more double bonds and is used as a measure of lipid peroxidation together with ethane from the terminal 2-carbon of ω3 fatty acids and pentane from the terminal 5-carbon of ω6 fatty acids.

initiation, and (2) chain-breaking antioxidants, which interfere with chain propagation. Preventive antioxidants include catalase and other peroxidases that react with ROOH, and chelators of metal ions such as DTPA (diethylenetriaminepentaacetate) and EDTA (ethylenediaminetetraacetate). Chain-breaking antioxidants are often phenols or aromatic amines. In vivo, the principal chain-breaking antioxidants are superoxide dismutase (see p 114), which acts in the aqueous phase to trap superoxide free radicals ($O_2^{-\cdot}$); perhaps urate; and vitamin E, which acts in the lipid phase to trap ROO· radicals.

Peroxidation is also catalyzed in vivo by heme compounds and by **lipoxygenases** found in platelets and leukocytes, etc.

CHROMATOGRAPHIC METHODS ARE USED FOR SEPARATING & IDENTIFYING LIPIDS IN BIOLOGIC MATERIAL

The older methods of separation and identification of lipids, based on classic chemical procedures of crystallization, distillation, and solvent extraction, have now been largely supplanted by chromatographic procedures. Particularly useful for the separation of the various lipid classes is **thin-layer chromatography** (TLC) and for the separation of the individual fatty acids, **gas-liquid chromatography** (GLC; Fig 16–29). Before these techniques are applied to wet tissues, the lipids are extracted by a solvent system based usually on a mixture of chloroform and methanol (2 : 1).

Gas-liquid chromatography involves the physical separation of a moving gas phase by adsorption onto a stationary phase consisting of an inert solid such as silica gel or inert granules of ground firebrick coated with a nonvolatile liquid (eg, lubricating grease or silicone oils). In practice, a glass or metal column is packed with the inert solid, and a mixture of the methyl esters of fatty acids is evaporated at one end of the column, the entire length of which is kept at tem-

peratures of 170–225 °C (Fig 16–29). A constantly flowing stream of an inert gas such as argon or helium keeps the volatilized esters moving through the column. As with other types of chromatography, separation of the vaporized fatty acid esters is dependent upon the different affinities of the components of the gas mixture for the stationary phase. Gases that are strongly attracted to the stationary phase move through the column at a slower rate and therefore emerge at the end of the column later than those that are relatively less attracted. As the individual fatty acid esters emerge from the column, they are detected by physical or chemical means and recorded automatically as a series of peaks that appear at different times according to the tendency of each fatty ester to be retained by the stationary phase (Fig 16–29). The area under each peak is proportionate to the concentration of a particular component of the mixture. The identity of each component is established by comparison with the gas chromatographic pattern of

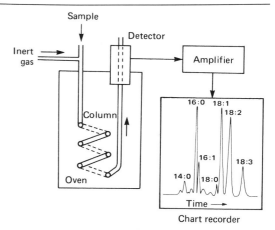

Figure 16–29. Diagrammatic representation of a gas-liquid chromatography apparatus and the separation of long-chain fatty acids (as methyl esters). (A section of the record of a chromatogram is shown at right.)

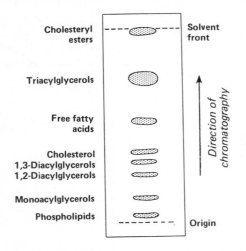

Figure 16–30. Separation of major lipid classes by thin-layer chromatography. A suitable solvent system for the above would be hexane-diethyl ether-formic acid (80:20:2 v/v/v).

a related standard mixture of known composition. A detector of radioactivity may also be incorporated into the gas stream, together with the mass detector. Thus, a measure of the specific radioactivity of each component separated is obtained.

The advantages of gas-liquid chromatography are its extreme **sensitivity,** which allows very small quantities of mixtures to be separated, and the fact that the columns **may be used repeatedly.** Application of the technique has shown that natural fats contain a wide variety of hitherto undetected fatty acids.

Thin-layer chromatography (TLC) is carried out on glass plates coated with a thin slurry of adsorbent, usually silica gel. This is allowed to dry and is then heated in an oven at a standard temperature and for a standard time. After cooling, the "activated" plate is "spotted" with the lipid mixture contained in a suitable solvent. The solvent is evaporated, the edge of the plate nearest the spots is dipped in an appropriate solvent mixture, and the plate is run inside a closed tank until the solvent front arrives near the top edge of the plate. The plate is dried of solvent, and the position of

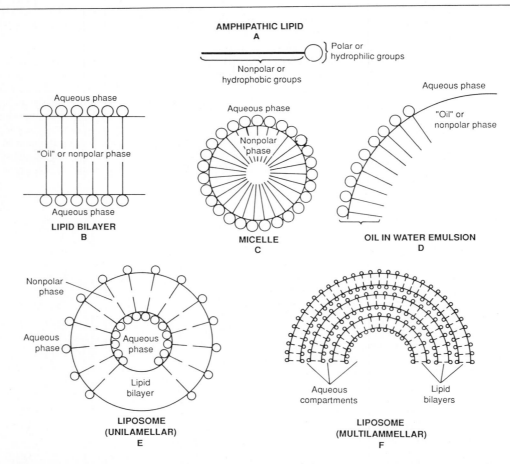

Figure 16–31. Formation of lipid membranes, micelles, emulsions, and liposomes from amphipathic lipids, eg, phospholipids.

the spots is determined by "charring" (spraying with sulfuric acid followed by heating) or by fluorescence (with dichlorofluorescein) or by reacting with iodine vapor (Fig 16–30).

AMPHIPATHIC LIPIDS SELF-ORIENT AT OIL: WATER INTERFACES

They Form Membranes, Micelles, Liposomes, & Emulsions

In general, lipids are insoluble in water, since they contain a predominance of nonpolar (hydrocarbon) groups. However, fatty acids, phospholipids, sphingolipids, bile salts, and, to a lesser extent, cholesterol contain polar groups. Therefore, part of the molecule is **hydrophobic,** or water-insoluble, and part is **hydrophilic,** or water-soluble. Such molecules are described as **amphipathic** (Fig 16–31). They become oriented at oil-water interfaces with the **polar group in the water phase** and the **nonpolar group in the oil phase.** A **bilayer** of such polar lipids has been regarded as a basic structure in biologic membranes (see Chapter 42). When a critical concentration of polar lipids is present in an aqueous medium, they form **micelles.** Aggregations of bile salts into micelles and **liposomes,** and the formation of mixed micelles with the products of fat digestion are important in facilitating absorption of lipids from the intestine. Liposomes may be formed by sonicating an amphipathic lipid in an aqueous medium. They consist of spheres of lipid bilayers that enclose part of the aqueous medium. They are of potential clinical use, particularly when combined with tissue-specific antibodies, as carriers of drugs in the circulation, targeted to specific organs. **Emulsions** are much larger particles, formed usually by nonpolar lipids in an aqueous medium. These are stabilized by emulsifying agents such as polar lipids (eg, lecithin), which form a surface layer separating the main bulk of the nonpolar material from the aqueous phase (Fig 16–31).

REFERENCES

Christie WW: *Lipid Analysis,* 2nd ed. Pergamon Press, 1982.

Cotgreave IA et al: Host biochemical defence mechanisms against prooxidants. *Annu Rev Pharmacol Toxicol* 1988; **28:**189.

Frankel EN: Chemistry of free radical and singlet oxidation of lipids. *Prog Lipid Res* 1985;**23:**197.

Gunstone FD, Harwood JL, Padley FB: *The Lipid Handbook.* Chapman & Hall, 1986.

Gurr AI, James AT: *Lipid Biochemistry: An Introduction,* 3rd ed. Wiley, 1980.

Hawthorne JN, Ansell GB (editors): *Phospholipids.* Elsevier, 1982.

Johnson AR, Davenport JB: *Biochemistry and Methodology of Lipids.* Wiley, 1971.

Vance DE, Vance JE (editors): *Biochemistry of Lipids and Membranes.* Benjamin/Cummings, 1985.

17

Overview of Intermediary Metabolism

Peter A. Mayes, PhD, DSc

INTRODUCTION

The fate of dietary components after digestion and absorption constitutes intermediary metabolism. Thus, it encompasses a wide field that not only seeks to describe the metabolic pathways taken by individual molecules but also attempts to understand their interrelationships and the mechanisms that regulate the flow of metabolites through the pathways. Metabolic pathways fall into 3 categories (Fig 17–1): (1) **Anabolic pathways** are those involved in the synthesis of the compounds constituting the body's structure and machinery. Protein synthesis is such a pathway. The free energy required for these processes comes from the next category. (2) **Catabolic pathways** involve oxidative processes that release free energy, usually in the form of high-energy phosphate or reducing equivalents, eg, the respiratory chain and oxidative phosphorylation. (3) **Amphibolic pathways** have more than one function and occur at the "crossroads" of metabolism, acting as links between the anabolic and catabolic pathways, eg, the citric acid cycle.

BIOMEDICAL IMPORTANCE

A knowledge of metabolism in the normal animal is a prerequisite to a sound understanding of many diseases. Normal metabolism includes the variations and adaptation in metabolism due to periods of starvation, exercise, pregnancy, and lactation. Abnormal metabolism results from, for example, nutritional deficiency, enzyme deficiency, or abnormal secretion of hormones. An important example of a disease resulting from abnormal metabolism (a "metabolic disease") is **diabetes mellitus.**

THE BASIC METABOLIC PATHWAYS PROCESS THE MAJOR PRODUCTS OF DIGESTION

The nature of the diet sets the basic pattern of metabolism in the tissues. Mammals such as humans need to process the absorbed products of digestion of dietary carbohydrate, lipid, and protein. These are mainly

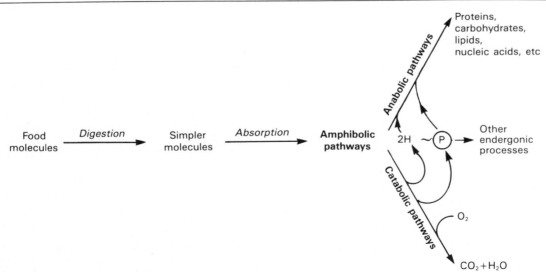

Figure 17–1. The 3 major categories of metabolic pathways. Catabolic pathways release free energy in the form of reducing equivalents (2H) or high-energy phosphate (~P) to power the anabolic pathways. Amphibolic pathways act as links between the other 2 categories of pathways.

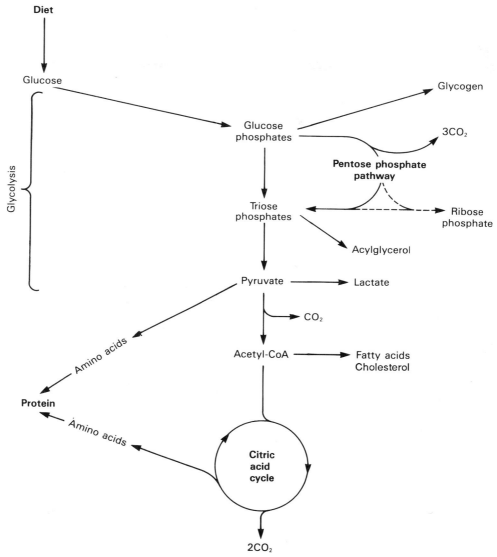

Figure 17-2. Overview of carbohydrate metabolism showing the major end products.

glucose, fatty acids and glycerol, and amino acids, respectively. In ruminants (and to a lesser extent other herbivores), cellulose in the diet is digested by symbiotic microorganisms to lower fatty acids (acetic, propionic, butyric), and tissue metabolism in these animals is adapted to utilize lower fatty acids as major substrates. All of these products of digestion are processed by their respective metabolic pathways to **a common product, acetyl CoA,** which is then completely oxidized by the **citric acid cycle** (see Fig 14-1 and Figs 17-2, 17-3, and 17-4).

Carbohydrate Metabolism Is Concerned With the Fate of Glucose (See Fig 17-2.)

Glucose is metabolized to pyruvate and lactate in all mammalian cells by the pathway of **glycolysis.**

Phosphorylation is necessary for glucose to enter this pathway. Glycolysis can occur in the absence of oxygen (anaerobic), when the end product is lactate only. Tissues that can utilize oxygen (aerobic) are able to metabolize pyruvate to acetyl-CoA, which can enter the **citric acid cycle** for complete oxidation to CO_2 and H_2O, with liberation of much free energy as ATP in the process of **oxidative phosphorylation** (see Fig 18-2). Thus, glucose is a major fuel of many tissues. But it (and certain of its metabolites) also takes part in other processes, as follows: (1) Conversion to its storage polymer, **glycogen,** particularly in skeletal muscle and liver. (2) The **pentose phosphate pathway,** which arises from intermediates of glycolysis. It is a source of reducing equivalents (2H) for biosynthesis—eg, of fatty acids—and it is also the source of ribose, which is important for nucleotide and nucleic acid formation. (3) Triose phosphate gives rise to the **glycerol moiety**

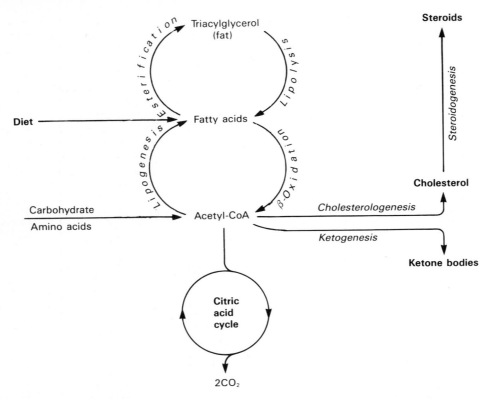

Figure 17–3. Overview of fatty acid metabolism showing the major end products. Ketone bodies comprise the substances acetoacetate, 3-hydroxybutyrate, and acetone.

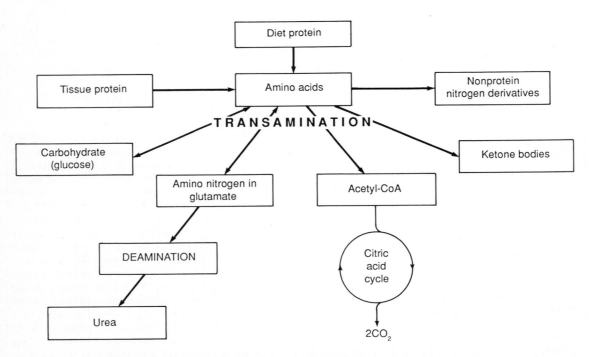

Figure 17–4. Overview of amino acid metabolism showing the major end products.

of acylglycerols (fat). (4) Pyruvate and intermediates of the citric acid cycle provide the carbon skeletons for the synthesis of **amino acids,** and acetyl-CoA is the building block for long-chain **fatty acids** and **cholesterol,** the precursor of all steroids synthesized in the body.

Lipid Metabolism Is Concerned Mainly With Fatty Acids & Cholesterol (See Fig 17–3.)

The source of long-chain fatty acids is either de novo synthesis from acetyl-CoA derived from carbohydrate, or from dietary lipid. In the tissues, fatty acids may be oxidized to acetyl-CoA (**β-oxidation**) or esterified to acylglycerols, where as triacylglycerol (fat) they constitute the body's main caloric reserve. Acetyl-CoA formed by β-oxidation has several important fates.

(1) As in the case of acetyl-CoA derived from carbohydrate, it is **oxidized completely** to CO_2 + H_2O via the **citric acid cycle.** Fatty acids yield considerable energy both in β-oxidation and in the citric acid cycle and are therefore very effective tissue fuels.

(2) It is a source of the carbon atoms in **cholesterol** and other **steroids.**

(3) In the liver it forms acetoacetate, the parent **ketone body.** Ketone bodies are alternative water-soluble tissue fuels, which become important sources of energy under certain conditions (eg, starvation).

Much of Amino Acid Metabolism Involves Transamination (See Fig 17–4.)

The amino acids are necessary for protein synthesis. Some must be supplied specifically in the diet (the **essential amino acids**), since the tissues are unable to synthesize them. The remainder, or **nonessential amino acids,** are also supplied in the diet, but they can be formed from intermediates by **transamination** using the amino nitrogen from other surplus amino acids. After **deamination,** excess amino nitrogen is removed as **urea,** and the carbon skeletons that remain after transamination (1) are oxidized to CO_2 via the citric acid cycle, (2) form glucose (gluconeogenesis), or (3) form ketone bodies.

In addition to their requirement for protein synthesis, the amino acids are also the precursors of many other important compounds, eg, purines, pyrimidines, and hormones such as epinephrine and thyroxine.

Metabolic Pathways May Be Studied at Different Levels of Organization

So far we have looked at metabolism as it occurs in the whole organism. The location and integration of metabolic pathways is revealed by studies at lower levels of organization, namely: (1) **At the tissue and organ level**—the nature of the substrates entering and metabolites leaving tissues and organs is defined, and their overall fate is described. (2) **At the subcellular**

level—each cell organelle (eg, the mitochondrion) or compartment (eg, the cytosol) carries out specific biochemical roles that form part of a subcellular pattern of metabolic pathways.

At the Tissue & Organ Level the Blood Circulation Integrates Metabolism

Amino acids resulting from the digestion of dietary protein and **glucose** resulting from the digestion of carbohydrate share a common route of absorption via the **hepatic portal vein.** This ensures that both of these metabolites and other water-soluble products of digestion are initially directed to the **liver** (Fig 17–5). The liver has the primary metabolic function of regulating the blood concentration of most metabolites, particularly glucose and amino acids. In the case of glucose this is achieved by taking up excess glucose and converting it to glycogen (**glycogenesis**) or to fat (**lipogenesis**). Between meals, it can draw upon its glycogen stores to replenish glucose in the blood (**glycogenolysis**) or, in company with the kidney, to convert non-carbohydrate metabolites such as lactate, glycerol, and amino acids to glucose (**gluconeogenesis**). The maintenance of an adequate concentration of blood glucose is vital for certain tissues in which it is an obligatory fuel, eg, brain and erythrocytes. The liver also has the task of **synthesizing the major plasma proteins** (eg, albumin) and of **deaminating amino acids** that are in excess of requirements, with formation of urea, which is transported via the blood to the kidney and excreted.

Skeletal muscle utilizes glucose as a fuel, forming both lactate and CO_2. It stores glycogen as a fuel for its use in muscular contraction and synthesizes muscle protein from plasma amino acids. Muscle accounts for approximately 50% of body mass and consequently represents a considerable store of protein that can be drawn upon to supply plasma amino acids, particularly during dietary shortage.

Lipids (Fig 17–6) upon digestion form monoacylglycerols and fatty acids. These are recombined in the intestinal cells with protein and secreted initially into the lymphatic system and then into the circulation as a **lipoprotein** known as a **chylomicron.** All hydrophobic, lipid-soluble products of digestion (eg, cholesterol) form lipoproteins, which facilitates their transport between tissues in an aqueous environment—the plasma. Unlike glucose and amino acids, chylomicron triacylglycerol is not taken up by the liver. It is metabolized by extrahepatic tissues possessing the enzyme **lipoprotein lipase,** which hydrolyzes the triacylglycerol, releasing fatty acids that are incorporated into tissue lipids or oxidized as fuel. The other major source of long-chain fatty acid is synthesis (**lipogenesis**) from carbohydrate, mainly in adipose tissue and the liver.

Adipose tissue triacylglycerol is the main fuel reserve of the body. Subsequent to its hydrolysis (**lipolysis**), fatty acids are released into the circulation as free fatty acids. These are taken up by most tissues

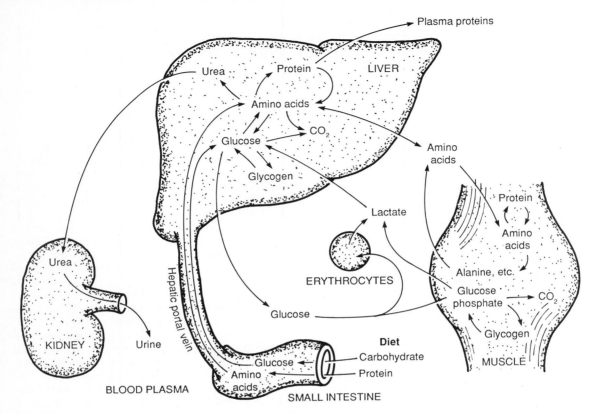

Figure 17-5. Transport and fate of major carbohydrate and amino acid substrates and metabolites. Note that there is little free glucose in muscle, since it is rapidly phosphorylated upon entry.

(but not brain or erythrocytes) and esterified to acylglycerols or oxidized as a major fuel to CO_2. Two pathways of additional importance occur in liver: (1) Surplus triacylglycerol, arising from both lipogenesis and free fatty acids, is secreted into the circulation as **very low density lipoprotein** (VLDL). This triacylglycerol undergoes a fate similar to that of chylomicrons. (2) Partial oxidation of free fatty acid leads to ketone body production (**ketogenesis**). Ketone bodies are transported to extrahepatic tissues where they act as another major fuel source.

At the Subcellular Level, Glycolysis Is Found in the Cytosol & the Citric Acid Cycle in the Mitochondria

A summary of the main biochemical functions of the subcellular components and organelles of the cell is given in Table 2–4. However, most cells are specialized in their functions and tend to emphasize certain metabolic pathways and relegate others. Fig 17–7 depicts the major metabolic pathways and their integration in a hepatic parenchymal cell, with special emphasis on their intracellular location.

The central role of the **mitochondrion** is immedi-

ately apparent, since it acts as the focus and crossroad of carbohydrate, lipid, and amino acid metabolism. In particular, it houses the enzymes of the citric acid cycle, of the respiratory chain and ATP synthase, of β-oxidation of fatty acids, and of ketone body production. In addition, it is the collecting point for the carbon skeletons of amino acids after transamination and for providing these skeletons for the synthesis of the nonessential amino acids.

Glycolysis, the pentose phosphate pathway, and fatty acid synthesis all occur in the **cytosol.** It will be noticed that in gluconeogenesis, even substances such as lactate and pyruvate that are formed in the cytosol **must enter the mitochondrion and form oxaloacetate** before conversion to glucose.

The membranes of the **endoplasmic reticulum** contain the enzyme system for acylglycerol synthesis, and the ribosomes are responsible for protein synthesis.

It will be appreciated that the transport of metabolites of varying size, charge, and solubility through the membranes separating organelles involves complex mechanisms. Some have been discussed in relation to the mitochondrial membrane (see Chapter 14), and others will be discussed in succeeding chapters.

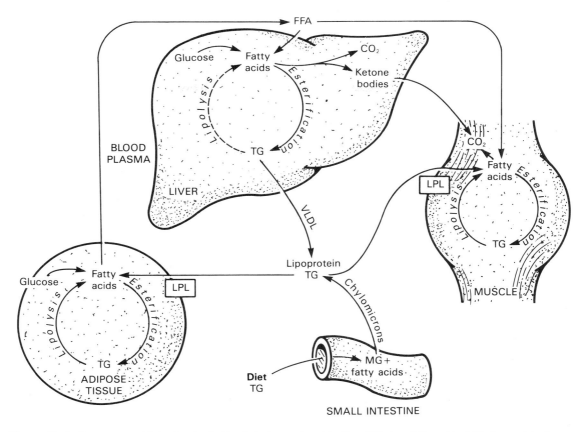

Figure 17–6. Transport and fate of major lipid substrates and metabolites. FFA, free fatty acids; LPL, lipoprotein lipase; MG, monoacylglycerol; TG, triacylglycerol; VLDL, very low density lipoprotein.

THE FLUX OF METABOLITES IN METABOLIC PATHWAYS MUST BE REGULATED IN A CONCERTED MANNER

Regulation of the overall flux along a metabolic pathway is often concerned with the control of only one or perhaps 2 key reactions in the pathway, catalyzed by **"regulatory enzymes."** The physicochemical factors that control the rate of an enzyme-catalyzed reaction, eg, substrate concentration (see Chapter 10), are of primary importance in the control of the overall rate of a metabolic pathway. However, temperature and pH, factors that can influence enzyme activity, are held constant in warm-blooded vertebrates and have little regulatory significance. (Note, however, the variation in pH in the gastrointestinal tract and its effects on digestion [see Chapter 56].)

Nonequilibrium Reactions Give Direction and Are Potential Control Points in Metabolic Pathways

In a reaction at equilibrium, the forward and reverse reactions take place at equal rates, and there is therefore no net flux in either direction. Many reactions in metabolic pathways are of this type, ie, "equilibrium reactions":

$$A \leftrightarrow B \leftrightarrow C \leftrightarrow D$$

In vivo, under "steady-state" conditions, there would probably be a net flux from left to right due to continuous supply of A and continuous removal of D. Such a pathway could function, but there would be little scope for control of the flux via regulation of enzyme activity, since an increase in activity would only serve to speed up attainment of the equilibrium.

In practice, there are invariably one or more "nonequilibrium" type reactions in a metabolic pathway, where the reactants are present in concentrations that are far from equilibrium. In attempting to reach equilibrium, large losses of free energy occur as heat, which cannot be reutilized, making this type of reaction essentially nonreversible, eg,

Heat

$$A \longleftrightarrow B \overset{\nearrow}{\longrightarrow} C \longleftrightarrow D$$

Nonequilibrium reaction

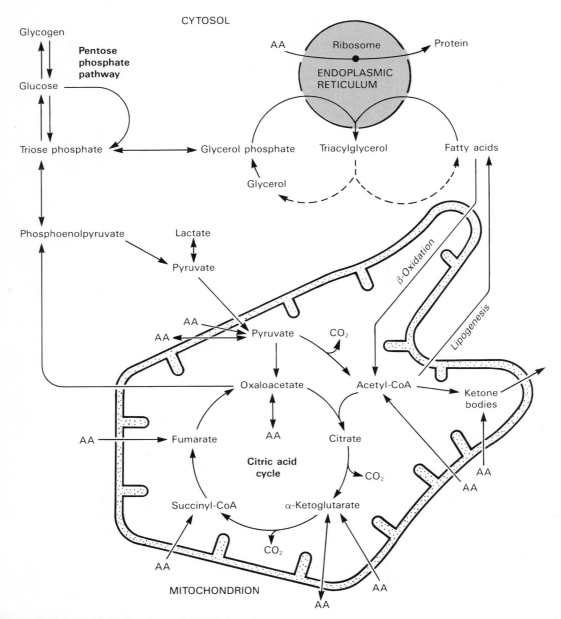

CYTOSOL

Figure 17–7. Intracellular location and integration of major metabolic pathways in a liver parenchymal cell. AA →, metabolism of one or more essential amino acids; AA ↔, metabolism of one or more nonessential amino acids.

Such a pathway has both flow and direction but would exhaust itself if control were not exerted. The enzymes catalyzing nonequilibrium reactions are usually low in concentration and are subject to other controlling mechanisms. This is similar to the opening and shutting of a "one-way" valve, making it possible to control the net flow.

The Flux-Generating Reaction Is the First Reaction In a Pathway That Is Saturated With Substrate

It may be identified as a nonequilibrium reaction in which the K_m of the enzyme is considerably lower than the normal substrate concentration. The first reaction in glycolysis catalyzed by **hexokinase** (see Fig 19–2) is such a flux-generating step.

ALLOSTERIC AND HORMONAL MECHANISMS ARE IMPORTANT IN THE METABOLIC CONTROL OF ENZYME-CATALYZED REACTIONS

A hypothetical metabolic pathway, A,B,C,D, is shown in Fig 17–8, in which reactions A ↔ B and C ↔ D are equilibrium reactions and B → C is a non-

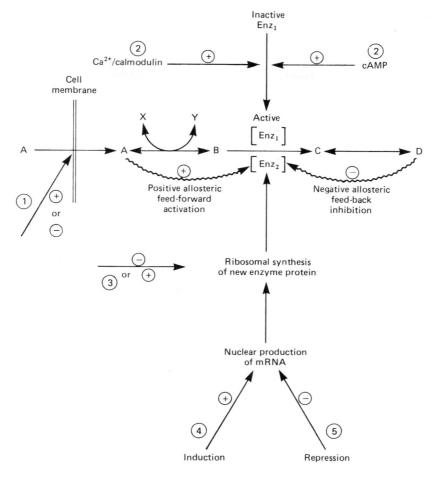

Figure 17–8. Mechanisms of control of an enzyme-catalyzed reaction. Circled numbers indicate possible sites of action of hormones. 1, Alteration of membrane permeability; 2, conversion of an inactive to an active enzyme; 3, alteration of the rate of translation of mRNA at the ribosomal level; 4, induction of new mRNA formation; and 5, repression of mRNA formation.

equilibrium reaction. The flux through such a pathway can be regulated by the availability of substrate A. This depends on its supply from the blood, which in turn depends on adequate food intake to the gut or on certain key reactions that maintain and release major substrates to the blood, eg, the flux-generating reactions catalyzed by phosphorylase in liver (see Fig 20–1), which provides blood glucose, and hormone-sensitive lipase in adipose tissue (see Fig 27–8), which supplies free fatty acids. It also depends on the ability of substrate A to permeate the cell membrane. The flux will also be determined by the efficiency of removal of the end product D and on the availability of cosubstrate or cofactors represented by X and Y.

Enzymes catalyzing nonequilibrium reactions are often allosteric proteins subject to the rapid action of "feed-back" or "feed-forward" control by **allosteric modifiers,** often in immediate response to the needs of the cell (see Chapter 10). Frequently the product of a biosynthetic pathway, eg, long-chain acyl-CoA, will

inhibit the enzyme catalyzing the first reaction in the pathway, eg, acetyl-CoA carboxylase. Other control mechanisms depend on the action of **hormones** responding to the needs of the body as a whole. These act by several different mechanisms (see Chapter 44). One is **covalent modification** of the enzyme by **phosphorylation** and **dephosphorylation.** This is **rapid** and is often mediated through the formation of the second messenger **cAMP,** which in turn causes the conversion of an inactive enzyme into an active enzyme. This change is brought about via the activity of a **cAMP-dependent protein kinase** that phosphorylates the enzyme or of specific **phosphatases** that dephosphorylate the enzyme. The active form of the enzyme can be either the phosphorylated enzyme, as in enzymes catalyzing **degradative** pathways (eg, phosphorylase a), or the dephosphorylated enzyme, as in enzymes catalyzing **synthetic** processes (eg, glycogen synthase a).

Some regulatory enzymes can be phosphorylated

without the mediation of cAMP and cAMP-dependent protein kinase. These enzymes respond to other metabolic signals such as the [ATP]/[ADP] ratio, eg, pyruvate dehydrogenase (Fig 19–6), or Ca^{2+}/calmodulin-dependent protein kinase, eg, phosphorylase kinase (see Fig 20–6).

The synthesis of rate-controlling enzymes can be affected by hormones. Because this involves new protein synthesis, it is not a rapid change but is often a response to a change in nutritional state. Hormones can act as inducers or repressors of mRNA formation in the nucleus or as stimulators of the translation stage of protein synthesis at the ribosomal level (see Chapters 41 and 44).

A significant feature that aids metabolic control is that pathways catalyzing degradation of a substance are not the simple reversal of synthesis. Usually two entirely separate pathways are involved, allowing separate control of each, eg, glucogen synthesis and breakdown (Fig 20–1).

REFERENCES

Cohen P: *Control of Enzyme Activity,* 2nd ed. Chapman & Hall, 1983.

Hue L, Van de Werve G (editors): *Short-Term Regulation of Liver Metabolism.* Elsevier/North Holland, 1981.

Newsholme EA, Crabtree B: Flux-generating and regulatory steps in metabolic control. *Trends Biochem Sci* 1981;**6:**53.

Newsholme EA, Start C: *Regulation in Metabolism.* Wiley, 1973.

The Citric Acid Cycle: The Catabolism of Acetyl-CoA

18

Peter A. Mayes, PhD, DSc

INTRODUCTION

The citric acid cycle (Krebs cycle, tricarboxylic acid cycle) is a series of reactions in mitochondria that bring about the catabolism of acetyl residues, liberating hydrogen equivalents, which, upon oxidation, lead to the release of most of the free energy of tissue fuels. The acetyl residues are in the form of **acetyl-CoA** (CH_3–$CO\sim S$–CoA, active acetate), an ester of coenzyme A. CoA contains the vitamin pantothenic acid.

BIOMEDICAL IMPORTANCE

The major function of the citric acid cycle is to act as the final common pathway for the oxidation of carbohydrate, lipids, and protein. This is because glucose, fatty acids, and many amino acids are all metabolized to acetyl-CoA or intermediates of the cycle. It also plays a major role in gluconeogenesis, transamination, deamination, and lipogenesis. Several of these processes are carried out in many tissues, but the liver is the only tissue in which all occur to a significant extent. The repercussions are therefore profound when, for example, large numbers of hepatic cells are damaged or replaced by connective tissue, as in acute **hepatitis** and **cirrhosis,** respectively. A mute testimony to the vital importance of the citric acid cycle is the fact that very few if any genetic abnormalities of its enzymes have been reported in humans; such abnormalities are presumably incompatible with normal development.

THE CITRIC ACID CYCLE MAKES AVAILABLE THE SUBSTRATE FOR THE RESPIRATORY CHAIN

Essentially, the cycle comprises the combination of a molecule of acetyl-CoA with the 4-carbon dicarboxylic acid oxaloacetate, resulting in the formation of a **6-carbon tricarboxylic acid, citrate.** There follows a series of reactions in the course of which 2 molecules of CO_2 are released and oxaloacetate is regenerated (Fig 18–1). Since only a small quantity of oxaloacetate is needed to facilitate the conversion of a large quantity of acetyl units to CO_2, oxaloacetate may be considered to play a **catalytic role.**

The citric acid cycle is an integral part of the process by which much of the free energy liberated during the oxidation of carbohydrate, lipids, and amino acids is made available. During the course of oxidation of acetyl-CoA in the cycle, **reducing equivalents** in the form of hydrogen or of electrons are formed as a result of the activity of specific dehydrogenases. These reducing equivalents then enter the respiratory chain, where large amounts of ATP are generated in the process of oxidative phosphorylation (Fig 18–2; see also Chapter 14). This process is **aerobic,** requiring oxygen as the final oxidant of the reducing equivalents. Therefore, **absence (anoxia) or partial deficiency (hypoxia) of O_2 causes total or partial inhibition of the cycle.**

The enzymes of the citric acid cycle are located in the **mitochondrial matrix,** either free or attached to the inner surface of the inner mitochondrial membrane, which facilitates the **transfer of reducing equivalents to the adjacent enzymes of the respiratory chain,** situated also in the inner mitochondrial membrane.

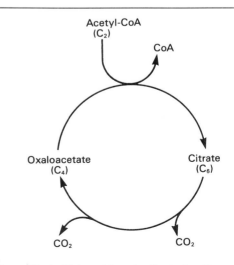

Figure 18–1. Citric acid cycle, illustrating the catalytic role of oxaloacetate.

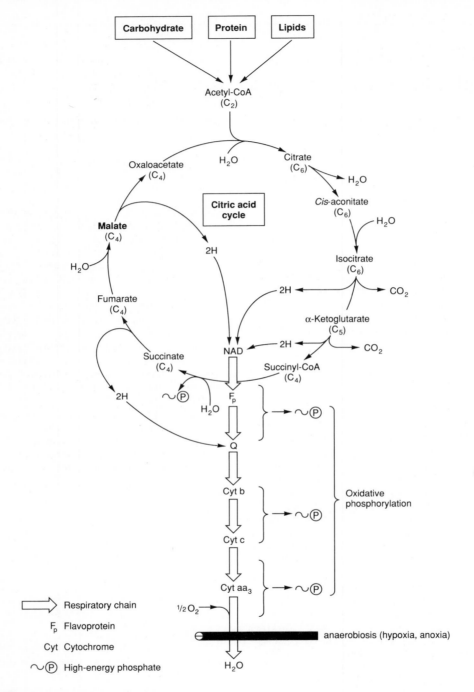

Figure 18–2. The citric acid cycle: the major catabolic pathway for acetyl-CoA in aerobic organisms. Acetyl-CoA, the product of carbohydrate, protein, and lipid catabolism, is taken up into the cycle, together with H_2O, and oxidized to CO_2 with the release of reducing equivalents (2H). Subsequent oxidation of 2H in the respiratory chain leads to coupled phosphorylation of ADP to ATP. For one turn of the cycle, 11 ~ Ⓟ are generated via oxidative phosphorylation and one ~ Ⓟ arises at substrate level from the conversion of succinyl-CoA to succinate.

REACTIONS OF THE CITRIC ACID CYCLE LIBERATE BOTH REDUCING EQUIVALENTS AND CO$_2$ (See Fig 18–3.)*

The initial condensation of acetyl-CoA with oxaloacetate to form citrate is catalyzed by a condensing enzyme, **citrate synthase,** which effects synthesis of a carbon-to-carbon bond between the methyl carbon of acetyl-CoA and the carbonyl carbon of oxaloacetate. The condensation reaction, which forms citryl-CoA, is followed by hydrolysis of the thioester bond of CoA, accompanied by considerable loss of free energy as heat, ensuring that the reaction goes to completion.

$$\text{Acetyl–CoA} + \text{Oxaloacetate} + H_2O \rightarrow \text{Citrate} + \text{CoA}$$

Citrate is converted to isocitrate by the enzyme **aconitase** (aconitate hydratase), which contains iron in the Fe^{2+} state in the form of an iron-sulfur protein (Fe:S). This conversion takes place in 2 steps: dehydration to *cis*-aconitate, some of which remains bound to the enzyme, and rehydration to isocitrate.

$$\text{Citrate} \leftrightarrow \underset{H_2O}{\overset{}{\text{Cis-aconitate}}} \leftrightarrow \underset{H_2O}{\overset{}{\text{Isocitrate}}}$$
$$\text{(enzyme bound)}$$

The reaction is inhibited by **fluoroacetate,** which, in the form of fluoroacetyl-CoA, condenses with oxaloacetate to form fluorocitrate. The latter inhibits aconitase, causing citrate to accumulate.

Experiments using ^{14}C-labeled intermediates indicate that aconitase reacts with citrate in an asymmetric manner, with the result that aconitase always acts on that part of the citrate molecule that is derived from oxaloacetate. This was puzzling, since citric acid appeared to be a symmetric compound. However, it is now realized (when the molecule is viewed in 3 dimensions) that the two –CH$_2$COOH groups are not identical in space with respect to the –OH and –COOH groups. The consequences of the asymmetric action of aconitase may be appreciated by reference to the fate of labeled acetyl-CoA in the citric acid cycle as shown in Fig 18–3. It is possible that *cis*-aconitate may not be an obligatory intermediate between citrate and isocitrate but may in fact be a side branch from the main pathway.

Isocitrate undergoes dehydrogenation in the presence of **isocitrate dehydrogenase** to form oxalosucci-nate. Three different isocitrate dehydrogenases have been described. One, which is NAD$^+$-specific, is found only in mitochondria. The other 2 enzymes are NADP$^+$-specific and are found in the mitochondria and the cytosol, respectively. Respiratory chain-linked oxidation of isocitrate proceeds almost completely through the NAD$^+$-dependent enzyme.

$$\text{Isocitrate} + \text{NAD}^+ \leftrightarrow \underset{\text{(enzyme bound)}}{\text{Oxalosuccinate}} \leftrightarrow$$
$$\alpha\text{-Ketoglutarate} + CO_2 + \text{NADH} + H^+$$

There follows a decarboxylation to α-ketoglutarate, also catalyzed by isocitrate dehydrogenase. Mn^{2+} (or Mg^{2+}) is an important component of the decarboxylation reaction. It would appear that oxalosuccinate remains bound to the enzyme as an intermediate in the overall action.

Next, α-ketoglutarate undergoes **oxidative decarboxylation** in a manner analogous to the oxidative decarboxylation of pyruvate (see Fig 19–5), both substrates being α-keto acids.

$$\alpha\text{-Ketoglutarate} + \text{NAD}^+ + \text{CoA} \rightarrow \text{Succinyl-CoA} + CO_2 + \text{NADH} + H^+$$

The reaction, catalyzed by an **α-ketoglutarate dehydrogenase** complex, also requires identical cofactors—eg, thiamin diphosphate, lipoate, NAD$^+$, FAD, and CoA—and results in the formation of succinyl-CoA, a thioester containing a high-energy bond. The equilibrium of this reaction is so much in favor of succinyl-CoA formation that the reaction must be considered as physiologically unidirectional. As in the case of pyruvate oxidation (see p 169), arsenite inhibits the reaction, causing the substrate, **α-ketoglutarate,** to accumulate.

To continue the cycle, succinyl-CoA is converted to succinate by the enzyme **succinate thiokinase (succinyl-CoA synthetase).**

$$\text{Succinyl-CoA} + P_i + \text{GDP} \leftrightarrow \text{Succinate} + \text{GTP} + \text{CoA}$$

This reaction requires GDP or IDP, which is converted in the presence of inorganic phosphate to either GTP or ITP. This is the only example in the citric acid cycle of the **generation of a high-energy phosphate at the substrate level** and arises because the release of free energy from the oxidative decarboxylation of α-ketoglutarate is sufficient to generate a high-energy bond in addition to the formation of NADH (equivalent to 3 ~℗). By means of a nucleoside diphosphate kinase (see p 104), ATP may be formed from either GTP or ITP,

$$\text{eg, GTP} + \text{ADP} \leftrightarrow \text{GDP} + \text{ATP}$$

An alternative reaction in extrahepatic tissues, which is catalyzed by **succinyl-CoA-acetoacetate-CoA**

*From Circular No. 200 of the Committee of Editors of Biochemical Journals Recommendations (1975): "According to standard biochemical convention, the ending *ate* in, eg, palmitate, denotes any mixture of free acid and the ionized form(s) (according to pH) in which the cations are not specified." The same convention is adopted in this text for all carboxylic acids.

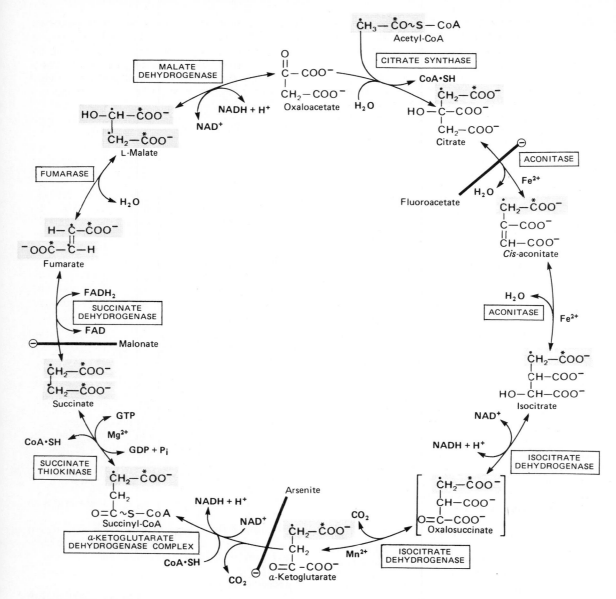

Figure 18–3. The citric acid (Krebs) cycle. Oxidation of NADH and $FADH_2$ in the respiratory chain leads to the generation of ATP via oxidative phosphorylation. In order to follow the passage of acetyl-CoA through the cycle, the 2 carbon atoms of the acetyl radical are shown labeled on the carboxyl carbon (using the designation [·]) and on the methyl carbon (using the designation [•]). Although 2 carbon atoms are lost as CO_2 in one revolution of the cycle, these atoms are not derived from the acetyl-CoA that has immediately entered the cycle but from that portion of the citrate molecule which derived from oxaloacetate. However, on completion of a single turn of the cycle, the oxaloacetate that is regenerated is now labeled, which leads to labeled CO_2 being evolved during the second turn of the cycle. Because succinate is a symmetric compound and because succinate dehydrogenase does not differentiate between its 2 carboxyl groups, "randomization" of label occurs at this step such that all 4 carbon atoms of oxaloacetate appear to be labeled after one turn of the cycle. During gluconeogenesis, some of the label in oxaloacetate is incorporated into glucose and glycogen (see Fig 21–1). For a discussion of the stereochemical aspects of the citric acid cycle, see Greville (1968). The sites of inhibition ($\ominus$) by fluoroacetate, malonate, and arsenite are indicated.

transferase (**thiophorase**), is the conversion of succinyl-CoA to succinate coupled with the conversion of acetoacetate to acetoacetyl-CoA (see p 214). In liver there is also deacylase activity, causing some hydrolysis of succinyl-CoA to succinate plus CoA.

Succinate is metabolized further by undergoing a dehydrogenation followed by the addition of water, and subsequently by a further dehydrogenation which regenerates oxaloacetate.

$$\text{Succinate} + \text{FAD} \leftrightarrow \text{Fumarate} + \text{FADH}_2$$

The first dehydrogenation reaction is catalyzed by **succinate dehydrogenase,** which is bound to the inner surface of the inner mitochondrial membrane unlike the other enzymes of the cycle, which are found in the matrix. It is the only dehydrogenation in the citric acid cycle that involves the **direct transfer of hydrogen from the substrate to a flavoprotein without the participation of NAD$^+$**. The enzyme contains FAD and iron-sulfur (Fe : S) protein. Fumarate is formed as a result of the dehydrogenation. Isotopic experiments have shown that the enzyme is stereospecific for the *trans* hydrogen atoms of the methylene carbons of succinate. Addition of malonate or oxaloacetate inhibits succinate dehydrogenase competitively, resulting in succinate accumulation.

Fumarase (fumarate hydratase) catalyzes the addition of water to fumarate to give malate.

$$\text{Fumarate} + \text{H}_2\text{O} \leftrightarrow \text{L-Malate}$$

In addition to being specific for the L-isomer of malate, fumarase catalyzes the addition of the elements of water to the double bond of fumarate in the *trans* configuration. Malate is converted to oxaloacetate by **malate dehydrogenase,** a reaction requiring NAD$^+$.

$$\text{L-Malate} + \text{NAD}^+ \leftrightarrow \text{Oxaloacetate} + \text{NADH} + \text{H}^+$$

Although the equilibrium of this reaction strongly favors malate, the net flux is toward the direction of oxaloacetate because this compound, together with the other product of the reaction (NADH), is removed continuously in further reactions.

The enzymes of the citric acid cycle, except for the α-ketoglutarate and succinate dehydrogenases, are also found outside the mitochondria. While they may catalyze similar reactions, some of the enzymes, eg, malate dehydrogenase, may not in fact be the same proteins as the mitochondrial enzymes of the same name.

12 ATP MOLECULES ARE FORMED PER TURN OF THE CITRIC ACID CYCLE

As a result of oxidation catalyzed by dehydrogenase enzymes of the citric acid cycle, **3 molecules of NADH** and **one of FADH$_2$** are produced for each mol-

ecule of acetyl-CoA catabolized in one revolution of the cycle. These reducing equivalents are transferred to the respiratory chain in the inner mitochondrial membrane (Fig 18–2). During passage along the chain, reducing equivalents from NADH generate 3 high-energy phosphate bonds by the esterification of ADP to ATP in the process of oxidative phosphorylation (see Chapter 14). However, FADH$_2$ produces only 2 high-energy phosphate bonds because it transfers its reducing power to Q, thus bypassing the first site for oxidative phosphorylation in the respiratory chain (see Fig 14–6). A further high-energy phosphate is generated at the level of the cycle itself (ie, at substrate level) during the conversion of succinyl-CoA to succinate. Thus, **12 ATP molecules are generated for each turn of the cycle** (Table 18–1).

SEVERAL VITAMINS PLAY KEY ROLES IN THE CITRIC ACID CYCLE

Four of the soluble vitamins of the B complex have precise roles in the functioning of the citric acid cycle. They are (1) **Riboflavin** in the form of **flavin adenine dinucleotide (FAD)**, a cofactor in the α-ketoglutarate dehydrogenase complex and in succinate dehydrogenase. (2) **Niacin** in the form of **nicotinamide adenine dinucleotide (NAD)**, the coenzyme for 3 dehydrogenases in the cycle, **isocitrate dehydrogenase, α-ketoglutarate dehydrogenase,** and **malate dehydrogenase.** (3) **Thiamin (B$_1$)**, as thiamin diphosphate, the coenzyme for decarboxylation in the α-ketoglutarate dehydrogenase reaction. (4) **Pantothenic acid** as part of **coenzyme-A,** the cofactor attached to "active" carboxylic acid residues such as acetyl-CoA and succinyl-CoA.

THE CITRIC ACID CYCLE PLAYS A PIVOTAL ROLE IN METABOLISM

Some metabolic pathways end in a constituent of the citric acid cycle, while other pathways originate from the cycle. These pathways concern the processes of

Table 18–1. Generation of ATP by the citric acid cycle.

Reaction Catalyzed By	Method of ~ ℗ Production	ATP Molecules Formed
Isocitrate dehydrogenase	Respiratory chain oxidation of NADH	3
α-Ketoglutarate dehydrogenase	Respiratory chain oxidation of NADH	3
Succinate thiokinase	Oxidation at substrate level	1
Succinate dehydrogenase	Respiratory chain oxidation of FADH$_2$	2
Malate dehydrogenase	Respiratory chain oxidation of NADH	3
		Net 12

gluconeogenesis, transamination, deamination, and fatty acid synthesis. Therefore **the citric acid cycle plays roles in both oxidative and synthetic processes;** ie, it is **amphibolic.** These roles are summarized below.

The Citric Acid Cycle Takes Part in Gluconeogenesis, Transamination, & Deamination

All major members of the cycle, from citrate to oxaloacetate, are potentially glucogenic, since they can give rise to a net production of glucose in the liver or kidney, the organs that contain a complete set of enzymes necessary for gluconeogenesis (see p 179). The key enzyme that facilitates the net transfer out of the cycle into the main pathway of gluconeogenesis is **phosphoenolpyruvate carboxykinase,** which catalyzes the decarboxylation of oxaloacetate to phosphoenolpyruvate, GTP acting as the source of high-energy phosphate (Fig 18–4).

Oxaloacetate + GTP →
$\qquad$ Phosphoenolpyruvate + CO$_2$ + GDP

Net transfer into the cycle occurs as a result of several different reactions. Among the most significant is the formation of oxaloacetate by the carboxylation of pyruvate, catalyzed by **pyruvate carboxylase.**

ATP + CO$_2$ + H$_2$O + Pyruvate →
$\qquad$ Oxaloacetate + ADP + P$_i$

This reaction is considered important in maintaining adequate concentrations of oxaloacetate for the condensation reaction with acetyl-CoA. If acetyl-CoA accumulates, it acts as an allosteric activator of pyruvate carboxylase, thereby ensuring a supply of oxaloacetate. Lactate, an important substrate for gluconeogenesis, enters the cycle via conversion to pyruvate and oxaloacetate.

Transaminase (aminotransferase) reactions produce pyruvate from alanine, oxaloacetate from aspartate, and α-ketoglutarate from glutamate. Because these reactions are reversible, **the cycle also serves as a source of carbon skeletons for the synthesis of nonessential amino acids,** eg,

Aspartate + Pyruvate ↔ Oxaloacetate + Alanine

Glutamate + Pyruvate ↔ α-Ketoglutarate + Alanine

Other amino acids contribute to gluconeogenesis because all or part of their carbon skeletons enter the

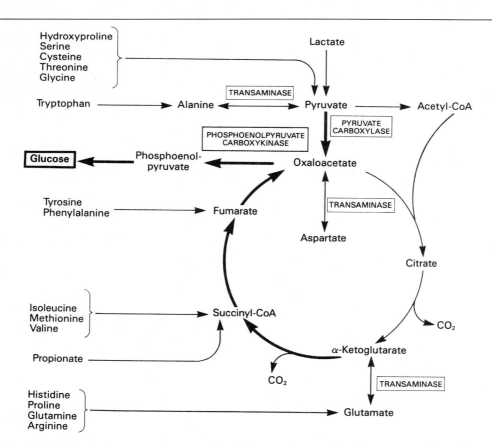

Figure 18–4. Involvement of the citric acid cycle in transamination and gluconeogenesis. The bold arrows indicate the main pathway of gluconeogenesis.

citric acid cycle after deamination or transamination. Examples are alanine, cysteine, glycine, hydroxyproline, serine, threonine, and tryptophan, which form pyruvate; arginine, histidine, glutamine, and proline, which form α-ketoglutarate via glutamate; isoleucine, methionine, and valine, which form succinyl-CoA; and tyrosine and phenylalanine, which form fumarate (see Fig 18–4). Substances forming pyruvate have the option of complete oxidation to CO_2 if they follow the pyruvate dehydrogenase pathway to acetyl-CoA, or they may follow the gluconeogenic pathway via carboxylation to oxaloacetate.

Of particular significance to ruminants is the conversion of propionate, the major glucogenic product of rumen fermentation, to succinyl-CoA via the methylmalonyl-CoA pathway (see Fig 21–2).

The Citric Acid Cycle Takes Part in Fatty Acid Synthesis (See Fig 18–5.)

Acetyl-CoA, formed from pyruvate by the action of pyruvate dehydrogenase, is the major building block for long-chain fatty acid synthesis in nonruminants. (In ruminants, acetyl-CoA is derived directly from acetate.) As pyruvate dehydrogenase is a mitochondrial enzyme and the enzymes responsible for fatty acid synthesis are extramitochondrial, the cell needs to transport acetyl-CoA through the mitochondrial membrane, which is impermeable to acetyl-CoA. This is accomplished by allowing **acetyl-CoA to form citrate** in the citric acid cycle, **transporting citrate** out of the mitochondria, and finally making acetyl-CoA available in the cytosol by **cleaving citrate** in a reaction catalyzed by the enzyme **ATP-citrate lyase.**

Citrate + ATP + CoA →
    Acetyl-CoA + Oxaloacetate + ADP + P_i

Regulation of the Citric Acid Cycle Depends Primarily on a Supply of Oxidized Cofactors

In most tissues, where the primary function of the citric acid cycle is to provide energy, there is little doubt that **respiratory control** via the respiratory chain and oxidative phosphorylation is the overriding control on citric acid cycle activity. Thus, activity is immediately dependent on the supply of oxidized dehydrogenase cofactors (eg, NAD), which in turn, because of the tight coupling between oxidation and phosphorylation, is dependent on the availability of ADP and ultimately, therefore, on the rate of utilization of ATP. Therefore, provided there is adequate O_2, the rate of doing work via utilization of ATP determines both the rate of respiration and the activity of the citric acid cycle. In addition to this overall or coarse control, the properties of some of the enzymes of the cycle indicate that control might also be exerted at the level of the cycle itself. In a tissue such as brain, which is largely dependent on carbohydrate to supply acetyl-CoA, control of the citric acid cycle may occur at the pyruvate dehydrogenase step. In the cycle proper, several enzymes are responsive to the energy status as expressed by the [ATP]/[ADP] and [NADH]/[NAD] ratios. Thus, there is allosteric inhibition of citrate synthase by ATP and long-chain fatty acyl-CoA. Allosteric activation of mitochondrial NAD-dependent isocitrate dehydrogenase by ADP is counteracted by ATP and NADH. The α-ketoglutarate dehydrogenase complex appears to be under control analogous to that

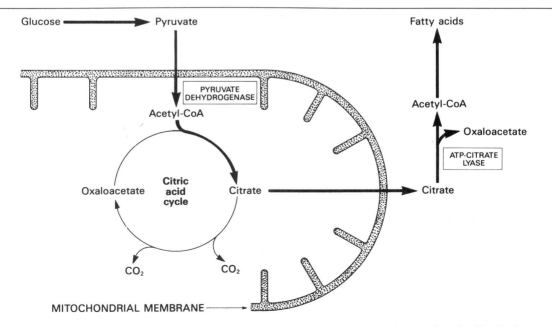

Figure 18–5. Participation of the citric acid cycle in fatty acid synthesis from glucose. See also Fig 23–5.

of pyruvate dehydrogenase. Succinate dehydrogenase is inhibited by oxaloacetate, and the availability of oxaloacetate, as controlled by malate dehydrogenase, depends on the [NADH]/[NAD$^+$] ratio. Since the K_m for oxaloacetate of citrate synthase is of the same order of magnitude as the intramitochondrial concentration, it would appear that the concentration of oxaloacetate could play a part in controlling the rate of citrate formation. Which (if any) of these mechanisms operates in vivo has still to be resolved.

REFERENCES

Baldwin JE, Krebs HA: The evolution of metabolic cycles. *Nature* 1981;**291**:381.

Boyer PD (editor): *The Enzymes*, 3rd ed. Academic Press, 1971.

Goodwin TW (editor): *The Metabolic Roles of Citrate*. Academic Press, 1968.

Greville GD: Vol 1, p 297, in: *Carbohydrate Metabolism and Its Disorders*. Dickens F, Randle PJ, Whelan WJ (editors). Academic Press, 1968.

Kay J, Weitzman PDJ (editors): *Kreb's Citric Acid Cycle—Half a Century and Still Turning*. Biochemical Society of London, 1987.

Lowenstein JM (editor): *Citric Acid Cycle: Control and Compartmentation*. Dekker, 1969.

Lowenstein JM (editor): *Citric Acid Cycle*. Vol 13 in: *Methods in Enzymology*. Academic Press, 1969.

Srere PA: The enzymology of the formation and breakdown of citrate. *Adv Enzymol* 1975;**43**:57.

Glycolysis & the Oxidation of Pyruvate

19

Peter A. Mayes, PhD, DSc

INTRODUCTION

There is a minimal requirement for glucose in all tissues, and in some (eg, brain and erythrocytes) the requirement is substantial. Glycolysis is the major pathway for the utilization of glucose and is found in all cells. It is a unique pathway, since **it can utilize oxygen if available (aerobic), or it can function in the absence of oxygen (anaerobic).**

BIOMEDICAL IMPORTANCE

Glycolysis is not only the principal route for glucose metabolism leading to the production of acetyl-CoA and oxidation in the citric acid cycle, but it also provides the main pathway for the metabolism of fructose and galactose derived from the diet. Of crucial biomedical significance is the ability of glycolysis to provide ATP in the absence of oxygen, because this allows skeletal muscle to perform at very high levels when aerobic oxidation becomes insufficient and it allows tissues with significant glycolytic ability to survive anoxic episodes. Conversely, heart muscle, which is adapted for aerobic performance, has relatively poor glycolytic ability and poor survival under conditions of **ischemia.** A small number of diseases occur in which enzymes of glycolysis (eg, pyruvate kinase) are deficient in activity; these conditions are mainly manifested as **hemolytic anemias.** In fast-growing **cancer** cells, glycolysis proceeds at a much higher rate than is required by the citric acid cycle. Thus, more pyruvate is produced than can be metabolized. This in turn results in excessive production of lactate, which favors a relatively acid local environment in the tumor, a situation that may have implications for certain types of cancer therapy. **Lactic acidosis** results from several causes, including pyruvate dehydrogenase deficiency.

GLYCOLYSIS CAN FUNCTION UNDER ANAEROBIC CONDITIONS

At an early period in the course of investigations on glycolysis it was realized that the process of fermentation in yeast was similar to the breakdown of glycogen in muscle. It was noted that when a muscle contracts in an anaerobic medium, ie, one from which oxygen is excluded, **glycogen disappears** and **lactate appears** as the principal end product. When oxygen is admitted, aerobic recovery takes place and glycogen reappears, while lactate disappears. However, if contraction takes place under aerobic conditions, lactate does not accumulate and pyruvate becomes the major end product of glycolysis; it does not accumulate, however, because it is oxidized further to CO_2 and water (Fig 19–1). As a result of these observations, it has been customary to separate carbohydrate metabolism into anaerobic and aerobic phases. However, this distinction is arbitrary, since the reactions in glycolysis are the same in the presence of oxygen as in its absence, except in extent and end products. When oxygen is in short supply, reoxidation of NADH formed from NAD during glycolysis is impaired. Under these circumstances, NADH is reoxidized by coupling to the reduction of pyruvate to lactate, and the NAD so formed allows further glycolysis to proceed (Fig 19–1). Thus, glycolysis can take place under anaerobic conditions, but this has a price, for it limits the amount of energy liberated per mole of glucose oxidized. Consequently, **to provide a given amount of energy, more glucose must undergo glycolysis under anaerobic as compared with aerobic conditions.**

THE SEQUENCE OF REACTIONS IN GLYCOLYSIS IS THE MAIN PATHWAY OF GLUCOSE UTILIZATION

The overall equation for glycolysis to lactate is

Glucose + 2ADP + 2P$_i$ →
2L(+)-Lactate + 2ATP + 2H$_2$O

All of the enzymes of the glycolysis pathway (Fig 19–2) are found in the extramitochondrial soluble fraction of the cell, the cytosol. They catalyze the reactions involved in the glycolysis of glucose to pyruvate and lactate, as follows:

Glucose enters into the glycolytic pathway by phosphorylation to glucose 6-phosphate. This is accomplished by the enzyme **hexokinase** and in liver parenchymal cells by **glucokinase,** whose activity is inducible and affected by changes in the nutritional state. ATP is required as phosphate donor, and as in

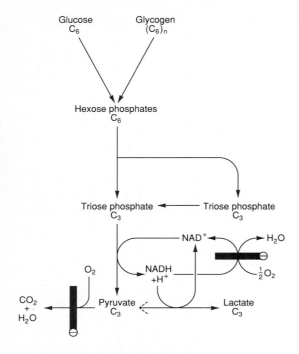

Figure 19–1. Summary of glycolysis. ~$\ominus$, blocked by anaerobic conditions or absence of mitochondria, eg, as in erythrocytes.

many reactions involving phosphorylation, it reacts as the Mg-ATP complex. One high-energy phosphate bond of ATP is utilized, and ADP is produced. The reaction is accompanied by considerable loss of free energy as heat and therefore, under physiologic conditions, may be regarded as irreversible. Hexokinase is inhibited in an allosteric manner by the product, glucose 6-phosphate.

$$\text{Glucose} + \text{ATP} \xrightarrow{\text{Mg}^{2+}} \text{Glucose 6-phosphate} + \text{ADP}$$

Hexokinase, present in all cells except those of the liver parenchyma, has a high affinity (low K_m) for its substrate, glucose. Its function is to ensure a supply of glucose for the tissues, even in the presence of low blood glucose concentrations, by phosphorylating all the glucose that enters the cell, thereby maintaining a large glucose concentration gradient between the blood and the intracellular environment. It acts on both the α and β anomers of glucose and will also catalyze the phosphorylation of other hexoses but at a much slower rate than glucose.

The function of glucokinase is to remove glucose from the blood following a meal. In contrast to hexokinase, it has a high K_m for glucose and operates optimally at blood glucose concentrations above 5 mmol/L (see Fig 21–5). It is specific for glucose.

Glucose 6-phosphate is an important compound at the junction of several metabolic pathways (glycolysis, gluconeogenesis, the pentose phosphate pathway,

glycogenesis, and glycogenolysis). In glycolysis it is converted to fructose 6-phosphate by **phosphohexose isomerase,** which involves an aldose-ketose isomerization. Only the α anomer of glucose 6-phosphate is acted upon.

$$\alpha\text{-D-Glucose 6-phosphate} \leftrightarrow \alpha\text{-D-Fructose 6-phosphate}$$

This reaction is followed by another phosphorylation with ATP catalyzed by the enzyme **phosphofructokinase (phosphofructokinase-1)** to produce fructose 1,6-bisphosphate. Phosphofructokinase is both an allosteric and an inducible enzyme whose activity is considered to play a major role in the regulation of the rate of glycolysis. The phosphofructokinase reaction is another that may be considered to be functionally irreversible under physiologic conditions.

$$\text{D-Fructose 6-phosphate} + \text{ATP} \rightarrow \text{D-Fructose 1,6-bisphosphate}$$

Fructose 1,6-bisphosphate is split by **aldolase** (fructose 1,6-bisphosphate aldolase) into 2 triose phosphates, glyceraldehyde 3-phosphate and dihydroxyacetone phosphate.

$$\text{D-Fructose 1,6-bisphosphate} \leftrightarrow \text{D-Glyceraldehyde 3-phosphate} + \text{Dihydroxyacetone phosphate}$$

Several different aldolases have been described, all of which contain 4 subunits. Aldolase A occurs in most tissues, and, in addition, aldolase B occurs in liver and kidney. The fructose phosphates exist in the cell mainly in the furanose form, but they react with phosphohexose isomerase, phosphofructokinase, and aldolase in the open-chain configuration.

Glyceraldehyde 3-phosphate and dihydroxyacetone phosphate are interconverted by the enzyme **phosphotriose isomerase.**

$$\text{D-Glyceraldehyde 3-phosphate} \leftrightarrow \text{Dihydroxyacetone phosphate}$$

Glycolysis proceeds by the oxidation of glyceraldehyde 3-phosphate to 1,3-bisphosphoglycerate, and because of the activity of phosphotriose isomerase, the dihydroxyacetone phosphate is also oxidized to 1,3-diphosphoglycerate via glyceraldehyde 3-phosphate.

$$\text{D-Glyderaldehyde 3-phosphate} + \text{NAD}^+ + \text{P}_i \leftrightarrow \text{1,3-Bisphosphoglycerate} + \text{NADH} + \text{H}^+$$

The enzyme responsible for the oxidation, **glyceraldehyde-3-phosphate dehydrogenase,** is NAD dependent. Structurally, it consists of 4 identical polypeptides (monomers) forming a tetramer. Four —SH groups are present on each polypeptide, derived from cysteine residues within the polypeptide chain.

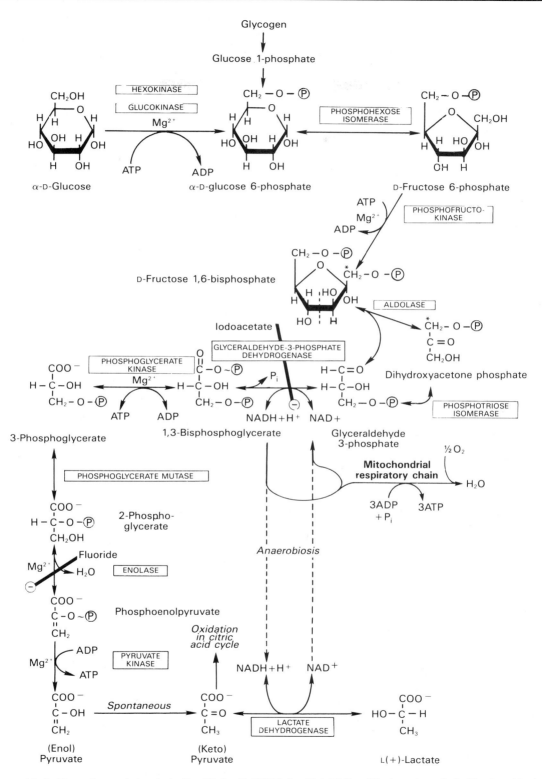

Figure 19–2. The pathway of glycolysis. $Ⓟ$, $-PO_3^{2-}$; P_i, $HOPO_3^{2-}$; $⊖$, inhibition. *Carbon atoms 1–3 of fructose bisphosphate form dihydroxyacetone phosphate, whereas carbons 4–6 form glyceraldehyde 3-phosphate. The term bis- as in bisphosphate indicates that the phosphate groups are separated, whereas diphosphate, as in adenosine diphosphate, indicates they are joined.

One of the −SH groups is found at the active site of the enzyme. The substrate initially combines with this −SH group, forming a thiohemiacetal that is converted to a high-energy thiol ester by oxidation; the hydrogens removed in this oxidation are transferred to NAD bound to the enzyme. The NADH produced on the enzyme is not so firmly bound to the enzyme as is NAD. Consequently, NADH is easily displaced by another molecule of NAD. Finally, by phosphorolysis, inorganic phosphate (P_i) is added, forming 1,3-bisphosphoglycerate, and the free enzyme with a reconstituted −SH group is liberated (Fig 19–3). Energy released during the oxidation is conserved by the formation of a high-energy sulfur bond that becomes, after phosphorolysis, a high-energy phosphate bond in position 1 of 1,3-bisphosphoglycerate. This high-energy phosphate is captured as ATP in a further reaction with ADP catalyzed by **phosphoglycerate kinase,** leaving 3-phosphoglycerate.

1,3-Bisphosphoglycerate + ADP ↔
3-Phosphoglycerate + ATP

Since 2 molecules of triose phosphate are formed per molecule of glucose undergoing glycolysis, 2 molecules of ATP are generated at this stage per molecule of glucose, an example of phosphorylation "at the substrate level."

If arsenate is present, it will compete with inorganic phosphate (P_i) in the above reactions to give 1-arseno-3-phosphoglycerate, which hydrolyzes spontaneously to give 3-phosphoglycerate plus heat, without generating ATP. This is an important example of the ability of arsenate to accomplish uncoupling of oxidation and phosphorylation.

3-Phosphoglycerate arising from the above reactions is converted to 2-phosphoglycerate by the enzyme **phosphoglycerate mutase.** It is likely that 2,3-bisphosphoglycerate (diphosphoglycerate, DPG) is an intermediate in this reaction.

3-Phosphoglycerate ↔ 2-Phosphoglycerate

The subsequent step is catalyzed by **enolase** and involves a dehydration and redistribution of energy within the molecule, raising the phosphate on position 2 to the high-energy state, thus forming phosphoenolpyruvate. Enolase is inhibited by **fluoride,** a property that can be made use of when it is required to prevent glycolysis prior to the estimation of blood glucose. The enzyme is also dependent on the presence of either Mg^{2+} or Mn^{2+}.

2-Phosphoglycerate ↔
Phosphoenolpyruvate + H_2O

The high-energy phosphate of phosphoenolpyruvate is transferred to ADP by the enzyme **pyruvate**

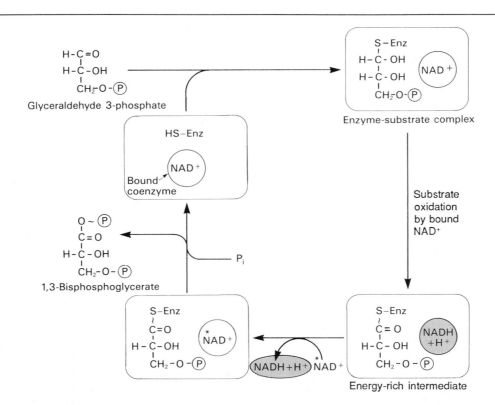

Figure 19–3. Mechanism of oxidation of glyceraldehyde 3-phosphate. Enz, glyceraldehyde-3-phosphate dehydrogenase. The enzyme is inhibited by the −SH poison iodoacetate, which is thus able to inhibit glycolysis.

kinase to generate, at this stage, 2 mol of ATP per mole of glucose oxidized. Enolpyruvate formed in this reaction is converted spontaneously to the keto form of pyruvate. This is another nonequilibrium reaction that is accompanied by considerable loss of free energy as heat and must be regarded as physiologically irreversible.

Phosphoenolpyruvate + ADP → Pyruvate + ATP

The redox state of the tissue now determines which of 2 pathways is followed. If **anaerobic** conditions prevail, the reoxidation of NADH by transfer of reducing equivalents through the respiratory chain to oxygen is prevented. Pyruvate is reduced by the NADH to lactate, the reaction being catalyzed by **lactate dehydrogenase.** Several isozymes of this enzyme have been described and have clinical significance (see p 58).

Pyruvate + NADH + H$^+$ ↔ L(+)-Lactate + NAD$^+$

The reoxidation of NADH via lactate formation allows glycolysis to proceed in the absence of oxygen by **regenerating sufficient NAD$^+$** for another cycle of the reaction catalyzed by glyceraldehyde-3-phosphate dehydrogenase. Thus, tissues that function under **hypoxic circumstances tend to produce lactate** (Fig 19–2). This is particularly true of skeletal muscle, where the rate at which the organ performs work is not limited by its capacity for oxygenation. The additional quantities of lactate produced may be detected in the tissues and in the blood and urine. **Glycolysis in erythrocytes,** even under aerobic conditions, **always terminates in lactate,** because mitochondria that contain the enzymatic machinery for the aerobic oxidation of pyruvate are absent. The mammalian erythrocyte is unique in that about 90% of its total energy requirement is provided by glycolysis. Besides skeletal muscle and erythrocytes, other tissues that normally derive most of their energy from glycolysis and produce lactate include brain, gastrointestinal tract, renal medulla, retina, and skin. The liver, kidneys, and heart usually take up lactate and oxidize it but will produce it under hypoxic conditions.

Glycolysis Is Regulated at 3 Steps Involving Nonequilibrium Reactions

Although most of the glycolytic reactions are reversible, 3 of them are markedly exergonic and must therefore be considered physiologically irreversible. These reactions are catalyzed by **hexokinase** (and glucokinase), **phosphofructokinase,** and **pyruvate kinase** and are the major sites of regulation of glycolysis. Cells that are capable of effecting a net movement of metabolites in the synthetic direction of the glycolytic pathway (gluconeogenesis) do so because of the presence of different enzyme systems which provide alternative routes around the irreversible reactions catalyzed by the above-mentioned enzymes.

These, together with the regulation of glycolysis, are discussed with the regulation of gluconeogenesis (Chapter 21).

In Erythrocytes the Second Site in Glycolysis for ATP Generation May Be By-Passed

In the erythrocytes of many mammalian species, the step catalyzed by **phosphoglycerate kinase** is bypassed by a process that effectively dissipates as heat the free energy associated with the high-energy phosphate of 1,3-bisphosphoglycerate (Fig 19–4). An additional enzyme, **bisphosphoglycerate mutase,** catalyzes the conversion of 1,3-bisphosphoglycerate to 2,3-bisphosphoglycerate. The latter is converted to 3-phosphoglycerate by **2,3-bisphosphoglycerate phosphatase,** an activity also attributed to phosphoglycerate mutase. The loss of a high-energy phosphate, which means that there is no net production of ATP when glycolysis takes this route, may be of advantage to the economy of the red cell, since it would allow glycolysis to proceed when the need for ATP was minimal. However, 2,3-bisphosphoglycerate, which is present in high concentration, combines with hemoglobin, causing a decrease in affinity for oxygen and a

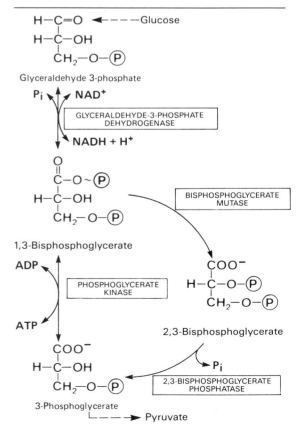

Figure 19–4. 2,3-Bisphosphoglycerate pathway in erythrocytes.

displacement of the oxyhemoglobin dissociation curve to the right. Thus, its **presence in the red cells helps oxyhemoglobin to unload oxygen** (see Chapter 7).

THE OXIDATION OF PYRUVATE TO ACETYL-CoA IS THE IRREVERSIBLE ROUTE FROM GLYCOLYSIS TO THE CITRIC ACID CYCLE

Before pyruvate can enter the citric acid cycle, it must be transported into the mitochondrion via a special **pyruvate transporter** that aids its passage across the inner mitochondrial membrane. This involves a symport mechanism whereby one proton is cotransported (see Fig 14–15). Within the mitochondrion, pyruvate is oxidatively decarboxylated to acetyl-CoA. This reaction is catalyzed by several different enzymes

working sequentially in a multienzyme complex. They are collectively designated as the **pyruvate dehydrogenase complex** and are analogous to the α-ketoglutarate dehydrogenase complex of the citric acid cycle (see p 156). Pyruvate is decarboxylated to a hydroxyethyl derivative of the thiazole ring of enzyme-bound **thiamin diphosphate,** which in turn reacts with oxidized lipoamide to form acetyl lipoamide (Fig 19–5). In the presence of **dihydrolipoyl transacetylase,** acetyl lipoamide reacts with coenzyme A to form acetyl-CoA and reduced lipoamide. The cycle of reaction is completed when the latter is reoxidized by a flavoprotein in the presence of **dihydrolipoyl dehydrogenase.** Finally, the reduced flavoprotein is oxidized by NAD, which in turn transfers reducing equivalents to the respiratory chain.

Pyruvate + NAD$^+$ + CoA →
    Acetyl-CoA + NADH + H$^+$ + CO$_2$

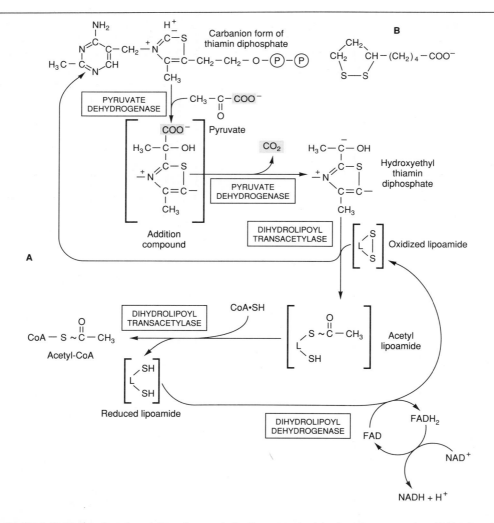

Figure 19–5. *A:* Oxidative decarboxylation of pyruvate by the pyruvate dehydrogenase complex. *B:* Lipoic acid. Lipoic acid is joined by an amide link to a lysine residue of the transacetylase component of the enzyme complex.

The pyruvate hydrogenase complex consists of a number of polypeptide chains of each of the 3 component enzymes, all organized in a regular spatial configuration. Movement of the individual enzymes appears to be restricted, and the metabolic intermediates do not dissociate freely but remain bound to the enzymes. Such a complex of enzymes, in which the substrates are handed on from one enzyme to the next, increases the reaction rate and eliminates side reactions, increasing overall efficiency.

It is to be noted that the pyruvate dehydrogenase system is sufficiently electronegative with respect to the respiratory chain that, in addition to generating a reduced coenzyme (NADH), it also generates a high-energy thio ester bond in acetyl-CoA.

Pyruvate Dehydrogenase Is Regulated by End-Product Inhibition and Covalent Modification

Pyruvate dehydrogenase is inhibited by its products, acetyl-CoA and NADH (Fig 19–6). It is also regulated by phosphorylation involving an ATP-specific kinase that causes a decrease in activity, and by dephosphorylation by a phosphatase that causes an increase in ac-

tivity of the dehydrogenase. The kinase is activated by increases in the [acetyl-CoA]/[CoA], [NADH]/[NAD$^+$], or [ATP]/[ADP] ratios. Thus, **pyruvate dehydrogenase—and therefore glycolysis—is inhibited not only by a high energy potential, but also under conditions of fatty acid oxidation,** which leads to increases in these ratios. In starvation, there is a decrease in the proportion of the enzyme in the active form, and an increase in activity occurs after administration of insulin in adipose tissue but not in the liver.

Inhibition of Pyruvate Metabolism Leads to Lactacidosis

Arsenite or mercuric ions complex the $-SH$ groups of lipoic acid and inhibit pyruvate dehydrogenase, as does a dietary deficiency of thiamin, allowing pyruvate to accumulate. Nutritionally deprived alcoholics are thiamin deficient and if administered glucose exhibit rapid accumulation of pyruvate and **lactacidosis,** which is frequently lethal. Patients with inherited pyruvate dehydrogenase deficiency present with a similar lactacidosis, particularly after glucose load. Mutations have been reported for virtually all of the enzymes of carbohydrate metabolism, each associated

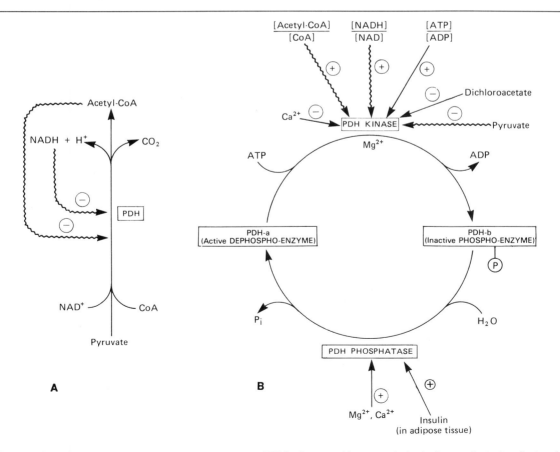

Figure 19–6. Regulation of pyruvate dehydrogenase (PDH). Arrows with wavy shafts indicate allosteric effects. *A:* Regulation by end-product inhibition. *B:* Regulation by interconversion of active and inactive forms.

Table 19–1. Generation of high-energy phosphate in the catabolism of glucose.

Pathway	Reaction Catalyzed by	Method of ~ (P) Production	Number of ~ (P) Formed per Mole of Glucose
Glycolysis	Glyceraldehyde-3-phosphate dehydrogenase	Respiratory chain oxidation of 2 NADH	6*
	Phosphoglycerate kinase	Oxidation at substrate level	2
	Pyruvate kinase	Oxidation at substrate level	2
			10
Allow for consumption of ATP by reactions catalyzed by hexokinase and phosphofructokinase			−2
			Net 8
Citric acid cycle	Pyruvate dehydrogenase	Respiratory chain oxidation of 2 NADH	6
	Isocitrate dehydrogenase	Respiratory chain oxidation of 2 NADH	6
	α-Ketoglutarate dehydrogenase	Respiratory chain oxidation of 2 NADH	6
	Succinate thiokinase	Oxidation at substrate level	2
	Succinate dehydrogenase	Respiratory chain oxidation of 2 $FADH_2$	4
	Malate dehydrogenase	Respiratory chain oxidation of 2 NADH	6
			Net 30
Total per mole of glucose under aerobic conditions			38
Total per mole of glucose under anaerobic conditions			2

*It is assumed that NADH formed in glycolysis is transported into mitochondria via the malate shuttle (see Fig 14–15). If the glycerophosphate shuttle is used, only 2 ~ (P) would be formed per mole of NADH, the total net production being 36 instead of 38. The calculation ignores the small loss of ATP due to a transport of H^+ into the mitochondrion with pyruvate and a similar transport of H^+ in the operation of the malate shuttle, totaling about 1 mol of ATP.

with human disease. Inherited aldolase A deficiency and pyruvate kinase deficiency in erythrocytes cause **hemolytic anemia.**

Oxidation of Glucose Yields up to 38 mol of ATP Under Aerobic Conditions but Only 2 When O_2 is Absent

When 1 mol of glucose is combusted in a calorimeter to CO_2 and water, approximately 2780 kJ are liberated as heat. When oxidation occurs in the tissues, some of this energy is not lost immediately as heat but is "captured" as high-energy phosphate. On the order of 38 mol of ATP are generated per molecule of glucose oxidized to CO_2 and water. Assuming each high-energy bond to be equivalent to 30.5 kJ, the total energy captured in ATP per mole of glucose oxidized is 1159 kJ, or approximately 41.7% of the energy of combustion. Most of the ATP is formed as a consequence of oxidative phosphorylation resulting from the reoxidation of reduced coenzymes by the respiratory chain. The remainder is generated by phosphorylation at the "substrate level." (See Chapter 14.) Table 19–1 indicates the reactions responsible for the generation of high-energy phosphate during oxidation of glucose and the net production under aerobic and anaerobic conditions.

REFERENCES

Boiteux A, Hess B: Design of glycolysis. *Phil Trans R Soc London B* 1981;**293**:5.

Blass JP: Disorders of pyruvate metabolism. *Neurology* 1979;**29**:280.

Boyer PD (editor): *The Enzymes,* 3rd ed. Vols 5–9. Academic Press, 1972.

Dickens, F, Randle PJ, Whelan WJ (editors): *Carbohydrate Metabolism and Its Disorders.* 2 vols. Academic Press, 1968.

Greenberg DM (editor): *Metabolic Pathways,* 3rd ed. Vol 1. Academic Press, 1967.

Randle PJ, Steiner DF, Whelan WJ (editors): *Carbohydrate Metabolism and Its Disorders.* Vol 3. Academic Press, 1981.

Scriver CR (editor): *Metabolic Basis of Inherited Disease.* McGraw Hill, 6th ed. 1989.

Sols A: Multimodulation of enzyme activity. *Curr Top Cell Reg* 1981;**19**:77.

Veneziale CM (editor): *The Regulation of Carbohydrate Formation and Utilization in Mammals.* University Park Press, 1981.

Metabolism of Glycogen

20

Peter A. Mayes, PhD, DSc

INTRODUCTION

Glycogen is the major storage form of carbohydrate in animals and corresponds to starch in plants. It occurs mainly in liver (up to 6%) and muscle, where it rarely exceeds 1%. However, because of its greater mass, muscle represents some 3–4 times as much glycogen store as liver (Table 20–1). Like starch, it is a branched polymer of α-glucose (see Fig 15–15).

BIOMEDICAL IMPORTANCE

The function of muscle glycogen is to act as a readily available source of hexose units for glycolysis **within the muscle itself.** Liver glycogen is largely concerned with storage and export of hexose units for maintenance of the **blood glucose,** particularly between meals. After 12–18 hours of fasting, the liver becomes almost totally depleted of glycogen, whereas muscle glycogen is only depleted significantly after prolonged vigorous exercise. **Glycogen storage diseases** are a group of inherited disorders characterized by deficient mobilization of glycogen and deposition of abnormal forms of glycogen, leading to muscular weakness or even death.

GLYCOGENESIS OCCURS MAINLY IN MUSCLE AND LIVER

The Pathway of Glycogen Biosynthesis Involves a Special, Active Nucleotide of Glucose (See Fig 20–1.)

Glucose is phosphorylated to glucose 6-phosphate, a reaction that is common to the first reaction in the pathway of glycolysis from glucose. This reaction is catalyzed by **hexokinase** in muscle and **glucokinase** in liver. Glucose 6-phosphate is converted to glucose 1-phosphate in a reaction catalyzed by the enzyme **phosphoglucomutase.** The enzyme itself is phosphorylated, and the phospho- group takes part in a reversible reaction in which glucose 1,6-bisphosphate is an intermediate.

$$\text{Enz-P} + \text{Glucose 6-phosphate} \leftrightarrow$$
$$\text{Enz} + \text{Glucose 1,6-bisphosphate} \leftrightarrow$$
$$\text{Enz-P} + \text{Glucose 1-phosphate}$$

Next, glucose 1-phosphate reacts with uridine triphosphate (UTP) to form the active nucleotide **uridine diphosphate glucose (UDPGlc)*** (Fig 20–2).

The reaction between glucose 1-phosphate and uridine triphosphate is catalyzed by the enzyme **UDPGlc pyrophosphorylase.**

$$\text{UTP} + \text{Glucose 1-phosphate} \leftrightarrow \text{UDPGlc} + \text{PP}_i$$

The subsequent hydrolysis of inorganic pyrophosphate by **inorganic pyrophosphatase** pulls the reaction to the right of the equation.

By the action of the enzyme **glycogen synthase,** the C_1 of the activated glucose of UDPGlc forms a glycosidic bond with the C_4 of a terminal glucose residue of glycogen, liberating uridine diphosphate (UDP). A preexisting glycogen molecule, or "primer," must be present to initiate this reaction. The glycogen primer may in turn be formed on a protein backbone, which may be a process similar to the synthesis of other glycoproteins (see Chapter 57).

$$\text{UDPGlc} + (C_6)_n \rightarrow \text{UDP} + (C_6)_{n+1}$$
$$\text{glycogen} \qquad\qquad \text{glycogen}$$

The Mechanism of Branching Involves Detachment of Existing Glycogen Chains

The addition of a glucose residue to a preexisting glycogen chain, or "primer," occurs at the nonreducing, outer end of the molecule so that the "branches" of the glycogen "tree" become elongated as successive $1 \rightarrow 4$-**linkages** occur (Fig 20–3). When the chain has been lengthened to a minimum of 11 glucose residues, a second enzyme, the **branching enzyme (amylo[$1 \rightarrow 4$]→[$1 \rightarrow 6$]-transglucosidase)** transfers a part of the $1 \rightarrow 4$-chain (minimum length 6 glucose resi-

---

*Other nucleoside diphosphate sugar compounds are known, eg, UDPGal. In addition, the same sugar may be linked to different nucleotides. For example, glucose may be linked to uridine (as shown above) as well as to guanosine, thymidine, adenosine, or cytidine nucleotides.

Table 20–1. Storage of carbohydrate in postabsorptive normal adult humans (70 kg).

Liver glycogen	4.0% =	72 g*
Muscle glycogen	0.7% =	245 g†
Extracellular glucose	0.1% =	10 g‡
		327 g

*Liver weight, 1800 g.
†Muscle mass, 35 kg.
‡Total volume, 10 L.

dues) to a neighboring chain to form a **1→6- linkage,** thus establishing a **branch point** in the molecule. The branches grow by further additions of 1→4- glucosyl units and further branching. When the number of non-reducing terminal residues increases, the total number of reactive sites in the molecule increases, speeding up both glycogenesis and glygenolysis.

The action of the branching enzyme has been studied in the living animal by feeding ^{14}C-labeled glucose and examining the liver glycogen at intervals thereafter (Fig 20–3).

GLYCOGENOLYSIS IS NOT THE REVERSE OF GLYCOGENESIS BUT IS A SEPARATE PATHWAY

The Pathway of Degradation Involves a Debranching Mechanism
(See Fig 20–1.)

It is the step catalyzed by **phosphorylase** that is rate-limiting in glycogenolysis.

$$(C_6)_n + P_i \rightarrow (C_6)_{n-1} + \text{Glucose 1-phosphate}$$
glycogen   glycogen

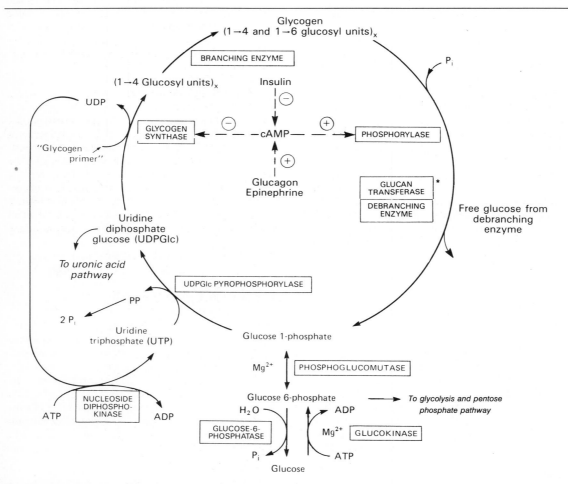

Figure 20–1. Pathway of glycogenesis and of glycogenolysis in the liver. Two high-energy phosphates are used in the incorporation of 1 mol of glucose into glycogen. ⊕, Stimulation; ⊖, inhibition. Insulin decreases the level of cAMP only after it has been raised by glucagon or epinephrine: ie, it antagonizes their action. Glucagon is active in heart muscle but not in skeletal muscle. * Glucan transferase and debranching enzyme appear to be 2 separate activities of the same enzyme.

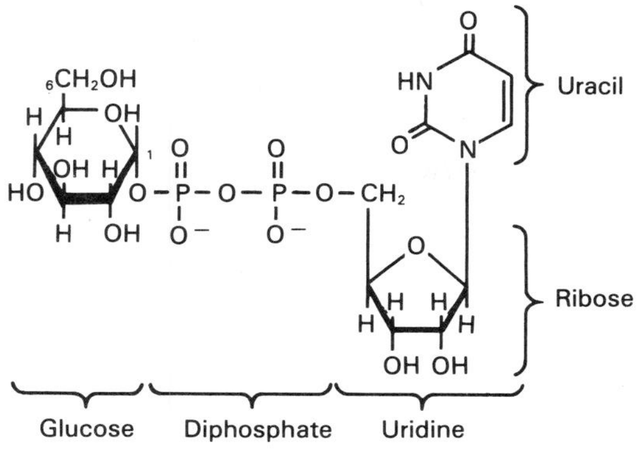

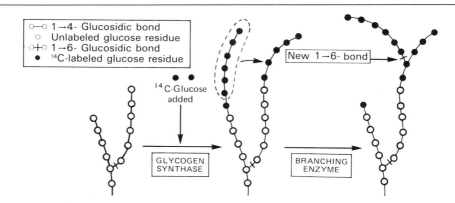

Figure 20–2. Uridine diphosphate glucose (UDPGlc).

This enzyme is specific for the phosphorylytic breaking (phosphorolysis) of the 1→4- linkages of glycogen to yield glucose 1-phosphate. Glucosyl residues from the outermost chains of the glycogen molecule are removed until approximately 4 glucose residues remain on either side of a 1→6- branch (Fig 20–4). Another enzyme (α-[1→4]→α-[1→4] **glucan transferase**) transfers a trisaccharide unit from one branch to the other, exposing the 1→6- branch points. The **hydrolytic** splitting of the 1→6- linkages requires the action of a specific **debranching enzyme (amylo-[1→6]-glucosidase)**. With the removal of the branch, further action by phosphorylase can proceed. The combined action of phosphorylase and these other enzymes leads to the complete breakdown of glycogen. The reaction catalyzed by phosphoglucomutase is reversible, so that glucose 6-phosphate can be formed from glucose 1-phosphate. In **liver** and **kidney** (but not in muscle), there is a specific enzyme, **glucose-6-phosphatase,** that removes phosphate from glucose 6-phosphate, enabling glucose to diffuse from the cell into the blood. This is the final step in hepatic glycogenolysis, which is reflected by a rise in the blood glucose.

CYCLIC AMP INTEGRATES THE REGULATION OF GLYCOGENOLYSIS AND GLYCOGENESIS

The principal enzymes controlling glycogen metabolism—glycogen phosphorylase and glycogen synthase—are regulated by a complex series of reactions involving both allosteric mechanisms (see p 87) and covalent modifications due to reversible phosphorylation and dephosphorylation of enzyme protein (see p 90).

Many covalent modifications are due to the action of cAMP (3′,5′-cyclic adenylic acid; cyclic AMP) (see Fig 20–5 and p 475). cAMP is the intracellular intermediate compound or **second messenger** through which many hormones act. It is formed from ATP by an enzyme, **adenylate cyclase,** occurring in the inner surface of cell membranes. Adenylate cyclase is activated by hormones such as **epinephrine** and **norepinephrine** acting through β-adrenergic receptors on the cell membrane and additionally in liver by **glucagon** acting through an independent **glucagon receptor.** cAMP is destroyed by a **phosphodiesterase,** and it is the activity of this enzyme that maintains the normally low level of cAMP. Insulin has been reported to increase its activity in liver, thereby lowering the concentration of cAMP.

Phosphorylase Differs between Liver and Muscle

In liver, the enzyme exists in both an active and an inactive form. Active phosphorylase (**phosphorylase a**) has one of its serine hydroxyl groups phosphorylated in an ester linkage. By the action of a specific phosphatase (**protein phosphatase-1**), the enzyme is inactivated to **phosphorylase b** in a reaction that involves hydrolytic removal of the phosphate from the serine residue. Reactivation requires rephosphorylation with ATP and a specific enzyme, **phosphorylase kinase.**

Muscle phosphorylase is immunologically and ge-

Figure 20–3. The biosynthesis of glycogen. The mechanism of branching as revealed by the addition of ^{14}C-labeled glucose.

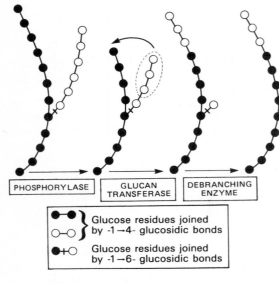

Figure 20-4. Steps in glycogenolysis.

netically distinct from that of liver. It is a dimer, each monomer containing 1 mol of pyridoxal phosphate. It is present in 2 forms: **phosphorylase a,** which is phosphorylated and active in either the presence or absence of AMP (its allosteric modifier), and **phosphorylase b,** which is dephosphorylated and active only in the presence of AMP. This occurs during exercise when the level of AMP rises. Phosphorylase a is the normal physiologically active form of the enzyme.

Activation of Muscle Phosphorylase Is Via cAMP

Phosphorylase in **muscle** is activated by epinephrine (Fig 20-6). However, this occurs not as a direct effect but rather by way of the action of cAMP.

Increasing the concentration of cAMP activates an enzyme of rather wide specificity, **cAMP-dependent protein kinase.** This kinase catalyzes the phosphoryl-

ation by ATP of inactive **phosphorylase kinase b** to active **phosphorylase kinase a,** which in turn, by means of a further phosphorylation, activates phosphorylase b to phosphorylase a.

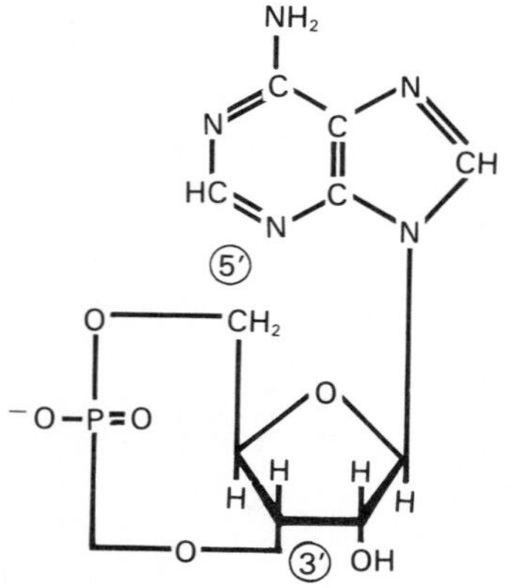

Inactive cAMP-dependent protein kinase comprises 2 pairs of subunits, each pair consisting of a regulatory subunit (R), which binds 2 mol of cAMP, and a catalytic subunit (C), which contains the active site. Combination with cAMP causes the R_2C_2 complex to dissociate, releasing active C monomers (see p 476).

$$R_2C_2 + 4cAMP \leftrightarrow 2C + 2(R-cAMP_2)$$

| Inactive enzyme | Active enzyme |

Ca²⁺ Synchronizes the Activation of Phosphorylase With Muscle Contraction

Glycogenolysis increases in muscle several hundred-fold immediately after the onset of contraction. This involves the rapid activation of phosphorylase owing to activation of phosphorylase kinase by Ca^{2+}, the same signal that initiates contraction. Muscle phosphorylase kinase has 4 types of subunits, α, β, γ, and δ, in a structure represented as $(\alpha\beta\gamma\delta)_4$. The α and β subunits contain serine residues that are phosphorylated by cAMP-dependent protein kinase. The β subunit binds $4 Ca^{2+}$ and is identical to the Ca^{2+} binding protein **calmodulin** (see p 479). The binding of Ca^{2+} activates the catalytic site of the γ subunit while the molecule remains in the dephosphorylated b configuration. However, the phosphorylated a form is only fully activated in the presence of Ca^{2+}. It is of significance that calmodulin is similar in structure to TpC, the Ca^{2+} binding protein in muscle. A second molecule of calmodulin or TpC can interact with the phosphorylase kinase, causing further activation. Thus, activation of muscle contraction and glycogenolysis are carried out by the same Ca^{2+} binding protein, ensuring their synchronization.

Glycogenolysis in Liver Can be cAMP Independent

In addition to the major action of **glucagon** in causing formation of cAMP and activation of phosphorylase in liver, studies have shown that α_1 **receptors** are the major mediators of catecholamine stimulation of glycogenolysis. This involves a **cAMP-independent** mobilization of Ca^{2+} from mitochondria into the cytosol, followed by the stimulation of a **Ca²⁺/calmodulin-sensitive phosphorylase kinase.** cAMP-independent glycogenolysis is also caused by vasopressin, oxytocin, and angiotensin II acting through calcium or the phosphatidylinositol bisphosphate pathway (see p 480).

Inactivation of Phosphorylase Is by Protein Phosphatase-1

Both phosphorylase **a** and phosphorylase kinase **a** are dephosphorylated and inactivated by **protein phosphatase-1.** Protein phosphatase-1 is inhibited by

Figure 20-5. 3′.5′-Adenylic acid (cyclic AMP; cAMP).

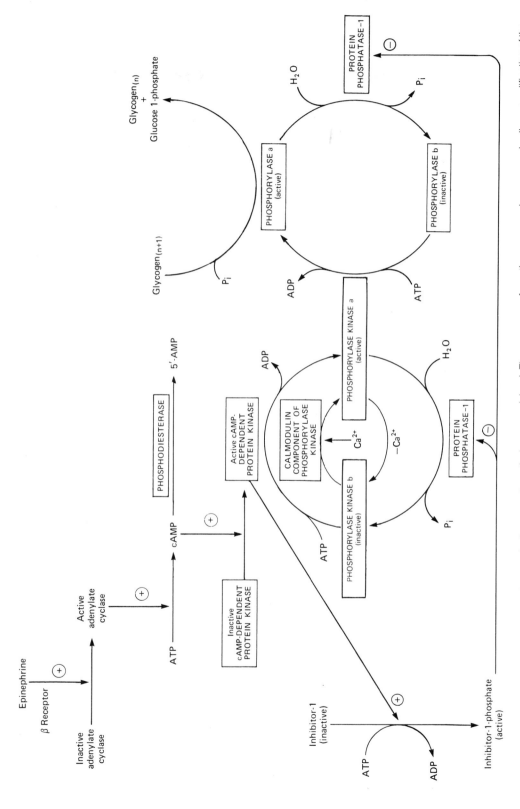

Figure 20–6. Control of phosphorylase in muscle (n = number of glucose residues). The sequence of reactions arranged as a cascade allows amplification of the hormonal signal at each step.

a protein called **inhibitor-1,** which is active only after it has been phosphorylated by cAMP-dependent protein kinase. Thus, **cAMP controls both the activation and inactivation of phosphorylase** (Fig 20–6).

Glycogen Synthase and Phosphorylase Activity are Reciprocally Regulated (See Fig 20–7.)

Like phosphorylase, glycogen synthase exists in either a phosphorylated or nonphosphorylated state. However, unlike phosphorylase, the active form is dephosphorylated (**glycogen synthase a**) and may be inactivated to **glycogen synthase b** by phosphorylation on 7 serine residues by no fewer than 6 different protein kinases. All 7 phosphorylation sites are contained on each of 4 identical subunits. Two of the protein kinases are Ca^{2+}/calmodulin-dependent (one of these is phosphorylase kinase). Another kinase is cAMP-dependent protein kinase, **which allows cAMP-mediated hormonal action to inhibit glycogen synthesis synchronously with the activation of glycogenolysis.** The remaining kinases are known as glycogen synthase kinase -3, -4, and -5.

Glucose 6-phosphate is an allosteric activator of glycogen synthase b, causing a decrease in K_m for UDP-glucose and allowing glycogen synthesis by the phosphorylated enzyme. Glycogen also exerts an inhibition on its own formation, and **insulin** also stimulates glycogen synthesis in muscle by promoting dephosphorylation and activation of glycogen synthase b. Normally, dephosphorylation of glycogen synthase b is carried out by protein phosphatase-1, which is under the control of cAMP-dependent protein kinase (Fig 20–7).

REGULATION OF GLYCOGEN METABOLISM IS EFFECTED BY A BALANCE IN ACTIVITIES BETWEEN GLYCOGEN SYNTHASE AND PHOSPHORYLASE (See Fig 20–8.)

Glycogen synthase and phosphorylase are under substrate control (through allostery) as well as hormonal control. Not only is phosphorylase activated by a rise in concentration of cAMP (via phosphorylase

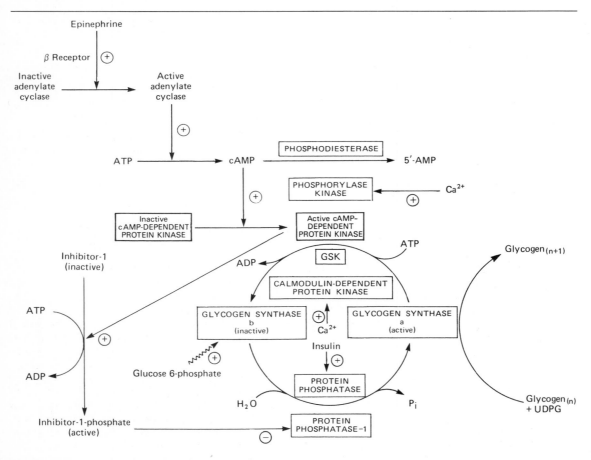

Figure 20–7. Control of glycogen synthase in muscle (n = number of glucose residues). The sequence of reactions arranged in a cascade causes amplification at each step, allowing only nanomole quantities of hormone to cause major changes in glycogen concentration. GSK, glycogen synthase kinase-3, -4, and -5; wavy arrow, allosteric activation.

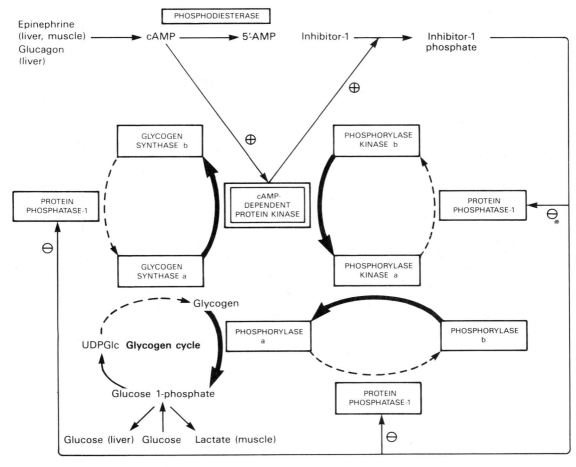

Figure 20–8. Coordinated control of glycogenolysis and glycogenesis by cAMP-dependent protein kinase. The reactions that lead to glycogenolysis as a result of an increase in cAMP concentrations are shown with bold arrows, and those that are inhibited are shown as broken arrows. The reverse occurs when cAMP concentrations decrease as a result of phosphodiesterase activity, leading to glycogenesis.

kinase), but glycogen synthase is at the same time converted to the inactive form; both effects are mediated via **cAMP-dependent protein kinase.** Thus, inhibition of glycogenolysis enhances net glycogenesis, and inhibition of glycogenesis enhances net glycogenolysis. Of further significance in the regulation of glycogen metabolism is the finding that the dephosphorylation of phosphorylase a, phosphorylase kinase, and glycogen synthase b is accomplished by a single enzyme of wide specificity—**protein phosphatase-1.** In turn, protein phosphatase-1 is inhibited by cAMP-dependent protein kinase via inhibitor-1 (Fig 20–8). Thus, glycogenolysis can be terminated and glycogenesis can be stimulated synchronously, or vice versa, because both processes are keyed to the activity of cAMP-dependent protein kinase. Both phosphorylase kinase and glycogen synthase may be reversibly phosphorylated in more than one site by separate kinases and phosphatases. These secondary phosphorylations modify the sensitivity of the primary sites to phosphorylation and dephosphorylation. This is known as **multisite phosphorylation.**

The Major Factor That Controls Glycogen Metabolism In the Liver Is the Concentration of *Phosphorylase a*

Not only does this enzyme control the rate-limiting step in glycogenolysis, but it also inhibits the activity of protein phosphatase-1 and thereby controls glycogen synthesis (Fig 20–8). Inactivation of phosphorylase occurs as a result of allosteric inhibition by glucose as it rises in concentration after a meal. Activation is caused by 5′-AMP responding to depletion of ATP. Administration of insulin causes an immediate inactivation of phosphorylase followed by activation of glycogen synthase. The effects of insulin require the presence of glucose.

Regulation of the branching and debranching enzymes does not occur.

GLYCOGEN STORAGE DISEASES ARE INHERITED

The term "glycogen storage disease" is a generic one intended to describe a group of inherited disorders characterized by deposition of an abnormal type or quantity of glycogen in the tissues.

In **type I glycogenosis (von Gierke's disease)**, both the liver cells and the cells of the renal convoluted tubules are characteristically loaded with glycogen. However, these glycogen stores are unavailable, as evidenced by the occurrence of hypoglycemia and a lack of glucose release under stimulus by epinephrine or glucagon. Ketosis and hyperlipemia are also present in these patients, as would be characteristic of an organism deprived of carbohydrate. In liver, kidney, and intestinal tissue, **the activity of glucose-6-phosphatase is either extremely low or entirely absent.**

Type II (Pompe's disease) is fatal and is characterized by a deficiency of lysosomal α-1→4- and 1→6-glucosidase (acid maltase), whose function is to degrade glycogen, which otherwise accumulates in the lysosomes.

Type III (limit dextrinosis; Forbes', or Cori's, disease) is characterized by the absence of debranching enzyme, which causes the accumulation of a characteristic branched polysaccharide.

Type IV (amylopectinosis; Andersen's disease) is characterized by the absence of branching enzyme, with the result that a polysaccharide having few branch points accumulates. Death due to cardiac or liver failure usually occurs in the first year of life.

An absence of muscle phosphorylase (myophosphorylase) is the cause of **type V glycogenosis (myo-** phosphorylase deficiency glycogenosis; **McArdle's syndrome).** Patients with this disease exhibit a markedly diminished tolerance to exercise. Although their skeletal muscles have an abnormally high content of glycogen (2.5–4.1%), little or no lactate is detectable in their blood after exercise.

Also described among the glycogen storage diseases are phosphorylase deficiency in the liver (**type VI; Hers' disease**), a deficiency of phosphofructokinase in the muscles and erythrocytes (**type VII; Tarui's disease**), and a glycogenosis in which liver phosphorylase kinase is deficient (**type VIII).** Deficiencies of **adenylate kinase** and **cAMP-dependent protein kinase** have also been reported.

REFERENCES

Cohen P: *Control of Enzyme Activity,* 2nd ed. Chapman & Hall, 1983.

Cohen P: The role of protein phosphorylation in the hormonal control of enzyme activity. *Eur J Biochem* 1985;**151:**439.

Exton JH: Molecular mechanisms involved in α-adrenergic responses. *Mol Cell Endocrinol* 1981;**23:**233.

Geddes R: Glycogen: A metabolic viewpoint. *Bioscience Rep* 1986;**6:**415.

Hers HG: The control of glycogen metabolism in the liver. *Annu Rev Biochem* 1976;**45:**167.

Randle PJ, Steiner DF, Whelan WJ (editors): *Carbohydrate Metabolism and Its Disorders.* Vol 3. Academic Press, 1981.

Scriver CR et al (editor): *The Metabolic Basis of Inherited Disease,* 6th ed. McGraw-Hill, 1989.

Sperling O, de Vries A (editors): *Inborn Errors of Metabolism in Man.* Karger, 1978.

Gluconeogenesis & Control of the Blood Glucose

21

Peter A. Mayes, PhD, DSc

INTRODUCTION

Gluconeogenesis includes all mechanisms and pathways responsible for converting noncarbohydrates to glucose or glycogen. The major substrates for gluconeogenesis are the glucogenic amino acids, lactate, glycerol, and (important in ruminants) propionate. Liver and kidney are the major tissues involved, since they contain a full complement of the necessary enzymes.

BIOMEDICAL IMPORTANCE

Gluconeogenesis meets the needs of the body for glucose when carbohydrate is not available in sufficient amounts from the diet. A continual supply of glucose is necessary as a source of energy, especially for the **nervous system and the erythrocytes.** Below a critical blood glucose concentration, there is brain dysfunction, which under conditions of severe hypoglycemia can lead to coma and death. Glucose is also required in adipose tissue as a source of glyceride-glycerol, and it probably plays a role in maintaining the level of intermediates of the citric acid cycle in many tissues. It is clear that even under conditions where fat may be supplying most of the caloric requirement of the organism, **there is always a certain basal requirement for glucose.** In addition, glucose is the only fuel that will supply energy to skeletal muscle under anaerobic conditions. It is the precursor of milk sugar (lactose) in the mammary gland, and it is taken up actively by the fetus. In addition, gluconeogenic mechanisms are used to clear the products of the metabolism of other tissues from the blood, eg, lactate, produced by muscle and erythrocytes, and glycerol, which is continuously produced by adipose tissue. Propionate, the principal glucogenic fatty acid produced in the digestion of carbohydrates by ruminants, is a major substrate for gluconeogenesis in these species.

GLUCONEOGENESIS INVOLVES GLYCOLYSIS, THE CITRIC ACID CYCLE, PLUS SOME SPECIAL REACTIONS
(See Fig 21–1.)

Thermodynamic Barriers Prevent a Simple Reversal of Glycolysis

Krebs pointed out that energy barriers obstruct a simple reversal of glycolysis: (1) between pyruvate and phosphoenolpyruvate, (2) between fructose 1,6-bisphosphate and fructose 6-phosphate, (3) between glucose 6-phosphate and glucose, and (4) between glucose 1-phosphate and glycogen. These reactions are all nonequilibrium, releasing much free energy as heat and therefore physiologically irreversible. They are circumvented by special reactions.

Pyruvate & Phosphoenolpyruvate: Present in mitochondria is an enzyme, **pyruvate carboxylase,** which in the presence of ATP, the B vitamin biotin, and CO_2 converts pyruvate to oxaloacetate. The function of the biotin is to bind CO_2 from bicarbonate onto the enzyme prior to the addition of the CO_2 to pyruvate. A second enzyme, **phosphoenolpyruvate carboxykinase,** catalyzes the conversion of oxaloacetate to phosphoenolpyruvate. High-energy phosphate in the form of GTP or ITP is required in this reaction, and CO_2 is liberated. Thus, with the help of these 2 enzymes and lactate dehydrogenase, lactate can be converted to phosphoenolpyruvate.

In pigeon, chicken, and rabbit liver, phosphoenolpyruvate carboxykinase is a mitochondrial enzyme, and phosphoenolpyruvate is transported into the cytosol for conversion into fructose 1,6-bisphosphate by reversal of glycolysis. In the rat and the mouse the enzyme is in the cytosol, but oxaloacetate does not diffuse readily from mitochondria. Alternative means are available to achieve the same end by converting oxaloacetate into compounds that can diffuse from the mitochondria, followed by their reconversion to oxaloacetate in the extramitochondrial portion of the cell. Such a compound is malate, whose formation from

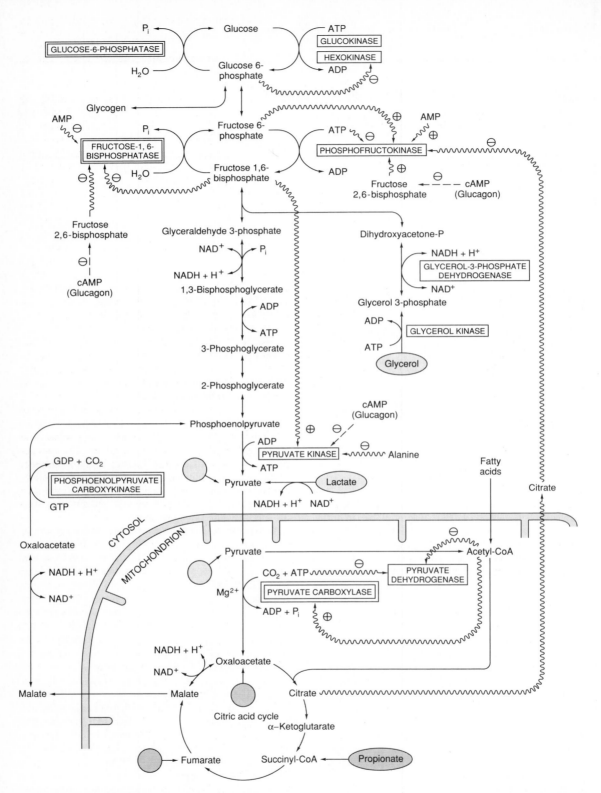

Figure 21–1. Major pathways and regulation of gluconeogenesis and glycolysis in the liver. Entry points of glucogenic amino acids after transamination are indicated by ○→. (See also Fig 18–7.) The key gluconeogenic enzymes are shown thus ▭. The ATP required for gluconeogenesis is supplied by the oxidation of long-chain fatty acids. Propionate is of quantitative importance only in ruminants. ∿→, allosteric effect; - - - →, covalent modification by reversible phosphorylation. High concentrations of alanine act as a "gluconeogenic signal" by inhibiting glycolysis at the pyruvate kinase step.

oxaloacetate within mitochondria and conversion back to oxaloacetate in the extramitochondrial compartment involves malate dehydrogenase. In humans, guinea pig, and cow, the enzyme is equally distributed between mitochondria and cytosol.

Fructose 6-Phosphate & Fructose 1,6-Bisphosphate: The conversion of fructose 1,6-bisphosphate to fructose 6-phosphate, necessary to achieve a reversal of glycolysis, is catalyzed by a specific enzyme, **fructose-1,6-bisphosphatase.** This is a key enzyme from another point of view in that its presence determines whether or not a tissue is capable of synthesizing glycogen not only from pyruvate but also from triosephosphates. It is present in liver and kidney and has been demonstrated in striated muscle. It is held to be absent from heart muscle and smooth muscle.

Glucose 6-Phosphate & Glucose: The conversion of glucose 6-phosphate to glucose is catalyzed by another specific phosphatase, **glucose-6-phosphatase.** It is present in liver and kidney but absent from muscle and adipose tissue. **Its presence allows a tissue to add glucose to the blood.**

Glucose 1-Phosphate & Glycogen: The breakdown of glycogen to glucose 1-phosphate is carried out by phosphorylase. The synthesis of glycogen involves an entirely different pathway through the formation of uridine diphosphate glucose and the activity of **glycogen synthase** (see Fig 20–1).

These key enzymes allow reversal of glycolysis to play a major role in gluconeogenesis. The relationships between gluconeogenesis and the glycolytic pathway are shown in Fig 21–1. After transamination or deamination, glucogenic amino acids form either pyruvate or members of the citric acid cycle. Therefore, the reactions described above can account for the conversion of both glucogenic amino acids and lactate to glucose or glycogen. Thus, lactate forms pyruvate and **enters the mitochondria before conversion to oxaloacetate and ultimate conversion to glucose.**

Propionate is a major source of glucose in ruminants, and enters the main gluconeogenic pathway via the citric acid cycle after conversion to succinyl-CoA. Propionate is first activated with ATP and CoA by an appropriate **acyl-CoA synthetase.** Propionyl-CoA, the product of this reaction, undergoes a CO_2 fixation reaction to form D-methylmalonyl-CoA, catalyzed by **propionyl-CoA carboxylase** (Fig 21–2). This reaction is analogous to the fixation of CO_2 in acetyl-CoA by acetyl-CoA carboxylase (see Chapter 23) in that it forms a malonyl derivative and requires the vitamin **biotin** as a coenzyme. D-Methylmalonyl-CoA must be converted to its stereoisomer, L-methylmalonyl-CoA, by **methylmalonyl-CoA racemase** before its final isomerization to succinyl-CoA by the enzyme **methylmalonyl-CoA isomerase,** which requires **vitamin B_{12}** as a coenzyme. Vitamin B_{12} deficiency in humans and animals results in the excretion of large amounts of methylmalonate (**methylmalonic aciduria**).

Although the pathway to succinate is its main route of metabolism, propionate may also be used as the priming molecule for the synthesis—in adipose tissue and mammary gland—of fatty acids that have an odd number of carbon atoms in the molecule. **C_{15} and C_{17} fatty acids** are found particularly in the lipids of ruminants.

Glycerol is a product of the metabolism of adipose tissue, and only tissues that possess the activating enzyme, **glycerol kinase,** can utilize it. This enzyme, which requires ATP, is found in liver and kidney, among other tissues. Glycerol kinase catalyzes the conversion of glycerol to glycerol 3-phosphate. This pathway connects with the triosephosphate stages of the glycolysis pathway, because glycerol 3-phosphate may be oxidized to dihydroxyacetone phosphate by NAD^+ in the presence of **glycerol-3-phosphate dehydrogenase.** Liver and kidney are able to convert glycerol to blood

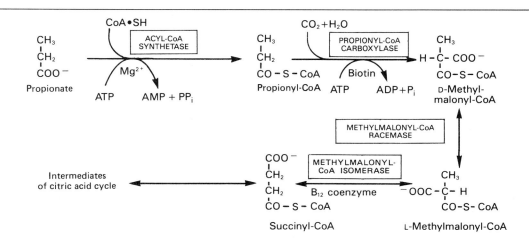

Figure 21–2. Metabolism of propionate.

glucose by making use of the above enzymes, some enzymes of glycolysis, and the specific enzymes of the gluconeogenic pathway, fructose-1,6-bisphosphatase and glucose-6-phosphatase (Fig 21–1).

GLYCOLYSIS & GLUCONEOGENESIS SHARE THE SAME PATHWAY: THEREFORE, THEY MUST BE RECIPROCALLY CONTROLLED

Changes in availability of substrates are either directly or indirectly responsible for most changes in metabolism. Fluctuations in their blood concentrations due to changes in dietary availability may alter the rate of secretion of hormones that influence, in turn, the pattern of metabolism in metabolic pathways—often by affecting the activity of key enzymes which attempt to compensate for the original change in substrate availability. Three types of mechanism can be identified as responsible for regulating the activity of enzymes concerned in carbohydrate metabolism and may be identified in Table 21–1: (1) changes in the rate of enzyme synthesis, (2) covalent modification by reversible phosphorylation, and (3) allosteric effects.

Induction & Repression of Key Enzyme Synthesis Is Not Rapid but Takes Place Over Several Hours

Some of the better documented changes in enzyme activity that are considered to occur under various metabolic conditions are listed in Table 21–1. The information in this table applies mainly to the liver. The enzymes involved catalyze nonequilibrium reactions that may be regarded physiologically as "one-way" rather than balanced reactions. Often the effects are reinforced because the activity of the enzymes catalyzing the changes in the opposite direction vary reciprocally (Fig 21–2). It is of importance that the key enzymes involved in a metabolic pathway are all activated or depressed in a coordinated manner. Table 21–1 shows that this is clearly the case. The enzymes involved in the utilization of glucose (ie, those of glycolysis and lipogenesis) all become more active when there is a superfluity of glucose, and under these conditions the enzymes responsible for producing glucose by the pathway of gluconeogenesis are all low in activity. The secretion of insulin, which is responsive to the blood glucose concentration, enhances the synthesis of the enzymes responsible for glycolysis. Likewise, it antagonizes the effect of the glucocorticoids and glucagon-stimulated cAMP in stimulating the key enzymes responsible for gluconeogenesis. All of these effects, which can be explained on the basis of enzyme induction or repression, can be prevented by agents that block the synthesis of protein, such as puromycin and ethionine. The regulation of the mRNA species of these enzymes and modulation of the expression of their genes have been demonstrated.

Both dehydrogenases of the pentose phosphate pathway can be classified as adaptive enzymes, since they increase in activity in the well-fed animal and when insulin is given to a diabetic animal. Activity is low in diabetes or fasting. "Malic enzyme" and ATP-citrate lyase behave similarly, indicating that these 2 enzymes are involved in lipogenesis rather than gluconeogenesis.

Covalent Modification by Reversible Phosphorylation Is Rapid

Glucagon, and to a lesser extent **epinephrine, inhibit glycolysis and stimulate gluconeogenesis** in the liver by increasing the concentration of cAMP, which in turn activates cAMP-dependent protein kinase, leading to the phosphorylation and inactivation of **pyruvate kinase.** They also affect the concentration of fructose 2,6-bisphosphate and therefore glycolysis and gluconeogenesis, as explained below.

Allosteric Modification is Also Rapid

Several examples are available from carbohydrate metabolism to illustrate allosteric control of the activity of an enzyme. In gluconeogenesis, the synthesis of oxaloacetate from bicarbonate and pyruvate, which is catalyzed by the enzyme **pyruvate carboxylase,** requires the presence of acetyl-CoA as an **allosteric activator.** The addition of acetyl-CoA results in a change in the tertiary structure of the protein, lowering the K_m value for bicarbonate. This effect has important implications for the self-regulation of intermediary metabolism, for, as acetyl-CoA is formed from pyruvate, it automatically ensures the provision of oxaloacetate, and, therefore, its further oxidation in the citric acid cycle, by activating pyruvate carboxylase. The activation of pyruvate carboxylase and the reciprocal inhibition of pyruvate dehydrogenase by acetyl-CoA formed from the oxidation of fatty acids helps to explain the sparing action of fatty acid oxidation on the oxidation of pyruvate and the stimulation of gluconeogenesis in the liver. The **reciprocal relationship between the activity of pyruvate dehydrogenase and pyruvate carboxylase** in both liver and kidney alters the metabolic fate of pyruvate as the tissue changes from carbohydrate oxidation, via glycolysis, to gluconeogenesis during transition from a fed to a starved state (Fig 21–1). A major role of fatty acid oxidation in promoting gluconeogenesis is to supply ATP required in the pyruvate carboxylase and phosphoenolpyruvate carboxykinase reactions.

Another enzyme that is subject to feedback control is **phosphofructokinase (phosphofructokinase-1).** It occupies a key position in regulating glycolysis. Phosphofructokinase-1 is **inhibited by citrate and by ATP** and is **activated by AMP.** AMP acts as an indicator of the energy status of the cell. The presence of **adenylate kinase** in liver and many other tissues allows rapid equilibration of the reaction:

$$ATP + AMP \leftrightarrow 2ADP$$

Table 21–1. Regulatory and adaptive enzymes of the rat (mainly liver).

	Activity In		Inducer	Repressor	Activator	Inhibitor
	Carbo-hydrate Feeding	Starva-tion and Diabetes				
Enzymes of glycogenesis, glycolysis & pyruvate oxidation						
Glycogen synthase system	↑	↓	Insulin		Insulin	Glucagon (cAMP), phosphorylase, glycogen
Hexokinase						*Glucose 6-phosphate
Glucokinase	↑	↓	Insulin	Glucagon (cAMP)		
Phosphofructokinase-1	↑	↓	Insulin		*AMP, *fructose 6-P, *P_i, *fructose 2,6-bisphosphate	*Citrate (fatty acids, ketone bodies), *ATP, glucagon (cAMP)
Pyruvate kinase	↑	↓	Insulin, fructose	Glucagon (cAMP)	*Fructose 1,6-bisphosphate	ATP, alanine, glucagon (cAMP), epinephrine
Pyruvate dehydrogenase	↑	↓			CoA, NAD, insulin†, ADP, pyruvate	Acetyl-CoA, NADH, ATP (fatty acids, ketone bodies)
Enzymes of gluconeogenesis						
Pyruvate carboxylase	↓	↑	Glucocorticoids, glucagon, epinephrine (cAMP)	Insulin	*Acetyl-CoA	*ADP
Phosphoenolpyruvate carboxykinase	↓	↑	Glucocorticoids, glucagon, epinephrine (cAMP)	Insulin	Glucagon?	
Fructose-1,6-bisphosphatase	↓	↑	Glucocorticoids, glucagon, epinephrine (cAMP)	Insulin	Glucagon (cAMP)	*Fructose 1,6-bisphosphate, *AMP, fructose 2,6-bisphosphate*
Glucose-6-phosphatase	↓	↑	Glucocorticoids, glucagon, epinephrine (cAMP)	Insulin		
Enzymes of the pentose phosphate pathway and lipogenesis						
Glucose-6-phosphate dehydrogenase	↑	↓	Insulin			
6-Phosphogluconate dehydrogenase	↑	↓	Insulin			
"Malic enzyme"	↑	↓	Insulin			
ATP-citrate lyase	↑	↓	Insulin			ADP
Acetyl-CoA carboxylase	↑	↓	Insulin?		*Citrate, insulin	Long-chain acyl-CoA, cAMP, glucagon
Fatty acid synthase	↑	↓	Insulin?			

*Allosteric.
†In adipose tissue but not in liver.

Thus, when ATP is used in energy-requiring processes resulting in formation of ADP, [AMP] rises. As [ATP] may be 50 times [AMP] at equilibrium, a small fractional decrease in [ATP] will cause a several-fold increase in [AMP]. Thus, **a large change in [AMP] acts as a metabolic amplifier of a small change in [ATP].** This mechanism allows the activity of phosphofructokinase-1 to be **highly sensitive to even small changes in energy status of the cell** and to **control the quantity of carbohydrate undergoing glycolysis prior to its entry into the citric acid cycle.** The increase in [AMP] can also explain why glycolysis is increased during hypoxia when [ATP] decreases. Simultaneously, AMP activates phosphorylase, increasing glycogenolysis. The inhibition of phosphofructokinase-1 by citrate and ATP could be another explanation of the sparing action of fatty acid oxidation on glucose oxidation and also of the **Pasteur effect** whereby aerobic oxidation (via the citric acid cycle) inhibits the anaerobic degradation of glucose. A consequence of the inhibition of phosphofructokinase-1 is an accumulation of glucose 6-phosphate which, in turn, inhibits further uptake of glucose in extrahepatic tissues by allosteric inhibition of hexokinase.

Fructose 2,6-Bisphosphate Has a Unique Role in the Regulation of Glycolysis & Gluconeogenesis

The most potent positive allosteric effector of phosphofructokinase-1 and inhibitor of fructose-1,6-bisphosphatase in liver is **fructose 2,6-bisphosphate.** It relieves inhibition of phosphofructokinase-1 by ATP and increases affinity for fructose 6-phosphate. It inhibits fructose-1,6-bisphosphatase by increasing the K_m for fructose 1,6-bisphosphate. Its concentration is under both substrate (allosteric) and hormonal control (covalent modification) (Fig 21–3).

Fructose 2,6-bisphosphate is formed by phosphorylation of fructose 6-phosphate by **phosphofructokinase-2.** The same enzyme protein is also responsible for its breakdown, since it contains **fructose-2,6-bisphosphatase** activity. This **bifunctional enzyme** is under the allosteric control of fructose 6-phosphate, which when raised in concentration owing to an abundance of glucose, ie, in the well-fed state, stimulates the kinase and inhibits the phosphatase. On the other hand, when glucose is short, glucagon stimulates the production of cAMP, activating cAMP-dependent protein kinase, which in turn inactivates phosphofructokinase-2 and activates fructose-2,6-bisphosphatase by phosphorylation. Thus, **under a superfluity of glucose, fructose 2,6-bisphosphate increases in concentration, stimulating glycolysis by activating phosphofructokinase-1 and inhibiting fructose-1,6-bisphosphatase.** Under conditions of glucose shortage, gluconeogenesis is stimulated by a decrease in the concentration of fructose 2,6-bisphosphate, which deactivates phosphofructokinase-1 and deinhibits fructose-1,6-bisphosphatase. **This mechanism also ensures that glucagon stimulation of glycogenolysis in liver results in glucose release rather than glycolysis.**

Substrate (Futile) Cycles Allow Fine Tuning

It will be apparent that many of the control points in glycolysis and glycogen metabolism involve a cycle of phosphorylation and dephosphorylation, eg, glucokinase/glucose-6-phosphatase; phosphofructokinase-1/fructose-1,6-bisphosphatase; pyruvate kinase/ pyruvate carboxylase/phosphoenolpyruvate carboxykinase; glycogen synthase/phosphorylase. If these were allowed to cycle unchecked, they would amount to futile cycles whose net result was hydrolysis of ATP. That this does not occur extensively is due to the various control mechanisms, which ensure that one limb of the cycle is inhibited as the other is stimulated, according to

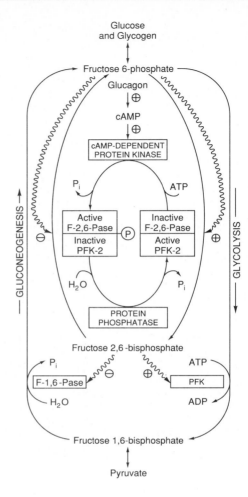

Figure 21–3. Control of glycolysis and gluconeogenesis in the liver by fructose 2,6-bisphosphate and the bifunctional enzyme PFK-2/F-2,6-Pase (6-phosphofructo-2-kinase/fructose 2,6-bisphosphatase). PFK, phosphofructokinase (6-phosphofructo-1-kinase); F-1,6-Pase, fructose-1,6-bisphosphatase. Arrows with wavy shafts indicate allosteric effects.

the need of the tissue and of the body. However, there may be a physiologic advantage in allowing some cycling. For example, in the phosphofructokinase/fructose-1,6-bisphosphatase cycle, an amplification of the effect of an allosteric modifier, eg, fructose 2,6-bisphosphate, would occur, causing a larger change in net flux of metabolites in either direction than would occur in the absence of substrate cycling. This "fine tuning" of metabolic control only occurs at the expense of some loss of ATP.

THE CONCENTRATION OF BLOOD GLUCOSE IS REGULATED WITHIN FINE LIMITS

In the postabsorptive state, the concentration of blood glucose in individual humans and many mammals is set within the range 4.5–5.5 mmol/L. After the ingestion of a carbohydrate meal, it may rise to 6.5–7.2. During fasting, the levels fall to around 3.3–3.9. The blood glucose level in birds is considerably higher (14.0) and in ruminants considerably lower (approximately 2.2 in sheep and 3.3 in cattle). These lower normal levels appear to be associated with the fact that ruminants ferment virtually all dietary carbohydrate to lower (volatile) fatty acids, and these largely replace glucose as the main metabolic fuel of the tissues in the fed condition. A sudden decrease in blood glucose will cause convulsions, as in insulin overdose, due to the immediate dependence of the brain on a supply of glucose. However, much lower concentrations can be tolerated, provided progressive adaptation is allowed; eg, rats adapted to high-fat diets appear normal with a blood glucose concentration as low as 1.1 mmol/L.

BLOOD GLUCOSE IS DERIVED FROM THE DIET, GLUCONEOGENESIS, & GLUCOGENOLYSIS

Glucose From Carbohydrates in the Diet: Most carbohydrates in the diet form glucose, galactose, or fructose upon digestion. These are transported to the liver via the **hepatic portal vein.** Galactose and fructose are readily converted to glucose in the liver (see Chapter 22).

Glucose From Various Glucogenic Compounds That Undergo Gluconeogenesis: (Figs 18–7 and 21–1.) These compounds fall into 2 catego-

ries: (1) those which involve a direct net conversion to glucose without significant recycling, such as some **amino acids** and **propionate;** and (2) those which are the products of the partial metabolism of glucose in certain tissues and which are conveyed to the liver and kidney, where they are resynthesized to glucose. Thus, **lactate,** formed by the oxidation of glucose in skeletal muscle and by erythrocytes, is transported to the liver and kidney where it re-forms glucose, which again becomes available via the circulation for oxidation in the tissues. This process is known as the **Cori cycle** or lactic acid cycle (Fig 21–4). **Glycerol** for the synthesis of triacylglycerols of adipose tissue is derived from the blood glucose. Acylglycerols of adipose tissue are continuously undergoing hydrolysis to form free glycerol, which cannot be utilized by adipose tissue and therefore diffuses out into the blood. It is converted back to glucose by gluconeogenetic mechanisms in the liver and kidney (Fig 21–1). Thus, a continuous cycle exists in which glucose is transported from the liver and kidney to adipose tissue and glycerol is returned from adipose tissue to be resynthesized into glucose by the liver and kidney.

Of the amino acids transported from muscle to the liver during starvation, alanine predominates. This has led to the postulation of a **glucose-alanine cycle,** as shown in Fig 21–4, which has the effect of cycling glucose from liver to muscle with formation of pyruvate, followed by transamination to alanine, then transport of alanine to liver, followed by gluconeogenesis back to glucose. A net transfer of amino nitrogen from muscle to liver and of free energy from liver

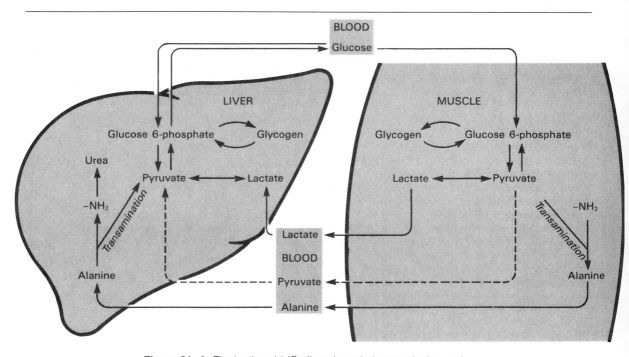

Figure 21–4. The lactic acid (Cori) cycle and glucose-alanine cycle.

to muscle is effected. The energy required for the hepatic synthesis of glucose from pyruvate is derived from the oxidation of fatty acids.

Glucose From Liver Glycogen by Glycogenolysis: (See Chapter 20.)

Metabolic & Hormonal Mechanisms Control the Concentration of Blood Glucose

The maintenance of stable levels of glucose in the blood is one of the most finely regulated of all homeostatic mechanisms and one in which the liver, the extrahepatic tissues, and several hormones play a part. **Liver cells appear to be freely permeable to glucose,** whereas cells of **extrahepatic tissues are relatively impermeable.** As a result, the passage through the cell membrane is the rate-limiting step in the uptake of glucose in extrahepatic tissues, and glucose is rapidly phosphorylated by hexokinase on entry into the cells. On the other hand, it is probable that the activity of certain enzymes and the concentration of key intermediates exert a much more direct effect on the uptake or output of glucose from liver. Nevertheless, the concentration of glucose in the blood is an important factor controlling the rate of uptake of glucose in both liver and extrahepatic tissues.

Glucokinase Is Important in Regulating Blood Glucose After a Meal: It is to be noted that hexokinase is inhibited by glucose 6-phosphate, so that some feedback control may be exerted on glucose uptake in extrahepatic tissues dependent on hexokinase for glucose phosphorylation. The liver is not subject to this constraint because glucokinase is not affected by glucose 6-phosphate. Glucokinase, which has a higher K_m (lower affinity) for glucose than does hexokinase, increases in activity over the physiologic range of glucose concentrations (Fig 21–5) and seems to be specifically concerned with glucose uptake into the liver at the higher concentrations found in the hepatic portal vein after a carbohydrate meal. Its absence in

ruminants, which have little glucose entering the portal circulation from the intestines, is compatible with this function.

At normal systemic blood glucose concentrations (4.5–5.5 mmol/L), the liver appears to be a net producer of glucose. However, as the glucose level rises, the output of glucose ceases, so that at high levels there is a net uptake. In the rat, it has been estimated that the rate of uptake of glucose and the rate of output are equal at a hepatic portal vein blood glucose concentration of 8.3 mmol/L.

Insulin Plays a Central Role in Regulating Blood Glucose: In addition to the direct effects of hyperglycemia in enhancing the uptake of glucose into both the liver and peripheral tissues, the hormone insulin plays a central role in regulating the blood glucose concentration. It is produced by the B cells of the islets of Langerhans in the pancreas and is secreted into the blood as a direct response to hyperglycemia. Its concentration in the blood parallels that of the blood glucose, and its administration results in prompt **hypoglycemia.** Substances causing release of insulin include also amino acids, free fatty acids, ketone bodies, glucagon, secretin, and the drug tolbutamide. Epinephrine and norepinephrine block the release of insulin. Insulin has an immediate effect of increasing glucose uptake in tissues such as adipose tissue and muscle. This action is due to an enhancement of glucose transport through the cell membrane by recruitment of insulin transporters from the interior of the cell to the plasma membrane. In contrast, there is no direct effect of insulin on glucose penetration of hepatic cells; this finding agrees with the fact that glucose metabolism by liver cells is not rate-limited by their permeability to glucose. However, insulin does indirectly enhance uptake of glucose by the liver as a result of its actions on the enzymes controlling glycolysis and glycogenesis.

Glucagon is the hormone produced by the A cells of the islets of Langerhans of the pancreas. Its secretion is stimulated by hypoglycemia. When it reaches the liver (via the portal vein), it causes glycogenolysis by activating phosphorylase. Most of the endogenous glucagon is cleared from the circulation by the liver. Unlike epinephrine, glucagon does not have an effect on muscle phosphorylase. Glucagon also enhances gluconeogenesis from amino acids and lactate. Both hepatic glycogenolysis and gluconeogenesis contribute to the **hyperglycemic** effect of glucagon, whose actions oppose those of insulin.

The **anterior pituitary gland** secretes hormones that tend to elevate the blood glucose and therefore antagonize the action of insulin. These are growth hormone, ACTH (corticotropin), and possibly other "diabetogenic" principles. Growth hormone secretion is stimulated by hypoglycemia. Growth hormone decreases glucose uptake in certain tissues, eg, muscle. Some of this effect may not be direct, since it mobilizes free fatty acids from adipose tissue which themselves inhibit glucose utilization. Chronic administration of growth hormone leads to diabetes. By producing hy-

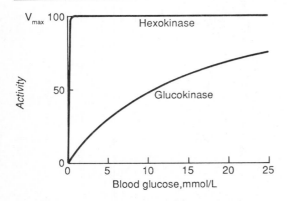

Figure 21–5. Variation in glucose phosphorylating activity of hexokinase and glucokinase with increase of blood glucose concentration. The K_m for glucose of hexokinase is 0.05 mmol/L and of glucokinase is 10 mmol/L.

perglycemia, it stimulates secretion of insulin, eventually causing B cell exhaustion.

The **glucocorticoids** (11-oxysteroids) are secreted by the adrenal cortex and are important in carbohydrate metabolism. Administration of these steroids causes increased gluconeogenesis. This is a result of increased protein catabolism in the tissues, increased hepatic uptake of amino acids, and increased activity of transaminases and other enzymes concerned with gluconeogenesis in the liver. In addition, glucocorticoids **inhibit the utilization of glucose in extrahepatic tissues.** In all these actions, glucocorticoids act in a manner **antagonistic to insulin.**

Epinephrine is secreted by the adrenal medulla as a result of stressful stimuli (fear, excitement, hemorrhage, hypoxia, hypoglycemia, etc) and leads to glycogenolysis in liver and muscle owing to stimulation of phosphorylase. In muscle, as a result of the absence of glucose-6-phosphatase, glycogenolysis ensues with the formation of lactate, whereas in liver, glucose is the main product leading to increase in blood glucose.

Thyroid hormone should also be considered as affecting the blood glucose. There is experimental evidence that thyroxine has a diabetogenic action and that thyroidectomy inhibits the development of diabetes. It has also been noted that there is a complete absence of glycogen from the livers of thyrotoxic animals. In humans, the fasting blood glucose is elevated in hyperthyroid patients and decreased in hypothyroid patients. However, hyperthyroid patients apparently utilize glucose at a normal or increased rate, whereas hypothyroid patients have a decreased ability to utilize glucose. In addition, hypothyroid patients are much less sensitive to insulin than are normal or hyperthyroid individuals.

Glycosuria Occurs When the Renal Threshold for Glucose is Exceeded

When the blood glucose rises to relatively high levels, the kidney also exerts a regulatory effect. Glucose is continuously filtered by the glomeruli but is ordinarily returned completely to the blood by the reabsorptive system of the renal tubules. The reabsorption of glucose against its concentration gradient is linked to the provision of ATP in the tubular cells. The capacity of the tubular system to reabsorb glucose is limited to a rate of about 350 mg/min. When the blood levels of glucose are elevated, the glomerular filtrate may contain more glucose than can be reabsorbed; the excess passes into the urine to produce **glycosuria.** In normal individuals, glycosuria occurs when the venous blood glucose concentration exceeds 9.5–10.0 mmol/L. This is termed the **renal threshold** for glucose.

Glycosuria may be produced in experimental animals with **phlorhizin,** which inhibits the glucose reabsorptive system in the tubule. This is known as renal glycosuria. Glycosuria of renal origin may result from inherited defects in the kidney, or it may be acquired as a result of disease processes. **The presence of glycosuria is frequently an indication of diabetes mellitus.**

Fructose-1,6-Bisphosphatase Deficiency Causes Lactacidosis and Hypoglycemia

Blockage of gluconeogenesis by deficiency of this enzyme prevents lactate and other glucogenic substrates from being converted to glucose in the liver. The condition may be controlled by feeding high-carbohydrate diets.

THE BODY'S ABILITY TO UTILIZE GLUCOSE MAY BE ASCERTAINED BY MEASURING ITS GLUCOSE TOLERANCE

Glucose tolerance is indicated by the nature of the blood glucose curve following the administration of a test amount of glucose (Fig 21–6). **Diabetes mellitus** ("sugar" diabetes) is characterized by decreased glucose tolerance due to decreased secretion of insulin (type I, or insulin-dependent diabetes mellitus, IDDM). This is manifested by elevated blood glucose levels (hyperglycemia) and glycosuria and may be accompanied by changes in fat metabolism. Tolerance to glucose declines not only in type I diabetes but also in conditions where the liver is damaged; in some infections; in type II diabetes mellitus (non-insulin-depen-

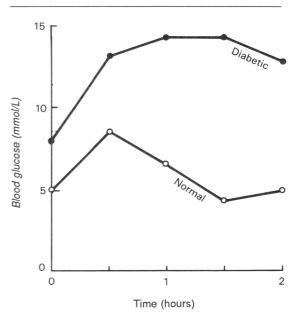

Figure 21–6. Glucose tolerance test. Blood glucose curves of a normal and a diabetic individual after oral administration of 50 g of glucose. Note the initial raised concentration in the diabetic. A criterion of normality is the return of the curve to the initial value within 2 hours.

dent diabetes mellitus, NIDDM), which is often associated with obesity and raised levels of plasma free fatty acids; under the influence of some drugs; and sometimes in atherosclerosis. It can also be expected to occur in the presence of hyperactivity of the pituitary or adrenal cortex, because of the antagonism of the hormones of these endocrine glands to the action of insulin.

Insulin increases glucose tolerance. Injection of insulin lowers the content of the glucose in the blood and increases its utilization and its storage in the liver and muscle as glycogen. An excess of insulin may cause severe **hypoglycemia,** resulting in convulsions and even in death unless glucose is administered promptly. Increased tolerance to glucose is observed in pituitary or adrenocortical insufficiency, attributable to a decrease in the normal antagonism to insulin which results in a relative excess of that hormone.

REFERENCES

Krebs HA: Gluconeogenesis. *Proc R Soc London (Biol)* 1964;**159**:545.

Newsholme EA, Chaliss RAJ, Crabtree B: Substrate cycles: Their role in improving sensitivity in metabolic control. *Trends Biochem Sci* 1984;**9**:277.

Newsholme EA, Start C: *Regulation in Metabolism.* Wiley, 1973.

Pagliara AS et al: Hepatic Fructose-1,6-Diphosphatase Deficiency, a cause of lactic acidosis and hypoglycemia in infancy. *J Clin Invest* 1972;**51**:2115.

Pilkis SJ, El-Maghrabi MR, Claus TH: Hormonal regulation of hepatic gluconeogenesis and glycolysis. *Annu Rev Biochem* 1988;**57**:755.

Watford M: What is the metabolic fate of dietary glucose? *Trends Biochem Sci* 1988;**13**:329.

The Pentose Phosphate Pathway & Other Pathways of Hexose Metabolism

22

Peter A. Mayes, PhD, DSc

INTRODUCTION

The pentose phosphate cycle does not generate ATP but has 2 major functions: (1) The generation of **NADPH** for reductive syntheses such as fatty acid and steroid biosynthesis, and (2) the provision of **ribose** for nucleotide and nucleic acid biosynthesis.

Glucose, fructose, and galactose are quantitatively the most important hexoses absorbed from the gastrointestinal tract. They are derived from dietary starch, sucrose, and lactose, respectively. Specialized pathways have been developed, particularly in the liver, for the conversion of fructose and galactose to glucose.

BIOMEDICAL IMPORTANCE

Deficiencies of certain enzymes of the pentose phosphate pathway are major causes of hemolysis of red blood cells, resulting in one type of **hemolytic anemia.** The principal enzyme involved is glucose-6-phosphate dehydrogenase. As many as 100 million people in the world may have genetically determined low levels of this enzyme.

The major metabolic routes for the utilization of glucose are glycolysis and the pentose phosphate pathway. Of minor quantitative importance but of major significance for the excretion of metabolites and foreign chemicals (xenobiotics) as **glucuronides** is the elaboration of glucuronic acid via the uronic acid pathway from glucose. A deficiency in the pathway leads to the condition of **essential pentosuria.** The total absence of one particular enzyme of the pathway in all primates accounts for the fact that **ascorbic acid** (vitamin C) is required in the diet of humans but not most other mammals. Deficiencies in the enzymes of fructose and galactose metabolism lead to metabolic diseases such as **essential fructosuria** and the **galactosemias.** Fructose has been used for parenteral nutrition, but at high concentration it can cause depletion of adenine nucleotides in liver and hepatic necrosis.

THE PENTOSE PHOSPHATE PATHWAY GENERATES NADPH & RIBOSE

The pentose phosphate pathway (hexose monophosphate shunt) is an alternative route for the oxidation of glucose. It is a multicyclic process in which 3 molecules of glucose 6-phosphate give rise to 3 molecules of CO_2 and three 5-carbon residues. The latter are rearranged to regenerate 2 molecules of glucose 6-phosphate and one molecule of the glycolytic intermediate, glyceraldehyde 3-phosphate. Since 2 molecules of glyceraldehyde 3-phosphate can regenerate glucose 6-phosphate, glucose may be completely oxidized by this pathway.

$$3 \text{ Glucose 6-phosphate} + 6\text{NADP}^+ \rightarrow$$
$$3\text{CO}_2 + 2 \text{ Glucose 6-phosphate} +$$
$$\text{Glyceraldehyde 3-phosphate} + 6\text{NADPH} + 6\text{H}^+$$

THE SEQUENCE OF REACTIONS IN THE PENTOSE PHOSPHATE PATHWAY OCCURS IN THE CYTOSOL

The enzymes of the pentose phosphate pathway like in glycolysis are found in the cytosol. As in glycolysis, oxidation is achieved by dehydrogenation; but in the case of the pentose phosphate pathway, **NADP** and not NAD is used as a hydrogen acceptor.

The sequence of reactions of the pathway may be divided into 2 phases: **oxidative** and **nonoxidative.** In the first, glucose 6-phosphate undergoes dehydrogenation and decarboxylation to give a pentose, ribulose 5-phosphate. In the second phase, ribulose 5-phosphate is converted back to glucose 6-phosphate by a series of reactions involving mainly 2 enzymes: **transketolase** and **transaldolase** (Fig 22–1).

The Oxidative Phase Generates NADPH

Dehydrogenation of glucose 6-phosphate to 6-phosphogluconate occurs via the formation of 6-phos-

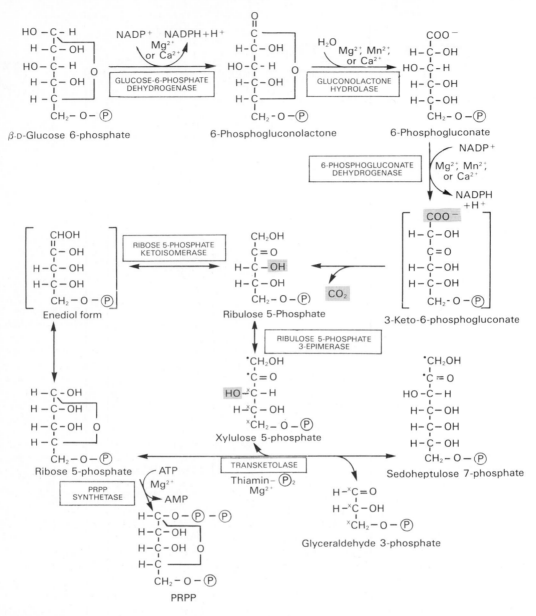

Figure 22–1. The pentose phosphate pathway. P, $-PO_3^{2-}$, PRPP, 5-phosphoribosyl-1-pyrophosphate.

phogluconolactone catalyzed by **glucose-6-phosphate dehydrogenase,** an NADP-dependent enzyme. The hydrolysis of 6-phosphogluconolactone is accomplished by the enzyme **gluconolactone hydrolase.** A second oxidative step is catalyzed by **6-phosphogluconate dehydrogenase,** which also requires NADP$^+$ as hydrogen acceptor. Decarboxylation follows with the formation of the ketopentose, ribulose 5-phosphate. The reaction probably takes place in 2 steps through the intermediate 3-keto-6-phosphogluconate.

The Nonoxidative Phase Generates Ribose Precursors

Ribulose 5-phosphate now serves as substrate for 2 different enzymes. **Ribulose 5-phosphate 3-epimerase** alters the configuration about carbon 3, forming the epimer xylulose 5-phosphate, another ketopentose. **Ribose 5-phosphate ketoisomerase** converts ribulose 5-phosphate to the corresponding aldopentose, ribose 5-phosphate.

Transketolase transfers the 2-carbon unit comprising carbons 1 and 2 of a ketose to the aldehyde carbon

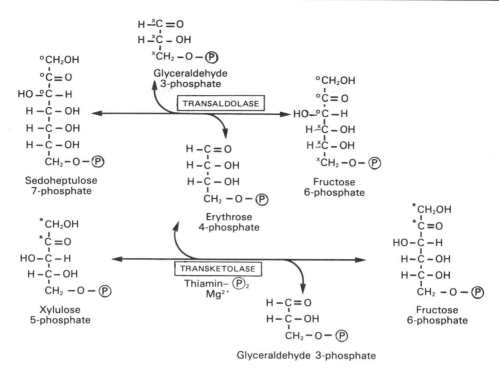

Figure 22–1 (cont'd). The pentose phosphate pathway.

of an aldose sugar. It therefore effects the conversion of a ketose sugar into an aldose with 2 carbons less, and simultaneously converts an aldose sugar into a ketose with 2 carbons more. The reaction requires a B vitamin, **thiamin,** as the coenzyme thiamin diphosphate in addition to Mg^{2+} ions. The 2-carbon moiety transferred is probably glycolaldehyde bound to thiamin diphosphate, ie, "active glycolaldehyde." Thus, transketolase catalyzes the transfer of the 2-carbon unit from xylulose 5-phosphate to ribose 5-phosphate, producing the 7-carbon ketose sedoheptulose 7-phosphate and the aldose glyceraldehyde 3-phosphate. These 2 products then enter another reaction known as transaldolation. **Transaldolase** allows the transfer of a 3-carbon moiety, "active dihydroxyacetone" (carbons 1–3), from the ketose sedoheptulose 7-phosphate to the aldose glyceraldehyde 3-phosphate to form the ketose fructose 6-phosphate and the 4-carbon aldose erythrose 4-phosphate.

A further reaction takes place, again involving **transketolase,** in which xylulose 5-phosphate serves as a donor of "active glycolaldehyde." In this case the erythrose 4-phosphate formed above acts as acceptor, and the products of the reaction are fructose 6-phosphate and glyceraldehyde 3-phosphate.

In order to oxidize glucose completely to CO_2 via the pentose phosphate pathway, it is necessary that the enzymes be present in the tissue to convert glyceraldehyde 3-phosphate to glucose 6-phosphate. This in-

volves enzymes of the glycolysis pathway working in a reverse direction and, in addition, the gluconeogenic enzyme **fructose-1,6-bisphosphatase.** A summary of the pathway is shown in Fig 22–2.

The Two Major Pathways for the Catabolism of Glucose Have Little in Common

Although some metabolites are common eg, glucose 6-phosphate, the pentose phosphate pathway is markedly different from glycolysis. Oxidation occurs in the first reactions utilizing NADP rather than NAD, and CO_2, which is not produced at all in the glycolysis pathway, is a characteristic product. ATP is not generated in the pentose phosphate pathway whereas it is a major function of glycolysis. Ribose phosphates are generated in the pentose phosphate pathway but not in glycolysis.

Reducing Equivalents Are Generated in Those Tissues Specializing in Reductive Syntheses

Estimates of the activity of the pentose phosphate pathway in various tissues indicate its metabolic significance. It is active in liver, adipose tissue, adrenal cortex, thyroid, erythrocytes, testis, and lactating mammary gland. It is not active in nonlactating mammary gland, and its activity is low in skeletal muscle.

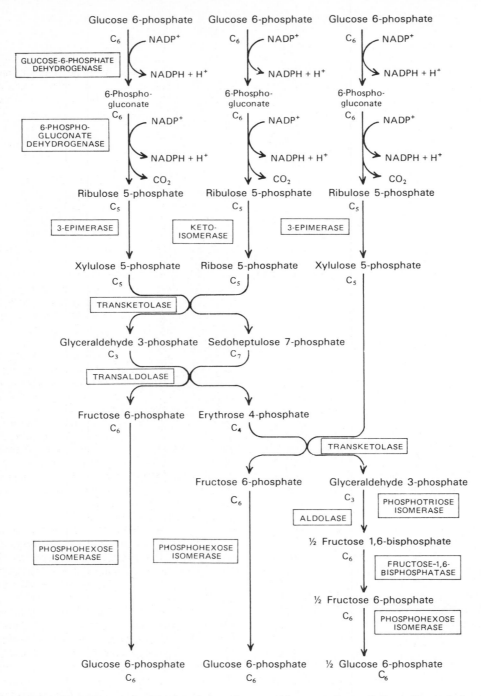

Figure 22–2. Flow chart of pentose phosphate pathway and its connections with the pathway of glycolysis.

All of the tissues in which the pathway is active use NADPH in reductive syntheses, eg, synthesis of fatty acids, steroids, amino acids via glutamate dehydrogenase, or reduced glutathione in erythrocytes. It is probable that the presence of active lipogenesis or of a system which utilizes NADPH stimulates an active degradation of glucose via the pentose phosphate pathway. This provides NADP, which is normally very low in concentration because of the nonequilibrium nature of the first reactions in the pathway. The synthesis of glucose-6-phosphate dehydrogenase and 6-phosphogluconate dehydrogenase may also be induced by insulin during conditions associated with the "fed state" (Table 21–1).

Ribose Can Be Synthesized in Virtually All Tissues

The pentose phosphate pathway provides ribose for nucleotide and nucleic acid synthesis (Fig 22–1). The source is the ribose 5-phosphate intermediate that reacts with ATP to form PRPP used in nucleotide biosynthesis (see p 344). Muscle tissue contains only very small amounts of glucose-6-phosphate dehydrogenase and 6-phosphogluconate dehydrogenase. Nevertheless, skeletal muscle, like most other tissues, is capable of synthesizing ribose 5-phosphate for nucleotide synthesis. This is accomplished by a reversal of the nonoxidative phase of the pentose phosphate pathway utilizing fructose 6-phosphate. **Thus, it is not necessary to have a completely functioning pentose phosphate pathway in order that a tissue may synthesize ribose phosphates.**

Ribose is not a significant constituent of systemic blood. Therefore tissues must satisfy their own requirement for this important precursor of nucleotides.

THE PENTOSE PHOSPHATE PATHWAY ASSISTS GLUTATHIONE PEROXIDASE IN PROTECTING ERYTHROCYTES AGAINST HEMOLYSIS

The pentose phosphate pathway in the erythrocyte provides NADPH for the reduction of oxidized glutathione (G–S–S–G) to reduced glutathione (2G–SH), catalyzed by **glutathione reductase,** a flavoprotein enzyme containing FAD. In turn, reduced glutathione removes H_2O_2 from the erythrocyte in a reaction catalyzed by **glutathione peroxidase,** an enzyme that contains the trace element **selenium.**

$$G–S–S–G + NADPH + H^+ \rightarrow 2G–SH + NADP^+$$

GLUTHATHIONE REDUCTASE

FAD

$$2G–SH + H_2O_2 \rightarrow G–S–S–G + 2H_2O$$

GLUTHATHIONE PEROXIDASE

Selenium

This reaction is important, since accumulation of H_2O_2 may decrease the life span of the erythrocyte by increasing the rate of oxidation of hemoglobin to methemoglobin. A mutation present in some populations causes a deficiency in glucose-6-phosphate dehydrogenase, with consequent impairment of the generation of NADPH. This impairment is manifested as red cell hemolysis when the susceptible individual is subjected to oxidants, such as the antimalarial primaquine, aspirin, or sulfonamides, or when the susceptible individual has eaten fava beans (*Vicia fava*—favism).

Gluthathione peroxidase is a natural antioxidant found in many tissues. It will attack organic peroxides in addition to H_2O_2. Together with vitamin E, it is part of the body's defense against lipid peroxidation (see p 138). An association between the incidence of some cancers and low levels of blood selenium and glutathione peroxidase activity has been reported.

Measurement of **transketolase** activity in blood reflects the degree of thiamin deficiency. The only condition in which its activity is raised is pernicious anemia.

GLUCURONATE, A PRECURSOR OF PROTEOGLYCANS & CONJUGATED GLUCURONIDES, IS A PRODUCT OF THE URONIC ACID PATHWAY

Besides the major pathways of metabolism of glucose 6-phosphate that have been described, there exists a pathway for the conversion of glucose to glucuronic acid, ascorbic acid, and pentoses that is referred to as the **uronic acid pathway.** It is also an alternative oxidative pathway for glucose, but like the pentose phosphate pathway, it does not lead to the generation of ATP.

In the uronic acid pathway, glucuronate is formed from glucose by the reactions shown in Fig 22–3. Glucose 6-phosphate is converted to glucose 1-phosphate, which then reacts with uridine triphosphate (UTP) to form the active nucleotide, uridine diphosphate glucose (UDPGlc). This latter reaction is catalyzed by the enzyme **UDPGlc pyrophosphorylase.** All of the steps up to this point are those previously indicated as in the pathway of glycogenesis in the liver. UDPGlc is oxidized at carbon 6 by a 2-step process to glucuronate. The product of the oxidation, which is catalyzed by an NAD-dependent **UDPGlc dehydrogenase, is UDP-glucuronate.**

UDP-glucuronate is the "active" form of glucuronate for reactions involving incorporation of glucuronic acid into proteoglycans or for reactions in which glucuronate is conjugated to such substrates as steroid hormones, certain drugs, or bilirubin (see Fig 34–13).

In an NADPH-dependent reaction, glucuronate is reduced to L-gulonate (Fig 22–3). This latter compound is the direct precursor of **ascorbate** in those animals capable of synthesizing this vitamin. In humans and other primates, as well as in guinea pigs, ascorbic acid cannot be synthesized because of the absence of the enzyme L-**gulonolactone oxidase.**

Gulonate is oxidized to 3-keto-L-gulonate, which is then decarboxylated to the pentose, L-xylulose. D-Xylulose is a constituent of the pentose phosphate pathway, but in the reactions shown in Fig 22–3, the L isomer of xylulose is formed from ketogulonate. If the 2 pathways are to connect, it is necessary to convert L-xylulose to the D isomer. This is accomplished by an NADPH-dependent reduction to xylitol, which is then oxidized in an NAD-dependent reaction to D-xylulose; this latter compound, after conversion to D-xylulose 5-

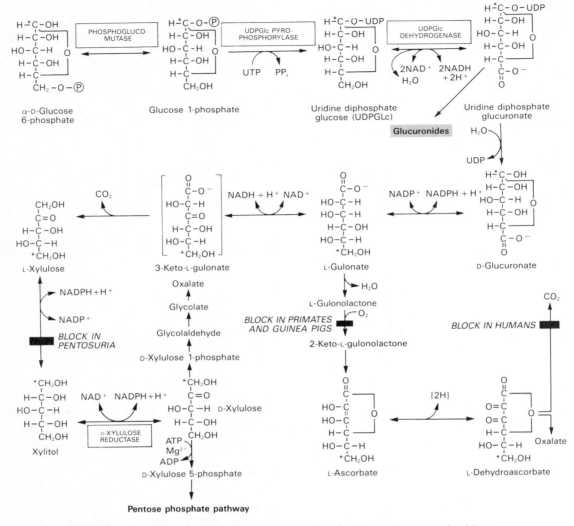

Figure 22–3. Uronic acid pathway. *Indicates the fate of carbon 1 of glucose; $\textcircled{P}$,$-PO_3^{2-}$.

phosphate, is further metabolized in the pentose phosphate pathway.

Disruption of the Uronic Acid Pathway Is Caused by Enzyme Defects & Some Drugs

In the rare hereditary disease **essential pentosuria,** considerable quantities of **L-xylulose** appear in the urine. It is now believed that this may be explained by the absence in pentosuric patients of the enzyme necessary to accomplish reduction of L-xylulose to xylitol. Parenteral administration of xylitol may lead to **oxalosis** involving calcium oxalate deposition in brain and kidneys. This results from the conversion of D-xylulose to oxalate via xylulose 1-phosphate, glycolaldehyde, and glycolate formation.

Various drugs markedly increase the rate at which glucose enters the uronic acid pathway. For example, administration of barbital or of chlorobutanol to rats

results in a significant increase in the conversion of glucose to glucuronate, L-gulonate, and ascorbate. Aminopyrine and antipyrine have been reported to increase the excretion of L-xylulose in pentosuric subjects.

INGESTION OF LARGE QUANTITIES OF FRUCTOSE HAS PROFOUND METABOLIC CONSEQUENCES

Diets high in sucrose content lead to large amounts of fructose (and glucose) entering the hepatic portal vein.

Fructose is more rapidly glycolyzed by the liver than glucose. This is due to the fact that it bypasses the step in glucose metabolism catalyzed by **phosphofructokinase,** at which point metabolic control is exerted on the rate of catabolism of glucose. **This allows**

fructose to flood the pathways in the liver leading to enhanced fatty acid synthesis, esterification of fatty acids, and VLDL secretion, which may raise serum triacylglycerols. The extra glucose taken into the blood stream stimulates more insulin secretion, which enhances all these effects.

Fructose may be phosphorylated to form fructose 6-phosphate, catalyzed by the same enzyme, hexokinase, that accomplishes the phosphorylation of glucose (or mannose) (see Fig 22–4). However, the affinity of the enzyme for fructose is very low compared with its affinity for glucose. It is unlikely, therefore, that this is a major pathway for fructose utilization.

Another enzyme, **fructokinase,** is present in liver and effects the transfer of phosphate from ATP to fructose, forming fructose 1-phosphate. It has also been demonstrated in kidney and intestine. This enzyme will not phosphorylate glucose, and, unlike glucoki-

nase, its activity is not affected by fasting or by insulin, which may explain why fructose disappears from the blood of diabetic patients at a normal rate. The K_m for fructose of the enzyme in liver is very low, indicating a very high affinity of the enzyme for its substrate. It seems probable that this is the major route for the phosphorylation of fructose. **Essential fructosuria** results from a lack of hepatic fructokinase.

Fructose 1-phosphate is split into D-glyceraldehyde and dihydroxyacetone phosphate by **aldolase B,** an enzyme found in the liver. The enzyme also attacks fructose 1,6-bisphosphate. Absence of this enzyme leads to a **hereditary fructose intolerance,** D-Glyceraldehyde gains entry to the glycolysis sequence of reaction via another enzyme present in liver, **triokinase,** which catalyzes its phosphorylation to glyceraldehyde 3-phosphate. The 2 triose phosphates, dihydroxyacetone phosphate and glyceraldehyde 3-phosphate, may be degraded via the glycolysis pathway or they may

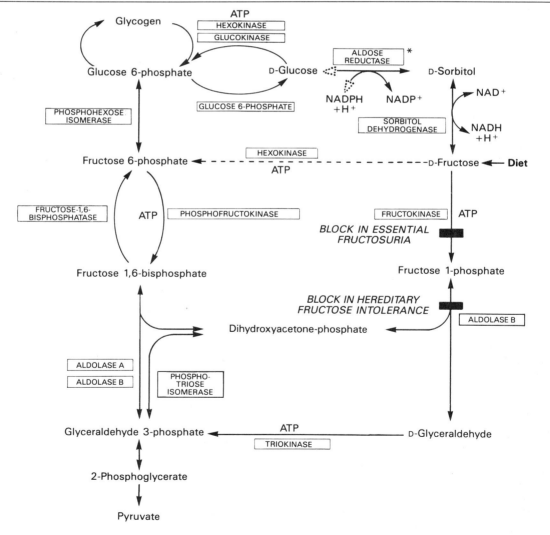

Figure 22–4. Metabolism of fructose. Aldolase A is found in all tissues except the liver, where only aldolase B is present.

combine under the influence of aldolase and be converted to glucose. The latter is the fate of much of the fructose metabolized in the liver.

One consequence of hereditary fructose intolerance and of another condition due to **fructose-1,6-bisphosphatase deficiency** is a fructose-induced **hypoglycemia** despite the presence of high glycogen reserves. Apparently, the accumulation of fructose 1-phosphate and fructose 1,6-bisphosphate inhibits the activity of liver phosphorylase by allosteric mechanisms.

If the liver and intestines of an experimental animal are removed, the conversion of injected fructose to glucose does not take place and the animal succumbs to hypoglycemia unless glucose is administered. However, it is reported that the human kidney can convert fructose to glucose and lactate. In humans but not in the rat, a significant amount of the fructose resulting from the digestion of sucrose is converted to glucose in the intestinal wall prior to passage into the portal circulation.

Free fructose is found in seminal plasma and is secreted in quantity into the fetal circulation of ungulates and whales, where it accumulates in the amniotic and allantoic fluids.

Fructose & Sorbitol in the Lens Are Associated With Diabetic Cataract

Both fructose and **sorbitol** are found in the human lens, where they increase in concentration in diabetes and may be involved in the pathogenesis of **diabetic cataract.** The **sorbitol (polyol) pathway** (not found in liver) is responsible for fructose formation from glucose (Fig 22–4) and increases in activity as the glucose concentration rises in diabetes in those tissues that are not insulin sensitive, ie, the lens, peripheral nerves, and renal glomoruli. Glucose undergoes reduction by NADPH to sorbitol catalyzed by **aldose reductase,** followed by oxidation of sorbitol to fructose in the presence of NAD and **sorbitol dehydrogenase** (polyol dehydrogenase). Sorbitol does not diffuse through cell membranes easily and therefore accumulates, causing osmotic damage. Simultaneously, myoinositol levels fall. Sorbitol accumula-

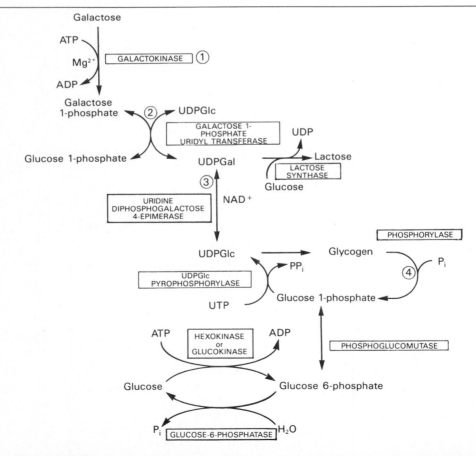

Figure 22–5. The pathway for conversion of galactose to glucose and for the synthesis of lactose.

tion, myoinositol depletion, and diabetic cataract can be prevented by aldose reductase inhibitors in diabetic rats.

Aldose reductase is found in the placenta of the ewe and is responsible for the secretion of sorbitol into the fetal blood. The presence of sorbitol dehydrogenase in the liver, including the fetal liver, is responsible for the conversion of sorbitol into fructose. The pathway is also responsible for the occurrence of fructose in seminal fluid. When sorbitol is administered intravenously, it is converted to fructose rather than to glucose, although if given by mouth, much escapes absorption from the gut and is fermented in the colon by bacteria to products such as acetate and H_2. Abdominal pain (**sorbitol intolerance**) may be caused by "sugar-free" sweeteners containing sorbitol.

GALACTOSE IS NEEDED FOR THE SYNTHESIS OF LACTOSE, GLYCOLIPIDS, PROTEOGLYCANS & GLYCOPROTEINS

Galactose is derived from intestinal hydrolysis of the disaccharide **lactose,** the sugar of milk. It is readily converted in the liver to glucose. The ability of the liver to accomplish this conversion may be used as a test of hepatic function in the **galactose tolerance test.** The pathway by which galactose is converted to glucose is shown in Fig 22–5.

In reaction 1, galactose is phosphorylated with the aid of **galactokinase,** using ATP as phosphate donor. The product, galactose 1-phosphate, reacts with **uridine diphosphate glucose** (UDPGlc) to form

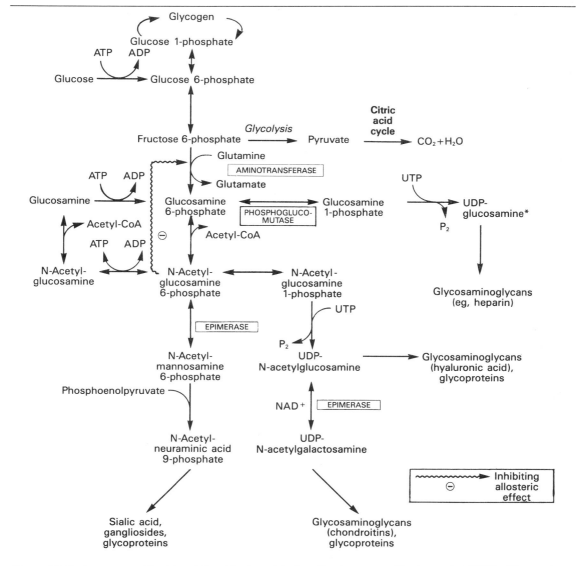

Figure 22–6. A summary of the interrelationships in metabolism of amino sugars. *Analogous to UDPGlc. Other purine or pyrimidine nucleotides may be similarly linked to sugars or amino sugars. Examples are thymidine diphosphate (TDP)-glucosamine and TDP-N-acetylglucosamine.

uridine diphosphate galactose (UDPGal) and glucose 1-phosphate. In this step (reaction 2), which is catalyzed by an enzyme called **galactose 1-phosphate uridyl transferase,** galactose is transferred to a position on UDPGlc, replacing glucose. The conversion of galactose to glucose takes place (reaction 3) in a reaction of the galactose-containing nucleotide that is catalyzed by an **epimerase.** The product is UDPGlc. Epimerization probably involves an oxidation and reduction at carbon 4 with NAD as coenzyme. Finally (reaction 4), glucose is liberated from UDPGlc as glucose 1-phosphate, probably after incorporation into glycogen followed by phosphorolysis.

Reaction 3 is freely reversible. In this manner, glucose can be converted to galactose, so that preformed galactose is not essential in the diet. Galactose is required in the body not only in the formation of lactose but also as a constituent of glycolipids (cerebrosides), proteoglycans, and glycoproteins.

In the synthesis of lactose in the mammary gland, glucose is converted to UDPGal by the enzymes described above. UDPGal condenses with glucose to yield lactose, catalyzed by **lactose synthase.**

Enzyme Deficiencies in the Galactose Pathway cause Galactosemia

Inability to metabolize galactose occurs in the **galactosemias,** which may be caused by inherited defects in any of the 3 enzymes marked 1, 2, and 3 in Fig 22–5, although a deficiency in the **uridyl transferase** (2) is the best known. Galactose, which increases in concentration in the blood, is reduced by aldose reductase in the eye to the corresponding polyol (galactitol), which accumulates, causing cataract. The general condition is more severe if it is due to a defect in the uridyl transferase, since galactose 1-phosphate accumulates and depletes the liver of inorganic phosphate. Ultimately, liver failure and mental deterioration result.

In inherited **galactose 1-phosphate uridyl transferase deficiency** affecting the liver and red blood cells (reaction 2), the epimerase (reaction 3) is, however, present in adequate amounts, so that the galactosemic individual can still form UDPGal from glucose. This explains how it is possible for normal growth and development of affected children to occur regardless of the galactose-free diets used to control the symptoms of the disease. Several different genetic defects have been described that cause reduced rather than total transferase deficiency. As the enzyme is normally present in excess, a reduction in activity to 50% or even less does not cause clinical disease, which manifests itself only in homozygotes. The epimerase has been found deficient in erythrocytes but present in liver and elsewhere, and this third condition appears to be symptom-free.

Glucose is the Precursor of all Amino Sugars (Hexosamines) (See Fig 22–6.)

Amino sugars are important components of **glycoproteins** (see Chapter 57), of certain **glycosphingolipids** (eg, gangliosides) (see Chapter 16), and of **glycosaminoglycans** (see Chapter 57). The major amino sugars are **glucosamine, galactosamine,** and **mannosamine** (these are all hexosamines) and the 9-carbon compound **sialic acid.** The principal sialic acid found in human tissues is N-acetylneuraminic acid (NeuAc). A summary of the interrelationships among the amino sugars is shown in Fig 22–6; the following are its important features: (1) **Glucosamine** is the major amino sugar. It is formed as glucosamine 6-phosphate from fructose 6-phosphate, using glutamine as the donor of the amino group. (2) The amino sugars occur mainly in the **N-acetylated** form. The acetyl donor is acetyl-CoA. (3) N-acetylmannosamine 6-phosphate is formed by epimerization of glucosamine 6-phosphate. (4) **NeuAc** is formed by the condensation of mannosamine 6-phosphate with phosphoenolpyruvate. (5) **Galactosamine** is formed by the epimerization of UDP-N-acetylglucosamine (UDPGlcNAc) to UDP-N-acetylgalactosamine (UDPGalNAc). (6) **Nucleotide sugars** are the forms in which amino sugars are used for the biosynthesis of glycoproteins and other complex compounds; the important amino sugar-containing nucleotide sugars are UDPGlcNAc, UDPGalNAc, and CMP-NeuAc.

REFERENCES

Huijing F: Textbook errors: Galactose metabolism and galactosemia. *Trends Biochem Sci* 1978;**3:**N129.

James HM et al: Models for the metabolic production of oxalate from xylitol in humans. *Aust J Exp Biol Med Sci* 1982;**60:**117.

Kador PF: The role of aldose reductase in the development of diabetic complications. Med Res Rev 1988;**8:**325.

Macdonald I, Vrana A (editors): *Metabolic Effects of Dietary Carbohydrates.* Karger, 1986.

Randle PJ, Steiner DF, Whelan WJ (editors): *Carbohydrate Metabolism and Its Disorders.* Vol 3. Academic Press, 1981.

Sperling O, de Vries A (editors): *Inborn Errors of Metabolism in Man.* Karger, 1978.

Various authors: In: *The Metabolic Basis of Inherited Disease,* 6th ed. Scriver CR et al (editors). McGraw-Hill, 1989.

Wood T: *The Pentose Phosphate Pathway.* Academic Press, 1985.

Biosynthesis of Fatty Acids

23

Peter A. Mayes, PhD, DSc

INTRODUCTION

Like many other degradative and synthetic processes (eg, glycogenolysis and glycogenesis), fatty acid synthesis (lipogenesis) was formerly considered to be merely the reversal of oxidation. However, it now seems clear that a mitochondrial system for fatty acid synthesis, involving some modification of the β-oxidation sequence, is responsible only for elongation of existing fatty acids of moderate chain length, whereas a radically different and highly active **extramitochondrial** system is responsible for the complete synthesis of palmitate from acetyl-CoA. An active system for **chain elongation** is also present in liver endoplasmic reticulum.

BIOMEDICAL IMPORTANCE

There are wide variations among species both in disposition of the principal lipogenic pathways between the tissues and in the main substrates for fatty acid synthesis. In the rat, a species that has provided most information about lipogenesis, the pathway is well represented in adipose tissue and liver, whereas in humans adipose tissue may not be an important site, and liver has only low activity. In birds, lipogenesis is confined to the liver, where it is particularly important in providing lipids for egg formation. In most mammals, glucose is the primary substrate for lipogenesis, but in ruminants acetate, which is the main fuel molecule produced by the diet, takes over this role. As the lipogenesis pathway may be of reduced importance in humans it is not surprising that critical diseases of the pathway have not been reported. However, variations in its activity between individuals may have a bearing on the nature and extent of obesity.

THE MAIN PATHWAY FOR DE NOVO SYNTHESIS OF FATTY ACIDS (LIPOGENESIS) OCCURS IN THE CYTOSOL

This system is present in many tissues, including liver, kidney, brain, lung, mammary gland, and adipose tissue. Its cofactor requirements include NADPH, ATP, Mn^{2+}, biotin, and HCO_3^- (as a source of CO_2). Acetyl-CoA is the immediate substrate, and free palmitate is the end product. These characteristics contrast markedly with those of β-oxidation.

Production of Malonyl-CoA Is the Initial & Controlling Step in Fatty Acid Synthesis

Bicarbonate as a source of CO_2 is required in the initial reaction for the carboxylation of acetyl-CoA to **malonyl-CoA** in the presence of ATP and **acetyl-CoA carboxylase.** Acetyl-CoA carboxylase has a requirement for the vitamin **biotin** (Fig 23–1). The enzyme contains a variable number of identical subunits, each containing biotin, biotin carboxylase, biotin carboxyl carrier protein, and transcarboxylase, as well as a regulatory allosteric site. It is therefore a **multienzyme protein.** The reaction takes place in 2 steps: (1) carboxylation of biotin (involving ATP; see Fig 53–13) and (2) transfer of the carboxyl to acetyl-CoA to form malonyl-CoA.

Fatty Acid Synthase Complex Is a Polypeptide Containing Six Enzymes

There appear to be 2 types of **fatty acid synthase** systems found in the soluble portion of the cell. In bacteria, plants, and lower forms, the individual enzymes of the system are separate, and the acyl radicals are found in combination with a protein called the **acyl carrier protein** (**ACP**). However, in yeast, mammals, and birds, the synthase system is a multienzyme complex that may not be subdivided without loss of activity, and ACP is part of this complex. ACP of both bacteria and the multienzyme complex contain the vitamin pantothenic acid in the form of 4'-phosphopantetheine (see Fig 53–6). In this system, ACP takes over the role of CoA.

The aggregation of all the enzymes of a particular pathway into one multienzyme functional unit offers great efficiency and freedom from interference by competing processes, thus achieving the effect of compartmentalization of the process within the cell, without the erection of permeability barriers. Another advantage of the single multienzyme polypeptide is that synthesis of all enzymes in the complex is coordinated.

The fatty acid synthase complex is a dimer (Fig 23–2). In mammals, each monomer is identical, consisting of one remarkable polypeptide chain containing all

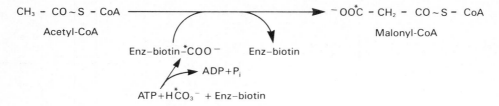

$$CH_3 - CO \sim S - CoA \longrightarrow {}^-OO\overset{*}{C} - CH_2 - CO \sim S - CoA$$

Acetyl-CoA   Malonyl-CoA

Enz–biotin–$\overset{*}{C}OO^-$ Enz–biotin

ADP + P$_i$

ATP + H$\overset{*}{C}O_3^-$ + Enz–biotin

Figure 23–1. Biosynthesis of malonyl-CoA. Enz, acetyl-CoA carboxylase.

the 6 enzymes of fatty acid synthase and an ACP with a 4'-phosphopantetheine –SH group. In close proximity is another thiol of a cysteine residue attached to **3-ketoacyl synthase (condensing enzyme)** of the other monomer (Fig 23–2). Since both thiols participate in the synthase activity, **only the dimer is active.**

Initially, a priming molecule of acetyl-CoA combines with the cysteine –SH group catalyzed by **acetyl transacylase** (Fig 23–3). Malonyl-CoA combines with the adjacent –SH on the 4'-phosphopantetheine of ACP of the other monomer, catalyzed by **malonyl transacylase,** to form **acetyl (acyl)-malonyl enzyme.** Recent work has shown that acetyl transacylase and malonyl transacylase are probably the same enzyme (as shown in Fig 23–2 and Fig 23–3). The acetyl group attacks the methylene group of the malonyl residue, catalyzed by **3-ketoacyl synthase,** and liberates CO_2, forming 3-ketoacyl enzyme (acetoacetyl enzyme).

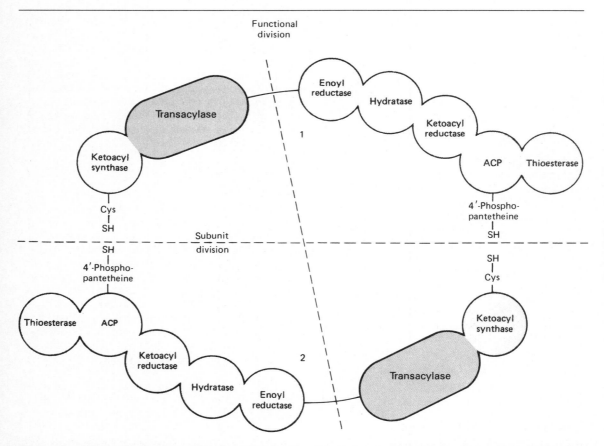

Figure 23–2. Fatty acid synthase multienzyme complex. The complex is a dimer of 2 identical polypeptide monomers, 1 and 2, each consisting of 6 separate enzyme activities and the acyl carrier protein (ACP). Cys–SH cysteine thiol. The –SH of the 4'-phosphopantetheine of one monomer is in close proximity to the –SH of the cysteine residue of the ketoacyl synthase of the other monomer, suggesting a "head-to-tail" arrangement of the 2 monomers. The detailed sequence of the enzymes in each monomer is tentative (based on Tsukamoto). Though each monomer contains all the partial activities of the reaction sequence, the actual functional unit consists of one-half of a monomer interacting with the complementary half of the other. Thus, 2 acyl chains are produced simultaneously.

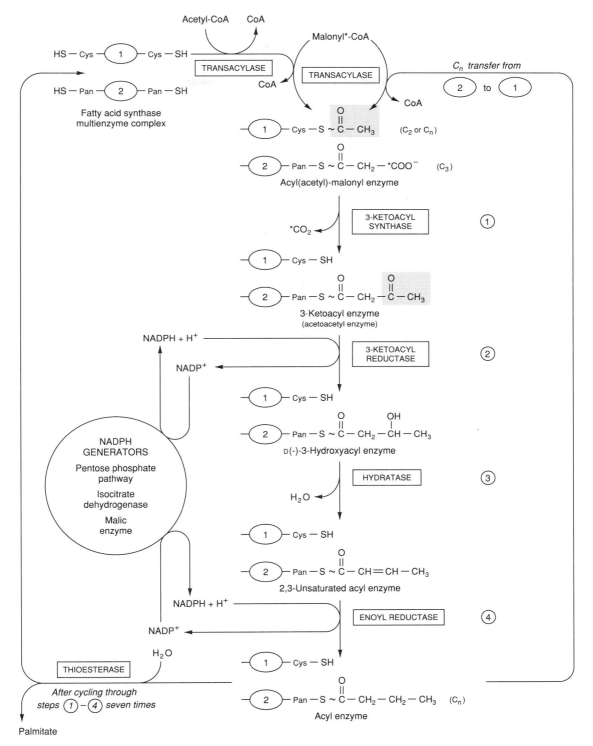

Figure 23–3. Biosynthesis of long-chain fatty acids. Details of how addition of a malonyl residue causes the acyl chain to grow by 2 carbon atoms. Cys-, cystine residue; pan, 4'-phosphopantetheine. Details of the fatty acid synthase dimer are shown in Fig 23–2. ① and ② represent the individual monomers of fatty acid synthase. Two acyl chains are synthesized simultaneously on one dimer using each pair of Cys/pan –SH groups.

This frees the cysteine –SH group, hitherto occupied by the acetyl group. Decarboxylation allows the reaction to go to completion, acting as a pulling force for the whole sequence of reactions. The 3-ketoacyl group is reduced, dehydrated, and reduced again to form the corresponding saturated acyl-S-enzyme. These reactions are analogous to those in β-oxidation, except that the 3-hydroxy acid is the D(−) isomer instead of the L(+) isomer and NADPH rather than NADH serves as hydrogen donor for both reductions. A new malonyl-CoA molecule combines with the –SH of 4′-phosphopantetheine, displacing the saturated acyl residue onto the free cysteine –SH group. The sequence of reactions is repeated 6 more times, a new malonyl residue being incorporated during each sequence, until a saturated 16-carbon acyl radical (palmityl) has been assembled. It is liberated from the enzyme complex by the activity of a sixth enzyme in the complex, **thioesterase** (deacylase). The free palmitate must be activated to acyl-CoA before it can proceed via any other metabolic pathway. Its usual fate is esterification into acylglycerols (Fig 23–4).

In mammary gland, there is a separate thioesterase specific for acyl residues of C_8, C_{10}, or C_{12}, which are subsequently found in milk lipids. In ruminant mammary gland, this enzyme is part of the fatty acid synthase complex.

There would appear to be 2 centers of activity in one dimer complex that function independently and simultaneously to form 2 molecules of palmitate. The equation for the overall synthesis of palmitate from acetyl-CoA and malonyl-CoA is shown below:

$$CH_2CO\cdot S\cdot CoA + 7HOOC\cdot CH_2CO\cdot S\cdot CoA$$
$$+ 14NADPH + 14H^+ \rightarrow$$
$$CH_3(CH_2)_{14}COOH + 7CO_2 + 6H_2O$$
$$+ 8CoA\cdot SH + 14NADP^+$$

The acetyl-CoA used as a primer forms carbon atoms 15 and 16 of palmitate. The addition of all the subsequent C_2 units is via malonyl-CoA formation. Butyryl-CoA may act as a primer molecule in mammalian liver and mammary gland. If propionyl-CoA acts as primer, long-chain fatty acids having an odd number of carbon atoms result. These are found particularly in ruminants, where propionate is formed by microbial action in the rumen.

The Main Source of Reducing Equivalents (NADPH) Is the Pentose Phosphate Pathway

NADPH is involved as coenzyme in both the reduction of the 3-ketoacyl and of the 2,3-unsaturated acyl derivatives. The oxidative reactions of the pentose phosphate pathway (see p. 175) are the chief source of the hydrogen required for the reductive synthesis of fatty acids. It is significant that tissues which possess an active pentose phosphate pathway are also the tissues specializing in active lipogenesis, ie, liver, adipose tissue, and the lactating mammary gland. Moreover, both metabolic pathways are found in the extramitochondrial region of the cell, so there are no membranes or permeability barriers for the transfer of NADPH/NADP from one pathway to the other. Other sources of NADPH include the reaction that converts malate to pyruvate catalyzed by the **"malic enzyme"** (NADP malate dehydrogenase) (Fig 23–5) and the extramitochondrial **isocitrate dehydrogenase** reaction (probably not a substantial source).

Acetyl-CoA Is the Main Building Block of Fatty Acids

It is formed from carbohydrate via the oxidation of pyruvate within the mitochondria. However, acetyl-CoA does not diffuse readily into the extramitochondrial compartment, the principal site of fatty acid synthesis. The activity of the extramitochondrial **ATP-citrate lyase,** like the "malic enzyme," increases in activity in the well-fed state, closely paralleling the activity of the fatty acid synthesizing system (Table 21–1). It is now believed that utilization of pyruvate for lipogenesis is by way of citrate. The pathway involves glycolysis followed by the oxidative decarboxylation of pyruvate to acetyl-CoA within the mitochondria and subsequent condensation with oxaloacetate to form citrate, as part of the citric acid cycle. This is followed by the **translocation of citrate** into the extramitochondrial compartment, where in the presence of CoA and ATP, it undergoes cleavage to acetyl-CoA and oxaloacetate catalyzed by ATP-citrate lyase. The acetyl-CoA is then available for malonyl-CoA formation and synthesis to palmitate (Fig 23–5). The oxaloacetate can form malate via NADH-linked malate dehydrogenase, followed by the generation of NADPH via the malic enzyme. In turn, the NADPH

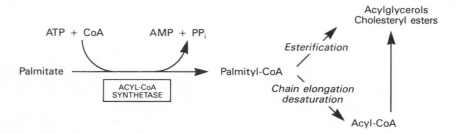

Figure 23–4. Fate of palmitate after biosynthesis.

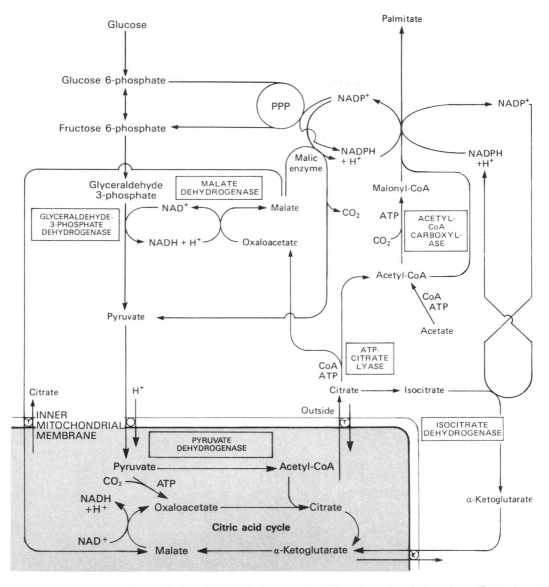

Figure 23–5. The provision of acetyl-CoA and NADPH for lipogenesis. PPP, pentose phosphate pathway; T, tricarboxylate transporter; K, α-ketoglutarate transporter.

becomes available for lipogenesis. This pathway is a means of transferring reducing equivalents from extramitochondrial NADH to NADP. Alternatively, malate can be transported into the mitochondrion, where it is able to reform oxaloacetate. It is to be noted that the citrate (tricarboxylate) transporter in the mitochondrial membrane requires malate to exchange with citrate (see Fig 14–16).

There is little ATP-citrate lyase or malic enzyme in ruminants, probably because in these species acetate (derived from the rumen) is the main source of acetyl-CoA. Since the acetate is activated to acetyl-CoA extramitochondrially, there is no necessity for it to enter mitochondria and form citrate prior to incorporation into long-chain fatty acids. Generation of NADPH via

extramitochondrial isocitrate dehydrogenase is more important in these species because of the deficiency in malic enzyme.

Chain Elongation of Fatty Acids Takes Place in the Endoplasmic Reticulum

The pathway (the "microsomal system") converts acyl-CoA compounds of fatty acids to acyl derivatives having 2 carbons more, using malonyl-CoA as acetyl donor and NADPH as reductant. Intermediates in the process are the CoA thioesters. The acyl groups that may act as a primer molecule include the saturated series from C_{10} upward, as well as unsaturated fatty acids. Fasting largely abolishes chain elongation.

Elongation of stearyl-CoA in brain increases rapidly during myelination in order to provide C_{22} and C_{24} fatty acids that are present in sphingolipids (Fig 23–6).

THE NUTRITIONAL STATE REGULATES LIPOGENESIS

Many animals, including humans, take their food as spaced meals and must therefore store much of the

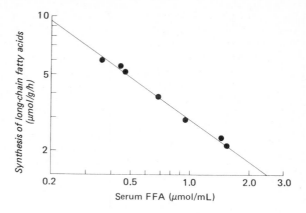

Figure 23–7. Direct inhibition of hepatic lipogenesis by free fatty acids. Lipogenesis was determined from the incorporation of ^3H_2O into long-chain fatty acids in the perfused rat liver. FFA, free fatty acids. (Experiments from the author's laboratory with DL Topping.)

energy of their diet for use between meals. The process of lipogenesis is concerned with the conversion of glucose and intermediates such as pyruvate, lactate, and acetyl-CoA to fat, which constitutes the anabolic phase of this cycle. The **nutritional state** of the organism and tissues is the **main factor controlling the rate of lipogenesis.** Thus, the rate is high in the well-fed animal whose diet contains a high proportion of carbohydrate. It is depressed under conditions of restricted caloric intake, on a high-fat diet, or when there is a deficiency of insulin, as in diabetes mellitus. All of these conditions are associated with increased concentrations of plasma free fatty acids. The regulation of the mobilization of free fatty acids from adipose tissue is described in Chapter 27.

There is an inverse relationship between hepatic lipogenesis and the concentration of serum free fatty acids (Fig 23–7). The greatest inhibition of lipogenesis occurs over the range of free fatty acids (0.3–0.8 μmol/mL of plasma) through which the plasma free fatty acids increase during transition from the fed to the starved state. Fat in the diet also causes depression of lipogenesis in the liver, and when there is more than 10% of fat in the diet, there is little conversion of dietary carbohydrate to fat. Lipogenesis is higher when sucrose is fed instead of glucose because fructose bypasses the phosphofructokinase control point in glycolysis and floods the lipogenic pathway (see Fig 22–4).

SHORT & LONG-TERM MECHANISMS ARE INVOLVED IN REGULATING LIPOGENESIS

Long-chain fatty acid synthesis is controlled in the short term by allosteric and covalent modification of enzymes and in the long term by changes in rates of synthesis and degradation of enzymes.

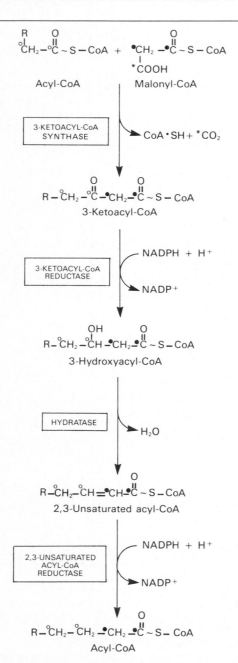

Figure 23–6. Microsomal system for chain elongation (elongase).

The Citrate and Acyl-CoA Concentration Regulate Acetyl-CoA Carboxylase

The rate-limiting reaction in the lipogenic pathway is at the **acetyl-CoA carboxylase step.** Acetyl-CoA carboxylase is activated by **citrate,** which increases in concentration in the well fed state and is an indicator of a plentiful supply of acetyl-CoA. However, it is inhibited by long-chain acyl-CoA molecules, an example of metabolic negative feedback inhibition by a product of a reaction sequence. Thus, if acyl-CoA accumulates because it is not esterified quickly enough, it will automatically reduce the synthesis of new fatty acid. Likewise, if acyl-CoA accumulates as a result of increased lipolysis or an influx of free fatty acids into the tissue, this will also inhibit synthesis of new fatty acid. Acyl-CoA may also inhibit the mitochondrial **tricarboxylate transporter,** thus preventing egress of citrate from the mitochondria into the cytosol.

Pyruvate Dehydrogenase Is Also Regulated by Acyl-CoA

There is also an inverse relationship between free fatty acid concentration and the proportion of active to inactive pyruvate dehydrogenase which regulates the availability of acetyl-CoA for lipogenesis. Acyl-CoA causes an inhibition of pyruvate dehydrogenase by inhibiting the ATP-ADP exchange transporter of the inner mitochondrial membrane, which leads to increased intramitochondrial $[ATP]/[ADP]$ ratios and therefore to conversion of active to inactive pyruvate dehydrogenase (see Fig 24–4). Also, oxidation of acyl-CoA due to increased levels of free fatty acids may increase the ratio of $[acetyl-CoA]/[CoA]$ and $[NADH]/[NAD^+]$ in mitochondria, inhibiting pyruvate dehydrogenase.

Hormones Also Regulate Lipogenesis

Insulin stimulates lipogenesis by several mechanisms. It increases the transport of glucose into the cell (eg, in adipose tissue) and thereby increases the availability of both pyruvate for fatty acid synthesis and glycerol 3-phosphate for esterification of the fatty acids. Insulin converts the inactive form of pyruvate dehydrogenase to the active form in adipose tissue but not in liver. In addition, acetyl-CoA carboxylase is an enzyme that can be regulated by reversible phosphorylation. **Insulin activates acetyl-CoA carboxylase,** possibly by activation of a protein phosphatase. Also, insulin, by its ability to depress the level of intracellular cAMP, **inhibits lipolysis** and thereby reduces the concentration of long-chain acyl-CoA, an inhibitor of lipogenesis. By this same mechanism insulin antagonizes the actions of **glucagon** and **epinephrine,** which inhibit acetyl-CoA carboxylase, and therefore lipogenesis, by increasing **cAMP,** allowing cAMP-dependent protein kinase to inactivate the enzyme by phosphorylation.

Recently, an **AMP-dependent protein kinase** has been described that detects a low energy status of the cell by responding to increased concentrations of AMP (see p. 191). This new kinase also inactivates acetyl-CoA carboxylase by phosphorylation. In addition, catecholamines inhibit the enzyme through α-adrenergic receptors and a Ca^{2+}/calmodulin-dependent protein kinase.

In ruminants, **acetate**—not glucose—is the starting material for lipogenesis. It follows that, in these species, many of the control mechanisms involving mitochondria are bypassed and thus do not apply.

Both the Fatty Acid Synthase Complex and Acetyl-CoA Carboxylase Are Adaptive Enzymes

They adapt to the body's physiologic needs by increasing in total amount in the fed state and decreasing in fasting, feeding of fat, and diabetes. **Insulin** is an important hormone causing induction of enzyme biosynthesis, and glucagon antagonizes this effect. These effects on lipogenesis take several days to become fully manifested and augment the direct and immediate effect of free fatty acids and hormones such as insulin and glucagon.

REFERENCES

Goodridge AG: Fatty acid synthesis in eukaryotes. Page 143 in: *Biochemistry of Lipids and Membranes.* Vance DE, Vance JE (editors). Benjamin/Cummings, 1985.

Goodridge AG: Dietary regulation of gene expression: Enzymes involved in carbohydrate and lipid metabolism. *Annu Rev Nutr* 1987;**7**:157.

Hardie DG, Carling D, Sim ATR: The AMP-activated protein kinase: a multisubstrate regulator of lipid metabolism. *Trends Biochem Sci* 1989;**14**:20.

Singh N, Wakil SJ, Stoops JK: On the question of half- or full-site reactivity of animal fatty acid synthetase. *J Biol Chem* 1984;**259**:3605.

Tsukamoto Y et al: The architecture of the fatty acid synthetase complex. *J Biol Chem* 1983;**258**:15312.

Wakil SJ, Stoops JK, Joshi VC: Fatty acid synthesis and its regulation. *Annu Rev Biochem* 1983;**52**:537.

24 Oxidation of Fatty Acids: Ketogenesis

Peter A. Mayes, PhD, DSc

INTRODUCTION

Fatty acids are both oxidized to acetyl-CoA and synthesized from acetyl-CoA. Although the starting material of one process is identical to the product of the other and the chemical stages involved are comparable, fatty acid oxidation is not the simple reverse of fatty acid biosynthesis but an entirely different process taking place in a separate compartment of the cell. The separation of fatty acid oxidation from biosynthesis allows each process to be individually controlled and integrated with tissue requirements.

Fatty acid oxidation takes place in mitochondria; each step involves acyl-CoA derivatives catalyzed by separate enzymes, utilizes NAD and FAD as coenzymes, and generates ATP. In contrast, fatty acid biosynthesis (lipogenesis) takes place in the cytosol, involves acyl derivatives continuously attached to a multienzyme complex, utilizes NADP as coenzyme, and requires both ATP and bicarbonate ion. Fatty acid oxidation is an aerobic process, requiring the presence of oxygen.

BIOMEDICAL IMPORTANCE

Increased fatty acid oxidation is characteristic of starvation and of diabetes mellitus, leading to **ketone body** production by the liver (**ketosis**). Ketone bodies are acidic and when produced in excess over long periods, as in diabetes, cause **ketoacidosis,** which is ultimately fatal. Because gluconeogenesis is dependent upon fatty acid oxidation, any impairment in fatty acid oxidation leads to hypoglycemia. This occurs in various states of carnitine deficiency or deficiency of essential enzymes in fatty acid oxidation (eg, carnitine palmitoyltransferase) or inhibition of fatty acid oxidation by poisons (eg, hypoglycin).

OXIDATION OF FATTY ACIDS OCCURS IN THE MITOCHONDRIA

Fatty Acids Are Transported in the Blood in the Form of Free Fatty Acids (FFA)

The term "free fatty acid" refers to fatty acids that are in the **unesterified state.** Alternative nomenclature is UFA (unesterified fatty acids) or NEFA (non-esterified fatty acids). In plasma, FFA of longer-chain fatty acids are combined with **albumin,** and in the cell they are attached to a **fatty acid binding protein,** or Z-protein, so that in fact they are never really "free." Shorter chain fatty acids are more water soluble and exist as the un-ionized acid or as a fatty acid anion.

Fatty Acids Must Be Activated Before They Can Be Metabolized

As in the metabolism of glucose, fatty acids must first be converted in a reaction with ATP to an active intermediate before they will react with the enzymes responsible for their further metabolism. This is the only step in the complete degradation of a fatty acid that requires energy from ATP. In the presence of ATP and coenzyme A, the enzyme **acyl-CoA synthetase** (**thiokinase**) catalyzes the conversion of a fatty acid (or free fatty acid) to an "active fatty acid" or acyl-CoA, accompanied by the expenditure of one high-energy phosphate bond.

$$\text{Fatty acid} + \text{ATP} + \text{CoA} \xrightarrow{\text{ACYL-CoA SYNTHETASE}} \text{Acyl-CoA} + \text{PP}_i + \text{AMP}$$

The presence of **inorganic pyrophosphatase** ensures that activation goes to completion by facilitating the loss of the additional high-energy phosphate bond of pyrophosphate. Thus, in effect, **2 high-energy phosphates are expended during the activation of each fatty acid molecule.**

$$PP_i + H_2O \rightarrow 2\ P_i$$

INORGANIC PYROPHOSPHATASE

Acyl-CoA synthetases are found in the endoplasmic reticulum and inside and on the outer membrane of mitochondria. Several acyl-CoA synthetases have been described, each specific for fatty acids of different chain length.

Long Chain Fatty Acids Penetrate the Inner Mitochondrial Membrane Only in Combination With Carnitine

Carnitine (β-hydroxy-γ-trimethylammonium butyrate), $(CH_3)_3N^+-CH_2-CH(OH)-CH_2-COO^-$, is widely distributed and is particularly abundant in muscle. It is synthesized from lysine and methionine in liver and kidney. Activation of

Acyl-CoA + Carnitine ↔ Acylcarnitine + CoA

CARNITINE PALMITOYLTRANSFERASE

lower fatty acids, and their oxidation may occur within the mitochondria, independently of carnitine, but long-chain acyl-CoA (or FFA) **will not penetrate mitochondria** and become oxidized unless they form acylcarnitines. An enzyme, **carnitine palmitoyltransferase I,** present on the inner side of the outer mitochondrial membrane, converts long-chain acyl CoA to acylcarnitine, which is able to penetrate mitochondria and gain access to the β-oxidation system of enzymes. **Carnitine-acylcarnitine translocase** acts as a membrane carnitine exchange transporter. Acylcarnitine is transported in, coupled with the transport out of one molecule of carnitine. The acylcarnitine then reacts with CoA, catalyzed by **carnitine palmitoyltransferase II,** attached to the inside of the inner membrane. Acyl-CoA is re-formed in the mitochondrial matrix, and carnitine is liberated (Fig 24–1).

Another enzyme, **carnitine acetyltransferase,** is present within mitochondria and catalyzes the transfer of short-chain acyl groups between CoA and carnitine. The function of this enzyme is obscure, but it may facilitate transport of acetyl groups through the mitochondrial membrane.

Acetyl-CoA + Carnitine ↔ Acetylcarnitine + CoA

CARNITINE ACETYLTRANSFERASE

β-OXIDATION OF FATTY ACIDS INVOLVES SUCCESSIVE CLEAVAGE & RELEASE OF ACETYL-CoA

In β-oxidation (Fig 24–2), 2 carbons are cleaved at a time from acyl-CoA molecules, starting at the carboxyl end. The chain is broken between the α(2)- and β(3)-carbon atoms, hence the name β-oxidation. The

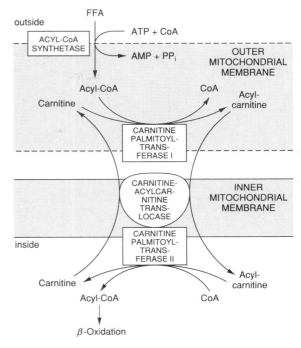

Figure 24–1. Role of carnitine in the transport of long-chain fatty acids through the inner mitochondrial membrane. Long-chain acyl-CoA cannot pass through the inner mitochondrial membrane, but its metabolic product, acylcarnitine, can.

2-carbon units formed are acetyl-CoA; thus, palmitoyl-CoA forms 8 acetyl-CoA molecules.

The Cyclic Reaction Sequence Generates NADH & FADH$_2$

Several enzymes, known collectively as "fatty acid oxidase," are found in the mitochondrial matrix adjacent to the respiratory chain (which is found in the inner membrane). These catalyze the oxidation of acyl-CoA to acetyl-CoA, the system being coupled with the phosphorylation of ADP to ATP (Fig 24–3).

After the penetration of the acyl moiety through the mitochondrial membrane via the carnitine transporter system and the re-formation of acyl-CoA, there follows the removal of 2 hydrogen atoms from the 2(α)- and 3(β)-carbon atoms, catalyzed by **acyl-CoA dehydrogenase.** This results in the formation of Δ²-*trans*-enoyl-CoA. The coenzyme for the dehydrogenase is a flavoprotein, containing FAD as prosthetic group, whose reoxidation by the respiratory chain requires the mediation of another flavoprotein, termed **electron-transferring flavoprotein** (see p 103). Water is added to saturate the double bond and form 3-hydroxyacyl-CoA, catalyzed by the enzyme **Δ²-enoyl-CoA hydratase.** The 3-hydroxy derivative undergoes further dehydrogenation on the 3-carbon (L(+)-**3-hydroxyacyl-CoA dehydrogenase**) to form the corresponding 3-ketoacyl-CoA compound. In this case, NAD is the coenzyme involved in the dehydrogenation. Finally, 3-

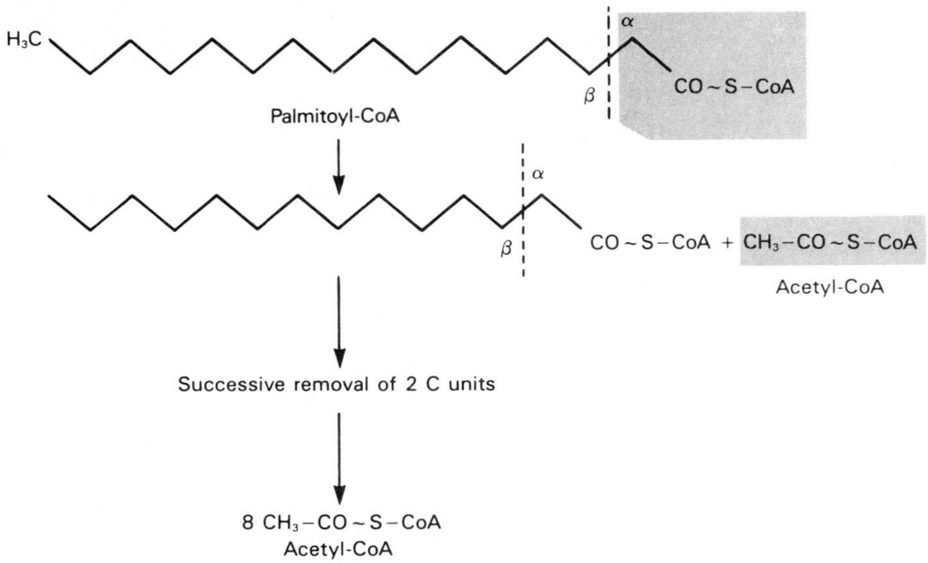

Figure 24–2. Overview of β-oxidation of fatty acids.

ketoacyl-CoA is split at the 2,3-position by **thiolase** (3-ketoacylthiolase or acetyl-CoA acyltransferase), which catalyzes a thiolytic cleavage involving another molecule of CoA. The products of this reaction are acetyl-CoA and an acyl-CoA derivative containing 2 carbons less than the original acyl-CoA molecule that underwent oxidation. The acyl-CoA formed in the cleavage reaction reenters the oxidative pathway at reaction 2 (Fig 24–3). In this way, a long-chain fatty acid may be degraded completely to acetyl-CoA (C_2 units). As acetyl-CoA can be oxidized to CO_2 and water via the citric acid cycle (which is also found within the mitochondria), the complete oxidation of fatty acids is achieved.

Oxidation of a Fatty Acid With an Odd Number of Carbon Atoms Yields a Molecule of Propionyl-CoA

Fatty acids with an odd number of carbon atoms are oxidized by the pathway of β-oxidation producing acetyl-CoA until a 3-carbon (propionyl-CoA) residue remains. This compound is converted to succinyl-CoA, a constituent of the citric acid cycle (see also Fig 21–2). Hence, **the propionyl residue from an odd-chain fatty acid is the only part of a fatty acid that is glucogenic.**

Oxidation of Fatty Acids Produces a Large Number of ATP Molecules

Transport in the respiratory chain of electrons from reduced flavoprotein and NAD will lead to the synthesis of 5 high-energy phosphate bonds (see Chapter 14) for each of the first 7 acetyl-CoA molecules formed by β-oxidation of palmitate ($7 \times 5 = 35$). A total of 8 mol of acetyl-CoA is formed, and each will give rise to 12 high-energy bonds on oxidation in the citric acid cycle,

making $8 \times 12 = 96$ high-energy bonds derived from the acetyl-CoA formed from palmitate, minus 2 for the initial activation of the fatty acid, yielding a net gain of 129 high-energy bonds/ mol, or $129 \times 30.5 = 3935$ kJ. As the free energy of combustion of palmitic acid is 9791 kJ/mol, the process captures as high-energy phosphate on the order of 40% of the total energy of combustion of the fatty acid.

Very Long Chain Fatty Acids Are Oxidized in Peroxisomes

A modified form of β-oxidation is found in peroxisomes and leads to the formation of acetyl-CoA and H_2O_2 (from the flavoprotein-linked dehydrogenase step). The system is **not linked directly to phosphorylation and the generation of ATP** but aids the oxidation of very long chain fatty acids (eg, C_{20}, C_{22}) and is induced by high-fat diets and hypolipidemic drugs such as clofibrate.

The enzymes in peroxisomes do not attack shorter-chain fatty acids; the β-oxidation sequence ends at octanoyl-CoA. Octanoyl and acetyl groups are subsequently removed from the peroxisomes in the forms of octanoyl and acetyl carnitine, and both are further oxidized in mitochondria.

α- & ω-OXIDATION OF FATTY ACIDS ARE SPECIALIZED PATHWAYS

Quantitatively, β-oxidation in mitochondria is the most important pathway for fatty acid oxidation. However, α-oxidation, ie, the removal of one carbon at a time from the carboxyl end of the molecule, has been detected in brain tissue. It does not require CoA intermediates and does not generate high-energy phosphates.

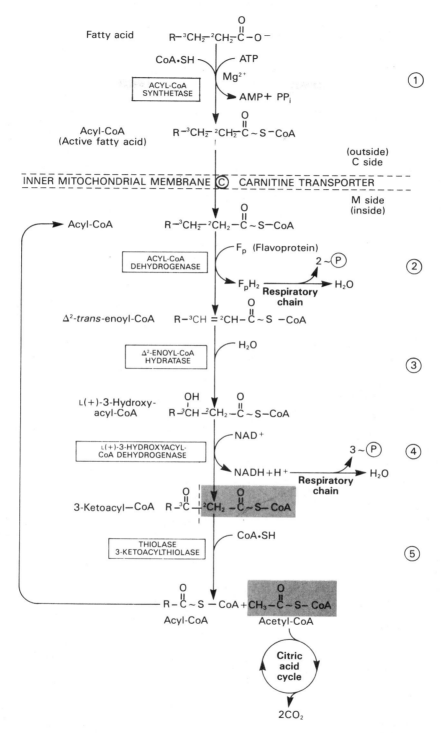

Figure 24–3. β-Oxidation of fatty acids. Long-chain acyl-CoA is cycled through reactions 2–5, acetyl-CoA being split off each cycle by thiolase (reaction 5). When the acyl radical is only 4 carbon atoms in length, 2 acetyl-CoA molecules are formed in reaction 5.

ω-Oxidation is normally a very minor pathway and is brought about by hydroxylase enzymes involving cytochrome P-450 in the endoplasmic reticulum (see p 103). The $-CH_3$ group is converted to a $-CH_2OH$ group that subsequently is oxidized to $-COOH$, thus forming a dicarboxylic acid. This is β-oxidized usually to adipic (C_6) and suberic (C_8) acids, which are excreted in the urine.

IMPAIRED OXIDATION OF FATTY ACIDS GIVES RISE TO METABOLIC DISEASES

Carnitine deficiency can occur particularly in the newborn—and especially in preterm infants—owing to inadequate biosynthesis or renal leakage. Losses can also occur in hemodialysis; patients with organic aciduria have large losses of carnitine, which is excreted conjugated to the organic acids. This indicates a vitaminlike dietary requirement for carnitine in some individuals. Signs and symptoms of deficiency include episodic periods of hypoglycemia owing to reduced gluconeogenesis resulting from impaired fatty acid oxidation, impaired ketogenesis in the presence of raised plasma FFA, muscular weakness, and lipid accumulation. Treatment is by oral supplementation of carnitine. The symptoms are similar to Reye's syndrome, in which carnitine is adequate, but the cause of Reye's syndrome is unknown.

Hepatic carnitine palmitoyltransferase deficiency results in hypoglycemia and low plasma ketone bodies, whereas **muscle carnitine palmitoyltransferase deficiency** leads to impaired fatty acid oxidation that results in recurrent muscle weakness and myoglobinuria. In a similar manner, the hypoglycemic sulfonylureas (**glyburide** and **tolbutamide**) inhibit fatty acid oxidation by inhibiting carnitine palmitoyltransferase.

Jamaican vomiting sickness is caused by eating the unripe fruit of the akee tree, which contains a toxin, **hypoglycin,** that inactivates acyl-CoA dehydrogenase, inhibiting β-oxidation and causing hypoglycemia.

Dicarboxylic aciduria is characterized by the excretion of C_6-C_{10} ω-dicarboxylic acids and by nonketotic hypoglycemia. It is caused by a lack of mitochondrial medium-chain acyl-CoA dehydrogenase. This impairs β-oxidation but increases ω-oxidation of long- and medium-chain fatty acids, which are then shortened by β-oxidation to medium-chain dicarboxylic acids, which are excreted.

Refsum's disease is a rare neurologic disorder caused by accumulation of phytanic acid, formed from phytol, a constituent of chlorophyll found in plant foodstuffs. Phytanic acid contains a methyl group on carbon 3 that blocks β-oxidation. Normally, an initial α-oxidation removes the methyl group, but persons with Refsum's disease have an inherited defect in α-oxidation that allows accumulation of phytanic acid.

Zellweger's (cerebrohepatorenal) syndrome occurs in individuals with a rare inherited absence of peroxisomes in all tissues. They accumulate $C_{26}-C_{38}$ polyenoic acids in brain tissue owing to inability to oxidize long-chain fatty acids in peroxisomes.

OXIDATION OF UNSATURATED FATTY ACIDS OCCURS BY A MODIFIED β-OXIDATION PATHWAY

The CoA esters of these acids are degraded by the enzymes normally responsible for β-oxidation until either a Δ^3-*cis*-acyl-CoA compound or a Δ^4-*cis*-acyl-CoA compound is formed, depending upon the position of the double bonds (Fig 24–4). The former compound is isomerized (Δ^3-*cis*→Δ^2-*trans*-**enoyl-CoA isomerase**) to the corresponding Δ^2-*trans*-CoA stage of β-oxidation for subsequent hydration and oxidation. Any Δ^4-*cis*-acyl-CoA either remaining, as in the case of linoleic acid (Fig 24–4), or entering the pathway at this point is converted by acyl-CoA dehydrogenase to Δ^2-*trans*-Δ^4-*cis*-dienoyl-CoA. This is converted to Δ^3-*trans*-enoyl-coA by an NADP-dependent enzyme, **Δ^2-*trans*-Δ^4-*cis*-dienoyl-CoA reductase.** Δ^3-*cis*→Δ^2-*trans*-enoyl-CoA isomerase will also attack the Δ^3-*trans* double bond to produce Δ^2-*trans*-enoyl-CoA, an intermediate in β-oxidation.

Microsomal Peroxidation of Polyunsaturated Fatty Acids Is NADPH Dependent

NADPH-dependent peroxidation of unsaturated fatty acids is catalyzed by microsomal enzymes (see p 106). The antioxidants BHT (butylated hydroxytoluene) and α-tocopherol (vitamin E) inhibit microsomal lipid peroxidation.

KETOGENESIS OCCURS WHEN THERE IS A HIGH RATE OF FATTY ACID OXIDATION IN THE LIVER

Under certain metabolic conditions associated with a high rate of fatty acid oxidation, the liver produces considerable quantities of **acetoacetate** and **D(–)-3-hydroxybutyrate** (β-hydroxybutyrate). Acetoacetate continually undergoes spontaneous decarboxylation to yield **acetone.** These 3 substances are collectively known as the **ketone bodies** (also called acetone bodies or [incorrectly*] "ketones") (Fig 24–5). Acetoacetate and 3-hydroxybutyrate are interconverted by the mitochondrial enzyme **D(–)-3-hydroxybutyrate dehydrogenase;** the equilibrium is controlled by the mitochondrial ratio of [NAD$^+$] to [NADH], ie, the **redox state.** The ratio [3-hydroxybutyrate]/[acetoacetate] in blood varies between 1:1 and 10:1.

The concentration of total ketone bodies* in the

---

*The term "ketones" should not be used because 3-hydroxybutyrate is not a ketone and there are many ketones in blood that are not ketone bodies, eg, pyruvate, fructose.

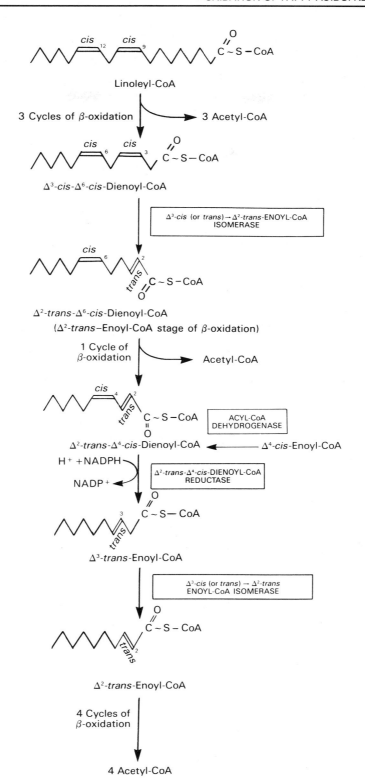

Figure 24–4. Sequence of reactions in the oxidation of unsaturated fatty acids, eg, linoleic acid. Δ^4-*cis*-fatty acids or fatty acids forming Δ^4-*cis*-enoyl-CoA enter the pathway at the position shown.

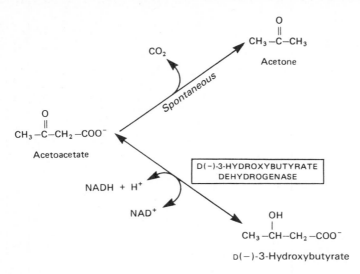

Figure 24–5. Interrelationships of the ketone bodies. D(−)-3-hydroxybutyrate dehydrogenase is a mitochondrial enzyme.

blood of well-fed mammals does not normally exceed 0.2 mmol/L. It is somewhat higher than this in ruminants, owing to 3-hydroxybutyrate formation from butyric acid (a product of ruminal fermentation) in the rumen wall. Loss via the urine is usually less than 1 mg/24 h in humans. Higher than normal quantities present in the blood or urine constitute **ketonemia** (hyperketonemia) or **ketonuria,** respectively. The overall condition is called **ketosis.** Acetoacetic and 3-hydroxybutyric acids are both moderately strong acids and are buffered when present in blood or the tissues. However, their continual excretion in quantity entails some loss of buffer cation (in spite of ammonia production by the kidney) that progressively depletes the alkali reserve, causing **ketoacidosis.** This may be fatal in uncontrolled **diabetes mellitus.**

The simplest form of ketosis occurs in **starvation** and involves depletion of available carbohydrate coupled with mobilization of free fatty acids. No other condition in which ketosis occurs seems to differ qualitatively from this general pattern of metabolism, but quantitatively it may be exaggerated to produce the pathologic states found in **diabetes mellitus, pregnancy toxemia in sheep,** and **ketosis in lactating cattle.** Nonpathologic forms of ketosis are found under conditions of high-fat feeding and after severe exercise in the postabsorptive state.

In vivo, the liver appears to be the only organ in nonruminants to add significant quantities of ketone bodies to the blood. Extrahepatic tissues utilize them as respiratory substrates. Extrahepatic sources of ketone bodies, as in fed ruminants, do not contribute significantly to the occurrence of ketosis in these species.

The net flow of ketone bodies from the liver to the extrahepatic tissues results from an active enzymatic mechanism in the liver for the production of ketone bodies coupled with very low activity of enzymes re-

sponsible for their utilization. The reverse situation occurs in extrahepatic tissues (Fig 24–6).

3-Hydroxy-3-Methylglutaryl-CoA (HMG-CoA) Is an Intermediate in the Pathway of Ketogenesis in Liver

Enzymes responsible for ketone body formation are associated mainly with the mitochondria. Originally it was thought that only one molecule of acetoacetate was formed from the terminal 4 carbons of a fatty acid upon oxidation. Later, to explain both the production of more than one equivalent of acetoacetate from a long-chain fatty acid and the formation of ketone bodies from acetic acid, it was proposed that C_2 units formed in β-oxidation condensed with one another to form acetoacetate. This may occur by a reversal of the **thiolase** reaction whereby 2 molecules of acetyl-CoA condense to form acetoacetyl-CoA. Thus, acetoacetyl-CoA, which is the starting material for ketogenesis, arises either directly during the course of β-oxidation or as a result of the condensation of acetyl-CoA (Fig 24–7). Two pathways for the formation of acetoacetate from acetoacetyl-CoA have been proposed. The first is by simple deacylation. The second pathway (Fig 24–8) involves the condensation of acetoacetyl-CoA with another molecule of acetyl-CoA to form **HMG-CoA,** catalyzed by **3-hydroxy-3-methylglutaryl-CoA synthase.** The presence of another enzyme in the mitochondria, **3-hydroxy-3-methylglutaryl-CoA lyase,** causes acetyl-CoA to split off from the HMG-CoA, leaving free acetoacetate. The carbon atoms split off in the acetyl-CoA molecule are derived from the original acetoacetyl-CoA molecule (Fig 24–8). **Both of these enzymes must be present in mitochondria for ketogenesis to take place.** This occurs solely in liver and rumen epithelium.

Present opinion favors the HMG-CoA pathway as

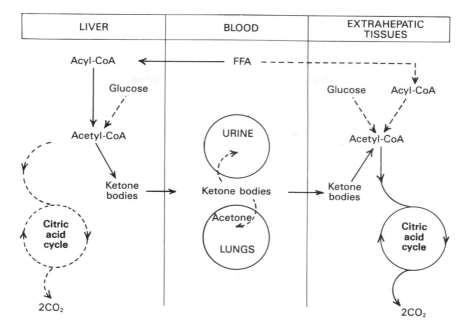

Figure 24–6. Formation, utilization, and excretion of ketone bodies. (The main pathway is indicated by the solid arrows.)

the major route of ketone body formation. Although there is a marked increase in activity of HMG-CoA lyase in fasting, evidence does not suggest that this enzyme is rate-limiting in ketogenesis.

Acetoacetate is converted to D(−)-3-hydroxybutyrate by D(−)-**3-hydroxybutyrate dehydrogenase,** which is present in mitochondria of many tissues, including the liver. D(−)-3-hydroxybutyrate is quan-

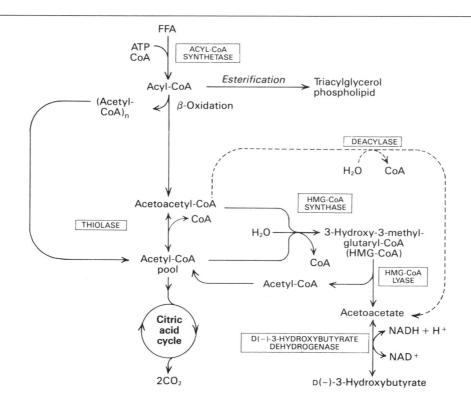

Figure 24–7. Pathways of ketogenesis in the liver. FFA, free fatty acids; HMG, 3-hydroxy-3-methylglutaryl.

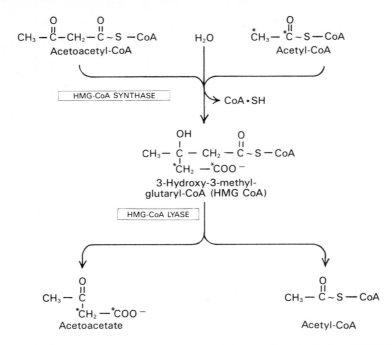

Figure 24–8. Formation of acetoacetate through intermediate production of HMG-CoA.

titatively the predominant ketone body present in the blood and urine in ketosis.

Ketone Bodies Are Utilized as a Fuel by Extrahepatic Tissues

While the liver is equipped with an active enzymatic mechanism for the production of acetoacetate from acetoacetyl-CoA, acetoacetate once formed cannot be reactivated directly in the liver except in the cytosol, where it is a precursor in cholesterol synthesis, a much less active pathway. This accounts for the net production of ketone bodies by the liver.

Two reactions take place in extrahepatic tissues that activate acetoacetate to acetoacetyl-CoA. One mechanism involves succinyl-CoA and the enzyme **succinyl-CoA-acetoacetate-CoA transferase.** Acetoacetate reacts with succinyl-CoA, the CoA being transferred to form acetoacetyl-CoA and succinate.

CH₃COCH₂COO⁻ → CH₂COO⁻ CH₂CO·S·CoA
Acetoacetate → Succinyl-CoA

CoA TRANSFERASE

CH₃COCH₂CO·S·CoA → CH₂COO⁻ CH₂COO⁻
Acetoacetyl-CoA → Succinate

The other reaction involves the activation of ace-

toacetate with ATP in the presence of CoA catalyzed by **acetoacetyl-CoA synthetase.**

ACETOACETYL-CoA SYNTHETASE

$CH_3COCH_2COO^- + ATP + CoA \cdot SH \longrightarrow$
Acetoacetate

$CH_3COOH_2CO \cdot S \cdot CoA + AMP + PP_i$
Acetoacetyl-CoA

D(−)-3-hydroxybutyrate may be activated directly in extrahepatic tissues by a synthetase; however, conversion to acetoacetate with D(−)-3-hydroxybutyrate dehydrogenase and NAD⁺, followed by activation to acetoacetyl-CoA, is the more important route leading to its further metabolism. The acetoacetyl-CoA formed by these reactions is split to acetyl-CoA by thiolase and oxidized in the citric acid cycle as shown in Fig 24–7.

Ketone bodies are oxidized in extrahepatic tissues proportionately to their concentration in the blood. They are also oxidized in preference to glucose and to FFA. If the blood level is raised, oxidation of ketone bodies increases until, at a concentration of approximately 12 mmol/L, they saturate the oxidative machinery. When this occurs, a large proportion of the oxygen consumption of the animal may be accounted for by the oxidation of ketone bodies.

Most of the evidence suggests that **ketonemia is due to increased production of ketone bodies** by the liver rather than to a deficiency in their utilization by extrahepatic tissues. However, the results of experiments on depancreatized rats support the possibility that ketosis in the severe diabetic may be enhanced by a reduced ability to catabolize ketone bodies.

In moderate ketonemia, the loss of ketone bodies via the urine is only a few percent of the total ketone body production and utilization. Since there are renal threshold-like effects (there is not a true threshold) that vary between species and individuals, measurement of the ketonemia, not the ketonuria, is the preferred method of assessing the severity of ketosis.

While acetoacetate and D(−)-3-hydroxybutyrate are readily oxidized by extrahepatic tissues, acetone is difficult to oxidize in vivo.

KETOGENESIS IS REGULATED AT THREE CRUCIAL STEPS

1. Control is exercised initially in adipose tissue. Ketosis does not occur in vivo unless there is a concomitant rise in the level of circulating free fatty acids that arise from lipolysis of triacylglycerol in adipose tissue. Fatty acids are the precursors of ketone bodies in the liver. The liver, both in fed and in fasting conditions, has the ability to extract about 30% or more of the free fatty acids passing through it, so that at high concentrations of free fatty acids the flux passing into the liver is substantial. **Therefore, the factors regulating mobilization of free fatty acids from adipose tissue are important in controlling ketogenesis** (Fig 24–9).

2. One of 2 fates awaits the free fatty acids upon uptake by the liver and after they are activated to acyl-CoA: They are **esterified** mainly to triacylglycerol and phospholipid, or they are **β-oxidized** to CO_2 or ketone bodies. The capacity for esterification as an antiketogenic factor depends on the availability of precursors in the liver to supply sufficient glycerol 3-phosphate. However, the availability of glycerol 3-phosphate does not limit esterification in fasting perfused livers. Whether its availability in the liver is ever rate-limiting on esterification is not clear; neither is there critical information on whether the in vivo activities of the enzymes involved in esterification are rate-limiting. It does not seem that they are, since neither free fatty acids nor any intermediates in their pathway of esterification to triacylglycerol (see Fig 26–1) ever accumulate in the liver.

By use of the perfused liver, it has been shown that livers from fed rats esterify considerably more ^{14}C-free fatty acids than livers from fasted rats, and the balance not esterified in the livers from fasted rats is oxidized to either [14]CO_2 or ^{14}C-ketone bodies. These results may be explained by the fact that **carnitine palmitoyltransferase I** activity in the outer mitochondrial membrane regulates the entry of long-chain acyl groups into mitochondria prior to β-oxidation (see Fig 24–1). Its activity is low in the fed state, when fatty acid oxidation is depressed, and high in fasting, when fatty acid oxidation increases. **Malonyl-CoA,** the ini-

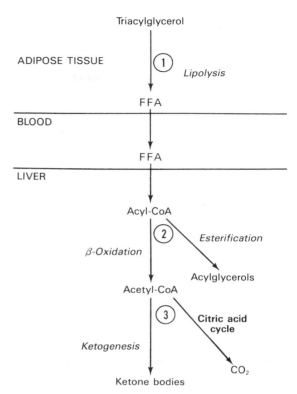

Figure 24–9. Regulation of ketogenesis. 1–3, three crucial steps in the pathway of metabolism of free fatty acids (FFA) that determine the magnitude of ketogenesis.

tial intermediate in fatty acid biosynthesis (see Fig 23–1), which increases in concentration in the fed state, inhibits this enzyme, switching off β-oxidation. Thus, in the fed condition there is active lipogenesis and high [malonyl-CoA], which inhibits carnitine palmitoyltransferase I (Fig 24–10). Free fatty acids entering the liver cell in low concentrations are nearly all esterified to acylglycerols and transported out of the liver in very low density lipoproteins (VLDL). However, as the concentration of free fatty acids increases with the onset of starvation, acetyl-CoA carboxylase is inhibited, and [malonyl-CoA] decreases, releasing the inhibition of carnitine palmitoyl-transferase and allowing more acyl-CoA to be oxidized. These events are reinforced in starvation by the **[insulin]/[glycagon] ratio,** which decreases, causing increased lipolysis in adipose tissue, the release of free fatty acids, and inhibition of acetyl-CoA carboxylase in the liver. (See Fig 24-10)

3. In turn, the acetyl-CoA formed in β-oxidation is oxidized in the citric acid cycle, or it enters the pathway of ketogenesis to form ketone bodies. As the level of serum free fatty acids is raised, proportionately more free fatty acid is converted to ketone bodies and less is oxidized via the citric acid cycle to CO_2. The partition of acetyl-CoA between the ketogenic pathway and the pathway of oxidation to CO_2 is so regu-

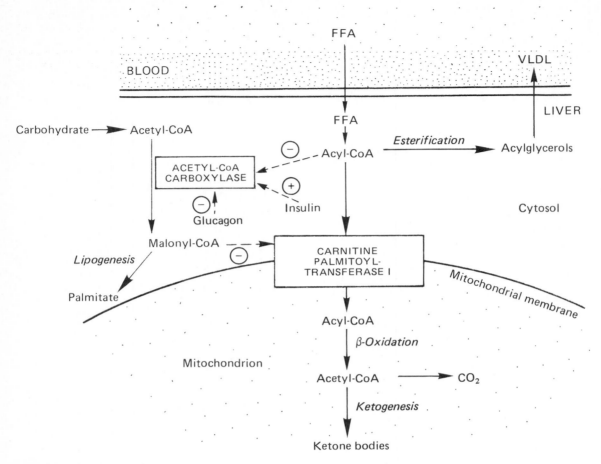

Figure 24–10. Regulation of long-chain fatty acid oxidation in the liver. FFA, free fatty acids; VLDL, very low density lipoprotein. Positive (⊕) and negative (⊖) regulatory effects are represented by broken lines and substrate flow by solid lines.

lated that the total free energy trapped in ATP which results from the oxidation of free fatty acids remains constant. **Complete oxidation of 1 mol of palmitate involves a net production of 129 mol of ATP via β-oxidation and CO_2 production in the citric acid cycle (see above), whereas only 33 mol of ATP is produced when acetoacetate is the end product and only 21 mol when 3-hydroxybutyrate is the end product.** Thus, ketogenesis may be regarded as a mechanism that allows the liver to oxidize increasing quantities of fatty acids within a tightly coupled system of oxidative phosphorylation, without increasing its total energy expenditure.

Several other hypotheses have been advanced to account for the diversion of fatty acid oxidation from CO_2 formation to ketogenesis. Theoretically, a fall in concentration of oxaloacetate, particularly within the mitochondria, could impair the ability of the citric acid cycle to metabolize acetyl-CoA. This fall may occur because of an increase in the [NADH]/[NAD+] ratio caused by increased β-oxidation. Krebs has suggested that since oxaloacetate is also on the main pathway of

gluconeogenesis, enhanced gluconeogenesis leading to a fall in the level of oxaloacetate may be the cause of the severe forms of ketosis found in diabetes and the ketosis of cattle. However, Utter and Keech have shown that pyruvate carboxylase, which catalyzes the conversion of pyruvate to oxaloacetate, is activated by acetyl-CoA. Consequently, when there are significant amounts of acetyl-CoA, there should be sufficient oxaloacetate to initiate the condensing reaction of the citric acid cycle.

REFERENCES

Boyer PD (editor): *The Enzymes,* 3rd ed. Vol 16 of *Lipid Enzymology.* Academic Press, 1983.

Debeer LJ, Mannaerts GP: The mitochondrial and peroxisomal pathways of fatty acid oxidation in rat liver. *Diabete Metab (Paris)* 1983;**9:**134.

Mayes PA, Laker ME: Regulation of ketogenesis in the liver. *Biochem Soc Trans* 1981;**9:**339.

McGarry JD, Foster DW: Regulation of hepatic fatty acid

oxidation and ketone body production. *Annu Rev Biochem* 1980;**49**:395.

Pande SV, Parvin R: Page 143 in: *Carnitine Biosynthesis, Metabolism, and Functions*. Frenkel RA, McGarry JD (editors). Academic Press, 1980.

Schulz H: Oxidation of fatty acids. Page 116 in: *Biochemistry of Lipids and Membranes*. Vance DE, Vance JE (editors). Benjamin/Cummings, 1985.

Various authors: In: *The Metabolic Basis of Inherited Disease*, 6th ed. Scriver CR et al (editors). McGraw-Hill, 1989.

Vianey-Liaud C et al: The inborn errors of mitochondrial fatty acid oxidation. *J Inher Metab Dis* 1987;**10**: Suppl 1:159.

25

Metabolism of Unsaturated Fatty Acids & Eicosanoids

Peter A. Mayes, PhD, DSc

INTRODUCTION

Compared with plants, animal tissues have limited ability in desaturating fatty acids. This necessitates dietary intake of certain polyunsaturated fatty acids derived ultimately from a plant source. These **essential fatty acids** give rise to eicosanoic (C_{20}) fatty acids, from which are derived families of compounds known as **eicosanoids.** These make up the prostaglandins, thromboxanes, and leukotrienes.

BIOMEDICAL IMPORTANCE

The content of unsaturated fatty acids in a fat determines its melting point and therefore its fluidity. Similarly, phospholipids of the cell membrane contain unsaturated fatty acids important in maintaining membrane fluidity. A high ratio of polyunsaturated fatty acids to saturated fatty acids(P:S ratio) in the diet is a major factor in lowering plasma cholesterol concentrations by dietary means and is considered to be beneficial in preventing coronary heart disease. The prostaglandins and thromboxanes are local hormones that are synthesized rapidly when required and act near their sites of synthesis. Nonsteroidal anti-inflammatory drugs, such as **aspirin,** act by inhibiting prostaglandin synthesis. The major physiologic roles played by prostaglandins are as modulators of adenyl cyclase activity, eg, (1) in controlling platelet aggregation, and (2) in inhibiting the effect of antidiuretic hormone in the kidney. Leukotrienes have muscle contractant and chemotactic properties, suggesting that they could be important in allergic reactions and inflammation. A mixture of leukotrienes has been identified as the slow-reacting substance of anaphylaxis (SRS-A). By varying the proportions of the different polyunsaturated fatty acids in the diet, it is possible to influence the type of eicosanoids synthesized, indicating that it might be possible to influence the course of disease by dietary means.

SOME POLYUNSATURATED FATTY ACIDS CANNOT BE SYNTHESIZED BY MAMMALS & ARE NUTRITIONALLY ESSENTIAL

Some long-chain unsaturated fatty acids of metabolic significance in mammals are shown in Fig 25–1. (For a review of the nomenclature of fatty acids, see Chapter 16.)

Other C_{20}, C_{22}, and C_{24} polyenoic fatty acids may be detected in the tissues. These may be derived from oleic, linoleic, and α-linolenic acids by chain elongation. It is to be noted that all double bonds present in naturally occurring unsaturated fatty acids of mammals are of the *cis* configuration.

Palmitoleic and oleic acids are not essential in the diet, because the tissues are capable of introducing a double bond at the Δ^9 position into the corresponding saturated fatty acid. Experiments with labeled palmitate have demonstrated that the label enters freely into palmitoleic and oleic acids but is absent from **linoleic, α-linolenic,** and **arachidonic acids.** These are **the only fatty acids known to be essential for the complete nutrition of many species** of animals, including the human, and must therefore be supplied in the diet; as a consequence, they are known as the **nutritionally essential fatty acids.** Although linoleic acid cannot be synthesized and therefore must be supplied preformed in the diet, arachidonic acid can be formed from linoleic acid in most mammals (see Fig 25–4). In animals, double bonds can be introduced at the Δ^4, Δ^5, Δ^6, and Δ^9 positions (counting from the carboxyl terminal; see Chapter 16) but never beyond the Δ^9 position. In contrast, plants are able to introduce new double bonds at the Δ^6, Δ^9, Δ^{12}, and Δ^{15} positions and can therefore synthesize the nutritionally essential fatty acids.

MONOUNSATURATED FATTY ACIDS ARE SYNTHESIZED BY A Δ^9 DESATURASE SYSTEM

As far as the nonessential monounsaturated fatty

Palmitoleic acid ($\omega 7$, 16:1, Δ^9)

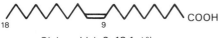

Oleic acid ($\omega 9$, 18:1, Δ^9)

*Linoleic acid ($\omega 6$, 18:2, $\Delta^{9,12}$)

*α-Linolenic acid ($\omega 3$, 18:3, $\Delta^{9,12,15}$)

*Arachidonic acid ($\omega 6$, 20:4, $\Delta^{5,8,11,14}$)

Eicosapentaenoic acid ($\omega 3$, 20:5, $\Delta^{5,8,11,14,17}$)

Figure 25–1. Structure of some unsaturated fatty acids. Although the carbon atoms in the molecules are conventionally numbered, ie, numbered from the carboxyl terminal, the ω numbers (eg, $\omega 7$ in palmitoleic acid) are calculated from the reverse end (the methyl terminal) of the molecules. The information in parentheses shows, for instance, that α-linolenic acid contains double bonds starting at the third carbon from the methyl terminal, has 18 carbons and 3 double bonds, and has these double bonds at the ninth, 12th, and 15th carbons from the carboxyl terminal. *Classified as "essential fatty acids."

acids are concerned, several tissues including the liver are considered to be responsible for their formation from saturated fatty acids. The first double bond introduced into a saturated fatty acid is nearly always in the Δ^9 position. An enzyme system, **Δ^9 desaturase** (Fig 25–2), in the endoplasmic reticulum will catalyze the conversion of palmitoyl-CoA or stearoyl-CoA to palmitoleyl-CoA or oleyl-CoA, respectively. Oxygen and either NADH or NADPH are necessary for the reaction. The enzymes appear to be those of a typical monooxygenase system involving cytochrome b_5 (hydroxylase) (see p 106).

SYNTHESIS OF POLYUNSATURATED FATTY ACIDS INVOLVES DESATURASE & ELONGASE ENZYME SYSTEMS

Additional double bonds introduced into existing monounsaturated fatty acids are always separated from each other by a methylene group (methylene interrupted), except in bacteria. In animals, the additional double bonds are **all introduced between the existing double bond and the carboxyl group,** but in plants they may also be introduced between the existing double bond and the ω (methyl terminal) carbon. Thus, since animals have a Δ^9 desaturase, they are able to synthesize the $\omega 9$ (oleic acid) family of unsaturated fatty acids completely by a combination of chain elongation and desaturation (Fig 25–3). However, since they are unable to synthesize either linoleic ($\omega 6$) or α-linolenic ($\omega 3$) acids, the required desaturases being absent, **these acids must be supplied in the diet to accomplish the synthesis of the other members of the $\omega 6$ and $\omega 3$ families of polyunsaturated fatty**

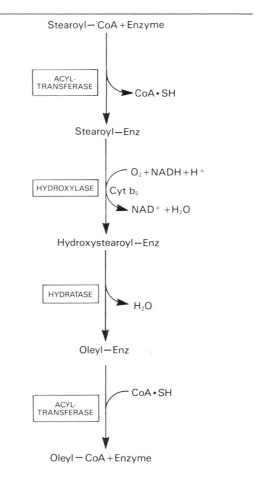

Figure 25–2. Microsomal Δ^9 desaturase system.

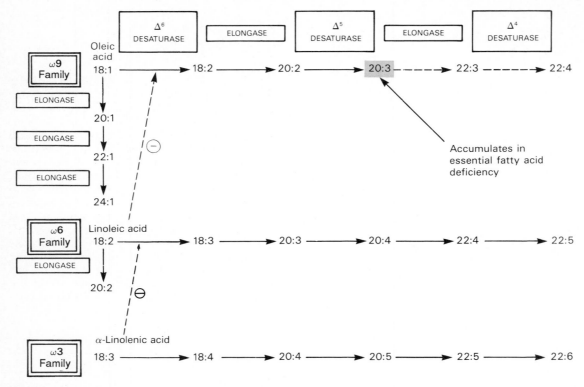

Figure 25–3. Biosynthesis of the ω9, ω6, and ω3 families of polyunsaturated fatty acids. Each step is catalyzed by the microsomal chain elongation or desaturase systems. ω9 Polyunsaturated fatty acids only become quantitatively significant when linoleic and α-linolenic acids are withheld from the diet. This is because each series competes for the same enzyme systems, and affinities decrease from the ω3 to ω9 series. ⊖, Inhibition.

acids. Linoleate may be converted to arachidonate (Fig 25–4). The pathway is first by dehydrogenation of the CoA ester through **γ-linolenate** followed by the addition of a 2-carbon unit via malonyl-CoA in the microsomal system for chain elongation (see p 203), to give eicosatrienoate (dihomo γ-linolenate). The latter forms arachidonate by a further dehydrogenation. The dehydrogenating system is similar to that described above for saturated fatty acids. **The nutritional requirement for arachidonate may thus be dispensed with if there is adequate linoleate in the diet.**

The desaturation and chain elongation system is greatly diminished in the fasting state and in the absence of insulin.

DEFICIENCY SYMPTOMS ARE PRODUCED WHEN THE ESSENTIAL FATTY ACIDS (EFA) ARE ABSENT FROM THE DIET

In 1928, Evans and Burr noticed that rats fed on a purified nonlipid diet to which vitamins A and D were added exhibited a reduced growth rate and a reproductive deficiency. Later work showed that the deficiency syndrome was cured by the addition of **linoleic, α-linolenic,** and **arachidonic acids** to the diet. Further diagnostic features of the syndrome include scaly skin, necrosis of the tail, and lesions in the urinary system, but the condition is not fatal. These fatty acids are found in high concentrations in various vegetable oils (see p 131) and in small amounts in animal carcasses.

The functions of the essential fatty acids appear to be various, though not well defined, apart from prostaglandin and leukotriene formation (see below). Essential fatty acids are found in the structural lipids of the cell and are concerned with the structural integrity of the mitochondrial membrane.

Arachidonic acid is present in membranes and accounts for 5–15% of the fatty acids in phospholipids. Docosahexaenoic acid (DHA; ω3,20:6), which is synthesized from α-linolenic acid or obtained direct from fish oils, is present in high concentrations in retina, cerebral cortex, testis, and sperm. DHA is particularly needed for brain development and is supplied via the placenta and milk.

In many of their structural functions, essential fatty acids are present in phospholipids, mainly in the 2 position. **In essential fatty acid deficiency,** nonessential polyenoic acids of the ω9 family replace the essential fatty acids in phospholipids, other complex lipids,

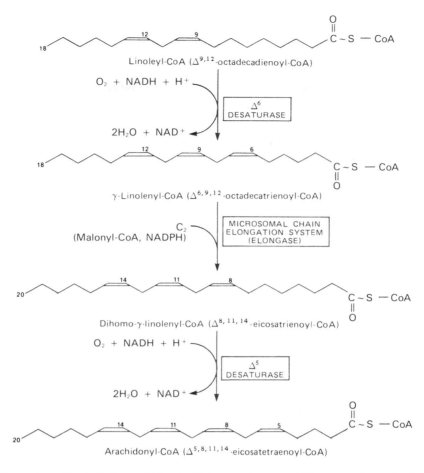

Figure 25–4. Conversion of linoleate to arachidonate. Cats cannot carry out this conversion owing to absence of Δ^6 desaturase and must obtain arachidonate in their diet.

and membranes, particularly $\Delta^{5,8,11}$ eicosatrienoic acid (Fig. 25-3). The triene:tetraene (arachidonate) ratio in plasma lipids can be used to diagnose the extent of essential fatty acid deficiency.

Skin symptoms and impairment of lipid transport have been noted in human subjects ingesting a diet lacking in essential fatty acids. In adults subsisting on ordinary diets, no signs of essential fatty acid deficiencies have been reported. However, infants receiving formula diets low in fat developed skin symptoms that were cured by giving linoleate. Deficiencies attributable to a lack of essential fatty acids, including α-linolenic acid, also occur among patients maintained for long periods exclusively by intravenous nutrition low in essential fatty acids. Deficiency can be prevented by an essential fatty acid intake of 1–2% of the total caloric requirement.

Trans-Fatty Acids May Compete With *Cis*-Fatty Acids

Traces of *trans*-unsaturated fatty acids are found in ruminant fat, where they arise from the action of microorganisms in the rumen, but the presence of large amounts of *trans*-unsaturated fatty acids in partially hydrogenated vegetable oils (eg, margarine) raises the question of their safety as food additives. Their long-term effects in humans are difficult to assess but they have been in the human diet for many years. Up to 15% of tissue fatty acids have been found at autopsy to be in the *trans* configuration. To date, no serious effects have been substantiated. They are metabolized more like saturated than like the *cis*-unsaturated fatty acids. This may be due to their similar straight-chain conformation (see Chapter 16). *Trans*-polyunsaturated fatty acids do not possess essential fatty acid activity and may antagonize the metabolism of essential fatty acids and exacerbate essential fatty acid deficiency.

Abnormal Metabolism of Essential Fatty Acids Occurs in Several Diseases

Apart from essential fatty acid deficiency and changes in unsaturated fatty acid patterns in chronic malnutrition, abnormal metabolism of essential fatty acids, which may be connected with dietary insufficiency, has been noted in cystic fibrosis, acroder-

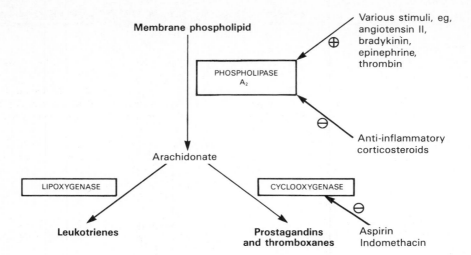

Figure 25–5. Conversion of arachidonic acid to prostaglandins and thromboxanes via the cyclooxygenase pathway and to leukotrienes via the lipoxygenase pathway. The figure indicates why steroids, which inhibit total eicosanoid production, are better antiinflammatory agents than aspirinlike drugs, which only inhibit the cyclooxygenase pathway. Antiinflammatory steroids are thought to inhibit phospholipase A_2 through induction of an inhibitory protein named lipocortin.

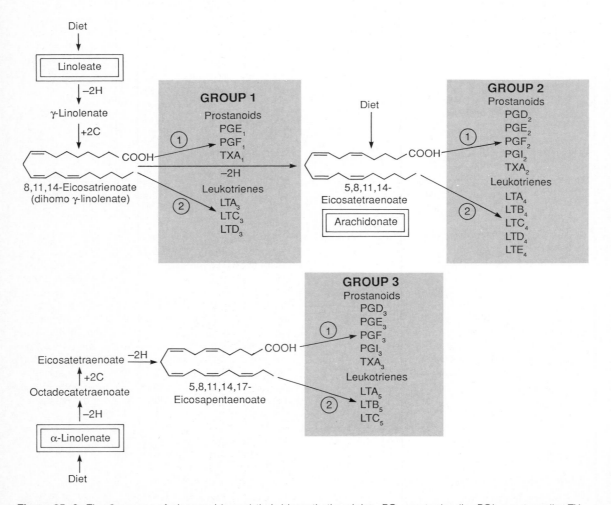

Figure 25–6. The 3 groups of eicosanoids and their biosynthetic origins. PG, prostaglandin; PGI, prostacyclin; TX, thromboxane; LT, leukotriene; 1, cyclooxygenase pathway; 2, lipoxygenase pathway. The subscript denotes the total number of double bonds in the molecule and the series to which the compound belongs.

matitis enteropathica, hepatorenal syndrome, Sjögren-Larsson syndrome, multisystem neuronal degeneration, Crohn's disease, cirrhosis and alcoholism, and Reye's syndrome. Elevated levels of very long chain polyenoic acids have been found in the brains of patients with Zellweger's Syndrome (see p 202). Diets with a high P:S (polyunsaturated:saturated fatty acid) ratio lower serum cholesterol levels, particularly in low-density lipoproteins. This is considered to be beneficial in view of the relationship between serum cholesterol level and coronary heart disease.

EICOSANOIDS ARE FORMED FROM C_{20} POLYUNSATURATED FATTY ACIDS

Arachidonate and some other C_{20} fatty acids with methylene-interrupted bonds give rise to **eicosanoids,** physiologically and pharmacologically active compounds known as **prostaglandins (PG)**, **thromboxanes (TX)**, and **leukotrienes (LT)**. (See Chapter 16).

Arachidonate, usually derived from the 2-position of phospholipids in the plasma membrane, as a result of phospholipase A_2 activity (see Fig 26–5), is the substrate for the synthesis of the PG_2, TX_2, and LT_4 compounds. The pathways of metabolism are divergent, the synthesis of the PG, and TX_2 series (**prostanoids**) competing with the synthesis of LT_4 for the arachidonate substrate. These 2 pathways are known as the **cyclooxygenase** and **lipoxygenase pathways,** respectively (Fig 25–5).

There are 3 groups of eicosanoids (each comprising PG, TX, and LT) that are synthesized from each of the essential fatty acids, respectively, **linoleate, arachidonate,** and **α-linolenate** (Fig 25–6).

THE CYCLOOXYGENASE PATHWAY IS RESPONSIBLE FOR PROSTANOID SYNTHESIS

Prostanoid synthesis (Fig 25–7) involves the consumption of 2 molecules of O_2 catalyzed by **prostaglandin endoperoxide synthase,** which possesses 2 separate enzyme activities, **cyclooxygenase** and **peroxidase.** The product of the cyclooxygenase pathway, an endoperoxide (PGH), is converted to prostaglandins D, E, and F as well as to the thromboxane (TXA_2) and prostacyclin (PGI_2). Each cell type pro-

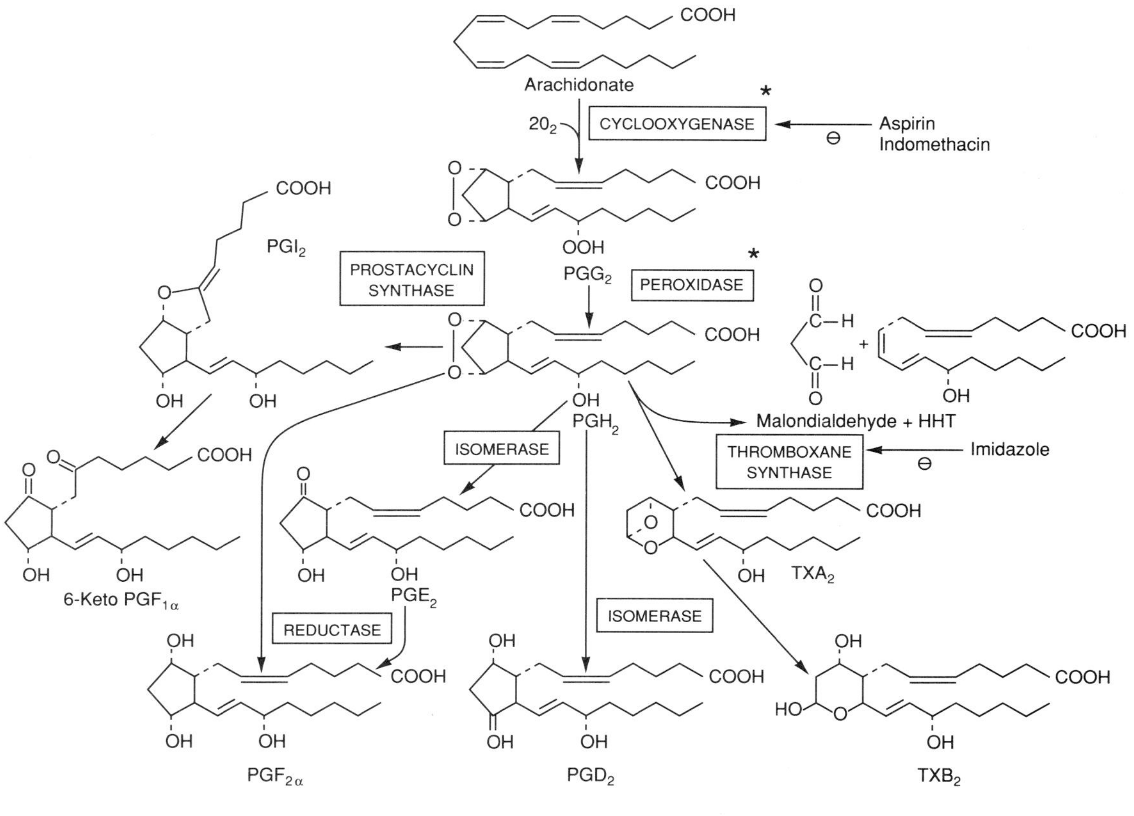

Figure 25–7. Conversion of arachidonic acid to prostaglandins and thromboxanes of series 2. PG, prostaglandin; TX, thromboxane; PGI, prostacyclin; HHT, hydroxyheptadecatrienoate. *Both of these activities are attributed to one enzyme—prostaglandin endoperoxide synthase. Similar conversions occur in prostaglandins and thromboxanes of series 1 and 3.

duces only one type of prostanoid. **Aspirin** inhibits the cyclooxygenase, as does indomethacin.

Essential Fatty Acid Activity & Production of Prostaglandins Is Correlated

Although there is a marked correlation between essential fatty acid activity of various fatty acids and their ability to be converted to prostaglandins, it does not seem that essential fatty acids exert all of their physiologic effects via prostaglandin synthesis. The role of essential fatty acids in membrane formation is unrelated to prostaglandin formation. Prostaglandins do not relieve symptoms of essential fatty acid deficiency, and an essential fatty acid deficiency syndrome is not caused by chronic inhibition of prostaglandin synthesis.

Cyclooxygenase Is a "Suicide Enzyme"

"Switching off" of prostaglandin formation is partly achieved by a remarkable property of cyclooxygenase—that of self-catalyzed destruction; ie, it is a "suicide enzyme." The inactivation of prostaglandins, once formed, is rapid. The presence of the enzyme **15-hydroxyprostaglandin dehydrogenase** in most mammalian tissues is probably the principal cause. It has been shown that blocking the action of this enzyme with sulfasalazine or indomethacin can prolong the half-life of prostaglandins in the body.

Prostanoids Are Potent Biologically Active Substances

Thromboxanes are synthesized in platelets and upon release cause vasoconstriction and platelet aggregation. **Prostacyclins (PGI$_2$)** are produced by blood vessel walls and are potent inhibitors of platelet aggregation. Thus, thromboxanes and prostacyclins are antagonistic. The low incidence of heart disease, diminished platelet aggregation, and prolonged clotting times in Greenland Eskimos have been attributed to their high intake of fish oils containing 20:5 ω3 (EPA, or eicosapentaenoic acid), which gives rise to the series 3 prostaglandins (PG$_3$) and thromboxane TX$_3$ (Fig 25–6). PG$_3$ and TX$_3$ inhibit the release of arachidonate from phospholipids and the formation of PG$_2$ and TX$_2$. PGI$_3$ is as potent an antiaggregator of platelets as PGI$_2$, but TXA$_3$ is a weaker aggregator than TXA$_2$; thus, the balance of activity is shifted toward nonaggregation. In addition, the plasma concentrations of cholesterol, triacylglycerol, and low-density and very low density lipoproteins are all low in

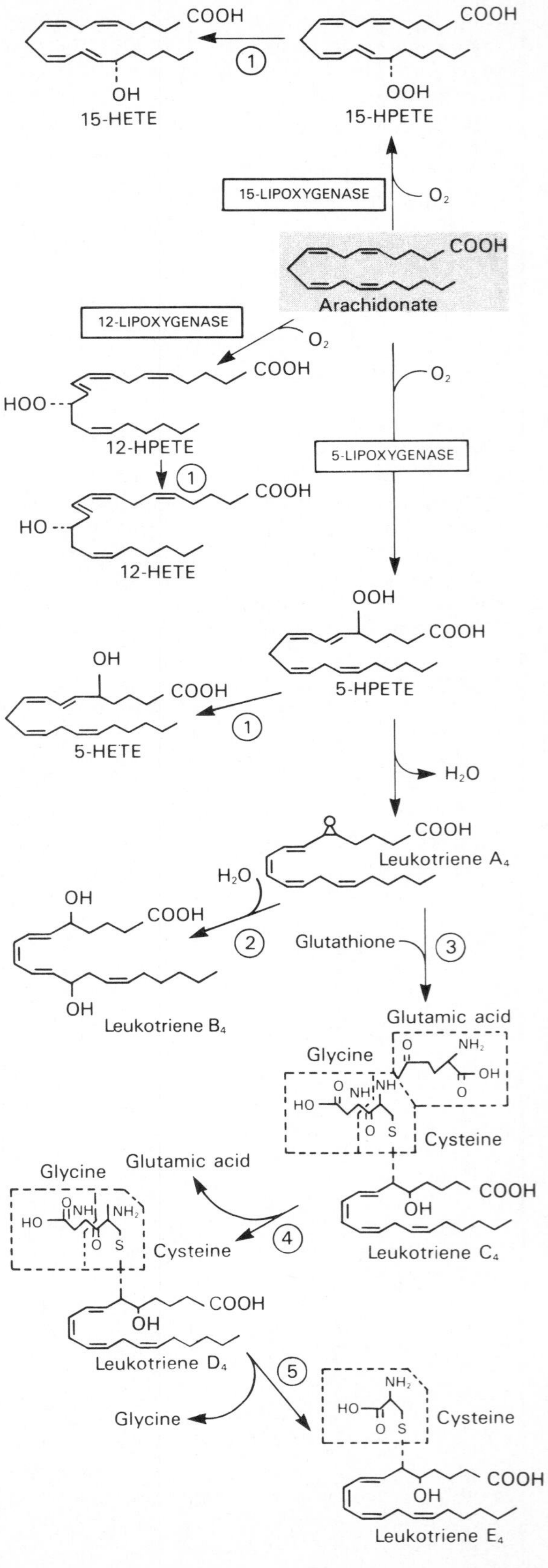

Figure 25–8. Conversion of arachidonic acid to leukotrienes of series 4 via the lipoxygenase pathway. HPETE, hydroperoxyeicosatetraenoate; HETE, hydroxyeicosatetraenoate. Some similar conversions occur in series 3 and 5 leukotrienes. 1, Peroxidase; 2, leukotriene A$_4$ epoxide hydrolase; 3, glutathione S-transferase; 4, γ-glutamyl-transferase; 5, cysteinyl-glycine dipeptidase.

Eskimos, whereas the high-density lipoprotein concentration is raised—all factors considered to militate against atherosclerosis and myocardial infarction.

As little as 1 ng/mL of prostaglandins causes contraction of smooth muscle in animals. Potential therapeutic uses include prevention of conception, induction of labor at term, termination of pregnancy, prevention or alleviation of gastric ulcers, control of inflammation and of blood pressure, and relief of asthma and nasal congestion.

Prostaglandins increase cAMP in platelets, thyroid, corpus luteum, fetal bone, adenohypophysis, and lung but lower cAMP in renal tubule cells and adipose tissue (see p 247).

LEUKOTRIENES ARE FORMED BY THE LIPOXYGENASE PATHWAY

The leukotrienes are a family of conjugated trienes formed from eicosanoic acids in leukocytes, mastocytoma cells, platelets, and macrophages by the **lipoxygenase pathway,** in response to both immunologic and nonimmunologic stimuli. Three different lipoxygenases (dioxygenases) insert oxygen into the 5, 12, and 15 positions of arachidonic acid, giving rise to hydroperoxides (HPETE). Only **5-lipoxygenase** forms leukotrienes. The first formed is leukotriene A_4, which in turn is metabolized to either leukotriene B_4 or leukotriene C_4 (Fig 25–8). Leukotriene C_4 is formed by the addition of the peptide glutathione via a thioether bond. The subsequent removal of glutamate and glycine generates leukotriene D_4 and leukotriene E_4, sequentially.

Leukotrienes Are Potent Regulators of Many Disease Processes

The slow-reacting substance of anaphylaxis (**SRS-A**) is a mixture of leukotrienes C_4, D_4, and E_4. This mixture of leukotrienes is 100–1000 times more potent than histamine or prostaglandins as a constrictor of the bronchial airway musculature. These leukotrienes together with leukotriene B_4 also cause vascular permeability and attraction and activation of leukocytes and seem to be important regulators in many diseases involving inflammatory or immediate hypersensitivity reactions, such as asthma. Leukotrienes are vasoactive and 5-lipoxygenase has been found in arterial walls.

REFERENCES

Hammarström S: Leukotrienes. *Annu Rev Biochem* 1983;**52:** 355.
Holman RT: Control of polyunsaturated acids in tissue lipids. *J Am Coll Nutr* 1986;**5:**183.
Kinsella JE: Food components with potential therapeutic benefits: The n-3 polyunsaturated fatty acids of fish oils. *Food Technology* 1986;**40:**89.
Lagarde M, Gualde N, Rigaud M: Metabolic interactions between eicosanoids in blood and vascular cells. *Biochem J* 1989;**257:**313.
Moncada S (editor): Prostacyclin, thromboxane and leukotrienes. *Br Med Bull* 1983;**39:**209.
Neuringer M, Anderson GJ, Connor WE: The essentiality of N-3 fatty acids for the development and function of the retina and brain. *Annu Rev Nutr* 1988;**8:**517.
Piper P: Formation and actions of leukotrienes. *Physiol Rev* 1984;**64:**744.
Smith WL, Borgeat P: The eicosanoids. In: *Biochemistry of Lipids and Membranes.* Vance DE, Vance JE (editors). Benjamin/Cummings, 1985.

26 Metabolism of Acylglycerols & Sphingolipids

Peter A. Mayes, PhD, DSc

INTRODUCTION

Acylglycerols constitute the majority of lipids in the body. Triaglycerols are the major lipids in fat deposits and in food. In addition, acylglycerols, particularly phospholipids, are major components of the plasma and other membranes. Phospholipids also take part in the metabolism of many lipids. Glycosphingolipids, which contain sphingosine and sugar residues as well as fatty acids, account for 5–10% of the lipids of the plasma membrane.

BIOMEDICAL IMPORTANCE

The role of triacylglycerol in lipid transport and storage and in various diseases such as obesity, diabetes, and hyperlipoproteinemia will be described in detail in subsequent chapters. Phosphoglycerols, phosphosphingolipids, and glycosphingolipids are all amphipathic lipids and consequently ideally suited as the main lipid constituents of the plasma membrane. Some phospholipids have specialized functions; eg, dipalmitoyl lecithin is a major component of **lung surfactant,** the lack of which in premature infants is responsible for respiratory distress syndrome of the newborn. Inositol phospholipids act as precursors of **hormone second messengers,** and **platelet-activating factor** is an alkylphospholipid. Glycosphingolipids, found in the outer leaflet of the plasma membrane with their oligosaccharide chains facing outward, form part of the glycocalyx of the cell surface and are considered to be important (1) in intercellular communication and contact; (2) as receptors for bacterial toxins (eg, the toxin that causes cholera); and (3) as ABO blood group substances. A dozen or so **glycolipid storage diseases** have been described (eg, Gaucher's disease, Tay-Sachs disease); these are due to deficiencies in glycolipid hydrolases present in lysosomes.

CATABOLISM OF ACYLGLYCEROLS IS NOT THE REVERSAL OF BIOSYNTHESIS

Hydrolysis Initiates Catabolism of Triacylglycerol

Triacylglycerols must be hydrolyzed by a suitable **lipase** to their constituent fatty acids and glycerol before further catabolism can proceed. Much of this hydrolysis (lipolysis) occurs in adipose tissue with release of free fatty acids into the plasma, where they are found combined with serum albumin. This is followed by free fatty acid uptake into tissues and subsequent oxidation or reesterification. Many tissues (including liver, heart, kidney, muscle, lung, testis, brain, and adipose tissue) have the ability to oxidize long-chain fatty acids, although brain cannot extract them readily from the blood. The utilization of glycerol depends upon whether such tissues possess the necessary activating enzyme, **glycerol kinase** (Fig 26–1). The enzyme has been found in significant amounts in liver, kidney, intestine, brown adipose tissue, and lactating mammary gland.

PHOSPHATIDATE IS THE COMMON PRECURSOR IN THE BIOSYNTHESIS OF TRIACYLGLYCEROLS & PHOSPHOGLYCEROLS

Although reactions involving the hydrolysis of triacylglycerols by lipase can be reversed in the laboratory, this is not the mechanism by which acylglycerols are synthesized in tissues. Both glycerol and fatty acids must be activated by ATP before they can be incorporated into acylglycerols. Glycerol kinase will catalyze the activation of glycerol to *sn*-glycerol 3-phosphate. If this enzyme is absent—or low in activity, as it is in muscle or adipose tissue—most of the glycerol 3-phosphate **must be derived from an intermediate of the glycolytic system, dihydroxyacetone**

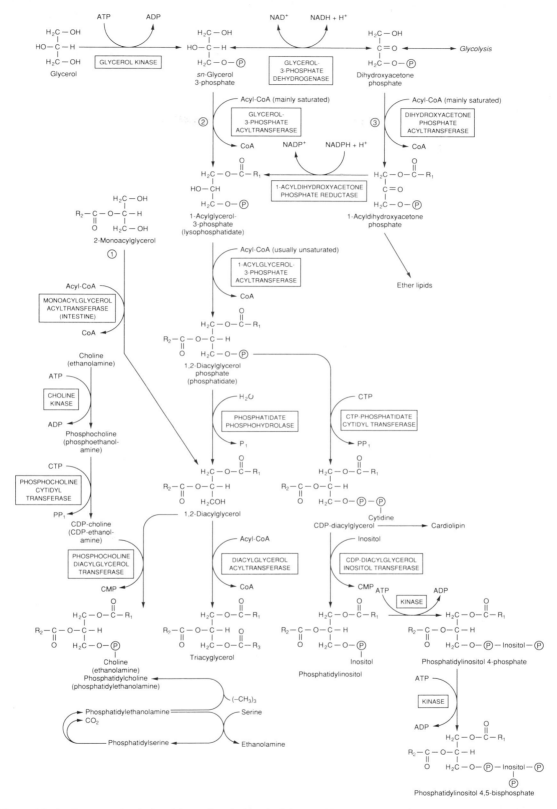

Figure 26–1. Biosynthesis of triacylglycerol and phospholipids. 1, Monoacylglycerol pathway; 2, glycerol phosphate pathway; 3, dihydroxyacetone pathway. Phosphoethanolamine-diacylglycerol transferase is not present in liver.

phosphate, which forms glycerol 3-phosphate by reduction with NADH catalyzed by **glycerol-3-phosphate dehydrogenase** (Fig 26–1).

Biosynthesis of Triacylglycerols: Fatty acids are activated to acyl-CoA by the enzyme **acyl-CoA synthetase,** utilizing ATP and CoA (see p 206). Two molecules of acyl-CoA combine with glycerol 3-phosphate to form **phosphatidate** (1,2-diacylglycerol phosphate). This takes place in 2 stages via lysophosphatidate, catalyzed first by **glycerol-3-phosphate acyltransferase** and then by **1-acylglycerol-3-phosphate acyltransferase** (lysophosphatidate acyltransferase). Phosphatidate is converted by **phosphatidate phosphohydrolase** to a 1,2-diacylglycerol. In intestinal mucosa, a **monoacylglycerol pathway** exists whereby monoacylglycerol is converted to 1,2-diacylglycerol as a result of the presence of **monoacylglycerol acyltransferase.** A further molecule of acyl-CoA is esterified with the diacylglycerol to form a triacylglycerol, catalyzed by **diacylglycerol acyltransferase.** Most of the activity of these enzymes resides in the endoplasmic reticulum of the cell, but some is found in mitochondria, eg, glycerol 3-phosphate acyltransferase. Phosphatidate phosphohydrolase activity is found mainly in the particle-free supernatant fraction but also is membrane bound. Dihydroxyacetone phosphate may be acylated and converted to lysophosphatidate after reduction by NADPH. The quantitative significance of this pathway remains controversial. The pathway appears to be more important in peroxisomes, where it is involved in **ether lipid synthesis.**

Biosynthesis of Phosphoglycerols: These phospholipids are synthesized either from phosphatidate, eg, phosphatidylinositol, or from 1,2-diacylglycerol, eg, phosphatidylcholine and phosphatidylethanolamine. In the synthesis of phosphatidylinositol, cytidine triphosphate (CTP), a high-energy phosphate formed from ATP (see p 104), reacts with phosphatidate to form a cytidine-diphosphate-diacylglycerol (CDP-diacylglycerol). Finally, this compound reacts with inositol, catalyzed by the enzyme **CDP-diacylglycerol inositol transferase,** to form phosphatidylinositol (Fig 26–1). By successive phosphorylations, phosphatidylinositol is transformed first to phosphatidylinositol 4-phosphate and then to phosphatidylinositol 4,5-bisphosphate. The latter is broken down into diacylglycerol and inositol trisphosphate by hormones that increase [Ca^{2+}], eg, vasopressin. These 2 products act as second messengers in the action of the hormone (see p 480).

In the biosynthesis of phosphatidylcholine and phosphatidylethanolamine (lecithins and cephalins), choline or ethanolamine must first be converted to "active choline" or "active ethanolamine," respectively. This is a 2-stage process involving, first, a reaction with ATP to form the corresponding monophosphate, followed by a further reaction with CTP to form either cytidine diphosphocholine (CDP-choline) or cytidine diphosphoethanolamine (CDP-ethanolamine). In this form, choline or ethanolamine reacts with 1,2-diacylglycerol so that a phosphorylated base (either phosphocholine or phosphoethanolamine) is transferred to the diacylglycerol to form either phosphatidylcholine or phosphatidylethanolamine, respectively. The cytidyl transferase appears to be the regulatory enzyme of the phosphatidylcholine pathway.

Phosphatidylserine is formed from phosphatidylethanolamine directly by reaction with serine. Phosphatidylserine may re-form phosphatidylethanolamine by decarboxylation. An alternative pathway in liver enables phosphatidylethanolamine to give rise directly to phosphatidylcholine by progressive methylation of the ethanolamine residue utilizing S-adenosylmethionine as the methyl donor. In turn, the methyl group in methionine can be derived from methyl-H$_4$ folate (see Fig 53–15). In spite of these sources of choline, it is considered to be an essential nutrient in many mammalian species, but this has not been established in humans.

A phospholipid present in mitochondria is **cardiolipin** (diphosphatidylglycerol, see p 138). It is formed from phosphatidylglycerol, which in turn is synthesized from CDP-diacylglycerol (Fig 26–1) and glycerol 3-phosphate according to the scheme shown in Fig 26–2.

Lung surfactant is a secretion with marked surface active properties, composed mainly of lipid with some proteins and carbohydrate, that prevent the alveoli from collapsing. Surfactant activity is largely attributed to the presence of a phospholipid, **dipalmitoylphosphatidylcholine,** which is synthesized shortly before parturition in full-term infants. Deficiency of lung surfactant in the lungs of many preterm newborns gives rise to the **respiratory distress syndrome.** Administration of either natural or artificial surfactant has been of therapeutic benefit.

Biosynthesis of Glycerol Ether Phospholipids & Plasmalogens: A plasmalogenic diacylglycerol is one in which the 1 (or 2) position has an alkenyl resi-

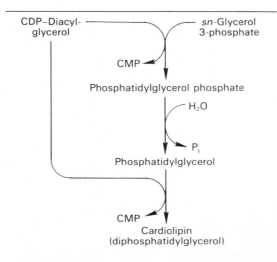

Figure 26–2. Biosynthesis of cardiolipin.

due containing the vinyl ether aldehydogenic linkage ($-CH_2-O-CH=CH-R'$). Dihydroxyacetone phosphate is the precursor of the glycerol moiety (Fig 26–3). This compound combines with acyl-CoA to give 1-acyldihydroxyacetone phosphate. An exchange reaction takes place between the acyl group and a long-chain alcohol to give a 1-alkyldihydroxyacetone phosphate (containing the ether link) which in the presence of NADPH is converted to 1-alkylglycerol 3-phosphate. After further acylation in the 2 position, the resulting 1-alkyl-2-acylglycerol 3-phosphate (analogous to phosphatidate in Fig 26–1) is hydrolyzed to give the free glycerol derivative. **Plasmalogens** are formed by desaturation of the analogous 3-phosphoethanolamine derivative (Fig 26–3). Much of the phospholipid in mitochondria consists of plasmalogens. **Platelet activating factor (PAF)** is synthesized from the corresponding 3-phosphocholine derivative and has been identified as 1-alkyl-2-acetyl-sn-glycerol-3-phosphocholine. It is formed by many blood cells and other tissues and aggregates platelets at concentrations as low as 10^{-11} mol/L. It also has hypotensive and ulcerogenic properties.

Phospholipases Allow Degradation & Remodeling of Glycerophospholipids

Degradation of many complex molecules in tissues procedes to completion, eg, proteins to amino acids. Thus, a turnover time can be determined for such a molecule. Although phospholipids are actively degraded, each portion of the molecule turns over at a different rate; eg, the turnover time of the phosphate group is different from that of the 1-acyl group. This is due to the presence of enzymes that allow partial degradation followed by resynthesis (Fig 26–4). **Phospholipase A_2** catalyzes the hydrolysis of the ester bond in position 2 of glycerophospholipids to form a

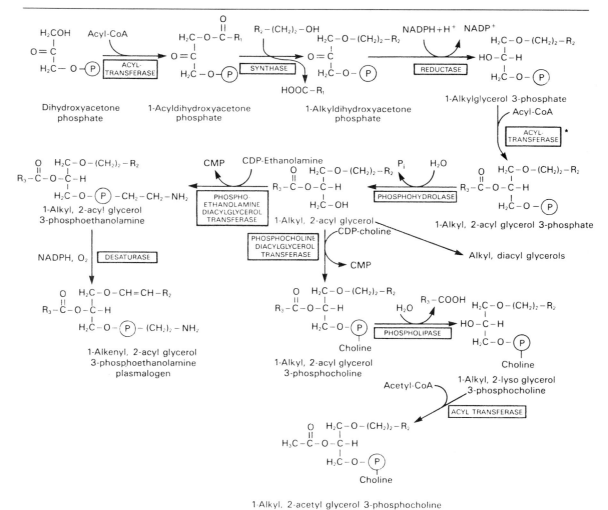

Figure 26–3. Biosynthesis of ether lipids, plasmalogens, and platelet-activating factor (PAF). In the de novo pathway for PAF synthesis, acetyl-CoA is incorporated at stage,* avoiding the last 2 steps in the pathway shown here.

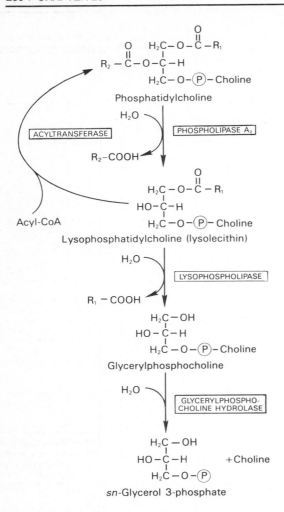

Figure 26–4. Metabolism of phosphatidylcholine (lecithin).

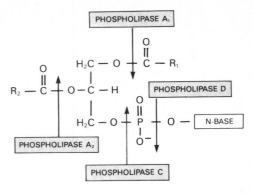

Figure 26–5. Sites of the hydrolytic activity of phospholipases on a phospholipid substrate.

form cholesteryl ester and is considered to be responsible for much of the cholesteryl ester in plasma lipoproteins. The consequences of **LCAT deficiency** are discussed on p 255.

LECITHIN:
CHOLESTEROL
ACYLTRANSFERASE

Lecithin + Cholesterol ⟶ Lysolecithin + Cholesteryl ester

nantly in the 1 position of phospholipids, whereas the polyunsaturated acids (eg, the precursors of prostaglandins) are incorporated more into the 2 position. The incorporation of fatty acids into lecithin occurs by complete synthesis of the phospholipid, by transacylation between cholesteryl ester and lysolecithin, and by direct acylation of lysolecithin by acyl-CoA. Thus, a continuous exchange of the fatty acids is possible, particularly with regard to **introducing essential fatty acids into phospholipid molecules.**

ALL SPHINGOLIPIDS ARE FORMED FROM CERAMIDE

The amino-alcohol **Sphingosine** (Fig 26–6) is synthesized in the endoplasmic reticulum. Following activation by combination with pyridoxal phosphate, the amino acid serine combines with palmitoyl-CoA to form 3-ketosphinganine after loss of CO_2. Sphingosine itself is formed after a reductive step, which is known to utilize NADPH as H donor. This is followed by an oxidative step which involves a flavoprotein enzyme, analogous to the acyl-CoA dehydrogenase step in β-oxidation.

Ceramide (N-acylsphingosine) is formed by a combination of acyl-CoA and sphingosine (Fig 26–7). The acyl group is represented frequently by long-chain saturated or monoenoic acids.

Sphingomyelins are phospholipids (see p 140) and are formed when ceramide reacts with either CDP-choline or phosphatidylcholine; the former reaction is

free fatty acid and lysophospholipid, which in turn may be reacylated by acyl-CoA in the presence of an acyltransferase. Alternatively, lysophospholipid (eg, lysolecithin) is attacked by **lysophospholipase** (phospholipase B), removing the remaining 1-acyl group and forming the corresponding glyceryl phosphoryl base, which in turn may be split by a hydrolase liberating glycerol 3-phosphate plus base. **Phospholipase A₁** attacks the ester bond in position 1 of phospholipids (Fig 26–5). **Phospholipase C** attacks the ester bond in position 3, liberating 1,2-diacylglycerol plus a phosphoryl base. It is one of the major toxins secreted by bacteria. **Phospholipase D** is an enzyme, described mainly in plants, that hydrolyzes the nitrogenous base from phospholipids.

Lysolecithin may be formed by an alternative route that involves **lecithin:cholesterol acyltransferase** (**LCAT**). This enzyme, found in plasma and synthesized in liver, catalyzes the transfer of a fatty acid residue from the 2 position of lecithin to cholesterol to

Long-chain saturated fatty acids are found predomi-

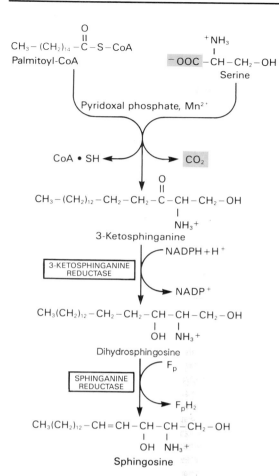

Figure 26–6. Biosynthesis of sphingosine. Fp, flavoprotein.

analogous to that employed in the biosynthesis of phosphatidylcholine (Fig 26–1).

Glycosphingolipids Are a Combination of Ceramide With One or More Sugar Residues

Characteristically, C_{24} fatty acids occur in many glycosphingolipids, particularly those in brain (lignoceric, cerebronic, and nervonic acids). Lignoceric acid ($C_{23}H_{47}COOH$) is completely synthesized from acetyl-CoA. Cerebronic acid, the 2-hydroxy deriva-

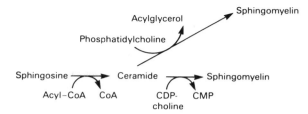

Figure 26–7. Biosynthesis of ceramide and sphingomyelin.

tive of lignoceric acid, is formed from it. Nervonic acid ($C_{23}H_{45}COOH$), a monounsaturated acid, is formed by elongation of oleic acid.

The simplest glycosphingolipids (**cerebrosides**) are **galactosylceramide** (**GalCer**) and **glucosylceramide** (**GlcCer**). GalCer is a major lipid of myelin, whereas GlcCer is the major glycosphingolipid of extraneural tissues and a precursor of most of the more complex glycosphingolipids. **Uridine diphosphogalactose epimerase** (Fig 26–8) utilizes uridine diphosphate glucose (UDPGlc) as substrate and accomplishes epimerization of the glucose moiety to galactose, thus forming uridine diphosphate galactose (UDPGal). The reaction in brain is similar to that described in Fig 22–5 for the liver and mammary gland. Galactosylceramide is formed in a reaction between ceramide and UDPGal. **Sulfogalactosylceramide** is formed after further reaction with 3′-phosphoadenosine-5′-phosphosulfate (PAPS; "active sulfate"). PAPS is also involved in the biosynthesis of the other sulfolipids, ie, the **sulfo(galacto)glycerolipids** and the **steroid sulfates**.

Gangliosides are synthesized from ceramide by the stepwise addition of activated sugars (eg, UDPGlc and UDPGal) and a **sialic acid,** usually **N-acetylneuraminic acid** (Fig 26–9). A large number of gangliosides of increasing molecular weight may be formed. Most of the enzymes transferring sugars from nucleotide sugars (glycosyl transferases) are found in the Golgi apparatus.

Glycosphingolipids are constituents of the outer leaflet of plasma membranes, and as such may be important in **intercellular communication and contact.** Some are antigens, eg, the Forssman antigen and ABO blood group substances. Similar oligosaccharide chains are found in glycoproteins in the plasma mem-

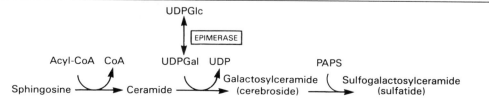

Figure 26–8. Biosynthesis of galactosylceramide and its sulfo derivative. PAPS, "active sulfate," phosphoadenosine-5′-phosphosulfate.

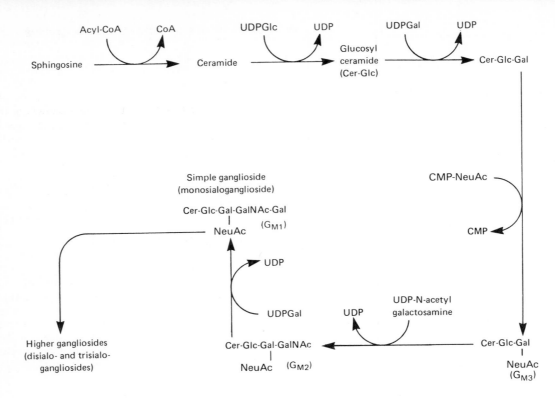

Figure 26–9. Biosynthesis of gangliosides. NeuAc, N-acetylneuraminic acid.

Table 26–1. Summary of the sphingolipidoses.

Disease	Enzyme Deficiency	Lipid Accumulating	Clinical Symptoms
Fucosidosis	α-Fucosidase	Cer–Glc–Gal–GalNAc–Gal$\div$Fuc H-Isoantigen	Cerebral degeneration, muscle spasticity, thick skin.
Generalized gangliosidosis	G_{M1}-β-galactosidase	Cer–Glc–Gal(NeuAc)–GalNAc$\div$Gal G_{M1} Ganglioside	Mental retardation, liver enlargement, skeletal deformation.
Tay-Sachs disease	Hexosaminidase A	Cer–Glc–Gal(NeuAc)$\div$GalNAc G_{M2} Ganglioside	Mental retardation, blindness, muscular weakness.
Tay-Sachs variant or Sandhoff's disease	Hexosaminidase A and B	Cer–Glc–Gal–Gal$\div$GalNAc Globoside plus G_{M2} ganglioside	Same as Tay-Sachs but progressing more rapidly.
Fabry's disease	α-Galactosidase	Cer–Glc–Gal$\div$Gal Globotriaosylceramide	Skin rash, kidney failure (full symptoms only in males; X-linked recessive).
Ceramide lactoside lipidosis	Ceramide lactosidase (β-galactosidase)	Cer–Glc$\div$Gal Ceramide lactoside	Progressing brain damage, liver and spleen enlargement.
Metachromatic leukodystrophy	Arylsulfatase A	Cer–Gal$\div$OSO$_3$ 3-Sulfogalactosylceramide	Mental retardation and psychologic disturbances in adults; demyelination.
Krabbe's disease	β-Galactosidase	Cer$\div$Gal Galactosylceramide	Mental retardation; myelin almost absent.
Gaucher's disease	β-Glucosidase	Cer$\div$Glc Glucosylceramide	Enlarged liver and spleen, erosion of long bones, mental retardation in infants.
Niemann-Pick disease	Sphingomyelinase	Cer$\div$P–choline Sphingomyelin	Enlarged liver and spleen, mental retardation; fatal in early life.
Farber's disease	Ceramidase	Acyl$\div$Sphingosine Ceramide	Hoarseness, dermatitis, skeletal deformation, mental retardation; fatal in early life.

NeuAc, N-acetylneuraminic acid; Cer, ceramide; Glc, glucose; Gal, galactose; Fuc, fucose. $\div$, site of deficient enzyme reaction.

brane. Certain gangliosides function as receptors for bacterial toxins (eg, for cholera toxin, which subsequently activates adenylate cyclase).

PHOSPHOLIPIDS & SPHINGOLIPIDS ARE INVOLVED IN MULTIPLE SCLEROSIS AND LIPIDOSES

Certain diseases are characterized by abnormal quantities of these lipids in the tissues, often in the nervous system. They may be classified into 3 groups: (1) true demyelinating diseases (2) sphingolipidoses, and (3) leukodystrophies.

In **multiple sclerosis,** which is a demyelinating disease, there is loss of both phospholipids (particularly ethanolamine plasmalogen) and of sphingolipids from white matter. Thus, the composition of white matter resembles that of gray matter. Cholesteryl esters may be found in white matter although they are normally absent. The cerebrospinal fluid shows raised phospholipid levels.

The **sphingolipidoses** are a group of inherited diseases that are often manifested in childhood. These diseases are part of a larger group of lysosomal disorders.

Lipid storage diseases exhibit several constant features: (1) In various tissues, there is an accumulation of complex lipids that have a portion of their structure in common—**ceramide.** (2) The rate of **synthesis** of the stored lipid is comparable to that in normal humans. (3) The enzymatic defect in each of these diseases is **a deficiency of a specific lysosomal hydrolytic enzyme necessary to break down the lipid.** (4) The extent to which the activity of the affected enzyme is decreased is similar in all of the tissues of the affected individual. As a result of these unifying basic considerations, procedures for the diagnosis of patients with these disorders have been developed. It has also become possible to detect heterozygous carriers of the genetic abnormalities responsible for these diseases as well as to discover in the unborn fetus the fact that a sphingolipodystrophy is present. A summary of the more important lipidoses is shown in Table 26–1.

Multiple sulfatase deficiency results in accumulation of sulfogalactosylceramide, steroid sulfates, and proteoglycans, owing to a combined deficiency of arylsulfatases A, B, and C and steroid sulfatase (see p 606).

REFERENCES

Boyer PD (editor): *The Enzymes,* 3rd ed. Vol 16: *Lipid Enzymology.* Academic Press, 1983.
Brady RO: Sphingolipidoses. *Annu Rev Biochem* 1978;**47:** 687.
Hawthorne JN, Ansell GB (editors): *Phospholipids.* Elsevier, 1982.
Snyder F: Biochemistry of platelet-activating factor. *PSEBM* 1989;**190:**125.
Various authors: In: *The Metabolic Basis of Inherited Disease,* 6th ed. Scriver CR et al (editors). McGraw-Hill, 1989.
Various authors: Metabolism of triacylglycerols; phospholipid metabolism; ether-linked glycerolipids; sphingolipids. In: *Biochemistry of Lipids and Hormones.* Vance DE, Vance JE (editors). Benjamin/Cummings, 1985.
VanGolde LMG, Batenburg JJ, Robertson B: The pulmonary surfactant system: biochemical aspects and functional significance. *Physiol Rev* 1988;**68:**374.
Watts RWE, Gibbs DA: *Lysosomal Storage Diseases: Biochemical and Clinical Aspects.* Taylor & Francis, London, 1986.

27

Lipid Transport & Storage

Peter A. Mayes, PhD, DSc

INTRODUCTION

Fat absorbed from the diet and lipids synthesized by the liver and adipose tissue must be transported between the various tissues and organs for utilization and storage. Since lipids are insoluble in water, the problem arises of how to transport them in an aqueous environment—the blood plasma. This is solved by associating nonpolar lipids (triacylglycerol and cholesteryl esters) with amphipathic lipids (phospholipids and cholesterol) and proteins to make water-miscible lipoproteins.

BIOMEDICAL IMPORTANCE

In a meal-eating omnivore such as the human, excess calories are ingested in the anabolic phase of the feeding cycle, followed by a period of negative caloric balance when the organism draws upon its carbohydrate and fat stores. Lipoproteins mediate this cycle by transporting lipids from the intestines as chylomicrons, and from the liver as very low density lipoproteins (VLDL), to most tissues for oxidation and to adipose tissue for storage. Lipid is mobilized from adipose tissue as free fatty acids (FFA) attached to serum albumin. Abnormalities of lipid metabolism occur at the sites of production or utilization of lipoproteins, causing various **hypo-** or **hyperlipoproteinemias.** The most common of these is **diabetes mellitus,** where insulin deficiency causes excessive mobilization of FFA and underutilization of chylomicrons and VLDL, leading to **hypertriacylglycerolemia.** Most other pathologic conditions affecting lipid transport primarily are due to inherited defects in synthesis of the apoprotein portion of the lipoprotein, of key enzymes, or of lipoprotein receptors. Some of these defects cause **hypercholesterolemia** and premature **atherosclerosis.** Excessive fat deposits constitute **obesity,** one form of which may be due to defective diet-induced thermogenesis in brown adipose tissue.

LIPIDS ARE TRANSPORTED IN THE PLASMA AS LIPOPROTEINS

Four Major Groups of Lipoproteins Have Been Identified

Extraction of the plasma lipids with a suitable lipid solvent and subsequent separation of the extract into various classes of lipids show the presence of **triacylglycerols, phospholipids, cholesterol,** and **cholesteryl esters** and, in addition, the existence of a much smaller fraction of unesterified long-chain fatty acids (free fatty acids) that accounts for less than 5% of the total fatty acid present in the plasma. This latter fraction, the **free fatty acids (FFA)**, is now known to be metabolically the most active of the plasma lipids. An analysis of blood plasma showing the major lipid classes is given in Table 27–1.

Pure fat is less dense than water; it follows that as the proportion of lipid to protein in lipoproteins increases, the density decreases (Table 27–2). Use is made of this property in separating the various lipoproteins in plasma by **ultracentrifugation.** The rate at which each lipoprotein floats through a solution of NaCl (specific gravity 1.063) may be expressed in Svedberg (Sf) units of flotation. One Sf unit is equal to 10^{-13} cm/s/dyne/g at 26 °C. The composition of the different lipoprotein fractions obtained by centrifugation is shown in Table 27–2. The various chemical classes of lipids are seen to occur in varying amounts in most of the lipoprotein fractions. Since the density fractions represent the physiologic entities present in the plasma, mere chemical analysis of the plasma lipids (apart from FFA) yields little information on their physiology.

In addition to the use of techniques depending on their density, lipoproteins may be separated according to their electrophoretic properties into **α-, β, and pre-β-lipoproteins** (Fig 27–1) and may be identified more accurately by means of immunoelectrophoresis.

Apart from FFA, **4 major groups of lipoproteins have been identified that are important physiologically and in clinical diagnosis.** These are (1)

Table 27–1. Lipids of the blood plasma in humans.

Lipid	mmol/L	
	Mean	**Range**
Triacylglycerol	1.6	0.9–2.0
Total phospholipid†	3.1	1.8–5.8
Total cholesterol	5.2	2.8–8.3
Free cholesterol (nonesterified)	1.4	0.7–2.7
Free fatty acids (nonesterified)	0.4	0.2–0.6*

Of total fatty acids, 45% are triacylglycerols, 35% phospholipids, 15% cholesteryl ester, and less than 5% free fatty acids.

*Varies with nutritional state.
†Analyzed as lipid phosphorus.

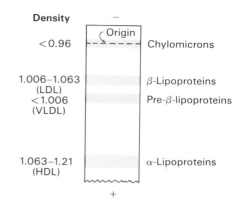

Figure 27–1. Separation of plasma lipoproteins by electrophoresis.

chylomicrons, derived from intestinal absorption of triacylglycerol; (2) **very low density lipoproteins** (VLDL, or pre-β-lipoproteins), derived from the liver for the export of triacylglycerol; (3) **low-density lipoproteins** (LDL, or β-lipoproteins), representing a final stage in the catabolism of VLDL; and (4) **high-density lipoproteins** (HDL, or α-lipoproteins), involved in VLDL and chylomicron metabolism and also in cholesterol metabolism. Triacylglycerol is the predominant lipid in chylomicrons and VLDL, whereas cholesterol and phospholipid are the predominant lipids in LDL and HDL, respectively (Table 27–2).

Amphipathic Lipids Are Essential Components of Lipoproteins

A typical lipoprotein—such as chylomicron or VLDL—consists of a **lipid core** of mainly **nonpolar triacylglycerol and cholesteryl ester** surrounded by a **single surface layer** of **amphipathic phospholipid and cholesterol** molecules. These are oriented so that their polar groups face outward to the aqueous medium, as in the cell membrane (see p 145).

The protein moiety of lipoproteins is known as an **apolipoprotein** or **apoprotein,** constituting nearly 60% of some HDL and as little as 1% of chylomicrons.

Table 27–2. Composition of the lipoproteins in plasma of humans.

Fraction	Source	Diameter (nm)	Density	Sf	Protein (%)	Total Lipid (%)	Triacylglycerol	Phospholipid	Cholesteryl Ester	Cholesterol (Free)	Free Fatty Acids
Chylomicrons	Intestine	90–1000	<0.95	>400	1–2	98–99	88	8	3	1	. . .
Very low density lipoproteins (VLDL)	Liver and intestine	30–90	0.95–1.006	20–400	7–10	90–93	56	20	15	8	1
Intermediate-density lipoproteins (IDL)	VLDL and chylomicrons	25–30	1.006–1.019	12–20	11	89	29	26	34	9	1
Low-density lipoproteins (LDL)	VLDL	20–25	1.019–1.063	2–12	21	79	13	28	48	10	1
High-density lipoproteins HDL₂	Liver and intestine VLDL? Chylomicrons?	10–20	1.063–1.125		33	67	16	43	31	10	. . .
HDL₃		7.5–10	1.125–1.210		57	43	13	46	29	6	6
Albumin-FFA	Adipose tissue		>1.2810		99	1	0	0	0	0	100

FFA, free fatty acids. VHDL (very high density lipoprotein) is a minor fraction occurring at density 1.21–1.25.

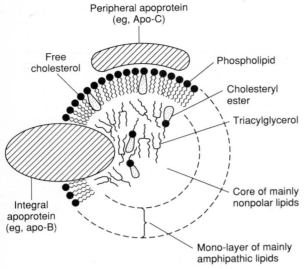

Figure 27-2. Generalized structure of a plasma lipoprotein. The similarities with the structure of the plasma membrane are to be noted. Recent work indicates that a small amount of cholesteryl ester and triacylglycerol are to be found in the surface layer and a little free cholesterol in the core.

Some apolipoproteins are integral and cannot be removed, whereas others are free to transfer to other lipoproteins (Fig 27–2).

The Distribution of Apolipoproteins Characterizes the Lipoprotein

One or more apolipoproteins (proteins or polypeptides) are present in each lipoprotein. According to the ABC nomenclature, the major apolipoprotein of HDL (α-lipoprotein) is designated A. The main apolipoprotein of LDL (β-lipoprotein) is apolipoprotein B, which is found also in VLDL and chylomicrons. However,

apo B of chylomicrons (B-48) is smaller than apo B of LDL or VLDL (B-100). B-48 is synthesized in the intestine and B-100 in the liver. (In the rat, the liver appears to form B-48 in addition to B-100.)

Apo B-100 is possibly the longest single polypeptide chain known, having 4536 amino acids. Apo B-48 (48% as large as B-100) is formed from the same mRNA as Apo B-100. Apparently, in intestine, a stop codon that is not present in genomic DNA is introduced by an RNA-editing mechanism that stops translation at amino acid residue 2153. Apolipoproteins C-I, C-II, and C-III are smaller polypeptides freely transferable between several different lipoproteins (Table 27–3). Carbohydrates account for approximately 5% of apo B and include mannose, galactose, fucose, glucose, glucosamine, and sialic acid. Thus, some lipoproteins are also glycoproteins. Several other apolipoproteins have been found in plasma lipoproteins. One is the arginine-rich apolipoprotein E isolated from VLDL and HDL; it contains arginine to the extent of 10% of the total amino acids and accounts for 5–10% of total VLDL apolipoproteins in normal subjects but is present in excess in the broad β-VLDL of patients with type III hyperlipoproteinemia.

Apolipoproteins carry out several roles: (1) They are enzyme cofactors, eg, C-II for lipoprotein lipase, A-I for lecithin:cholesterol acyltransferase. (2) They can act as lipid transfer proteins, eg, apo D in HDL. (3) They act as ligands for interaction with lipoprotein receptors in tissues, eg, apo B-100, apo E for the LDL receptor, apo E for the remnant receptor, and apo A-I for the HDL receptor.

FREE FATTY ACIDS ARE VERY RAPIDLY METABOLIZED

The free fatty acids (FFA, nonesterified fatty acids, unesterified fatty acids) arise in the plasma from

Table 27–3. Apolipoproteins of human plasma lipoproteins.

Apolipoprotein	Lipoprotein	Molecular Mass (Da)	Additional Remarks
A-I	HDL, chylomicrons	28,000	Activator of lecithin:cholesterol acyltransferase (LCAT).
A-II	HDL, chylomicrons	17,000	Structure is 2 identical monomers joined by a disulfide bridge. Inhibitor of LCAT?
B-100	LDL, VLDL, IDL	550,000	Synthesized in liver. Ligand for LDL receptor.
B-48	Chylomicrons, chylomicron remnants	260,000	Synthesized in intestine.
C-I	VLDL, HDL	7600	Possible activator of LCAT.
C-II	VLDL, HDL, chylomicrons	8800	Activator of extrahepatic lipoprotein lipase.
C-III	VLDL, HDL, chylomicrons	8750	Several polymorphic forms depending on content of sialic acids.
D	Subfraction of HDL	20,000	Possibly identical to the cholesteryl ester transfer protein.
E (arginine-rich)	VLDL, HDL, chylomicrons, chylomicron remnants	34,000	Present in excess in the β-VLDL of patients with type III hyperlipoproteinemia. It is the sole apoprotein found in HDL$_c$ of diet-induced hypercholesterolemic animals. Ligand for chylomicron remnant receptor in liver and LDL receptor.

lipolysis of triacylglycerol in adipose tissue or as a result of the action of lipoprotein lipase during uptake of plasma triacylglycerols into tissues. They are found **in combination with serum albumin** in concentrations varying between 0.1 and 2 µeq/mL of plasma and comprise the long-chain fatty acids found in adipose tissue, ie, palmitic, stearic, oleic, palmitoleic, linoleic, and other polyunsaturated acids, and smaller quantities of other long-chain fatty acids. Binding sites on albumin of varying affinity for the fatty acids have been described. Low levels of free fatty acids are recorded in the fully fed condition, rising to about 0.5 µeq/mL in the postabsorptive and between 0.7 and 0.8 µeq/mL in the fully fasting state. In uncontrolled **diabetes mellitus,** the level may rise to as much as 2 µeq/mL. The level falls just after eating and rises again prior to the next meal; however, in such continuous feeders as ruminants—where there is uninterrupted influx of nutrient from the intestine—the free fatty acids remain at a low level.

The rate of removal of free fatty acids from the blood is extremely rapid. Some of the uptake is oxidized and supplies about 25–50% of the energy requirements in fasting. The remainder of the uptake is esterified. In starvation, the respiratory quotient (RQ) indicates that considerably more fat is being oxidized than can be traced to the oxidation of free fatty acids. This difference is accounted for by the oxidation of esterified lipids from the circulation or of those present in tissues. The latter is thought to occur particularly in heart and skeletal muscle, where considerable stores of lipid are to be found in the muscle cells.

The free fatty acid turnover is related directly to free fatty acid concentration. Thus, the rate of free fatty acid production in adipose tissue controls the free fatty acid concentration in plasma, which in turn determines the free fatty acid uptake by other tissues. The nutritional condition does not appear to have a great effect on the fractional uptake of free fatty acids by tissues. It does, however, alter the proportion of the uptake which is oxidized compared to the fraction which is esterified, more being oxidized in the fasting than in the fed state.

The presence of a **fatty acid-binding protein,** or **Z-protein,** in the cytosol of many of the major tissues has been reported. The role of this protein in intracellular transport is thought to be similar to the role of serum albumin in extracellular transport of long-chain fatty acids.

TRIACYLGLYCEROL IS TRANSPORTED FROM THE INTESTINES IN CHYLOMICRONS & FROM THE LIVER IN VERY LOW DENSITY LIPOPROTEINS

By definition, **chylomicrons** are found in **chyle formed only by the lymphatic system draining the intestine.** They are responsible for the transport of all dietary lipids into the circulation. Smaller and denser particles having the physical characteristics of VLDL are also to be found in chyle. However, their composi-

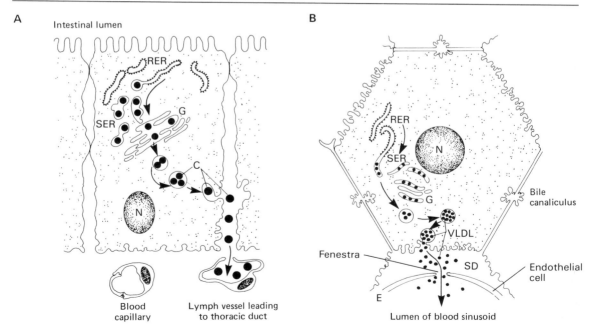

Figure 27–3. The formation and secretion of (A) chylomicrons by an intestinal cell and (B) very low density lipoproteins by a hepatic cell. RER, rough endoplasmic reticulum; SER, smooth endoplasmic reticulum; G, Golgi apparatus; N, nucleus; C, chylomicrons; VLDL, very low density lipoproteins; E, endothelium; SD, space of Disse, containing blood plasma. The figure is a diagrammatic representation of events that can be seen with electron microscopy.

tion resembles chylomicrons rather than VLDL, indicating that they should be regarded as small chylomicrons. Their formation occurs even in the fasting state, their lipids originating mainly from bile and intestinal secretions. On the other hand, chylomicron formation increases with the load of triacylglycerol absorbed. Most of the plasma **VLDL** are of hepatic origin. They are the **vehicles of transport of triacylglycerol from the liver to the extrahepatic tissues.**

There are many similarities in the mechanism of formation of chylomicrons by intestinal cells and of VLDL by hepatic parenchymal cells (Fig 27–3). Apolipoprotein B is synthesized by ribosomes in the rough endoplasmic reticulum and is incorporated into lipoproteins in the smooth endoplasmic reticulum, which is the main site of synthesis of triacylglycerol. Lipoproteins pass through the Golgi apparatus, where, it is thought, carbohydrate residues are added to the lipoprotein. The chylomicrons and VLDL are released from either the intestinal or hepatic cell by fusion of the secretory vacuole with the cell membrane (reverse pinocytosis). Chylomicrons pass into the spaces between the intestinal cells, eventually making their way into the lymphatic system (lacteals) draining the intestine. VLDL are secreted by hepatic parenchymal cells into the space of Disse and then into the hepatic sinusoids through fenestrae in the endothelial lining. The similarities between the 2 processes and the anatomic mechanisms are striking, for—apart from the mammary gland—the intestine and liver are the only tissues

from which particulate lipid is secreted. The inability of particulate lipid of the size of chylomicrons and VLDL to pass through endothelial cells of the capillaries without prior hydrolysis is probably the reason dietary fat enters the circulation via the lymphatics (thoracic duct) and not via the hepatic portal system.

Although both chylomicrons and VLDL isolated from blood contain apolipoproteins C and E, the newly secreted or "nascent" lipoproteins contain little or none, and it would appear that the full complement of apo C and E polypeptides is taken up by transfer from HDL once the chylomicrons and VLDL have entered the circulation (Figs 27–4 and 27–5). A more detailed account of the factors controlling hepatic VLDL secretion is given below.

Apo B is essential for chylomicron and VLDL formation. In **abetalipoproteinemia** (a rare disease), apo B is not synthesized; lipoproteins containing this apolipoprotein are not formed, and lipid droplets accumulate in the intestine and liver.

CATABOLISM OF CHYLOMICRONS & VERY LOW DENSITY LIPOPROTEINS IS A RAPID PROCESS

The clearance of labeled chylomicrons from the blood is rapid, the half-time of disappearance being on the order of minutes in small animals (eg, rats) but longer in larger animals (eg, humans), in whom it is

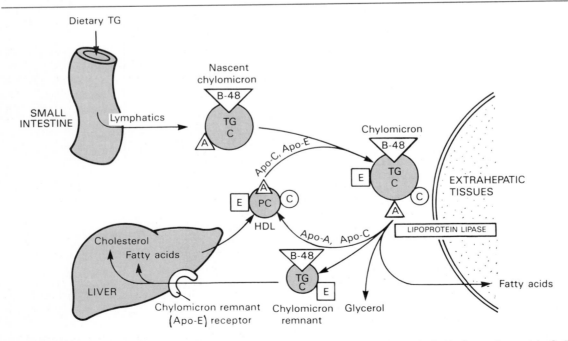

Figure 27–4. Metabolic fate of chylomicrons. A, apolipoprotein A; B-48, apolipoprotein B-48; ©, apolipoprotein C; E, apolipoprotein E; HDL, high-density lipoprotein; TG, triacylglycerol; C, cholesterol and cholesteryl ester; P, phospholipid. Only the predominant lipids are shown.

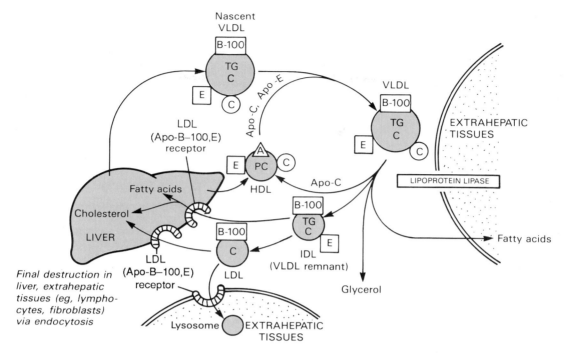

Figure 27–5. Metabolic fate of very low density lipoproteins (VLDL) and production of low-density lipoproteins (LDL). A, apolipoprotein A; B-100, apolipoprotein B-100; ©, apolipoprotein C; E, apolipoprotein E; HDL, high-density lipoprotein; TG, triacylglycerol; IDL, intermediate-density lipoprotein; C, cholesterol and cholesteryl ester; P, phospholipid. Only the predominant lipids are shown.

still under 1 hour. Larger particles are catabolized more quickly than smaller ones. When chylomicrons labeled in the triacylglycerol fatty acids are administered intravenously, some 80% of the label is found in adipose tissue, heart, and muscle and approximately 20% in the liver. As experiments with the perfused organ have shown that the **liver does not metabolize native chylomicrons or VLDL significantly,** the label in the liver must result secondarily from their metabolism in extrahepatic tissues.

Triacylglycerol of Chylomicrons & VLDL Is Hydrolyzed by Lipoprotein Lipase

There is a significant correlation between the ability of a tissue to incorporate lipoprotein triacylglycerol fatty acids and the activity of the enzyme **lipoprotein lipase.** It is located on the **walls of blood capillaries,** anchored by proteoglycan chains of heparan sulfate, and has been found in extracts of heart, adipose tissue, spleen, lung, renal medulla, aorta, diaphragm, and lactating mammary gland. Normal blood does not contain appreciable quantities of the enzyme; however, following injection of **heparin,** lipoprotein lipase is released from its heparan sulfate binding into the circulation and is accompanied by the clearing of lipemia. A lipase is also released from the liver by large quantities of heparin (**heparin-releasable hepatic lipase**), but this enzyme has properties different

from those of lipoprotein lipase and does not react readily with chylomicrons.

Both **phospholipids** and **apolipoprotein C-II** are required as cofactors for lipoprotein lipase activity. Apo C-II contains a specific phospholipid binding site through which it is attached to the lipoprotein. Thus, chylomicrons and VLDL provide the enzyme for their metabolism with both its substrate and cofactors. Hydrolysis takes place while the lipoproteins are attached to the enzyme on the endothelium. The triacylglycerol is hydrolyzed progressively through a diacylglycerol to a monoacylglycerol that is finally hydrolyzed to free fatty acid plus glycerol. Some of the released free fatty acids return to the circulation, attached to albumin, but the bulk are transported into the tissue (Figs 27–4 and 27–5). Heart lipoprotein lipase has a low K_m for triacylglycerol, whereas the K_m of the enzyme in adipose tissue is 10 times greater. As the concentration of plasma triacylglycerol decreases in the transition from the fed to the starved condition, the heart enzyme remains saturated with substrate but the saturation of the enzyme in adipose tissue diminishes, thus **redirecting uptake from adipose tissue toward the heart.** A similar redirection occurs during lactation, in which adipose tissue activity diminishes and mammary gland activity increases, allowing uptake of lipoprotein triacylglycerol long-chain fatty acid for milk fat synthesis.

In adipose tissue insulin enhances lipoprotein lipase

synthesis in adipocytes and its translocation to the luminal surface of the capillary endothelium.

The Action of Lipoprotein Lipase Results in Formation of Remnant Lipoproteins

Reaction with lipoprotein lipase results in the loss of approximately 90% of the triacylglycerol of chylomicrons and in the loss of the apo C (which returns to HDL) but not apo E (which is retained). The resulting lipoprotein or **chylomicron remnant** is about half the diameter of the parent chylomicron and in terms of the percentage composition becomes relatively enriched in cholesterol and cholesteryl esters because of the loss of triacylglycerol (Fig 27–4). Similar changes occur to VLDL, with the formation of VLDL remnants or IDL (intermediate-density lipoprotein) (Fig 27–5).

The Liver Is Responsible for the Uptake of Remnant Lipoproteins

Chylomicron remnants are taken up by the liver, and the cholesteryl esters and triacylglycerols are hydrolyzed and metabolized. Uptake appears to be mediated by a **receptor specific for apo E** (Fig 27–4).

Studies using apo B-100-labeled VLDL have shown that VLDL is the precursor of IDL and IDL the precursor of LDL. Only one molecule of apo B-100 is present in each of these lipoprotein particles and these are conserved during the transformations. Thus, each LDL particle is derived from only one VLDL particle (Fig 27–5). Two possible fates await IDL. It can be taken up by the liver directly via the LDL (Apo B-100, E) receptor, or it is converted to LDL. In the rat, most of the apo B from VLDL appears in the liver, and only a small percentage appears in LDL. This may be due to the fact that a proportion of the VLDL in the rat contains apo B-48 rather than B-100. If the hepatic apo E receptor excludes apo B-100 but not apo B-48, this would account for the extensive hepatic removal of IDL in the rat and for the low production of LDL in this species.

LDL IS METABOLIZED VIA THE LDL RECEPTOR

Most LDL appears to be formed from VLDL, as described above, but there is evidence for some production directly by the liver. The half-time of disappearance from the circulation of apoprotein B-100 in LDL is approximately 2½ days.

Studies on cultured fibroblasts, lymphocytes, and arterial smooth muscle cells, and liver have shown the existence of specific binding sites, or receptors, for LDL, the B-100, E receptor. It is so designated because it is specific for apo B-100 but not B-48, and under some circumstances it will take up lipoproteins rich in apo E. Apo B-48 lacks the carboxy-terminal domain of B-100 which contains the ligand for the LDL receptor. These receptors are defective in **familial hypercholesterolemia.** Approximately 50% of LDL is degraded in extrahepatic tissues and 50% in the liver. A positive correlation exists between the incidence of **coronary atherosclerosis** and the plasma concentration of LDL. For further discussion of the regulation of the LDL receptor, see p 253.

HDL TAKES PART IN BOTH TRIACYLGLYCEROL AND CHOLESTEROL METABOLISM

HDL is synthesized and secreted from both liver and intestine (Fig 27–6). However, nascent (newly secreted) HDL from intestine does not contain apolipoprotein C or E but only apolipoprotein A. Thus, apo C and E are synthesized in the liver and transferred to intestinal HDL when the latter enters the plasma. **A major function of HDL is to act as a repository for apolipoproteins C and E that are required in the metabolism of chylomicrons and VLDL** (see Figs 27–4 and 27–5).

Nascent HDL consists of discoid phospholipid bilayers containing apolipoprotein and free cholesterol. These lipoproteins are similar to the particles found in the plasma of patients with a deficiency of the plasma enzyme **lecithin:cholesterol acyltransferase (LCAT)** and in the plasma of patients with obstructive jaundice. LCAT—and the LCAT activator apolipoprotein A-I—bind to the disk. Catalysis by LCAT converts surface phospholipid and free cholesterol into cholesteryl esters and lysolecithin. The nonpolar cholesteryl esters move into the hydrophobic interior of the bilayer, whereas lysolecithin is transferred to plasma albumin. The reaction continues, generating a nonpolar core that pushes the bilayer apart until a spherical, pseudomicellar HDL is formed, covered by a surface film of polar lipids and apolipoproteins. The esterified cholesterol can be transferred from HDL to the lower density lipoproteins, eg, chylomicrons, VLDL, and LDL, by means of the **cholesteryl ester transfer protein** (apo D), which is another protein component of HDL. Thus, the cholesteryl ester transfer protein allows cholesteryl ester of HDL to be transported to the liver via the remnants of chylomicrons and VLDL or via hepatic uptake of LDL. Thus, the LCAT system is involved in the removal of excess unesterified cholesterol from lipoproteins and from the tissues. The liver and possibly the intestines seem to be the final sites of degradation of HDL apolipoproteins. It is not clear whether a true HDL or Apo A receptor is involved.

An HDL cycle has been proposed to account for the transport of cholesterol from the tissues to the liver (see Fig 27–6 for details). This explains why **HDL$_2$ concentrations in the plasma vary reciprocally with the chylomicron and VLDL concentration and directly with the activity of lipoprotein lipase.** HDL (HDL$_2$) concentrations are **inversely related to the incidence of coronary atherosclerosis,** possibly be-

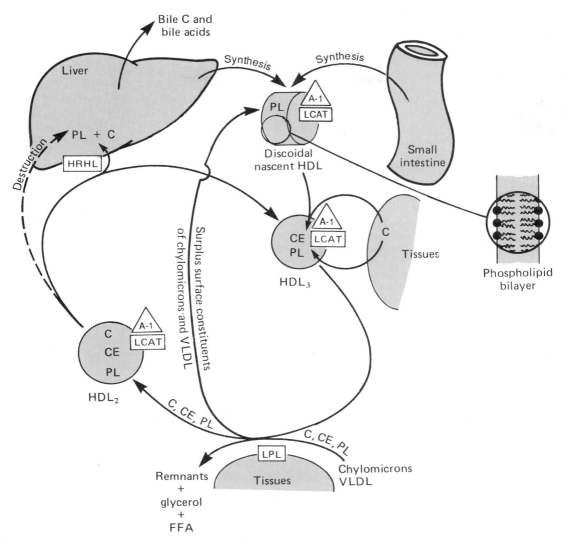

Figure 27–6. Metabolism of high-density lipoprotein (HDL). HRHL, heparin-releasable hepatic lipase; LCAT, lecithin cholesterol acyltransferase; LPL, lipoprotein lipase; C, cholesterol; CE, cholesteryl ester, PL, phospholipid; FFA, free fatty acids; A-I, apoprotein A-I. The figure illustrates the role of the 3 enzymes HRHL, LCAT, and LPL in the postulated HDL cycle for the transport of cholesterol from the tissues to the liver. HDL_2, HDL_3—see Table 27–2. In addition to triacylglycerol, HRHL hydrolyzes phospholipid on the surface of HDL_2, releasing cholesterol for uptake into the liver, allowing formation of smaller and more dense HDL_3. HRHL activity is increased by androgens and decreased by estrogens, which may account for higher concentrations of plasma HDL_2 in women.

cause they reflect the efficiency of cholesterol scavenging from the tissues. HDL_c is found in the blood of diet-induced hypercholesterolemic animals. It is rich in cholesterol, and its sole apolipoprotein is apo E. It is taken up by the liver via the apo E remnant receptor but also by LDL receptors. It is for this reason that the latter are sometimes designated apo B-100, E receptors. Atherosclerotic plaques contain scavenger cells that have taken up so much cholesterol that they are converted into cholesteryl ester-laden foam cells. Most of the cells arise from macrophages that ingest the more abnormal cholesterol-rich lipoproteins such as

chemically modified LDL or β-VLDL (see p 253). Macrophages secrete both cholesterol (to a suitable recipient such as HDL) and apo E. This apo E, after suitable processing in the presence of LCAT, may be the source of cholesterol-rich HDL_c. Thus, HDL_c could be an important component in the movement of cholesterol from the tissues to the liver ("reverse cholesterol transport").

It appears that all plasma lipoproteins are interrelated components of one or more metabolic cycles that together are responsible for the complex process of plasma lipid transport.

THE LIVER PLAYS A CENTRAL ROLE IN LIPID TRANSPORT & METABOLISM

Much of the lipid metabolism of the body was formerly thought to be the prerogative of the liver. The discovery that most tissues have the ability to oxidize fatty acids completely and the knowledge which has accumulated showing that adipose tissue is extremely active metabolically have tended to modify the former emphasis on the role of the liver. Nonetheless, the concept of a central and unique role for the liver in lipid metabolism is still an important one. The liver carries out the following major functions in lipid metabolism: (1) It facilitates the digestion and absorption of lipids by the production of bile, which contains cholesterol and bile salts synthesized within the liver (see p 256). (2) The liver has active enzyme systems for synthesizing and oxidizing fatty acids (see Chapters 23 and 24) and for synthesizing triacylglycerols, phospholipids (see Chapter 26), and cholesterol (see Chapter 27). (3) It synthesizes plasma lipoproteins (this chapter). (4) It converts fatty acids to ketone bodies (ketogenesis) (see Chapter 24). (5) It plays an integral part in the metabolism of plasma lipoproteins (this chapter).

Hepatic VLDL Secretion Is Related to the Dietary Status

The cellular events involved in VLDL formation and secretion have been described (p 238). Hepatic triacylglycerols are the immediate precursors of triacylglycerols contained in plasma VLDL (Fig 27–7). The synthesis of triacylglycerol provides the immediate stimulus for the formation and secretion of VLDL. The fatty acids used in the synthesis of hepatic triacylglycerols are derived from 2 possible sources: (1) synthesis within the liver from acetyl-CoA derived mainly from carbohydrate and (2) uptake of free fatty acids from the circulation. The first source is predominant in the well-fed condition, when fatty acid synthesis is high and the level of circulating free fatty acids is low. As triacylglycerol does not normally accumulate in the liver under this condition, it must be inferred that it is transported from the liver in VLDL as rapidly as it is synthesized. On the other hand, during fasting, during the feeding of high-fat diets, or in diabetes mellitus, the level of circulating free fatty acids is raised and more is abstracted into the liver. Under these conditions, lipogenesis is inhibited and free fatty acids are the main source of triacylglycerol fatty acids in the liver and in VLDL. The enzyme mechanisms responsible for the synthesis of triacylglycerols and phospholipids have been described on pp 226 and 228. Factors that enhance both the synthesis of triacylglycerol and the secretion of VLDL by the liver include (1) the fed state rather than the fasting state; (2) the feeding of diets high in carbohydrate (particularly if they contain sucrose or fructose), leading to high rates of lipogenesis and esterification of fatty acids; (3) high levels of circulating free fatty acids; (4) ingestion of ethanol; and (5) the presence of high concentrations of insulin and low concentrations of glucagon, which enhance fatty acid synthesis and esterification and inhibit their oxidation.

Imbalance in the Rate of Triacylglycerol Formation & Export Causes Fatty Liver (Fig 27-7)

For a variety of reasons, lipid—mainly as triacylglycerol—can accumulate in the liver. Extensive accumulation is regarded as a pathologic condition. When accumulation of lipid in the liver becomes chronic, fibrotic changes occur in the cells that progress to **cirrhosis** and impaired liver function.

Fatty livers fall into 2 main categories. (1) The first type is associated with **raised levels of plasma free fatty acids** resulting from mobilization of fat from adipose tissue or from the hydrolysis of lipoprotein or chylomicron triacylglycerol by lipoprotein lipase in extrahepatic tissues. Increasing amounts of free fatty acids are taken up by the liver and esterified. The production of plasma lipoprotein does not keep pace with the influx of free fatty acids, allowing triacylglycerol to accumulate, causing a fatty liver. The quantity of triacylglycerol present in the liver is significantly increased during **starvation** and the feeding of **high-fat diets**. In many instances (eg, in starvation), the ability to secrete VLDL is also impaired. This may be due to low levels of insulin and impaired protein synthesis. In uncontrolled **diabetes mellitus, pregnancy toxemia of ewes,** and **ketosis in cattle,** fatty infiltration is sufficiently severe to cause visible pallor or fatty appearance and enlargement of the liver.

(2) The second type of fatty liver is usually due to a **metabolic block in the production of plasma lipoproteins,** thus allowing triacylglycerol to accumulate. Theoretically, the lesion may be due to (a) a block in apolipoprotein synthesis, (b) a block in the synthesis of the lipoprotein from lipid and apolipoprotein, (c) a failure in provision of phospholipids that are found in lipoproteins, or (d) a failure in the secretory mechanism itself.

One type of fatty liver that has been studied extensively in rats is due to a deficiency of **choline,** which has therefore been called a **lipotropic factor.** As choline may be synthesized using labile methyl groups donated by methionine in the process of **transmethylation** (see Chapters 32 and 33), the deficiency is basically due to a shortage of the type of methyl group donated by methionine. Several mechanisms have been suggested to explain the role of choline as a lipotropic agent, including its absence, causing an impairment in synthesis of lipoprotein phospholipids.

The antibiotic puromycin inhibits protein synthesis and causes a fatty liver and a marked reduction in concentration of VLDL in rats. Other substances that act similarly include ethionine (α-amino-γ-mercaptobutyric acid), carbon tetrachloride, chloroform, phosphorus, lead, and arsenic. Choline will not protect the organism against these agents but appears to aid in recovery. It is very likely that carbon tetrachloride also affects the secretory mechanism itself or the conjuga-

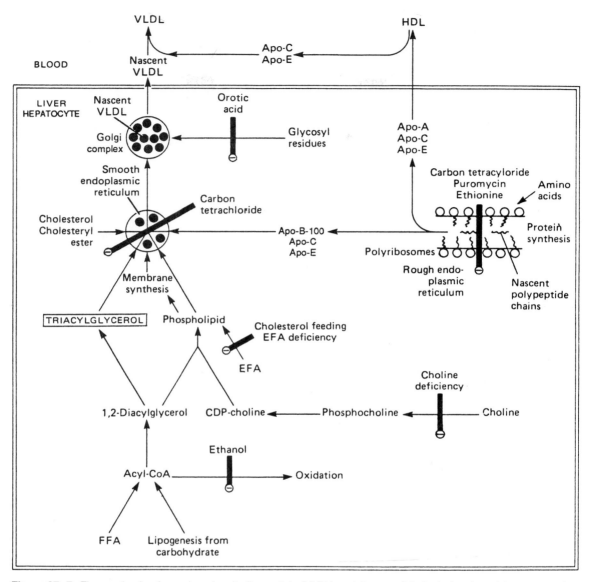

Figure 27–7. The synthesis of very low density lipoprotein (VLDL) and the possible loci of action of factors causing accumulation of triacylglycerol and a fatty liver. EFA, essential fatty acids; FFA, free fatty acids; HDL, high-density lipoproteins; Apo A, apolipoprotein A; Apo B, apolipoprotein B; Apo C, apolipoprotein C; Apo E, apolipoprotein E. The pathways indicated form a basis for events depicted in Fig 27–3B.

tion of the lipid with apolipoprotein. Its effect is not direct but depends on further transformation of the molecule. This probably involves formation of free radicals that disrupt lipid membranes in the endoplasmic reticulum by formation of lipid peroxides. Some protection against carbon tetrachloride-induced lipid peroxidation is provided by vitamin E-supplemented diets. The action of ethionine is thought to be due to a reduction in availability of ATP. This results when ethionine, replacing methionine in S-adenosylmethionine, traps available adenine and prevents synthesis of ATP. Orotic acid also causes fatty livers; as VLDL accumulate in the Golgi apparatus, it is consid-

ered that orotic acid interferes with glycosylation of the lipoprotein, thus inhibiting its release and accounting for the marked decrease in plasma lipoproteins containing apo B.

A deficiency of vitamin E enhances the hepatic necrosis of the choline deficiency type of fatty liver. Added vitamin E or a source of selenium has a protective effect by combatting lipid peroxidation. In addition to protein deficiency, essential fatty acid and vitamin deficiencies (eg, linoleic acid, pyridoxine and pantothenic acid) can cause fatty infiltration of the liver. A deficiency of essential fatty acids is thought to depress the synthesis of phospholipids; therefore, other substances such as cholesterol that compete for available essential fatty acids for esterification can also cause fatty livers.

Ethanol Also Causes Fatty Liver

Alcoholism leads to fat accumulation in the liver, hyperlipidemia, and ultimately **cirrhosis.** The exact mechanism of action of ethanol in the long term is still uncertain. Whether or not extra free fatty acid mobilization plays some part in causing the accumulation of fat is not clear, but several studies have demonstrated elevated levels of free fatty acids in the rat after administration of a single intoxicating dose of ethanol. However, ethanol consumption over a long period leads to the accumulation of fatty acids in the liver that are derived from endogenous synthesis rather than from adipose tissue. There is no impairment of hepatic synthesis of protein after ethanol ingestion. There is good evidence of increased hepatic triacylglycerol synthesis, decreased fatty acid oxidation, and decreased citric acid cycle activity, caused by oxidation of ethanol in the hepatic cytosol by **alcohol dehydrogenase** leading to excess production of NADH.

$$CH_3-CH_2-OH + NAD^+ \xrightarrow{\text{ALCOHOL DEHYDROGENASE}} CH_3-CHO + NADH + H^+$$

Ethanol   Acetaldehyde

The NADH generated competes with reducing equivalents from other substrates for the respiratory chain, inhibiting their oxidation. The increased [NADH]/[NAD$^+$] ratio causes a shift to the left in the equilibrium malate $\rightleftharpoons$ oxaloacetate, which may reduce activity of the citric acid cycle. The net effect of inhibiting fatty acid oxidation is to cause increased esterification of fatty acids in triacylglycerol, which appears to be the cause of the fatty liver. Oxidation of ethanol leads to the formation of acetaldehyde, which is oxidized by **aldehyde dehydrogenase** in mitochondria, acetate being the end product. Other effects of ethanol may include increased lipogenesis and cholesterol synthesis from acetyl-CoA. The increased [NADH]/[NAD$^+$] ratio also causes an increased [lactate]/[pyruvate] ratio that results in hyperlactacidemia, which in turn decreases the capacity of the kidney to excrete uric acid. The latter is probably the cause of aggravation of gout by drinking alcohol. Although the major route for ethanol metabolism is via the alcohol dehydrogenase pathway, some metabolism takes place via a cytochrome P-450-dependent microsomal ethanol oxidizing system involving NADPH and O_2. This system increases in activity in **chronic alcoholism** and may account for the increased metabolic clearance in this condition as indicated by increased blood levels of both acetaldehyde and acetate.

$$CH_2-CH_2-OH + NADPH + H^+ + O_2 \longrightarrow$$
Ethanol

$$CH_3-CHO + NADP^+ + 2H_2O$$
Acetaldehyde

ADIPOSE TISSUE IS THE MAIN STORE OF TRIACYLGLYCEROL IN THE BODY

The triacylglycerol stores in adipose tissue are continually undergoing lipolysis (hydrolysis) and reesterification (Fig 27–8). These 2 processes are not the forward and reverse phases of the same reaction. Rather, they are entirely different pathways involving different reactants and enzymes. Many of the nutritional, metabolic, and hormonal factors that regulate the metabolism of adipose tissue **act either upon the process of esterification or on lipolysis.** The resultant of these 2 processes determines the magnitude of the free fatty acid pool in adipose tissue, which in turn is the source and determinant of the level of free fatty acids circulating in the plasma. Since the level of plasma free fatty acids has most profound effects upon the metabolism of other tissues, particularly liver and muscle, **the factors operating in adipose tissue that regulate the outflow of free fatty acids exert an influence far beyond the tissue itself.**

The Provision of Glycerol 3-Phosphate Regulates Esterification: Lipolysis is Controlled by Hormone Sensitive Lipase

In adipose tissue, triacylglycerol is synthesized from acyl-CoA and glycerol 3-phosphate according to the mechanism shown in Fig 26–1. Because the enzyme **glycerol kinase** is low in activity in adipose tissue, glycerol cannot be utilized to any great extent in the esterification of acyl-CoA. **For the provision of glycerol 3-phosphate, the tissue is dependent on glycolysis and a supply of glucose.**

Triacylglycerol undergoes hydrolysis by a **hormone-sensitive lipase** to form free fatty acids and glycerol. **This lipase is distinct from lipoprotein lipase** that catalyzes lipoprotein triacylglycerol hydrolysis prior to its uptake into extrahepatic tissues (see p 239). Since glycerol cannot be utilized readily in this tissue, it diffuses out into the plasma, from which it is utilized by such tissues as liver and kidney, which possess an active glycerol kinase. The free fatty acids formed by lipolysis can be reconverted in the tissue to acyl-CoA by **acyl-CoA synthetase** and reesterified with glycerol 3-phosphate to form triacylglycerol. Thus, **there is a continuous cycle of lipolysis and reesterification within the tissue.** However, when the rate of reesterification is not sufficient to match the rate of lipolysis, free fatty acids accumulate and diffuse into the plasma, where they bind to albumin and raise the concentration of plasma free fatty acids. These are a most important source of fuel for many tissues.

Increased Glucose Metabolism Reduces the Output of Free Fatty Acids

When the utilization of glucose by adipose tissue is increased, the free fatty acid outflow decreases. How-

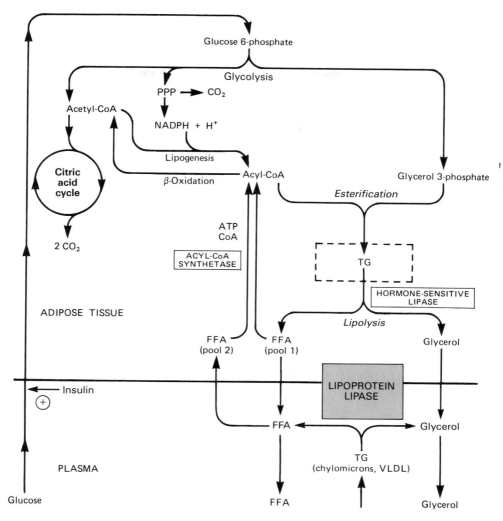

Figure 27–8. Metabolism of adipose tissue. Hormone-sensitive lipase is activated by ACTH, TSH, glucagon, epinephrine, norepinephrine, and vasopressin and inhibited by insulin, prostaglandin E_1, and nicotinic acid. Details of the formation of glycerol 3-phosphate from intermediates of glycolysis are shown in Fig 26–1. PPP, pentose phosphate pathway; TG, triacylglycerol; FFA, free fatty acids; VLDL, very low density lipoprotein.

ever, the release of glycerol continues, demonstrating that the effect of glucose is not mediated by reducing the rate of lipolysis. It is believed that the effect is due to the provision of glycerol 3-phosphate, which enhances esterification of free fatty acids via acyl-CoA.

Glucose can take several pathways in adipose tissue, including oxidation to CO_2 via the citric acid cycle, oxidation in the pentose phosphate pathway, conversion to long-chain fatty acids, and formation of acylglycerol via glycerol 3-phosphate. When glucose utilization is high, a larger proportion of the uptake is oxidized to CO_2 and converted to fatty acids. However, as total glucose utilization decreases, the greater proportion of the glucose is directed to the formation of glycerol 3-phosphate for the esterification of acyl-

CoA, which helps to minimize the efflux of free fatty acids.

Free Fatty Acids Are Taken Up as a Result of Lipoprotein Lipase Activity

There is more than one free fatty acid pool within adipose tissue. It has been shown that the free fatty acid pool (Fig 27–8, pool 1) formed by lipolysis of triacylglycerol is the same pool that supplies fatty acids for reesterification; also, it releases them into the external medium (plasma). Fatty acids taken up from the external medium as a result of the action of lipoprotein lipase on the triacylglycerol of chylomicrons and VLDL do not label pool 1 before they are incorporated into triacylglycerol but travel through a small pool 2 of high turnover.

HORMONES REGULATE FAT MOBILIZATION

Insulin Reduces the Output of Free Fatty Acids

The rate of release of free fatty acids from adipose tissue is affected by many hormones that influence either the rate of esterification or the rate of lipolysis. Insulin inhibits the release of free fatty acids from adipose tissue, which is followed by a fall in circulating plasma free fatty acids. It enhances lipogenesis and the synthesis of acylglycerol and increases the oxidation of glucose to CO_2 via the pentose phosphate pathway. All of these effects are dependent on the presence of glucose and can be explained, to a large extent, on the basis of the ability of insulin to enhance the uptake of glucose into adipose cells. This is achieved by insulin causing the translocation of glucose transporters from the Golgi apparatus to the plasma membrane (see

p 529). Insulin has also been shown to increase the activity of **pyruvate dehydrogenase, acetyl-CoA carboxylase, and glycerol phosphate acyltransferase,** which would reinforce the effects arising from increased glucose uptake on the enhancement of fatty acid and acylglycerol synthesis. These 3 enzymes are now known to be regulated in a coordinate manner by covalent modification, ie, by phosphorylation-dephosphorylation mechanisms.

A principal action of insulin in adipose tissue is to inhibit the activity of the **hormone-sensitive lipase,** reducing the release not only of free fatty acids but of glycerol as well. Adipose tissue is much more sensitive to insulin than are many other tissues, **which points to adipose tissue as a major site of insulin action in vivo.**

A Large Number of Hormones Are Capable of Promoting Lipolysis

Other hormones accelerate the release of free fatty acids from adipose tissue and raise the plasma free fatty acid concentration by increasing the rate of lipolysis of the triacylglycerol stores (Fig 27–9).

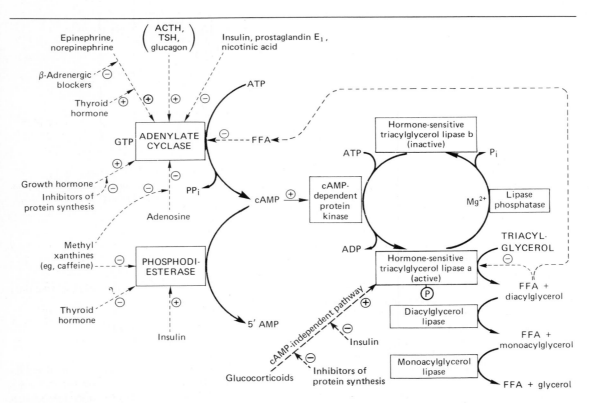

Figure 27–9. Control of adipose tissue lipolysis. TSH, thyroid-stimulating hormone; FFA, free fatty acids. Note the cascade sequence of reactions affording amplification at each step. The lipolytic stimulus is "switched off" by (1) removal of the stimulating hormone; (2) the action of lipase phosphatase; (3) the inhibition of the lipase and adenylate cyclase by high concentrations of FFA; (4) the inhibition of adenylate cyclase by adenosine; and (5) the removal of cAMP by the action of phosphodiesterase. ACTH, TSH, and glucagon may not activate adenylate cyclase in vivo, since the concentration of each hormone required in vitro is much higher than is found in the circulation. Positive (⊕) and negative (⊖) regulatory effects are represented by broken lines and substrate flow by solid lines.

These include epinephrine, norepinephrine, glucagon, adrenocorticotropic hormone (ACTH), α- and β-melanocyte-stimulating hormones (MSH), thyroid-stimulating hormone (TSH), growth hormone (GH), and vasopressin. Many of these activate the hormone-sensitive lipase. For an optimal effect, most of these lipolytic processes require the presence of **glucocorticoids** and **thyroid hormones**. On their own, these particular hormones do not increase lipolysis markedly but act in a **facilitatory** or **permissive** capacity with respect to other lipolytic endocrine factors.

The hormones that act rapidly in promoting lipolysis, ie, catecholamines, do so by stimulating the activity of **adenylate cyclase,** the enzyme that converts ATP to cAMP. The mechanism is analogous to that responsible for hormonal stimulation of glycogenolysis (see Chapter 20). It appears that cAMP, by stimulating **cAMP-dependent protein kinase,** converts inactive hormone-sensitive triacylglycerol lipase into active lipase. Lipolysis is controlled largely by the amount of cAMP present in the tissue. It follows that processes which destroy or preserve cAMP have an effect on lipolysis. cAMP is degraded to 5′-AMP by the enzyme **cyclic 3′,5′-nucleotide phosphodiesterase.** This enzyme is inhibited by methyl xanthines such as **caffeine** and **theophylline.** It is significant that the drinking of coffee, containing caffeine, causes marked and prolonged elevation of plasma FFA in humans.

Insulin antagonizes the effect of the lipolytic hormones. It is now considered that lipolysis may be more sensitive to changes in concentration of insulin than are glucose utilization and esterification. The anti-lipolytic effects of insulin, nicotinic acid, and prostaglandin E_1 may be accounted for by inhibition of the synthesis of cAMP at the adenylate cyclase site. Also, insulin stimulates phosphodiesterase. Possible mechanisms for the action of thyroid hormones include an augmentation of the level of cAMP by facilitation of the passage of the stimulus from the receptor site on the outside of the cell membrane to the adenylate cyclase site on the inside of the membrane and an inhibition of phosphodiesterase activity. The effect of growth hormone in promoting lipolysis is slow. It is dependent on synthesis of proteins involved in the formation of cAMP. Glucocorticoids promote lipolysis via synthesis of new lipase protein by a cAMP-independent pathway, which may be inhibited by insulin. These findings help to explain the role of the pituitary gland and the adrenal cortex in enhancing fat mobilization.

The sympathetic nervous system, through liberation of norepinephrine in adipose tissue, plays a central role in the mobilization of free fatty acids by exerting a tonic influence even in the absence of augmented nervous activity. Thus, the increased lipolysis caused by many of the factors described above can be reduced or abolished by denervation of adipose tissue, by ganglionic blockade with hexamethonium, or by depleting norepinephrine stores with reserpine.

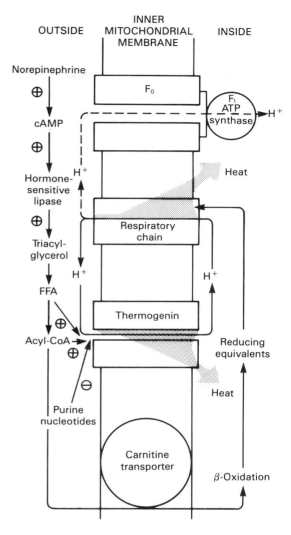

Figure 27–10. Thermogenesis in brown adipose tissue. Activity of the respiratory chain produces heat in addition to translocating protons (see p 115). These protons dissipate heat when returned to the inner mitochondrial compartment via thermogenin, instead of generating ATP when returning via the F_1 ATP synthase. The passage of H^+ via thermogenin is inhibited by purine nucleotides when brown adipose tissue is unstimulated. Under the influence of norepinephrine, the inhibition is removed by the production of free fatty acids (FFA) and acyl-CoA. Note the dual role of acyl-CoA in both facilitating the action of thermogenin and supplying reducing equivalents for the respiratory chain. Positive (⊕) or negative (⊖) regulatory effects.

IN THE VARIOUS SPECIES STUDIED, A VARIETY OF MECHANISMS HAVE EVOLVED FOR FINE CONTROL OF ADIPOSE TISSUE METABOLISM

Human adipose tissue may not be an important site of lipogenesis. This is indicated by the observation that there is not significant incorporation of label into long-chain fatty acids from labeled glucose or pyruvate; ATP-citrate lyase, a key enzyme in lipogenesis, does not appear to be present and has extremely low activity in liver. Other enzymes—eg, glucose-6-phosphate dehydrogenase and the malic enzyme—which in the rat undergo adaptive changes coincident with increased lipogenesis, do not undergo similar changes in human adipose tissue. Indeed, it has been suggested that in humans there is a "carbohydrate excess syndrome" due to a unique limitation in ability to dispose of excess carbohydrate by lipogenesis. In birds, lipogenesis is confined to the liver, where it is particularly important in providing lipids for egg formation, stimulated by estrogens.

Human adipose tissue is unresponsive to most of the lipolytic hormones apart from the catecholamines. Of further interest is the lack of lipolytic response to epinephrine in the rabbit, guinea pig, pig, and chicken; the pronounced lipolytic effect of glucagon in birds, together with an absence of any antilipolytic effect of insulin; and the lack of acylglycerol-glycerol synthesis from glucose in the pigeon.

On consideration of the profound derangement of metabolism in diabetes mellitus (which is due mainly to increased release of free fatty acids from the depots) and the fact that insulin to a large extent corrects the condition, it must be concluded that **insulin plays a prominent role in the regulation of adipose tissue metabolism.**

BROWN ADIPOSE TISSUE PROMOTES THERMOGENESIS

Brown adipose tissue is involved in metabolism particularly at times when heat generation is necessary. Thus, the tissue is extremely active in some species in arousal from hibernation, in animals exposed to cold (nonshivering thermogenesis), and in heat production in the newborn animal. Though not a prominent tissue in humans, recently it has been shown to be active in normal individuals, where it appears to be responsible for **"diet-induced thermogenesis,"** which may account for how some persons can "eat and not get fat."

It is noteworthy that brown adipose tissue is reduced or absent in obese persons. Brown adipose tissue is characterized by a well-developed blood supply and a high content of mitochondria and cytochromes but low activity of ATP synthase. Metabolic emphasis is placed on oxidation of both glucose and fatty acids.

Norepinephrine liberated from sympathetic nerve endings is important in increasing lipolysis in the tissue. Oxidation and phosphorylation are not coupled in mitochondria of this tissue, since dinitrophenol has no effect and there is no respiratory control by ADP. The phosphorylation that does occur is at the substrate level, eg, at the succinate thiokinase step and in glycolysis. Thus, **oxidation produces much heat, and little free energy is trapped in ATP.** In terms of the chemiosmotic theory, it would appear that the proton gradient, normally present across the inner mitochondrial membrane of coupled mitochondria, is continually dissipated in brown adipose tissue by a thermogenic protein, **thermogenin,** which acts as a proton conductance pathway through the membrane. This would explain the apparent lack of effect of uncouplers (Fig 27–10).

REFERENCES

Borensztajn J (editor): *Lipoprotein Lipase.* Evener Publishers, Chicago, 1987.

Brewer HB, et al: Apolipoproteins and lipoproteins in human plasma: An overview. *Clin Chem* 1988;**34:**B4.

Brown MS, Goldstein JL: Lipoprotein metabolism in the macrophage: Implication for cholesterol deposition in atherosclerosis. *Annu Rev Biochem* 1983;**52:**223.

Eisenberg S: High density lipoprotein metabolism. *J Lipid Res* 1984;**25:**1017.

Fielding CJ, Fielding PE: Metabolism of cholesterol and lipoproteins. Page 404 in: *Biochemistry of Lipids and Membranes.* Vance DE, Vance JE (editors). Benjamin/Cummings, 1985.

Himms-Hagen J: Brown adipose tissue metabolism and thermogenesis. *Annu Rev Nutr* 1985;**5:**69.

Lieber CS: Biochemical and molecular basis of alcohol-induced injury to liver and other tissues. *New Eng J Med* 1988;**319:**1639.

Sparks JD, Sparks CE: Apolipoprotein B and lipoprotein metabolism. *Adv Lipid Res* 1985;**21:**1.

Cholesterol Synthesis, Transport, & Excretion

28

Peter A. Mayes, PhD, DSc

INTRODUCTION

Cholesterol is present in tissues and in plasma lipoproteins either as free cholesterol or, combined with a long-chain fatty acid, as cholesteryl ester. It is synthesized in many tissues from acetyl-CoA and is ultimately eliminated from the body in the bile as cholesterol or bile salts. Cholesterol is the precursor of all other steroids in the body such as corticosteroids, sex hormones, bile acids, and vitamin D. It is typically a product of animal metabolism and therefore occurs in foods of animal origin such as egg yolk, meat, liver, and brain.

BIOMEDICAL IMPORTANCE

Cholesterol is an amphipathic lipid and as such is an essential structural component of membranes and of the outer layer of plasma lipoproteins. Additionally, lipoproteins transport free cholesterol in the circulation, where it readily equilibrates with cholesterol in other lipoproteins and in membranes. Cholesteryl ester is a storage form of cholesterol found in most tissues. It is transported as cargo in the core of lipoproteins. LDL is the mediator of cholesterol and cholesteryl ester uptake into many tissues. Free cholesterol is removed from tissues by HDL and transported to the liver for conversion to bile acids. Cholesterol is a major constituent of **gallstones**. However, its chief role in pathologic processes is as a factor in the genesis of **atherosclerosis** of vital arteries, causing cerebrovascular, coronary, and peripheral vascular disease. Coronary atherosclerosis correlates with a high plasma LDL:HDL cholesterol ratio.

CHOLESTEROL IS DERIVED FROM THE DIET & FROM BIOSYNTHESIS ABOUT EQUALLY

Approximately half the cholesterol of the body arises by synthesis (about 500 mg/d), and the remainder is provided by the average diet. The liver accounts for approximately 50% of total synthesis, the gut for about 15%, and the skin for a large proportion of the remainder.

Virtually all tissues containing nucleated cells are capable of synthesizing cholesterol. The microsomal (endoplasmic reticulum) and cytosol fraction of the cell is responsible for cholesterol synthesis.

Acetyl-CoA Is the Source of All the Carbon Atoms in Cholesterol

The biosynthesis of cholesterol may be divided into 5 stages. (1) The synthesis of mevalonate, a 6-carbon compound, from acetyl-CoA (Fig 28–1). (2) Isoprenoid units are formed from mevalonate by loss of CO_2 (Fig 28–2). (3) Six isoprenoid units condense to form the intermediate, squalene. (4) Squalene cyclizes to give rise to the parent steroid, lanosterol. (5) Cholesterol is formed from lanosterol after several further steps, including the loss of 3 methyl groups (Fig 28–3).

Step 1. Acetyl-CoA Forms HMGCoA and Mevalonate: The pathway through HMG-CoA (3-hydroxy-3-methylglutaryl-CoA) follows the same sequence of reactions described in Chapter 24 for the synthesis in mitochondria of ketone bodies. However, since cholesterol synthesis is extramitochondrial, the 2 pathways are distinct. Initially, 2 molecules of acetyl-CoA condense to form acetoacetyl-CoA catalyzed by a cytosolic **thiolase** enzyme. Alternatively, in liver, acetoacetate made inside the mitochondrion in the pathway of ketogenesis (see Chapter 24) diffuses into the cytosol and may be activated to acetoacetyl-CoA by **acetoacetyl-CoA synthase,** requiring ATP and CoA. Acetoacetyl-CoA condenses with a further molecule of acetyl-CoA catalyzed by **HMG-CoA synthase** to form HMG-CoA.

HMG-CoA is converted to **mevalonate** in a 2-stage reduction by NADPH catalyzed by **HMG-CoA reductase,** a microsomal enzyme considered to catalyze the **rate-limiting step in the pathway of cholesterol synthesis** (Fig 28–1).

Step 2. Mevalonate Forms Active Isoprenoid Units: Mevalonate is phosphorylated by ATP to form several active phosphorylated intermediates (Fig 28–2). By means of a decarboxylation, the active isoprenoid unit, **isopentenylpyrophosphate,** is formed.

Step 3. Six Isoprenoid Units Form Squalene: This stage involves the condensation of 3 molecules of

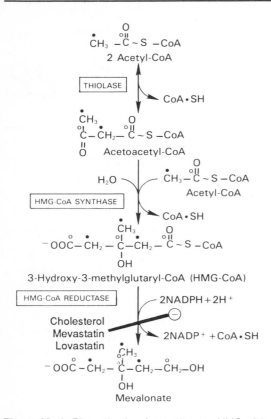

Figure 28–1. Biosynthesis of mevalonate. HMG, 3-hydroxy-3-methylglutaryl. HMG-CoA reductase is inhibited by cholesterol and the fungal metabolites mevastatin (compactin) and lovastatin (mevinolin), which are competitive with HMG-CoA.

isopentenylpyrophosphate to form **farnesyl pyrophosphate.** This occurs via an isomerization of isopentenylpyrophosphate involving a shift of the double bond to form **dimethylallyl pyrophosphate,** followed by condensation with another molecule of isopentenylpyrophosphate to form the 10-carbon intermediate, **geranyl pyrophosphate** (Fig 28–2). A further condensation with isopentenylpyrophosphate forms farnesyl pyrophosphate. Two molecules of farnesyl pyrophosphate condense at the pyrophosphate end in a reaction involving first an elimination of pyrophosphate to form pre-squalene pyrophosphate, followed by a reduction with NADPH with elimination of the remaining pyrophosphate radical. The resulting compound is **squalene.** An alternative pathway known as the "*trans*-methylglutaconate shunt" may be present. This pathway removes a significant proportion (5% in fed livers, rising to 33% in fasted livers) of the dimethylallyl pyrophosphate and returns it, via *trans*-3-methylglutaconate-CoA, to HMG-CoA. This pathway may have regulatory potential with respect to the overall rate of cholesterol synthesis.

Step 4. Squalene Is Converted to Lanosterol: Squalene has a structure that closely resembles the steroid nucleus (Fig 28–3). Before ring closure occurs,

squalene is converted to squalene 2,3-oxide by a mixed-function oxidase in the endoplasmic reticulum, **squalene epoxidase.** The methyl group on C_{14} is transferred to C_{13} and that on C_8 to C_{14} as cyclization occurs, catalyzed by **oxidosqualene : lanosterolcyclase.**

Step 5. Lanosterol Is Converted to Cholesterol: In this last stage (Fig 28–3), the formation of cholesterol from **lanosterol,** takes place in the membranes of the endoplasmic reticulum and involves changes in the steroid nucleus and side chain. The methyl group on C_{14} is oxidized to CO_2 to form 14-desmethyl lanosterol. Likewise, 2 more methyl groups on C_4 are removed to produce zymosterol. $\Delta^{7,24}$-Cholestadienol is formed from zymosterol by the double bond between C_8 and C_9 moving to a position between C_8 and C_7. **Desmosterol** is formed at this point by a further shift in the double bond in ring B to take up a position between C_5 and C_6, as in cholesterol. Finally, cholesterol is produced when the double bond of the side chain is reduced, although this can occur at any stage of the overall conversion to cholesterol. The exact order in which the steps described actually take place is not known with certainty.

It is probable that the intermediates from squalene to cholesterol are attached to a special carrier protein known as the **squalene and sterol carrier protein.** This protein binds sterols and other insoluble lipids, allowing them to react in the aqueous phase of the cell. In addition, it seems likely that it is in the form of cholesterol-sterol carrier protein that cholesterol is converted to steroid hormones and bile acids and participates in the formation of membranes and of lipoproteins.

Farnesyl Pyrophosphate Gives Rise to Other Important Isoprenoid Compounds

Farnesyl pyrophosphate is the branch point for the synthesis of the other polyisoprenoids, **dolichol** and **ubiquinone.** The polyisoprenyl alcohol dolichol (see p 141 and p 596) is formed by the further addition of up to 16 isopentenylpyrophosphate residues, whereas the side chain of ubiquinone (see p 112) is formed by the addition of a further 3–7 isoprenoid units.

CHOLESTEROL SYNTHESIS IS REGULATED AT THE HMG-CoA REDUCTASE STEP

Regulation of cholesterol synthesis is exerted near the beginning of the pathway, at the HMG-CoA reductase step. There is a marked decrease in the activity of HMG-CoA reductase in fasting animals, which explains the **reduced synthesis of cholesterol during fasting.** There is a feedback mechanism whereby HMG-CoA reductase in liver is inhibited by cholesterol, the main product of the pathway. Since a direct inhibition of the enzyme by cholesterol cannot be demonstrated, cholesterol (or a metabolite, eg, oxygenated

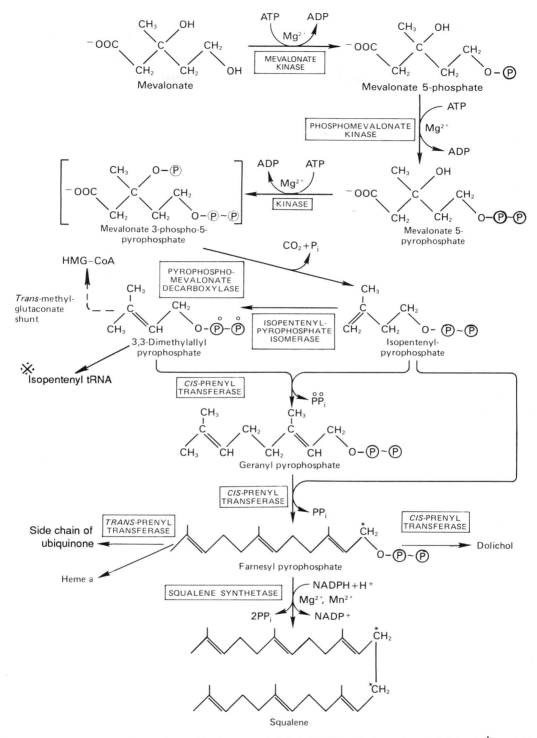

Figure 28–2. Biosynthesis of squalene, ubiquinone, and dolichol. HMG, 3-hydroxy-3-methylglutaryl; ※, cytokinin. A farnesyl residue is present in heme a of cytochrome oxidase. The carbon marked * becomes C_{11} or C_{12} in squalene. Squalene synthetase is a microsomal enzyme; all other enzymes indicated are soluble cytosolic proteins.

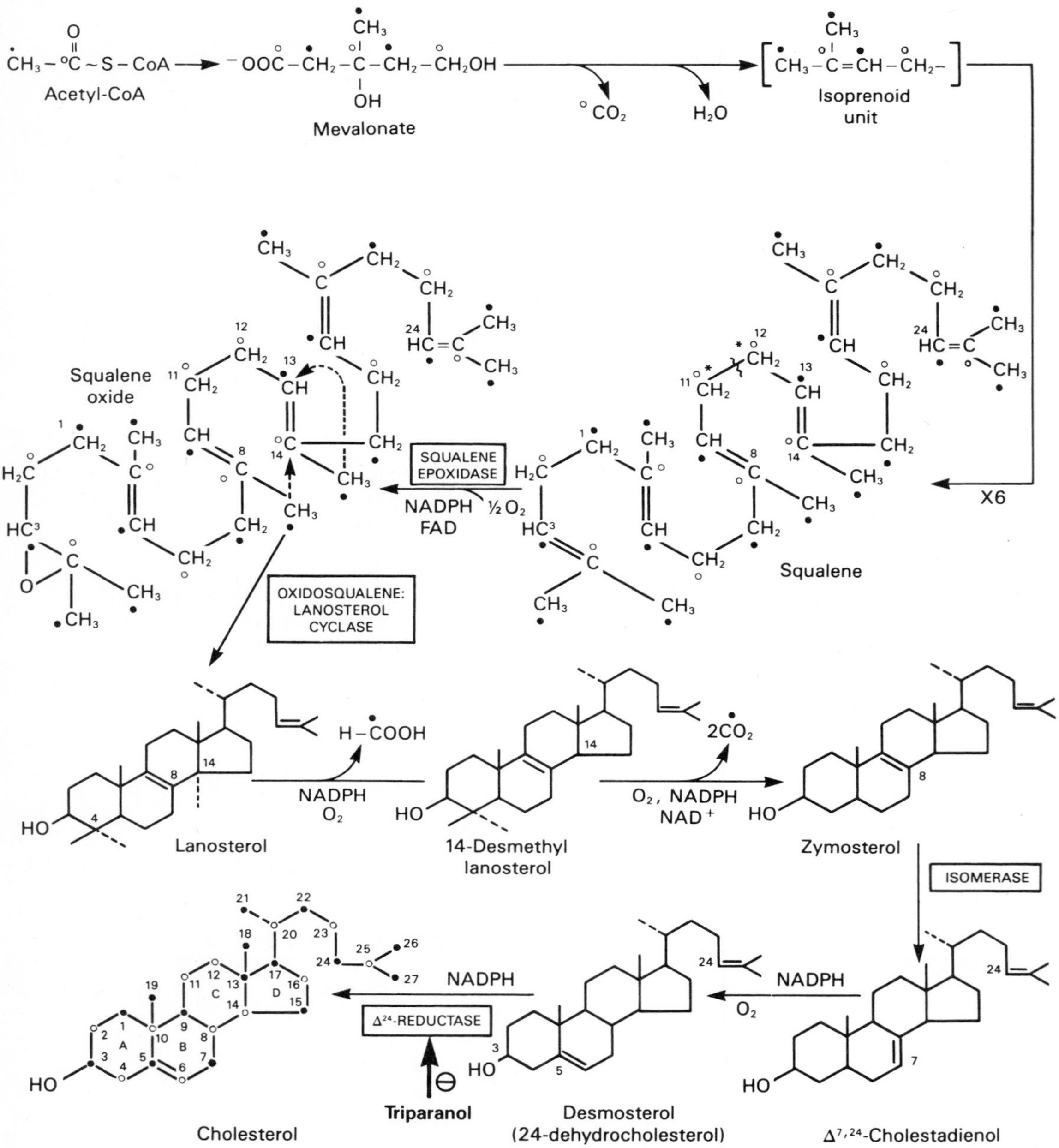

Figure 28–3. Biosynthesis of cholesterol. The numbered positions are those of the steroid nucleus. *Refers to labeling of squalene in Fig 28–2.

sterol) may act either by repression of the synthesis of new reductase or by inducing the synthesis of enzymes that degrade existing reductase. Cholesterol synthesis is also inhibited by LDL-cholesterol taken up via LDL receptors (apo-B-100,E receptors). A **diurnal variation** occurs in both cholesterol synthesis and reductase activity. However, there are more rapid effects of cholesterol on reductase activity than can be explained solely by changes in the rate of protein synthesis. Administration of insulin or thyroid hormone increases

HMG-CoA reductase activity, whereas glucagon or glucocorticoids decrease it. The enzyme exists in both active and inactive forms that may be reversibly modified by phosphorylation-dephosphorylation mechanisms, some of which may be cAMP dependent and therefore immediately responsive to glucagon and possibly insulin (Fig 28–4).

The effect of variations in the amount of cholesterol in the diet on the endogenous production of cholesterol has been studied in rats. When there was only 0.05%

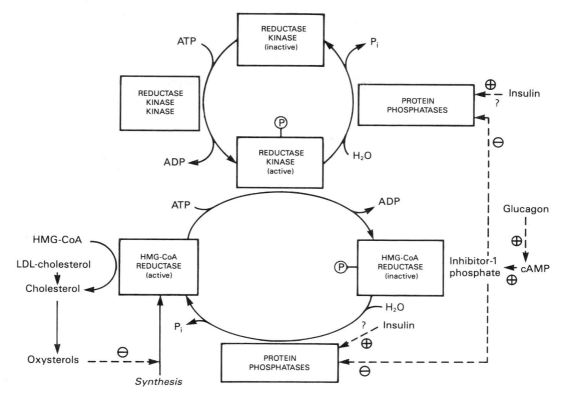

Figure 28–4. Possible mechanisms in the regulation of cholesterol synthesis by HMG-CoA reductase. Insulin has a dominant role compared with glucagon.

cholesterol in the diet, 70–80% of the cholesterol of the liver, small intestine, and adrenal gland was synthesized within the body, whereas when the dietary intake was raised to 2%, the endogenous production fell. It appears that it is only hepatic synthesis which is inhibited. Experiments with the perfused liver have demonstrated that cholesterol-rich chylomicron remnants, which are taken up by the liver (see p 240) inhibit sterol synthesis.

Attempts to lower plasma cholesterol in humans by reducing the amount of cholesterol in the diet produces variable results. Generally, a decrease of 100 mg in dietary cholesterol causes a decrease of approximately 0.13 mmol/L serum.

MANY FACTORS INFLUENCE THE CHOLESTEROL BALANCE IN TISSUES

At the tissue level, the following processes are considered to govern the cholesterol balance of cells (Fig 28–5).

Increase is due to (1) uptake of cholesterol-containing lipoproteins by receptors, eg, the LDL receptor; (2) uptake of cholesterol-containing lipoproteins by a non-receptor-mediated pathway; (3) uptake of free cholesterol from cholesterol-rich lipoproteins to the cell membrane; (4) cholesterol synthesis; and (5) hydrolysis of cholesteryl esters by the enzyme **cholesteryl ester hydrolase.**

Decrease is due to (1) efflux of cholesterol from the membrane to lipoproteins of low cholesterol potential, particularly to HDL$_3$ or nascent HDL, promoted by **LCAT** (lecithin:cholesterol acyltransferase); (2) esterification of cholesterol by **ACAT** (acyl-CoA : cholesterol acyltransferase); and (3) utilization of cholesterol for synthesis of other steroids, such as hormones, or bile acids in the liver.

The LDL Receptor Is Highly Regulated

The LDL (Apo B-100,E) receptors occur on the cell surface in pits that are coated on the cytosolic side of the cell membrane with a protein called clathrin. The receptor is a glycoprotein that spans the membrane, the B-100 binding region being at the exposed N-terminal end. It reacts with the ligand on LDL, apo B-100, and the LDL is taken up intact by endocytosis of the pit. It is broken down in the lysosomes, which involves hydrolysis of the apoprotein and cholesteryl ester followed by translocation of cholesterol into the cell. The receptors are not destroyed but return to the cell surface. This influx of cholesterol inhibits HMG-CoA reductase and cholesterol synthesis and stimulates ACAT activity. It also appears that the number of LDL

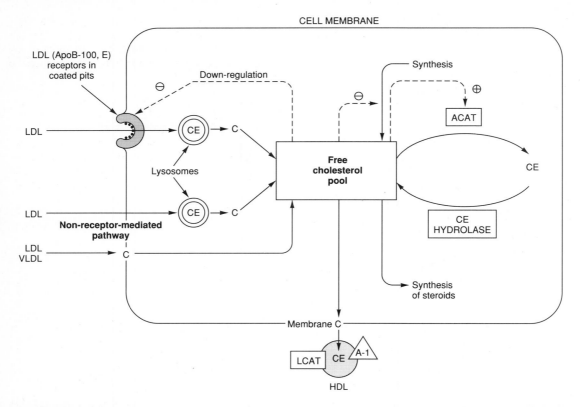

Figure 28–5. Factors affecting cholesterol balance at the cellular level. C, cholesterol; CE, cholesteryl ester; ACAT, acyl-CoA:cholesterol acyltransferase; LCAT, lecithin:cholesterol acyltransferase; A-I, apoprotein A-I; LDL, low-density lipoprotein; VLDL, very low density lipoprotein.

receptors on the cell surface is regulated by the cellular need of cholesterol for membranes and steroid hormone synthesis. Thus, influx of cholesterol down-regulates the number of LDL receptors (Fig 28–5).

The apo B-100,E receptor is a "high-affinity" LDL receptor, which may be saturated under most circumstances. Other "low-affinity" LDL receptors also appear to be present in addition to the non-receptor-mediated pathway, or scavenger pathway, which is not regulated.

CHOLESTEROL IS TRANSPORTED BETWEEN TISSUES IN THE PLASMA LIPOPROTEINS
(See Fig 28–6.)

In humans, the total plasma cholesterol is about 5.2 mmol/L, rising with age, although there are wide variations between individuals. The greater part is found in the esterified form. It is transported in lipoproteins in the plasma, and the highest proportion of cholesterol is found in the LDL (β-lipoproteins). However, under conditions where the VLDL are quantitatively predominant, an increased proportion of the plasma cholesterol will reside in this fraction.

Dietary cholesterol takes several days to equilibrate with cholesterol in the plasma and several weeks to equilibrate with cholesterol of the tissues. The turnover of cholesterol in the liver is relatively fast compared with the half-life of the total body cholesterol, which is several weeks. Free cholesterol in plasma and liver equilibrates in a matter of hours.

Cholesteryl ester in the diet is hydrolyzed to free cholesterol, which mixes with dietary free cholesterol and biliary cholesterol before absorption from the intestine in company with other lipids. It mixes with cholesterol synthesized in the intestines and is incorporated into chylomicrons. Of the cholesterol absorbed, 80–90% is esterified with long-chain fatty acids in the intestinal mucosa. The plant sterols (sitosterols) are poorly absorbed. When chylomicrons react with lipoprotein lipase to form chylomicron remnants, only about 5% of the cholesteryl ester is lost. The rest is taken up by the liver when the remnant reacts with the apo-E receptor and is hydrolyzed to free cholesterol. VLDL formed in the liver transport cholesterol into the plasma. However, in humans this is mainly free cholesterol because of low ACAT activity in liver. The cholesteryl ester in VLDL is derived mainly from the action of LCAT in plasma. Most of the cholesterol in VLDL is retained in the VLDL remnant that is taken up by the liver or converted to LDL, which in turn is taken up by the LDL receptor in liver and extrahepatic tissues.

LCAT Is Responsible for Most of the Plasma Cholesteryl Ester

The activity of plasma LCAT is responsible for virtually all plasma cholesteryl ester in humans. (This is not so in other species such as the rat, where there is appreciable ACAT activity in liver, allowing significant export of cholesteryl ester in nascent VLDL.) LCAT activity is associated with a species of HDL containing apo A-I. As cholesterol in HDL becomes esterified, it creates a concentration gradient and draws in cholesterol from tissues and from other lipoproteins (Fig 28–6). It becomes less dense, form-

ing HDL_2, which is thought to deliver cholesterol to the liver (see Fig 28–6). Thus, HDL features prominently in **reverse cholesterol transport,** the process whereby tissue cholesterol is transported to the liver.

Cholesteryl Ester Transfer Protein Facilitates Transfer of Cholesteryl Ester Between Lipoproteins

This protein, found in plasma of the human but not the rat, is also associated with HDL and seems identical to apo D. It facilitates transfer of cholesteryl ester from HDL to VLDL, LDL, and to a lesser extent, chylomicrons in exchange for triacylglycerol. It therefore relieves product inhibition of LCAT activity in HDL. Thus, in humans, much of the cholesteryl ester

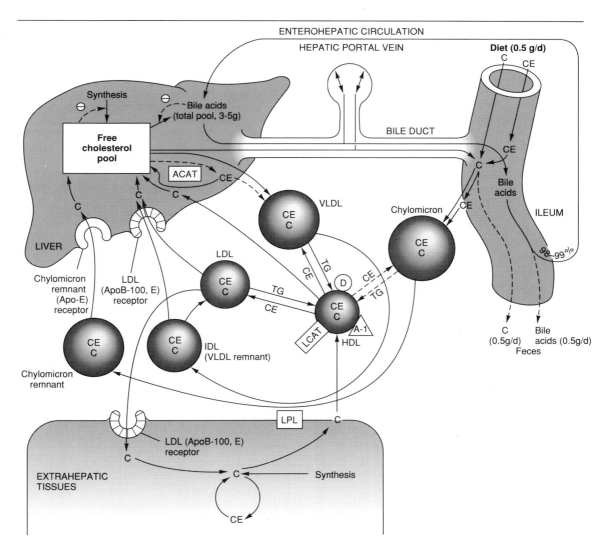

Figure 28–6. Transport of cholesterol between the tissues in humans. C, free cholesterol; CE, cholesteryl ester; VLDL, very low density lipoprotein; IDL, intermediate-density lipoprotein; LDL, low-density lipoprotein; HDL, high-density lipoprotein; ACAT, acyl-CoA:cholesterol acyltransferase; LCAT, lecithin:cholesterol acyltransferase; A-I, apoprotein A-I; D, apoprotein D; LPL, lipoprotein lipase.

formed by LCAT in HDL finds its way to the liver via VLDL remnants (IDL) or LDL (see Fig 28–6).

ULTIMATELY, CHOLESTEROL DESTINED FOR EXCRETION FROM THE BODY MUST ENTER THE LIVER AND BE EXCRETED IN THE BILE, EITHER AS CHOLESTEROL OR AS BILE ACIDS (SALTS).

About 1 g of cholesterol is eliminated from the body per day. Approximately half is excreted in the feces after conversion to bile acids. The remainder is excreted as neutral steroids. Much of the cholesterol secreted in the bile is reabsorbed, and it is believed that at least some of the cholesterol that serves as precursor for the fecal sterols is derived from the intestinal mucosa. **Coprostanol** is the principal sterol in the feces; it is formed from cholesterol in the lower intestine by the bacterial flora therein. A large proportion of the biliary excretion of bile salts is reabsorbed into the portal circulation, taken up by the liver, and reexcreted in the bile. This is known as the **enterohepatic circulation.** The bile salts not reabsorbed, or their derivatives, are excreted in the feces. Bile salts undergo changes brought about by intestinal bacteria to form secondary bile acids.

Bile Acids Are Formed From Cholesterol

The **primary bile acids** are synthesized in the liver from cholesterol by several intermediate steps. **Cholic acid** is the bile acid found in the largest amount in the bile itself. Both cholic acid and **chenodeoxycholic acid** are formed from a common precursor, itself derived from cholesterol (Fig 28–7).

The 7α-hydroxylation of cholesterol is the first committed step in the biosynthesis of bile acids, and it is this reaction that is rate-limiting in the pathway for synthesis of the acids. The reaction is catalyzed by **7α-hydroxylase,** a microsomal enzyme. It requires oxygen, NADPH, and cytochrome P-450 and appears to be a typical monooxygenase, as are subsequent hydroxylation steps. Vitamin C deficiency interferes with bile acid formation at the 7α-hydroxylation step and leads to cholesterol accumulation and atherosclerosis in scorbutic guinea pigs.

The pathway of bile acid biosynthesis divides early into one subpathway leading to **cholic acid,** characterized by an extra α-OH group on position 12, and another pathway leading to **chenodeoxycholic acid.** Apart from this difference, both pathways involve similar hydroxylation reactions and shortening of the side chain (Fig 28–7) to give the typical bile acid structures of α-OH groups on positions 3 and 7 and full saturation of the steroid nucleus.

The bile acids normally enter the bile as glycine or taurine conjugates. The newly synthesized primary bile acids are considered to exist within the liver cell as esters of CoA, ie, cholyl- or chenodeoxycholyl-CoA

(Fig 28–7). The CoA derivatives are formed with the aid of an activating enzyme occurring in the microsomes of the liver. A second enzyme catalyzes conjugation of the CoA derivatives with glycine or taurine to form glycocholic or glycochenodeoxycholic and taurocholic or taurochenodeoxycholic acids. These are the primary bile acids. In humans, the ratio of the glycine to the taurine conjugates is normally 3:1.

Since bile contains significant quantities of sodium and potassium and the pH is alkaline, it is assumed that the bile acids and their conjugates are actually in a salt form—hence the term "bile salts."

A portion of the primary bile acids in the intestine may be subjected to some further changes by the activity of the intestinal bacteria. These include deconjugation and 7α-dehydroxylation, which produce the **secondary bile acids,** deoxycholic acid from cholic acid, and lithocholic acid from chenodeoxycholic acid (Fig 28–7).

Nearly All of the Bile Acids Are Returned to the Liver in the Enterohepatic Circulation

Although fat digestion products, including cholesterol, are absorbed in the first 100 cm of small intestine, the primary and secondary bile acids are absorbed almost exclusively in the ileum, returning to the liver by way of the portal circulation about 98–99% of the bile acids secreted into the intestine. This is known as the **enterohepatic circulation.** However, lithocholic acid, because of its insolubility, is not reabsorbed to any significant extent.

A small fraction of the bile salts—perhaps only as little as 500 mg/d—escapes absorption and is therefore eliminated in the feces. Even though this is a very small amount, it nonetheless represents a major pathway for the elimination of cholesterol. The enterohepatic circulation of the bile salts is so efficient that each day the relatively small pool of bile acids (about 3–5 g) can be cycled through the intestine 6–10 times with only a small amount lost in the feces; ie, approximately 1–2% per pass through the enterohepatic circulation. However, **each day, an amount of bile acid equivalent to that lost in the feces is synthesized from cholesterol** by the liver, so that a pool of bile acids of constant size is maintained. This is accomplished by a system of feedback control.

Bile Acid Synthesis Is Regulated at the 7α-Hydroxylase Step

The principal rate-limiting step in the biosynthesis of bile acids is at the **7α-hydroxylase reaction,** and in the biosynthesis of cholesterol it is at the HMG-CoA reductase step (Fig 28–1). The activities of these 2 enzymes often change in parallel, and consequently it has been difficult to ascertain whether inhibition of bile acid synthesis takes place primarily at the HMG-CoA reductase step or at the 7α-hydroxylase reaction. Both enzymes undergo similar diural variation in activity. However, the stimulatory effect of cholesterol feeding on 7α-hydroxylase activity seems to be an effect on the

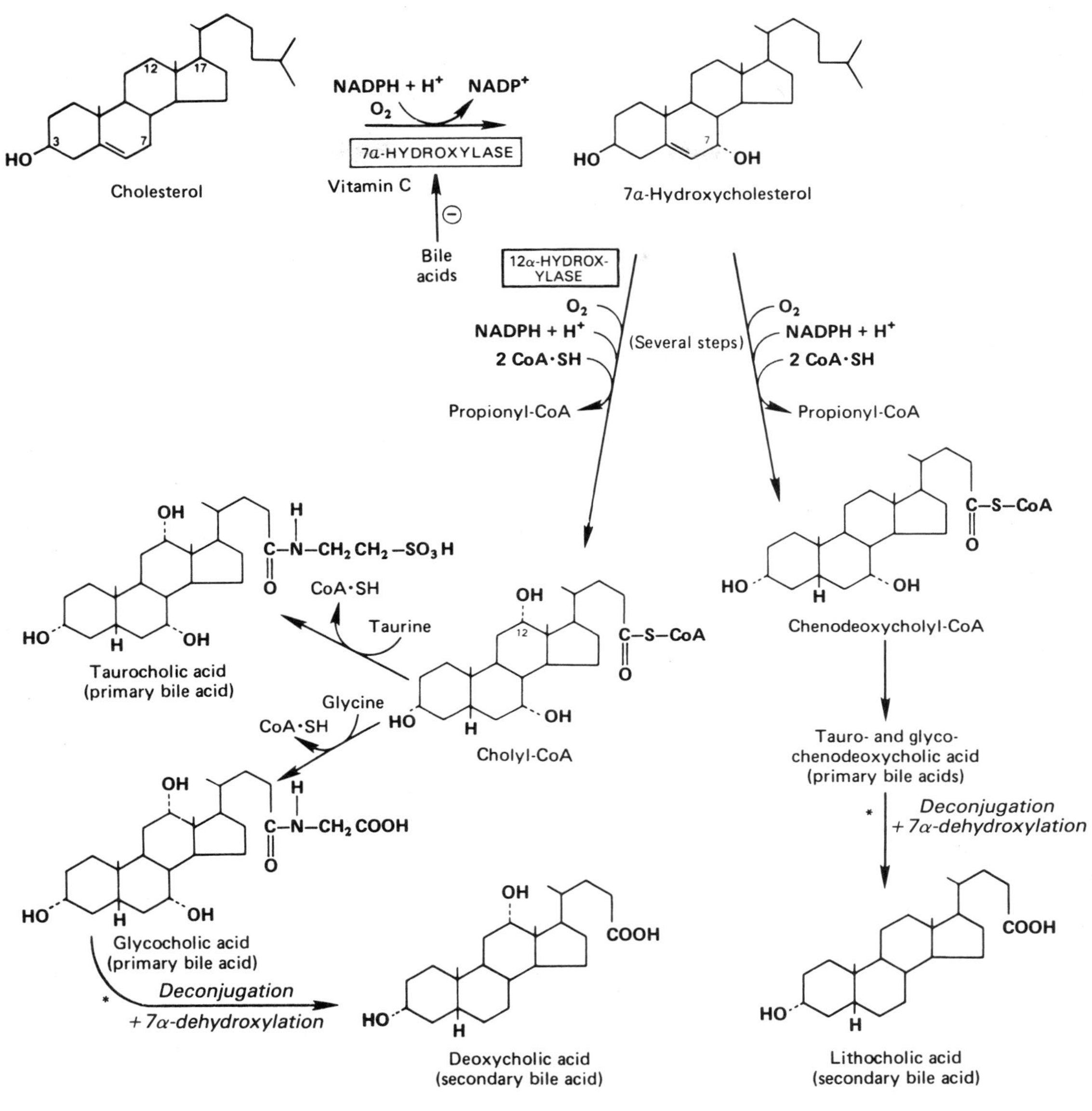

Figure 28–7. Biosynthesis and degradation of bile acids. *Catalyzed by microbial enzymes.

activity of the enzyme itself rather than being due to substrate availability. Bile acids certainly exert feedback inhibition on 7α-hydroxylase, but it does not seem to be a direct allosteric mechanism. In this regard, the return of bile salts to the liver via the enterohepatic circulation is an important control that, if interrupted, leads to activation of 7α-hydroxylase. Recent work indicates that 7α-hydroxylase (as well as HMG-CoA reductase) can be controlled by covalent phosphorylation-dephosphorylation. In contrast to HMG-CoA reductase, it is the phosphorylated form that results in increased activity of 7α-hydroxylase.

Hypercholesterolemia May Be Treated by Interrupting the Enterohepatic Circulation of Bile Acids

Significant reductions of plasma cholesterol can be effected by the use of cholestyramine resin (Questran), or surgically by the ileal exclusion operations. Both procedures cause a block in the reabsorption of bile acids. Then, because of release from feedback regulation normally exerted by bile acids, the conversion of cholesterol to bile acids is greatly enhanced in an effort to maintain the pool of bile acids. Consequently, LDL receptors in the liver are up-regulated, causing increased uptake of LDL with consequent lowering of plasma cholesterol.

The Serum Cholesterol Is Correlated With the Incidence of Atherosclerosis & Coronary Heart Disease

Of the serum lipids, cholesterol has been the one most often singled out as being chiefly concerned in the relationship. However, other parameters—such as serum triacylglycerol concentration—show similar correlations. Patients with arterial disease can have any one of the following abnormalities: (1) elevated concentrations of VLDL with normal concentrations of LDL; (2) elevated LDL with normal VLDL; (3) elevation of both lipoprotein fractions. There is also an inverse relationship between HDL (HDL$_2$) concentrations and coronary heart disease, and some consider that the most predictive relationship is the **LDL:HDL cholesterol ratio.** This relationship is explainable in terms of the proposed roles of LDL in transporting cholesterol to the tissues and of HDL acting as the scavenger of cholesterol in reverse cholesterol transport.

Atherosclerosis is characterized by the deposition of cholesterol and cholesteryl ester of lipoproteins containing apo B-100 in the connective tissue of the arterial walls. Diseases in which prolonged elevated levels of VLDL, IDL, or LDL occur in the blood (eg, diabetes mellitus, lipid nephrosis, hypothyroidism, and other conditions of hyperlipidemia) are often accompanied by premature or more severe atherosclerosis.

Experiments on the induction of atherosclerosis in animals indicate a wide species variation in susceptibility. The rabbit, pig, monkey, and humans are species in which atherosclerosis can be induced by feeding cholesterol. The rat, dog, and cat are resistant. Thyroidectomy or treatment with thiouracil drugs will allow induction of atherosclerosis in the dog and rat. Low blood cholesterol is a characteristic of hyperthyroidism.

Changes in Diet Play an Important Role in Reducing Serum Cholesterol Concentrations

Hereditary factors play the greatest role in determining individual blood cholesterol concentrations, but of the dietary and environmental factors that lower blood cholesterol, the substitution in the diet of **polyunsaturated and monounsaturated fatty acids** for some of the saturated fatty acids is most beneficial. Naturally occurring oils that contain a high proportion of polyunsaturated fatty acids include sunflower, cottonseed, corn, soybean oil, and olive oil contains high concentrations of monounsaturated fatty acids. On the other hand, butterfat, beef fat, and coconut oil contain a high proportion of saturated fatty acids. Sucrose and fructose have a greater effect in raising blood lipids, particularly triacylglycerols, than do other carbohydrates.

The reason for the cholesterol-lowering effect of polyunsaturated fatty acids is still not clear. However, several hypotheses have been advanced to explain the effect, including the stimulation of cholesterol excretion into the intestine and the stimulation of the oxidation of cholesterol to bile acids. It is possible that cholesteryl esters of polyunsaturated fatty acids are more rapidly metabolized by the liver and other tissues, which might enhance their rate of turnover and excretion. There is other evidence that the effect is largely due to a shift in distribution of cholesterol from the plasma into the tissues because of increased catabolic rate of LDL due to up-regulation of the LDL receptor by poly- and mono-unsaturated fatty acids and down-regulation by saturated fatty acids. Saturated fatty acids cause the formation of smaller VLDL particles that contain relatively more cholesterol, and they are utilized by extrahepatic tissues at a slower rate than are larger particles, tendencies that may be regarded as atherogenic.

Life Style Affects the Serum Cholesterol Level

Additional factors considered to play a part in coronary heart disease include high blood pressure, smoking, obesity, lack of exercise, and drinking soft as opposed to hard water. Elevation of plasma free fatty acids will also lead to increased VLDL secretion by the liver, involving extra triacylglycerol and cholesterol output into the circulation. Factors leading to higher or fluctuating levels of free fatty acids include emotional stress, nicotine from cigarette smoking, coffee drinking, and partaking of a few large meals rather than more continuous feeding. Premenopausal women appear to be protected against many of these deleterious factors, possibly because they have higher concentrations of HDL than do men and postmenopausal women.

When Diet Fails, Hypolipidemic Drugs Will Reduce Serum Cholesterol

Several drugs are known to block the formation of cholesterol at various stages in the biosynthetic pathway. Many of these drugs have harmful effects, but the fungal inhibitors of HMG-CoA reductase, **mevastatin** and **lovastatin,** reduce LDL cholesterol levels with few adverse effects. **Sitosterol** is a hypocholesterolemic agent that acts by blocking the absorption of cholesterol in the gastrointestinal tract. Resins such as **colestipol** and **cholestyramine** (Questran) prevent the reabsorption of bile salts by combining with them, thereby increasing their fecal loss. **Clofibrate** and **gemfibrozil** exert at least part of their hypolipidemic effect by diverting the hepatic inflow of free fatty acids from the pathways of esterification into those of oxidation, thus decreasing the secretion of triacylglycerol and cholesterol containing VLDL by the liver. In addition, they facilitate hydrolysis of VLDL triacylglycerols by lipoprotein lipase. **Probucol** appears to increase LDL catabolism via receptor-independent pathways. **Nicotinic acid** reduces the flux of FFA by inhibiting adipose tissue lipolysis, thereby inhibiting VLDL production by the liver.

Primary Disorders of the Plasma Lipoproteins (Dyslipoproteinemias) Are Inherited

A few individuals in the population exhibit inherited defects in lipoprotein metabolism, leading to the primary condition of either **hypo-** or **hyperlipoproteinemia.** Many others having defects such as diabetes mellitus, hypothyroidism, and atherosclerosis show secondary abnormal lipoprotein patterns that are very similar to one or another of the primary inherited conditions. Virtually all of these primary conditions are due to a defect at a stage in the course of lipoprotein formation, transport, or destruction (see Figs 27–4, 28–5, and 28–6). Not all of the abnormalities are harmful.

A. Hypolipoproteinemia:

1. Abetalipoproteinemia—This is a rare inherited disease characterized by absence of β-lipoprotein (LDL) in plasma. The blood lipids are present in low concentrations—especially acylglycerols, which are virtually absent, since **no chylomicrons or VLDL are formed.** Both the intestine and the liver accumulate acylglycerols. Abetalipoproteinemia is due to a defect in apoprotein B synthesis.

2. Familial hypobetalipoproteinemia—In hypobetalipoproteinemia, LDL concentration is between 10 and 50% of normal, but chylomicron formation occurs. It must be concluded that apo B is essential for triacylglycerol transport. Most individuals are healthy and long-lived.

3. Familial alpha-lipoprotein deficiency (Tangier disease)—In the homozygous individual, there is near absence of plasma HDL and accumulation of cholesteryl esters in the tissues. There is no impairment of chylomicron formation or secretion of VLDL by the liver. However, on electrophoresis, there is no pre-β-lipoprotein, but a broad β-band is found containing the endogenous triacylglycerol. This is because the normal pre-β-band contains other apolipoproteins normally provided by HDL. Patients tend to develop hypertriacylglycerolemia as a result of the absence of apo C-II, which normally activates lipoprotein lipase.

B. Hyperlipoproteinemia:

1. Familial lipoprotein lipase deficiency (type I)—This condition is characterized by very slow clearing of chylomicrons from the circulation, leading to abnormally raised levels of chylomicrons. VLDL may be raised, but there is a decrease in LDL and HDL. Thus, the condition is fat-induced. It may be corrected by reducing the quantity of fat and increasing the proportion of complex carbohydrate in the diet. A variation of this disease is caused by a deficiency in apo C-II, required as a cofactor for lipoprotein lipase.

2. Familial hypercholesterolemia (type II)—Patients are characterized by hyperbetalipoproteinemia (LDL), which is associated with increased plasma total cholesterol. There may also be a tendency for the VLDL to be elevated in type IIb. Therefore, the patient may have somewhat elevated triacylglycerol levels but the plasma—as is not true in the other types of hyperlipoproteinemia—remains clear. Lipid deposition in the tissue (eg, xanthomas, atheromas) is common. A type II pattern may also arise as a secondary result of hypothyroidism. The disease is due to reduced rates of clearance of LDL from the circulation because of **defective LDL receptors** and is associated with an **increased incidence of atherosclerosis.** Reduction of dietary cholesterol and saturated fats may be of use in treatment. A disease producing hypercholesterolemia but due to a different cause is **Wolman's disease** (cholesteryl ester storage disease). This is due to a deficiency of cholesteryl ester hydrolase in lysosomes of cells such as fibroblasts that normally metabolize LDL.

3. Familial type III hyperlipoproteinemia (broad beta disease, remnant removal disease, familial dysbetalipoproteinemia)—This condition is characterized by an increase in both chylomicron and VLDL remnants; these are lipoproteins of density less than 1.019 but appear as a broad β-band on electrophoresis (β-VLDL). They cause hypercholesterolemia and hypertriacylglycerolemia. Xanthomas and atherosclerosis of both peripheral and coronary arteries are present. Treatment by weight reduction and diets containing complex carbohydrates, unsaturated fats, and little cholesterol is recommended. The disease is due to a **deficiency in remnant metabolism by the liver caused by an abnormality in apo E,** which is normally present in 3 isoforms, E2, E3, and E4. Patients with type III hyperllipoproteinemia possess only E2, which does not react with the E receptor.

4. Familial hypertriacylglycerolemia (type IV)—This condition is characterized by high levels of endogenously produced triacylglycerol (VLDL). Cholesterol levels rise in proportion to the hypertriacylglycerolemia, and glucose intolerance is frequently present. Both LDL and HDL are subnormal in quantity. This lipoprotein pattern is also commonly associated with coronary heart disease, type II non-insulin-dependent diabetes mellitus, obesity, and many other conditions, including alcoholism, and the taking of progestational hormones. Treatment of primary type IV hyperlipoproteinemia is by weight reduction; replacement of soluble diet carbohydrate with complex carbohydrate, unsaturated fat, low-cholesterol diets; and also hypolipidemic agents.

5. Familial type V hyperlipoproteinemia—The lipoprotein pattern is complex, since both chylomicrons and VLDL are elevated, causing both triacylglycerolemia and cholesterolemia. Concentrations of LDL and HDL are low. Xanthomas are frequently present, but the incidence of atherosclerosis is apparently not striking. Glucose tolerance is abnormal and frequently associated with obesity and diabetes. The reason for the condition, which is familial, is not clear. Treatment has consisted of weight reduction followed by a diet not too high in either carbohydrate or fat.

It has been suggested that a further cause of hyperlipoproteinemia is overproduction of apo B, which can raise plasma concentrations of VLDL and LDL.

6. Familial hyperalphalipoproteinemia—This is a rare condition associated with increased concentrations of HDL apparently beneficial to health.

C. Familial Lecithin : Cholesterol Acyltransferase (LCAT) Deficiency: In affected subjects, the plasma concentration of cholesteryl esters and lysolecithin is low, whereas the concentration of cholesterol and lecithin is raised. The plasma tends to be turbid. Abnormalities are also found in the lipoproteins. One HDL fraction contains disk-shaped structures in stacks or rouleaux that are clearly nascent HDL unable to take up cholesterol owing to the absence of LCAT. Also present as an abnormal LDL subfraction is lipoprotein-X, otherwise found only in patients with **cholestasis.** VLDL are also abnormal, migrating as β-lipoproteins upon electrophoresis (β-VLDL). Patients with **parenchymal liver disease** also show a decrease of LCAT activity and abnormalities in the serum lipids and lipoproteins.

REFERENCES

Björkhem I, Akerlund JE: Studies on the link between HMG-CoA reductase and cholesterol 7α-hydroxylase in rat liver. *J Lipid Res* 1988;**29:**136.

Brown MS, Goldstein JL: Lipoprotein metabolism in the macrophage: Implications for cholesterol deposition in atherosclerosis. *Annu Rev Biochem* 1983;**52:**223.

Fears R, Sabine JR (editors): *Cholesterol 7α-Hydroxylase (7α-Monooxygenase)*. CRC Press, 1986,

Fielding CJ, Fielding PE: Metabolism of cholesterol and lipoproteins. Page 404 in: *Biochemistry of Lipids and Membranes.* Vance DE, Vance JE (editors). Benjamin/Cummings, 1985.

Kane JB, Havel RJ: Treatment of hypercholesterolemia. *Annu Rev Med* 1986;**37:**427.

Mahley RW, Innerarity TL: Lipoprotein receptors and cholesterol homeostasis. *Biochim Biophys Acta* 1983;**737:**197.

NIH Publication No 88-2925: *Report of the Expert Panel on Detection, Evaluation and Treatment of High Blood Cholesterol in Adults.* January, 1988.

Rudney H, Sexton RC: Regulation of Cholesterol Biosynthesis. *Annu Rev Nutr* 1986;**6:**245.

Integration of Metabolism & the Provision of Tissue Fuels

29

Peter A. Mayes, PhD, DSc

INTRODUCTION

Carbohydrates and lipids play many structural and metabolic roles, but it is as the provider of a large proportion of dietary calories that they have their greatest impact on metabolism and health. The regulation of this fuel influx and the manner in which it is integrated with other tissue fuels are of central interest, since they impinge on many other metabolic processes and are concerned with metabolic disease.

BIOMEDICAL IMPORTANCE

Under positive caloric balance, a significant proportion of the food energy intake is stored as either glycogen or fat. However, in many tissues, even under fed conditions, fatty acids are oxidized in preference to glucose, but particularly under conditions of caloric deficit or starvation. The purpose is to spare glucose for those tissues (eg, brain and erythrocytes) that require it under all conditions. Thus, regulatory mechanisms, often hormone-mediated, ensure a supply of suitable fuel for all tissues, at all times, from the fully fed to the totally starved state. Breakdown of these mechanisms occurs due to hormone imbalance (eg, insulin deficiency in diabetes mellitus), to metabolic imbalance due to heavy lactation (eg, ketosis of cattle), or to high metabolic demands in pregnancy (eg, pregnancy toxemia in sheep). All of these conditions are pathologic aberrations of the **starvation syndrome** which is a complication of many medical situations when appetite is diminished.

NOT ALL THE MAJOR FOODSTUFFS ARE INTERCONVERTIBLE
(See Fig 29–1.)

That animals may be fattened on a predominantly carbohydrate diet demonstrates the ease of conversion of carbohydrate into fat. However, as has been pointed out, humans may be limited in the extent to which glucose can be converted to fatty acids. A most significant reaction in this respect is the conversion of pyruvate to acetyl-CoA, as acetyl-CoA is the starting material for the synthesis of long-chain fatty acids. However, with respect to the reverse process, the conversion of fatty acids to glucose, **the pyruvate dehydrogenase reaction is essentially nonreversible,** which prevents the direct conversion of acetyl-CoA to pyruvate. In addition, there cannot be a net conversion of acetyl-CoA to oxaloacetate via the citric acid cycle, since one molecule of oxaloacetate is required to condense with acetyl-CoA and only one molecule of oxaloacetate is regenerated. For similar reasons, **there cannot be a net conversion of fatty acids having an even number of carbon atoms** (which form acetyl-CoA) **to glucose or glycogen.** Only the terminal 3-carbon portion of a fatty acid having an odd number of carbon atoms is glucogenic, as this portion of the molecule will ultimately form **propionyl-CoA** upon β-oxidation. Nevertheless, it is possible for labeled carbon atoms of fatty acids to be found ultimately in glycogen after traversing the citric acid cycle: This is because oxaloacetate is an intermediate both in the citric acid cycle and in the pathway of gluconeogenesis. The glycerol moiety of triacylglycerol will form glucose after activation to glycerol 3-phosphate.

Many of the carbon skeletons of the nonessential amino acids can be produced from carbohydrate via the citric acid cycle and transamination. By reversal of these processes, glucogenic amino acids yield carbon skeletons that are either members or precursors of the citric acid cycle. They are therefore readily converted by gluconeogenic pathways to glucose and glycogen. The ketogenic amino acids give rise to acetoacetate, which will in turn be metabolized as ketone bodies, forming acetyl-CoA in extrahepatic tissues.

For the same reasons that it is not possible for a net conversion of fatty acids to carbohydrate to occur, it is not possible for a net conversion of fatty acids to glucogenic amino acids to take place. Neither is it possible to reverse the pathways of breakdown of ketogenic and other amino acids, which fall into the category of **nutritionally essential amino acids.** Conversion of the carbon skeletons of glucogenic amino acids

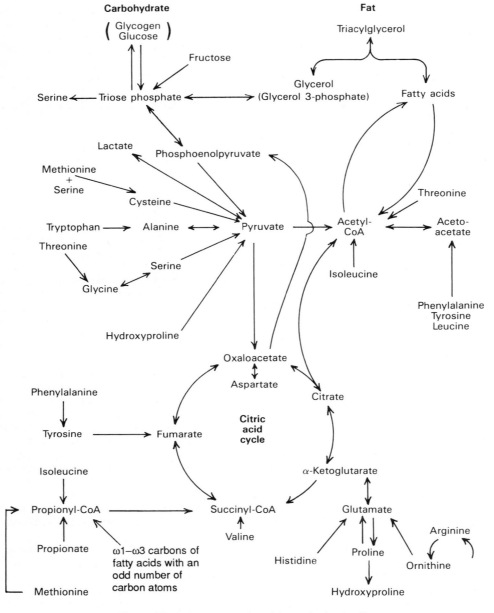

Figure 29–1. Interconversion of the major foodstuffs.

to fatty acids is possible, either by formation of pyruvate and acetyl-CoA or by reversal of nonmitochondrial reactions of the citric acid cycle from α-ketoglutarate to citrate followed by the action of ATP-citrate lyase to give acetyl-CoA (see Chapter 23). However, under most natural conditions, eg, starvation, a net breakdown of protein and amino acids is usually accompanied by a net breakdown of fat. The net conversion of amino acids to fat is therefore not a significant process except possibly in animals receiving a high-protein diet.

THE ECONOMICS OF CARBOHYDRATE & LIPID METABOLISM ENCOMPASS THE WHOLE BODY

Glucose Is a Metabolic Necessity in All Nutritional States

Many of the details of the interplay between carbohydrate and lipid metabolism in various tissues have been described, particularly the ready conversion of many glucogenic substances to glucose and glycogen by

gluconeogenesis. Gluconeogenesis is particularly important because certain tissues and cell types, including the central nervous system and the erythrocytes, are much more dependent upon a continual supply of glucose than others. A minimal supply of glucose is probably necessary in extrahepatic tissues to maintain oxaloacetate concentrations and the integrity of the citric acid cycle. In addition, glucose appears to be the main source of glycerol 3-phosphate in tissues devoid of glycerol kinase. There is, therefore, **a minimal and obligatory rate of glucose oxidation under all conditions.** Large quantities of glucose are also a necessity for fetal nutrition and the synthesis of lactose in milk. Certain mechanisms, in addition to gluconeogenesis, safeguard essential supplies of glucose in times of shortage **by allowing other substrates to spare its oxidation.**

The Preferential Utilization of Ketone Bodies & Free Fatty Acids Spare Glucose for Its Essential Functions

Ketone bodies and free fatty acids spare the oxidation of glucose in muscle by impairing its entry into the cell, its phosphorylation to glucose 6-phosphate, the phosphofructokinase reaction, and the oxidative decarboxylation of pyruvate. Oxidation of free fatty acids and ketone bodies causes an increase in the intracellular concentration of citrate that in turn inhibits phosphofructokinase allosterically. These observations and others, which demonstrate that acetoacetate is oxidized in the perfused heart preferentially to free fatty acids, justify the conclusion that under conditions of carbohydrate shortage available **fuels are oxidized in the following order of preference:** (1) **ketone bodies** (and probably other short-chain fatty acids, eg, acetate), (2) **free fatty acids,** and (3) **glucose.** This does not imply that any particular fuel is oxidized to the total exclusion of any other (Fig 29–2). However, these mechanisms are more important in tissues having a high capacity for aerobic oxidation of fatty acids, eg, heart and slow-twitch muscle, than in tissues with a low capacity, eg, fast-twitch muscle.

The combination of the effects of free fatty acids in sparing glucose utilization in muscle and heart and the effect of the spared glucose in inhibiting free fatty acid mobilization in adipose tissue has been called the **glucose-fatty acid cycle.**

DURING STARVATION A CONTINUAL SUPPLY OF FUEL FOR THE TISSUES IS PROVIDED

In animals fed high-carbohydrate diets, fatty acid oxidation is spared. This is because lipolysis in adipose tissue is inhibited due to high blood glucose and insulin concentrations and therefore, free fatty acid levels remain low (Fig 29–3). As the animal passes from the fed to the fasting condition, glucose availability from food becomes less, and liver glycogen is drawn upon in an attempt to maintain the blood glucose. The concentration of insulin in the blood decreases, and glucagon increases. As glucose utilization diminishes in adipose tissue and the inhibitory effect of insulin on lipolysis becomes less, fat is mobilized as free fatty acids and glycerol. The free fatty acids are transported to the tissues, where they are either oxidized or esterified. Glycerol joins the carbohydrate pool after activation to glycerol 3-phosphate, mainly in the liver and kidney. During this transition phase from the fully fed to the fully fasting state, endogenous glucose production (from amino acids and glycerol) does not keep pace with its utilization and oxidation, since the liver glycogen stores become depleted and blood glucose tends to fall. Thus, fat is mobilized at an ever-increasing rate, but in several hours the plasma free fatty acids and blood glucose stabilize at the fasting level (0.7–0.8 and 3.3–3.9 mmol/L, respectively). At this point, it must be presumed that in the whole animal the supply of glucose balances the obligatory demands for glucose utilization and oxidation. This is achieved by the increased oxidation of free fatty acids and ketone bodies, sparing the nonobligatory oxidation of glucose. This fine balance is disturbed in conditions that demand more glucose or in which glucose utilization is impaired and which therefore lead to further mobilization of fat. The provision of carbohydrate by adipose tissue, in the form of **glycerol,** is an important function, for it is only this source of carbohydrate together with that provided by **gluconeogenesis from protein** that can supply the starving organism with the glucose needed for those processes which must utilize glucose. In prolonged starvation in humans, **gluconeogenesis from protein is diminished** owing to reduced release of amino acids, particularly alanine, from muscle. This coincides with adaptation of the brain to **replace approximately half of the glucose oxidized with ketone bodies.**

Ketosis Is a Metabolic Adaptation to Starvation

The primary function of ketogenesis is to remove excess fatty acid carbon from the liver in a form that is readily oxidized by extrahepatic tissues in place of glucose. Ketosis arises as a result of a deficiency in available carbohydrate. This has the following actions in fostering ketogenesis (see Figs 24–9 and 24–10). (1) It causes an imbalance between esterification and lipolysis in adipose tissue, with consequent release of free fatty acids into the circulation. Free fatty acids are the principal substrates for ketone body formation in the liver, and therefore all factors, metabolic or endocrine, affecting the release of free fatty acids from adipose tissue influence ketogenesis. (2) Upon entry of free fatty acids into the liver, the balance between their esterification and oxidation is governed by carnitine palmitoyltransferase I, whose activity is increased indirectly by the concentration of free fatty acids and the increased glucagon : insulin ratio. (3) As more fatty

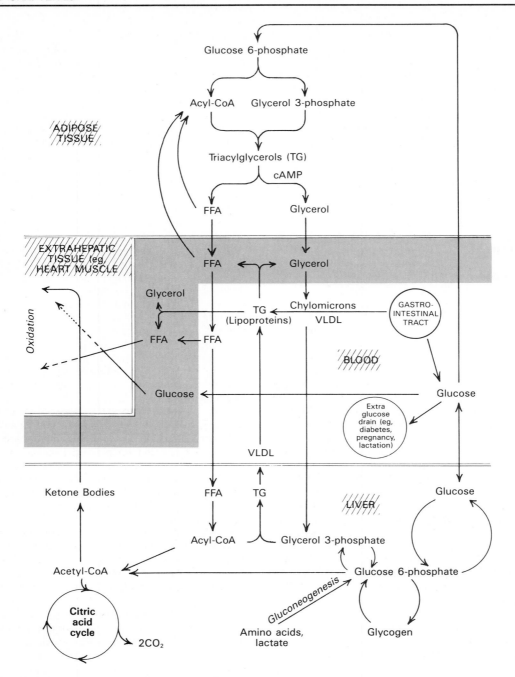

Figure 29–2. Metabolic interrelationships between adipose tissue, the liver, and extrahepatic tissues. Stippled areas, lipoprotein lipase region of capillary wall; FFA, free fatty acids, VLDL, very low density lipoproteins.

acid is oxidized, more forms ketone bodies and less forms CO_2, regulated in such a manner that the total ATP production of the liver remains constant.

A feedback mechanism for controlling free fatty acid output from adipose tissue in starvation may operate as a result of the action of ketone bodies and free fatty acids to directly stimulate the pancreas to produce insulin.

Under most conditions free fatty acids are mobilized in excess of oxidative requirements, since a large proportion are esterified, even during fasting. As the liver takes up and esterifies a considerable proportion of the free fatty acid output, it plays a regulatory role in removing excess free fatty acids from the circulation. When carbohydrate supplies are adequate, most of the influx is esterified and ultimately retransported from

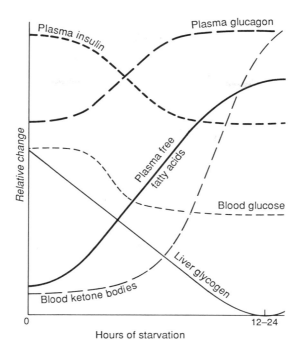

Figure 29–3. Relative changes in metabolic parameters during the onset of starvation.

the liver as VLDL to be utilized by other tissues. However, in the face of an increased influx of free fatty acids, an alternative route, **ketogenesis,** is available that enables the liver to continue to retransport much of the influx of free fatty acids in a form readily utilized by extrahepatic tissues under all nutritional conditions.

Most of these principles are depicted in Fig 29–2. It will be noted that there is a carbohydrate cycle involving release of glycerol from adipose tissue and its conversion in the liver to glucose, followed by its transport back to adipose tissue to complete the cycle. The other cycle, a lipid cycle, involves release of free fatty acid by adipose tissue, its transport to and esterification in the liver, and retransport as VLDL back to adipose tissue.

Pathological Ketosis Is Caused by an Amplification of the Factors Causing Starvation Ketosis

The ketosis that occurs in starvation and fat feeding is relatively mild compared with the condition encountered in uncontrolled **diabetes mellitus, pregnancy toxemia of ewes,** or **ketosis of lactating cattle.** The main reason appears to be that in the severe conditions carbohydrate is even less available to the tissues than in the mild conditions. Thus, in the milder forms of diabetes mellitus, in fat feeding, and in chronic starvation, glycogen is present in the liver in variable amounts, and free fatty acid levels are lower, which probably accounts for the less severe ketosis associated with these conditions.

In type I diabetes mellitus, the lack (or relative lack) of insulin probably affects adipose tissue more than

any other tissue, because of its extreme sensitivity to this hormone. As a result, free fatty acids are released in quantities that give rise to plasma free fatty acid levels more than twice those in fasting normal subjects. Many changes also occur in the activity of enzymes within the liver, and these changes enhance both the rate of gluconeogenesis and the rate of transfer of glucose to the blood despite high levels of circulating glucose.

In ketosis of ruminants, there is a severe drain of glucose from the blood owing to excessive fetal demands of twins or the demands of heavy lactation (Fig 29–2). Extreme hypoglycemia results, coupled with negligible amounts of glycogen in the liver. Ketosis in these conditions tends to be severe. As hypoglycemia develops, the secretion of insulin diminishes, allowing not only less glucose utilization but also enhancement of lipolysis in adipose tissue. Pregnant women often exhibit mild ketosis.

In untreated type I diabetes mellitus death occurs as a result of complications of acidosis caused by long-term depletion of base needed to neutralize acidic ketone bodies excreted in the urine (see Chapter 62, Case no. 8: Diabetes mellitus with ketoacidosis). In pregnancy toxemia of ewes, death occurs rapidly due to the severe hypoglycemia.

REFERENCES

Caprio S et al: Oxidative fuel metabolism during mild hypoglycemia: critical role of free fatty acids. *Am J Physiol* 1989;**256:**E413.

Cohen P: *Control of Enzyme Activity,* 2nd ed. Chapman & Hall, 1983.

Hue L, Van de Werve G (editors): *Short-Term Regulation of Liver Metabolism.* Elsevier/North Holland, 1981.

Knopp RH et al: Lipoprotein metabolism in pregnancy, fat transport to the fetus and the effects of diabetes. *Biol Neonate* 1986;**50:**297.

Mayes PA, Laker ME: Regulation of ketogenesis in the liver. *Biochem Soc Trans* 1981;**9:**339.

McGarry JD, Foster DW: Regulation of hepatic fatty acid oxidation and ketone body production. *Annu Rev Biochem* 1980;**49:**395.

Siess EA, Kientsch-Engel RI, Wieland OH: Concentration of free oxaloacetate in the mitochondrial compartment of isolated liver cells. *Biochem J* 1984;**218:**171.

Zorzano A et al: Effects of starvation and exercise on concentrations of citrate, hexose phosphates and glycogen in skeletal muscle and heart: Evidence for selective operation of the glucose-fatty acid cycle. *Biochem J* 1985;**232:**585.

Section III.
Metabolism of Proteins & Amino Acids

Biosynthesis of the Nutritionally Nonessential Amino Acids

30

Victor W. Rodwell, PhD

INTRODUCTION

We often refer to nutritionally essential amino acids as "essential" or "indispensable" and to nutritionally nonessential amino acids as "nonessential" or "dispensable" (Table 30–1). While in a nutritional context these terms are correct, they obscure the biologically essential nature of all 20 amino acids. It might be argued that the nutritionally nonessential amino acids are more important to the cell than the nutritionally essential ones, since organisms (eg, humans) have evolved that lack the ability to manufacture the latter but not the former group.

Since this book emphasizes metabolic processes of human tissues, we will discuss only biosynthesis of the nutritionally nonessential amino acids, not of the nutritionally essential amino acids by plants and microorganisms.

Table 30–1. Amino acid requirements of humans.

Nutritionally Essential	Nutritionally Nonessential
Arginine*	Alanine
Histidine*	Asparagine
Isoleucine	Aspartate
Leucine	Cysteine
Lysine	Glutamate
Methionine	Glutamine
Phenylalanine	Glycine
Threonine	Hydroxyproline†
Tryptophan	Hydroxylysine†
Valine	Proline
	Serine
	Tyrosine

*"Nutritionally semiessential." Synthesized at rates inadequate to support growth of children.
†Not necessary for protein synthesis but formed during posttranslational processing of collagen.

BIOMEDICAL IMPORTANCE

Medical implications of the material in this chapter relate to amino acid deficiency states that can result if any of the nutritionally essential amino acids are omitted from the diet or are present in inadequate amounts. Since certain grains are relatively poor sources of tryptophan and lysine, in regions where the diet relies heavily on these grains for total protein and is unsupplemented by protein sources such as milk, fish, or meat, dramatic deficiency states may be observed. **Kwashiorkor** and **marasmus** are endemic in certain regions of West Africa. Kwashiorkor results when a child is weaned onto a starchy diet poor in protein. In marasmus, both caloric intake and specific amino acids are deficient.

DURING EVOLUTION, HIGHER ANIMALS LOST THE ABILITY TO BIOSYNTHESIZE AMINO ACIDS WHOSE FORMATION REQUIRES PROTRACTED REACTION SEQUENCES

The existence of nutritional requirements suggests that dependence on an external supply of a required intermediate can be of greater survival value than the ability to biosynthesize it. If a specific intermediate is present in the food, an organism that can synthesize it is reproducing and transferring to future generations genetic information of negative survival value. The survival value is negative rather than nil because ATP and nutrients are used to synthesize unnecessary DNA. The number of enzymes required by prokaryotic cells to synthesize the nutritionally essential amino acids is large relative to the number of enzymes required to synthesize the nutritionally nonessential amino acids (Table 30–2). This suggests that there is a survival

Table 30–2. Enzymes required for the synthesis of amino acids from amphibolic intermediates.

Number of Enzymes Required to Synthesize			
Nutritionally Essential		**Nutritionally Nonessential**	
Arg*	7	Ala	1
His	6	Asp	1
Thr	6	Asn†	1
Met	5 (4 shared)	Glu	1
Lys	8	Gln*	1
Ile	8 (6 shared)	Hyl¶	1
Val	1 (7 shared)	Hyp#	1
Leu	3 (7 shared)	Pro*	3
Phe	10	Ser	3
Trp	5 (8 shared)	Gly‡	1
	59	Cys§	2
		Tyr‖	1
			17

*From Glu. †From Asp. ¶From Lys. #From Pro. ‡From Ser. §From Ser plus S^{2-}. ‖From Phe.

advantage in retaining the ability to manufacture "easy" amino acids while losing the ability to make "difficult" amino acids.

MOST NUTRITIONALLY NONESSENTIAL AMINO ACIDS ARE FORMED FROM AMPHIBOLIC INTERMEDIATES BY SHORT ANABOLIC PATHWAYS

Of the 12 nutritionally nonessential amino acids (Table 30–1), 9 are formed from amphibolic intermediates. The remaining 3 (Cys, Tyr, Hyl) are formed from nutritionally essential amino acids.

Glutamate dehydrogenase, glutamine synthetase, and transaminases occupy central positions in amino acid biosynthesis. Their combined effect is to catalyze transformation of inorganic ammonium ion into the organic α-amino nitrogen of various amino acids.

Glutamate: Reductive amination of α-ketoglutarate is catalyzed by glutamate dehydrogenase

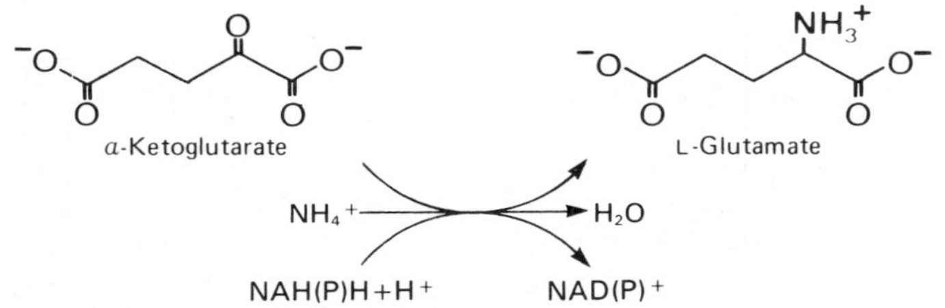

Figure 30–1. The glutamate dehydrogenase reaction. Reductive amination of α-ketoglutarate by NH_4^+ proceeds at the expense of NAD(P)H.

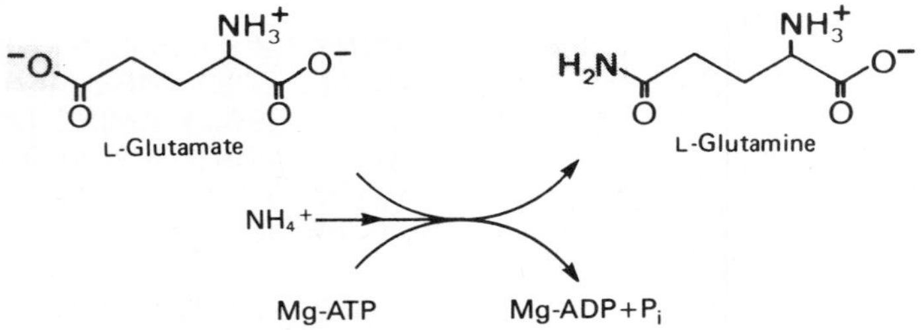

Figure 30–2. The glutamine synthetase reaction.

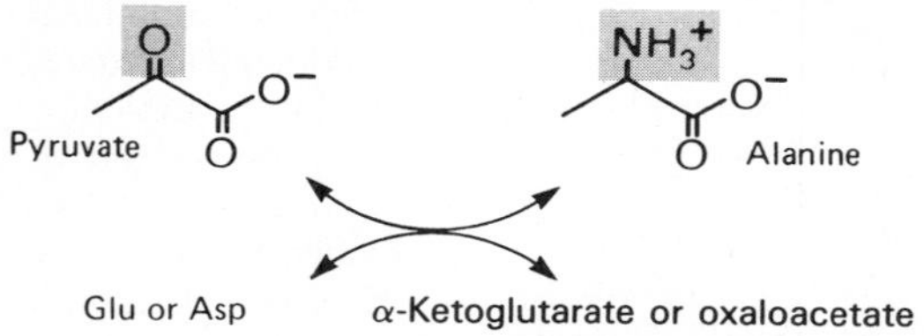

Figure 30–3. Formation of alanine by transamination of pyruvate. The amino donor may be glutamate or aspartate. The other product thus is α-ketoglutarate or oxaloacetate.

(Fig 30–1). In addition to forming L-glutamate from the amphibolic intermediate α-ketoglutarate, this reaction constitutes a key first step in the biosynthesis of many additional amino acids.

Glutamine: Biosynthesis of glutamine from glutamate is catalyzed by glutamine synthetase (Fig 30–2). The reaction exhibits both similarities to and differences from the glutamate dehydrogenase reaction. Both "fix" inorganic nitrogen—one into amino and the other into amide linkage. Both reactions are coupled to highly exergonic reactions—for glutamate dehydrogenase the oxidation of NAD(P)H and for glutamine synthetase the hydrolysis of ATP.

Alanine & Aspartate: Transamination of pyruvate forms L-alanine, and transamination of oxaloacetate forms L-aspartate (Fig 30–3). Transfer of the α-amino group of glutamate to these amphibolic intermediates illustrates the ability of a transaminase to channel ammonium ion, via glutamate, into the α-amino nitrogen of amino acids.

Asparagine: Formation of asparagine from aspartate, catalyzed by asparagine synthetase (Fig 30–4), resembles glutamine synthesis (Fig 30–2). However, since the mammalian enzyme uses glutamine rather than ammonium ion as the nitrogen source, mammalian asparagine synthetase does not "fix" inorganic nitrogen. By contrast, bacterial asparagine synthetases do use ammonium ion and hence do "fix" nitrogen. As for other reactions in which PP$_i$ is formed, hydrolysis of PP$_i$ to P$_i$ by pyrophosphatase ensures that the reaction is strongly favored energetically.

Serine: Serine is formed from the glycolytic intermediate D-3-phosphoglycerate (Fig 30–5). The α-hydroxyl group is reduced to an oxo group by NAD$^+$, then transaminated, forming phosphoserine. This is then dephosphorylated, forming serine.

Glycine: Synthesis of glycine in mammalian tissues can occur in several ways. Liver cytosol contains glycine transaminases that catalyze the synthesis of glycine from glyoxylate and glutamate or alanine. Unlike most transaminase reactions, this strongly favors glycine synthesis. Two additional important mammalian routes for glycine formation are from choline (Fig 30–6) and from serine via the serine hydroxymethyl-transferase reaction (Fig 30–7).

Proline: In mammals and some other life forms, proline is formed from glutamate by reversal of the reactions of proline catabolism (Fig 30–8).

Cysteine: Cysteine, while not itself nutritionally essential, is formed from methionine (nutritionally essential) and serine (nutritionally nonessential). Methionine is first converted to homocysteine via S-adenosylmethionine and S-adenosylhomocysteine (see Chapter 32). Conversion of homocysteine and serine to cysteine and homoserine is shown in Fig 30–9.

Tyrosine: Tyrosine is formed from phenylalanine by the reaction catalyzed by phenylalanine hydroxylase (Fig 30–10). Thus, whereas phenylalanine is a nutritionally essential amino acid, tyrosine is not—provided the diet contains adequate quantities of phenylalanine. The reaction is not reversible, so tyrosine

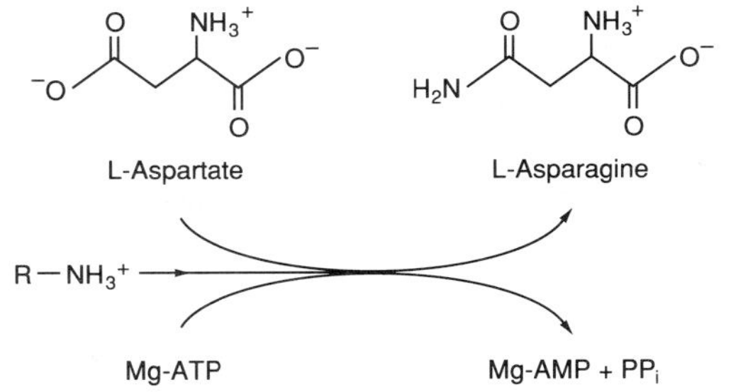

Figure 30–4. The asparagine synthetase reaction. Note similarities to and differences from the glutamine synthetase reaction (Fig 30–2).

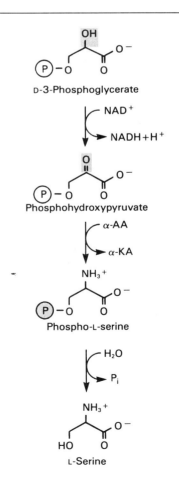

Figure 30–5. Serine biosynthesis α-AA, α-amino acids; α-KA, α-keto acids.

cannot replace the nutritional requirement for phenylalanine. The **phenylalanine hydroxylase complex** is a mixed-function oxygenase present in mammalian liver but absent from other tissues. The reaction involves incorporation of one atom of molecular oxygen into the para position of phenylalanine while the other atom is reduced, forming water (Fig 30–11). The reducing power, supplied ultimately by NADPH, is immediately provided as **tetrahydrobiopterin,** a pteridine which resembles folic acid.

Hydroxyproline: Since proline serves as a precursor of hydroxyproline, proline and hydroxyproline belong to the glutamate family of amino acids. Although both 3- and 4-hydroxyprolines occur in mammalian tissues, what follows refers solely to *trans*-4-hydroxyproline.

Hydroxyproline, like hydroxylysine, is almost exclusively associated with collagen, the most abundant protein of mammalian tissues. Collagen contains about one-third glycine and one-third proline and hydroxyproline. Hydroxyproline, which accounts for many of the amino acid residues of collagen, stabilizes the collagen triple helix to digestion by proteases. Unlike the hydroxyl groups of hydroxylysine, which serve as sites for attachment of galactosyl and glucosyl residues, the hydroxyl groups of collagen hydroxyproline are unsubstituted.

A unique feature of hydroxyproline and hydroxylysine metabolism is that the preformed amino acids, as they may occur in ingested food protein, are not incorporated into collagen. There is no tRNA capable of accepting hydroxyproline or hydroxylysine and inserting them into an elongating polypeptide chain. Dietary proline, however, is a precursor of collagen hydroxyproline, and dietary lysine a precursor of collagen hydroxylysine. Hydroxylation of proline or lysine is catalyzed by prolyl hydroxylase or by lysyl hydroxylase, enzymes associated with the microsomal fraction of many tissues (skin, liver, lung, heart, skeletal muscle, and granulating wounds). These enzymes are peptidyl hydroxylases, since hydroxylation only occurs subsequent to incorporation of proline or lysine into polypeptide linkage.

Both hydroxylases are mixed function oxygenases that require, in addition to substrate, molecular O_2, ascorbate, Fe^{2+}, and α-ketoglutarate. Prolyl hydroxylase has been more extensively studied, but lysyl hydroxylase appears to be an entirely analogous enzyme. For every mole of proline hydroxylated, 1 mol of α-ketoglutarate is decarboxylated to succinate. During this process, one atom of molecular O_2 is incorporated into proline and one into succinate (Fig 30–11).

Hydroxylysine: 5-Hydroxylysine (α,ε-diamino-δ-hydroxycaproate) is present in collagen but absent from most other mammalian proteins. Collagen hydroxylysine arises directly from dietary lysine, not dietary hydroxylysine. Before lysine is hydroxylated, it must first be incorporated into peptide linkage. Hydroxylation of the lysyl peptide is then catalyzed by lysyl hydroxylase, a mixed-function oxidase analogous to prolyl hydroxylase (Fig 30–9).

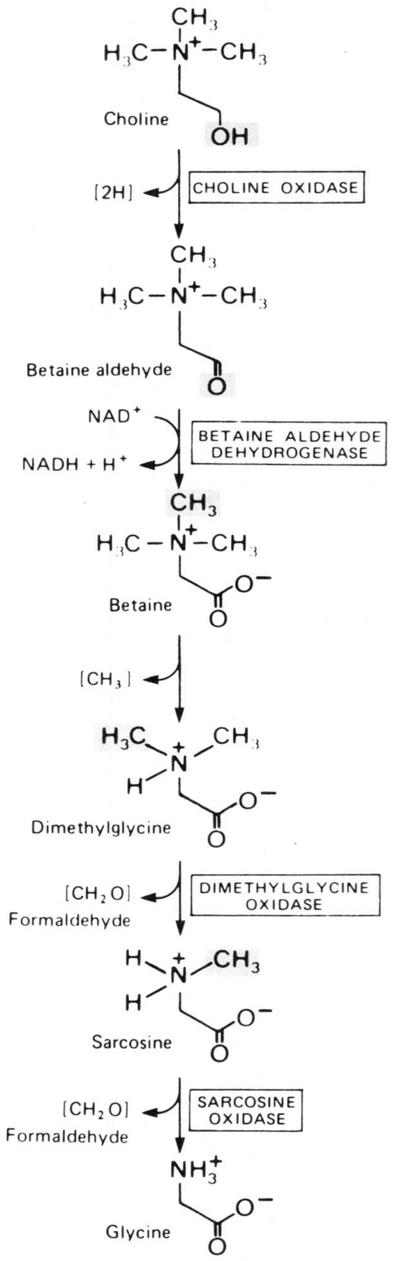

Figure 30–6. Formation of glycine from choline.

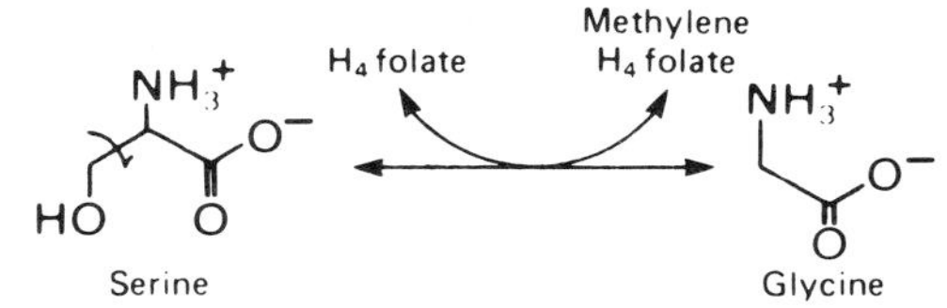

Figure 30–7. The serine hydroxymethyltransferase reaction. The reaction is freely reversible. H_4 folate, tetrahydrofolate.

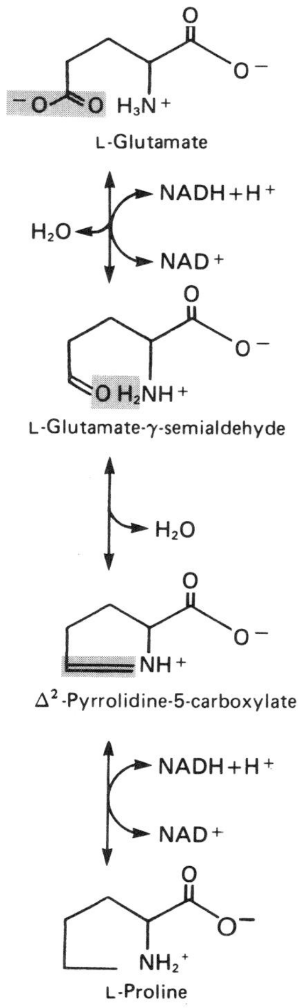

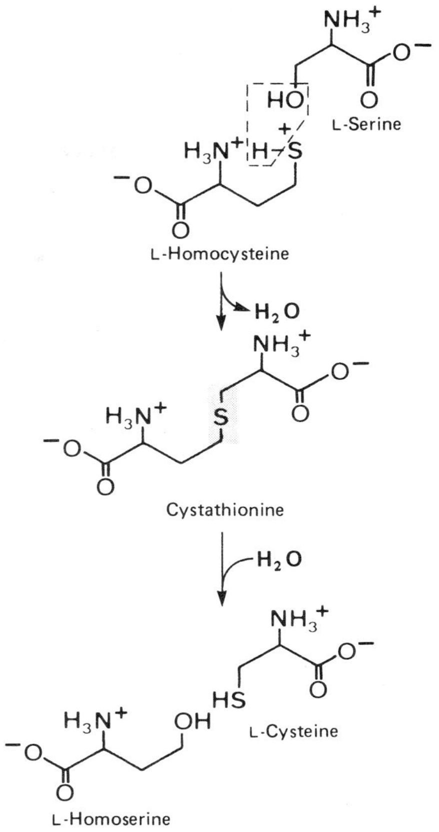

Figure 30–9. Conversion of homocysteine and serine to homoserine and cysteine. Note that while the sulfur of cysteine derives from methionine by transsulfuration, the carbon skeleton is provided by serine.

Figure 30–8. Biosynthesis of proline from glutamate by reversal of the reactions of proline catabolism.

The α-Keto Acids of the Three Branched-Chain Amino Acids Can Replace Their Corresponding Amino Acids in the Human Diet

While leucine, valine, and isoleucine are all nutritionally essential amino acids for humans and other higher animals, mammalian tissue transaminases reversibly interconvert all 3 amino acids and their corresponding α-keto acids. These α-keto acids thus can replace their amino acids in the diet.

Histidine and Arginine are Regarded as Nutritionally Semiessential

Arginine, a nutritionally essential amino acid for growing humans, can be synthesized by rats but not in quantities sufficient to permit normal growth.

Histidine, like arginine, is nutritionally semiessential. Adult humans and adult rats have been maintained in nitrogen balance for short periods in the absence of histidine. The growing animal does, however, require

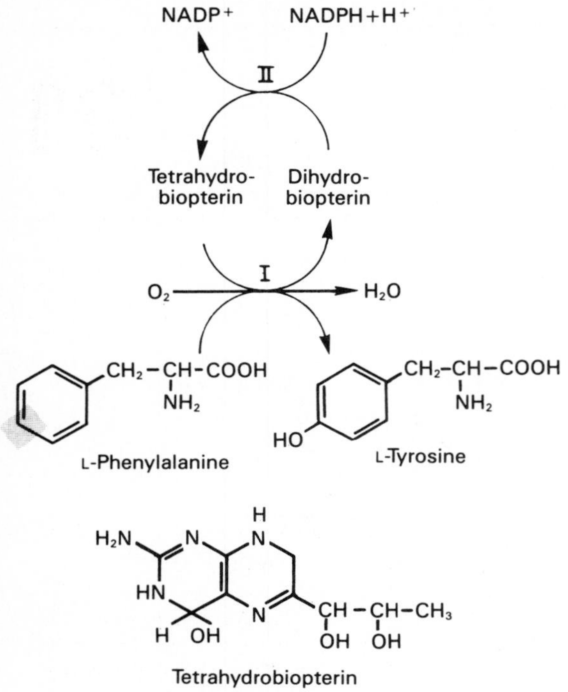

NADP⁺   NADPH+H⁺

II

Tetrahydro-   Dihydro-
biopterin   biopterin

I

O_2 ————————→ H_2O

CH₂–CH–COOH
      |
      NH₂

L-Phenylalanine

CH₂–CH–COOH
      |
      NH₂

HO ◯ L-Tyrosine

Tetrahydrobiopterin

Figure 30–10. The phenylalanine hydroxylase reaction. Two distinct enzymatic activities are involved. Activity II catalyzes reduction of dihydrobiopterin by NADPH, and activity I the reduction of O_2 to H_2O and of phenylalanine to tyrosine. This reaction is associated with several defects of phenylalanine metabolism discussed in Chapter 32.

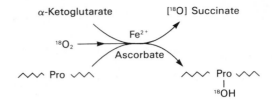

Figure 30–11. The prolyl hydroxylase reaction. The substrate is a proline-rich peptide. During the course of the reaction, molecular oxygen is incorporated into both succinate and proline (shown by the use of heavy oxygen, $^{18}O_2$).

histidine in the diet. If studies were to be carried on for longer periods, it is probable that a requirement for histidine in adult human subjects would also be elicited.

REFERENCES

Cardinale GJ, Udenfriend S: Prolyl hydroxylase. *Adv Enzymol* 1974;**41**:245.

Mercer LP, Dodds SJ, Smith DI: Dispensable, indispensable, and conditionally indispensable amino acid ratios in the diet. Chapter 1 in Vol. 1 of: *Adsorption and Utilization of Amino Acids.* Friedman M (editor). CRC Press. 1989.

Rosenberg LE, Scriver CR: Disorders of amino acid metabolism. Chapter 11 in: *Metabolic Control and Disease.* Bondy PK, Rosenberg LE (editors). Saunders, 1980.

Tyler B: Regulation of the assimilation of nitrogen compounds. *Annu Rev Biochem* 1978;**47**:1127.

Catabolism of Proteins and of Amino Acid Nitrogen

31

Victor W. Rodwell, PhD

INTRODUCTION

We shall consider how nitrogen is removed from amino acids and converted to urea and the medical problems that arise when there are defects in these reactions.

BIOMEDICAL IMPORTANCE

Ammonia, derived mainly from the α-amino nitrogen of amino acids, is potentially toxic to humans. The mechanisms by which ammonia causes toxicity are not fully understood. The body disposes of ammonia by converting it to the nontoxic compound **urea.** Normal operation of the metabolic pathway that converts ammonia to urea—the urea cycle—is essential for maintenance of health. In conditions in which liver function is seriously compromised—eg, in individuals with massive cirrhosis (where normal liver cells are replaced by fibroblasts and collagen) or severe hepatitis—ammonia accumulates in the blood and results in clinical signs and symptoms. In those few infants born with a deficiency in the activity of one of the enzymes of the urea cycle, proper treatment requires an understanding of the biochemistry of the formation of urea.

PROTEIN TURNOVER OCCURS IN ALL FORMS OF LIFE

The proteins of living cells are constantly renewed by **protein turnover,** the continuous process of degradation into, and subsequent resynthesis from, free amino acids. The physiologic importance of protein turnover is indicated by its presence in all forms of life, even bacteria growing logarithmically in rich media. While turnover involves both synthesis and degradation of proteins, **this chapter discusses the degradation of proteins and amino acids.** Protein synthesis is discussed in Chapter 40.

Clinicians & Nutritionists Employ Specific Terms to Characterize Different States of Nitrogen Balance

Specific terms describe the status of nitrogen metabolism in human subjects. **Nitrogen balance** refers to the difference between total nitrogen intake and total nitrogen loss. Ingestion of more nitrogen than is excreted constitutes a **positive nitrogen balance,** a state typical of a growing infant or pregnant woman. In **nitrogen equilibrium,** typical for an adult human subject, nitrogen intake matches nitrogen output in feces and urine. **Negative nitrogen balance,** in which nitrogen output exceeds intake, characterizes certain postsurgical patients, patients with advanced cancer, and individuals who ingest inadequate nitrogen (eg, in kwashiorkor) or low-quality dietary protein.

Each Day, Adults Degrade One to Two Percent of Their Body Protein

Turnover of 1–2% of total body protein per day in a normal human subject results principally from degradation of muscle protein. From 75 to 80% of the liberated amino acids are then reutilized for new protein synthesis. The nitrogen of the remainder is catabolized to urea and the carbon skeletons to amphibolic intermediates (Fig 31–1). In Western society, this daily loss amounts to 30–40 g of protein, or 5–7 g of nitrogen.

Rates of Protein Degradation Vary Widely Between Proteins & in Different Physiologic States

Individual proteins are degraded at different rates. The mean rate of protein degradation in a given cell, tissue, or intact organism thus reflects a mean value for numerous proteins. The degradation rates, or half-lives, of proteins in specific tissues respond to physiologic demand. The mean rate may be extremely high when a tissue undergoes major structural rearrangement (eg, uterine tissue during pregnancy or tadpole tail tissue during metamorphosis). Degradation of skeletal muscle proteins also increases in severe starvation.

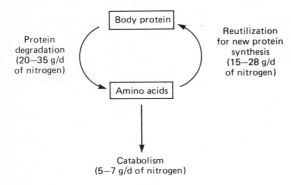

Figure 31–1. Quantitative relationships for protein and amino acid turnover.

The Half-Lives of Proteins Range From 30 Minutes to Over 150 Hours

Since the in vivo rate of degradation of a protein obeys first-order kinetics, **an enzyme's susceptibility to degradation is expressed by its half-life, $t_{1/2}$, the time required to reduce its concentration to 50% of its initial value.** Half-lives for liver enzymes range from less than 40 minutes to over 150 hours. Enzymes exhibiting short half-lives include many key regulatory enzymes. For example, tryptophan oxygenase and tyrosine transaminase have a $t_{1/2}$ of about 2 hours, and the $t_{1/2}$ for HMG-CoA reductase can be as short as 30 min. These values contrast sharply with half-lives of over 100 hours for aldolase, lactate dehydrogenase, and cytochromes. Domains rich in Pro, Ser, Asp, and Glu, termed PEST sequences, are present in proteins with short half-lives. In response to physiologic demand, the rates of degradation of key regulated enzymes may be accelerated or retarded, altering enzyme levels, and hence partitioning metabolites between different metabolic pathways.

Amino Acids Ingested in Excess of Need Are Degraded, Not Stored

To maintain health, a typical Western adult requires 30–60 g of protein per day, or its equivalent in free amino acids. However, *protein quality* (**the proportion of essential amino acids in a food relative to their proportion in proteins being synthesized**) is of critical importance. **Excess amino acids are not stored.** Regardless of their source, those not immediately incorporated into new protein are rapidly degraded. Consumption of excess amino acids thus serves no useful purpose that cannot equally well be served, and at a lower cost, by carbohydrates and lipids.

PROTEASES & PEPTIDASES DEGRADE PROTEINS TO AMINO ACIDS

Like parallel processes in the gastrointestinal tract, the proteolytic digestion of endogenous proteins within cells involves proteases and peptidases. **Intracellular proteases** hydrolyze the internal peptide bonds of proteins, forming peptides. These peptides are then degraded to free amino acids by **peptidases. Endopeptidases** cleave internal bonds in peptides, forming shorter peptides. **Aminopeptidases** and **carboxypeptidases** then remove amino acids from the N- and C-termini of peptides, respectively. The ultimate products are free amino acids.

Proteins Are Degraded by ATP-Dependent & ATP-Independent Pathways

Two major pathways degrade intracellular proteins of eukaryotic cells. Extracellular, membrane-associated, and long-lived intracellular proteins are degraded in cellular organelles termed **lysosomes** by ATP-independent processes. By contrast, degradation of abnormal and other short-lived proteins requires ATP and ubiquitin, and occurs in the **cytosol.**

Liver Cell Asialoglycoprotein Receptors Bind & Internalize Glycoproteins Targeted for Degradation

What factors target a protein for rapid degradation? For proteins in the circulation (eg, certain peptide hormones), loss of a sialic acid moiety from the nonreducing ends of their oligosaccharide chains signals their degradation. These asialated glycoproteins are recognized by the asialoglycoprotein receptor of liver cells and are internalized. They are then degraded in lysosomes by proteases termed **cathepsins.**

Ubiquitin Targets Many Intracellular Proteins for Degradation

Many intracellular proteins are targeted for degradation by **ubiquitin,** a small (8.5 kd) protein present in all eukaryotic cells. The primary structure of ubiquitin is highly conserved. Only 3 of 76 residues differ between yeast and human ubiquitin. Proteins destined for degradation via ubiquitin-dependent reactions are derivatized by several ubiquitin molecules. These are attached via 3 reactions that result in formation of non-α-peptide bonds between the carboxy terminus of ubiquitin and ε-amino groups of lysyl residues in a protein (Fig 31–2). Whether a protein is derivatized by ubiquitin depends on which aminoacyl residue is present at its N-terminus. Reaction with ubiquitin is retarded by methionyl or seryl, and accelerated by aspartyl or arginyl N-terminal residues.

1.　$UB-\overset{\overset{O}{\|}}{C}-O^- + E_1-SH + ATP \rightarrow AMP + PP_i + UB-\overset{\overset{O}{\|}}{C}-S-E1$

2.　$UB-\overset{\overset{O}{\|}}{C}-S-E_1 + E_2-SH \rightarrow E_1-SH + UB-\overset{\overset{O}{\|}}{C}-S-E_2$

3. $UB-\overset{\overset{O}{\|}}{C}-S-E_2 + H_2N-\varepsilon-Protein \overset{E_3}{\rightarrow} E_2-SH + UB-\overset{\overset{O}{\|}}{C}-\overset{\overset{H}{|}}{N}-\varepsilon-Protein$

Figure 31–2. Partial reactions in the attachment of ubiquitin (UB) to proteins. (1) The terminal COOH of ubiquitin forms a thioester bond with an —SH of E_1 in a reaction driven by conversion of ATP to AMP and PP_1. (2) A thioester exchange reaction transfers activated ubiquitin to E_2. (3) E_3 catalyzes transfer of ubiquitin to ε-lysyl groups of the target protein.

DEPENDING ON THEIR ECOLOGICAL NICHE, ANIMALS CONVERT NITROGEN TO DIFFERENT END PRODUCTS

Animals convert nitrogen from amino acids and other sources to one of three end products: **ammonia, uric acid,** or **urea.** Which product prevails in different animals depends upon the availability of water in their specific ecological niche. **Ammonotelic organisms,** the teleostean fish, excrete nitrogen as ammonia. Their aqueous niche, which compels them to excrete water continuously, facilitates continuous excretion of the highly toxic compound, ammonia. By contrast, land animals convert nitrogen either to **uric acid** (**uricotelic** life style) or to **urea** (**ureotelic** organisms). Birds, which must both conserve water and maintain low weight, are uricotelic. The relatively insoluble end-product uric acid is then excreted as semisolid guano. Land animals, including man, are **ureotelic,** converting nitrogen to the highly soluble, nontoxic compound, urea. The extremely low toxicity of urea is sometimes obscured by the elevated levels of blood urea in patients with renal disease. This elevation in blood urea is a consequence, not a cause, of malaise.

THE BIOSYNTHESIS OF UREA OCCURS IN SEVERAL STAGES

The biosynthesis of urea will be divided for discussion into 4 stages: (1) transamination, (2) oxidative deamination, (3) ammonia transport, and (4) reactions of the urea cycle. Fig 31–3 relates these areas to overall catabolism of amino acid nitrogen. Although each stage also plays a role in amino acid biosynthesis, what follows is discussed from the perspective of amino acid catabolism.

THE INITIAL REACTION IN AMINO ACID CATABOLISM IS REMOVAL OF THE α-AMINO GROUP

Free amino acids derived from exogenous dietary proteins or from degradation of endogenous proteins by the processes described above are metabolized in identical ways. Their α-amino nitrogen is first removed, either by transamination or by oxidative de-

amination. The resulting carbon "skeleton" is then degraded by pathways discussed in the next chapter. This chapter discusses the fate of the nitrogen itself.

Pyridoxal Phosphate Is Present at the Catalytic Site of All Transaminases

Pyridoxal phosphate forms an essential part of the catalytic site of transaminases and of many other enzymes with amino acid substrates. In all pyridoxal phosphate-dependent reactions of amino acids, the initial step is formation of an enzyme-bound Schiff base intermediate. This intermediate, stabilized by interaction with a cationic region of the active site, can be rearranged in ways that include release of a keto acid with formation of enzyme-bound pyridoxamine phosphate. The bound amino form of the coenzyme can then form an analogous Schiff base intermediate with a keto acid. During transamination, bound coenzyme thus serves as a carrier of amino groups. Since the equilibrium constant for most transaminase reactions is close to unity, transamination is a freely reversible process. This permits transaminases to function in both amino acid catabolism and biosynthesis.

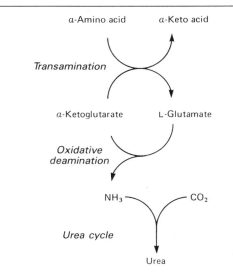

Figure 31–3. Overall flow of nitrogen in amino acid catabolism.

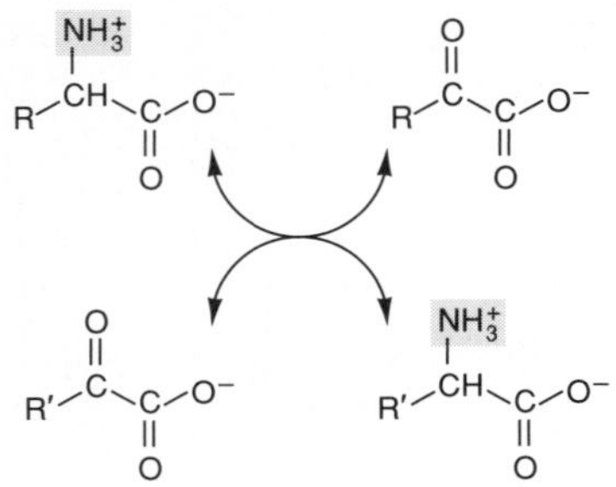

Figure 31–4. Transamination. The reaction is shown for 2 α-amino and 2 α-keto acids. Non-α-amino or carbonyl groups also participate in transamination, although this is relatively uncommon. The reaction is freely reversible with an equilibrium constant of about 1.

Transamination Channels α-Amino Acid Nitrogen Into Glutamate

Transamination, catalyzed by **transaminases** or **aminotransferases,** interconverts a pair of amino acids and a pair of keto acids. These generally are α-amino acid and α-keto acids (Fig 31–4).

Two transaminases, alanine-pyruvate transaminase (**alanine transaminase**) and glutamate-α-keto-glutarate transaminase (**glutamate transaminase**), present in most mammalian tissues, catalyze transfer of amino groups from most amino acids to form alanine (from pyruvate) or glutamate (from α-ketoglutarate) (Fig 31–5).

Transaminases Are Specific for Only One Pair of α-Amino & α-Keto Acids: Each transaminase is specific for the specified pair of amino and keto acids as one pair of substrates but nonspecific for the other pair, which may be any of a wide variety of amino acids and their corresponding keto acids. Since alanine is also a substrate for glutamate transaminase, all of the amino nitrogen from amino acids that can undergo transamination can be concentrated in glutamate. This is important, because **L-glutamate is the only amino**

acid in mammalian tissues that undergoes oxidative deamination at an appreciable rate. The formation of ammonia from α-amino groups thus occurs mainly via conversion to the α-amino nitrogen of L-glutamate.

Transamination Is Not Restricted to α-Amino Groups: Most (but not all) amino acids are substrates for transamination. Exceptions include lysine, threonine, and the cyclic imino acids, proline and hydroxyproline. Transamination is not restricted to α-amino groups. The δ-amino group of ornithine (but not the ε-amino group of lysine) is readily transaminated, forming glutamate-γ-semialdehyde (see Fig 32–3). Serum levels of transaminases are elevated in some disease states (see the Appendix).

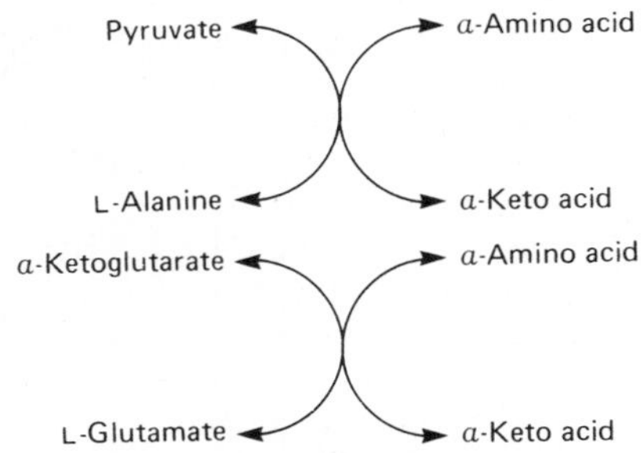

L-AMINO ACID OXIDASE ALSO CONVERTS α-AMINO ACIDS TO α-KETO ACIDS

Oxidative conversion of many amino acids to their corresponding α-keto acids occurs in mammalian liver and kidney. Although most of the activity toward L-α-amino acids is due to the coupled action of transaminases plus **L-glutamate dehydrogenase,** both L- and D-amino acid oxidase activities occur in mammalian liver and kidney tissue and are widely distributed in other animals and microorganisms.

Amino Acid Oxidases Are Autoxidizable Flavoproteins

The reduced FMN or FAD of amino acid oxidases is reoxidized directly by molecular oxygen, forming hydrogen peroxide (H_2O_2) without participation of cytochromes or other electron carriers (Fig 31–6). The toxic product H_2O_2 is then split to O_2 and H_2O by **catalase,** which occurs widely in tissues, especially liver. Although the amino acid oxidase reactions are reversible, if catalase is absent the α-keto acid product

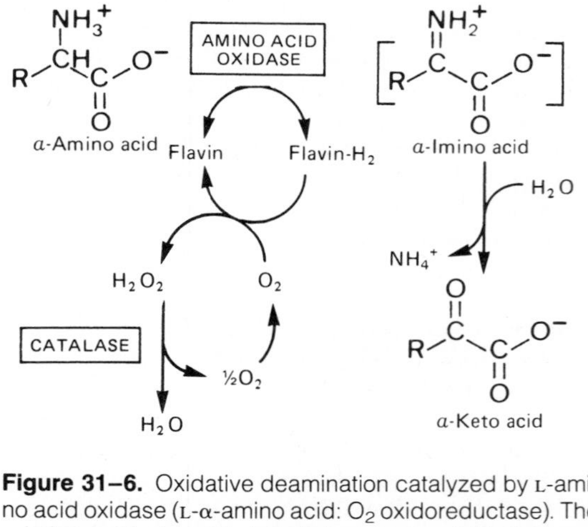

Figure 31–5. Alanine (*top*) and glutamate (*bottom*) transaminases.

Figure 31–6. Oxidative deamination catalyzed by L-amino acid oxidase (L-α-amino acid: O_2 oxidoreductase). The α-imino acid, shown in brackets, is not a stable intermediate.

is nonenzymatically decarboxylated by H_2O_2, forming a carboxylic acid with one less carbon atom. It is doubtful, however, whether this decarboxylation occurs to any great extent in intact human tissues.

In the amino acid oxidase reaction (Fig 31–6), the amino acid is first dehydrogenated by the flavoprotein of the oxidase, forming an α-imino acid. This spontaneously adds water, then decomposes to the corresponding α-keto acid with loss of the α-imino nitrogen as ammonium ion.

L-GLUTAMATE DEHYDROGENASE OCCUPIES A CENTRAL POSITION IN NITROGEN METABOLISM

The amino groups of most amino acids ultimately are transferred to α-ketoglutarate by transamination, forming L-glutamate (Fig 31–3). Release of this nitrogen as ammonia is catalyzed by **L-glutamate dehydrogenase,** an enzyme of high activity widely distributed in mammalian tissues (Fig 31–7). Liver glutamate dehydrogenase is a regulated enzyme whose activity is affected by allosteric modifiers such as ATP, GTP, and NADH, which inhibit the enzyme, and ADP, which activates it. Certain hormones appear also to influence glutamate dehydrogenase activity in vitro.

Glutamate dehydrogenase uses either NAD^+ or $NADP^+$ as cosubstrate. The reaction is reversible and functions in both amino acid catabolism and biosynthesis. It therefore functions not only to funnel nitrogen from glutamate to urea (catabolism) but also to catalyze **amination** of α-ketoglutarate by free ammonia (see Chapter 30).

BLOOD AMMONIA ARISES ALSO FROM AMMONIA FORMED BY ENTERIC BACTERIA

Ammonia* is absorbed from the intestine into the portal venous blood, which thus contains higher levels of ammonia than does systemic blood. Under normal circumstances, the liver promptly removes the ammonia from the portal blood, so that blood leaving the liver (and indeed all of the peripheral blood) is virtually ammonia-free. This is essential, since even minute quantities of ammonia are toxic to the central nervous system. With severely impaired hepatic function or development of collateral communications between the portal and systemic veins (as may occur in cirrhosis), portal blood may bypass the liver. Ammonia may thus rise to toxic levels in the systemic blood.

Ammonia Intoxication Is Life-Threatening

Symptoms of **ammonia intoxication** include tremor, slurring of speech, blurring of vision, and, in severe cases, coma and death. These symptoms resemble those of the syndrome of hepatic coma which occurs

*Present at physiologic pH almost exclusively as ammonium ion, NH_4^+.

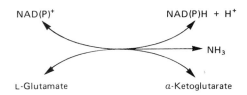

Figure 31–7. The L-glutamate dehydrogenase reaction. NAD(P)+ means that either NAD+ or NADP+ can serve as cosubstrate. The reaction is reversible, but the equilibrium constant favors glutamate formation.

when blood and, presumably, brain ammonia levels are elevated. Ammonia intoxication is assumed to be a factor in the etiology of hepatic coma. Therefore, treatment includes measures designed to reduce blood ammonia levels.

FORMATION & SECRETION OF AMMONIA BY THE KIDNEYS MAINTAINS ACID-BASE BALANCE

The ammonia content of the blood in renal veins exceeds that in renal arteries, indicating that the kidneys produce ammonia and add it to the blood. However, the excretion into the urine of the ammonia produced by renal tubular cells constitutes a more significant aspect of renal ammonia metabolism. Ammonia production, an important renal tubular mechanism for regulation of acid-base balance and conservation of cations, is markedly increased in metabolic acidosis and depressed in alkalosis. This ammonia is derived, not from urea, but from intracellular amino acids, particularly glutamine. Ammonia release is catalyzed by renal **glutaminase** (Fig 31–8).

Although ammonia may be excreted as ammonium salts—particularly in metabolic acidosis—the vast majority is excreted as urea, the principal nitrogenous component of urine. Ammonia, constantly produced in the tissues but present only in traces in peripheral blood (10–20 µg/dL), is rapidly removed from the circulation by the liver and converted to glutamate, to glutamine, or to urea.

Glutamine Synthetase Converts Ammonia to Nontoxic Glutamine For Transport Between Tissues

Removal of ammonia via glutamate dehydrogenase was mentioned above. Formation of glutamine is catalyzed by **glutamine synthetase** (Fig 31–9), a mitochondrial enzyme present in highest quantities in renal tissue. Synthesis of the amide bond of glutamine is accomplished at the expense of hydrolysis of one equivalent of ATP to ADP and P_i. The reaction is thus strongly favored in the direction of glutamine synthesis.

Glutaminase & Asparaginase Release Ammonia From Glutamine & Asparagine

Liberation of the amide nitrogen of glutamine as

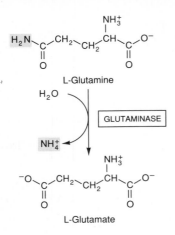

Figure 31–8. The glutaminase reaction proceeds essentially irreversibly in the direction of glutamate and NH_4^+ formation. Note that the amide nitrogen, not the α-amino nitrogen is removed.

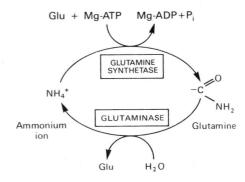

Figure 31–10. Interconversion of ammonia and of glutamine catalyzed by glutamine synthetase and glutaminase. Both reactions are strongly favored in the directions indicated by the arrows. Glutaminase thus serves solely for glutamine deamidation and glutamine synthetase solely for synthesis of glutamine from glutamate. Glu, glutamate.

ammonia occurs by hydrolytic removal of ammonia catalyzed by **glutaminase** (Fig 31–8). The glutaminase reaction, unlike the glutamine synthetase reaction, does not involve adenine nucleotides, strongly favors glutamate formation, and does not function in glutamine synthesis. Glutamine synthetase and glutaminase thus catalyze interconversion of free ammonium ion and glutamine (Fig 31–10) in a manner reminiscent of the interconversion of glucose and glucose 6-phosphate by glucokinase and glucose-6-phosphatase (see Chapter 18).

A reaction analogous to that catalyzed by glutaminase is catalyzed by L-**asparaginase** of animal, plant, and microbial tissue. Asparaginase and glutaminase have both been investigated as antitumor agents, since certain tumors exhibit abnormally high requirements for glutamine and asparagine.

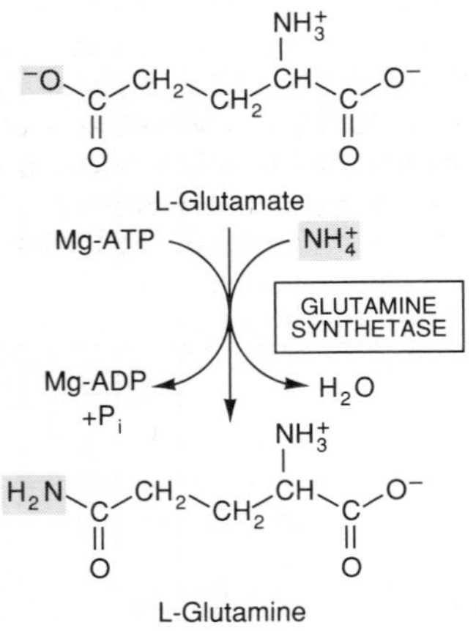

The Liver Converts Ammonia to Urea

Whereas in brain the major mechanism for removal of ammonia is glutamine formation, in the liver the most important pathway is urea formation. Brain tissue can form urea, although this does not play a significant role in ammonia removal. Formation of glutamine in the brain must be preceded by synthesis of glutamate in the brain, because the supply of blood glutamate is inadequate in the presence of high levels of blood ammonia. The immediate precursor of glutamate is α-ketoglutarate. Thus, formation of glutamine from ammonia would rapidly deplete citric acid cycle intermediates unless they could be replaced by CO_2 fixation with conversion of pyruvate to oxaloacetate (see Chapter 18). A significant fixation of CO_2 into amino acids does indeed occur in the brain, presumably by way of the citric acid cycle, and after infusion of ammonia more oxaloacetate is diverted to the synthesis of glutamine (rather than to aspartate) via α-ketoglutarate.

EXCHANGE BETWEEN ORGANS MAINTAINS CIRCULATING LEVELS OF AMINO ACIDS

The maintenance of steady-state concentrations of circulating plasma amino acids between meals depends upon the net balance between release from endogenous protein stores and utilization by various tissues. Muscle generates greater than 50% of the total body pool of free amino acids, while liver is the site of

Figure 31–9. The glutamine synthetase reaction. The reaction strongly favors glutamine synthesis.

the urea cycle enzymes necessary for disposal of nitrogenous waste. Thus, muscle and liver play a major role in determining the circulating levels and the turnover of amino acids.

Muscle: Alanine and glutamine account for over 50% of the total α-amino acid nitrogen **released** from muscle tissue. By contrast, muscle consistently takes up small quantities of serine, cysteine, and glutamate from the circulation.

Liver & Gut: The liver and gut (the splanchnic tissues) consistently **take up** from the plasma large quantities of alanine and glutamine, the predominant amino acids released by muscle. The liver is the primary site of uptake of alanine, and the gut of glutamine. In the gut, the majority of the amino groups of glutamine are released from that tissue as alanine or free ammonia. Serine is also extracted by these splanchnic tissues as well as by muscle.

Kidney: The kidney is the major source of release of serine; in addition, it releases small but significant quantities of alanine. The kidney takes up glutamine, proline, and glycine from the circulation.

There thus is a fairly close correspondence between the output of most amino acids from peripheral muscle and their uptake by the splanchnic tissues.

Brain: The uptake of valine by the brain exceeds that of all other amino acids, and the capacity of the rat brain to oxidize the branched-chain amino acids (leucine, isoleucine, and valine) is at least 4-fold greater than that of muscle and liver. Although in the postabsorptive state significant quantities of these branched-chain amino acids are released from muscle, they are not extracted by liver, and thus it is likely that the brain is the primary site for their utilization.

A Complex Set of Interorgan Exchanges Characterize the Postabsorptive State

Fig 31–11 summarizes the postabsorptive state. Free amino acids, particularly alanine and glutamine, are released from muscle into the circulation. Alanine, which appears to be the vehicle of nitrogen transport in the plasma, is extracted primarily by the liver. Glutamine is extracted by the gut and the kidney, both of which convert a significant portion to alanine. Glutamine also serves as a source of ammonia for excretion by the kidney. The kidney provides a major source of serine for uptake by peripheral tissues, including liver and muscle. Branched-chain amino acids, particularly valine, are released by muscle and taken up predominantly by brain.

Alanine serves as a key protein-derived glucose precursor, ie, a key **gluconeogenic amino acid** (Fig 31–12). In the liver, the rate of glucose synthesis from alanine and serine is far higher than that observed from all other amino acids. The capacity of the liver for gluconeogenesis from alanine is enormous; it does not reach saturation until the alanine concentration is 9 mmol/L, some 20–30 times its physiologic level. The predominance of alanine in the outflow of α-amino acids from muscle reflects its **synthesis in muscle** by transamination of pyruvate.

Interorgan Exchange of Branched Amino Acids Follows Feeding

After a protein-rich meal is ingested, the splanchnic tissues release amino acids, predominantly the branched-chain amino acids (Fig 31–13), while the peripheral muscles extract amino acids, also predominantly the branched-chain amino acids. The branched-chain amino acids, which are also oxidized in muscle in response to feeding, probably serve as the major donors of amino groups for the transamination of pyruvate to alanine.

Thus, the branched-chain amino acids have a special role in nitrogen metabolism, both in the fasting state, when they provide the brain with an energy source, and after feeding, when they are extracted predominantly by muscles, having been spared by the liver. In muscle, they seem to provide an important source of energy as well as of nitrogen.

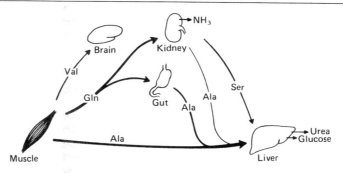

Figure 31–11. Interorgan amino acid exchange in normal postabsorptive humans. The key role of alanine in amino acid output from muscle and gut and uptake by the liver is shown. (Reproduced, with permission, from Felig P: Amino acid metabolism in man. *Annu Rev Biochem* 1975;**44**:937. Copyright © 1975 by Annual Reviews, Inc.)

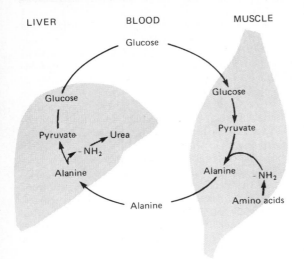

Figure 31-12. The glucose-alanine cycle. Alanine is synthesized in muscle by transamination of glucose-derived pyruvate, released into the bloodstream, and taken up by the liver. In the liver, the carbon skeleton of alanine is reconverted to glucose and released into the bloodstream, where it is available for uptake by muscle and resynthesis of alanine. (Reproduced, with permission, from Felig P: Amino acid metabolism in man. *Annu Rev Biochem* 1975; **44**:938. Copyright © 1975 by Annual Reviews, Inc.)

UREA IS THE MAJOR END PRODUCT OF NITROGEN CATABOLISM IN HUMAN SUBJECTS

A moderately active man consuming about 300 g of carbohydrate, 100 g of fat, and 100 g of protein daily must excrete about 16.5 g of nitrogen daily. Ninety-five percent is eliminated by the kidneys and the remaining 5% in the feces. The **major pathway of nitrogen excretion in humans is as urea** synthesized in the liver, released into the blood, and cleared by the kidney. In humans eating an occidental diet, urea constitutes 80–90% of the nitrogen excreted.

The Reactions of the Urea Cycle Form Urea From Ammonia, Carbon Dioxide, & Aspartate

The reactions and intermediates in biosynthesis of 1 mol of urea from 1 mol each of ammonium ion, of carbon dioxide (activated with Mg^{2+} and ATP), and of the α-amino nitrogen of aspartate are shown in Fig 31–14. The overall process requires 3 mol of ATP (2 of which are converted to ADP + P_i and 1 to AMP + PP_i) and the successive participation of 5 enzymes catalyzing the numbered reactions of Fig 31–14. Of the 6 amino acids involved in urea synthesis, one (N-acetylglutamate) functions as an enzyme activator rather than as an intermediate. The remaining 5—aspartate, arginine, ornithine, citrulline, and argininosuccinate— all function as carriers of atoms which ultimately become urea. Two (aspartate and arginine) occur in proteins, while the remaining 3 (ornithine, citrulline, and argininosuccinate) do not. The major metabolic role of these latter 3 amino acids in mammals is urea synthesis. Note that urea formation is in part a **cyclical process.** The ornithine used in reaction 2 is regenerated in reaction 5. There is thus no net loss or gain of ornithine, citrulline, argininosuccinate, or arginine during urea synthesis; however, ammonium ion, CO_2, ATP, and aspartate are consumed.

Synthesis of Carbamoyl Phosphate From Ammonium Ion & Carbon Dioxide Requires 2 Molecules of ATP

Formation of carbamoyl phosphate is catalyzed by **carbamoyl phosphate synthetase,** an enzyme present in liver **mitochondria** of all ureotelic organisms, including humans. The 2 mol of ATP hydrolyzed during this reaction provide the driving force for synthesis of 2 covalent bonds—the amide bond and the mixed carboxylic acid-phosphoric acid anhydride bond of carbamoyl phosphate. In addition to Mg^{2+}, a dicarboxylic acid, preferably N-acetylglutamate, is required. Its

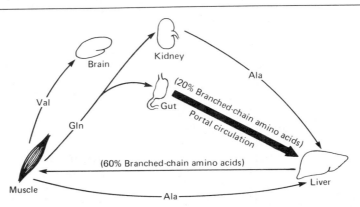

Figure 31-13. Summary of amino acid exchange between organs immediately after feeding.

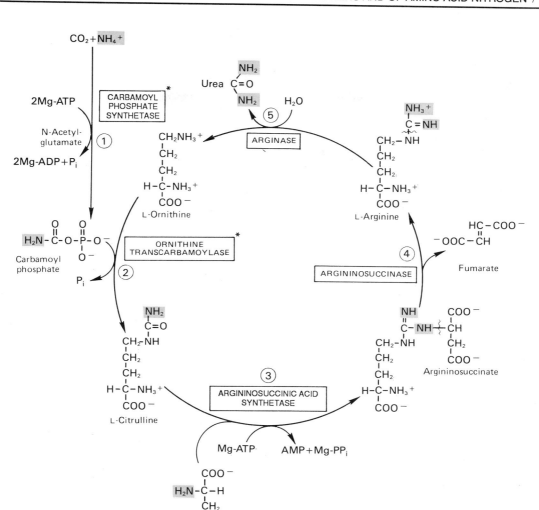

Figure 31–14. Reactions and intermediates of urea biosynthesis. The amines contributing to the formation of urea are shaded. *Mitochondrial enzymes.

presence brings about a profound conformational change in the structure of carbamoyl phosphate synthetase that exposes certain sulfhydryl groups, conceals others, and affects the affinity of the enzyme for ATP.

Carbamoyl phosphate synthetase acts with mitochondrial glutamate dehydrogenase to channel nitrogen from glutamate (and hence from all amino acids; see Fig 31–3) into carbamoyl phosphate and thus into urea. While the equilibrium constant of the glutamate dehydrogenase reaction favors glutamate rather than ammonia formation, removal of ammonia by carbamoyl phosphate synthetase and oxidation of α-ketoglutarate by citric acid cycle enzymes in the mitochondrion serve to favor glutamate catabolism. This effect is enhanced by ATP, which, in addition to being a substrate for carbamoyl phosphate synthesis, stimulates glutamate dehydrogenase activity unidirectionally, favoring ammonia formation.

Condensation of Carbamoyl Phosphate with Ornithine Forms Citrulline

Transfer of a carbamoyl moiety from carbamoyl phosphate to ornithine, forming citrulline + P_i, is catalyzed by L-ornithine transcarbamoylase of liver mitochondria. The reaction is highly specific for ornithine, and the equilibrium strongly favors citrulline synthesis. (Fig 31–14)

Condensation of Citrulline With Aspartate Forms Argininosuccinate

In the argininosuccinate synthetase reaction, aspartate and citrulline are linked together via the amino group of aspartate. The reaction requires ATP, and the equilibrium strongly favors argininosuccinate synthesis. (Fig 31–14)

282 / CHAPTER 31

Cleavage of Argininosuccinate Forms Arginine & Fumarate

Reversible cleavage of argininosuccinate to arginine plus fumarate is catalyzed by **argininosuccinase,** an enzyme of mammalian liver and kidney tissues. The reaction proceeds via a *trans* elimination mechanism. The fumarate formed may be converted to oxaloacetate via the fumarase and malate dehydrogenase reactions and then transaminated to regenerate aspartate.

Cleavage of Arginine Releases Urea & Re-Forms Ornithine

This reaction completes the urea cycle and regenerates ornithine, a substrate for reaction 2. Hydrolytic cleavage of the guanidino group of arginine is catalyzed by **arginase,** present in the livers of all ureotelic organisms. Smaller quantities of arginase also occur in renal tissue, brain, mammary gland, testicular tissue, and skin. Mammalian liver arginase is activated by Co^{2+} or Mn^{2+}. Ornithine and lysine are potent inhibitors competitive with arginine.

KNOWN METABOLIC DISORDERS ARE ASSOCIATED WITH EACH REACTION OF THE UREA CYCLE

The rate-limiting reactions of urea synthesis appear to be those catalyzed by carbamoyl phosphate synthetase I (reaction 1), ornithine transcarbamoylase (reaction 2), and arginase (reaction 5) (Fig 31–14). Since the urea cycle converts toxic ammonia to nontoxic urea, all disorders of urea synthesis cause ammonia intoxication. This intoxication is more severe when the metabolic block occurs at reaction 1 or 2, since some covalent linking of ammonia to carbon has already occurred if citrulline can be synthesized. Clinical symptoms common to all urea cycle disorders include vomiting in infancy, avoidance of high-protein foods, intermittent ataxia, irritability, lethargy, and mental retardation.

The clinical features and the treatment of all 5 of the disorders discussed below are similar. Significant improvement is noted on a low-protein diet, and much of the brain damage may thus be prevented. Food intake should be in frequent small meals to avoid sudden increases in blood ammonia levels.

Hyperammonemia Type I: About 24 cases of **carbamoyl phosphate synthetase** deficiency (reaction 1, Fig 31–14) have been reported. This probably is a familial disorder.

Hyperammonemia Type II: Ornithine transcarbamoylase (reaction 2, Fig 31–14) deficiency produces this condition, which is X chromosome-linked. The mothers also exhibit hyperammonemia and an aversion to high-protein foods. The only consistent clinical finding is an elevation of glutamine in blood, cerebrospinal fluid, and urine. This probably reflects enhanced synthesis of glutamine by glutamine synthetase consequent to elevated tissue levels of ammonia.

Citrullinemia: This rare disorder probably is recessively inherited. Large quantities (1–2 g/d) of citrulline are excreted in the urine, and both plasma and cerebrospinal fluid citrulline levels are markedly elevated. In one patient, complete absence of **argininosuccinate synthetase** activity (reaction 3, Fig 31–14) was noted. In another, the K_m for citrulline was 25 times normal. This suggests a mutation causing a significant but not "lethal" modification of the catalytic site of the synthetase.

Citrulline (and argininosuccinate; see below) may serve as a carrier of waste nitrogen, since it contains nitrogen intended for urea synthesis. Feeding arginine enhances the excretion of citrulline in these patients. Similarly, feeding benzoate diverts ammonium nitrogen to hippurate via glycine (see Fig 32–3).

Argininosuccinic Aciduria: This rare recessive inherited disease, characterized by elevated levels of argininosuccinate in the blood, cerebrospinal fluid, and urine, frequently is associated with the occurrence of friable, tufted hair (trichorrhexis nodosa). While both early- and late-onset types are known, the disease is always manifest by age 2 and usually terminates fatally early in life.

Argininosuccinicaciduria reflects the absence of **argininosuccinase** (reaction 4, Fig 31–14). Cultured skin fibroblasts from normal subjects contain this enzyme whereas those from patients with argininosuccinicacidemia do not. Argininosuccinase is also absent from brain, liver, kidney, and erythrocytes of patients with this disease. While the diagnosis is readily made by 2-dimensional paper chromatography of the urine, additional abnormal spots appear in urine on standing owing to the tendency of argininosuccinate to form cyclic anhydrides. Confirmatory diagnosis is by measurement of erythrocyte levels of argininosuccinase. This test can be performed on umbilical cord blood for early detection. Since argininosuccinase is present in amniotic fluid cells, diagnosis by amniocentesis is also possible. For reasons discussed relevant to citrullinemia, feeding arginine and benzoate promotes nitrogen waste excretion in these patients also.

Hyperargininemia: This defect in urea synthesis is characterized by elevated blood and cerebrospinal fluid arginine levels, low erythrocyte levels of **arginase** (reaction 5, Fig 31–14), and a urinary amino acid pattern resembling that of lysine-cystinuria. Possibly this pattern reflects competition by arginine with lysine and cystine for reabsorption in the renal tubule. In patients, a low-protein diet lowers plasma ammonia levels and causes disappearance of the urinary lysine-cystinuria pattern.

REFERENCES

Adams E, Frank L: Metabolism of proline and the hydroxyprolines. *Annu Rev Biochem* 1980;**49**:1005.

Batshaw ML et al: Treatment of inborn errors of urea synthesis: Activation of alternative pathways of waste nitrogen synthesis and excretion. *N Engl J Med* 1982;**306**: 1387.

Felig P: Amino acid metabolism in man. *Annu Rev Biochem* 1975;**44:**933.

Msall M et al: Neurologic outcome in children with inborn errors of urea synthesis: Outcome of urea-cycle enzymopathies. *N Engl J Med* 1984;**310:**1500.

Rosenberg LE, Scriver CR: Disorders of amino acid metabolism. Chapter 11 in: *Metabolic Control and Disease.* Bondy PK, Rosenberg LE (editors). Saunders, 1980.

Stanbury JB et al: *The Metabolic Basis of Inherited Disease,* 5th ed. McGraw-Hill, 1983.

Torchinsky YM: Transamination: Its discovery, biological and clinical aspects (1937–1987). *Trends Biochem Sci* 1987;**12:**115.

Wellner D, Meister A: A survey of inborn errors of amino acid metabolism and transport in man. *Annu Rev Biochem* 1981;**50:**911.

32

Catabolism of the Carbon Skeletons of Amino Acids

Victor W. Rodwell, PhD

INTRODUCTION

This section deals with conversions of the carbon skeletons of common L-amino acids to amphibolic intermediates and with the metabolic diseases or "inborn errors of metabolism" associated with these catabolic pathways.

BIOMEDICAL IMPORTANCE

Historically, certain disorders of amino acid metabolism in humans played key roles in elucidating the pathways by which amino acids are metabolized in normal human subjects. Most of these diseases are rare and so are unlikely to be encountered by most practicing physicians. These disorders nevertheless present a formidable challenge to the psychiatrist, pediatrician, genetic counselor, or biochemist. They are detected most frequently at infancy, often are fatal at an early age, and often result in irreversible brain damage if untreated. Early detection and rapid initiation of appropriate treatment, if available, is essential. Since several of the enzymes concerned are detectable in cultures of amniotic fluid cells, prenatal diagnosis of these disorders by amniocentesis is possible. Current treatment consists primarily of feeding diets low in the amino acids whose catabolism is impaired. Recombinant DNA technology may eventually provide a means of correcting genetic defects by "gene therapy."

These metabolic disorders reflect genetic mutations that cause production of proteins with modified primary structures. While some changes in the primary structures of enzymes may have little or no effect, others may profoundly modify the 3-dimensional structure of catalytic or regulatory sites. The modified or mutant enzyme may possess altered catalytic efficiency (low V_{max} or high K_m) or altered ability to bind an allosteric regulator of its catalytic activity. A variety of mutations may cause the same clinical disease. For example, any mutation that significantly lowers the catalytic activity of argininosuccinase (see Fig 31–14) will cause the metabolic disorder known as argininosuccinicacidemia. It is extremely unlikely, however, that all cases of argininosuccinicacidemia represent the same alteration in primary structure of argininosuccinase. At a molecular level these are

therefore distinct molecular diseases. Some known disorders of amino acid metabolism are discussed in this chapter. For further examples, the reader should consult major reference works that specialize in this subject, eg, Scriver et al. *The Metabolic Basis of Inherited Disease,* 6th ed, 1989.

AMINO ACIDS ARE CATABOLIZED TO SUBSTRATES FOR CARBOHYDRATE & LIPID BIOSYNTHESIS

Nutritional studies in the period 1920–1940, reinforced by studies using isotopically labeled amino acids conducted from 1940 to 1950, substantiated the interconvertibility of fat, carbohydrate, and protein carbons and established that each amino acid is convertible either to carbohydrate (13 amino acids), fat (one amino acid), or both (5 amino acids) (Table 32–1). Fig 32–1 outlines how these interconversions occur.

FOR MANY AMINO ACIDS, THE INITIAL REACTION IS REMOVAL OF THE α-AMINO NITROGEN

Removal of the α nitrogen usually (but not always, eg, proline, hydroxyproline, lysine) involves transamination. Once removed, the nitrogen may be re-

Table 32–1. Fates of the carbon skeletons of the common L-α-amino acids.

Converted to amphibolic intermediates forming:			
Glycogen ("Glycogenic")		Fat ("Ketogenic")	Glycogen and Fat ("Glycogenic" and "Ketogenic")
Ala	Hyp	Leu	Ile
Arg	Met		Lys
Asp	Pro		Phe
Cys	Ser		Trp
Glu	Thr		Tyr
Gly	Val		
His			

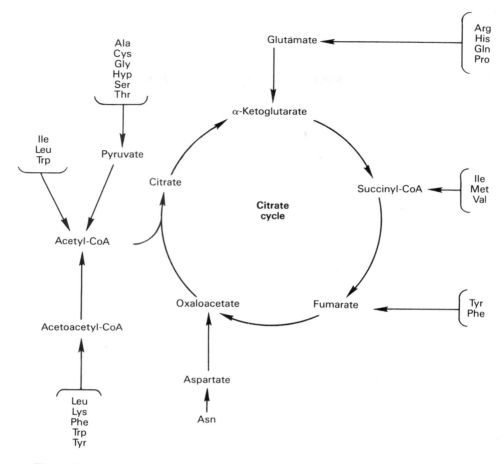

Figure 32–1. Amphibolic intermediates formed from the carbon skeleton of amino acids.

utilized for anabolic processes (eg, protein synthesis) or converted to urea and excreted. The nitrogen-free carbon skeleton that remains is, in most instances, an oxidized hydrocarbon, which is degraded to amphibolic intermediates by reactions similar to those by which other oxidized hydrocarbons are catabolized.

Deamidation of Asparagine Yields Aspartate, Which Transaminates, Forming Oxaloacetate

All 4 carbons of asparagine and of aspartate are converted to oxaloacetate via asparaginase and a transaminase (Fig 32–2, top).

No known metabolic defect is associated with this short catabolic pathway. This possibly is because a defect in the transaminase may have consequences incompatible with life, since transaminases fulfill central anabolic as well as catabolic functions.

Deamidation of Glutamine Yields Glutamate, Which Transaminates, Forming α-Ketoglutarate

Catabolism of glutamine and of glutamate proceeds like that of asparagine and aspartate but with formation of α-ketoglutarate (Fig 32–2, bottom). While both glutamate and aspartate are substrates for the same transaminase, deamidation of asparagine and glutamine is catalyzed by distinct enzymes, glutaminase and asparaginase.

Possibly for the reasons given above for asparagine and aspartate, there are no known metabolic defects of the glutamine-glutamate catabolic pathway.

All Five Carbons of Proline Form α-Ketoglutarate

Proline is oxidized to a dehydroproline which, on addition of water, forms glutamate-γ-semialdehyde. This is then oxidized to glutamate and transaminated to α-ketoglutarate (Fig 32–3, left).

Two genetically distinct hyperprolinemias have been described, both apparently autosomal recessive traits. Despite mental retardation in half of the known cases, neither type is life threatening.

Hyperprolinemia Type I: The site of the metabolic block in hyperprolinemia type I is proline dehydrogenase (Fig 32–3). In contrast to hyperprolinemia type II, there is no associated impairment of **hydroxy**proline catabolism. Type I heterozygotes exhibit only a mild

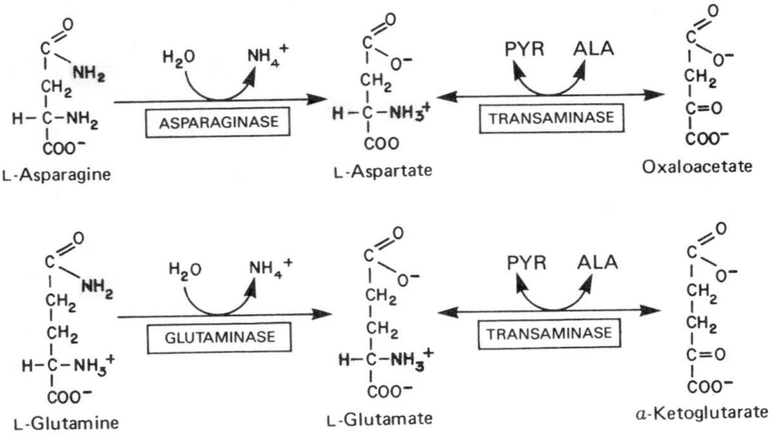

Figure 32–2. Catabolism of L-asparagine (*top*) and of L-glutamine (*bottom*) to amphibolic intermediates. PYR, pyruvate; ALA, L-alanine. In this and subsequent figures, shading on functional groups highlights portions of the molecules undergoing chemical change.

hyperprolinemia. An animal model, the Pro/Re mouse, has only 10% of normal hepatic proline dehydrogenase activity.

Hyperprolinemia Type II: The site of the metabolic block is the dehydrogenase that catalyzes oxidation of glutamate-γ-semialdehyde to glutamate (Fig 32–3). The same dehydrogenase functions in hydroxyproline catabolism (see below). Therefore, both proline and hydroxyproline catabolism are affected, and the urine contains Δ¹-pyrroline-3-hydroxy-5-carboxylate, a hydroxyproline catabolite (see Fig 32–12). Unlike type I heterozygotes, type II heterozygotes exhibit no hyperprolinemia.

Arginase Converts Arginine to Ornithine, Which Is Then ε-Transaminated, Forming α-Ketoglutarate

While arginine and histidine also form α-ketoglutarate, one carbon and either 2 (histidine) or 3 (arginine) nitrogens must first be removed from these 6-carbon amino acids. With arginine, this requires but a single step: hydrolytic removal of the guanidino group catalyzed by arginase. The product, ornithine, then undergoes transamination of the δ-amino group, forming glutamate-γ-semialdehyde, which is converted to α-ketoglutarate as described above for proline (Fig 32–3).

Hyperargininemia, a metabolic disorder caused by a deficiency of liver arginase, is discussed in Chapter 31 in conjunction with metabolic disorders of the urea cycle.

Catabolism of Histidine Ultimately Yields α-Ketoglutarate

For histidine, removal of the extra carbon and nitrogens requires 4 reactions (Fig 32–4). Deamination of histidine produces urocanate. Conversion of urocanate

to 4-imidazolone-5-propionate, catalyzed by urocanase, involves both addition of H_2O and an internal oxidation-reduction. Although 4-imidazolone-5-propionate may undergo additional fates, conversion to α-ketoglutarate involves hydrolysis to N-formiminoglutamate followed by transfer of the formimino group on the α carbon to tetrahydrofolate, forming N^5-formiminotetrahydrofolate. In patients with folic acid deficiency, this last reaction is partially or totally blocked and N-formiminoglutamate (Figlu) is excreted in the urine. This provides a test for folic acid deficiency in which N-formiminoglutamate is detected in the urine following a large dose of histidine.

Histidinemia is inherited as an autosomal recessive trait. Over half of the affected individuals are mentally retarded and exhibit a characteristic speech defect. In addition to increased levels of histidine in blood and urine, there is increased excretion of imidazolepyruvate (which in a color test with ferric chloride may be mistaken for phenylpyruvate, so that a mistaken diagnosis of phenylketonuria could be made). The metabolic defect in histidinemia is inadequate activity of liver histidase, which impairs conversion of histidine to urocanate (Fig 32–4). Transamination to imidazolepyruvate, is then favored. The excess imidazolepyruvate, plus imidazoleacetate and imidazolelactate, the reduction products of imidazolepyruvate, are excreted in the urine.

A conspicuous increase in histidine excretion typifies normal pregnancy. This reflects changes in renal function.

SIX AMINO ACIDS FORM PYRUVATE

All of the carbons of glycine, alanine, cysteine, and serine—but only 2 of the carbons of threonine—form

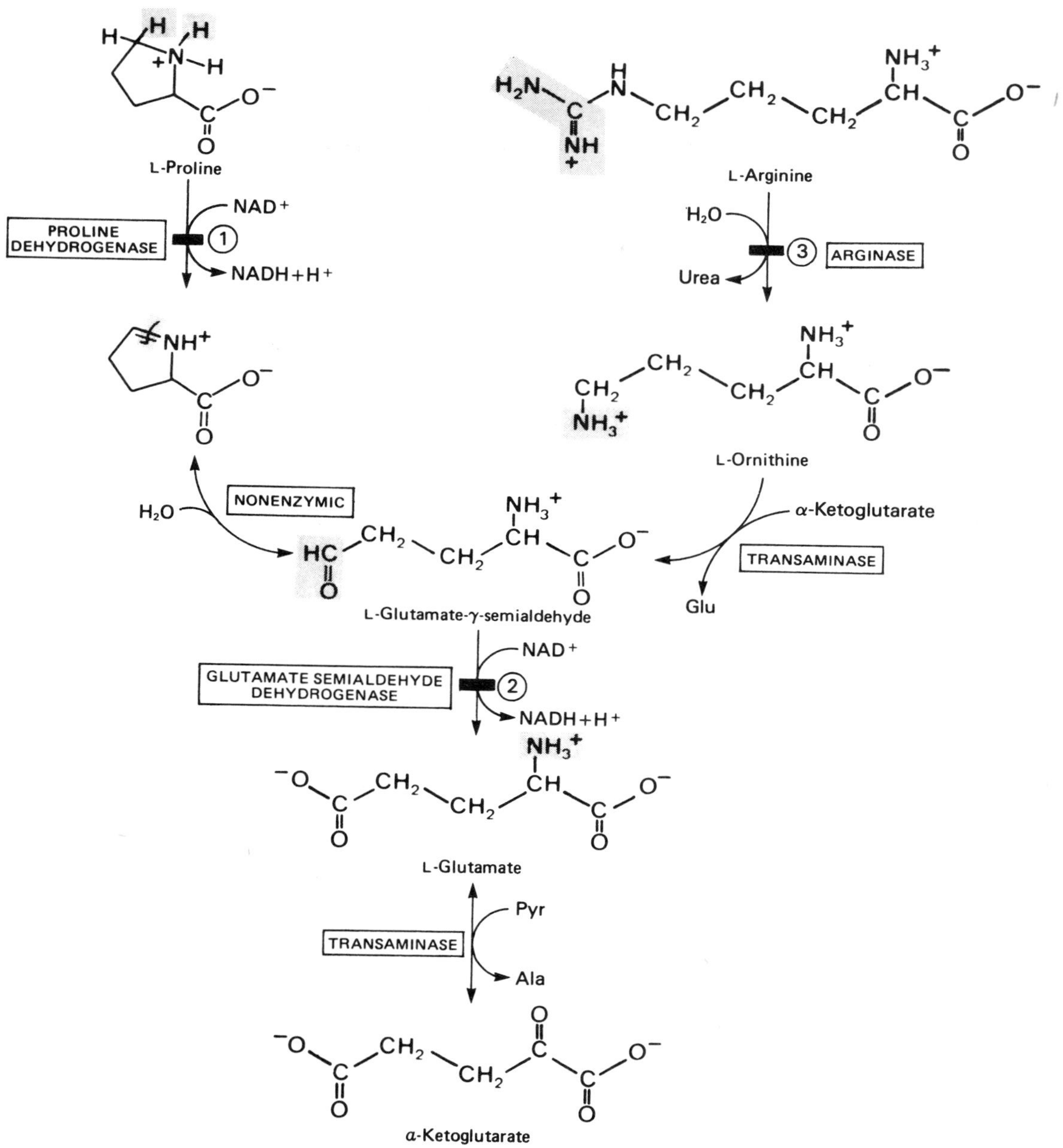

Figure 32–3. Catabolism of L-proline (*left*) and of L-arginine (*right*) to α-ketoglutarate. Circled numerals mark the sites of the metabolic defects in 1, type I hyperprolinemia; 2, type II hyperprolinemia; and 3, hyperargininemia (see Chapter 31).

pyruvate. Pyruvate may then be converted to acetyl-CoA.

L-Threonine
↓
Glycine
↓
L-Serine   L-Cystine
↓   ↓
L-Alanine → Pyruvate ← L-Cysteine
↓
Acetyl-CoA

Catabolism of Glycine Proceeds Via the Glycine-Cleavage System

While glycine can form pyruvate via initial conversion to serine (Fig 32–5), the major pathway for glycine catabolism in vertebrates involves conversion to CO_2, NH_4^+, and N^5,N^{10}-methylenetetrahydrofolate catalyzed by the glycine synthase complex. This reversible reaction (Fig 32–6) in some respects resembles conversion of pyruvate to acetyl-CoA by the pyruvate dehydrogenase complex. Both complexes comprise macromolecular aggregates in liver mitochon-

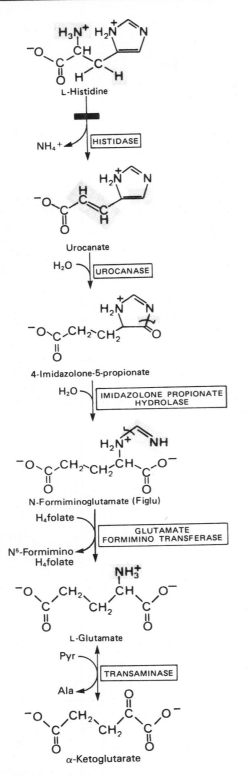

Figure 32–4. Catabolism of L-histidine to α-ketoglutarate. H_4 folate, tetrahydrofolate. The reaction catalyzed by histidase represents the site of the probable metabolic defect in histidinemia.

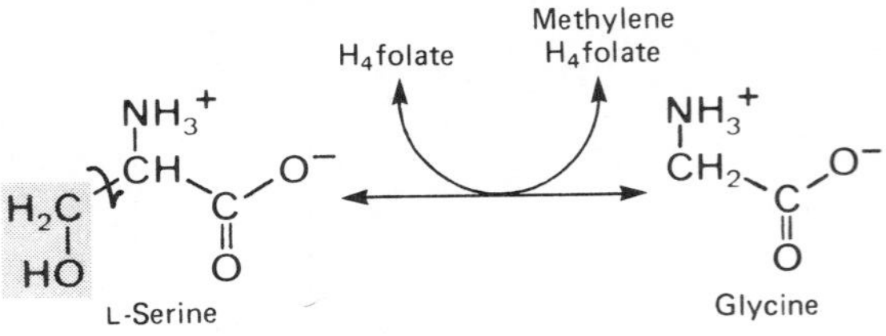

Figure 32–5. The freely reversible serine hydroxymethyltransferase reaction. H_4 folate, tetrahydrofolate.

dria. The reactions of the glycine cleavage system probably constitute the major route not only for glycine but also for serine catabolism in humans and many other vertebrates.

Glycinuria: Glycinuria, a rare disorder, is characterized by excess urinary excretion of glycine and a tendency to form oxalate renal stones. Glycinuria appears to be a dominant, possibly X-linked, trait. Since plasma glycine levels are normal while urinary excretion of glycine ranges from 600 to 1000 mg/d, glycinuria is attributed to a defect in renal tubular reabsorption of glycine.

Primary Hyperoxaluria: Primary hyperoxaluria is characterized by continuous high urinary excretion of oxalate unrelated to dietary intake of oxalate. Progressive bilateral calcium oxalate urolithiasis, nephrocalcinosis, and recurrent infection of the urinary tract are followed by death in childhood or early adult life from renal failure or hypertension. The excess oxalate apparently arises from glycine, which may be deaminated to form glyoxylate, a precursor of oxalate. The metabolic defect is a disorder of glyoxylate metabolism associated with failure to catabolize glyoxylate. The excessive glyoxylate is therefore oxidized to oxalate.

Transamination of Alanine Forms Pyruvate

Pyruvate, formed by transamination of L-alanine (Fig 32–7), may then be decarboxylated to acetyl-CoA by the pyruvate dehydrogenase complex.

Possibly for the reasons advanced under glutamate and aspartate catabolism, there is no known metabolic defect of α-alanine catabolism.

Serine Is Catabolized Via Conversion to Glycine

Conversion of serine to pyruvate by serine dehydratase, a pyridoxal phosphate protein, involves elimination of water and hydrolytic loss of ammonia from an amino acid intermediate (Fig 32–7). While rat and guinea pig liver convert serine to pyruvate by serine dehydratase, humans and many other vertebrates degrade serine primarily to glycine and N^5,N^{10}-methylenetetrahydrofolate. The initial reaction is catalyzed by serine hydroxymethyltransferase (Fig 32–5). Further catabolism of serine then merges with glycine catabolism (Fig 32–6).

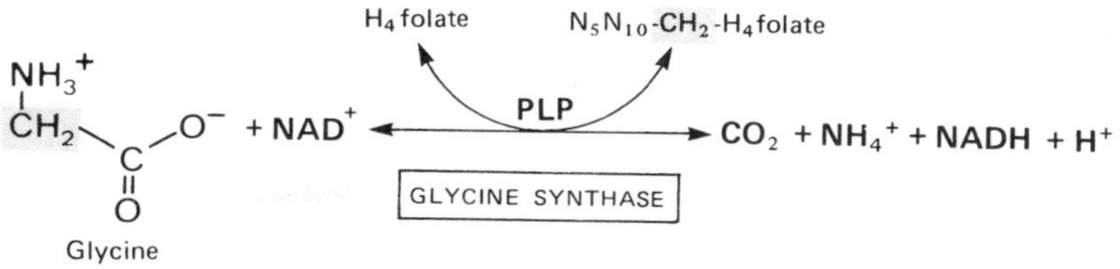

H_4 folate $N_5 N_{10}$-CH_2-H_4 folate

PLP

GLYCINE SYNTHASE

Glycine + NAD^+ ⟶ CO_2 + NH_4^+ + NADH + H^+

Figure 32–6. Reversible cleavage of glycine by the mitochondrial glycine synthase complex. PLP, pyridoxal phosphate.

Cystine Reductase Reduces Cystine to Cysteine

Like carbon and nitrogen, sulfur is continuously recycled through the biosphere through the combined metabolic activities of prokaryotic and eukaryotic organisms. Mammals participate in this cycle by catabolism of organic sulfur compounds to inorganic sulfur compounds. For example, human subjects excrete 20–30 mmol of sulfur per day, mostly as inorganic sulfate.

The major catabolic fate of cystine in mammals is conversion to cysteine, catalyzed by cystine reductase (Fig 32–8). Catabolism of cystine then merges with that of cysteine.

Conversion of Cysteine to Pyruvate Can Involve an Initial Oxidative Reaction

Cysteine is catabolized via 2 catabolic pathways: the direct oxidative (cysteine sulfinate) pathway and the transamination (3-mercaptopyruvate) pathway. Conversion of cystine to cysteine sulfinate (Fig 32–9) is catalyzed by cysteine dioxygenase, an enzyme that requires Fe^{2+} and NAD(P)H. Further catabolism of cysteine sulfinate probably involves its transamination to β-sulfinylpyruvate, although β-sulfinylpyruvate has yet to be isolated as a catabolite. Conversion of β-sulfinylpyruvate to pyruvate and sulfite may not be enzyme catalyzed. Desulfination is extremely rapid even in the absence of enzymatic catalysis.

Initial Transamination of Cysteine Forms 3-Mercaptopyruvate

Reversible transamination of cysteine to 3-mercaptopyruvate (thiolpyruvate) is catalyzed by specific cysteine transaminases or by glutamate or asparagine transaminases of mammalian liver and kidney (Fig 32–9). 3-Mercaptopyruvate may then be reduced by L-lactate dehydrogenase. The product, 3-mercaptolactate, a normal constituent of human urine in the form of its mixed disulfide with cysteine, is excreted in increased amounts in the urine of patients with **mercaptolactate-cysteine disulfiduria.** Alternatively, 3-mercaptopyruvate undergoes desulfuration, forming pyruvate and H_2S (Fig 32–9).

Table 32–2 summarizes known defects of sulfur-containing amino acid catabolism.

Cystinuria (Cystine-Lysinuria): In this inherited metabolic disease, urinary excretion of cystine is 20–30 times normal. Excretion of lysine, arginine, and ornithine is also markedly increased, suggesting a defect in the renal reabsorptive mechanisms for these 4 amino acids. "Cystinuria" is therefore a misnomer. Cystine-lysinuria may now be the preferred descriptive term for this disease.

Because cystine is relatively insoluble, in cystinuric patients it may precipitate in the kidney tubules and form cystine calculi. Were it not for this possibility, cystinuria would be an entirely benign anomaly.

The mixed disulfide of L-cysteine and L-homocysteine (Fig 32–10), present in the urine of cystinuric patients, is somewhat more soluble than cystine. To the extent that it may be formed at the expense of cystine, it reduces the tendency to formation of cystine crystals and calculi.

Cystinosis (Cystine Storage Disease): In cystinosis, which is also inherited, cystine crystals are deposited in many tissues and organs (particularly the reticuloendothelial system). Cystinosis is usually accompanied by a generalized aminoaciduria. Other renal functions are also seriously impaired, and affected patients usually die at an early age with all of the manifestations of acute renal failure.

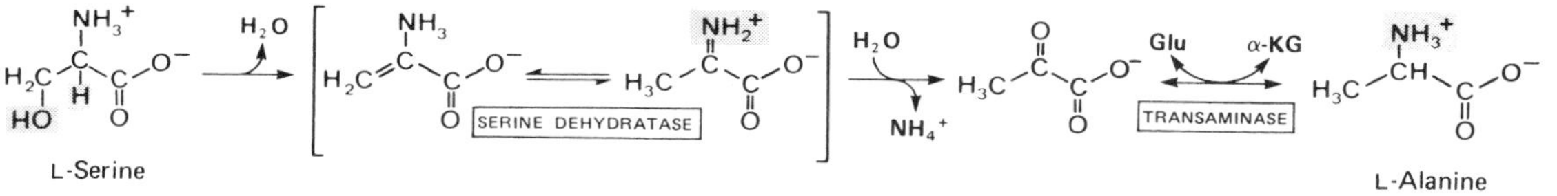

L-Serine   SERINE DEHYDRATASE   TRANSAMINASE   L-Alanine

Figure 32–7. Conversion of alanine and serine to pyruvate. Both the alanine transaminase and serine dehydratase reactions require pyridoxal phosphate. The serine dehydratase reaction proceeds via elimination of H_2O from serine, forming an unsaturated amino acid. This rearranges to an α-imino acid that is spontaneously hydrolyzed to pyruvate plus ammonia. Glu, glutamate; α-KG, α-ketoglutarate.

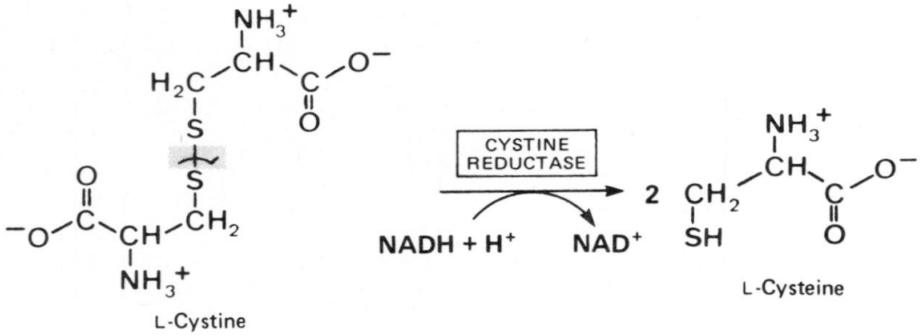

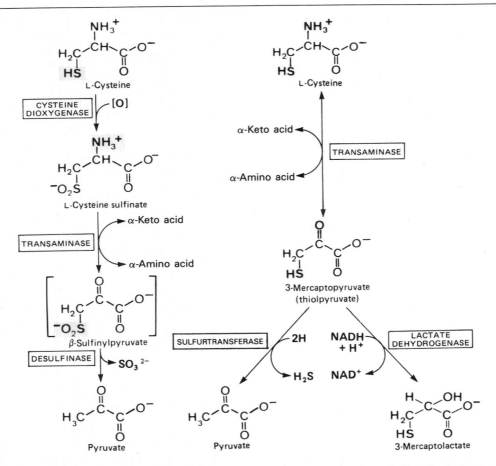

Figure 32–8. The cystine reductase reaction.

Figure 32–9. Catabolism of L-cysteine via the direct oxidative (cysteine sulfinate) pathway (*left*) and by the transamination (3-mercaptopyruvate) pathway (*right*). β-Sulfinylpyruvate is a putative intermediate. Oxidation of the sulfite produced in the last reaction of the direct oxidative pathway is catalyzed by sulfite oxidase, α-KA, α-keto acid, α-AA, α-amino acid.

Table 32–2. Inborn errors of sulfur-containing amino acid metabolism.

Name	Defect	Reference
Homocystinuria I	Cystathionine β-synthase	Fig 30–9, reaction 1
Homocystinuria II	N^5, N^{10}-methylenetetrahydrofolate reductase	
Homocystinuria III	Low N^5-methyltetrahydrofolate-homocysteine transmethylase owing to inability to synthesize methylcobalamin	
Homocystinuria IV	Low N^5-methyltetrahydrofolate-homocysteine transmethylase owing to defective intestinal absorption of cobalamin	
Hypermethioninemia	Liver methionine adenosyltransferase*	Fig 32–22
Cystathioninuria	Cystathionase	Fig 30–9, reaction 2
Sulfituria (sulfocysteinuria)	Sulfite oxidase	Fig 32–9, legend
Cystinosis	Defect in lysosomal function	
3-Mercaptopyruvate-cysteine disulfiduria	3-Mercaptopyruvate sulfurtransferase	Fig 32–9
Methionine malabsorption syndrome	Inability to absorb methionine from gut	

*May also occur in cystathioninuria, tyrosinemia, and fructose intolerance.

Homocystinurias: The incidence of these heritable defects of methionine catabolism is about one per 160,000 births. Homocystine (up to 300 mg/d), together with S-adenosylmethionine in some cases, is excreted in the urine, and plasma methionine levels are elevated.

At least 4 metabolic defects cause homocystinuria (Table 32–2). In homocystinuria type I, clinical findings include thromboses, osteoporosis, dislocated lenses in the eyes, and frequently mental retardation. Two forms of this disease are known: a vitamin B_6-responsive form and a vitamin B_6-unresponsive form. Feeding a diet low in methionine and high in cystine prevents pathologic changes if initiated early in life. Other types of homocystinuria reflect defects in the remethylation cycle (Table 32–2).

Threonine Aldolase Initiates Threonine Catabolism

Threonine is cleaved to acetaldehyde and glycine by **threonine aldolase.** Acetaldehyde then forms acetyl-CoA (Fig 32–11). Catabolism of glycine is discussed above.

Hydroxyproline Catabolism Resembles That of Proline

4-Hydroxy-L-proline is converted to pyruvate and

glyoxylate (Fig 32–12). A mitochondrial dehydrogenase catalyzes conversion of hydroxyproline to L-Δ^1-pyrroline-3-hydroxy-5-carboxylate. This is in nonenzymatic equilibrium with γ-hydroxy-L-glutamate-γ-semialdehyde, formed by addition of water. The semialdehyde is oxidized to the corresponding carboxylic acid, erythro-γ-hydroxy-L-glutamate, and transaminated to α-keto-γ-hydroxyglutarate. An aldol type cleavage then forms glyoxylate plus pyruvate.

Hyperhydroxyprolinemia is characterized by high plasma levels of 4-hydroxyproline. The site of the metabolic defect in this autosomal recessive trait is 4-hydroxyproline dehydrogenase (Fig 32–12). In contrast to type II hyperprolinemia, there is no accompanying impairment of proline catabolism, since the affected enzyme functions solely in hydroxyproline catabolism. The condition has no effect on collagen metabolism and, like the hyperprolinemias, appears to be harmless.

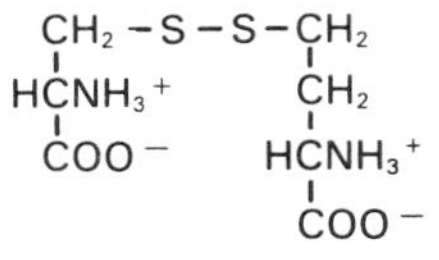

TWELVE AMINO ACIDS FORM ACETYL-CoA

All amino acids that form pyruvate (alanine, cysteine, cystine, glycine, hydroxyproline, serine, and threonine) form acetyl-CoA (recall pyrvate dehydrogenase). In addition, 5 amino acids form acetyl-CoA without first forming pyruvate. These include phenylalanine, tyrosine, tryptophan, lysine, and leucine.

Five Sequential Reactions Convert Tyrosine to Fumarate & Acetoacetate

Several intermediates of tyrosine metabolism (Fig 32–13) were discovered during studies of the human genetic disease alkaptonuria. Patients with alkaptonuria excrete homogentisate in the urine, and much useful information was obtained by feeding suspected precursors of homogentisate to these patients.

Figure 32–10. Mixed disulfide of cysteine and homocysteine.

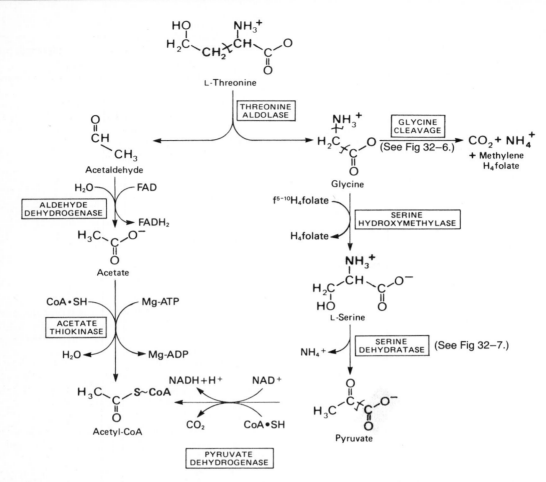

Figure 32–11. Conversion of threonine and glycine to serine, pyruvate, and acetyl-CoA. $f^{5-10}\cdot H_4$folate, formyl [5–10] tetrahydrofolic acid.

Transamination of Tyrosine Forms *p*-Hydroxyphenylpyruvate: Transamination of tyrosine to *p*-hydroxyphenylpyruvate is catalyzed by **tyrosine-α-ketoglutarate transaminase,** an inducible enzyme of mammalian liver.

Hydroxyphenylpyruvate Oxidase, a Copper Metalloprotein, Oxidizes *p*-Hydroxyphenylpyruvate to Homogentisate: Although the reaction (Fig 32–13) appears to involve hydroxylation of *p*-hydroxyphenylpyruvate in the ortho position accompanied by oxidative loss of the carboxyl carbon, it actually involves migration of the side chain. Ring hydroxylation and side-chain migration occur in a concerted manner. Since the physiologic reductant is ascorbate, scorbutic patients excrete incompletely oxidized products of tyrosine metabolism.

Homogentisate Oxidase, an Iron Metalloprotein, Opens the Aromatic Ring: The benzene ring of homogentisate is ruptured, forming maleylacetoacetate in an oxidative reaction catalyzed by **homogentisate oxidase** of mammalian liver.

Isomerization Followed by Hydrolysis Forms Fumarate Plus Acetoacetate: Conversion of mal-

eylacetoacetate to fumarylacetoacetate, a *cis* to *trans* isomerization, is catalyzed by **maleylacetoacetate *cis*, *trans* isomerase.** Subsequent hydrolysis of fumarylacetoacetate by **fumarylacetoacetate hydrolase** forms fumarate and acetoacetate. Acetoacetate can then be converted to acetyl-CoA plus acetate by β-ketothiolase (see Chapter 23).

Several Metabolic Disorders Are Characterized by Tyrosinemia, Tyrosinuria, & Phenolaciduria

Tyrosinemia Type I (Tyrosinosis): In tyrosinosis, accumulation of metabolites adversely affects the activities of several enzymes and transport systems. The pathophysiology is therefore complex. The proposed metabolic defect is in fumarylacetoacetate hydrolase (Fig 32–13) and possibly also in maleylacetoacetate hydrolase. Both acute and chronic forms of tyrosinosis are known. In acute tyrosinosis, infants exhibit diarrhea, vomiting, a "cabbagelike" odor, and failure to thrive. Death from liver failure in untreated acute tyrosinosis ensues within 6–8 months. In chron-

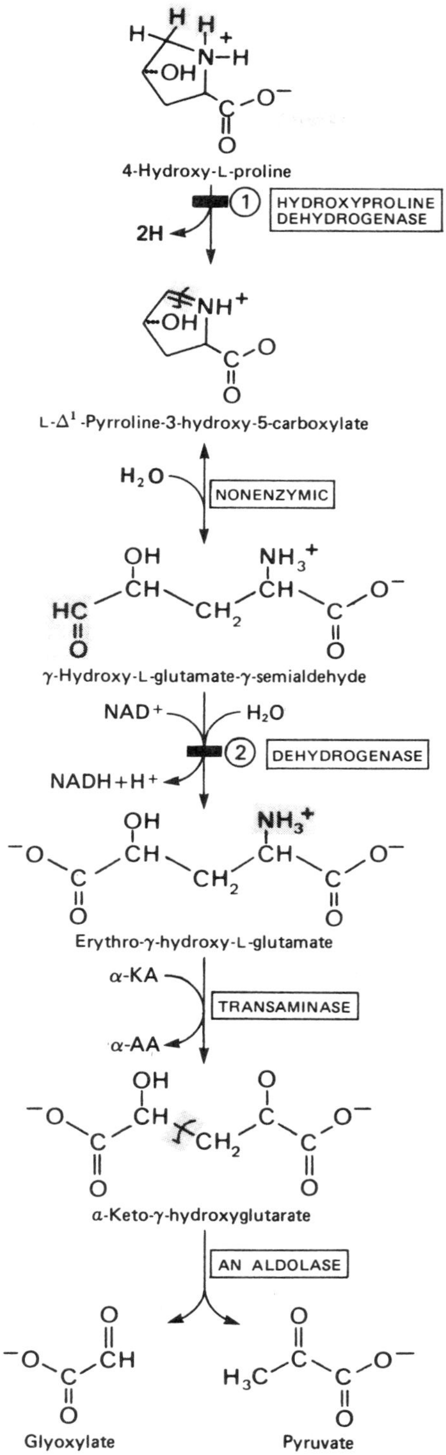

Figure 32–12. Intermediates in L-hydroxyproline catabolism. α-KA, α-keto acid; α-AA, α-amino acid. The circled numerals represent sites of metabolic defects in 1, hyperhydroxyprolinemia; and 2, type II hyperprolinemia.

ic tyrosinemia, similar but milder symptoms lead to death by the age of 10 years. Plasma tyrosine levels are elevated (6–12 mg/dL), as are those of additional amino acids, notably methionine. Treatment involves a diet low in tyrosine and phenylalanine and, on occasion, low in methionine also.

Tyrosinemia Type II (Richner-Hanhart Syndrome): The probable site of the metabolic defect in tyrosinemia type II is hepatic tyrosine transaminase (Fig 32–13). Clinical findings include elevated plasma tyrosine levels (4–5 mg/dL), eye and skin lesions, and moderate mental retardation. Tyrosine is the only amino acid whose urinary concentration is elevated. However, renal clearance and reabsorption of tyrosine fall within normal limits. Urinary metabolites include p-hydroxyphenylpyruvate, p-hydroxyphenyllactate, p-hydroxyphenylacetate, N-acetyltyrosine, and tyramine (Fig 32–14).

Neonatal Tyrosinemia: The disorder is thought to result from a relative deficiency of p-hydroxyphenylpyruvate hydroxylase (Fig 32–13). Blood levels of tyrosine and phenylalanine are elevated, as are urinary levels of tyrosine, p-hydroxyphenylacetate, N-acetyltyrosine, and tyramine. Therapy involves feeding a diet low in protein.

Alkaptonuria: This inherited metabolic disorder, noted as early as the 16th century, was characterized in 1859. The disease is of considerable historic interest, because it formed the basis for Garrod's ideas concerning heritable metabolic disorders. Its most striking clinical manifestation is the occurrence of dark urine on standing in air. Late in the disease, there occur generalized pigmentation of connective tissues (ochronosis) and a form of arthritis. The metabolic defect is lack of **homogentisate oxidase** (Fig 32–13). The substrate, homogentisate, is excreted in the urine, where it is oxidized in air to a brownish-black pigment. Over 600 cases have been reported; the estimated incidence of alkaptonuria is 2–5 per million live births. Alkaptonuria is inherited as an autosomal recessive trait. At present, no diagnostic procedure for the detection of heterozygotes is available. The mechanism of the ochronosis involves oxidation of homogentisate by polyphenol oxidase, forming benzoquinoneacetate, which polymerizes and binds to connective tissue macromolecules.

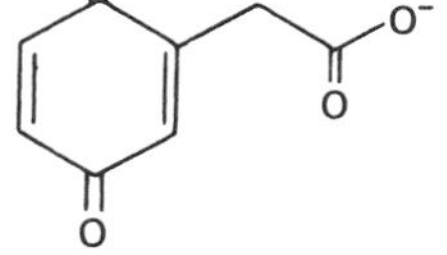

Benzoquinoneacetate

Hydroxylation of Phenylalanine Forms Tyrosine

Phenylalanine is first converted to tyrosine by phenylalanine hydroxylase (see Fig 30–12). The labeling pattern in the amphibolic products fumarate and

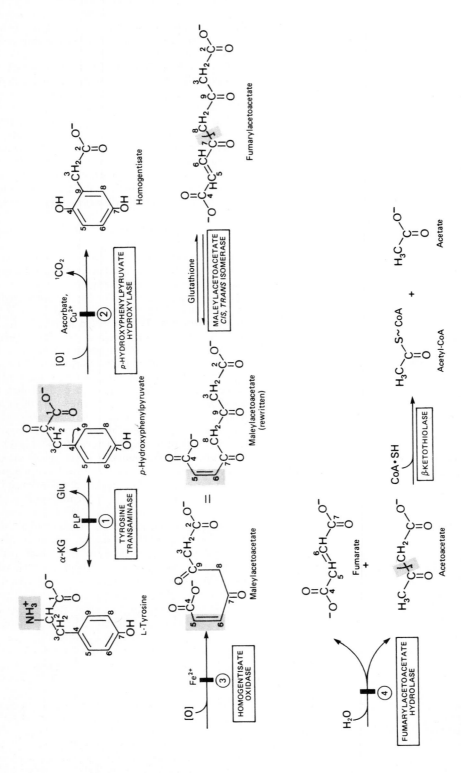

Figure 32–13. Intermediates in tyrosine catabolism. With the exception of β-ketothiolase, reactions are discussed in the text. Carbon atoms of intermediates are numbered to assist readers in determining the ultimate fate of each carbon (see also Fig 32–15). α-KG, α-ketoglutarate; Glu, glutamate; PLP, pyridoxal phosphate. The circled numerals represent the probable sites of the metabolic defects in 1, type II tyrosinemia; 2, neonatal tyrosinemia; 3, alkaptonuria; and 4, type I tyrosinemia, or tyrosinosis.

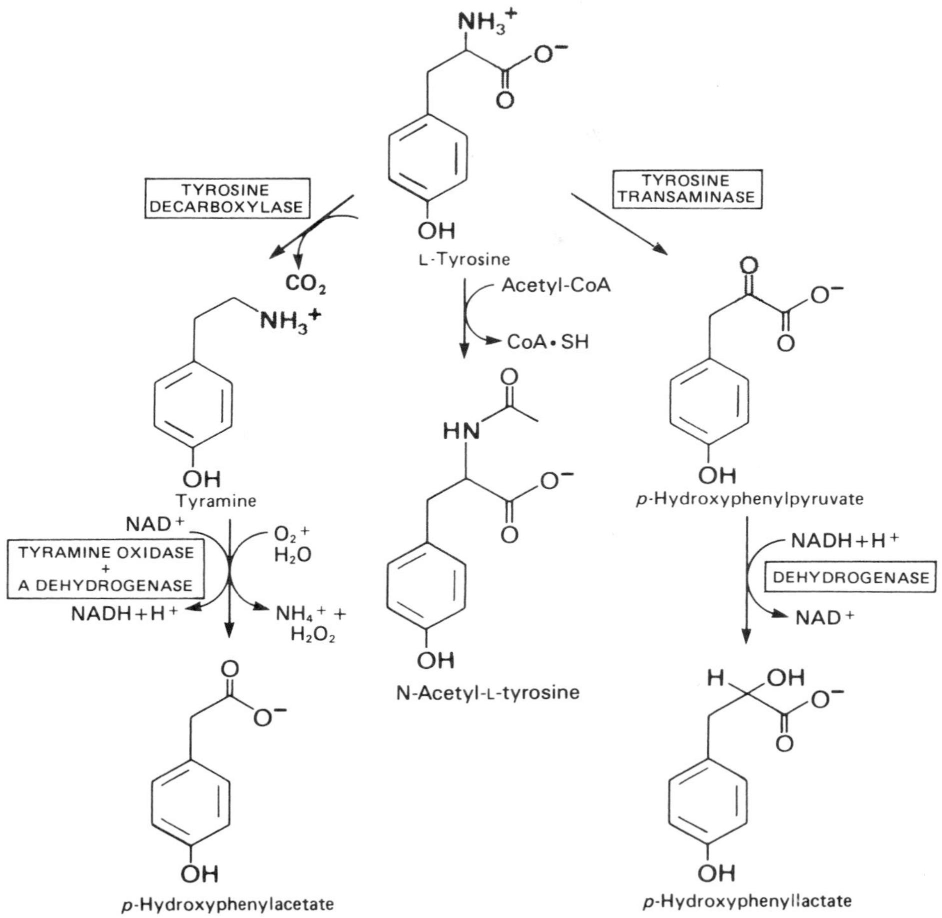

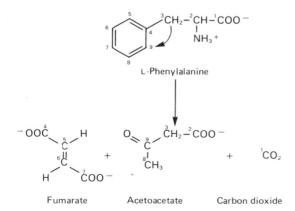

Figure 32–14. Alternative catabolites of tyrosine. *p*-Hydroxyphenylacetaldehyde is formed as an intermediate during oxidation of tyramine to *p*-hydroxyphenylacetate.

acetoacetate (Fig 32–15) thus is identical to that for tyrosine (Fig 32–13).

Figure 32–15. Ultimate catabolic fate of each carbon atom of phenylalanine. Pattern of isotopic labeling in the ultimate catabolites of phenylalanine (and tyrosine).

Numerous Metabolic Disorders Characterize the Pathway of Phenylalanine Catabolism

Major metabolic disorders associated with impaired ability to convert phenylalanine to tyrosine (see Fig 30–12) may be classified into 3 broad groups: defects in phenylalanine hydroxylase (hyperphenylalaninemia type I, or classic phenylketonuria), defects in dihydrobiopterin reductase (hyperphenylalaninemia types II and III), and defects in dihydrobiopterin biosynthesis (hyperphenylalaninemia types IV and V). Additional types, however, have been identified (Table 32–3).

The major consequence of untreated **hyperphenylalaninemia type I (classic phenylketonuria; PKU)** is mental retardation. Additional clinical signs include seizures, psychoses, eczema, and a "mousy" odor. However, if diagnosis and initiation of appropriate treatment are prompt, these symptoms may be avoided. Because of the availability of animal models and because prompt dietary intervention can ameliorate the

Table 32–3. Hyperphenylalaninemias.*

Type	Condition	Defect	Treatment
I	Phenylketonuria	Phe hydroxylase absent	Low Phe diet
II	Persistent hyperphenylalaninemia	Decreased Phe hydroxylase	None, or temporary dietary therapy
III	Transient mild hyperphenylalaninemia	Maturational delay of hydroxylase	Same as type II
IV	Dihydropteridine reductase deficiency	Deficient or absent dihydropteridine reductase	Dopa, 5-hydroxytryptophan, carbidopa
V	Abnormal dihydrobiopterin function	Dihydrobiopterin synthesis defect	Dopa, 5-hydroxytryptophan, carbidopa
VI	Persistent hyperphenylalaninemia and tyrosinemia	? Catabolism of tyrosine	Reduced Phe intake
VII	Transient neonatal tyrosinemia	p-Hydroxyphenylpyruvic oxidase inhibition	Vitamin C
VIII	Hereditary tyrosinemia	Deficiency: 1. p-Hydroxyphenylpyruvate deoxygenase 2. Cytoplasmic tyrosine aminotransferase 3. Fumarylacetoacetate hydrolase	Low Tyr diet Low Tyr diet plus glutathione injections

*Modified and reproduced, with permission, from Tourian A, Sidbury JB: Phenylketonuria and hyperphenylalaninemia. Page 273 in: Stanbury JB et al (editors): *The Metabolic Basis of Inherited Disease,* 5th ed. McGraw-Hill, 1983.

otherwise inevitable mental retardation, PKU has served as a model for study of the mental retardation associated with metabolic diseases. In classic PKU, a heritable disorder with a frequency of about 1:10,000 live births, levels of component I of liver phenylalanine hydroxylase (see Fig 30–12) average approximately 25% of normal, and the hydroxylase is insensitive to regulation by phenylalanine. Since patients cannot convert phenylalanine to tyrosine, alternative catabolites are produced (Fig 32–16). These include phenylpyruvic acid (from deamination of phenylalanine), phenyllactic acid (from reduction of phenylpyruvic acid), and phenylacetic acid (from decarboxylation and oxidation of phenylpyruvic acid). Much of the phenylacetate is excreted as phenylacetylglutamine. Table 32–4 illustrates the chemical pattern in the blood and urine of a phenylketonuric patient. While phenylpyruvate can be detected by a simple biochemical spot test, definitive diagnosis requires detection of elevated plasma phenylalanine levels.

Deterioration of mental performance of phenylketonuric children can be prevented if they are maintained on a diet containing very low levels of phenylalanine. The diet can be terminated at 6 years of age, when high concentrations of phenylalanine and its derivatives no longer are injurious to the brain.

While plasma phenylalanine may be measured by an automated micro method that requires as little as 20 μL of blood, abnormally high blood phenylalanine levels may not occur in phenylketonuric infants until the third or fourth day of life, because of their initially low intake of dietary protein. Furthermore, false-positive results may occur in premature infants owing to delayed maturation of the enzymes of phenylalanine catabolism. A useful but less reliable screening test depends on detecting elevated urinary levels of phenylpyruvate with ferric chloride.

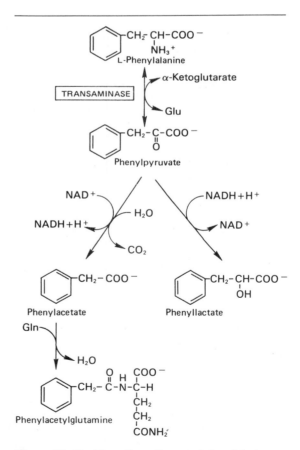

Figure 32–16. Alternative pathways of phenylalanine catabolism in phenylketonuria. The reactions also occur in the liver tissue of normal individuals but are of minor significance if a functional phenylalanine hydroxylase is present. Glu, glutamate; Gln, glutamine.

Table 32–4. Metabolites of phenylalanine which accumulate in the plasma and urine of phenylketonuric patients.

Metabolite	Plasma (mg/dL)		Urine (mg/dL)	
	Normal	Phenylketonuric	Normal	Phenylketonuric
Phenylalanine	1–2	15–63	30	300–1000
Phenylpyruvate		0.3–1.8		300–2000
Phenyllactate				290–550
Phenylacetate				Increased
Phenylacetylglutamine			200–300	2400

Administration of phenylalanine to a phenylketonuric subject should result in prolonged elevation of the level of this amino acid in the blood, indicating diminished tolerance to phenylalanine. However, abnormally low tolerance to injected phenylalanine and a high fasting level of phenylalanine are also characteristic of the parents of phenylketonurics. The defective gene responsible for phenylketonuria thus can be detected biochemically in the phenotypically normal, heterozygous parents.

Neither Nitrogen of Lysine Participates in Transamination

Mammals convert the intact carbon skeleton of L-lysine to α-aminoadipate and α-ketoadipate (Fig 32–17) via saccharopine (Fig 32–18), an intermediate in lysine biosynthesis by fungi. L-Lysine first condenses with α-ketoglutarate, forming a Schiff base. This is reduced to saccharopine by a dehydrogenase and then oxidized by a second dehydrogenase. Addition of water forms L-glutamate and L-α-aminoadipate-δ-semialdehyde. The net effect of this reaction sequence is equivalent to removal of the ε nitrogen of lysine by transamination. However, NAD$^+$ and NADH are required as cofactors.

Further catabolism of α-aminoadipate involves transamination to α-ketoadipate, probably followed by oxidative decarboxylation to glutaryl-CoA. While lysine is both glycogenic and ketogenic, the nature of the subsequent catabolites of glutaryl-CoA in mammalian systems is not known.

Two rare metabolic abnormalities result from impaired conversion of L-lysine and α-ketoglutarate to saccharopine (Fig 32–18).

Periodic Hyperlysinemia With Associated Hyperammonemia: In periodic hyperlysinemia, ingestion of normal levels of protein triggers hyperlysinemia. Hyperammonemia then results from competitive inhibition of liver arginase activity by elevated levels of tissue lysine. Fluid therapy and restriction of lysine intake relieve both the hyperammonemia and its clinical manifestations. Conversely, administration of a lysine load precipitates severe crises and coma.

Persistent Hyperlysinemia Without Hyperammonemia: Some patients are mentally retarded. There is no associated hyperammonemia, even in response to a lysine load. Lysine catabolites may or may not accumulate in biologic fluids. Persistent hyperlysinemia is believed to be inherited as an autosomal recessive trait. In addition to impaired conversion of lysine and α-ketoglutarate to saccharopine, some patients appear to have an additional deficiency in the conversion of saccharopine to L-glutamate and α-aminoadipate-δ-semialdehyde (Fig 32–18).

The Iron Porphyrin Metalloprotein Tryptophan Oxygenase Initiates Tryptophan Catabolism

The carbon atoms both of the side chain and of the aromatic ring of tryptophan may be completely degraded to amphibolic intermediates via the **kynurenine-anthranilate pathway** (Fig 32–19), important both for tryptophan degradation and for conversion of tryptophan to **nicotinamide.**

Tryptophan oxygenase (**tryptophan pyrrolase**) catalyzes cleavage of the indole ring with incorporation of 2 atoms of molecular oxygen, forming N-formylkynurenine. The oxygenase, an iron porphyrin metalloprotein, is inducible in liver by adrenal corticosteroids and by tryptophan. A considerable portion of newly synthesized enzyme is in a latent form that

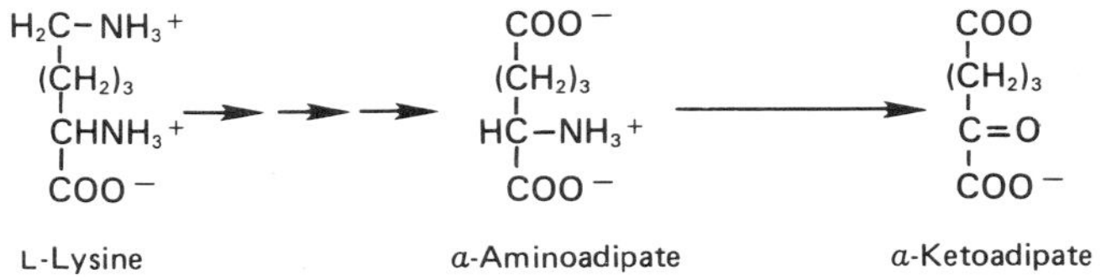

Figure 32–17. Conversion of L-lysine to α-aminoadipate and α-ketoadipate. Multiple arrows represent multiple reactions.

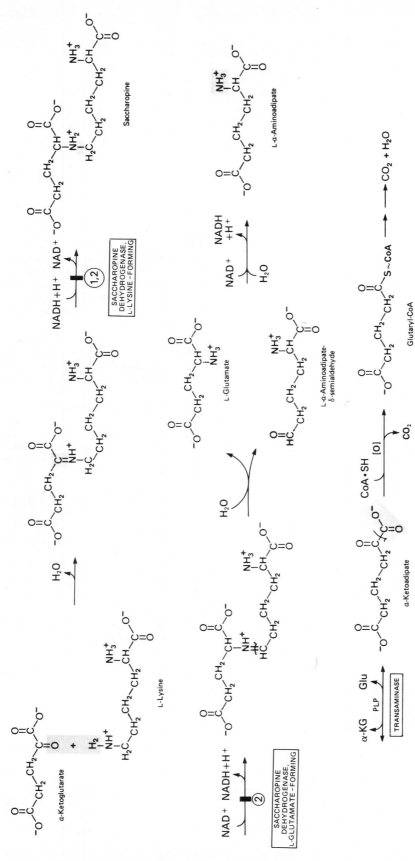

Figure 32–18. Catabolism of L-lysine. (α-KG, α-ketoglutarate; Glu, glutamate; PLP, pyridoxal phosphate.) The circled numerals indicate the probable sites of the metabolic defects in 1, periodic hyperlysinemia with associated hyperammonemia; and 2, persistent hyperlysinemia without associated hyperammonemia.

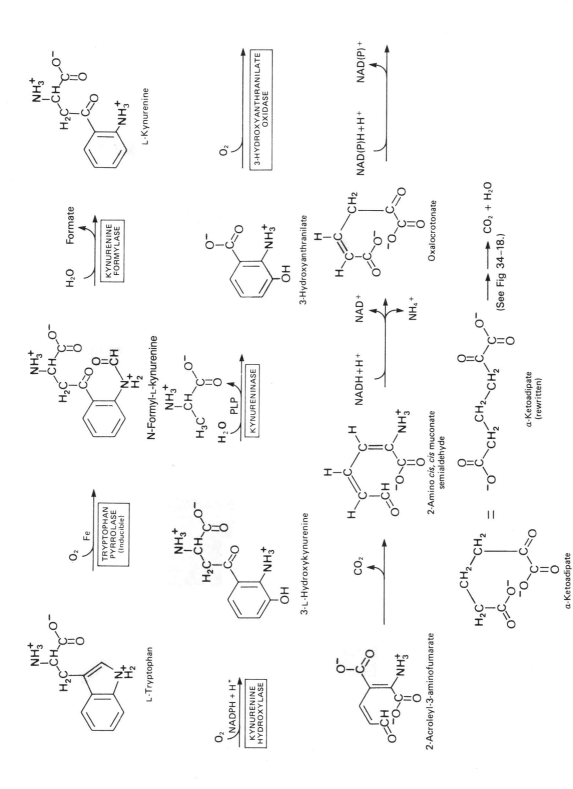

Figure 32–19. Catabolism of L-tryptophan. PLP, pyridoxal phosphate.

requires activation. Tryptophan also stabilizes the oxygenase toward proteolytic degradation. Tryptophan oxygenase is feedback-inhibited by nicotinic acid derivatives, including NADPH.

Hydrolytic removal of the formyl group of N-formylkynurenine, catalyzed by **kynurenine formylase** of mammalian liver produces **kynurenine** (Fig 32–19). The enzyme catalyzes similar reactions with various arylformylamines.

Kynurenine may be deaminated by transamination of the amino group of the side chain to ketoglutarate. The resulting keto derivative, 2-amino-3-hydroxybenzoyl pyruvate, loses water, and spontaneous ring closure forms **kynurenic acid,** a by-product not formed in the main pathway of tryptophan breakdown (Fig 32–19).

Further metabolism of kynurenine involves conversion to **3-hydroxykynurenine,** then **3-hydroxyanthranilate.** Hydroxylation requires molecular oxygen in an NADPH-dependent reaction similar to that for hydroxylation of phenylalanine (see Chapter 30).

The Alternative Tryptophan Metabolite Xanthurenate Accumulates in Vitamin B_6 Deficiency

Kynurenine and hydroxykynurenine are converted to hydroxyanthranilate by **kynureninase,** a pyridoxal phosphate enzyme. A deficiency of vitamin B_6 results in partial failure to catabolize these kynurenine derivatives, which thus reach extrahepatic tissues where they are converted to **xanthurenate** (Fig 32–20). This abnormal metabolite occurs in the urine when dietary

vitamin B_6 is inadequate. Feeding excess tryptophan induces excretion of xanthurenate in vitamin B_6 deficiency.

In many animals, conversion of tryptophan to nicotinic acid makes a supply of the vitamin in the diet unnecessary. In the rat, rabbit, dog, and pig, tryptophan can completely replace the vitamin in the diet; in humans and other animals, tryptophan increases the urinary excretion of nicotinic acid derivatives (eg, N-methylnicotinamide). In vitamin B_6 deficiency, synthesis of NAD^+ and $NADP^+$ may be impaired, a result of inadequate conversion of tryptophan to nicotinic acid for pyridine nucleotide synthesis. If an adequate supplement of nicotinic acid is supplied, pyridine nucleotide synthesis proceeds normally even in the absence of vitamin B_6.

Hartnup disease, a hereditary abnormality in metabolism of tryptophan, is characterized by a pellagralike skin rash, intermittent cerebellar ataxia, and mental deterioration. The urine contains increased amounts of indoleacetate (α-N[indole-3-acetyl]glutamine) and tryptophan.

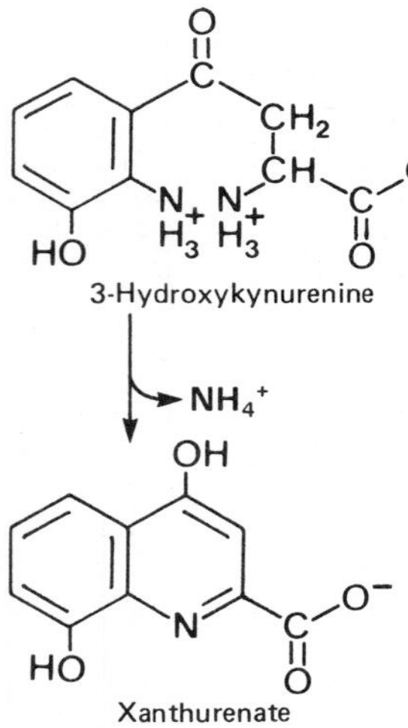

METHIONINE, ISOLEUCINE, & VALINE ARE CATABOLIZED TO SUCCINYL-CoA

While succinyl-CoA is the amphibolic end product for catabolism of methionine, isoleucine, and valine, only portions of the skeletons are converted (Fig 32–21). Four-fifths of the carbons of valine, three-fifths of those of methionine, and half of those of isoleucine form succinyl-CoA. The carboxyl carbons of all 3 form CO_2. The terminal 2 carbons of isoleucine form acetyl-CoA, and the methyl group of methionine is removed as such.

What follows relates only to conversion of methionine and isoleucine to propionyl-CoA and of valine to methylmalonyl-CoA. The reactions leading from propionyl-CoA through methylmalonyl-CoA to succinyl-CoA are discussed in Chapter 23 in connection

Figure 32–20. Formation of xanthurenate in vitamin B_6 deficiency. Conversion of the tryptophan metabolite 3-hydroxykynurenine to 3-hydroxyanthranilate is impaired (see Fig 32–19). A large portion is therefore converted to xanthurenate.

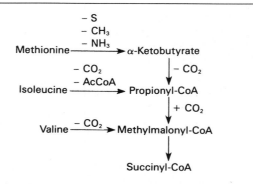

Figure 32–21. Overall catabolism of methionine, isoleucine, and valine.

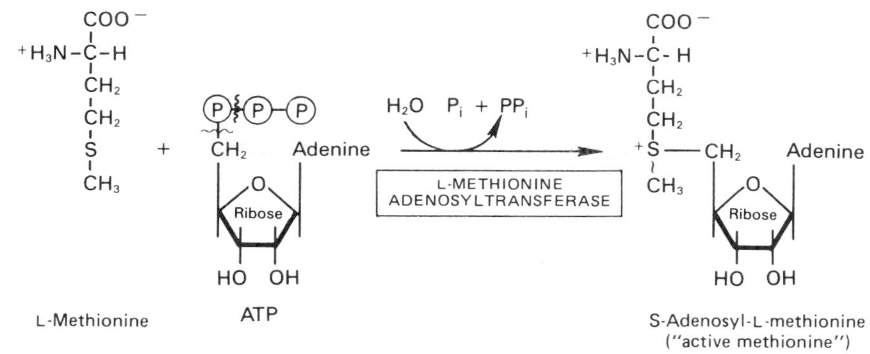

Figure 32–22. Formation of S-adenosylmethionine. ~CH_3 represents the high transfer potential of "active methionine."

with catabolism of propionate and of fatty acids containing an odd number of carbon atoms.

Methionyl Carbons Form Propionyl-CoA Via Homocysteine, Cystathionine & α-Ketobutyrate

L-Methionine first condenses with ATP, forming S-adenosylmethionine,* "active methionine" (Fig 32–22). The activated S-methyl group may transfer to various acceptors. Removal of the methyl group forms S-adenosylhomocysteine. Hydrolysis of the S–C bond yields L-homocysteine plus adenosine. Homocysteine then condenses with serine, forming cystathionine (Fig 32–23). Hydrolytic cleavage of cystathionine forms L-homoserine plus cysteine, so that the net effect is conversion of homocysteine to homoserine and of serine to cysteine. These 2 reactions are therefore also involved in biosynthesis of cysteine from serine (see Chapter 30). Homoserine is converted to α-ketobutyrate by homoserine deaminase (Fig 32–24). Conversion of α-ketobutyrate to propionyl-CoA then occurs in the usual manner for oxidative carboxylation of α-keto acids (eg, pyruvate, α-ketoglutarate) to form acyl-CoA derivatives.

For the metabolic disorders of methionine catabolism, see Table 32–2.

THE TWO INITIAL CATABOLIC REACTIONS ARE COMMON TO ALL THREE BRANCHED-CHAIN AMINO ACIDS

Catabolism of leucine, valine, and isoleucine initially involves the same reactions. Subsequently, each amino acid skeleton follows a unique pathway to am-phibolic intermediates (Figs 32–25 and 32–26). The nature of these amphibolic end products determines whether an amino acid is glycogenic (valine), ketogenic (leucine), or both (isoleucine). **Many of the reactions involved are analogous to reactions of straight- and branched-chain fatty acid catabolism.** In what follows, reaction numbers refer to reactions of Figs 32–26 through 32–29.

A Single Enzyme Appears To Transaminate All Three Branched Amino Acids

Reversible transamination (reaction 1, Fig 32–26) of all 3 branched amino acids probably involves a single transaminase. Reversibility of this reaction accounts for the ability of the corresponding α-keto acids to replace the amino acids in the diet.

Oxidative Decarboxylation of the Branched α-Keto Acids Is Analogous to Conversion of Pyruvate to Acetyl-CoA

The branched-chain α-keto acid dehydrogenase, (reaction 2, Fig 32–26) an intramitochondrial multi-enzyme complex, catalyzes oxidative decarboxylation of α-ketoisocaproate (from leucine), α-keto-β-methylvalerate (from isoleucine), and α-ketoisovalerate (from valine). The α-keto acid dehydrogenase complex subunits are analogous to those of pyruvate dehydrogenase: an α-keto acid decarboxylase, a transacylase, and a dihydrolipoyl dehydrogenase. As for pyruvate dehydrogenase, the complex is inactivated when phosphorylated by ATP and a protein kinase. A Ca^{2+}-independent phosphoprotein phosphatase catalyzes its dephosphorylation and accompanying reactivation. The phosphorylation state thus can regulate the catabolism of the branched-chain amino acids. The protein kinase is inhibited by ADP, the branched-chain α-keto acid products, the hypolipidemic agents clofibrate and dichloroacetate, and coenzyme A thioesters (eg, acetoacetyl-CoA). Of the branched-chain α-keto acids, α-ketoisocaproate (α-ketoleucine) is the most potent inhibitor.

---

*Compounds whose methyl groups derive from S-adenosylmethionine include betaines, choline, creatine, epinephrine, melatonin, sarcosine, N-methylated amino acids, nucleotides, and many plant alkaloids.

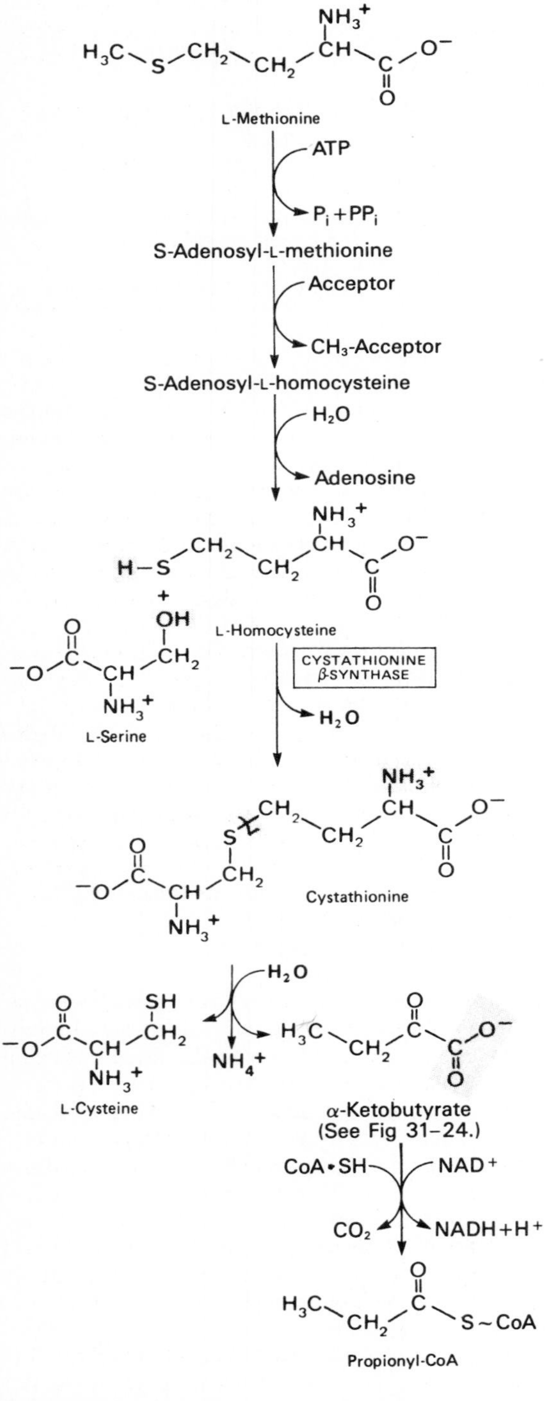

Figure 32–23. Conversion of methionine to propionyl-CoA.

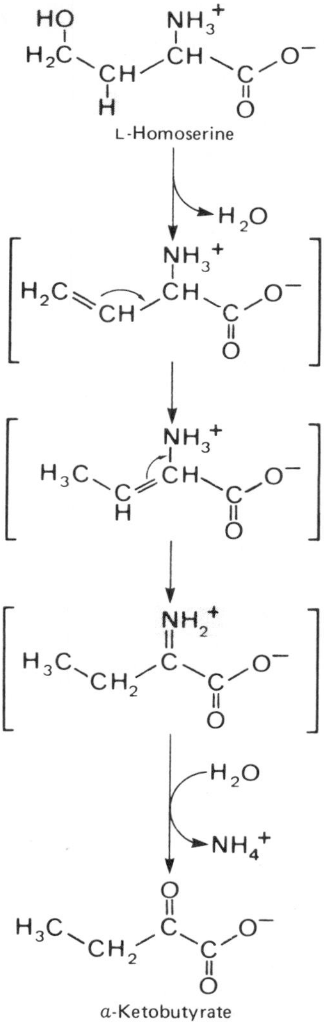

Figure 32–24. Conversion of L-homoserine to α-keto-butyrate, catalyzed by homoserine deaminase.

The Dehydrogenation of Branched Acyl-CoA Thioesters Is Analogous to a Reaction of Fatty Acid Catabolism

Reaction 3 is analogous to dehydrogenation of straight-chain acyl-CoA thioesters in fatty acid catabolism. While a single enzyme may catalyze dehydrogenation of all 3 branched acyl-CoA thioesters, indirect evidence implicates at least 2 enzymes. In **isovaleric acidemia,** following ingestion of protein-rich foods, isovalerate (a deacylation product of iso-valeryl-CoA) accumulates in the blood. An increase in other branched α-keto acids does not occur.

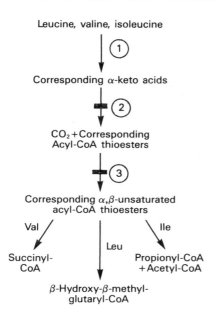

Figure 32–25. Catabolism of the branched-chain amino acids. Reactions 1–3 are common to all 3 amino acids. Double lines intersecting arrows mark sites of metabolic blocks in 2 rare human diseases; at 2, maple syrup urine disease, a defect in catabolism of all 3 amino acids; and at 3, isovaleric acidemia, a defect of leucine catabolism.

Three Reactions Are Specific to Leucine Catabolism

Reaction 4L, Carboxylation of β-Methylcrotonyl-CoA: The key to understanding the ketogenic action of leucine was the discovery that 1 mol of CO_2 was "fixed" (ie, covalently bound) per mole of isopropyl groups (from the terminal isopropyl group of leucine) converted to acetoacetate. This CO_2 fixation (reaction 4L, Fig 32–27) requires biotinyl-CO_2, and forms β-methylglutaconyl-CoA.

Reaction 5L, Hydration of β-Methylglutaconyl-CoA: The product, β-hydroxy-β-methylglutaryl-CoA, is a precursor both of ketone bodies (reaction 6L, Fig 32–27) and of mevalonate, and hence of cholesterol and other polyisoprenoids (see Chapter 29).

Reaction 6L, Cleavage of β-Hydroxy-β-Methylglutaryl-CoA: Cleavage of β-hydroxy-β-methylglutaryl-CoA to acetyl-CoA and acetoacetate occurs in liver, kidney, and heart mitochondria. It explains the strong ketogenic effect of leucine, since not only is 1 mol of acetoacetate formed per mole of leucine, but another ½ mole of ketone bodies may be formed from the remaining product, acetyl-CoA (see Chapter 29).

Four Reactions Are Specific to Valine Catabolism

Reaction 4V, Hydration of Methylacrylyl-CoA: This reaction, which occurs nonenzymatically at a relatively rapid rate, is catalyzed by crotonase, a hydrolase of broad specificity for L-β-hydroxyacyl-CoA thioesters having 4–9 carbon atoms.

Reaction 5V, Deacylation of β-Hydroxyisobutyryl-CoA: Since the CoA thioester is not a substrate for the subsequent reaction (reaction 6V, Fig 32–28), it must first be deacylated to β-hydroxyisobutyrate (reaction 5V, Fig 32–28) by a deacylase, whose only other substrate is β-hydroxypropionyl-CoA.

Reaction 6V, Oxidation of β-Hydroxyisobutyrate: The reversible, NAD^+-dependent oxidation of the primary alcohol group of β-hydroxyisobutyrate to an aldehyde (reaction 6V, Fig 32–28) forms methylmalonate semialdehyde.

Reaction 7V, Fate of Methylmalonate Semialdehyde: Possible fates for methylmalonate semialdehyde include transamination and conversion to succinyl-CoA. Transamination forms α-aminoisobutyrate, a normal urinary amino acid (reaction 7V, Fig 32–28). The second major fate involves oxidative acylation to methylmalonyl-CoA and isomerization to succinyl-CoA (reactions 8V and 9V, Fig 32–28). Isomerization (reaction 9V, Fig 32–28) requires adenosylcobalamin coenzyme and is catalyzed by methylmalonyl-CoA mutase. This reaction is important not only for valine catabolism but also for that of propionyl-CoA, a catabolite of isoleucine (Fig 32–29). In cobalamin (vitamin B_{12}) deficiency, mutase activity is impaired. This produces a "dietary metabolic defect" in ruminants that utilize propionate (from fermentation in the rumen) as an energy source. Rearrangement to succinyl-CoA occurs via an intramolecular shift of the CoA-carboxyl group.

Three Reactions Are Specific to Isoleucine Catabolism

Dietary studies in intact animals initially identified isoleucine as glycogenic and weakly ketogenic. Glycogen synthesis from isoleucine was later confirmed using D_2O. Use of ^{14}C-labeled intermediates and liver slice preparations revealed that the isoleucine skeleton was cleaved, forming acetyl-CoA and propionyl-CoA (Fig 32–29).

Reaction 4I, Hydration of Tiglyl-CoA: This reaction, like the analogous reaction in valine catabolism (reaction 4V, Fig 32–28), is catalyzed by crotonase.

Reaction 5I, Dehydrogenation of α-Methyl-β-Hydroxybutyryl-CoA: This reaction is analogous to reaction 5V of valine catabolism (Fig 32–28).

Reaction 6I, Thiolysis of α-Methylacetoacetyl-CoA: Thiolytic cleavage of the bond linking carbons 2 and 3 of α-methylacetoacetyl-CoA resembles thiolysis of acetoacetyl-CoA by β-ketothiolase. The products, acetyl-CoA (ketogenic) and propionyl-CoA (glyco-

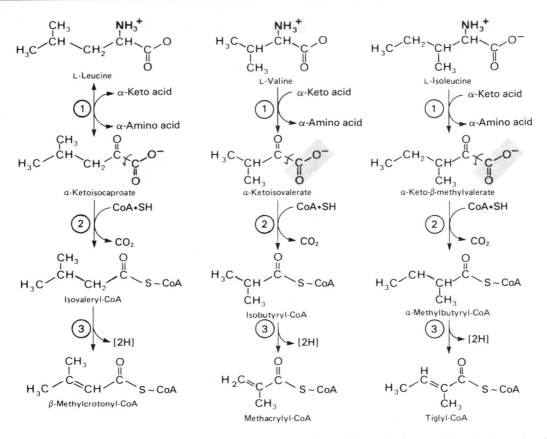

Figure 32–26. The analogous first 3 reactions in the catabolism of leucine, valine, and isoleucine. Note also the analogy in reactions 2 and 3 to the catabolism of fatty acids. This analogy continues, as shown in subsequent figures. α-KA, α-keto acid; α-AA, α-amino acid.

genic), account for the ketogenic and glycogenic properties of isoleucine.

Several Metabolic Disorders Are Associated With the Catabolic Pathways for the Branched-Chain Amino Acids

Hypervalinemia: This rare metabolic disease, characterized by elevated plasma levels of valine (but not of leucine or isoleucine), reflects the inability to transaminate valine to α-ketoisovalerate (reaction 1, Fig 32–26). However, transamination of leucine and isoleucine is unimpaired.

Maple Syrup Urine Disease (Branched Chain Ketonuria): The most striking feature of this hereditary disease (incidence 5 to 10 per million live births) is the odor of the urine, which resembles that of maple syrup or burnt sugar. Plasma and urinary levels of leucine, isoleucine, valine, and their α-keto acids are greatly elevated. Smaller quantities of branched-chain α-hydroxy acids, formed by reduction of the α-keto acids, also are present in the urine.

The disease is evident by the end of the first week of extrauterine life. In addition to the biochemical abnormalities described above, the infant is difficult to feed, may vomit, and may be lethargic. Diagnosis prior to 1 week of age is possible only by enzymatic analysis. Extensive brain damage occurs in surviving children. Without treatment, death usually occurs by the end of the first year of life.

The biochemical defect is the absence or greatly reduced activity of the **α-keto acid decarboxylase** that catalyzes conversion of all 3 branched-chain α-keto acids to CO_2 plus acyl-CoA thioesters (reaction 2, Fig 32–26). The mechanism of toxicity is unknown.

Providing that treatment is initiated in the first week of life, the dire consequences of this disease may be largely averted. Dietary protein is replaced by a mixture of purified amino acids from which leucine, isoleucine, and valine are omitted. When plasma levels of these amino acids fall within the normal range, they are restored to the diet in the form of milk and other foods in amounts adequate to supply—but not to exceed—the requirements for branched-chain amino acids. There is no indication when, if ever, dietary restrictions may be eased.

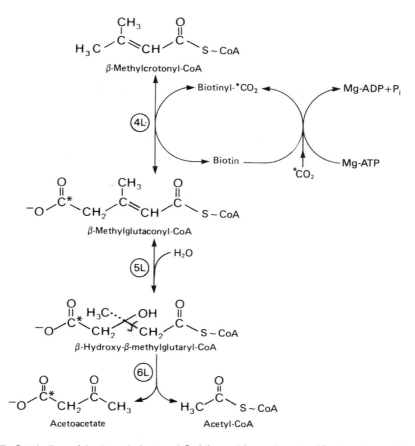

Figure 32–27. Catabolism of the β-methylcrotonyl-CoA formed from L-leucine. *Carbon atoms derived from CO_2.

Intermittent Branched-Chain Ketonuria: This variant of maple syrup urine disease probably reflects a less severe structural modification of the **α-keto acid decarboxylase.** Since affected individuals possess an impaired but nevertheless distinct capability for catabolism of leucine, valine, and isoleucine, symptoms of maple syrup urine disease occur later in life and only intermittently. The prognosis for dietary therapy is more favorable in these individuals.

Maple syrup urine disease and intermittent branched-chain ketonuria illustrate mutations causing different changes in the primary structure of the same enzyme. It is probable that a spectrum of activities ranging from frank disease through intermittent manifestations to normal values occurs in individual subjects.

Isovaleric Acidemia: Relevant findings include a "cheesy" odor of the breath and body fluids, vomiting, acidosis, and coma precipitated by excessive ingestion of protein. The impaired enzyme is isovaleryl-CoA dehydrogenase (reaction 3, Fig 32–26). Isovaleryl-CoA thus accumulates, is hydrolyzed to isovalerate, and is excreted in the urine and sweat.

Disorders of Methylmalonyl-CoA Catabolism Reflect Failure to Convert Vitamin B_{12} to Its Coenzyme

The conversion to amphibolic intermediates of propionyl-CoA (derived from isoleucine, methionine, the side chain of cholesterol, and fatty acids with odd numbers of carbon atoms) involves biotin-dependent carboxylation to methylmalonyl-CoA. Methylmalonyl-CoA also is formed directly (ie, without prior formation of propionyl-CoA) from valine. A vitamin B_{12} coenzyme-dependent reaction then isomerizes methylmalonyl-CoA to succinyl-CoA. Patients with acquired vitamin B_{12} deficiency excrete methylmalonate in their urine. This methylmalonic aciduria disappears when sufficient vitamin B_{12} is administered.

Propionic Acidemia: Propionyl-CoA carboxylase deficiency is characterized by high serum propionate levels. Treatment involves a low-protein diet and measures to counteract metabolic acidosis.

Methylmalonic Aciduria: Two forms are known. One responds to pharmacologic levels of vitamin B_{12}. The second type, requires therapy by massive doses (1 g/d) of vitamin B_{12}.

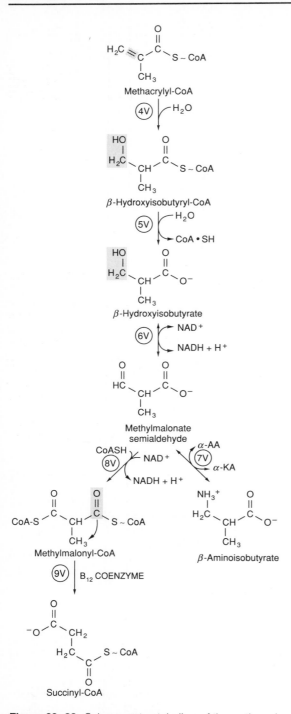

Methacrylyl-CoA

(4V) — H$_2$O

β-Hydroxyisobutyryl-CoA

(5V) — H$_2$O, CoA·SH

β-Hydroxyisobutyrate

(6V) — NAD$^+$, NADH + H$^+$

Methylmalonate semialdehyde

CoASH (8V) NAD$^+$ / NADH + H$^+$ α-AA (7V) α-KA

Methylmalonyl-CoA   β-Aminoisobutyrate

(9V) B$_{12}$ COENZYME

Succinyl-CoA

Figure 32–28. Subsequent catabolism of the methacryl-yl-CoA formed from L-valine (see Fig 32–26). α-KA, α-keto acid; α-AA, α-amino acid.

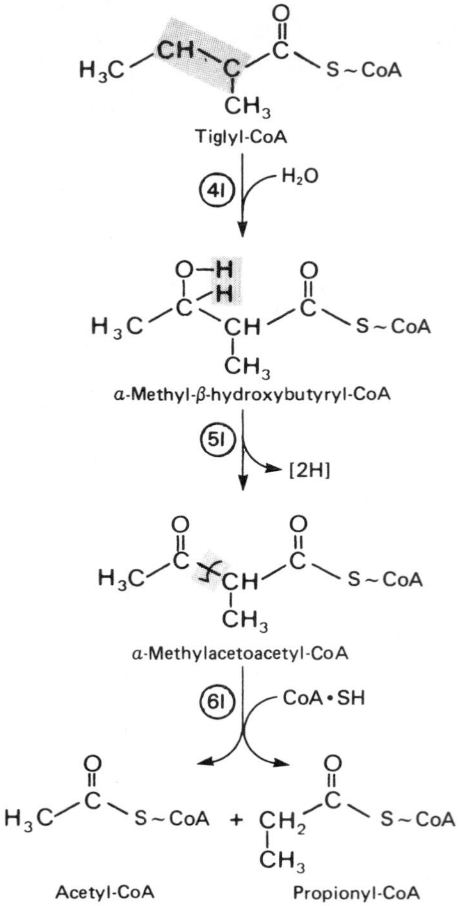

Tiglyl-CoA

(4I) — H$_2$O

a-Methyl-β-hydroxybutyryl-CoA

(5I) — [2H]

a-Methylacetoacetyl-CoA

(6I) — CoA·SH

Acetyl-CoA   +   Propionyl-CoA

Figure 32–29. Subsequent catabolism of the tiglyl-CoA formed from L-isoleucine.

REFERENCES

Cooper AJL: Biochemistry of the sulfur-containing amino acids. *Annu Rev Biochem* 1983;**52**:187.

Paxton R, Harris RA: Isolation of rabbit liver branched chain α-ketoacid dehydrogenase and regulation by phosphorylation. *J Biol Chem* 1982;**257**:14433.

Paxton R, Harris RA: Regulation of branched-chain α-ketoacid dehydrogenase kinase. *Arch Biochem Biophys* 1984;**231**:48.

Rosenberg LE, Scriver CR: Disorders of amino acid metabolism. Chapter 11 in: *Metabolic Control and Disease.* Bondy PK, Rosenberg LE (editors). Saunders, 1980.

Scriver CR et al (editors): *The Metabolic Basis of Inherited Disease,* 6th ed. McGraw-Hill, 1989.

Wellner D, Meister A: A survey of inborn errors of amino acid metabolism and transport in man. *Annu Rev Biochem* 1981;**50**:911.

Conversion of Amino Acids to Specialized Products

33

Victor W. Rodwell, PhD

INTRODUCTION

Amino acids are the primary source of nitrogen for animals and serve as precursors for other nitrogenous compounds. Since most are not themselves amino acids, this discussion merges with metabolic pathways discussed elsewhere in this book. Physiologically important products derived from amino acids include heme, purines, pyrimidines, hormones, and neurotransmitters including biologically active peptides. In addition, many proteins contain amino acids that have been modified for a specific function, eg, calcium binding or cross-linking, and thus the amino acid residues in those proteins serve as precursors for these modified residues. Finally, there are small peptides or peptide-like molecules not synthesized on ribosomes that carry out specific functions in cells.

BIOMEDICAL IMPORTANCE

The biologically active amine histamine, formed by decarboxylation of histidine, plays a central role in many allergic reactions. Specific neurotransmitters derived from amino acids include γ-aminobutyrate from glutamate; 5-hydroxytryptamine (serotonin) from tryptophan; and dopamine, norepinephrine, and epinephrine from tyrosine. Many drugs used to treat neurologic and psychiatric conditions affect the metabolism of the neurotransmitters mentioned above.

GLYCINE PARTICIPATES IN THE BIOSYNTHESIS OF HEME, PURINES, GLYCINE CONJUGATES, & CREATINE

Synthesis of Heme: The α-carbon and the nitrogen atom of glycine are used for synthesis of the porphyrin moiety of hemoglobin (see Chapter 34). The pyrrole nitrogen is derived from glycine nitrogen, and an adjoining carbon from the α carbon of glycine. The α carbon is also the source of the methylene bridge atoms that link the pyrrole rings.

Metabolic disorders of heme metabolism are discussed in Chapter 34.

Synthesis of Purine Nucleotides: The entire glycine molecule forms positions 4, 5, and 7 of the purine skeleton (see Chapter 36).

Formation of Glycine Conjugates: Glycine conjugates with cholic acid, forming glycocholic acid. With benzoate, glycine forms hippurate (Fig 33–1). The quantitative ability of liver to convert a measured dose of benzoate to hippurate formerly was used as a test of liver function.

Synthesis of Creatine: The sarcosine (N-methylglycine) component of creatine is derived from glycine and S-adenosylmethionine.

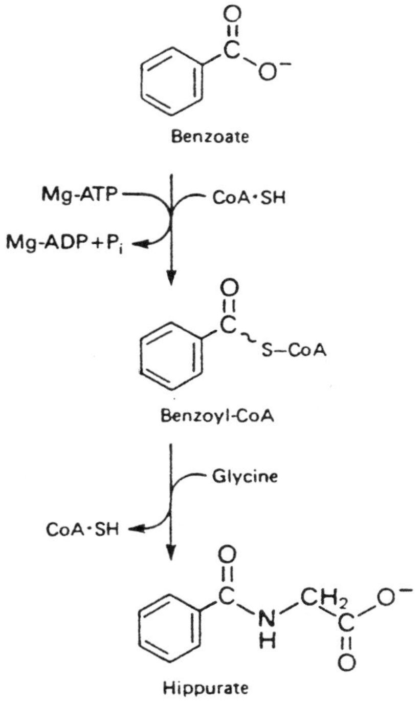

Figure 33–1. Hippurate biosynthesis.

α-ALANINE IS A MAJOR PLASMA AMINO ACID

Alanine, with glycine, makes up a considerable fraction of the amino nitrogen in human plasma. Alanine is also a major component of bacterial cell walls, partly as the D isomer; 39–50% in *Streptococcus faecalis;* 67% in *Staphylococcus aureus*.

MAMMALS FORM β-ALANINE FROM URACIL & CATABOLIZE IT VIA MALONATE SEMIALDEHYDE

Little free β-alanine is present in tissues. Considerably more is present as β-alanyl dipeptides (see below) and as coenzyme A (see Fig 18–6).

While microorganisms form β-alanine by α-decarboxylation of aspartate, mammalian tissue β-alanine arises principally from catabolism of uracil, carnosine, and anserine (Fig 33–2).

Catabolism of β-alanine in mammals involves transamination to malonate semialdehyde, which is oxidized to acetate and thence to CO_2.

In the rare metabolic disorder **hyper-β-alaninemia,** free β-alanine levels are elevated in body fluids and tissues. Taurine and β-aminoisobutyrate levels also are elevated.

CARNOSINE IS A β-ALANYL DIPEPTIDE

β-Alanine is present primarily as the human skeletal dipeptide **carnosine** (Fig 33–2). The closely related β-alanyl dipeptide **anserine** (N-methylcarnosine Fig 33–2) is absent from human muscle but present in species whose skeletal muscle is characterized by rapid contractile activity (rabbit limb and bird pectoral muscle). It thus may fulfill physiologic functions distinct from carnosine.

β-Alanyl-imidazole buffers the pH of anaerobically contracting skeletal muscle. Carnosine and anserine activate myosin ATPase activity in vitro. Both dipeptides also chelate copper and enhance copper uptake.

β-Alanyl Dipeptides Are Synthesized & Degraded by Short Pathways

Carnosine is formed from β-alanine and L-histidine in an ATP-requiring reaction catalyzed by carnosine synthetase:

ATP + L-Histidine + β-Alanine →
AMP + PP$_i$ + Carnosine

Anserine is formed from carnosine (the methyl group donor is S-adenosylmethionine) in a reaction catalyzed by carnosine N-methyltransferase:

S-Adenosylmethionine + Carnosine →
S-Adenosylhomocysteine + Anserine

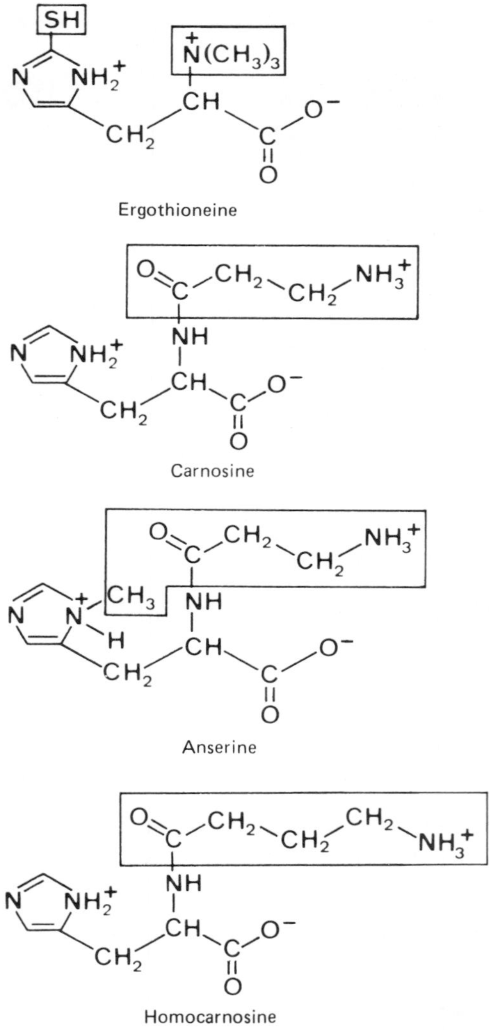

Figure 33–2. Compounds related to histidine. The boxes surround the components not derived from histidine.

The overall reaction involves enzyme-bound β-alanyl-adenylate.

Carnosine is hydrolyzed to β-alanine and L-histidine by the serum zinc metalloenzyme **carnosinase** (carnosine hydrolase). A presumably autosomal recessive, heritable disorder, **serum carnosinase deficiency** is characterized by persistent carnosinuria and occasionally also by carnosinemia. Carnosinuria persists even if carnosine is excluded from the diet.

The Response to Dietary Carnosine Facilitates Diagnosis of Hartnup Disease

Kidney tissue and intestinal enterocytes take up carnosine and β-alanine by membrane carriers that discriminate one substrate from the other, and both from other dipeptides. This ability to differentiate β-alanine from carnosine uptake has been applied to the definition of Hartnup disease, a heritable disorder in the transport of certain neutral α-amino acids (see Chapter 32). In patients with Hartnup disease, the blood histidine response is normal if carnosine is fed but is attenuated following administration of L-histidine.

Homocarnosine Is a Central Nervous System Dipeptide

The physiologic function of homocarnosine (γ-aminobutyryl-L-histidine; Fig 33–2), a central nervous system dipeptide related structurally and metabolically to carnosine, is not known. This **dipeptide of γ-aminobutyrate and L-histidine** is present in human brain tissue, where its concentration varies with the region examined. Biosynthesis of homocarnosine in human brain appears to be catalyzed by carnosine synthetase. However, serum carnosinase does not hydrolyze homocarnosine.

PHOSPHORYLATED SERYL & THREONYL RESIDUES ARE PRESENT IN MANY PROTEINS

Much of the serine in phosphoproteins is present as O-phosphoserine. Serine is involved in synthesis of sphingosine (see Chapter 26) and in purine and pyrimidine synthesis. The β carbon is a source of the methyl groups of thymine (and of choline) and of the carbon in positions 2 and 8 of the purine nucleus (see Chapter 36).

Since threonine does not participate in transamination, the D isomer and the α-keto acid are not utilized by mammals. Threonine is present in certain proteins as O-phosphothreonine.

S-ADENOSYLMETHIONINE IS THE PRINCIPAL SOURCE OF METHYL GROUPS FOR BIOSYNTHESIS

Methionine as a methyl group donor is discussed in Chapter 32. In the form of S-adenosylmethionine, it is the principal source of methyl groups in the body. In addition to direct utilization, the methyl group is also oxidized. The methyl carbon may be used to produce the one-carbon moiety that conjugates with glycine in synthesis of serine. As S-adenosylmethionine, methionine serves as precursor to the 1,3-diaminopropane portions of the polyamines spermine and spermidine (see below).

URINARY SULFATE DERIVES PRINCIPALLY FROM CYSTEINE

Urinary sulfate arises almost entirely from oxidation of L-cysteine. The sulfur of methionine (as homocysteine) is transferred to serine (see Fig 30–9) and thus contributes to the urinary sulfate indirectly (ie, via cysteine). L-Cysteine serves as a precursor of the thio-

ethanolamine portion of coenzyme A and is a precursor of the taurine that conjugates with bile acids, forming, for example taurocholic acid.

DECARBOXYLATION OF HISTIDINE FORMS HISTAMINE

Histamine is derived from histidine by decarboxylation, a reaction catalyzed in mammalian tissues by an **aromatic L-amino acid decarboxylase.** This enzyme also catalyzes decarboxylation of dopa, 5-hydroxytryptophan, phenylalanine, tyrosine, and tryptophan (see below). The decarboxylase is inhibited by α-methyl amino acids, which thus have clinical application as antihypertensive agents. A different enzyme, **histidine decarboxylase,** present in most cells, catalyzes decarboxylation of histidine.

Histidine compounds present in the body include **ergothioneine,** in red blood cells and liver; carnosine; and anserine (Fig 33–2). 1-Methylhistidine in human urine probably is derived from anserine. 3-Methylhistidine, identified in human urine in amounts of about 50 mg/dL, is unusually low in the urine of patients with **Wilson's disease** (see Chapter 58).

ARGININE, VIA ORNITHINE, IS A PRECURSOR OF POLYAMINES

Arginine serves as a formamidine donor for creatine synthesis in primates (see Fig 33–10) and for streptomycin synthesis in *Streptomyces*. Other fates include conversion, via ornithine, to putrescine, spermine, and spermidine (Fig 33–3) and synthesis of arginine phosphate (functionally analogous to creatine phosphate) in invertebrate muscle.

In addition to its role in urea biosynthesis (see Chapter 31), ornithine (with methionine) serves as a precursor of the ubiquitous mammalian (and bacterial) polyamines spermidine and spermine (Fig 33–4). Normal humans biosynthesize approximately 0.5 mmol of spermine per day. Pharmacologic doses of polyamines are hypothermic and hypotensive.

Spermidine and spermine are implicated in diverse physiologic processes that share as a common thread a close relationship to cell proliferation and growth. They are growth factors for cultured mammalian and bacterial cells and have been implicated in the stabilization of intact cells, subcellular organelles, and membranes. As a consequence of their multiple positive charges, polyamines associate readily with polyanions such as DNA and RNAs and have been implicated in such fundamental processes as stimulation of DNA and RNA biosynthesis, DNA stabilization, and packaging of DNA in bacteriophage. Polyamines also exert diverse effects on protein synthesis and act as inhibitors of enzymes that include protein kinases.

While it is not presently possible to explain (in precise mechanistic terms) the mode of action of poly-

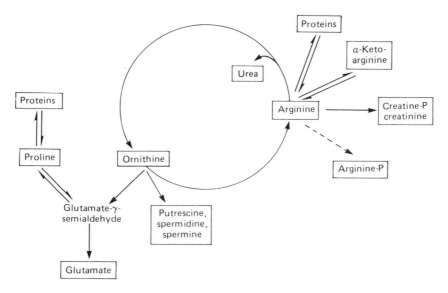

Figure 33–3. Arginine, ornithine, and proline metabolism. Reactions with solid arrows all occur in mammalian tissues. Putrescine and spermine synthesis occurs in both mammals and bacteria. Arginine phosphate occurs in invertebrate muscle, where it functions as a phosphagen analogous to creatine phosphate in mammalian tissues.

amines on any specific metabolic process, the essential nature of polyamines in mammalian metabolism is convincingly documented by experiments of the following type. The initial reaction in polyamine biosynthesis is catalyzed by ornithine decarboxylase (Fig 33–5). Addition to cultured mammalian cells of inhibitors of ornithine decarboxylase activity (eg, α-methylornithine or difluoromethylornithine) triggers overproduction of ornithine decarboxylase. This suggests

an essential physiologic role for this enzyme, whose only known function is polyamine biosynthesis.

Biosynthesis of Polyamines Is Important in Cell & Tissue Growth

Fig 33–5 summarizes polyamine biosynthesis in mammalian tissues. Note that the putrescine portion of spermidine and spermine derives from L-ornithine and the diaminopropane portion from L-methionine via in-

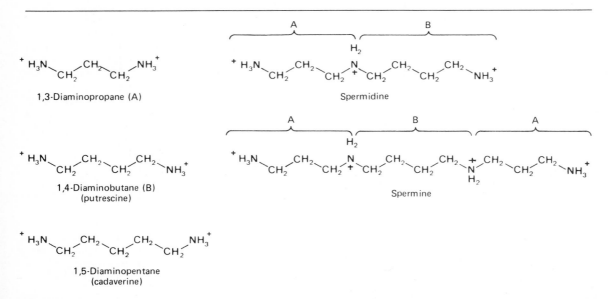

Figure 33–4. Structures of the natural polyamines. Note that spermidine and spermine are polymers of diaminopropane (A) and diaminobutane (B). Diaminopentane (cadaverine) also occurs in mammalian tissues.

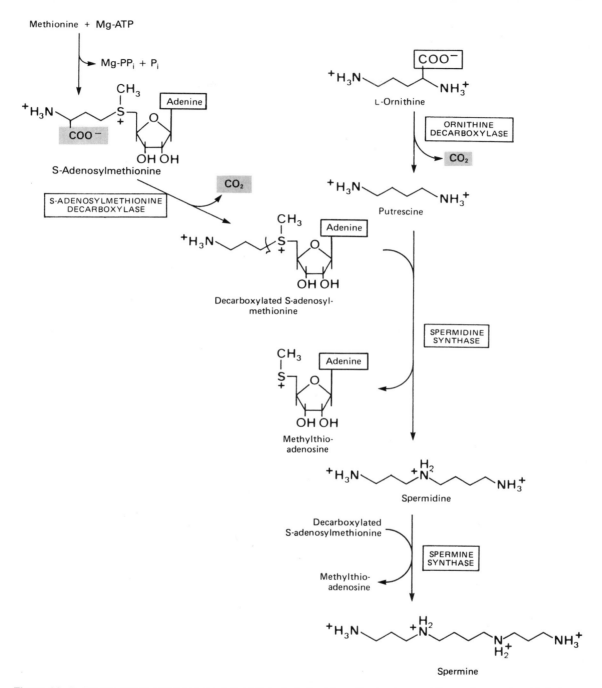

Figure 33–5. Intermediates and enzymes that participate in the biosynthesis of spermidine and spermine. Methylene groups are abbreviated to facilitate visualization of the overall process.

termediate formation of S-adenosylmethionine. Ornithine decarboxylase and S-adenosylmethionine decarboxylase both are inducible enzymes with short half-lives. Spermine and spermidine synthases are, by contrast, neither inducible nor unusually labile enzymes.

Of the enzymes of mammalian polyamine biosynthesis, 2 (ornithine decarboxylase and S-adenosylmethionine decarboxylase) are of interest with respect to both their regulation and their potential for enzyme-directed chemotherapy. The half-life of ornithine decarboxylase (approximately 10 minutes) is shorter than that of any other known mammalian enzyme, and its activity responds rapidly and dramatically to many stimuli. Increases in ornithine decarboxylase activity of 10- to 200-fold rapidly follow administration to cultured mammalian cells of growth hormone, corticosteroids, testosterone, or epidermal growth factor. Polyamines added to cultured cells induce synthesis of a protein antizyme that binds to ornithine decarboxylase and inhibits its activity. The activity of ornithine decarboxylase thus appears also to be controlled by a protein-protein interaction reminiscent of the regulation of trypsin activity by protein trypsin inhibitors. Difluoromethylornithine, a "suicide inhibitor" of ornithine decarboxylase, has been used both to isolate mutant cell lines that overproduce ornithine decarboxylase and to inhibit cell replication by enzyme-directed chemotherapy.

S-Adenosylmethionine decarboxylase, the only known eukaryotic enzyme that contains bound pyruvate as an essential cofactor (decarboxylases normally contain pyridoxal phosphate, which is absent from S-adenosylmethionine decarboxylase), has a short half-life (1–2 hours) and responds to promoters of cell growth in a manner qualitatively similar to ornithine decarboxylase. Both the rapidity and the extent of the response are, however, less dramatic. S-Adenosylmethionine decarboxylase (Fig 33–5) is inhibited by decarboxylated S-adenosylmethionine and activated by putrescine.

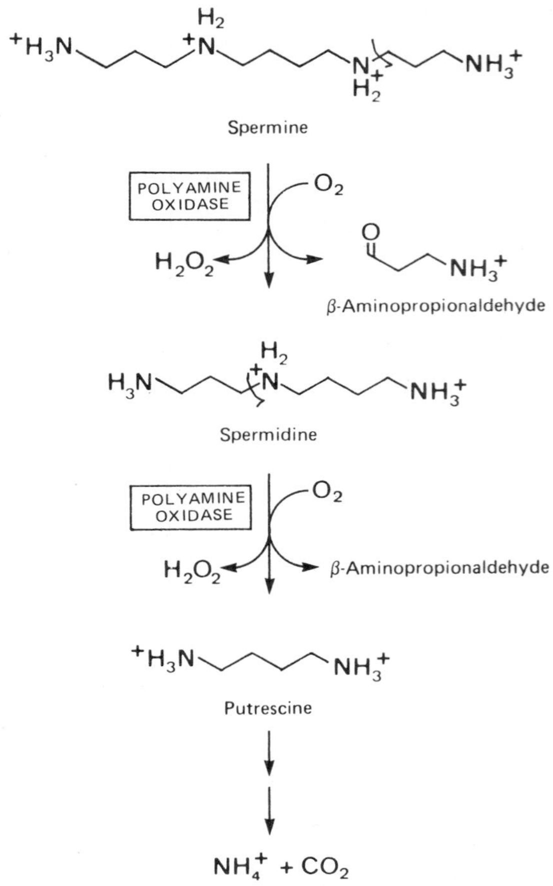

Catabolic Products of Polyamines Are Excreted in Urine

Fig 33–6 summarizes the catabolism of polyamines in mammalian tissues. The enzyme polyamine oxidase, present in liver peroxisomes, oxidizes spermine to spermidine and subsequently oxidizes spermidine to putrescine. Both diaminopropane moieties are converted to β-aminopropionaldehyde. Subsequently, putrescine is partially oxidized to NH_4^+ and CO_2 by mechanisms that remain to be elucidated. However, major portions of putrescine and spermidine are excreted in urine as conjugates, principally as acetyl derivatives.

TRYPTOPHAN FORMS SEROTONIN

A secondary pathway for the metabolism of tryptophan involves hydroxylation to 5-hydroxytryptophan. Oxidation of tryptophan to the hydroxy

Figure 33–6. Catabolism of polyamines. Structures are abbreviated to facilitate presentation.

derivative is analogous to conversion of phenylalanine to tyrosine (Fig 30–10), and liver phenylalanine hydroxylase also catalyzes hydroxylation of tryptophan. Decarboxylation of 5-hydroxytryptophan forms **5-hydroxytryptamine (serotonin)** (Fig 33–7) a potent vasoconstrictor and stimulator of smooth muscle contraction.

The 5-hydroxytryptophan decarboxylase that forms serotonin from hydroxytryptophan is present in the kidney (hog and guinea pig), liver, and stomach. However, the widely distributed aromatic L-amino acid decarboxylase will also catalyze decarboxylation of 5-hydroxytryptophan.

Most serotonin is metabolized by oxidative deamination to 5-hydroxyindoleacetate. The enzyme that catalyzes this reaction is **monoamine oxidase** (Fig 33–7). Inhibitors of this enzyme include **iproniazid.** The psychic stimulation that follows the administration of this drug is attributed to its ability to prolong the stimulating action of serotonin through inhibition of monoamine oxidase. In normal human urine, 2–8 mg of 5-hydroxyindoleacetate is excreted per day.

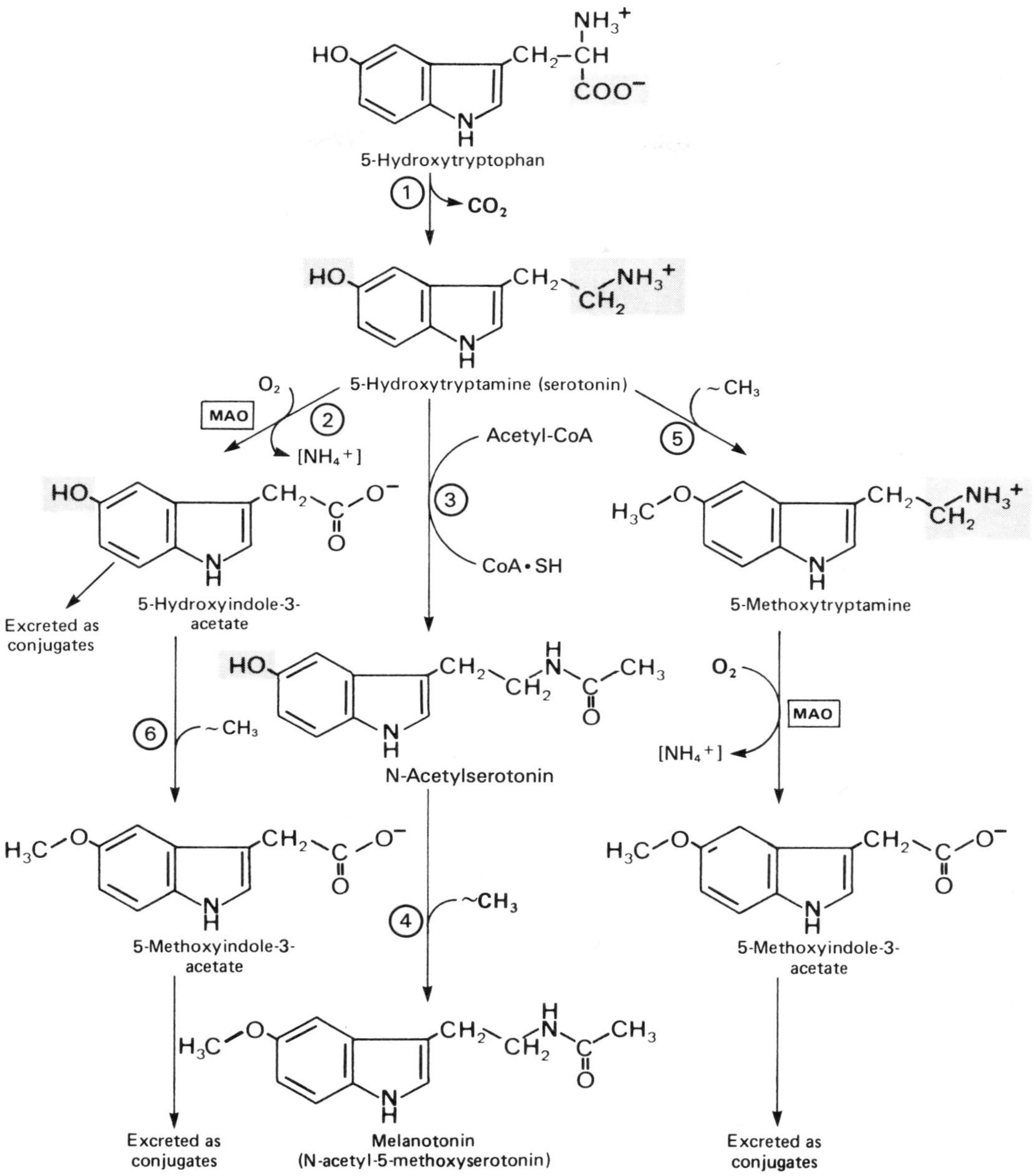

Figure 33–7. Biosynthesis and metabolism of melatonin. [NH_4^+], by transamination; MAO, monoamine oxidase.

Increased Serotonin Production Occurs in Malignant Carcinoid

Greatly increased production of serotonin occurs in malignant **carcinoid** (argentaffinoma), a disease characterized by widespread serotonin-producing tumor cells in the argentaffin tissue of the abdominal cavity. Carcinoid has been considered an abnormality in tryptophan metabolism in which a much greater proportion of tryptophan than normal is metabolized by way of hydroxyindole. One percent of tryptophan is normally converted to serotonin, but in the carcinoid patient as much as 60% may follow this pathway. This metabolic diversion markedly reduces production of nicotinic acid from tryptophan; consequently, symptoms of pellagra as well as negative nitrogen balance may occur. Other metabolites of serotonin identified in the urine of patients with carcinoid include 5-hydroxyindoleaceturate (the glycine conjugate of 5-hydroxyindoleacetate) and N-acetylserotonin conjugated with glucuronic acid.

N-Acetylation of Serotonin Forms Melanotonin

Melatonin is derived from serotonin by N-acetyla-

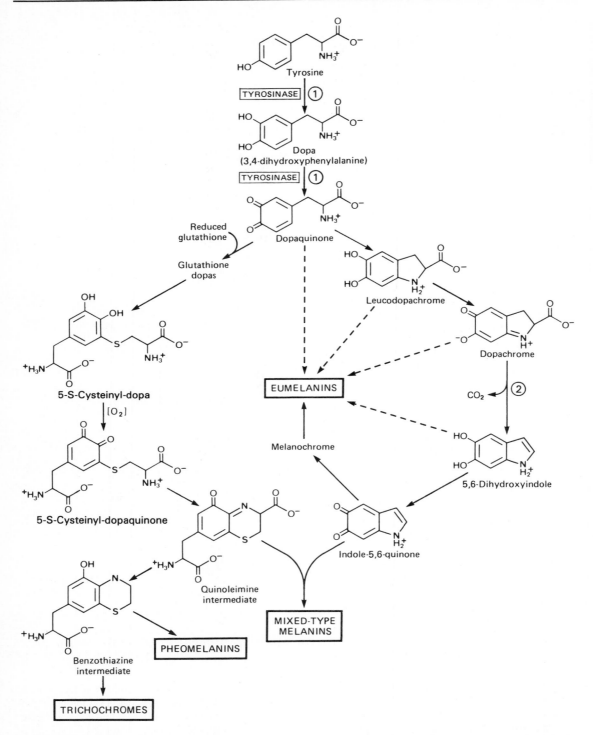

Figure 33–8. Known intermediates and reactions of eumelanin and pheomelanin biosynthesis. Melanin polymers contain both eumelanin and pheomelanin in varying proportions. Dotted arrows indicate that intermediates contribute toward synthesis of eumelanins in varying proportions. Circled numerals indicate probable regulated reactions of the biosynthetic pathway. Reaction 1, that catalyzed by tyrosinase, is defective in tyrosinase-negative oculocutaneous albinism.

tion followed by methylation of the 5-hydroxy group (Fig 33–7). Methylation is localized in pineal body tissue. In addition to methylation of N-acetylserotonin, direct methylation of serotonin and of 5-hydroxyindoleacetate (Fig 33–7), the serotonin metabolite, also occurs.

Serotonin and 5-methoxytryptamine are metabolized to the corresponding acids by monoamide oxidase. Circulating melatonin is taken up by all tissues, including brain, but is rapidly metabolized by hydroxylation at position 6, followed by conjugation with sulfate (70%) and with glucuronic acid (6%). A portion is also converted to nonindolic reacting compounds.

Tryptophan Metabolites Are Excreted in Urine & Feces

Tryptophan may be converted to several indole derivatives (Fig 33–7). The end products of these conversions that appear in the urine are principally 5-hydroxyindoleacetate, the major end product of the hydroxytryptophan-to-serotonin pathway, and indole-3-acetate, from decarboxylation and oxidation of indolepyruvate, the keto acid of tryptophan.

Mammalian kidney and liver and bacteria from human feces decarboxylate tryptophan to tryptamine, which can then be oxidized to indole-3-acetate. Patients with phenylketonuria excrete increased quantities of indoleacetate (and indolelactate, formed by reduction of indolepyruvate).

MELANINS ARE POLYMERS OF TRYPTOPHAN CATABOLITES

Melanin biosynthesis is complicated by the protracted, branched biosynthetic pathway, the complex chemical structures of melanin heteropolymers, and their insolubility, which hinders their structural determination. Melanins are synthesized in **melanosomes**—membrane-bound particles within melanocytes. The developing eumelanin polymer is thought to entrap free radicals and to undergo partial degradation by H_2O_2 generated during the auto-oxidative process. Pheomelanins and eumelanins then complex with proteins of the melanosomal matrix, forming **melanoprotein.**

Fig 33–8 summarizes the known intermediates and reactions of eumelanin and pheomelanin biosynthesis. The initial reaction is catalyzed by tyrosinase, a copper-dependent enzyme. **The tyrosinase reaction is defective in tyrosinase-negative oculocutaneous albinism** (see below).

Several Defects in Melanin Biosynthesis Can Cause Albinism

The term "albinism" encompasses a spectrum of clinical syndromes characterized by **hypomelanosis** arising from heritable defects in the pigment cells

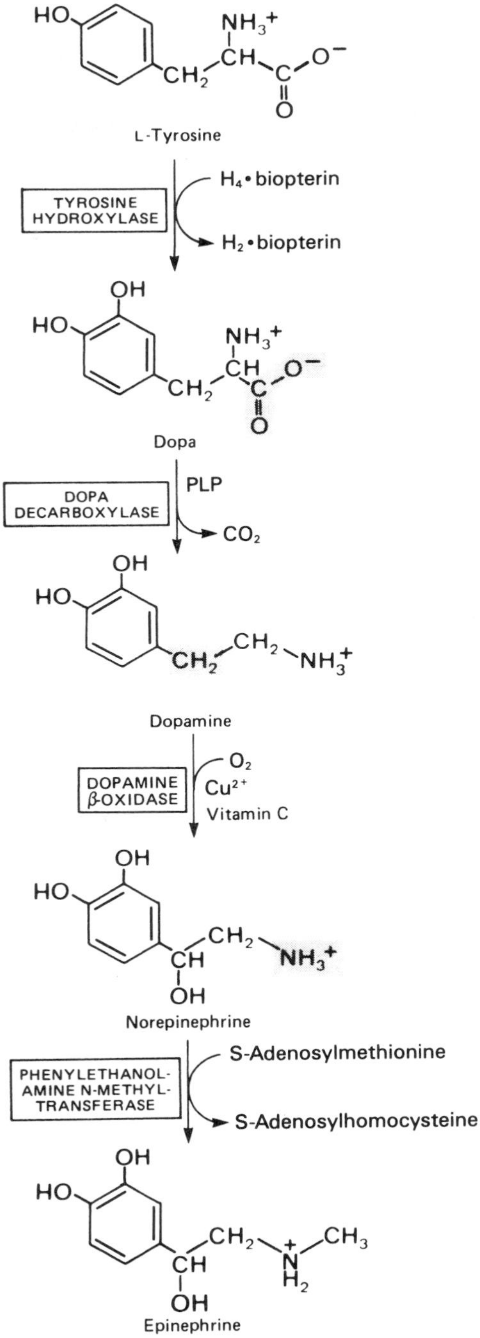

Figure 33–9. Conversion of tyrosine to epinephrine and norepinephrine in neuronal and adrenal cells. PLP, pyridoxal phosphate.

(melanocytes) of the eye and skin. Several useful rodent models of albinism are known.

Clinical signs common to all 10 human forms of **oculocutaneous albinism** include decreased pigmentation of the skin and eye. All 10 forms can be differentiated on the basis of their clinical, biochemical, ultrastructural, and genetic characteristics.

Tyrosinase-negative albinos completely lack visual pigment. Hair bulbs from these patients fail to convert added tyrosine to pigment in vitro, and their melanocytes contain unpigmented melanosomes. **Tyrosinase-positive albinos** have some visible pigment, although this may not be evident in white infants. Hair color ranges from white-yellow to light tan, and lightly pigmented nevi may be present. Hair bulb melanocytes may contain lightly pigmented melanosomes, which convert tyrosine to black eumelanin in vitro.

Ocular albinism occurs both as an autosomal recessive and as an X-linked trait. The melanocytes of X-linked and heterozygous (but not autosomal recessive) ocular albinos contain macromelanosomes. The retinas of females heterozygous for X-linked ocular albinism (Nettleship variety) exhibit a mosaic pattern of pigment distribution due to random X-chromosome inactivation. The precise metabolic defects leading to hypomelanosis in ocular albinism are unknown.

Oculocutaneous albinoidism is inherited as an autosomal recessive trait. With rare exceptions, patients lack associated nystagmus, photophobia, and decreased visual acuity.

TYROSINE FORMS EPINEPHRINE & NOREPINEPHRINE

Tyrosine is a precursor of **epinephrine** and **norepinephrine,** which are formed in cells of neural origin. Although dopa is an intermediate in the formation of both melanin in melanocytes and norepinephrine in neuronal cells, different enzymes carry out the tyrosine hydroxylation reactions in different cell types. The enzyme **tyrosine hydroxylase** forms dopa in neuronal and adrenal cells on the pathway to norepinephrine and epinephrine production (Fig 33–9). **Dopa decarboxylase,** a pyridoxal phosphate-dependent enzyme, forms dopamine. The latter is subjected to further hydroxylation by dopamine β-oxidase, a copper-dependent enzyme that seems to utilize vitamin C to generate norepinephrine. In the **adrenal medulla,** phenylethanolamine-N-methyltransferase utilizes S-adenosylmethionine to methylate the primary amine of norepinephrine to form **epinephrine** (Fig 33–9).

Tyrosine is also a precursor of the thyroid hormones triiodothyronine and thyroxine (see Chapter 47).

CREATININE EXCRETED IS A FUNCTION OF MUSCLE MASS

Creatine is present in muscle, brain, and blood both as phosphocreatine and in the free state. Traces of creatine are also normally present in urine. **Creatinine,** the anhydride of creatine, is formed

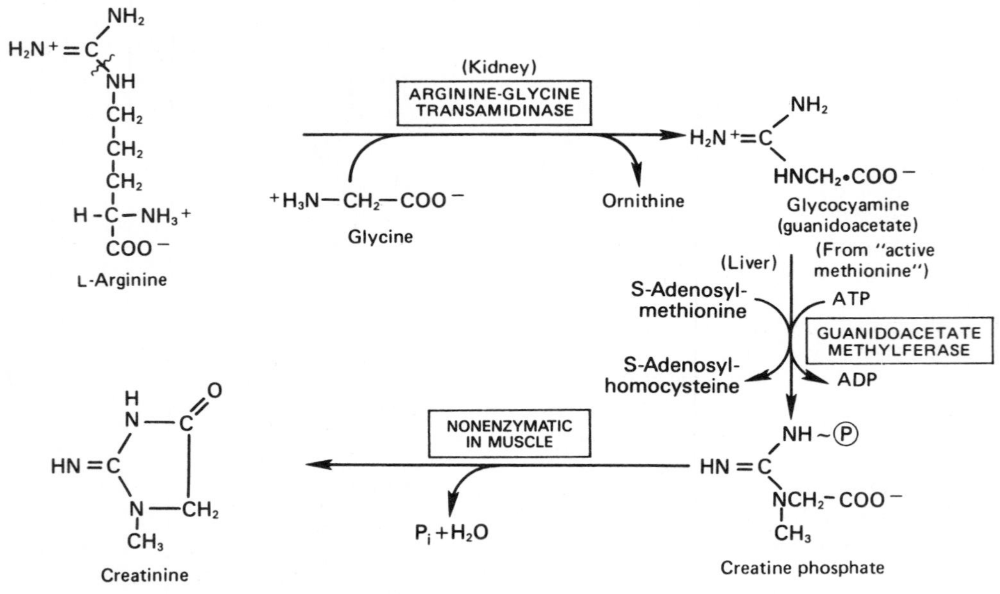

Figure 33–10. Biosynthesis of creatine and creatinine.

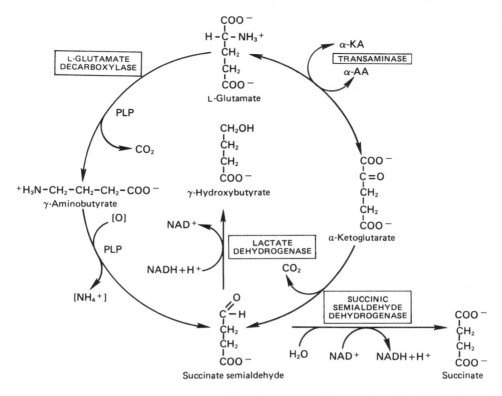

Figure 33–11. Metabolism of γ-aminobutyrate. α-KA, α-keto acids; α-AA, α-amino acids; PLP, pyridoxal phosphate.

largely in muscle by irreversible nonenzymatic dehydration of creatine phosphate (Fig 33–10). The 24-hour excretion of creatinine in the urine of a given subject is remarkably constant from day to day and proportionate to muscle mass.

For synthesis of creatine, 3 amino acids—**glycine, arginine,** and **methionine**—are directly involved. The first reaction is transamidination from arginine to glycine to form guanidoacetate (glycocyamine). This occurs in the kidney, but not in the liver or in heart muscle. Synthesis of creatine is completed by methylation of glycocyamine by "active methionine" in the liver (Fig. 33–10).

γ-AMINOBUTYRATE, FORMED FROM GLUTAMATE, IS CATABOLIZED TO SUCCINATE SEMIALDEHYDE

γ-Aminobutyrate (GABA) is formed by decarboxylation of L-glutamate, a reaction catalyzed by the pyridoxal phosphate-dependent enzyme L-glutamate decarboxylase (Fig 33–11). This decarboxylase is present in the tissues of the central nervous system, principally in the gray matter. Two minor reaction sequences also convert putrescine (Fig 33–4) to γ-aminobutyrate. One involves deamination by diamine oxidase; the other utilizes N-acetylated intermediates.

The relative importance of these 3 routes of γ-aminobutyrate biosynthesis varies among tissues and with developmental stage. For example, ornithine (Fig 33–5) is efficiently converted to γ-aminobutyrate in embryonic chick retinal tissue and in adult brain nerve terminals.

Catabolism of γ-aminobutyrate (Fig 33–11) involves transamination, catalyzed by γ-aminobutyrate transaminase, to succinate semialdehyde. Succinate semialdehyde may undergo reduction to γ-hydroxybutyrate, a reaction catalyzed by L-lactate dehydrogenase, or oxidation to the citric acid cycle intermediate succinate and thence to CO_2 and H_2O.

γ-Aminobutyrate, in common with the anions of other ω-amino acids, is poorly transported across plasma cell membranes. The urinary levels of γ-aminobutyrate vary directly with the serum levels of this compound. While the precise biochemical defect is not known, it may result from impaired transamination of γ-aminobutyrate to succinate semialdehyde.

REFERENCES

Scriver CR et al (editors): *The Metabolic Basis of Inherited Disease,* 6th ed. McGraw-Hill, 1989.
Tabor CW, Tabor H: Polyamines. *Annu Rev Biochem* 1984; **53**:749.

34

Porphyrins & Bile Pigments

Robert K. Murray, MD, PhD

INTRODUCTION

The biochemistry of the porphyrins and of the bile pigments is presented in this chapter. These topics are closely related, because heme is synthesized from porphyrins and iron, and the products of its degradation are the bile pigments and iron.

BIOMEDICAL IMPORTANCE

Knowledge of the biochemistry of the porphyrins and of heme is basic to understanding the varied **functions of hemoproteins** (involvement in oxygen transport, electron transport, drug metabolism, etc) in the body. The **porphyrias** are a group of diseases caused by abnormalities in the pathway of biosynthesis of the various porphyrins. They are uncommon, but medical practitioners must be aware of them and dermatologists, hepatologists, and psychiatrists will encounter patients with these conditions. A much more common clinical condition is **jaundice,** due to elevation of bilirubin in the plasma. This elevation is due to overproduction of bilirubin or to failure of its excretion and is seen in numerous diseases, ranging from viral hepatitis to cancer of the pancreas.

METALLOPORPHYRINS ARE IMPORTANT IN NATURE

Porphyrins are cyclic compounds formed by the linkage of 4 pyrrole rings through methenyl bridges (Fig 34–1). A characteristic property of the porphyrins is the formation of complexes with metal ions bound to the nitrogen atom of the pyrrole rings. Examples are the iron porphyrins such as **heme** of hemoglobin and the magnesium-containing porphyrin **chlorophyll,** the photosynthetic pigment of plants.

In nature, the metalloporphyrins are conjugated to proteins to form many compounds important in biologic processes. These include the following:

Hemoglobins: Iron porphyrins attached to the protein, globin. These conjugated proteins possess the ability to combine reversibly with oxygen. They serve as the transport mechanism for oxygen within the

Figure 34–1. The porphin molecule. Rings are labeled I, II, III, IV. Substituent positions on rings are labeled 1, 2, 3, 4, 5, 6, 7, 8. Methenyl bridges (−HC =) are labeled α, β, γ, δ.

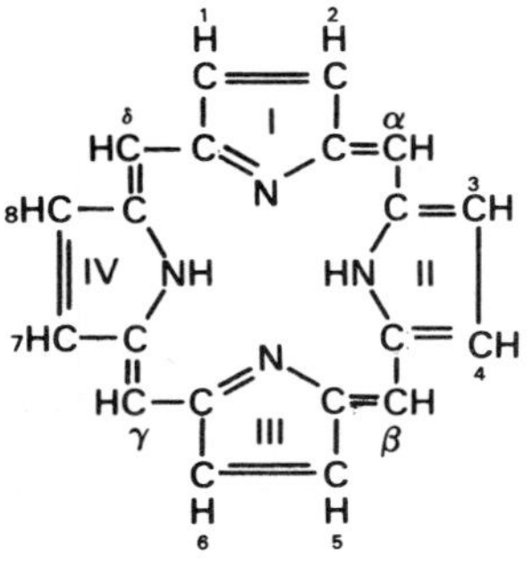

blood (see Chapter 7). The structure of heme is shown in Fig 7–2.

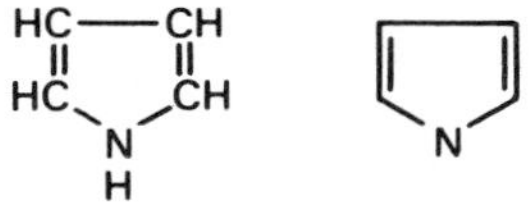

Erythrocruorins: Iron porphyrinoproteins occurring in the blood and in the tissue fluids of some invertebrates. They correspond in function to hemoglobin.

Myoglobins: Respiratory proteins that occur in the **muscle cells** of vertebrates and invertebrates. A myoglobin molecule is similar to a subunit of hemoglobin.

Cytochromes: Compounds that act as electron transfer agents in oxidation-reduction reactions. An important example is **cytochrome c,** which has a molecular weight of about 13,000 and contains 1 gram-atom of iron per mole.

Catalases: Iron porphyrin enzymes that degrade hydrogen peroxide; several have been obtained in crystalline form. In plants, catalase activity is minimal, but the iron porphyrin enzyme peroxidase performs similar functions.

Tryptophan Pyrrolase: This enzyme catalyzes the oxidation of tryptophan to formyl kynurenine. It is an iron porphyrinoprotein.

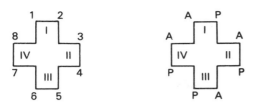

Figure 34–2. Uroporphyrin III.

Natural Porphyrins Have Substituent Side Chains on the Porphin Nucleus

The porphyrins found in nature are compounds in which various **side chains** are substituted for the 8 hydrogen atoms numbered in the porphin nucleus shown in Fig 34–1. As a simple means of showing these substitutions, Fischer proposed a shorthand formula in which the methenyl bridges are omitted and each pyrrole ring is shown as a bracket with the 8 substituent positions numbered as shown (Fig 34–2). Various porphyrins are represented in Figs 34–2 through 34–4 (A [acetate] = –CH₂COOH; P [propionate] = –CH₂CH₂COOH; M [methyl] = –CH₃; V [vinyl] = –CH=CH₂).

The arrangement of the A and P substituents in the uroporphyrin shown in Fig 34–2 is asymmetric (in ring IV, the expected order of the acetate and propionate substituents is reversed). A porphyrin with this type of **asymmetric substitution** is classified as a type III porphyrin. A porphyrin with a completely symmetric arrangement of the substituents is classified as a type I porphyrin. Only types I and III are found in nature, and the **type III series** is by far the more abundant (Fig 34–3) and more important, because it contains heme.

Heme and its immediate precursor, protoporphyrin

IX (Fig 34–4), are both type III porphyrins (ie, the methyl groups are asymmetrically distributed, as in type III coproporphyrin). However, they are sometimes identified as belonging to series IX, because they were designated ninth in a series of isomers postulated by Hans Fischer, the pioneer worker in the field of porphyrin chemistry.

PORPHYRINS HEME IS SYNTHESIZED FROM SUCCINYL-CoA & GLYCINE

Heme, the iron protoporphyrin of hemoglobin is synthesized in living cells by a pathway that has been studied. The 2 starting materials are **succinyl-CoA,** derived from the citric acid cycle in mitochondria, and the amino acid **glycine.** Pyridoxal phosphate is also necessary in this reaction to "activate" glycine. The product of the condensation reaction between succinyl-CoA and glycine is α-amino-β-ketoadipic acid, which is rapidly decarboxylated to form δ-aminolevulinate (ALA) (Fig 34–5). This step is catalyzed by the enzyme **ALA synthase.** This appears to be the **rate-controlling** enzyme in porphyrin biosynthesis in mammalian liver. Synthesis of ALA occurs in the mitochondria. In the cytosol, 2 molecules of ALA are condensed by the enzyme ALA dehydratase to form 2 molecules of water and one of porphobilinogen (PBG) (Fig 34–5). ALA dehydratase is a Zn-containing enzyme and is sensitive to inhibition by lead.

The formation of a tetrapyrrole, ie, a porphyrin, occurs by condensation of 4 molecules of PBG (Fig 34–6). These 4 molecules condense in a head-to-tail manner to form a linear tetrapyrrole, hydroxymethylbilane. The reaction is catalyzed by uroporphyrinogen I synthase, also known as porphobilinogen deaminase.

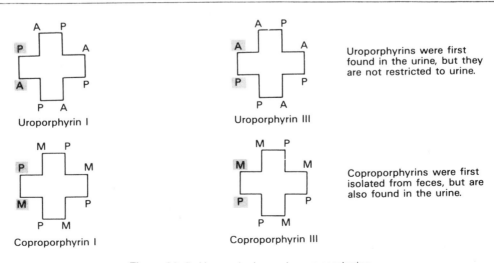

Uroporphyrins were first found in the urine, but they are not restricted to urine.

Coproporphyrins were first isolated from feces, but are also found in the urine.

Figure 34–3. Uroporphyrins and coproporphyrins.

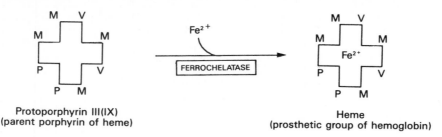

Figure 34–4. Addition of iron to protoporphyrin to form heme.

Hydroxymethylbilane cyclizes spontaneously to form **uroporphyrinogen I** (left-hand side of Fig 34–6), or is converted to **uroporphyrinogen III** by the combined action of uroporphyrinogen I synthase and uroprophyrinogen III cosynthase (right-hand side of Fig 34–6). Under normal conditions, the uroporphyrinogen formed is almost exclusively the III isomer, but in certain of the porphyrias (discussed below) the type I isomers of porphyrinogens are also formed in excess.

Note that both of these uroporphyrinogens have the pyrrole rings connected by methylene bridges $(-CH_2 =)$, which do not form a conjugated ring system. Thus, these compounds (as are all porphyrinogens) are colorless. However, the porphyrinogens are readily auto-oxidized to their respective porphyrins.

These oxidations are catalyzed by light and by the porphyrins that are formed.

Uroporphyrinogen III is converted to coproporphyrinogen III by decarboxylation of all of the acetate (A) groups, which changes them to methyl (M) substituents. The reaction is catalyzed by **uroporphyrinogen decarboxylase,** which is also capable of converting uroporphyrinogen I to coproporphyrinogen I (Fig 34–7). Coproporphyrinogen III then enters the mitochondria, where it is converted to **protoporphyrinogen III** and then to **protoporphyrin III.** Several steps seem to be involved in this conversion. The mitochondrial enzyme **coproporphyrinogen oxidase** catalyzes the decarboxylation and oxidation of 2 propionic side chains to form protoporphyrinogen. This enzyme is able to act only on type III coproporphyrinogen,

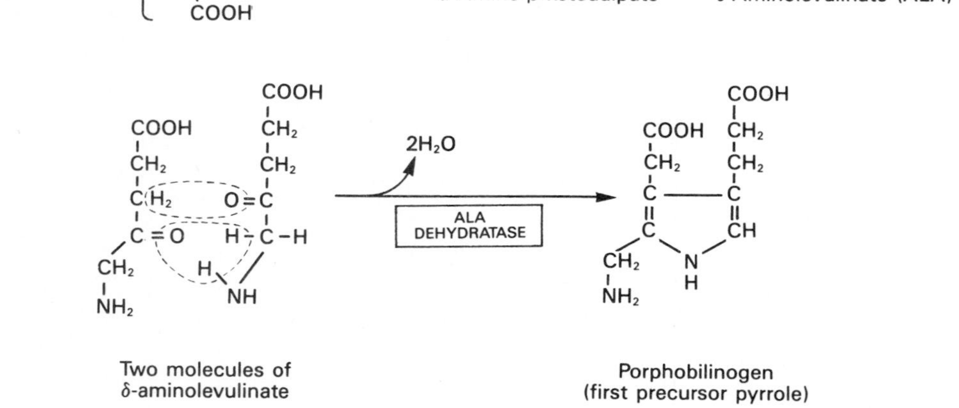

Figure 34–5. Biosynthesis of porphobilinogen. ALA synthase occurs in the mitochondria, whereas ALA dehydratase is present in the cytosol.

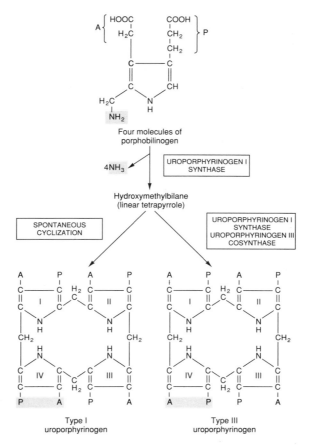

Figure 34–6. Conversion of porphobilinogen to uroporphyrinogens.

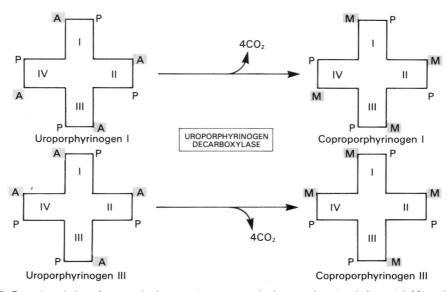

Figure 34–7. Decarboxylation of uroporphyrinogens to coproporphyrinogens in cytosol. A, acetyl; M, methyl; P, propyl.

which would explain why a type I protoporphyrin has not been identified in natural materials. The oxidation of protoporphyrinogen to protoporphyrin is catalyzed by another mitochondrial enzyme, **protoporphyrinogen oxidase.** In mammalian liver, the conversion of coproporphyrinogen to protoporphyrin requires molecular oxygen.

Formation of Heme Involves Incorporation of Fe Into Protoporphyrin

The final step in heme synthesis involves the incorporation of ferrous iron into protoporphyrin in a reaction catalyzed by **heme synthase** or **ferrochelatase,** another mitochondrial enzyme (Fig 34–4).

A summary of the steps in the biosynthesis of the porphyrin derivatives from PBG is given in Fig 34–8. Heme biosynthesis occurs in most mammalian tissues with the exception of mature erythrocytes, which do not contain mitochondria.

The porphyrinogens that have been described above are colorless, containing 6 extra hydrogen atoms as compared to the corresponding colored porphyrins. It is now apparent that these **reduced porphyrins** (the porphyrinogens) and not the corresponding porphyrins are the actual intermediates in the biosynthesis of protoporphyrin and of heme.

ALA Synthase Is the Key Regulatory Enzyme in Heme Biosynthesis

The rate-limiting reaction for the synthesis of heme occurs at the condensation of succinyl-CoA and glycine to form ALA (Fig 34–5), a reaction catalyzed by the enzyme δ-aminolevulinic acid synthase (ALA synthase). ALA synthase is a regulated enzyme. It appears that heme, probably acting through an aporerepressor molecule, acts as a negative regulator of the accumulation of ALA synthase. This repression and derepression mechanism is depicted diagrammatically in Fig 34–9. It is possible that there is also significant feedback inhibition at this step, but the major regulatory effect of heme appears to be one in which the rate of accumulation of ALA synthase increases greatly in the absence of heme and is diminished in its presence. The rate of ALA synthase turn-

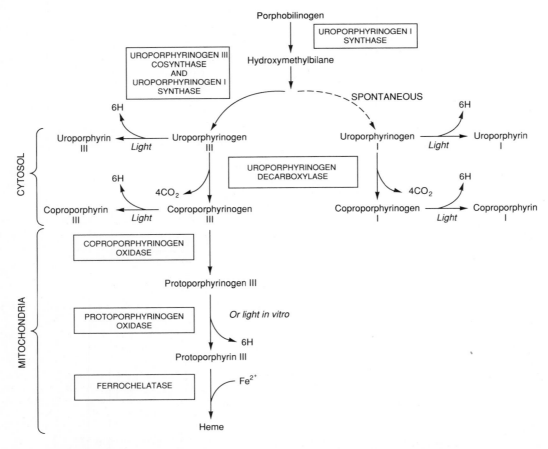

Figure 34–8. Steps in the biosynthesis of the porphyrin derivatives from porphobilinogen.

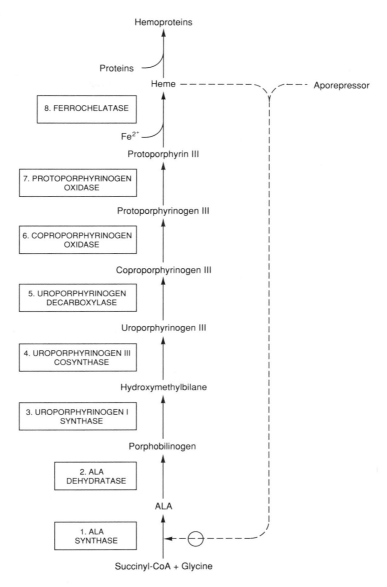

Figure 34–9. Intermediates, enzymes, and regulation of heme synthesis. The enzyme numbers are those referred to in Table 34–1. Lesions of enzymes 2–8 cause the porphyrias. Regulation of heme synthesis occurs at ALA synthase by a repression-derepression mechanism mediated by heme and its hypothetical aporepressor. The dotted lines indicate the negative ($\ominus$) regulation by repression.

over is normally rapid (half-life is about 1 hour) in mammalian liver, not a surprising property for an enzyme catalyzing a rate-limiting reaction.

Many xenobiotics (see Chapter 60), when administered to humans, can result in a marked increase in hepatic ALA synthase. Most of these compounds are metabolized by a system in the liver that utilizes a specific hemoprotein, cytochrome P-450. During their metabolism, the utilization of heme by cytochrome P-450 is greatly increased, which in turn diminishes the intracellular heme concentration. This latter event effects a derepression of ALA synthase with a corre-

sponding increased rate of heme synthesis to meet the needs of the cells.

Several other factors affect the induction of ALA synthase in the liver. **Glucose** can prevent the induction of ALA synthase; **iron** in chelated form exerts a synergistic effect on the induction of hepatic ALA synthase; and **steroids** play at least a permissive role in the drug-mediated derepression of ALA synthase in vivo. The administration of **hematin** in vivo can prevent the drug-mediated derepression of ALA synthase, as well as that of other hemoproteins in liver.

The importance of some of these regulatory mechanisms is further discussed below when the porphyrias are described.

PORPHYRINS ARE COLORED & FLUORESCE

The various porphyrinogens are colorless, whereas the various **porphyrins are all colored.** In the study of porphyrins or porphyrin derivatives, the characteristic absorption spectrum that each exhibits, in both the visible and the ultraviolet regions of the spectrum, is of great value. An example is the absorption curve for a solution of porphyrin in 5% hydrochloric acid (Fig 34–10). Note particularly the sharp absorption band near 400 nm. This is a distinguishing feature of the porphin ring and is characteristic of all porphyrins regardless of the side chains present. This band is termed the **Soret band,** after its discoverer.

When porphyrins dissolved in strong mineral acids or in organic solvents are illuminated by ultraviolet light, they emit a strong red fluorescence. This **fluorescence** is so characteristic that it is frequently used to detect small amounts of free porphyrins. The double bonds in the porphyrins are responsible for the characteristic absorption and fluorescence of these compounds, and, as previously noted, the reduction (by addition of hydrogen) of the methenyl (–HC=) bridges to methylene (–CH$_2$–) leads to the formation of colorless compounds termed porphyrinogens.

Spectrophotometry is Used to Test for Porphyrins & Their Precursors

The presence of coproporphyrins or of uroporphyrins is of clinical interest, since these 2 types of compounds are excreted in increased amounts in the porphyrias. These compounds, when present in urine or feces, can be separated from each other by extraction with appropriate solvent mixtures. They can then be identified and quantified using spectrophotometric methods.

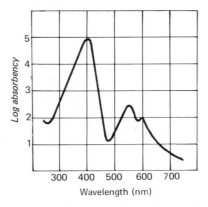

Figure 34–10. Absorption spectrum of hematoporphyrin (0.01% solution in 5% HCl).

ALA and PBG can also be measured in urine by appropriate colorimetric tests.

THE PORPHYRIAS ARE GENETIC DISORDERS OF HEME METABOLISM

The **porphyrias** are a group of inborn errors of metabolism due to mutations in the genes directing the synthesis of the enzymes involved in the biosynthesis of heme. They do not occur frequently, but it is important to consider them in certain circumstances (eg, in the differential diagnosis of abdominal pain and of a variety of neuropsychiatric findings); otherwise patients will be subjected to inappropriate treatments. It has been speculated that King George III had variegate porphyria, which may account for his periodic confinements in Windsor Castle and possibly some of his views regarding American colonists. Also, the photosensitivity (favoring nocturnal activities) and severe disfigurement exhibited by some victims of congenital erythropoietic porphyria have led to the suggestion that these individuals may have been the prototypes of werewolves.

Biochemistry Underlies the Causes, Diagnoses & Treatments of the Porphyrias

Six types of porphyria have been described, resulting from depressions in the activities of each of the enzymes 3–8 shown in Fig 34–9 (see Table 34–1). Assay of the activity of one or more of these enzymes using an appropriate source (eg, red blood cells) is thus important in making a definitive diagnosis in a suspected case of porphyria. Individuals with **low activities of enzyme 1 (ALA synthase)** have not been reported. Subjects exhibiting depressions of the activity of enzyme 2 (ALA dehydratase) have been reported but are very rare.

In general, the porphyrias described are **inherited** in an autosomal dominant manner, with the exception of congenital erythropoietic porphyria, which is inherited in a recessive mode. The precise abnormalities in the genes directing synthesis of the enzymes involved in heme biosynthesis are being determined by the methods of recombinant DNA technology.

As is the case of most inborn errors, the **clinical signs and symptoms** in cases of porphyria result from either a **deficiency** of metabolic products beyond the enzymatic block or from an **accumulation** of metabolites behind the block.

Where the **enzyme lesion occurs early in the pathway prior to the formation of porphyrinogens** (eg, enzyme 3 of Fig 34–9, in intermittent acute porphyria), ALA and PBG will accumulate in body tissues and fluids (Fig 34–11). One or both of these compounds can cause toxic effects in abdominal nerves and in the central nervous system, resulting in the **abdominal pain and neuropsychiatric symptoms** seen in this type of porphyria. Possible biochemical bases for these symptoms are that ALA may inhibit an ATPase

Table 34–1. Summary of major findings in the porphyrias.

*Enzyme Involved	Porphyria Type and [Class]	Major Symptoms	Results of Lab Tests
3. Uroporphyrinogen I cosynthetase	Acute intermittent porphyria [Hepatic]	Abdominal pain Neuropsychiatric No photosensitivity	Urinary PBG + Urinary uroporphyrin +
4. Uroporphyrinogen III cosynthetase	Congenital erythropoietic [Erythropoietic]	Photosensitivity	Urinary uroporphyrin + Urinary PBG −
5. Uroporphyrinogen decarboxylase	Porphyria cutanea tarda [Hepatic]	Photosensitivity	Urinary uroporphyrin + Urinary PBG −
6. Coproporphyrinogen oxidase	Hereditary coproporphyria [Hepatic]	Photosensitivity Abdominal pain Neuropsychiatric	Urinary PBG + Urinary uroporphyrin + Fecal coproporphyrin +
7. Protoporphyrinogen oxidase	Variegate porphyria [Hepatic]	Photosensitivity Abdominal pain Neuropsychiatric	Urinary PBG + Urinary uroporphyrin + Fecal protoporphyrin +
8. Ferrochelatase	Protoporphyria [Erythrohepatic]	Photosensitivity	Fecal protoporphyrin + Red cell protoporphyrin +

Only the biochemical findings in the active stages of these diseases are indicated. Certain biochemical abnormalities are detectable in the latent stages of some of the above conditions.
*The numbering of the enzymes in this table corresponds to that used in Fig 34–9.

in nervous tissue and/or that ALA may be taken up by brain and somehow cause a conduction paralysis.

On the other hand, **enzyme blocks later in the pathway** result in the accumulation of the porphyrinogens indicated in Figs 34–9 and 34–11. Their oxidation products, the corresponding porphyrin derivatives, cause **photosensivity,** a reaction to visible light of about 400 nm. The porphyrins, when exposed to light of this wavelength, are thought to become "excited" and then react with molecular oxygen to form oxygen radicals. These latter species injure lysosomes and other organelles. Damaged lysosomes release their degradative enzymes, causing variable degrees of skin damage, including scarring.

The porphyrias can be **classified on the basis of the organ or cells that are most affected.** These are generally organs or cells in which synthesis of heme is particularly active. The bone marrow synthesizes considerable hemoglobin per day and the liver is very active in the synthesis of another hemoprotein, cytochrome P-450. Thus one classification of the porphyrias is to designate them as **erythropoietic, hepatic, and erythrohepatic** (mixed); the types of porphyria that fall into these classes are indicated in Table 34–1.

The question arises as to why specific types of porphyria affect certain organs more markedly than others. A partial answer is that the levels of metabolites

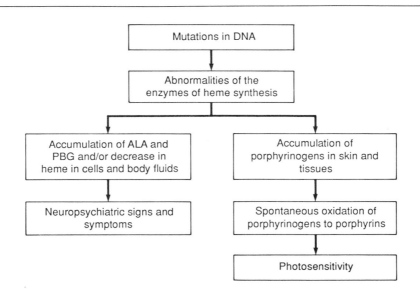

Figure 34–11. Biochemical causes of the major signs and symptoms of the porphyrias.

that cause damage (eg, ALA, PBG or specific porphyrins) can vary markedly in different organs or cells, depending upon the differing activities of their heme-forming enzymes.

As described above, ALA synthase is the key regulatory enzyme of the heme biosynthetic pathway. Although this enzyme is not directly implicated as a cause of the six types of porphyria discussed here, it is important to understand its regulation in order to comprehend some features of these diseases. ALA synthase is subject to both induction and repression and its activity can increase markedly (up to 50-fold) under certain conditions. A large number of **drugs** (eg, barbiturates, griseofulvin) induce the enzyme. Most of these drugs do so by inducing cytochrome P-450 (see Chapter 60), which uses up heme and thus derepresses (induces) ALA synthase. In patients with porphyria, increased activities of ALA synthase result in increased level of potentially harmful heme precursors prior to the metabolic block. Thus, taking drugs that cause induction of cytochrome P-450 (microsomal inducers) can precipitate attacks of porphyria.

The **diagnosis** of a specific type of porphyria can generally be established by consideration of the **clinical and family history,** of the **physical examination** and of appropriate **laboratory tests.** The major findings in the six types of porphyria are listed in Table 34–1. High levels of **lead** can affect heme metabolism, by combining with SH groups in enzymes such as ferrochelatase and ALA dehydrase. This affects porphyrin metabolism. Elevated levels of protoporphyrin are found in red blood cells and elevated levels of ALA and of coproporphyrin in urine.

It is hoped that appropriate **treatment of the porphyrias at the gene level** may become possible. In the meantime, **treatment is essentially symptomatic.** It is important for patients to avoid any anesthetics and drugs, including alcohol, that cause induction of cytochrome P-450. Ingestion of large amounts of food rich in carbohydrates (glucose loading) or administration of hematin (a hydroxide of heme) may repress ALA synthase, resulting in diminished production of harmful heme precursors. Patients exhibiting photosensitivity may benefit from administration of β-carotene; this compound appears to lessen production of free radicals, thus diminishing photosensitivity. Sunscreens that filter out visible light can also be helpful to such patients.

CATABOLISM OF HEME PRODUCES BILIRUBIN

Under physiologic conditions in the human adult, $1–2 \times 10^8$ erythrocytes are destroyed per hour. Thus, in 1 day, a 70-kg human turns over approximately 6 g of hemoglobin. When hemoglobin is destroyed in the body, the protein portion, **globin,** may be reutilized either as such or in the form of its constituent amino acids, and the **iron** of heme enters the iron pool, also for reuse. However, the iron-free **porphyrin** portion of heme is degraded, mainly in the reticuloendothelial cells of the liver, spleen, and bone marrow.

The catabolism of heme from all of the heme proteins appears to be carried out in the microsomal fractions of the reticuloendothelial cells by a complex enzyme system called **heme oxygenase.** By the time the heme of heme proteins reaches the heme oxygenase system, the iron has usually been oxidized to the ferric form, constituting **hemin,** and may be loosely bound to albumin as methemalbumin. The heme oxygenase system is substrate-inducible. It is located in close proximity to the microsomal electron transport system. As depicted in Fig 34–12, the hemin is reduced with NADPH, and, with the aid of more NADPH, oxygen is added to the α-methenyl bridge between pyrroles I and II of the porphyrin. The ferrous iron is again oxidized to the ferric form. With the further addition of oxygen, **ferric ion** is released, **carbon monoxide** is produced, and an equimolar quantity of **biliverdin IX-α** results from the splitting of the tetrapyrrole ring. The heme itself participates in this reaction as a catalyst.

In birds and amphibia, the green biliverdin IX-α is excreted; in mammals, a soluble enzyme called **biliverdin reductase** reduces the methenyl bridge between pyrrole III and pyrrole IV to a methylene group to produce **bilirubin IX-α**, a yellow pigment (Fig 33–12).

It is estimated that 1 g of hemoglobin yields 35 mg of bilirubin. The daily bilirubin formation in human adults is approximately 250–350 mg.

The chemical conversion of heme to bilirubin by the reticuloendothelial cells can be observed in vivo as the purple color of the heme in a hematoma is slowly converted to the yellow pigment of bilirubin.

The further metabolism of bilirubin occurs primarily in the liver. It can be divided into 3 processes: (1) uptake of bilirubin by liver parenchymal cells, (2) conjugation of bilirubin in the smooth endoplasmic reticulum, and (3) secretion of conjugated bilirubin into the bile. Each of these processes will be considered separately.

Uptake of Bilirubin by the Liver

Bilirubin is only sparingly soluble in plasma and water, but in the plasma it is protein-bound, specifically to albumin. Each molecule of albumin appears to have one high-affinity site and one low-affinity site for bilirubin. In 100 mL of plasma, approximately 25 mg of bilirubin can be **tightly bound to albumin** at its high-affinity site. Bilirubin in excess of this quantity can be bound only loosely and thus can easily be detached and diffused into tissues. A number of compounds such as antibiotics and other **drugs** compete with bilirubin for the high-affinity binding site on albumin. Thus, these compounds can displace bilirubin from albumin and have significant clinical effects.

In the liver, the bilirubin seems to be removed from the albumin and taken up at the sinusoidal surface of the hepatocytes by a carrier-mediated saturable system. This **facilitated transport system** has a very

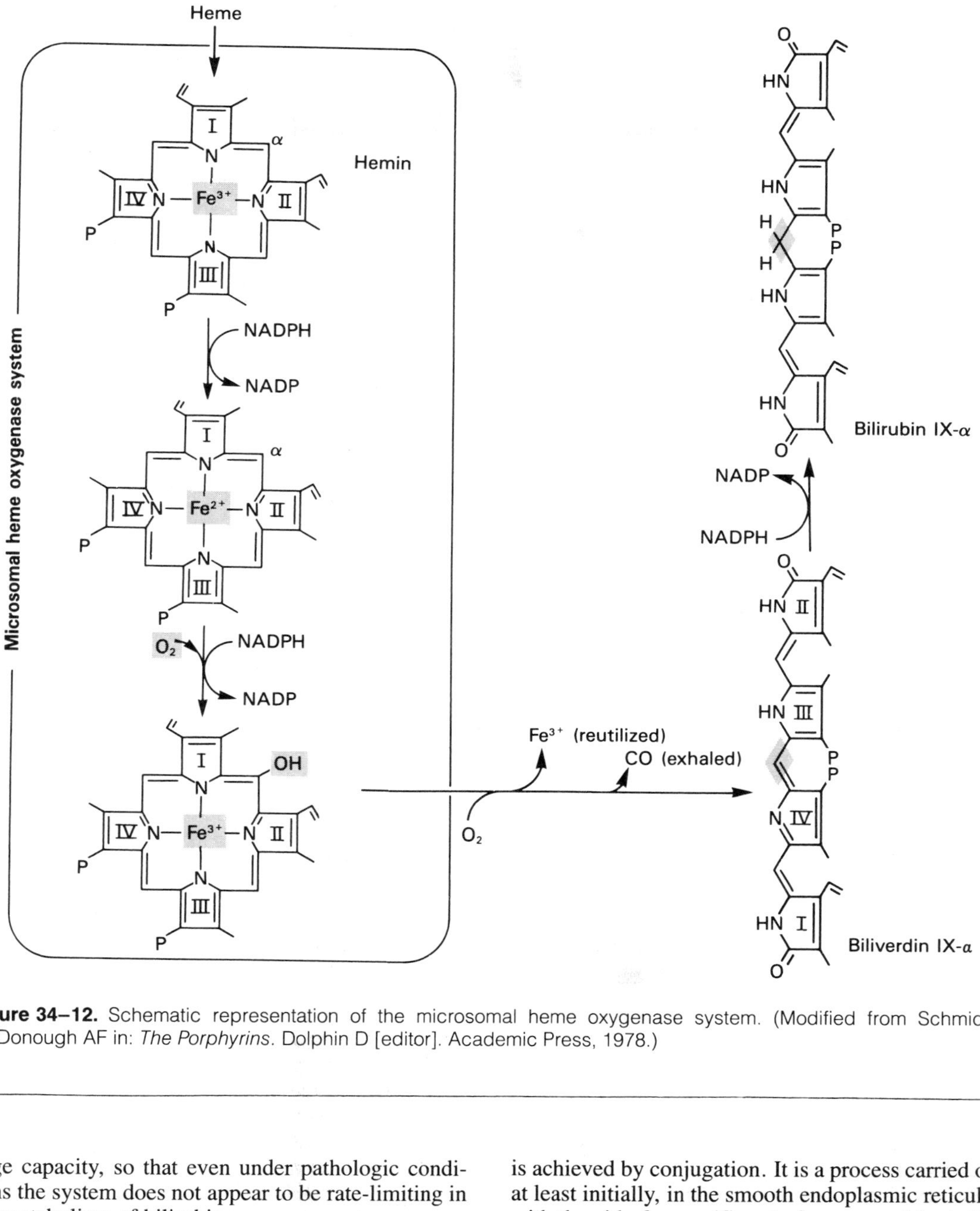

Figure 34–12. Schematic representation of the microsomal heme oxygenase system. (Modified from Schmid R, McDonough AF in: *The Porphyrins.* Dolphin D [editor]. Academic Press, 1978.)

large capacity, so that even under pathologic conditions the system does not appear to be rate-limiting in the metabolism of bilirubin.

Since this facilitated transport system allows the equilibrium of bilirubin across the sinusoidal membrane of the hepatocyte, the net uptake of bilirubin will be dependent upon the removal of bilirubin by subsequent metabolic pathways.

Conjugation of Bilirubin Occurs in the Liver

By adding polar groups to bilirubin, the liver converts bilirubin to a water-soluble form that can subsequently be secreted into the bile. This process of **increasing the water solubility** or polarity of bilirubin

is achieved by conjugation. It is a process carried out, at least initially, in the smooth endoplasmic reticulum with the aid of a specific set of enzymes. Most of the bilirubin excreted in the bile of mammals is in the form of a **bilirubin diglucuronide** (Fig 34–13). The formation of the intermediate monoglucuronide of bilirubin is catalyzed by uridine diphosphate glucuronate glucuronyltransferase (**UDP-glucuronyltransferase**), an enzyme that exists in the smooth endoplasmic reticulum and is probably composed of more than a single entity. The reaction is depicted in Fig 34–14. It occurs chiefly in the liver but also in the kidney and the intestinal mucosa. When bilirubin conjugates exist abnormally in human serum, they are predominantly monoglucuronides.

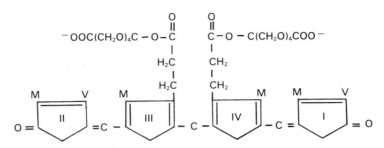

Figure 34–13. Structure of bilirubin diglucuronide (conjugated, "direct-reacting" bilirubin). Glucuronic acid is attached via ester linkage to the 2 propionic acid groups to form an acylglucuronide.

The formation of the diglucuronide of bilirubin may occur in the bile canalicular region of the hepatocyte membrane by a similar UDP-glucuronyltransferase (Fig 34–14) or by an enzyme-catalyzed dismutation of 2 mol of bilirubin monoglucuronide to 1 mol of bilirubin diglucuronide and 1 mol of unconjugated bilirubin (Fig 34–14). More will be said about this conjugation system in the discussion of the inherited disorders of bilirubin conjugation.

UDP-glucuronyltransferase activity can be **induced** by a number of clinically useful drugs, including phenobarbital.

Bilirubin Is Secreted Into Bile

Secretion of conjugated bilirubin into the bile occurs against a large concentration gradient and must be carried out by an active transport mechanism. The **active transport** is probably **rate-limiting** for the entire process of hepatic bilirubin metabolism. The hepatic transport of conjugated bilirubin into the bile is inducible by those same drugs that are capable of inducing the conjugation of bilirubin. Thus, the conjugation and excretion systems for bilirubin behave as a coordinated functional unit.

Under physiologic conditions, essentially all (<97%) of the bilirubin secreted into the bile is conjugated. Only after phototherapy can significant quantities of unconjugated bilirubin be found in bile.

In the liver, there are multiple systems for secreting naturally occurring and pharmaceutical compounds into the bile after their metabolism. Some of these secreting systems are shared by the bilirubin diglucuronides, but others operate independently.

Conjugated Bilirubin Is Reduced to Urobilinogen by Intestinal Bacteria

As the conjugated bilirubin reaches the terminal ileum and the large intestine, the glucuronides are removed by specific bacterial enzymes (β-**glucuronidases**), and the pigment is subsequently reduced by the fecal flora to a group of colorless tetrapyrrolic com-

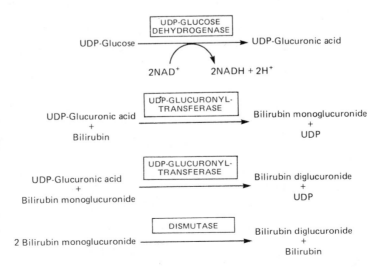

Figure 34–14. Conjugation of bilirubin with glucuronic acid. The glucuronate donor, UDP-glucuronic acid, is formed from UDP-glucose as depicted.

pounds called **urobilinogens** (Fig 34–15). In the terminal ileum and large intestine, a small fraction of the urobilinogens is reabsorbed and reexcreted through the liver to constitute the **intrahepatic urobilinogen cycle.** Under abnormal conditions, particularly when excessive bile pigment is formed or liver disease interferes with this intrahepatic cycle, urobilinogen may also be excreted in the urine.

Normally, most of the **colorless urobilinogens** formed in the colon by the fecal flora are oxidized there to urobilins (colored compounds) and are excreted in the feces (Fig 34–15). Darkening of feces upon standing in air is due to the oxidation of residual urobilinogens to urobilins.

HYPERBILIRUBINEMIA CAUSES JAUNDICE

When bilirubin in the blood exceeds 1 mg/dL (17.1 μmol/L), hyperbilirubinemia exists. Hyperbilirubinemia may be due to the **production of more bilirubin** than the normal liver can excrete, or it may result from the **failure of a damaged liver to excrete bilirubin** produced in normal amounts. In the absence of hepatic damage, obstruction to the excretory ducts of the liver—by preventing the excretion of bilirubin—will also cause hyperbilirubinemia. In all of these situations, bilirubin accumulates in the blood, and when it reaches a certain concentration, it diffuses into the tissues, which then become yellow. The condition is called **jaundice** or **icterus.**

In clinical studies of jaundice, measurement of bilirubin in the serum is of great value. A method for quantitatively assaying the bilirubin content of the serum was first devised by **Van den Bergh** by application of Ehrlich's test for bilirubin in urine. The Ehrlich reaction is based on the coupling of diazotized sulfanilic acid (Ehrlich's diazo reagent) and bilirubin to produce a reddish-purple azo compound. In the original procedure as described by Ehrlich, methanol was used to provide a solution in which both bilirubin and

the diazo reagent were soluble. Van den Bergh inadvertently omitted the methanol on an occasion when assay of bile pigment in human bile was being attempted. To his surprise, normal development of the color occurred "directly." This form of bilirubin that would react without the addition of methanol was thus termed **"direct-reacting."** It was then found that this same direct reaction would also occur in serum from cases of jaundice due to biliary obstruction. However, it was still necessary to add methanol to detect bilirubin in normal serum or that which was present in excess in serum from cases of hemolytic jaundice where no evidence of obstruction was to be found. To that form of bilirubin which could be measured only after the addition of methanol, the term **"indirect-reacting"** was applied.

It has now been demonstrated that the **indirect** bilirubin is **"free" (unconjugated) bilirubin** en route to the liver from the reticuloendothelial tissues where the bilirubin was originally produced by the breakdown of heme porphyrins. Since this bilirubin is not water soluble, it requires methanol to initiate coupling with the diazo reagent. In the liver, the free bilirubin becomes conjugated with glucuronic acid, and the conjugate, bilirubin glucuronide, can then be excreted into the bile. Furthermore, conjugated bilirubin, being water soluble, can react directly with the diazo reagent, so that the **"direct bilirubin"** of Van den Bergh is actually a **bilirubin conjugate** (bilirubin glucuronide).

Depending on the type of bilirubin present in plasma, ie, unconjugated bilirubin or conjugated bilirubin, the hyperbilirubinemia may be classified as **retention** hyperbilirubinemia or **regurgitation** hyperbilirubinemia, respectively.

Only unconjugated bilirubin can cross the blood-brain barrier into the central nervous system; thus, encephalopathy due to hyperbilirubinemia (kernicterus) can occur only in connection with retention bilirubin or unconjugated hyperbilirubinemia. On the other hand, **only conjugated bilirubin can appear in urine.** Accordingly, **choluric jaundice** occurs only in

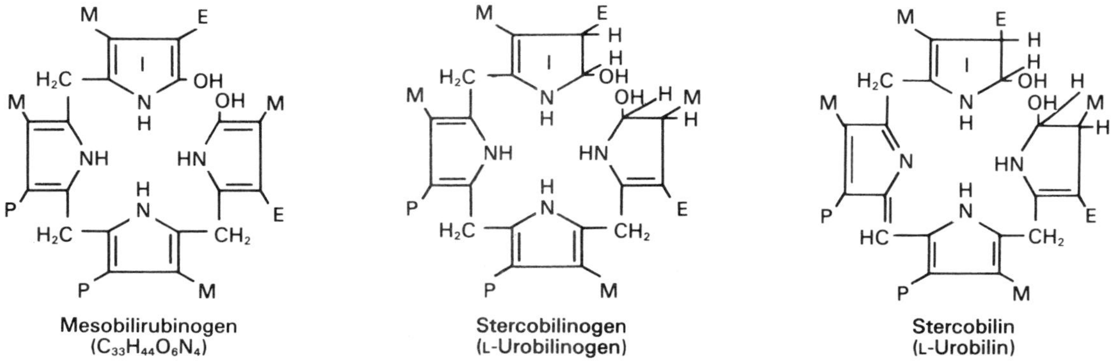

Mesobilirubinogen
($C_{33}H_{44}O_6N_4$)

Stercobilinogen
(L-Urobilinogen)

Stercobilin
(L-Urobilin)

Figure 34–15. Structure of some bile pigments.

regurgitation hyperbilirubinemia, and **acholuric jaundice** occurs only in the presence of an excess of unconjugated bilirubin.

Unconjugated Hyperbilirubinemia

Even in the event of extensive hemolysis, unconjugated hyperbilirubinemia is usually only slight (<4 mg/dL; <68.4 μmol/L), because of the liver's large capacity for handling bilirubin. However, if the handling of bilirubin is defective owing to either an acquired defect or an inherited abnormality, unconjugated hyperbilirubinemia may occur. The following are the commonest causes of this condition:

Neonatal "Physiologic Jaundice": This transient condition is the most common cause of unconjugated hyperbilirubinemia. It results from an accelerated hemolysis and an immature hepatic system for the uptake, conjugation, and secretion of bilirubin. Not only is the **UDP-glucuronyltransferase** activity reduced, but there probably is reduced synthesis of the substrate for that enzyme, UDP-glucuronic acid. Since the increased bilirubin is unconjugated, it is capable of penetrating the blood-brain barrier when its concentration in plasma exceeds that which can be tightly bound by albumin (20–25 mg/dL). This can result in a hyperbilirubinemic toxic encephalopathy, or **kernicterus,** which can cause mental retardation. Because of the recognized inducibility of this bilirubin metabolizing system, phenobarbital has been administered to jaundiced neonates and is effective in this disorder. In addition, exposure to visible light (phototherapy) can promote (by a mechanism that is not understood) the hepatic excretion of unconjugated bilirubin by converting some of the bilirubin to other derivatives such as maleimide fragments and geometric isomers that are excreted in the bile.

Crigler-Najjar Syndrome, Type I; Congenital Nonhemolytic Jaundice: Type I Crigler-Najjar syndrome, a rare autosomal recessive disorder of humans, is due to a primary metabolic defect in the conjugation of bilirubin. It is characterized by severe congenital jaundice due to the **inherited absence of bilirubin UDP-glucuronyltransferase activity** in hepatic tissues. The disease is usually fatal within the first 15 months of life, but a few teenagers have been reported who did not develop difficulties until puberty. These children have been treated with phototherapy with some reduction in plasma bilirubin levels. Phenobarbital has no effect on the formation of bilirubin glucuronides in patients with type I Crigler-Najjar syndrome. Serum bilirubin usually exceeds 20 mg/dL when untreated.

Crigler-Najjar Syndrome, Type II: This rare inherited disorder seems to result from a **milder defect** in the bilirubin conjugating system and has a more benign course. The serum bilirubin concentrations usually do not exceed 20 mg/dL, but all of the bilirubin accumulated is of the unconjugated type. Surprisingly, the bile in these patients does contain bilirubin monoglucuronide, and it has been proposed that the genetic defect **may involve the hepatic UDP-glucuronyltransferase that adds the second glucuronyl group** to bilirubin monoglucuronide. Patients with this syndrome can respond to treatment with large doses of phenobarbital.

Gilbert's Disease: Gilbert's disease is a heterogeneous group of disorders, many of which are now recognized to be due to a compensated hemolysis associated with unconjugated hyperbilirubinemia. There also appears to be a defect in the hepatic clearance of bilirubin, possibly due to a **defect in the uptake of bilirubin** by the liver parenchymal cells. However, **bilirubin UDP-glucuronyltransferase activities** in the livers of those patients studied with this disease were found to be reduced.

Toxic Hyperbilirubinemia: Unconjugated hyperbilirubinemia can result from toxin-induced liver dysfunction such as that caused by chloroform, arsphenamines, carbon tetrachloride, acetaminophen, hepatitis virus, cirrhosis, and *Amanita* mushroom poisoning. Although most of these acquired disorders are due to hepatic parenchymal cell damage, there is frequently a component of obstruction of the biliary tree within the liver that results in the presence of some conjugated hyperbilirubinemia.

Conjugated Hyperbilirubinemia Is Detected in Urine

Because conjugated bilirubin is water-soluble, it is detectable in the urine of most patients with conjugated hyperbilirubinemia; thus, they are frequently said to have **choluric jaundice.** Following are the commonest causes of this condition.

Chronic Idiopathic Jaundice (Dubin-Johnson Syndrome): This autosomal recessive disorder consists of conjugated hyperbilirubinemia in childhood or during adult life. The hyperbilirubinemia is apparently caused by a **defect in the hepatic secretion of conjugated bilirubin into the bile.** This secretory defect of conjugated compounds is not restricted to bilirubin but also involves secretion of conjugated estrogens and test compounds such as the dye sulfobromophthalein. Characteristically, in patients with Dubin-Johnson syndrome, the hepatocytes in the centrilobular area contain an abnormal pigment that has not been identified.

Biliary Tree Obstruction: Conjugated hyperbilirubinemia also results from blockage of the hepatic or common bile ducts. The bile pigment is believed to pass from the blood into the liver cells as usual but fails to be excreted. As a consequence of this, the conjugated bilirubin is absorbed into the hepatic veins and lymphatics.

The term **cholestatic jaundice** may be used to include all forms of extrahepatic obstructive jaundice in addition to some forms of parenchymal jaundice characterized by conjugated hyperbilirubinemia.

Urobilinogen in Urine Is a Clinical Indicator

Normally, there are mere traces of urobilinogen in

Table 34–2. Lab results in normal patients and patients with 3 different causes of jaundice.

Condition	Serum Bilirubin	Urine Urobilinogen	Urine Bilirubin	Fecal Urobilinogen
Normal	Direct: 0.1–0.4 mg/dL Indirect: 0.2–0.7 mg/dL	0–4 mg/24 h	Absent	40–280 mg/24 h
Hemolytic anemia	Elevation of indirect	Increased	Absent	Increased
Hepatitis	Elevations of direct and indirect	Increased	Present	Decreased
*Obstructive jaundice	Elevation of direct	Absent	Present	Trace to absent

*The commonest causes of obstructive (posthepatic) jaundice are cancer of the head of the pancreas and a gallstone lodged in the common bile duct.

the urine. In **complete obstruction of the bile duct,** no urobilinogen is found in the urine, since bilirubin has no access to the intestine where it can be converted to urobilinogen. In this case, the presence of bilirubin in the urine without urobilinogen suggests **obstructive jaundice,** either intrahepatic or posthepatic.

In **hemolytic jaundice,** the increased production of bilirubin leads to increased production of urobilinogen, which appears in the urine in large amounts. Bilirubin is not usually found in the urine in hemolytic jaundice (because unconjugated bilirubin does not pass into the urine), so that the combination of **increased urobilinogen and absence of bilirubin** is suggestive of hemolytic jaundice. Increased blood destruction from any cause (eg, pernicious anemia) will, of course, also bring about an increase in urine urobilinogen.

Table 34–2 summarizes the different lab results obtained on patients with 3 different causes of jaundice—hemolytic anemia (a prehepatic cause), hepatitis (a hepatic cause), and obstruction of the common bile duct (a posthepatic cause).

REFERENCES

Billett, HH. Porphyrias: Inborn errors in heme production. *Hosp Pract* 1988;.**41**:60.

Goldberg A et al: Porphryin metabolism and the porphyrias. In: *Oxford Textbook of Medicine,* 2nd ed. Weatherall DJ et al (editors). Oxford University Press, 1987.

Kappas A, Sassa S, Anderson KE: The porphyrias. In: *The Metabolic Basis of Inherited Disease,* 5th ed. Stanbury JB et al (editors). McGraw-Hill, 1983.

Meyer UA. Porphyrias. In: *Harrison's Principles of Internal Medicine,* 11th ed. Braunwald E et al (editors). McGraw-Hill, 1987.

Wolkoff AW, Chowdhury JR, Arias IM. Hereditary jaundice and disorders of bilirubin metabolism. In: *The Metabolic Basis of Inherited Disease,* 5th ed. Stanbury JB et al (editors). McGraw-Hill, 1983.

Section IV.
Structure, Function, & Replication of Informational Macromolecules

Nucleotides

35

Victor W. Rodwell, PhD

INTRODUCTION

Nucleotides participate in a wide variety of biochemical processes. Perhaps the best known role of the purine and pyrimidine nucleotides is to serve as the monomeric precursors of RNA and DNA. However, the **purine ribonucleotides** serve also as the ubiquitous high-energy source, ATP; as regulatory signals (cyclic AMP [cAMP] and cyclic GMP [cGMP]); and as components of the coenzymes FAD, NAD+, and NADP+ and of the methyl group donor S-adenosylmethionine. The **pyrimidine nucleotides,** in addition to providing monomeric precursors for nucleic acid synthesis, also serve as high-energy intermediates, such as UDP-glucose and UDP-galactose in carbohydrate metabolism and CDP-acylglycerol in lipid synthesis.

BIOMEDICAL IMPORTANCE

The heterocyclic bases purine and pyrimidine are the parent molecules of nucleosides and nucleotides. Nucleotides are ubiquitous in living cells, where they perform numerous key functions. Examples include incorporation, as their ribose (RNA) or deoxyribose (DNA) monophosphates, into nucleic acids, energy transduction (ATP), parts of coenzymes (AMP), acceptors for oxidative phosphorylation (ADP), allosteric regulators of enzyme activity, and "second messengers" (cAMP, cGMP). Synthetic analogs of naturally occurring nucleotides find application in cancer chemotherapy as enzyme inhibitors and can replace the naturally occurring nucleotides in nucleic acids. Therapeutic attempts to inhibit the growth of cancer cells or certain viruses have often employed administration of analogs of bases, nucleosides, or nucleotides that inhibit the synthesis of either DNA or RNA. Such compounds include 5-fluorouracil, 5'-iodo-2'-deoxyuridine, 6-thioguanine, 6-mercaptopurine, 6-azauridine, and arabinosyl cytosine, as described below. **Allopurinol,** a purine analog, is widely used in the treatment of gout.

THE HETEROCYCLIC BASES OF NUCLEOTIDES ARE DERIVATIVES OF PURINE & PYRIMIDINE

The purine and pyrimidine bases (purines and pyrimidines) of nucleotides arise by substitution of atoms on the aromatic rings of the parent heterocycles, purine and pyrimidine (Fig 35–1). Note that **the direction in which atoms of the six-membered ring are numbered is opposite in purine (counterclockwise) and pyrimidine (clockwise).** However, atom 5 (carbon) is the same in both compounds. Purines and pyrimidines are **planar** molecules, a property that has major significance for nucleic acid structure (see Chapter 38).

The Terms "Major" & "Minor" Bases Refer to Relative Abundance, Not Physiologic Importance

Since certain purines and pyrimidines are present in cells in much greater quantities than others, we customarily distinguish between major and minor purines and pyrimidines. However, since additional purines and pyrimidines also play key roles in metabolism, this distinction rests solely on relative abundance, not on their relative physiologic importance.

Purine   Pyrimidine

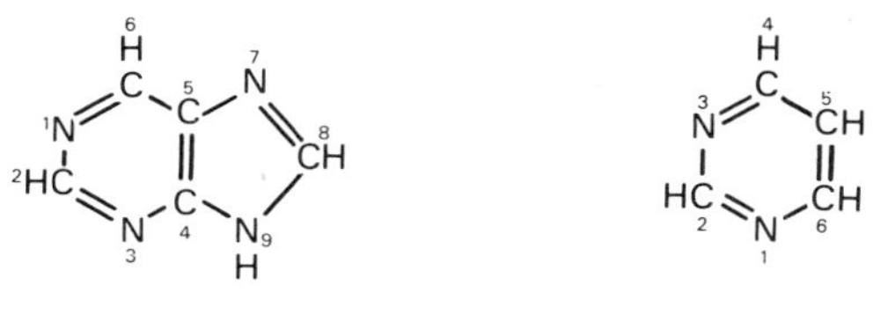

Figure 35–1. Structures of purine and pyrimidine with the positions of the elements numbered according to the international system.

The major purines and pyrimidines of nucleic acids of both prokaryotes and eukaryotes are the purines **adenine** and **guanine** and the pyrimidines **cytosine, thymine,** and **uracil** (Figs 35–2 and 35–3). The purines **hypoxanthine** and **xanthine** (Fig 35–3) are intermediates in the metabolism of adenine and guanine, and human subjects excrete the oxidized purine **uric acid** as the end product of purine catabolism.

Additional bases are present in the DNA and transfer RNAs (tRNAs) of both prokaryotes and eukaryotes. Others, whose functions are in many cases poorly understood, occur only in the nucleic acids of bacteria and viruses. For example, bacterial and human DNA contains **5-methylcytosine,** and bacteriophage DNA contains **5-hydroxymethylcytosine** (Fig 35–4). Minor bases present in mRNAs of mammalian cells include **N6-methyladenine, N6,N6-dimethyladenine,** and **N7-methylguanine** (Fig 35–5).

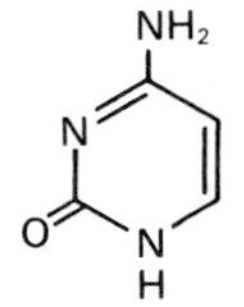

Methylated Purines of Plants Exhibit Pharmacologic Properties

Plants* and their methylated purines with pharmacologic properties include coffee (contains **caffeine,** or 1,3,7-trimethylxanthine), tea (contains **theophylline,** or 1,3-dimethylxanthine), and cocoa (contains **theobromine,** or 3,7-dimethylxanthine) (Fig 35–6).

Purines & Pyrimidines Exhibit Keto-Enol Tautomerism

While purines and pyrimidines can exist in a **lactim** (–OH) or **lactam** (=O) form (Fig 35–7), the lactam is by far the predominant tautomer of guanine or thymine under physiologic conditions. Chapters 39 and 41 discuss the importance of lactam/lactim tautomerism in base pairing and mutagenesis.

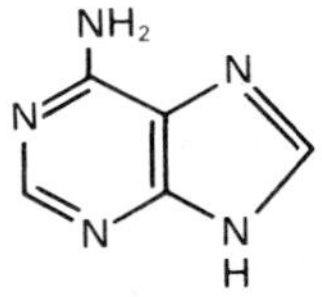

LOW SOLUBILITY OF PURINE BASES AT ACID pH CAN CAUSE MEDICAL PROBLEMS

At neutral pH, guanine and xanthine are the least soluble bases. Guanine, however, is not normally present in human urine. While salts of uric acid (**urates**) are relatively soluble at a neutral pH, uric acid is highly insoluble in solutions of low pH (eg, urine). **Xanthine** and **uric acid** may occur as constituents of urinary tract stones.

FREE PURINES AND PYRIMIDINES ARE FAR LESS ABUNDANT THAN THEIR CORRESPONDING NUCLEOSIDES AND NUCLEOTIDES

Nucleosides (Fig 35–8) consist of a **sugar** (usually **D-ribose** or **2-deoxy-D-ribose**) attached to a purine or a pyrimidine by the relatively acid-labile **β-N-glycosidic linkage** at **N9** (purine) or **N1** (pyrimidine). The major

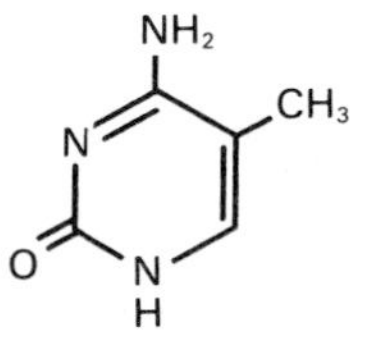

Cytosine
(2-oxy-4-aminopyrimidine)

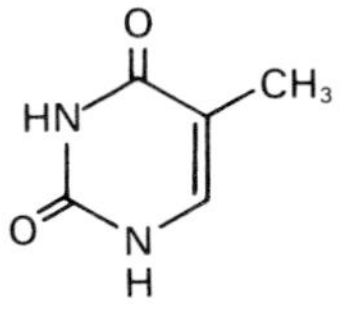

Thymine
(2,4-dioxy-5-methylpyrimidine)

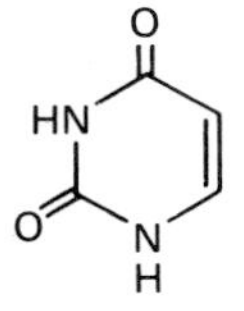

Uracil
(2,4-dioxypyrimidine)

Figure 35–2. The 3 major pyrimidine bases found in nucleotides.

Adenine
(6-aminopurine)

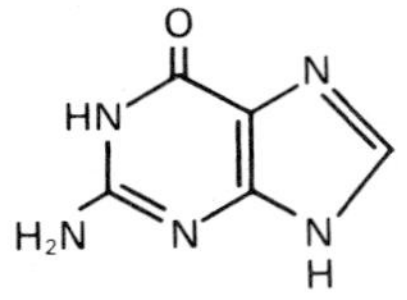

Guanine
(2-amino-6-oxypurine)

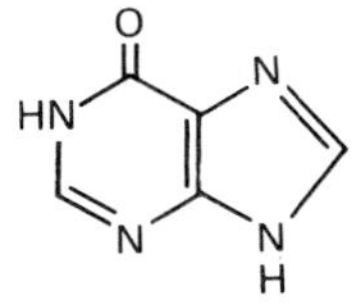

Hypoxanthine
(6-oxypurine)

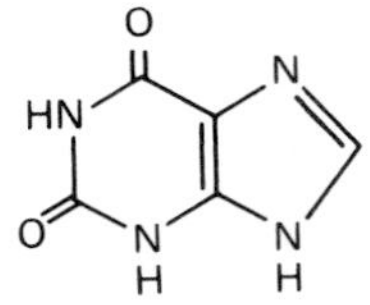

Xanthine
(2,6-dioxypurine)

Figure 35–3. The major purine bases of nucleotides.

5-Methylcytosine

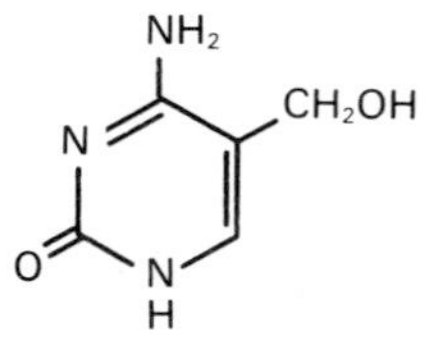

5-Hydroxymethylcytosine

Figure 35–4. The structures of 2 uncommon naturally occurring pyrimidine bases.

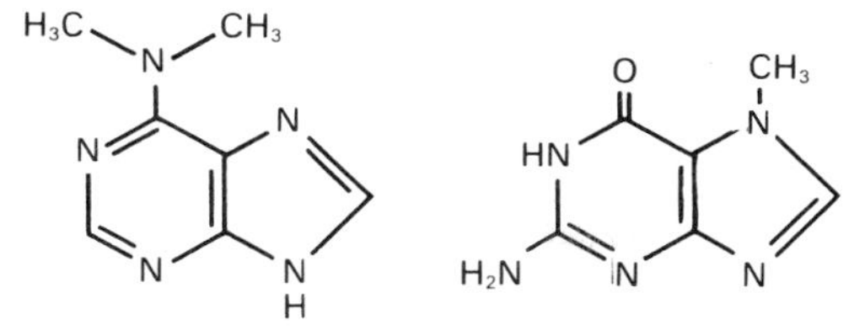

N⁶,N⁶-Dimethyladenine   N⁷-Methylguanine

Figure 35–5. The structures of 2 uncommon naturally occurring purine bases.

ribonucleotides are **adenosine** (D-ribose linked to N^9 of adenine), **guanosine** (D-ribose linked to N^9 of guanine), **cytidine** (D-ribose linked to N^1 of cytosine), and **uridine** (D-ribose linked to N^1 of uracil). The 2'-deoxyribonucleosides consist of 2-deoxy-D-ribose attached to purines or pyrimidines at the positions described above, again via a β-N-glycosidic bond.

Steric factors hinder rotation about the N-glycosidic bond, and the **anti** conformation (Fig 35–9) predominates in naturally occurring nucleosides. As discussed in Chapter 38, the anti form is essential for alignment of complementary purine and pyrimidine bases in double-stranded DNA. However, the conventional representation of D-ribose dictates that most figures of this and other chapters show nucleosides and nucleotides in the less favored **syn** conformation.

Nucleotides Are Phosphorylated Nucleosides

Mononucleotides are nucleosides singly phosphory-

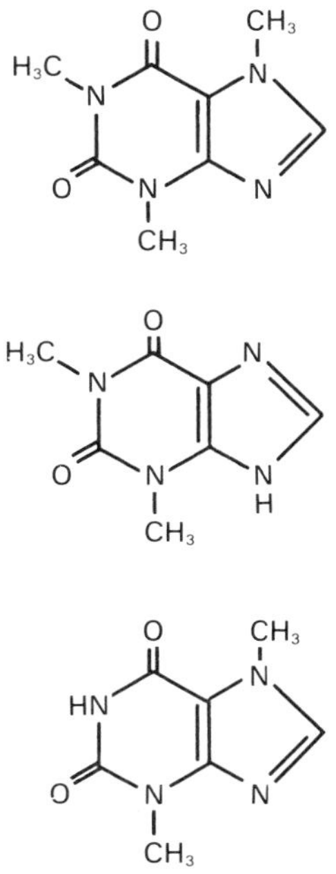

Caffeine
(1,3,7-trimethyl-xanthine)

Theophylline
(1,3-dimethyl-xanthine)

Theobromine
(3,7-dimethyl-xanthine)

Figure 35–6. Methyl xanthines present in foods.

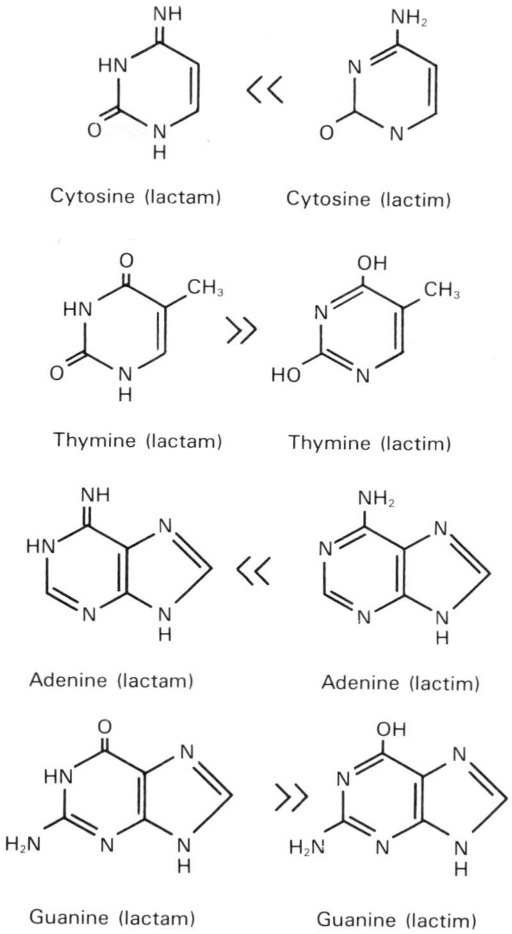

Cytosine (lactam)   Cytosine (lactim)

Thymine (lactam)   Thymine (lactim)

Adenine (lactam)   Adenine (lactim)

Guanine (lactam)   Guanine (lactim)

Figure 35–7. Tautomers of cytosine, thymine, adenine, and guanine with the predominant forms indicated.

lated on hydroxyl groups of the sugar (generally ribose or 2'-deoxyribose) (Fig 35–10). For example, **adenosine monophosphate** (AMP or adenylate) is adenine + ribose + phosphate. The only sugars commonly linked to uracil and thymine are D-ribose and 2-deoxy-D-ribose, respectively. Table 35–1 illustrates the relationships between the major purines and pyrimidines and their corresponding 5'-oxy- and 5'-deoxyribonucleoside monophosphates. DNA is a polymer of dTMP, dCMP, dAMP, and dGMP, and RNA a polymer of UMP, CMP, AMP, and GMP.

There are exceptions to the above structures of nucleotides. For example, **in tRNA, ribose occasionally is attached to atom 5 of uracil by a carbon-to-carbon rather than the usual N-to-C bond.** This unusual compound is **pseudouridine** (ψ). tRNAs contain additional unusual nucleotides, ie, **TMP, or thymine attached to ribose monophosphate** (Fig 35–11). TMP is formed after synthesis of the tRNA by methylation of UMP by S-adenosylmethionine. Similarly, **pseudouridylic acid** (ψMP) is formed by rearrange-

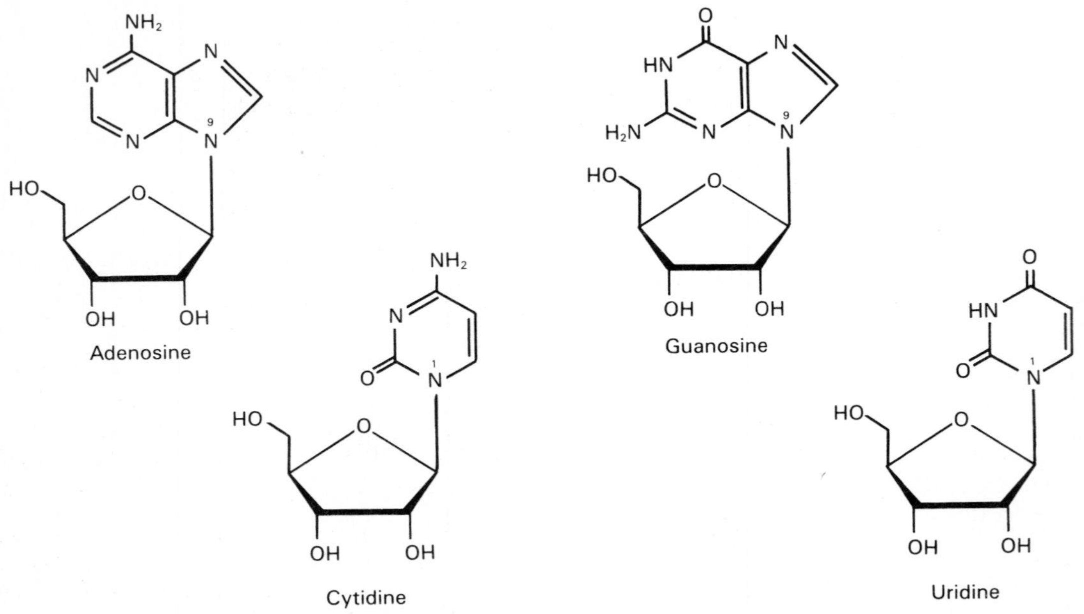

Figure 35–8. Structures of ribonucleosides.

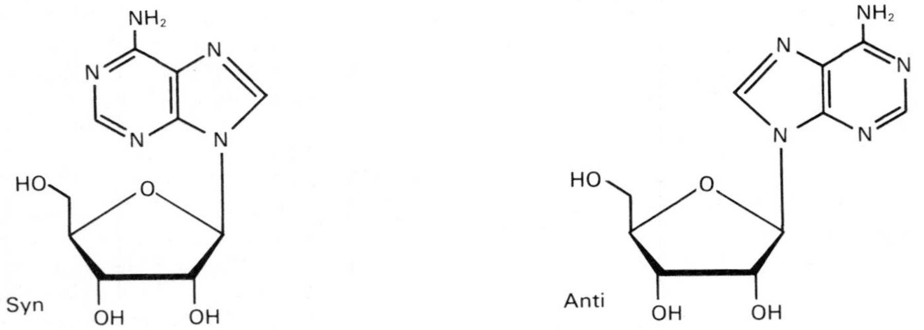

Figure 35–9. The syn and anti conformations of adenosine.

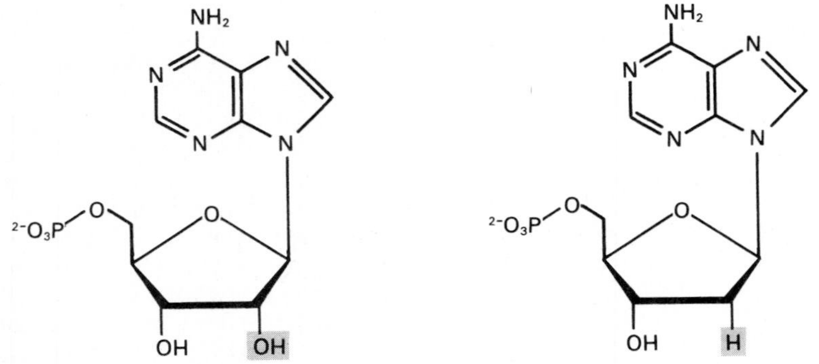

Figure 35–10. The structures of adenylic acid (AMP) (*left*) and 2'-deoxyadenylic acid (dAMP) (*right*).

Table 35–1. Principal bases, nucleosides, and nucleotides.

Base	Ribonucleoside	Ribonucleotide (5′-monophosphate)
Adenine (A)	Adenosine	Adenosine monophosphate (AMP)
Guanine (G)	Guanosine	Guanosine monophosphate (GMP)
Cytosine (C)	Cytidine	Cytidine 5′-monophosphate (CMP)
Uracil (U)	Uridine	Uridine 5′-monophosphate (UMP)
Base	**Deoxyribonucleoside**	**Deoxyribonucleotide (5′-monophosphate)**
Adenine (A)	Deoxyadenosine	Deoxyadenosine 5′-monophosphate (dAMP)
Guanine (G)	Deoxyguanosine	Deoxyguanosine 5′-monophosphate (dGMP)
Cytosine (C)	Deoxycytidine	Deoxycytidine 5′-monophosphate (dCMP)
Thymine (T)	Thymidine	Thymidine 5′-monophosphate (dTMP)

ment of uridylic acid after the tRNA molecule has been synthesized.

The position of the phosphate in the nucleotide is indicated by a numeral with a prime. For example, adenosine with the phosphate attached to carbon 3 of the sugar ribose is **adenosine 3′-monophosphate.** The prime differentiates atoms of the sugar from those of the purine or pyrimidine, which are not followed by a prime (Fig 35–12).

A, G, C, T, and U denote the nucleosides of adenine, guanine, cytosine, thymine, and uracil, respectively. The prefix **d** indicates that the sugar is 2′-deoxy-D-ribose. The abbreviation "MP" (monophosphate) is added to the abbreviation designating the nucleoside. The 5′ is omitted when phosphate is esterified to carbon 5′ of ribose or 2′-deoxyribose. For example, guanosine 5′-monophosphate is abbreviated **GMP.**

Additional phosphates attached to the sugar of the mononucleotide by **acid anhydride bonds** form **nucleoside di-** and **triphosphates** such as ADP (adenosine diphosphate) and GTP (guanosine triphosphate) (Fig 35–13). Recall that acid anhydrides have a high group transfer potential, and that $\Delta G°$ for hydrolysis of ATP to ADP is about 7 kcal/mol.

Nucleoside Triphosphates Participate in Covalent Bond Formation

Since all purine and pyrimidine triphosphates have a high group transfer potential, they participate in nu-merous reactions that form covalent bonds. A prominent example is polymerization of the major triphosphates to form DNA and RNA. In addition, specific nucleoside di- and triphosphates fulfill specific physiologic functions in different tissues and life forms.

Adenosine Derivatives: ADP and ATP are substrates and products, respectively, for oxidative phosphorylation, and ATP serves as the major intracellular transducer of free energy. The mean intracellular concentration of ATP, the most abundant free nucleotide in mammalian cells, is about 1 mM.

Cyclic AMP (**cAMP,** or adenosine 3′,5′-monophosphate), formed from ATP in a reaction catalyzed by **adenylate cyclase** (Fig 35–14), mediates diverse intracellular processes. Adenylate cyclase activity is regulated by complex interactions, many of which involve hormone receptors (see Chapter 44). Intracellular concentrations of cAMP (about 1 μM) are about three orders of magnitude below those of ATP. Hydrolysis of cAMP, catalyzed by **cAMP phosphodiesterase** (Fig 35–14), forms 5-′AMP.

The incorporation of sulfate into ester linkages in compounds such as sulfated proteoglycans (see Chapter 43) requires prior "activation" of the sulfate by reaction with ATP to form **adenosine 3′-phosphate-5′-phosphosulfate** (**PAPS,** or phosphoadenosine phosphosulfate) (Fig 35–15). PAPS is also the substrate for sulfate conjugation reactions.

S-Adenosylmethionine (Fig 35–16), a form of "active" methionine, serves as a methyl donor in methyla-

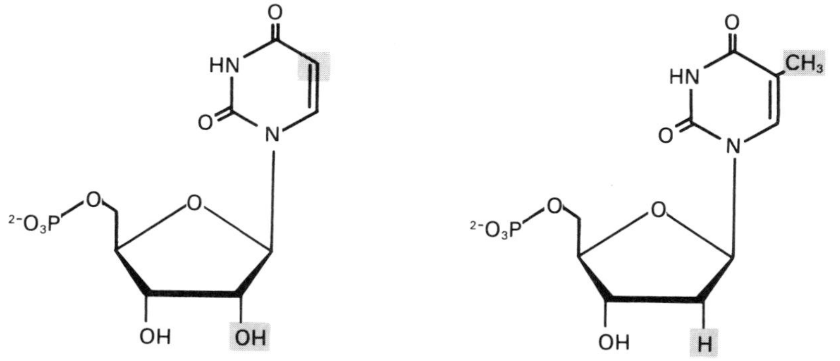

Figure 35–11. Uridylic acid (UMP) (*left*) and thymidylic acid (TMP) (*right*).

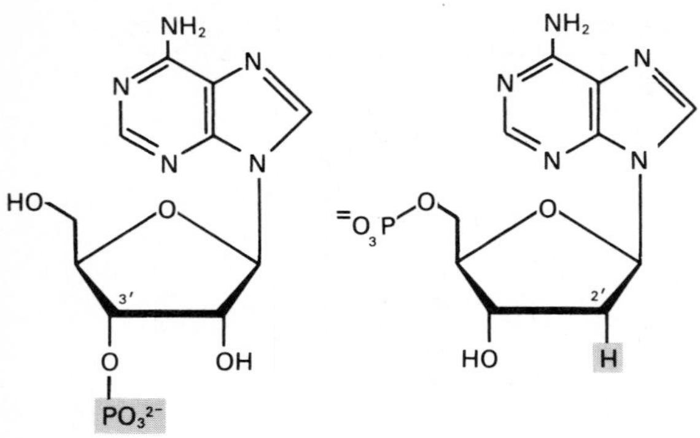

Figure 35–12. Adenosine 3'-monophosphate (*left*) and 2'-deoxyadenosine-5'-monophosphate (*right*).

tion reactions and as a source of propylamine for the synthesis of polyamines (see Chapter 33).

Guanosine Derivatives: Oxidation of α-keto-glutarate to succinyl-CoA involves phosphorylation of GDP to GTP. GTP is required for activation of adenylate cyclase by some hormones, and serves both as an allosteric regulator and as an energy source for protein synthesis on polyribosomes. GTP thus plays an important role in maintenance of the internal milieu.

Cyclic GMP (**cGMP,** or guanosine 3',5'-monophosphate) (Fig 35–17) is also an intracellular signal, or second messenger, of extracellular events. cGMP may act antagonistically to cAMP. cGMP is formed from GTP by **guanylate cyclase.** Like adenylate cyclase, guanylate cyclase is regulated by various effectors, including hormones. Like cAMP, cGMP is also catabolized by a **phosphodiesterase** to its 5'-monophosphate, GMP.

Hypoxanthine Derivatives: Hypoxanthine ribonucleotide (**IMP,** or inosinate), is a precursor of all purine ribonucleotides synthesized *de novo*. IMP also

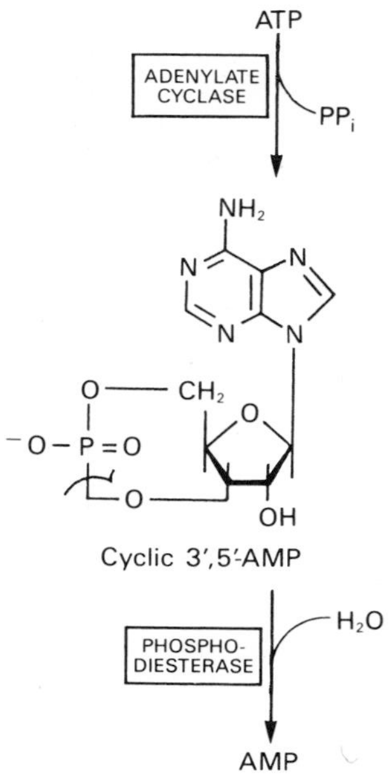

Figure 35–14. Formation of cAMP from ATP by adenylate cyclase and hydrolysis of cAMP by cAMP phosphodiesterase.

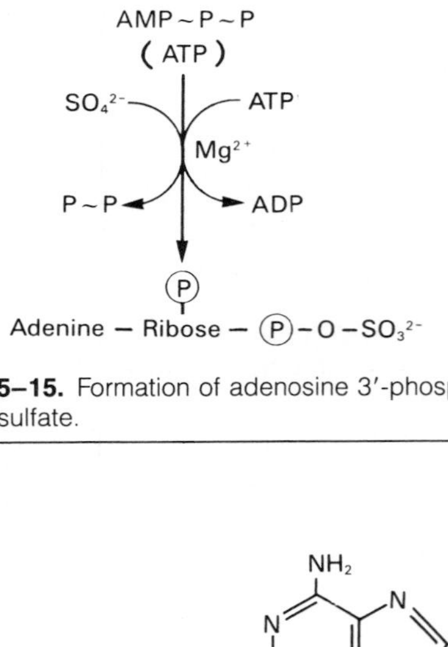

Figure 35–15. Formation of adenosine 3'-phosphate-5'-phosphosulfate.

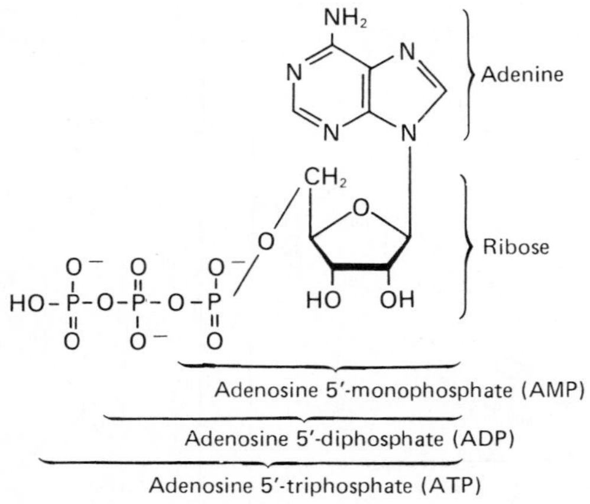

Figure 35–13. ATP, its diphosphate, and its monophosphate.

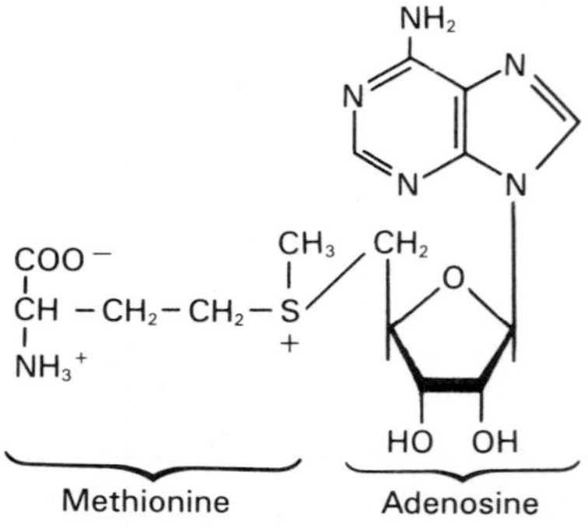

Figure 35–16. S-Adenosylmethionine.

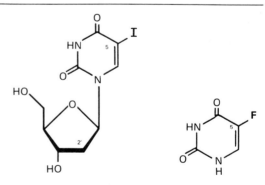

Figure 35–17. Cyclic 3′,5′-guanosine monophosphate (cyclic GMP; cGMP).

arises by deamination of AMP, a reaction that in muscle tissue forms part of the purine nucleotide cycle (Fig 35–18). Amination of IMP, reforming AMP, results in the net production of ammonia from aspartate. Dephosphorylation of IMP forms the nucleoside **inosine** (hypoxanthine riboside), an intermediate in the purine salvage cycle (see Chapter 36).

Uracil Derivatives: UDP-sugar derivatives participate in sugar epimerizations (eg, interconversion of glucose- and galactose-1-phosphate), and **UDP-glucose** is the glycosyl donor for biosynthesis of glycogen and glucosyl disaccharides. UDP-sugars also participate in biosynthesis of the oligosaccharides of glycoproteins and proteoglycans (see Chapter 55). Finally, **UDP-glucuronic acid** is the glycosidic acid donor for conjugation reactions such as the formation of bilirubin glucuronide (see Chapter 34).

Cytosine Derivatives: CTP is required for the biosynthesis of some phosphoglycerides in animal tissues. Reactions involving ceramide and CDP-choline are responsible for the formation of sphingomyelin and other substituted sphingosines. Cyclic nucleotide derivatives of cytidine, analogous to those of adenosine and guanosine, have been described.

Many Coenzymes Are Nucleotide Derivatives

Many coenzymes incorporate nucleotides as well as structures similar to purine and pyrimidine nucleotides (Table 35–2).

Table 35–2. Many coenzymes and related compounds are derivatives of adenosine monophosphate.

Coenzyme	R	R′	R″	n
Active methionine	Methionine*	H	H	0
Amino acid adenylates	Amino acid	H	H	1
Active sulfate	SO_3^{2-}	H	PO_3^{2-}	1
3′,5′-Cyclic AMP	H	H	PO_3^{2-}	
NAD+	†	H	H	2
NADP+	†	PO_3^{2-}	H	2
FAD	†	H	H	2
CoA·SH	†	H	PO_3^{2-}	2

*Replaces phosphate group.
†R is a B-vitamin derivative.

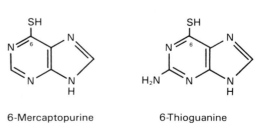

5-Iodo-2′-deoxyuridine   5-Fluorouracil

6-Mercaptopurine   6-Thioguanine

Figure 35–19. Synthetic pyrimidine analogs (*above*) and synthetic purine analogs (*below*).

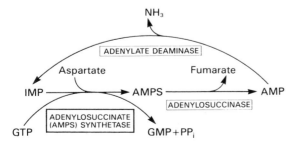

Figure 35–18. The purine nucleotide cycle.

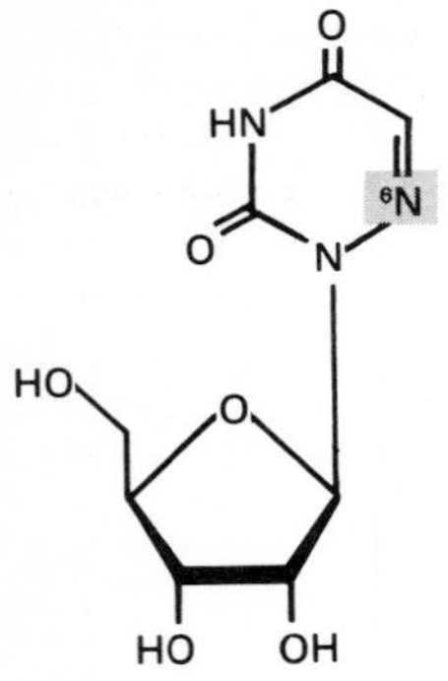

6-Azauridine

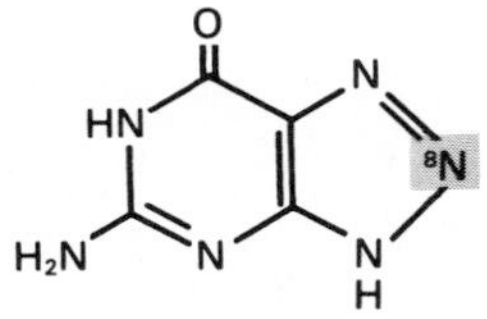

8-Azaguanine

Figure 35–20. 6-Azauridine (*left*) and 8-azaguanine (*right*).

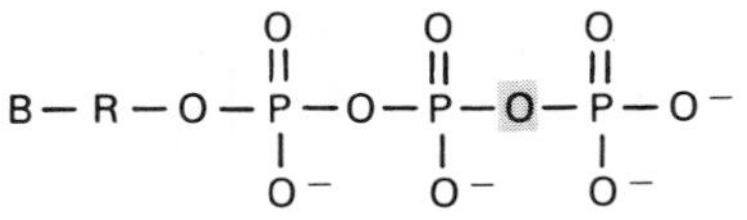

Parent nucleoside triphosphate

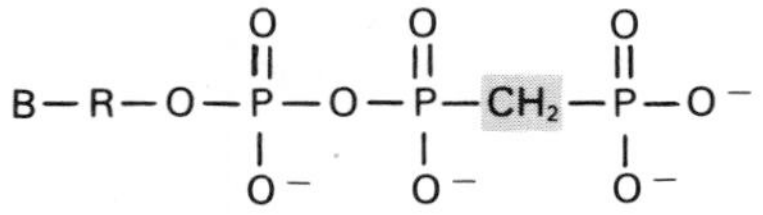

β,γ-Methylene derivative

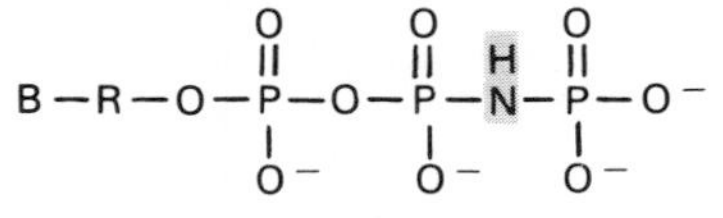

β,γ-Imino derivative

Figure 35–22. Synthetic derivatives of nucleoside triphosphates incapable of undergoing hydrolytic release of the terminal phosphate group. B, a purine or pyrimidine base; R, ribose or deoxyribose. Shown are the parent (hydrolyzable) nucleoside triphosphate (*top*) and the unhydrolyzable β,γ-methylene (*center*) and β,γ-imino derivatives (*bottom*).

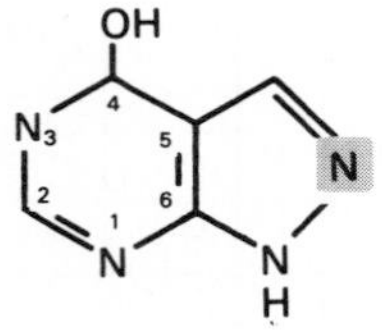

Allopurinol (lactim)

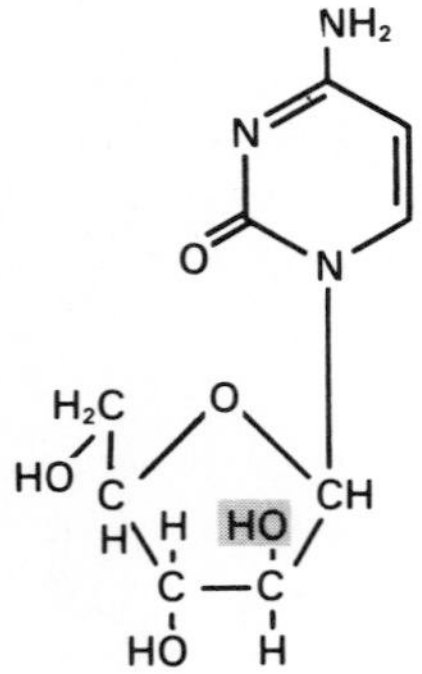

Arabinosyl cytosine

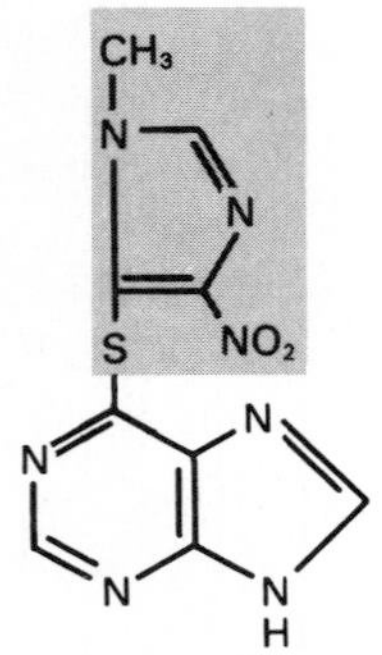

Azathioprine

Figure 35–21. 4-Hydroxypyrazolopyrimidine (allopurinol), arabinosyl cytosine (cytarabine), and azathioprine.

SYNTHETIC NUCLEOTIDE ANALOGS ARE USED IN CLINICAL MEDICINE

Synthetic analogs of purines and pyrimidines, nucleosides, and nucleotides are widely used in the medical sciences and clinical medicine. Typically, most uses exploit the role of nucleotides as components of the nucleic acids essential for cellular growth and division. For a cell to divide, its DNA must be replicated. All the precursors of DNA—the normal purine and pyrimidine deoxyribonucleotides—must therefore be available. One of the most important components of an oncologist's pharmacopeia is the group of synthetic analogs of purines, pyrimidines, and their nucleosides.

The pharmacologic approach uses **an analog in which either the heterocyclic ring structure or the sugar moiety has been altered so as to induce toxic effects** when the analog is incorporated into specific cellular constituents. Many of these effects reflect one of two processes: (1) inhibition by the drug of specific enzymes essential for nucleic acid synthesis, or (2) incorporation of metabolites of the drug into nucleic acids where they affect the base pairing essential to accurate transfer of information. Examples include the **5-fluoro** or **5-iodo derivatives** of uracil or deoxyuridine, which serve as thymine or thymidine analogs, respectively (Fig 35–19). Both **6-thioguanine** and **6-mercaptopurine,** in which thiol groups replace the hydroxyl groups at the 6 position, are widely used clinically. Analogs such as **5-** or **6-azauridine, 5-** or **6-**

azacytidine, and **8-azaguanine** (Fig 35–20), analogs in which a heterocyclic ring carbon has been replaced by a nitrogen atom, also have been employed clinically.

The purine analog 4-hydroxypyrazolopyrimidine (**allopurinol**), used in treatment of hyperuricemia and gout, inhibits de novo purine biosynthesis and xanthine oxidase. Nucleosides containing **arabinose** rather than ribose as the sugar moieties, eg, **cytarabine** (arabinosyl cytosine, Ara-C), are used in the chemotherapy of cancer and viral infections (Fig 35–21).

Azathioprine, which is catabolized to 6-mercaptopurine, is employed in organ transplantation to suppress events involved in immunologic rejection. Among several nucleoside analogs with antiviral activities, **5-iododeoxyuridine** (see above), is effective in the local treatment of herpetic keratitis, an infection of the cornea by herpes virus.

Numerous analogs of purine and pyrimidine ribonucleotides with nonhydrolyzable di- or triphosphates have been synthesized for in vitro use. These analogs allow the investigator to determine whether given biochemical effects of nucleoside di- or triphosphates require hydrolysis, or whether their effects are mediated by occupying specific nucleotide-binding sites on enzymes or regulatory proteins. Fig 35–22 illustrates two nonhydrolyzable analogs of GTP.

REFERENCES

Mainwaring WIP, Parrish JH, Pickering JD, Mann NH: Nucleic Acid Biochemistry and Molecular Biology. Blackwell Scientific Publications, 1982.

36

Metabolism of Purine & Pyrimidine Nucleotides

Victor W. Rodwell, PhD

INTRODUCTION

This chapter will discuss the digestion, biosynthesis, and catabolism of purine and pyrimidine nucleotides and selected diseases associated with genetic defects in these processes.

BIOMEDICAL IMPORTANCE

Neither dietary nucleotides nor their parent purine and pyrimidine bases are incorporated into human tissue nucleic acids or into purine or pyrimidine derivatives such as ATP, NAD$^+$, or coenzyme A. Even when a diet rich in nucleoproteins is ingested, human subjects form the constituents of tissue nucleic acids from amphibolic intermediates. This de novo synthesis permits purine and pyrimidine analogs with potential as **anticancer drugs** to be incorporated into DNA. The rates of synthesis of purine and pyrimidine oxy- and deoxyribonucleotides are precisely regulated by mechanisms that ensure production of these compounds in the quantities, and at the times, appropriate to varying physiologic demand. These mechanisms include "salvage" pathways for reutilization of purine or pyrimidine bases released in vivo by degradation of nucleic acids.

Human diseases that involve abnormalities in purine or pyrimidine metabolism include **gout, Lesch-Nyhan syndrome, adenosine deaminase deficiency,** and **purine nucleoside phosphorylase deficiency.**

PURINES & PYRIMIDINES ARE NOT ESSENTIAL IN THE HUMAN DIET

While animals ingest nucleic acids and nucleotides in their food, survival does not depend on their absorption and utilization, since man and most other vertebrates readily synthesize purine and pyrimidine nucleotides de novo (ie, from amphibolic intermediates). Humans convert most of the purines derived from ingested nucleoproteins to uric acid (Fig 36–1). While little or no dietary purine or pyrimidine is incorporated into tissue nucleic acids, parenterally administered compounds are incorporated. Consequently, incorporation of injected [^{3}H]thymidine directly into newly synthesized DNA provides a widely used technique for measuring the rate of DNA synthesis both in vivo and in vitro.

ENZYMES OF THE GASTRO-INTESTINAL TRACT DEGRADE INGESTED NUCLEIC ACIDS TO PURINES & PYRIMIDINES

Nucleic acids released from ingested **nucleoproteins** by proteolysis in the intestinal tract are degraded to **mononucleotides** by ribonucleases, deoxyribonucleases, and polynucleotidases. Nucleotidases and phosphatases hydrolyze the mononucleotides to **nucleosides,** which are either absorbed or are further degraded by intestinal phosphorylase to **purine and pyrimidine bases.** Purine bases oxidized to **uric acid** (Fig 36–1) may be absorbed and subsequently excreted in the urine.

ANIMALS FORM NUCLEOTIDES FROM AMPHIBOLIC INTERMEDIATES

Humans and other mammals synthesize purine nucleotides from amphibolic intermediates in quantities sufficient both for nucleic acid synthesis and for the additional functions mentioned in Chapter 35. Nucleotide biosynthesis serves additional purposes in some vertebrates. Although mammals and other **ureotelic** animals excrete nitrogenous waste as **urea, uricotelic** animals (birds, amphibians, reptiles) excrete nitrogenous wastes as **uric acid.** While the reactions of de novo purine nucleotide synthesis are analogous in ureotelic and uricotelic organisms, uricotelic life forms must therefore synthesize purine nucleotides at a relatively greater rate.

Inosine Monophosphate (IMP), The Parent Purine Mononucleotide, Is Formed From Amphibolic Intermediates

Long before the details of the protracted pathway of IMP biosynthesis were known, tracer studies identified the source of each atom of a purine base (Fig 36–2). These early isotopic studies prepared the way for

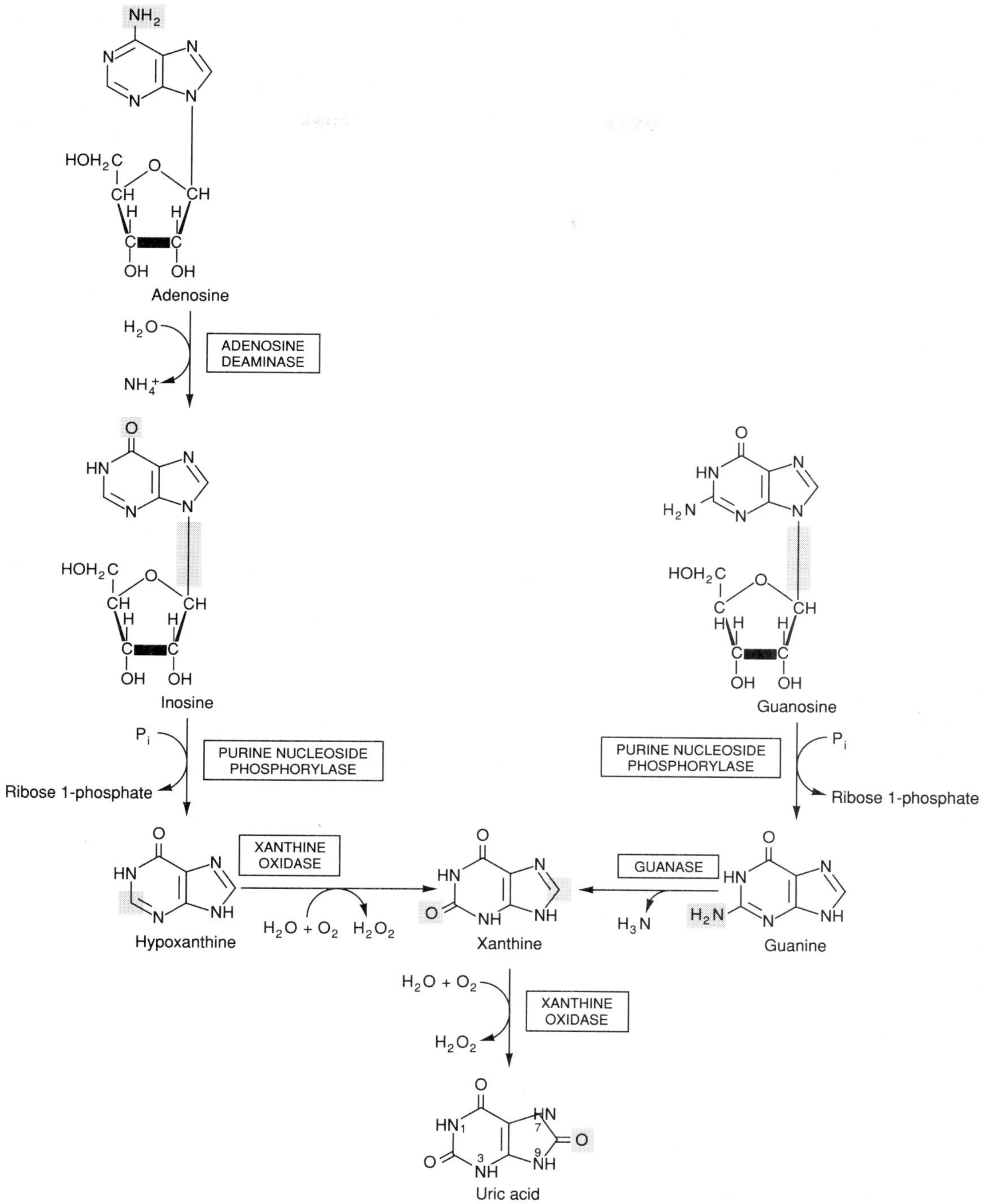

Figure 36–1. Formation of uric acid from purine nucleosides by way of the purine bases hypoxanthine, xanthine, and guanine. Purine deoxyribonucleosides are degraded by the same catabolic pathway and enzymes, all of which exist in the mucosa of the mammalian gastrointestinal tract.

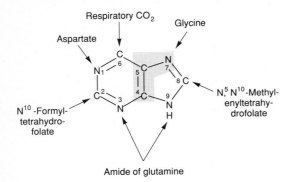

Figure 36–2. The sources of the nitrogen and carbon atoms of the purine ring. Atoms 4, 5, and 7 (shaded) derive from glycine.

discovery of the reactions and intermediates discussed below. In what immediately follows, Arabic numerals of paragraphs correspond to numbered reactions of Fig 36–3, and Roman numerals of compounds to the structures of intermediates in that figure.

(1) Transfer of pyrophosphate from ATP to C-1 of D-ribose 5-phosphate (I) forms **5-phosphoribosyl-1-pyrophosphate**, PRPP (II), an intermediate also in NAD+, NADP+, and pyrimidine nucleotide biosynthesis.

(2) Conversion of PRPP (II) and glutamine to **5-phospho-β-D-ribosylamine** (III) involves displacement of pyrophosphate by the amide nitrogen of glutamine with **inversion of configuration at C-1.** This forms the β-N-glycosidic bond of the ultimate nucleotide.

(3) Condensation of glycine with (III) to form **glycinamide ribosyl-5-phosphate** (IV) adds 3 atoms, which ultimately become purine atoms C-4, C-5, and N-7.

(4) Transfer to (IV) of a formyl group from N5,N10-methenyl tetrahydrofolate (N5,N10-methenyl-THF) forms **formylglycinamide ribosyl-5-phosphate** (V) and adds C-8 of the purine base.

(5) Amidation of (V) by the amide nitrogen of a second glutamine forms **formylglycinamidine ribosyl-5-phosphate** (VI) and adds what will become N-3 of the purine ring.

(6) Elimination of water accompanied by closure of the imidazole ring forms **aminoimidazole ribosyl-5-phosphate** (VII). The initial event, which requires ATP, is phosphoryl group transfer from ATP to the oxo function of (VI). Nucleophilic attack by the adjacent amino nitrogen then displaces P_i, with accompanying closure of the imidazole ring.

(7) Addition to (VII) of CO_2, a reaction that requires neither ATP nor biotin, forms **aminoimidazole carboxylate ribosyl-5-phosphate** (VIII). The added carbon will become purine atom C-6.

(8) Reactions 8 and 9 resemble conversion of ornithine to arginine in the urea cycle (see Fig 31–13). Condensation of aspartate with (VIII) forms **aminoimidazole succinyl carboxamide ribosyl-5-phosphate** (IX) and adds N-1 of the purine ring.

(9) The succinyl group of (IX) departs as fumarate,

forming **aminoimidazole carboxamide ribosyl-5-phosphate** (X).

(10) Formylation of (X) by **N10-formyl-THF** forms **amidoimidazole carboxamide ribosyl-5-phosphate** (XI). The newly added carbon will be C-2 of the purine nucleus.

(11) Ring closure of (XI) forms the first purine nucleotide, **inosinic acid** (XII) (inosine monophosphate, or IMP).

Multifunctional Polypeptides, Evolved by Gene Fusion, Catalyze Multiple Reactions of Purine Nucleotide Biosynthesis

In procaryotes, each reaction of Fig 36–3 is catalyzed by a different polypeptide. However, in eukaryotes, gene fusion has resulted in the evolution of **single polypeptides with multiple catalytic functions.** This development confers two major advantages. The first concerns metabolite channeling. Proximity of the catalytic sites of contiguous activities ensures that a given product is physically close to the catalytic activity for which it is a substrate. Secondly, gene fusion ensures production of equal quantities of different catalytic activities. Three multifunctional polypeptides in eukaryotes catalyze more than one reaction of Fig 36–3. These are **5′-phosphoribosyl-aminoimidazole synthase** (catalyzes reactions 3, 4, and 6), **5′-phosphoribosyl-aminoimidazole synthase** (catalyzes reactions 7 and 8), and **IMP synthase** (catalyzes reactions 10 and 11).

Drugs That Interfere With Tetrahydrofolate (THF) Metabolism Can Block Purine Nucleotide Biosynthesis

The two carbons inserted in reactions 4 and 10, which become atoms 8 and 2 of the purine ring, arise from N5,N10-methenyl-THF and N10-formyl-THF. N5,N10-Methenyl-THF is formed by oxidation of N5,N10-methylene-THF. However, once formed, N5,N10-methenyl-THF is committed to purine synthesis, either directly or after conversion to N10-formyl-THF. Inhibiting formation of these tetrahydrofolate compounds thus can inhibit de novo purine synthesis.

Oxidation & Amination of IMP Forms AMP & GMP

Arabic numerals refer to reactions of Fig 36–4, which outlines the reactions and intermediates whereby IMP is converted to AMP and GMP.

(12) Addition of aspartate to IMP forms **adenylosuccinate.** While this reaction, catalyzed by **adenylosuccinate synthase,** superficially resembles reaction 8, it specifically requires GTP, a requirement that provides a potential focus for regulation of adenine nucleotide synthesis (see below).

(13) Release of fumarate forms **adenosine 5′-monophosphate** (AMP). This reaction is catalyzed by **adenylosuccinase,** the same enzyme that catalyzes reaction 9.

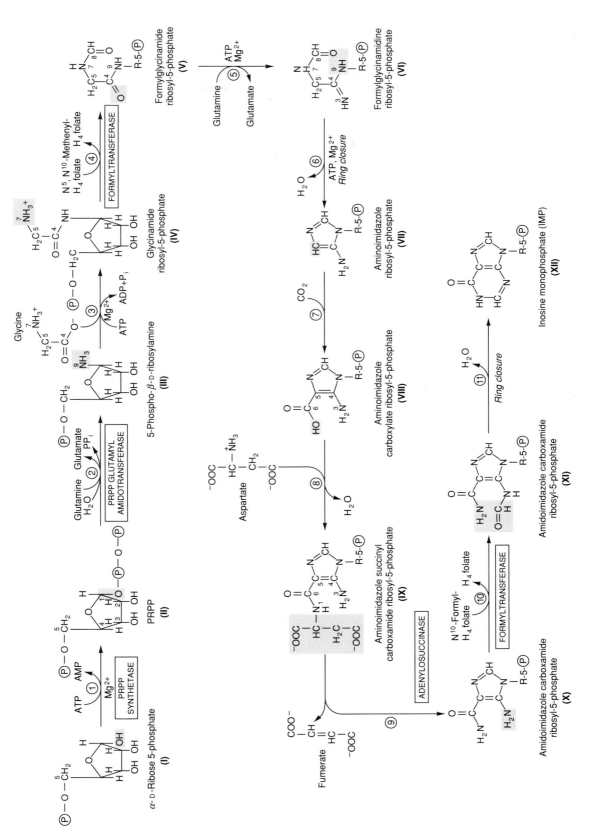

Figure 36–3. The pathway of de novo purine biosynthesis from ribose 5-phosphate and ATP. (See text for explanation.) Ⓟ, PO_3^{2-} or PO_2^-.

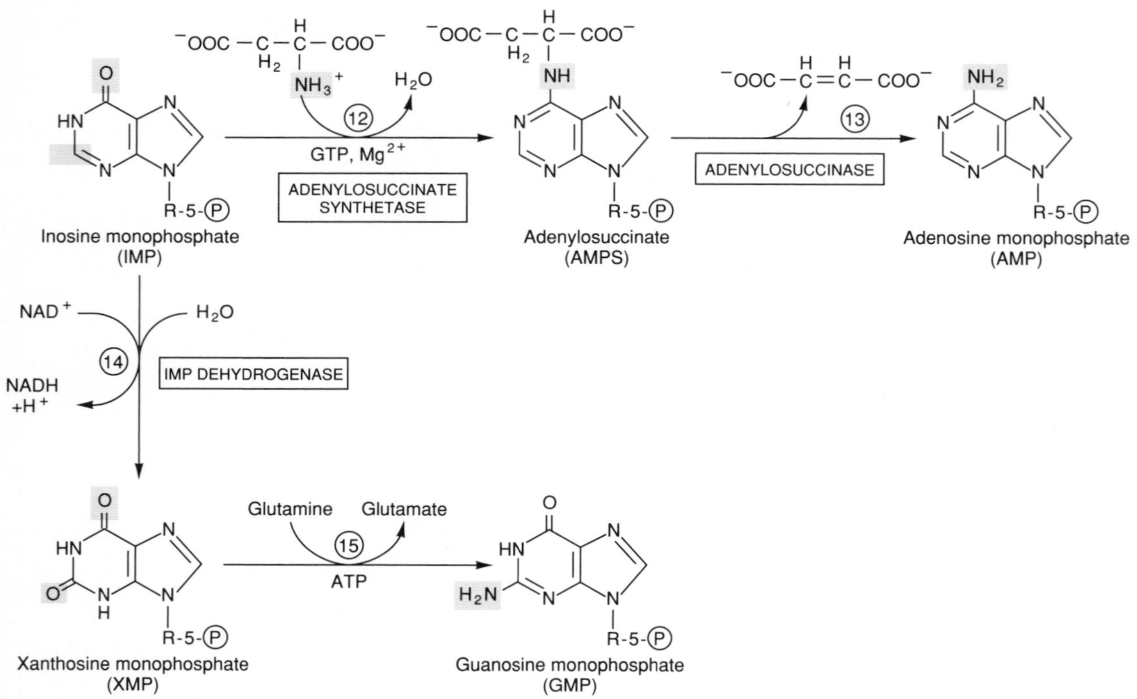

Figure 36-4. Conversion of IMP to AMP and GMP.

(14) Oxidation of IMP by NAD$^+$, catalyzed by **IMP dehydrogenase,** forms **xanthosine monophosphate (XMP).**

(15) Amination by the amide nitrogen of glutamine proceeds by analogy to reaction 5.

Certain Glutamine Analogs Inhibit Purine Nucleotide Biogenesis

Several antimetabolites that are glutamine analogs are effective inhibitors of purine nucleotide biosynthesis. **Azaserine** (0-diazoacetyl-L-serine) is an antagonist to glutamine, particularly at reaction 5. **Diazanorleucine** [(6-diazo-5-oxo)-L-norleucine] blocks reaction 2, and **6-mercaptopurine,** among other actions, inhibits reactions 13 and 14. **Mycophenolic acid** specifically inhibits reaction 14.

Conversion of AMP & GMP to Their Di- & Triphosphates Occurs Stepwise

Conversion of AMP and GMP to their respective nucleoside di- and triphosphates occurs in 2 stages (Fig 36–5). Successive phosphoryl transfers from ATP are catalyzed by **nucleoside monophosphate kinase** and **nucleoside diphosphate kinase,** respectively. The enzyme that phosphorylates adenylate is also called **myokinase.**

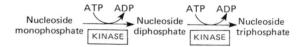

Figure 36–5. Conversion of nucleoside monophosphates to nucleoside diphosphates and nucleoside triphosphates.

PURINES & THEIR NUCLEOSIDES MAY BE CONVERTED DIRECTLY TO MONONUCLEOTIDES BY "SALVAGE" REACTIONS

Conversion of purines and purine nucleosides directly to mononucleotides by "salvage" reactions requires considerably less energy than does de novo synthesis. The quantitatively more important mechanism is **phosphoribosylation of a free purine (Pu) by PRPP,** forming a 5'-mononucleotide (Pu-RP).

$$Pu + PP-RP \rightarrow PP_i + Pu-RP$$

Two enzymes catalyze PRPP-dependent phosphoribosylation of a purine base: **adenine phosphoribosyl transferase,** which converts adenine to AMP (Fig 36–6), and **hypoxanthine-guanine phosphoribosyltransferase,** which adds PRPP to hypoxanthine or guanine, forming IMP or GMP (Fig 36–7).

A second salvage mechanism involves **direct phos-**

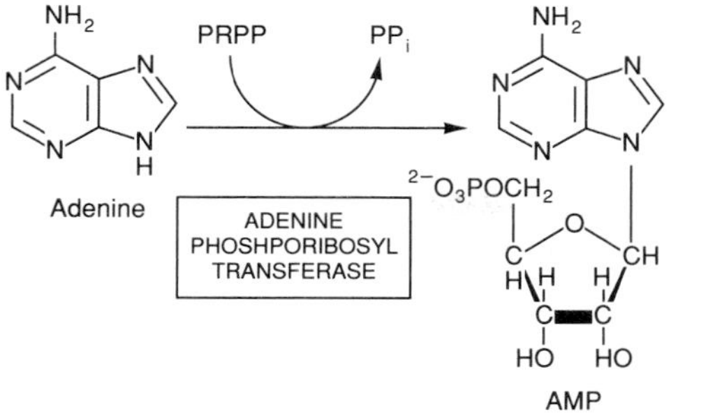

Figure 36–6. Phosphoribosylation of adenine catalyzed by adenine phosphoribosyltransferase.

phorylation of a purine ribonucleoside (PuR) by ATP:

$$PuR + ATP \rightarrow ADP + PuR–P$$

Adenosine kinase catalyses phosphorylation of adenosine or deoxyadenosine to AMP or dAMP, while **deoxycytidine kinase** phosphorylates deoxycytidine, deoxyadenosine, and 2′-deoxyguanosine to dCMP, dAMP, and dGMP, respectively.

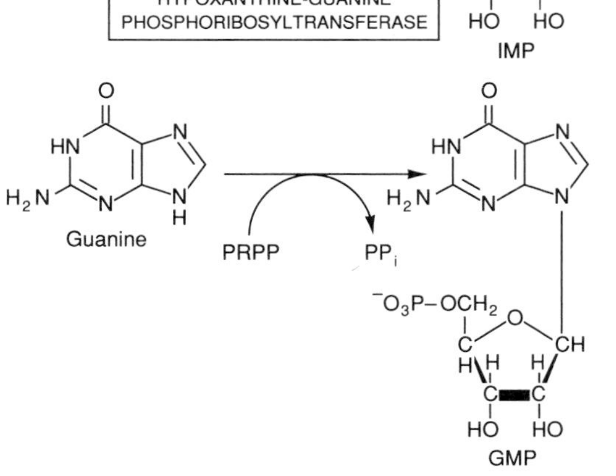

Figure 36–7. Phosphoribosylation of hypoxanthine and guanine to form IMP and GMP, respectively. Both reactions are catalyzed by hypoxanthine-guanine phosphoribosyltransferase.

HEPATIC PURINE NUCLEOTIDE BIOSYNTHESIS IS STRINGENTLY REGULATED

Mammalian liver, the major site of de novo purine nucleotide synthesis, provides purine bases and their nucleosides to be salvaged and utilized by tissues incapable of their de novo synthesis. For example, brain tissue has a low level of PRPP amidotransferase, and human brain may at least partially depend on exogenous purines. Erythrocytes and polymorphonuclear leukocytes cannot synthesize 5-phosphoribosylamine and hence must utilize exogenous purines for formation of purine nucleotides. Peripheral lymphocytes, however, possess some ability to synthesize purines de novo.

The Availability of PRPP Is the Major Determinant of the Rate of Biosynthesis of Purine Nucleotides

De novo synthesis of IMP consumes 6 mol ATP equivalent plus glycine, glutamine, methenyl-THF, and aspartate. To achieve efficient resource utilization, it is therefore imperative that cells regulate de novo purine biosynthesis. **The major determinant of the overall rate of de novo synthesis of purine nucleosides is the concentration of PRPP.** This, in turn, is determined by the relative rates of PRPP synthesis, utilization, and degradation. The rate of PRPP **synthesis** depends both on the availability of ribose 5-phosphate and on the activity of PRPP synthase. **PRPP synthase** activity is sensitive both to phosphate concentration and to the purine ribonucleotides that act as allosteric regulators (Fig 36–8). PRPP utilization reflects primarily activity of the salvage pathway for hypoxanthine and guanine, and only secondarily the rate of de novo purine synthesis. This conclusion arises from the observation that levels of PRPP in the erythrocytes and cultured fibroblasts of males with an inherited deficiency of hypoxanthine-guanine phosphoribosyl transferase were elevated severalfold.

AMP & GMP Feedback Regulate PRPP Glutamyl Amidotransferase

PRPP glutamyl amidotransferase (reaction 2, Fig 36–3), the first enzyme uniquely committed to de novo purine synthesis, **is sensitive to feedback inhibition by purine nucleotides, particularly AMP and GMP,** which inhibit competitively with PRPP (Fig 36–8). However, regulation of de novo purine synthesis via the amidotransferase is probably of less physiologic importance than regulation of PRPP synthetase.

AMP & GMP Feedback Regulate Their Formation From IMP

Two mechanisms regulate conversion of IMP to GMP and AMP (Fig 36–9). **AMP feedback regulates adenylosuccinate synthetase, and GMP feedback inhibits IMP dehydrogenase.** Furthermore, the con-

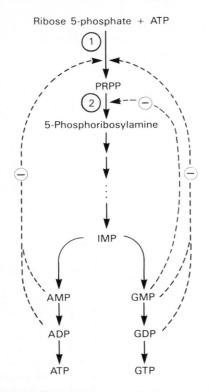

Figure 36–8. Control of the rate of de novo purine nucleotide synthesis. Solid lines represent chemical flow, and broken lines represent feedback inhibition (⊖) by end products of the pathway. Reactions ① and ② are catalyzed by PRPP synthetase and by PRPP glutamyltransferase (see Fig 35–3), respectively.

version of IMP to adenylosuccinate en route to AMP requires GTP, and conversion of xyanthinylate (XMP) to GMP requires ATP. **Cross-regulation between the pathways of IMP metabolism thus serves to decrease synthesis of one purine nucleotide when there is a deficiency of the other nucleotide.** Hypoxanthine-guanine phosphoribosyltransferase, which converts hypoxanthine and guanine to IMP and GMP, respectively (Fig 36–7), is also inhibited by these same nucleotides.

REDUCTION OF NDP'S FORMS dNDP'S

Reduction of purine and pyrimidine deoxyribonucleotides at the 2'-carbon of the ribose of the corresponding ribonucleotide diphosphate (NDP) is catalyzed by **ribonucleotide reductase complex** (Fig 36–10). This system is active only in cells that are actively synthesizing DNA preparatory to cell division. Reduction requires **thioredoxin** (a protein cofactor), **thioredoxin reductase** (a flavoprotein), and **NADPH.** In certain bacteria, but not in mammals, this reaction also requires cobalamin (vitamin B_{12}). Reduced thioredoxin, produced from oxidized thioredoxin in a reaction catalyzed by **NADPH:thioredoxin reductase,** is the immediate reductant of the NDP (Fig 36–10).

Reduction of ribonucleoside diphosphates (NDPs) to deoxyribonucleoside diphosphates (dNDPs) is subject to complex regulation (Fig 36–11), which achieves balanced production of deoxyribonucleotides for synthesis of DNA.

HUMANS CATABOLIZE PURINES TO URIC ACID

While lower primates and other mammals further catabolize uric acid by uricase-catalyzed hydrolysis to

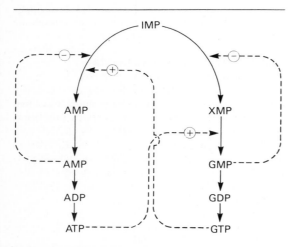

Figure 36–9. Regulation of the interconversion of IMP to adenosine nucleotides and guanosine nucleotides. Solid lines represent chemical flow, and broken lines represent both positive (⊕) and negative (⊖) feedback regulation.

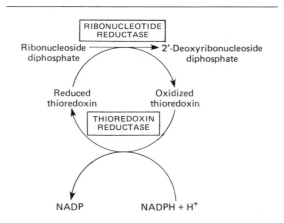

Figure 36–10. Reduction of ribonucleoside diphosphates to 2'-deoxyribonucleoside diphosphates.

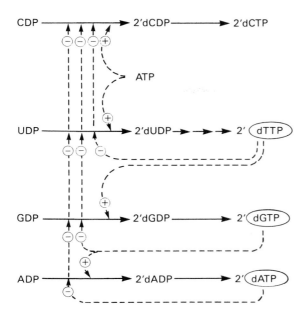

Figure 36–11. Regulation of the reduction of purine and pyrimidine ribonucleotides to their respective 2'-deoxyribonucleotides. Solid lines represent chemical flow, and broken lines represent negative (⊖) or positive (⊕) feedback regulation.

the highly water-soluble end product allantoin (Fig 36–12), the end product of purine catabolism in human subjects is uric acid. Amphibians, birds, and reptiles, which like humans lack uricase, excrete uric acid and guanine as the end products of both purine and protein catabolism.

Guanine and hypoxanthine form uric acid via xanthine in reactions catalyzed by **guanase** and **xanthine oxidase** of liver, small intestine, and kidney

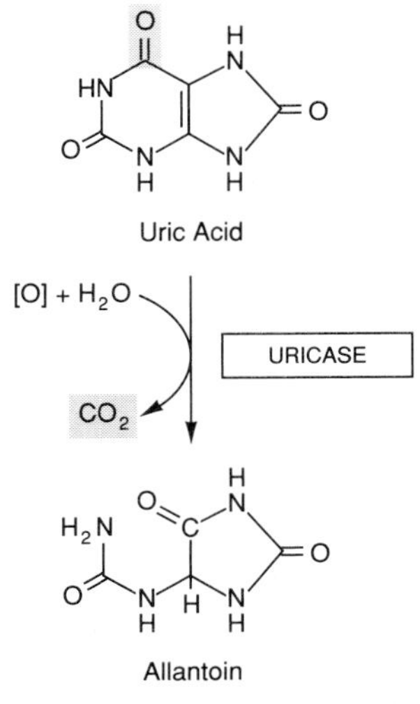

Uric Acid

[O] + H_2O

URICASE

CO_2

Allantoin

Figure 36–12. Conversion of uric acid to allantoin.

(Fig 36–1). Xanthine oxidase thus constitutes a potential site for pharmacologic intervention in patients with hyperuricemia and gout (see below).

FORMATION OF THE β-N-GLYCOSIDIC BOND OCCURS LATE IN THE BIOSYNTHESIS OF PYRIMIDINE NUCLEOTIDES

The de novo biosynthesis of pyrimidine and purine nucleosides involves several common precursors: **PRPP, glutamine, CO_2, aspartate,** and for thymidine nucleotides, **tetrahydrofolate** derivatives. A striking **difference** between these biosynthetic processes is that, while ribose phosphate is an integral part of the earliest precursor in purine nucleotide synthesis (see Fig 36–3), **attachment of the ribose phosphate moiety to N-3 of the pyrimidine base occurs late in the biosynthetic pathway** (Fig 36–13).

In what follows, Arabic numerals of paragraphs correspond to numbered reactions of Fig 36–13.

(1) Synthesis of the pyrimidine ring commences with formation of **carbamoyl phosphate** from glutamine, ATP, and CO_2 in a reaction catalyzed by **cytosolic carbamoyl phosphate synthase.** The carbamoyl phosphate synthase functional in urea synthesis is, by contrast, mitochondrial. This compartmentation ensures independent supplies of carbamoyl phosphate for these processes.

(2) Condensation of carbamoyl phosphate with aspartate forms **carbamoyl aspartate** in a reaction catalyzed by **aspartate transcarbamoylase.**

(3) Ring closure via loss of water, catalyzed by **dihydroorotase,** forms **carbamoyl aspartate.**

(4) Abstraction of hydrogens from carbons 5 and 6 by NAD⁺ introduces a double bond, forming **orotic acid.** This reaction is catalyzed by **dihydroorotate dehydrogenase,** a **mitochondrial** enzyme. All other enzymes of de novo pyrimidine synthesis are **cytosolic.**

(5) Addition of a ribose phosphate moiety derived from PRPP forms the β-N-glycosidic bond of **orotidine monophosphate** (OMP). This reaction, which is analogous to the *trans*-ribosylation reactions of Fig 36–7, is catalyzed by **orotate phosphoribosyltransferase.**

(6) Decarboxylation of orotidylate forms **uridine monophosphate** (UMP), the first true pyrimidine ribonucleotide. Thus, only at the penultimate step in the formation of UMP is the pyrimidine ring phosphoribosylated.

(7 & 8) Phosphate transfer from ATP yields **UDP** and **UTP** in reactions analogous to those for phosphorylation of purine nucleoside monophosphates (Fig 36–5).

(9) UTP is aminated to **CTP** by glutamine and ATP.

(10) Reduction of ribonucleotide diphosphates (NDPs) to their corresponding **dNDPs** occurs by reactions analogous to those for purine nucleotides (see Figs 36–10 and 36–11).

(11) dUMP may accept a phosphate from ATP forming **dUTP** (not shown). Alternatively, and since the substrate for TMP synthesis is dUMP, dUDP is dephosphorylated to **dUMP.**

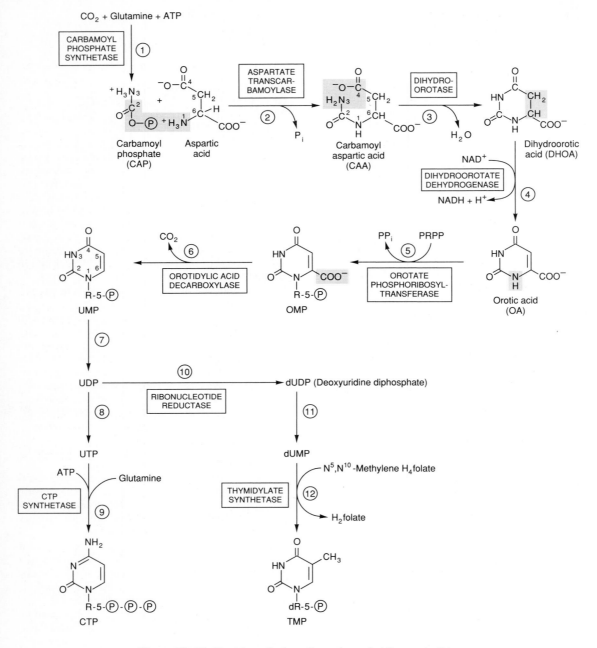

Figure 36–13. The biosynthetic pathway for pyrimidine nucleotides.

(12) Methylation of dUMP at C-5 by N^5,N^{10}-methylene-THF, catalyzed by **thymidylate synthase,** forms **thymidylate** (thymidine monophosphate, TMP).

The Anticancer Drug Methotrexate Blocks Reduction of Dihydrofolate

Reaction 12 of Fig 36–13 is the sole reaction of pyrimidine nucleotide biosynthesis which requires a tetrahydrofolate derivative. During the transfer process, the methylene group of N^5,N^{10}-methylene-THF is reduced to a methyl group, and the tetrahydrofolate carrier is oxidized to **dihydrofolate.** For further synthesis to occur, dihydrofolate must be reduced to tetrahydrofolate in a reaction catalyzed by **dihydrofolate reductase.** Consequently, dividing cells, that are by necessity generating TMP and dihydrofolate, are especially sensitive to inhibitors of dihydrofolate reductase. One such inhibitor is **methotrexate** (amethopterin), a widely used anti-cancer drug.

URACIL & CYTOSINE RIBO- & DEOXYRIBONUCLEOTIDES ARE SUBSTRATES FOR PYRIMIDINE SALVAGE

While mammalian cells lack efficient means of salvaging free pyrimidine bases, salvage reactions convert the pyrimidine ribonucleosides uridine and cytidine and the deoxyribonucleosides thymidine and deoxycytidine to their respective nucleotides (Fig 36–14). 2'-Deoxycytidine is phosphorylated by **deoxycytidine kinase,** an enzyme that also phosphorylates deoxyguanosine and deoxyadenosine. **Orotate phosphoribosyltransferase** (reaction 5, Fig 36–13), an enzyme of the de novo synthesis of pyrimidine nucleotides, can salvage orotic acid to OMP.

PYRIMIDINE ANALOGS ARE SUBSTRATES FOR CERTAIN ENZYMES OF PYRIMIDINE NUCLEOTIDE BIOGENESIS

While **orotate phosphoribosyltransferase** cannot use normal pyrimidine bases as substrates, it does catalyze conversion of the drug **allopurinol** (4-hydroxypyrazolopyramidine) to a nucleotide in which the ribosyl phosphate is attached to N-1 of the pyrimidine ring of allopurinol. The anticancer drug **5-fluorouracil** is also phosphoribosylated by orotate phosphoribosyl transferase.

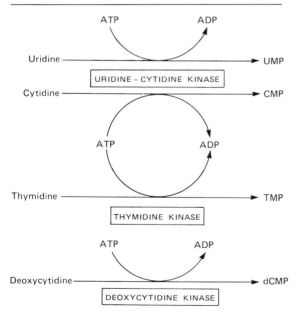

Figure 36–14. The pyrimidine nucleoside kinase reactions responsible for formation of the respective pyrimidine nucleoside monophosphates.

CATABOLISM OF PYRIMIDINES PRODUCES WATER-SOLUBLE PRODUCTS

The catabolism of pyrimidines, which occurs mainly in the liver, produces highly soluble end products (Fig 36–15). This contrasts with the production from purines of sparingly soluble uric acid and sodium urate. Release of respiratory CO_2 from the ureido carbon (C2) of the pyrimidine nucleus represents a major pathway for the catabolism of uracil, cytosine, and thymine. β-Alanine and β-aminoisobutyrate are end products of cytosine, uracil, and thymine catabolism.

Excretion of β-aminoisobutyrate increases in leukemia and after exposure to x-irradiation, due to increased destruction of cells and their DNA. Abnormally high excretion of β-aminoisobutyrate, traceable to a recessively expressed gene, occurs in heterozygous offspring (heterozygotes) of otherwise normal individuals. Approximately 25% of tested persons of Chinese or Japanese ancestry consistently excrete large amounts of β-aminoisobutyrate. Although little is known concerning how humans degrade β-aminoisobutyrate, pig kidney transaminates β-aminoisobutyrate to methylmalonate semialdehyde. Following oxidation to propionate, this semialdehyde forms succinyl-CoA (see Fig 21–2).

Pseudouridine Is Excreted Unchanged

Since no human enzyme catalyzes hydrolysis or phosphorolysis of **pseudouridine,** this unusual nucleoside, which was first detected in human urine, is excreted unchanged in the urine of normal subjects.

PYRIMIDINE NUCLEOTIDE BIOSYNTHESIS IS REGULATED AT THE LEVELS OF BOTH GENE EXPRESSION & ENZYME ACTIVITY

The first two enzymes of pyrimidine nucleotide biosynthesis are sensitive to **allosteric regulation.** With respect to the regulation of gene expression, the first three and last two enzymes of the pathway are regulated by apparently coordinate **repression and derepression. Carbamoyl phosphate synthase** is inhibited by UTP and purine nucleotides, but is activated by PRPP (Fig 36–16). **Aspartate transcarbamoylase** is particularly sensitive to inhibition by CTP. The allosteric properties of procaryotic aspartate transcarbamoylase constitute classic studies in allostery.

BALANCED PRODUCTION OF NUCLEIC ACID PRECURSORS REQUIRES COORDINATE CONTROL OF PURINE & PYRIMIDINE NUCLEOTIDE BIOSYNTHESIS

On a molar basis, the rate of pyrimidine biosynthesis parallels that of purine biosynthesis, sug-

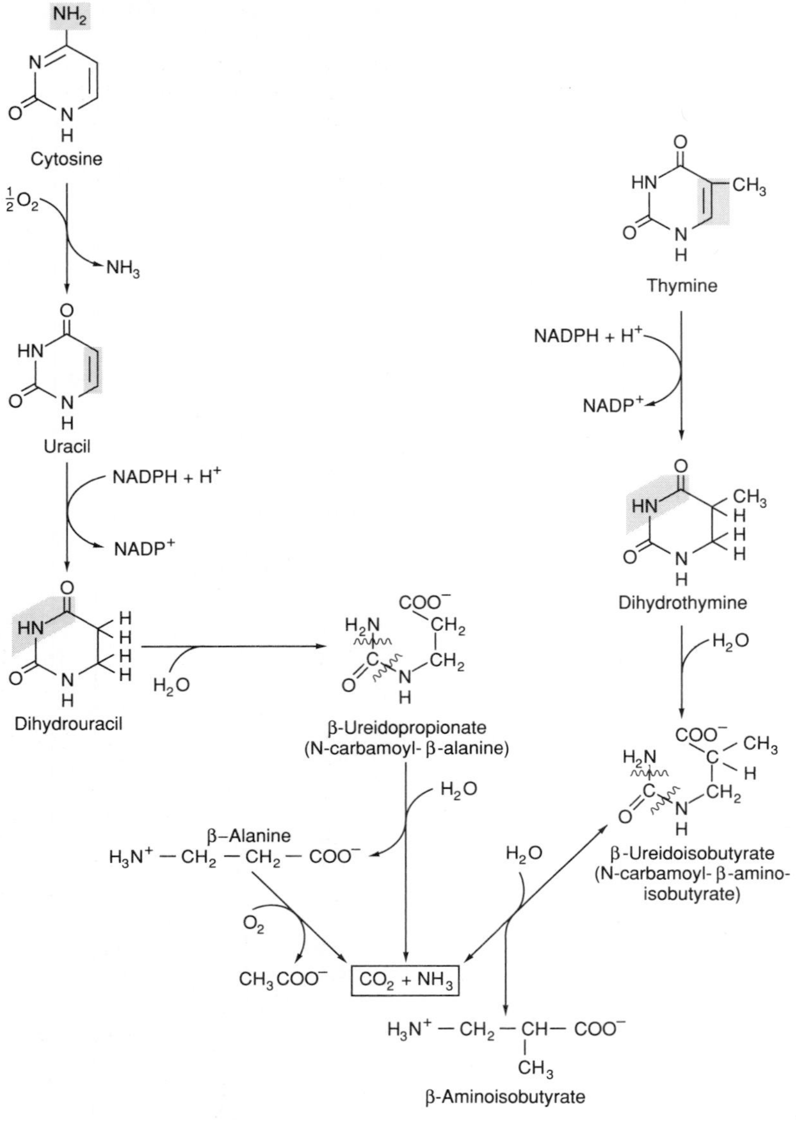

Figure 36–15. Catabolism of pyrimidines.

gesting coordinate control of these two processes. **PRPP synthetase** (reaction 1, Fig 36–3), an enzyme that catalyzes a reaction which forms an essential precursor for both processes, is subject to feedback inhibition by both purine and pyrimidine nucleotides and to activation by PRPP. Thus, there are several sites at which there is significant cross-regulation between purine and pyrimidine nucleotide synthesis.

GOUT IS A METABOLIC DISORDER OF PURINE CATABOLISM

As for any weak acid, the relative proportions of undissociated acid (uric acid) and its conjugate base (sodium urate) depend upon pH. For pH values of physiologic interest, we need consider only dissociation from uric acid of the first proton ($pK = 5.75$), since the second proton ($pK = 10.3$) dissociates at pH values well above that of any physiologic fluid. Thus, only uric acid and its monosodium salt are present in body fluids. In **hyperuricemia,** serum levels of sodium urate exceed the solubility limit. Sodium urate crystals may then form in soft tissues and joints, forming deposits termed **tophi.** This process causes an acute inflammatory reaction, **acute gouty arthritis,** which can progress to **chronic gouty arthritis.**

While urine at pH 5 becomes saturated with urates at 15 mg/dL, at pH 7, urine will dissolve 150–200 mg/dL. However, in normal urine, typically below pH

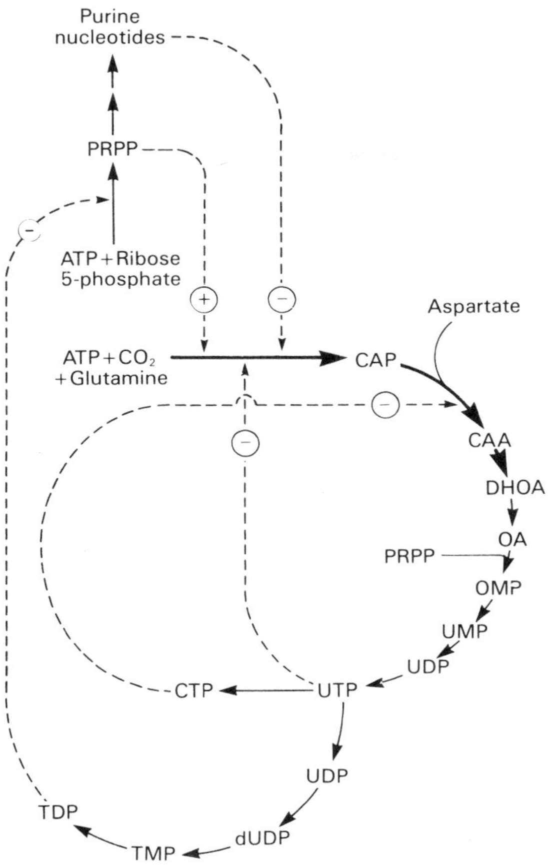

Figure 36–16. Control of pyrimidine nucleotide synthesis. Solid lines represent chemical flow. Broken lines represent positive (⊕) and negative (⊖) feedback regulation.

5.75, uric acid predominates. Crystals in the urinary system thus are sodium urate at any site proximal to, the site of acidification in the distal tubule and collecting ducts, but uric acid will be present at any site distal to that site. Since most stones of the urinary collecting system thus are uric acid, stone formation can be reduced by alkalinization of the urine.

Isotopic Methods Measure Miscible Urate Pools

Following intravenous injection of [^{15}N]uric acid into normal human subjects and into patients suffering from gout, the extent of dilution of the isotope was used to calculate the **miscible urate pool.** This pool averaged 1200 mg in normal adult males and 600 mg in normal adult females, from 2000 to 4000 mg in gouty patients without tophi, and as high as 31,000 mg in patients with severe tophaceous gout.

Sodium urate is extensively reabsorbed and partially secreted in the proximal tubule, further secreted in the loop of Henle, and again partially reabsorbed in the distal convoluted tubule. Net excretion of total uric

acid in normal men averages 400–600 mg/24 h. Many pharmacologic and naturally occurring compounds influence renal absorption and secretion of sodium urate. For example, aspirin in high doses competitively inhibits both urate excretion and reabsorption.

Urate Crystals Are Diagnostic of Gout

Visualization under a polarizing light microscope of needle-shaped, intensively **negatively birefringent** crystals of sodium urate in joint fluid polymorphonuclear leukocytes is diagnostic of gout. The crystals appear yellow when their long axis is parallel to the plane of polarized light, and blue when perpendicular to it.

DISORDERS OF PURINE METABOLISM INCLUDE HYPER- & HYPOURICEMIAS

Disorders of purine metabolism (Table 36–1) include **hyperuricemias, hypouricemias,** and immunodeficiency diseases. Hyperuricemias can be further subdivided on the basis of excretion of normal or excessive quantities (over 600 mg/24 h) of total urates (Table 36–2). Some disorders reflect specific enzyme defects. In others, hyperuricemia is secondary to disease processes such as cancer or psoriasis, which enhance tissue turnover.

Lesch-Nyhan Syndrome Reflects a Total Absence of Hypoxanthine-Guanine Phosphoribosyl Transferase

Lesch-Nyhan syndrome is a consequence of nonfunctional **hypoxanthine-guanine phosphoribosyl transferase** (HGPRTase), an enzyme of the purine salvage pathway (Fig 36–7). This inherited, X-linked, recessive disorder is characterized by cerebral palsy with choreoathetosis and spasticity, severe overproduction hyperuricemia with frequent uric acid lithiasis, and a bizarre syndrome of self-mutilation. Less deleterious mutations, which result in partial deficiency of HGPRTase, are accompanied in males by severe hyperuricemia, but without significant neurologic signs or symptoms. Purine overproduction in HGPRTase deficiency reflects sparing of PRPP from purine salvage, with consequent increased intracellular concentrations of PRPP.

Purine Overproduction and Hyperuricemia Characterize Von Gierke's Disease

Purine overproduction and hyperuricemia in **von Gierke's disease** (glucose-6-phosphatase deficiency) occurs secondary to enhanced generation of the PRPP precursor ribose 5-phosphate. In addition, however, the associated lactic acidosis elevates the renal thresh-

Table 36–1. Inherited disorders of purine metabolism and their associated enzyme abnormalities.

Clinical Disorder	Defective Enzyme	Nature of the Defect	Characteristics of Clinical Disorder	Inheritance Pattern
Gout	PRPP synthetase	Superactive (increased V_{max})	Purine overproduction and overexcretion	X-linked recessive
Gout	PRPP synthetase	Resistance to feedback inhibition	Purine overproduction and overexcretion	X-linked recessive
Gout	PRPP synthetase	Low K_m for ribose 5-phosphate	Purine overproduction and overexcretion	Probably X-linked recessive
Gout	HGPRTase*	Partial deficiency	Purine overproduction and overexcretion	X-linked recessive
Lesch-Nyhan syndrome	HGPRTase*	Complete deficiency	Purine overproduction and overexcretion; cerebral palsy and self-mutilation.	X-linked recessive
Immune deficiency	Adenosine deaminase	Severe deficiency	Combined (T cell and B cell) immunodeficiency, deoxyadenosinuria	Autosomal recessive
Immune deficiency	Purine nucleoside phosphorylase	Severe deficiency	T cell deficiency, inosinuria, deoxyinosinuria, guanosinuria, deoxyguanosinuria, hypouricemia	Autosomal recessive
Renal lithiasis	Adenine phosphoribosyltransferase	Complete deficiency	2,8-Dihydroxyadenine renal lithiasis	Autosomal recessive
Xanthinuria	Xanthine oxidase	Complete deficiency	Xanthine renal lithiasis, hypouricemia	Autosomal recessive

*HGPRTase = hypoxanthine-guanine phosphoribosyltransferase.

old for secretion of urate, contributing to accumulation of total body urates.

Hypouricemias Result From Enhanced Secretion or From Decreased Production of Uric Acid

Dalmatian dogs, which incompletely reabsorb filtered uric acid, excrete urate and uric acid in amounts that are excessive with respect to their serum urate levels. Similar defects have been noted in humans with **hypouricemia.**

Hypouricemia Is Associated With Xanthine Oxidase Deficiency

Hypouricemia and increased excretion of hypoxanthine and xanthine is associated with **xanthine oxidase** deficiency, due either to a genetic defect or to severe liver damage. In severe xanthine oxidase deficiency, patients may exhibit **xanthinuria** and **xanthine lithiasis.**

Table 36–2. Classification of patients with hyperuricemia.

I. Normal excretion of urate; renal disorder responsible for elevated serum urate.
II. Excessive excretion of urate because of overproduction.
  A. Secondary to other disease, eg, cancer, psoriasis.
  B. Known enzyme defects responsible for overproduction.
    1. PRPP synthetase abnormalities.
    2. Hypoxanthine-guanine phosphoribosyltransferase deficiencies.
    3. Glucose-6-phosphatase deficiencies.
  C. Unrecognized defects.

Genetic Defects Cause Adenosine Deaminase & Purine Nucleoside Phosphorylase Deficiency

Adenosine deaminase deficiency is associated with a severe combined immunodeficiency disease in which both thymus-derived lymphocytes (T cells) and bone marrow-derived lymphocytes (B cells) are sparse and dysfunctional. **Purine nucleoside phosphorylase deficiency** is associated with a severe thymus-derived lymphocyte deficiency with apparently normal B cell function. Both are autosomal recessive disorders. Immune dysfunctions appear to result from accumulation of dGTP and dATP, which allosterically inhibit ribonucleotide reductase and thereby deplete cells of DNA precursors, particularly dCTP.

Purine Deficiency Is Rare in Humans

Purine deficiency states in humans are limited to circumstances attributable primarily to deficiencies of folic acid and perhaps of vitamin B_{12} when the latter results in secondary deficiency of folate derivatives.

OVERPRODUCTION OF PYRIMIDINE CATABOLITES IS RARELY ASSOCIATED WITH CLINICALLY SIGNIFICANT ABNORMALITIES

The products of pyrimidine catabolism, unlike those of purine catabolism, are highly water soluble. Thus, even where pyrimidine overproduction occurs, clinically detectable abnormalities are rare (Table 36–

Table 36–3. Inherited disorders of pyrimidine metabolism and their associated enzyme abnormalities.

Clinical Disorder	Defective Enzyme	Characteristics of Clinical Disorder	Inheritance Pattern
β-Aminoisobutyric aciduria	Transaminase	No symptoms; frequent in Orientals.	Autosomal recessive
Orotic aciduria, type I	Orotate phosphoribosyl-transferase and orotidy-late decarboxylase	Orotic acid crystalluria, failure to thrive, and megaloblastic anemia. Immune deficiency. Remission with oral uridine.	Autosomal recessive
Orotic aciduria, type II	Orotidylate decarboxy-lase	Orotidinuria and orotic aciduria, mega-loblastic anemia. Remission with oral uridine.	Autosomal recessive
Ornithine transcarba-moylase deficiency	Ornithine transcarba-moylase	Protein intolerance, hepatic encepha-lopathy, and mild orotic aciduria.	X-linked recessive.

3). In hyperuricemia associated with severe over-production of PRPP, there is overproduction of pyrimidine nucleotides and increased excretion of β-alanine. Since N^5,N^{10}-methylene-THF is required for thymidylate synthesis, disorders of folate and vitamin B_{12} metabolism result in deficiencies of TMP (for vitamin B_{12} deficiency, by an indirect mechanism). β-Aminoisobutyric aciduria was discussed above in connection with pyrimidine catabolism. The orotic aciduria which accompanies **Reye's syndrome** is probably secondary to the inability of severely damaged mitochondria to utilize carbamoyl phosphate, which then may be available for cytosolic overproduction of orotic acid.

Patients With Orotic Aciduria Are Pyrimidine Auxotrophs Who Respond to Dietary Pyrimidine Nucleosides

In the more common type I disease, both **orotate phosphoribosyltransferase** and **orotidylate decarboxylase** (reactions 5 and 6, Fig 36–13) are deficient. In the more rare type II form, orotic aciduria results from deficiency of **orotidylate decarboxylase** (reaction 6, Fig 36–13). The major abnormal excretory product is orotic acid. Both types of patients are pyrimidine auxotrophs who respond to administered uridine. The greatly increased activities of **aspartate transcarbamoylase** and **dihydroorotase** in patients with type I orotic aciduria return to normal following treatment with oral uridine.

Deficiency of a Urea Cycle Enzyme Is Associated With Excretion of Pyrimidine Precursors

Increased excretion of orotic acid, uracil, and uridine has been described in patients deficient in the liver mitochondrial enzyme **ornithine transcarbamoylase.** Accumulated carbamoyl phosphate, a substrate of this enzyme, therefore enters the cytosol, where it is utilized for pyrimidine nucleotide syn-thesis. The resulting mild **orotic aciduria** increases when patients eat high-nitrogen foods.

Drugs May Precipitate Orotic Aciduria

The purine analog **allopurinol** (see Fig 35–21) is an inhibitor of xanthine oxidase used to treat gout. A sub-strate for orotate phosphoribosyltransferase (reaction 5, Fig 36–13), allopurinol competitively inhibits phos-phoribosylation of the natural substrate, orotic acid. In addition, the resulting nucleotide product inhibits **orotidylate decarboxylase** (reaction 6, Fig 36–13), producing **orotic aciduria** and **orotidinuria.** Since the de novo pathway for pyrimidine nucleotide bio-synthesis adjusts to this inhibition, human subjects are only transiently starved for pyrimidine nucleotides during early stages of treatment.

6-Azauridine, following its conversion to **6-azauridylate,** competitively inhibits **orotidylate de-carboxylase** (reaction 6, Fig 36–13), greatly enhanc-ing excretion of orotic acid and orotidine.

REFERENCES

Ames BN, Cathcart R, Schwiers E, Hochstein P: Uric acid provides an antioxidant defense in humans against oxi-dant- and radical-caused aging and cancer: A hypothesis. *Proc Natl Acad Sci USA* 1981;**78**:6858.

Benkovic SJ: The transformylase enzymes in de novo purine biosynthesis. *Trends Biochem Sci* 1984;**9**:320.

Holmgren A: Thioredoxin. *Annu Rev Biochem* 1985;**54**:237.

Jones M: Pyrimidine nucleotide biosynthesis in animal cells. *Annu Rev Biochem* 1980;**49**:253.

Martin DW Jr, Gelfand EW: Biochemistry of diseases of immunodevelopment. *Annu Rev Biochem* 1981;**50**:845.

Schinke RT: Methotrexate resistance and gene amplification. Mechanisms and implications. *Cancer* 1986;**57**:1912.

Seegmiller JE: Overview of the possible relation of defects in purine metabolism to immune deficiency. *Ann NY Acad Sci* 1985;**45**:9.

Stanbury JB et al (editors): *The Metabolic Basis of Inherited Disease,* 5th ed. McGraw-Hill, 1983.

37

Nucleic Acid Structure & Function

Daryl K. Granner, MD

INTRODUCTION

The discovery that genetic information is coded along the length of a polymeric molecule composed of only 4 types of monomeric units is one of the major scientific achievements of this century. This polymeric molecule, **DNA, is the chemical basis of heredity** and is organized into **genes,** the fundamental units of genetic information. Genes control the synthesis of various types of RNA, most of which are involved in protein synthesis. Genes do not function autonomously; their replication and function are controlled, in still vaguely understood ways, by feedback loops in which gene products themselves play a critical role. Knowledge of the structure and function of nucleic acids is essential in understanding genetics and provides the basis for future research.

BIOMEDICAL IMPORTANCE

The chemical basis of heredity and of genetic diseases is found in the structure of DNA. The basic information pathway (ie, DNA directs the synthesis of RNA, which in turn directs protein synthesis) has been elucidated. This knowledge is being used to define normal cellular physiology and the pathophysiology of disease at the molecular level.

DNA CONTAINS THE GENETIC INFORMATION

The demonstration that DNA contained the genetic information was first made in 1944 in a series of experiments by Avery, MacLeod, and McCarty, who showed that the genetic determination of the character (type) of the capsule of a specific pneumococcus could be transmitted to another of a distinctly different capsular type by introducing purified DNA from the former coccus into the latter. These authors referred to the agent (later shown to be DNA) accomplishing the change as "transforming factor." Subsequently, this type of genetic manipulation has become commonplace. Similar experiments have recently been performed utilizing yeast, cultured mammalian cells, and insect and rodent embryos as recipients, and cloned DNA as the donor of genetic information.

DNA Consists of 4 Deoxynucleotides

The chemical nature of the monomeric deoxynucleotide units of DNA—**deoxyadenylate, deoxyguanylate, deoxycytidylate,** and **thymidylate**—is described in Chapter 35. These monomeric units of DNA are held in polymeric form by 3',5'-phosphodiester bridges constituting a single strand, as depicted in Fig 37–1. The informational content of DNA (the genetic code) resides in the sequence in which these monomers—purine and pyrimidine deoxyribonucleotides—are ordered. **The polymer as depicted possesses a polarity;** one end has a 5'-hydroxyl or phosphate terminus while the other has a 3'-phosphate or hydroxyl moiety. The importance of this polarity will become evident. Since the genetic information resides in the order of the monomeric units within the polymers, there must exist a mechanism of reproducing or replicating this specific information with a high degree of fidelity. That requirement, together with x-ray diffraction data from the DNA molecule and the observation of Chargaff that in DNA molecules the concentration of deoxyadenosine (A) nucleotides equals that of thymidine (T) nucleotides (A = T), while the concentration of deoxyguanosine (G) nucleotides equals that of deoxycytidine (C) nucleotides (G = C), led Watson, Crick, and Wilkins to propose in the early 1950s a model of a double-stranded DNA molecule. The model they proposed is depicted in Fig 37–2. The 2 strands of this right-handed, double-stranded molecule are held in register by **hydrogen bonds** between the purine and pyrimidine bases of the respective linear molecules. The pairings between the purine and pyrimidine nucleotides on the opposite strands are very specific and are dependent upon hydrogen bonding of **A with T,** and **G with C** (Fig 37–3).

In the double-stranded molecule, restrictions imposed by the rotation about the phosphodiester bond, the favored **anti** configuration of the glycosidic bond (see Fig 35–9), and the predominant tautomers (see Fig 35–4) of the 4 bases (A, G, T, and C) allow A to pair only with T, and G only with C, as depicted in Fig 37–3. This base-pairing restriction explains the earlier observation that in a double-stranded DNA molecule the content of A equals that of T and the content of G equals that of C. The 2 strands of the double-helical

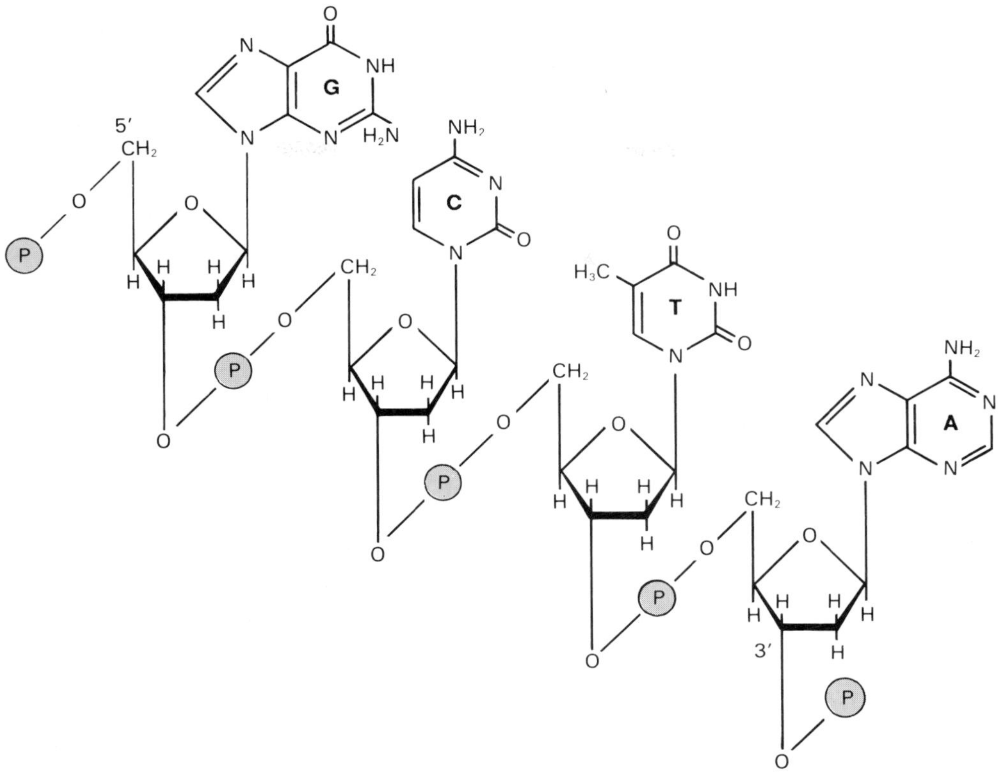

Figure 37–1. A segment of one strand of a DNA molecule in which the purine and pyrimidine bases adenine (A), thymine (T), cytosine (C), and guanine (G) are held together by a phosphodiester backbone between 2′-deoxyribosyl moieties attached to the nucleobases by an N-glycosidic bond. Note that the backbone has a polarity (ie, a direction).

molecule, each of which possesses a polarity, are **antiparallel;** ie, one strand runs in the 5′ to 3′ direction and the other in the 3′ to 5′ direction. This is analogous to 2 parallel streets, each running one way but carrying traffic in opposite directions. In the double-stranded DNA molecules the genetic information resides in the sequence of nucleotides on one strand, the **template strand;** the opposite strand is considered the **coding strand** because it matches the RNA transcript that encodes the protein.

As depicted in Fig 37–3, 3 hydrogen bonds hold the deoxyguanosine nucleotide to the deoxycytidine nucleotide, whereas the other pair, the A-T pair, is held together by 2 hydrogen bonds. Thus, the G-C bond is stronger by approximately 50%. Because of this added strength and also because of stacking interactions, regions of DNA that are rich in G-C bonds are much more resistant to denaturation, or "melting," than A-T-rich regions.

DNA Exists in Several Double-Helical Structures

To date 6 forms (A to E and Z) have been described, but most of these have been found only under rigidly controlled experimental conditions. These forms are distinguished by (1) the number of base pairs that occupy each turn of the helix, (2) the pitch or angle between each base pair, (3) the helical diameter of the molecule, and (4) the handedness (right or left) of the double helix (Table 37–1). Some of these forms interconvert if salt and hydration conditions are manipulated. It is possible that interconversion might happen in vivo.

The **B form,** the overwhelmingly dominant form of DNA under physiologic conditions (low salt, high degree of hydration), has a pitch of 3.4 nm per turn (Fig 37–2). Within a single turn, 10 base pairs (bp) exist (this can vary between 10.0 and 10.6 bp per turn), each planar base being stacked to resemble 2 winding stacks of coins side by side. The 2 stacks are held together by hydrogen bonding at each level between the 2 coins on opposite stacks and by 2 ribbons wound in a right-hand turn about the 2 stacks and representing the phosphodiester backbone.

A variation of this basic structure is seen in the **A form,** which is favored by an environment slightly less hydrous and richer in Na⁺ and K⁺ ions. This right-handed structure is bulkier than the B form, has more base pairs per turn, and resembles the structure of double-stranded RNA and DNA-RNA hybrid duplexes. The C-E forms, also right-handed, are seen in very special experimental circumstances and are not thought to exist in vivo.

Z-DNA forms a left-handed double helix in which

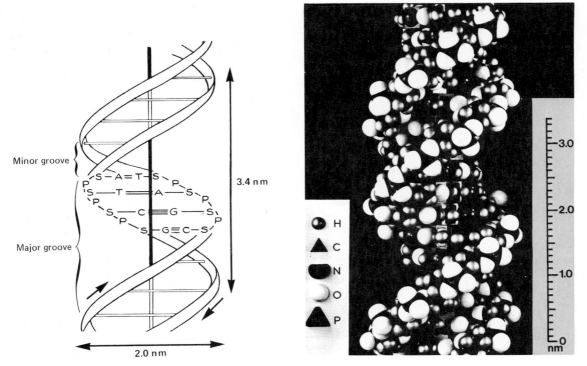

Figure 37–2. The Watson and Crick model of the double-helical structure of the B form of DNA. *Left:* Diagrammatic representation of structure. The horizontal arrow indicates the width of the double helix (2.0 nm) and the vertical arrow indicates the distance spanned by one complete turn of the double helix (3.4 nm). The central axis of the double helix is indicated by the vertical rod. The short arrows designate the polarity of the antiparallel strands. (A, adenine; C, cytosine; G, guanine; T, thymine; P, phosphate; S, sugar [deoxyribose].) *Right:* Space-filling model of DNA structure. (Photograph from James D. Watson, *Molecular Biology of the Gene,* 3rd ed. Copyright © 1976, 1970, 1965, by W. A. Benjamin, Inc., Menlo Park, Calif.)

the phosphodiester backbone zigzags along the molecule; hence, the name Z-DNA. Z-DNA is the least twisted (12 bp per turn) and thinnest DNA helix known to exist, and it has only one groove (see below). Z-DNA occurs in repeated sequences of alternating purine and pyrimidine deoxynucleotides (GC or AC) but also requires one or more stabilizing influences. These stabilizing influences include (1) the presence of high-salt or specific cations such as spermine or spermidine, (2) a high degree of negative supercoiling of the DNA (see Chapter 38), (3) the binding of Z-DNA-specific proteins, and (4) the methylation of the 5-carbon of some of the deoxycytidine nucleotides in the alternating sequence.

Z-DNA could exert regulatory effects both proximal and distal to the site of its existence. For instance, some proteins that bind in the minor or major groove of B-form DNA could probably not bind to the Z form. In addition, the reversion of Z form to a B form of DNA, an event that might occur as a consequence of loss of methyl groups from 5-methyldeoxycytidine, would likely result in torsional differences of DNA distal to the actual site of Z-DNA. As discussed below, torsional winding and unwinding are thought to affect gene activity, as is the methylation of deoxycytidine.

The existence of Z-DNA in *Drosophila* (fruit fly) chromosomes has been demonstrated utilizing antibodies that recognize and bind specifically to Z-DNA. Human DNA contains potential Z-DNA-forming regions dispersed throughout the genome, and the stabilizing influences may also exist.

The Denaturation (Melting) of DNA Is Used To Analyze Its Structure

The double-stranded structure of DNA can be melted in solution by increasing the temperature or decreasing the salt concentration. Not only do the 2 stacks of bases pull apart but the bases themselves unstack while still connected in the polymer by the phosphodiester backbone. Concomitant with this denaturation of the DNA molecule is an increase in the optical absorbance of the purine and pyrimidine bases— a phenomenon referred to as **hyperchromicity** of denaturation. Because of the stacking of the bases and the hydrogen bonding between the stacks, the double-stranded DNA molecule exhibits properties of a rigid rod and in solution is a viscous material that loses its viscosity upon denaturation.

The strands of a given molecule of DNA separate over a temperature range. The midpoint is called the

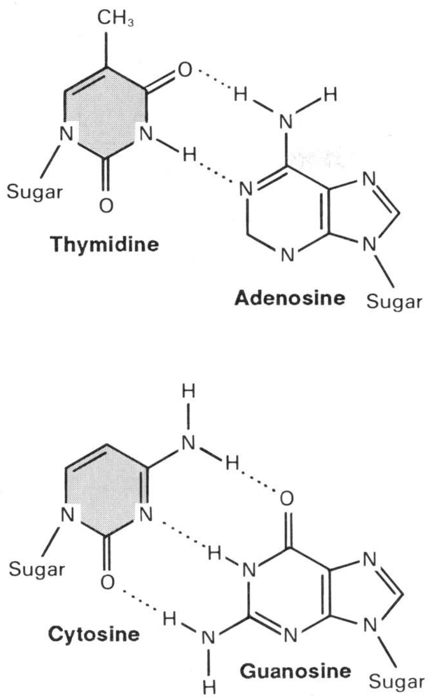

Figure 37–3. Base pairing between deoxyadenosine and thymidine involves the formation of 2 hydrogen bonds. Three such bonds form between deoxycytidine and deoxyguanosine. The broken lines represent hydrogen bonds. In DNA, the sugar moiety is 2-deoxyribose, whereas in RNA it is D-ribose.

melting temperature, or T_m. The T_m is influenced by the base composition of the DNA and by the salt concentration of the solution. DNA rich in G-C pairs, which have 3 hydrogen bonds, melts at a higher temperature than that rich in A-T pairs, which have 2 hydrogen bonds. A 10-fold increase of monovalent cation concentration increases the T_m by 16.6 °C. Formamide, which is commonly used in recombinant DNA experiments, destabilizes hydrogen bonding between bases, thereby lowering the T_m. This allows the strands of DNA or DNA-RNA hybrids to be separated at much lower temperatures and minimizes the strand breakage that occurs at high temperatures.

There Are Grooves in the DNA Molecule

Careful examination of the model depicted in Fig 37–2 reveals a **major groove** and a **minor groove**

Table 37–1. Features of some structures of DNA.

Type	Handedness	Base Pairs per Turn	Pitch per Base Pair	Helical Diameter
A	Right	11	0.256 nm	2.3 nm
B	Right	10	0.338 nm	1.9 nm
Z	Left	12	0.371 nm	1.8 nm

winding along the molecule parallel to the phosphodiester backbones. In these grooves, proteins can interact specifically with exposed atoms of the nucleotides (usually by H bonding) and thereby recognize and bind to specific nucleotide sequences without disrupting the base pairing of the double-helical DNA molecule. As discussed in Chapters 39 and 41, regulatory proteins can control the expression of specific genes via such interactions.

DNA Exists in Relaxed & Supercoiled Forms

In some organisms such as bacteria, bacteriophages, and many DNA-containing animal viruses, the ends of the DNA molecules are joined to create a **closed circle** with no terminus. This of course does not destroy the polarity of the molecules, but it eliminates all free 3′ and 5′ hydroxyl and phosphoryl groups. Closed circles exist in relaxed or supercoiled forms. Supercoils are introduced when a closed circle is twisted around its own axis or when a linear piece of duplex DNA, whose ends are fixed, is twisted. This energy-requiring process puts the molecule under stress, and the greater the number of supercoils, the greater the stress or torsion (test this with a rubber band). **Negative supercoils** are formed when the molecule is twisted in the direction opposite from the clockwise turns of the right-handed double helix found in B-DNA. Such DNA is said to be **underwound.** The energy required to achieve this state is, in a sense, stored in the supercoils. The transition to another form that requires energy is thereby facilitated by the underwinding. One such transition is strand separation, which is a prerequisite for DNA replication and transcription. Supercoiled DNA is therefore a preferred form in biologic systems. Enzymes that catalyze topologic changes of DNA are called **topoisomerases.** Topoisomerases can relax or insert supercoils. The best characterized is **bacterial gyrase,** which induces negative supercoiling in DNA using ATP as an energy source.

THE FUNCTION OF DNA IS TO PROVIDE A TEMPLATE FOR REPLICATION & TRANSCRIPTION

The genetic information stored in the nucleotide sequence of DNA serves 2 purposes. It is the source of information for the synthesis of all protein molecules of the cell and organism, and it provides the information inherited by daughter cells or offspring. Both of these functions require that the DNA molecule serve as a template—in the first case for the **transcription** of the information into RNA and in the second case for the **replication** of the information into daughter DNA molecules.

The complementarity of the Watson and Crick double-stranded model of DNA strongly suggests that **replication of the DNA molecule occurs in a semiconservative manner.** Thus, when each strand of the double-stranded parental DNA molecule separates from its complement during replication, each serves as a template on which a new complementary strand is synthesized (Fig 37–4). The 2 newly formed double-stranded daughter DNA molecules, each containing one strand (but complementary rather than identical) from the parent double-stranded DNA molecule, are then sorted between the 2 daughter cells (Fig 37–5). Each daughter cell contains DNA molecules with information identical to that which the parent possessed; yet in each daughter cell the DNA molecule of the parent cell has been only semiconserved.

The **semiconservative nature of DNA replication** in the bacterium *Escherichia coli* was unequivocally demonstrated by Meselson and Stahl in a classic experiment using the heavy isotope of nitrogen and centrifugal equilibrium techniques. The DNA of *E coli* is chemically identical to that of humans, although the sequences of nucleotides are, of course, different, and the human cell contains about 1000 times more DNA per cell than does the bacterium. Furthermore, the chemistry of replication of DNA in prokaryotes such as *E coli* appears to be identical to that in eukaryotes, including humans, even though the enzymes carrying out the reactions of DNA synthesis and replication are different. Thus, any observations on the chemical nature or chemical reactions of nucleic acids of prokaryotes are very likely applicable to eukaryotic organisms. Indeed, the Meselson and Stahl type of experiment has now been performed in mammalian cells and has yielded results comparable to those obtained with *E coli*.

THE CHEMICAL NATURE OF RNA DIFFERS FROM THAT OF DNA

Ribonucleic acid (RNA) is a polymer of purine and pyrimidine ribonucleotides linked together by 3′,5′-phosphodiester bridges analogous to those in DNA (Fig 37–6). Although sharing many features with DNA, RNA possesses several specific differences:

(1) In RNA, the sugar moiety to which the phosphates and purine and pyrmidine bases are attached is ribose rather than the 2′-deoxyribose of DNA.

(2) The pyrimidine components of RNA differ from that of DNA. Although RNA contains the ribonucleotides of adenine, guanine, and cytosine, it does not possess thymine except in the rare case mentioned below. Instead of thymine, **RNA contains the ribonucleotide of uracil.**

(3) RNA exists as a single strand, whereas DNA exists as a double-stranded helical molecule. However, given the proper complementary base sequence with opposite polarity, the single strand of RNA, as demonstrated in Fig 37–7, is capable of folding back on itself like a hairpin and thus acquiring double-stranded characteristics.

(4) Since the RNA molecule is a single strand complementary to only one of the 2 strands of a gene, its

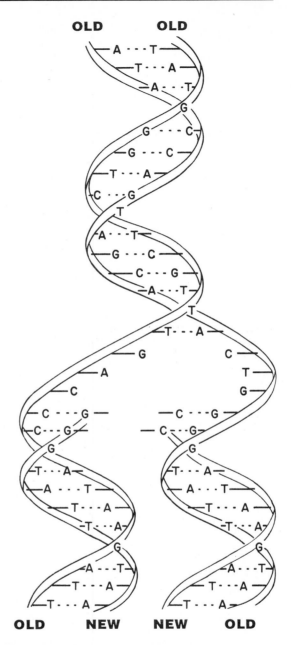

Figure 37–4. The double-stranded structure of DNA and the template function of each old strand on which a new complementary strand (shaded) is synthesized. (From James D. Watson. *Molecular Biology of the Gene,* 3rd ed. Copyright © 1976, 1970, 1965, by W. A. Benjamin, Inc., Menlo Park, Calif.)

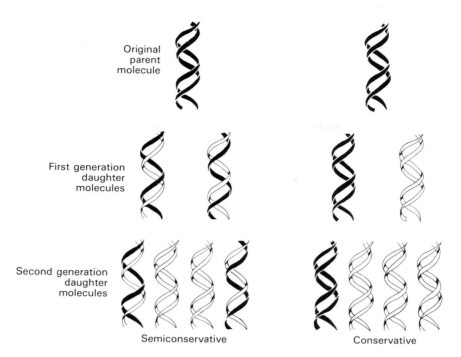

Figure 37–5. The expected distributions of parental DNA strands if a semiconservative or conservative mechanism of replication was used. The parental strands are solid, and the newly synthesized strands are open. DNA replication occurs by the semiconservative mechanism. (Redrawn and reproduced, with permission, from Lehninger AL: *Biochemistry,* 2nd ed. Worth, 1975.)

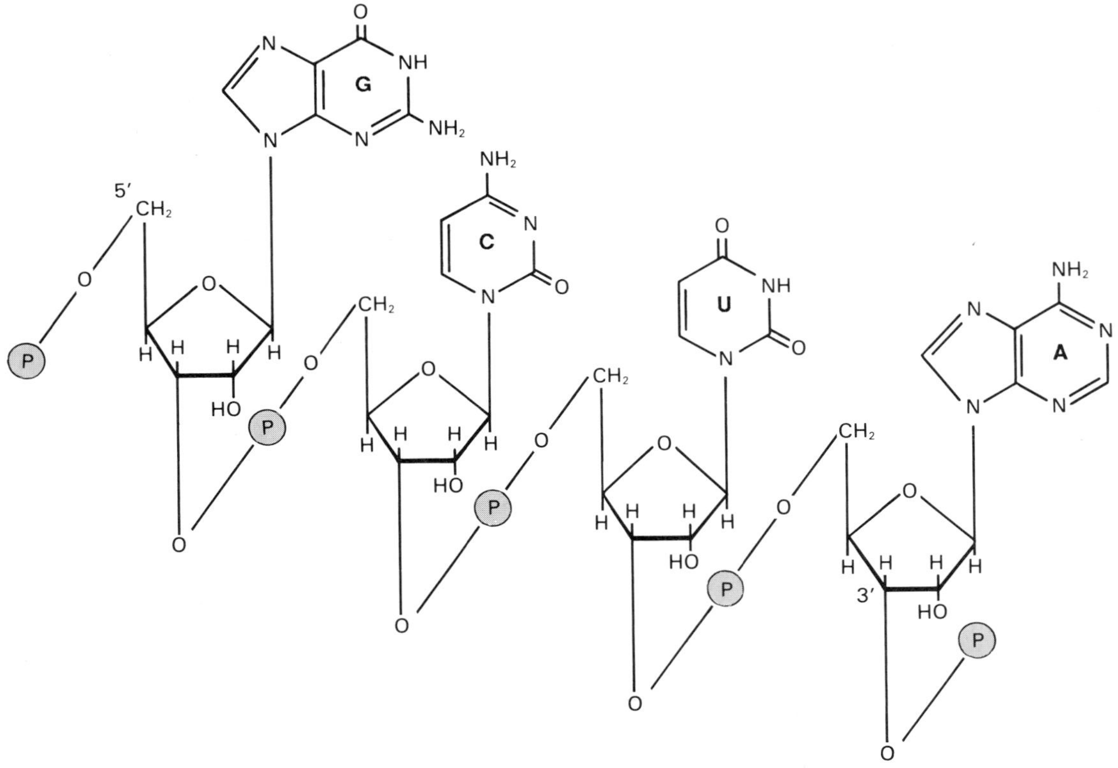

Figure 37–6. A segment of a ribonucleic acid (RNA) molecule in which the purine and pyrimidine bases—adenine (A), uracil (U), cytosine (C), and guanine (G)—are held together by phosphodiester bonds between ribosyl moieties attached to the nucleobases by N-glycosidic bonds. Note that the polymer has a polarity as indicated by the labeled 3′- and 5′-attached phosphates.

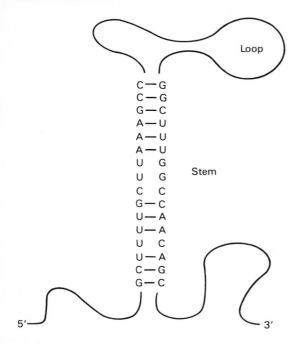

Figure 37–7. Diagrammatic representation of the secondary structure of a single-stranded RNA molecule in which a stem loop, or "hairpin," has been formed and is dependent upon the intramolecular base pairing.

guanine content does not necessarily equal its cytosine content, nor does its adenine content necessarily equal its uracil content.

(5) **RNA can be hydrolyzed by alkali** to 2′,3′ cyclic diesters of the mononucleotides, compounds that cannot be formed from alkali-treated DNA because of the absence of a 2′-hydroxyl group. The alkali lability of RNA is useful both diagnostically and analytically.

Information within the single strand of RNA is contained in its sequence ("primary structure") of purine and pyrimidine nucleotides within the polymer. The sequence is complementary to the template strand of the gene from which it was transcribed. Because of this complementarity, an RNA molecule can bind specifically via the base-pairing rules to its template DNA strand; it will not bind ("hybridize") with the other (coding) strand of its gene. The sequence of the RNA molecule (except for U replacing T) is the same as that of the coding strand of the gene (Fig 37–8).

Nearly All of the Several Species of RNA Are Involved in Some Aspect of Protein Synthesis

Those cytoplasmic RNA molecules that serve as templates for protein synthesis (ie, that transfer genetic information from DNA to the protein-synthesizing machinery) are designated **messenger RNA, or mRNA.** Many other cytoplasmic RNA molecules (**ribosomal RNA, or rRNA**) have structural roles wherein they contribute to the formation of ribosomes (the organellar machinery for protein synthesis) or serve as adapter molecules (**tRNA**) for the translation of RNA information into specific sequences of polymerized amino acids.

Much of the RNA synthesized from DNA templates in eukaryotic cells, including mammalian cells, is **degraded within the nucleus,** and it never serves as either a structural or an informational entity within the cellular cytoplasm.

In human cells there are **small nuclear RNA (snRNA)** species that are not directly involved in protein synthesis but that may have roles in RNA processing and the cellular architecture. These relatively small molecules vary in size from 90 to about 300 nucleotides (Table 37–3).

The genetic material for some animal and plant viruses is RNA rather than DNA. Although some RNA viruses do not ever have their information transcribed into a DNA molecule, many animal RNA viruses—specifically the retroviruses (the HIV or AIDS virus, for example)—are transcribed by an **RNA-dependent DNA polymerase,** the so-called **reverse transcriptase,** to produce a double-stranded DNA copy of their RNA genome. In many cases, the resulting double-stranded DNA transcript is integrated into the host genome and subsequently serves as a template for gene expression and from which new viral RNA genomes can be transcribed.

RNA Is Organized in Several Unique Structures

In all prokaryotic and eukaryotic organisms, 3 main classes of RNA molecules exist: messenger RNA

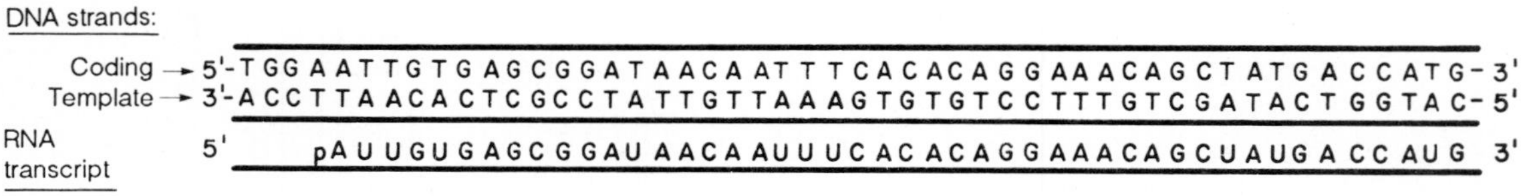

DNA strands:

Coding → 5′-T GG A A T T G T G A G C G G A T A A C A A T T T C A C A C A G G A A A C A G C T A T G A C C A T G- 3′
Template → 3′-A C C T T A A C A C T C G C C T A T T G T T A A A G T G T G T C C T T T G T C G A T A C T G G T A C- 5′

RNA transcript   5′   pA U U G U G A G C G G A U A A C A A U U U C A C A C A G G A A A C A G C U A U G A C C A U G   3′

Figure 37–8. The relationship between the sequences of an RNA transcript and its gene, in which the coding and noncoding strands are shown with their polarities. The RNA transcript with a 5′ to 3′ polarity is complementary to the coding strand with its 3′ to 5′ polarity. Note that the sequence in the RNA transcript and its polarity is the same as that in the noncoding strand, except that the U of the transcript replaces the T of the gene.

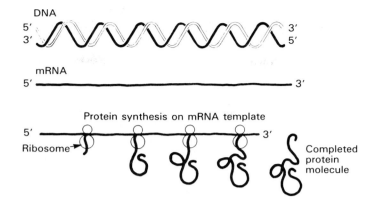

Figure 37–9. The expression of genetic information in DNA into the form of an mRNA transcript. This is subsequently translated by ribosomes into a specific protein molecule.

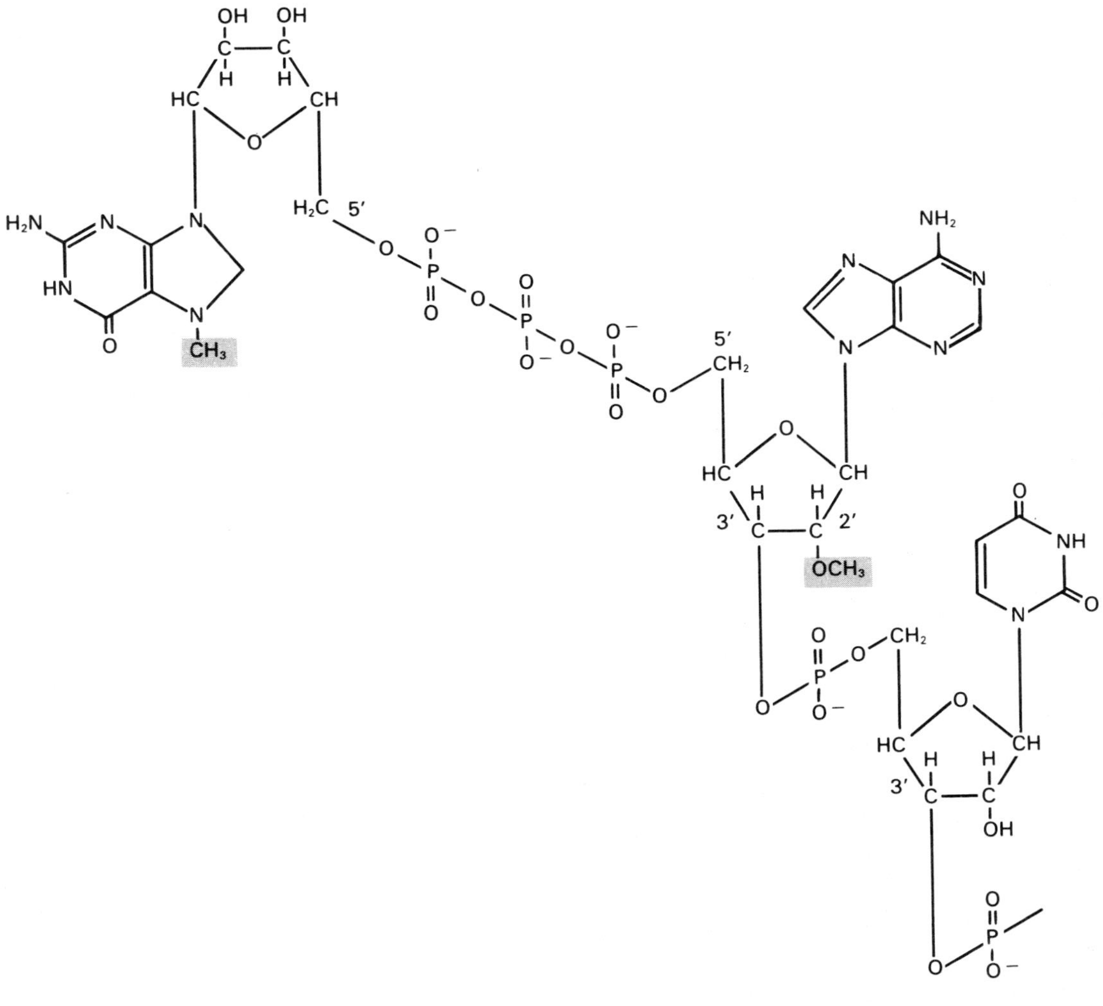

Figure 37–10. The cap structure attached to the 5′ terminus of most eukaryotic messenger RNA molecules. A 7-methyl-guanosine triphosphate is attached at the 5′ terminus of the mRNA, which usually contains a 2′-O-methylpurine nucleotide.

(mRNA), transfer RNA (tRNA), and ribosomal RNA (rRNA). Each class differs from the others by size, function, and general stability.

Messenger RNA (mRNA): This class is the most heterogeneous in size and stability. All of the members of the class function as messengers conveying the information in a gene to the protein-synthesizing machinery, where each serves as a template on which a specific sequence of amino acids is polymerized to form a specific protein molecule, the ultimate gene product (Fig 37–9).

Messenger RNAs, particularly in eukaryotes, have some unique chemical characteristics. The 5′ terminus of mRNA is **"capped"** by a 7-methylguanosine triphosphate that is linked to an adjacent 2′-O-methyl ribonucleoside at its 5′-hydroxyl through the 3 phosphates (Fig 37–10). The mRNA molecules frequently contain internal 6-methyladenylates and other 2′-O-ribose methylated nucleotides. The cap is probably involved in the recognition of mRNA by the translating machinery and it helps stabilize the mRNA by preventing the attack of 5′-exonucleases. The protein-synthesizing machinery begins translating the mRNA into proteins at the 5′ or capped terminus. The other end of most mRNA molecules, the 3′-hydroxyl terminus, has attached a polymer of adenylate residues 20–250 nucelotides in length. The specific function of the **poly(A) "tail"** at the 3′-hydroxyl terminus of mRNAs is not fully understood, but it seems that it maintains the intracellular stability of the specific mRNA by preventing the attack of 3′-exonucleases. Some mRNAs, including those for some histones, do not contain poly(A). The poly(A) tail, because it will form a base pair with oligodeoxythymidine polymers attached to a solid substrate like cellulose, can be used to separate mRNA from other species of RNA.

In **mammalian cells,** including cells of humans, the mRNA molecules present in the cytoplasm are not the RNA products immediately synthesized from the DNA template but must be formed by **processing** from a precursor molecule before entering the cytoplasm. Thus, in mammalian nuclei, the immediate products of gene transcription constitute a fourth class of RNA molecules. These nuclear RNA molecules are very heterogeneous in size and are quite large. The **heterogeneous nuclear RNA (hnRNA)** molecules may have a MW in excess of 10^7, whereas the MW of mRNA molecules is generally less than 2×10^6. As is discussed in Chapter 39, hnRNA molecules are processed to generate the mRNA molecules which then enter the cytoplasm to serve as templates for protein synthesis.

Transfer RNA (tRNA): tRNA molecules consist of approximately 75 nucleotides. They also are generated by nuclear processing of a precursor molecule (see Chapter 39). The tRNA molecules serve as adapters for the translation of the information in the sequence of nucleotides of the mRNA into specific amino acids. There are at least 20 species of tRNA molecules in every cell, at least one (and often several) corresponding to each of the 20 amino acids required for protein

synthesis. Although each specific tRNA differs from the others in its sequence of nucleotides, the tRNA molecules as a class have many features in common. The primary structure—ie, the nucleotide sequence—of all tRNA molecules allows extensive folding and intrastrand complementarity to generate a secondary structure that appears like a **cloverleaf** (Fig 37–11).

All tRNA molecules contain 4 main arms. The **acceptor arm** consists of a base-paired stem that terminates in the sequence CCA (5′ to 3′). It is through an ester bond to the 3′-hydroxyl group of the adenosyl moiety that the carboxyl groups of amino acids are attached. The other arms have base-paired stems and unpaired loops (Fig 37–7). The **anticodon arm** at the end of a base-paired stem recognizes the triplet nucleotide or codon (discussed in Chapter 40) of the template mRNA. It has a nucleotide sequence complementary to the codon, and is responsible for the specificity of the tRNA. The **D arm** is named for the presence of the base dihydouridine, and the **TψC arm** for the sequence T, pseudouridine, and C. The **extra arm** is the most variable feature of tRNA, and it provides a basis for classification. **Class 1** tRNAs (about 75% of all tRNAs) have an extra arm that is 3–5 bp

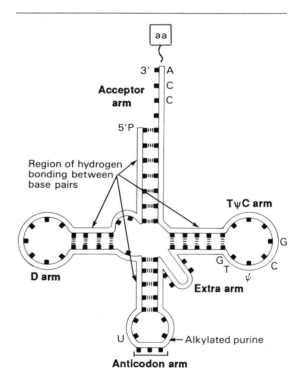

Figure 37–11. Typical aminoacyl tRNA in which the amino acid (aa) is attached to the 3′ CCA terminus. The anticodon, TφC, and dihydrouracil (DHU) arms are indicated, as are the positions of the intramolecular hydrogen bonding between these base pairs. (From James D. Watson, *Molecular Biology of the Gene,* 3rd ed. Copyright © 1976, 1970, 1965, by W. A. Benjamin, Inc., Menlo Park, Calif.)

Table 37–2. Components of mammalian ribosomes.*

Component	Mass (mw)	Protein Number	Protein Mass	RNA Size	RNA Mass	Bases
40S subunit	1.4×10^6	~35	7×10^5	18S	7×10^5	1900
60S subunit	2.8×10^6	~50	1×10^6	5S	35,000	120
				5.8S	45,000	160
				28S	1.6×10^6	4700

*The ribosomal subunits are defined according to their sedimentation velocity in Svedberg units (40S or 60S). This table illustrates the total mass (MW) of each. The number of unique proteins and their total mass (MW) and the RNA components of each subunit in size (Svedberg units), mass, and base composition are listed.

long. **Class 2** tRNAs have an extra arm 13–21 bp long and often have a stem-loop structure.

The secondary structure of tRNA molecules is maintained by the base pairing in these arms, and this is a consistent feature. The acceptor arm has 7 bp, the TψC and anticodon arms 5 bp, and the D arm 3 (or 4) bp.

Although tRNAs are quite stable in prokaryotes, they are somewhat less stable in eukaryotes. The opposite is true for mRNAs, which are quite unstable in prokaryotes but generally stable in eukaryotic organisms.

Ribosomal RNA (rRNA): A ribosome is a cytoplasmic nucleoprotein structure that acts as the machinery for the synthesis of proteins from the mRNA templates. On the ribosomes, the mRNA and tRNA molecules interact to translate into a specific protein molecule information transcribed from the gene.

The components of the mammalian ribosome, which has a molecular weight of about 4.2×10^6 and a sedimentation velocity of 80S (Svedberg units), are shown in Table 37–2. The mammalian ribosome contains 2 major nucleoprotein subunits, a larger one of MW 2.8×10^6 (60S) and a smaller subunit of 1.4×10^6 MW (40S). The **60S subunit** contains a **5S ribosomal RNA** (rRNA), a **5.8S rRNA,** and a **28S rRNA;** there are also probably more than 50 specific polypeptides. The smaller, or **40S, subunit** contains a single **18S rRNA** and approximately 30 polypeptide chains. All of the ribosomal RNA molecules except the 5S rRNA are processed from a single 45S precursor RNA molecule in the nucleolus (see Chapter 40). The 5S rRNA apparently has its own precursor that is independently transcribed. The highly methylated ribosomal RNA molecules are packaged in the nucleolus with the specific ribosomal proteins. In the cytoplasm, the ribosomes remain quite stable and capable of many translation cycles. The functions of the ribosomal RNA molecules in the ribosomal particle are not fully understood, but they are necessary for ribosomal assembly and seem to play key roles in the binding of mRNA to ribosomes and its translation.

Small Stable RNA: A large number of discrete, highly conserved, and small stable RNA species are found in eukaryotic cells. The majority of these molecules exist as ribonucleoproteins and are distributed in the nucleus, in the cytoplasm, or in both. They range in size from 90 to 300 nucleotides and are present in 100,000–1,000,000 copies per cell.

Small nuclear ribonucleoprotein particles, often called **snurps,** may be significantly involved in gene regulation. The U7 snurp appears to be involved in the production of correct 3′ ends of histone mRNA. The U4 and U6 RNAs may be required for poly(A) processing, and the U1 snurp is implicated in intron removal and mRNA processing (see Chapter 39).

Table 37–3 summarizes some characteristics of small stable RNAs.

Table 37–3. Some of the species of small stable RNAs found in mammalian cells.

Name	Length (nucleotides)	Molecules per Cell	Localization
U1	165	1×10^6	Nucleoplasm/hnRNA
U2	188	5×10^5	Nucleoplasm
U3	216	3×10^5	Nucleolus
U4	139	1×10^5	Nucleoplasm
U5	118	2×10^5	Nucleoplasm
U6	106	3×10^5	Perichromatin granules
4.5S	91–95	3×10^5	Nucleus and cytoplasm
7S	280	5×10^5	Nucleus and cytoplasm
7-2	290	1×10^5	Nucleus and cytoplasm
7-3	300	2×10^5	Nucleus

REFERENCES

Hunt T: *DNA Makes RNA Makes Protein.* Elsevier, 1983.
Rich A et al: The chemistry and biology of left-handed Z-DNA. *Annu Rev Biochem* 1984;**53**:847.
Turner P: Controlling roles for snurps. *Nature* 1985;**316**:105.
Watson JD: *The Double Helix.* Atheneum, 1968.
Watson JD, Crick FHC: Molecular structure of nucleic acids. *Nature* 1953;**171**:737.
Zieve GW: Two groups of small stable RNAs. *Cell* 1981;**25**:296.

38

DNA Organization & Replication

Daryl K. Granner, MD

INTRODUCTION

The DNA in prokaryotic organisms is generally not combined with proteins other than those involved in DNA replication or transcription. Much of the DNA in eukaryotic organisms is covered with a variety of proteins. These proteins and DNA form a complex structure, chromatin, that allows for numerous configurations of the DNA molecule and types of control unique to the eukaryotic organism.

The genetic information in the DNA of a chromosome can be transmitted by exact replication, or it can be exchanged by a number of processes, including crossing over, recombination, transposition, and conversion. These provide a means of ensuring adaptability and diversity for the organism but can also result in disease.

DNA replication, a highly complex and ordered process, follows the 5' to 3' polarity typical of the synthesis of RNA and protein described in other chapters. In eukaryotic cells replication of the DNA in a chromosome begins at multiple sites and proceeds in both directions simultaneously. A number of enzymes are required for the synthesis and repair of DNA, both of which follow the rules of Watson-Crick base pairing.

BIOMEDICAL IMPORTANCE

Mutations are due to a change in the base sequence of DNA. These may result from the faulty replication, movement, or repair of DNA and occur with a frequency of about one in every 10^6 cell divisions. An abnormal gene product can be the result of mutations that occur in coding or regulatory-region DNA. A mutation in a germ cell will be transmitted to offspring (so-called vertical transmission of hereditary disease). A number of factors, including viruses, chemicals, ultraviolet light, and ionizing radiation, increase the rate of mutation. Mutations often affect somatic cells and so are passed on to successive generations of cells within an organism. It is becoming apparent that a number of diseases, and perhaps most cancers, are due to horizontal transmission of induced mutations.

CHROMATIN IS THE CHROMOSOMAL MATERIAL EXTRACTED FROM NUCLEI OF CELLS OF EUKARYOTIC ORGANISMS

Chromatin* consists of very long double-stranded **DNA molecules** and a nearly equal mass of rather small basic proteins termed **histones** as well as a smaller amount of **nonhistone proteins** (most of which are acidic and larger than histones) and a small quantity of **RNA.** Electron microscopic studies of chromatin have demonstrated dense spherical particles called **nucleosomes,** which are approximately 10 nm in diameter and connected by DNA filaments (Fig 38–1).

Histones Are the Most Abundant Chromatin Proteins

The **histones** are somewhat heterogeneous, consisting of a series of closely related basic proteins. Among the histones, **H1** histones are the least tightly bound to chromatin and are, therefore, easily removed with a salt solution, after which chromatin becomes soluble. The isolated core nucleosomes contain 4 classes of histones: **H2A, H2B, H3,** and **H4.** The structures of slightly lysine-rich histones—H2A and H2B—appear to have been significantly conserved between species, while the structures of arginine-rich histones—H3 and H4—have been highly conserved between species. This severe conservation implies that the function of histones is identical in all eukaryotes and that the entire molecule is involved quite specifically in carrying out this function. The C-terminal two-thirds of the molecules have a usual amino acid composition, while their N-terminal thirds contain most of the basic amino acids. **These 4 core histones are subject to 5 types of**

---

*So far as possible, the discussion of this chapter and of Chapters 39, 40, and 41 will pertain to mammalian organisms, which are, of course, among the higher eukaryotes. At times it will be necessary to refer to observations made in prokaryotic organisms such as bacteria and viruses, but when such occurs it will be acknowledged as being information that may be extrapolated to mammalian organisms. The division of the material presented in Chapters 37–41 is somewhat arbitrary and should not be taken to mean that the processes described are not fully integrated and interdependent.

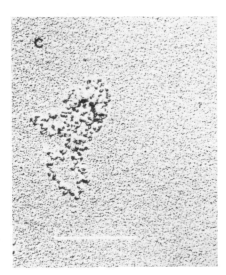

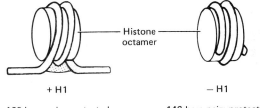

Histone octamer

+ H1

166 base pairs protected
(2 superhelical turns) and
exposed (unprotected)
linker DNA

− H1

146 base pairs protected
(1.75 superhelical turns)

Figure 38–2. Model for the structure of the nucleosome (*left*) and nucleosome core (*right*), in which DNA is wrapped around the surface of a flat protein cylinder consisting of 2 each of histones H2A, H2B, H3, and H4, Histone H1 (shaded area) expands the number of base pairs protected. (Reproduced, with permission, from Laskey RA, Earnshaw WC: Nucleosome assembly. *Nature* 1980; **286**:763.)

Figure 38–1. Electron micrograph of nucleosomes attached by strands of nucleic acid. (White bar represents 2.5 μm.) (Reproduced, with permission, from Oudet P, Gross-Bellard M, Chambon P: Electron microscopic and biochemical evidence that chromatin structure is a repeating unit. *Cell* 1975;**4**:281.)

covalent modifications: acetylation, methylation, phosphorylation, ADP-ribosylation, and covalent linkage (H2A only) to ubiquitin, the nuclear protein. These histone modifications likely play some role in chromatin structure and function, but little is currently understood.

When removed from chromatin, the histones interact with each other in very specific ways. **H3 and H4 aggregate to form a tetramer** containing 2 molecules of each $(H3/H4)_2$, while **H2A and H2B form dimers** (H2A-H2B) and higher oligomeric complexes $([H2A-H2B]_n)$. In low salt concentrations, the tetrameric H3-H4 does not associate with the H2A-H2B dimer or oligomer, and H1 does not associate directly with any of the other histones in solution.

The Nucleosome Is a Structure Containing Histone and DNA

However, when the $H3_2$-$H4_2$ tetramer and H2A-H2B dimers are mixed with purified, double-stranded DNA, the same x-ray diffraction pattern is formed as that observed in freshly isolated chromatin. Electron microscopic studies confirm the existence of reconstituted **nucleosomes.** Furthermore, the reconstitution of nucleosomes from DNA and histones H2A, H2B, H3, and H4 is independent of the organismal or cellular origin of the various components. The histone H1 and the nonhistone proteins are not necessary for the reconstitution of the nucleosome core.

In the nucleosome, the DNA is supercoiled in a left-handed helix over the surface of the disk-shaped histone octamer consisting of a central H3-H4 tetramer $(H3/H4)_2$ and two H2A-H2B dimers (Fig 38–2). The

core histones interact with the DNA on the inside of the supercoil without protruding.

The $(H3/H4)_2$ itself can confer nucleosome-like properties on DNA and thus has a central role in the formation of the nucleosome. The addition of two H2A-H2B dimers stabilizes the primary particle and binds firmly 2 additional half-turns of DNA previously bound only loosely to the $(H3/H4)_2$. Thus, **1.75 superhelical turns of DNA** are wrapped around the surface of the histone octamer, **protecting 146 base pairs of DNA** and forming the **nucleosome core** (Fig 38–2). As the DNA wraps around the surface of the histone octamer to form the nucleosome, it comes in contact with the histones in the order

H2A–H2B–H4–H3–H3–H4–H2B–H2A

Histone H1 binds to the DNA, where it enters and leaves the nucleosome core to seal a **2-turn, 166-base-pair DNA superhelix** generating the **nucleosome** (Fig 38–2).

The assembly of nucleosomes is probably mediated by the anionic nuclear protein **nucleoplasmin.** Histones, which are strongly cationic, can bind nonspecifically to the strongly anionic DNA by forming salt bridges. Clearly, such a nonspecific interaction of histones and DNA would be detrimental to nucleosome formation and chromatin function. Nucleoplasmin is an anionic pentameric protein that binds neither to DNA nor to chromatin, but it can interact reversibly with a histone octamer in such a way that the histones no longer adhere nonspecifically to negatively charged surfaces such as DNA. It seems that nucleoplasmin thereby maintains in the nucleus an ionic environment conducive to the specific interaction of histones and DNA and the assembly of nucleosomes. As the nucleosome is assembled, nucleoplasmin must be released from the histones. Nucleosomes appear to exhibit preference for certain regions on specific DNA molecules, but the basis for this nonrandom distribution, termed **phasing,** is unknown, although it is probably related to the relative physical flexibility of certain nucleotide sequences that are able to accommodate the regions of kinking within the supercoil.

The super-packing of nucleosomes in nuclei is seemingly dependent upon the interaction of the H1 histones with the double-stranded DNA connecting the nucleosomes. The topology of the interaction of the double-stranded DNA with the H1 histones to form the internucleosome spacer regions is not well delineated.

HIGHER ORDER STRUCTURES PROVIDE FOR THE COMPACTION OF CHROMATIN

Electron microscopy of chromatin reveals 2 higher orders of structure—the 10-nm fibril and the 25- to 30-nm chromatin fiber—beyond that of the nucleosome itself. The disklike nucleosome structure has a 10-nm diameter and a height of 5 nm. The **10-nm fibril** seems to consist of nucleosomes arranged with their edges touching and their flat faces parallel with the fibril axis (Fig 38–3). The 10-nm fibril is probably further supercoiled with 6–7 nucleosomes per turn to form the **30-nm chromatin fiber** (Fig 38–3). Each turn of the supercoil is relatively flat, and the faces of the nucleosomes of successive turns would be nearly parallel to each other. H1 histones appear to stabilize the 30-nm fiber, but their position and that of the variable length spacer DNA are not clear. It is probable that nucleosomes can form a variety of packed structures. In order to form a mitotic chromosome, the 30-nm fiber must be compacted in length another 100-fold (see below).

In **interphase chromosomes,** chromatin fibers appear to be organized into 30,000–100,000 base-pair **loops or domains** anchored in a scaffolding (or supporting matrix) within the nucleus. Within these domains, some DNA sequences may be located nonrandomly. It has been suggested that each looped domain of chromatin corresponds to a separate genetic function, containing both coding and noncoding regions of the gene.

SOME REGIONS OF CHROMATIN ARE "ACTIVE"; OTHERS ARE "INACTIVE"

Generally, every cell of an individual metazoan organism contains the same genetic information in the form of the same DNA sequences. Thus, the differences between different cell types within an organism must be explained by differential expression of the common genetic information. Chromatin containing active genes (ie, transcriptionally active chromatin) has been shown to differ in several ways from that of nonactive regions. The nucleosome structure of active chromatin appears to be altered or even absent in highly active regions. DNA in active chromatin contains large regions (about 100,000 bases long) that are **sensitive to digestion by a nuclease** such as DNase I. The sensitivity to DNase I of chromatin regions being actively transcribed reflects only a potential for transcription rather than transcription itself and in several systems can be correlated with a relative lack of 5-methyldeoxycytidine in the DNA.

Within the large regions of active chromatin there exist shorter stretches of 100–300 nucleotides that exhibit an even greater (another 10-fold) sensitivity to DNase I. These **hypersensitive sites** probably result from a structural conformation that favors access of the nuclease to the DNA. These regions are generally located immediately upstream from the active gene and are the location of interrupted nucleosomal structure caused by the binding of non-histone proteins. (See Chapters 39 and 41.) In many cases, it seems that if a gene is capable of being transcribed, it must have a DNase hypersensitive site in the chromatin immediately upstream. The proteins involved in transcription, and those involved in maintaining access to the template strand, lead to the formation of hypersensitive sites. Hypersensitive sites often provide the first clue about the presence and location of a transcription control element.

Transcriptionally inactive chromatin is densely packed during interphase as observed by electron microscopic studies and is referred to as **heterochromatin;** transcriptionally active chromatin stains less densely and is referred to as **euchromatin.** Generally, euchromatin is replicated earlier in the mammalian cell cycle (see below) than is heterochromatin.

There are 2 types of heterochromatin, constitutive heterochromatin and facultative heterochromatin. **Constitutive heterochromatin** is always condensed and, thus, inactive. Constitutive heterochromatin is found in the regions near the chromosomal centromere and at chromosomal ends (telomeres). **Facultative heterochromatin** is at times condensed, but at other

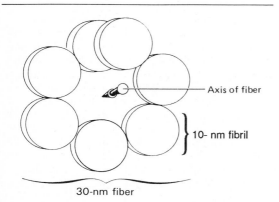

Figure 38–3. Proposed structure of the 30-nm chromatin fiber consisting of superhelixes of 10-nm fibrils of nucleosomes. The axis of the 30-nm fiber is perpendicular to the plane of the page.

times it is actively transcribed and, thus, uncondensed and appearing as euchromatin. Of the 2 members of the X chromosome pair in mammalian females, one X chromosome is almost completely inactive transcriptionally and is heterochromatic. However, the heterochromatic X chromsome decondenses during gametogenesis and becomes transcriptionally active during early embryogenesis; thus, it is facultative heterochromatin.

Certain cells of insects, eg, *Chironomus*, contain giant chromosomes that have been replicated for 10 cycles without separation of daughter chromatids. These copies of DNA line up side by side in precise register and produce a banded chromosome containing regions of condensed chromatin and lighter bands of more extended chromatin. Transcriptionally active regions of these **polytene chromosomes** are especially decondensed into **"puffs,"** which can be shown to contain the enzymes responsible for transcription and to be the sites of RNA synthesis (Fig 38–4).

DNA IS ORGANIZED INTO CHROMOSOMES

At metaphase, mammalian **chromosomes** possess a 2-fold symmetry, with identical **sister chromatids** connected at a **centromere,** the relative position of which is characteristic for a given chromosome (Fig 38–5). Each sister chromatid contains one double-stranded DNA molecule. During interphase, the packing of the DNA molecule is less dense than it is in the condensed chromosome during the metaphase. Metaphase chromosomes are transcriptionally **inactive.**

The human haploid genome consists of 3.5×10^9 base pairs or pairs of nucleotides and about 1.7×10^7 nucleosomes. Thus, each of the 23 chromatids in the human haploid genome would contain on the average 1.5×10^8 nucleotides in one double-stranded DNA molecule. The length of each DNA molecule must be **compressed about 8000-fold** to generate the structure of a condensed metaphase chromosome! In metaphase chromosomes, the 25- to 30-nm chromatin fibers are also folded into a series of **looped domains,** the proximal portions of which are anchored to a nonhistone proteinaceous scaffolding. The packing ratios of each of the orders of DNA structure are summarized in Table 38–1.

The packaging of nucleoproteins within chromatids is not random, as evidenced by the characteristic patterns observed when chromosomes are

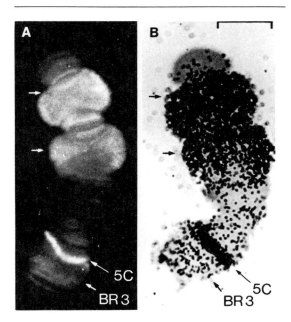

Figure 38–4. The correlation between RNA polymerase II activity and RNA synthesis. A number of genes are activated when *Chironomus tentans* larvae are subjected to heat shock (39 °C for 30 min). *A:* Distribution of RNA polymerase B (also called type II) in isolated chromosome IV from the salivary gland. The enzyme was detected by immunofluorescence using an antibody directed against the polymerase. The 5C and BR3 are specific bands of chromosome IV, and the arrows indicate puffs. *B:* Autoradiogram of a chromosome IV that was incubated in ^{3}H-uridine to label the RNA. Note the correspondence of the immunofluorescence and presence of the radioactive RNA (dots). Bar = 7 μm. (Reproduced, with permission, from Sass H: RNA polymerase B in polytene chromosomes. *Cell* 1982;**28:**274. Copyright © 1982 by the Massachusetts Institute of Technology.)

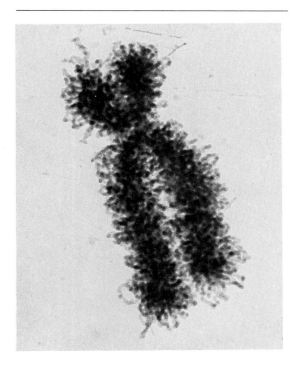

Figure 38–5. The 2 sister chromatids of human chromosome 12. × 27,850. (Reproduced, with permission, from DuPraw EJ: *DNA and Chromosomes.* Holt, Rinehart, & Winston, 1970.)

Table 38–1. The packing ratios of each of the orders of DNA structure.

Chromatin Form	Packing Ratio
Bare double-helix DNA	~1.0
10-nm fibril of nucleosomes	7–10
25- to 30-nm chromatin fiber of superhelical nucleosomes	40–60
Condensed metaphase chromosome of loops	8000

stained with specific dyes such as quinacrine or Giemsa's stain (Fig 38–6).

From individual to individual within a single species, the pattern of staining (banding) of the entire chromosome complement is highly reproducible; nonetheless, it differs significantly from other species, even those closely related. Thus, the packaging of the nucleoproteins in chromosomes of higher eukaryotes must in some way be dependent upon species-specific characteristics of the DNA molecules.

MUCH OF THE MAMMALIAN GENOME IS NOT TRANSCRIBED

The haploid genome of each human cell consists of 3.5×10^9 base pairs of DNA, subdivided into 23 chromosomes. The entire haploid genome contains sufficient DNA to code for nearly 1.5 million pairs of genes. However, studies of mutation rates and of the complexities of the genomes of higher organisms

strongly suggest that humans have only about 100,000 essential proteins. This implies that most of the DNA is noncoding; ie, its information is never translated into an amino acid sequence of a protein molecule. Certainly, some of the excess DNA sequences serve to regulate the expression of genes during development, differentiation, and adaptation to the environment. Some excess clearly makes up the intervening sequences that split the coding regions of genes, but much of the excess appears to be composed of many families of repeated sequences for which no functions have been clearly defined.

The DNA in a eukaryotic genome can be divided into different "sequence classes." These are **unique-sequence, or nonrepetitive, DNA** and **repetitive-sequence DNA.** In the haploid genome unique-sequence DNA generally includes the single copy genes that code for proteins. The repetitive DNA in the haploid genome includes sequences that vary in copy number from 2 to as many as 10^7 copies per cell.

More Than Half the DNA in Eukaryotic Organisms Is in Unique or Nonrepetitive Sequences

This estimation (and the distribution of repetitive-sequence DNA) is based on a variety of DNA-RNA hybridization techniques. Similar techniques are used to estimate the number of active genes in a population of unique-sequence DNA. In yeast, a lower eukaryote, about 4000 genes are expressed. In typical tissues in a higher eukaryote (eg, mammalian liver and kidney),

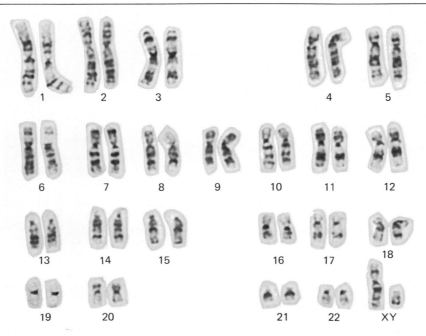

Figure 38–6. A human karyotype (of a man with a normal 46 XY constitution), in which the chromosomes have been stained by the Giemsa method and aligned according to the Paris Convention (Courtesy of H. Lawce and F. Conte).

between 10,000 and 15,000 genes are expressed. Different combinations of genes are expressed in each tissue, of course, and how this is accomplished is one of the major unanswered questions in biology.

Coding Regions Are Often Interrupted by Intervening Sequences

The coding regions of DNA, the transcripts of which ultimately appear in the cytoplasm as single mRNA molecules, **are usually interrupted in the genome by large intervening sequences of noncoding DNA.** Accordingly, the primary transcripts of DNA—hnRNA—contain noncoding intervening sequences of RNA that must be removed in a process which also joins together the appropriate coding segments to form the mature mRNA. Most coding sequences for a single mRNA are interrupted in the genome (and thus in the primary transcript) by at least one—and as many as 50 in some cases—noncoding intervening sequences (**introns**). In most cases the introns are much longer than the continuous coding regions (**exons**). The processing of the primary transcript, which involves removal of introns and splicing of adjacent exons, is described in detail in Chapter 39.

The function of the intervening sequences, or introns, is not clear. They may serve to separate functional domains (exons) of coding information in a form that permits genetic rearrangement by recombination to occur more rapidly than if all coding regions for a given genetic function were contiguous. Such an enhanced rate of genetic rearrangement of functional domains might allow more rapid evolution of biologic function.

In Human DNA, at Least 20–30% of the Genome Consists of Repetitive Sequences

Repetitive-sequence DNA can be broadly classified as moderately repetitive or as highly repetitive. The **highly repetitive sequences** consist of 5–500 base-pair lengths repeated many times in tandem. These sequences are usually **clustered** in centromeres and telomeres of the chromosome and are present in about 1–10 million copies per haploid genome. These sequences are transcriptionally inactive and may play a structural role in the chromosome.

The **moderately repetitive sequences,** which are defined as being present at less than 10^6 copies per haploid genome, are not clustered but are interspersed with unique sequences and can be categorized as being long or short in length. The **long interspersed repeats** are 5000–7000 base pairs in length and are present at 1000–100,000 copies per haploid genome. These long interspersed repeats are flanked on either side by 300–600 base-pair **direct repeats** (Fig 38–7) that closely resemble the **long terminal repeats** (LTR) at the ends of integrated retrovirus DNAs. In many cases, these long interspersed repeats are transcribed by **RNA polymerase II** and contain caps indistinguishable from those on mRNA.

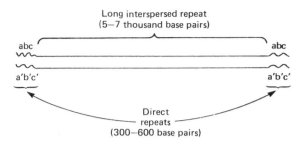

Figure 38–7. Representation of a long interspersed repeated sequence with its short direct repeat sequences (abc) and complements (a′b′c′) indicated at the termini.

The **short interspersed repeats** are families of related but individually distinct members that consist of a few to several hundred nucleotide pairs. The short interspersed repeats are actively transcribed either as integral components of introns or by DNA-dependent RNA polymerase III (see Chapter 39) as discrete elements. Of the short interspersed repeats in the human genome, one family, the **Alu family,** is present in about 500,000 copies per haploid genome and accounts for at least 3–6% of the human genome. Members of the human Alu family and their closely related analogs in other animals are transcribed as integral components of hnRNA or as discrete RNA molecules, including the well-studied 4.5S RNA and 7S RNA. These particular family members are highly conserved within a species as well as between mammalian species. The structures of the short interspersed repeats, including the members of the Alu family, resemble the retroviral long terminal repeat itself and may be mobile elements, capable of jumping into and out of various sites within the genome (see below).

GENETIC MATERIAL CAN BE ALTERED AND REARRANGED

An alteration in the sequence of purine and pyrimidine bases in a gene due to a change, a removal, or an insertion of one or more bases may result in an altered gene product that in most instances ultimately is a protein. Such alteration in the genetic material results in a **mutation,** the consequences of which are discussed in detail in Chapter 40.

Chromosomal Recombination Is One Way of Rearranging Genetic Material

Prokaryotic and eukaryotic organisms are capable of exchanging genetic information between similar or homologous chromosomes. The exchange or **recombination** event occurs primarily during meiosis in mammalian cells and requires alignment of homologous chromosomes, an alignment that almost always occurs with great exactness. A process of crossing-over occurs as shown in Fig 38–8. This usually results

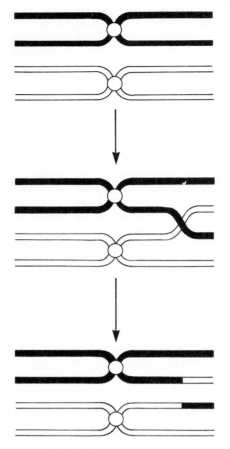

Figure 38–8. The process of crossing over between homologous chromosomes to generate recombinant chromosomes.

in an equal and reciprocal exchange of genetic information between homologous chromosomes. If the homologous chromosomes possess different alleles of the same genes, the crossover may produce noticeable and heritable genetic linkage differences. In the rare case where the alignment of homologous chromosomes is not exact, the crossing over or recombination event may result in an unequal exchange of information. One chromosome may receive less genetic material and thus a deletion, while the other partner of the chromosome pair receives more genetic material and thus an insertion or duplication (Fig 38–8). Unequal crossing over does occur in humans, as evidenced by the existence of hemoglobins designated Lepore and anti-Lepore. **Unequal crossover** affects tandem arrays of repeated DNAs whether they are related globin genes, as in Fig 38–9, or more abundant repetitive DNA. The unequal crossover through slippage in the pairing can result in expansion or contraction in the copy number of the repeat family and may contribute to the expansion and fixation of variant members throughout the array.

Chromosomal Integration Occurs With Some Viruses

Some bacterial viruses (bacteriphages) are capable of recombining with the DNA of a bacterial host in such a way that the genetic information of the bacteriophage is incorporated in a linear fashion into the genetic information of the host. This integration, which is a form of recombination, occurs by the mechanism simplified in Fig 38–10. The backbone of the circular bacteriophage genome is broken, as is that of the DNA molecule of the host; the appropriate ends are resealed with the proper polarity. The bacteriophage DNA is figuratively straightened out ("linearized") as it is integrated into the bacterial DNA molecule—frequently a closed circle as well. The site at which the bacteriophage genome integrates or recombines with the bacterial genome is chosen by one of 2 mechanisms. If the

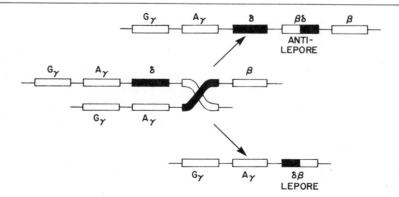

Figure 38–9. The process of unequal crossover in the region of the mammalian genome that harbors the structural genes for hemoglobin and the generation of the unequal recombinant products hemoglobin delta-beta Lepore and beta-delta anti-Lepore. The examples given show the locations of the crossover regions between amino acid residues. (Redrawn and reproduced, with permission, from Clegg JB, Weatherall DJ: β° Thalassemia: Time for a reappraisal? *Lancet* 1974;**2**:133.)

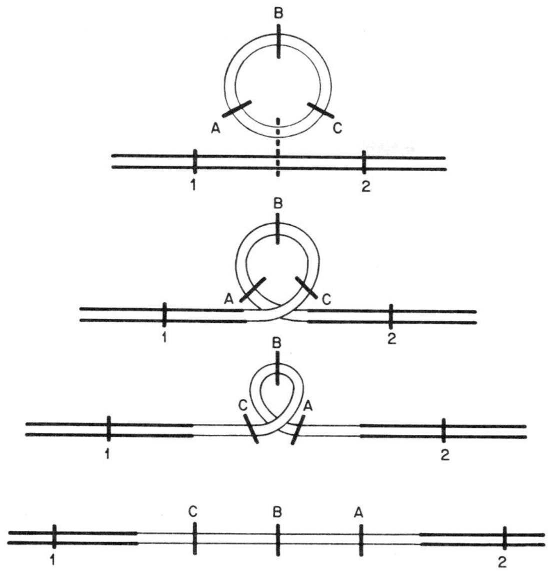

Figure 38–10. The integration of a circular genome (with genes A, B, and C) into the DNA molecule of a host (with genes 1 and 2) and the consequent ordering of the genes.

bacteriophage contains a DNA sequence **homologous** to a sequence in the host DNA molecule, then a recombination event analogous to that occurring between homologous chromosomes can occur. However, some bacteriophages synthesize proteins that bind specific sites on bacterial chromosomes with a nonhomologous site specifically of the bacteriophage DNA molecule. Integration occurs at the site and is said to be **"site specific."**

Many animal viruses, particularly the oncogenic viruses—either directly or, in the case of RNA viruses, their DNA transcripts—can be integrated into chromosomes of the mammalian cell. The integration of the animal virus DNA into the animal genome generally is not "site specific."

Transposition Can Produce Processed Genes

In eukaryotic cells, small DNA elements that clearly are not viruses are capable of transposing themselves in and out of the host genome in ways that affect the function of neighboring DNA sequences. These mobile elements, sometimes called "jumping DNA," can carry flanking regions of DNA and, therefore, profoundly affect evolution. As mentioned above, the Alu family of moderately repeated DNA sequences has structural characteristics similar to the termini of retroviruses, which would account for the ability of the latter to move into and out of the mammalian genome.

Direct evidence for the transposition of other small DNA elements into the human genome has been provided by the discovery of "processed genes" for immunoglobulin molecules, α-globin molecules, and

several others. These **processed genes** consist of DNA sequences identical or nearly identical to those of the messenger RNA for the appropriate gene product. That is, the 5' nontranscribed region, the coding region without intron representation, and the 3' poly(A) tail are all present contiguously. This particular DNA sequence arrangement must have resulted from the reverse transcription of an appropriately processed messenger RNA molecule from which the intron regions had been removed and the poly(A) tail added. The only recognized mechanism that this reverse transcript could have used to integrate into the genome would have been a transposition event. In fact, these "processed genes" have short terminal repeats at each end, as do known transposed sequences in lower organisms. Some of the processed genes have been randomly altered through evolution so that they now contain nonsense codons that preclude their expression (see Chapter 40). Thus, they are referred to as **"pseudogenes."**

Gene Conversion Produces Rearrangements

Besides unequal crossover and transposition, a third mechanism can effect rapid changes in the genetic material. Similar sequences on homologous or nonhomologous chromosomes may occasionally pair up and eliminate any mismatched sequences between them. This may lead to the accidental fixation of one variant or another throughout a family of repeated sequences and thereby homogenize the sequences of the members of repetitive DNA families. This latter process is referred to as **gene conversion.**

In diploid eukaryotic organisms such as humans, after cells progress through the S phase they contain a tetraploid content of DNA. This is in the form of sister chromatids of chromosome pairs. Each of these sister chromatids contains identical genetic information, since each is a product of the semiconservative replication of the original parent DNA molecule of that chromosome. Crossing over occurs between these genetically identical sister chromatids. Of course, these **sister chromatid exchanges** (Fig 38–11) have no genetic consequence so long as the exchange is the result of an equal crossover.

In mammalian cells, some interesting gene rearrangements occur normally during development and differentiation. For example, in mice the V_L and C_L genes for a single immunoglobulin molecule (see Chapter 41) are widely separated in the germ line DNA. In the DNA of a differentiated immunoglobulin-producing (plasma) cell, the same V_L and C_L genes have been moved physically closer together in the genome, and into the same transcription unit. However, even then, this rearrangement of DNA during differentiation does not bring the V_L and C_L genes into contiguity in the DNA. Instead, the DNA contains an interspersed or interruption sequence of about 1200 base pairs at or near the junction of the V and C regions. The interspersed sequence is transcribed into RNA along with the V_L and C_L genes, and the in-

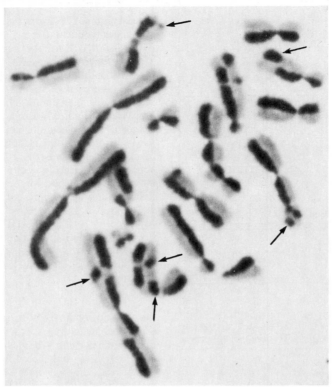

Figure 38–11. Sister chromatid exchanges between human chromosomes. These are detectable by Giemsa staining of the chromosomes of cells replicated for 2 cycles in the presence of bromodeoxyuridine. The small arrows indicate some regions of exchange. (Courtesy of S Wolff and J Bodycote.)

terspersed information is removed from the RNA during its nuclear processing (see Chapters 39 and 41).

DNA SYNTHESIS & REPLICATION ARE RIGIDLY CONTROLLED

The primary function of DNA replication is understood to be the provision of progeny with the genetic information possessed by the parent. Thus, the replication of DNA must be complete and carried out with **high fidelity to maintain genetic stability** within the organism and the species. The process of DNA replication is complex and involves many cellular functions and several verification procedures to ensure fidelity in replication. The first enzymologic observations on DNA replication were made in *Escherichia coli* by Arthur Kornberg, who described in that organism the existence of an enzyme now called DNA polymerase I. This enzyme has multiple catalytic activities, a complex structure, and a requirement for the triphosphates of the 4 deoxyribonucleosides of adenine, guanine, cytosine, and thymine. The polymerization reaction catalyzed by DNA polymerase I of *E coli* has served as a prototype for all DNA polymerases of both prokaryotes and eukaryotes, even though it is now recognized that the major role of this

polymerase is to ensure fidelity and to repair rather than to replicate DNA.

Initiation of DNA Synthesis Requires RNA Priming

The initiation of DNA synthesis (Fig 38–12) requires priming by a **short length of RNA,** about 10–200 nucleotides long. This priming process involves the nucleophilic attack by the 3′-hydroxyl group of the RNA primer to the α phosphate of the deoxynucleoside triphosphate with the splitting off of pyrophosphate. The 3′-hydroxyl group of the recently attached deoxyribonucleoside monophosphate is then free to carry out a nucleophilic attack on the next entering deoxyribonucleoside triphosphate, again at its α phosphate moiety, with the splitting off of pyrophosphate. Of course, the selection of the proper deoxyribonucleotide whose terminal 3′-hydroxyl group is to be attacked is dependent upon **proper pairing with the other strand** of the DNA molecule according to the rules proposed originally by Watson and Crick (Fig 38–13). When an adenine deoxyribonucleoside monophosphoryl moiety is in the template position, a thymidine triphosphate will enter and its α phosphate will be attacked by the 3′-hydroxyl group of the deoxyribonucleoside monophosphoryl most recently added to the polymer. By this stepwise process, the template dictates which deoxyribonucleoside triphosphate is

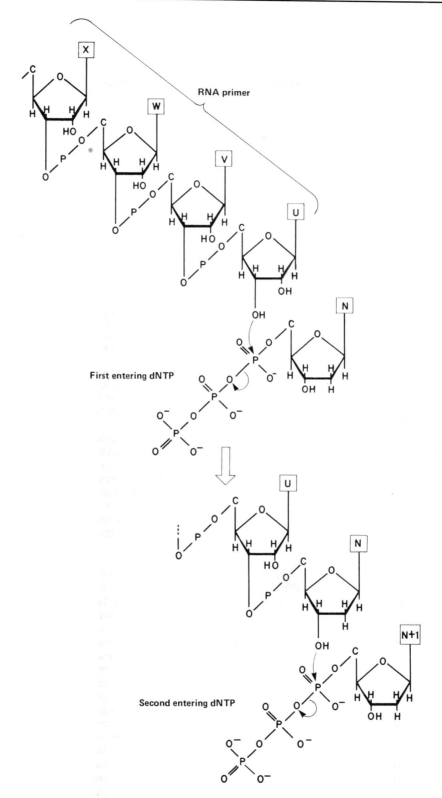

Figure 38–12. The initiation of DNA synthesis upon a primer of RNA and the subsequent attachment of the second deoxyribonucleoside triphosphate.

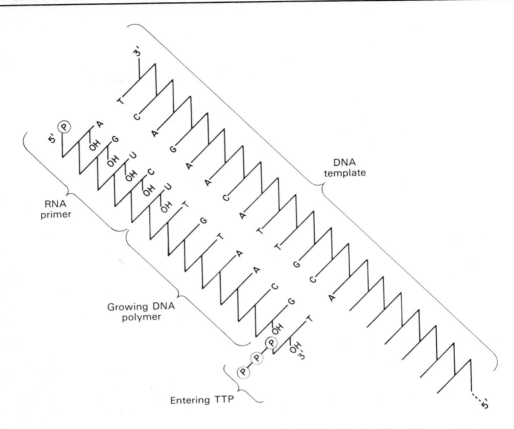

Figure 38–13. The RNA-primed synthesis of DNA demonstrating the template function of the complementary strand of parental DNA.

complementary and by hydrogen bonding holds it in place while the 3′-hydroxyl group of the growing strand attacks and incorporates the new nucleotide into the polymer. These fragments of DNA attached to an RNA initiator component, discovered by Okazaki, are referred to as **Okazaki pieces** (Fig 38–14). In mammals, after many Okazaki pieces are generated, the replication complex begins to remove the RNA primers, to fill in the gaps left by their removal with the proper base-paired deoxynucleotide, and then to seal the fragments of newly synthesized DNA by enzymes referred to as **DNA ligases.**

There is a Polarity of Replication

As has already been noted, DNA molecules are double-stranded and the 2 strands are antiparallel, ie, running in opposite directions. The replication of DNA in prokaryotes and eukaryotes occurs on **both strands simultaneously.** However, an enzyme capable of polymerizing DNA in the 3′ to 5′ direction does not exist in any organism, so that both of the newly replicated DNA strands cannot grow in the same direction simultaneously. Nevertheless, the same enzyme does replicate both strands at the same time. The single enzyme replicates **one strand ("leading strand") in a**

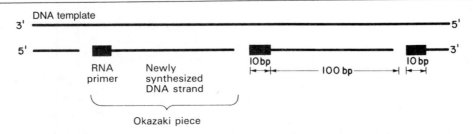

Figure 38–14. The discontinuous polymerization of deoxyribonucleotides and formation of Okazaki pieces.

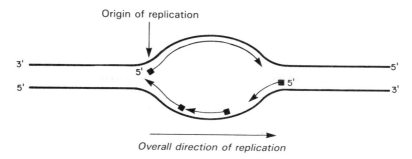

Figure 38–15. The process of semidiscontinuous, simultaneous replication of both strands of double-stranded DNA.

continuous manner in the 5' to 3' direction, with the same overall forward direction. It replicates the **other strand ("lagging strand") discontinuously** while polymerizing the nucleotides in short spurts of 150–250 nucleotides again in the 5' to 3' direction, but at the same time it faces toward the back end of the preceding RNA primer rather than toward the unreplicated portion. This process of **semidiscontinuous DNA synthesis** is shown diagrammatically in Fig 38–15.

In the mammalian nuclear genome, most of the RNA primers are eventually removed as part of the replication process, whereas after replication of the mitochondrial genome the small piece of RNA remains as an integral part of the closed circular DNA structure.

Several Enzymes Are Involved in DNA Polymerization & Repair

In mammalian cells, there is one class of DNA polymerase enzymes, called **polymerase alpha,** which is present in the nucleus and responsible for chromosome replication. One polymerase alpha molecule is capable of polymerizing about 100 nucleotides per second, a rate 10-fold less than the rate of polymerization of deoxynucleotides by the bacterial DNA polymerase. This reduced rate may result from the interference by

nucleosomes. It is not known how DNA polymerase negotiates nucleosomes. However, after DNA replication, the assembling nucleosomes are distributed randomly to either daughter strand. Newly synthesized core histones in the octameric form would then attach to the other strand as the replication fork progresses.

A lower-molecular-weight polymerase, **polymerase beta,** is also present in mammalian nuclei but is not responsible for the usual DNA replication. It may function in DNA repair (see below). Mitochondrial DNA polymerase, **polymerase gamma,** is responsible for replication of the mitochondrial genome, another DNA molecule that exists in circular form.

The entire mammalian genome replicates in 9 hours, the average period required for formation of a tetraploid genome from a diploid genome in a replicating cell. This requires the presence of **multiple origins** of DNA replication that occur in clusters of up to 100 of these replication units. Replication occurs in **both directions** along the chromosome, and both strands are replicated simultaneously. This replication process generates "replication bubbles" (Fig 38–16).

The multiple sites that serve as origins for DNA replication in eukaryotes are poorly defined except in a few animal viruses and in yeast. However, it is clear that initiation is regulated both spatially and temporally, since clusters of adjacent sites initiate syn-

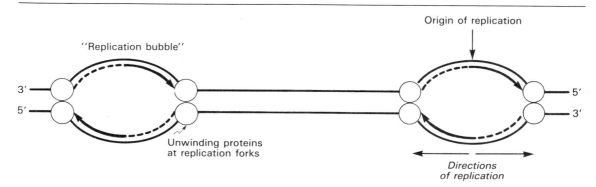

Figure 38–16. The generation of "replication bubbles" during the process of DNA synthesis. The bidirectional replication and the proposed positions of unwinding proteins at the replication forks are depicted.

chronously. There are suggestions that functional domains of chromatin replicate as intact units, implying that the origins of replication are specifically located with respect to transcription untis.

During the replication of DNA, there must be a separation of the 2 strands to allow each to serve as a template by hydrogen bonding its nucleotide bases to the incoming deoxynucleoside triphosphate. The separation of the DNA double helix is promoted by specific protein molecules that **stabilize the single-stranded structure** as the replication fork progresses. These stabilizing proteins bind stoichiometrically to the single strands without interfering with the abilities of the nucleotides to serve as templates (Fig 38–17). In addition to separating the 2 strands of the double helix, there must be an **unwinding** of the molecule (once every 10 nucleotide pairs) to allow strand separation. Given the time during which DNA replication must occur this must happen in segments. There are multiple "swivels" interspersed in the DNA molecules of all organisms. The swivel function is provided by specific enzymes that introduce **"nicks" in one strand of the unwinding double helix,** thereby allowing the unwinding process to proceed. The nicks are quickly resealed without requiring energy input, because of the formation of a high-energy covalent bond between the nicked phosphodiester backbone and the nicking-sealing enzyme. The nicking-resealing enzymes are called DNA **topoisomerases.** This process is depicted diagrammatically in Fig 38–18 and there compared to the ATP-dependent resealing carried out by the DNA ligases. Topoisomerases are also capable of unwinding supercoiled DNA. Supercoiled DNA is a higher ordered structure occurring in circular DNA molecules wrapped around a core, as depicted in Fig 38–19.

There exists in one species of animal viruses (retroviruses) a class of enzymes capable of synthesizing a single-stranded and then a double-stranded DNA molecule from a single-stranded RNA template. This polymerase, RNA-dependent DNA polymerase or **"reverse transcriptase,"** first synthesizes a DNA-RNA hybrid molecule utilizing the RNA genome as a template. A specific enzyme, RNase H, degrades the RNA strand, and the remaining DNA strand in turn serves as a template to form a double-stranded DNA molecule containing the information originally present in the RNA genome of the animal virus.

DNA Synthesis Occurs During the S Phase of the Cell Cycle

In animal cells, including human cells, the replication of the DNA genome occurs only at a specified time during the life span of the cell. This period is referred to as the synthetic or S phase. This is usually temporally separated from the mitotic phase by nonsynthetic periods referred to as gap 1 (G_1) and gap 2 (G_2), occurring before and after the S phase, respectively (Fig 38–20). The cell regulates its DNA synthesis grossly by allowing it to occur only at specific times and mostly in cells preparing to divide by a mitotic process. The regulation of the entry of a cell into an S phase may involve cyclic purine nucleotides and perhaps the substrates for DNA synthesis, but the mechanisms are unknown. Many of the cancer-causing viruses (oncoviruses) are capable of alleviating or disrupting the apparent restriction that normally controls the entry of mammalian cells from G_1 into the S phase. Again, this mechanism is currently unknown but probably involves the phosphorylation of specific host protein molecules.

During the S phase, mammalian cells contain greater quantities of DNA polymerase alpha than during the nonsynthetic phases of the cell cycle. Furthermore, those enzymes responsible for the formation of the substrates for DNA synthesis, ie, deoxyribonucleoside triphosphates, are also increased in activity, and their activity will diminish following the synthetic phase until the reappearance of the signal for renewed DNA synthesis. During S phase, the nuclear DNA is **completely replicated once and only once.** It seems that once chromatin has been replicated, it is marked so as to prevent its further replication until it again passes through mitosis. It has been suggested that DNA methylation may serve as such a covalent marker.

In general, a given pair of chromosomes will replicate simultaneously and within a fixed portion of the S phase upon every replication. On a chromosome, clusters of replication units replicate coordinately. The nature of the signals that regulate DNA synthesis at these levels is unknown, but the regulation does appear to be an intrinsic property of each individual chromosome.

Damaged DNA Is Repaired by Enzymes

The maintenance of the integrity of the information in DNA molecules is of utmost importance to the survival of a particular organism as well as to survival of the species. Thus, it can be concluded that surviving species have evolved mechanisms for repairing DNA damage incurred as a result of either replication errors

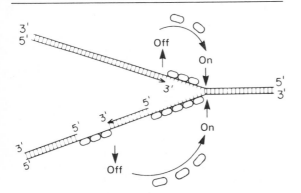

Figure 38–17. Hypothetical scheme for single-strand binding protein action at a replicating fork. The protein is recycled after binding single-stranded regions of the template and facilitating replication. (Courtesy of B Alberts.)

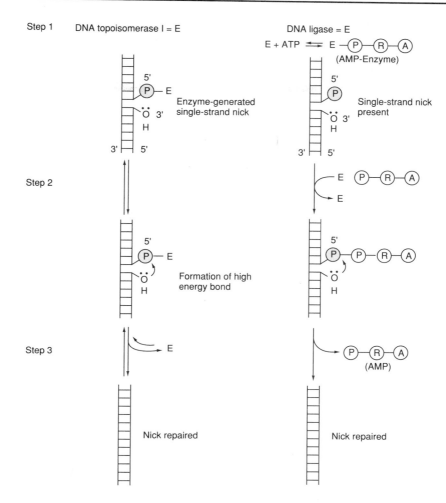

Figure 38–18. Comparison of 2 types of nick-sealing reactions on DNA. The series of reactions at left is catalyzed by DNA ligase; that at right by DNA topoisomerase I. (Slightly modified and reproduced, with permission, from Lehninger AL: *Biochemistry*, 2nd ed. Worth, 1975.)

or environmental insults. It has been estimated that DNA replication and environmentally induced DNA damage result in an average of about 6 nucleotide changes per year in the germ line cells of an individual. Presumably, at least that number of nucleotide changes or mutations must occur per year in the somatic cells as well.

As described in Chapter 37, the major responsibility for the fidelity of replication resides in the specific pairing of nucleotide bases. Proper pairing is dependent upon the presence of the favored tautomers of the purine and pyrimidine nucleotides (see Fig 35–7), but the equilibrium wherein one tautomer is more stable than another is only about 10^4 or 10^5 in favor of that with the greater stability. Although this is not sufficiently favorable to ensure the high fidelity that is necessary, the favoring of the preferred tautomers, and thus of the proper base pairing, could be ensured by monitoring the base pairing twice. Such double monitoring does appear to occur in both bacterial and

mammalian systems: once at the time of insertion of the deoxyribonucleoside triphosphates, and later by a follow-up, energy-requiring mechanism which removes all improper bases that may occur in the newly formed strand. This double monitoring does not permit errors of mispairing due to the presence of the unfavored tautomers to occur more frequently than once every 10^8–10^{10} base pairs. The molecule responsible for this monitoring mechanism in *E coli* is the built-in $3' \rightarrow 5'$ exonuclease activity of DNA polymerase, but mammalian DNA polymerases do not seem to possess such a nuclease proofreading function. Other enzymes provide this repair function.

Damage to DNA by environmental, physical, and chemical agents may be classified into 4 types (Table 38–2). The damaged regions of DNA may be **repaired, replaced** by recombination, or **retained.** Retention leads to mutations and, potentially, cell death. Repair and replacement exploit the redundancy of information inherent in the double helical DNA

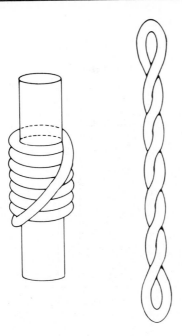

Figure 38–19. Supercoiling of DNA. A left-handed toroidal (solenoidal) supercoil, at left, with convert to a right-handed interwound supercoil, at right, when the cylindric core is removed. Such a transition is analogous to that which occurs when nucleosomes are disrupted by the high salt extraction of histones from chromatin.

Table 38–2. Types of damage to DNA.

I. Single-base alteration
  A. Depurination
  B. Deamination of cytosine to uracil
  C. Deamination of adenine to hypoxanthine
  D. Alkylation of base
  E. Insertion or deletion of nucleotide
  F. Base-analog incorporation
II. Two-base alteration
  A. UV light-induced thymine-thymine dimer
  B. Bifunctional alkylating agent cross-linkage
III. Chain breaks
  A. Ionizing radiation
  B. Radioactive disintegration of backbone element
IV. Cross-linkage
  A. Between bases in same or opposite strands
  B. Between DNA and protein molecules (eg, histones)

structure. The defective region in one strand can be returned to its original form by relying on the complementary information stored in the unaffected strand.

The key to all of the repair or recombinational processes is the initial **recognition of the defect** and either repairing it during the recognition step or marking it for future attention. The **depurination** of DNA, which happens spontaneously owing to the thermal lability of the purine N-glycosidic bond, occurs at a rate of 5000–10,000/cell/d at 37 °C. Specific enzymes recognize a depurinated site and replace the appropriate purine directly, without interruption of the phosphodiester backbone.

Both cytosine and adenine bases in DNA spontaneously **deaminate** to form uracil and hypoxanthine, respectively. Since neither uracil nor hypoxanthine normally exists in DNA, it is not surprising that specific **N-glycosylases** can recognize these abnormal bases and remove the base itself from the DNA. This removal marks the site of the defect and allows an **apurinic or apyrimidinic endonuclease** to incise the appropriate backbone near the defect. Subsequently, the sequential actions of an **exonuclease,** a repair DNA polymerase, and a **ligase** return the DNA to its original state (Fig 38–21). This series of events is called **excision-repair.** By a similar series of steps involving initially the recognition of the defect, alkylated bases and base analogs can be removed from DNA and the DNA returned to its original informational content.

The repair of insertions or deletions of nucleotides normally occurs by recombinational mechanisms either with or without replication.

Ultraviolet light induces the formation of pyrimidine-pyrimidine dimers, predominantly the dimerization of 2 juxtaposed thymines in the same strand (Fig 38–22). There are apparently 2 mechanisms for removing or repairing these thymine-thymine dimers. One is **excision-repair** analogous to that described above. The second mechanism involves **visible light photoactivation** of a specific enzyme that directly reverses the dimer formation in situ.

Single-strand breaks induced by ionizing radiation can be repaired by direct ligation or by recombination.

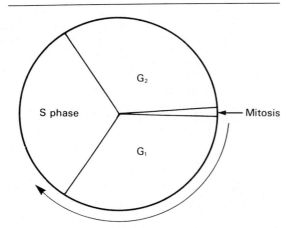

Figure 38–20. Mammalian cell cycle. The DNA synthetic phase (S phase) is separated from mitosis by gap 1 (G_1) and gap 2 (G_2). (Arrow indicates direction of cell progression.)

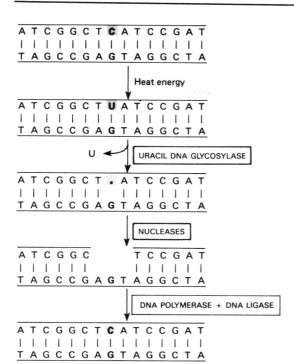

Figure 38–21. Excision-repair of DNA. The enzyme uracil DNA glycosylase removes the uracil created by spontaneous deamination of cytosine in the DNA. An endonuclease cuts the backbone near the defect; then, after an endonuclease removes a few bases, the defect is filled in by the action of a repair polymerase and the strand is rejoined by a ligase. (Courtesy of B Alberts.)

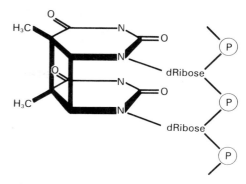

Figure 38–22. A thymine-thymine dimer formed via a cyclobutane moiety between juxtaposed thymine residues of DNA.

The mechanisms responsible for the repair of cross-linkages between bases on opposite strands of the DNA double helix or between the DNA and protein molecules are poorly understood.

In general, damage caused by ionizing radiation and by alkylation of bases is repaired in short patches of excision and resynthesis. Ultraviolet light damage and strand cross-linkages are repaired by long patches of excision and resynthesis. In mammalian cells, repair replication can be observed as **unscheduled DNA synthesis,** ie, incorporation of DNA precursors (radioactive thymidine) into DNA when a cell is not in S phase.

Associated with the increased excision-repair activity in response to DNA damaging agents, mammalian cells exhibit increased activity of the enzyme **poly(ADP-ribose) polymerase.** This enzyme uses the coenzyme NAD^+ to ADP-ribosylate chromatin proteins. It adds mostly mono(ADP-ribose), but to some extent homopolymeric chains of ADP-ribose are added. It is not evident what function poly(ADP-ribose) polymerase or its product, (ADP-ribose)$_n$, has in the excision-repair process. There is a temporal rela-

tionship between the increased repair activity and the increased specific enzyme. Furthermore, inhibition of the enzyme by specific inhibitors prevents the rejoining of broken DNA strands. The increased activity of poly(ADP-ribose) polymerase appears to be a response to DNA fragmentation in the nucleus. This fragmentation might be induced primarily by physical agents such as x-ray, or secondarily by the incision mechanism responding to other chemical or physical agents such as ultraviolet light or alkylating agents. The activity of the polymerase is sufficiently great to cause a depletion of intracellular NAD^+ following environmentally induced DNA damage.

Xeroderma pigmentosum is an autosomal recessive genetic disease. The clinical syndrome includes marked sensitivity to sunlight (ultraviolet) with subsequent formation of multiple skin cancers and premature death. The inherited defect seems to involve the repair of damaged DNA. Cells cultured from patients with xeroderma pigmentosum exhibit low activity for the photoactivated thymine dimer cleavage process. However, the involved DNA repair processes in this disease are quite complex; there are at least 7 genetic complementation groups.

In cells from most if not all complementation groups of xeroderma pigmentosum, there is an abnormal temporal or quantitative response of the poly(ADP-ribose) polymerase to ultraviolet light exposure. It seems that the abnormal response in at least one complementation group is due to the inability to incise the DNA strand at the site of damage, since the addition of deoxyribonuclease to permeabilized defective cells is followed by a normal or nearly normal increase in poly(ADP-ribose) polymerase activity.

In patients with **ataxia-telangiectasia,** an autosomal recessive disease in humans resulting in the development of cerebellar ataxia and lymphoreticular neoplasms, there appears to exist an increased sensitivity to damage by x-ray. Patients with **Fanconi's anemia,** an autosomal recessive anemia characterized also by an increased frequency of cancer and by chromosomal instability, probably have defective repair of cross-linking damage. All 3 of these clinical syndromes are associated with increased frequency of cancer. It is likely that other human diseases resulting from disordered DNA repair capabilities will be found in the future.

REFERENCES

Kornberg, A: *J Biol Chem* 1988;**263**:1.

Igo-Kemenes T, Horz W, Zachau HG: Chromatin. *Annu Rev Biochem* 1982;**51**:89.

Jelinek WR, Schmid CW: Repetitive sequences in eukaryotic DNA and their expression. *Annu Rev Biochem* 1982;**51**:813.

McGhee JD, Felsenfeld G: Nucleosome structure. *Annu Rev Biochem* 1980;**49**:1115.

Sancar A, Sancar G: DNA repair enzymes. *Annu Rev Biochem* 1988;**57**:29.

Campbell JL: Eucaryotic DNA replication. *Annu Rev Biochem* 1986;**55**:733.

Gross DS, Garrard WT: *Annu Rev Biochem* 1988;**57**:159.

RNA Synthesis, Processing, & Metabolism

39

Daryl K. Granner, MD

INTRODUCTION AND BIOMEDICAL IMPORTANCE

The synthesis of an RNA molecule from DNA is a very complex process involving one of the group of RNA polymerase enzymes and a number of associated proteins. The general steps required to synthesize the primary transcript are initiation, elongation, and termination. Most is known about initiation. A number of DNA regions (generally located upstream from the initiation site) and protein factors that bind to these sequences to regulate the initiation of transcription have been identified. This process is best understood in prokaryotes and viruses, but considerable progress has been made in deciphering mammalian cell transcription in recent years. Certain RNAs, mRNAs in particular, have very different life spans in a cell. It is important to understand the basic principles of RNA metabolism, for modulation of this process results in altered rates of protein synthesis and hence a variety of metabolic changes. This is how all organisms adapt to changes of environment. It is also how differentiated cell structures and functions are established and maintained in higher metazoans.

The RNA molecules synthesized in mammalian cells are often very different from those made in prokaryotic organisms, particularly the mRNA-encoding transcripts. Prokaryotic mRNA can be translated as it is being synthesized, whereas in mammalian cells most RNAs are made as precursor molecules that have to be processed into mature, active RNA. Erroneous processing and splicing of mRNA transcripts are a cause of disease, eg, certain types of thalassemia (see Chapter 42).

RNA IS SYNTHESIZED FROM A DNA TEMPLATE BY AN RNA POLYMERASE

The process of synthesizing RNA from a DNA template has been characterized best in prokaryotes. Although in mammalian cells the regulation of RNA synthesis and the processing of the RNA transcripts are different from that in prokaryotes, the process of RNA synthesis per se is quite similar in these 2 classes of organisms. Therefore, the description of RNA synthesis in prokaryotes will be applicable to eukaryotes even though the enzymes involved and the regulatory signals are different.

The sequence of ribonucleotides in an RNA molecule is complementary to the sequence of deoxyribonucleotides in one strand of the double-stranded DNA molecule (see Fig 37–8). The strand that is transcribed into an RNA molecule is referred to as the **template strand** strand of the DNA. The other DNA strand is frequently referred to as the **coding strand** of that gene. It is called this because, with the exception of T for U changes, it corresponds exactly to the sequence of the primary transcript, which encodes the protein product of the gene. In the case of a double-stranded DNA molecule containing many genes, the template strand for each gene will not necessarily be the same strand of the DNA double helix (Fig 39–1). Thus, a given strand of a double-stranded DNA molecule will serve as the template strand for some genes and the coding strand of other genes. Note that the nucleotide sequence of an RNA transcript will be the same (except for U replacing T) as that of the coding strand. The information in the template strand is read out in the 3' to 5' direction.

DNA-dependent RNA polymerase is the enzyme responsible for the polymerization of ribonucleotides into a sequence complementary to the template strand of the gene (see Figs 39–2 and 39–3). The enzyme attaches at a specific site, the **promoter,** on the tem-

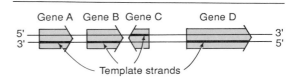

Figure 39–1. This figure illustrates that genes can be transcribed off both strands of DNA. The arrowheads indicate the direction of transcription (polarity). Note that the template strand is always read in the 3' to 5' direction. The opposite strand is called the coding strand because it is identical (except for T for U changes) to the mRNA transcript (the primary transcript in eukaryotic cells) that encodes the protein product of the gene.

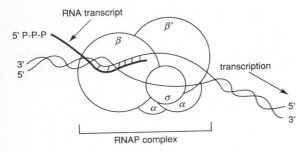

Figure 39-2. RNA polymerase (RNAP) catalyzes the polymerization of ribonucleotides into an RNA sequence that is complementary to the template strand of the gene. The RNA transcript has the same polarity (5' to 3') as the coding strand but contains U rather than T. *Escherichia coli* RNAP consists of a core complex of two α subunits and two β subunits (β and β'). The holoenzyme contains the σ subunit when near the transcription start site (within ~10 bp) but then transcription proceeds with just the core complex. The transcription "bubble" is a 17-bp area of melted DNA and the entire complex covers 30–75 bp, depending on the conformation of RNAP.

plate strand. This is followed by initiation of RNA synthesis at the starting point, and the process continues until a termination sequence is reached (Fig 39–3). A **transcription unit** is defined as that region of DNA that extends between the promoter and the terminator. The RNA product, which is synthesized in the 5' to 3' direction, is the **primary transcript.** In prokaryotes, this can represent the product of several contiguous genes; in mammalian cells, it usually represents the product of a single gene. The 5' termini of the primary RNA transcript and the mature cytoplasmic RNA are identical. Thus, the **start point of transcription corresponds to the 5' nucleotide of the mRNA.** The primary transcripts generated by RNA polymerase II are promptly capped by 7-methylguanosine triphosphate (see Fig 37–10)—caps that persist and eventually appear on the 5' end of mature cytoplasmic mRNA. These caps are presumably necessary for the subsequent processing of the primary transcript to mRNA, for the translation of the mRNA, and for protection of the mRNA against 5' → 3' exonucleolytic attack.

The DNA-dependent RNA polymerase (RNAP) of

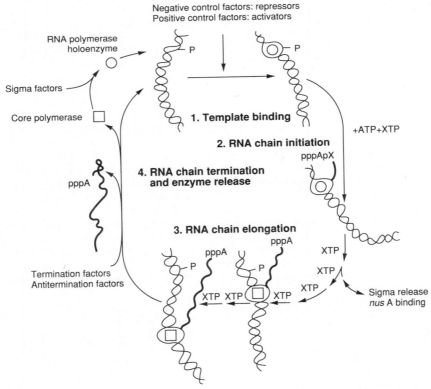

Figure 39-3. The transcription cycle in bacteria. Bacterial RNA transcription is described in 4 steps: (1) **Template binding:** RNA polymerase (RNAP) binds to DNA and locates a promoter. (2) **Chain initiation:** RNAP holoenzyme (core + sigma factors) catalyzes the coupling of the first base (usually ATP or GTP) to a second ribonucleoside triphosphate to form a dinucleotide. (3) **Chain elongation:** successive residues are added to the 3'-OH terminus of the nascent RNA molecule; sigma factor dissociates from the holoenzyme after a chain length of about 10 is achieved. (4) **Chain termination and release:** the completed RNA chain and RNAP are released from the template. The RNAP holoenzyme reforms, finds a promoter, and the cycle is repeated. Reproduced with slight modifications, from M. Chamberlin, *The Enzymes,* vol. XV, 1982.

the bacterium *Escherichia coli* exists as a core molecule composed of 4 subunits; 2 of these are identical to each other (the α subunits), and 2 are similar in size to each other but not identical (the β subunit and β′ subunit) (Figure 39–2). RNAP also contains 2 zinc molecules. The core RNA polymerase utilizes a specific protein factor (the sigma [σ] factor) that helps the core enzyme attach more tightly to the specific deoxynucleotide sequence of the promoter region. Bacteria contain multiple σ factors, each of which acts as a regulatory protein that modifies the **promoter recognition specificity** of the RNA polymerase. The appearance of different σ factors can be correlated temporally with various programs of gene expresson in prokaryotic systems, such as bacteriophage development, sporulation, and the response to heat shock.

RNA SYNTHESIS INVOLVES INITIATION, ELONGATION, AND TERMINATION

The process of RNA synthesis, depicted in Fig 39–3, involves first the binding of the RNA holopolymerase molecule to the template at the promoter site. Binding is followed by a conformational change of the RNAP, then the first nucleotide (almost always a purine) associates with the initiation site on the β subunit of the enzyme. In the presence of the 4 nucleotides, the RNAP moves to the second base in the template, a phosphodiester bond forms, and the nascent chain is now attached to the polymerization site on the β subunit of RNAP. (The analogy to the A and P sites on the ribosome should be noted here; see Figs 40–7 and 40–8).

Initiation of formation of the RNA molecule at its 5′ end then follows with the release of the σ factor, while the elongation of the RNA molecule from the 5′ to its 3′ end continues antiparallel to its template. The enzyme polymerizes the ribonucleotides in a specific sequence that is dictated by the template strand and interpreted by Watson-Crick base-pairing rules. Pyrophosphate is released in the polymerization reaction. In both prokaryotes and eukaryotes, a purine ribonucleotide is usually the first to be polymerized into the RNA molecule.

As the **elongation** complex containing the core RNA polymerase progresses along the DNA molecule, **DNA** unwinding must occur in order to provide access for the appropriate base pairing to the nucleotides of the coding strand. The extent of DNA unwinding is constant throughout transcription and has been estimated to be about 17 base pairs per polymerase molecule. Thus, it appears that the size of the unwound DNA region is dictated by the polymerase and is independent of the DNA sequence in the complex. This suggests that RNA polymerase has associated with it an "unwindase" activity that opens the DNA helix. The fact that the DNA double helix must unwind and the strands part at least transiently for transcription implies some disruption of the nucleosome structure of eukaryotic cells.

Termination of the synthesis of the RNA molecule is signaled by a sequence in the template strand of the DNA molecule, a signal that is recognized by a termination protein, the rho (ρ) factor. After termination of synthesis of the RNA molecule, the core enzyme separates from the DNA template. With the assistance of another σ factor, the core enzyme then recognizes a promoter at which the synthesis of a new RNA molecule commences. More than one RNA polymerase molecule may transcribe the same template strand of a gene simultaneously, but the process is phased and spaced in such a way that at any one moment each is transcribing a different portion of the DNA sequence. An electron micrograph of RNA synthesis is shown in Fig 39–4.

MAMMALIAN CELLS POSSESS SEVERAL DNA-DEPENDENT RNA POLYMERASES

The properties of mammalian polymerases are described in Table 39–1. Each of these DNA-dependent

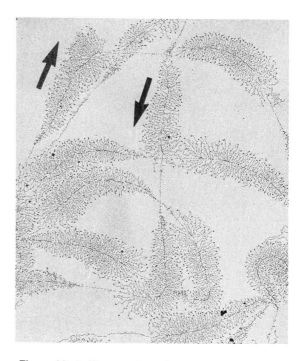

Figure 39–4. Electron photomicrograph of multiple copies of amphibian ribosomal RNA genes in the process of being transcribed. The magnification is ~6000×. Note that the length of the transcripts increases as the RNA polymerase molecules progress along the individual ribosomal RNA genes. Thus, the proximal end of the transcribed gene has short transcripts attached to it, while much longer transcripts are attached to the distal end of the gene. The arrows indicate the direction (5′ to 3′) of transcription from the sites of initiation (i) to the sites of termination (t). (Reproduced with permission, from Miller OL Jr, Beatty BR: Portrait of a gene. *J Cell Physiol* 1969;**74**[Suppl 1]:225).

Table 39–1. Nomenclature of animal DNA-dependent RNA polymerases.

Class of Enzyme	Sensitivity to α-Amanitin	Major Products
I (A)	Insensitive	rRNA
II (B)	Sensitive to low concentrations (10^{-8} to 10^{-9} mol/L)	hnRNA (mRNA)
III (C)	Sensitive to high concentrations	tRNA and 5S RNA

RNA polymerases seems to be responsible for the transcription of different sets of genes. The sizes of the RNA polymerases for the 3 major classes of eukaryotic RNA range from MW 500,000 to 600,000. All of these enzymes share the basic subunit structural organization of bacterial RNA polymerase. They all have 2 large subunits and a number of smaller subunits. Recent DNA cloning and sequencing work indicates that eukaryotic RNA polymerases have extensive amino acid homologies with prokaryotic RNA polymerases. The functions of each of the subunits are not yet understood. Many could have regulatory functions, such as serving to assist the polymerase in the recognition of specific sequences like promoters and termination signals.

One toxin from the mushroom *Amanita phalloides,* α-amanitin, is a specific inhibitor of the eukaryotic nucleoplasmic DNA-dependent RNA polymerase (RNA polymerase II) and as such has proved to be a powerful research tool (Table 39–1).

CERTAIN DNA SEQUENCES ARE IMPORTANT AS TRANSCRIPTION SIGNALS

The DNA sequence analysis of specific genes obtained by recombinant DNA technology has allowed the recognition of a number of sequences important in gene transcription. From the large number of bacterial genes studied it is possible to construct consensus models of transcription initiation and termination signals.

The question "How does RNAP find the correct site to initiate transcription?" is not trivial when the complexity of the genome is considered. *Escherichia coli* has $\sim2 \times 10^3$ transcription initiation sites in $\sim4 \times 10^6$ bp of DNA. The situation is even more complex in humans, where some 10^5 transcription initiation sites are scattered in 3×10^9 bp of DNA. RNAP can bind to many regions of DNA, but it scans the DNA sequence, at a rate of $\sim10^3$ bp/s, until it recognizes certain specific regions of DNA to which it binds with higher affinity. This region is called the **promoter,** and it is the association of RNAP with the promoter that assures accurate initiation of transcription.

Bacterial promoters are approximately 40 nucleotides (40 base pairs, or 4 turns of the DNA double helix) in length, a region sufficiently small to be covered by an *E coli* RNA holopolymerase molecule. In this consesus promoter regions are 2 short, conserved sequence elements. Approximately 35 base pairs upstream of the transcription start site there is a consensus sequence of 8 nucleotide pairs 5'-TGTTGACA-3', shown in Fig 39–5. More proximal to the transcription start site, about 10 nucleotides upstream, is a 6-nucleotide-pair AT-rich sequence (5'-TATAAT-3'). The latter sequence has a low melting temperature because of its deficiency of GC nucleotide pairs. Thus, the **TATA or Pribnow box** is thought to ease the dissociation between the coding and noncoding strands so that RNA polymerase bound to the promoter region can have access to the nucleotide sequence of its immediately downstream coding strand. Other bacteria have different combinations (boxes) but all generally have two components to the promoter; these tend to be in the same position relative to that transcription start site, and in all cases the sequences between the boxes have no similarity.

Rho-dependent transcription **termination signals**

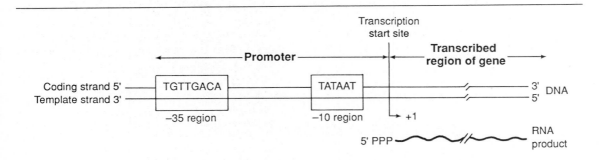

Figure 39–5. Bacterial promoters, such as that from *E coli* shown here, share 2 regions of highly conserved nucleotide sequence. These regions are located 35 and 10 bp upstream (in the 5' direction of the coding strand) from the start site of transcription, which is indicated as +1. By convention, all nucleotides upstream of the transcription initiation site (at +1) are numbered in a negative sense. Also by convention, the DNA regulatory sequence elements (TATA box, etc) are described in the 5' to 3' direction and as being on the coding strand. These elements only function in double-stranded DNA, however.

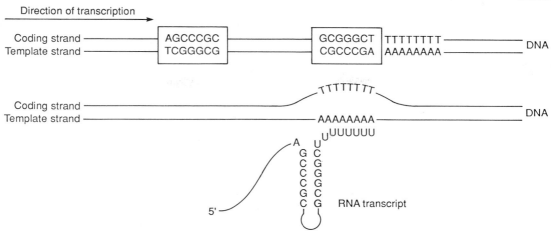

Figure 39–6. The bacterial transcription termination signal in the gene contains an inverted, hyphenated repeat (the two boxed areas) followed by a stretch of AT base pairs (top figure). The inverted repeat, when transcribed into RNA, can generate the secondary structure in the RNA transcript shown in the bottom figure.

in *E coli* also appear to have a distinct consensus sequence, as shown in Fig 39–6. The conserved consensus sequence, which is about 40 nucleotide pairs in length, can be seen to contain a hyphenated or interrupted inverted repeat, followed by a series of AT base pairs. As transcription proceeds through the hyphenated, inverted repeat, the generated transcript can form the intramolecular hairpin structure, also depicted in Fig 39–6. Transcription continues into the AT region, and with the aid of a termination protein factor called rho (ρ) the RNA polymerase stops and dissociates, releasing the primary transcript.

The Transcription Signals in Mammalian Cells Are, Not Unexpectedly, More Complex

It is clear that the signals in DNA which control transcription in eukaryotic cells are of several types. Two types of sequence elements are promoter-proximal. One of these defines **where transcription is to commence** along the DNA, and the other determines **how frequently** this event is to occur. For example, in the thymidine kinase gene of herpes simplex, which utilizes transcription factors of its mammalian host for gene expression, there is a unique transcription start site, and accurate transcription from this start site de-

pends upon a nucleotide sequence located −32 nucleotides upstream from the start site (Figure 39–7). This region has the sequence of **TATAAAAG** and bears remarkable homology to the functionally related **Pribnow box** (TATAAT) which is located about 10 base pairs upstream from prokaryotic mRNA start points. RNA polymerase II probably binds to DNA in the region of the TATA box and then commences transcription of the template strand about 32 nucleotides downstream. Therefore, the TATA box seems to provide the "where" signal.

Sequences farther upstream from the start site determine how frequently this transcription event occurs. Mutations in these regions reduce the frequency of transcriptional starts 10- to 20-fold. These DNA elements are referred to as GC and CAAT boxes because of the DNA sequences involved (Table 39–2). As illustrated in Fig 39–7, each of these boxes binds a unique protein, Spl in the case of the GC box, and CTF (or C/EPB, NF1, NFY) by the CAAT box. The frequency of transcription initiation is a consequence of these protein-DNA interactions, whereas the protein-DNA interaction at the TATA box ensures the fidelity of initiation. **These DNA elements are said to be *cis* acting because they are located on the same molecule of DNA as the gene being regulated. The pro-**

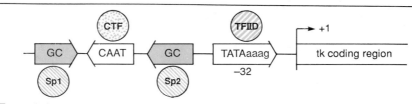

Figure 39–7. Transcription elements and binding factors in the thymidine kinase (tk) gene. DNA-dependent RNA polymerase II binds downstream from in the region of the TATA box (which binds transcription factor TfIID) to initiate transcription at a single nucleotide. The frequency of this event is increased by the presence of upstream *cis*-acting elements (the GC and CAAT boxes). These elements bind *trans*-acting transcription factors Sp1 and CTF, respectively, and can function in the reverse orientation (arrows).

Table 39–2. Some of the transcription control elements and the factors that bind to them that are found in mammalian genes transcribed by RNA polymerase II. The consensus DNA sequences of the elements are in parentheses. The small n means that any nucleotide will suffice.

Element	Factor
TATA box (TATAAT)	TFIID
CAAT box (CCAAT)	CAAT binding proteins
TGG(n)$_{6-7}$GCCAA	NF1 (nuclear factor 1)
GC box (GGGCGG)	Sp1
Ig Octamer (ATTTCGAT)	NFκB (immunoglobulin κ gene factor b)
Heat-shock element (CnnGAAnnTTCnnG)	Heat-shock transcription factor

tein factors are said to be *trans* acting because they originate from genes presumably located on different chromosomes.

These upstream elements confer fidelity and frequency of initiation and have rigid requirements for both position and orientation. Single base changes have dramatic effects on function; the spacing of these elements with respect to the start site is critical, and they generally do not work if the 5' to 3' orientation is reversed (Fig 39–8).

A third class of sequence elements can either increase or decrease the rate of transcription initiation of eukaryotic genes. These elements are called **enhancers** or **silencers,** depending on which effect they have. These elements have been found in a variety of locations both upstream and downstream from the transcription start site. In contrast to proximal and upstream promoter elements, enhancers and silencers can exert their effects when located hundreds or thousands of bases away from transcription units located on the same chromosome (*cis* linked). Surprisingly, enhancers and silencers can function in an orientation-independent fashion.

Hormone response elements (for steroids, T$_3$, TRH, cAMP, prolactin, etc) act as, or in conjunction with, enhancers or silencers (see Chapter 44). Other processes that enhance or silence gene expression, such as the response to heat shock, metals (Cd^{2+} and Zn^{2+}), and some toxic chemicals (eg, dioxin) are mediated through specific regulatory elements. Tissue-specific expression of genes, eg, the albumin gene in liver, is also mediated by specific DNA sequences.

A common feature of all the elements, basal and regulatory, is that the interaction of specific proteins with the DNA sequences is involved. A number of these factors have been identified (Table 39–2), and a great deal of research is directed at analyzing how these protein-DNA interactions influence gene transcription.

The **signals for the termination of transcription** by eukaryotic RNA polymerase II are very poorly understood. However, it appears that the termination signals exist far downstream of the coding sequence of eukaryotic genes. For example, the transcription termination signal for mouse β-globin occurs at several positions 1000–2000 bases beyond the site at which

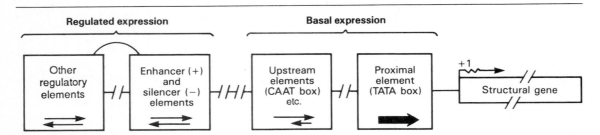

Figure 39–8. Schematic diagram showing the transcription control regions in a hypothetical class II (mRNA producing) eukaryotic gene. Such a gene can be divided into its structural and regulatory regions, as defined by the transcription start site (arrow). The structural gene contains the DNA sequence that is transcribed into mRNA, which is ultimately translated into protein. The regulatory region consists of 2 elements responsible for ensuring basal expression. The proximal component, generally the TATA box, directs RNA polymerase II to the correct site (fidelity). Another component, the upstream element, specifies the frequency of initiation. The best studied of these is the CAAT box, but several other elements (Sp1, NF1, AP1, etc) may be used in various genes. Regulated expression consists of elements that enhance or silence expression and of others that mediate the response to various signals, including hormones, heat shock, metals, and chemicals. Tissue-specific expression also involves specific sequences of this sort. It is possible that these 2 regulatory regions overlap in function (the connecting line). The orientation dependence of all the elements is indicated by the arrows within the boxes. For example, the proximal element must be in the 5' to 3' orientation. The upstream elements work best in the 5' to 3' orientation, but some of them can be reversed. The broken lines indicate that some elements are not fixed with respect to the transcription start site. Indeed, some elements responsible for regulated expression can be located either interspersed with the upstream elements, or they can be located downstream from the start site.

the poly(A) tail will eventually be added. Little is known about the termination process, or whether specific termination factors similar to the bacterial ρ factor are involved. However, it is known that the mRNA 3' terminus is generated posttranscriptionally, and it appears to involve 2 steps. After RNA polymerase II has traversed the region of the transcription unit encoding the 3' end of the transcript, an RNA endonuclease cleaves the primary transcript at a position about 15 bases 3' to the consensus sequence **AAUAAA** that seems to serve in eukaryotic transcripts as a cleavage signal. Finally, this newly formed 3' terminus is polyadenylated in the nucleoplasm, as described below.

THE LOCATION OF THE *CIS* ACTING DNA ELEMENTS IS INTRAGENIC IN CLASS III GENES

DNA-dependent RNA polymerase III, which transcribes tRNA genes and low molecular weight RNA genes (see Chapter 37), recognizes a promoter that is **internal to the gene** to be expressed, rather than upstream of the transcription starting point. In the case of eukaryotic tRNA genes, 2 internal, separated blocks (A and B) of sequences exist that act as an intragenic promoter. The sequences within the A and B blocks exist in the mature tRNA molecule in regions that are highly conserved and participate in the formation of the DHU loop and the TψC loop, respectively (Fig 37–11). By manipulating the structure of tRNA genes, it has been shown that for promoter function the optimal distance between the A and B blocks is 30–40 base pairs and that the transcription start point occurs between 10 and 16 base pairs upstream from the A block. For the 5S RNA gene, which is also transcribed by RNA polymerase III, there appears to be a specific transcription protein factor which, once bound to the intragenic promoter for that gene, probably interacts with an RNA polymerase III molecule to position its catalytic sites on the transcription start point of the DNA. A similar mechanism, employing a specific *trans*-acting transcription factor(s), may be involved in tRNA gene transcription.

RNA MOLECULES ARE OFTEN PROCESSED BEFORE THEY BECOME FUNCTIONAL

In prokaryotic organisms the primary transcripts of mRNA-encoding genes begin to serve as translation templates even before their transcription has been completed. This is presumably because the site of transcription is not compartmentalized into a nucleus as it is in eukaryotic organisms. Thus, transcription and translation are coupled in prokaryotic cells. Consequently, prokaryotic mRNAs are subjected to little modification and processing prior to carrying out their intended function in protein synthesis. Prokaryotic rRNA and tRNA molecules are transcribed in units considerably longer than the ultimate molecule. In fact, many of the tRNA transcription units contain more than one molecule. Thus, in prokaryotes the processing of these rRNA and tRNA precursor molecules is required for the generation of the mature functional molecules.

Nearly all eukaryotic RNA primary transcripts undergo extensive processing between the time they are synthesized and the time at which they serve their ultimate function, whether it be as mRNA or as a structural molecule such as rRNA, 5S RNA, or tRNA. The processing occurs primarily within the nucleus. The processing includes **capping, nucleolytic and ligation reactions, terminal additions** of nucleotides, and **nucleoside modifications.** However, it is clear that, for mammalian cells, 50–75% of the nuclear RNA, including those RNAs with capped 5' termini, do not contribute to the cytoplasmic mRNA. This nuclear RNA loss is significantly greater than can be reasonably accounted for by the loss of intervening sequences alone (see below). Thus, the exact function of the seemingly excessive transcripts in the nucleus of a mammalian cell is not known.

THE CODING PORTIONS (EXONS) OF MOST EUKARYOTIC GENES ARE INTERRUPTED BY INTRONS

As a result of advances in techniques for DNA cloning and DNA sequencing, it is now apparent that interspersed within the amino acid-coding portions (**exons**) of many genes are long sequences of DNA that do not contribute to the genetic information ultimately translated into the amino acid sequence of a protein molecule (see Chapter 38). In fact these sequences actually interrupt the coding region of structural genes. These **intervening sequences (introns)** exist within most but not all genes of higher eukaryotes. The primary transcripts of the structural genes contain RNA complementary to the interspersed sequences. However, the intron RNA sequences are cleaved out of the transcript, and the exons of the transcript are appropriately spliced together in the nucleus before the resulting mRNA molecule appears in the cytoplasm for translation (Figs 39–9 and 39–10).

The mechanisms whereby the introns are removed from the primary transcript in the nucleus, the exons are ligated to form the mRNA molecule, and the mRNA molecule is transported to the cytoplasm are being elucidated. Although the sequences of nucleotides in the introns of the various eukaryotic transcripts, and even those within a single transcript, are very heterogeneous, there are reasonably conserved sequences at each of the 2 exon-intron (splice) junctions (see concensus sequences in Fig 39–10). A special structure, the **spliceosome,** is involved in converting the primary transcript into mRNA. Spliceosomes consist of the primary transcript, 4 small nuclear RNAs (U1, U2, U5, and U4/U6), and an undetermined number of proteins. U1 snRNA is complemen-

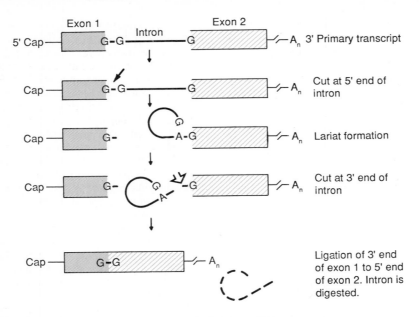

Figure 39–9. The processing of the primary transcript (hnRNA) to mRNA. In this hypothetical transcript the left end of the intron is cut (↓) and a lariat forms between the G at the 5' end of the intron and an A near the 3' end, in the consensus sequence UACUAAC. The right end of the intron is then cut (◇). This releases the lariat, which is digested, and exon 1 is joined to exon 2 at G residues.

tary to the consensus sequence at the 5' splice site, U2 to the branch point, and U5 to the 3' splice site. Catalysis may be provided by these snRNA molecules.

Recently, it has been discovered that during the process of removing the intron sequence from premRNA, there is formed an unusual RNA molecule resembling a lariat. It appears that the 5' end of the intervening sequence is joined via a 2'–5' phosphodiester linkage to an adenylate residue 28–37 nucleotides upstream from the 3' end of the intervening sequence. This process and structure are diagrammed in Fig 39–9.

It seems that the mystery of the relationship between hnRNA and the corresponding mature mRNA in eukaryotic cells is solved. The hnRNA molecules are the primary transcripts plus their early processed products, which, after the addition of caps and poly(A) tails and removal of the portion corresponding to the introns, are transported to the cytoplasm as mature mRNA molecules.

The processing of hnRNA molecules is a potential site for regulation of gene expression. In fact, it has been demonstrated that alternative patterns of RNA splicing are subject to developmental control.

Numerous examples could be cited, but three will make the point. For example, the cytoplasmic mRNAs for α-amylase in rat salivary gland and in rat liver differ in their 5' nucleotide sequence, while the remainder of the mRNA genes containing the coding region and poly(A) addition sites are identical. Further analysis has revealed that although the primary transcripts are huge and overlapping, different splice sites are used to join 2 different cap and leader sequences to the same mRNA "body." In addition, alternative patterns of RNA splicing are used to generate 2 different immunoglobulin heavy-chain mRNAs—one that codes for a membrane-bound heavy chain protein and another that codes for a secreted heavy chain protein (see Chapter 41). RNA splicing is usually necessary for the generation of messenger RNA molecules, and this splicing requirement provides an additional mechanism of differential regulation of gene expression.

At least one form of β-thalassemia, a disease in which the β-globin gene of hemoglobin is severely underexpressed, appears to result from a nucleotide change at an exon-intron junction, precluding removal of the intron and therefore leading to diminished or

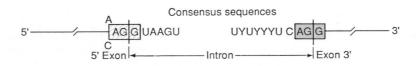

Figure 39–10. Consensus sequences at splice junctions.

absent synthesis of the β chain. This is a consequence of the fact that the normal translation reading frame of the mRNA is disrupted. Clearly a defect in this fundamental process (splicing) underscores the accuracy which the process of RNA-RNA splicing must attain.

RNA CAN ACT AS A CATALYST

In addition to the catalytic action served by the snRNAs in the formation of mRNA, at least three other RNA processing reactions are catalyzed by RNA. These observations, made in organelles from plants, yeast, viruses, and higher eukaryotic cells, show that **RNA can act as an enzyme.** This has revolutionized thinking about enzyme action and the origin of life itself.

MESSENGER RNA (mRNA) IS MODIFIED AT THE 5′ AND 3′ ENDS

As mentioned above, mammalian mRNA molecules contain a capped structure at their 5′ terminus, and most have a poly(A) tail at the 3′ terminus. The cap structure is added to the 5′ end of the newly transcribed mRNA precursor in the nucleus prior to transport of the mRNA molecule to the cytoplasm. Poly(A) tails (when present) appear to be added either in the nucleus or in the cytoplasm. The secondary methylations of mRNA molecules, those on the 2′-hydroxy groups and the N_6 of adenylate residues, occur after the mRNA molecule has appeared in the cytoplasm. The **5′ cap** of the RNA transcript appears to be required for the formation of the ribonucleoprotein complex necessary for the splicing reactions, may be involved in mRNA transport and translation initiations, and it protects the 5′ end of mRNA from attack by 5′ → 3′ exonucleases.

The function of the **poly(A) tail** is unknown but it does appear to protect the 3′ end of mRNA from 3′ → 5′ exonuclease attack. In any event, the presence or absence of the poly(A) tail does not determine whether a precursor molecule in the nucleus appears in the cytoplasm, because all poly(A)-tailed hnRNA molecules do not contribute to cytoplasmic mRNA, nor do all cytoplasmic mRNA molecules contain poly(A) tails (the histones are most notable in this regard). Cytoplasmic processes in mammalian cells can both add and remove adenylate residues from the poly(A) tails; this has been associated with an alteration of mRNA stability.

The size of the cytoplasmic mRNA molecules even after the poly(A) tail is removed is still considerably greater than the size required to code for the specific protein for which it is template, often by a factor of 2 or 3. **The extra nucleotides occur in untranslated (noncoding) regions** both 5′ and 3′ to the coding region; the longest untranslated sequences are usually at the 3′ end. The exact function of these sequences is unknown, but they have been implicated in RNA processing, transport, degradation, and translation.

Transfer RNA (tRNA) Is Extensively Modified

The tRNA molecules, as described in Chapters 37 and 40, serve as adapter molecules for the translation of mRNA into protein sequences. The tRNAs contain many modifications of the standard bases A, U, G, and C. Some are simply methylated derivatives and some possess rearranged glycosidic bonds. The tRNA molecules are transcribed in both prokaryotes and eukaryotes as large precursors, frequently containing the sequence for more than one tRNA, which are then subjected to **nucleolytic processing** and reduced in size by a specific class of ribonucleases. In addition, the genes of some tRNA molecules contain—very near the portion corresponding to the anticodon loop—a single intron 10–40 nucleotides long. These introns in the tRNA genes are transcribed; therefore, the processing of the precursor transcripts of many tRNA molecules must include removal of the introns and proper splicing of the anticodon region to generate an active adapter molecule for protein synthesis. The nucleolytic processing enzymes of tRNA precursors apparently recognize 3-dimensional structure and not just linear RNA sequences. This enzyme system thereby processes only molecules capable of folding into functionally competent products.

The further modification of the tRNA molecules includes nucleotide **alkylations** and the **attachment of the characteristic C · C · A terminus** at the 3′ end of the molecule. This C · C · A terminus is the point of attachment for the specific amino acid that is to enter into the polymerization reaction of protein synthesis. The methylation of mammalian tRNA precursors probably occurs in the nucleus, whereas the cleavage and attachment of **C · C · A** are cytoplasmic functions, since the termini turn over more rapidly than do the tRNA molecules themselves. Enzymes within the cytoplasm of mammalian cells are required for the attachment of amino acids to the C · C · A residues.

Ribosomal RNA (tRNA) Is Synthesized as a Large Precursor

In mammalian cells, the 2 major rRNA molecules and one minor rRNA molecule are transcribed as part of a single large precursor molecule (Fig 39–11). The precursor is subsequently processed in the **nucleolus** to provide the RNA component for the ribosome subunits found in the cytoplasm. The rRNA genes are located in the nucleoli of mammalian cells. Hundreds of copies of these genes are present in every cell. The rRNA genes are transcribed as units, each of which encodes (5′ to 3′) an 18S, a 5.8S, and a 28S ribosomal RNA. The primary transcript is a 45S molecule that is highly methylated in the nucleolus. In the **45S precursor,** the eventual 28S segment contains 65 ribose-methyl groups and 5- base-methyl groups. Only those portions of the precursor that eventually become stable rRNA molecules are methylated. The 45S precursor is nucleolytically processed, but the processing signals are clearly distinct from those in hnRNA. Therefore, the nucleolytic processing likely is mediated by a

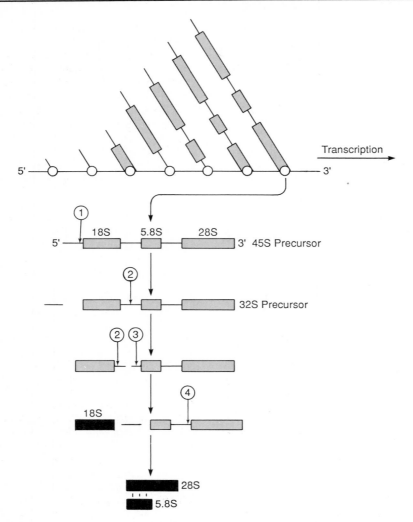

Figure 39–11. Diagrammatic representation of the processing of ribosomal RNA from precursor RNA molecules. The final products are indicated by the solid bars. A complex series of posttranscriptional processes, involving endo- and exonucleolytic processing by rRNA-specific ribonucleases, generates 18S, 5.8S, and 28S rRNAs. The 5.8S and 28S are associated through hydrogen bonding. Fig 39–4 shows an electron micrograph of the transcription of ribosomal genes.

mechanism unique from that responsible for processing hnRNA to mRNA.

Nearly half of the original primary transcript is degraded, as shown in Fig 39–11. During the processing of rRNA, further methylation occurs, and eventually, in the nucleoli, the 28S chains self-assemble with ribosomal proteins to form the larger 60S subunit. The 5.8S rRNA molecule also formed from the 45S precursor RNA in the nucleolus becomes an integral part of the larger ribosomal subunit. The smaller ribosomal subunits (40S) are formed by the association of appropriate ribosome proteins with the 18S rRNA molecule.

SPECIFIC NUCLEASES DIGEST NUCLEIC ACIDS

Enzymes capable of degrading nucleic acids have been recognized for many years. These can be classified in several ways. Those which exhibit specificity for deoxyribonucleic acid are referred to as **deoxyribonucleases.** Those which specifically hydrolyze ribonucleic acids are **ribonucleases.** Within both of these classes are enzymes capable of cleaving internal phosphodiester bonds to produce either 3'-hydroxyl and 5'-phosphoryl termini or 5'-hydroxyl and 3'-phosphoryl termini. These are referred to as **endonucleases.** Some are capable of hydrolyzing both strands of a **double-stranded** molecule, whereas others can only cleave **single strands** of nucleic acids. Some nucleases can hydrolyze only unpaired single strands, while others are capable of hydrolyzing single strands participating in the formation of a double-stranded molecule. There exist classes of endonucleases that recognize specific sequences in DNA; the majority of these are the **restriction endonucleases,** which have in recent years become important tools in molecular genetics and medical sciences. A list of some currently recognized restriction endonucleases is presented in Table 42–1.

Some nucleases are capable of hydrolyzing a nucleotide only when it is present at a terminus of a molecule; these are referred to as **exonucleases.** Exonucleases act in one direction ($3' \rightarrow 5'$ or $5' \rightarrow 3'$) only. In bacteria, a $3' \rightarrow 5'$ exonuclease is an integral part of the DNA replication machinery and there serves to edit the most recently added deoxynucleotide for base-pairing errors.

REGULATION OF DEGRADATION PROVIDES ANOTHER MECHANISM FOR REGULATING THE AMOUNT OF MESSENGER RNA

Although most mRNAs in mammalian cells are very stable (half-lives measured in hours), some turn over very rapidly (half-lives of 10–30 min). In certain instances mRNA stability is subject to regulation. This has important implications since there is usually a direct relationship between mRNA amount and the translation of that mRNA into its cognate protein. Changes in the stability of a specific mRNA can therefore have major effects on biologic processes.

Messenger RNAs exist in the cytoplasm as ribonucleoprotein particles (RNPs). Some of these proteins protect the mRNA from digestion by nucleases, while others may, under certain conditions, promote nuclease attack. It is thought that mRNAs are stabilized, or destabilized, by the interaction of proteins with these various structures or sequences. Certain effectors, such as hormones, may regulate mRNA stability by increasing or decreasing the amount of these proteins.

It appears that the ends of mRNAs molecules are involved in mRNA stability (see Fig 39–12). The 5′ cap structure in eukaryotic mRNA prevents attack by 5′ exonucleases and the poly(A) tail prohibits the action of 3′ exonucleases. In mRNA molecules with those structures it is presumed that a single endonucleolytic cut allows exonucleases to attack and digest the entire molecule. Other structures (sequences) in the 5′ noncoding sequence (5′ NCS), the coding region, and the 3′ NCS are thought to promote or prevent this initial endonucleolytic action (Fig 39–12). A few illustrative examples will be cited.

Deletion of the 5′ NCS results in a 3- to 5-fold prolongation of the half-life of c-*myc* mRNA. Shortening the coding region of histone mRNA results in a prolonged half-life. A form of autoregulation of mRNA stability indirectly involves the coding region. Free tubulin binds to the first 4 amino acids of a nascent chain of tubulin as it emerges from the ribosome. This appears to activate an RNase associated with the ribosome (RNP) which then digests the tubulin mRNA.

Structures at the 3′ end, including the poly(A) tail, enhance or diminish the stability of specific mRNAs. The absence of a poly(A) tail is associated with rapid degradation of mRNA and the removal of poly(A) from some RNAs results in their destabilization. Histone mRNAs lack a poly(A) tail but have a sequence near the 3′ terminus that can form a stem-loop structure, and this appears to provide resistance to exonucleolytic attack. Histone H4 mRNA, for example, is degraded in the 3′ to 5′ direction, but only after a single endonucleolytic cut occurs about 9 nucleotides from the 3′ end, in the region of the putative stem-loop structure. Stem-loop structures in the 3′ noncoding sequence are also critical for the regulation, by iron, of the mRNA encoding the transferrin receptor. Stem-loop structures are also associated with mRNA stability in bacteria, which suggests that this mechanism may be commonly employed.

Other sequences in the 3′ end of certain eukaryotic mRNAs appear to be involved in the destabilization of these molecules. Of particular interest are AU-rich regions, many of which contain the sequence AUUUA. This sequence appears in mRNAs that have a very short half life, including several oncogenes and cytokines. The importance of this region is underscored by an experiment in which a sequence corre-

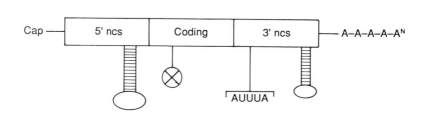

Figure 39–12. Structure of a typical eukaryotic mRNA showing elements that are involved in regulating mRNA stability. The typical eukaryotic mRNA has a 5′ noncoding sequence (5′ NCS), a coding region, and a 3′ NCS. All are capped at the 5′ end and most have a polyadenylate sequence at the 3′ end. The 5′ cap and 3′ poly(A) tail protect the mRNA against exonuclease attack. Stem loop structures in the 5′ and 3′ NCS, features in the coding sequence, and the AU-rich region in the 3′ NCS are thought to play roles in mRNA stability.

sponding to the 3' noncoding region of the short half-life colony stimulating factor (CSF) mRNA, which contains the AUUUA motif, was added to the 3' end of the β-globin mRNA. Instead of becoming very stable, this hybrid β-globin mRNA now had the short half-life characteristic of CSF mRNA.

From the few examples cited it is clear that a number of mechanisms are used to regulate mRNA stability, just as several mechanisms are used to regulate the synthesis of mRNA. Coordinate regulation of these two processes affords the cell remarkable adaptability.

REFERENCES

Breathnach R, Chambon P: Organization and expression of eucaryotic split genes coding for proteins. *Annu Rev Biochem* 1981;**50:**349.

Dynan WS, Tjian R: Control of eucaryotic mRNA synthesis by sequence specific DNA binding proteins. *Nature (London)* 1985;**316:**774.

Maniatis T, Read R: The role of small nuclear ribonucleoprotein particles in pre-mRNA splicing. *Nature (London)* 1987;**325:**673.

McClure WR: Protein-nucleic acid interactions in transcription: A molecular analysis. *Annu Rev Biochem* 1985;**54:** 171.

McKnight S, Tjian R: Transcriptional selectivity of viral genes in mammalian cells. *Cell* 1986;**46:**795.

Nevins JR: The pathway of eukaryotic mRNA formation. *Annu Rev Biochem* 1983;**52:**441.

Padgett RA et al: Splicing of messenger RNA precursors. *Annu Rev Biochem* 1986;**55:**1119.

Ross J: The turnover of messenger RNA. *Sci Am* 1989;**260:** 48.

Ruskin B et al: Excision of an intact intron as a novel lariat structure during pre-mRNA splicing in vitro. *Cell* 1984; **38:**317.

Sentenec A: Eucaryotic RNA polymerases. *Crit Rev Biol* 1985;**18:**31.

Shapiro DJ et al: Regulation of mRNA stability in eukaryotic cells. *Bioessays* 1987;**6:**221.

Sharp PA: On the origin of RNA splicing and introns. *Cell* 1985;**42:**397.

Protein Synthesis & the Genetic Code

40

Daryl K. Granner, MD

INTRODUCTION

The letters A, G, T, and C correspond to the nucleotides found in DNA. They are organized into 3-letter code words called **codons,** and the collection of these codons comprises the **genetic code**. A linear array of codons (a **gene**) specifies the synthesis of various RNA molecules, most of which are involved in some aspect of protein synthesis. Protein synthesis occurs in 3 major steps: initiation, elongation, and termination. This process resembles DNA replication and transcription in its general features and in the fact that it, too, follows a 5′ to 3′ polarity.

BIOMEDICAL IMPORTANCE

It was impossible to understand protein synthesis, or to explain mutations, before the genetic code was elucidated. The genetic code provides a foundation for explaining the way in which protein defects may cause genetic disease and for the diagnosis and perhaps treatment of these disorders. In addition, the pathophysiology of many viral infections is related to the ability of these agents to disrupt host cell protein synthesis.

GENETIC INFORMATION FLOWS FROM DNA TO RNA TO PROTEIN

The genetic information within the nucleotide sequence of DNA is transcribed in the nucleus into the specific nucleotide sequence of an RNA molecule. The sequence of nucleotides in the RNA transcript is complementary to the nucleotide sequence of the coding strand of its gene in accordance with the base-pairing rules. Several different classes of RNA combine to direct the synthesis of proteins.

In prokaryotes there is a linear correspondence between the gene, the **messenger RNA (mRNA)** transcribed from the gene, and the polypeptide product. The situation is more complicated in higher eukaryotic cells, in which the primary transcript, **heterogeneous nuclear RNA (hnRNA)**, is much larger than the mature mRNA. The large hnRNA contains coding regions (**exons**) that will form the mature mRNA and long intervening sequences (**introns**) that separate the exons. The hnRNA is processed within the nucleus, and the introns, which often make up much more of the hnRNA than the exons, are removed. Exons are spliced to form mature mRNA, which is transported to the cytoplasm, where it is translated into protein.

The cell must possess the machinery necessary to translate information accurately and efficiently from the nucleotide sequence of an mRNA into the sequence of amino acids of the corresponding specific protein. Clarification of our understanding of this process, which is termed **translation**, awaited deciphering of the genetic code. It was realized early that mRNA molecules in themselves have no affinity for amino acids and, therefore, that the translation of the information in the mRNA nucleotide sequence into the amino acid sequence of a protein requires an intermediate adapter molecule. This adapter molecule must recognize a specific nucleotide sequence on the one hand as well as a specific amino acid on the other. With such an adapter molecule, the cell can direct a specific amino acid into the proper sequential position of a protein as dictated by the nucleotide sequence of the specific mRNA. In fact, the functional groups of the amino acids do not themselves actually come into contact with the mRNA template.

THE NUCLEOTIDE SEQUENCE OF AN mRNA MOLECULE CONSISTS OF A SERIES OF CODONS THAT CORRESPOND TO EACH AMINO ACID

The adaptor molecules that translate the codons into the amino acid sequence of a protein are the **transfer RNA (tRNA)** molecules. The **ribosome** is the cellular component on which these various functional entities interact to assemble the protein molecule. Many of these subcellular units (ribosomes) can assemble to translate simultaneously a single mRNA molecule and, in so doing, form a **polyribosome**. The **rough endoplasmic reticulum** is a compartment of membrane-attached polyribosomes that provides for the synthesis of integral membrane proteins and proteins to be exported. Polyribosomal structures also exist free in the cytoplasm, where they synthesize proteins that remain within the cell.

Twenty different amino acids are required for the synthesis of the cellular complement of proteins; thus, there must be at least 20 distinct codons that comprise the genetic code. Since there are only 4 different nucleotides in mRNA, each codon must consist of more than a single purine or pyrimidine nucleotide. Codons consisting of 2 nucleotides each could provide for only 16 (4^2) specific codons, whereas codons of 3 nucleotides could provide 64 (4^3) specific codons.

As a result of the initial observations of Matthaei and Nirenberg, it is now known that each codon consists of a sequence of 3 nucleotides; ie, **it is a triplet code.** The deciphering of the genetic code depended heavily on the chemical synthesis of nucleotide polymers, particularly triplets.

THE GENETIC CODE IS DEGENERATE, UNAMBIGUOUS, NONOVERLAPPING, WITHOUT PUNCTUATION AND UNIVERSAL

Three codons do not code for specific amino acids; these have been termed **nonsense codons**. At least 2 of these nonsense codons are utilized in the cell as **termination signals**; they specify where the polymerization of amino acids into a protein molecule is to stop. The remaining 61 codons code for 20 amino acids. Thus, there must be **"degeneracy"** in the genetic code; ie, multiple codons must decode the same amino acid. An examination of the genetic code reveals that the 64 possible codons may be arranged in **16 families,** a family of codons being those which have the **same first 2 bases.** Each family occupies a single column between the horizontal lines. For example, the codons CCN, where N can be U, C, A, or G, define a family located in the second column of the second box from the top. In some families, all 4 codons code for the same amino acid, as do members of the CC family described immediately above. These are referred to as **unmixed** families. Eight of the 16 families of codons are unmixed (Table 40–1). Those families of codons which code for more than one amino acid are said to be **mixed** families. In 6 of the mixed families, codons with pyrimidines (U or C) at the third position code for one amino acid, while members with purines (A or G) at the third position code for another amino acid or chain termination signal (Table 40–1). The 2 remaining families—the UG family and the AU family—do not exhibit either pattern and are unique. Thus, in general, the third nucleotide in a codon is less important than the other 2 in determining the specific amino acid to be incorporated, and this accounts for most of the degeneracy of the code. However, for any specific codon only a single amino acid is indicated; with rare exceptions the genetic code is **unambiguous**—ie, given a specific codon, only a single amino acid is indicated. **The distinction between ambiguity and degeneracy is an important concept to be emphasized.**

The unambiguous but degenerate code can be ex-

plained in molecular terms. The recognition of specific codons in the mRNA by the tRNA adapter molecules is dependent upon their **anticodon region** and specific base-pairing rules. Each tRNA molecule contains a specific sequence, complementary to a codon, which is termed its anticodon. For a given codon in the mRNA, only a single species of tRNA molecule possesses the proper anticodon. Since each tRNA molecule can be charged with only one specific amino acid, each codon therefore specifies only one amino acid. However, some tRNA molecules can utilize the anticodon to recognize more than one codon. As can be seen from the example of unmixed families, the nucleotide in the anticodon that recognizes the third (3′-) base of the codon could be less discriminating (mixed families) or nondiscriminating (unmixed families) and still manage to insert the proper amino acid when called for. This reduced stringency between the third base of the codon and the complementary nucleotide in the anticodon is referred to as **wobble. With few exceptions, given a specific codon, only a specific amino acid will be incorporated—although, given a specific amino acid, more than one codon may call for it.**

As discussed below, the reading of the genetic code during the process of protein synthesis does not involve any overlap of codons. **Thus, the genetic code is nonoverlapping.** Furthermore, once the reading is commenced at a specific codon, there is **no punctuation** between codons, and the message is read in a continuing sequence of nucleotide triplets until a nonsense codon is reached.

Until recently, the genetic code was thought to be universal. It has now been shown that the set of tRNA molecules in mitochondria (which contain their own separate and distinct set of translation machinery) from lower and higher eukaryotes, including humans, reads 4 codons differently from the tRNA molecules in the cytoplasm of even the same cells. As noted in Table 40–1, the codon AUA is read as Met, and UGA codes for Trp in mammalian mitochondria. These 2 codons reside in the 2 codon families noted to be unique: the UG family and the AU family. Apparently, in order to minimize the number of tRNA molecules necessary to translate the genetic code, mitochondria have managed to convert the UG family and the AU family to simple mixed types. In addition, the codons AGA and AGG are read as stop or chain terminator codons rather than as Arg. As a result, mitochondria require only 22 tRNA molecules to read their genetic code, whereas the cytoplasmic translation system possesses a full complement of 31 tRNA species. These exceptions noted, **the genetic code is universal.** The frequency of use of each amino acid codon varies considerably between species and in different tissues within a species. Tables of codon usage are becoming more accurate as more genes are sequenced. This is of considerable importance because investigators often need to deduce mRNA structure from the primary sequence of a portion of protein in order to synthesize an oligonucleotide probe and initiate a recombinant DNA cloning project.

Table 40–1. The genetic code (codon assignments in messenger RNA).*

First Nucleotide	Second Nucleotide				Third Nucleotide
	U	C	A	G	
U	Phe	Ser	Tyr	Cys	U
	Phe	Ser	Tyr	Cys	C
	Leu	Ser	Term	Term†	A
	Leu	Ser	Term	Trp	G
C	Leu	Pro	His	Arg	U
	Leu	Pro	His	Arg	C
	Leu	Pro	Gln	Arg	A
	Leu	Pro	Gln	Arg	G
A	Ile	Thr	Asn	Ser	U
	Ile	Thr	Asn	Ser	C
	Ile†	Thr	Lys	Arg†	A
	Met	Thr	Lys	Arg†	G
G	Val	Ala	Asp	Gly	U
	Val	Ala	Asp	Gly	C
	Val	Ala	Glu	Gly	A
	Val	Ala	Glu	Gly	G

*The terms first, second, and third nucleotide refer to the individual nucleotides of a triplet codon. U, uridine nucleotide; C, cytosine nucleotide; A, adenine nucleotide; G, guanine nucleotide; Met, chain initiator codon; Term, chain terminator codon. Aug, which codes for Met, serves as the initiator codon in mammalian cells. (Abbreviations of amino acids are explained in Chapter 3.)

†In mammalian mitochondria, AUA codes for Met and UGA for Trp, and AGA and AGG serve as chain terminators.

AT LEAST ONE SPECIES OF TRANSFER RNA (tRNA) EXISTS FOR EACH OF THE 20 AMINO ACIDS

tRNA molecules have extraordinarily similar functions and 3-dimensional structures. The adapter function of the tRNA molecules requires the charging of each specific tRNA with its specific amino acid. Since there is no affinity of nucleic acids for specific functional groups of amino acids, this recognition must be carried out by a protein molecule capable of recognizing both a specific tRNA molecule and a specific amino acid. At least 20 specific enzymes are required for these specific recognition functions and for the proper attachment of the 20 amino acids to specific tRNA molecules. The process of recognition and attachment (charging) is carried out in 2 steps by one enzyme for each of the 20 amino acids. These enzymes are termed **aminoacyl-tRNA synthetases.** They form an activated intermediate of aminoacyl-AMP-enzyme complex as depicted in Fig 40–1. The specific aminoacyl-AMP-enzyme complex then recognizes a specific tRNA to which it attaches the aminoacyl moiety at the 3′-hydroxyl adenosine terminus. The amino acid remains attached to its specific tRNA in an ester linkage until it is polymerized at a specific position in the fabrication of a polypeptide precursor of a protein molecule.

The regions of the tRNA molecule referred to in Chapter 37 (and illustrated in Fig 37–11) now become important. The thymidine-pseudouridine-cytidine (TψC) arm is involved in binding of the aminoacyl-tRNA to the ribosomal surface at the site of protein synthesis. The D arm is one of the sites important for the proper recognition of a given tRNA species by its proper aminoacyl-tRNA synthetase. The acceptor arm, located at the 3′-hydroxyl adenosyl terminus, is the site of attachment of the specific amino acid.

The anticodon region consists of 7 nucleotides, and it recognizes the 3-letter codon in mRNA (Fig 40–2). The sequence read from the 3′ to 5′ direction in that anticodon loop consists of a variable base · modified purine · X · Y · Z · pyrimidine · pyrimidine-5′. Note that this direction of reading the anticodon is 3′ to 5′, whereas the genetic code in Table 40–1 is read 5′ to 3′, since the codon and the anticodon loop of the mRNA and tRNA molecules, respectively, are **antiparallel** in their complementarity.

The degeneracy of the genetic code resides mostly in the last nucleotide of the codon triplet, suggesting that the base pairing between this last nucleotide and the corresponding nucleotide of the anticodon is not strict. As referred to above, this is called **wobble**; the pairing of the codon and anticodon can "wobble" at this specific nucleotide-to-nucleotide pairing site. For example, the 2 codons for arginine, A · G · A and A · G · G, can bind to the same anticodon having a uracil at its 5′ end. Similarly, 3 codons for glycine, G · G · U, G

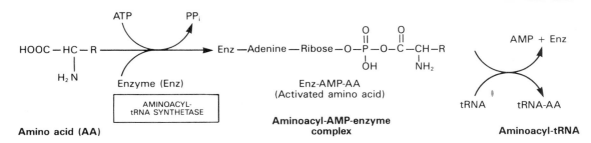

Figure 40–1. Formation of aminoacyl-tRNA. A-2 step reaction, involving the enzyme aminoacyl-tRNA synthetase, results in the formation of aminoacyl-tRNA. The first reaction involves the formation of an AMP-amino acid-enzyme complex. This activated amino acid is next transferred to the corresponding tRNA molecule. The AMP and enzyme are released, and the latter can be reutilized.

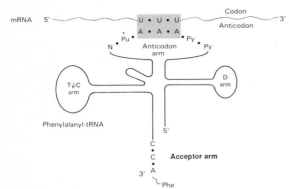

Figure 40–2. Recognition of the codon by the anticodon. One of the codons for phenylalanine is U · U · U. tRNA charged with phenylalanine (Phe) has the complementary sequence A · A · A; hence, it forms a base-pair complex with the codon. The anticodon region typically consists of a sequence of 7 nucleotides: variable (N), modified purine (Pu*), X, Y, Z, and 2 pyrimidines (Py) in the 3′ to 5′ direction.

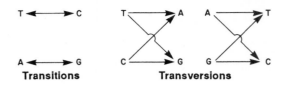

Figure 40–3. Diagrammatic representation of transition mutations and transversion mutations.

· G · C, and G · G · A, can form a base pair from one anticodon, C · C · I. I is an inosine nucleotide, another of the peculiar bases appearing in tRNA molecules.

The codon recognition by a tRNA molecule does not depend upon the amino acid that is attached at its 3′-hydroxyl terminus. This has been ingeniously demonstrated by charging a tRNA specific for cysteine (tRNA$_{cys}$) with radioactively labeled cysteine. By chemical means, the cysteinyl residue was then altered to generate a tRNA molecule specific for cysteine but charged instead with alanine. The chemical transformation of the cysteinyl to the alanyl moiety did not alter the anticodon portion of the cysteine-specific tRNA molecule. When this alanyl-tRNA$_{cys}$ was used in the translation of a hemoglobin mRNA, a radioactive alanine was incorporated at what was normally a cysteine site in the hemoglobin protein molecule. The experiment demonstrated that the aminoacyl derivative of an aminoacyl-tRNA molecule does not play a role in the codon recognition. As already noted, the aminoacyl moiety never comes in contact with the template mRNA containing the codons.

A MUTATION RESULTS WHEN ONE OF A NUMBER OF TYPES OF CHANGES OCCURS IN THE NUCLEOTIDE SEQUENCE

Although the initial change may not occur in the coding strand of the double-stranded DNA molecule for that gene, after replication, daughter DNA molecules with mutations in the coding strand will segregate and appear in the population of organisms.

Some Mutations Occur by Base Substitution

Single base changes (**point mutations**) may be **transitions** or **transversions**. In the former, a given

pyrimidine is changed to the other pyrimidine or a given purine is changed to the other purine. Transversions are changes from a purine to either of the 2 pyrimidines or the change of a pyrimidine into either of the 2 purines, as shown in Fig 40–3.

If the nucleotide sequence of the gene containing the mutation is transcribed into an RNA molecule, then the RNA molecule will possess a complementary base change at this corresponding locus.

Single base changes in the mRNA molecules may have one of several effects when translated into protein:

(1) There may be **no detectable effect** because of the degeneracy of the code. This would be more likely if the changed base in the mRNA molecule were to be at the third nucleotide of a codon. Because of wobble, the translation of a codon is least sensitive to a change at the third position.

(2) A **missense** effect will occur when a different amino acid is incorporated at the corresponding site in the protein molecule. This mistaken amino acid, or missense, depending upon its location in the specific protein, might be **acceptable, partially acceptable,** or **unacceptable** to the function of that protein molecule. From a careful examination of the genetic code, one can conclude that most single base changes would result in the replacement of one amino acid by another with rather similar functional groups. This is an effective mechanism to avoid drastic change in the physical properties of a protein molecule. If an acceptable missense effect occurs, the resulting protein molecule may not be distinguishable from the normal one. A partially acceptable missense will result in a protein molecule with partial but abnormal function. If an unacceptable missense effect occurs, then the protein molecule will not be capable of functioning in its assigned role.

(3) A **nonsense** codon may appear that would then result in the **premature termination** of amino acid incorporation into a peptide chain and the production of only a fragment of the intended protein molecule. The probability is high that a prematurely terminated protein molecule or peptide fragment would not function in its assigned role.

The Hemoglobin Molecule Can Be Used to Demonstrate the Effects of Single Base Changes in the Hemoglobin Structural Gene

Some mutations have no apparent effect. The lack of effect of a single base change would be demonstrable only by sequencing the nucleotides in the mRNA molecules or structural genes for hemoglobin from a large number of humans with normal hemoglobin molecules. However, it can be deduced that the codon for valine at position 67 of the β chain of hemoglobin is not identical in all persons possessing the normal β chain of hemoglobin. Hemoglobin Milwaukee has at position 67 a glutamic acid; hemoglobin Bristol contains aspartic acid at position 67. In order to account for the amino acid change by the change of a single nucleotide residue in the codon for amino acid 67, one must infer that the mRNA encoding hemoglobin Bristol possessed a G · U · U or G · U · C codon prior to a later change to G · A · U or G · A · C, both codons for aspartic acid (Fig 40–4). However, the mRNA encoding hemoglobin Milwaukee would have to possess at position 67 a codon G · U · A or G · U · G in order that a single nucleotide change could provide for the appearance of the glutamic acid codons G · A · A or G · A · G. Hemoglobin Sydney, which contains an alanine at position 67, could have arisen by the change of a single nucleotide in any of the 4 codons for valine (G · U · U, G · U · C, G · U · A, or G · U · G) to the alanine codons (G · C · U, G · C · C, G · C · A, or G · C · G, respectively).

Substitution of Amino Acids Causes Missense Mutations

Acceptable Missense Mutations: An example of an acceptable missense mutation (Fig 40–5, top) in the structural gene for the β chain of hemoglobin could be detected by the presence of an electrophoretically altered hemoglobin in the red cells of an apparently healthy individual. Hemoglobin Hikari has been found in at least 2 families of Japanese people. This hemoglobin has asparagine substituted for lysine at the 61

position in the β chain. The corresponding transversion might be either A · A · A or A · A · G changed to either A · A · U or A · A · C. The replacement of the specific lysine with asparagine apparently does not alter the normal function of the β chain in these individuals.

Partially Acceptable Missense Mutations: A partially acceptable missense mutation (Fig 40–5, center) is best exemplified by **hemoglobin S,** sickle hemoglobin, in which the normal amino acid in position 6 of the β chain, glutamic acid, has been replaced by valine. The corresponding single nucleotide change within the codon would be G · A · A or G · A · G of glutamic acid to G · U · A or G · U · G of valine. Clearly, this missense mutation hinders normal function and results in sickle cell anemia when the mutant gene is present in the homozygous state. The glutamate-to-valine-change may be considered to be partially acceptable because hemoglobin S does bind and release oxygen, although abnormally.

Unacceptable Missense Mutations: An unacceptable missense mutation (Fig 40–5, bottom) in a hemoglobin gene generates a nonfunctioning hemoglobin molecule. For example, the hemoglobin M mutations generate molecules that allow the Fe^{2+} of the heme moiety to be oxidized to Fe^{3+}, producing methemoglobin. Methemoglobin cannot transport oxygen (see Chapter 7).

Frame Shift Mutations Result From the Deletion or Insertion of Nucleotides in the Gene and Thus Generate Altered Nucleotide Sequences of mRNA Molecules

The deletion of a single nucleotide from the coding strand of a gene results in an altered reading frame in

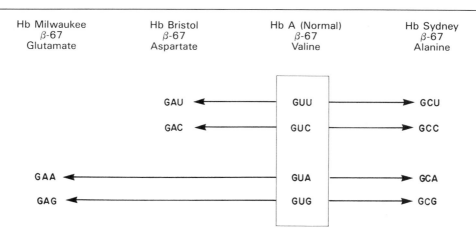

Figure 40–4. The normal valine at position 67 of the β chain of hemoglobin A can be coded for by one of the 4 codons shown in the box. In abnormal hemoglobin Milwaukee, the amino acid at position 67 of the β chain contains glutamate, coded for by G·A·A or G·A·G, either one of which could have resulted from a single-step transversion from the valine codons G·U·A or G·U·G. Similarly, the alanine present at position 67 of the β chain of hemoglobin Sydney could have resulted from a single-step transition from any one of the 4 valine codons. However, the aspartate residue at position 67 of hemoglobin Bristol could have resulted from a single-step transversion only from the G·U·U or G·U·C valine codons.

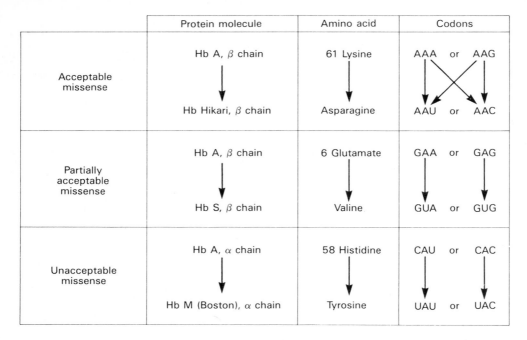

	Protein molecule	Amino acid	Codons
Acceptable missense	Hb A, β chain ↓ Hb Hikari, β chain	61 Lysine ↓ Asparagine	AAA or AAG (crossed) AAU or AAC
Partially acceptable missense	Hb A, β chain ↓ Hb S, β chain	6 Glutamate ↓ Valine	GAA or GAG ↓ ↓ GUA or GUG
Unacceptable missense	Hb A, α chain ↓ Hb M (Boston), α chain	58 Histidine ↓ Tyrosine	CAU or CAC ↓ ↓ UAU or UAC

Figure 40–5. Examples of 3 types of missense mutations resulting in abnormal hemoglobin chains. The amino acid alterations and possible alterations in the respective codons are indicated. The hemoglobin Hikari β-chain mutation has apparently normal physiologic properties but is electrophoretically altered. Hemoglobin S has a β-chain mutation and partial function; hemoglobin S binds oxygen but precipitates when deoxygenated. Hemoglobin M Boston, an α-chain mutation, permits the oxidation of the heme ferrous iron to the ferric state and so will not bind oxygen at all.

the mRNA. The machinery translating the mRNA does not recognize that a base was missing, since there is no punctuation in the reading of codons. Thus, a major alteration in the sequence of polymerized amino acids, as depicted in example 1, Fig 40–6, results. Altering the reading frame results in a garbled translation of the mRNA distal to the single nucleotide deletion. Not only is the sequence of amino acids distal to this deletion garbled, but reading of the message can also result in the appearance of a nonsense codon and thus the production of a polypeptide both garbled and prematurely terminated near its carboxyl terminus (example 3, Fig 40–6).

If 3 nucleotides or a multiple of 3 were deleted from a gene, the corresponding messenger when translated would provide a protein from which was missing the corresponding number of amino acids (example 2, Fig 40–6). Because the reading frame is a triplet, the reading phase would not be disturbed for those codons distal to the deletion. If, however, deletion of one or 2 nucleotides occurs just prior to or within the normal termination codon (nonsense codon), the reading of the normal termination signal is disturbed. Such a deletion might result in reading through a termination signal until another nonsense codon was encountered (example 1, Fig 40–6). Excellent examples of this phenomenon are described in discussions of hemoglobinopathies.

Insertions of one or 2 or nonmultiples of 3 nucleotides into a gene result in an mRNA in which the read-

ing frame is distorted upon translation, and the same effects that occur with deletions are reflected in the mRNA translation. This may be **garbled amino acid sequences** distal to the insertion, and the generation of a **nonsense codon** at or distal to the insertion, or perhaps **reading through** the normal termination codon. Following a deletion in a gene, an insertion (or vice versa) can reestablish the proper reading frame (example 4, Fig 40–6). The corresponding mRNA, when translated, would contain a garbled amino acid sequence between the insertion and deletion. Beyond the reestablishment of the reading frame, the amino acid sequence would be correct. One can imagine that different combinations of deletions, of insertions, or of deletions and insertions would result in formation of a protein wherein a portion is abnormal, but this portion is surrounded by the normal amino acid sequences. Such phenomena have been demonstrated convincingly in the bacteriophage T4, a finding which contributed significantly to evidence that the reading frame was a triplet.

Suppressor Mutations Reduce the Effects of Missense, Nonsense, and Frame Shift Mutations

The above discussion of the altered protein products of gene mutations is based on the presence of normally functioning tRNA molecules. However, in prokaryotic and lower eukaryotic organisms abnormally functioning tRNA molecules have been discovered that are

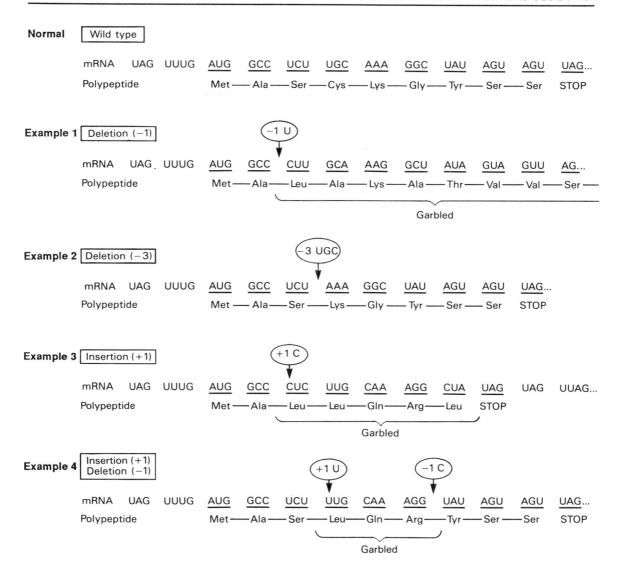

Figure 40–6. Demonstration of the effects of deletions and insertions in a gene on the sequence of the mRNA transcript and of the polypeptide chain translated therefrom. The arrows indicate the sites of deletions of insertions, and the numbers in the circles indicate the number of nucleotide residues deleted or inserted.

themselves the results of mutations. Some of these abnormal tRNA molecules are capable of suppressing the effects of mutations in distant structural genes. These **suppressor tRNA molecules,** usually formed as the result of alterations in their anticodon regions, are capable of suppressing missense mutations, nonsense mutations, and frame shift mutations. However, since the suppressor tRNA molecules are not capable of distinguishing between a normal codon and one resulting from a gene mutation, their presence in a cell usually results in decreased viability. For instance, the nonsense suppressor tRNA molecules can suppress the normal termination signals to allow a read-through when it is not desirable. Frame shift suppressor tRNA molecules may read a normal codon plus a component of a juxtaposed codon to provide a frame shift, also

when it is not desirable. Suppressor tRNA molecules may exist in mammalian cells, since read-through transcription occurs.

THE PROCESS OF PROTEIN SYNTHESIS, LIKE THAT OF DNA REPLICATION AND GENE TRANSCRIPTION, CAN BE DESCRIBED IN 3 PHASES: INITIATION, ELONGATION, AND TERMINATION

The general structural characteristics of ribosomes and their self-assembly process have been discussed in Chapter 39. These particulate entities serve as the machinery on which the mRNA nucleotide sequence is translated into the sequence of amino acids of the specified protein. The translation of the mRNA commences near its 5′-terminus with the formation of the corresponding amino terminus of the protein molecule. The message is read from 5′ to 3′, concluding with the formation of the carboxyl terminus of the protein. Again, the concept of **polarity** is apparent. As described in Chapter 39, the transcription of a gene into the corresponding mRNA or its precursor first forms the 5′ terminus of the RNA molecule. In prokaryotes, this allows for the beginning of mRNA translation before the transcription of the gene is completed. In eukaryotic organisms, the process of transcription is a nuclear one; mRNA translation occurs in the cytoplasm. This precludes simultaneous transcription and translation in eukaryotic organisms and makes possible the processing necessary to generate mature mRNA from the primary transcript— hnRNA.

Initiation Involves a Complex of Structures and Proteins (See Fig 40–7.)

The 5′ termini of most mRNA molecules in eukaryotes are "capped" as described in Chapter 39. This methyl-guanosyl triphosphate cap facilitates the binding of mRNA molecules to the 40S ribosomal subunit. A number of mRNA binding proteins, eIF-4A, eIF-4B, eIF-4E, and eIF-4F, modify mRNA before it binds to the 40S ribosome. Presence of the cap-binding complex, eIF-4F, consisting of four polypeptides (24 kd, 46 kd, 73 kd, and 220 kd), is necessary for protein synthesis to be initiated from the capped mRNAs characteristic of eukaryotic organisms. The first codon to be translated is usually A · U · G. The 18S ribosomal RNA (rRNA) of the 40S ribosomal subunit binds to a region of the mRNA that precedes the first translated codon. This binding of the mRNA to the 40S ribosomal subunit requires the presence of a protein factor, **initiation factor 3 (IF-3).**

The aminoacyl-tRNA called for by the first codon then interacts with GTP and **initiation factor 2 (IF-2)** to form a complex. This complex in the presence of **initiation factor 1 (IF-1)** attaches the anticodon of the tRNA to the first codon of the message to form an initiation complex with the 40S ribosomal subunit. Upon release of the initiation factors (IF-1, IF-2, and IF-3), the 60S ribosomal subunit attaches and the GTP is hydrolyzed. The formation of the 80S ribosome is thus complete.

The complete ribosome contains 2 sites for tRNA molecules. The peptidyl site (**P site**) contains the peptidyl-tRNA attached to its codon on the mRNA. The aminoacyl site (**A site**) contains the aminoacyl-tRNA attached to its respective codon on the mRNA. With the formation of the initiation complex for the first codon, the aminoacyl-tRNA molecule enters at what will become the P site, leaving the A site free. Thus, the reading frame is defined by attachment of the

tRNA to the first codon to be translated in the mRNA. The recognition of this specific initiating codon is apparently dependent upon the secondary structure of the mRNA molecule and in addition involves in prokaryotes, and perhaps in eukaryotes, a specific sequence of nucleotides complementary to a segment of the 16S (18S) ribosomal RNA.

In prokaryotes, a specific aminoacyl-tRNA is involved in the initiation of synthesis of most, if not all, protein molecules. N-Formylmethionyl-tRNA initiates most proteins in prokaryotes. Although methionine is the N-terminal amino acid in many eukaryotic proteins, the methionyl-tRNA is not formylated in eukaryotes. In prokaryotes, the N-formylation of the methionyl moiety on the tRNA seems to deceive the P site of the ribosome by appearing to be a peptide bond. There exists also in prokaryotes an enzyme capable of removing the N-terminal formyl moiety or N-terminal methionyl residue (or both) from proteins, in many cases even before the complete protein molecule has been formed.

Elongation Is by Translocation (See Fig 40–8.)

In the complete 80S ribosome formed during the process of initiation, the A site is free. The binding of the proper aminoacyl-tRNA in the A site requires proper codon recognition. **Elongation factor 1 (EF-1)** forms a complex with GTP and the entering aminoacyl-tRNA (Fig 40–8). This complex then allows the aminoacyl-tRNA to enter the A site with the release of EF-1 · GDP and phosphate. As shown in Fig 40–8, EF-1 · GDP then recycles to EF-1 · GTP with the aid of other soluble protein factors and GTP.

The α-amino group of the new aminoacyl-tRNA in the A site carries out a nucleophilic attack on the esterified carboxyl group of the peptidyl-tRNA occupying the P site. This reaction is catalyzed by a protein component, **peptidyltransferase,** of the 60S ribosomal subunit. Because the amino acid on the aminoacyl-tRNA is already "activated," no further energy source is required for this reaction. The reaction results in attachment of the growing peptide chain to the tRNA in the A site.

Upon removal of the peptidyl moiety from the tRNA in the P site, the discharged tRNA quickly dissociates from the P site. **Elongation factor 2 (EF-2)** and GTP are responsible for the **translocation** of the newly formed peptidyl-tRNA at the A site into the empty P site. The GTP required for EF-2 is hydrolyzed to GDP and phosphate during the translocation process. The translocation of the newly formed peptidyl-tRNA and its corresponding codon into the P site then frees the A site for another cycle of aminoacyl-tRNA codon recognition and elongation.

The charging of the tRNA molecule with the aminoacyl moiety requires the hydrolysis of an ATP to an AMP, equivalent to the hydrolysis of 2 ATPs to 2 ADPs and phosphates. The entry of the aminoacyl-tRNA into

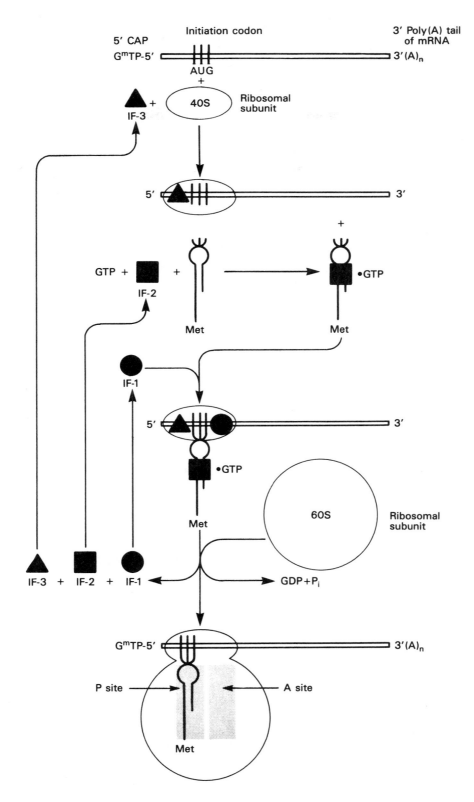

Figure 40–7. Diagrammatic representation of the initiation of protein synthesis on the mRNA template containing a 5′ cap and 3′ poly(A) terminus. IF-1, IF-2, and IF-3 represent initiation factor 1, initiation factor 2, and initiation factor 3, respectively, and the hairpinlike structure with Met at one end represents the methionyl-tRNA. The P site and the A site represent the peptidyl-tRNA and aminoacyl-tRNA binding sites of the ribosome, respectively.

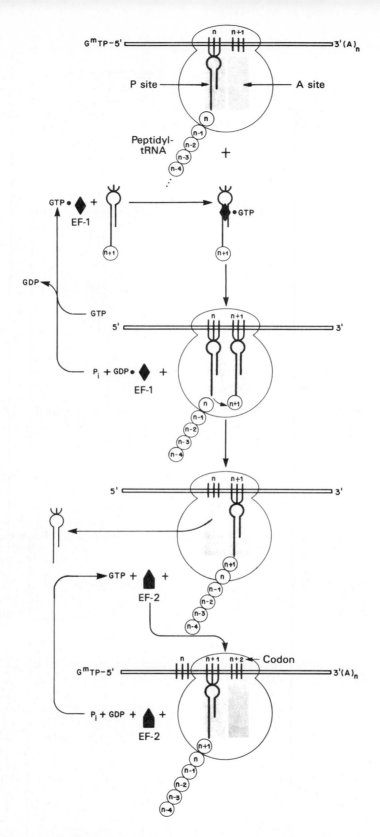

Figure 40–8. Diagrammatic representation of the peptide elongation process of protein synthesis. The small circles labeled n − 1, n, n + 1, etc represent the amino acid residues of the newly formed protein molecule. EF-1 and EF-2 represent elongation factors 1 and 2, respectively. The peptidyl-tRNA and aminoacyl-tRNA sites on the ribosome are represented by P site and A site, respectively.

the A site results in the hydrolysis of one GTP to GDP. The translocation of the newly formed peptidyl-tRNA in the A site into the P site by EF-2 similarly results in the hydrolysis of GTP to GDP and phosphate. Thus, the energy requirements for the formation of one peptide bond include the equivalent of the hydrolysis of 2 ATP molecules to ADP and 2 GTP molecules to GDP, or the hydrolysis of 4 high-energy phosphate bonds.

Termination Occurs When a Nonsense Codon Is Recognized (See Fig 40–9.)

After multiple cycles of elongation culminating in polymerization of the specific amino acids into a protein molecule, the nonsense or terminating codon of mRNA appears in the A site. Normally, there is no tRNA with an anticodon capable of recognizing such a termination signal. **Releasing factors** are capable of recognizing that a termination signal resides in the A site (Fig 40–9). The releasing factor, in conjunction with GTP and the peptidyl transferase, promotes the hydrolysis of the bond between the peptide and the tRNA occupying the P site. This hydrolysis releases the protein and the tRNA from the P site. Upon hydrolysis and release, the **80S ribosome dissociates** into its 40S and 60S subunits, which are then recycled. Therefore, the releasing factors are proteins that hydrolyze the peptidyl-tRNA bond when a nonsense codon occupies the A site.

Many ribosomes can translate the same mRNA molecule simultaneously. Because of their relatively large size, the ribosome particles cannot attach to an mRNA any closer than 80 nucleotides apart. Multiple ribosomes on the same mRNA molecule form a **polyribosome,** or "polysome." In an unrestricted system, the number of ribosomes attached to an mRNA (and thus the size of polyribosomes) correlates positively with the length of the mRNA molecule. The mass of the mRNA molecule is, of course, quite small compared to the mass of even a single ribosome.

A single mammalian ribosome is capable of synthesizing about 100 peptide bonds each minute. Polyribosomes actively synthesizing proteins can exist as free particles in the cellular cytoplasm or may be attached to sheets of membranous cytoplasmic material referred to as **endoplasmic reticulum.** The attachment of the particulate polyribosomes to the endoplasmic reticulum is responsible for its "rough" appearance as seen by electron microscopy. The proteins synthesized by the attached polyribosomes are extruded into the cisternal space between the sheets of rough endoplasmic reticulum and are exported from there. Some of the protein products of the rough endoplasmic reticulum are packaged by the Golgi apparatus into zymogen particles for eventual export (see Chapter 42). The polyribosomal particles free in the cytosol are responsible for the synthesis of proteins required for intracellular functions.

The Complex Machinery of Protein Synthesis Can Be Exploited to Respond to Environmental Threats or It Can Be Coopted To Become a Part of a Disease Mechanism

Ferritin, an iron-binding protein, prevents ionized iron (Fe^{2+}) from reaching toxic levels within cells. Iron stimulates ferritin synthesis by activating one or more cytoplasmic proteins that then can bind to a specific region in the 5′ nontranslated region of ferritin mRNA. This protein-mRNA interaction activates ferritin mRNA and results in its translation. This mechanism provides for rapid control of the synthesis of a protein that sequesters Fe^{2+}, a potentially toxic molecule.

The protein synthesis machinery can also be modified in deleterious ways. Viruses replicate by using host cell processes, including those involved in protein synthesis. Some viral mRNAs are translated much more efficiently than those of the host cell (eg, mengovirus and encephalomycarditis virus). Others, such as reovirus and vesicular stomatitis virus, replicate abundantly and their mRNAs have a competitive advantage over host cell mRNAs for limited translation factors. Other viruses inhibit host cell protein synthesis by preventing the association of mRNA with the 40S ribosome. Polio virus accomplishes this by activating a cellular protease that degrades the 220-kd component of the eIF-4F cap-binding complex described above.

POSTTRANSLATIONAL PROCESSING AFFECTS THE ACTIVITY OF MANY PROTEINS

Some animal viruses, notably poliovirus (an RNA virus), synthesize long polycistronic proteins from one long mRNA molecule. These protein molecules are subsequently cleaved at specific sites to provide the several specific proteins required for viral function. In animal cells, many proteins are synthesized from the mRNA template as a precursor molecule, which then must be modified to achieve the active protein. The prototype is insulin, which is a low-molecular-weight protein having 2 polypeptide chains with interchain and intrachain disulfide bridges. The molecule is synthesized as a single chain precursor, or **prohormone,** which folds to allow the disulfide bridges to form. A specific protease then clips out the segment that connects the 2 chains that form the functional insulin molecule (see Fig 52–3).

Many other peptides are synthesized as proproteins that require modification before attaining biologic activity. Many of the posttranslational modifications involve the removal of N-terminal amino acid residues by specific aminopeptidases. Collagen, an abundant protein in the extracellular spaces of higher eukaryotes, is synthesized as procollagen. Three procollagen polypeptide molecules, frequently not identical in se-

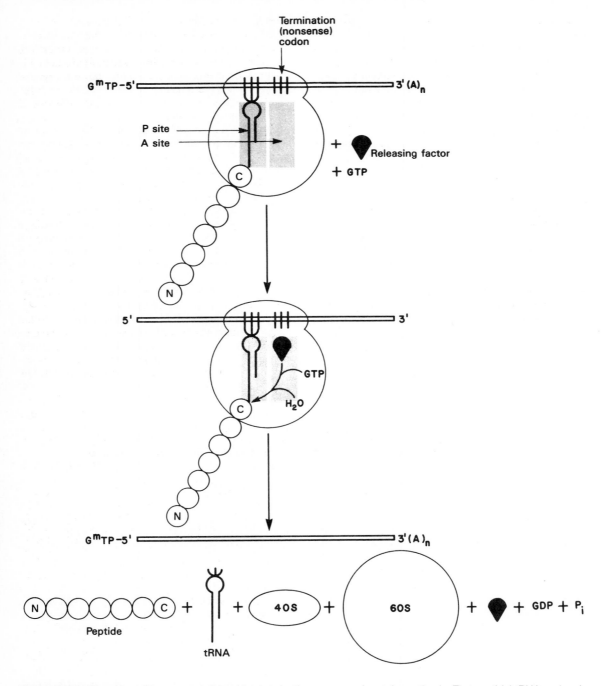

Figure 40–9. Diagrammatic representation of the termination process of protein synthesis. The peptidyl-tRNA and aminoacyl-tRNA sites are indicated as P site and A site, respectively. The hydrolysis of the peptidyl-tRNA complex is shown by the entry of H_2O. N and C indicate the NH_2- and carboxy-terminal amino acids, respectively, and illustrate the polarity of protein synthesis.

quence, align themselves in a way dependent upon the existence of specific amino-terminal peptides. Specific enzymes then carry out hydroxylations and oxidations of specific amino acid residues within the procollagen molecules to provide cross links for greater stability. Amino-terminal peptides are cleaved off the molecule to form the final product, a strong, insoluble collagen molecule (see Chapter 58). Many other posttranslational modifications of proteins occur. Covalent modification by acetylation, phosphorylation, and glycosylation is common, for example.

MANY ANTIBIOTICS WORK BECAUSE THEY SELECTIVELY INHIBIT PROTEIN SYNTHESIS IN BACTERIA

Ribosomes in bacteria and in the mitochondria of higher eukaryotic cells differ from the mammalian ribosome described in Chapter 37. The bacterial ribosome is smaller (70S rather than 80S) and has a different, somewhat simpler, complement of RNA and protein molecules. This difference has been exploited for clinical purposes, because many effective antibiotics interact specifically with the proteins of prokaryotic ribosomes and thus inhibit protein synthesis. This results in growth arrest or death of the bacterium. The most useful of this class of antibiotics (eg, tetracyclines, lincocin, erythromycin, and chloramphenicol) do not interact with the specific proteins of eukaryotic ribosomal particles and are thus not toxic to eukaryotes.

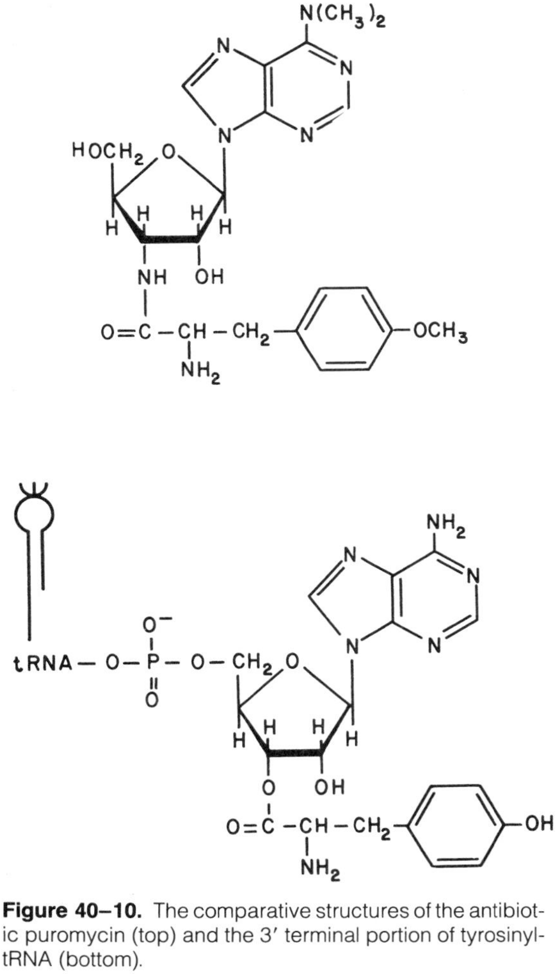

Other antibiotics inhibit protein synthesis on all ribosomes (**puromycin**) or only on those of eukaryotic cells (**cycloheximide**). Puromycin, the structure of which is shown in Fig 40–10, is a structural analog of tyrosinyl-tRNA. Puromycin is incorporated via the A site on the ribosome into the carboxy-terminal position of a peptide but causes the premature release of the polypeptide. Puromycin, as a tyrosinyl-tRNA anlog, effectively inhibits protein synthesis in both prokaryotes and eukaryotes.

Diphtheria toxin, an exotoxin of *Corynebacterium diphtheriae* infected with a specific lysogenic phage, catalyzes the ADP ribosylation of EF-2 in mammalian cells. This modification inactivates EF-2 and thereby specifically inhibits mammalian protein synthesis. Many animals (eg, mice) are resistant to diphtheria toxin. This resistance is due to inability of diphtheria toxin to cross the cell membrane rather than to insensitivity of mouse EF-2 to diphtheria toxin-catalyzed ADP ribosylation by NAD.

Many of these compounds, puromycin and cycloheximide in particular, are not clinically useful but have been important in elucidating the role of protein synthesis in the regulation of metabolic processes, particularly enzyme induction by hormones.

Figure 40–10. The comparative structures of the antibiotic puromycin (top) and the 3' terminal portion of tyrosinyl-tRNA (bottom).

REFERENCES

Banerjee AK: 5'-Terminal cap structure in eucaryotic messenger ribonucleic acids. *Microbiol Rev* 1980;**44**:175.

Barrell BG et al: Different pattern of codon recognition by mammalian mitochonrial tRNAs. *Proc Natl Acad Sci USA* **77**:3164.

Caskey CT: Peptide chain termination. *Trends Biochem Sci* 1980;**5**:234.

Drake JW, Baltz RH: The biochemistry of mutagenesis. *Annu Rev Biochem* 1976;**45**:11.

Forget BG: Molecular genetics of human hemoglobin synthesis. *Ann Int Med* 1979;**91**:605.

Kozak M: Comparison of initiation of protein synthesis in procaryotes, eucaryotes and organelles. *Microbiol Rev* 1983;**47**:1.

Maitra U, Stringer EA, Chaudhuri, A: Initiation factors in protein biosynthesis. *Annu Rev Biochem* 1982;**51**:869.

Pelletier J, Sonenberg N: Internal initiation of translation of eukaryotic mRNA directed by a sequence derived from picornavirus mRNA. *Nature (London)* 1988;**334**:320.

Schlessinger D: Genetic and antibiotic modification of protein synthesis. *Annu Rev Genet* 1974;**8**:135.

Schneider RJ, Shenk T: Impact of virus infection on host cell protein synthesis. *Annu Rev Biochem* 1987;**56**:317.

Shatkin AJ: mRNA cap binding proteins: Essential factors for initiating translation. *Cell* 1985;**40**:223.

Wool I: The structure and function of eukaryotic ribosomes. *Annu Rev Biochem* 1979;**48**:719.

41

Regulation of Gene Expression

Daryl K. Granner, MD

INTRODUCTION

Organisms adapt to environmental changes by altering gene expression. The process of gene expression alteration has been studied in detail in bacteria and viruses, and it generally involves the interaction of specific binding proteins with various regions of DNA in the immediate vicinity of the transcription start site. This can have a positive or negative effect on transcription. Eukaryotic cells use this basic paradigm but employ other mechanisms as well to regulate transcription. Such processes as enhancement/silencing; tissue-specific expression; regulation by hormones, metals, and chemicals; gene amplification; gene rearrangement; and posttranscriptional modifications are also used to control gene expression.

BIOMEDICAL IMPORTANCE

Many of the mechanisms that control gene expression are used to respond to hormones and therapeutic agents. An understanding of these processes may lead to development of agents that inhibit the function or arrest the growth of pathogenic organisms.

REGULATED EXPRESSION OF GENES IS REQUIRED FOR DEVELOPMENT, DIFFERENTIATION AND ADAPTATION

The genetic information present in each somatic cell of a metazoan organism is practically identical. The exceptions are found in those few cells that have amplified or rearranged genes in order to carry out specialized cellular functions. The expression of the genetic information must be regulated during ontogeny and differentiation of the organism and its cellular components. Furthermore, in order for the organism to adapt to its environment and to conserve energy and nutrients, the expression of genetic information must be responsive to extrinsic signals. As organisms have evolved, more sophisticated regulatory mechanisms have appeared to provide the organism and its cells with the responsiveness necessary for survival in its complex environment. Mammalian cells possess about 1000 times more genetic information than does the bacterium *Escherichia coli*. Much of this additional genetic information is probably involved in the regulation of gene expression during the differentiation of tissues and biologic processes in the multicellular organism and to ensure that the organism can respond to complex environmental challenges.

In simple terms, there are only 2 types of gene regulation: **positive regulation** and **negative regulation** (Table 41–1). When the expression of genetic information is quantitatively **increased** by the presence of a specific regulatory element, regulation is said to be **positive;** whereas when the expression of genetic information is **diminished** by the presence of a specific regulatory element, regulation is said to be **negative.** The element or molecule mediating the negative regulation is said to be a negative regulator; that mediating positive regulation is a positive regulator. However, a **double negative** has the effect of acting as a **positive.** Thus, an effector that inhibits the function of a negative regulator will appear to bring about a positive regulation. In many regulated systems that appear to be induced, they are, in fact, derepressed at the molecular level. (See Chapter 11 for a description of these terms.)

IN BIOLOGIC SYSTEMS, THERE ARE 3 TYPES OF TEMPORAL RESPONSES TO A REGULATORY SIGNAL

These 3 responses are depicted diagrammatically in Fig 41–1 as rate of gene expression in temporal response to an inducing signal.

A **type A response** is characterized by an increased rate of gene expression that is **dependent** upon the continued presence of the inducing signal. When the inducing signal is removed, the rate of gene expression

Table 41–1. Effects of positive and negative regulation on gene expression.

	Rate of Gene Expression	
	Negative Regulation	**Positive Regulation**
Regulator present	Decreased	Increased
Regulator absent	Increased	Decreased

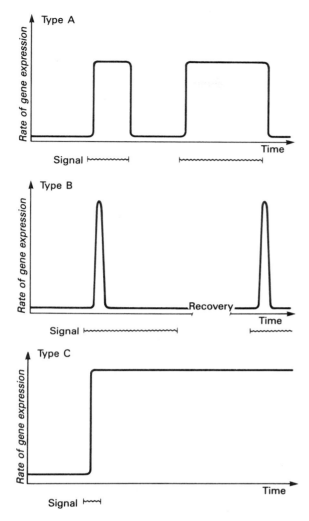

Figure 41–1. Diagrammatic representations of the responses of the rate of expression of a gene to specific regulatory signals such as a hormone.

diminishes to its basal level, but the rate repeatedly increases in response to the reappearance of the specific signal. This type of response is commonly observed in many higher organisms after exposures to inducers such as steroid hormones (see Chapter 45).

A **type B response** exhibits an increased rate of gene expression that is **transient** even in the continued presence of the regulatory signal. After the regulatory signal has terminated and the cell has been allowed to recover, a second transient response to a subsequent regulatory signal may be observed. This phenomenon of response–desensitization–recovery characterizes the action of many pharmacologic agents, but it is also a feature of many naturally occurring processes. This type of response may commonly occur during development of an organism when only the transient appearance of a specific gene product is required although the signal persists.

The **type C response** pattern exhibits, in response to the regulatory signal, an increased rate of gene expression that persists **indefinitely** even after the termination of the signal. The signal acts as a trigger in this pattern. Once the gene expression is initiated in the cell, it cannot be terminated even in the daughter cells; it is therefore an irreversible and inherited alteration.

Prokaryotes Provide Models for the Study of the Regulation of Gene Expression in Mammalian Cells

In the last 20 years, with the understanding of how information flows from the gene through a messenger RNA to a specific protein molecule, there has developed sophisticated knowledge of the regulation of gene expression in prokaryotic cells. Most of the detailed knowledge about molecular mechanisms has been limited until recent years to prokaryotic and lower eukaryotic systems. This was due to the more advanced genetic analyses first available in these primitive organisms. Recent advances in recombinant DNA technology have allowed the sophisticated analysis of mammalian gene expression to begin. In this chapter, the initial discussion will center on prokaryotic systems. The impressive genetic studies will not be described, but rather what may be termed the physiology of gene expression will be discussed. However, nearly all of the conclusions about this physiology have been derived from genetic studies.

Before the physiology can be explained, a few specialized genetic terms must be defined for prokaryotic systems.

The **cistron** is the smallest unit of genetic expression. As described in Chapter 11, some enzymes and other protein molecules are composed of 2 or more nonidentical subunits. Thus, the "one gene, one enzyme" concept is now known to be not necessarily valid. The cistron is the genetic unit coding for the structure of the subunit of a protein molecule, acting as it does as the smallest unit of genetic expression. Thus, the one gene, one enzyme idea might more accurately be regarded as a **one cistron, one subunit concept.**

An **inducible gene** is a gene whose expression increases in response to an **inducer,** a specific regulatory signal.

The expression of some genes is **constitutive,** meaning that they are expressed at a reasonably constant rate and not known to be subject to regulation. As the result of mutation, some inducible gene products become constitutively expressed. A mutation resulting in constitutive expression of what was formerly a regulated gene is called a constitutive mutation.

ANALYSIS OF LACTOSE METABOLISM IN *E COLI* LED TO THE OPERON HYPTHESIS

François Jacob and Jacques Monod in 1961 described their **operon** model in a classic paper. Their hypothesis was to a large extent based on observations on the regulation of lactose metabolism by the intestinal bacterium *Escherichia coli*. The molecular mechanisms responsible for the regulation of the genes involved in the metabolism of lactose are now among the best understood in any organism. β-Galactosidase hydrolyzes the β-galactoside lactose to galactose and glucose (Fig 41–2). The structural gene for β-galactosidase (the lac Z gene) is clustered with the genes responsible for the permeation of galactose into the cell (Y) and for galactoside acetylase (A), whose function is not understood. The structural genes for these 3 enzymes are physically associated to constitute the **lac operon** as depicted in Fig 41–3. This genetic arrangement of the structural genes and their regulatory genes allows for the **coordinate expression** of the 3 enzymes concerned with lactose metabolism. Each of these linked genes is transcribed into one large mRNA molecule that contains multiple, independent translation start (AUG) and stop (UAA) codons for each cistron. This type of mRNA molecule is called a **polycistronic mRNA.** Polycistronic mRNAs are predominantly found in prokaryotic organisms.

When *E coli* are presented with lactose or some specific lactose analogs, the expression of the activities of β-galactosidase, galactoside permease, and galactoside acetylase is increased 10-fold to 100-fold. This is a type A response, as depicted in Fig 41–1. Upon removal of the signal, ie, the inducer, the rate of synthesis of these 3 enzymes declines. Since there is no significant degradation of these enzymes in bacteria, the level of β-galactosidase as well as that of the other 2 enzymes will remain the same unless they are diluted out by cell division.

When *E coli* are exposed to both lactose and glucose as sources of carbon, the organisms first metabolize the glucose and then temporarily cease growing until the genes of the lac operon become induced to provide the ability to metabolize lactose. Although lactose is present from the beginning of the bacterial growth phase, the cell does not induce those enzymes necessary for catabolism of lactose until the glucose has been exhausted. This phenomenon was first thought to

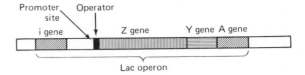

Figure 41–3. The positional relationships of the structural and regulatory genes of the lac operon. The Z gene encodes β-galactosidase, the Y gene encodes a permease, and the A gene encodes an acetylase. The i gene encodes the lac operon repressor protein.

be attributable to the repression of the lactose operon by some catabolite of glucose; hence, it was termed catabolite repression. It is now known that "catabolite repression" is in fact mediated by a **catabolite gene activator protein (CAP)** in conjunction with **cyclic AMP (cAMP**; see p. 174). The expression of many inducible enzyme systems or operons in *E coli* and other prokaryotes is sensitive to catabolite repression, as discussed below.

The physiology of the induction of the lac operon is well understood at the molecular level (Fig 41–4). The expression of the normal **i gene** of the lac operon is constitutive; it is expressed at a constant rate, resulting in the formation of the subunits of the **lac repressor.** Four identical subunits of MW 38,000 assemble into a lac repressor molecule. The repressor protein molecule, the product of the i gene, has a high affinity (K_d about 10^{-12} mol/L) for the operator locus. The **operator locus** is a region of double-stranded DNA 27 base pairs long with a 2-fold rotational symmetry (indicated by solid lines about the dotted axis) in a region that is 21 base pairs long, as shown below:

$$
\begin{array}{l}
: \\
5'\text{-}\underline{\text{AAT TGTGAGC}}\ \text{G}\ \underline{\text{GATAACAATT}} \\
3'\text{-}\underline{\text{TTA ACACTCG}}\ \text{C}\ \underline{\text{CTATTGTTAA}} \\
:
\end{array}
$$

The minimum effective size of an operator for lac repressor binding is 17 base pairs (boldface letters in

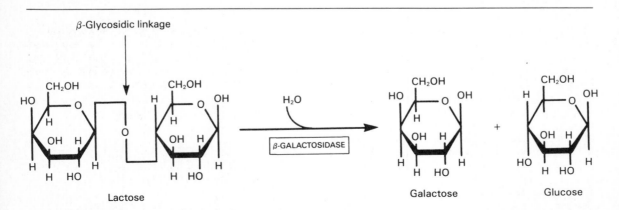

Figure 41–2. The hydrolysis of lactose to galactose and glucose by the enzyme β-galactosidase.

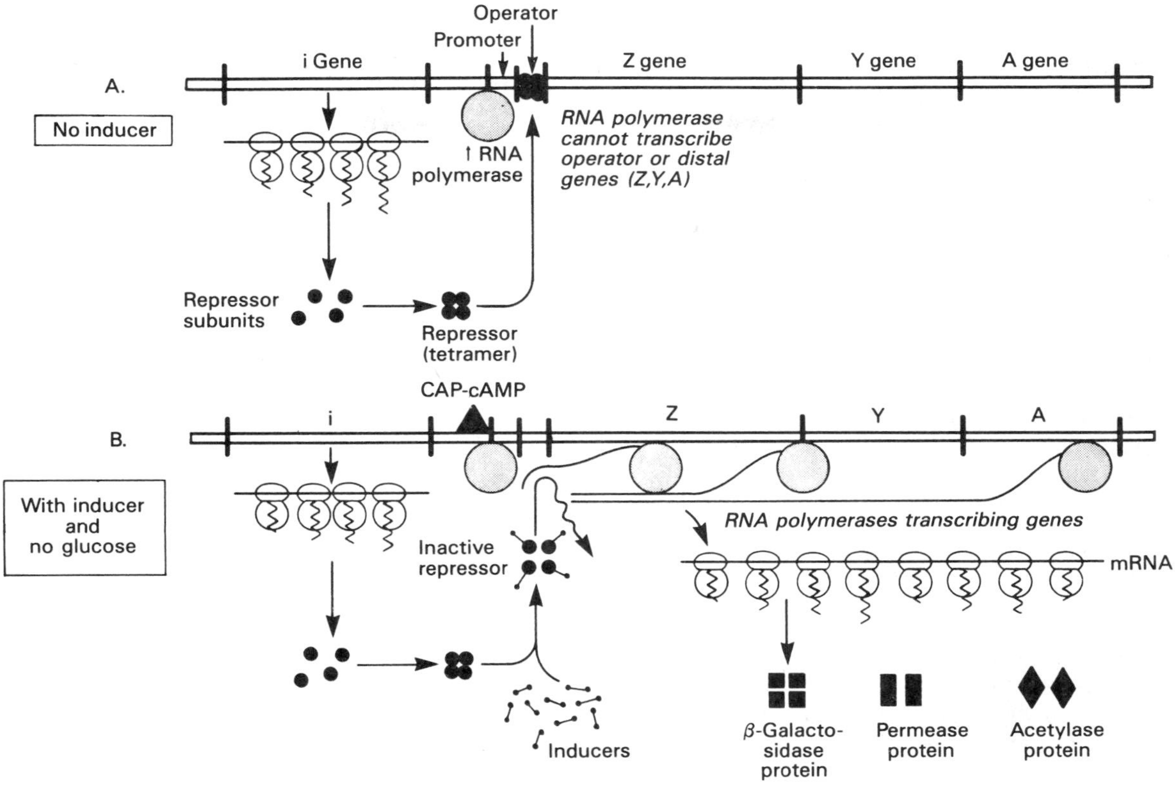

Figure 41–4. The mechanism of repression and derepression of the lactose operon. When no inducer is present (*A*), the i gene products that are synthesized constitutively form a repressor molecule which binds at the operator locus to prevent the binding of RNA polymerase at the promoter locus and thus to prevent the subsequent transcription of the Z, Y, and A structural genes. When inducer is present (*B*), the constitutively expressed i gene forms repressor molecules that are inactivated by the inducer and cannot bind to the operator locus. In the presence of cAMP and its binding protein (CAP), the RNA polymerase can transcribe the structural genes Z, Y, and A, and the polycistronic mRNA molecule formed can be translated into the corresponding protein molecules β-galactosidase, permease, and acetylase, allowing for the catabolism of lactose.

above sequence). At any one time, only 2 subunits of the repressors appear to bind to the operator, and within the 17-base-pair region at least one base of each base pair is involved in the lac repressor recognition and binding. The binding occurs mostly in the **major groove** without interrupting the base-paired, double helical nature of the operator DNA. The **operator locus** is between the **promoter site,** at which the DNA-dependent RNA polymerase attaches to commence transcription, and the transcription initiation site of the **Z gene,** the structural gene for β-galactosidase (Fig 41–3). When attached to the operator locus, the repressor molecule prevents the transcription of the operator locus as well as of the distal structural genes, Z, Y, and A. Thus, the repressor molecule is a **negative regulator;** in its presence (and in the absence of inducer, see below) the expression of the Z, Y, and A genes is prevented. There are normally 20–40 repressor tetramer molecules and one operator locus per cell.

A lactose analog that is capable of inducing the lac operon while not itself serving as a substrate for β-galactosidase is an example of a **gratuitous inducer.** The addition of lactose or of a gratuitous inducer to bacteria growing on a poorly utilized carbon source (such as succinate) results in the prompt induction of the lac operon enzymes. Small amounts of the gratuitous inducer or of lactose are able to enter the cell even in the absence of permease. The repressor molecules, both those attached to the operator loci and those free in the cytosol, have a high affinity for the inducer. The binding of the inducer to a repressor molecule attached to the operator locus will induce a comformational change in the structure of the repressor and cause it to dissociate from the DNA. If DNA-dependent RNA polymerase has already attached to the coding strand at the promoter site, transcription will begin. The polymerase generates a polycistronic mRNA, the 5′ terminus of which is complementary to the template strand of the operator. In such a manner, an inducer derepresses the lac operon and allows the transcription of the structural genes for β-galactosidase, galactoside permease, and galactoside acetylase. The translation of the polycistronic mRNA can occur even before transcription is completed. Derepression of the lac operon allows the cell to synthesize the enzymes necessary to catabolize lactose as an energy source.

In order for the RNA polymerase to attach at the promoter site, there must also be present the catabolite gene activator protein (CAP) to which cAMP is bound. By an independent mechanism, the bacterium accumulates cAMP only when it is starved for a source of carbon. In the presence of glucose, or glycerol in concentrations sufficient for growth, the bacteria will lack sufficient cAMP to bind to CAP. Thus, in the presence of glucose or glycerol, cAMP-saturated CAP is lacking, so that the DNA-dependent RNA polymerase cannot initiate transcription of the lac operon. In the presence of the CAP-cAMP complex, which binds to DNA just upstream of the promoter site, transcription then occurs (Fig 41–4). Thus, the CAP-cAMP regulator is acting as a **positive regulator,** because its presence is required for gene expression. Hence, the lac operon is subject to both positive and negative regulation.

When the i gene has been mutated so that its product, the lac repressor, is not capable of binding to operator DNA, the organism will exhibit **constitutive expression** of the lac operon. In a contrary manner, an organism with an i gene mutation that prevents the binding of an inducer to the repressor will remain repressed even in the presence of the inducer molecule, because the inducer cannot bind to the repressor on the operator locus in order to derepress the operon.

Bacteria harboring mutations in their operator locus such that the operator sequence will not bind a normal repressor molecule constitutively express the lac operon genes.

The Genetic Switch of *Bacteriophage Lambda* (λ) Provides a Paradigm for Protein-DNA Interactions in Eukaryotic Cells

Some bacteria harbor viruses that can reside in a dormant state within the bacterial chromosome or can replicate within the bacterium and eventually lead to lysis and killing of the bacterial host. Some *E coli* harbor such a "temperate" virus, bacteriophage lambda (λ). When a lambda infects a sensitive *E coli,* it injects its 45,000-base-pair, double-stranded, linear DNA genome into the cell (Fig 41–5). Depending upon the nutritional state of the cell, the lambda DNA will either **integrate** into the host genome (**lysogenic pathway**) and remain dormant until activated (see below), or it will commence **replicating** until it has made about 100 copies of complete, protein-packaged virus at which point it effects lysis of its host (**lytic pathway**). The newly generated virus particles can then infect other sensitive hosts.

When integrated into the host genome in its dormant state, lambda will remain in such a state until activated by exposure of its lysogenic bacterial host to DNA-damaging agents. In response to such a noxious stimulus, the dormant bacteriophage becomes "induced"

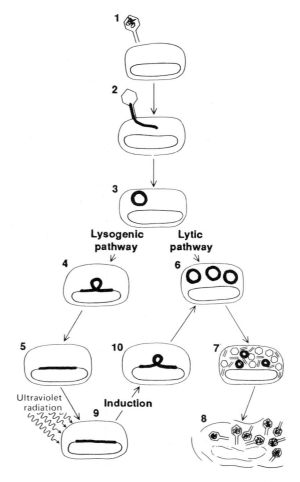

Figure 41–5. Infection of the bacterium *E coli* by phage lambda begins when a virus particle attaches itself to the bacterial cell (1) and injects its DNA (shaded line) into the cell (2, 3). Infection can take either of 2 courses depending on which of 2 sets of viral genes is turned on. In the lysogenic pathway, the viral DNA becomes integrated into the bacterial chromosome (4, 5), where it replicates passively as the bacterial cell divides. The dormant virus is called a prophage, and the cell that harbors it is called a lysogen. In the alternative lytic mode of infection, the viral DNA replicates itself (6) and directs the synthesis of viral proteins (7). About 100 new virus particles are formed. The proliferating viruses lyse, or burst, the cell (8). A prophage can be "induced" by an agent such as ultraviolet radiation (9). The inducing agent throws a switch, so that a different set of genes is turned on. Viral DNA loops out of the chromosome (10) and replicates; the virus proceeds along the lytic pathway. (Reproduced, with permission, from Ptashne M, Johnson AD, Pabo CO: A genetic switch in a bacterial virus. *Sci Am* [Nov] 1982;**247**:128.)

and begins to transcribe and subsequently translate those genes of its own genome which are necessary for its excision from the host chromosome, its DNA replication, and its protein coat and lysis enzymes. This event acts like a trigger or type C (Fig 41–1) response; that is, once lambda has committed itself to induction, there is no turning back until the cell is lysed and the replicated bacteriophage released. This **switch** from a dormant or **prophage state** to a **lytic infection** is well understood at the genetic and molecular levels and will be described in detail here.

The switching event in lambda is centered around an 80-base-pair region in its double-stranded DNA molecule referred to as the "right operator" (O_R) (Fig 41–6A). The **right operator** is flanked on its left side by the structural gene for the lambda repressor and on its right side by the structural gene for another regulatory protein called **cro**. When lambda is in its prophage state, ie, integrated into the host genome, the **repressor gene** is the *only* lambda gene that is expressed. When the bacteriophage is undergoing lytic growth, the repressor gene is not expressed, but the cro gene, as well as many other genes in lambda, is expressed. That is, when the **repressor gene is on,** the **cro gene is off,** and when the **cro gene is on,** the **repressor gene is off.** As we shall see, these 2 genes regulate each other's expression and thus, ultimately, the decision between lytic and lysogenic growth of lambda. **This decision between repressor gene transcription and cro gene transcription is an example of a molecular switch.**

The operator region can be subdivided into 3 discrete sites, each consisting of 17 base pairs of similar but not identical DNA sequence joined to one another (Fig 41–6B). Each of these 3 subregions, O_R1, O_R2, and O_R3, can bind either repressor or cro proteins predominantly through major groove contacts between

repressor and the DNA double helix. The DNA region between the cro and repressor genes also contains 2 promoter sequences that direct the binding of RNA polymerase in a specified orientation, where it commences transcribing the adjacent genes. One promoter directs RNA polymerase to transcribe in the **rightward direction** and, thus, to transcribe cro and other distal genes, while the other promoter directs the transcription of the **repressor** gene in the **leftward direction** (Fig 41–6B).

The product of the repressor gene, the 236-amino-acid **repressor protein,** exists as a **2-domain** molecule in which the **amino-terminal domain binds to operator DNA** and the **carboxy-terminal domain promotes the association** of one repressor protein with another to form a dimer. A **dimer** of repressor molecules binds to **operator DNA** much more tightly than does the monomeric form (Fig 41–7A to C).

The product of the cro gene, the 66-amino-acid **cro protein,** has a single domain but also binds the operator DNA more tightly as a **dimer** (Fig 41–7D). Obviously, the cro protein's single domain mediates both operator binding and dimerization.

In a lysogenic bacterium, ie, a bacterium containing a lambda prophage, the lambda repressor dimer binds **preferentially to O_R1** but in so doing, by a cooperative interaction, **enhances** the binding (by a factor of 10) of another repressor dimer to O_R2 (Fig 41–8). The affinity of repressor for O_R3 is the least of the 3 operator subregions. The binding of repressor to O_R1 has 2 major effects. The occupation of O_R1 by repressor **blocks the binding of RNA polymerase to the rightward promoter** and thereby prevents the expression of the cro gene. Second, as mentioned above, repressor dimer bound to O_R1 enhances the binding of repressor dimer to O_R2. The binding of repressor to O_R2 has the important added effect of **enhancing the**

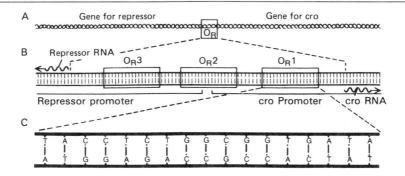

Figure 41–6. Right operator (O_R) is shown in increasing detail in this series of drawings. The operator is a region of the viral DNA some 80 base pairs long (*A*). To its left lies the gene encoding lambda repressor, to its right the gene (cro) encoding the regulator protein cro. When the operator region is enlarged (*B*), it is seen to include 3 subregions, O_R1, O_R2, and O_R3, each 17 base pairs long. They are recognition sites to which both repressor and cro can bind. The recognition sites overlap 2 promoters: sequences of bases to which the enzyme RNA polymerase binds in order to transcribe a gene into mRNA (wavy lines), which is translated into protein. Site O_R1 is enlarged (*C*) to show its base sequence. Note that in this region of the λ chromosome, both strands of DNA act as a template for transcription (see Chapter 39). (Reproduced, with permission, from Ptashne M, Johnson AD, Pabo CO: A genetic switch in a bacterial virus. *Sci Am* [Nov] 1982;**247**:128.)

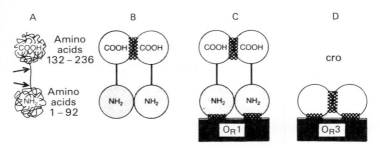

Figure 41–7. Schematic molecular structures of cI (lambda repressor, shown in A, B, and C) and cro. Lambda repressor protein is a polypeptide chain 236 amino acids long. The chain folds itself into a dumbbell shape with 2 substructures: an amino-terminal (NH_2) domain and a carboxy-terminal (COOH) domain. The 2 domains are linked by a region of the chain that is susceptible to cleavage by proteases (indicated by the 2 arrows in A). Single repressor molecules (monomers) tend to associate to form dimers (B); a dimer can dissociate to form monomers again. A dimer is held together mainly by contact between the carboxy-terminal domains (hatching). Repressor dimers bind to (and can fall off) the recognition sites in the operator region; their greatest affinity is for site O_R1 (C). It is the amino-terminal domain of the repressor molecule that makes contact with the DNA (hatching). Cro (D) has a single domain with sites that promote dimerization and other sites that promote binding of dimers to operator, preferentially to O_R3. (Reproduced, with permission, from Ptashne M, Johnson AD, Pabo CO: A genetic switch in a bacterial virus. *Sci Am* [Nov] 1982;**247**:128.)

binding of RNA polymerase to the leftward promoter that overlaps O_R2 and thereby enhances the transcription and subsequent expression of the repressor gene. This enhancement of transcription is apparently mediated through direct protein-protein interactions between promoter-bound RNA polymerase and O_R2-bound repressor. Thus, the lambda repressor is both a **negative regulator,** by preventing transcription of the cro gene, and a **positive regulator**, by enhancing the transcription of its own gene, the repressor gene. This dual effect of repressor is responsible for the stable state of the dormant lambda bacteriophage; not only does the repressor prevent the expression of the genes necessary for lysis, but it also promotes the expression of itself to stabilize this state of differentiation. In the event that the repressor protein concentration becomes very high, repressor can bind to O_R3 and by so doing diminish the transcription of the repressor gene from the leftward promoter, until the repressor concentration drops and repressor dissociates itself from O_R3.

When a DNA-damaging signal, such as ultraviolet light, strikes the lysogenic host bacterium, fragments of single-stranded DNA are generated that activate a specific **protease** coded by a bacterial gene and referred to as **recA** (Fig 41–8). The activated recA protease hydrolyzes the portion of the repressor protein that connects the amino-terminal and carboxy-terminal domains of that molecule. Such cleavage of the repressor domains causes the **repressor dimers to dissociate,** which in turn causes a **dissociation of the repressor molecules from O_R2** and eventually from O_R1. The effects of removal of repressor from O_R1 and O_R2 are predictable. RNA polymerase immediately has access to the rightward promoter and commences transcribing the **cro gene,** and the enhancement effect of the repressor at O_R2 on leftward transcription is lost (Fig 41–8).

The cro protein translated from the newly transcribed cro gene also binds to the operator region as dimers, but its order of preference is the opposite of that of repressor (Fig 41–8). That is, **cro binds most tightly to O_R3,** but there is no cooperative effect of cro at O_R3 on the binding of cro to O_R2. At increasingly higher concentrations of cro, the protein will bind to O_R2 and eventually to O_R1.

The occupancy of O_R3 by cro immediately turns off the transcription from the leftward promoter and, hence, **prevents any further expression of the repressor gene.** Thereby the switch is completely effected: the cro gene is now expressed, and the repressor gene is fully turned off. This event is irreversible, and the expression of other lambda genes begins as part of the lytic cycle. When cro repressor concentration becomes quite high, it will eventually occupy O_R1 and in so doing turn down the expression of its own gene, a process that is necessary in order to effect the final stages of the lytic cycle.

The 3-dimensional structure of the cro protein and that of the lambda repressor protein have been determined by x-ray crystallography, and models for their binding and effecting the above-described molecular and genetic events have been proposed and tested. **To date, this system provides the best understanding of the molecular events involved in gene regulation.**

GENE REGULATION IN PROKARYOTES AND EUKARYOTES DIFFERS IN SEVERAL IMPORTANT RESPECTS

In addition to transcription, eukaryotic cells employ a variety of mechanisms to regulate gene expression (see Table 41–2). The nuclear membrane of eukaryotic

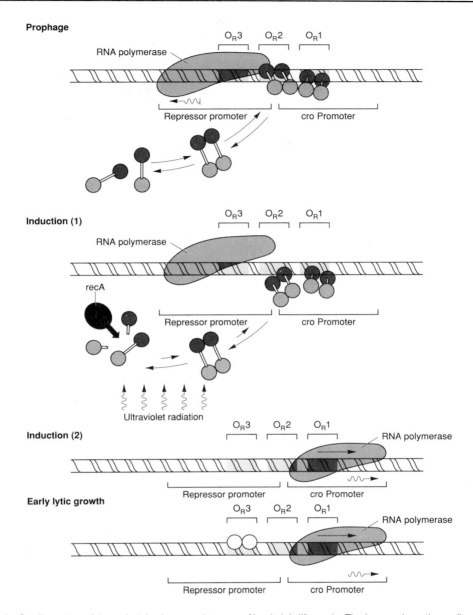

Figure 41–8. Configuration of the switch is shown at 4 stages of lambda's life cycle. The lysogenic pathway (in which the virus remains dormant as a prophage) is selected when a repressor dimer binds to O_R1, thereby making it likely that O_R2 will be filled immediately by another dimer. In the prophage (top), the repressor dimers bound at O_R1 and O_R2 prevent RNA polymerase from binding to the rightward promoter and so block the synthesis of cro (negative control). The repressors also enhance the binding of polymerase to the leftward promoter (positive control), with the result that the repressor gene is transcribed into RNA (wavy line) and more repressor is synthesized, maintaining the lysogenic state. The prophage is induced when ultraviolet radiation activates the protease recA, which cleaves repressor monomers. The equilibrium of free monomers, free dimers, and bound dimers is thereby shifted, and dimers leave the operator sites. Polymerase is no longer encouraged to bind to the leftward promoter, so that repressor is no longer synthesized. As induction proceeds, all the operator sites become vacant, and so polymerase can bind to the rightward promoter and cro is synthesized. During early lytic growth, a single cro dimer binds to O_R3, the site for which it has the highest affinity. Now polymerase cannot bind to the leftward promoter, but the rightward promoter remains accessible. Polymerase continues to bind there, transcribing cro and other early lytic genes. Lytic growth ensues. (Reproduced, with permission, from Ptashne M, Johnson AD, Pabo CO. A genetic switch in a bacterial virus. *Sci Am* [Nov] 1982;**247**:128.)

Table 41–2. Gene expression is regulated by transcription and in numerous other ways in eukaryotic cells.

Other Methods of Gene Regulation
Gene amplification
Gene rearrangement
RNA processing
Alternate mRNA splicing
Transport of mRNA from nucleus to cytoplasm
Regulation of mRNA stability

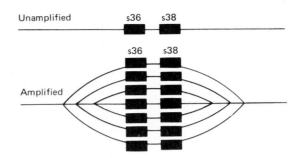

Figure 41–9. Schematic representation of the amplification of chorion protein genes s36 and s38. (Reproduced, with permission, from Chisholm R: Gene amplification during development. *Trends Biochem Sci* 1982;**7**:161.)

cells physically segregates gene transcription from translation, since ribosomes exist only in the cytoplasm. There are many more steps, especially in RNA processing, involved in the expression of eukaryotic genes than of prokaryotic genes, and these steps provide additional sites for regulatory influences that cannot exist in prokaryotes. These RNA processing steps in eukaryotes include capping of the 5′ end of the primary transcript, addition of a polyadenylate tail to the 3′ end of transcripts, and excision of intron regions to generate spliced exons in the mature mRNA molecule. To date, the analyses of eukaryotic gene expression have provided evidence that regulation occurs at the level of **transcription, nuclear RNA processing, and mRNA stability.** In addition, **gene amplification** and **rearrangement** have been shown to occur and to influence gene expression.

Owing to the advent of recombinant DNA technology, much progress has been made in recent years in the understanding of eukaryotic gene expression. However, because most eukaryotic organisms contain so much more genetic information than do prokaryotes and the manipulation of their genes is so much more limited, molecular aspects of eukaryotic gene regulation are less well understood than the examples discussed earlier in this chapter. This section briefly describes a few different types of eukaryotic gene regulation.

Eukaryotic Genes Can Be Amplified During Development, or in Response to Drugs

During early development of metazoans, there is an abrupt increase in the need for specific molecules such as ribosomal RNAs and messenger RNA molecules for proteins that make up such organs as the eggshell. One way to increase the rate at which such molecules can be formed is to increase the number of genes available for transcription of these specific molecules. Among the repetitive DNA sequences are hundreds of copies of ribosomal RNA genes and tRNA genes. These genes preexist repetitively in the genomic material of the gametes and, thus, are transmitted in high copy number from generation to generation. In some specific organisms such as the fruit fly (*Drosophila*), there occurs during oogenesis an amplification of a few preexisting genes, such as those for the chorion (eggshell) proteins. Subsequently, these amplified genes, presumably generated by a process of repeated initiations during DNA synthesis (compounded replication bub-

bles) provide multiple sites for gene transcription (Figs 38-15 and 41-9).

In recent years, it has been possible to promote the amplification of specific genetic regions in cultured mammalian cells. In some cases, a several thousandfold increase in the copy number of specific genes can be achieved over a period of time involving increasing doses of selective drugs. In fact, it has been demonstrated in patients receiving methotrexate for treatment of cancer that malignant cells can develop **drug resistance** by increasing the number of genes for dihydrofolate reductase, the target of methotrexate. Gene amplification events such as these occur spontaneously in vivo, ie, in the absence of exogenously supplied selective agents, and these unscheduled extra rounds of replication can become "frozen" in the genome under appropriate selective pressures.

The Formation of Active Immunoglobulin Genes Involves Selective DNA Rearrangement

Some of the most interesting and perplexing questions raised by biologists in recent decades concern the genetic and molecular basis of antibody diversity (see Chapter 55). In addition, advances in immunology have made it apparent that as cells of the humoral immunity system differentiate, they produce antibodies with the same specificity but different effector functions. Within the last several years, many laboratories have contributed greatly to the understanding of the genetic basis of antibody diversity and regulation of the expression of immunoglobulin genes during development and differentiation.

As described in Chapter 39, the coding segments responsible for the generation of specific protein molecules are frequently not contiguous in the mammalian genome. The coding segments for the variable and the constant domains of the immunoglobulin (antibody) light chain were the first recognized to be separated in the genome. As described in more detail in Chapter 55, immunoglobulin molecules are composed of 2 types of polypeptide chains, light (L) and heavy (H) chains (see Fig 55–6). The L and H chains are each divided into N-terminal variable (V) and carboxy-terminal constant (C) regions. The V regions are responsible for the recognition of antigens (foreign molecules) and the constant regions for effector functions that determine how the antibody molecule will dispense with the antigen.

There are 3 unlinked families of genes responsible for immunoglobulin molecule structure. Two families are responsible for the light chains (λ and κ chains) and one family for heavy chains.

Each **light chain** is encoded by 3 distinct segments: the variable (V_L), the joining (J_L), and the constant (C_L) segments. The mammalian haploid genome contains over 500 V_L segments, five or six J_L segments, and perhaps ten or twenty C_L segments. During the differentiation of a lymphoid B cell, a V_L segment is brought from a distant site on the same chromosome to a position closer to the region of the genome containing the J_L and C_L segments. This **DNA rearrangement** then allows the V_L, J_L, and C_L segments to be transcribed as a single mRNA precursor and subse-

quently processed to generate the mRNA for a specific antibody light chain. By rearrangement of the various V_L, J_L, and C_L segments in the genome, the immune system can generate an immensely diverse library (millions) of antigen-specific immunoglobulin molecules. This DNA rearrangement is referred to as **V-J joining** of the light chain.

The **heavy chain** is encoded by 4 gene segments: the V_H, the D (diversity), the J_H, and the C_H DNA segments. The variable region of the heavy chain is generated by joining the V_H with a D and a J_H segment. The resulting V_H-D-J_H DNA region is in turn linked to a C_H gene, of which there are 8. These C_H genes (C_μ, C_δ, $C_\gamma 3$, $C_\gamma 1$, $C_\gamma 2b$, $C_\gamma 2a$, C_α, and C_ϵ) determine the immunoglobulin class or subclass—IgM, IgG, IgA, etc—of the immunoglobulin molecule (see Chapter 55). An example of the rearrangements and processing events that result in the formation of a $C_\gamma 2b$ heavy chain gene is shown in Fig 41–10.

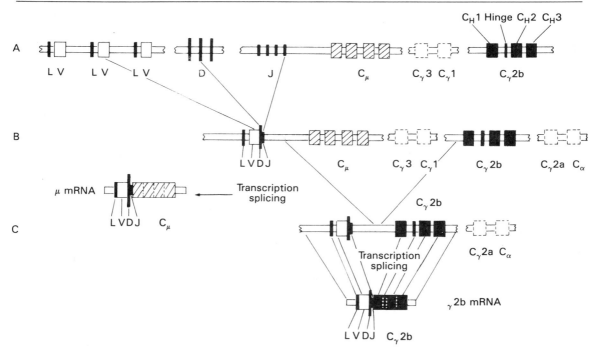

Figure 41–10. Recombination events leading to a complete immunoglobulin heavy chain ($\gamma 2b$) gene. *A:* The germ line DNA before rearrangement. There is a cluster of at least 50 genes (each with a short leader sequence L) coding for part of the variable (V) region, a cluster of D gene segments coding for most of the third hypervariable region, and some distance away there are four J segments that complete the V region coding sequence. The J segments lie about 8000 bases from the C_μ gene that lies at the start of a cluster containing all the C region genes. The C region gene sequences are interrupted by noncoding sequences to give a series of exons that coincide with the domains and the hinge region in the C region amino acid sequence. *B:* In the first translocation event, one of each of the V, D, and J segments are recombined to give a complete μ chain transcription unit. The transcript is a copy of the gene as shown, but the noncoding sequences (introns) are removed by splicing events that lead to a continuous coding sequence in the μmRNA. *C:* A second translocation event, the heavy-chain switch, deletes the C_μ, $C_\gamma 3$, and $C_\gamma 1$ gene segments and places the V-D-J segment and part of the J-C_μ intron near the $C_\gamma 2b$ gene. Following transcription, the introns are removed by splicing, leading to a continuous coding sequence in the $\gamma 2b$ mRNA. (Reproduced, with permission, from Molgaard HV: Assembly of immunoglobulin heavy chain genes. *Nature* 1980;**286**:659.)

DIFFERENTIAL EFFECTS ON CHROMATIN STRUCTURE MAY BE INVOLVED IN DEVELOPMENT AND DIFFERENTIATION

Most of the DNA in prokaryotic cells is organized into genes and the templates can always be transcribed. A very different situation exists in mammalian cells. Here relatively little of the total DNA is organized into genes and their associated regulatory regions. The function of the extra DNA is unknown (this is one of the reasons there is such interest in sequencing the entire human genome).

Chromatin structure provides an additional level of control. As discussed in Chapter 38, there are large regions of chromatin that are transcriptionally inactive, while others are either active or potentially active. With few exceptions, each cell contains the same complement of genes (antibody-producing cells are a notable exception). The development of specialized organs, tissues, and cells and their function in the intact organism depend upon the differential expression of genes.

Some of this differential expression is accomplished by having different regions of chromatin available for transcription in cells from various tissues. For example, the DNA containing the β-globin gene cluster is in "active" chromatin in the reticulocyte but is in "inactive" chromatin in a muscle cell. The mechanisms that determine "active" versus "inactive" chromatin are not known, but protein-DNA interactions are probably involved.

Additionally, as described in Chapter 38, there is evidence that the methylation of deoxycytidine residues (in the sequence $5'-^{m}CpG-3'$) in DNA may effect gross changes in chromatin so as to preclude its active transcription. For example, in mouse liver only the unmethylated ribosomal genes can be expressed, and there is evidence that many animal viruses are not transcribed when their DNA is methylated. However, it is *not* possible to generalize that methylated DNA is transcriptionally inactive, that all inactive chromatin is methylated, or that active DNA is not methylated.

Eukaryotic DNA that is in an "active" region of chromatin can be transcribed. As in procaryotic cells, a promoter dictates where the RNA polymerase will initiate transcription, but this promoter cannot be neatly defined as a -35 and -10 box, particularly in mammalian cells (see Chapter 39). Also, the transacting factors generally come from other chromosomes (so act in *trans*), whereas this consideration is moot in the case of the single chromosome-containing prokaryotic cells. Additional complexity is added by elements/factors that enhance or silence transcription, that define tissue specific expression, and that modulate the actions of many effector molecules.

CERTAIN DNA ELEMENTS ENHANCE OR SILENCE TRANSCRIPTION OF EUKARYOTIC GENES

In addition to gross changes in chromatin affecting transcriptional activity, there is increasing evidence that there are DNA elements which facilitate or enhance initiation at the promoter. For example, in simian virus 40 (SV40) there exists about 200 bp upstream from the promoter of the early genes a region of 2 identical, tandem 72-bp lengths that can greatly increase the expression of genes in vivo. Each of these 72-bp elements can be subdivided into a series of smaller elements; hence, some enhancers have a very complex structure. Enhancer elements differ from the promoter in 2 remarkable ways. They can exert their positive influence on transcription even when separated by thousands of base pairs from a promoter, they work when oriented in either direction, and they can work upstream (5′) or downstream (3′) from the promoter. Enhancers are promiscuous; they can stimulate any promoter in the vicinity. The SV40 enhancer element can exert an influence on, for example, the transcription of β-globin by increasing its transcription 200-fold in cells containing both the enhancer and the β-globin gene on the same plasmid (see below and Fig 41–11). The enhancer element does not seem to be producing a product that in turn acts on the promoter, since it is active only when it exists within the same DNA molecule as (ie, *cis* to) the promoter. Enhancer binding proteins are now being isolated, and these should help elucidate how these elements work. Enhancer elements do appear to convey nuclease hypersensitivity to those regions where they reside (see Chapter 38). A summary of the properties of enhancers is given in Table 41–3.

The *cis*-acting elements that decrease or **silence** the expression of specific genes have also been identified. Fewer of these elements have been studied, so it is not possible to state generalizations about their mechanism of action.

Tissue-Specific Expression May Result from the Action of Enhancers or Silencers

Many genes have now been recognized to harbor enhancer elements in various locations relative to their coding regions. In addition to being able to enhance gene transcription, some of these enhancer elements clearly possess the ability to do so in a tissue-specific manner. Thus, the enhancer element associated with the immunoglobulin genes between the J and C regions enhances the expression of those genes preferentially in lymphoid cells. Enhancer elements associated with the genes for pancreatic enzymes are capable of enhancing even unrelated but physically linked genes preferentially in the pancreatic cells of mice into which the specifically engineered gene constructions were introduced microsurgically at the single-cell embryo stage. This **transgenic animal** approach has proved

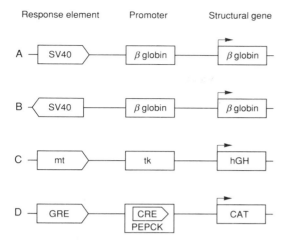

Figure 41–11. A schematic explanation of the action of enhancers and other *cis*-acting regulatory elements. This model, chimeric gene consists of a reporter (structural) gene that encodes a protein that can be readily assayed, a promoter that ensures initiation of transcription, and the putative regulatory element(s). Examples A and B illustrate the fact that enhancers (eg, SV40) work in either orientation, and upon a heterologous promoter. Example C illustrates that the metallothionein (mt) regulatory element (which under the influence of cadmium or zinc induces transcription of the endogenous mt gene and hence the metal-binding mt protein) will work through the thymidine kinase (tk) promoter to enhance transcription of the human growth hormone (hGH) gene. The engineered genetic constructions were introduced into the male pronuclei of single-cell mouse embryos and the embryos placed into the uterus of a surrogate mother to develop as transgenic animals. Offspring have been generated under these conditions, and in some the addition of zinc ions to their drinking water will effect an increase in liver growth hormone. In this case, these transgenic animals have responded to the high levels of growth hormone by becoming twice as large as their normal litter mates. Example D illustrates that a glucocorticoid response element (GRE) will work through homologous (PEPCK gene) or heterologous promoters and that the PEPCK gene promoter also contains an element which functions as both a basal level enhancer and a cyclic AMP response element (CRE).

Table 41–3. A summary of the properties of enhancers.

Properties of Enhancers
Work when located long distances from the promoter
Work when upstream or downstream from the promoter
Work when oriented in either direction
Work through heterologous promoters
Work by binding one or more proteins

very useful in studying tissue-specific gene expression. For example, DNA containing a pancreatic β-cell tissue-specific enhancer (from the insulin gene), when ligated in a vector to polyoma large-T antigen, produced β-cell tumors in transgenic mice. Tumors did not develop in any other tissue. Tissue-specific gene expression may therefore be mediated by enhancers or enhancerlike elements.

Fusion Genes Are Used To Define Enhancers and Other Regulatory Elements

By ligating regions of DNA suspected of harboring regulatory sequences to various reporter genes (the **fusion** or **chimeric gene approach**) (see Figs 41–11, 41–12) one can determine which regions in the vicinity of structural genes have an influence on their expression. Pieces of DNA thought to harbor regulatory elements are ligated to a suitable reporter gene and introduced into a host cell (Fig 41–11). Basal ex-

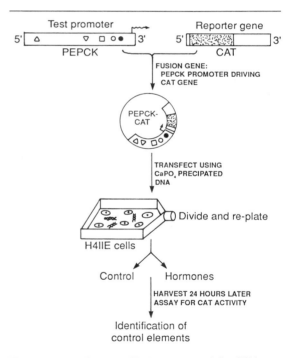

Figure 41–12. The use of fusion genes to define DNA regulatory elements. A DNA fragment thought to contain one or more regulatory elements is ligated into a plasmid vector that contains a suitable reporter gene, the bacterial enzyme chloramphenicol transferase (CAT). CAT is not present in mammalian cells; hence detection of this activity in a cell extract means the cell was successfully transfected by the plasmid. An increase of CAT activity over the basal level, eg, after addition of a glucocorticoid hormone, means the region of DNA inserted contains a functional glucocorticoid hormone response element (GRE). Progressively shorter pieces of DNA, regions with internal deletions, or regions with point mutations can be constructed and inserted to pinpoint the response element.

pression of the reporter gene will be increased if the DNA contains an enhancer. Addition of a hormone or heavy metal to the culture medium will increase expression of the reporter gene if the DNA contains a hormone or metal response element (Fig 41–12). The location of the element can be pinpointed by using progressively shorter pieces of DNA, deletions, or point mutations (Fig 41–13).

This strategy, **using transfected cells in culture and transgenic animals,** has led to the identification of dozens of enhancers, silencers, tissue-specific elements, hormone/metal/drug response elements, etc. The activity of a gene at any moment reflects the interaction of these numerous *cis*-acting DNA elements with their respective *trans*-acting factors. The challenge is to figure out how this occurs.

SEVERAL MOTIFS MEDIATE THE BINDING OF REGULATORY PROTEINS TO DNA

The specificity involved in the control of transcription requires that regulatory proteins bind, with high affinity, to the correct region of DNA. It is known that three unique motifs, the **helix-turn-helix,** the **zinc finger,** and the **leucine zipper,** account for many of these specific protein-DNA interactions. Examples of proteins containing these motifs are given in Table 41–4.

A comparison of the binding activities of the proteins that contain these motifs leads to several important generalizations. These are as follows:

1. Binding must be of high affinity to the specific site and of low affinity to other DNA.

2. Small regions of the protein make direct contact

Table 41–4. Examples of transcription regulatory proteins that contain the various binding motifs.

Binding Motif	Organism	Regulatory Protein
Helix-turn-helix	E coli	lac repressor CAP
	Phage	cro, λ, tryptophan and 434 repressors
	Mammals	homeo box proteins ? pou proteins
Zinc finger	E coli	Gene 32 protein
	Yeast	Gal4
	Drosophila	Serendipity, Hunchback
	Xenopus	TFIIIA
	Mammals	steroid receptor family, Sp1
Leucine zipper	Yeast	GCN4
	Mammals	C/EBP, fos, Jun/Ap1, CRE binding protein, c-myc, n-myc, l-myc

with DNA; the rest of the protein may ensure the proper information of the DNA recognition site, or be involved in the dimerization of monomers of the binding protein.

3. The protein-DNA interactions are maintained by hydrogen bonds and Van der Waals forces.

4. The motifs found in these proteins are unique; their presence in a protein of unknown function suggests the protein may bind to DNA.

5. Proteins with the helix-turn-helix or leucine zipper motifs form symmetric dimers and their respective DNA binding sites are symmetric palindromes. In proteins with the zinc finger motif the binding site is repeated 2–9 times. These features allow for cooperative interactions between binding sites and enhance the degree and affinity of binding.

The Helix-Turn-Helix Motif

The first motif described, and the one studied most extensively, is the helix-turn-helix. Analysis of the 3-dimensional structure of cro has revealed that each monomer consists of three antiparallel β sheets and three α helices (Fig 41–14). The dimer forms by the association of the antiparallel β_3 sheets. The α_3 helices form the DNA recognition surface, and the rest of the molecule appears to be involved in stabilizing these structures. The average diameter of an α helix is 12 Å, which is the approximate width of the major groove in the B form of DNA. The DNA recognition domain of each cro monomer interacts with 5 bp and the dimer binding sites span 34 Å, which allows it to fit into successive half turns of the major groove, on the same surface (Fig 41–14). x-Ray analysis of λ repressor, CRP (the cyclic AMP receptor protein of *Escherichia coli*), tryptopham repressor, and phage 434 repressor confirm this dimeric helix-turn-helix structure.

The Zinc Finger Motif

The zinc finger was the second DNA binding motif elucidated. It was known that the protein TFIIIA, which is a positive regulator of 5S RNA transcription, required zinc for activity. Structural and biophysical

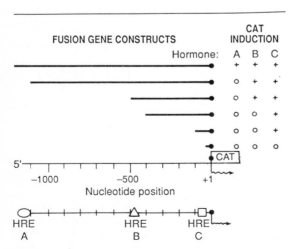

Figure 41–13. The location of hormone responsive elements A, B, and C using the fusion gene—transfection approach. A fusion gene, constructed as described in Figure 41–12, can be transfected into a recipient cell. By analysing when certain hormone responses are lost in comparison to the 5′ deletion, specific hormone responsive elements can be located.

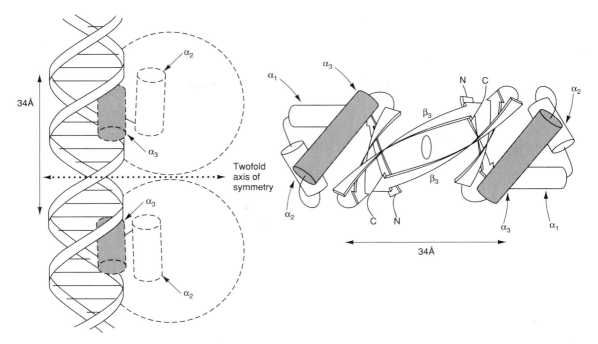

Figure 41–14. A schematic representation of the 3-dimensional structure of cro protein and its binding to DNA by the helix-turn-helix motif. The cro monomer consists of 3 antiparallel β sheets (β_1–β_3) and 3 α helices (α_1–α_3). The helix-turn-helix motif is formed because the α_3 and α_2 helices are held at 90° to each other by a turn of 4 amino acids. The α_3 helix of cro is the DNA recognition surface (shaded). Two monomers associate through the antiparallel β_3 sheets to form a dimer that has a 2-fold axis of symmetry (right). A cro dimer binds to DNA through its α_3 helices, each of which contacts about 5 bp on the same surface of the major groove. The distance between comparable points on the two DNA α helices is 34 Å, which is the distance required for one complete turn of the double helix. (Courtesy of Dr Brian Mathews.)

analyses revealed that each TFIIIA molecule contains 9 zinc ions in a repeating coordination complex formed by closely spaced cysteine-cysteine residues followed 12–13 amino acids later by a histidine-histidine pair (Fig 41–15). In some instances, notably the steroid-thyroid receptor family, the his-his doublet is replaced by a second cys-cys pair. The protein containing zinc fingers appears to lie on one face of the DNA helix, with successive fingers alternatively positioned in one turn in the major groove. As in the case with the recognition domain in the helix-turn-helix protein, each TFIIIA zinc finger contacts about 5 bp of DNA. The importance of this motif in the action of steroid hormones is underscored by an "experiment of nature." A single amino acid mutation in either of the two zinc fingers of the calcitriol receptor protein results in resistance to the action of this hormone and the clinical syndrome of rickets (Chapter 48).

The Leucine Zipper Motif

Careful analysis of a 30-amino acid sequence in the carboxyl-terminal region of the enhancer binding protein C/EBP revealed a novel structure. As illustrated in Fig 41–16, this region of the protein forms an α helix in which there is a periodic repeat of leucine residues at every seventh position. This occurs for eight helical turns and four leucine repeats. Similar structures have

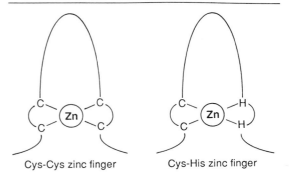

Cys-Cys zinc finger   Cys-His zinc finger

Figure 41–15. Zinc fingers are a series of repeated domains (2–9) in which each is centered on a tetrahedral coordination with zinc. In the case of TFIIIA the coordination is provided by a pair of cysteine residues (C) separated by 12–13 amino acids from a pair of histidine (H) residues. In other zinc finger proteins the second pair also consists of C residues. Zinc fingers bind in the major groove, with adjacent fingers making contact with 5 bp along the same face of the helix.

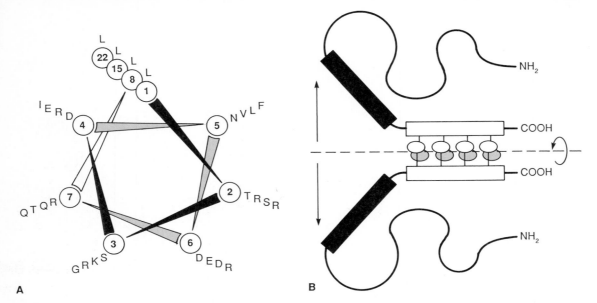

Figure 41–16. The leucine zipper motif. The figure on the left shows a helical wheel analysis of a carboxyl-terminal portion of the DNA binding protein C/EBP. The amino acid sequence is displayed end-to-end down the axis of schematic α-helix. The helical wheel consists of seven spokes which correspond to the seven amino acids that comprise every two turns of the α-helix. Note that leucine residues (L) occur at every seventh position. Other proteins with "leucine zippers" have a similar helical wheel pattern.

A schematic model of the DNA binding domain of C/EBP is shown on the right. Two identical C/EBP polypeptide chains are held in dimer formation by the leucine zipper domain of each polypeptide (denoted by the rectangles and attached ovals). This association is apparently required to hold the DNA binding domains of each polypeptide (the shaded rectangles) in the proper conformation for DNA binding. (Figure courtesy of Steven McKnight.)

been found in a number of other proteins associated with the regulation of transcription in mammalian and yeast cells. It is thought that this structure allows two identical monomers, or perhaps a heterodimer (eg, fos with Jun/AP1) to "zip" together and form a tight dimeric complex (Fig 41–16). This protein-protein interaction may serve to enhance the association of the separate DNA binding domains with their target (Fig 41–16).

The DNA Binding & Transactivation Domains of These Regulatory Proteins Are Separate & Noninteractive

DNA binding could result in a general conformational change that allows the bound protein to activate transcription, or these two functions could be served by separate and independent domains. Domain-swap experiments suggest the latter is the case.

The GAL1 gene product is involved in galactose metabolism in yeast. This gene is positively regulated by the GAL4 protein, which binds to an upstream activator sequence (UAS) through an amino terminal domain. The 73 amino acid DNA-binding terminus of GAL4 was removed and replaced with the DNA-binding domain of lex A, an *E coli* protein. This resulted in a molecule that did not bind to the GAL 1 UAS and, of course, did not activate the GAL1 gene (Fig 41–17). If, however, the lex A operator was inserted into the promoter region of the GAL gene, the hybrid protein

bound to this promoter (at the lex A operator) and it activated transcription of GAL1. This experiment, which has been repeated a number of times (including lex A-glucocorticoid receptor fusion proteins that transactivate glucocorticoid-responsive genes), affords solid evidence that the carboxyl-terminal region of GAL4 causes transcriptional activation. The DNA binding and transactivation domains appear to be independent and noninteractive. The carboxyl-terminal regions appear to have dense concentrations of negatively charged amino acids. These domains, often referred to as "acid blobs" or "negative noodles" presumably interact with positively charged regions of some component of the transcription complex.

Alternative RNA Processing Is Another Control Mechanism

In addition to affecting the efficiency of promoter utilization, eukaryotic cells employ alternative RNA processing to control gene expression. This can result when alternative promoters, intron-exon splice sites, or polyadenylation sites are used. Occasionally this results in heterogeneity within a cell, but more commonly the same primary transcript is processed differently in different tissues. A few examples of each of these types of regulation are presented below.

The use of alternate **transcription start sites** results in a different 5' exon on mRNAs corresponding to mouse amylase and myosin light chain, rat gluco-

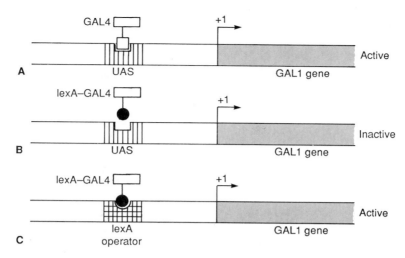

Figure 41–17. Domain-swap experiments show the separateness of DNA binding and transcription activation. The GAL1 gene promoter contains an upstream activating sequence (UAS) that binds the regulatory protein GAL4 (example A). This interaction results in a stimulation of GAL1 gene transcription. A fusion protein, in which the NH_2-terminal DNA-binding domain of GAL4 is removed and replaced with the DNA-binding region of the E coli protein lex A, fails to stimulate GAL1 transcription because the lexA domain cannot bind to the UAS (example B). The lex A-GAL4 fusion protein does increase GAL1 transcription when the lexA operator (its natural target) is inserted into the GAL1 promoter region (example C).

kinase, and *Drosophila* alcohol dehydrogenase and actin. **Alternative polyadenylation sites** in the μ immunoglobulin heavy-chain primary transcript result in mRNAs that are either 2700 bases long (μ_m) or 2400 bases long (μ_s). This results in a different carboxyl-terminal region of the encoded proteins such that the μ_m protein remains attached to the membrane of the β lymphocyte and the μ_s immunoglobulin is secreted. **Alternative splicing and processing** results in the formation of seven unique α-tropomysin mRNAs in seven different tissues. It is not clear how these processing/splicing decisions are made, or whether these steps can be regulated.

Regulation of Messenger RNA Stability Provides Another Control Mechanism

The stability of messenger RNA molecules in the cytoplasm can clearly affect the level of gene expression in a positive or negative direction. Stabilization of mRNA, given a fixed rate of transcription, would lead to increased accumulation, and vice versa. The mechanisms involved in the regulation of mRNA stability were discussed in an earlier chapter. Each of these is a potential control site that can be influenced by hormones and other effectors. Some hormones influence the synthesis and degradation of specific mRNAs. For example, estradiol prolongs the half-life of vitellogenin mRNA from a few hours to more than 200 hours. This, coupled with the fact that estrogens

enhance the rate of transcription of this gene by 4- to 6-fold, results in a tremendous increase of vitellogenin mRNA.

REFERENCES

Breitbart RE, Andreadis A, Wadal-Ginard B: Alternative splicing: A ubiquitous mechanism for the generation of multiple protein isoforms from single genes. *Annu Rev Biochem* 1987;**56:**467.

Jacob F, Monod J: Genetic regulatory mechanisms in protein synthesis. *J Mol Biol* 1961;**3:**318.

Klug A, Rhodes D: "Zinc fingers:" A novel protein motif for nucleic acid recognition. *Trends Biochem Sci* 1987;**12.**

Landschulz WH, Johnson PF, McKnight SL: The leucine zipper: A hypothetical structure common to a new class of DNA binding proteins. *Science* 1989;**240:**1759.

McKnight S, Tjian R: Transcriptional selectivity of viral genes in mammalian cells. *Cell* 1986;**46:**795.

Ptasne M: Gene regulation by proteins acting nearby and at a distance. *Nature (London)* 1986;**322:**697.

Ptasne M: *A genetic switch.* Cell Press and Blackwell Scientific Publications, 1986.

Shimizu A, Honjo T: Immunoglobulinn class switching. *Cell* 1984;**36:**801.

Schlief R: DNA binding by proteins. *Science* 1988;**241:** 1182.

Struhl K: Promoters, activator proteins and the mechanism of transcriptional initiation in yeast. *Cell* 1987;**49:**295.

Wu R, Bahl CP, Narang SA: Lactose operator-repressor interaction. *Curr Top Cell Reg* 1978;**13:**137.

Recombinant DNA Technology

Daryl K. Granner, MD

INTRODUCTION

Recombinant DNA technology, often referred to as genetic engineering, has revolutionized biology and is having an ever-increasing impact on clinical medicine. Much has been learned about human genetic disease from pedigree analysis and study of affected proteins, but in many cases where the specific genetic defect is unknown, these approaches cannot be used. The new technology circumvents these limitations by going directly to the DNA molecule for information.

This chapter is aimed at clarifying this rather complex topic. It presents the basic concepts of recombinant DNA technology, its applications to clinical medicine, and a glossary. In order for the chapter to be complete in itself, some repetition of subjects discussed in other chapters will be found.

BIOMEDICAL IMPORTANCE

Understanding recombinant DNA technology is important for several reasons. (1) The information explosion occurring in this area is truly staggering. To understand and keep up with this field, one must have an appreciation of the fundamental concepts involved. (2) There is now a rational approach to understanding the molecular basis of a number of diseases (eg, familial hypercholesterolemia, sickle cell disease, the thalassemias, cystic fibrosis, muscular dystrophy). (3) Using recombinant DNA technology, human proteins can be produced in abundance for therapy (eg, insulin, growth hormone, plasminogen activator). (4) Proteins for vaccines (eg, hepatitis B) and for diagnostic tests (eg, AIDS test) can be obtained. (5) Recombinant DNA technology is used to diagnose existing diseases and predict the risk of developing a given disease. (6) Special techniques have led to remarkable advances in forensic medicine. (7) Gene therapy for sickle cell disease, the thalassemias, adenosine deaminase deficiency, and other diseases may be devised.

ELUCIDATION OF THE BASIC FEATURES OF DNA LED TO RECOMBINANT DNA TECHNOLOGY

DNA Is a Complex Biopolymer That Is Organized as a Double Helix

The fundamental organizational element is the sequence of purine (adenine [A] or guanine [G]) and pyrimidine (cytosine [C] or thymine [T]) bases. These bases are attached to the C-1' position of the sugar deoxyribose, and the bases are linked together through joining of the sugar moieties at their 3' and 5' positions via a phosphodiester bond (see Fig 37–1). The alternating deoxyribose and phosphate groups form the backbone of the double helix (see Fig 37–2). These 3'–5' linkages also define the orientation of a given strand of the DNA molecule, and since the 2 strands run in opposite directions, they are said to be antiparallel.

Base Pairing Is One of the Most Fundamental Concepts of DNA Structure and Function

Adenine and thymine always pair, by hydrogen bonding, as do guanine and cytosine (see Fig 37–3). These base pairs are said to be **complementary,** and the guanine content of a fragment of double-stranded DNA will always equal its cytosine content; likewise the thymine and adenine contents are equal. Base pairing and hydrophobic base-stacking interactions hold the 2 DNA strands together. These interactions can be reduced by heating the DNA to denature it. The laws of base pairing predict that 2 complementary DNA strands will reanneal exactly in register upon renaturation, as happens when the temperature of the solution is slowly reduced to normal. Indeed, the degree of base-pair matching (or mismatching) can be estimated from the temperature required for denaturation-renaturation. Segments of DNA with high degrees of base-pair matching require more energy input (heat) to accomplish denaturation, or, to put it another way, a closely matched segment will withstand more heat before the strands separate. This reaction is used to determine whether there are significant differences between 2 DNA sequences, and it underlies the concept of **hybridization,** which is fundamental to the processes described below.

There are about 3×10^9 base pairs (bp) in each human haploid genome. If an average gene length is 3×10^3 bp (3 kilobases [kb]), the genome could consist of 10^6 genes, assuming that there is no overlap and that transcription proceeds in only one direction. It is thought that there are only about 10^5 genes in the human and that only 10% of the DNA codes for proteins. The function of the remaining 90% of the human genome has not yet been defined.

The double-helical DNA is packaged into a more compact structure by a number of proteins, most notably the basic proteins called histones. This condensation may serve a regulatory role and certainly has a practical purpose. The DNA present within the nucleus of a cell, if simply extended, would be about a meter long. The chromosomal proteins compact this long length of DNA so that it can be packaged into a nucleus with a volume of a few cubic microns.

DNA Is Organized Into Genes

In general, prokaryotic genes consist of a small regulatory region (100–500 bp) and a large protein-coding segment (500–10,000 bp). Several genes are often controlled by a single regulatory unit. Most mammalian genes are more complicated, in that the coding regions are interrupted by noncoding regions that are eliminated when the primary RNA transcript is processed into mature **messenger RNA (mRNA).** The **coding regions** (those regions that appear in the mature RNA species) are called **exons,** and the **noncoding regions,** which interpose or intervene between the exons, are called **introns** (Fig 42–1). Introns are always removed from precursor RNA before transport into the cytoplasm occurs. The process by which introns are removed from precursor RNA and by which exons are ligated together is called **RNA splicing.** Incorrect processing of the primary transcript into the

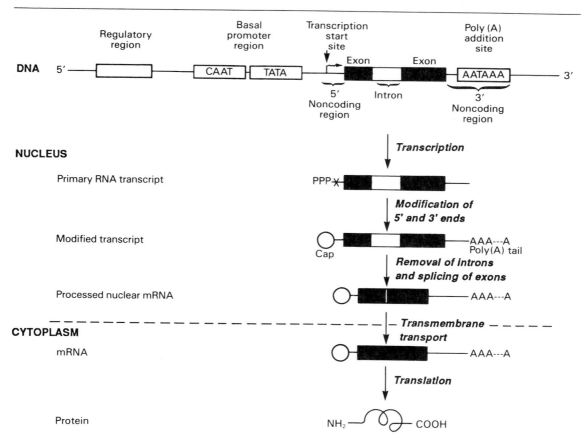

Figure 42–1. Organization of a eukaryotic transcription unit and the pathway of eukaryotic gene expression. Eukaryotic genes have structural and regulatory regions. The structural region consists of the coding DNA and 5' and 3' noncoding DNA sequences. The coding regions are divided into 2 parts: (1) exons, which eventually become mature RNA, and (2) introns, which are processed out of the primary transcript. The structural region is bounded at its 5' end by the transcription initiation site and at its 3' end by the polyadenylate addition or termination site. The promoter region, which contains specific DNA sequences that interact with various protein factors to regulate transcription, is discussed in detail in Chapters 39 and 41. The primary transcript has a special structure, a cap, at the 5' end and a stretch of A's at the 3' end. This transcript is processed to remove the introns, and the mature mRNA is then transported to the cytoplasm, where it is translated into protein.

mature mRNA can result in disease in humans (see below); this underscores the importance of these post-transcriptional processing steps. Regulatory regions for specific eukaryotic genes are usually located in the DNA that flanks the transcription initiation site at its 5′ end (**5′ flanking-sequence DNA**). Occasionally, such sequences are found within the gene itself or in the region that flanks the 3′ end of the gene. In mammalian cells, each gene has its own regulatory region. Many eukaryotic genes (and some viruses that replicate in mammalian cells) have special regions, called **enhancers,** that increase the rate of transcription. Some genes also have DNA sequences, known as **silencers,** that diminish transcription. Mammalian genes are obviously complicated, multicomponent structures.

Genes Are Transcribed Into RNA

Information generally flows from DNA to mRNA to protein, as illustrated in Fig 42–1 and discussed in more detail in Chapter 41. This is a rigidly controlled process involving a number of complex steps, each of which no doubt is regulated by one or more enzymes or factors; faulty function at any of these steps can cause disease.

RECOMBINANT DNA TECHNOLOGY INVOLVES ISOLATION & MANIPULATION OF DNA TO MAKE CHIMERIC MOLECULES

Isolation and manipulation of DNA, including end-to-end joining of sequences from very different sources to make chimeric molecules (eg, molecules containing both human and bacterial DNA sequences in a sequence-independent fashion), is the essence of recombinant DNA research. This involves several unique techniques and reagents.

Restriction Enzymes Cut DNA Chains at Specific Locations

Certain endonucleases, enzymes that cut DNA at specific DNA sequences within the molecule (as opposed to exonucleases, which digest from the ends of DNA molecules), are a key tool in recombinant DNA research. These enzymes were originally called **restriction enzymes** because their presence in a given bacterium restricted the growth of certain bacterial viruses called bacteriophages. Restriction enzymes cut DNA into short pieces in a sequence-specific manner, in contrast to most other enzymatic, chemical, or physical methods, which break DNA randomly. These defensive enzymes (over 200 have been discovered) protect the host bacterial DNA from DNA from foreign organisms (primarily infective phages). However, they are only present in cells that also have a companion enzyme that methylates the host DNA, rendering it an unsuitable substrate for digestion by the restriction enzyme. Thus, **site-specific DNA methylases** and restriction enzymes always exist in pairs in a bacterium.

Restriction enzymes are named after the bacterium from which they are isolated. For example, Eco RI is from *Escherichia coli,* and Bam HI is from *Bacilius amyloliquefaciens* (Table 42–1). The first 3 letters in the restriction enzyme name consist of the first letter of the genus (E) and the first 2 letters of the species (co). These may be followed by a strain designation (R) and a roman numeral (I) to indicate the order

Table 42–1. Selected restriction endonucleases and their sequence specificities.*

Endonuclease	Sequence Cleaved	Bacterial Source
Bam HI	↓ G G A T C C C C T A G G 　　　　　↑	*Bacillus amylolique-faciens* H
Bgl II	↓ A G A T C T T C T A G A 　　　　　↑	*Bacillus globigii*
Eco RI	↓ G A A T T C C T T A A G 　　　　　↑	*Escherichia coli* RY13
Eco RII	↓ C C T G G G G A C C 　　　　↑	*Escherichia coli* R245
Hind III	↓ A A G C T T T T C G A A 　　　　　↑	*Haemophilus influenzae* R$_d$
Hha I	↓ G C G C C G C G ↑	*Haemophilus haemolyticus*
Hpa I	↓ G T T A A C C A A T T G 　　　↑	*Haemophilus para-influenzae*
Mst II	↓ C C T N A G G G G A N T C C 　　　　　↑	*Microcoleus* strain
Pst I	↓ C T G C A G G A C G T C 　↑	*Providencia stuartii* 164
Taq I	↓ T C G A A G C T 　　　↑	*Thermus aquaticus* YTI

*A, adenine; C, cytosine; G, guanine; T, thymine. Arrows show the site of cleavage; depending on the site, sticky ends (Bam HI) or blunt ends (Hpa I) may result. The length of the recognition sequence can be 4 bp (Taq I), 5 bp (Eco RII), 6 bp (Eco RI), or 7 bp (Mst II). By convention, these are written in the 5′ to 3′ direction for the upper strand of each recognition sequence, and the lower strand is shown with the opposite (ie, 3′ to 5′) polarity. Note that most recognition sequences are palindromes (ie, the sequence reads the same in opposite directions on the 2 strands). A residue designated N means that any nucleotide is permitted.

A. Sticky or staggered ends

```
——— G G A T C C ———              — G         G A T C C ———
                    Bam HI                  +
——— C C T A G G ———              — C C T A G        G ———
```

B. Blunt ends

```
——— G T T A A C ———              — G T T     A A C ———
                    Hpa I                  +
——— C A A T T G ———              — C A A     T T G ———
```

Figure 42–2. Results of restriction endonuclease digestion. Digestion with a restriction endonuclease can result in the formation of DNA fragments with sticky, or cohesive, ends (A) or blunt ends (B). This is an important consideration in devising cloning strategies.

of discovery (eg, Eco RI, Eco RII). Each enzyme recognizes and cleaves a specific double-stranded DNA sequence that is 4–7 bp long. These DNA cuts result in **blunt ends** (Hpa I) or overlapping (**sticky**) **ends** (Bam HI) (Fig 42–2), depending on the mechanism used by the enzyme. Sticky ends are particularly useful in constructing hybrid or chimeric DNA molecules (see below). If the nucleotides are distributed randomly in a given DNA molecule, one can calculate how frequently a given enzyme would cut a length of DNA. For each position in the DNA molecule there are 4 possibilities (A, C, G, or T); therefore, a restriction enzyme that recognizes a 4-bp sequence will cut, on average, once every 256 bp (4^4), whereas another enzyme that recognizes a 6-bp sequence will cut once every 4096 bp (4^6). A given piece of DNA will have a characteristic linear array of sites for the various enzymes; hence, a **restriction map** can be constructed. When DNA is digested with a given enzyme, the ends

of all the fragments will have the same DNA sequence. The fragments produced can be isolated by electrophoresis on agarose or polyacrylamide (see the discussion of blot transfer, below); this is an essential step in cloning and a major use of these enzymes.

A number of other enzymes that act on DNA and RNA are an important part of recombinant DNA technology. Many of these are referred to in this and subsequent chapters (Table 42–2).

Digestion and Religation are Used to Prepare Chimeric DNA Molecules

Sticky-end ligation is technically easy, but some special techniques are often required to overcome problems inherent in this approach. Sticky ends of a vector may reconnect with themselves, with no net gain of DNA. Sticky ends of fragments can also anneal, so that tandem heterogeneous inserts form. Also,

Table 42–2. Enzymes used in recombinant DNA research.*

Enzyme	Reaction	Primary Use
Alkaline phosphatase	Dephosphorylates 5′ ends of RNA and DNA.	Removal of 5′–PO_4 groups prior to kinase labeling to prevent self-ligation.
BAL 31 nuclease	Degrades both the 3′ and 5′ ends of DNA.	Progressive shortening of DNA molecules.
DNA ligase	Catalyzes bonds between DNA molecules.	Joining of DNA molecules.
DNA polymerase I	Synthesizes double-stranded DNA from single-stranded DNA.	Synthesis of double-stranded cDNA; nick translation.
DNase I	Under appropriate conditions, produces single-stranded nicks in DNA.	Nick translation; mapping of hypersensitive sites.
Exonuclease III	Removes nucleotides from 3′ ends of DNA.	DNA sequencing; mapping of DNA-protein interactions.
λ Exonuclease	Removes nucleotides from 5′ ends of DNA.	DNA sequencing.
Polynucleotide kinase	Transfers terminal phosphate (γ position) from ATP to 5′–OH groups of DNA or RNA.	^{32}P labeling of DNA or RNA.
Reverse transcriptase	Synthesizes DNA from RNA template.	Synthesis of cDNA from mRNA; RNA (5′ end) mapping studies.
SI nuclease	Degrades single-stranded DNA.	Removal of "hairpin" in synthesis of cDNA; RNA mapping studies (both 5′ and 3′ ends).
Terminal transferase	Adds nucleotides to the 3′ ends of DNA.	Homopolymer tailing.

*Adapted and reproduced, with permission, from Emery AEH: Page 41 in: *An Introduction to Recombinant DNA*. Wiley, 1984.

sticky-end sites may not be available or in a convenient position. To circumvent these problems, an enzyme that generates blunt ends is used, and new ends are added using the enzyme terminal transferase. If poly d(G) is added to the 3′ ends of the vector and poly d(C) is added to the 3′ ends of the foreign DNA, the 2 molecules can only anneal to each other, thus circumventing the problems listed above. This procedure, called **homopolymer tailing,** also generates an Sma I restriction site, and so it is easy to retrieve the fragment. Sometimes, synthetic oligonucleotide linkers with a convenient restriction enzyme sequence are ligated to the blunt-ended DNA. Direct blunt-end ligation is accomplished using the enzyme bacteriophage T4 DNA ligase. This technique, though more difficult than sticky-end ligation, has the advantage of joining together any pairs of ends. The disadvantages are that there is no control over the orientation of insertion or the number of molecules annealed together, and there is no easy way of retrieving the insert.

Cloning Amplifies Chimeric DNA
A clone is a large population of identical mole-

cules, bacteria, or cells that arise from a common ancestor. Cloning allows for the production of a large number of identical DNA molecules, which can then be characterized or used for other purposes. This technique is based on the fact that chimeric or hybrid DNA molecules can be constructed in **cloning vectors,** typically bacterial plasmids, phages, or cosmids, which then continue to replicate in a host cell under their own control systems. In this way, the chimeric DNA is amplified. The general procedure is illustrated in Fig 42–3.

Bacterial **plasmids** are small, circular, duplex DNA molecules whose natural function is to confer antibiotic resistance to the host cell. Plasmids have several properties that make them extremely useful as cloning vectors. They exist as single or multiple copies within the bacterium and replicate independently from the bacterial DNA. The complete DNA sequence of many plasmids is known; hence, the precise location of restriction enzyme cleavage sites for inserting the foreign DNA is available. Plasmids are smaller than the host chromosome and are therefore easily separated from the latter, and the desired DNA is readily re-

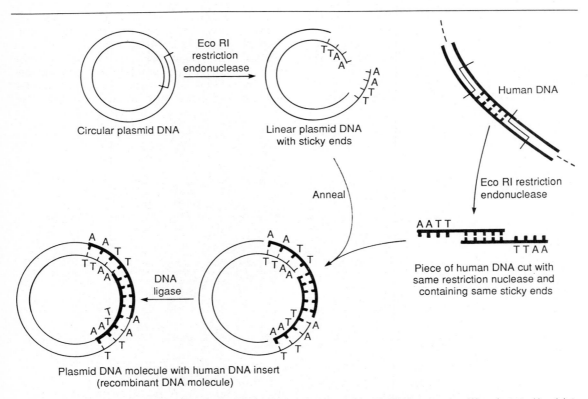

Figure 42–3. Use of restriction nucleases to make new recombinant or chimeric DNA molecules. When inserted back into a bacterial cell (by the process called transformation), the plasmid DNA replicates not only itself but also the physically linked new DNA insert. Since recombining the sticky ends, as indicated, regenerates the same DNA sequence recognized by the original restriction enzyme, the cloned DNA insert can be cleanly cut back out of the recombinant plasmid circle with this endonuclease. If a mixture of all of the DNA pieces created by treatment of total human DNA with a single restriction nuclease is used as the source of human DNA, a million or so different types of recombinant DNA molecules can be obtained, each pure in its own bacterial clone. (Modified and reproduced, with permission, from Cohen SN: The manipulation of genes. *Sci AM* [July] 1975;**233**:34.)

moved by cutting the plasmid with the enzyme specific for the restriction site into which the original piece of DNA was inserted.

Phages usually have linear DNA molecules into which foreign DNA can be inserted at several restriction enzyme sites. The chimeric DNA is collected after the phage proceeds through its lytic cycle and produces mature, infective phage particles. A major advantage of phage vectors is that while plasmids accept DNA pieces about 6–10 kb long, phages can accept DNA fragments 10–20 kb long, a limitation imposed by the amount of DNA that can be packed into the phage head.

Even larger fragments of DNA can be cloned in **cosmids,** which combine the best features of plasmids and phages. Cosmids are plasmids that contain the DNA sequences, so-called **cos sites**, required for packaging lambda DNA into the phage particle. These vectors grow in the plasmid form in bacteria, but since much of the unnecessary lambda DNA has been removed, more chimeric DNA can be packaged into the particle head. It is not unusual for cosmids to carry inserts of chimeric DNA that are 35–50 kb long. A comparison of these vectors is shown in Table 42-3.

Insertion of DNA into a functional region of the vector will interfere with the action of this region, and so care must be taken not to interrupt an essential function of the vector. This concept can be exploited, however, to provide a selection technique. The common plasmid vector **pBR322** has both **tetracycline** (tet) and **ampicillin** (amp) **resistance genes.** A single Pst I site within the amp-resistance gene is commonly used as

Table 42–3. Common cloning vectors.

Vector	DNA Insert Size
Plasmid pBR322	0.01–10 kb
Lambda charon 4A	10–20 kb
Cosmids	35–50 kb

the insertion site for a piece of foreign DNA. In addition to having sticky ends (Table 42–1 and Fig 42–2), the DNA inserted at this site disrupts the amp-resistance gene and makes the bacterium carrying this plasmid amp sensitive (Fig 42–4). Thus, the parental plasmid, which provides resistance to both antibiotics, can be readily separated from the chimeric plasmid, which is resistant only to tetracycline. Additional confirmation that insertion has taken place comes from sizing the plasmid DNA obtained from the putative recombinant on an agarose gel, since the chimeric DNA molecule is now larger than the host vector DNA.

The Recombinant Clones of an Organism's Genome Are Its Library

The combination of restriction enzymes and various cloning vectors allows the entire genome of an organism to be packed into a vector. A collection of these different recombinant closes is called a library. A **genomic library** is prepared from the total DNA of a cell line or tissue. A **cDNA library** represents the population of mRNAs in a tissue. Genomic libraries are prepared by performing partial digestion of total DNA

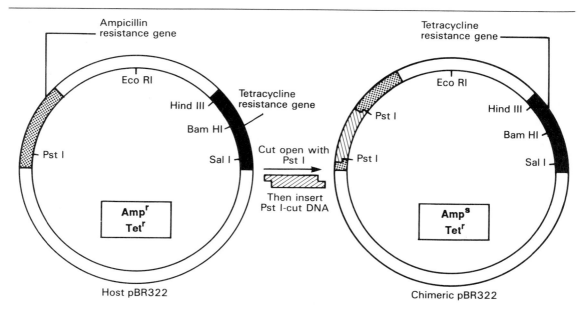

Figure 42–4. A method of screening recombinants for inserted DNA fragments. Using the plasmid pBR322, a piece of DNA is inserted into the unique Pst I site. This insertion disrupts the gene coding for a protein that provides ampicillin resistance to the host bacterium. Hence, the chimeric plasmid will no longer survive when plated on a substrate medium that contains this antibiotic. The differential sensitivity to tetracycline and ampicillin can therefore be used to distinguish clones of plasmid that contain an insert.

with a restriction enzyme that cuts DNA frequently (eg, Sau IIIA). The idea is to generate rather large fragments, so that most genes will be left intact. Phage vectors are preferred for these libraries because they accept large pieces of DNA (up to 20 kb). The goal is to achieve a complete library. The number of fragments required to attain this objective is inversely related to fragment size and directly related to genome size (Table 42–4). A human library that contains 10^6 recombinant fragments of large size has a 99% probability of being complete. Thus, the chances of finding any single-copy gene are excellent.

cDNA libraries are prepared by first isolating all the mRNAs in a tissue and then copying these molecules into double-stranded DNA, using (sequentially) the enzymes reverse transcriptase and DNA polymerase. For technical reasons, full-length cDNA copies are seldom obtained, and so smaller DNA fragments are cloned. Plasmids are often the favored vectors for cDNA libraries because they are much more convenient to work with than are phages or cosmids, although various lambda phage vectors have special advantages for cDNA cloning (see below).

A vector in which the protein coded by the gene introduced by recominant DNA technology is actually synthesized is known as an **expression vector.** Such vectors are now commonly used to detect specific cDNA molecules in libraries and to produce proteins by genetic engineering techniques. These vectors are specially constructed to contain very active inducible promoters, proper in-phase translation initiation codons, both transcription and translation termination signals, and appropriate protein processing signals, if needed. Some expression vectors even contain genes that code for protease inhibitors, so that the final yield of product is enhanced. The vector λgt11 is popular for library construction because it accepts larger cDNA molecules that are replicated and translated into proteins; thus, λgt11 recombinant libraries can be screened with either cDNA or antibody probes.

Probes Search Libraries for Specific Genes or cDNA Molecules

A variety of molecules can be used to "probe" libraries in search of a specific gene or cDNA molecule or to define and quantitate DNA or RNA separated by electrophoresis through various gels. Probes are generally pieces of DNA or RNA labeled with a ^{32}P-containing nucleotide. The probe must recognize a complementary sequence to be effective. A cDNA synthesized from a specific mRNA can be used to screen either a cDNA library for a longer cDNA or a genomic library for a complementary sequence in the coding region of a gene. A popular technique for finding specific genes entails taking a short amino acid sequence and, using the codon usage for that species (see Chapter 40), making an oligonucleotide probe that will detect the corresponding DNA fragment in a genomic library. If the sequences match exactly, probes 15–20 nucleotides long will hybridize. cDNA probes are used to detect DNA fragments on Southern blot transfers and to detect and quantitate RNA on Northern blot transfers.

Table 42–4. The composition of complete genomic libraries.*

Source	Complete Genomic Library
E. coli	1500 fragments
Yeast	4500 fragments
Drosophila	50,000 fragments
Mammals	800,000 fragments

*The number of random fragments (unique clones) that a library should have to ensure that any single gene is represented is inversely related to the average fragment size used to construct the library and directly related to the number of genes in the organism. The numbers given above represent the number of fragments (independent clones) necessary to achieve a 99% probability of finding a given DNA sequence in a recombinant DNA library with an average insert size of 2×10^4 nucleotides. The differences represent the variations in genomic complexity between the creatures.

The number of clones necessary is calculated from the following formula:

$$N = \frac{\ln(1 - P)}{\ln(1 - f)}$$

where P is the probability desired and f is the fraction of the total genome in a single clone. In the case of the mammalian genomic library cited above, given the presence of 3×10^9 nucleotides in the haploid genome, the equation is as follows:

$$N = \frac{\ln(1 - 0.99)}{\ln\left(1 - \left[\dfrac{2 \times 10^4}{3 \times 10^9}\right]\right)}$$

The advantage of having a library composed of large DNA inserts is immediately apparent if this equation is solved using an average fragment size of 5×10^3 nucleotides rather than 2×10^4.

Blotting & Hybridization Techniques Allow Vizualization of Specific Fragments

Visualization of a specific DNA or RNA fragment among the many thousands of "contaminating" molecules requires the convergence of a number of techniques, which are collectively termed **blot transfer**. Fig 42–5 illustrates the **Southern** (DNA), **Northern** (RNA), and **Western** (protein) **blot transfer** procedures. (The first is named for the person who devised the technique, and the other names began as laboratory jargon but are now accepted terms.) These procedures are useful in determining how many copies of a gene are in a given tissue or whether there are any gross alterations in a gene (deletions, insertions, or rearrangements). Occasionally, if a specific base is changed and a restriction site is altered, these procedures can detect a point mutation. The Northern and Western blot transfer techniques are used to size and quantitate specific RNA and protein molecules, respectively.

Colony or **plaque hybridization** is the method by which specific clones are identified and purified. Bacteria are grown on colonies on an agar plate and over-

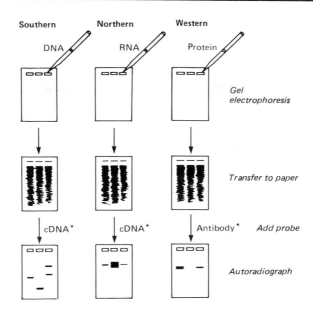

Figure 42–5. The blot transfer procedure. In a Southern, or DNA, blot transfer, DNA isolated from a cell line or tissue is digested with one or more restriction enzymes. This mixture is pipetted into a well in an agarose or polyacrylamide gel and exposed to a direct electrical current. DNA, being negatively charged, migrates toward the cathode; the smaller fragments move the most rapidly. After a suitable time, the DNA is denatured by exposure to mild alkali and transferred to nitrocellulose paper, in an exact replica of the pattern on the gel, by the blotting technique devised by Southern. The DNA is annealed to the paper by exposure to heat, and the paper is then exposed to the labeled cDNA probe, which hybridizes to complementary fragments on the filter. After thorough washing, the paper is exposed to x-ray film, which is developed to reveal several specific bands corresponding to the DNA fragment that recognized the sequences in the cDNA probe. The RNA, or Northern, blot is conceptually similar. RNA is subjected to electrophoresis before blot transfer. This requires some different steps from those of DNA transfer, primarily to ensure that the RNA remains intact, and is generally somewhat more difficult. In the protein, or Western, blot, proteins are electrophoresed and transferred to nitrocellulose and then probed with a specific antibody or other probe molecule.

laid with a nitrocellulose filter paper. Cells from each colony stick to the filter and are permanently fixed thereto by heat, which with NaOH treatment also lyses the cells and denatures the DNA so that it will hybridize with the probe. A radioactive probe is added to the filter, and after washing, the hybrid complex is localized by exposing the filter to x-ray film. By matching the spot on the autoradiograph to a colony, the latter can be picked from the plate. A similar strategy is used to identify fragments in phage libraries. Successive rounds of this procedure result in a clonal isolate (bacterial colony) or individual phage plaque.

All of the hybridization procedures discussed in this section depend on the specific base-pairing properties of complementary nucleic acid strands described above. Perfect matches hybridize readily and withstand high temperatures in the hybridization and washing reactions. These complexes also form in the presence of low salt concentrations. Less than perfect matches do not tolerate these **stringent conditions** (ie, elevated temperatures and low salt concentrations); thus, hybridization either never occurs or is disrupted during the washing step. Gene families, in which there is some degree of homology, can be detected by varying the stringency of the hybridization and washing

steps. Cross-species comparisons of a given gene can also be made using this approach.

DNA Sequencing is Used to Analyze Recombinant DNA Molecules

The segments of specific DNA molecules obtained by recombinant DNA technology can be analyzed for their nucleotide sequence. This method depends upon having a large number of identical DNA molecules. This requirement can be satisfied by cloning the fragment of interest, using the techniques described above. The enzymatic method (Sanger's) employs specific dideoxynucleotides that terminate DNA strand synthesis at specific nucleotides as the strand is synthesized on purified template nucleic acid. The reactions are adjusted so that a population of DNA fragments representing termination at every nucleotide is obtained. By having a radioactive label incorporated at the end opposite the termination site one can separate the fragments according to size using polyacrylamide gel electrophoresis. An autoradiograph is made and each of the fragments produces an image (band) on an x-ray film. These are read in order to give the DNA sequence (see Fig 42–6). Semiautomated DNA sequencing is

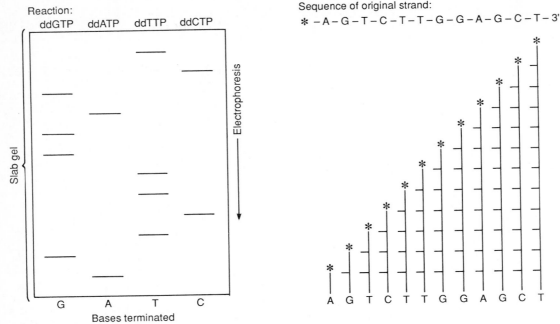

Figure 42–6. Sequencing of DNA by the method devised by Sanger. The ladderlike arrays represent from bottom to top all of the successively longer fragments of the original DNA strand. Knowing which specific dideoxynucleotide reaction was conducted to produce each mixture of fragments, one can determine the sequence of nucleotides from the labeled end toward the unlabeled end by reading up the gel. The base-pairing rules of Watson and Crick (A–T, G–C) dictate the sequence of the other (complementary) strand.

now possible. Another method, that of Maxam and Gilbert, employs chemical methods to cleave the DNA molecules where they contain the specific nucleotides.

Oligonucleotide Synthesis Is Now Routine

The automated, chemical synthesis of moderately long (~100 nt) oligonucleotides of precise sequence is now a routine laboratory procedure. Each synthetic cycle takes a few minutes so an entire molecule can be made in hours to days depending on its length. Oligonucleotides are now indispensable for DNA sequencing, library screening, DNA mobility shift assays, the polymerase chain reaction (see below), and numerous other applications.

The Polymerase Chain Reaction (PCR) Amplifies DNA Sequences

The polymerase chain reaction (PCR) is a method of amplifying a target sequence of DNA. Specificity is based on the use of two oligonucleotide primers that hybridize to complementary sequences on opposite strands of DNA and flank the target sequence (Fig 42–7). The DNA sample is first heated to separate the 2 strands, the primers are allowed to bind to the DNA, and each strand is copied by a DNA polymerase, starting at the primer site. The 2 DNA strands each serve as

a template for the synthesis of new DNA from the two primers. Repeated cycles of heat denaturation, annealing of the primers to their complementary sequences, and extension of the annealed primers with DNA polymerase result in the exponential amplification of DNA segments of defined length. Early PCR reactions used an *E coli* DNA polymerase that was destroyed by each heat denaturation cycle. Substitution of a heat-stable DNA polymerase from *Thermus aquaticus,* an organism that lives and replicates at 70–80 °C, obviates this problem and has allowed for automation of the reaction, since the polymerase reactions can be run at 70 °C. This has also improved the specificity and the yield of DNA.

DNA sequences as short as 50–100 bp and as long as 2.5 kbp can be amplified. Twenty cycles provides an amplification of ~10^6 and 30 cycles of ~10^9. The PCR allows the DNA in a single cell, hair follicle, or sperm to be amplified and analyzed. Thus, the applications of PCR to forensic medicine are obvious. The PCR is also used (1) to detect infectious agents, especially latent viruses; (2) to make prenatal genetic diagnoses, (3) to detect allelic polymorphisms; (4) to establish precise tissue types for transplants; and (5) to study evolution, using DNA from archeological samples. There are an equal number of applications of PCR to problems in basic science, and new uses will no doubt ensue.

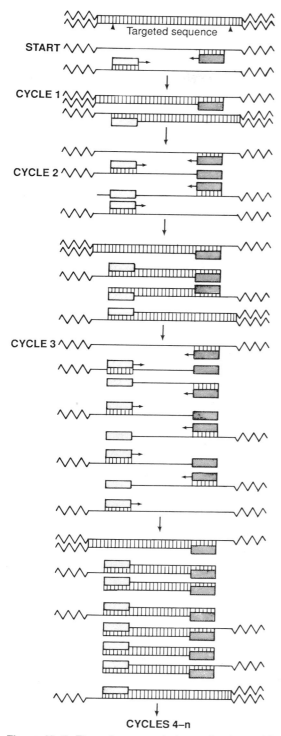

Figure 42–7. The polymerase chain reaction is used to amplify specific gene sequences. Double-stranded DNA is heated to separate it into individual strands. These bind 2 distinct primers that are directed at specific sequences on opposite strands and that define the segment to be amplified. DNA polymerase extends the primers in each direction and synthesizes 2 strands complementary to the original 2. This cycle is repeated several times, giving an amplified product of defined length and sequence.

PRACTICAL APPLICATIONS OF RECOMBINANT DNA TECHNOLOGY ARE GROWING

The isolation of a specific gene from an entire genome requires a technique that will discriminate one part in a million. The identification of a regulatory region that may be only 10 bp in length requires a sensitivity of one part in 3×10^8; a disease such as sickle cell anemia is caused by a single base change, or one part in 3×10^9. Recombinant DNA technology is powerful enough to accomplish all of these things.

Gene Mapping Localizes Specific Genes to Distinct Chromosomes

Gene localizing thus can define a map of the human genome. This is already yielding useful information in the definition of human disease. Somatic cell hybridization and in situ hybridization are 2 techniques used to accomplish this. In **in situ hybridization,** the simpler and more direct procedure, a radioactive probe is added to a metaphase spread of chromosomes on a glass slide. The exact area of hybridization is localized by layering photographic emulsion over the slide and, after exposure, lining up the grains with some histologic identification of the chromosome. This often places the gene at a location on a given band or region on the chromosome. Some of the human genes localized using this technique are listed in Table 42–5.

This table represents only a sampling, since hundreds of genes have been mapped. This human map will become more complete in ensuing years, and there are plans to sequence the entire human genome. The following conclusions can already be drawn: (1) Genes that code for proteins with similar functions can be located on separate chromosomes (α- and β-globin). (2) Genes that form part of a family can also be on separate chromosomes (growth hormone and prolactin). (3) The genes involved in many hereditary disorders known to be due to specific protein deficiencies, including X chromosome-linked conditions, are indeed located at specific sites. Of most interest, perhaps, is the fact that because of the availability of defined and cloned restriction fragments, the chromosomal location for many disorders for which the protein deficiency is unknown is being defined, eg, Huntington's chorea, chromosome 4; cystic fibrosis, chromosome 7; adult polycystic kidney disease, chromosome 16; and Duchenne-type muscular dystrophy, chromosome X. Once the defect is localized to a region of DNA that has the characteristic structure of a gene (Fig 42–1), a synthetic gene can be constructed and expressed in an appropriate vector and its function can be assessed, or the putative peptide, deduced from the open reading frame in the coding region, can be synthesized. Antibodies directed against this peptide can be used to assess whether this peptide is expressed in normal persons and whether it is absent in those with the genetic syndrome.

Table 42–5. Localization of human genes.*

Gene	Chromosome	Disease
Insulin	<u>11</u>p15	
Prolactin	<u>6</u>p23-q12	
Growth hormone	<u>17</u>q21-qter	Growth hormone deficiency
α-Globin	<u>16</u>p12-pter	α-Thalassemia
β-Globin	<u>11</u>p12	β-Thalassemia, sickle cell
Adenosine deaminase	<u>20</u>q13-qter	Adenosine deaminase deficiency
Phenylalanine hydroxylase	<u>12</u>q24	Phenylketonuria
Hypoxanthine-guanine phosphoribosyltransferase	<u>X</u>q26-q27	Lesch-Nyhan syndrome
DNA segment G8	<u>4</u>p	Huntington's chorea

*This table indicates the chromosomal location of several genes and the diseases associated with deficient or abnormal production of the gene products. The chromosome involved is indicated by the first (underlined) number or letter. The other numbers and letters refer to precise localizations, as defined in McKusick VA: *Mendelian Inheritance in Man,* 6th ed. Johns Hopkins Univ Press, 1983.

Proteins Can Be Produced for Research & Diagnosis

A practical goal of recombinant DNA research is the production of materials for biomedical application. This technology has 2 distinct merits: (1) It can supply large amounts of material that could not be obtained by conventional purification methods (eg, interferon, plasminogen activating factor). (2) It can provide human material (eg, insulin, growth hormone). The advantages in both cases are obvious. Although the primary aim is to supply products, generally proteins, for treatment (insulin) and diagnosis (AIDS test) of human and other animal diseases and for disease prevention (hepatitis B vaccine), there are other real and potential commercial applications, especially in agriculture. An example of the latter is the attempt to engineer plants that are more resistant to drought or temperature extremes or more efficient at fixing nitrogen.

Recombinant DNA Technology Is Used in the Molecular Analysis of Disease

Normal Gene Variations: There is a normal variation of DNA sequence just as there is with more obvious aspects of human structure. Variations of DNA sequence, **polymorphisms,** occur approximately once in every 500 nucleotides, or about 10^7 times per genome. There are, no doubt, deletions and insertions of

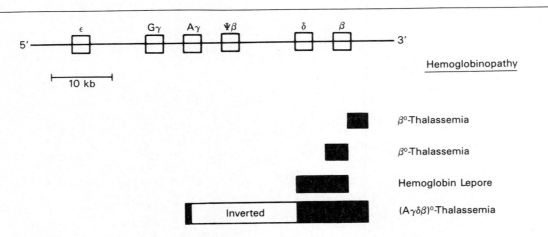

Figure 42–8. Schematic representation of the β-globin gene cluster and some genetic disorders. The β-globin gene is located on chromosome 11 in close association with the 2 γ-globin genes and the δ-globin gene. The β-gene family is arranged in the order 5'-ξ-Gγ-Aγ-ψβ-δ-β-3'. The ξ locus is expressed in early embryonic life ($\alpha_2\xi_2$). The γ genes are expressed in fetal life, making fetal hemoglobin (HbF, $\alpha_2\gamma_2$). Adult hemoglobin consists of HbA ($\alpha_2\beta_2$) or HbA₂($\alpha_2\delta_2$). The ψβ is a pseudogene that has sequence homology with β but contains mutations that prevent its expression. Deletions (solid bar) of the β locus cause β-thalassemia (deficiency or absence [β°] of β-globin). A deletion of δ and β causes hemoglobin Lepore (only hemoglobin α is present). An inversion (Aγδβ)° in this region (open bar) disrupts gene function and also results in thalassemia (type III). Each type of thalassemia tends to be found in a certain group of people, eg, the (Aγδβ)° deletion inversion occurs in persons from India. Many more deletions in this region have been mapped, and each causes some type of thalassemia.

DNA as well as single-base substitutions. In healthy people, these alterations obviously occur in noncoding regions of DNA or at sites that cause no change in function of the encoded protein. This polymorphism of DNA structure can be associated with certain diseases and can be used to search for the specific gene involved, as is illustrated below. It can also be used in a variety of applications in forensic medicine.

Gene Variations Causing Disease: Classical genetics taught that most genetic diseases were due to point mutations that resulted in an impaired protein. This may still be true, but if on reading the initial sections of this chapter one predicted that genetic disease could result from derangement of any of the steps illustrated in Fig 42–1, one would have made a proper assessment.

This point is nicely illustrated by an examination of the β-globin gene. This gene is located in a cluster on chromosome 11 (Fig 42–8), and an expanded version of the gene is illustrated in Fig 42–9. Defective production of β-globin results in a variety of diseases and is due to many different lesions in and around the β-globin gene (Table 42–6).

Point Mutations: The classic example is **sickle cell disease,** which is caused by mutation of a single base out of the 3×10^9 in the genome, a T-to-A DNA substitution, which in turn results in an A-to-U change in the mRNA corresponding to the sixth codon of the β-globin gene (see Fig 7–20). The altered codon specifies a different amino acid (valine rather than glutamic acid), and this causes a structural abnormality of the β-globin molecule. Other point mutations in and around the β-globin gene result in decreased or, in some instances, no production of β-globin; β-thalassemia is the result of these mutations. (The thalassemias are characterized by defects in the synthesis of hemoglobin subunits, and so β-thalassemia results when there is insufficient production of β-globin.) Fig 42–9 illustrates that point mutations affecting each of the many processes involved in generating a normal mRNA (and therefore a normal protein) have been implicated as a cause of β-thalassemia.

Table 42–6. Structural alterations of the β-globin gene.

Alteration	Function Affected	Disease
Point mutations	Protein folding Transcriptional control Frameshift and non-sense mutations RNA processing	Sickle cell disease β-Thalassemia β-Thalassemia β-Thalassemia
Deletion	mRNA production	β°-Thalassemia Hemoglobin Lepore
Rearrangement	mRNA production	β-Thalassemia type III

Deletions, Insertions, & Rearrangements of DNA: Studies of bacteria, viruses, yeasts, and fruit flies show that pieces of DNA can move from one place to another within a genome. The deletion of a critical piece of DNA, the rearrangement of DNA within a gene, or the insertion of a piece of DNA within a coding or regulatory region can all cause changes in gene expression resulting in disease. Again, a molecular analysis of β-thalassemia produces numerous examples of these processes, particularly deletions, as a cause of disease (Fig 42–8). The globin gene clusters seem particularly prone to this lesion. Deletions in the α-globin cluster, located on chromosome 16, cause α-thalassemia. There is a strong ethnic association for many of these deletions, so that northern Europeans, Filipinos, blacks, and Mediterranean peoples have different lesions all resulting in the absence of hemoglobin A and α-thalassemia.

A similar analysis could be made for a number of other diseases. Point mutations are usually defined by sequencing the gene in question, though occasionally, if the mutation destroys or creates a restriction enzyme site, the technique of restriction fragment analysis can be used to pinpoint the lesion. Deletions or insertions of DNA larger than 50 bp can often be detected by the Southern blotting procedure.

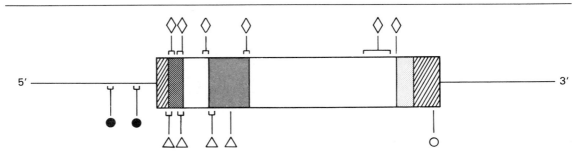

Figure 42–9. Mutations in the β-globin gene causing β-thalassemia. The β-globin gene is shown in the 5′ to 3′ orientation. The cross-hatched areas indicate the 5′ and 3′ nontranslated regions. Reading from the 5′ to 3′ direction, the shaded areas are exons 1–3 and the clear spaces are introns 1 and 2. Mutations that affect transcription control (●) are located in the 5′ flanking-region DNA. Examples of nonsense mutations (△), mutations in RNA processing (◇), and RNA cleavage mutations (○) have been identified and are indicated. In some regions, many mutations have been found. These are indicated by a bracket (⌐⌐).

Pedigree Analysis: Sickle cell disease again provides an excellent example of how recombinant DNA technology can be applied to the study of human disease. The substitution of T for A in the template strand of DNA in the β-globin gene changes the sequence in the region that corresponds to the sixth codon from

```
      ↓
C C T G A G G        coding strand
G G A C Ⓣ C C        template strand
          ↑
```

to

```
C C T G T G G        coding strand
G G A C Ⓐ C C        template strand
```

and destroys a recognition site for the restriction enzyme Mst II (CCTNAGG; denoted by the small vertical arrows; Table 42–1). Other Mst II sites 5' and 3' from this site (Fig 42–10) are not affected and so will be cut. Therefore, incubation of DNA from normal (AA), heterozygous (AS), and homozygous (SS) individuals results in 3 different patterns on Southern blot transfer (Fig 42–10). This illustrates how a DNA pedigree can be established using the principles discussed in this chapter. Pedigree analysis has been applied to a number of genetic diseases and is most useful in those caused by deletions and insertions or the rarer instances in which a restriction endonuclease cleavage site is affected, as in the example cited in this paragraph. The analysis is facilitated by the PCR reaction, which can provide sufficient DNA for analysis from just a few red blood cells.

Prenatal Diagnosis: If the genetic lesion is understood and a specific probe is available, prenatal diagnosis is possible. DNA from cells collected from as little as 10 mL of amniotic fluid (or by chorionic villus biopsy) can be analyzed by Southern blot transfer. A fetus with the restriction pattern AA in Fig 42–10 does not have sickle cell disease, nor is it a carrier. A fetus with the SS pattern will develop the disease. Probes are now available for this type of analysis of many genetic diseases.

Restriction Fragment Length Polymorphism (RFLP): The differences in DNA sequence cited above can result in variations of restriction sites and thus in the length of restriction fragments. Inherited differences in the pattern of restriction (eg, a DNA variation occurring in more than 1% of the general population) are known as restriction fragment length polymorphism, or RFLP, RFLPs result from single-base changes (eg, sickle cell disease) or from deletions or insertions of DNA into a restriction fragment (eg, the thalassemias) and are proving to be a useful diagnostic tool. They have been found at known gene loci and in sequences that have no known function; thus, RFLPs may disrupt the function of the gene or may have no biologic consequences.

RFLPs are inherited, and they segregate in a Mendelian fashion. A major use of RFLPs (nearly 350 are now known) is in the definition of inherited diseases in which the functional deficit is unknown. RFLPs can be used to establish linkage groups, which in turn, by the process of **chromosome walking,** will eventually define the disease locus. In chromosome walking (Fig 42–11), a fragment representing one end of a long piece of DNA is used to isolate another that overlaps but extends the first. The direction of extension is determined by restriction mapping, and the procedure is repeated sequentially until the desired sequence is obtained. The X chromosome-linked disorders are particularly amenable to this approach, since only a single allele is expressed. Hence, 20% of the defined RFLPs are on the X chromosome, and a reasonably complete linkage map of this chromosome exists. The gene for the X-linked disorder, Duchenne-type muscular dystrophy, has been found using RFLPs. Likewise, the defect in Huntington's chorea has been localized to the terminal region of the short arm of chromosome 4, and the defect that causes polycystic kidney disease is linked to the α-globin locus on chromosome 16.

Gene Therapy: Diseases caused by deficiency of a gene product (Table 42–5) are amenable to replacement therapy. The strategy is to clone a gene (eg, the gene that codes for adenosine deaminase) into a vector that will readily be taken up and incorporated into the genome of a host cell. Bone marrow precursor cells are being investigated for this purpose because they presumably will resettle in the marrow and replicate there. The introduced gene would begin to direct the expression of its protein product, and this would correct the deficiency in the host cell.

This somatic cell gene replacement would obviously not be passed on to offspring. Other strategies to alter germ cell lines have been devised but have been tested only in experimental animals. A certain percentage of genes injected into a fertilized mouse ovum will be incorporated into the genome and found in both somatic and germ cells. These **transgenic animals** are proving to be useful for analysis of tissue-specific effects on gene expression and effects of overproduction of gene products (eg, those from the growth hormone gene or oncogenes) and in discovering genes involved in development, a process that heretofore has been difficult to study. The transgenic approach has recently been used to correct a genetic deficiency in mice. Fertilized ova obtained from mice with genetic hypogonadism were injected with DNA containing the coding sequence for the gonadotropin-releasing hormone (GnRH) precursor protein. This gene was expressed and regulated normally in the hypothalamus of a certain number of the resultant mice, and these animals were in all respects normal. Their offspring also showed no evidence of GnRH deficiency. This is, therefore, evidence of somatic cell expression of the transgene and of its maintenance in germ cells.

A. Mst II restriction sites around and in the β-globin gene

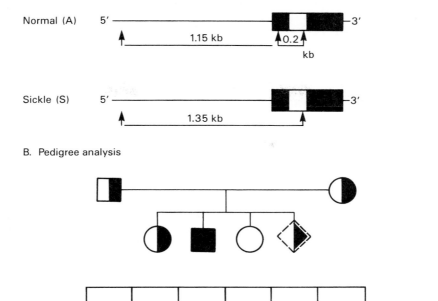

B. Pedigree analysis

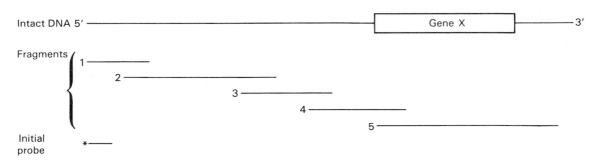

Figure 42–10. Pedigree analysis of sickle cell disease. The top part of the figure (*A*) shows the first part of the β-globin gene and the Mst II restriction enzyme sites (↑) in the normal (A) and sickle cell (S) β-globin genes. Digestion with the restriction enzyme Mst II results in DNA fragments 1.15 kb and 0.2 kb long in normal individuals. The T-to-A change in individuals with sickle cell disease abolishes one of the 3 Mst II sites around the β-globin gene, hence, a single restriction fragment 1.35 kb in length is generated in response to Mst II. This size difference is easily detected on a Southern blot (*B*). (The 0.2-kb fragment would run off the gel in this illustration.) Pedigree analysis shows 3 possibilities: AA - normal (○), AS = heterozygous (◑,◨); SS = homozygous (■). This approach allows for prenatal diagnosis of sickle cell disease or trait (◈).

Figure 42–11. The technique of chromosome walking. Gene X is to be isolated from a large piece of DNA. The exact location of this gene is not known, but a probe (■––) directed against a fragment of DNA (shown at the 5′ end in this representation) is available, as is a library containing a series of overlapping DNA fragments. For the sake of simplicity, only 5 of these are shown. The initial probe will hybridize only with clones containing fragment 1, which can then be isolated and used as a probe to detect fragment 2. This procedure is repeated until fragment 4 hybridizes with fragment 5, which contains the entire sequence of gene X.

GLOSSARY

Autoradiography: The detection of radioactive molecules (eg, DNA, RNA, protein) by visualization of their effects on photographic film.

Bacteriophage: A virus that infects a bacterium.

Blunt-ended DNA: Two strands of a DNA duplex having ends that are flush with each other.

cDNA: A single-stranded DNA molecule that is complementary to an mRNA molecule and is synthesized from it by the action of reverse transcriptase.

Chimeric molecule: A molecule (eg, DNA, RNA, protein) containing sequences derived from 2 different species.

Clone: A large number of cells or molecules that are identical with a single parental cell or molecule.

Cosmid: A plasmid into which the DNA sequences from bacteriophage lambda that are necessary for the packaging of DNA (cos sites) have been inserted; this permits the plasmid DNA to be packaged in vitro.

Endonuclease: An enzyme that cleaves internal bonds in DNA or RNA.

Exon: The sequence of a gene that is represented (expressed) as mRNA.

Exonuclease: An enzyme that cleaves nucleotides from either the 3' or 5' ends of DNA or RNA.

Hairpin: A double-helical stretch formed by base pairing between neighboring complementary sequences of a single strand of DNA or RNA.

Hybridization: The specific reassociation of complementary strands of nucleic acids (DNA with DNA, DNA with RNA, or RNA with RNA).

Insert: An additional length of base pairs in DNA, generally introduced by the techniques of recombinant DNA technology.

Intron: The sequence of a gene that is transcribed but excised before translation.

Library: A collection of cloned fragments that represents the entire genome. Libraries may be either genomic DNA (in which both introns and exons are represented) or cDNA (in which only exons are represented).

Ligation: The enzyme-catalyzed joining in phosphodiester linkage of 2 stretches of DNA or RNA into one; the respective enzymes are DNA and RNA ligases.

Nick translation: A technique for labeling DNA based on the ability of the DNA polymerase from *E coli* to degrade a strand of DNA that has been nicked and then to resynthesize the strand; if a radioactive nucleoside triphosphate is employed, the rebuilt strand becomes labeled and can be used as a radioactive probe.

Northern blot: A method for transferring RNA from an agarose gel to a nitrocellulose filter, on which the RNA can be detected by a suitable probe.

Oligonucleotide: A short, defined sequence of nucleotides joined together in the typical phosphodiester linkage.

Palindrome: A sequence of duplex DNA that is the same when the 2 strands are read in opposite directions.

Plasmid: A small, extrachromosomal, circular molecule of DNA that replicates independently of the host DNA.

Polymerase Chain Reaction (PCR): An enzymatic method for the repeated copying of the 2 strands of DNA that make up a particular gene sequence.

Probe: A molecule used to detect the presence of a specific fragment of DNA or RNA in, for instance, a bacterial colony that is formed from a genetic library or during analysis by blot transfer techniques; common probes are cDNA molecules, synthetic oligodeoxynucleotides of defined sequence, or antibodies to specific proteins.

Pseudogene: An inactive segment of DNA arising by mutation of a parental active gene.

Recombinant DNA: The altered DNA that results from the insertion of a sequence of deoxynucleotides not previously present into an existing molecule of DNA by enzymatic or chemical means.

Restriction enzyme: An endodeoxynuclease that causes cleavage of both strands of DNA at highly specific sites dictated by the base sequence.

Reverse transcription: RNA-directed synthesis of DNA, catalyzed by reverse transcriptase.

Signal: The end product observed when a specific sequence of DNA or RNA is detected by autoradiography or some other method. Hybridization with a complementary radioactive polynucleotide (eg, by Southern or Northern blotting) is commonly used to generate the signal.

Southern blot: A method for transferring DNA from an agarose gel to nitrocellulose filter, on which the DNA can be detected by a suitable probe (eg, complementary DNA or RNA).

Splicing: The removal of introns from RNA accompanied by the joining of its exons.

Sticky-ended DNA: Complementary single strands of DNA that protrude from opposite ends of a DNA duplex or from the ends of different duplex molecules (see also Blunt-ended DNA, above).

Tandem: Used to describe multiple copies of the same sequence (eg, DNA) that lie adjacent to one another.

Terminal transferase: An enzyme that adds nucleotides of one type (eg, deoxyadenonucleotidyl residues) to the 3' end of DNA strands.

Transcription: DNA-directed synthesis of RNA.

Transgenic: Describing the introduction of new DNA into germ cells by its injection into the nucleus of the ovum.

Translation: Synthesis of protein using mRNA as template.

Vector: A plasmid or bacteriophage into which foreign DNA can be introduced for the purposes of cloning.

Western blot: A method for transferring protein to a nitrocellulose filter, on which the protein can be detected by a suitable probe (eg, an antibody).

REFERENCES

Beaudet AL: Bibliography of cloned human and other selected DNAs. *Am J Hum Genet* 1985;**37**:386.

Berger SL, Kimmel AR: *Guide to Molecular Cloning Techniques,* Vol 152, *Methods in Enzymology,* Academic Press, New York, 1987.

DNA in medicine. *Lancet* 1984;**2**:853, 908, 966, 1022, 1086, 1138, 1194, 1257, 1329, 1380, 1440.

Gusella JF: Recombinant DNA techniques in the diagnosis and treatment of inherited disorders. *J Clin Invest* 1986;**77**: 1723.

Kan YW et al: Pages 275–283 in: *Thalassemia: Recent Advances in Detection and Treatment.* Cao A, Carcassi U, Rowley P (editors). AR Liss, 1982.

Lewin B: *Genes III.* Wiley, 1987.

Maniatis T, Fritsch EF, Sambrook J: *Molecular Cloning,* 2nd ed. Cold Spring Harbor Laboratory, 1989.

Martin JB, Gusella JF: Huntington's disease: Pathogenesis and management. *N Engl J Med* 1986:**315:**1267.

Orkin SH et al: Improved detection of the sickle mutation by DNA analysis: Application to prenatal diagnosis. *N Engl J Med* 1982;**307:**32.

Watson JD, Tooze J, Kurtz DT: *Recombinant DNA: A Short Course.* Freeman, 1983.

Weatherall DJ: *The New Genetics and Clinical Practice,* 2nd ed. Oxford Univ Press, 1986.

Yuan R: Structure and mechanism of multifunctional restriction endonucleases. *Annu Rev Biochem* 1981;**50:**285.

Section V.
Biochemistry of Extracellular & Intracellular Communication

Membranes: Structure, Assembly, & Function

43

Daryl K. Granner, MD

INTRODUCTION

Membranes are highly viscous yet plastic structures. Plasma membranes form closed compartments around cellular protoplasm to separate one cell from another and thus permit cellular individuality. The plasma membrane has selective permeabilities and acts as a barrier, thereby maintaining differences in composition between the inside and the outside of the cell. The selective permeabilities are provided by channels and pumps for ions and substrates and by specific receptors for signals, eg, hormones. Plasma membranes also exchange material with the extracellular environment by exocytosis and endocytosis, and there are special areas of membrane structures—the gap junctions—through which adjacent cells exchange material.

Membranes also form specialized compartments within the cell. Such intracellular membranes form many of the morphologically distinguishable structures (organelles), eg, mitochondria, endoplasmic reticulum, sarcoplasmic reticulum, Golgi complexes, secretory granules, lysosomes, and the nuclear membrane. Membranes localize enzymes, function as integral elements in excitation-response coupling, and provide sites of energy transduction, such as in photosynthesis and oxidative phosphorylation.

BIOMEDICAL IMPORTANCE

Gross alterations of membrane structure can affect water balance and ion flux and therefore every process within the cell. Specific deficiencies or alterations of certain membrane components lead to a variety of diseases. Examples include the lysosomal absence of acid maltase, causing type II glycogen storage disease; the lack of an iodide transporter, causing congenital goiter (see Fig 47–2); and defective endocytosis of low-density lipoproteins, resulting in accelerated hypercholesterolemia and coronary artery disease. Normal cellular function obviously begins with normal membranes.

THE MAINTENANCE OF A NORMAL ENVIRONMENT AROUND AND IN A CELL IS A FUNDAMENTAL REQUIREMENT

Life originated in an aqueous environment; enzyme reactions, cellular and subcellular processes, and so forth have therefore evolved to work in this milieu. Since most mammals live in a gaseous environment, how is the aqueous state maintained? Membranes accomplish this by internalizing and compartmentalizing body water.

The Body's Internal Water Is Compartmentalized

Water makes up about 56% of the lean body mass of the human body and is distributed in 2 large compartments.

Intracellular Fluid (ICF): This compartment constitutes two-thirds of total water and provides the environment for the cell to (1) make, store, and utilize energy; (2) repair itself; (3) replicate; and (4) perform special functions.

Extracellular Fluid (ECF): This compartment contains about one-third of total water and is distributed between the plasma and interstitial compartments. The extracellular fluid is a delivery system. It brings to the cells nutrients (eg, glucose, fatty acids, amino acids), oxygen, various ions and trace minerals, and a variety of regulatory molecules (hormones) that coordinate the functions of widely separated cells. Extracellular fluid removes CO_2, waste products, and toxic or detoxified materials from the immediate cellular environment.

The Ionic Composition of Intracellular Fluid Differs Greatly From That of Extracellular Fluid

As illustrated in Table 43–1, the internal environment is rich in K^+ and Mg^{2+}, and phosphate is its major anion. Extracellular fluid is characterized by high Na^+ and Ca^{2+} content and Cl^- is the major anion. Note also that glucose is higher in extracellular fluid than in the cell, whereas the opposite is true for proteins. Why is there such a difference? It is thought that the primordial sea in which life originated was rich in K^+ and Mg^{2+}. It therefore follows that enzyme reactions and other biologic processes evolved to function best in that environment, hence the high concentration of these ions within cells. Cells were faced with strong selection pressure as the sea gradually changed to a composition rich in Na^+ and Ca^{2+}. Vast changes would have been required for evolution of a completely new set of biochemical and physiologic machinery; instead, as it happened, cells developed barriers—membranes with associated "pumps"—to maintain the internal microenvironment.

MEMBRANES ARE COMPLEX STRUCTURES COMPOSED OF LIPIDS, PROTEINS, AND CARBOHYDRATES

Different membranes within the cell and between cells have different compositions, as reflected in the ratio of protein to lipid (Fig 43–1). The difference is not surprising given the very different functions of membranes. Membranes are asymmetric sheetlike enclosed structures with an inside and an outside surface. These sheetlike structures are noncovalent assemblies that are thermodynamically stable and metabolically active. Specific protein molecules are anchored in membranes, where they carry out specific functions of the organelle, the cell, or the organism.

The Major Lipids in Mammalian Membranes Are Phospholipids, Glycosphingolipids, and Cholesterol

Phospholipids: Of the 2 major phospholipid groups present in membranes, **phosphoglycerides are**

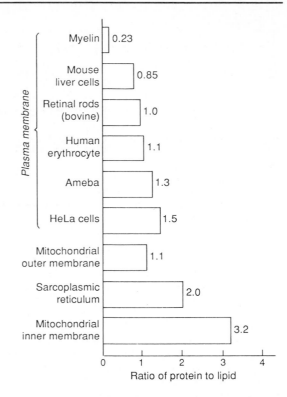

Figure 43–1. Ratio of protein to lipid in different membranes. Proteins equal or exceed the quantity of lipid in nearly all membranes. The outstanding exception is myelin, an electrical insulator found on many nerve fibers.

the more common and consist of a glycerol backbone to which are attached 2 fatty acids in ester linkage and a phosphorylated alcohol (Fig 43–2). The fatty acid constituents are usually even-numbered carbon molecules, most commonly containing 14 or 16 carbons. They are unbranched and can be saturated or unsaturated. The simplest phosphoglyceride is phosphatidic acid, which is 1,2-diacylglcerol 3-phosphate, a key intermediate in the formation of all other phospho-

Table 43–1. Comparison of the mean concentration of various substances outside and inside a mammalian cell.

Substance	Extracellular Fluid	Intracellular Fluid
Na^+	140 mmol/L	10 mmol/L
K^+	4 mmol/L	140 mmol/L
Ca^{2+} (free)	2.5 mmol/L	0.1 μmol/L
Mg^{2+}	1.5 mmol/L	30 mmol/L
Cl^-	100 mmol/L	4 mmol/L
HCO_3^-	27 mmol/L	10 mmol/L
PO_4^{3-}	2 mmol/L	60 mmol/L
Glucose	5.5 mmol/L	0–1 mmol/L
Protein	2 g/dL	16 g/dL

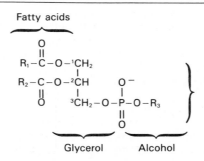

Figure 43–2. A phosphoglyceride showing the fatty acids (R_1 and R_2), glycerol, and phosphorylated alcohol components. In phosphatidic acid, R_3 is hydrogen.

lipids (see Chapter 25). In other phospholipids, the 3-phosphate is esterified to an alcohol such as ethanolamine, choline, serine, glycerol, or inositol (see pp 134, 135).

The second class of phospholipids is composed of **sphingomyelins**, which contain a sphingosine backbone rather than glycerol. A fatty acid is attached by an amide linkage to the amino group of sphigosine. The primary hydroxyl group of sphingosine is esterified to phosphorylcholine. Sphingomyelins, as the name implies, are prominent in myelin sheaths.

Glycosphingolipids: The glycosphingolipids are **sugar-containing lipids** such as cerebrosides and gangliosides and are also derived from sphingosine. The **cerebrosides** and **gangliosides** differ from sphingomyelin in the moiety attached to the primary hydroxyl group of sphingosine. In sphingomyelin, a phosphorylcholine is attached to the alcohol group. A cerebroside contains a single hexose moiety, glucose or galactose, at that site. A ganglioside contains a chain of 3 or more sugars, at least one of which is a sialic acid, attached to the primary alcohol of sphingosine.

Sterols: The most common sterol in membranes is **cholesterol,** which exists almost exclusively in the plasma membranes of mammalian cells but can also be found in lesser quantity in mitochondria, Golgi complexes, and nuclear membranes. Cholesterol is generally more abundant toward the outside of the plasma membrane. Cholesterol intercalates among the phospholipids of the membrane, with its hydroxyl group at the aqueous interface and the remainder of the molecule within the leaflet. At temperatures above the transition temperature (see discussion of fluid mosaic model, below), its rigid sterol ring interacts with the acyl chains of the phospholipids, limits their movement, and thus decreases membrane fluidity. On the other hand, when the temperature approaches the transition temperature, the interaction of cholesterol with the acyl chains interferes with their alignment with each other; this phenomenon lowers the temperature at which the fluid → gel transition occurs, thus assisting in keeping the membrane fluid at lower temperatures.

Membranes Are Amphipathic

All major lipids in membranes contain both hydrophobic and hydrophilic regions and are therefore termed **amphipathic**. Membranes themselves are thus amphipathic. If the hydrophobic regions were separated from the rest of the molecule, it would be insoluble in water but soluble in oil. Conversely, if the hydrophilic region were separated from the rest of the molecule, it would be insoluble in oil but soluble in water. The amphipathic membrane lipids have a polar head group and nonpolar tails, as represented in Fig 43–3. Saturated fatty acids make straight tails, whereas unsaturated fatty acids, which generally exist in the *cis* form in membranes, make kinked tails. As more kinks are inserted in the tails, the membrane becomes less tightly packed and therefore more fluid. Detergents are amphipathic molecules that are important

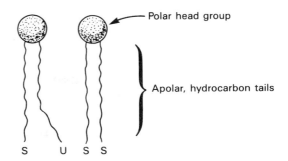

Figure 43–3. Diagrammatic representation of a phospholipid or other membrane lipid. The polar head group is hydrophilic, and the hydrocarbon tails are hydrophobic or lipophilic. The fatty acids in the tails are saturated (S) or unsaturated (U); the former are usually attached to carbon 1 of glycerol and the latter to carbon 2.

in biochemistry and in the household. The molecular structure of a detergent is not unlike that of a phospholipid.

Membrane Lipids Are Organized Into Bilayers

The amphipathic character of phospholipids suggests that the 2 regions of the molecule have incompatible solubilities; however, in a solvent such as water, phospholipids organize themselves into a form that thermodynamically satisfies both regions. A micelle (Fig 43–4) is such a structure; the hydrophobic regions are shielded from water, while the hydrophilic polar groups are immersed in the aqueous environment.

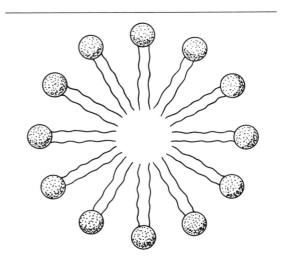

Figure 43–4. Diagrammatic cross section of a micelle. The polar head groups are bathed in water, whereas the hydrophobic hydrocarbon tails are surrounded by other hydrocarbons and thereby protected from water. Micelles are spherical structures.

The Lipid Bilayer: As recognized nearly 60 years ago by Gorter and Grendel, a bimolecular layer, or bilayer, can also satisfy the thermodynamic requirements of amphipathic molecules in an aqueous environment. A bilayer exists as a sheet in which the hydrophobic regions of the phospholipids are protected from the aqueous environment, while the hydrophilic regions are immersed in water (Fig 43–5). Only the ends or edges of the bilayer sheet are exposed to an unfavorable environment, but even these exposed edges can be eliminated by folding the sheet back upon itself to form an enclosed vesicle with no edges. The closed bilayer provides one of the essential properties of membranes. It is impermeable to most water-soluble molecules, since they would be insoluble in the hydrophobic core of the bilayer.

Two questions immediately arise. First, **how many biologic materials are lipid soluble and can therefore readily enter the cell?** Gases such as oxygen, CO_2, and nitrogen—small molecules with little interaction with solvents—readily diffuse through the hydrophobic regions of the membrane. Lipid-derived molecules, eg, steroid hormones, readily traverse the bilayer. Organic nonelectrolyte molecules exhibit diffusion rates that are dependent upon their oil-water partition coefficients (Fig 43–6); the greater the lipid solubility of a molecule, the greater its diffusion rate across the membrane.

The second question concerns molecules that are not lipid soluble. How are the transmembrane concentration gradients for non-lipid-soluble molecules maintained? The answer is that membranes contain proteins, and proteins are also amphipathic molecules that can be inserted into the correspondingly amphipathic lipid bilayer. Proteins form channels for the movement of ions and small molecules and serve as transporters for larger molecules that otherwise could not pass the bilayer. These processes are described below.

Membrane Proteins Are Associated With the Lipid Bilayer

Membrane phospholipids act as a solvent for mem-

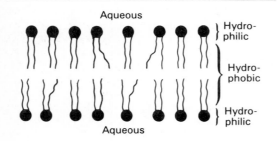

Figure 43–5. Diagram of a section of a bilayer membrane formed from phospholipid molecules. The unsaturated fatty acid tails are kinked and lead to more spacing between the polar head groups, hence more room for movement. Thus under the region more "fluid." (Slightly modified and reproduced, with permission, from Stryer L: *Biochemistry,* 2nd ed. Freeman, 1981.)

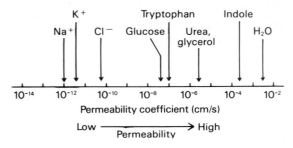

Figure 43–6. Permeability coefficients of water, some ions, and other small molecules in lipid bilayer membranes. Molecules that move rapidly through a given membrane are said to have a high permeability coefficient. (Slightly modified and reproduced, with permission, from Stryer L: *Biochemistry,* 2nd ed. Freeman, 1981.)

brane proteins, creating an environment in which the latter can function. Of the 20 amino acids contributing to the primary structure of proteins, the functional groups attached to the α carbon are strongly hydrophobic in 6, weakly hydrophobic in a few, and hydrophilic in the remainder. As described in Chapter 6, the α-helical structure of proteins minimizes the hydrophilic character of the peptide bonds themselves. Thus, proteins can be amphipathic and form an integral part of the membrane by having hydrophilic regions protruding at the inside and outside faces of the membrane but connected by a hydrophobic region traversing the hydrophobic core of the bilayer. In fact, those portions of membrane proteins that traverse membranes do contain substantial numbers of hydrophobic amino acids and a high α-helical or β-pleated sheet content.

The number of different proteins in a membrane varies from 6 to 8 in the sarcoplasmic reticulum to over 100 in the plasma membrane. The proteins consist of enzymes, transport proteins, structural proteins, antigens (eg, for histocompatibility), and receptors for various molecules. Because every membrane possesses a different complement of proteins, there is no such thing as a typical membrane structure. The enzymatic properties of several different membranes are shown in Table 43–2.

Table 43–2. Enzymatic markers of different membranes.*

Membrane	Enzyme
Plasma	5' nucleotidase Adenylate cyclase Na$^+$/K$^+$-ATPase
Endoplasmic reticulum	Glucose-6-phosphatase
Golgi complex	Galactosyltransferase
Inner mitochondrial membrane	ATP synthase

*Membranes contain many proteins, some of which have enzymatic activity. Some of these enzymes are located only in certain membranes and can therefore be used as markers to follow the purification of these membranes.

Membranes and their components are dynamic structures. The lipids and proteins in membranes turn over, just as they do in other compartments of the cell. Different lipids have different turnover rates, and the turnover rates of individual species of membrane proteins may vary widely. The membrane itself can turn over even more rapidly than any of its constituents. This is discussed in more detail in the section of endocytosis.

Asymmetry Is an Important Feature of Membranes

This asymmetry can be partially attributed to the irregular distribution of proteins within the membranes. An inside-outside asymmetry is also provided by the external location of the carbohydrates attached to membrane proteins. In addition, specific enzymes are located exclusively on the outside or inside of membranes, as in the mitochondrial and plasma membranes.

There are regional asymmetries in membranes. Some, such as occur at the villous border of mucosal cells, are almost macroscopically visible. Others, such as those at gap junctions, tight junctions, and synapses, occupy much smaller regions of the membrane and generate correspondingly smaller local asymmetries.

There is also inside-outside (transverse) asymmetry of the phospholipids. The choline-containing phospholipids (phosphatidylcholine and sphingomyelin) are located mainly in the outer molecular layer; the aminophospholipids (phosphatidylserine and ethanolamine) are preferentially located in the inner layer. Cholesterol is generally present in larger amounts on the outside than on the inside. Obviously, if this asymmetry is to exist at all, there must be limited transverse mobility (flip-flop) of the membrane phospholipids. In fact, phospholipids in synthetic bilayers exhibit an extraordinarily slow rate of flip-flop; the half-life of the asymmetry can be measured in days or weeks. However, when certain membrane proteins such as the erythrocyte protein glycophorin are inserted artificially into synthetic bilayers, the frequency of phospholipid flip-flop may increase as much as 100-fold.

The mechanisms involved in the establishment of lipid asymmetry are not understood. The enzymes involved in the synthesis of phospholipids are located on the cytoplasmic side of microsomal membrane vesicles. It has therefore been postulated that translocases exist which transfer certain phospholipids from the inner leaflet to the outer. Also, specific proteins that preferentially bind individual phospholipids may be present in the 2 leaflets, leading to the asymmetric distribution of these lipid molecules.

There Are Integral and Peripheral Membrane Proteins

Most membrane proteins are integral components of the membrane (they interact with the phospholipids), and in fact all of those which have been adequately studied span the entire 5- to 10-nm transverse distance of the bilayer. These integral proteins are usually globular and are themselves amphipathic. They consist of 2 hydrophilic ends separated by an intervening hydrophobic region that traverses the hydrophobic core of the bilayer. As the structures of integral membrane proteins are being elucidated, it is apparent that some (notably transporter molecules) may span the bilayer many times, as illustrated in Fig 43–9.

Integral proteins are asymmetrically distributed across the membrane bilayer (Fig 43–7).

Some proteins must be given their asymmetric orientation in the membrane at the time of their insertion in the lipid bilayer. The hydrophilic external region of an amphipathic protein, which is clearly synthesized inside the cell, must traverse the hydrophobic core of the membrane and eventually be found outside the membrane. The molecular mechanisms of membrane assembly are discussed below.

Peripheral proteins do not interact directly with the phospholipids in the bilayer but are instead by definition weakly bound to the hydrophilic regions of specific integral proteins. For example, ankyrin, a peripheral protein, is bound to the integral protein "band III" of erythrocyte membrane. Spectrin, a cytoskeletal structure within the erythrocyte, is in turn bound to ankyrin and thereby plays an important role in the maintenance of the biconcave shape of the erythrocyte. The immunoglobulin molecules on the plasma membranes of lymphocytes are integral membrane proteins and can be released by the shedding of small fragments of the membrane. Many hormone receptor molecules are integral proteins, and the specific polypeptide hormones that bind to these receptor molecules may therefore be considered peripheral proteins. Peripheral proteins, such as peptide hormones, may even organize the distribution of integral proteins, such as their receptors, within the plane of the bilayer (see below).

ARTIFICIAL MEMBRANES CAN BE USED TO STUDY MEMBRANE FUNCTION

Artificial membrane systems can be prepared by appropriate techniques. These systems generally consist of mixtures of one or more phospholipids of natural or synthetic origin that can be treated (eg, by using mild sonication) to form spherical vesicles in which the lipids form a bilayer. Such vesicles, surrounded by a lipid bilayer, are termed liposomes.

Some of the advantages and uses of artificial membrane systems in the study of membrane function follow.

(1) The lipid content of the membranes can be varied, allowing systematic examination of the effects of varying lipid composition on certain functions. For instance, vesicles can be made that are composed solely of phosphatidylcholine or, alternatively, of known mixtures of different phospholipids, glycolipids, and cholesterol. The fatty acid moieties of the lipids used can also be varied by employing synthetic

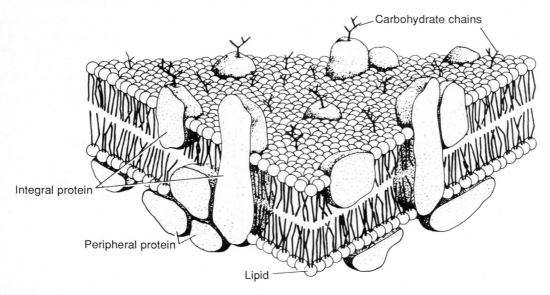

Figure 43–7. The fluid mosaic model of membrane structure. The membrane consists of a bimolecular lipid layer with proteins inserted in it or bound to the cytoplasmic surface. Integral membrane proteins are firmly embedded in the lipid layers. Some of these proteins completely span the bilayer and are called transmembrane proteins, while others are embedded in either the outer or inner leaflet of the lipid bilayer. Loosely bound to the inner surface of the membrane are the peripheral proteins. Many of the proteins and lipids have externally exposed oligosaccharide chains. (Reproduced, with permission, from Junqueira LC, Carneiro J, Long JA: *Basic Histology,* 5th ed. Appleton & Lange, 1986.)

lipids of known composition to permit systematic examination of the effects of fatty acid composition on certain membrane functions (eg, transport).

(2) Purified membrane proteins or enzymes can be incorporated into these vesicles in order to assess what factors (eg, specific lipids or ancillary proteins) the proteins require to reconstitute their function,. Investigations of purified proteins, eg, the Ca^{2+}/ATPase of the sarcoplasmic reticulum, have in certain cases suggested that only a single protein and a single lipid are required to reconstitute an ion pump.

(3) The environment of these systems can be rigidly controlled and systematically varied (eg, ion concentrations). The systems can also be exposed to known ligands if, for example, the liposomes contain specific receptor proteins.

(4) When liposomes are formed, they can be made to entrap certain compounds inside themselves, eg, drugs and isolated genes. There is interest in using liposomes to distribute drugs to certain tissues, and if components (eg, antibodies to certain cell surface molecules) could be incorporated into liposomes so that they would be targeted to specific tissues or tumors, the therapeutic impact could be considerable. DNA entrapped inside liposomes appears to be less sensitive to attack by nucleases; this approach may prove useful in attempts at gene therapy.

FUNCTIONAL MEMBRANES ARE 2-DIMENSIONAL SOLUTIONS OF GLOBULAR INTEGRAL PROTEINS DISPERSED IN A FLUID PHOSPHOLIPID MATRIX

This fluid mosaic model of membrane structure was proposed in 1972 by Singer and Nicolson (Fig 43–7). Early evidence for the model was the rapid and random redistribution of species-specific integral proteins in the plasma membrane of an interspecies hybrid cell formed by the artificially induced fusion of 2 different parent cells. It has subsequently been demonstrated that phospholipids also undergo rapid redistribution in the plane of the membrane. This diffusion within the plane of the membrane, termed translational diffusion, can be quite rapid for a phospholipid; in fact, within the plane of the membrane, one molecule of phospholipid can move several micrometers per second.

The phase changes, and thus the fluidity of membranes, are highly dependent upon the lipid composition of the membrane. In a lipid bilayer, the hydrophobic chains of the fatty acids can be highly aligned or ordered to provide a rather stiff structure. As the temperature increases, the hydrophobic side chains undergo a transition from the ordered state to a disordered one, taking on a more liquidlike or fluid arrangement. The temperature at which the structure undergoes the transition from ordered to disordered is called the transition temperature. The longer and more saturated fatty acid chains exhibit higher transition temperatures, ie, higher temperatures are required to increase the fluidity of the structure. Unsaturated bonds that exist in the *cis* configuration tend to increase the fluidity of a bilayer by decreasing the compactness of the side chain packing without diminishing hydrophobicity (Fig 43–3). The phospholipids of cellular membranes generally contain at least one unsaturated fatty acid with at least one *cis* double bond.

Cholesterol also acts as a moderator molecule in membranes, producing intermediate states of fluidity. If the acyl side chains exist in a disordered phase, cholesterol will have a condensing effect; if the acyl side chains are ordered or in a crystalline phase, cholesterol will induce disorder. At high cholesterol: phospholipid ratios, transition temperatures are abolished altogether.

The fluidity of a membrane significantly affects its functions. As membrane fluidity increases, so does its permeability to water and other small hydrophilic molecules. The lateral mobility of integral proteins increases as the fluidity of the membrane increases. If the active site of an integral protein involved in some given function resides exclusively in its hydrophilic regions, changing lipid fluidity will probably have little effect on the activity of the protein; however, if the protein is involved in a transport function in which transport components span the membrane, lipid phase effects may significantly alter the transport rate. The insulin receptor is an excellent example of altered function with changes in fluidity (see Chapter 52). As the concentration of unsaturated fatty acids in the membrane is increased (by growing cultured cells in a medium rich in such molecules), fluidity increases. This alters the receptor so that it binds more insulin.

A state of fluidity and thus of translational mobility in a membrane may be confined to certain regions of membranes under certain conditions. For example, protein-protein interaction may take place within the plane of the membrane, such that the integral proteins form a rigid matrix, in contrast to the more usual situation, where the lipid acts as the matrix. Such regions of rigid protein matrix can exist side by side in the same membrane with the usual lipid matrix. Gap junctions, tight junctions, and bacteriorhodopsin-containing regions of the purple membranes of halobacteria are clear examples of such side-by-side coexistence of different matrices.

Some of the protein-protein interactions taking place within the plane of the membrane may be mediated by interconnecting peripheral proteins, such as cross linking antibodies or lectins that are known to patch or cap on membrane surfaces. Thus, peripheral proteins, by their specific attachments, may restrict the mobility of integral proteins within the membrane.

MEMBRANES ARE ASSEMBLED

Lipids and proteins must both be considered in models of membrane assembly. Coverage of the former will be brief, as little is known of how lipids are assembled into membranes. Assembly of proteins into the membranes of the endoplasmic reticulum (ER), the

Golgi apparatus (GA) and the cell surface (plasma membrane, PM) of animal cells is given special attention. Some points that have emerged with regard to the biosynthesis of proteins of other membranes (eg, mitochondrial) will also be discussed. The observation that specific amino acid sequences (or mannose 6-phosphate in the case of the lysosome) target certain proteins to their unique cellular destinations is of great importance. The locating of such proteins in their appropriate organelles is said to be **signal mediated.** The elucidation of targeting sequences has been greatly facilitated by recombinant DNA technology, through use of which specific sequences in a protein may be deleted or added. The cDNAs for such altered proteins can be introduced into suitable cells by transfection; alterations in cellular distribution are then detected by the use of suitable fluorescent antibodies or other techniques.

Secreted proteins generally follow the route ER → GA → PM, prior to exiting cells. Thus, some aspects of their biosynthesis will be covered. It has become apparent that such proteins must also carry specific targeting sequences that mark them for secretion, or alternatively they must lack signals that target other proteins to become residents of membranes. Evidence that will permit distinction between these two possibilities is just becoming available.

Membrane Asymmetry Is Maintained During Assembly

The Golgi apparatus and the ER vesicles exhibit transverse asymmetries of both lipid and protein, and these asymmetries are maintained during fusion with the plasma membrane. The inside of the vesicle after fusion becomes the outside of the plasma membrane, and the cytoplasmic side of the vesicles remains the cytoplasmic side of the membrane (Fig 43–8). Since the transverse asymmetry of the membranes already exists in the vesicles of the ER well before they are fused to the PM, a major problem of membrane assembly becomes understanding how the integral proteins are inserted asymmetrically into the lipid bilayer of the ER.

Phospholipids are the major class of lipid in membranes. The enzymes responsible for the synthesis of phospholipids reside on the cytoplasmic surface of the cisternae of the endoplasmic reticulum (ER). As phospholipids are synthesized at that site, they probably self-assemble into thermodynamically stable bimolecular layers, thereby expanding the membrane and promoting the detachment of **lipid vesicles** from it. This process is known as **membrane budding.** These lipid vesicles then appear to migrate and fuse with other membranes such as those of the Golgi apparatus, which in turn eventually fuse with the plasma membrane. Cytosolic proteins that take up phospholipids from one membrane and release them to another, **phospholipid exchange proteins,** probably also play a role in contributing to the specific lipid composition of various membranes.

The variety of ways in which proteins are inserted

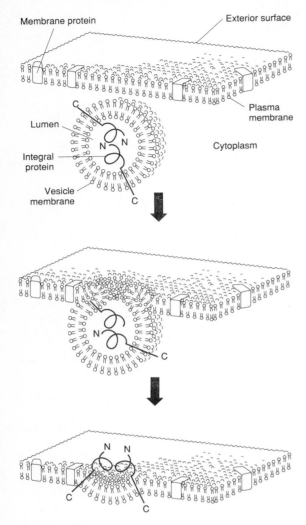

Figure 43–8. Fusion of a vesicle with the plasma membrane preserves the orientation of any integral proteins embedded in the vesicle bilayer. Initially, the N terminus of the protein faces the lumen, or inner cavity, of such a vesicle. After fusion, the N terminus is on the exterior surface of the plasma membrane. That the orientation of the protein has not been reversed can be perceived by noting that the other end of the molecule, the C terminus, is always immersed in the cytoplasm. The lumen of a vesicle and the outside of the cell are topologically equivalent. (Redrawn and modified, with permission, from Lodish HF, Rothman JE: The assembly of cell membranes. *Sci Am* [Jan] 1979; **240**:43.)

into membranes is shown in Fig 43–9. Models for each of these types of insertion will not be presented, but an indication of the major features known about insertion of proteins into membranes of the ER and GA and into the PM will be discussed.

The "Signal Hypothesis" Describes Membrane Assembly

The signal hypothesis was proposed by Sabatini and Blobel. These investigators found that proteins synthesized on **membrane-bound polyribosomes** (ie, secreted proteins, some integral plasma membrane proteins, lysosomal enzymes, proteins of the membranes of the GA, some integral proteins of membranes of the ER) contained **a peptide extension** (signal peptide) at their N-termini which mediated their attachment to the membranes of the ER. Proteins whose entire synthesis occurs on **free polyribosomes** (eg, cytosolic proteins, extrinsic proteins of the inner face of the plasma membrane, mitochondrial proteins encoded by nuclear DNA, peroxisomal proteins) lack this signal peptide. An important aspect of the signal hypothesis was that it suggested—as appears to be the case—that all cytoplasmic ribosomes have the same structure, and that the distinction between membrane-bound and free ribosomes depended solely on the former carrying proteins that have signal peptides. A variety of experimental evidence has confirmed the original hypothesis. Because many membrane proteins are synthesized on membrane-bound polyribosomes, the signal hypothesis plays an important role in concepts of membrane assembly. Some characteristics of signal peptides are summarized in Table 43–3.

Fig 43–10 illustrates the principal features of the signal hypothesis in relation to the passage of a secreted protein through the membrane of the ER. The mRNA for such a protein encodes for a signal peptide at the N-terminus. The signal hypothesis proposes that the protein is inserted into the membrane simultaneously with the translation of its mRNA on polyribosomes, so-called **cotranslational insertion.** As the leader sequence of the protein emerges from the ribosome, it is recognized by a signal recognition particle (SRP) that blocks further translation after about 70 amino acids have been polymerized (40 buried in the large ribosomal complex and 30 exposed). The SRP contains 6 proteins and has associated with it a 7S RNA that is closely related to the "Alu family" of highly repeated DNA sequences (see Chapter 38). The SRP-imposed block is not released until the SRP-leader sequence-ribosome complex has bound to the so-called docking protein (a receptor for the SRP) on the ER. Cotranslational insertion of the protein into the ER then commences at that site. The process of elongation of the remaining portion of the protein molecule drives the nascent protein across the lipid bilayer as the ribosomes remain attached to the ER. Thus, the rough (or ribosome-studded) ER is formed. Ribosomes remain attached to the ER during synthesis of the membrane protein but are released and dissociated into their respective subunits as the protein is completed. The leader sequence is cleaved off by signal peptidase and carbohydrate is attached as the early synthesized portion of the protein enters the lumen of the ER.

Integral ER membrane proteins, such as cytochrome P-450, do not completely cross the membrane. Instead they reside in the membrane of the ER with the signal peptide intact. Apparently their passage through that membrane is prevented by a sequence of amino acids called a **stop signal.**

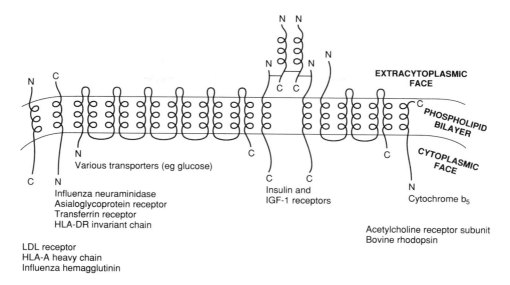

Various transporters (eg glucose)

Influenza neuraminidase
Asialoglycoprotein receptor
Transferrin receptor
HLA-DR invariant chain

LDL receptor
HLA-A heavy chain
Influenza hemagglutinin

Insulin and
IGF-1 receptors

Cytochrome b₅

Acetylcholine receptor subunit
Bovine rhodopsin

Figure 43–9. Variations in the way proteins are inserted into membranes. This schematic representation, which illustrates a number of possible orientations, shows the portion of the peptide within the membrane as α helices and the other portions as lines. N, NH₂ terminus; C, COOH terminus. (Adapted, with permission, from Wickner WT, Lodish HF; Multiple mechanisms of protein insertion into and across membranes. *Science* 1985;**230**:400. Copyright © 1985 by the American Association for the Advancement of Science.)

Secreted proteins completely traverse the membrane bilayer and are discharged into the lumen of the ER. Carbohydrate moieties are attached (generally from dolichol-pyrophosphate oligosaccharide; see Chapter 57) as these proteins traverse the inner part of the ER membrane, a process called **cotranslational glycosylation.** Subsequently, the secretory proteins are found in the lumen of the GA, where their carbohydrate chains are modified prior to secretion.

There is strong evidence that the leader sequence is involved in the process of protein insertion into ER membranes. Mutant proteins containing altered leader sequences in which a hydrophobic amino acid is replaced by a hydrophilic one are not inserted into ER membranes. Nonmembrane proteins (eg, hemoglobin), to which leader sequences have been attached by genetic engineering, can be inserted into membranes or even secreted.

Proteins Can Be Inserted Into the Membranes of the ER by a Number of Different Mechanisms

The mechanisms used to insert proteins into membranes include:

1. Cotranslational insertion, mediated by a stop signal—eg, cytochrome P-450 (see above).

2. Synthesis on free polyribosomes and subsequent attachment to the ER membrane—eg, cytochrome b₅.

3. Retention at the luminal aspect of the ER by specific amino acid sequences. Recent work has indicated that a number of proteins possess the amino acid sequence KDEL (Lys-Asp-Glu-Leu) at their C-terminus. This sequence specifies that such proteins will be

attached to the inner aspect of the cisternae of the ER, in a relatively loose manner.

4. Certain proteins destined for the membranes of the ER may pass to the Golgi by vesicular transport and then return to the ER to be inserted therein.

The above list indicates that a variety of mechanisms are involved in the biosynthesis of the proteins of ER membranes; a similar situation probably holds for other membranes (eg, mitochondrial membranes and the plasma membrane). Targeting sequences have only been identified for a few of the above mechanisms (eg, KDEL sequences). It has been shown that the turnover of the lipids of the ER membranes of rat liver is generally more rapid than that of its proteins, so that lipid and protein turnover are independent. Indeed, different lipids have been found to exhibit different turnovers. Also, the turnover rates of the proteins of these membranes vary quite widely, some exhibiting rapid (hours) and others quite slow (days) turnovers. Thus, individual lipids and proteins of the ER membranes appear to be inserted into these membranes relatively independently; this appears to be the case for most membranes.

Table 43–3. Some properties of signal peptides.

Usually, but not always, located at the N-terminus.
Contain approximately 12–35 amino acids.
Methionine is usually the N-terminal amino acid.
Contain a central cluster of hydrophobic amino acids.
Contain at least one positively charged amino acid near their N-terminus.
Usually cleaved off at the C-terminal end of an Ala residue by signal peptidase.

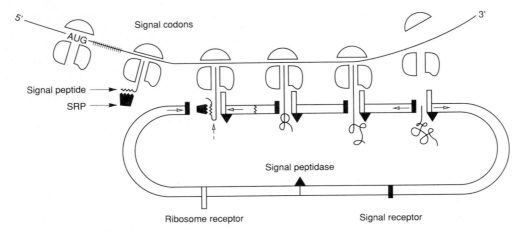

Figure 43–10. Diagram of the signal hypothesis for the transport of secreted proteins across the endoplasmic reticulum membrane. The ribosomes synthesizing a protein move along the messenger RNA specifying the amino acid sequence of the protein. (The messenger is represented by the line between 5' and 3'.) The codon AUG marks the start of the message for the protein; the hatched lines that follow AUG represent the codons for the signal sequence. As the protein grows out from the larger ribosomal subunit, the signal sequence is exposed and bound by the signal recognition particle (SRP). Translation is blocked until the complex binds to the "docking protein" (represented by the solid bar) on the endoplasmic reticulum membrane. There is also a receptor (open bar) for the ribosome itself. The interaction of the ribosome and growing peptide chain with the endoplasmic reticulum membrane results in the opening of a pore through which the protein is transported to the interior space of the endoplasmic reticulum. During transport, the signal sequence of most proteins is removed by an enzyme called the signal peptidase. The completed protein is eventually released by the ribosome, which then separates into its 2 components, the large and small ribosomal subunits. The protein ends up inside the endoplasmic reticulum. (Slightly modified and reproduced, with permission, from: Newly made proteins zip through the cell. *Science* 1980;**207**:154. Copyright © 1980 by the American Association for the Advancement of Science.)

Proteins Move Through Cellular Compartments to Specific Membranes

A scheme representing the possible flow of membrane proteins along the ER → GA → PM route is shown in Fig 43–11. The horizontal arrows denote transport steps that may be independent of targeting signals. Thus flow of certain membrane proteins from the ER to the PM (designated **bulk flow,** as it is non-selective) may occur without any targeting sequences being involved—ie, by default. On the other hand, insertion of **resident** proteins into the ER and Golgi membranes may be dependent upon specific signals (eg, KDEL or stop sequences for the ER). Similarly, transport to lysosomes and into secretory storage granules may also be signal mediated (mannose 6-phosphate in the case of certain lysosomal enzymes).

The major mechanism for proteins that are synthesized on membrane-bound polyribosomes to reach the GA or PM by bulk flow appears to involve **vesicular transport.** How proteins that are synthesized in the rough ER are inserted into appropriate vesicles is not known. The vesicles involved in bulk flow appear to be clathrin free, whereas those involved in targeting pro-

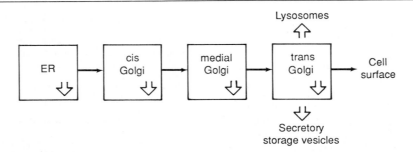

Figure 43–11. Flow of membrane proteins from the ER to the cell surface. Horizontal arrows denote steps that have been proposed to be signal independent and thus represent bulk flow. The open vertical arrows in the boxes indicate denote retention of proteins that are resident in the membranes of the organelle indicated. The open vertical arrows outside the boxes indicate signal-mediated transport to lysosomes and secretory storage granules. (From Pfeffer & Rothman, 1987.)

teins to lysosomes and into secretory storage granules appear to be coated with this protein. Specific sequences that encode targeting signals indicating Golgi and PM destinations have not yet been revealed. It is likely that some proteins destined for insertion into the PM by vesicular transport may not contain specific signals, and that their final destination is determined by default. Whether individual vesicles (easily visualized by electron microscopy) contain only one protein, or a multitude of proteins, remains to be determined. The GA plays two important roles in membrane synthesis: (i) it is involved in the **modification** (processing) of the oligosaccharide chains of membrane and other N-linked glycoproteins (Chap 57); (ii) it is involved in the **sorting** of various proteins for their appropriate intracellular destination. All parts of the GA participate in the first role, whereas the *trans* Golgi is particularly involved in the second and is very rich in vesicles. Certain proteins of the inner face of the PM may be synthesized on free polyribosomes and and thereafter become incorporated into ER membranes. The recent development of *in vitro* systems for studying vesicular transport and the fate of their transported proteins should clarify many problems in this area.

Mitochondrial Proteins Come From Inside & Outside This Organelle

Mitochondria contain many proteins. Certain proteins (eg, some integral membrane proteins) are synthesized in mitochondria using the mitochondrial protein synthesizing system. However, the majority are synthesized outside the mitochondria and must be imported. Yeast cells have proved to be a particularly useful system for studying the mechanisms of import of such mitochondrial proteins. Most progress has been made in the study of the proteins present in the **mitochondrial matrix** (eg, the F_1-ATPase subunits). These proteins are synthesized on free polyribosomes. Subsequently, they pass through the outer and inner mitochondrial membranes to reach their destination. Evidence suggests that they must be in an unfolded state to pass through these membranes. ATP must be present for protein import to occur. It is not clear whether ATP affects the electric potential or the proton-motive force across the inner membrane, or whether it plays a role in unfolding the imported proteins. It has been found that these proteins contain a leader sequence, up to 70 amino acids in length, which must also contain some positively charged amino acids. This sequence is equivalent to a signal peptide, directing these proteins into the matrix; if the peptide extension is cleaved off, these proteins will not enter. There are probably receptors for the imported proteins on the surface of the outer mitochondrial membrane. The presequence is split off by a metalloprotease present in the matrix. Certain other mitochondrial proteins do not contain presequences, so that a variety of mechanisms and routes are employed by proteins to enter mitochondria.

Table 43–4. Sequences or compounds that direct proteins to specific organelles.

Targeting Sequence or Compound	Organelle Targeted
Signal peptide sequence	Membrane of the ER
KDEL sequence (Lys·Asp·Glu·Leu)	Luminal surface ER
N-terminal sequence (70-residue positive region)	Mitochondrion
Short, basic amino acid sequences	Nucleus
Mannose-6-phosphate	Lysosome

Lysosomes & Other Organelles Are Special Cases

As described in Chapter 57, certain hydrolases destined for lysosomes contain mannose 6-phosphate residues that mark them for that specific destination. Evidence is accumulating that distinct sequences exist that mark proteins for other organelles such as nuclei and peroxisomes; this is an active area of research. Some of the sequences or signals that have been elucidated for targeting various proteins for their correct intracellular membrane sites are summarized in Table 43–4.

THEIR SELECTIVITY ALLOWS SPECIALIZED FUNCTIONS OF MEMBRANES

If the plasma membrane is relatively impermeable, how do most molecules enter a cell? How is selectivity of this movement established? Answers to such questions are important in understanding how cells adjust to a constantly changing extracellular environment. Metazoan organisms also must have means of communicating between adjacent and distant cells, so that complex biologic processes can be coordinated. These signals must arrive at and be transmitted by the membrane, or they must be generated as a consequence of some interaction with the membrane. Some of the major mechanisms used to accomplish these different objectives are listed in Table 43–5.

Table 43–5. Transfer of material and information across membranes.

Cross-membrane movement of small molecules
  Diffusion (passive and facilitated)
  Active transport
Cross-membrane movement of large molecules
  Endocytosis
  Exocytosis
Signal transmission across membranes
  Cell surface receptors
    1. Signal transduction (eg, glucagon → cAMP)
    2. Signal internalization (coupled with endocytosis, eg, the LDL receptor)
    Movement to intracellular receptors (steroid hormones; a form of diffusion)
Intracellular contact and communication

Some Small Molecules Move Across Membranes

Molecules can passively traverse the bilayer down electrochemical gradients by simple or facilitated diffusion. This spontaneous movement toward equilibrium contrasts with active transport, which requires energy because it constitutes movement against an electrochemical gradient. Fig 43–12 provides a schematic representation of these mechanisms.

Passive Diffusion: As described above, some solutes such as gases can enter the cell by diffusing down an electrochemical gradient across the membrane and do not require metabolic energy. The simple diffusion of a solute across the membrane is limited by the thermal agitation of that specific molecule, by the concentration gradient across the membrane, and by the solubility of that solute (the permeability coefficient, Fig 43–6) in the hydrophobic core of the membrane bilayer. Solubility is inversely proportionate to the number of hydrogen bonds that must be broken in order for a solute in the external aqueous phase to become incorporated in the hydrophobic bilayer. Electrolytes, poorly soluble in lipid, do not form hydrogen bonds with water, but they do acquire a shell of water from hydration by electrostatic interaction. The size of the shell is directly proportionate to the charge density of the electrolyte. Electrolytes with a large charge density have a larger shell of hydration and thus a slower diffusion rate. Na^+, for example, has a higher charge density than K^+. Hydrated Na^+ is therefore larger

than hydrated K^+; hence, the latter tends to move more easily through the membrane.

In natural membranes, as opposed to synthetic membrane bilayers, there are transmembrane channels, porelike structures composed of proteins that constitute selective ion-conductive pathways. Cation-conductive channels have an average diameter of about 5–8 nm and are negatively charged within the channel. The permeability of a channel depends upon the size, extent of hydration, and extent of charge density on the ion. Specific channels for Na^+, K^+, and Ca^{2+} have been identified.

The membranes of nerve cells contain well-studied ion channels that are responsible for the action potentials generated across the membrane. The activity of some of these channels is controlled by neurotransmitters; hence, channel activity can be regulated. One ion can regulate the activity of the channel of another ion. For example, a decrease of Ca^{2+} concentration in the extracellular fluid increases membrane permeability and increases the diffusion of Na^+. This depolarizes the membrane and triggers nerve discharge. This may explain the numbness, tingling, and muscle cramps symptomatic of a low level of serum Ca^{2+}.

Channels are open transiently and thus are gated. Gates can be controlled by opening or closing. In ligand-gated channels, a specific molecule binds to a receptor and opens the channel. **Voltage-gated channels** open (or close) in response to a change in membrane potential.

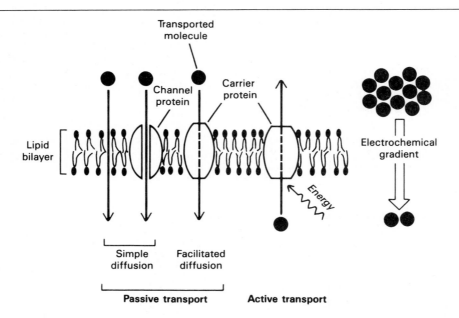

Figure 43–12. Many small uncharged molecules pass freely through the lipid bilayer. Charged molecules, larger uncharged molecules, and some small uncharged molecules are transferred through channels or pores or by specific carrier proteins. Passive transport is always down an electrochemical gradient, toward equilibrium. Active transport is against an electrochemical gradient and requires an input of energy, whereas passive transport does not. (Redrawn and reproduced, with permission, from Alberts B et al: *Molecular Biology of the Cell.* Garland, 1983.)

Some microbes synthesize small organic molecules, **ionophores**, that function as shuttles for the movement of ions across membranes. These ionophores contain hydrophilic centers that bind specific ions and are surrounded by peripheral hydrophobic regions; this arrangement allows the molecules to dissolve effectively in the membrane and diffuse transversely therein. Others, like the well-studied polypeptide gramicidin, form channels. Microbial toxins such as diphtheria toxin and activated serum complement components can produce large pores in cellular membranes and thereby provide macromolecules with direct access to the internal milieu.

In summary, net diffusion of a substance depends upon the following: (1) Its concentration gradient across the membrane. Solutes move from high to low concentration. (2) The electrical potential across the membrane. Solutes move toward the solution that has the opposite charge. The inside of the cell usually has a negative charge. (3) The permeability coefficient of the substance for the membrane. (4) The hydrostatic pressure gradient across the membrane. Increased pressure will increase the rate and force of the collision between the molecules and the membrane. (5) Temperature. Increased temperature will increase particle motion and thus increase the frequency of collisions between external particles and the membrane.

FACILITATED DIFFUSION & ACTIVE TRANSPORT

Transport systems can be described in a functional sense according to the number of molecules moved and the direction of movement (Fig 43–13) or according to whether movement is toward or away from equilibrium. A **uniport** system moves one type of molecule bidirectionally. In **cotransport** systems, the transfer of one solute depends upon the stoichiometric simultaneous or sequential transfer of another solute. A **symport** moves these solutes in the same direction.

Examples are the proton-sugar transporter in bacteria and the Na^+-sugar transporters (glucose, mannose, galactose, xylose, and arabinose) and the Na^+-amino acid transporters in mammalian cells. **Antiport** systems move 2 molecules in opposite directions (eg, Na^+ in and Ca^{2+} out).

Molecules that cannot pass freely through the lipid bilayer membrane by themselves do so in association with carrier proteins. This involves 2 processes, facilitated diffusion and active transport, and highly specific transport systems.

Facilitated diffusion and active transport share many features. Both appear to involve carrier proteins, and both show specificity for ions, sugars, and amino acids. Mutations in bacteria and mammalian cells (including some that result in human disease) have supported these conclusions. Facilitated diffusion and active transport resemble a substrate-enzyme reaction except that no covalent interaction occurs. These points of resemblance are as follows: (1) There is a specific binding site for the solute. (2) The carrier is saturable, so it has maximum rate of transport (V_{max}; see Fig 43–14). (3) There is a binding constant (K_m) for the solute, and so the whole system has a K_m (Fig 43–14). (4) Structurally similar competitive inhibitors block transport.

Major differences are the following: (1) Facilitated diffusion can operate bidirectionally, whereas active transport is usually unidirectional. (2) Active transport always occurs against an electrical or chemical gradient, and so it requires energy.

Facilitated Diffusion: Some specific solutes diffuse down electrochemical gradients across membranes more rapidly than might be expected from their size, charge, or partition coefficients. This facilitated diffusion exhibits properties distinct from those of simple diffusion. The rate of facilitated diffusion, a uniport system, can be saturated; ie, the number of sites involved in diffusion of the specific solutes appears finite. Many facilitated diffusion systems are stereospecific but, like simple diffusion, require no metabolic energy.

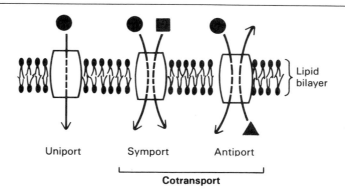

Figure 43–13. Schematic representation of types of transport systems. Transporters can be classified with regard to the direction of movement and whether one or more unique molecules are moved. (Redrawn and reproduced, with permission, from Alberts B et al: *Molecular Biology of the Cell.* Garland, 1983.)

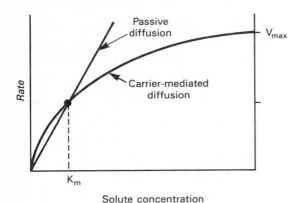

Figure 43–14. A comparison of the kinetics of carrier-mediated (facilitated) diffusion with passive diffusion. The rate of movement in the latter is directly proportionate to solute concentration, whereas the process is saturable when carriers are involved. V_{max}, maximal rate. The concentration at half-maximal velocity is equal to the binding constant (K_m) of the carrier for the solute.

which solutes enter a cell by facilitated diffusion is determined by the following factors: (1) The concentration gradient across the membrane. (2) The amount of carrier available (this is a key control step). (3) The rapidity of the solute-carrier interaction. (4) The rapidity of the conformational change for both the loaded and the unloaded carrier.

Hormones regulate facilitated diffusion by changing the number of transporters available. Insulin increases glucose transport in fat and muscle by recruiting transporters from an intracellular reservoir (see Fig 52–9). Insulin also enhances amino acid transport in liver and other tissues. One of the coordinated actions of glucocorticoid hormones is to enhance transport of amino acids into liver, where the amino acids then serve as a substrate for gluconeogenesis. Growth hormone increases amino acid transport in all cells, and estrogens do this in the uterus. There are at least 5 different carrier systems for amino acids in animal cells. Each is specific for a group of closely related amino acids, and most operate as Na^+-symport systems (Fig 43–13).

Active Transport: The process of active transport differs from diffusion in that molecules are transported away from thermodynamic equilibrium; hence, energy is required. This energy can come from the hydrolysis of ATP, from electron movement, or from light. The maintenance of electrochemical gradients in biologic systems is so important that it consumes perhaps 30–40% of the total energy expenditure in a cell.

In general, cells maintain a low intracellular Na^+ concentration and a high intracellular K^+ concentration (Table 43–1), along with a net negative electrical potential inside. The pump that maintains these gradients is an ATPase that is activated by Na^+ and K^+ (Fig 43–16). The ATPase is an integral membrane protein and requires phospholipids for activity. The ATPase has catalytic centers for both ATP and Na^+ on the cytoplasmic side of the membrane, but the K^+ binding site is located on the extracellular side of the membrane. Ouabain (or digitalis) inhibits this ATPase by

As described earlier, the inside-outside asymmetry of membrane proteins is stable, and mobility of proteins across (rather than in) the membrane is rare; therefore, transverse mobility of specific carrier proteins is not likely to account for facilitated diffusion processes except those for microbial ionophores, described earlier.

A **"ping-pong" mechanism (Fig 43–15) explains facilitated diffusion.** In this model, the carrier protein exists in 2 principal conformations. In the "pong" state it is exposed to high concentrations of solute, and molecules of the solute bind to specific sites on the carrier protein. Transport occurs when a conformational change exposes the carrier to a lower concentration of solute ("ping" state). This process is completely reversible, and net flux across the membrane depends upon the concentration gradient. The rate at

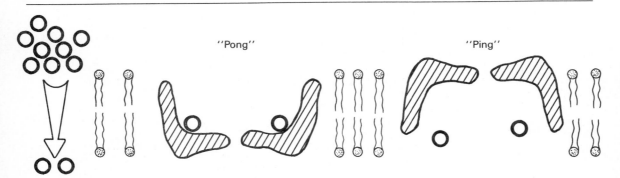

Figure 43–15. The "ping-pong" model of facilitated diffusion. A protein carrier (shaded structure) in the lipid bilayer associates with a solute in high concentration on one side of the membrane. A conformational change ensues ("pong" to "ping"), and the solute is discharged on the side favoring the new equilibrium. The empty carrier then reverts to the original conformation ("ping" to "pong") to complete the cycle.

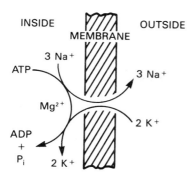

Figure 43–16. Stoichiometry of the Na^+/K^+-ATPase pump. This pump moves $3 Na^+$ ions from inside the cell to the outside and brings $2 K^+$ ions from the outside to the inside for every molecule of ATP hydrolyzed to ADP by the membrane-associated ATPase. Ouabain and other cardiac glycosides inhibit this pump by acting on the extracellular surface of the membrane. (Courtesy of R Post.)

binding to the extracellular domain. Inhibition of the ATPase by ouabain can be antagonized by extracellular K^+.

Transmission of Nerve Impulses Occurs Up & Down Membranes

The membrane forming the surface of neuronal cells maintains an asymmetry of inside-outside voltage (electrical potential) and is electrically excitable. When appropriately stimulated by a chemical signal mediated by a specific synaptic membrane receptor (see discussion of the transmission of biochemical signals, below), gates in the membrane are opened to allow the rapid influx of Na^+ or Ca^{2+} (with or without the efflux of K^+), so that the voltage difference rapidly collapses and that segment of the membrane is depolarized. However, as a result of the action of the ion pumps in the membrane, the gradient is quickly restored.

When large areas of the membrane are depolarized in this manner, the electrochemical disturbance propagates in wavelike form down the membrane, generating a nerve impulse. Myelin sheets, formed by Schwann cells, wrap around nerve fibers and provide an electrical insulator that surrounds most of the nerve and greatly speeds up the propagation of the wave (signal) by allowing ions to flow in and out of the membrane only where the membrane is free of the insulation. The myelin membrane is composed of phospholipids, including sphingomyelin, cholesterol, proteins, and glycosphingolipids. There are relatively few integral and peripheral proteins associated with the myelin membrane; those present appear to hold together multiple membrane bilayers to form the hydrophobic, insulating structure that is impermeable to ions and water. Certain diseases, eg, multiple sclerosis and the Guillain-Barré syndrome, are characterized by demyelination and impaired nerve conduction.

Glucose Transport Involves Several Mechanisms

A discussion of the transport of glucose summarizes many of the points made in this chapter. Glucose must enter cells as the first step in energy utilization. In adipocytes and muscle, glucose enters by a specific transport system that is enhanced by insulin. Changes in transport are primarily due to alterations of V_{max} (presumably from more or fewer active transporters) but changes in K_m may also be involved. Glucose transport involves different aspects of the principles of transport discussed above. Glucose and Na^+ bind to different sites on the glucose transporter. Na^+ moves into the cell down its electrochemical gradient and "drags" glucose with it (Fig 43–17). Therefore, the greater the Na^+ gradient, the more glucose enters, and if Na^+ in extracellular fluid is low, glucose transport stops. To maintain a steep Na^+ gradient, this Na^+-glucose symport is dependent on gradients generated by an Na^+/K^+ pump that maintains a low intracellular Na^+ concentration. Similar mechanisms are used to transport other sugars as well as amino acids.

The transcellular movement of sugars involves one additional component, a uniport that allows the glucose accumulated within the cell to move across a different surface toward a new equilibrium; this occurs in intestinal and renal cells, for example.

Cells Also Transport Macromolecules Across the Plasma Membrane

Endocytosis is the process by which cells take up

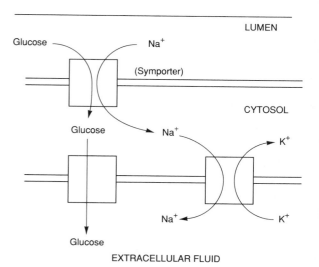

Figure 43–17. The transcellular movement of glucose in an intestinal cell. Glucose follows Na^+ across the luminal epithelial membrane. The Na^+ gradient that drives this symport is established by Na^+/K^+ exchange, which occurs at the basal membrane facing the extracellular fluid compartment. Glucose at high concentration within the cell moves "downhill" into the extracellular fluid by facilitated diffusion (a uniport mechanism).

large molecules. Some of these (eg, polysaccharides, proteins, and polynucleotides) can be sources of nutritional elements. Endocytosis provides a mechanism for regulating the content of certain membrane components, hormone receptors being a case in point. Endocytosis can be used to learn more about how cells function. DNA from one cell type can be used to transfect a different cell and alter the latter's function or phenotype. A specific gene is often employed in these experiments, and this provides a unique way to study and analyze the regulation of that gene. DNA transfection depends upon endocytosis; endocytosis is responsible for the entry of DNA into the cell. Such experiments commonly use calcium phosphate, since Ca^{2+} stimulates endocytosis and precipitates DNA, which makes the DNA a better object for endocytosis. Cells also release macromolecules by exocytosis. Endocytosis and exocytosis both involve vesicle formation with or from the plasma membrane.

Endocytosis: All eukaryotic cells are continuously ingesting parts of their plasma membranes. Endocytotic vesicles are generated when segments of the plasma membrane invaginate, enclosing a minute volume of extracellular fluid and its contents. The vesicle then pinches off as the fusion of plasma membranes seals the neck of the vesicle at the original site of invagination (Fig 43–18). This vesicle fuses with other membrane structures and thus achieves the transport of its contents to other cellular compartments or even back to the cell exterior. Most endocytotic vesicles fuse with primary lysosomes to form secondary lysosomes, which contain hydrolytic enzymes and are

therefore specialized organelles for intracellular disposal. The macromolecular contents are digested to yield amino acids, simple sugars, and nucleotides, and they diffuse out of the veislces to be reused in the cytoplasm. Endocytosis requires (1) energy, usually from the hydrolysis of ATP: (2) Ca^{2+} in extracellular fluid; and (3) contractile elements in the cell (probably the microfilament system).

There are 2 general types of endocytosis. **Phagocytosis** occurs only in specialized cells such as macrophages and granulocytes. Phagocytosis involves the ingestion of large particles such as viruses, bacteria, cells, or debris. Macrophages are extremely active in this regard and may ingest 25% of their volume per hour. In so doing, a macrophage may internalize 3% of its plasma membrane each minute or the entire membrane every 30 minutes.

Pinocytosis is a property of all cells and leads to the cellular uptake of fluid and fluid contents. There are 2 types. **Fluid-phase pinocytosis** is a nonselective process in which the uptake of a solute by formation of small vesicles is simply proportionate to its concentration in the surrounding extracellular fluid. The formation of these vesicles is an extremely active process. Fibroblasts, for example, internalize their plasma membrane at about one-third the rate of macrophages. This process occurs more rapidly than membranes are made. The surface area and volume of a cell do not change much, so membranes must be replaced by exocytosis or by being recycled as fast as they are removed by endocytosis.

The other type of pinocytosis, **absorptive pi-**

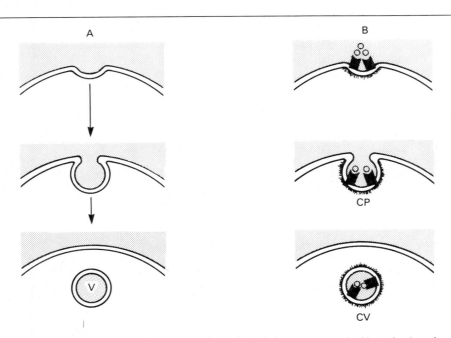

Figure 43–18. Two types of endocytosis. An endocytotic vesicle (V) forms as a result of invagination of a portion of the plasma membrane. Fluid-phase endocytosis (A) is random and nondirected. Receptor-mediated endocytosis (B) is selective and occurs in coated pits (CP) lined with the protein clathrin (the fuzzy material). Targeting is provided by receptors (■) specific for a variety of molecules. This results in the formation of a coated vesicle (CV).

nocytosis, is a receptor-mediated selective process primarily responsible for the uptake of macromolecules for which there are a finite number of binding sites on the plasma membrane. These high-affinity receptors permit the selective concentration of ligands from the medium, minimize the uptake of fluid or soluble unbound macromolecules, and markedly increase the rate at which specific molecules enter the cell. The vesicles formed during absorptive pinocytosis are derived from invaginations (pits) that are coated on the cytoplasmic side with a filamentous material. In many systems, **clathrin** is the filamentous material; it is probably a peripheral membrane protein. **Coated pits** may constitute as much as 2% of the surface of some cells.

For example, the low-density lipoprotein (LDL) molecule and its receptor (see Chapter 26) are internalized by means of coated pits containing the LDL receptor. These endocytotic vesicles containing LDL and its receptor fuse to lysosomes in the cell. The receptor is released and recycled back to the cell surface membrane, but the apoprotein of LDL is degraded and the cholesteryl esters metabolized. Synthesis of the LDL receptor is regulated by secondary or tertiary consequences of pinocytosis, eg, by metabolic products, such as cholesterol, released during the degradation of LDL. Disorders of the LDL receptor and its internalization are medically important and are discussed in Chapter 26.

Other macromolecules, including several hormones, are subject to **adsorptive pinocytosis** and form **receptosomes,** vesicles that avoid lysosomes and deliver their contents to other intracellular sites, such as the Golgi system.

Adsorptive pinocytosis of extracellular glycoproteins requires that the glycoproteins carry specific carbohydrate recognition signals. These recognition signals are bound by membrane receptor molecules, which play a role analogous to that of the LDL receptor. A galactosyl receptor on the surface of hepatocytes is instrumental in the adsorptive pinocytosis of asialoglycoproteins from the circulation. Acid hydrolases taken up by adsorptive pinocytosis in fibroblasts are recognized by their mannose 6-phosphate moieties. Interestingly, the mannose 6-phosphate moiety also seems to play an important role in the intracellular targeting of the acid hydrolases to the lysosomes of the cells in which they are synthesized (see Chapter 57).

There is a dark side to receptor-mediated endocytosis, for viruses that cause such diseases as hepatitis (affecting liver cells), poliomyelitis (affecting motor neurons), and AIDS (affecting T cells) initiate their damage by this mechanism. Iron toxicity also begins with excessive uptake due to endocytosis.

Exocytosis: Most cells also release macromolecules to the exterior by exocytosis. This process is also involved in membrane remodeling when the components synthesized in the Golgi apparatus are carried in vesicles to the plasma membrane. The signal for exocytosis is often a hormone which, when it binds to a cell-surface receptor, induces a local and transient change in Ca^{2+} concentration. Ca^{2+} triggers exocytosis. Fig 43–19 provides a comparison of the mechanisms of exocytosis and endocytosis.

Molecules released by exocytosis fall into 3 categories: (1) They can attach to the cell surface and become peripheral proteins, eg, antigens. (2) They can become part of the extracellular matrix, eg, collagen and glycosaminoglycans. (3) They can enter extracellular fluid and signal other cells. Insulin, parathyroid hormone, and the catecholamines are all packaged in granules and processed within cells, to be released upon appropriate stimulation (see Chapters 48, 50, and 52).

Some Signals Are Transmitted Across Membranes

Biochemical Signals: Specific biochemical signals such as neurotransmitters, hormones, and immunoglobulins bind to specific receptors (integral proteins) exposed to the outside of cellular membranes and transmit information through these membranes to the cytoplasm. This mechanism involves the generation of a number of signals, including cyclic nucleotides, calcium, phosphosites, and diacylglycerol. It is discussed in detail in Chapter 45.

Information Can Be Communicated by Intercellular Contact

There are many areas of intercellular contact in a metazoan organism. This necessitates contact between the plasma membranes of the individual cells. Cells have developed specialized regions on their membranes for intracellular communication in close proximity. Gap junctions mediate and regulate the passage

Figure 43–19. A comparison of the mechanisms of endocytosis and exocytosis. Exocytosis involves the contact of 2 inside surface (cytoplasmic side) monolayers, whereas endocytosis results from the contact of 2 outer surface monolayers.

of ions and small molecules through a narrow hydrophilic core connecting the cytoplasm of adjacent cells. It is through this central opening that ions and small molecules can pass from one cell to another in a regulated fashion.

REFERENCES

Blobel G et al: Translocation of proteins across membranes: The signal hypothesis and beyond. *Symp Soc Exp Biol* 1979;**33**:9.

Dautry-Varsat A, Lodish HF: How receptors bring proteins and particles into cells. *Sci Am* (May) 1984;**250**:52.

Dawidowicz EA: Dynamics of membrane lipid metabolism and turnover. *Annu Rev Biochem* 1987;**56**:43.

Ellers M, Schatz G: Protein Unfolding and the Energetics of Protein Translocation across Biological Membranes. *Cell* 1988;**52**:481.

Goldstein J et al: Receptor-mediated endocytosis. *Annu Rev Cell Biol* 1985;**1**:1.

Lodish HF: Transport of secretory and membrane glycoproteins from the rough endoplasmic reticulum to the Golgi: A rate-limiting step in protein maturation and secretion. *J Biol Chem* 1988;**263**:2107.

Matlin, KS: The sorting of proteins to the plasma membrane in epithelial cells. *J Cell Biol* 1986;**103**:2565.

Pfeffer SR, Rothman JE: Biosynthetic protein transport and sorting by the endoplasmic reticulum and Golgi. *Annu Rev Biochem* 1987;**56**:829.

Roise D, Schatz G: Mitochondrial presequences. *J Biol Chem* 1988;**263**:4509.

Singer SJ, Nicolson GL: The fluid mosaic model of the structure of cell membranes. *Science* 1972;**175**:720.

Stahl P, Schwartz AL: Receptor-mediated endocytosis. *J Clin Invest* 1986;**77**:657.

Stein WD: *Transport and Diffusion Across Cell Membranes.* Academic Press, 1986.

Unwin N, Henderson R: The structures of proteins in biological membranes. *Sci Am* (Feb) 1984:**250**:78.

Walter P, Gilmore R, Blobel G: Protein translocation across the endoplasmic reticulum. *Cell* 1984;**38**:5.

Wickner WT, Lodish HF: Multiple mechanisms of protein insertion into and across membranes. *Science* 1985;**230**:400.

Characteristics of Hormone Systems

44

Daryl K. Granner, MD

INTRODUCTION

The objective of this chapter is to impart an appreciation for the varied nature of the endocrine system and to define several of the fundamental concepts that will reappear in subsequent chapters.

BIOMEDICAL IMPORTANCE

The advances made in understanding the general features of the endocrine system—such as why some glands are located in proximity to others, how hormones are produced, and the concepts of target cells, feedback control, and receptors—have direct clinical application. We are now able to provide some precise definitions of the causes of endocrine diseases. Most of these involve the identification of specific hormone receptor defects, but other types of causes will surely be found.

THE SPECIALIZED TISSUES OF MULTICELLULAR ORGANISMS REQUIRE INTRICATE COORDINATION

A distinguishing characteristic of multicellular organisms is the presence of differentiated tissues that perform the specialized functions necessary for the survival of the organism. Mechanisms are required for intercellular communication to ensure the coordination of the responses necessary for adjusting to a constantly changing external and internal environment. Two general systems have evolved to serve these functions. These are the **nervous system,** often viewed as conducting signals or messages through a fixed, structural system; and the **endocrine system,** in which various **hormones** secreted by specific glands are transported as mobile messages to act on adjacent and distant tissue.

It is now clear that there is an exquisite convergence of these regulatory systems. Neural regulation of the endocrine system is very important; eg, epinephrine is produced and secreted by postganglionic cells in the adrenal medulla, and vasopressin is synthesized in the hypothalamus and transported by axons to the posterior pituitary, from which it is released. Likewise, many **neurotransmitters** (catecholamines, dopamine, acet-

ylcholine, etc) are similar to hormones with regard to their synthesis, release, transport, and mechanism of action. In fact, catecholamines are neurotransmitters in one tissue and hormones in other tissues. The recent demonstration that certain metabolites of adrenal steroid hormones are barbiturate-like modulators of the γ-aminobutyrate (GABA) receptor in brain is another example of this interaction. Finally, many hormones such as insulin, ACTH, vasoactive intestinal polypeptide (VIP), somatostatin, thyrotropin-releasing hormone (TRH), and cholecystokinin have recently been found in brain. It remains to be established whether all of these molecules are synthesized in brain and whether they act there as neuromodulators or as neurotransmitters. Since specific receptors for many of these hormones are found in brain, the possibility exists that these molecules act in brain.

The word hormone is derived from a Greek term meaning "to arouse to activity." By the classic definition, a hormone is a substance that is synthesized in one tissue and transported by the circulatory system to act on another organ. This original description, which defines an endocrine function, is too restrictive; it is now appreciated that hormones act on adjacent cells in a given tissue (**paracrine function**) as well as on the cells in which they are synthesized (**autocrine function**).

THE HALLMARK OF THE ENDOCRINE SYSTEM IS DIVERSITY

One of the most remarkable features of the endocrine system is that it provides an organism with a number of different ways for solving problems. The purpose of this section is to present a very brief discussion of selected examples that highlight this diversity.

Endocrine Glands Are Mostly Derived From Epithelium & Are Located Strategically

The endocrine glands are mostly derived from epithelial cells. Notable exceptions include the connective tissue origin of the testosterone-producing Leydig cells in the testis and the estrogen-producing granulosa cells in the ovary, and the neuronal origin of the secretory cells in the neurohypophysis. The **neural crest** has been suggested as the embryologic origin of a number of endocrine cell types. If true, this would provide a rational link between the central nervous system and the endocrine system. Since neural crest tissue can appear in any organ, this may explain why some hormones appear to be made in brain and in tissues predominantly derived from the midgut and foregut. It may also explain the **ectopic hormone syndromes,** in which there is production of hormones by the "wrong tissue" (eg, production of parathyroid hormone [PTH] and ACTH by malignant cells in the case of lung cancer). These syndromes generally involve a rather restricted number of peptide hormones but a large and apparently diverse number of tissues. Although commonly thought to represent the activation of silent genes within a given cell, these syndromes could represent the activation of silent cells of common embryologic ancestry within a tissue. Another curious example is afforded by the **multiple endocrine neoplasia (MEN) syndromes,** in which there is a peculiar familial clustering of neoplasia of several endocrine glands. The production of excessive amounts of peptide or catecholamine hormones, often with one tissue making several, is a feature of these syndromes.

Hormone-producing cells are not randomly distributed; they are present in various tissues for specific reasons. Locally high concentrations of some hormones (ie, values in excess of plasma hormone levels) are often required for specific biologic processes. For example, a level of testosterone higher than that available in the plasma is required for spermatogenesis; thus, the testosterone-secreting Leydig cells and the seminiferous tubules are in juxtaposition. A very high concentration of estrogen is required for corpus luteum formation; hence, there is close proximity of this structure and the granulosa cells. A major action of insulin and glucagon is to regulate hepatic production of glucose; thus, there is close association of the pancreatic islets with the hepatic portal circulation. Cortisol, which is required in high concentration in the adrenal medulla for the induction of phenylethanolamine-N-methyltransferase (a rate-limiting enzyme is catecholamine biosynthesis), reaches this site by a portal vascular system that originates in the adrenal cortex. There is an intimate association of the hypothalamus and anterior pituitary, so that high concentrations of the very labile hypothalamic releasing hormones can easily reach the pituitary target via another special portal vascular system. Finally, there is a unique anatomic relationship between the various cells of the pancreatic islets in which locally high gradients of somatostatin, pancreatic polypeptide, glucagon, and insulin interact to regulate the secretion of one another.

Hormone Biosynthesis & Modification Are Finely Tuned to Specific Functions

There is great diversity in the chemical nature of and in the biosynthetic and postsynthetic mechanisms employed in the generation of active hormones. Hormones are derived from lipid precursors, from modifications of the amino acid tyrosine, and from the synthesis of proteins (simple and complex peptides and carbohydrage-containing glycoproteins).

Hormones may be synthesized and secreted in final form; examples include aldosterone, hydrocortisone, triiodothyronine (T_3), estradiol, and the catecholamines. Others must be modified within the cell before they are secreted or before they have full biologic activity. Examples include insulin, which is synthesized as proinsulin, the prototype of **precursor proteins,** and parathyroid hormone (PTH), which has at least 2 precursor peptides (a prepro segment) that must be removed to achieve full biologic activity. A description of precursor proteins, their synthesis, and intracellular processing into the final product can be found in Chapter 43. Pro-opiomelanocortin (POMC), a 285-amino-acid peptide that is the product of a single gene, represents an even more complicated case. POMC is cleaved to form ACTH, β-lipotropin, β-endorphin, α-MSH, and β-MSH, and the parent or precursor molecule may contain sequences of peptide hormones as yet unidentified. The **processing** of the precursor molecule is tissue specific (see Chapter 46).

Perhaps the most exaggerated example of a large precursor of a hormone is thyroglobulin. This is a large protein (MW 660,000) found in the lumen of the thyroid follicle. Among the 5000 amino acids of the thyroglobulin molecule, there are 120 tyrosyl residues, some of which are iodinated in the process of thyroid hormone biosynthesis (see Chapter 47). The entire thyroglobulin molecule must be degraded to release the few tetraiodothyronine (T_4) and T_3 molecules present.

Some hormones are converted into more active molecules in peripheral tissues. This can occur in target tissues, as is the case in conversion of T_4 to T_3 in liver and pituitary and in conversion of testosterone to dihydrotestosterone in secondary sex tissues. **Peripheral conversion** can also occur in nontarget tissues; dehydroepiandrosterone is synthesized in the adrenal and converted to androstenedione in the liver. This latter molecule can then be converted to testosterone or to estrone and estradiol in fat cells, liver, or skin. Combined target and nontarget tissue peripheral conversion of an inactive molecule to an active hormone occurs. This is illustrated by the conversion of vitamin D_3 (from skin) to 25-hydroxycholecalciferol in liver, with subsequent conversion to 1,25-dihydroxycholecalciferol in kidney (see Chapter 48).

Hormones that are secreted from very different tissues and have different cell specificity may have structural similarities. The glycoprotein hormones from the pituitary and placenta (TSH, LH, FSH, and hCG) are heterodimers consisting of α and β subunits in which the α subunits are identical.

HORMONES MAY BE TARGETED TO MORE THAN ONE TISSUE: THE TARGET GLAND CONCEPT

There are about 200 types of differentiated cells in humans. Only a few of these produce hormones, but the 75 trillion cells in a human probably are all targets for one or more of the approximately 50 known hormones. A hormone may have one target tissue or it may affect a number of tissues. A target tissue was classically defined as having a unique biochemical or physiologic response to a hormone. For example, the thyroid is a specific target gland of TSH; TSH increases the number and size of the thyroid acinar cells and enhances all of the enzymatic steps involved in thyroid hormone biosynthesis. In contrast, insulin affects many tissues. Insulin enhances glucose uptake and oxidation in muscle, lipogenesis in fat, amino acid transport in liver and lymphocytes, and protein synthesis in liver and muscle, to name a few effects. More recently, with the delineation of specific cell surface and intracellular hormone receptors, the definition of a target has been expanded to include any tissue in which the hormone can be demonstrated to bind to its receptor, whether or not a classic biochemical or physiologic response has been determined (eg, insulin binding to endothelial cells). This definition is also incomplete, but it has heuristic merit, since it recognizes that not all actions of hormones have been elucidated.

Several factors determine the overall response of a target tissue to a hormone. The local concentration of a hormone around the target tissue depends upon (1) the rate of synthesis and secretion of the hormone; (2) the proximity of target and source; (3) the association-dissociation constants of the hormone with specific carrier proteins in the plasma, if such exist; (4) the rate of conversion of an inactive or suboptimally active form of the hormone into the active form; and (5) the rate of clearance of the hormone from blood by degradation or excretion, primarily accomplished by the liver and kidneys. The actual response then depends upon (1) the relative activity and state of occupancy, or both, of the specific hormone receptors on the plasma membrane or within the cytoplasm or nucleus; and (2) the postreceptor sensitization-desensitization of the cell. Alterations of any of these processes can result in a change of the hormonal activity on a given target tissue and must be considered as an addition to the classic feedback loops.

NEGATIVE & POSITIVE FEEDBACK PROVIDE CONTROL

Physiologic hormone levels in blood are maintained by a variety of homeostatic mechanisms that entail precise signaling between the hormone-secreting gland and the target tissue, and this often involves one or more intermediate glands. **Negative feedback control** is commonly employed, especially by the hypothalamic-pituitary target gland systems. An example is illustrated in Fig 44–1. A hypothalamic releasing hormone stimulates the synthesis and release of an anterior pituitary hormone, which in turn stimulates the production of the target organ hormone. High levels of the latter inhibit the system by decreasing hypothala-

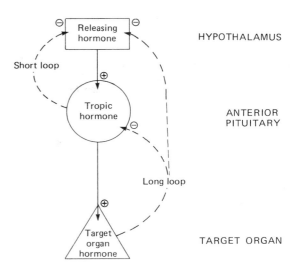

Figure 44–1. Example of the negative feedback control system used to regulate the function of the thyroid, adrenal, ovary, and testes.

mic hormone synthesis and action, while low levels result in the system being activated at the level of the hypothalamus. A unique feature of this particular axis is that the pituitary hormone may itself blunt the system by **"short-loop" feedback inhibition** of its own synthesis. This tonic system provides for exquisite control of the plasma hormone level and also illustrates that there are several hormones and several target tissues within one endocrine system. Such control loops are recognized for the adrenal, thyroid, testicular, and ovarian systems.

In other instances, negative feedback regulation is accomplished through various metabolites or substrates whose plasma concentration is changed as a result of a hormone acting on a target cell. For example, an increased plasma glucose concentration (hyperglycemia) provokes a measured release of insulin, which enhances glucose uptake and utilization by a number of tissues, thereby decreasing the plasma glucose to a normal level and, in turn, diminishing insulin release. Under certain pathologic conditions, the insulin response may be excessive, in which case hypoglycemia ensues. The physiologic response to this life-threatening condition is the release of catecholamines, growth hormone, glucagon, ACTH, vasopressin, and angiotensin II, all of which act to increase the plasma glucose concentration. Thus, a complex network has evolved to regulate a critical metabolite (in this example, glucose) which is required for brain function.

In other cases, hormones exert **positive feedback control.** For example, estrogen and progesterone are required for the acute burst of LH secretion that results in ovulation and follicular luteinization and the further production of these steroid hormones. In many instances, the feedback loops have not been established, usually because the end products of the action of the hormone are not known.

A number of **pathophysiologic events,** including shock, trauma, hypoglycemia, pain, and stress, affect the hypothalamic-pituitary target gland axis through higher brain centers. These stimuli have profound effects on catecholamine and growth hormone metabolism and on the function of the adrenal cortex, thyroid, and gonads, but the precise components of these circuits are still poorly defined.

Endocrine and metabolic diseases result from the disruption of these normal feedback control mechanisms, and diagnostic perturbations of these systems (eg, the metyrapone test), are used to distinguish normal from pathologic conditions.

HORMONE RECEPTORS ARE
OF CENTRAL IMPORTANCE

Receptors Discriminate Precisely

One of the major challenges faced in making the hormone-based communication system work is depicted in Fig 44–2. Hormones are present at very low concentrations in the extracellular fluid, generally in the range of 10^{-15}–10^{-9} mol/L. This is a much lower concentration than that of the many structurally similar molecules (sterols, amino acids, peptides, proteins) and other molecules that circulate at concentrations in the 10^{-5}–10^{-3} mol/L range. Target cells, therefore, must distinguish not only between different hormones present in small amounts but also between a given hormone and the 10^6- to 10^9-fold excess of other mole-

cules. This high degree of discrimination is provided by cell-associated recognition molecules called receptors. Hormones initiate their biologic effects by binding to specific receptors, and since any effective control system also must provide a means of stopping a response, hormone-induced actions generally terminate when the effector dissociates from the receptor.

A target cell is defined by its ability to bind selectively a given hormone via such a receptor, an interaction that is often quantitated using radioactive ligands that mimic hormone binding. Several features of this interaction are important: (1) the radioactivity must not alter the biologic activity of the ligand; (2) the binding should be specific, ie, displaceable by unlabeled agonist or antagonist; (3) binding should be saturable; and (4) binding should occur within the concentration range of the expected biologic response.

Both Recognition & Coupling
Domains Occur on Receptors

All receptors, whether for polypeptides or steroids, have at least 2 functional domains. A recognition domain binds the hormone, and a second region generates a signal that couples hormone recognition to some intracellular function. The binding of hormone by receptor implies that some region of the hormone molecule has a conformation that is complementary to a region of the receptor molecule. The degree of similarity, or fit, determines the tightness of the association; this is measured as the affinity of binding (K). If the native hormone has a relative K value of 1, other natural molecules range between 0 and 1. In absolute terms, this actually spans a binding affinity range of more than a trillion. Ligands with a relative K value of

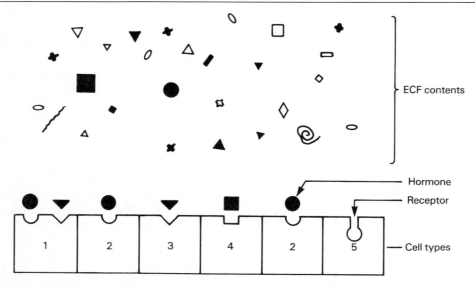

Figure 44–2. The specificity and selectivity of hormone receptors. Many different molecules circulate in the extracellular fluid (ECF), but only a few are recognized by hormone receptors. Receptors must select these molecules from high concentrations of the other molecules. There can be one class of receptor on each cell, or a given cell can have receptors for several hormones.

> 1 for some receptors have been synthesized and are used to study various aspects of receptor biology.

Coupling (signal transduction) occurs in 2 general ways. Polypeptide and protein hormones and the catecholamines bind to receptors located in the plasma membrane and thereby generate a signal that regulates various intracellular functions, often by changing the activity of an enzyme. Steroid and thyroid hormones interact with intracellular receptors, and this complex provides the signal (see below).

The amino acid sequences of these 2 domains in many polypeptide hormone receptors have now been identified. Hormone analogs with specific amino acid substitutions were used to change binding and to alter the biologic activity of the hormone. Steroid hormone receptors have several functional domains: one site binds the hormone, another binds to specific DNA regions, a third activates (or represses) gene transcription, and a fourth may specify high-affinity binding to DNA.

The dual functions of binding and coupling ultimately define a receptor, and it is the coupling of hormone binding to signal transduction, so-called **receptor-effector coupling,** that provides the first step in the amplification of the hormonal response. This dual purpose also distinguishes the target cell receptor from the plasma carrier proteins that bind hormone but do not generate a signal.

Receptors & Transport Proteins Show Complementary Action

It is important to distinguish the binding of hormones to receptors from the association hormones have with various transport (carrier) proteins. Table 44–1 illustrates several comparative features. Thousands of receptor molecules per cell bind the ligand, and there is high binding affinity and specificity. Receptors are capable of recognizing and selecting specific molecules against a concentration gradient of 10^6 or 10^7, and this binding is saturable at physiologic concentrations of the hormone. Receptor-hormone interactions exhibit a salt, temperature, and pH dependency that is characteristic for each hormone. Binding

Table 44–1. Comparison of hormone receptors with proteins that transport hormones in the plasma.

Feature	Receptors	Plasma Transport Proteins
Concentration	Very low (thousands/cell)	Very high (billions/μL)
Binding affinity	Very high (10^{-11}–10^{-9} mol/L)	Low (10^{-7}–10^{-5} mol/L)
Binding specificity	High	Low
Saturability at physiologic concentrations	Yes	No
Reversibility	Yes	Yes
Signal transduction	Yes	No

is by **hydrophobic** and **electrostatic** mechanisms and thus is readily reversible except in special cases.

The hydrophobic steroid and thyroid hormones circulate bound to specific transport proteins. Transport proteins are much more abundant that the intracellular receptor proteins, but they bind hormones with lower affinity and lower specificity. Transport proteins provide a circulating reservoir of hormones, which when bound cannot be metabolized or excreted. Only the unbound, or free, hormone has biologic activity. Peptide and protein hormones do not have plasma transport proteins and thus have a much shorter plasma half-life than do the steroid hormones (seconds and minutes as opposed to hours).

The Relationship Between Receptor Occupancy and Biologic Effect Is Complex

The concentrations of hormone required for occupancy of the receptor and for elicitation of a specific biologic response are often very similar (Fig 44–3A). This is especially true for steroid hormones, but some peptide hormones also exhibit the same characteristics. This is remarkable, considering the many steps that must occur between hormone binding and complex responses, such as enzyme induction, cell lysis, and amino acid transport. In other instances, there is a marked dissociation of these 2 processes, so that a maximal biologic effect occurs when only a small percentage of the receptors are occupied (Fig 44–3B, effect 2). Those receptors not involved in the elicitation of the response are said to be **spare receptors.**

Spare receptors are observed in the response of several polypeptide hormones and are thought to provide a means of increasing the sensitivity of a target cell to activation by low concentrations of hormone and to provide a reservoir of receptors. The concept of spare receptors is an operational one and may depend on which aspect of the response is examined and which tissue is involved. For example, there is excellent agreement between LH binding and cAMP production in granulosa cells (there generally are no spare receptors when any hormone activates adenylate cyclase), but steroidogenesis in these cells, which is cAMP-dependent, occurs when fewer than 1% of the receptors are occupied (compare effect 1 with effect 2, Fig 44–3B). Transcription of the phosphoenolpyruvate carboxykinase gene is repressed when far fewer than 1% of liver cell insulin receptors are occupied, whereas in thymocytes there is a high correlation between insulin binding and amino acid transport. Other examples of the dissociation of receptor binding and biologic effects include the effects of catecholamines on muscle contraction, lipolysis, and ion transport. The presumption is that these end responses reflect a cascade or multiplier effect of the hormone. Variable sensitivities of different responses even within the same cell have been noted. Successively greater degrees of occupancy of the adipose cell insulin receptor increase (in sequence) lipolysis, glucose oxidation, amino acid transport, and protein synthesis.

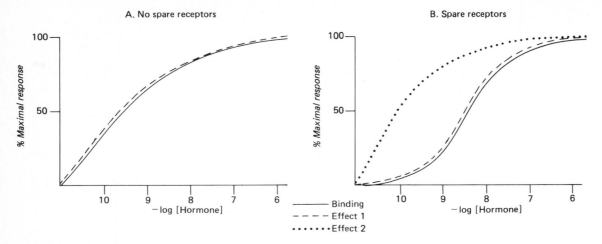

Figure 44–3. A comparison of hormone binding with biologic effect in the absence (*A*) or presence (*B*, effect 2) of spare receptors. In some cases, one biologic effect may be tightly coupled to binding in a tissue, whereas another effect shows the spare receptor phenomenon (compare effects 1 and 2 in *B*).

Receptor Numbers Are Up- & Down-Regulated

The number of receptors on or in a cell is in a dynamic state and can be regulated physiologically or be influenced by diseases or therapeutic measures. Most is known about the plasma membrane receptors. Both the receptor concentration and the affinity of hormone binding can be regulated in plasma membrane. These changes can be acute and can significantly affect hormone responsiveness of the cell. For instance, cells exposed to β-adrenergic agonists for minutes to hours no longer activate adenylate cyclase in response to the addition of more agonist, and the biologic response is lost. This **desensitization** occurs by 2 mechanisms. The first involves a loss of receptors from the plasma membrane. This **down-regulation** involves the internal sequestration of receptors, thereby segregating them from the other components of the response system including the regulatory and catalytic subunits of adenylate cyclase (see Chapter 45). Removal of the agonist results in the return of receptors to the cell surface and restoration of hormonal sensitivity. A second form of desensitization of the β-adrenergic system involves the covalent modification of receptor by phosphorylation. This cAMP-dependent process entails no change in receptor number and no translocation. Reconstitution experiments show that the phosphorylated receptor is unable to activate cyclase, so that the activation and hormone binding functions are uncoupled. Other examples of physiologic adaptation accomplished through down-regulation of receptor number by the homologous hormone include insulin, glucagon, TRH, growth hormone, LH, FSH, and catecholamines. A few hormones, such as angiotensin II and prolactin, **"up-regulate"** their receptors. These changes in receptor number can occur rapidly (minutes to hours) and are probably an important means of regulating biologic responses.

How the loss of receptors affects the biologic response elicited at a given hormone concentration depends on whether or not there are spare receptors. Fig 44–4 illustrates the effects of a 5-fold loss of receptors on the concentration-response curve in both conditions. In condition A (with no spare receptors), the maximal response obtained is 20% that of control; hence, the effect is on the "V_{max}." In condition B (with spare receptors), the maximal response is obtained but at 5 times the originally effective hormone concentration, analogous to a "K_m" effect.

Receptors Are Proteins

Most is known about the acetylcholine receptor, which was easy to purify, since it exists in relatively large amounts in the electric organ of the eel *Torpedo californica*. The acetylcholine receptor consists of 4 subunits in the configuration α_2, β, γ, δ. The two α subunits bind acetylcholine; the technique of site-directed mutagenesis has been used to show which regions of this subunit are involved in the formation of the transmembrane ion channel, which performs the major function of the acetylcholine receptor.

Other receptors are present in very small amounts; thus, purification and characterization have been difficult. Recombinant DNA techniques can now provide the requisite amounts of material for such studies, and so this has become an active area of investigation. The insulin receptor is a heterotetramer ($\alpha_2\beta_2$) linked by multiple disulfide bonds, in which the extramembrane α subunit binds insulin and the membrane-spanning β subunit transduces the signal, presumably through the tyrosine kinase component of the cytoplasmic portion of this polypeptide. The receptors for insulinlike growth factor I (IGF-I), epidermal growth factor (EGF), and low-density lipoprotein (LDL) are generally similar to the insulin receptor (see Fig 52–12). Other polypeptide hormone receptors are less well characterized but are assumed to have a fundamental protein component because of their sensitivity to a va-

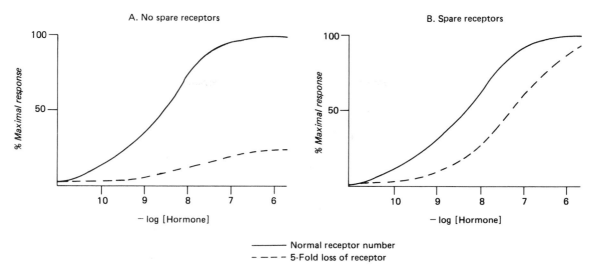

Figure 44–4. Effect of a 5-fold loss of receptors on a biologic effect in a system that lacks (*A*) or has (*B*) spare receptors.

riety of peptidases and proteolytic enzymes. Hormone binding appears to require disulfide bonds, phospholipid, and carbohydrates moieties in many instances.

Steroid hormone receptors also are proteins. The function of this group of molecules has been defined in recent years, and we now are beginning to learn about their structure. The glucocorticoid receptor is a good example (Fig 49–6). This molecule has several functional domains: (1) a hormone-binding region in the carboxy-terminal portion; (2) an adjacent DNA-binding region; (3) a specifier region, located in the amino-terminal half of the molecule, which is necessary for high-affinity binding to the proper region of DNA (and which contains most of the antigenic sites in the molecule); and (4) one or more regions that activate or repress gene transcription. These functional regions are being confirmed by analyses of receptor synthesized in cDNA-driven expression systems. It appears that the various types of steroid receptors share this general structure and that there is a great deal of amino acid sequence homology in these regions in the different receptors. There is also a curious homology between this class of receptors and the v-*erb* A oncogene.

Hormone Receptor Abnormalities Are Involved in Disease

The realization that **certain diseases involve abnormalities of receptor function** followed the basic elucidation of the role that receptors play in hormone action. Three general categories of receptor-related defects are shown in Table 44–2. In the first category, antibodies directed against a specific hormone receptor are responsible for the disease. These antibodies of the IgG class can block hormone binding (acanthosis nigricans with insulin resistance; asthma), mimic hormone binding (Graves' disease), or enhance receptor turnover (myasthenia gravis).

In the second category, no hormone binding to re-

ceptor can be detected. Whether receptors are absent in these diseases or present but defective cannot be determined as yet, since the assays for most receptors depend upon hormone binding.

The third category consists of diseases involving abnormal receptor regulation. Patients with obesity and those with type II diabetes mellitus and obesity often manifest glucose intolerance and insulin re-

Table 44–2. Hormone receptors and disease.

Disease	Receptor	Problem
Graves' disease (hyperthyroidism)	TSH	Antibody stimulates TSH receptor
Acanthosis nigricans with insulin resistance	Insulin	Antibody blocks insulin binding to receptor
Myasthenia gravis	Acetylcholine	Antibody enhances turnover of the acetylcholine receptor
Asthma	β-Adrenergic receptor	Antibody blocks β-adrenergic binding
Congenital nephrogenic diabetes insipidus	ADH	Receptor deficiency
Testicular feminization syndrome	Androgen	Receptor deficiency
Pseudohypoparathyroidism	PTH	Receptor deficiency
Vitamin D-resistant rickets type II	Calcitriol	Receptor deficiency
Obesity	Insulin	Hormone binding decreased
Diabetes mellitus type II (non-insulin-dependent diabetes mellitus [NIDDM])	Insulin	Hormone binding decreased

sistance in spite of elevated plasma insulin levels. These individuals have fewer insulin receptors (owing to down-regulation) on target cells such as fat, liver, and muscle. With weight reduction, the plasma insulin level decreases, the number of receptors increases, insulin sensitivity improves, and glucose intolerance is reduced. Recent studies of the molecular basis of cancer strongly suggest that abnormalities of growth factor receptor-effector coupling can account for the uncontrolled growth of malignant cells. These examples are illustrative of the many hormone receptor-mediated diseases.

REFERENCES

Ginsberg BH: Synthesis and regulation of receptors for polypeptide hormones. Pages 59–97 in: *Biological Regulation and Development*. Vol 3B. Yamamoto K (editor). Plenum Press, 1985.

Granner DK, Lee F: The multiple endocrine neoplasia syndromes. Chapter 76 in: *Comprehensive Textbook of Oncology*. Moosa AR, Robson MC, Schimpff SC (editors). Williams & Wilkins, 1984.

Mishina M et al: Expression of functional acetylcholine receptor from cloned cDNAs. *Nature* 1984;**307**:604.

Roth J, Taylor SI: Receptors for peptide hormones: Alterations in disease of humans. *Annu Rev Physiol* 1982;**44**:639.

Roth J et al: The evolutionary origins of hormones, neurotransmitters, and other extracellular messengers. *N Engl J Med* 1982;**306**:523.

Sibley DR, Lefkowitz RJ: Molecular mechanisms of receptor desensitization using the β-adrenergic receptor-coupled adenylate cyclase system as a model. *Nature* 1985;**317**:124.

Hormone Action

45

Daryl K. Granner, MD

INTRODUCTION

Hormone action at the cellular level begins with the association of the hormone and its specific receptor. Hormones can be classified by the location of the receptor and by the nature of the signal or second messenger used to mediate hormone action within the cell. A number of these second messengers have been defined, but for several hormones the intracellular signal has not been discovered. Considerable progress has been made in elucidating how hormones work intracellularly, particularly in regard to the regulation of expression of specific genes.

BIOMEDICAL IMPORTANCE

The rational diagnosis and therapy of a disease depend upon understanding the pathophysiology involved and the ability to quantitate it. Diseases of the endocrine system, which are generally due to excessive or deficient production of hormones, are an excellent example of the application of basic principles to clinical medicine. Knowing the general aspects of hormone action and understanding the physiologic and biochemical effects of the individual hormones enable one to recognize endocrine disease syndromes that result from hormone imbalance and to apply effective therapy.

HORMONES CAN BE CLASSIFIED IN A NUMBER OF WAYS

Hormones can be classified according to chemical composition, solubility properties, location of receptors, and nature of the signal used to mediate hormone action within the cell. A classification based on the last 2 properties is illustrated in Table 45–1, and general features of each group are illustrated in Table 45–2.

The hormones in group I are lipophilic and, with the exception of T_3 and T_4, are derived from cholesterol. After secretion, these hormones associate with transport proteins, a process that circumvents the problem of solubility while prolonging the plasma half-life. The free hormone readily traverses the plasma mem-

brane of all cells and encounters receptors in either the cytosol or nucleus of target cells. The ligand-receptor complex is assumed to be the intracellular messenger in this group.

The second major group consists of water-soluble hormones that bind to the plasma membrane of the target cell. Hormones that bind to the surface of cells communicate with intracellular metabolic processes through intermediary molecules, so-called **second messengers** (the hormone itself is the first messenger), which are generated as a consequence of the ligand-receptor interaction. The second-messenger concept arose from Sutherland's observation that epinephrine binds to the plasma membrane of pigeon erythrocytes and increases intracellular cAMP. This was followed by a series of experiments in which cAMP was found to mediate the metabolic effects of many hormones. Hormones that clearly employ this mechanism are shown in group II.A. To date only one hormone, atrial natriuretic factor (ANF), uses cGMP as its second messenger, but other hormones will probably be added to group II.B. Several hormones, many of which were previously thought to affect cAMP, appear to use calcium or metabolites of complex phosphoinositides (or both) as the intracellular signal. These are shown in group I.C. The intracellular messenger has not been definitively identified for group I.D., a large and very interesting class of hormones. It would not be surprising if a number of mediators or different mechanisms are involved in the action of this group of hormones. A few hormones fit into more than one category, and assignments change with new information.

GROUP I HORMONES HAVE INTRACELLULAR RECEPTORS AND AFFECT GENE EXPRESSION

The general features of the action of this group of hormones are illustrated in Fig 45–1. These lipophilic molecules diffuse through the plasma membrane of all cells but only encounter their specific, high-affinity receptor in target cells. The hormone-receptor complex next undergoes a temperature- and salt-dependent **"activation" reaction** that results in size, conformation, and surface charge changes that render it able to

Table 45–1. Classification of hormones by mechanism of action.

Group I. Hormones that bind to intracellular receptors

Estrogens	Calcitriol (1,25[OH]$_2$-D$_3$)
Glucocorticoids	Androgens
Mineralocorticoids	Thyroid hormones (T$_3$ and T$_4$)
Progestins	

Group II. Hormones that bind to cell surface receptors
A. The second messenger is cAMP.

Adrenocorticotropic hormone (ACTH)	Parathyroid hormone (PTH)
Angiotensin II	Opioids
Antidiuretic hormone (ADH)	Acetylcholine
Follicle-stimulating hormone (FSH)	Glucagon
Human chorionic gonadotropin (hCG)	α_2-Adrenergic catecholamines
Lipotropin (LPH)	Corticotropin-releasing hormone (CRH)
Luteinizing hormone (LH)	Calcitonin
Melanocyte-stimulating hormone (MSH)	Somatostatin
Thyroid-stimulating hormone (TSH)	β-Adrenergic catecholamines

B. The second messenger is cGMP.

Atrial natriuretic factor (ANF)	

C. The second messenger is calcium or phosphatidylinositides (or both):

α_1-Adrenergic catecholamines	Acetylcholine (muscarinic)
Cholecystokinin	Oxytocin
Gastrin	Gonadotropin-releasing hormone (GnRH)
Substance P	Angiotensin II
Thyrotropin-releasing hormone (TRH)	
Vasopressin	

D. The intracellular messenger is unknown:

Chorionic somatomammotropin (CS)	
Growth hormone (GH)	Nerve growth factor (NGF)
Insulin	Epidermal growth factor (EGF)
Insulinlike growth factors (IGF-I, IGF-II)	Fibroblast growth factor (FGF)
Prolactin (PRL)	Platelet-derived growth factor

Table 45–2. General features of hormone classes.

	Group I	Group II
Types	Steroids, iodothyronines, calcitriol	Polypeptides, proteins, glycoproteins, catecholamines
Solubility	Lipophilic	Hydrophilic
Transport proteins	Yes	No
Plasma half-life	Long (hours to days)	Short (minutes)
Receptor	Intracellular	Plasma membrane
Mediator	Receptor-hormone complex	cAMP, Ca^{2+}, metabolites of complex phosphoinositides, others

bind to chromatin. Whether this association and "activation" process occurs in the cytoplasm or nucleus is debatable but not crucial to understanding the whole process. The hormone receptor complex binds to a specific region of DNA, called the "hormone response element," and activates or inactivates specific genes. By selectively affecting gene transcription and the production of the respective mRNAs, the amounts of specific proteins are changed and metabolic processes are influenced. The effect of each of these hormones is quite specific; generally, the hormone affects less than 1% of the proteins or mRNA in a target cell. This discussion has concentrated on nuclear actions of steroid and thyroid hormones because these are quite well defined. Direct actions in the cytoplasm and upon various organelles and membranes have also been described. Most evidence suggests that steroid hormones exert their predominent effect on gene transcription,

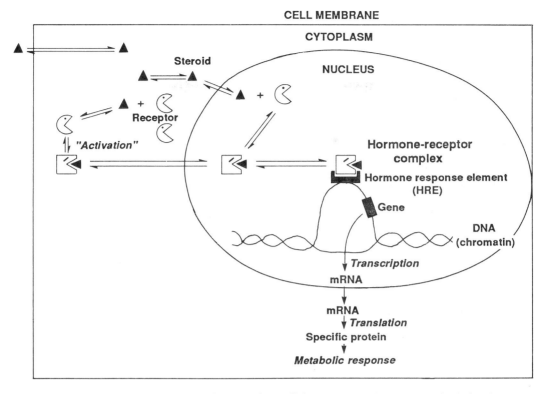

CELL MEMBRANE

Figure 45–1. The steroid or thyroid hormone binds to an intracellular receptor and causes a conformational change of the latter. This complex then binds to a specific DNA region, the hormone response element, and this interaction results in the activation or repression of a restricted number of genes.

but these hormones, and many of those found in the other classes discussed below, can act at any step of the "information pathway" illustrated in Fig 45–2. Although the biochemistry of gene transcription in mammalian cells is not well understood, a general model of the structural requirements for steroid and thyroid regulation of gene transcription can be drawn (Fig 45–3). These genes must be in regions of "open" or transcriptionally active chromatin (depicted as the bubble in Fig 45–1), as defined by their susceptibility to digestion by the enzyme DNase I. The genes studied to date have at least 2 separate regulatory elements (control sites) in the DNA sequence immediately 5' of the transcription initiation site (Fig 45–3). The first of these, the **promoter element (PE)**, is generic, since it is present in some form or other in all genes. This element specifies the site of RNA polymerase II attachment to DNA and therefore the accuracy of transcript initiation (see Chapter 41).

A second element, the **hormone response element (HRE)**, has been identified in many genes regulated by steroid hormones. This is located slightly farther 5' than the PE and may consist of several discrete elements. The HRE presumably modulates the frequency of transcript initiation and is less dependent on position and orientation; in these respects, it resembles the transcription **enhancer elements** found in other genes (see

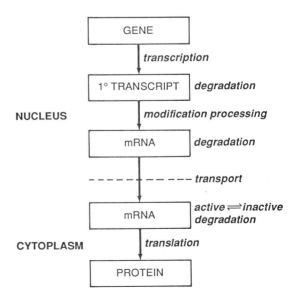

Figure 45–2. The "information pathway." Hormones can affect any of these steps.

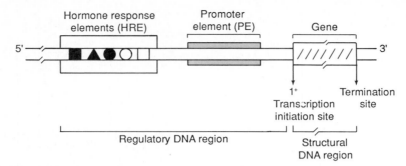

Figure 45–3. Structural requirements for hormonal regulation of gene transcription.

Chapter 41). Generally, the HRE is found within a few hundred nucleotides upstream of the transcription initiation site, but the precise location of the HRE varies from gene to gene. In some instances, it is located within the gene. Genes controlled by several hormones have a corresponding number of HREs. Although the initial reactions are different, peptide hormones also exert their effects on transcription through HREs. For example, many of the hormones that use cAMP as a second messenger affect transcription. A special protein, the cAMP response element binding (CREB) protein, is the *trans*-acting factor (analogous to the steroid/thyroid hormone receptor) in these instances. The consensus DNA sequences of several HREs have been defined (Table 45–3).

The identification of an HRE requires that it bind the hormone-receptor complex more avidly than does sur-

rounding DNA or DNA from another source. In the cases just cited, such specific binding has been demonstrated. The HRE must also confer hormone responsiveness. Putative regulatory sequence DNA can be ligated to reporter genes to assess this point. Usually, these **"fusion genes"** contain reporter genes not ordinarily influenced by the hormone, and often these genes are not normally expressed in the tissue being tested. Commonly used reporter genes are globin, thymidine kinase, bacterial chloramphenicol acetyltransferase and luciferase. The "fusion gene" is transfected into a target cell, and if the hormone now regulates the transcription of the reporter gene, one has functionally defined an HRE. Position, orientation, and base substitution effects can be precisely defined using this technique. Exactly how the hormone-receptor interaction with the HRE affects transcription is an area of active investigation. Transcript initiation is a probable control site, but effects on elongation and termination might also occur. Control sites farther 5' from the initiation site, or 3' downstream, either within or beyond the gene, have been proposed. Finally, *trans*-acting control mechanisms (eg, from another chromosome) may also be operative.

Table 45–3. The DNA sequences of several hormone response elements (HRE).

Hormone/Effector	HRE	DNA Sequence*
Glucocorticoids	GRE	$\overrightarrow{GGTACA}$ nnn $\overleftarrow{TGTTCT}$
Progestins	PRE	"
Mineralocorticoids	MRE	"
Androgens	ARE	"
Estrogens	ERE	$\overrightarrow{AGGTCA}$ nnn $\overleftarrow{TGACCT}$
Thyroid hormone	TRE	$\overrightarrow{GATCA}$ nnnnn $\overleftarrow{TGACC}$
Retinoic acid	RRE	"
cAMP	CRE	$\overrightarrow{TGACGTCA}$

*Letters indicate nucleotide triphosphates; n means any one of the four can be used in that position. The arrows pointing in opposite directions illustrate the slightly imperfect inverted palindromes present in many HREs; in some cases these are called "half-binding sites" because each binds one monomer of the receptor. The GRE, PRE, MRE, and ARE consist of the same DNA sequence. Specificity may be conferred by the intracellular concentration of hormone receptor, or by flanking DNA sequences not included in the consensus. A second group of HREs includes those for thyroid hormones, estrogens, and retinoic acid. These HREs are very similar except for the spacing between the half palindromes. All these HREs are consensus sequences except that for the thyroid hormones; this sequence represents the TRE for the growth hormone gene. The retinoic acid receptor may bind to the same sequence as the thyroid hormone receptor. Peptide hormones that change intracellular cAMP concentrations affect gene transcription through the CRE.

GROUP II (PEPTIDE) HORMONES HAVE MEMBRANE RECEPTORS AND USE INTRACELLULAR MESSENGERS

The largest number of hormones are water-soluble, have no transport proteins (and therefore a short plasma half-life), and initiate a response by binding to a receptor located in the plasma membrane (Tables 45–1 and 45–2). The mechanism of action of this group of hormones can best be discussed in terms of their intracellular messengers.

Many Hormones Use cAMP as the Second Messenger

cAMP (cyclic AMP, 3'.5'-adenylic acid; see p 169), a ubiquitous nucleotide derived from ATP through the action of the enzyme adenylate cyclase, plays a crucial role in the action of a number of hormones. The intracellular level of cAMP is increased or decreased by

45–4. Subclassification of group II.A hormones.

Hormones That Stimulate Adenylate Cyclase (Hs)	Hormones That Inhibit Adenylate Cyclase (Hi)
ACTH	Acetylcholine
ADH	α_2-Adrenergics
βAdrenergics	Angiotensin II
Calcitonin	Opioids
CRH	Somatostatin
FSH	
Glucagon	
hCG	
LH	
LPH	
MSH	
PTH	
TSH	

decreased by various hormones (Table 45–4), and this effect varies from tissue to tissue. Epinephrine causes large increases of cAMP in muscle and relatively small changes in liver. The opposite is true of glucagon. Tissues that respond to several hormones of this group do so through unique receptors converging upon a single adenylate cyclase molecule. The best example of this is the adipose cell, in which epinephrine, ACTH, TSH, glucagon, MSH, and vasopressin (ADH) stimulate adenylate cyclase and increase cAMP. Combinations of maximally effective concentrations are not additive, and treatments that destroy one receptor have no effect on the cellular response to other hormones.

Adenylate Cyclase System: The components of this system in mammalian cells are illustrated in Fig 45–4. The interaction of the hormone with its receptor results in the activation or inactivation of adenylate cyclase. This process is mediated by at least 2 GTP-dependent regulatory proteins, designated G_s (stimulatory) and G_i (inhibitory), each of which is composed of 3 subunits, α, β, and γ. Adenylate cyclase, located on the inner surface of the plasma membrane, catalyzes the formation of cAMP from ATP in the presence of magnesium (see Fig 34–14).

What was originally conceived of as a single protein with 2 functional domains is now viewed as a system of extraordinary complexity. Over the past 15 years, a number of studies have established the biochemical uniqueness of the hormone receptor and GTP regulatory and catalytic domains of the adenylate cyclase complex, a current mode of which is illustrated in Fig 45–4. This model explains how different peptide hormones can either stimulate (s) or inhibit (i) the production of cAMP (Table 45–4).

Two parallel systems, a stimulatory (s) one and an inhibitory (i) one, converge upon a single catalytic molecule (C). Each consists of a receptor, R_s or R_i, and regulatory complex, G_s and G_i. G_s and G_i are each trimers composed of α, β, and γ subunits. The β and γ subunits in G_s appear to be identical to their respective counterparts in G_i. Because the α subunit in G_s differs from that in G_i, the proteins are designated α_s (MW 45,000) and α_i (MW 41,000). The binding of a hor-

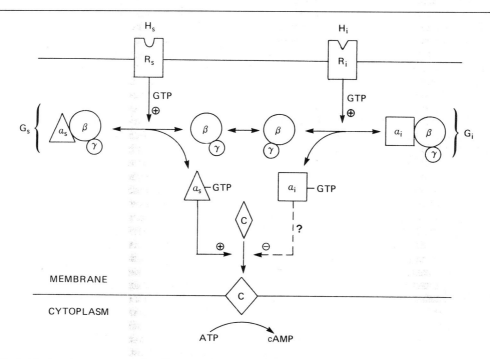

Figure 45–4. The hormone-receptor signal is transmitted through a stimulatory (s) or inhibitory (i) regulatory complex (G_s or G_i) to either stimulate (s) or inhibit (i) the activity of adenylate cyclase (C). Adenylate cyclase catalyzes the formation of cAMP from ATP. (Modified, with permission, from Gilman AG: G proteins and dual control of adenylate cyclase. *Cell* 1984;**36**:577. Copyright © the Massachusetts Institute of Technology.)

mone to R_s or R_i results in a receptor-mediated activation of G, which entails Mg^{2+}-dependent binding of GTP by α and the concomitant dissociation of β and γ from α.

$$\alpha\beta\gamma \underset{\text{GTPase}}{\overset{\text{GTP}}{\rightleftharpoons}} \alpha\cdot\text{GTP} + \beta\gamma$$

The α_s has intrinsic GTPase activity, and the active form, $\alpha_s \cdot$ GTP, is inactivated upon hydrolysis of the GTP to GDP, and the trimeric G_s complex is reformed. **Cholera toxin, known to be an irreversible activator of cyclase, causes ADP ribosylation of α_s and in so doing inactivates the GTPase;** therefore, α_s is frozen in the active form. The α_i also has a GTPase activity; however, GDP does not freely dissociate from $\alpha_i \cdot$ GDP. The α_i is reactivated by an exchange of GTP for GDP. **Pertussis toxin irreversibly activates adenylate cyclase by promoting the ADP ribosylation of α_i, which prevents the α_i subunit from being activated.** NaF, another irreversible activator of cyclase, presumably acts on the α_s or α_i subunit, because it affects G_s and G_i similarly. The exact role of each of the α, β, and γ subunits has not been defined. Two possibilities have been tested. The α_s and α_i could interact noncompetitively with C, causing opposite effects; the net effect would depend on the balance of active α_s and active α_i. Unfortunately, active α_i has little inhibitory effect on C in isolated systems. The more likely possibility, therefore, is that the β subunit of G_i inhibits α_s. In this model, α_i would be an anti-inhibitor of adenylate cyclase by binding β and, as such, would have no direct effect.

Many of the components of the cyclase system have been purified, including the catalytic subunit, and it is now apparent that there is a family of G proteins. Transducin, the protein that is important in coupling light to photoactivation in the retina, is closely related to the G protein of adenylate cyclase, as are the products of the *ras* oncogenes. A unique protein, G_o, has been isolated from brain. Others, not yet well-characterized, appear to be involved in calcium and potassium flux or phosphoinositide metabolism.

The importance of these components is underscored by an "experiment of nature." Pseudohypoparathyroidism is a syndrome characterized by hypocalcemia and hyperphosphatemia, the biochemical hallmarks of hypoparathyroidism, and by a number of congenital defects. Affected individuals do not have defective parathyroid function; in fact, they secrete large amounts of biologically active PTH. Some have target organ resistance on the basis of a postreceptor defect. They are partially deficient in G protein (probably only the α_s subunit) and thus fail to couple binding to adenylate cyclase stimulation. Others appear to generate cAMP in response to PTH but fail to respond to the cAMP signal. The observation that patients with pseudohypoparathyroidism often show evidence of defective responses to other hormones, including TSH, glucagon, and β-adrenergic agents, is not surprising.

Protein Kinase: In **prokaryotic cells,** cAMP binds to a specific protein, called **catabolite regulatory protein (CRP),** which binds directly to DNA and influences gene expression. The analogy of this to steroid hormone action described above is apparent. In **eukaryotic cells,** cAMP binds to a protein kinase that is a heterotetrameric molecule consisting of 2 **regulatory** subunits (R) and 2 **catalytic** subunits (C). cAMP binding results in the following reaction:

$$4 \text{ cAMP} + R_2C_2 \rightleftharpoons R_2 \cdot (4 \text{ cAMP}) + 2 \text{ C}$$

The R_2C_2 complex has no enzymatic activity, but the binding of cAMP by R dissociates R from C, thereby activating the latter (Fig 45–5). The active C subunit catalyzes the transfer of the γ phosphate of ATP (Mg^{2+}) to a serine or threonine residue in a variety of proteins. The consensus phosphorylation sites are -Arg-Arg-X-Ser- and -Lys-Arg-X-X-Ser-, where X can be any amino acid.

Protein kinase activities were originally described as being **cAMP-dependent** or **cAMP-independent.** Table 45–5 shows that this too has become considerably more complex, as protein phosphorylation is now recognized as being an important regulatory mechanism. The kinases listed are all unique molecules and show considerable variability with respect to subunit composition, molecular weight, autophosphorylation, K_m for ATP, and substrate specificity.

The cAMP-dependent protein kinases I and II have been studied in greatest detail. These kinases share a common C subunit and have different R subunits. They were originally classified by type on the basis of their different surface charges and hence differential elution from ion exchange chromatography columns (type I is less acidic and elutes at a lower salt concentration than does type II). Most tissues have both forms, but there is a marked species-to-species and tissue-to-tissue variation in the distribution of the 2 isozymes. Recent studies suggest that types I and II respond differently to

Table 45–5. Purified protein kinases.

Hormone responsive
  Calcium/calmodulin-dependent kinase
  Calcium/phospholipid-dependent kinase
  Casein kinase II
  cAMP-dependent kinase I and II
  cGMP-dependent kinase
  Epidermal growth factor-dependent tyrosine kinase
  Insect cyclic nucleotide-dependent kinase
  Insulin-dependent tyrosine kinase
  Myosin light chain kinase
  Phosphorylase kinase
  Pyruvate dehydrogenase kinase
Not known to be hormone responsive
  dsRNA-dependent eIF-2α kinase
  Hemin-dependent eIF-2α kinase
  Protease-activated kinase I
  Rhodopsin kinase
  Viral tyrosine kinases I, II, and III
May be hormone responsive
  Casein kinase I
  Protease-activated kinase II

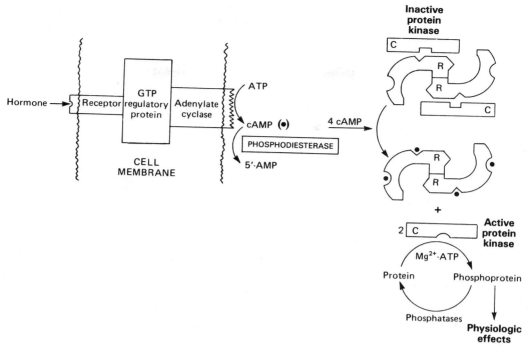

Figure 45–5. Hormonal regulation of cellular processes through cAMP-dependent protein kinases. The cAMP (•) generated by the action of adenylate cyclase binds to the regulatory (R) subunit of cAMP-dependent protein kinase. This results in the release and activation of the catalytic (C) subunit. (Courtesy of J Corbin.)

various combinations of cAMP analogs, and this approach may be useful in defining which isozyme mediates a specific biologic response. There is also some evidence that hormone stimulation selectively enhances either type I or type II kinase activity.

Several other protein kinases are involved in hormone action. The role some of these play is illustrated in Fig 45–6 and in the later parts of this chapter. The epidermal growth factor and insulin-dependent kinases are unique in that the enzymatic activity resides within the hormone receptor and depends upon ligand-receptor binding for activation (see Chapter 52). Another distinguishing feature is that these kinases preferentially phosphorylate tyrosine residues, and tyrosine phosphorylation is infrequent in mammalian cells. The role these receptor-associated protein tyrosine kinases play in hormone action is not clear, but it is possible that the hormone initiates a phosphorylation cascade and that one or more products of the cascade are the intracellular messenger.

Phosphoproteins: The effects of cAMP in eukaryotic cells are all thought to be mediated by protein phosphorylation-dephosphorylation. The control of any of the effects of cAMP, including such diverse processes as steroidogenesis, secretion, ion transport, carbohydrate and fat metabolism, enzyme induction, gene regulation, and cell growth and replication, could be conferred by a specific protein kinase, a specific phosphatase, or by specific substrates for phosphorylation. In some instances a phosphoprotein that

is a known participant in a metabolic pathway has been identified; however, in most processes cited above, the phosphoproteins involved have not been identified. These substrates may help define a target tissue and certainly are involved in defining the extent of the response within a given cell. Many proteins can be phosphorylated, including casein, histones, and protamine; such phosphorylations may be epiphenomena, although they are useful for assaying protein kinase activity. Until recently, the only actions of cAMP that had been defined were actions that occurred outside the nucleus. Effects of cAMP on the transcription of several genes have been described. These effects appear to be mediated by the CREB protein described above.

Phosphodiesterases: Actions caused by hormones that increase cAMP concentration can be terminated in a number of ways, including the hydrolysis of cAMP by phosphodiesterases. The presence of these hydrolytic enzymes ensures a rapid turnover of the signal (cAMP) and hence a rapid termination of the biologic process once the hormonal stimulus is removed. cAMP phosphodiesterases exist in low and high K_m forms and are themselves subject to regulation by hormones as well as by intracellular messengers such as calcium, probably acting through calmodulin. Inhibitors of phosphodiesterase, most notably methylated xanthine derivatives such as caffeine, increase intracellular cAMP and mimic or prolong the actions of hormones.

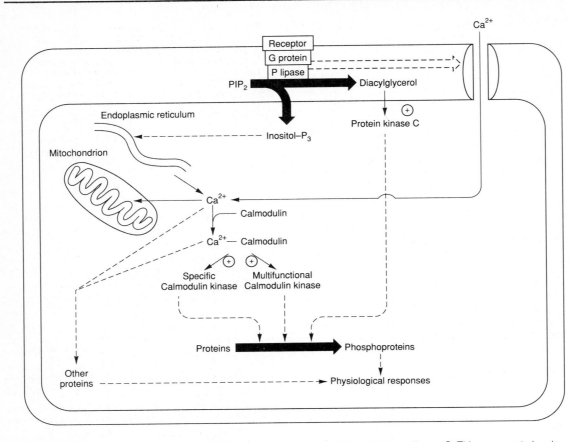

Figure 45–6. Certain hormone receptor interactions result in the activation of phospholipase C. This appears to involve a specific G protein, which also may activate a calcium channel. Phospholipase C results in the generation of IP$_3$, which liberates stored intracellular Ca^{2+}, and diacylglycerol (DAG), which activates protein kinase C. In this scheme the activated protein kinase C phosphorylates specific substrates, which then alter physiologic processes. Likewise, the Ca^{2+}-calmodulin (Cam) complex can activate specific kinases. These actions result in the modification of substrates and this leads to altered physiologic responses.

Phosphoprotein Phosphatases: Another means of controlling hormone action is the regulation of the protein dephosphorylation reaction. The phosphoprotein phosphatases are themselves subject to regulation by phosphorylation-dephosphorylation and by a variety of other substances. Most is known about the role of phosphatase in regulation of glycogen metabolism in muscle. In this tissue, 2 types of phosphoprotein phosphatases have been described. Type I preferentially dephosphorylates the β subunit of phosphorylase kinase, whereas type II dephosphorylates the α subunit. Two heat-stable protein inhibitors regulate type I phosphatase activity. Inhibitor-1 is phosphorylated and activated by cAMP-dependent protein kinase, and inhibitor-2, which may be a subunit of the inactive phosphatase, is also phosphorylated, possibly by glycogen synthase kinase-3. Phosphorylation of inhibitor-2 also results in activation of the phosphatase. Certain phosphatases may attack specific residues; eg, there may be a phosphatase that removes phosphate from tyrosine residues.

Extracellular cAMP: Some cAMP leaves cells and can be readily detected in extracellular fluids. The action of glucagon on liver and vasopressin or PTH on kidney is reflected in elevated levels of cAMP in plasma and urine, respectively; this has led to diagnostic tests of target organ responsiveness. Extracellular cAMP has little if any biologic activity in mammals, but it is an extremely important intercellular messenger in lower eukaryotes and prokaryotes.

One Hormone Uses cGMP as the Second Messenger

Cyclic GMP is made from GTP by the enzyme guanylate cyclase, which exists in soluble and membrane-bound forms. Each of these isozymes has unique kinetic, physiochemical, and antigenic properties. For some time, cGMP was thought to be the functional counterpart of cAMP. It now appears that cGMP has its unique place in hormone action. The atriopeptins, a family of peptides produced in cardiac atrial tissues, cause natriuresis, diuresis, vasodilation, and inhibition of aldosterone secretion. These peptides (eg, atrial natriuretic factor) bind to and activate the membrane-bound form of guanylate cyclase. This results in an increase of cGMP of as much as 50-fold in some cases, which is thought to mediate these effects. Other evidence links cGMP to vasodilation. A series of compounds, including nitroprusside, nitroglycerin, sodium nitrite, and sodium azide, all cause smooth muscle relaxation and are potent vasodilators. These agents increase cGMP by activating the soluble form of guanylate cyclase, and inhibitors of cGMP phosphodiesterase enhance and prolong these responses. The increased cGMP activates cGMP-dependent protein kinase, which in turn phosphorylates a number of smooth muscle proteins, including the myosin light chain. Presumably, this is involved in relaxation of smooth muscle and vasodilation.

Several Hormones Act Through Calcium or Phosphoinositides

Ionized calcium is an important regulator of a variety of cellular processes including muscle contraction, stimulus-secretion coupling, the blood clotting cascade, enzyme activity, and membrane excitability. It is also an intracellular messenger of hormone action.

Calcium Metabolism: The extracellular calcium (Ca^{2+}) concentration is about 5 mmol/L and is very rigidly controlled (see Chapter 48). The intracellular concentration of this free ion is much lower, 0.1–10 μmol/L, and the concentration associated with intracellular organelles such as mitochondria and endoplasmic reticula is in the range of 1–20 μmol/L. In spite of this 5000- to 10,000-fold concentration gradient and a favorable transmembrane electrical gradient, Ca^{2+} is restrained from entering the cell. There are 3 ways of changing cytosolic Ca^{2+}. Certain hormones (class II.C.) enhance membrane permeability to Ca^{2+} and thereby increase Ca^{2+} influx. This is probably accomplished by an Na^+/Ca^{2+} exchange mechanism that has a high capacity but low affinity for Ca^{2+}. There also is a $Ca^{2+}/2H^+$-ATPase-dependent pump that extrudes Ca^{2+} in exchange for H^+. This has a high affinity for Ca^{2+} but a low capacity and is probably responsible for fine-tuning cytosolic Ca^{2+}. Finally, Ca^{2+} can be mobilized (or deposited) from (or into) the mitochondrial and endoplasmic reticulum pools.

Two observations led to the current understanding of how Ca^{2+} serves as an intracellular messenger of hormone action. First was the ability to quantitate the rapid changes of intracellular Ca^{2+} concentration that are implicit in a role for Ca^{2+} as an intracellular messenger. Such evidence was provided by a variety of techniques, including the use of Quin 2 or Fura 2, fluorescent Ca^{2+} chelators. Rapid changes of Ca^{2+} in the submicromolar range can be quantitated using these compounds. The second important observation linking Ca^{2+} to hormone action involved the definition of the intracellular targets of Ca^{2+} action. The discovery of a Ca^{2+}-dependent regulator of phosphodiesterase activity provided the basis for under-

Table 45–6. Enzymes regulated by calcium/calmodulin.

Adenylate cyclase	Glycogen synthase
Ca^{2+}-dependent protein kinase	Guanylate cyclase
Ca^{2+}/Mg^{2+}-ATPase	Myosin kinase
Ca^{2+}/phospholipid-dependent protein kinase	NAD kinase
Cyclic nucleotide phosphodiesterase	Phospholipase A_2
	Phosphorylase kinase
Glycerol-3-phosphate dehydrogenase	Phosphoprotein phosphatase 2B
	Pyruvate carboxylase
	Pyruvate dehydrogenase
	Pyruvate kinase

standing how Ca^{2+} and cAMP interact within cells.

Calmodulin: The calcium-dependent regulatory protein is now referred to as calmodulin, a 17,000-MW protein that is homologous to the muscle protein troponin C in structure and function. Calmodulin has four Ca^{2+} binding sites, and full occupancy of these sites leads to a marked conformational change, so that most of the molecule assumes an alpha-helical structure. This conformational change is presumably linked to calmodulin's ability to activate or inactivate enzymes. The interaction of Ca^{2+} with calmodulin (with the resultant change of activity of the latter) is conceptually similar to the binding of cAMP to protein kinase and the subsequent activation of this molecule. Calmodulin is often one of numerous subunits of complex proteins and is particularly involved in regulating various kinases and enzymes of cyclic nucleotide generation and degradation. A partial list of the enzymes regulated directly or indirectly by Ca^{2+}, probably through calmodulin, is given in Table 45–6.

In addition to its effects on enzymes and ion transport, Ca^{2+}/calmodulin regulates the activity of many structural elements in cells. These include the actin-myosin complex of smooth muscle, which is under β-adrenergic control, and various microfilament-mediated processes in noncontractile cells including cell motility, conformation changes, mitosis, granule release, and endocytosis.

Calcium is a Mediator of Hormone Action

A role for ionized calcium in hormone action is suggested by the observations that the effect of many hormones (1) is blunted by Ca^{2+}-free media or when intracellular calcium is depleted; (2) can be mimicked by agents that increase cytosolic Ca^{2+}, such as the Ca^{2+} ionophore A23187; and (3) influences cellular calcium flux. These processes have been studied in some detail in pituitary, smooth muscle, platelets, and salivary gland, but most is probably known about how vasopressin and α-adrenergic catecholamines regulate glycogen metabolism in liver. This is shown schematically in Figs 19–5 and 19–7.

Addition of α_1 agonists or vasopressin to isolated hepatocytes results in a 3-fold increase of cytosolic Ca^{2+} (from 0.2 to 0.6 μmol/L) within a few seconds. This change precedes and equals the increase in phosphorylase a activity, and the hormone concentrations required for both processes are comparable. This effect on Ca^{2+} is inhibited by α_1 antagonists, and removal of the hormone results in a prompt decline of both cytosolic Ca^{2+} and phosphorylase a. The initial source of the Ca^{2+} appears to be the intracellular organelle reservoirs, which seem to be sufficient for the early effects of the hormones. More prolonged action appears to require enhanced influx or inhibition of Ca^{2+} efflux through the Ca^{2+} pump. The latter may depend upon concomitant increases of cAMP.

Phosphorylase activation results from the conversion of phosphorylase b to phosphorylase a through the action of the enzyme phosphorylase b kinase. This enzyme contains calmodulin as its δ subunit, and its activity is increased through a Ca^{2+} concentration range of $0.1-1$ μmol/L, the range through which hormones increase Ca^{2+} in liver. The link between Ca^{2+} and phosphorylase activation is definite.

A number of critical metabolic enzymes are regulated by Ca^{2+}, phosphorylation, or both, including glycogen synthase, pyruvate kinase, pyruvate carboxylase, glycerol-3-phosphate dehydrogenase, and pyruvate dehydrogenase. It is uncertain whether calmodulin is directly involved or whether the newly discovered Ca^{2+}/calmodulin-dependent or Ca^{2+}/phospholipid-dependent protein kinases are responsible.

Phosphoinositide Metabolism Plays a Role in Ca^{2+}-Dependent Hormone Action

Some signal must provide communication between the hormone receptor on the plasma membrane and the intracellular Ca^{2+} reservoirs. The best candidates appear to be products of phosphoinositide metabolism. Cell surface receptors such as those for acetylcholine, antidiuretic hormone, and α_1-type catecholamines are, when occupied by their respective ligands, potent activators of phospholipase C. Receptor binding and activation of phospholipase C are coupled by a unique G protein (Fig 45–6). Phospholipase C catalyzes the hydrolysis of phosphatidylinositol 4,5-bisphosphate to inositol trisphosphate and 1,2-diacylglycerol (Fig 45–7). The diacylglycerol is itself capable of activating protein kinase C, the activity of which also depends upon free ionic calcium. Inositol trisphosphate is an effective releaser of calcium from intracellular storage sites such as the sarcoplasmic reticulum and mitochondria. Thus, the hydrolysis of phosphatidylinositol 4,5-bisphosphate leads to activation of protein kinase C and promotes an increase of cytoplasmic calcium ion.

Steroidogenic agents, including ACTH and cAMP in the adrenal cortex; angiotensin II, K^+, serotonin, ACTH, and dibutyryl cAMP in the zona glomerulosa of the adrenal; LH in the ovary; and LH and cAMP in the Leydig cells of the testes, have been associated with increased amounts of phosphatidic acid, phosphoinositol, and polyphosphoinositides in the respective target tissues. Several other examples could be cited.

The roles that Ca^{2+} and polyphosphoinositide breakdown products might play in hormone action are presented in Fig 45–6. In this scheme the activated protein kinase C can phosphorylate specific substrates, which then alter physiologic processes. Likewise, the Ca^{2+}-calmodulin (Cam) complex can activate specific kinases. These then modify substrates and thereby alter physiologic responses.

The Intracellular Messenger Is Unknown for Some Hormones

A large number of important hormones have no identified intracellular messenger. It is curious that these hormones cluster into 2 groups. One group consists of insulin, the insulinlike growth factors (IGF-I and IGF-II), and a variety of other growth factors, all of which may share a common ancestor. The other major group consists of proteins from the growth hormone gene family (growth hormone, prolactin,

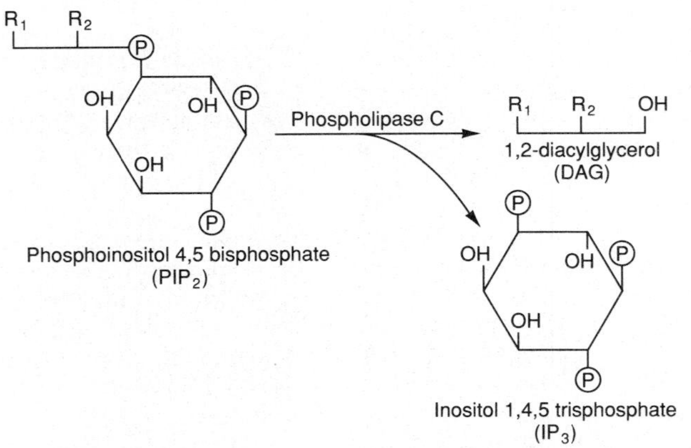

Figure 45–7. Phospholipase C cleaves PIP_2 into diacylglycerol and inositol trisphosphate. R_1 generally is stearate and R_2 is usually arachidonate. IP_3 can be dephosphorylated (to the inactive I-1,4-P_2) or phosphorylated (to the potentially active I-1,3,4,5-P_4).

chorionic somatomammotropin), which clearly are related to one another (see Chapter 46). There is some overlap between these 2 groups, since many of the actions of growth hormone appear to be mediated by IGF-I. Oxytocin appears to stand alone.

Much effort has been directed toward finding the intracellular mediator of insulin action. A variety of candidates including cAMP, cGMP, H_2O_2, Ca^{2+}, and insulin itself have been proposed. Various "mediator" substances of peptide or phospholipid derivation have been found in tissue extracts, but to date these have not been purified or characterized. The recent observation that the insulin receptor has intrinsic tyrosine kinase activity has spurred interest in finding a phosphorylation cascade that might explain the actions of this hormone. This is not an isolated observation, since the epidermal growth factor receptor is also a tyrosine kinase; indeed, this observation led to the studies of the insulin receptor. Finally, the platelet-derived growth factor is also a tyrosine kinase that closely resembles v-*sis* and c-*sis,* specific virus-associated and cell-associated oncogene products. Stimulation of platelet-derived growth factor target cells (fibroblasts, glial cells, smooth muscle cells) results in the production of several gene products that appear to be involved in the replication of those cells.

It is probable that entirely different mechanisms of intracellular signaling are employed by this large group of hormones. The traditional messengers certainly do not seem to be involved.

REFERENCES

Anderson JE: The effect of steroid hormones on gene transcription. In: *Biological Regulation and Development.* Goldberger RF, Yamamoto KR (editors). Vol 3B: Hormone Action. Plenum Press, 1985.

Blackmore PF, Exton JH: Mechanisms involved in the actions of calcium dependent hormones. In: *Biochemical Action of the Hormones.* Vol 12. Litwack G (editor). Academic Press, 1985

Beato, M: Gene regulation by steroid hormones. *Cell* 1989; **56:**335.

Enhancers and eukaryotic gene expression. In: *Current Communications in Molecular Biology.* Gluzman Y, Shenk T (editors). Cold Spring Harbor Press, 1983.

Gilman A: G proteins and dual control of adenylate cyclase. *Cell* 1984;**36:**577.

Rasmussen H: The calcium messenger system. (2 parts.) *N Engl J Med* 1986;**314:**1094, 1164.

46

Pituitary & Hypothalamic Hormones

Daryl K. Granner, MD

INTRODUCTION

The anterior pituitary, under control of hypothalamic hormones, secretes a number of hormones (trophic hormones) that regulate the growth and function of other endocrine glands or influence metabolic reactions in other target tissues. The posterior pituitary produces hormones that regulate water balance and milk ejection from the lactating mammary gland.

BIOMEDICAL IMPORTANCE

The loss of anterior pituitary function (panhypopituitarism) results in atrophy of the thyroid, adrenal cortex, and gonads. Secondary effects due to the absence of the hormones secreted by these target glands affect most body organs and tissues and many general processes such as protein, fat, carbohydrate, and fluid and electrolyte metabolism. The loss of posterior pituitary function results in diabetes insipidus, the inability to concentrate the urine.

HYPOTHALAMIC HORMONES REGULATE THE ANTERIOR PITUITARY

The release (and in some cases production) of each of the pituitary hormones listed in Table 46–1 is under tonic control by at least one hypothalamic hormone. The hypothalamic hormones are released from the hypothalamic nerve fiber endings around the capillaries of the hypothalamic-hypophyseal system in the pituitary stalk and reach the anterior lobe through the special portal system that connects the hypothalamus and the anterior lobe. The structures of several hypothalamic hormones are illustrated in Table 46–2.

The hypothalamic hormones are released in a pulsatile manner, and isolated anterior pituitary target cells respond better to pulsatile administration of these hormones than to continuous exposure. The release of LH and FSH is controlled by the concentration of one releasing hormone, GnRH; this in turn is primarily a function of circulating levels of gonadal hormones that reach the hypothalamus (see the feedback loop in Fig 44–1. Similar feedback loops exist for all of the hypo-

Acronyms Used in This Chapter	
ACTH	Adrenocorticotropic hormone
ADH	Antidiuretic hormone
CG	Chorionic gonadotropin
CLIP	Corticotropinlike intermediate lobe peptide
CRH	Corticotropin-releasing hormone
CS	Chorionic somatomammotropin; placental lactogen
FSH	Follicle-stimulating hormone
GAP	GnRH-associated peptide
GH	Growth hormone
GHRH or GRH	Growth hormone-releasing hormone
GHRIH	Growth hormone release-inhibiting hormone; somatostatin
GnRH	Gonadotropin-releasing hormone
IGF-I, -II	Insulinlike growth factors I and II
LH	Luteinizing hormone
LPH	Lipotropin
MSH	Melanocyte-stimulating hormone
NGF	Nerve growth factor
POMC	Pro-opiomelanocortin peptide family
PRIH or PIH	Prolactin release-inhibiting hormone
PRL	Prolactin
SRIH	Somatotropin release-inhibiting hormone
T$_3$	Triiodothyronine
T$_4$	Thyroxine; tetraiodothyronine
TRH	Thyrotropin-releasing hormone
TSH	Thyroid-stimulating hormone
VIP	Vasoactive intestinal polypeptide

thalamic-pituitary-target gland systems (Table 46–1).

The release of ACTH is primarily controlled by CRH, but a number of other hormones, including ADH, catecholamines, VIP, and angiotensin II, may be involved. CRH release is influenced by cortisol, a glucocorticoid hormone secreted by the adrenal. TSH release is primarily affected by TRH, which in turn is regulated by the thyroid hormones T$_3$ and T$_4$; but TSH release is also inhibited by somatostatin. Growth hormone release and production are under tonic control by both stimulating and inhibiting hypothalamic hormones. In addition, a peripheral feedback loop is in-

Table 46–1. Hypothalamic-hypohyseal-target gland hormones form integrated feedback loops.*

Hypothalamic Hormone	Acronym	Pituitary Hormone Affected†	Target Gland Hormone Affected
Corticotropin-releasing hormone	CRH	ACTH (LPH, MSH, endorphins)	Hydrocortisone
Thyrotropin-releasing hormone	TRH	TSH (PRL)	T_3 and T_4
Gonadotropin-releasing hormone	GnRH (LHRH, FSHRH)	LH, FSH	Androgens, estrogens, progestins
Growth hormone-releasing hormone	GHRH or GRH	GH	IGF-1; others (?)
Growth hormone release-inhibiting hormone; somatostatin; somatotropin release-inhibiting hormone	GHRIH or SHIH	GH (TSH, FSH, ACTH)	IGF-1; T_3 and T_4; others (?)
Prolactin release-inhibiting hormones; dopamine and GAP	PRIH or PIH	PRL	Neurohormones (?)

*The general features of each major feedback system can be deduced by substituting the corresponding hypothalamic, pituitary, or target gland hormone into the appropriate place in Fig 44–1.
†The hypothalamic hormone has a secondary or lesser effect on the hormones in parentheses.

volved in GH regulation. IGF-I (somatomedin C), which mediates some of the effects of GH, stimulates the release of somatostatin (GHRIH) while inhibiting the release of GHRH. The regulation of PRL synthesis and secretion is primarily under tonic inhibition by hypothalamic agents. It is unique because of the combined neural (nipple stimulation) and neurotransmitter/neurohormone link. Dopamine (Table 46–2) inhibits PRL synthesis (by inhibiting transcription of the PRL gene) and release, but does not account for overall PRL inhibition. Recently, a 56-amino-acid neuropeptide was discovered that has both GnRH and PRIH activities—thus the name GnRH-associated peptide (GAP). GAP is a potent inhibitor of PRL release and may be the elusive PRIH peptide. GAP may explain the curious link between GnRH and PRL secretion that is particularly obvious in some species.

Many of the hypothalamic hormones, in particular TRH, CRH, and somatostatin, are found in other portions of the nervous system and in a variety of peripheral tissues.

Although cAMP was originally thought to mediate the action of releasing hormones on the adenohypophysis, recent studies with GnRH and TRH suggest that a calcium-phospholipid mechanism, similar to that described above, is involved. Whether the releasing hormones also affect the synthesis of the corresponding pituitary hormone has been argued, but recently GHRH has been shown to stimulate the rate of transcription of the GH gene, and TRH has a similar effect on the prolactin gene.

THE ANTERIOR PITUITARY PRODUCES A LARGE NUMBER OF HORMONES THAT STIMULATE A VARIETY OF PHYSIOLOGIC & BIOCHEMICAL PROCESSES IN TARGET TISSUES

The anterior pituitary hormones have traditionally been discussed individually, but recent studies dealing

Table 46–2. Structures of hypothalamic releasing hormones.

Hormone	Structure
TRH	(pyro)Glu-His-Pro-NH$_2$
Somatostatin	Ala-Gly-Cys-Lys-Asn-Phe-Phe-Trp-Lys-Thr-Phe-Thr-Ser-Cys-NH$_2$ (S—S bridge)
GnRH	(pyro)Glu-His-Trp-Ser-Tyr-Gly-Leu-Arg-Pro-Gly-NH$_2$
PRIH	HO—(ring)—CH$_2$CH$_2$NH$_2$;GNRH-associated peptide (GAP)
Ovine CRH	Ser-Gln-Glu-Pro-Pro-Ile-Ser-Leu-Asp-Leu-Thr-Phe-His-Leu-Leu-Arg-Glu-Val-Leu-Glu-Met-Thr-Lys-Ala-Asp-Gln-Leu-Ala-Gln-Gln-Ala-His-Ser-Asn-Arg-Lys-Leu-Leu-Asp-Ile-Ala-NH$_2$
Human GHRH	Tyr-Ala-Asp-Ala-Ile-Phe-Thr-Asn-Ser-Tyr-Arg-Lys-Val-Leu-Gly-Gln-Leu-Ser-Ala-Arg-Lys-Leu-Leu-Gln-Asp-Ile-Met-Ser-Arg-Gln-Gln-Gly-Glu-Ser-Asn-Gln-Glu-Arg-Gly-Ala-Arg-Ala-Arg-Leu-NH$_2$

with the mechanism of synthesis and with the intra-cellular mediators of action (see Table 45–1) allow one to classify these hormones into 3 categories: (1) the growth hormone-prolactin-chorionic somatomammotropin group, (2) the glycoprotein hormone group, and (3) the pro-opiomelanocortin peptide family.

Growth Hormone, Prolactin, & Chorionic Somatomammotropin Are One Group of Hormones

Growth hormone (GH), prolactin (PRL), and chorionic somatomammotropin (CS; placental lactogen) are a family of protein hormones having considerable sequence homology. GH, CS, and PRL range in size from 190 to 199 amino acids in different species. Each has a single tryptophan residue (locus 85 in GH and CS; locus 91 in PRL), and each has 2 homologous disulfide bonds. The amino acid homology between hGH and hCS is 85%, whereas that between hGH and hPRL is 35%. In view of this homology, it is not surprising that these 3 hormones share common antigenic determinants and that all have growth-promoting and lactogenic activity. The hormones are produced in a tissue-specific manner, with GH and PRL produced in the anterior pituitary and CS in the syncytiotrophoblast cells of the placenta. Each appears to be under different regulation (see Table 46–1).

On the basis of these striking similarities, it was postulated several years ago that these hormones may have arisen by duplication of an ancestral gene. Recombinant DNA technology has revealed that there are multiple genes for GH and CS in primates and humans; that the single PRL gene, while encoding a very similar protein, is 5 times as large as those for GH and CS; that hCS is a variant of hGH; and that the GH-CS group in humans is located on chromosome 17, while PRL in humans is found on chromosome 6. There is marked evolutionary divergence of these genes. Rat and bovine tissues have a single copy of GH and PRL per haploid genome, and humans have a single PRL gene. Humans have one functional GH gene (GH-N) and a variant (GH-V), two CS genes that are expressed (CS-A and CS-B), and one CS gene that is not expressed (CS-L). Several simian species have at least 4 of the genes from the GH-CS family. The coding sequence of all of these genes is organized into 5 exons interrupted by 4 introns (Fig 46–1). The genes are highly homologous in the 5′ flanking regions and the coding sequence areas (~ 93% homology in the latter) and diverge in the 3′ flanking regions. The splice junctions

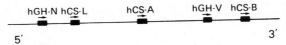

are highly conserved, even though the introns in the PRL gene are much longer.

The human GH-CS gene family is located on a linkage group in region q22–24 on the long arm of chromosome 17. Fig 46–2 indicates the relative positions of each of these genes in a 5′ to 3′ orientation. The genes are all transcribed in the 5′ to 3′ direction, and GH-N is separated from CS-B by about 45 kb.

The GH-N coding sequence matches the amino acid sequence for circulating GH, and the gene is DNase I-sensitive, signifying its location in a region of "**active chromatin.**" The GH-V gene, if expressed, would encode for a protein with 13 amino acid differences. This gene is DNase I resistant, and thus it may not be active. The GH-V gene is present in patients who lack the GH-N gene (inherited GH deficiency), but since these persons have complete GH deficiency, the GH-V gene either is silent or is producing an inactive GH molecule. The first possibility is most likely, because these individuals form antibodies in response to exogenous GH, an indication that this molecule has not previously been seen by the immune system.

The CS-A and CS-B genes are expressed in placenta; CS-L is silent.

Growth Hormone (GH)

Synthesis & Structure: Growth hormone is synthesized in **somatotropes,** a subclass of the pituitary acidophilic cells; somatotropes are the most abundant cells in the gland. The concentration of GH in the pituitary is 5–15 mg/g, which is much higher than the $\mu g/g$ quantities of other pituitary hormones. Growth hormone is a single polypeptide, with a molecular weight of about 22,000 in all mammalian species. The general structure of the 191-amino acid human growth hormone molecule is shown in Fig 46–3. Although there is a high degree of sequence homology between various mammalian growth hormones, only human growth hormone or that of other higher primates is active in humans. Human GH made by recombinant DNA techniques is now available for therapeutic use.

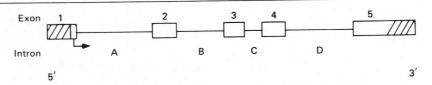

Figure 46–1. Schematic representation, drawn to scale, of the structure of the human growth hormone gene. The gene is about 45 kb in length and consists of 5 exons and 4 introns. Cross-hatching represents noncoding regions in exons 1 and 5. Arrow indicates direction of transcription.

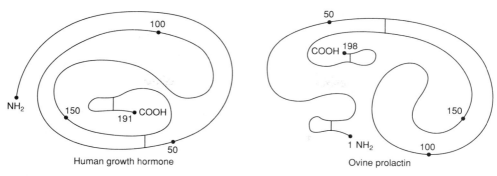

Figure 46–3. The structures of human growth hormone (left) and ovine prolactin (right) are compared. Growth hormone has disulfide bonds between residues 53–165 and 182–189. Prolactin has disulfide bonds between residues 4–11, 58–73, and 190–198.

Physiologic and Biochemical Actions: GH is essential for postnatal growth and for normal carbohydrate, lipid, nitrogen, and mineral metabolism. The growth-related effects are primarily mediated by **IGF-I,** a member of the insulinlike gene family. This was originally known as "sulfation factor" because of its ability to enhance the incorporation of sulfate into cartilage. It next was known as somatomedin C. Structurally, it is similar to proinsulin (see Chapter 52 and Fig 52–5). Another closely related peptide found in human plasma, **IGF-II,** has activity similar or identical to what is often referred to in the rat as multiplication-stimulating activity (MSA). IGF-I and IGF-II both bind to membrane receptors; however, they can be differentiated on the basis of specific radioimmunoassays. IGF-I has 70 amino acids, and IGF-II has 67. Plasma levels of IGF-II are twice those of IGF-I, but it is IGF-I that correlates most directly with GH effects. Individuals who lack sufficient IGF-I but have IGF-II (GH-deficient dwarfs and pygmies; see Table 46–3) fail to grow normally.

GH Has Several Actions

1. Protein synthesis–GH increases the transport of amino acids into muscle cells and also increases protein synthesis by a mechanism separate from the transport effect. Animals treated with GH show positive nitrogen balance, reflecting a generalized increase in protein synthesis and a decrease in plasma and urinary levels of amino acids and urea. This is accompanied by increased synthesis of RNA and DNA in some tissues.

Table 46–3. Relationship of GH, IGF-I, and IGF-II to dwarfism.

	Plasma Levels			Response to GH Stimulation
	GH	**IGF-I**	**IGF-II**	
GH-deficient dwarfs	Low	Low	Low to normal	Yes
Pygmies	Normal	Low	Normal	No
Laron type dwarfs	High	Low	Low	No

In these respects GH actions resemble some of the actions of insulin.

2. Carbohydrate metabolism–GH generally antagonizes the effects of insulin. Hyperglycemia after growth hormone administration is the combined result of decreased peripheral utilization of glucose and increased hepatic production via gluconeogenesis. In liver, GH increases liver glycogen, probably from activation of gluconeogenesis from amino acids. Impairment of glycolysis may occur at several steps, and the mobilization of fatty acids from triacylglycerol stores may also contribute to the inhibition of glycolysis in muscle. Prolonged administration of GH may result in diabetes mellitus.

3. Lipid metabolism–GH promotes the release of free fatty acids and glycerol from adipose tissue, increases circulating free fatty acids, and causes increased oxidation of free fatty acids in the liver. Under conditions of insulin deficiency (eg, diabetes), increased ketogenesis may occur. These effects and those on carbohydrate metabolism probably are not mediated by IGF-I.

4. Mineral metabolism–GH, or more likely IGF-I, promotes a positive calcium, magnesium, and phosphate balance and causes the retention of Na^+, K^+, and Cl^-. The first effect probably relates to the action of GH in bone, where it promotes growth of long bones at the epiphyseal plates in growing children and appositional or acral growth in adults. In children, GH also increases formation of cartilage.

5. Prolactinlike effects–GH binds to lactogenic receptors and thus has many of the properties of prolactin, such as stimulation of the mammary glands, lactogenesis, and stimulation of the pigeon crop sac.

Growth Hormone Pathophysiology: Deficient amounts of GH, whether from panhypopituitarism or isolated GH deficiency, are most serious in infancy because affected infants fail to grow properly. The other metabolic effects are less troublesome. Several types of dwarfism help illustrate the importance of the various steps in GH action (Table 46–3). **GH-deficient dwarfs** respond normally to exogenous GH. Two types of target organ resistance have been described.

Laron type dwarfs have excessive amounts of GH-N, but they lack hepatic GH receptors. **Pygmies** apparently have a post-GH receptor defect, and this may be limited to the action GH exerts through IGF-I.

GH excess, usually from an acidophilic tumor, causes **gigantism** if it occurs before the epiphyseal plates close, since there is accelerated growth of the long bones. **Acromegaly** results from excessive release of GH that begins after epiphyseal closure and the cessation of long bone growth. Acral bone growth causes the characteristic facial changes (protruding jaw, enlarged nose) and enlargement of the hands, feet, and skull. Other findings include enlarged viscera, thickening of the skin, and a variety of metabolic problems, including diabetes mellitus.

A knowledge of GH regulation allows one to understand the clinical tests used to confirm these diagnoses. GH-deficient patients fail to increase GH levels in response to induced hypoglycemia or administration of arginine or levodopa. Patients with increased GH from a tumor (gigantism or acromegaly) fail to suppress GH levels in response to glucose administration.

Prolactin (PRL: Lactogenic Hormone, Mammotropin, Luteotropic Hormone)

Synthesis & Structure: PRL is a protein hormone with a molecular weight of about 23,000; its general structure is compared to that of GH in Fig 46–3. It is secreted by **lactotropes,** which are acidophilic cells in the anterior pituitary. The number of these cells and their size increase dramatically during pregnancy. The similarities between the structures and functions of PRL, GH, and CS are noted above.

Prolactin Physiologic & Biochemical Actions: PRL is involved in the initiation and maintenance of lactation in mammals. Physiologic levels act only upon breast tissue primed by female sex hormones, but excessive levels can trigger breast development in ovariectomized females or in males. In rodents, PRL is capable of maintaining the corpus luteum–hence the name **luteotropic hormone.** Related molecules appear to be responsible for the adaptation of saltwater fish to fresh water, for the molting of reptiles, and for crop-sac milk production in birds. The intracellular mediator of PRL action is unknown. A peptide has been proposed, but this has not been verified.

Prolactin Pathophysiology: Tumors of prolactin-secreting cells cause **amenorrhea** (cessation of menses) and **galactorrhea** (breast discharge) in women. Excessive PRL has been associated with **gynecomastia** (breast enlargement) and **impotence** in men.

Chorionic Somatomammotropin (CS; Placental Lactogen)

The final member of the GH-PRL-CS family has no definite function in humans. In bioassays, CS has lactogenic and luteotropic activity and metabolic effects that are qualitatively similar to those of growth hormone, including inhibition of glucose uptake, stimulation of free fatty acid and glycerol release, enhance-

ment of nitrogen and calcium retention (despite increased urinary calcium excretion), and reduction in the urinary excretion of phosphorus and potassium.

The Glycoprotein Hormones Are Another Group

The most complex protein hormones yet discovered are the pituitary and placental glycoproteins: **thyroid-stimulating hormone (TSH), luteinizing hormone (LH), follicle-stimulating hormone (FSH),** and **chorionic gonadotropin (CG).** These hormones affect diverse biologic processes and yet have remarkable structural similarities. This class of hormones is found in all mammals. These molecules, like other peptide and protein hormones, interact with cell surface receptors and activate adenylate cyclase; thus, they employ cAMP as their intracellular messenger.

Each of these hormones consists of 2 subunits, α and β, joined by noncovalent bonding. The α subunits are identical for all of these hormones within a species, and there is considerable interspecies homology. The specific biologic activity is determined by the β subunit, which also is highly conserved between hormones but to a lesser extent than that noted for the α subunit. The β subunit is not active by itself, and receptor recognition involves the interaction of regions of both subunits. Interhormone and interspecies hybrid molecules are fully active; eg, $TSH_\alpha LH_\beta = LH$ activity and $hTSH_\alpha mTSH_\beta = $ mouse TSH activity. Thus, interspecies differences between α and β do not affect subunit association or the biologic function domain on β. Each subunit is synthesized from unique mRNAs from separate genes. It is thought that all hormones in this class evolved from a common ancestral gene that resulted in 2 molecules, α and β, and that the latter evolved further to provide the separate hormones.

A great deal is known about the structure of these molecules. For example, the carboxy-terminal pentapeptide of α is essential for receptor binding but not for $\alpha\beta$ association. The feature that distinguishes hormones in the glycoprotein group from hormones in other groups is their glycosylation. In each glycoprotein hormone, the α subunit contains 2 complex asparagine-linked oligosaccharides, and the β subunit has either one or 2. The glycosylation may be necessary for $\alpha\beta$ interaction. The α subunit has five S–S bridges, and the β moiety has 6.

Free α subunits are found in the pituitary and placenta. This finding and the observation that α and β are translated from separate mRNAs support the concept that the syntheses of α and β are under separate control and that β is limiting for the production of the complete hormone. All are synthesized as preprohormones and are subject to posttranslational processing within the cell to yield the glycosylated proteins.

The Gonadotropins (FSH, LH, & hCG): These hormones are responsible for gametogenesis and steroidogenesis in the gonads. Each is a glycoprotein with a molecular weight of about 25,000.

1. Follicle-Stimulating Hormone (FSH)–FSH

binds to specific receptors on the plasma membranes of its target cells, the **follicular cells** in the ovary and the **Sertoli cells** in the testis. This results in activation of adenylate cyclase and increased cAMP production. The actions of FSH are described in more detail in Chapter 51.

2. Luteinizing Hormone (LH)–LH binds to specific plasma membrane receptors and stimulates the production of progesterone by **corpus luteum cells** and of testosterone by the **Leydig cells.** The intracellular signal of LH action is cAMP. This nucleotide mimics the actions of LH, which include enhanced conversion of acetate to squalene (the precursor for cholesterol synthesis) and enhanced conversion of cholesterol to 2α-hydroxycholesterol, a necessary step in the formation of progesterone and testosterone. The actions of LH are described in more detail in Chapter 51. There is tight coupling between the binding of LH and the production of cAMP, but steroidogenesis occurs when very small increases of cAMP have occurred. There thus are spare receptors in this response (see Fig 43–3). Prolonged exposure to LH results in desensitization, perhaps owing to down-regulation of LH receptors. This phenomenon may be exploited as an effective means of birth control.

3. Human Chorionic Gonadotropin (hCG)–hCG is a glycoprotein synthesized in the **syncytiotrophoblast cells** of the placenta. It has the $\alpha\beta$ dimer structure characteristic of this class of hormones and most closely resembles LH. It increases in blood and urine shortly after implantation (see above); hence, its detection is the basis of many pregnancy tests.

Thyroid-Stimulating Hormone (TSH): TSH is a glycoprotein of $\alpha\beta$ dimer structure with a molecular weight of about 30,000. Like other hormones of this class, TSH binds to plasma membrane receptors and activates adenylate cyclase. The consequent increase of cAMP is responsible for the action of TSH in thyroid hormone biosynthesis. Its relationship to the trophic effects of TSH on the thyroid is less certain.

TSH has several acute effects on thyroid function. These occur in minutes and involve increases of all phases of T_3 and T_4 biosynthesis, including iodide concentration, organification, coupling, and thyroglobulin hydrolysis. TSH also has several chronic effects on the thyroid. These require several days and include increases in the synthesis of proteins, phospholipids, and nucleic acids and in the size and number of thyroid cells. Long-term metabolic effects of TSH are due to the production and action of the thyroid hormones.

Complex Processing Generates the Pro-Opiomelanocortin (POMC) Peptide Family

The POMC family consists of peptides that act as hormones (ACTH, LPH, MSH) and others that may serve as neurotransmitters or neuromodulators (endorphins). POMC is synthesized as a precursor molecule of ~ 285 amino acids and is processed differently in various regions of the pituitary.

Distribution, Processing, & Functions of the POMC Gene Products: The POMC gene is expressed in the anterior and intermediate lobes of the pituitary. The most conserved sequences between species are within the N-terminal fragment, the ACTH region, and the β-endorphin region. POMC or related products are found in several other vertebrate tissues, including the brain, placenta, gastrointestinal tract, reproductive tract, lung, and lymphocytes. This is presumably due to gene expression in these tissues (rather than to absorption from plasma). Related peptides have also been found in many invertebrate species.

The POMC protein is processed differently in the anterior lobe than in the intermediate lobe. The intermediate lobe is rudimentary in adult humans, but it is active in human fetuses and in pregnant women during late gestation and is also active in many animal species. Processing of the POMC protein in the peripheral tissues (gut, placenta, male reproductive tract) resembles that in the intermediate lobe. There are 3 basic peptide groups: (1) ACTH, which can give rise to α-MSH and corticotropinlike intermediate lobe peptide (CLIP); (2) β-lipotropin (β-LPH), which can yield γ-LPH, β-MSH, and β-endorphin (and thus α- and γ-endorphins); and (3) a large N-terminal peptide, which generates γ-MSH. The diversity of these products is due to the many dibasic amino acid clusters that are potential cleavage sites for trypsinlike enzymes. Each of the peptides mentioned is preceded by Lys-Arg, Arg-Lys, Arg-Arg, or Lys-Lys residues. The prehormone segment is cleaved, and modification by glycosylation, acetylation, and phosphorylation occurs after translation. The next cleavage, in both anterior and intermediate lobes, is between ACTH and β-LPH, resulting in an N-terminal peptide with ACTH and a β-LPH segment (Fig 46–4). $ACTH_{1-39}$ is subsequently cleaved from the N-terminal peptide, and in the anterior lobe essentially no further cleavages occur. In the intermediate lobe, $ACTH_{1-39}$ is cleaved into α-MSH (residues 1–13) and CLIP (18–39); β-LPH (42–134) is converted to γ-LPH (42–101) and β-endorphin (104–134). β-MSH (84–101) is derived from γ-LPH.

There are extensive additional modifications of these peptides. Much of the N-terminal peptide and $ACTH_{1-39}$ in the anterior pituitary is glycosylated. α-MSH is found predominantly in an N-acetylated and carboxy-terminal amidated form; deacetylated α-MSH is much less active. β-Endorphin is rapidly acetylated in the intermediate lobe; acetylated β-endorphin, in contrast to α-MSH, is 1000 times less active than the unmodified form. β-Endorphin may therefore be inactive in the pituitary. In the hypothalamus, these molecules are not acetylated and presumably are active. β-Endorphin is also trimmed at the C-terminal end to form α- and γ-endorphin (Fig 46–4). These form the 3 major endorphins in the rodent intermediate lobe. The large N-terminal fragment is probably also extensively cleaved but while γ-MSH has been found in rat and bovine pituitaries, less is known about this fragment. This structural information has come largely from

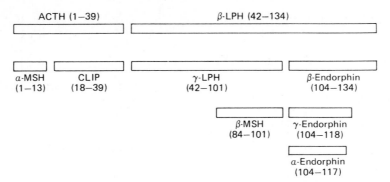

Figure 46–4. Products of pro-opiomelanocortin (POMC) cleavage. MSH, a melanocyte-stimulating hormone. CLIP, corticotropinlike intermediate lobe peptide; LPH, lipotropin.

studies of the rodent pituitary, but the general scheme is thought to apply to other species. Precise functions for most of the POMC peptides have not been established.

Action & Regulation of Specific Peptides:

1. Adrenocorticotropic hormone (ACTH) structure and mechanism of action–ACTH, a single-chain polypeptide consisting of 39 amino acids (Fig 46–5), regulates the growth and function of the adrenal cortex. The 24 N-terminal amino acids are required for full biologic activity and are invariant between species, whereas the 15 C-terminal amino acids are quite variable. A synthetic $ACTH_{1-24}$ analog is widely used in diagnostic testing.

ACTH increases the synthesis and release of adrenal steroids by enhancing the conversion of cholesterol to pregnenolone. This step entails the conversion from a C_{27} to a C_{21} steroid by removal of a 6-carbon side chain. Since pregnenolone is the precursor of all adrenal steroids (see Fig 49–3), prolonged ACTH stimulation results in excessive production of glucocorticoids, mineralocorticoids, and dehydroepiandrosterone (an androgen precursor). However, the contribution of ACTH to the last 2 classes of steroids is minimal under physiologic conditions. ACTH increases adrenal cortical growth (the trophic effect) by enhancing protein and RNA synthesis.

ACTH, like other peptide hormones, binds to a plasma membrane receptor. Within a few seconds of this interaction, intracellular cAMP levels increase markedly. cAMP analogs mimic the action of ACTH, but calcium also is involved.

2. ACTH pathophysiology–Excessive production of ACTH by the pituitary or by ectopic production from a tumor results in **Cushing's syndrome.** The weak MSH-like activity of ACTH or associated release of β- or α-MSH results in hyperpigmentation. The metabolic manifestations are due to excessive production of adrenal steroids and include (1) negative nitrogen, potassium, and phosphorus balance; (2) sodium retention, which can result in hypertension, edema, or both; (3) glucose intolerance or overt diabetes mellitus; (4) increased plasma fatty acids; and (5) decreased circulating eosinophils and lymphocytes, with increased polymorphonuclear leukocytes. Patients with Cushing's syndrome may have muscle atrophy and a peculiar redistribution of fat, ie, truncal obesity. Loss of ACTH owing to tumor, infection, or infarction of the pituitary results in an opposite constellation of findings.

β-Lipotropin (β-LPH): This peptide consists of the carboxy-terminal 91 amino acids of POMC (Fig 46–4). β-LPH contains the sequences of β-MSH, γ-LPH, Met-enkephalin, and β-endorphin. Of these, β-LPH, γ-LPH, and β-endorphin have been found in human pituitary but β-MSH has not been detected. β-LPH is found only in the pituitary, since it is rapidly converted to γ-LPH and β-endorphin in other tissues. β-LPH contains a 7-amino-acid sequence ($β$-LPH_{47-53}) that is identical to $ACTH_{4-10}$. β-LPH causes lipolysis and fatty acid mobilization, but its physiologic role is minimal. It probably serves only as the precursor for β-endorphin.

Endorphins: β-Endorphin consists of the carboxy-terminal 31 amino acids of β-LPH (Fig 46–4.) The α- and γ-endorphins are modifications of β-endorphin from which 15 and 14 amino acids, respectively, are removed from the C-terminal end. These peptides are found in the pituitary, but they are acetylated there (see above) and probably are inactive. In other sites (eg, central nervous system neurons), they are not modified and hence probably serve as neurotransmitters or neu-

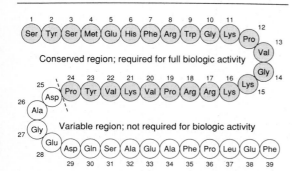

Figure 46–5. Structure of human ACTH.

romodulators. Endorphins bind to the same central nervous system receptors as do the morphine opiates and may play a role in endogenous control of pain perception. They have higher analgesic potencies (18–30 times on a molar basis) than morphine. The sequence for enkephalin is present in POMC, but it is not preceded by dibasic amino acids and presumably is not cleaved or expressed.

Melanocyte-Stimulating Hormone (MSH): MSH stimulates **melanogenesis** in some species by causing the dispersion of intracellular melanin granules, resulting in darkening of the skin. Three different MSH molecules, α, β, and γ, are contained within the POMC molecule, and 2 of these, α and β, are secreted in some nonhuman species. In humans, the actual circulating MSH activity is contained within the larger molecules γ- or β-LPH. α-MSH contains an amino acid sequence that is identical to $ACTH_{1-13}$, but it has an acetylated N terminus. α-MSH (and CLIP) are generally found in animals that have a well-developed intermediate lobe. These peptides are not found in postnatal humans.

Patients with insufficient production of glucocorticoids (**Addison's disease**) have hyperpigmentation associated with increased plasma MSH activity. This could be due to ACTH but is more likely the result of concomitant secretion of β- and γ-LPH, with their associated MSH activity.

THE POSTERIOR PITUITARY CONTAINS 2 ACTIVE HORMONES, VASOPRESSIN & OXYTOCIN

Vasopressin, originally named because of its ability to increase blood pressure when administered in pharmacologic amounts, is more appropriately called **antidiuretic hormone** (ADH) because its most important physiologic action is to promote reabsorption of water from the distal renal tubules. **Oxytocin** is also named for an effect of questionable physiologic significance, the acceleration of birth by stimulation of uterine smooth muscle contraction. Its probable physiologic role is to promote milk ejection from the mammary gland.

Both hormones are produced in the hypothalamus and transported by axoplasmic flow to nerve endings in the posterior pituitary where, upon appropriate stimulation, the hormones are released into the circulation. The probable reason for this arrangement is to escape the blood-brain barrier. ADH is primarily synthesized in the **supraoptic nucleus** and oxytocin in the **paraventricular nucleus.** Each is transported through axons in association with specific carrier proteins called **neurophysins.** Neurophysins I and II are synthesized with oxytocin and ADH, respectively, each as a part of a single protein (sometimes referred to as propressophysin) from a single gene. Neurophysins I and II are unique proteins with molecular weights of 19,000 and 21,000, respectively. ADH and oxytocin are secreted separately into the bloodstream along with their appropriate neurophysins. They circulate unbound to proteins and have very short plasma half-lives, on the order of 2–4 minutes. The structures for ADH and oxytocin are shown below.

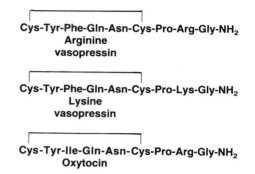

Cys-Tyr-Phe-Gln-Asn-Cys-Pro-Arg-Gly-NH₂
**Arginine
vasopressin**

Cys-Tyr-Phe-Gln-Asn-Cys-Pro-Lys-Gly-NH₂
**Lysine
vasopressin**

Cys-Tyr-Ile-Gln-Asn-Cys-Pro-Arg-Gly-NH₂
Oxytocin

Each is a nonapeptide containing cysteine molecules at positions 1 and 6 linked by an S–S bridge. Most animals have arginine vasopressin; however, the hormone in pigs and related species has a lysine substituted in position 8. Because of the close structural similarity, it is not surprising that ADH and oxytocin each exhibit some of the effects of the other molecule. These peptides are primarily metabolized in the liver, although renal excretion of ADH accounts for a significant part of its loss from blood.

Oxytocin

Regulation of Secretion: The neural impulses that result from stimulation of the nipples are the primary stimulus for oxytocin release. Vaginal and uterine distention are secondary stimuli. PRL is released by many of the stimuli that release oxytocin, and a fragment of oxytocin has been proposed as prolactin-releasing factor. Estrogen stimulates the production of oxytocin and of neurophysin I, and progesterone inhibits the production of these compounds.

Mechanism of Action: The mechanism of action of oxytocin is unknown. It causes contraction of uterine smooth muscle and thus is used in pharmacologic amounts to induce labor in humans. Interestingly, pregnant animals in which the hypothalamic-hypophyseal tract has been destroyed do not necessarily have trouble delivering their young. The most likely physiologic function of oxytocin is to stimulate contraction of myoepithelial cells surrounding the mammary alveoli. This promotes the movement of milk into the alveolar duct system and allows for milk ejection. Membrane receptors for oxytocin are found in both uterine and mammary tissues. These receptors are increased in number by estrogens and decreased by progesterone. The concomitant rise in estrogens and fall in progesterone occurring immediately before parturition probably explains the onset of lactation prior to delivery. Progesterone derivatives are commonly used to inhibit postpartum lactation in humans. Oxytocin and neurophysin I appear to be produced in the ovary, wherein oxytocin may inhibit steroidogenesis.

The chemical groups important for oxytocin action include the primary amino group of the N-terminal cysteine; the phenolic group of tyrosine; the 3 carboxy-amide groups of asparagine, glutamine, and glycina-mide; and the disulfide (S–S) linkage. By deleting or substituting these groups, numerous analogs of oxy-tocin have been produced. For example, deletion of the free primary amino group of the terminal half cysteine residue (position 1) results in desamino oxytocin, which has 4–5 times the antidiuretic activity of oxytocin.

Antidiuretic Hormone (ADH; Vasopressin)

Regulation of Secretion: The neural impulses that trigger ADH release are activated by a number of different stimuli. Increased osmolality of plasma is the primary physiologic stimulus. This is mediated by **osmoreceptors** located in the hypothalamus and by **baroreceptors** located in the heart and other regions of the vascular system. Hemodilution (decreased os-molality) has the opposite effect. Other stimuli include emotional and physical stress and pharmacologic agents including acetylcholine, nicotine, and mor-phine. Most of these effects involve increased syn-thesis of ADH and neurophysin II, since the depletion of stored hormone is not associated with this action. Epinephrine and agents that expand plasma volume inhibit ADH secretion, as does ethanol.

Mechanism of Action: The most important phys-iologic target cells of ADH in mammals are those of the distal convoluted tubules and collecting structures of the kidney. These ducts pass through the renal medulla, in which the extracellular solute pool has an osmolality gradient up to 4 times that of plasma. These cells are relatively impermeable to water, so that in the absence of ADH, the urine is not concentrated and may be excreted in amounts exceeding 2 L/d. ADH in-creases the permeability of the cells to water and per-mits osmotic equilibration of the collecting tubule urine with the hypertonic interstitium, resulting in urine volumes in the range of 0.5–1 L/d. ADH recep-tors exist on the mucosal (urinary) membrane of these epithelial cells. This receptor is linked to adenylate cyclase, and cAMP is thought to mediate the effects of ADH in the renal tubule. This physiologic action is the basis of the name "antidiuretic hormone." cAMP and inhibitors of phosphodiesterase activity (caffeine for example) mimic the actions of ADH. In vivo, an ele-vated level of calcium in the medium bathing the mucosal surface of the tubular cells inhibits the action of ADH on water movement, apparently by inhibiting the action of adenylate cyclase, since it does not dimin-ish the action of cAMP per se. This may account, in part, for the excessive volumes of urine that are charac-teristic of patients with hypercalcemia.

Pathophysiology: Abnormalities of ADH secre-tion or action lead to **diabetes insipidus,** which is characterized by the excretion of large volumes of di-lute urine. Primary diabetes insipidus, an insufficient amount of the hormone, is usually due to destruction of the hypothalamic-hypophyseal tract from a basal skull fracture, tumor, or infection, but it can be hereditary. In **hereditary nephrogenic diabetes insipidus,** ADH is secreted normally but the target cell is incapable of responding, presumably because of a receptor defect (see Table 43–2). This hereditary lesion is distin-guished from **acquired nephrogenic diabetes insipi-dus,** which most often is due to the pharmacologic administration of lithium for manic-depressive illness. The **inappropriate secretion of ADH** occurs in asso-ciation with ectopic production by a variety of tumors (usually tumors of the lung) but can also occur in con-junction with diseases of the brain, pulmonary infec-tions, or hypothyroidism. It is called inappropriate secretion because ADH is produced at a normal or increased rate in the presence of hypoosmolality, thus causing a persistent and progressive dilutional hypo-natremia with excretion of hypertonic urine.

REFERENCES

Anterior Pituitary Hormones

Douglass J, Civelli O, Herbert E: Polyprotein gene ex-pression: Generation of diversity of neuroendocrine pep-tides. *Annu Rev Biochem* 1984;**53**:665.

Frantz AG: Prolactin. *N Engl J Med* 1978;**298**:201.

Krieger DT: The multiple faces of pro-opiomelanocortin, a prototype precursor molecule. *Clin Res* 1983;**3**:342.

Nikolics K et al: A prolactin-inhibiting factor with the precur-sor for human gonadotropin-releasing hormone. *Nature* 1986;**316**:511.

Pierce JG, Parsons TF: Glycoprotein hormones: Structure and function. *Annu Rev Biochem* 1981;**50**:465.

Seeburg P: The human growth hormone gene family: Struc-ture and evolution of the chromosomal locus. *Nucleic Acids Res* 1983;**11**:3939.

Posterior Pituitary Hormones

Chord IT: The posterior pituitary gland. *Clin Endocrinol* 1975;**4**:89.

Robertson GL: Regulation of vasopressin function in health and disease. *Recent Prog Horm Res* 1977;**33**:333.

Hypothalamic Hormones

Imura H et al: Effect of CNS peptides on hypothalamic reg-ulation of pituitary secretion. *Adv Biochem Psychophar-macol* 1981;**28**:557.

Labrie F et al: Mechanism of action of hypothalamic hor-mones in the adenohypophysis. *Annu Rev Physiol* 1979; **41**:555.

Reichlin S: Systems for the study of regulation of neuropep-tide secretion. In: *Neurosecretion and Brain Peptides: Implications for Brain Function and Neurological Dis-ease.* Martin JB, Reichlin S, Bick KL (editors). Raven Press, 1981.

Thyroid Hormones

47

Daryl K. Granner, MD

INTRODUCTION

Thyroid hormones regulate gene expression, tissue differentiation, & general development. The thyroid gland produces 2 iodoamino acid hormones, **3,5,3′-triiodothyronine (T_3)** and **3,5,3′,5′-tetraiodothyronine (T_4, thyroxine)**, which have long been recognized for their importance in regulating general metabolism, development, and tissue differentiation. These hormones, whose structures are illustrated in Fig 47–1, regulate gene expression using mechanisms similar to those employed by steroid hormones.

BIOMEDICAL IMPORTANCE

Diseases of the thyroid are among the most common afflictions involving the endocrine system. Diagnosis and therapy are firmly based on the principles of thyroid hormone physiology and biochemistry. The availability of radioisotopes of iodine has greatly aided in the elucidation of these principles. Radioactive iodine, because it localizes in the gland, is widely used in the diagnosis and treatment of thyroid disorders. Radioiodine has a dangerous aspect as well, since excessive exposure, such as from nuclear fallout, is a major risk factor for thyroid cancer. This is especially true in infants and adolescents, whose thyroid cells are still actively dividing.

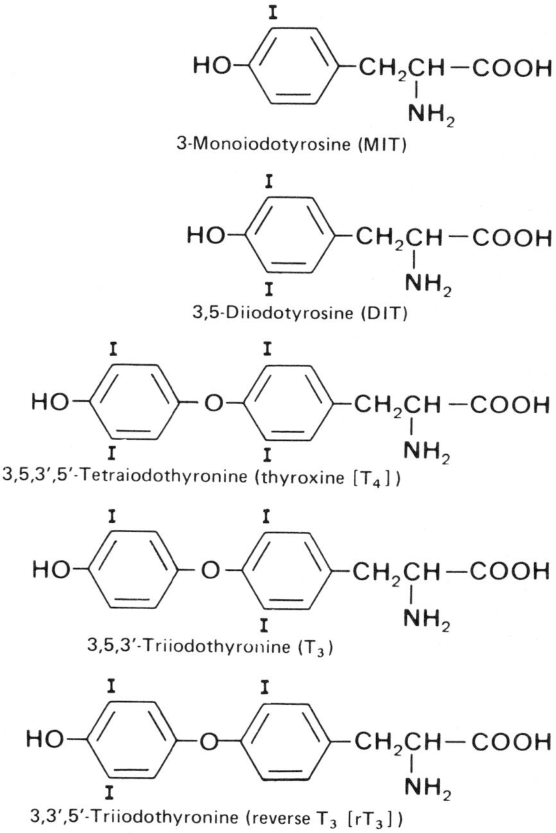

Figure 47–1. Structure of thyroid hormones and related compounds.

Acronyms Used in This Chapter

DIT	Diiodotyrosine
MIT	Monoiodotyrosine
T_3	Triiodothyronine
T_4	Thyroxine; tetraiodothyronine
TBG	Thyroxine-binding globulin
TBPA	Thyroxine-binding prealbumin
TRH	Thyrotropin-stimulating hormone
TSH	Thyroid-stimulating hormone
TSI	Thyroid-stimulating IgG

THYROID HORMONE BIOSYNTHESIS INVOLVES THYROGLOBULIN & IODIDE METABOLISM

Thyroid hormones are unique in that they require the trace element **iodine** for biologic activity. In most parts of the world, iodine is a scarce component of soil, and hence there is little in food. A complex mechanism has evolved to acquire and retain this crucial element and to convert it into a form suitable for incorporation into organic compounds. At the same time, the thyroid must synthesize thyronine, and this synthesis takes place in thyroglobulin. These processes will be discussed separately, although they occur concurrently.

Thyroglobulin is a Complex Protein

Biosynthesis: Thyroglobulin is the precursor of T_4 and T_3. It is a large, iodinated, glycosylated protein with a molecular weight of 660,000. Carbohydrate accounts for 8–10% of the weight of thyroglobulin and iodide for about 0.2–1%, depending upon the iodine content in the diet. Thyroglobulin is composed of 2 subunits. It contains 115 tyrosine residues, each of which is a potential site of iodination. About 70% of the iodide in thyroglobulin exists in the inactive precursors, **monoiodotyrosine (MIT)** and **diiodotyrosine (DIT)**, while 30% is in the **iodothyronyl** residues, T_4 and T_3. When iodine supplies are sufficient, the $T_4:T_3$ ratio is about 7:1. In **iodine deficiency,** this ratio decreases, as does the DIT:MIT ratio. The reason for synthesizing a molecule of 5000 amino acids to generate a few molecules of a modified diamino acid seems to be that the conformation of this large structure is required for tyrosyl coupling or iodide organification. Thyroglobulin is synthesized in the basal portion of the cell, moves to the lumen, where it is stored in the extracellular **colloid,** and reenters the cell and moves in an apical to basal direction during its hydrolysis into the active T_3 and T_4 hormones. All of these steps are enhanced by TSH, and this hormone (or cAMP) also enhances transcription of the thyroglobulin gene.

Hydrolysis: Thyroglobulin is a storage form of T_3 and T_4 in the colloid: a several weeks' supply of these hormones exists in the normal thyroid. Within minutes after the stimulation of the thyroid by TSH (or cAMP), there is a marked increase of microvilli on the apical membrane. This microtubule-dependent process entraps thyroglobulin, and subsequent pinocytosis brings it back into the follicular cell. These phagosomes fuse with lysosomes to form **phagolysosomes** in which various acid proteases and peptidases hydrolyze the thyroglobulin into amino acids, including the iodothyronines. T_4 and T_3 are discharged from the basal portion of the cell, perhaps by a facilitated process, into the blood. The $T_4:T_3$ ratio in this blood is lower than that in thyroglobulin, so that some selective deiodination of T_4 must occur in the thyroid. About 50 μg of thyroid hormone iodide is secreted each day. With an average uptake of iodide (25–30% of the iodide ingested), the daily iodide requirement is between 150 and 200 μg.

As mentioned above, most of the iodide in thyroglobulin is not in iodothyronine; about 70% is in the inactive compounds MIT and DIT. These amino acids are released when thyroglobulin is hydrolyzed and the iodide is scavenged by an environmentally conscious enzyme, **deiodinase.** This NADPH-dependent enzyme is also present in kidney and liver. The iodide removed from MIT and DIT constitutes an important pool within the thyroid, as distinguished from that I^- which enters from blood. Under steady-state conditions, the amount of iodide that enters the thyroid matches the amount that leaves. If one-third of the iodide in thyroglobulin leaves (as T_4 and T_3), it follows that two-thirds of the iodide available for biosynthesis comes from the deiodination of MIT and DIT within the thyroid.

Iodide Metabolism Involves a Number of Discrete Steps (See Fig 47–2)

1. Concentration of Iodide (I^-): The thyroid, along with several other epithelial tissues including mammary gland, chorion, salivary gland, and stomach, is able to concentrate I^- against a strong electrochemical gradient. This is an energy-dependent process and is linked to the ATPase-dependent Na^+/K^+ pump. The activity of the **thyroidal I^- pump** can be isolated from subsequent steps in hormone biosynthesis by inhibiting organification of I^- with drugs of the thiourea class (Fig 47–3). The ratio of iodide in thyroid to iodide in serum (T:S ratio) is a reflection of the activity of this pump or concentrating mechanism. This activity is primarily controlled by TSH and ranges from 500 in animals chronically stimulated with TSH to 5 or less in hypophysectomized animals. The T:S ratio in humans on a normal iodine diet is about 25.

A very small amount of iodide also enters the thyroid by diffusion. Any intracellular I^- that is not incorporated into MIT or DIT (generally < 10%) is free to leave by this mechanism.

The transport mechanism is inhibited by 2 classes of molecules. The first group consists of perchlorate (ClO_4^-), perrhenate (ReO_4^-), and pertechnetate (TcO_4^-), all anions with a similar partial specific volume to I^-. These anions compete with I^- for its carrier and are concentrated by the thyroid. A radioisotope of TcO_4^- is commonly used to study iodide transport in humans. The linear anion thiocyanate (SCN^-), an example of the second class, is a competitive inhibitor of I^- transport but is not concentrated by the thyroid. These inhibitors of I^- transport unmask the rapid diffusion of exchangeable I^- from the thyroid and are used to diagnose organification deficiencies. After the acute administration of a blocking concentration of a transport inhibitor, the amount of accumulated I^- (usually measured as the isotope ^{131}I) that leaves the thyroid is directly related to the unbound, or nonorganified, fraction. Individuals with incomplete organification will "discharge" more ^{131}I than will normal persons in response to ClO_4^-.

2. Oxidation of I^-: The thyroid is the only tissue that can oxidize I^- to a higher valence state, an obligatory step in I^- organification and thyroid hormone biosynthesis. This step involves a heme-containing peroxidase and occurs at the luminal surface of the follicular cell.

Thyroperoxidase, a tetrameric protein with a molecular weight of 60,000, requires hydrogen peroxide as an oxidizing agent. The H_2O_2 is produced by an NADPH-dependent enzyme resembling cytochrome c reductase. A number of compounds inhibit I^- oxidation and therefore its subsequent incorporation into MIT and DIT. The most important of these clinically are the thiourea drugs, some of which are shown in Fig 47–3. They are known as **antithyroid**

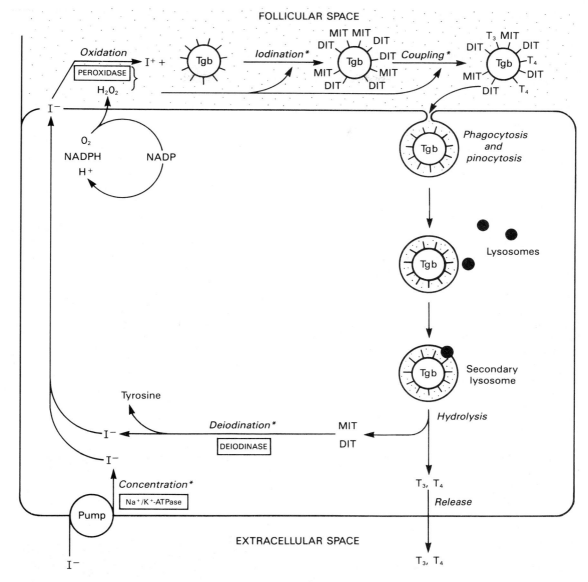

Figure 47–2. Model of iodide metabolism in the thyroid follicle. A follicular cell is shown facing the follicular lumen (stippled area) and the extracellular space (at bottom). Iodide enters the thyroid by a pump and by passive diffusion. Thyroid hormone synthesis occurs in the follicular space through a series of reactions, many of which are peroxidase-mediated. Thyroid hormones are released from thyroglobulin by hydrolysis. Tgb, thyroglobulin; MIT, monoiodotyrosine; DIT, diiodotyrosine; T_3, triiodothyronine; T_4, tetraiodothyronine. Asterisks indicate steps or processes that are inherited enzyme deficiencies which cause congenital goiter and often result in hypothyroidism.

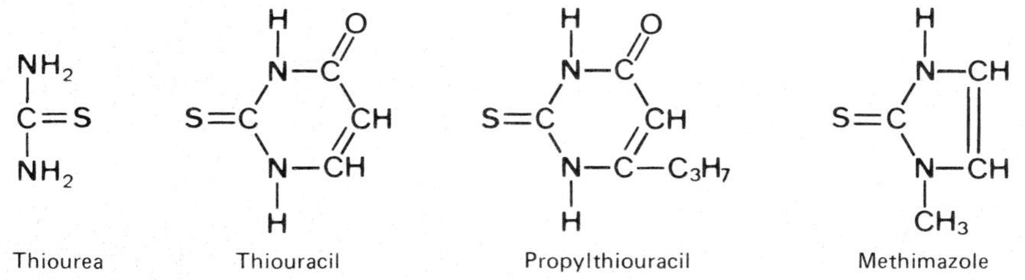

Thiourea   Thiouracil   Propylthiouracil   Methimazole

Figure 47–3. Thiourea class of antithyroid drugs.

drugs because of their ability to inhibit thyroid hormone biosynthesis at this step.

3. Iodination of Tyrosine: Oxidized iodide reacts with the tyrosyl residues in thyroglobulin in a reaction that probably also involves thyroperoxidase. The 3 position of the aromatic ring is iodinated first and then the 5 position to form MIT and DIT, respectively. This reaction, sometimes called **organification,** occurs within seconds in luminal thyroglobulin. Once iodination occurs, the iodine does not readily leave the thyroid. Free tyrosine can be iodinated, but it is not incorporated into proteins, since no tRNA recognizes iodinated tyrosine.

4. Coupling of Iodotyrosyls: The coupling of two DIT molecules to form T_4 or of an MIT and DIT to form T_3 occurs within the thyroglobulin molecule, although the addition of a free MIT or DIT to a bound DIT has not been conclusively excluded. A separate coupling enzyme has not been found, and since this is an oxidative process, it is assumed that the same thyroperoxidase catalyzes this reaction by stimulating free radical formation of iodotyrosine. This hypothesis is supported by the observation that the same drugs which inhibit I^- oxidation also inhibit coupling. The formed thyroid hormones remain as integral parts of thyroglobulin until the latter is degraded, as described above. Thyroglobulin hydrolysis is stimulated by TSH but is inhibited by I^-; this latter effect is occasionally exploited by using KI to treat hyperthyroidism.

THYROID HORMONES ARE TRANSPORTED BOUND & METABOLIZED

One-half to two-thirds of T_4 and T_3 in the body is extrathyroidal, and most of this circulates in bound form, ie, bound to 2 specific binding proteins, **thyroxine-binding globulin (TBG)** and **thyroxine-binding prealbumin (TBPA)**. TBG, a glycoprotein of 50,000 MW, is quantitatively the more important. It binds T_4 and T_3 with 100 times the affinity of TBPA and has the capacity to bind 20 μg/dL of plasma. Under normal circumstances, TBG binds, noncovalently, nearly all of the T_4 and T_3 in plasma (Table 47–1). The small, unbound (free) fraction is responsible for the biologic activity. In spite of the great difference in total amount, the free fraction of T_3 approximates that of T_4, but the plasma half-life of T_4 is 4–5 times that of T_3.

TBG is also subject to regulation, an important consideration in diagnostic testing of thyroid function, since most assays of T_4 or T_3 measure the total amount in plasma rather than the free hormone. TBG is produced in liver, and its synthesis is increased by estrogens (pregnancy and birth control pills). Decreased production of TBG occurs following androgen or glucocorticoid therapy and in certain liver diseases. Inherited increases or decreases of TBG also occur. All of these conditions result in changes of total T_4 and T_3 without a change of the free level. Phenytoin and salicylates compete with T_3 and T_4 for binding to TBG. This decreases the total level of hormone without changing the free fraction and must be considered when interpreting diagnostic tests.

Extrathyroidal deiodination converts T_4 to T_3. Since T_3 binds to the thyroid receptor in target cells with 10 times the affinity of T_4, T_3 is thought to be the preponderant metabolically active form of the molecule. About 80% of circulating T_4 is converted to T_3 or reverse T_3 (rT_3) in the periphery, and this conversion accounts for most of the production of T_3. Reverse T_3 is a very weak agonist that is made in relatively larger amounts in chronic disease, in carbohydrate starvation, and in the fetus. Propylthiouracil and propranolol decrease the conversion of T_4 to T_3.

Other forms of thyroid hormone metabolism include total deiodination and inactivation by deamination or decarboxylation. Hepatic glucuronidation and sulfation result in a more hydrophilic molecule that is excreted into bile, reabsorbed in the gut, deiodinated in the kidney, and excreted as the glucuronide conjugate in the urine.

THYROID HORMONES HAVE A NUCLEAR MECHANISM OF ACTION

Thyroid hormones bind to specific high-affinity receptors in the target cell nucleus; T_3 binds with approximately 10 times the affinity of T_4. The question as to whether all biologic activity of the thyroid hormones is mediated by T_3 is thus moot; both T_3 and T_4 are active. Thyroid hormones bind to low-affinity sites in cytoplasm, but this is apparently not the same protein as the nuclear receptor. The cytoplasmic binding may serve to keep thyroid hormones "in the neighborhood."

The general metabolic function of thyroid hormones is to increase oxygen consumption. Edelman and coworkers hypothesize that much of the energy utilized by a cell is for driving the **Na^+/K^+-ATPase pump.** Thyroid hormones enhance the function of this pump by increasing the number of pump units. Since all cells have this pump and virtually all cells respond to thyroid hormones, this increased utilization of ATP and the associated increase of oxygen consumption via oxidative phosphorylation could be the basic mechanism of thyroid hormone action.

Another major effect of T_3/T_4 is to enhance general protein synthesis and cause positive nitrogen balance. Thyroid hormones, like steroids, induce or repress proteins by increasing or decreasing gene transcription (see Fig 45–1). In the case of T_3/T_4 the *trans*-acting factor is the hormone-receptor complex, which always

Table 47–1. Comparison of T_4 and T_3 in plasma.

| Total Hormone (μg/dL) | Free Hormone | | | $t_{1/2}$ in Blood (days) |
	Percent of Total	ng/dL	Molarity	
T_4 8	0.03	~2.24	3.0×10^{-11}	6.5
T_3 0.15	0.3	~0.4	~0.6×10^{-11}	1.5

seems to reside in the nucleus. The *cis*-acting DNA hormone response element that binds this complex consists of the core sequence GATCAnnnnnnTGACC (Table 45–3).

There is a curious association between the 2 classes of hormones related to growth, the thyroid hormones and growth hormone itself. T_3 and glucocorticoids enhance transcription of the GH gene, so that more GH is produced. This explains a classic observation in which the pituitaries of T_3-deficient animals were found to lack GH, and it may account for some of the general anabolic effects of T_3. Very high concentrations of T_3 inhibit protein synthesis and cause negative nitrogen balance.

Thyroid hormones are known to be important modulators of developmental processes. This is most apparent in amphibian metamorphosis. Thyroid hormones are required for the conversion of a tadpole into a frog, a process that involves resorption of the tail, limb-bud proliferation, conversion from fetal to adult hemoglobin, stimulation of urea cycle enzymes (carbamoyl phosphate synthase) so that urea is excreted rather than ammonia, and epidermal changes. These effects probably result from the regulation of specific gene expression. Thyroid hormones are required for normal development in humans. Intrauterine or neonatal hypothyroidism results in **cretinism,** a condition characterized by multiple congenital defects and severe, irreversible mental retardation.

THE PATHOPHYSIOLOGY OF MANY THYROID DISEASES RELATES TO TSH & T_3,T_4

A Goiter Is An Enlarged Thyroid

Any enlargement of the thyroid is referred to as a goiter. Simple goiter represents an attempt to compensate for decreased thyroid hormone production; thus, in all of these situations, elevated TSH is the common denominator. Causes include iodide deficiency; iodide excess, when an autoregulatory mechanism fails; and a variety of rare inherited metabolic defects that illustrate the importance of various steps in thyroid hormone biosynthesis. These defects include (1) I^- transport defect; (2) iodination defect; (3) coupling defect; (4) deiodinase deficiency; and (5) production of abnormal iodinated proteins. Partial deficiencies of these functions may cause simple goiter in adults. Any of these causes of simple goiter can, when severe, cause hypothyroidism. Simple goiter is treated with exogenous thyroid hormone. Supplementation or restriction of iodide intake is appropriate for specific types of goiter.

Insufficient Amounts of Free T_3 or T_4 Result in Hypothyroidism

This is usually due to thyroid failure but can be due to disease of the pituitary or hypothalamus. In hypothyroidism, the basal metabolic rate is decreased, as are other processes dependent upon thyroid hormones. Prominent features include slow heart rate, diastolic hypertension, sluggish behavior, sleepiness, constipation, sensitivity to cold, dry skin and hair, and a sallow complexion. Other features depend upon the age at onset. Hypothyroidism later in childhood results in short stature but no mental retardation. The various kinds of hypothyroidism are treated with exogenous thyroid hormone replacement.

Hyperthyroidism, or Thyrotoxicosis, Is Due to the Excessive Production of Thyroid Hormone

There are many causes, but most cases in the USA are due to **Graves' disease,** which results from the production of **thyroid-stimulating IgG (TSI)** that activates the TSH receptor (see Table 43–2). This causes a diffuse enlargement of the thyroid and excessive, uncontrolled production of T_3 and T_4, since the production of TSI is not under feedback control. Findings are multisystemic and include rapid heart rate, widened pulse pressure, nervousness, inability to sleep, weight loss in spite of increased appetite, weakness, excessive sweating, sensitivity to heat, and red, moist skin. The hyperthyroidism of Graves' disease is treated by blocking hormone production with an antithyroid drug, by ablating the gland with a radioactive isotope of iodide (such as ^{131}I), or by a combination of these 2 methods. Occasionally, the gland is removed surgically.

REFERENCES

Chopra IJ et al: Pathways of metabolism of thyroid hormones. *Recent Prog Horm Res* 1978;**34:**531.

Larsen PR: Thyroid-pituitary interaction: Feedback regulation of thyrotropin secretion by thyroid hormones. *N Engl J Med* 1982;**396:**23.

Oppenheimer JH: Thyroid hormone action at the nuclear level. *Ann Intern Med* 1985;**102:**374.

Robins J et al: Thyroxine transport proteins of plasma: Molecular properties and biosynthesis. *Recent Prog Horm Res* 1978;**34:**477.

Samuels H: Regulation of gene expression by thyroid hormone. *J Clin Invest* 1988;**81:**957.

48 Hormones That Regulate Calcium Metabolism

Daryl K. Granner, MD

INTRODUCTION

Calcium ion regulates a number of important physiologic and biochemical processes. These include **neuromuscular excitability, blood coagulation, secretory processes, membrane integrity and plasma membrane transport, enzyme reactions,** the **release of hormones** and **neurotransmitters,** and the **intracellular action** of a number of hormones. In addition, the proper extracellular fluid (ECF) and periosteal concentrations of Ca^{2+} and PO_4^{3-} are required for **bone mineralization.** To ensure that these processes operate normally, the plasma Ca^{2+} concentration is maintained within *very* narrow limits. The purpose of this chapter is to explain how this is accomplished.

BIOMEDICAL IMPORTANCE

Deviations of the ionized calcium from the normal range cause many disorders and can be life-threatening. As many as 3% of hospitalized patients may have disorders of calcium homeostasis.

CALCIUM OCCURS IN BONE & EXTRACELLULAR FLUID

There is approximately 1 kg of calcium in the human body. Ninety-nine percent of this is located in bone where, with phosphate, it forms the **hydroxyapatite crystals** that provide the inorganic and structural component of the skeleton. Bone is a dynamic tissue, and it undergoes constant remodeling as stresses change; in the steady-state condition, there is a balance between new bone formation and bone resorption. Most of the calcium in bone is not freely exchangeable with **extracellular fluid (ECF) calcium.** Thus, in addition to its mechanical role, bone serves as a large reservoir of calcium. About 1% of skeletal Ca^{2+} is in a freely exchangeable pool and this, with another 1% of the total found in the periosteal space, constitutes the **miscible pool of Ca^{2+}.** The hormones discussed in this chapter regulate the amount of calcium in the ECF by influencing the transport of calcium across the membrane that separates the ECF space from the periosteal fluid space. This transport is primarily stimulated by parathyroid hormone (PTH) but calcitriol is also involved.

Plasma calcium exists in 3 forms: (1) **complexed** with organic acids, (2) **protein-bound,** and (3) **ionized.** About 6% of total calcium is complexed with citrate, phosphate, and other anions. The remainder is divided nearly equally between a protein-bound form (bound primarily to albumin) and an ionized (unbound) form. The ionized calcium (Ca^{2+}), which is maintained at concentrations between 1.1 and 1.3 mmol/L in most mammals, birds, and freshwater fish, is the biologically active fraction. The organism has very little tolerance for significant deviation from this normal range. If the ionized calcium level falls, the animal becomes increasingly hyperexcitable and may develop tetanic convulsions. A marked elevation of plasma calcium may result in death owing to muscle paralysis and coma.

Calcium ion and the counter-ion, phosphate, exist at or near their solubility product in plasma; hence, protein binding may protect against precipitation and **ectopic calcification.** An alteration of the plasma protein concentration (primarily albumin, but globulins also bind calcium) results in parallel changes in total plasma calcium. For example, hypoalbuminemia results in a decrease of total plasma calcium of approximately 0.8 mg/dL for each g/dL of albumin decrease. The converse is noted when the plasma albumin is increased. The association of calcium with plasma proteins is pH-dependent; acidosis favors the ionized form, whereas alkalosis enhances binding and causes a concomitant decrease in Ca^{2+}. The latter probably accounts for the numbness and tingling associated with the **hyperventilation syndrome,** which causes acute respiratory alkalosis.

2 MAIN HORMONES ARE INVOLVED IN CALCIUM HOMEOSTASIS

Parathyroid Hormone (PTH) Is An 84-Amino Acid Single-Chain Peptide

This hormone (MW 9500) contains no carbohydrate or other covalently bound molecules (Fig 48–1). Full

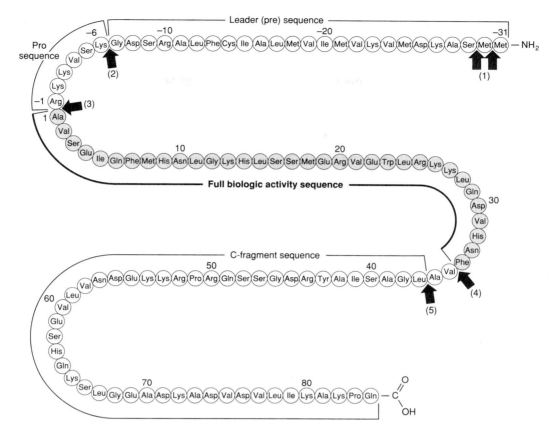

Figure 48–1. Structure of bovine preproparathyroid hormone. Arrows indicate sites cleaved by processing enzymes in the parathyroid gland (1–5) and in the liver after secretion of the hormone (4–5). The biologically active region of the molecule is flanked by sequence not required for activity on target receptors. (Slightly modified and reproduced, with permission, from Habener JF: Recent advances in parathyroid hormone research. *Clin Biochem* 1981;**14**:223.)

biologic activity resides in the N-terminal third of the molecule; PTH$_{1-34}$ has full biologic activity. The region 25–34 is primarily responsible for receptor binding.

PTH is synthesized as a 115-amino-acid precursor molecule (Fig 48–1). The immediate precursor of PTH is **proPTH,** which differs from the native hormone by having an N-terminal highly basic hexapeptide extension whose function is obscure. The primary gene product and the immediate precursor for proPTH is **preproPTH.** This differs from proPTH by having an additional 25-amino-acid N-terminal extension that, in common with the other leader or signal sequences characteristic of secreted proteins, is hydrophobic. The complete structure of preproPTH and the sequences of proPTH and PTH are illustrated in Fig 48–1. The sequence of events involved in the conversion of PTH preprohormone to PTH is shown in schematic form in Fig 48–2. PreproPTH is transferred to the cisternal space of the endoplasmic reticulum while the molecule is still being translated from PTH mRNA by the ribosomes. During this transfer, the 25-amino-acid prepeptide (signal or leader peptide) is removed to yield proPTH. ProPTH is then transported to the Golgi

apparatus, where an enzyme removes the pro-extension to yield the mature PTH molecule. The PTH released from the Golgi apparatus in secretory vesicles has 3 possible fates: (1) transport into a storage pool; (2) degradation; or (3) immediate secretion.

The Synthesis, Secretion, & Metabolism of PTH Is Regulated.

Regulation of Synthesis: The rate of synthesis and degradation of proPTH is unaffected by the ambient Ca^{2+} concentration, even though the rate of formation and secretion of PTH is always markedly enhanced at low Ca^{2+} concentrations. Indeed, 80–90% of the proPTH synthesized cannot be accounted for as intact PTH in cells or in the incubation medium of experimental systems. This led to the conclusion that most of the proPTH synthesized is quickly degraded. It was later discovered that this rate of degradation decreases when Ca^{2+} concentrations are low and increases when Ca^{2+} concentrations are high. This indicates that calcium affects PTH production through control of degradation and not synthesis. The constitutive synthesis of proPTH is reflected in PTH mRNA levels, which also do not change in spite of

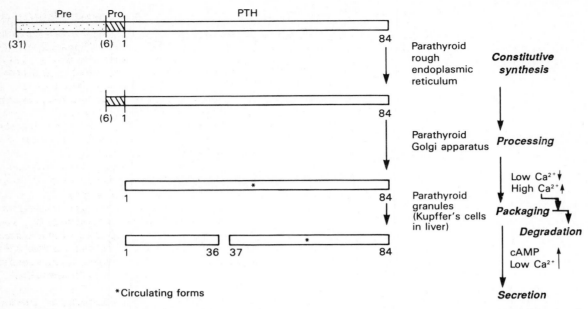

Figure 48–2. The precursors and cleavage products of PTH and the location of these steps in the parathyroid gland and liver. The numbers in parentheses indicate the number of amino acids in the pre (31) and pro (6) fragments.

wide fluctuations of extracellular Ca^{2+}. It appears that the only way that the organism can enhance PTH synthesis is to increase the size and number of PTH-producing chief cells in the parathyroid glands.

Regulation of Metabolism: The degradation of PTH begins about 20 minutes after proPTH is synthesized, is initially unaffected by Ca^{2+} concentration, and occurs after the hormone is in secretory vesicles. Newly formed PTH can either be secreted immediately or be placed in storage vesicles for subsequent secretion. Degradation occurs as soon as the secretory vesicle begins to enter the storage compartment.

Very specific fragments of PTH are generated during its proteolytic digestion (Figs 48–1 and 48–2), and large amounts of carboxy-terminal fragments of PTH are found in the circulation. These molecules, with molecular weights of about 7000, consist of PTH_{37-84} and lesser amounts of PTH_{34-84}. Most of the newly synthesized PTH is degraded. About 2 mol of the C-terminal fragments is secreted for each mole of intact PTH; hence, the bulk of circulating PTH consists of the C-terminal molecules. No biologic function for the C-terminal fragment of PTH has been defined, but it may prolong the half-life of the hormone in the circulation. A number of proteolytic enzymes, including **cathepsins B and D,** have been identified in parathyroid tissue. Cathepsin B cleaves PTH into 2 fragments, PTH_{1-36} and PTH_{37-84}. PTH_{37-84} is not further degraded; however, PTH_{1-36} is rapidly and progressively cleaved into di- and tripeptides. ProPTH has never been found in circulation, and little (if any) PTH_{1-34} escapes from the gland. PreproPTH was identified by deciphering the coding sequence of the PTH gene.

Most of the proteolysis of PTH occurs within the gland; however, a number of studies confirm that PTH, once secreted, is proteolytically degraded in other tissues. The exact contribution of extraglandular proteolysis has not been defined, nor is it clear whether the proteolytic enzymes in the 2 sites are similar or whether the patterns and products of cleavage are identical.

The liver and kidneys are involved in peripheral metabolism of secreted PTH. After hepatectomy, no 34–84 or 37–84 fragments are detected, indicating that the liver is the principal organ involved in the generation of these fragments. The role of the kidneys may be to remove and excrete these fragments. The principal site of **peripheral proteolysis** appears to be the **Kupffer cells** lining the intrasinusoidal passages of the liver. The endopeptidase responsible for the initial cleavage into the amino- and carboxy-terminal fragments is located on the surface of these macrophagelike cells, which are in intimate contact with plasma. This enzyme, also a cathepsin B, cleaves PTH between residues 36 and 37; as in the parathyroid, the resulting carboxy-terminal fragment continues to circulate, whereas the amino-terminal fragment is rapidly degraded.

Regulation of Secretion: PTH secretion is inversely related to the ambient concentration of ionized calcium and magnesium, as is the circulating level of immunoreactive PTH. Serum PTH declines in a rectilinear fashion in relation to serum calcium levels between 4 mg/dL and 10.5 mg/dL. The presence of biologically active PTH when the serum calcium level is 10.5 mg/dL or greater is an indication of **hyperparathyroidism.**

There is a linear relationship between PTH release

and the parathyroid intracellular level of cAMP. The intracellular Ca^{2+} level may be involved in this process, since there is an inverse relationship between the intracellular concentrations of calcium and cAMP. Calcium may exert this effect through its known action on phosphodiesterase (via Ca^{2+}/calmodulin-dependent protein kinase) or through a similar mechanism by inhibiting adenylate cyclase. Phosphate has no effect on PTH secretion.

Parathyroid glands have relatively few storage granules and contain enough hormone to maintain maximal secretion for only 1.5 hours. This is in contrast to the pancreatic islets, which contain insulin stores sufficient for several days, and to the thyroid, which contains hormone stores adequate for several weeks. PTH must therefore be continually synthesized and secreted.

The Mechanism of Action of PTH Involves a Membrane Receptor

PTH binds to a single membrane receptor protein of approximately 70,000 MW. This receptor appears to be identical in bone and kidney, and it is not found in nontarget cells. The hormone-receptor interaction initiates a typical cascade: activation of adenylate cyclase → increased intracellular cAMP → increased intracellular calcium → phosphorylation of specific intracellular proteins by kinases → activation of the intracellular enzymes or proteins that finally mediate the biologic actions of the hormone. The PTH response system, like that for many other peptide and protein hormones, is subject to **down-regulation** of receptor number and to "**desensitization**," which may involve a post-cAMP mechanism.

PTH Has a Major Effect on Calcium Homeostasis

The central role of PTH in calcium metabolism is underscored by the observation that the first evolutionary appearance of this hormone was in animals attempting to adapt to a terrestrial existence. The physiologic maintenance of calcium balance depends on the long-term effects of PTH acting on intestinal absorption through the formation of calcitriol. If in the face of prolonged dietary Ca^{2+} deficiency intestinal calcium absorption is inadequate, a complex regulatory system involving PTH is brought into play. PTH restores normal ECF calcium concentration by acting directly on bone and kidney and by acting indirectly on the intestinal mucosa (through stimulation of synthesis of calcitriol). PTH (1) increases the rate of dissolution of bone, including both organic and inorganic phases, which moves Ca^{2+} into ECF; (2) reduces the renal clearance or excretion of calcium, hence increasing the ECF concentration of this cation; and (3) increases the efficiency of calcium absorption from the intestine by promoting the synthesis of calcitriol. The most rapid changes occur through the action on the kidney, but the largest effect is from bone. Therefore, although PTH prevents hypocalcemia in the face of dietary calcium deficiency, it does so at the expense of bone substance.

PTH Also Affects Phosphate Homeostasis

The usual counterion for Ca^{2+} is phosphate, and the hydroxyapatite crystal in bone consists of calcium phosphate. Phosphate is released with calcium from bone whenever PTH increases dissolution of the mineral matrix. PTH increases renal phosphate clearance; thus, the net effect of PTH on bone and kidney is to increase the ECF calcium concentration and decrease the ECF phosphate concentration. Importantly, this prevents the development of a supersaturated concentration of calcium and phosphate in plasma.

Pathophysiology

Insufficient amounts of PTH result in **hypoparathyroidism.** The biochemical hallmarks of this condition are decreased serum ionized calcium and elevated serum phosphate levels. Symptoms include neuromuscular irritability which, when mild, causes muscle cramps and **tetany.** Severe, acute hypocalcemia results in tetanic paralysis of the respiratory muscles, laryngospasm, severe convulsions, and death. Long-standing hypocalcemia results in cutaneous changes, cataracts, and calcification of the basal ganglia of the brain. The usual cause of hypoparathyroidism is accidental removal or damage of the glands during neck surgery (secondary hypoparathyroidism), but the disorder occasionally results from **autoimmune destruction** of the glands (primary hypoparathyroidism).

Pseudohypoparathyroidism is discussed in Chapter 45. Biologically active PTH is produced in this inherited disorder, but there is end-organ resistance to its effects. The biochemical consequences are the same, however. There are usually associated developmental anomalies including short stature, short metacarpal or metatarsal bones, and mental retardation. There are several types of pseudohypoparathyroidism, and they have been attributed to (1) partial deficiency of the G_s adenylate cyclase regulatory protein, and (2) a defective step beyond the formation of cAMP.

Hyperparathyroidism, the excessive production of PTH, is usually due to the presence of a functioning **parathyroid adenoma** but can be due to **parathyroid hyperplasia** or to **ectopic production** of PTH in a malignant tumor. The biochemical hallmarks of hyperparathyroidism are elevated serum ionized calcium and PTH and depressed serum phosphate levels. In long-standing hyperparathyroidism, findings include extensive resorption of bone and a variety of renal effects, including kidney stones, nephrocalcinosis, frequent urinary tract infections, and (in severe cases) decreased renal function. **Secondary hyperparathyroidism,** characterized by hyperplasia of the glands and hypersecretion of PTH, may be seen in patients with progressive renal failure. Hyperparathyroidism in these patients is presumably due to the decreased conversion of 25OH-D_3 to 1,25(OH)$_2$-D_3 in the diseased renal parenchyma, which results in inefficient calcium absorption in the gut and the secondary release of PTH in a compensatory attempt to maintain normal ECF calcium levels.

CALCITRIOL [1,25(OH)$_2$ − D$_3$] AFFECTS SEVERAL ASPECTS OF CALCIUM HOMEOSTASIS

Historical Perspective: Rickets, a childhood disorder characterized by deficient mineralization of the skeleton and severe, crippling bone deformities, was epidemic in North America and Western Europe early in this century. Results of a series of studies suggested that rickets was due to a dietary deficiency. After the discovery that rickets could be prevented by ingestion of cod-liver oil and that the active ingredient in this agent was not vitamin A, the preventive factor was termed fat-soluble **vitamin D.** About the same time, it was found that ultraviolet light, either artificial or from sunlight, would also prevent the disorder. It was subsequently determined that there was an adult equivalent to rickets. **Osteomalacia,** in which there is a failure to mineralize bone, also responded to vitamin D. Clues to further developments resulted from the observation that patients with liver or kidney disease did not respond normally to vitamin D. For the last 50 years, efforts to elucidate the structure of vitamin D and to define its mechanism of action have proceeded, greatly accelerated during the last several years.

The Principal Biological Role of Calcitriol is To Stimulate Intestinal Absorption of Calcium & Phosphate

Calcitriol is the only hormone that can promote this translocation of calcium against the concentration gradient which exists across the intestinal cell membrane. Since the production of calcitriol is tightly regulated (Fig 48–3), a fine mechanism exists for controlling ECF Ca^{2+} in spite of marked fluctuations of the calcium content of food. This ensures a proper concentration of calcium and phosphate for deposition, as hydroxyapatite crystals, onto the collagen fibrils in bone. In vitamin D deficiency (calcitriol deficiency), new bone formation slows and bone remodeling is also impaired. These processes are primarily regulated by PTH acting on bone cells, but small concentrations of calcitriol are also required. Calcitriol may also augment the actions of PTH on renal calcium reabsorption.

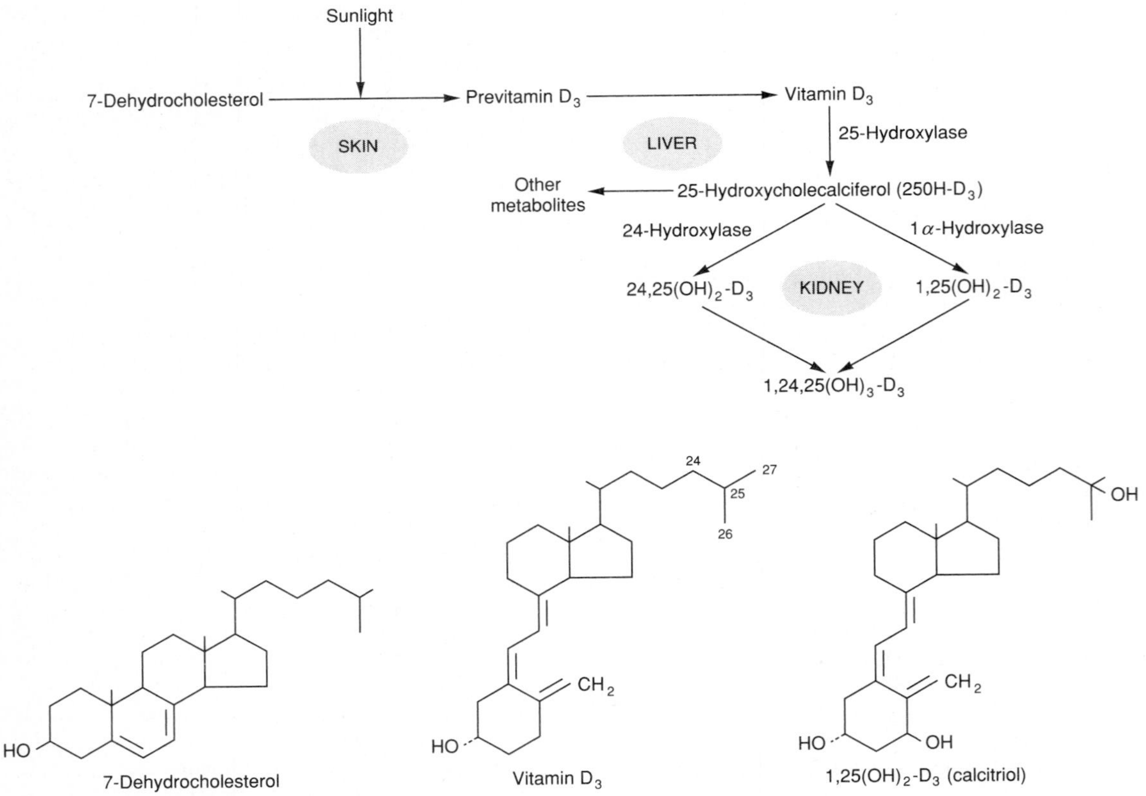

Figure 48–3. Formation and hydroxylation of vitamin D$_3$. 25-Hydroxylation takes place in the liver, and the other hydroxylations occur in the kidneys. 25,26(OH)$_2$-D$_3$ and 1,25,26(OH)$_3$-D$_3$ are probably formed as well. The formulas of 7-dehydrocholesterol, vitamin D$_3$, and 1,25(OH)$_2$-D$_3$ (calcitriol) are also shown. (Reproduced, with permission, from Ganong WF: *Review of Medical Physiology,* 14th ed. Appleton & Lange, 1989.)

The Synthesis & Metabolism of Calcitriol Involve Several Tissues & Are Highly Regulated Processes

Biosynthesis: Calcitriol is a hormone in every respect. It is produced by a complex series of enzymatic reactions that involve the plasma transport of precursor molecules to a number of different tissues (Fig 48–3). The active molecule, calcitriol, is transported to other organs where it activates biologic processes in a manner similar to that employed by the steroid hormones.

1. Skin—Small amounts of **vitamin D** occur in food (fish-liver oil, egg yolk), but most of the vitamin D available for calcitriol synthesis is produced in the malpighian layer of the epidermis from 7-dehydrocholesterol in an ultraviolet light-mediated, nonenzymatic **photolysis reaction.** The extent of this conversion is directly related to the intensity of the exposure and inversely related to the extent of pigmentation in the skin. There is an age-related loss of 7-dehydrocholesterol in the epidermis that may be related to the negative calcium balance associated with old age.

2. Liver—A specific transport protein called the **D-binding protein** binds vitamin D_3 and its metabolites and moves D_3 from the skin or intestine to the liver where it undergoes 25-hydroxylation, the first obligatory reaction in the production of calcitriol. 25-Hydroxylation occurs in the endoplasmic reticulum in a reaction that requires magnesium, NADPH, molecular oxygen, and an uncharacterized cytoplasmic factor. Two enzymes, an NADPH-dependent cytochrome P-450 reductase and a cytochrome P-450, are involved. This reaction is not regulated, and it also occurs with low efficiency in kidney and intestine. The $25OH-D_3$ enters the circulation, where it is the major form of vitamin D found in plasma, and is transported to the kidney by the D-binding protein.

3. Kidney—$25OH-D_3$ is a weak agonist and must be modified by hydroxylation at position C_1 for full biologic activity. This is accomplished in mitochondria of the renal proximal convoluted tubule in a complex, 3-component monooxygenase reaction that requires NADPH. Mg^{2+}, molecular oxygen, and at least 3 enzymes: (1) a flavoprotein, renal ferredoxin reductase; (2) an iron sulfur protein, renal ferredoxin; and (3) cytochrome P-450. This system produces $1.25(OH)_2$-D_3, which is the most potent naturally occurring metabolite of vitamin D.

4. Other tissues—The placenta has a 1α-hydroxylase that appears to be an important extrarenal source of calcitriol. Enzyme activity is found in a variety of other tissues, including bone; however, the physiologic significance of this appears to be minimal, since very little calcitriol is found in nonpregnant, nephrectomized animals.

Regulation of Metabolism and Synthesis: Like other steroid hormones, calcitriol is subject to tight feedback regulation (Fig 48–3 and Table 48–1). Low-calcium diets and hypocalcemia result in marked increases of 1α-hydroxylase activity in intact animals. This effect requires PTH, which is also released in

Table 48–1. Regulation of renal 1 α-hydroxylase.

Primary Regulators	Secondary Regulators
Hypocalcemia (↑) PTH (↑) Hypophosphatemia (↑) Calcitriol (↓)	Estrogens Androgens Progesterone Insulin Growth hormone Prolactin Thyroid hormone

response to hypocalcemia. The action of PTH is as yet unexplained, but it stimulates 1α-hydroxylase activity in both vitamin D-deficient and vitamin D-treated animals. Low-phosphorus diets and hypophosphatemia also induce 1α-hydroxylase activity, but this appears to be a weaker stimulus than that provided by hypocalcemia.

Calcitriol is an important regulator of its own production. High levels of calcitriol inhibit renal 1α-hydroxylase and stimulate the formation of a 24-hydroxylase that leads to the formation of $24,25(OH)_2$-D_3, an apparently inactive by-product. Estrogens, progestins, and androgens cause marked increases of 1α-hydroxylase in ovulating birds. The role that these hormones, along with insulin, growth hormone, and prolactin, play in mammals is uncertain.

The basic sterol molecule can be modified by **alternative metabolic pathways,** ie, by hydroxylation at positions 1, 23, 24, 25, and 26 and by the formation of a number of lactones. Over 20 metabolites have been found; none has unequivocally been shown to have biologic activity.

Calcitriol Acts at the Cellular Level in a Manner Similar to Other Steroid Hormones

Studies using radioactive calcitriol revealed localization in the nuclei of intestinal villus and crypt cells, osteoblasts, and distal renal tubular cells. There also is nuclear accumulation of this hormone in cells not previously suspected of being targets, including cells in the malpighian layer of the skin; pancreatic islet cells; some brain cells; some cells in the pituitary, ovary, testis, placenta, uterus, mammary gland, and thymus; and myeloid precursors. Calcitriol binding has also been noted in parathyroid cells, which leads to the intriguing possibility that it might be involved in PTH metabolism.

The Calcitriol Receptor: The calcitriol receptor is a member of the steroid receptor family (see Fig 49–7). The ligand-binding domain of this receptor binds calcitriol with high affinity and low capacity. This binding is saturable, specific, and reversible. The receptor has a DNA-binding domain that appears to contain the zinc finger motif characteristic of other steroid receptors.

Calcitriol-Dependent Gene Products: It has been known for several years that the response of intestinal transport to calcitriol requires RNA and protein synthesis. The observation of binding of the calcitriol receptor to chromatin in the nucleus suggests that calcitriol stimulates gene transcription and the formation of specific mRNAs. One such example, the induction of an mRNA that codes for a calcium-binding protein (CBP), has been reported.

There are several cytosolic proteins that bind Ca^{2+} with high affinity. One group, comprised of several proteins of different molecular weight, antigenicity, and tissue location (intestine, skin, and bone), is calcitriol-dependent. Of these, intestinal CBP has been studied most intensively. No CBP is found in the intestine of vitamin D-deficient rats, and the concentration of CBP is highly correlated with the extent of nuclear localization of calcitriol.

Effects of Calcitriol on Intestinal Mucosa: The transfer of Ca^{2+} or PO_4^{3-} across the intestinal mucosa requires (1) uptake across the brush border and microvillar membrane; (2) transport across the mucosal cell membrane; and (3) efflux across the basal lateral membrane into the ECF. It is clear that calcitriol enhances one or more of these steps, but the precise mechanism has not been established. CBP was thought to be actively involved until it was observed that Ca^{2+} translocation occurs within 1–2 hours after administration of calcitriol, well before CBP increases in response to calcitriol. CBP may bind Ca^{2+} and protect the mucosa cell against the large fluxes of Ca^{2+} coincident with the transport process. Several investigators are searching for other proteins that may be involved in Ca^{2+} transport, whereas others suggest that the process, particularly the early increase of Ca^{2+} flux, may be mediated by a membrane change. Metabolites of polyphosphoinositides have been implicated.

Pathophysiology: Rickets is a childhood disorder characterized by low plasma calcium and phosphorus levels and by poorly mineralized bone with associated skeletal deformities. Rickets is most commonly due to **vitamin D deficiency.** There are 2 types of **vitamin D-dependent rickets. Type I** is an inherited autosomal recessive trait characterized by a defect in the conversion of 25OH-D_3 to calcitriol. **Type II** is an autosomal recessive disorder in which there is a single amino acid change in one of the zinc fingers of the DNA-binding domain. This results in a nonfunctional receptor.

Vitamin D deficiency in the adult results in **osteo-malacia.** Calcium and phosphorus absorption are decreased, as are the ECF levels of these ions. Consequently mineralization of osteoid to form bone is impaired, and such undermineralized bone is structurally weak.

When substantial renal parenchyma is lost or diseased, the formation of calcitriol is reduced and calcium absorption decreases. When hypocalcemia ensues, there is a compensatory increase of PTH, which acts on bone in an attempt to increase ECF Ca^{2+}. The associated extensive bone turnover, structural changes, and symptoms are known as **renal osteodystrophy.** Early treatment with vitamin D will blunt this process.

THE ROLE OF CALCITONIN (CT) IN HUMAN CALCIUM HOMEOSTASIS IS UNCLEAR

CT is a 32-amino-acid peptide secreted by the parafollicular C cells of the human thyroid (less commonly, the parathyroid or thymus) or by similar cells located in the ultimobranchial gland of other species. These cells originate in the neural crest and are biochemically related to cells in a variety of other endocrine glands.

The entire CT molecule, including the 7-member N-terminal loop, formed by a Cys–Cys bridge, is required for biologic activity. There is tremendous interspecies variation of the amino acid sequence of CT (human and porcine CT share only 14 of 32 amino acids), but in spite of these differences there is cross-species bioactivity. The most potent naturally occuring CT is isolated from salmon.

CT has a history unmatched by any other hormone. Within a 7-year span (1962–1968), CT was discovered, isolated, sequenced, and synthesized, yet its role in human physiology is still uncertain.

REFERENCES

DeLuca HF, Schnoes HK: Vitamin D: Recent advances. *Annu Rev Biochem* 1983;**52**:411.

Norman AW, Roth J, Orci L: The vitamin D endocrine system: Steroid metabolism, hormone receptors, and biological response (calcium binding). *Endocr Rev* 1982;**3**:331.

Potts JT Jr, Kronenberg HM, Rosenblatt M: Parathyroid hormone: Chemistry, biosynthesis and mode of action. *Adv Protein Chem* 1982;**35**:323.

Hormones of the Adrenal Cortex

49

Daryl K. Granner, MD

INTRODUCTION

The adrenal cortex synthesizes dozens of different steroid molecules, but only a few of these have biologic activity. These sort into 3 classes of hormones: glucocorticoids, mineralocorticoids, and androgens. These hormones initiate their actions by combining with specific intracellular receptors, and this complex binds to specific regions of DNA to regulate gene expression. This results in altered rates of synthesis of a small number of proteins, which, in turn, affect a variety of metabolic processes, eg, gluconeogenesis and Na^+ and K^+ balance.

BIOMEDICAL IMPORTANCE

The hormones of the adrenal cortex, particularly the glucocorticoids, are an essential component of adaptation to severe stress. The mineralocorticoids are required for normal Na^+ and K^+ balance. Synthetic analogs of both classes are used therapeutically. In particular, many glucocorticoid analogs are potent anti-inflammatory agents. Excessive or deficient plasma levels of any of these 3 classes of hormones, whether due to disease or therapeutic use, result in serious, sometimes life-threatening, complications. A series of inherited enzyme deficiencies helps define the steps involved in steroidogenesis and illustrates the capacity of the adrenal cortex to alter the relative rates of production of these different hormones.

THE ADRENAL CORTEX MAKES 3 KINDS OF HORMONES

The adult cortex has 3 distinct layers or zones. The subcapsular area is called the **zona glomerulosa** and is associated with the production of mineralocorticoids. Next is the **zona fasciculata,** which, with the **zona reticularis,** produces glucocorticoids and androgens.

Some 50 steroids have been isolated and crystalized from adrenal tissue. Most of these are intermediates; only a small number are secreted in significant amounts; and few have significant hormonal activity. The adrenal cortex makes 3 general classes of steroid hormones, which are grouped according to their dominant action. There is an overlap of biologic activity, since all natural

Acronyms Used in This Chapter	
ACTH	Adrenocorticotropic hormone
ADH	Antidiuretic hormone
CBG	Corticosteroid-binding globulin
CRH	Corticotropin-releasing hormone
DHEA	Dehydroepiandrosterone
DOC	Deoxycorticosterone
GH	Growth hormone
3β-OHSD	3β-Hydroxysteroid dehydrogenase
PEPCK	Phosphoenolpyruvate carboxykinase
POMC	Pro-opiomelanocortin
PTH	Parathyroid hormone
TBG	Thyroid-binding globulin

glucocorticoids have mineralocorticoid activity and vice versa. This can now be understood on the basis of the commonality of the hormone response elements that mediate the effects of these hormones (and progestins) at the gene level (see Table 45–3).

The **glucocorticoids** are 21-carbon steroids with many actions, the most important of which is to promote gluconeogenesis. **Cortisol** is the predominant glucocorticoid in humans, and it is made in the zona fasciculata. **Corticosterone,** made in the zonae fasciculata and glomerulosa, is less abundant in humans, but it is the dominant glucocorticoid in rodents. **Mineralocorticoids** are also 21-carbon steroids. The primary action of these hormones is to promote retention of Na^+ and excretion of K^+ and H^+, particularly in the kidney. **Aldosterone** is the most potent hormone in this class, and it is made exclusively in the zona glomerulosa. The zonae fasciculata and reticularis of the adrenal cortex also produce significant amounts of the androgen precursor **dehydroepiandrosterone** and of the weak androgen **androstenedione.** These steroids are converted into more potent androgens in extraadrenal tissues and become pathologic sources of androgens when specific steroidogenic enzymes are deficient. Estrogens are not made in the normal adrenal in significant amounts, but in certain cancers of the adrenal they may be produced, and androgens of adrenal origin are important precursors of estrogen (converted by peripheral aromatization) in postmenopausal women.

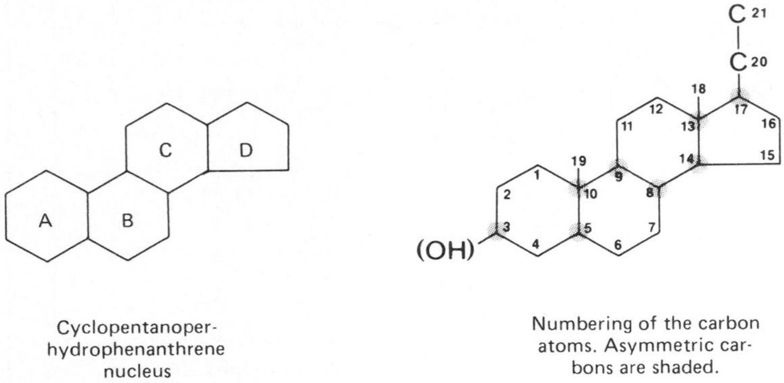

Figure 49–1. Structural features of steroid molecules.

A SPECIAL NOMENCLATURE DESCRIBES THE CHEMISTRY OF STEROIDS

All steroid hormones have in common the 17-carbon **cyclopentanoperhydrophenanthrene** structure with the 4 rings labeled A–D (Fig 49–1). Additional carbons can be added at positions 10 and 13 or as a side chain attached to C_{17}. Steroid hormones and their precursors and metabolites differ in number and type of substituted groups, number and location of double bonds, and stereochemical configuration. A precise nomenclature for designating these chemical formulations has been devised. The asymmetric carbon atoms (shaded on the C_{21} molecule in Fig 49–1) allow for **stereoisomerism.** The angular methyl groups (C_{19} and C_{18}) at positions 10 and 13 project in front of the ring system and serve as the point of reference. Nuclear substitutions in the same plane as these groups are designated *cis* or "β" and are represented in drawings by solid lines. Substitutions that project behind the plane of the ring system are designated *trans* or "α" and are represented as a dashed line. Double bonds are referred to by the number of the preceding carbon (eg, $\triangle^3$-$\triangle^4$). The steroid hormones are named according to whether they have one angular methyl group (estrane, 18 carbons), 2 angular methyl groups (androstane, 19 carbons), or 2 angular groups plus a 2-carbon side chain at C_{17} (pregnane, 21 carbons). This information (Fig 49–2), together with the glossary provided in Table 49–1, should allow one to understand the chemical names of the natural and synthetic hormones listed in Table 49–2.

SEVERAL ENZYMES ARE INVOLVED IN THE BIOSYNTHESIS OF ADRENAL STEROID HORMONES

The adrenal steroid hormones are synthesized from cholesterol that is mostly derived from the plasma, but a small portion is synthesized in situ from acetyl-CoA

via mevalonate and squalene. Much of the cholesterol in the adrenal is esterified and stored in cytoplasmic lipid droplets. Upon stimulation of the adrenal by ACTH (or cAMP), an esterase is activated, and the free cholesterol formed is transported into the mitochondrion, where a **cytochrome P-450 side chain cleavage enzyme** (P-450$_{scc}$) converts cholesterol to pregnenolone. Cleavage of the side chain involves sequential hydroxylations, first at C_{22} and then at C_{20}, followed by side chain cleavage (removal of the 6-carbon fragment isocaproaldehyde) to give the 21-carbon steroid (Fig 49–2). An ACTH-dependent protein may bind and activate cholesterol or P-450$_{scc}$. Aminoglutethimide is a very efficient inhibitor of P-450$_{scc}$ and of steroid biosynthesis.

All mammalian steroid hormones are formed from cholesterol via pregnenolone through a series of reactions that occur in either the mitochondria or endo-

Table 49–1. Nomenclature of steroids.

Prefix	Suffix	Chemical Nature
Hydroxy-	-ol	Alcohols
Dihydroxy-	-diol	
Oxo-	-one	Ketones (eg, -dione = 2 keto groups)
Cis-		Arrangement of 2 groups in same plane as C_{19}
Trans-		Arrangement of 2 groups in opposing plane to C_{19}
α-		A group *trans* to the 19-methyl
β-		A group *cis* to the 19-methyl
Deoxy-		Lacking a hydroxy group
Iso- or epi-		Isomerism at a C–C, C–OH, or C–H bond, eg, androsterone (5α) versus iso-androsterone (5β)
Dehydro-		Removal of 2 hydrogen atoms to form a double bond
Dihydro-		Addition of 2 hydrogen atoms to a double bond
Allo-		*Trans* configuration of the A and B rings

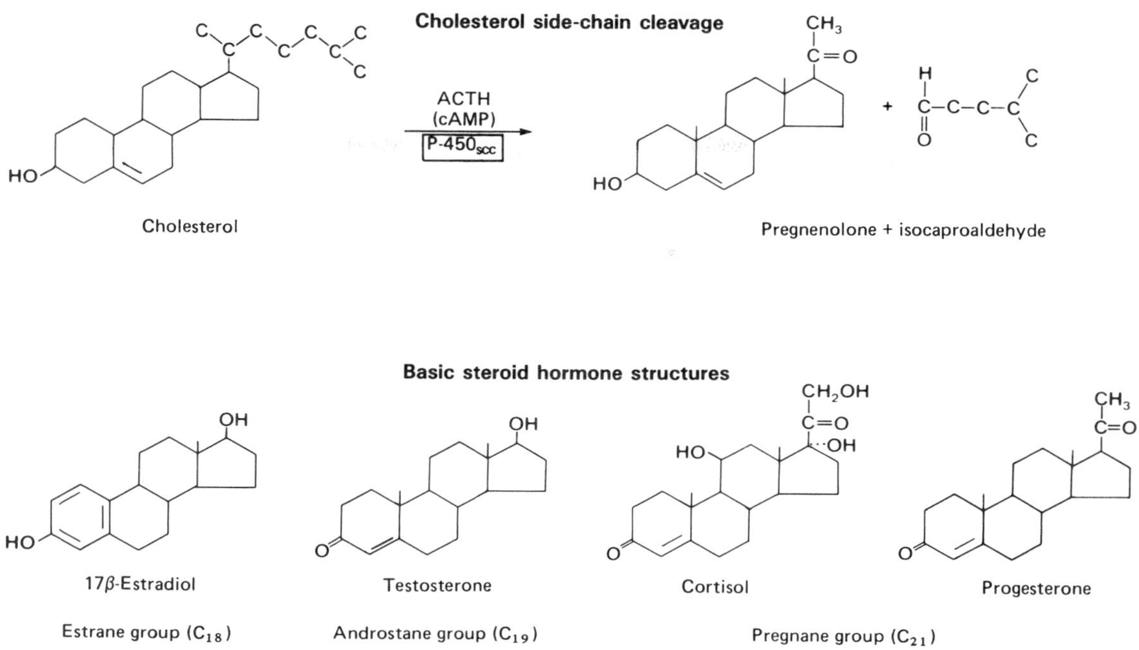

Figure 49–2. Cholesterol side-chain cleavage and basic steroid hormone structures.

Table 49–2. Trivial and chemical names of some steroids.

Trivial Name	Chemical Name
Aldosterone	11β,21-Dihydroxy-3,20-dioxo-4-pregnen-18-al
Androstenedione	4-Androstene-3,17-dione
Cholesterol	5-Cholesten-3β-ol
Corticosterone (compound B)	11β,21-Dihydroxy-4-pregnene-3,20-dione
Cortisol (compound F)	11β,17α,21-Trihydroxy-4-pregnene-3,20-dione
Cortisone (compound E)	17α,21-Dihydroxy-4-pregnene-3,11,20-trione
Dehydroepiandrosterone (DHEA)	3β-Hydroxy-5-androsten-17-one
11-Deoxycorticosterone (DOC)	21-Hydroxy-4-pregnene-3,20-dione
11-Deoxycortisol (compound S)	17,21-Dihydroxy-4-pregnene-3,20-dione
Dexamethasone	9α-Fluoro-16α-methyl-11β,17α,21-trihydroxypregna-1,4-diene-3,20-dione
Estradiol	1,3,5,(10)-Estratriene-3,17β-diol
Estriol	1,3,5,(10)-Estratriene-3,16α,17β-triol
Estrone	3-Hydroxy-1,3,5(10)-estratriene-3-ol-17-one
Etiocholanolone	3α-Hydroxy-5β-androstan-17-one
9α-Fluorocortisol	9α-Fluoro-11β,17α,21-trihydroxypregn-4-ene,3,20-dione
Prednisolone	11β,17α,21-Trihydroxypregna-1,4-diene-3,20-dione
Prednisone	17α,21-Dihydroxypregna-1,4-diene-3,11,20-trione
Pregnanediol	5β-Pregnane-3α,20α-diol
Pregnanetriol	5β-Pregnane-3α,17α,20α-triol
Pregnenolone	3β-Hydroxy-5-pregnen-20-one
Progesterone	4-Pregnene-3,20-dione
Testosterone	17β-Hydroxy-4-androsten-3-one
Triamcinolone	9α-Fluoro-11β,16α,17α,21-tetrahydroxypregna-1,4-diene-3,20-dione

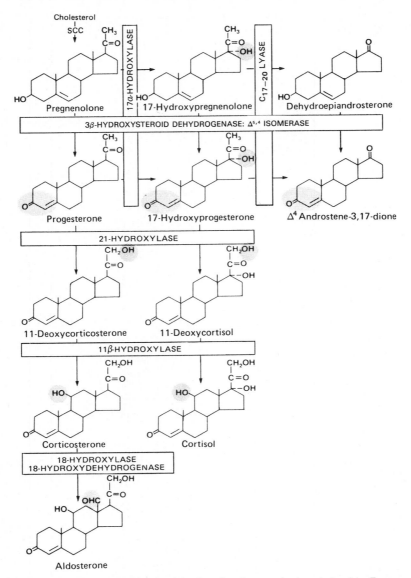

Figure 49–3. Pathways involved in the synthesis of the 3 major classes of adrenal steroids. Enzymes are shown in the rectangular boxes, and the modifications at each step are shaded. (Slightly modified and reproduced, with permission, from Harding BW. Page 1135 in: *Endocrinology* Vol 2. DeGroot LJ [editor]. Grune & Stratton, 1979.)

plasmic reticulum of the adrenal cell. **Hydroxylases** that require molecular oxygen and NADPH are essential, and **dehydrogenases,** an **isomerase,** and a **lyase** are also necessary for certain steps. There is some cellular specificity in steroidogenesis. For instance, 18-hydroxylase and 18-hydroxysteroid dehydrogenase, which are required for aldosterone synthesis, are found only in glomerulosa cells, so that the biosynthesis of this mineralocorticoid is confined to this region. A schematic representation of the pathways involved in the synthesis of the 3 major classes of adrenal steroids is presented in Fig 49–3. The enzymes are shown in the rectangular boxes, and the modifications at each step are shaded.

Mineralocorticoid Synthesis Occurs in the Zona Glomerulosa

Synthesis of aldosterone follows the mineralocorticoid pathway and occurs in the zona glomerulosa. Pregnenolone is converted to progesterone by the action of 2 smooth endoplasmic reticulum enzymes, **3β-hydroxysteroid dehydrogenase (3β-OHSD)** and $\Delta^{5,4}$ **isomerase.** Progesterone is hydroxylated at the C_{21} position to form 11-deoxycorticosterone (DOC), which is an active (Na$^+$-retaining) mineralocorticoid. The next hydroxylation, at C_{11}, produces corticosterone, which has glucocorticoid activity and is a weak mineralocorticoid (it has less than 5% of the potency of aldosterone). In some species (eg, rodents), it is the most potent glucocorticoid. C_{21} hydroxylation is necessary for both mineralocorticoid and glucocorticoid activity, but most steroids with a C_{17} hydroxyl group have more glucocorticoid and less mineralocorticoid action. In the zona glomerulosa, which does not have the smooth endoplasmic reticulum enzyme 17α-hydroxylase, a mitochondrial 18-hydroxylase is present. The **18-hydroxylase** acts on corticosterone to form 18-hydroxycorticosterone, which is changed to aldosterone by the conversion of the 18-alcohol to an aldehyde. This unique distribution of enzymes and the special regulation of the zona glomerulosa (see below) have led some investigators to suggest that, in addition to the adrenal being 2 glands, the adrenal cortex is actually 2 separate organs.

Glucocorticoid Synthesis: Cortisol synthesis requires 3 hydroxylases that act sequentially on the C_{17}, C_{21}, and C_{11} positions. The first 2 reactions are rapid, while C_{11} hydroxylation is relatively slow. If the C_{21} position is hydroxylated first, the action of 17α-hydroxylase is impeded and the mineralocorticoid pathway is followed (forming corticosterone or aldosterone, depending on the cell type). 17α-Hydroxylase is a smooth endoplasmic reticulum enzyme that acts upon either progesterone or, more commonly, pregnenolone. 17α-Hydroxyprogesterone is hydroxylated at C_{21} to form 11-deoxycortisol, which is then hydroxylated at C_{11} to form cortisol, the most potent natural glucocorticoid hormone in humans. The **21-hydroxylase** is a smooth endoplasmic reticulum enzyme, whereas the **11β-hydroxylase** is a mitochondrial enzyme. Steroidogenesis thus involves the repeated shuttling of substrates into and out of the mitochondria of the fasciculata and reticularis cells (Fig 49–4).

Androgen Synthesis: The major androgen or androgen precursor produced by the adrenal cortex is **dehydroepiandrosterone (DHEA)**. Most 17-hydroxy-pregnenolone follows the glucocorticoid pathway, but a small fraction is subjected to oxidative fission and removal of the 2-carbon side chain through the action of 17,20-lyase. This enzyme is found in the adrenals and gonads and acts exclusively on 17α-hydroxy-containing molecules. Adrenal androgen production increases markedly if glucocorticoid biosynthesis is impeded by the lack of one of the hydroxylases (see adrenogenital syndrome, below). Most DHEA is rapidly modified by the addition of sulfate, about half of which occurs in the adrenal and the rest in the liver. DHEA sulfate is inactive, but removal of the sulfate results in reactivation. DHEA is really a prohormone, since the actions of 3β-OHSD and $\triangle^{5,4}$ isomerase convert the weak androgen DHEA into the more potent **androstenedione.** Small amounts of androstenedione are also formed in the adrenal by the action of the lyase on 17α-hydroxyprogesterone. Reduction of androstenedione at the C_{17} position results in the formation of **testosterone,** the most potent adrenal androgen. Small amounts of testosterone are produced in the adrenal by this mechanism, but most of this conversion occurs in other tissues.

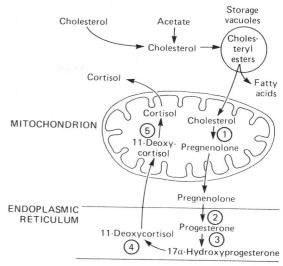

Figure 49–4. Subcellular compartmentalization of glucocorticoid biosynthesis. Adrenal steroidogenesis involves the shuttling of precursors between mitochondria and the endoplasmic reticulum. The enzymes involved are (1) C_{20-22} lyase, (2) 3β-hydroxysteroid dehydrogenase and $\triangle^{5,4}$ isomerase, (3) 17α-hydroxylase, (4) 21-hydroxylase, and (5) 11β-hydroxylase. (Slightly modified and reproduced, with permission, from Harding BW. Page 1135 in: *Endocrinology.* Vol 2. DeGroot LJ [editor]. Grune & Stratton, 1979.)

Small amounts of other steroids can be isolated from adrenal venous blood, including 11-deoxycorticosterone, progesterone, pregnenolone, 17α-hydroxyprogesterone, and a very small amount of estradiol (from the aromatization of testosterone). None of these amounts are important in relation to production from other glands, however.

THE SECRETION, TRANSPORT, & METABOLISM OF ADRENAL STEROID HORMONES AFFECT BIOAVAILABILITY

Secretion of Steroid Hormones

There is little, if any, storage of steroid hormones within the adrenal (or gonad) cell, since these hormones are released into the plasma when they are made. Cortisol release occurs with a periodicity that is regulated by the **diurnal rhythm** of ACTH release.

Plasma Transport

Glucocorticoids: Cortisol circulates in plasma in protein-bound and free forms. The main plasma binding protein is an α-globulin called **transcortin** or **corticosteroid-binding globulin (CBG)**. CBG is produced in the liver, and its synthesis, like that of thyroid-binding globulin (TBG), is increased by estrogens. CBG binds most of the hormone when plasma cortisol levels are within the normal range; much smaller amounts of cortisol are bound to albumin. The avidity of binding helps determine the biologic half-lives of various glucocorticoids. Cortisol binds tightly to CBG and has a $t_{1/2}$ of 1.5–2 hours, while corticosterone, which binds less tightly, has a $t_{1/2}$ of less than 1 hour. Binding to CBG is not restricted to glucocorticoids. Deoxycorticosterone and progesterone interact with CBG with sufficient affinity to compete for cortisol binding. **The unbound, or free, fraction constitutes about 8% of the total plasma cortisol and represents the biologically active fraction of cortisol.**

Mineralocorticoids: Aldosterone, the most potent natural mineralocorticoid, does not have a specific plasma transport protein, but it forms a very weak association with albumin. Corticosterone and 11-deoxycorticosterone, other steroids with mineralocorticoid effects, bind to CBG. These observations are important in understanding the mechanism of action of aldosterone (see below).

Metabolism & Excretion Rates Depend on the Presence or Absence of Carrier Proteins

Glucocorticoids: Cortisol and its metabolites constitute about 80% of the 17-hydroxycorticoids in plasma; the other 20% consist of cortisone and 11-deoxycortisol. About half of the cortisol (as well as cortisone and 11-deoxycortisol) circulates in the form of the reduced dihydro- and tetrahydro- metabolites that are produced from reduction of the A ring double bond by NADPH-requiring hydrogenases and from reduction of the 3-ketone group by a reversible dehydrogenase reaction. Substantial amounts of all of these compounds are also modified by conjugation at the C_3 position with glucuronide or, to a lesser extent, with sulfate. These modifications occur primarily in the liver and make the lipophilic steroid molecule water-soluble and excretable. In humans, most of the conjugated steroids that enter the intestine by biliary excretion are reabsorbed by the enterohepatic circulation. About 70% of the conjugated steroids are excreted in the urine, 20% leave in feces, and the rest exit through the skin.

Mineralocorticoids: Aldosterone is very rapidly cleared from the plasma by the liver, no doubt because it lacks a plasma carrier protein. The liver forms tetrahydroaldosterone 3-glucuronide, which is excreted in the urine.

Androgens: Androgens are excreted as 17-keto compounds including DHEA (sulfate) as well as androstenedione and its metabolites. Testosterone, secreted in small amounts by the adrenal, is not a 17-keto compound, but the liver converts about 50% of testosterone to androsterone and etiocholanolone, which are 17-keto compounds.

THE SYNTHESIS OF ADRENAL STEROID HORMONES IS REGULATED BY VERY DIFFERENT MECHANISMS

Glucocorticoid Hormones

The secretion of cortisol is dependent on ACTH, which in turn is regulated by corticotropin-releasing hormone (CRH). These hormones are linked by a classic negative feedback loop (see Fig 44–1 and Table 46–1).

Mineralocorticoid Hormones

The production of aldosterone by the glomerulosa cells is regulated in a completely different manner. The primary regulators are the renin-angiotensin system and potassium. Sodium, ACTH, and neural mechanisms are also involved.

The Renin-Angiotensin System: This system is involved in the regulation of blood pressure and electrolyte metabolism. The primary hormone in these processes is **angiotensin II,** an octapeptide made from **angiotensinogen** (Fig 49–5). Angiotensinogen, an α_2-globulin made in liver, is the substrate for renin, an enzyme produced in the **juxtaglomerular cells** of the renal afferent arteriole. The position of these cells makes them particularly sensitive to blood pressure changes, and many of the physiologic regulators of renin release act through renal **baroreceptors** (Table 49–3). The juxtaglomerular cells are also sensitive to changes of Na^+ and Cl^- concentration in the renal tubular fluid; therefore, any combination of factors that decreases fluid volume (dehydration, decreased blood pressure, fluid or blood loss) or decreases NaCl concentration stimulates renin release. Renal sympathetic nerves that terminate in the juxtaglomerular cells mediate the central nervous system and postural effects on renin release independent of the baroreceptor and salt effects, a mechanism that involves the β-adrenergic receptor.

Renin acts upon the substrate angiotensinogen to produce the decapeptide **angiotensin I.** The synthesis of angiotensinogen in liver is enhanced by glucocorticoids and estrogens. Hypertension associated with these hormones may be due in part to increased plasma levels of angiotensinogen. Since this protein circulates at about the K_m for its interaction with renin, small changes could markedly affect the generation of angiotensin II.

Angiotensin-converting enzyme, a glycoprotein found in lung, endothelial cells, and plasma, removes 2 carboxy-terminal amino acids from the decapeptide angiotensin I to form angiotensin II in a step that is not thought to be rate-limiting. Various nonapeptide analogs of angiotensin I and other compounds act as competitive inhibitors of converting enzyme and are used to

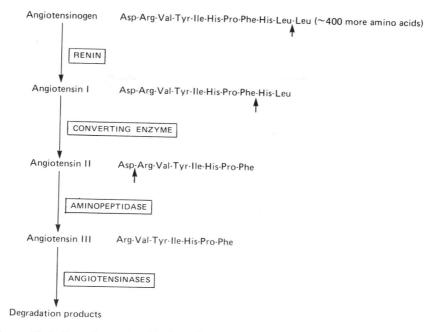

Angiotensinogen   Asp-Arg-Val-Tyr-Ile-His-Pro-Phe-His-Leu-Leu (~400 more amino acids)

RENIN

Angiotensin I   Asp-Arg-Val-Tyr-Ile-His-Pro-Phe-His-Leu

CONVERTING ENZYME

Angiotensin II   Asp-Arg-Val-Tyr-Ile-His-Pro-Phe

AMINOPEPTIDASE

Angiotensin III   Arg-Val-Tyr-Ile-His-Pro-Phe

ANGIOTENSINASES

Degradation products

Figure 49–5. Formation and metabolism of angiotensins. Small arrows indicate cleavage sites.

treat **renin-dependent hypertension.** Converting enzyme also degrades **bradykinin,** a potent vasodilator; thus, this enzyme increases blood pressure in 2 distinct ways.

Angiotensin II increases blood pressure by causing vasoconstriction of the arteriole and is a very potent vasoactive substance. It inhibits renin release from the juxtaglomerular cells and is a potent stimulator of aldosterone production. Although angiotensin II stimulates the adrenal directly, it has no effect on cortisol production.

In some species, angiotensin II is converted to the des-Asp[1] heptapeptide **angiotensin III** (Fig 49–5), an equally potent stimulator of aldosterone production. In humans, the plasma level of angiotensin II is 4 times greater than that of angiotensin III, so most effects are exerted by the octapeptide. Angiotensins II and III are rapidly inactivated by **angiotensinases.**

Angiotensin II binds to specific glomerulosa cell receptors. The hormone-receptor interaction does not activate adenylate cyclase, and cAMP does not appear to mediate the action of this hormone. The actions of angiotensin II, which are to stimulate the conversion of cholesterol to pregnenolone and of corticosterone to 18-hydroxycorticosterone and aldosterone, may involve changes in the concentration of intracellular calcium and of phospholipid metabolites by mechanisms similar to those described in Chapter 45.

Potassium: Aldosterone secretion is sensitive to changes in plasma potassium level; an increase as small as 0.1 meq/L stimulates production, whereas a similar decrease reduces aldosterone production and secretion. K^+ affects the same enzymatic steps as does angiotensin II, although the mechanism involved is obscure. Like angiotensin II, K^+ does not affect the biosynthesis of cortisol.

Other Effectors: In special circumstances **ACTH** and **sodium** may be involved in aldosterone production in humans.

ADRENAL STEROID HORMONES HAVE NUMEROUS & DIVERSE METABOLIC EFFECTS

Loss of adrenal cortical function results in death unless replacement therapy is instituted. In humans, treatment of adrenal insufficiency with mineralocorticoids is generally not sufficient; glucocorticoids seem to be more critical in this regard. Rats, in contrast, do quite well with mineralocorticoid replacement. Excessive or deficient plasma levels of either of these classes of hormones, whether due to disease or therapeutic use, cause a number of serious complications directly related to their metabolic actions.

Table 49–3. Factors that influence renin release.

Stimulators	Inhibitors
Decreased blood pressure	Increased blood pressure
Change from supine to erect posture	Change from erect to supine posture
Salt depletion	Salt loading
β-Adrenergic agents	β-Adrenergic antagonists
Prostaglandins	Prostaglandin inhibitors
	Potassium
	Vasopressin
	Angiotensin II

Table 49–4. The diverse effects of glucocorticoids.

I. Effects on Intermediary Metabolism
a. Increase glucose production in liver by:
   1. increasing the delivery of amino acids (the gluconeogenic substrate) from peripheral tissues
   2. increasing the rate of gluconeogenesis by increasing the amount (and activity) of several key enzymes
   3. "permitting" other key metabolic reactions to operate at maximal rates
b. Increase hepatic glycogen deposition by promoting the activation of glycogen synthetase.
c. Promote lipolysis (in extremities) but can cause lipogenesis in other sites (face and trunk) especially at higher than physiologic levels.
d. Promote protein and RNA metabolism. This is an anabolic effect at physiologic levels, but can be catabolic in certain conditions and at higher than physiologic levels.

II. Effects on Host Defense Mechanisms
a. Suppress the immune response. These hormones cause a species- and cell-type-specific lysis of lymphocytes.
b. Suppress the inflammatory response by:
   1. decreasing the number of circulating leukocytes and the migration of tissue leukocytes
   2. inhibiting fibroblast proliferation
   3. inducing lipocortins, which, by inhibiting phospholipase A_2, blunt the production of the potent anti-inflammatory molecules, the prostaglandins and leukotrienes.

III. Other Effects
a. Necessary for maintenance of normal blood pressure and cardiac output.
b. Required for maintenance of normal water and electrolyte balance, perhaps by restraining ADH release (H_2O) and by increasing angiotensinogen (Na^+). These effects contribute to the effect on blood pressure.
c. Necessary, with the hormones of the adrenal medulla, in allowing the organism to respond to stress.

Glucocorticoid Hormones Affect Basal Metabolism, Host Defense Mechanisms, Blood Pressure, & Response to Stress

A detailed discussion of the various metabolic effects of the glucocorticoid hormones is found in standard physiology texts. A brief description of the effects is presented in Table 49–4.

Mineralocorticoid Hormones Affect Electrolyte Balance & Ion Transport

Mineralocorticoid hormones act in the kidney to stimulate active Na^+ transport by the distal convoluted tubules and collecting tubules, the net result being Na^+ retention. These hormones also promote the secretion of K^+, H^+, and NH_4^+ by the kidney and affect ion transport in other epithelial tissues including sweat glands, intestinal mucosa, and salivary glands. Aldosterone is 30–50 times more potent than 11-deoxycorticosterone (DOC) and 1000 times more potent than cortisol or corticosterone. As the most potent naturally occurring mineralocorticoid, aldosterone accounts for most of this action in humans. Cortisol, although far less potent, has a much higher production

rate and thus has a significant effect on Na^+ retention and K^+ excretion. Since the amount of DOC produced is very small, it is much less important in this regard.

RNA and protein synthesis are required for the action of aldosterone, which appears to involve the production of specific gene products (see below).

ADRENAL STEROID HORMONES ACT BY SIMILAR MECHANISMS

Glucocorticoid hormones initiate their action in a target cell by interacting with a specific receptor. This step is necessary for entry into the nucleus and DNA binding. There is generally a high correlation between the association of a steroid with receptor and the elicitation of a given biologic response. This correlation holds true for a wide range of activities, so that a steroid with one-tenth the binding affinity evokes a correspondingly decreased biologic effect at a given steroid concentration. There are generally no "spare receptors" involved in steroid hormone action.

The biologic effect of a steroid depends upon both its ability to bind to the receptor and the concentration of free hormone in the plasma. Cortisol, corticosterone, and aldosterone all bind with high affinity to the glucocorticoid receptor, but in physiologic circumstances cortisol is the dominant glucocorticoid because of its much greater plasma concentration. Corticosterone is an important glucocorticoid in certain pathologic conditions (17α-hydroxylase deficiency), but aldosterone never reaches a concentration in plasma sufficient to exert glucocorticoid effects.

The Functional Domains of the Glucocorticoid Receptor Have Been Defined

A number of biochemical, immunologic, and genetic studies have led to the formulation of the glucocorticoid receptor illustrated in Fig 49–6. The amino-terminal half contains most of the antigenic sites and has a region that modulates promoter function (*trans*-activation). The carboxy terminal half contains the DNA- and hormone-binding domains. The DNA-binding domain is closer to the center of the molecule, while the hormone-binding domain is near the carboxyl terminus. Both of these domains are required for *trans*-activation of gene transcription. A sequence of amino acids in the carboxy terminal region is required for dimerization of two receptor molecules, a reaction thought to be required for binding to each of the two "half" binding sites in the glucocorticoid regulatory element (GRE) (see the arrows in Table 45–3). Two separate regions appear to be necessary for entry of the receptor into the nucleus (nuclear localization).

The amino acid sequence of the receptor, deduced by analysis of appropriate cDNA molecules, reveals two regions with an abundance of Cys-Lys-Arg residues in the DNA-binding domain. By comparing these regions with other known DNA-binding proteins (especially TFIIIA), it is possible to hypothesize a protein

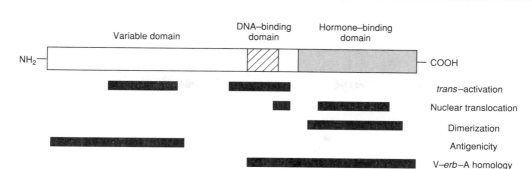

Figure 49–6. Schematic representation of the 777-amino acid human glucocorticoid receptor. The receptor can be generally divided into antigenic, DNA-binding, and hormone-binding domains. Other domains, and the homology to the oncogene v-*erb*-A, are shown.

structure that has two of the "fingers" (each with a coordinated zinc in its center) which are speculated to bind to a turn of DNA. This is one of the forms of protein-DNA interaction discussed in Chapter 41. This observation, coupled with others in which the carboxyl half of the glucocorticoid receptor was noted to be highly similar to the oncogene protein v-*erb*-A and in which the DNA-binding domains of the glucocorticoid, estrogen, and progesterone receptors were homologous, led to the discovery of the steroid-thyroid hormone receptor superfamily.

The Steroid-Thyroid Hormone Receptor Superfamily: Steroid and thyroid hormones regulate a variety of processes involved in development, differentiation, growth, reproduction, and adaptation to environmental changes. In recent years it has become obvious that a general mechanism could explain how these hormones work at the molecular level (see Chapter 45). An essential component in this mechanism is the hormone receptor. These molecules are not abundant, so structural analysis awaited the isolation of cDNA clones for each. The first structures deduced

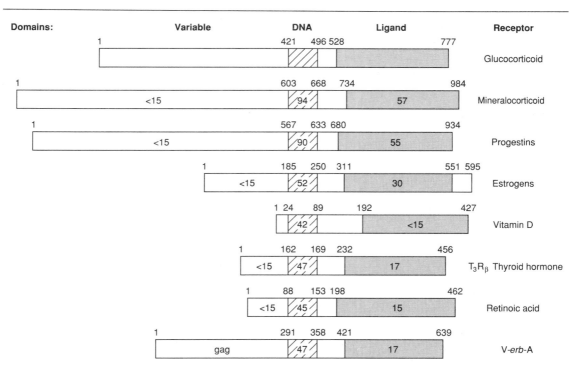

Figure 49–7. Schematic comparison of the steroid-thyroid hormone receptor superfamily. The sequences of the human receptors (v-*erb*-A is avian) are aligned by their DNA-binding domains, which show the highest amino acid similarity (numbers inside the rectangles are % similarity to the corresponding region of the glucocorticoid receptor). The numbers above the vertical lines that separate the domains show the amino acid positions. The amino-terminal position is designated as 1. Many other similar molecules have been isolated but the ligands/function for these have not been determined. Modified, with permission, from R. Evans.

were those for the glucocorticoid, estrogen, and progesterone receptors. The homology in the DNA-binding domains of these and the close similarity of each to v-*erb*-A, a DNA-binding oncogene protein, led to the hypothesis that these receptors might belong to a supergene family. If so, a corollary hypothesis was that other receptors should be isolated from cDNA libraries using probes directed against the common region (DNA domain) under conditions of reduced hybridization stringency (see Chapter 42). This hypothesis proved to be true. As illustrated in Fig 49–7, structures of all of the steroid receptors have been deduced, along with those of several receptors yet unidentified. The homology between the DNA-binding domains of these receptors is striking, and the general organization of each is the same. There is considerable variation in the total length of the receptors, most of which is due to the N-terminal half of the molecule. This observation has greatly accelerated understanding of how this class of hormones regulates gene transcription.

Glucocorticoid Hormones Generally Act by Regulating Gene Expression

The general features of glucocorticoid hormone action are described in Chapter 45 and illustrated in Fig 45–1. Numerous examples support the concept that this class of hormones affects specific cellular processes by influencing the amount of critical proteins, usually enzymes, within the cell. Glucocorticoids usually accomplish this by regulating the rate of transcription of specific genes in the target cell, but they also affect other steps in the "information flow" (see Fig 45–2). Regulation of transcription requires that the steroid-receptor complex bind to specific regions of DNA in the vicinity of the transcription initiation site and that such regions confer specificity to the response. How this interaction actually enhances or inhibits transcription, how tissue specificity is accomplished, and how a given gene can be stimulated in one tissue and inhibited in another are a few of the important questions that remain unanswered.

A brief description of how glucocorticoid hormones affect transcription of mouse mammary tumor virus DNA provides a good illustration of what is known about steroid hormone action. The mammary tumor virus system has been useful because the steroid effect is rapid and large and the molecular biology of the virus has been extensively studied. The glucocorticoid hormone receptor complex binds with high selectivity and specificity to a region, the glucocorticoid regulatory element (GRE), a few hundred base pairs upstream from the transcription initiation site. With the GRE are sequences closely related to the consensus sequence GGTACAnnnTGTTCT that is found in the regulatory elements of glucocorticoid-regulated genes. The receptor-charged glucocorticoid regulatory element enhances transcription initiation of the mouse mammary tumor virus genome and will also activate heterologous promoters. This *cis*-acting element also works when moved to different regions upstream and downstream, and it works in either the forward or the backward orientation. In these respects, the glucocorticoid regulatory element qualifies as a transcription enhancer. These features have also been demonstrated in several other glucocorticoid-regulated genes, and are summarized in Fig 49–8. Some genes regulated by glucocorticoids appear to require an additional DNA-binding protein and its cognate DNA element.

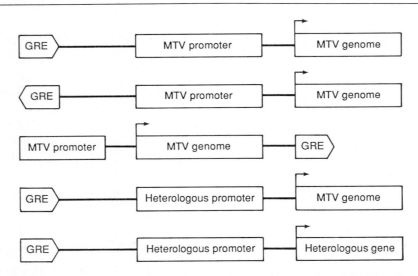

Figure 49–8. The GRE is a transcription enhancer. The wild-type mammary tumor virus (MTV) contains a region of DNA that is copied into the transcription unit (MTV genome), the promoter, and a GRE. The GRE is normally situated within a 200 bp 5′ of the transcription start site, but it works in either orientation and can be located downstream from the transcription start site. It will also work on heterologous promoters and coding units. These observations mean that this GRE, and others isolated from different genes, functions as a transcription enhancer.

Control of the rate of gene transcription appears to be the major action of the glucocorticoid hormones, but it is not the sole mechanism employed. The ability to measure specific processes has revealed that these hormones also regulate the rate of degradation of specific mRNAs (eg, growth hormone and phosphoenolpyruvate-carboxykinase), and posttranslational processing (various mammary tumor virus proteins). These and other classes of steroid hormones appear able to act at any level of the "information flow" from DNA to protein (see Fig 45–2) and the relative importance of each varies from system to system.

The General Features of Mineralocorticoid Hormone (Aldosterone) Action Are Similar to Those of Other Steroid Hormones (See Fig 45–1,2,3 & Table 45–3.)

Although specific gene products have not been isolated, protein and RNA synthesis are known to be required for aldosterone action, and it is presumed that specific proteins are involved in mediating the effects of aldosterone on ion transport.

Receptors That Bind Aldosterone With High Affinity ($K_d \sim 1$ nmol/L) Are Found in the Cytoplasm & Nuclei of Target Cells

These cells include those of the kidney, the parotid, and the colon and of other cells not thought to be targets of aldosterone action (hippocampus and heart). These receptors have equal affinity for aldosterone, cortisol, and corticosterone and are called type I receptors to distinguish them from the classical glucocorticoid receptor (type II).

Given the fact that the plasma level of aldosterone is much lower than that of either of the other 2 steroids, one might suppose that these would preferentially occupy the type I sites and that aldosterone would exert little effect. Recall that DOC and corticosterone are avidly bound to corticosteroid-binding globulin, the plasma glucocorticoid transport protein, while aldosterone has no specific carrier protein. Consequently, the effective "free" concentration of aldosterone in plasma is greater than that of either corticosterone or DOC. Aldosterone therefore is readily able to enter cells, and this ensures a competitive advantage for aldosterone with respect to occupying the type I receptor in vivo. The important action of aldosterone is assured by an additional "failsafe" mechanism. The receptor in mineralocorticoid target tissues has absolute selectivity for aldosterone because of the presence of the enzyme 11β-hydroxysteroid dehydrogenase. This enzyme converts cortisol and corticosterone to their 11β metabolites but it is not active on aldosterone. These metabolites cannot bind to the type I receptor, so aldosterone has unimpeded access.

The Major Actions of Aldosterone Are on Ion Transport

The molecular mechanisms of aldosterone action on Na^+ transport have not been elucidated, but several studies point to the following hypothesis.

Na^+ from the luminal fluid bathing the apical surface of the renal cell enters passively through Na^+ channels. Na^+ is then transported into the interstitial fluid through the serosal side of the cell by the Na^+/K^+-dependent ATPase pump. ATP provides the energy required for this active process.

Aldosterone increases the number of apical membrane Na^+ channels, and this presumably increases intracellular Na^+. Aldosterone also increases the activity of several mitochondrial enzymes, and this could result in the generation of the ATP required to drive the serosal membrane Na^+/K^+ pump. The NADH:NAD ratio increases as a result of aldosterone action, as do the activities of several mitochondrial enzymes including citrate synthase. The increased activity of citrate synthase involves a true induction (perhaps mediated by the gene transcription effects alluded to above), and the temporal increase of this protein correlates highly with the effect of aldosterone on Na^+ transport. Aldosterone has not been shown to have an effect on the Na^+ pump itself; therefore, it appears that the hormone increases the intracellular concentration of Na^+ and creates the energy source required for removal of this ion through the serosal pump. Other mechanisms, involving different aldosterone-regulated proteins, may be involved in the handling of K^+ and H^+.

PATHOPHYSIOLOGY OF THE ADRENAL CORTEX

Disorders of Glucocorticoid Hormone Insufficiency & Excess

Primary adrenal insufficiency (**Addison's disease**) results in hypoglycemia, extreme sensitivity to insulin, intolerance to stress, anorexia, weight loss, nausea, and severe weakness. Patients with Addison's disease have low blood pressure, decreased glomerular filtration rate, and decreased ability to excrete a water load. They often have a history of salt craving. Plasma Na^+ levels are low, K^+ levels are high, and blood lymphocyte and eosinophil counts are increased. Such patients often show increased pigmentation of skin and mucous membranes because of the exaggerated compensatory secretion of ACTH and associated products of the POMC gene. **Secondary adrenal insufficiency** is due to a deficiency of ACTH resulting from tumor, infarction, or infection. This results in a similar metabolic syndrome without hyperpigmentation.

Glucocorticoid excess, commonly called **Cushing's syndrome,** is usually due to the pharmacologic use of steroids, but it may result from an ACTH-secreting pituitary adenoma, from adrenal adenomas or carcinomas, or from the ectopic production of ACTH by a neoplasm. Patients with Cushing's syndrome typically lose the diurnal pattern of ACTH/cortisol secretion.

They have hyperglycemia or glucose intolerance (or both) because of accelerated gluconeogenesis. Related to this are severe protein catabolic effects, which result in thinning of the skin, muscle wasting, osteoporosis, extensive lymphoid tissue involution, and generally a negative nitrogen balance. There is a peculiar redistribution of fat, with truncal obesity and the typical "buffalo hump." Resistance to infections and inflammatory responses is impaired, as is wound healing. Several findings, including hypernatremia, hypokalemia, alkalosis, edema, and hypertension, are due to the mineralocorticoid actions of cortisol.

Disorders of Mineralocorticoid Excess

Small adenomas of the glomerulosa cells result in **primary aldosteronism (Conn's syndrome)**, the classic manifestations of which include hypertension, hypokalemia, hypernatremia, and alkalosis. Patients with primary aldosteronism do not have evidence of glucocorticoid hormone excess, and plasma renin and angiotensin II levels are suppressed.

Renal artery stenosis, with the attendant decrease in perfusion pressure, can lead to hyperplasia and hyperfunction of the juxtaglomerular cells and cause elevated levels of renin and angiotensin II. This action results in **secondary aldosteronism,** which resembles the primary form, except for the elevated renin and angiotensin II levels.

Congenital Adrenal Hyperplasia is due to Enzyme Deficiency

Insufficient amounts of steroidogenic enzymes result in the deficiency of end products, the accumulation of intermediates, and the exaggerated production of steroids from alternative pathways. A common feature of most of these syndromes, which develop in utero, is deficient cortisol production with ACTH overproduction and adrenal hyperplasia—hence, the term **congenital adrenal hyperplasia.** The overproduction of adrenal androgens is another common feature. This hormone excess results in increased body growth, virilization, and ambiguous external genitalia—hence, the alternative designation **adrenogenital syndrome.**

Two types of **21-hydroxylase deficiency** (partial, or simple virilizing, and complete, or salt wasting) account for more than 90% of cases of congenital adrenal hyperplasia, and most of the rest are due to **11β-hydroxylase deficiency.** Only a few cases of other deficiencies (3β-hydroxysteroid dehydrogenase, 17α-hydroxylase, cholesterol desmolase, 18-hydroxylase, and 18-dehydrogenase) have been described. The **18-hydroxylase and -dehydrogenase deficiencies** affect only aldosterone biosynthesis and so do not cause adrenal hyperplasia. The **cholesterol desmolase deficiency** prevents any steroid biosynthesis and so is usually incompatible with extrauterine life.

REFERENCES

Beanto M: Gene regulation by steroid hormones. *Cell* 1989; **56:**335.

Evans R: The steroid and thyroid hormone receptor superfamily. *Science* 1988;**240:**889.

Granner DK: The role of glucocorticoid hormones as biological amplifiers. In: *Glucocorticoid Hormone Action.* Baxter JD, Rousseau GG (editors). Springer-Verlag, 1979.

Gustafsson et al: Biochemistry, molecular biology and physiology of the glucocorticoid receptor. *Endocrine Rev* 1987;**8:**185.

Morris DJ: The metabolism and mechanism of action of aldosterone. *Endocr Rev* 1981;**2:**234.

Yamamoto KR: Steroids receptor-regulated transcription of specific genes and gene networks. *Ann Rev Genet* 1985; **19:**209.

Hormones of the Adrenal Medulla

50

Daryl K. Granner, MD

INTRODUCTION

The **sympathoadrenal** system consists of the **parasympathetic** nervous system with its cholinergic pre- and postganglionic nerves, the **sympathetic** nervous system with cholinergic preganglionic and adrenergic postganglionic nerves, and the **adrenal medulla.** The latter is actually an extension of the sympathetic nervous system, since preganglionic fibers from the splanchnic nerve terminate in the adrenal medulla, where they innervate the chromaffin cells that produce the catecholamine hormones **dopamine, norepinephrine,** and **epinephrine.** The adrenal medulla is thus a specialized ganglion without axonal extensions. Its chromaffin cells synthesize, store, and release products that act on distant sites, so that it also functions as an endocrine organ–a perfect illustration of the enmeshing of the nervous and endocrine systems alluded to in Chapter 45.

BIOMEDICAL IMPORTANCE

The hormones of the sympathoadrenal system, while not necessary to life, are required for adaptation to acute and chronic stress. Epinephrine, norepinephrine, and dopamine are the major elements in the response to severe stress. This response involves an acute, integrated adjustment of many complex processes in the organs vital to the response (brain, muscles, cardiopulmonary system, and liver) at the expense of other organs that are less immediately involved (skin, gastrointestinal system, and lymphoid tissue). Catecholamines do not facilitate the stress response alone but are aided by the glucocorticoids, growth hormone, vasopressin, angiotensin II, and glucagon.

THE CATECHOLAMINE HORMONES ARE 3,4-DIHYDROXY DERIVATIVES OF PHENYLETHYLAMINE

These amines, dopamine, norepinephrine, and epinephrine, are synthesized in the chromaffin cells of the adrenal medulla, so named because they contain granules that develop a red-brown color when exposed to

Acronyms Used in This Chapter	
COMT	Catechol-O-methyltransferase
DBH	Dopamine β-hydroxylase
MAO	Monoamine oxidase
PNMT	Phenylethanolamine-N-methyltransferase
VMA	Vanillylmandelic acid

potassium dichromate. Collections of these cells are also found in the heart, liver, kidney, gonads, adrenergic neurons of the postganglionic sympathetic system, and central nervous system.

The major product of the adrenal medulla is epinephrine. This compound constitutes about 80% of the catecholamines in the medulla, and it is not made in extramedullary tissue. In contrast, most of the norepinephrine present in organs innervated by sympathetic nerves is made in situ ($\sim$ 80% of the total), and most of the rest is made in other nerve endings and reaches the target sites via the circulation. Epinephrine and norepinephrine may be produced and stored in different cells in the adrenal medulla and other chromaffin tissues.

The conversion of tyrosine to epinephrine requires 4 sequential steps: (1) ring hydroxylation; (2) decarboxylation; (3) side chain hydroxylation; and (4) N-methylation. The biosynthetic pathway and the enzymes involved are illustrated in Fig 50–1, and a schematic representation is illustrated in Fig 50–2.

Tyrosine Hydroxylase Is the Rate-Limiting Enzyme in Catecholamine Biosynthesis

Tyrosine is the immediate precursor of catecholamines, and tyrosine hydroxylase is the rate-limiting enzyme in catecholamine biosynthesis. Tyrosine hydroxylase is found in both soluble and particle-bound forms only in tissues that synthesize catecholamines; it functions as an oxidoreductase, with tetrahydropteridine as a cofactor, to convert L-tyrosine to L-dihydroxyphenylalanine (L-dopa). As the rate-limiting enzyme, tyrosine hydroxylase is regulated in a variety of ways. The most important mechanism involves feedback inhibition by the catecholamines,

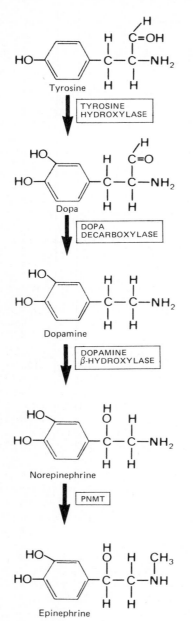

Figure 50–1. Biosynthesis of catecholamines. PNMT phenylethanolamine-N-methyltransferase. (Modified and reproduced, with permission, from Goldfien A: The adrenal medulla. In: *Basic & Clinical Endocrinology*, 2nd ed. Greenspan FS, Forsham PH [editors] Appleton & Lange, (1986.)

which compete with the enzyme for the pteridine cofactor by forming a Schiff base with the latter. Tyrosine hydroxylase is also competitively inhibited by a series of tyrosine derivatives, including α-methyltyrosine. This compound is occasionally used to treat catecholamine excess in pheochromocytoma, but other agents are more effective and have fewer side effects. A third group of compounds inhibit tyrosine hydrox-

ylase by chelating iron and thus removing available cofactor. An example is α,α ' -dipyridyl.

Catecholamines cannot cross the blood-brain barrier; hence, in the brain they must be synthesized locally. In certain central nervous system diseases, eg, Parkinson's disease, there is a local deficiency of dopamine synthesis. L-Dopa, the precursor of dopamine, readily crosses the blood-brain barrier and so is an important agent in the treatment of Parkinson's disease.

Dopa Decarboxylase Is Found in All Tissues

This soluble enzyme requires pyridoxal phosphate for the conversion of L-dopa to 3,4-dihydroxyphenylethylamine (dopamine). Compounds that resemble L-dopa, such as α-methyldopa, are competitive inhibitors of this reaction. Halogenated compounds form a Schiff base with L-dopa and also inhibit the decarboxylase reaction.

α-Methyldopa and other related compounds, such as 3-hydroxytyramine (from tyramine), α-methyltyrosine, and metaraminol, are effective in treating some kinds of hypertension.

Dopamine β-Hydroxylase (DBH) Catalyzes the Conversion of Dopamine to Norepinephrine

DBH is a mixed-function oxidase and uses ascorbate as an electron donor, copper at the active site, and fumarate as modulator. DBH is in the particulate fraction of the medullary cells, probably in the secretion granule; thus, the conversion of dopamine to norepinephrine occurs in this organelle. DBH is released from the adrenal medulla or nerve endings with norepinephrine, but (unlike norepinephrine) it cannot reenter nerve terminals via the reuptake mechanism.

Phenylethanolamine-N-Methyltransferase Catalyzes the Production of Epinephrine

The soluble enzyme phenylethanolamine-N-methyltransferase (PNMT) catalyzes the N-methylation of norepinephrine to form epinephrine in the epinephrine-forming cells of the adrenal medulla. Since PNMT is soluble, it is assumed that norepinephrine-to-epinephrine conversion occurs in the cytoplasm. The synthesis of PNMT is induced by glucocorticoid hormones that reach the medulla via the intraadrenal portal system. This system provides for a 100-fold steroid concentration gradient over systemic arterial blood, and this high intraadrenal concentration appears to be necessary for the induction of PNMT.

CATECHOLAMINES ARE STORED & RELEASED

Storage Is in Chromaffin Granules

The adrenal medulla contains the **chromaffin granules**—organelles capable of the biosynthesis, uptake,

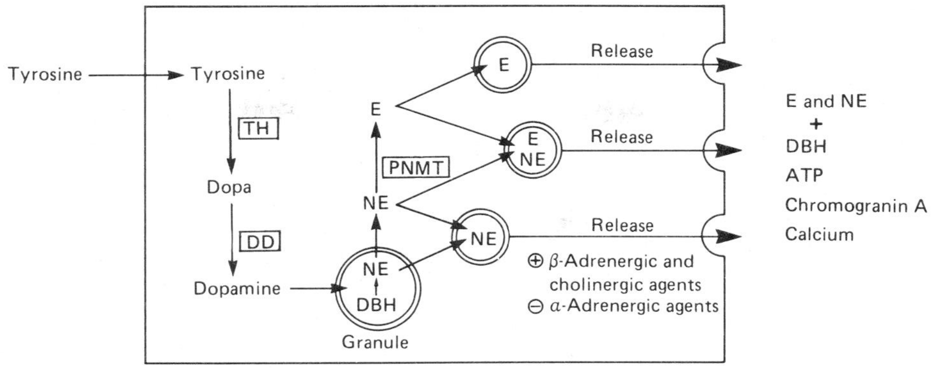

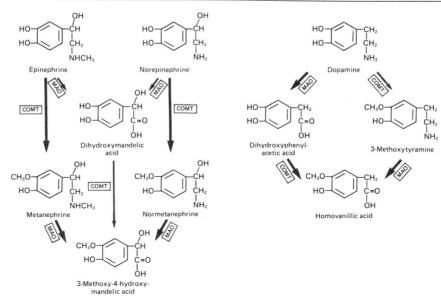

Figure 50–2. Schematic representation of catecholamine biosynthesis. TH, tyrosine hydroxylase; DD, dopa decarboxylase; PNMT, phenylethanolamine-N-methyltransferase; DBH, dopamine β-hydroxylase; ATP, adenosine triphosphate. The biosynthesis of catecholamines occurs within the cytoplasm and in various granules of the adrenal medullary cell. Some granules contain epinephrine (E), others have norepinephrine (NE), while still others have both hormones. Upon stimulation all the contents of the granules are released into the extracellular fluid (ECF).

storage, and secretion of catecholamines. These granules contain a number of substances in addition to the catecholamines, including ATP-Mg^{2+}, Ca^{2+}, DBH, and the protein chromagranin A. Catecholamines enter the granule via an ATP-dependent transport mechanism and bind this nucleotide in a 4:1 ratio (hormone:ATP). Norepinephrine is stored in these granules but can exit to be N-methylated; the epinephrine formed then enters a new population of granules.

Release Is Calcium Dependent

Neural stimulation of the adrenal medulla results in the fusion of the membranes of the storage granules with the plasma membrane, and this leads to the exocytotic release of norepinephrine and epinephrine. This is a calcium-dependent process and, like most exocytotic events, is stimulated by cholinergic and β-adrenergic agents and inhibited by α-adrenergic agents (Fig 50–2). Catecholamines and ATP are released in proportion to their intragranular ratio, as are the other contents including DBH, calcium, and chromagranin A.

Neuronal reuptake of catecholamines is an important mechanism for conserving these hormones and for quickly terminating hormonal or neurotransmitter activity. The adrenal medulla, unlike the sympathetic nerves, does not have a mechanism for the reuptake

Figure 50–3. Metabolism of catecholamines by catechol-O-methyltransferase (COMT) and monoamine oxidase (MAO). (Reproduced, with permission, from Goldfien A: The adrenal medulla. In: *Basic & Clinical Endocrinology,* 2nd ed. Greenspan FS, Forsham PH [editors]. Appleton & Lange, 1986.)

and storage of discharged catecholamines. The epinephrine discharged from the adrenal goes to the liver and skeletal muscle but then is rapidly metabolized. Very little adrenal norepinephrine reaches distal tissues. Catecholamines circulate in plasma in a loose association with albumin. They have an extremely short biologic half-life (10–30 seconds).

CATECHOLAMINES ARE RAPIDLY METABOLIZED

Very little epinephrine ($< 5\%$) is excreted in the urine. Catecholamines are rapidly metabolized by catechol-O-methyltransferase and monoamine oxidase to form the inactive O-methylated and deaminated metabolites (Fig 50–3). Most catecholamines are substrates for both of these enzymes, and these reactions can occur in any sequence.

Catechol-O-methyltransferase (COMT) is a cytosol enzyme found in many tissues. It catalyzes the addition of a methyl group, usually at the 3 position (meta) on the benzene ring, to a variety of catecholamines. The reaction requires a divalent cation, and S-adenosylmethionine is the methyl donor. The result of this reaction, depending on the substrate, is the production of homovanillic acid, normetanephrine, and metanephrine.

Monoamine oxidase (MAO) is an oxidoreductase that deaminates monoamines. It is located in many tissues, but it occurs in highest concentrations in the liver, stomach, kidney, and intestine. At least 2 isozymes of MAO have been described. MAO-A is found in neural tissue and deaminates serotonin, epinephrine, and norepinephrine, while MAO-B is found in extraneural tissues and is most active against 2-phenylethylamine and benzylamine. Dopamine and tyramine are metabolized by both forms. Much research effort is directed at correlating affective disorders with increases or decreases of the activity of these isozymes. MAO inhibitors have been used to treat hypertension and depression, but serious reactions with foods or drugs that contain sympathomimetic amines limit their usefulness.

O-Methoxylated derivatives are further modified by conjugation with glucuronic or sulfuric acid.

A bewildering number of metabolites of catecholamines are formed. Two classes of these have diagnostic significance, since they are found in readily measurable amounts in urine. **Metanephrines** represent the methoxy derivatives of epinephrine and norepinephrine, while the O-methylated deaminated product of epinephrine and norepinephrine is **3-methoxy-4-hydroxymandelic acid** (also called **vanillylmandelic acid [VMA]** (Fig 50–3). The concentration of metanephrines or VMA in urine is elevated in more than 95% of patients with pheochromocytoma. These tests have excellent diagnostic precision, particularly when coupled with a measurement of plasma or urine catecholamines.

THE SYNTHESIS OF CATECHOLAMINES IS REGULATED BY NERVE IMPULSES

Stimulation of the splanchnic nerve, which supplies the preganglionic fibers to the adrenal medulla, results in the exocytotic release of catecholamines, the granule carrier protein, and DBH. Such stimulation is controlled by the hypothalamus and brain stem, but the exact feedback loop has not been described.

Nerve stimulation also results in increased synthesis of catecholamines. Norepinephrine synthesis increases after acute stress, but the amount of tyrosine hydroxylase is unchanged even though tyrosine hydroxylase activity increases. Tyrosine hydroxylase is a substrate for cAMP-dependent protein kinase, and so this activation may involve phosphorylation. Prolonged stress accompanied by chronic sympathetic nerve activity results in an induction (increased amount) of tyrosine hydroxylase. A similar induction of DBH has also been reported. The induction of these enzymes of the catecholamine biosynthetic pathway is a means of adapting to physiologic stress and depends upon neural (tyrosine hydroxylase and DBH induction) and endocrine (PNMT induction) factors.

CATECHOLAMINES CAN BE CLASSIFIED BY THEIR MECHANISM OF ACTION

The mechanism of action of the catecholamines has attracted the attention of investigators for nearly a century. Indeed, many of the general concepts of receptor biology and hormone action can be traced to these early studies.

The catecholamines act through 2 major classes of receptors. These are designated α-adrenergic and β-adrenergic, and each consists of 2 subclasses, ie, alpha$_1$, alpha$_2$, beta$_1$, and beta$_2$. This classification is based on the relative order of binding of various agonists and antagonists. Epinephrine binds to and activates both alpha and beta receptors, so that its action in a tissue which has both depends on the relative affinity of these receptors for the hormone. Norepinephrine at physiologic concentrations primarily binds to alpha receptors.

The Structure of the β-Adrenergic Receptor is Known

Molecular cloning of the gene and cDNA for the mammalian β-adrenergic receptor revealed some surprising features. First, the gene has no introns and thus joins the histone and interferon genes as mammalian genes that lack these structures. Second, the β-adrenergic receptor is closely homologous to rhodopsin (in 3 peptide regions, at least), the protein that initiates the process which converts light into the visual response. Further similarities are discussed below.

Table 50–1. Actions mediated through various adrenergic receptors.

Alpha$_1$	Alpha$_2$	Beta$_1$	Beta$_2$

Increased glycogenolysis
Smooth muscle contraction
  Blood vessels
  Genitourinary tract
	Smooth muscle relaxation
	  Gastrointestinal tract
	Smooth muscle contraction
	  Some vascular beds
	Inhibition of:
	  Lipolysis
	  Renin release
	  Platelet aggregation
	  Insulin secretion
		Stimulation of lipolysis
		Myocardial contraction
		  Increased rate
		  Increased force
			Increased hepatic gluconeogenesis
			Increased hepatic glycogenolysis
			Increased muscle glycogenolysis
			Increased release of:
			  Insulin
			  Glucagon
			  Renin
			Smooth muscle relaxation
			  Bronchi
			  Blood vessels
			  Genitourinary tract
			  Gastrointestinal tract

Three of these adrenergic receptor subgroups are coupled to the adenylate cyclase system

Hormones that bind to the β_1 and β_2 receptors activate adenylate cyclase, whereas hormones that bind to α_2 receptors inhibit this enzyme (see Fig 45–4 and Table 45–4). Catecholamine binding induces the coupling of the receptor to a G protein that then binds GTP. This either stimulates (G_s) or inhibits (G_i) adenylate cyclase, thus stimulating or inhibiting the synthesis of cAMP. The response terminates when the α subunit-associated GTPase hydrolyzes the GTP (see Fig 45–4). α_1 Receptors are coupled to processes that alter intracellular calcium concentrations or modify phosphatidylinositide metabolism (or both) (see Chapter 45). A separate G protein complex is involved in this response.

THERE IS FUNCTIONAL SIMILARITY BETWEEN THE CATECHOLAMINE RECEPTOR AND THE VISUAL RESPONSE SYSTEM

The stimulation of rhodopsin by light couples it to transducin, a G protein complex whose α subunit also binds GTP. The activated G protein in turn activates a phosphodiesterase that hydrolyzes cGMP. This results in the closure of ion channels on the rod cell membrane and produces the visual response. The response terminates when the α subunit-associated GTPase hydrolyzes the bound GTP. A partial list of the biochemical and physiologic effects mediated by each of these receptors is provided in Table 50–1. Activation of phosphoproteins by cAMP-dependent protein kinase (see Fig 45–5) accounts for many of the biochemical effects of epinephrine.

PHEOCHROMOCYTOMAS ARE TUMORS OF THE ADRENAL MEDULLA

These tumors are usually not detected unless they produce and secrete enough epinephrine or norepinephrine to cause a severe hypertension syndrome. The ratio of norepinephrine to epinephrine is often increased in pheochromocytoma. This may account for differences in clinical presentation, since norepinephrine is thought to be primarily responsible for hypertension and epinephrine for hypermetabolism.

REFERENCES

Cryer PE: Diseases of the adrenal medulla and sympathetic nervous system. Pages 511-550 in: *Endocrinology and Metabolism.* Felig P et al (editors). McGraw-Hill, 1981.

Dixon RAF et al: Cloning of the gene and cDNA for mammalian β-adrenergic receptor and homology with rhodopsin. *Nature (London)* 1986;**321:**75.

Exton JH: Mechanisms involved in α-adrenergic phenomena: Role of calcium ion in actions of catecholamines in liver and other tissues. *Am J Physiol* 1980;**238:**E3.

Kaupp UB: Mechanism of photoreception in vertebrate vision. *Trends Biochem Sci* 1986;**11:**43.

Sibley DR, Lefkowitz RJ: Molecular mechanisms of receptor desensitization using the β-adrenergic receptor-coupled adenylate cyclase system as a model. *Nature (London)* 1985;**317:**124.

Stiles GL, Caron MG, Lefkowitz RK: The β-adrenergic receptor: Biochemical mechanisms of physiological regulation. *Physiol Rev* 1984;**64:**661.

51

Hormones of the Gonads

Daryl K. Granner, MD

INTRODUCTION

The gonads are bifunctional organs that produce germ cells and the sex hormones. These 2 functions are closely approximated, for high local concentrations of the sex hormones are required for germ cell development. The ovaries produce ova and the steroid hormones estrogen and progesterone; the testes produce spermatozoa and testosterone. As in the adrenal, a number of steroids are produced, but only a few are active as hormones. The production of these hormones is tightly regulated through a feedback loop that involves the pituitary and the hypothalamus. The gonadal hormones act by a nuclear mechanism similar to that employed by the adrenal steroid hormones.

BIOMEDICAL IMPORTANCE

Proper functioning of the gonads is crucial for reproduction and, hence, survival of the species. Conversely, an understanding of the endocrine physiology and biochemistry of the reproductive process is the basis of many approaches to contraception. The gonadal hormones have other important actions; for example, they are anabolic and thus required for maintenance of metabolism in skin, bone, and muscle.

THE TESTES PRODUCE TESTOSTERONE (THE MALE SEX HORMONE) AND SPERMATOZOA (THE MALE GERM CELLS)

These functions are carried out by 3 specialized cell types; (1) the **spermatogonia** and more differentiated germ cells, which are located in the seminiferous tubules; (2) the **Leydig cells** (also called interstitial cells), which are scattered in the connective tissue between the coiled seminiferous tubules and which produce testosterone in response to LH; and (3) the **Sertoli cells,** which form the basement membrane of the seminiferous tubules and provide the environment necessary for germ cell differentiation and maturation. Spermatogenesis is stimulated by FSH and LH from the pituitary. It requires an environment conducive to germ cell differentiation and a concentration of testos-

Acronyms Used in This Chapter	
ACTH	Adrenocorticotropic hormone
ABP	Androgen-binding protein
CBG	Corticosteroid-binding globulin
DHEA	Dehydroepiandrosterone
DHT	Dihydrotestosterone
E$_2$	Estradiol
FSH	Follicle-stimulating hormone
GnRH	Gonadotropin-releasing hormone
hCG	Human chorionic gonadotropin
hCS	Human chorionic somatomammotropin
LH	Luteinizing hormone
MIF	Müllerian inhibiting factor
3β-OHSD	3β-Hydroxysteroid dehydrogenase
17β-OHSD	17β-Hydroxysteroid dehydrogenase
PL	Placental lactogen
SHBG	Sex hormone-binding globulin
TBG	Thyroid-binding globulin
TEBG	Testosterone-estrogen-binding globulin

terone in excess of that found in the systemic circulation—a requirement that can be met because the Leydig cells and seminiferous tubules are in close approximation.

Although a Number of Enzymes Are Involved in the Synthesis of Gonadal Steroids, Cholesterol Side-Chain Cleavage Enzyme and 3β-Hydroxysteroid Dehydrogenase Catalyze the Critical Steps

Testicular androgens are synthesized in the interstitial tissue by the Leydig cells. The immediate precursor of the gonadal steroids, as with the adrenal steroids, is cholesterol. The rate-limiting step, as in the adrenal, is cholesterol side chain cleavage. The conversion of cholesterol to pregnenolone is identical in adrenal, ovary, and testis. In the latter 2 tissues, however, the reaction is promoted by LH rather than ACTH.

The conversion of pregnenolone to testosterone re-

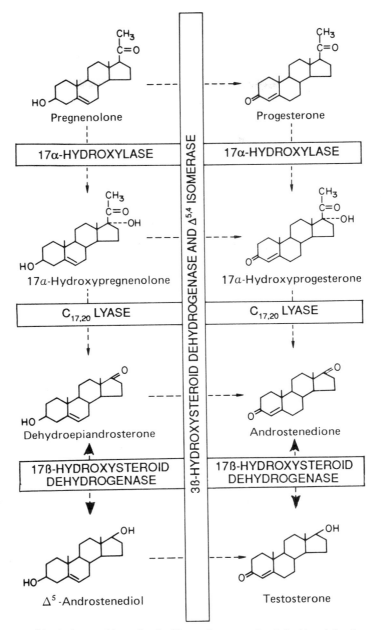

Figure 51–1. Pathways of testosterone biosynthesis. The pathway on the left side of the figure is called the $\triangle^5$ or dehydroepiandrosterone pathway; the pathway on the right side is called the $\triangle^4$ or progesterone pathway.

quires the action of 5 enzymes. These enzymes are: (1) 3β-hydroxysteroid dehydrogenase (3β-OHSD); (2) $\triangle^{5,4}$ isomerase; (3) 17α-hydroxylase; (4) C_{17-20} lyase; and (5) 17β-hydroxysteroid dehydrogenase (17β-OHSD). This sequence, referred to as the progesterone (or $\triangle^4$) pathway, is shown on the right side of Fig 51–1. Pregnenolone can also be converted to testosterone by the dehydroepiandrosterone (or $\triangle^5$) pathway, which is illustrated on the left side of Fig 51–1. The $\triangle^4$ route appears to be preferred in human testes; however, since these are seldom available for study, most

information about these pathways comes from studies in other animals. There may be significant species differences.

The 5 enzymes are localized in the microsomal fraction in rat testes, and there is a close functional association between the activities of 3β-OHSD and $\triangle^{5,4}$ isomerase and between those of 17α-hydroxylase and C_{17-20} lyase. These enzyme pairs are shown in the general reaction sequence in Fig 51–1 and in the schematic representation of androgen biosynthesis in the testicular microsomal membrane in Fig 51–2. The lat-

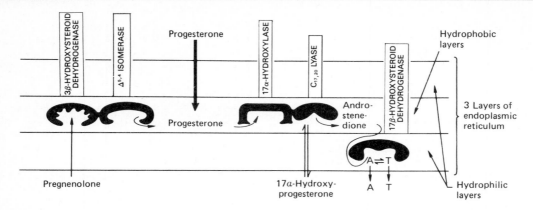

Figure 51–2. Schematic representation of androgen biosynthesis in the testicular microsomal membrane. The membrane is shown as horizontal, which may be the case in the cell; in microsomal preparations, however, it forms vesicles. A, androstenedione; T, testosterone. (Reproduced, with permission, from DeGroot LJ: *Endocrinology.* Vol 3. Grune & Stratton, 1979.)

ter shows how the various substrates for testosterone biosynthesis can enter the microsomal compartment and how they might proceed via the Δ^4 pathway from one reaction to the next. Since there are 4 potential substrates for what appears to be a single 3β-OHSD, multiple alternative pathways exist; thus, the route taken probably depends on the substrate concentration in the vicinity of the various enzymes. Partitioning in the microsomal membrane may provide these gradients.

Dihydrotestosterone (DHT) Is Formed From Testosterone by the Reduction of the A Ring Through the Action of The Enzyme 5α-reductase: Human testes secrete about 50–100 μg of DHT per day, but most DHT is derived from peripheral conversion (see below). The testes also make small but significant amounts of 17β-estradiol (E_2), the female sex hormone, but most of the E_2 produced by the male is derived from peripheral aromatization of testosterone and androstenedione. The Leydig cells, the Sertoli cells, and the seminiferous tubules are thought to be involved in E_2 production. The role of E_2 in the male has not been determined, but it may contribute to FSH regulation. Abnormally high plasma levels of E_2 and changes in the free E_2:testosterone ratio have been associated with pubertal or postpubertal gynecomastia (male breast enlargement), particularly in older individuals and in patients with chronic liver disease or hyperthyroidism.

There Are Remarkable Age-Related Changes in Testicular Hormone Production: Testosterone is the dominant hormone in the fetal and neonatal rat, but the testes make only androsterone soon after birth. The ability to produce testosterone is restored at puberty and continues throughout life. Similar observations have been made in other species, and these age-related changes may also occur in humans.

Testosterone Binds to a Specific Plasma Protein

Most mammals, humans included, have a plasma β-globulin that binds testosterone with specificity, relatively high affinity, and limited capacity (Table 51–1). This protein, usually called **sex hormone-binding globulin (SHBG), or testosterone-estrogen-binding globulin (TEBG),** is produced in the liver. Its production is increased by estrogens (women have twice the serum concentration of SHBG as men), certain types of liver disease, and hyperthyroidism; it is decreased by androgens, advancing age, and hypothyroidism. Many of these conditions also affect the production of CBG (see Chapter 44) and TBG (see Chapter 47). Since SHBG and albumin bind 97–99% of circulating testosterone, only a small fraction of the hormone in circulation is in the free (biologically active) form. The primary function of SHBG may be to restrict the free concentration of testosterone in the serum. Testosterone binds to SHBG with higher affinity than does estradiol (Table 51–2). Therefore, a change in the level of SHBG causes a greater change in the free testosterone level than in the free estradiol level. An increase of SHBG may contribute to the increased free E_2:testosterone ratio noted in aging, cirrhosis, and hyperthy-

Table 51–1. Hormone binding to sex hormone-binding globulin (SHBG).

Steroids Bound	Steroids Not Bound
Testosterone	Conjugated androgens
17β-Estradiol	17α-Testosterone
Dihydrotestosterone	Dehydroisoandrosterone
Other 17β-hydroxysteroids	Cortisol
Estrone	Progesterone

Table 51–2. Approximate affinities of steroids for serum-binding proteins.*

	SHBG†	CBG†
Estradiol	5	>10
Estrone	>10	>100
Androstenedione	. . .	. . .
Testosterone	2	>100
Dihydrotestosterone	1	>100
Progesterone	>100	2
Cortisol	>100	3

*Adapted from Siiteri PK, Febres F: Ovarian hormone synthesis, circulation and mechanisms of action. Page 1401 in: *Endocrinology.* Vol 3. DeGroot LJ (editor). Grune & Stratton, 1979.
†Affinity expressed as K_d of molar quantity × 10^9.

roidism and hence contribute to the attendant signs and symptoms of "estrogenization" alluded to above.

A number of steroids are present in testicular venous blood, but testosterone is the major steroid secreted by the adult testes.

The secretion rate of testosterone is about 5 mg/d in normal adult men. Like other steroid hormones, testosterone seems to be released as it is produced.

Many of the Metabolites of Testosterone Are Inactive; Others, Such as Dihydrotestosterone & Estradiol, Have Increased or Different Activities

Metabolic Pathways: Testosterone is metabolized by 2 pathways. One involves oxidation at the 17-position, and the other involves reduction of the A ring double bond and the 3-ketone. Metabolism via the first pathway occurs in many tissues, including liver, and produces 17-ketosteroids that are generally inactive or less active than the parent compound. Metabolism via the second pathway, which is less efficient, occurs primarily in target tissues and produces the potent metabolite DHT.

Metabolites of Testosterone: The most significant metabolic product of testosterone is DHT, since in many tissues, including seminal vesicles, prostate, external genitalia, and some areas of the skin, this is the active form of the hormone. The plasma content of

DHT in the adult male is about one-tenth that of testosterone, and approximately 400 μg of DHT is produced daily, as compared to about 5 mg of testosterone. The reaction is catalyzed by 5α-reductase, an NADPH-dependent enzyme (see below).

Testosterone can thus be considered a prohormone, since it is converted into a much more potent compound (dihydrotestosterone) and since most of this conversion occurs outside the testes. A small percentage of testosterone is also converted into estradiol by aromatization, a reaction that is especially important in the brain, where these hormones help determine the sexual behavior of the animal. Androstanediol, another potent androgen, is also produced from testosterone.

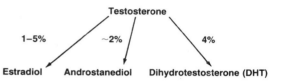

The major 17-ketosteroid metabolites, **androsterone** and **etiocholanolone,** are conjugated with glucuronide and sulfate in the liver to make water-soluble, excretable compounds.

REGULATION OF TESTICULAR FUNCTION IS MULTIHORMONAL

Testicular Steroidogenesis Is Stimulated by LH

LH stimulates steroidogenesis and testosterone production by binding to receptors on the plasma membrane of the Leydig cells (an analogous LH receptor is found in the ovary on cells of the corpus luteum) and activating adenylate cyclase, thus increasing intracellular cAMP. This action enhances the rate of cholesterol side-chain cleavage. The similarity between this action of LH and that of ACTH on the adrenal is apparent. Testosterone provides for feedback control at the hypothalamus through inhibition of GnRH release, GnRH production, or both (Fig 51–3).

Spermatogenesis Is Regulated by FSH and Testosterone

FSH binds to the Sertoli cells and promotes the synthesis of androgen-binding protein (ABP). ABP is

Testosterone   Dihydrotestosterone (DHT)

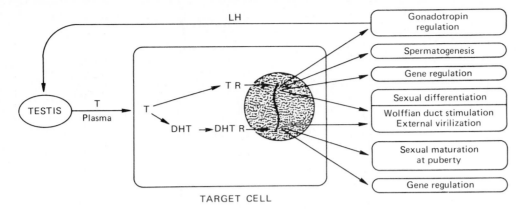

Figure 51–3. Mechanism of androgen action. LH, luteinizing hormone; T, testosterone; DHT, dihydrotestosterone; R, androgen receptor. (Modified and reproduced, with permission, from Wilson JD et al: The endocrine control of male phenotypic development. *Aust J Biol Sci* 1983;**36**:101.)

glycoprotein that binds testosterone. It is distinct from the intracellular androgen receptor but is homologous with SHBG. ABP is secreted into the lumen of the seminiferous tubule, and in this process testosterone produced by the Leydig cells is transported in very high concentration to the site of spermatogenesis. This appears to be a critical step, since normal systemic levels of testosterone, such as might be achieved by replacement therapy, do not support spermatogenesis.

Androgens Affect Several Complex Physiologic Prosesses

The androgens, principally testosterone and DHT, are involved in (1) sexual differentiation, (2) spermatogenesis, (3) development of secondary sexual organs and ornamental structures, (4) anabolic metabolism and gene regulation, and (5) male-pattern behavior (Fig 51–3). The numerous target tissues involved in these complex processes are defined according to whether they are affected by testosterone or DHT. The classic target cells for DHT (and those which coincidentally have the highest 5α-reductase activity) are the prostate, seminal vesicles, external genitalia, and genital skin. Targets for testosterone include the embryonic wolffian structures, spermatogonia, muscles, bone, kidney, and brain. The specific androgen involved in regulating the many other processes mentioned above has not been determined.

Androgens Act by a Nuclear Mechanism Similar to That Employed by the Adrenal Steroids

The current concept of androgen action is shown in Fig 51–3. Free testosterone enters cells through the plasma membrane by either passive or facilitated diffusion. Target cells retain testosterone, presumably because the hormone associates with a specific intracellular receptor. Although there is considerable tissue-to-tissue variability, most of the retained hormone is found in the cell nucleus. The cytoplasm of many (but not all) target cells contains the enzyme 5α-reductase, which converts testosterone to DHT. Whether there are

distinct receptors for testosterone and DHT has been a point of controversy, but the consensus is that while there is but a single class of receptors, the affinity of the receptor for DHT exceeds that for testosterone. Single gene mutations in mice result in loss of binding of both testosterone and DHT to the receptor in various tissues, suggesting that a single protein is involved. The affinity difference, coupled with the ability of a target tissue to form DHT from testosterone, may determine whether the testosterone-receptor complex or the DHT-receptor complex is active.

Nuclear localization of the testosterone/DHT-receptor complex is a prerequisite for androgen action. Binding of the receptor-steroid complex to chromatin may involve a prior activation step, and specificity is afforded by the androgen response element (Table 49–3).

In keeping with other steroid (and some peptide) hormones, the testosterone/DHT receptor activates specific genes. The protein products of these genes mediate many (if not all) of the effects of the hormone. Testosterone stimulates protein synthesis in male accessory organs, an effect that is usually associated with increased accumulation of total cellular RNA, including mRNA, tRNA, and rRNA. A more specific example involves the effect of testosterone on the synthesis of ABP. The hormone increases the rate of transcription of the ABP gene, which results in an increased amount of the mRNA that codes for this protein. Another well-studied example is α_{2u} globulin, the major protein excreted in the urine of male rats. The rate of synthesis of α_{2u} globulin is directly related to the amount of the cognate mRNA, which in turn is related to the rate of transcription of the α_{2u} globulin gene. All are stimulated by androgens.

The kidney is a major target tissue for androgens. These hormones cause a general enlargement of the kidney and induce the synthesis of a number of enzymes in various species.

Androgens also stimulate the replication of cells in some target tissues, an effect that is poorly understood. Testosterone or DHT, in combination with E_2, appears to be implicated in the extensive and uncontrolled divi-

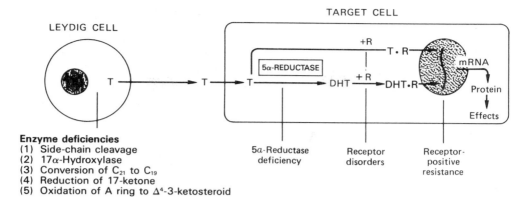

Figure 51–4. Steps involved in androgen resistance. Four stages at which mutations have been identified are shown. T, testosterone; DHT, dihydrotestosterone; R, androgen receptor. (Modified and reproduced, with permission, from Wilson JD et al: The endocrine control of male phenotypic development. *Aust J Biol Sci* 1983;**36**:101.

sion of prostate cells that results in **benign prostatic hypertrophy,** a condition that afflicts as many as 75% of men over the age of 60 years.

THE PATHOPHYSIOLOGY OF THE MALE REPRODUCTIVE SYSTEM RELATES TO HORMONAL DEFECTS

The lack of testosterone synthesis is called **hypogonadism.** If this occurs before puberty, secondary sex characteristics fail to develop, and if it occurs in adults, many of these features regress. **Primary hypogonadism** is due to processes that affect the testes directly and cause testicular failure, whereas **secondary hypogonadism** is due to defective secretion of the gonadotropins. Isolated genetic deficiencies help to establish the importance of specific steps in the biosynthesis and action of androgens. Fig 51–4 represents the pathway involved in androgen action from testosterone biosynthesis through postreceptor actions of testosterone and DHT. At least 5 distinct genetic defects in testosterone biosynthesis have been described.

In addition, a 5α-reductase deficiency is known; there are a number of instances in which either no testosterone/DHT receptor is detected or the receptor is abnormal in some manner; and there are a number of cases in which all measurable entities, including the receptor, are normal, but the patients (always genetic males) have variable degrees of feminization. The extent of the abnormality of differentiation is related to the severity of the deficit. Persons who completely lack a biosynthetic enzyme appear to be phenotypic females but have an XY (male) genotype, while the mildest cases may have only an abnormally located penile urethra. Genetic males who completely lack functioning receptors have testes and produce testosterone but have complete feminization of the external genitalia (the so-called **testicular feminization syndrome**). It is interesting that no comparable deficien-

cies in estrogen synthesis or action have been identified.

THE OVARIES PRODUCE THE FEMALE SEX HORMONES (ESTROGENS AND PROGESTINS) AND THE FEMALE GERM CELLS (OVA)

Biosynthesis & Metabolism of Ovarian Hormones Is Similar to those of Male Hormones

The estrogens are a family of hormones synthesized in a variety of tissues. 17β-Estradiol is the primary estrogen of ovarian origin. In some species, estrone, synthesized in numerous tissues, is more abundant. In pregnancy, relatively more estriol is produced and this comes from the placenta. The general pathway and the subcellular localization of the enzymes involved in the early steps of estradiol synthesis are the same as those involved in androgen biosynthesis. Features unique to the ovary are illustrated in Fig 51–5.

Estrogens are formed by the aromatization of androgens in a complex process that involves 3 hydroxylation steps, each of which requires O_2 and NADPH. The aromatase enzyme complex is thought to include a P-450 mixed-function oxidase. Estradiol is formed if the substrate of this enzyme complex is testosterone, whereas estrone results from the aromatization of androstenedione.

The cellular source of the various ovarian steroids has been difficult to unravel, but it now appears that 2 cell types are involved. These cells are the major sources of androstenedione (the principal androgen produced by the ovary), and of 17α-hydroxyprogesterone. Granulosa cells produce estradiol from the latter. Progesterone is produced and secreted by the corpus luteum, which also makes some estradiol.

Significant amounts of estrogens are produced by

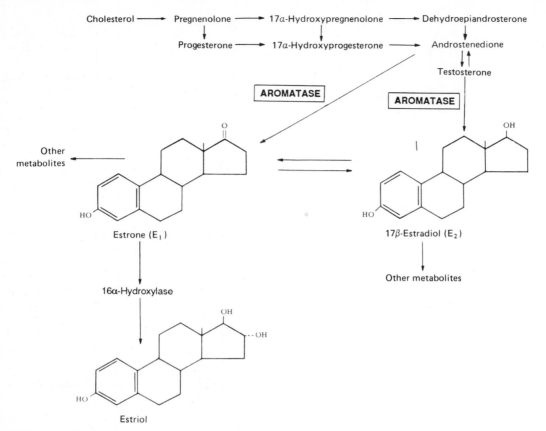

Figure 51–5. Biosynthesis of estrogens. (Slightly modified and reproduced, with permission, from Ganong WF: *Review of Medical Physiology,* 13th ed. Appleton & Lange, 1987.)

the peripheral aromatization of androgens. In human males, the peripheral aromatization of testosterone to estradiol (E_2) accounts for 80% of the production rate of the latter. In females, adrenal androgens are important substrates, since as much as 50% of the E_2 produced during pregnancy comes from the aromatization of androgens. Finally, the conversion of androstenedione to estrone is the major source of estrogens in postmenopausal women. Aromatase activity is present in adipose cells and also in liver, skin, and other tissues. Increased activity of this enzyme may contribute to the "estrogenization" that characterizes such diseases as cirrhosis of the liver, hyperthyroidism, aging, and obesity.

Estrogens & Progestins Are Bound in Varying Degrees to Plasma Transport Proteins

Estrogens are bound to SHBG and progestins to CBG. SHBG binds estradiol about 5 times less avidly than it binds testosterone or DHT, while progesterone and cortisol have little affinity for this protein (Table 51–2). In contrast, progesterone and cortisol bind with nearly equal affinity to CBG, which in turn has little avidity for estradiol and even less for testosterone, DHT, or estrone.

These plasma transport proteins play no apparent role in the mechanism of action of these hormones at the cellular level, and (as with other steroids) probably

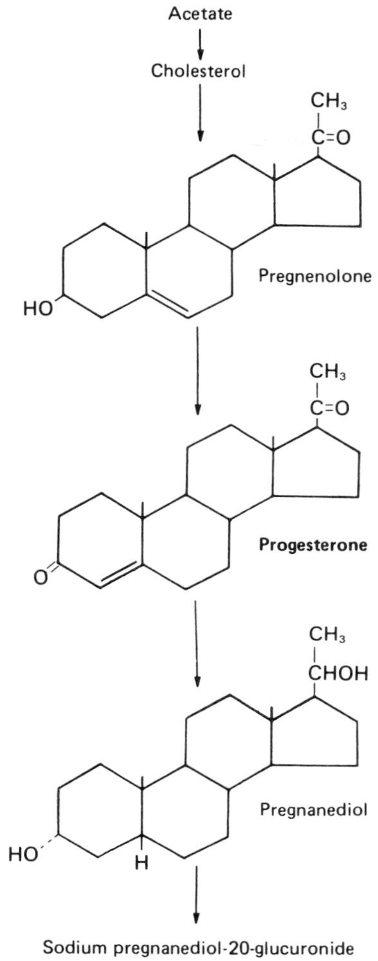

Figure 51–6. Biosynthesis of progesterone and major pathway for its metabolism. Other metabolites are also found. (Slightly modified and reproduced, with permission, from Ganong WF: *Review of Medical Physiology,* 13th ed. Appleton & Lange, 1987.)

only the free (unbound) hormone has biologic activity. The binding proteins do provide a circulating reservoir of hormone, and because of the relatively large binding capacity, they probably buffer against sudden changes in the plasma level. The metabolic clearance rates of these steroids are inversely related to the affinity of their binding to SHBG; hence, estrone is cleared more rapidly than estradiol, which in turn is cleared more rapidly than testosterone or DHT. In this regard, the conjugated derivatives of these hormones (see below) are not bound by either SHBG or CBG. The factors that regulate the production of SHBG are discussed above.

The rate of secretion of ovarian steroids varies considerably during the menstrual (or estrous) cycle and is directly related to rate of production in the ovary. There is no storage of these compounds; they are secreted when they are produced.

Estrogens & Progestins Are Actively Metabolized by the Liver

Estrogens: The liver converts estradiol and estrone to estriol by the pathways shown in Fig 51–5. Estradiol, estrone, and estriol are substrates for hepatic enzymes that add glucuronide or sulfate moieties. Activity of these conjugating enzymes varies among species. Rodents have such active metabolizing enzyme systems that estrogens are almost completely metabolized by the liver and thus are essentially without activity when given orally. These enzyme systems are less active in primates, so that oral estrogens are more effective. The conjugated steroids are water soluble and do not bind to transport proteins; thus, they are excreted readily in the bile, feces, and urine.

Progestins: Because the liver actively metabolizes progesterone to several compounds, **progesterone is ineffective when given orally.** Sodium pregnanediol-20-glucuronide is the major progestin metabolite found in human urine (Fig 51–6). Certain synthetic steroids, eg, derivatives of 17α-hydroxyprogesterone and 17α-alkyl-substituted 19-nortestosterone compounds, have progestational activity and avoid hepatic metabolism. Thus, they are widely used in oral contraceptives.

THE MATURATION & MAINTENANCE OF THE FEMALE REPRODUCTIVE SYSTEM IS THE MAJOR FUNCTION OF THE OVARIAN HORMONES

These hormones prepare the structural determinants of the female reproductive system (see below) for reproduction by (1) maturing the primordial germ cells; (2) developing the tissues that will allow for implantation of the blastocyst; (3) providing the "hormonal timing" for ovulation; (4) establishing the milieu required for the maintenance of pregnancy; and (5) providing the hormonal influences for parturition and lactation.

Estrogens stimulate the development of tissues involved in reproduction. In general, these hormones stimulate the size and number of cells by increasing the rate of synthesis of protein, rRNA, tRNA, mRNA, and DNA. Under estrogen stimulation, the vaginal epithelium proliferates and differentiates; the uterine endometrium proliferates and the glands hypertrophy and elongate; the myometrium develops an intrinsic, rhythmic motility; and breast ducts proliferate. Estradiol also has anabolic effects on bone and cartilage, and so it is growth promoting. By affecting peripheral blood vessels, estrogens typically cause vasodilation and heat dissipation.

Progestins reduce the proliferative activity of the estrogens on the vaginal epithelium and convert the uterine epithelium from proliferative to secretory (increased size and function of secretory glands and increased glycogen content), thus preparing the uterine epithelium for implantation of the fertilized ovum. Progestins enhance development of the acinar portions of breast glands after estrogens have stimulated ductal development. Progestins decrease peripheral blood flow, thereby decreasing heat loss, so that body temperature tends to increase during the luteal phase of the menstrual cycle, when these steroids are produced. This temperature increase, usually 0.5 °C, is used as an indicator of ovulation.

Progestins generally require the previous or concurrent presence of estrogens, perhaps because estrogens stimulate production of the progesterone receptor. The two classes of hormones often act synergistically, although they can be antagonists.

The number of oogonia in the human female ovary reaches a maximum of 6–7 million at about the fifth month of gestation. This decreases to about 2 million by birth and is further diminished to 100,000–200,000 by the onset of menarche. Some 400–500 of these develop into mature oocytes; the rest are gradually lost through a process that is not understood, although ovarian androgens have been implicated. Follicular maturation begins in infancy, and the ovaries gradually enlarge in prepubertal years owing to increased volume of the follicles because of the growth of granulosa

cells, to the accumulation of tissue from atretic follicles, and to the increased mass of medullary stromal tissue with the interstitial and theca cells that will produce the steroids.

The concentration of sex hormones is low in childhood, although exogenous gonadotropins increase production; hence, the immature ovary has the capacity to synthesize estrogen. It is thought that these low levels of sex steroids inhibit gonadotropin production in prepubertal girls and that at puberty the hypothalamic-pituitary system becomes less sensitive to suppression. At puberty, the pulsatile release of GnRH begins stimulating LH and this causes a dramatic increase of ovarian hormone production. FSH, the main stimulus for estrogen secretion, stimulates a follicle to ripen, and ovulation ensues.

The Menstrual Cycle Depends on a Complex Interaction Among Three Endocrine Glands

Hormones determine the frequency of ovulation and receptivity to mating. Monestrous species ovulate and mate once a year, whereas polyestrous species repeat

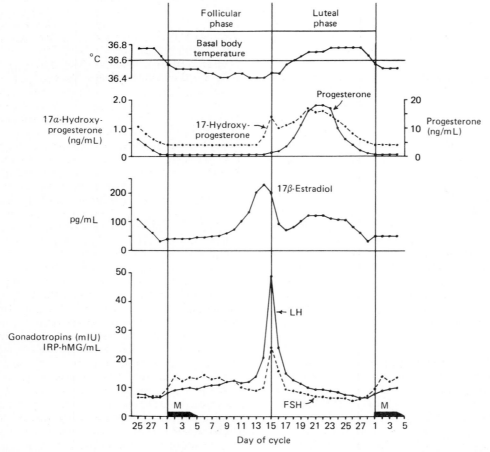

Figure 51–7. Hormonal and physiologic changes during a typical human menstrual cycle. M, menstruation; IRP-hMG, international reference standard for gonadotropins. (Reproduced, with permission, from Midgley AR in: *Human Reproduction*. Hafez ESE, Evans TN [editors]. Harper & Row, 1973.)

this cycle several times a year. Primates have menstrual cycles, with shedding of the endometrium at the end of each cycle, and mating behavior is not tightly coupled to ovulation. The human menstrual cycle results from a complex interaction between the hypothalamus, pituitary, and ovary. **The cycle normally varies between 25 and 35 days in length (average, 28 days).** It can be divided into a **follicular phase,** a **luteal phase,** and **menstruation** (Fig 51–7).

Follicular Phase: For reasons that are not clear, a particular follicle begins to enlarge under the general influence of FSH. E_2 levels are low during the first week of the follicular phase, but they begin to rise progressively as the follicle enlarges. E_2 reaches its maximal level 24 hours before the LH (FSH) peak and sensitizes the pituitary to GnRH. LH is released either in response to this high level of E_2 in a "positive feedback" manner or in response to a sudden decline of E_2 from this high level. Continual administration of high doses of estrogen (as in oral contraceptives) suppresses LH and FSH release and inhibits the action of GnRH on the pituitary. Progesterone levels are very low during the follicular phase. The LH peak heralds the end of the follicular phase and precedes ovulation by 16–18 hours.

Luteal Phase: After ovulation, the granulosa cells of the ruptured follicle luteinize and form the corpus luteum, a structure that soon begins to produce progesterone and some estradiol. Estradiol peaks about midway through the luteal phase and then declines to a very low level. **The major hormone of the luteal portion of the cycle is progesterone,** which (as noted above) is required for preparation and maintenance of the secretory endometrium that provides early nourishment for the implanted blastocyst. **LH is required for the early maintenance of the corpus luteum,** and the pituitary supplies it for about 10 days. If implantation occurs (day 22–24 of the average cycle), this LH function is assumed by chorionic gonadotropin (hCG), a placental hormone that is very similar to LH and is made by the cytotrophoblastic cells of the implanted early embryo. hCG stimulates progesterone synthesis by the corpus luteum until the placenta beings making large amounts of this steroid. In the absence of implantation (and hCG), the corpus luteum regresses and menstruation ensues; after the endometrium is shed, a new cycle commences. The luteal phase is always 14 ± 2 days in length. Variations in cycle length are almost always due to an altered follicular phase.

Pregnancy Activates Placental Hormones

The implanted blastocyst forms the trophoblast, which is subsequently organized into the placenta. The placenta provides the nutritional connection between the embryo and the maternal circulation and produces a number of hormones.

Human Chorionic Gonadotropin (hCG): The primary function of the glycoprotein hormone hCG (the structural similarity of hCG to LH is discussed in Chapter 46) is to support the corpus luteum until the placenta produces amounts of progesterone sufficient to support the pregnancy. hCG can be detected within a few days of implantation, and this provides the basis of early diagnostic tests for pregnancy. Peak hCG levels are reached in the middle of the first trimester, after which there is a gradual decline throughout the remainder of pregnancy. Changes in hCG and other hormone levels in pregnancy are illustrated in Fig 51–8.

Progestins: The corpus luteum is the major source of progesterone for the first 6–8 weeks of the pregnancy, and then the placenta assumes this function. The corpus luteum continues to function, but late in pregnancy the placenta makes 30–40 times more progesterone than does the corpus luteum. The placenta does not synthesize cholesterol and so depends upon a maternal supply.

Estrogens: Plasma concentrations of estradiol, estrone, and estriol gradually increase throughout pregnancy. **Estriol is produced in the largest amount, and its formation reflects a number of fetoplacental functions.** The fetal adrenal produces DHEA and DHEA sulfate, which are converted to 16α-hydroxy derivatives by the fetal liver. These are converted to estriol by the placenta; travel via the placental circulation to the maternal liver, where they are conjugated to glucuronides; and then are excreted in the urine (Fig 51–9). The measurement of urinary estriol levels is used to document the function of a number of maternal-fetal processes.

Another interesting exchange of substrates is required for fetal cortisol production. The fetal adrenal lacks the familiar 3β-hydroxysteroid dehydrogenase △^{5,4} isomerase complex and hence depends upon the placenta for the progesterone required for cortisol synthesis (Fig 51–9).

Placental Lactogens: The placenta makes a hormone called placental lactogen (PL). PL is also called chorionic somatomammotropin or placental growth hormone because it has biologic properties of prolactin and growth hormone. The genetic relationship of these hormones is discussed in Chapter 46. The physiologic function of PL is uncertain, since women who lack this hormone appear to have normal pregnancies and deliver normal babies.

The Trigger for Parturition Is Unknown

Pregnancy lasts a predetermined number of days for each species, but the factors responsible for its termination are unknown. Hormonal influences are suspected but unproved. Estrogens and progestins are candidates, since they affect uterine contractility, and there is evidence that catecholamines are involved in induction of labor. Since oxytocin stimulates uterine contractility, it is used to facilitate delivery, but it will not initiate labor unless the pregnancy is at term. There are 100 times more oxytocin receptors in the uterus at term than there are at the onset of pregnancy.

The increased amount of estrogen at term may increase the number of oxytocin receptors (see Chapter 46). Once labor begins, the cervix dilates, initiating a

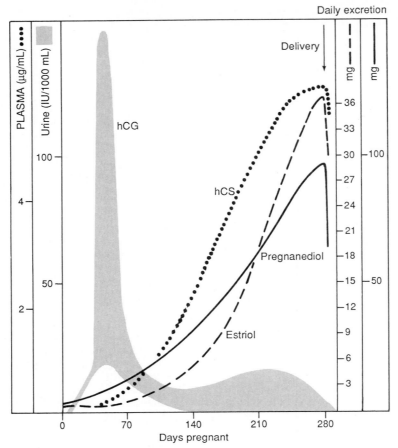

Figure 51–8. Hormone levels during normal pregnancy. hCG, human chorionic gonadotropin; hCS, human chorionic somatomammotropin. (Data from various authors.) (Reproduced, with permission, from Ganong WF: *Review of Medical Physiology,* 13th ed. Appleton & Lange, 1987.)

neural reflex that stimulates oxytocin release and hence further uterine contraction. Mechanical factors, such as the amount of stretch or force applied to the muscle, may be important. A sudden and dramatic change in the hormonal milieu of both the mother and newborn occurs with parturition, and plasma levels of progesterone (measured as pregnanediol) and estriol decline rapidly after the placenta is delivered (Fig 51–8).

Mammary Gland Development Is Stimulated by Estradiol & Progesterone & Lactation by Prolactin

The differentiation and function of the mammary gland are regulated by the concerted action of several hormones. The female sex hormones initiate this process, since estrogens are responsible for ductal growth and progestins stimulate alveolar proliferation. Some growth of glandular tissue occurs during puberty, along with deposition of adipose tissue, but extensive development occurs during pregnancy when glandular tissue is exposed to high concentrations of estradiol and progesterone. Complete differentiation, studied

mostly in rat mammary gland explants, requires the additional action of prolactin, glucocorticoids, insulin or a growth peptide, and an unidentified serum factor. Of these hormones, only the concentration of prolactin changes dramatically in pregnancy; it increase from < 2 ng/dL to > 200 ng/dL in late pregnancy. The effects of these hormones on the synthesis of various milk proteins, including lactalbumin, lactoglobulin, and casein, have been studied in detail. These hormones increase the rate of synthesis of these proteins by increasing amounts of the specific mRNAs, and in the case of casein at least, this is due to an increase in gene transcription and to stabilization of the mRNA.

Progesterone, required for alveolar differentiation, inhibits milk production and secretion in late pregnancy. Lactation commences when levels of this hormone decrease abruptly after delivery. Prolactin levels also fall rapidly postpartum but are stimulated with each episode of suckling (see Chapter 46), thereby ensuring continual lactation. Lactation gradually decreases if suckling is not allowed and can be rapidly terminated by administration of a large parenteral dose of an androgen before suckling is allowed.

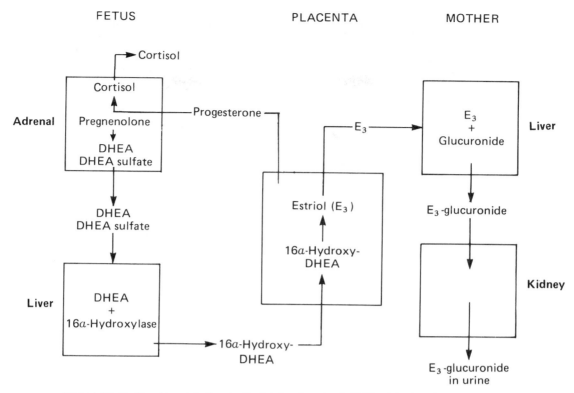

FETUS   PLACENTA   MOTHER

Figure 51–9. Steroid metabolism by the fetal-maternal unit DHEA, dehydroepiandrosterone.

Suckling also results in the release of oxytocin from the posterior pituitary. Oxytocin stimulates contraction of the myoepithelial cells that surround the alveolar ducts, thus expelling milk from the gland. The regulation of oxytocin synthesis and secretion is discussed in Chapter 46).

Menopause Is Complete With the Loss of Ovarian Estrogen Production

Women in the western hemisphere cease having regular menstrual cycles at about age 53, coincident with loss of all follicles and ovarian estrogen production. There is no alternative source of progesterone, but substantial amounts of a weak estrogen, estrone, are produced by the aromatization of androstenedione (Fig 51–5). The levels of estrone are not sufficient to suppress pituitary gonadotropin levels; thus, marked increases of LH and FSH are characteristic of the postmenopausal years. Postmenopausal women are particularly prone to 2 problems associated with tissue catabolism. Estrone is not always able to prevent the **atrophy of secondary sex tissues,** particularly the epithelium of the lower urinary tract and vagina. **Osteoporosis** is a major health problem in older individuals, and women with the most severe decrease in bone mass have lower than normal estrone levels.

Synthetic Agonists & Antagonists Are Used to Promote & Prevent Conception & to Inhibit Tumor Growth

Estrogens: Several synthetic compounds have estrogenic activity and one or more favorable pharmacologic features. Most modifications are designed to retard hepatic metabolism, so that the compounds can be given orally. One of the first developed was diethylstilbestrol. Other examples of modified steroids include 17α-ethinyl estradiol and mestranol, which are used in oral contraceptives.

Diethylstilbestrol

Numerous compounds with antiestrogenic activity have been synthesized, and several of these have clinical applications. Most of these antagonists act by competing with estradiol for its intracellular receptor (see

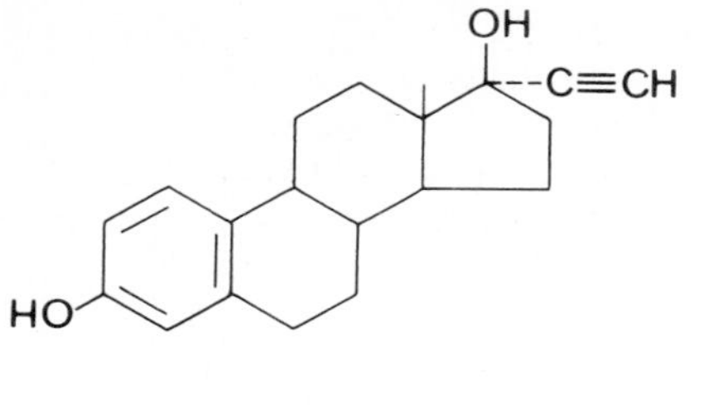

17α-Ethinyl estradiol

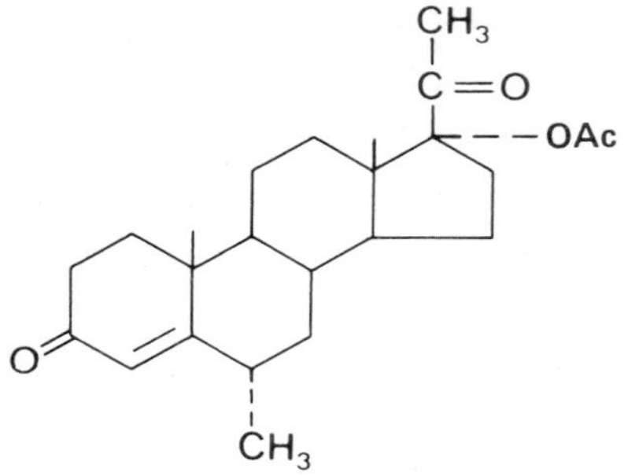

Medroxyprogesterone acetate

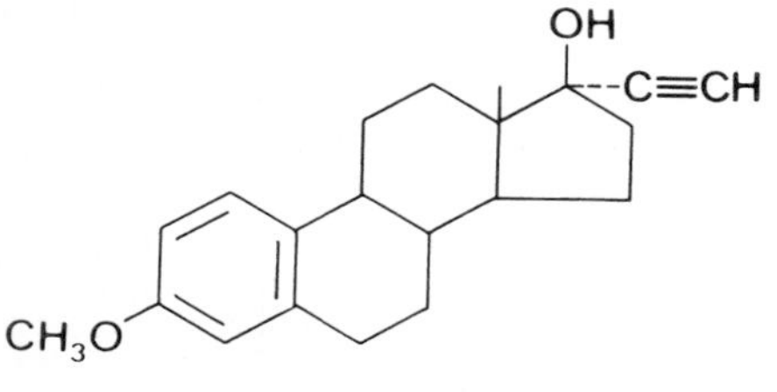

Mestranol

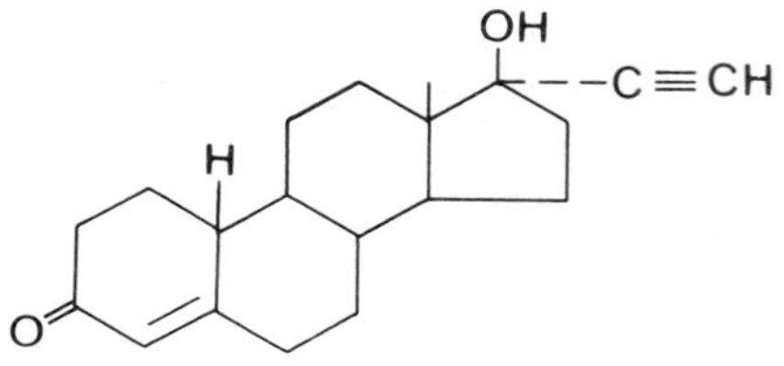

Norethindrone

below). Clomiphene citrate (Clomid) has a particular affinity for the estrogen receptor in the hypothalamus. Climophene was originally designed as an antifertility drug but, interestingly, it is now used for the opposite effect. Climophene competes with estradiol for hypothalamic receptor sites; thus, GnRH release is not restrained and excessive amounts of LH and FSH are released. Multiple follicles often mature simultaneously in response to clomiphene, and multiple pregnancies can ensue. Nafoxidine, a nonsteroidal compound, and tamoxifen combine with the estrogen receptor to form very stable complexes with chromatin; hence, the receptor cannot recycle and these agents inhibit the action of estradiol for prolonged periods. These antagonists are used in the treatment of estrogen-dependent breast cancer.

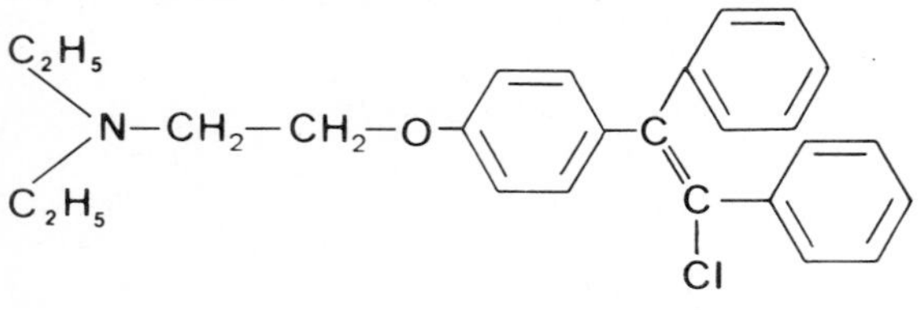

Clomiphene citrate

Progestins: It has been difficult to synthesize compounds that have progestin activity but no estrogenic or androgenic action. The 17α-alkyl-substituted 19-nortestosterone derivatives (eg, norethindrone) have minimal androgenic activity in most women and are used in oral contraceptives. Another potent progestin is medroxyprogesterone acetate (Provera). Medroxyprogesterone inhibits ovulation for several months when given as an intramuscular depot injection. However, since progestins inhibit cell growth, this compound is more frequently used for treating well-differentiated endometrial carcinoma.

Estrogens & Progestins Act by Regulating Gene Expression

These hormones act through their ability to combine with intracellular receptors that then bind to specific regions of chromatin or DNA (or both) to effect changes in the rate of transcription of specific genes. Much information has been learned from the analysis of how estradiol and progesterone stimulate transcription of the avian eggwhite protein genes, especially ovalbumin and conalbumin. The determination of exactly how these hormones activate gene transcription is under intense investigation.

The Estrogen & Progesterone Receptors Are Part of a Gene Family

The sequence of the receptors for estrogens (ER) and progestins (PR) has been deduced by analysis of the corresponding cDNA sequences. These receptors are part of the steroid and thyroid hormone receptor gene family discussed in Chapter 49. As illustrated in Fig 49–3, each receptor has several functional domains. The steroids bind to the ligand site in the carboxy-terminal portion of the receptor molecule. This causes a conformational change that allows the receptor to bind to DNA. The DNA-binding domain of the ER recognizes the sequence AGGTCAnnnTG$_\text{T}^\text{A}$CCT (the estrogen response element, or ERE), whereas the cognate domain in the PR recognizes the sequence GGTACAnnnTGTTCT, the PRE. The receptor-DNA interaction allows various *trans*-activating domains in each receptor to influence the activity of genes adjacent to the hormone response elements. The increased (or decreased) activity of specific genes results in altered rates of synthesis of specific proteins. This eventuates in altered metabolic responses.

Special Features: Some points, which may have

application to the mechanism of action of other hormones, bear noting: (1) There is considerable crosstalk between the sex hormone receptors. Progesterone binds to the androgen receptor and thus is a weak androgen; some androgens bind to the estrogen receptor and mimic the action of the latter in the uterus. Inspection of Fig 49–3 reveals that there is a close similarity in the core sequences of the hormone response elements. This could explain why some hormones exert "crossover" actions in certain instances. (2) Estrogens increase the concentration of both the estrogen and the progesterone receptor. (3) Progesterone appears to enhance the rate of turnover of its receptor. (4) So-called weak estrogens, such as estriol, act as potent estrogens when given frequently.

SOME PATHOPHYSIOLOGY OF THE FEMALE REPRODUCTIVE SYSTEM HAS A HORMONAL CONNECTION

A discussion of all of the disorders that affect the female reproductive system is beyond the scope of this chapter, but a few illustrative disorders follow. **Primary hypogonadism** is due to processes that directly involve the ovaries and thus cause ovarian deficiency (decreased ovulation, decreased hormone production, or both), whereas **secondary hypogonadism** is due to the loss of pituitary gonadotropin function. **Gonadal dysgenesis (Turner's syndrome)** is a relatively frequent genetic disorder in which individuals have an XO karyotype, female internal and external genitalia, several developmental abnormalities, and delayed puberty.

Several syndromes are related to abnormal amounts of hormones. The most frequent is **polycystic ovary syndrome** (Stein-Leventhal syndrome), in which overproduction of androgens causes hirsutism, obesity, irregular menses, and impaired fertility. The rare **Leydig cell and arrhenoblastoma tumors** produce testosterone; **granulosa-theca cell tumors** produce estrogens; and **intraovarian adrenal rests** produce

cortisol. Persistent trophoblastic tissue results in the benign **hydatidiform mole** or a malignant transformation of this—**choriocarcinoma;** both of these produce enormous quantities of hCG. The radioimmunoassay of hCG is a diagnostic test for these dangerous conditions and can also be used to monitor efficacy of therapy.

REFERENCES

General
Huggins C: Two principles in endocrine therapy of cancer. Hormone deprival and hormone interference. *Cancer Res* 1965;**25:**1163.
O'Malley BW: Steroid hormone action in eucaryotic cells. *J Clin Invest* 1984;**74:**307.
Wilson JD et al: The endocrine control of male phenotypic development. *Aust J Biol Sci* 1983;**36:**101.

Testicular Hormones
Chang C et al: Structural analysis of complementary DNA and amino acid sequences of human and rat androgen receptors. *Proc Natl Acad Sci USA* 1988;**85:**7211.
Hall PF: Testicular hormones: Synthesis and control. Pages 1511–1520 in: *Endocrinology,* Vol 3. DeGroot LJ (editor). Grune & Stratton, 1979.
Wilson J: Metabolism of testicular androgens. Chap 25, pp 491–508, in: *Handbook of Endocrinology. Section 7: Endocrinology,* Vol 5: *Male Reproductive System.* Hamilton DW, Greep RP (editors). American Physiological Society, Washington DC, 1975.

Ovarian Hormones
Green S et al: Human estrogen receptor DNA: Sequence, expression and homology to V-*erb*-A. *Nature (London)* 1986;**320:**134.
Siiteri PK, Febres F: Ovarian hormone synthesis, circulation and mechanisms of action. pages 1401–1417 in: *Endocrinology,* Vol 3. DeGroot LJ (editor). Grune & Stratton, 1979.
Toft D, Gorski J: A receptor molecule for estrogens. *Proc Natl Acad Sci USA* 1966;**55:**1574.

52

Hormones of the Pancreas and GI Tract

Daryl K. Granner, MD

INTRODUCTION

The pancreas is 2 very different organs contained within one structure. The acinar portion of the pancreas has an **exocrine** function, secreting into the duodenal lumen the enzymes and ions used for the digestive process. The **endocrine** portion consists of the islets of Langerhans. The 1–2 million islets of the human pancreas make up 1–2% of its weight and are collections of the several different cell types listed in Table 52–1.

The pancreatic islets secrete at least 4 hormones: insulin, glucagon, somatostatin, and pancreatic polypeptide. The hormones are released into the pancreatic vein, which empties into the portal vein—a convenient arrangement, since the liver is a primary site of action of insulin and glucagon. These 2 hormones are chiefly involved in regulating carbohydrate metabolism but affect many other processes. Somatostatin, first identified in the hypothalamus as the hormone that inhibits growth hormone secretion, is present in higher concentration in the pancreatic islets than in the hypothalamus and is involved in the local regulation of insulin and glucagon secretion. Pancreatic polypeptide affects gastrointestinal secretion.

The gastrointestinal tract secretes many hormones, perhaps more than any other single organ. The purpose of the gastrointestinal tract is to propel foodstuffs to sites of digestion, to provide the proper milieu (enzymes, pH, salt, etc) for the digestive process, to move the digested products across the intestinal mucosa through the mucosal cells and into the extracellular space, to move those products to distant cells via the circulation, and to expel waste products. The gastrointestinal hormones assist in all of these functions.

Table 52–1. Cell types in the islets of Langerhans.

Cell Type	Relative Abundance	Hormone Produced
A (or α)	~25%	Glucagon
B (or β)	~70%	Insulin
D (or δ)	<5%	Somatostatin
F	Trace	Pancreatic polypeptide

Acronyms Used in This Chapter

ACTH	Adrenocorticotropic hormone
EGF	Epidermal growth factor
FGF	Fibroblast growth factor
GIP	Gastric inhibitory polypeptide
IGF	Insulinlike growth factor
LDL	Low-density lipoproteins
IDDM	Insulin-dependent diabetes mellitus
NIDDM	Non-insulin-dependent diabetes mellitus
PDGF	Platelet-derived growth factor
PEPCK	Phosphoenolpyruvate carboxykinase
$PGF_{2\alpha}$	Prostaglandin $F_{2\alpha}$
PP	Pancreatic polypeptide
VLDL	Very low density lipoproteins

BIOMEDICAL IMPORTANCE

Insulin has been the model peptide hormone in many ways, being the first purified, crystallized, and synthesized by chemical and molecular biologic techniques. Studies of its biosynthesis led to the important concept of the propeptide. Insulin has important medical implications. Five percent of the population in developed countries has diabetes mellitus, and an equal number are liable to develop this disease. Diabetes mellitus is due to insufficient action of insulin, owing either to its absence or to resistance to its action. Glucagon, acting unopposed, aggravates this condition.

Disease syndromes due to excessive production of several of the gastrointestinal hormones have been described. Signs and symptoms often involve many organ systems, and accurate diagnosis can be difficult unless the physician is aware of these syndromes. The gastrointestinal hormones are of interest also because of their close link to neuropeptides.

THE PANCREATIC HORMONES ARE INSULIN, GLUCAGON, SOMATOSTATIN, & PANCREATIC POLYPEPTIDE

Insulin Production Is Regulated by Glucose Concentration

Historical Perspective: Langerhans identified the islets in the 1860s but did not understand their function—nor did von Mering and Minkowski, who demonstrated in 1889 that pancreatectomy produced diabetes. The link between the islets and diabetes was suggested by de Mayer in 1909 and by Sharpey-Schaffer in 1917, but it was Banting and Best who proved this association in 1921. These investigators used acid-ethanol to extract from the tissue an islet cell factor that had potent hypoglycemic activity. The factor was named insulin, and it was quickly learned that bovine and porcine islets contained insulin that was active in humans. Within a year, insulin was in widespread use for the treatment of diabetes and proved to be lifesaving.

Having large quantities of bovine or porcine insulin to study had an equally dramatic effect on biomedical research. Insulin was the first protein proved to have hormonal action, the first protein crystallized (Abel, 1926), the first protein sequenced (Sanger et al, 1955), the first protein synthesized by chemical techniques (Du et al; Zahn; Katsoyanis; ~ 1964), the first protein shown to be synthesized as a larger precursor molecule (Steiner et al, 1967), and the first protein prepared for commercial use by recombinant DNA technology. In spite of this impressive list of "firsts," less is known about how insulin works at the molecular level than about how most other hormones work at that level.

Insulin Is A Heterodimeric Polypeptide

Insulin is a polypeptide consisting of 2 chains, A and B, linked by 2 interchain disulfide bridges that connect A7 to B7 and A20 to B19. A third intrachain disulfide bridge connects residues 6 and 11 of the A chain. The location of these 3 disulfide bridges is invariant, and the A and B chains have 21 and 30 amino acids, respectively, in most species. The covalent

Table 52–2. Variations in the structure of insulin in mammalian species.*

Species	Variations From Human Amino Acid Sequence		
	A-Chain Position 8 9 10		B-Chain Position 30
Human	Thr-Ser-Ile		Thr
Pig, dog, sperm whale	Thr-Ser-Ile		Ala
Rabbit	Thr-Ser-Ile		Ser
Cattle, goat	Ala-Ser-Val		Ala
Sheep	Ala-Gly-Val		Ala
Horse	Thr-Gly-Ile		Ala
Sei whale	Ala-Ser-Thr		Ala

*Modified and reproduced, with permission, from Ganong WF: *Review of Medical Physiology*, 13th ed. Appleton & Lange, 1987.

structure of human insulin (MW 5734) is illustrated in Fig 52–1, and a comparison of the amino acid substitutions found in a variety of species is presented in Table 52–2. Substitutions occur at many positions within either chain without affecting bioactivity and are particularly common in positions 8, 9, and 10 of the A chain. Thus, this region is not crucial for bioactivity. **Several positions and regions are highly conserved,** however, including (1) the positions of the 3 disulfide bonds, (2) the hydrophobic residues in the C-terminal region of the B chain, and (3) the N- and C-terminal regions of the A chain. Chemical modification or substitution of specific amino acids in these regions has allowed investigators to formulate a composite active region (Fig 52–2). The C-terminal hydrophobic region of the B chain is also involved in the dimerization of insulin.

There Is a Close Similarity Among Human, Porcine, & Bovine Insulins as Shown in Table 52–2

Porcine insulin differs by a single amino acid, an alanine for threonine substitution at B30, while bovine insulin has this modification plus the substitutions of alanine for threonine at A8 and valine for isoleucine at A10. These modifications result in no appreciable change in biologic activity and very little antigenic

Figure 52–1. Covalent structure of human insulin. (Reproduced, with permission, from Ganong WF: *Review of Medical Physiology*, 13th ed. Appleton & Lange, 1987.)

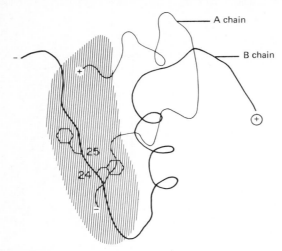

Figure 52–2. Region of the insulin molecule required for biologic activity. Diagrammatic structure of insulin as determined by x-ray crystallography. The shaded area illustrates the portion of insulin that is thought to be most important in conferring biologic activity to the hormone. The Phe residues B24 and B25 are the sites of mutations that affect insulin bioactivity. The N termini of the insulin A and B chains are indicated by ⊕, whereas the C termini are indicated by ⊖. (Redrawn and reproduced, with permission, from Tager HS: Abnormal products of the human insulin gene. *Diabetes* 1984;**33**:693.)

difference. Although all patients given heterologous insulin develop low titers of circulating antibodies against the molecule, few develop clinically significant titers. Porcine and bovine insulins were standard therapy for diabetes mellitus until human insulin was produced by recombinant DNA technology. Despite a wide variation in primary structure, biologic activity is about 25–30 IU/mg dry weight for all insulins.

Insulin forms very interesting complex structures. Zinc is present in high concentration in the B cell and forms complexes with insulin and proinsulin. Insulins from all vertebrate species form isologous dimers through hydrogen bonding between the peptide groups of the B24 and B26 residues of 2 monomers, and at high concentrations these are organized as hexamers, each with 2 atoms of zinc. This higher-order structure made studies of the crystalline structure of insulin feasible. Insulin is probably in the monomeric form at physiologic concentrations.

Insulin Is Synthesized as a Precursor Molecule

Insulin is synthesized as a **preprohormone** (MW ~ 11,500) and is the prototype for peptides that are processed from larger precursor molecules. The hydrophobic 23-amino-acid pre-, or leader, sequence directs the molecule into the cisternae of the endoplasmic reticulum and then is removed. This results in the 9000-MW proinsulin molecule that provides the conformation necessary for forming the proper disulfide bridges. As shown in Fig 52–3, the arrangement of

proinsulin, starting from the amino terminus, is B chain—connecting (C) peptide—A chain. The proinsulin molecule undergoes a series of site-specific peptide cleavages that result in the formation of equimolar amounts of mature insulin and C peptide. These enzymatic cleavages are summarized in Fig 52–3.

Other Islet Cell Hormones Are Also Synthesized as Precursor Molecules

The synthesis of other islet cell hormones also requires posttranslational enzymatic processing of higher molecular weight precursor molecules. Diagrammatic structures of pancreatic polypeptide, glucagon, and somatostatin are compared with that of insulin in Fig 52–4. Several combinations of endoproteolytic (trypsinlike) and exoproteolytic (carboxypeptidase B-like) cleavages are involved, since the hormone sequence may occur at the carboxyl terminus of the precursor (somatostatin), at the amino terminus (pancreatic polypeptide), at both ends (insulin), or in the middle (glucagon).

Insulin Synthesis & Granule Formation Occur in Subcellular Organelles

Proinsulin is synthesized by ribosomes on the rough endoplasmic reticulum, and the enzymatic removal of the leader peptide (pre- segment), disulfide bond formation, and folding (Fig 52–3), occur in the cisternae of this organelle. The proinsulin molecule is transported to the Golgi apparatus wherein proteolysis and packaging into the secretory granules begin. Granules continue to mature as they traverse the cytoplasm toward the plasma membrane. Proinsulin and insulin both combine with zinc to form hexamers, but since about 95% of the proinsulin is converted to insulin, it is the crystals of the latter that confer morphologic distinctness to the granules. Equimolar amounts of C peptide are present within these granules, but these molecules do not form a crystalline structure. Upon appropriate stimulation (see below), the mature granules fuse with the plasma membrane and discharge their contents into the extracellular fluid by **emiocytosis.**

The Properties of Proinsulin & C Peptide Differ From Those of Insulin

Proinsulins vary in length from 78 to 86 amino acids, with the variation occurring in the length of the C-peptide region. Proinsulin has the same solubility and isoelectric point as insulin; it also forms hexamers with zinc crystals, and it reacts strongly with insulin antisera. **Proinsulin has less than 5% of the bioactivity of insulin,** indicating that most of the active site of the latter is occluded in the precursor molecule. Some proinsulin is released with insulin and in certain conditions (islet cell tumors) in larger than usual amounts. Since the plasma half-life of proinsulin is significantly longer than that of insulin and since pro-

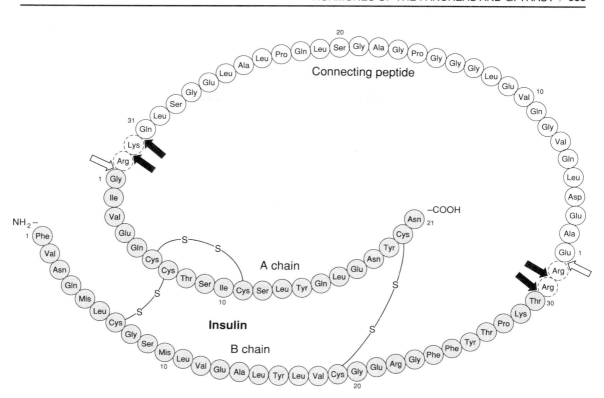

Figure 52–3. Structure of human proinsulin. Insulin and C-peptide molecules are connected at 2 sites by dipeptide links. An initial cleavage by a trypsin-like enzyme (⇨) followed by several cleavages by a carboxy peptidase-like enzyme (→) results in the production of the heterodimeric (AB) insulin molecule and the C-peptide.

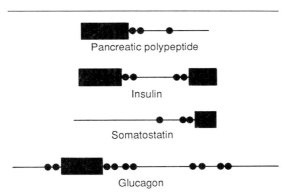

Figure 52–4. Diagrammatic structures of the precursors for the 4 major endocrine cell products of the pancreatic islet. Portions of the precursors that correspond to the named hormones are indicated by heavy bars, whereas portions that correspond to peptide extensions are indicated by lines; dibasic amino acid residues (arginine or lysine) corresponding to precursor conversion sites are shown as filled circles. Note that proinsulin has been drawn in an extended form that does not show disulfide bonds; the structure of proinsulin illustrated has the sequence B chain—C peptide—A chain. (Redrawn and reproduced, with permission, from Tager HS: Abnormal products of the human insulin gene. *Diabetes* 1984;**33**: 693.)

insulin is strongly cross-reactive with insulin antisera, a radioimmunoassay for "insulin" may occasionally overestimate the bioactivity of "insulin" in plasma.

The C peptide has no known biologic activity. It is a distinct molecule from an antigenic standpoint. Thus, C-peptide immunoassays can distinguish insulin secreted endogenously from insulin administered exogenously and can quantitate the former when antiinsulin antibodies preclude the direct measurement of insulin. The C peptides of different species have a high rate of amino acid substitution, an observation which underscores the statement that this fragment probably has no biologic activity.

Insulin-Related Peptides Have Precursors

The structural arrangement of the precursor molecule is not unique to insulin; very closely related peptide hormones (relaxin and the insulinlike growth factors) show the same arrangement (Fig 52–5). All of these hormones have highly homologous B- and A-chain regions at the amino and carboxyl termini of a precursor molecule, and these are joined by a connecting segment. In the relaxin and insulin precursor peptides, this connecting segment is bound on both ends by 2 basic amino acids. After the B and A chains are joined by disulfide bonds, this piece is removed by

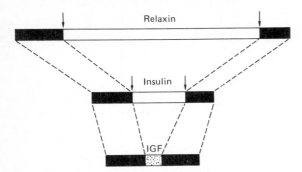

Figure 52–5. Diagrammatic structures of precursors for insulin-related peptides. Homologous regions of relaxin, insulin, and insulinlike growth factor are shown as solid bars. Amino acid sequences connecting B-chain and A-chain sequences in the precursors for relaxin and insulin are shown as open bars; these sequences are removed during processing of the precursors to their corresponding 2-chained products (vertical arrows). The amino acid sequence of insulinlike growth factor that corresponds to these connecting peptides, but is not removed by proteolytic processing events, is shown as a stippled bar; insulinlike growth factor is a single-chain peptide hormone. (Redrawn and reproduced, with permission, from Tager HS: Abnormal products of the human insulin gene. *Diabetes* 1984;**33**:693.)

endoproteolytic action, and these molecules are converted to 2-chain peptide hormones. The insulinlike growth factors, though highly homologous to insulin and relaxin in primary structure, lack these dibasic cleavage sites and thus remain single-chain peptide hormones.

The Human Insulin Gene Has Been Isolated

The human insulin gene (Fig 52–6) is located on the short arm of chromosome 11. Most mammals express a single insulin gene that is organized like the human gene, but rats and mice have 2 nonallelic genes. Each codes for a unique proinsulin that is processed into 2 distinct, active insulin molecules. The synthesis of human insulin in bacterial expression systems, using recombinant DNA technology, affords an excellent source of this hormone for diabetic patients.

Figure 52–6. Diagrammatic structure of the human insulin gene. Areas with diagonal stripes correspond to untranslated regions of the corresponding mRNA; open regions correspond to intervening sequences; and stippled regions correspond to coding sequences. L, B, C, and A identify coding sequences for the leader (or signal) peptide, the insulin B chain, the C peptide, and the insulin A chain, respectively. Note that the coding sequence for the C peptide is split by an intervening sequence. The diagrammatic structure is drawn to scale. (Redrawn and reproduced, with permission, from Tager HS: Abnormal products of the human insulin gene. *Diabetes* 1984;**33**: 693.)

Insulin Secretion Is Precisely Regulated

The human pancreas secrets 40–50 units of insulin daily, which represents about 15–20% of the hormone stored in the gland. Insulin secretion is an energy-requiring process that involves the microtubule-microfilament system in the B cells of the islets. A number of mediators have been implicated in insulin release.

Glucose: An increase in plasma glucose concentration is the most important physiologic regulator of insulin secretion. The threshold concentration for secretion is the fasting plasma glucose level (80–100 mg/dL), and the maximal response is obtained at glucose levels between 300 and 500 mg/dL. Two different mechanisms have been proposed to explain how glucose regulates insulin secretion. One hypothesis suggests that glucose combines with a receptor, possibly located on the B-cell membrane, that activates the release mechanism. The second hypothesis suggests that intracellular metabolites or the rate of metabolite flux through a pathway such as the pentose-phosphate shunt, the citric acid cycle, or the glycolytic pathway, is involved. There is experimental evidence to support both positions.

Hormonal Factors: Numerous hormones affect insulin release. α-Adrenergic agonists, principally epinephrine, inhibit insulin release even when this process has been stimulated by glucose. β-Adrenergic agonists stimulate insulin release, probably by increasing intracellular cAMP (see below).

Chronic exposure to excessive levels of growth hormone, cortisol, placental lactogen, estrogens, and progestins also increases insulin secretion. It is therefore not surprising that insulin secretion increases markedly during the later stages of pregnancy.

Pharmacologic Agents: Many drugs stimulate insulin secretion, but the **sulfonylurea compounds** are used most frequently for therapy in humans. Drugs such as tolbutamide stimulate insulin release by a mechanism different from that employed by glucose and have achieved widespread use in the treatment of type II (non-insulin-dependent) diabetes mellitus.

$$H_3C-\!\!\!\!\bigcirc\!\!\!\!-SO_2-NH-\underset{\underset{O}{\|}}{C}-NH-(CH_2)_3-CH_3$$

Tolbutamide

Insulin Is Rapidly Metabolized

Unlike the insulinlike growth factors, **insulin has no plasma carrier protein;** thus, its plasma half-life is less than 3–5 minutes under normal conditions. The major organs involved in insulin metabolism are the liver, kidneys, and placenta; about 50% of insulin is removed in a single pass through the liver. **Mechanisms involving 2 enzyme systems are responsible for the metabolism of insulin.** The first involves an insulin-specific **protease** found in many tissues but in highest concentration in those listed above. This protease has been purified from skeletal muscle and is known to be sulfhydryl-dependent and active at physiologic pH. The second mechanism involves hepatic **glutathione-insulin transhydrogenase.** This enzyme reduces the disulfide bonds, and then the individual A and B chains are rapidly degraded. It is not clear which of these mechanisms is most active under physiologic conditions, nor is it clear whether either process is regulated.

The Central Role of Insulin in Carbohydrate, Lipid, & Protein Metabolism Can Be Best Appreciated by Examining the Consequences of Insulin Deficiency in Humans

The cardinal manifestation of **diabetes mellitus** is **hyperglycemia,** which results from (1) decreased entry of glucose into cells, (2) decreased utilization of glucose by various tissues, and (3) increased production of glucose (gluconeogenesis) by the liver (see Fig 52–7). Each of these is discussed in more detail below.

Polyuria, polydipsia, and weight loss in spite of adequate caloric intake are the major symptoms of insulin deficiency. How is this explained? The plasma glucose level rarely exceeds 120 mg/dL in normal humans, but much higher levels are routinely found in patients with deficient insulin action (Fig 52–8). After a certain plasma glucose level is attained (generally > 180 mg/dL in humans), the maximum level of renal tubular reabsorption of glucose is exceeded, and sugar is excreted in the urine (glycosuria). The urine volume is increased owing to osmotic diuresis and coincident transporter translocation is temperature- and energy-dependent and is protein synthesis-independent (Fig 52–9).

The hepatic cell represents a notable exception to this scheme. Insulin does not promote the facilitated diffusion of glucose into hepatocytes, but it indirectly enhances net inward flux by converting intracellular glucose to glucose 6-phosphate through the action of glucokinase, an enzyme induced by insulin. This rapid phosphorylation keeps the free glucose concentration

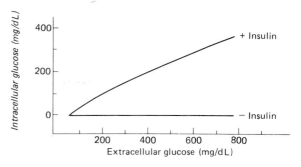

Figure 52–8. Entry of glucose into muscle cells.

very low in the hepatocyte, thus favoring entry by simple diffusion down a concentration gradient.

Insulin also promotes the entry of amino acids into cells, particularly in muscle, and enhances the movement of K^+, Ca^{2+}, nucleosides, and inorganic phosphate. These effects are independent of the action of insulin on glucose entry.

Effects on Glucose Utilization: Insulin influences the intracellular utilization of glucose in a number of ways, as illustrated below.

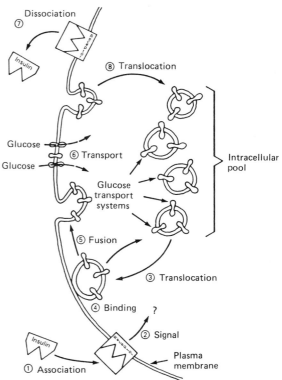

Figure 52–9. Translocation of glucose transporters by insulin. (Reproduced, with permission, from Karnieli E et al: Insulin-stimulated translocation of glucose transport systems in the isolated rat adipose cell. *J Biol Chem* 1981;**256:** 4772. Courtesy of S Cushman.)

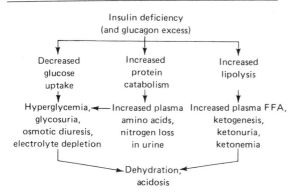

Figure 52–7. Pathophysiology of insulin deficiency. (Courtesy of RJ Havel.)

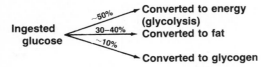

In a normal person, about half of the glucose ingested is converted to energy through the glycolytic pathway and about half is stored as fat or glycogen. Glycolysis decreases in the absence of insulin, and the anabolic processes of glycogenesis and lipogenesis are impeded. Indeed, only 5% of an ingested glucose load is converted to fat in an insulin-deficient diabetic.

Insulin increases hepatic glycolysis by increasing the activity and amount of several key enzymes including glucokinase, phosphofructokinase, and pyruvate kinase. Enhanced glycolysis increases glucose utilization and thus indirectly decreases glucose release into plasma. Insulin also decreases the activity of glucose-6-phosphatase, an enzyme found in liver but not in muscle. Since glucose 6-phosphate cannot exit from the plasma membrane, this action of insulin results in the retention of glucose within the liver cell.

Insulin stimulates lipogenesis in adipose tissue by (1) providing the acetyl-CoA and NADPH required for fatty acid synthesis; (2) maintaining a normal level of the enzyme acetyl-CoA carboxylase, which catalyzes the conversion of acetyl-CoA to malonyl-CoA; and (3) providing the glycerol involved in triacylglycerol synthesis. In insulin deficiency, all of these are decreased; thus, lipogenesis decreases. Another reason for the decreased lipogenesis in insulin deficiency is that fatty acids, released in large amounts by several hormones when unopposed by insulin, feedback-inhibit their own synthesis by inhibiting acetyl-CoA carboxylase. The net effect of insulin on fat is therefore anabolic.

The final action of insulin on glucose utilization involves another anabolic process. In liver and muscle, insulin stimulates the conversion of glucose to glucose 6-phosphate, which then undergoes isomerization to glucose 1-phosphate and is incorporated into glycogen by the enzyme glycogen synthase, the activity of which is stimulated by insulin. This action is indirect and dual in nature. Insulin decreases intracellular cAMP levels by activating a phosphodiesterase. Since cAMP-dependent phosphorylation inactivates glycogen synthase, low levels of this nucleotide allow the enzyme to stay in the active form. Insulin also activates a phosphatase that dephosphorylates glycogen synthase, thereby resulting in the activation of this enzyme. Finally, insulin inhibits phosphorylase by a mechanism involving cAMP and phosphatase as described above, and this decreases glucose liberation from glycogen. The net effect of insulin on glycogen metabolism is also anabolic.

Effects on Glucose Production (Gluconeogenesis):
The actions of insulin on glucose transport, glycolysis, and glycogenesis occur within seconds or minutes, since they primarily involve the activation or inactivation of enzymes by phosphorylation or dephosphorylation. A more long-term effect on plasma glucose involves **the inhibition of gluconeogenesis by insulin.** The formation of glucose from noncarbohydrate precursors involves a series of enzymatic steps, many of which are stimulated by glucagon (acting through cAMP), by glucocorticoid hormones, and to a lesser extent by α- and β-adrenergic agents, angiotensin II, and vasopressin. Insulin inhibits these same steps. The key gluconeogenic enzyme in the liver is phosphoenolpyruvate carboxykinase (PEPCK), which converts oxaloacetate to phosphoenolpyruvate. Recent studies (see below) show that insulin decreases the amount of this enzyme by selectively inhibiting transcription of the gene that codes for the mRNA for PEPCK.

Effects on Glucose Metabolism:
The net action of all of the above effects of insulin is to **decrease the blood glucose level.** In this action, insulin stands alone against an array of hormones that attempt to counteract this effect. This no doubt represents one of the organism's most important defense mechanisms, since prolonged hypoglycemia poses a potentially lethal threat to the brain and must be avoided.

Effects on Lipid Metabolism:
The lipogenic actions of insulin were discussed in the context of glucose utilization. Insulin also is a potent inhibitor of lipolysis in liver and adipose tissue and thus has an indirect anabolic effect. This is partly due to the ability of insulin to decrease tissue cAMP levels (which are increased in these tissues by the lipolytic hormones glucagon and epinephrine) but also to the fact that insulin inhibits hormone-sensitive lipase activity. This inhibition is presumably due to the activation of a phosphatase that dephosphorylates and thereby inactivates the lipase or cAMP-dependent protein kinase. Insulin therefore decreases circulating free fatty acids. This contributes to the action of insulin on carbohydrate metabolism, since fatty acids inhibit glycolysis at several steps and stimulate gluconeogenesis. This illustrates the point that one cannot discuss metabolic regulation in the context of a single hormone or metabolite. Regulation is a complex process in which the flux through a given pathway is the result of the interplay of a number of hormones and metabolites.

In patients with insulin deficiency, lipase activity increases, resulting in enhanced lipolysis and increased concentration of free fatty acids in plasma and liver. Glucagon levels also increase in these patients, and this enhances the release of free fatty acids. (Glucagon opposes most of the actions of insulin, and the metabolic state in the diabetic is a reflection of the relative levels of glucagon and insulin.) A portion of the free fatty acids is metabolized to acetyl-CoA (the reverse of lipogenesis) and then to CO_2 and H_2O via the citric acid cycle. In patients with insulin deficiency, the capacity of this process is rapidly exceeded and the acetyl-CoA is converted to acetoacetyl-CoA and then to acetoacetic and β-hydroxybutyric acids. Insulin reverses this pathway.

Insulin apparently affects the formation or clearance of VLDL and LDL, since levels of these particles, and consequently the level of cholesterol, are often elevated in poorly controlled diabetics. Accelerated ath-

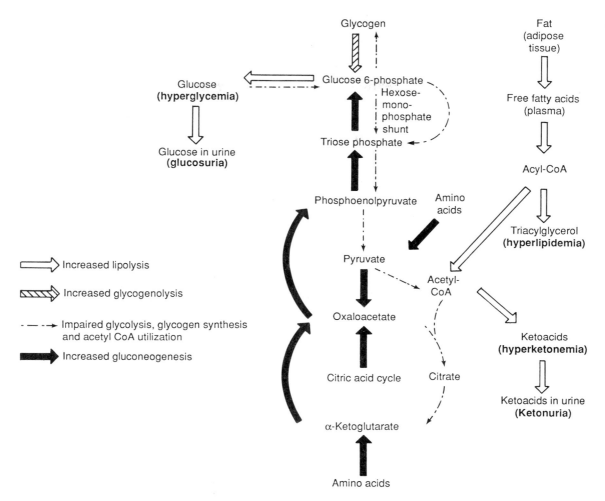

Figure 52–10. In severe insulin deficiency there is accelerated lipolysis. This results in elevated plasma triacylglycerol levels (hyperlipidemia). Little of the acetyl-CoA can be metabolized by the citric acid cycle, so the remainder is converted to ketoacids (ketonemia) and some is excreted ketonuria. Since glycolysis is inhibited, the G6P formed from accelerated glycogenolysis is converted to glucose. This, combined with accelerated gluconeogenesis, results in hyperglycemia (from increased availability of amino acids and increased amount of PEPCK enzyme). Insulin essentially reverses all these processes.

erosclerosis, a serious problem in many diabetics, is attributed to this metabolic defect.

The actions of insulin can be inferred by inspecting Fig 52–10, which depicts the flux through several critical pathways in the absence of the hormone.

Effects on Protein Metabolism: Insulin generally has an anabolic effect on protein metabolism in that it stimulates protein synthesis and retards protein degradation. Insulin stimulates the uptake of neutral amino acids into muscle, an effect that is not linked to glucose uptake or to subsequent incorporation of the amino acids into protein. The effects of insulin on general protein synthesis in skeletal and cardiac muscle and in liver are thought to be exerted at the level of mRNA translation.

In recent years, insulin has been shown to influence the synthesis of specific proteins by effecting changes in the corresponding mRNAs. This action of insulin, which may ultimately explain many of the effects the hormone has on the activity or amount of specific proteins, is discussed in more detail below.

Effects on Cell Replication: Insulin stimulates the proliferation of a number of cells in culture, and it may also be involved in the regulation of growth in vivo. Cultured fibroblasts are the most frequently used cells in studies of growth control. In such cells, insulin potentiates the ability of fibroblast growth factor (FGF), platelet-derived growth factor (PDGF), epidermal growth factor (EGF), tumor-promoting phorbol esters, prostaglandin $F_{2\alpha}$ ($PGF_{2-\alpha}$), vasopressin, and cAMP analogs to stimulate cell cycle progression of cells arrested in the G_1 phase of the cycle by serum deprivation.

An exciting new area of research involves the inves-

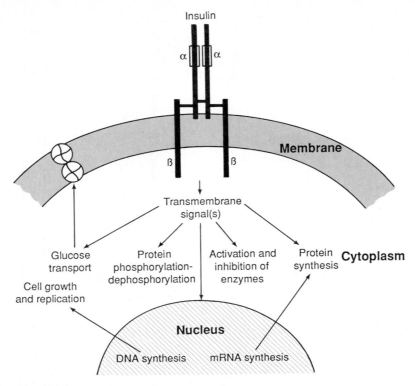

Figure 52–11. Relationship of the insulin receptor to insulin action. Insulin binds to its membrane receptor and this interaction generates one or more transmembrane signals. This signal (or signals) modulates a wide variety of intracellular events.

tigation of tyrosine kinase activity. The insulin receptor, along with receptors for many other growth-promoting peptides including those of PDGF and EGF, has tyrosine kinase activity. Interestingly, at least 10 oncogene products, many of which are suspected to be involved in stimulating malignant cell replication, are also tyrosine kinases. Mammalian cells contain analogs of these oncogenes (**proto-oncogenes**), which may be involved in the replication of normal cells. Support for the theory that they are involved comes from recent observations that the expression of at least 2 proto-oncogene products, c-*fos* and c-*myc,* increases following addition of serum PDGF or insulin growth-arrested cells.

The Mechanism of Action of Insulin Is Being Elucidated

Insulin action begins when the hormone binds to a specific glycoprotein receptor on the surface of the target cell. The diverse actions of the hormone (Fig 52–11) can occur within seconds or minutes (transport, protein phosphorylation, enzyme activation and inhibition, RNA synthesis) or after a few hours (protein and DNA synthesis and cell growth).

The insulin receptor has been studied in great detail using biochemical and recombinant DNA techniques. It is a heterodimer consisting of 2 subunits, designated α and β, in the configuration α_2-β_2, linked by di-

sulfide bonds (Fig 52–12). Both subunits are extensively glycosylated, and removal of sialic acid and galactose decreases insulin binding and insulin action. Each of these glycoprotein subunits has a unique structure and function. The α subunit (MW 135,000) is entirely extracellular, and it binds insulin, probably via a cysteine-rich domain. The β subunit (MW 95,000) is a transmembrane protein that performs the second major function of a receptor (see Chapter 45), ie, signal transduction. The cytoplasmic portion of the β subunit has tyrosine kinase activity and an autophosphorylation site. Both of these are thought to be involved in signal transduction and insulin action (see below). The striking similarity between 3 receptors with very different functions is illustrated in Fig 52–12. Indeed, several regions of the β subunit have sequence homology with the EGF receptor.

The insulin receptor is constantly being synthesized and degraded, and its half-life is 7–12 hours. The receptor is synthesized as a single-chain peptide in the rough endoplasmic reticulum and is rapidly glycosylated in the Golgi region. The precursor of the human insulin receptor has 1382 amino acids, has a molecular weight of 190,000, and is cleaved to form the mature α and β subunits. The human insulin receptor gene is located on chromosome 19.

Insulin receptors are found on most mammalian cells, in concentrations of up to 20,000 per cell, and

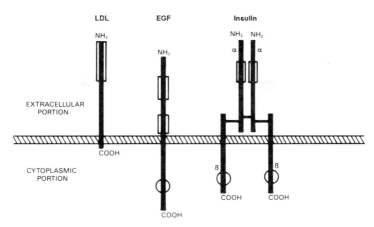

Figure 52–12. Schematic representation of the structure of the LDL, EGF, and insulin receptors. The amino terminus of each is in the extracellular portion of the molecule. The boxes represent cysteine-rich regions, which are thought to be involved in ligand binding. Each receptor has a short domain (~ 25 amino acids) that traverses the plasma membrane (the hatched line), and there is an intracellular domain of variable length. The EGF and insulin receptors have tyrosine kinase activity associated with the cytoplasmic domain (illustrated by the circles) and also have autophosphorylation sites in this region. The insulin receptor is a heterotetramer connected by disulfide bridges (vertical bars).

often on cells not typically thought of as being insulin targets. Insulin has a well-known set of effects on metabolic processes but also is involved in growth and replication of cells (see above) as well as in fetal organogenesis and differentiation and in tissue repair and regeneration. The structure of the insulin receptor and the ability of different insulins to bind to receptors and elicit biologic responses are virtually identical in all cells and all species. Thus, porcine insulin is always 10–20 times more effective than porcine proinsulin, which in turn is 10–20 times more effective than guinea pig insulin, even in the guinea pig. The insulin receptor has apparently been highly conserved, more so than even insulin itself.

When insulin binds to the receptor, several events occur. (1) There is a conformational change of the receptor; (2) the receptors crosslink and form microaggregates; (3) the receptor is internalized; and (4) one or more signals is generated. The significance of the conformational change is unknown, and internalization probably represents a means of controlling receptor concentration and turnover. In conditions in which plasma insulin levels are high, eg, obesity or acromegaly, the number of insulin receptors is decreased and target tissues become less sensitive to insulin. This "down-regulation" results from the loss of receptors by internalization, the process whereby insulin-receptor complexes enter the cell through endocytosis in clathrin-coated vesicles (see Chapter 41). Down-regulation explains part of the insulin resistance in obesity and type II diabetes mellitus.

The Intracellular Messenger(s) of Insulin Action Has Not Been Defined

Although the mechanism of insulin action has been under investigation for 60 years, certain critical points,

such as the nature of the intracellular signal, remain obscure. Insulin is not unique in this respect—the intracellular messenger has not been identified for a large number of hormones (see Table 45–1). A variety of different molecules have been proposed as the intracellular second messenger for insulin. These include insulin itself, calcium, cyclic nucleotides (cAMP, cGMP), H_2O_2, membrane-derived peptides, membrane phosphoinositol glycans, monovalent cations, and tyrosine kinase (the insulin receptor). None has withstood rigorous testing.

Current interest centers on the observation that the insulin receptor is itself an insulin-sensitive enzyme, since it undergoes autophosphorylation in response to insulin binding. This function is conferred by the β subunit, which acts as a protein kinase in transferring the γ-phosphate of ATP to a tyrosine residue in the β subunit (Fig 52–12). Insulin increases the V_{max} of this enzymatic reaction.

Tyrosine phosphorylation is unusual in mammalian cells (phosphotyrosine accounts for only 0.03% of the phosphoamino acid content of normal cells), and it is perhaps more than coincidental that the EGF, PDGF, and IGF-I receptors also have tyrosine kinase activity. Tyrosine kinase activity is thought to be an essential factor in the action of a number of viral oncogene products; the relationship of this and of cellular oncogene analogs with similar properties in malignant and normal cell growth was discussed above. As structures are elucidated, a high degree of homology between receptors and oncogenes is becoming apparent, eg, between the EGF receptor and *erb*-B, between the PDGF receptor and v-*sis,* and between the insulin receptor and v-*ros.*

Tyrosine kinase activity has not been proved to be involved in transduction of the insulin-receptor signal, but it could accomplish this by phosphorylating a

Table 52–3. Enzymes whose degree of phosphorylation and activity are altered by insulin.*

Enzyme	Change in Activity	Possible Mechanism
cAMP metabolism		
Phosphodiesterase (low K_m)	Increase	Phosphorylation
Protein kinase (cAMP-dependent)	Decrease	Assocation of R and C subunits
Glycogen metabolism		
Glycogen synthase	Increase	Dephosphorylation
Phosphorylase kinase	Decrease	Dephosphorylation
Phosphorylase	Decrease	Dephosphorylation
Glycolysis and gluconeo-genesis		
Pyruvate dehydrogenase	Increase	Dephosphorylation
Pyruvate kinase	Increase	Dephosphorylation
6-Phosphofructo-2-kinase	Increase	Dephosphorylation
Fructose-2,6-bisphos-phatase	Decrease	Dephosphorylation
Lipid metabolism		
Acetyl-CoA carboxylase	Increase	Phosphorylation
HMG-CoA reductase	Increase	Dephosphorylation
Triacylglycerol lipase	Decrease	Dephosphorylation
Other		
Tyrosine kinase (the insulin receptor)	?	Phosphorylation

*Modified and reproduced, with permission, from Denton RM et al: A partial view of the mechanism of insulin action. *Diabetologia* 1981;**21**:347.

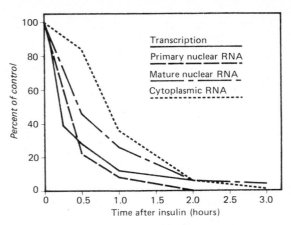

Figure 52–13. Effect of insulin on specific gene transcription. The addition of insulin to H4IIE hepatoma cells results in a rapid decrease in the rate of transcription of the PEPCK gene. This is followed by decreases in the amounts of primary transcript in the nucleus and mature mRNA[PEPCK]. The rate of synthesis of PEPCK protein declines after the amount of cytoplasmic mRNA[PEPCK] decreases. (Reproduced, with permission, from Sasaki K et al: Multihormonal regulation of phosphoenolpyruvate carboxykinase gene transcription. *J Biol Chem* 1984;**259**:15242.)

specific protein that initiates insulin action, by initiating a phosphorylation-dephosphorylation cascade, by changing some property of the cell membrane, or by generating a membrane-related product, eg, a phosphoinositol glycan.

Protein Phosphorylation-Dephosphorylation Is Involved in Some Actions of Insulin

Many of the metabolic effects of insulin, particularly those which occur rapidly, are mediated by influencing protein phosphorylation and dephosphorylation reactions that in turn alter the enzymatic activity of the protein. A list of enzymes affected in this way is presented in Table 52–3. In some instances, insulin decreases intracellular cAMP levels (by activating a cAMP-phosphodiesterase), thereby decreasing the activity state of cAMP-dependent protein kinase; examples of this action include glycogen synthase and phosphorylase. In other instances, this action is independent of cAMP and is exerted by activating other protein kinases (as is the case with the insulin receptor, tyrosine kinase); by inhibiting other protein kinases (see Table 45–5); or, more commonly, by stimulating the activity of phosphoprotein phosphatases. Dephosphorylation increases the activity of a number of key enzymes (Table 52–3). These

covalent modifications allow for almost immediate changes in the activity of enzymes.

Insulin Affects mRNA Translation

Insulin is known to affect the activity or amount of at least 50 proteins in a variety of tissues, and many of these effects involve covalent modification. A role for insulin in the translation of mRNA has been proposed, largely based on studies of ribosomal protein S6, a component of the 40S ribosomal subunit. Such a mechanism could account for the general effect insulin has on protein synthesis in liver, skeletal muscle, and cardiac muscle.

Effects on Gene Expression Are a Major Action of Insulin

The actions of insulin discussed heretofore all occur at the plasma membrane level or in the cytoplasm. In addition, insulin affects specific nuclear processes, presumably through its intracellular mediator. The enzyme **phosphoenolpyruvate carboxykinase (PEPCK)** catalyzes a rate-limiting step in gluconeogenesis. The synthesis of PEPCK is decreased by insulin; hence, gluconeogenesis decreases. Recent studies show that the rate of transcription of the PEPCK gene is selectively decreased within minutes after the addition of insulin to cultured hepatoma cells (Fig 52–13). The decrease in transcription accounts for the decreased amount of the primary transcript and of mature mRNA[PEPCK], which in turn is directly related to the decreased rate of PEPCK synthesis. This effect occurs at physiologic levels of insulin (10^{-12} to 10^{-9} mol/L), is mediated through the insulin receptor, and appears to be due to a decreased rate of mRNA[PEPCK] transcript initiation.

Table 52–4. Messenger RNAs regulated by insulin.

Intracellular enzymes
  Tyrosine aminotransferase*
  Phosphoenolpyruvate carboxykinase*
  Fatty acid synthase
  Pyruvate kinase*
  Glycerol-3-phosphate dehydrogenase*
  Glyceraldehyde-1-dehydrogenase*
  Glucokinase*
Secreted proteins and enzymes
  Albumin*
  Adipsin*
  Amylase*
  α_{2u} Globulin
  Growth hormone*
Proteins involved in reproduction
  Ovalbumin*
  Casein*
Structural proteins
  δ-Crystallin
Other proteins
  Liver (g33, etc)
  Adipose tissue
  Cardiac muscle
  Skeletal muscle

*Insulin regulates the rate of specific mRNA transcription.

Table 52–5. Comparison of insulin and the insulinlike growth factors. (Courtesy of CR Kahn.)

	Insulin	IGF-I	IGF-II
Other names	. . .	Somatomedin C	Multiplication-stimulating activity (MSA)
Number of amino acids	51	70	67
Source	Pancreatic B cells	Liver and other tissues	Diverse tissues
Level regulated by	Glucose	Growth hormone, nutritional status	Unknown
Plasma levels	0.3–2 ng/mL	ng/mL range	ng/mL range
Plasma binding protein	No	Yes	Yes
Major physiologic role	Control of metabolism	Skeletal and cartilage growth	Unknown; perhaps a role in embryonic development

Although studies of PEPCK regulation provided the first example of an effect of insulin on gene transcription, this case is no longer unique. Indeed, it appears that regulation of mRNA synthesis is a major action of insulin. A number of specific mRNAs are affected by insulin (Table 52–4), and a number of mRNAs in liver, adipose tissue, skeletal muscle, and cardiac muscle, as yet unidentified, are also affected by the hormone. In several instances, the hormone is known to affect gene transcription.

This effect of insulin involves enzymes retained in the cells, secreted enzymes and proteins, proteins involved in the reproductive process, and structural proteins (Table 52–4). A number of organs or tissues are involved, and the effect occurs in many species. The regulation of specific mRNA transcription by insulin is now well established, and as a means of modulating enzyme activity, it equals phosphorylation-dephosphorylation in importance. The effect of insulin on gene transcription may also explain its effect on embryogenesis, differentiation, and growth and replication of cells.

Pathophysiology Involving Insulin Is Mostly Expressed as Diabetes

Insulin deficiency or resistance to the action of insulin results in **diabetes mellitus**. About 90% of persons with diabetes have **non-insulin dependent (type II) diabetes mellitus (NIDDM)**. Such patients are usually obese, have elevated plasma insulin levels, and have down-regulated insulin receptors. The other 10% have **insulin-dependent (type I) diabetes mellitus (IDDM)**. The metabolic derangements discussed earlier most typically apply to the type I diabetic.

Certain rare conditions illustrate essential features about insulin action. A few individuals produce antibodies directed against their insulin receptors. These antibodies prevent insulin from binding to the receptor so that such persons develop a syndrome of severe insulin resistance (see Table 44–2). Tumors of B-cell origin cause hyperinsulinism and a syndrome characterized by severe hypoglycemia. The role of insulin (or perhaps of IGF-I or IGF-II) in organogenesis and development is illustrated by the rare cases of **leprechaunism.** This syndrome is characterized by low birth weight, decreased muscle mass, decreased subcutaneous fat, elfin facies, insulin resistance with markedly elevated plasma levels of biologically active insulin, and early death. Several individuals with leprechaunism have been shown to lack insulin receptors or to have defective receptors.

THE INSULINLIKE GROWTH FACTORS (IGF-I AND IGF-II) ARE NOT PANCREATIC HORMONES BUT ARE RELATED TO INSULIN IN STRUCTURE AND FUNCTION

It is difficult to separate the effects of insulin on cell growth and replication from similar actions exerted by IGF-I and IGF-II. Indeed, insulin and the IGFs may interact in this process. The structural similarity of these proteins was alluded to above and in Fig 52–5. A more detailed comparison is presented in Table 52–5. IGF-I and IGF-II are single-chain peptides of 70 and 67 amino acids, respectively. There is 62% homology between IGF-I and IGF-II, and these 2 hormones are identical with insulin in 50% of their residues. These molecules have unique antigenic sites and are regulated in different ways (Table 52–5). Insulin is the more potent metabolic hormone, whereas the IGFs are more potent in stimulating growth. Each hormone has a unique receptor. The IGF-I receptor, like the insulin receptor, is a hetero-

Table 52–6. Binding of insulin, IGF-I, and IGF-II to various receptors.

Hormone	Receptor		
	Insulin	IGF-I	IGF-II
Insulin	High	Low	Negligible
IGF-I	Moderate	High	Moderate
IGF-II	Negligible	Low	High

dimer of α_2-β_2 structure and is a tyrosine kinase. The IGF-II receptor, in contrast, is a single-chain polypeptide of MW 260,000 and is not a tyrosine kinase. It is closely related to, if not identical with, the mannose 6-phosphate receptor.

There is some crosstalk between these hormones and their receptors, and this probably accounts for the mixed biologic activity of these hormones (Table 52–6). In general, the growth-promoting effects of these hormones correlate best with their affinity for binding to the IGF-I or IGF-II receptor.

GLUCAGON IS AN INSULIN ANTAGONIST

The early commercial preparations of insulin increased the plasma glucose level before lowering it, owing to the presence of a contaminating peptide, glucagon, which was the second pancreatic islet cell hormone discovered.

Glucagon Is Also Synthesized as a Precursor Molecule

Glucagon, synthesized mainly in the A cells of the pancreatic islets, is a single-chain polypeptide (MW 3485) consisting of 29 amino acids (Fig 52–14). Glucagon is synthesized as a much larger (MW ~ 9000) proglucagon precursor. Molecules larger than this have been detected, but whether they represent glucagon precursors or closely related peptides is unclear. Only 30–40% of the immunoreactive "glucagon" in plasma is pancreatic glucagon; the rest consists of biologically inactive larger molecules.

Glucagon shares some immunologic and phys-

iologic properties with enteroglucagon, a peptide extracted from the duodenal mucosa, and 14 of the 27 amino acid residues of secretin are identical to those of glucagon.

Glucagon circulates in plasma in the free form. Since it does not associate with a transport protein, its plasma half-life is short (~5 minutes). Glucagon is inactivated by the liver, which has an enzyme that removes the first 2 amino acids from the N-terminal end by cleaving between Ser 2 and Gln 3. Since the liver is the first stop for glucagon after it is secreted and since the liver rapidly inactivates the hormone, the level of glucagon in the portal vein is much higher than that in the peripheral circulation.

The Secretion of Glucagon Is Inhibited by Glucose, an Action that Emphasizes the Opposing Metabolic Roles of Glucagon and Insulin

It is not clear whether glucose directly inhibits glucagon secretion or whether this is mediated through the actions of insulin or IGF-I, since both of these islet cell hormones directly inhibit glucagon release. Many other substances, including amino acids, fatty acids and ketones, gastrointestinal tract hormones, and neurotransmitters, affect glucagon secretion.

In General, the Actions of Glucagon Oppose Those of Insulin

Whereas insulin promotes energy storage by stimulating glycogenesis, lipogenesis, and protein synthesis, glucagon causes the rapid mobilization of potential energy sources into glucose by stimulating glycogenolysis and into fatty acids by stimulating lipolysis. Glucagon is also the most potent gluconeogenic hormone, and it is ketogenic.

The liver is the primary target of glucagon action. Glucagon binds to specific receptors in the hepatic cell plasma membrane, and this activates adenylate cyclase. The cAMP generated activates phosphorylase, which enhances the rate of glycogen degradation while inhibiting glycogen synthase and thus glycogen forma-

Glucagon NH₂— His — Ser — Gln — Gly — Thr — Phe — Thr — Ser — Asp — Tyr —
Ser — Lys — Tyr — Leu — Asp — Ser — Arg — Arg — Ala — Gln —
Asp — Phe — Val — Gln — Trp — Leu — Met — Asn — Thr — COOH

Somatostatin NH₂— Ala — Gly — Cys — Lys — Asn — Phe — Phe — Trp — Lys — Thr —
Phe — Thr — Ser — Cys — COOH

Figure 52–14. The amino acid sequences of glucagon and somatostatin.

Table 52–7. Enzymes induced or repressed by insulin or glucagon.*

Enzymes induced by a high insulin:glucagon ratio and repressed by a low insulin:glucagon ratio
Glucokinase
Citrate cleavage enzyme
Acetyl-CoA carboxylase
HMG-CoA reductase
Pyruvate kinase
6-Phosphofructo-1-kinase
6-Phosphofructo-2-kinase/fructose-2,6-bisphosphatase

Enzymes induced by a low insulin:glucagon ratio and repressed by a high insulin:glucagon ratio
Glucose-6-phosphatase
Phosphoenolpyruvate carboxykinase (PEPCK)
Fructose-1,6-bisphosphatase

*Slightly modified and reproduced, with permission, from Karam JH, Salber PR, Forsham PH: Pancreatic hormones and diabetes mellitus. In: *Basic & Clinical Endocrinology,* 2nd ed. Greenspan FS, Forsham PH (editors). Appleton & Lange, 1986.

tion (see Chapter 45). There is hormone and tissue specificity in this effect, since glucagon has no effect on glycogenolysis in muscle, whereas epinephrine is active in both muscle and liver.

The elevated cAMP level stimulates the conversion of amino acids to glucose by inducing a number of enzymes involved in the gluconeogenic pathway. Principal among these is PEPCK. Glucagon, through cAMP, increases the rate of transcription of mRNA from the PEPCK gene, and this stimulates the synthesis of more PEPCK. This is the opposite of the effect of insulin, which decreases PEPCK gene transcription. Other examples are illustrated in Table 52–7. The net action of glucagon in the liver is increased glucose production; since much of this glucose exits the liver, the plasma glucose concentration increases in response to glucagon.

Glucagon is a potent lipolytic agent, it increases adipose cell cAMP levels, and this activates the hormone-sensitive lipase. The increased fatty acids can be metabolized for energy or converted to the ketone bodies acetoacetate and β-hydroxybutyrate. This is an important aspect of metabolism in the diabetic, since glucagon levels are always increased in insulin deficiency.

SOMATOSTATIN INHIBITS GH SECRETION

Somatostatin, so-named because it was first isolated from the hypothalamus as the factor that inhibited growth hormone secretion, is a cyclic peptide synthesized as a large somatostatin prohormone (MW ~ 11,500) in the D cells of the pancreatic islets. The rate of transcription of the prosomatostatin gene is markedly enhanced by cAMP. The prohormone is first pro-

cessed into a 28-amino-acid peptide and finally into a molecule that has a molecular weight of 1640 and contains 14 amino acids (Fig 52–14). All forms have biologic activity.

In addition to its presence in the hypothalamus and pancreatic islets, somatostatin is found in many gastrointestinal tissues, where it is thought to regulate a variety of functions, and in multiple sites in the central nervous system, where it may be a neurotransmitter.

Somatostatin inhibits the release of the other islet cell hormones through a paracrine action. In pharmacologic amounts, somatostatin significantly blunts the ketosis associated with acute insulin deficiency. This is apparently due to its ability to inhibit the glucagon release that accompanies insulinopenia. It also decreases the delivery of nutrients from the gastrointestinal tract into the circulation, because it (1) prolongs gastric emptying, (2) decreases gastrin secretion and therefore gastric acid production, (3) decreases pancreatic exocrine (digestive enzyme) secretion, (4) decreases splanchnic blood flow, and (5) slows sugar absorption. Little is known about the biochemical and molecular actions of this hormone.

THE FUNCTION OF PANCREATIC POLYPEPTIDE IS UNKNOWN

Pancreatic polypeptide (PP), a 36-amino-acid peptide (MW ~ 4200), is a recently discovered product of the pancreatic F cells. Its secretion in humans is increased by a protein meal, fasting, exercise, and acute hypoglycemia and is decreased by somatostatin and intravenous glucose. The function of pancreatic polypeptide is unknown, but effects on hepatic glycogen levels and gastrointestinal secretion have been suggested.

THE GASTROINTESTINAL HORMONES

The Discipline of Endocrinology Began With the Discovery of a Gastrointestinal Hormone

In 1902, Bayliss and Starling instilled hydrochloric acid into a denervated loop of a dog's jejunum and showed that this increased the secretion of fluid from the pancreas. Intravenous injection of HCl did not mimic this effect, but the intravenous injection of an extract of the jejunal mucosa did. These investigators postulated that "secretin," released from the mucosa of the upper intestine in response to a stimulus, moved to the pancreas through the circulation, where it exerted its effect. Bayliss and Starling were the first to use the word "hormone," and secretin was the first hormone whose function was identified.

Although the activity of secretin was identified in

Table 52–8. Gastrointestinal hormones.

Hormone	Location	Major Action
Gastrin	Gastric antrum, duodenum	Gastric adid and pepsin secretion
Cholecystokinin (CCK)	Duodenum, jejunum	Pancreatic amylase secretion
Secretin	Duodenum, jejunum	Pancreatic bicarbonate secretion
Gastric inhibitory poly-peptide (GIP)	Small bowel	Enhances glucose-mediated insulin release; inhibits gastric acid secretion
Vasoactive intestinal polypeptide (VIP)	Pancreas	Smooth muscle relaxation; stimulates pancreatic bicarbonate secretion
Motilin	Small bowel	Initiates interdigestive intestinal motility
Somatostatin	Stomach, duodenum, pancreas	Numerous inhibitory effects
Pancreatic polypeptide (PP)	Pancreas	Inhibits pancreatic bicarbonate and protein secretion
Enkephalins	Stomach, duodenum, gallbladder	Opiate-like actions
Substance P	Entire gastrointestinal tract	Physiologic actions uncertain
Bombesin-like immuno-reactivity (BLI)	Stomach, duodenum	Stimulates release of gastrin and CCK
Neurotensin	Ileum	Physiologic actions unknown
Enteroglucagon	Pancreas, small intestine	Physiologic actions unknown

1902, it took 60 years before its chemical identity was proved. The reasons for the 60-year time span are now apparent; families of closely related gastrointestinal peptides have overlapping chemical structures and biologic functions, and most of these peptides exist in multiple forms.

Of the major gastrointestinal hormones, only secretin exists in a single form. The presence of multiple forms of gastrointestinal peptides in gastrointestinal tissues and in the circulation impeded the definition of the number and nature of these molecules. The concept of precursor molecules helped clarify this issue; much of tissue heterogeneity is due to this feature. Isolation techniques developed only recently also have helped to differentiate among them.

Gastrointestinal Hormones Have Some Special Features

More than a dozen peptides with uniques actions have been isolated from gastrointestinal tissues (Table 52–8). A unique feature of this group of hormones is that many fit the classic definition of a hormone, some have paracrine actions, and others act in a neurocrine fashion (as local neurotransmitters or neuromodulators).

Another unique aspect of the gastrointestinal endocrine system is that the cells are scattered throughout the gastrointestinal tract rather than collected in discrete organs as in more typical endocrine glands. The

distribution of the gastrointestinal hormones is outlined in Table 52–8.

Since many of the gastrointestinal peptides are found in the nerves in gastrointestinal tissues, it is not surprising that most of them are also present in the central nervous system.

There Are Families of GI Hormones

Many of these hormones can be placed in one of 2 families based on amino acid sequence and functional similarity. These are the **gastrin family,** which consists of gastrin and CCK, and the **secretin family,** which includes secretin, glucagon, GIP, VIP, and glicentin (which has glucagonlike immunoreactivity but is a distinct peptide). The neurocrine peptides neurotensin, bombesinlike peptides, substance P, and somatostatin bear no structural similarity to any other gastrointestinal peptide. A final general characteristic of this last group of molecules is that they have very short plasma half-lives and may play no physiologic role in plasma.

Relatively Little Is Known About the Mechanism of Action of the GI Hormones

Studies of the mechanism of action of the gastrointestinal peptide hormones have lagged behind those of other hormones, no doubt because most attention to date has been directed toward cataloging the various molecules and establishing their physiologic action. A notable exception to this statement involves the regulation of secretion of enzymes by the pancreatic acinar cell.

Six different classes of receptors on pancreatic acinar cells have been identified. These are for (1) muscarinic cholinergic agents, (2) the gastrin-CCK family, (3) bombesin and related peptides, (4) the physalaemin-substance P family, (5) secretin and VIP, and (6) cholera toxin. Classes 1–4 appear to act through the calcium-phosphoinositide mechanism, whereas groups 5 and 6 act through cAMP.

REFERENCES

Cohen P: The role of protein phosphorylation in neural and hormonal control of cellular activity. *Nature* 1982;**296:** 613.

Docherty K, Steiner D: Post-translational proteolysis in polypeptide hormone biosynthesis. *Annu Rev Physiol* 1982; **44:**625.

Granner DK, Andreone T: Insulin modulation of gene expression. In: *Diabetes and Metabolism Reviews.* Vol I. DeFronzo R (editor). Wiley, 1985.

Kahn CR: The molecular mechanism of insulin action. *Annu Rev Med* 1985;**36:**429.

Kono T: Action of insulin on glucose transport and cAMP phosphodiesterase in fat cells: Involvement of two distinct molecular mechanisms. *Recent Prog Horm Res* 1983;**30:** 519.

Rosen O: After insulin binds. *Science*

Ullrich A et al: Human insulin receptor and its relationship to the tyrosine kinase family of oncogenes. *Nature* 1985; **313:**756.

Unger RH, Orci L: Glucagon and the A cell. (2 parts.) *N Engl J Med* 1981;**304:**1518,1575.

Section VI.
Special Topics

Structure and Function of the Water-Soluble Vitamins

53

Peter A. Mayes, PhD, DSc

INTRODUCTION

Vitamins are organic nutrients that are required in small quantities for a variety of biochemical functions and which, generally, cannot be synthesized by the body and must therefore be supplied by the diet. The first discovered vitamins, A and B, were found to be fat and water soluble, respectively. As more vitamins were discovered they were also shown to be either fat or water soluble, and this property was used as a basis for their classification. The water-soluble vitamins were all designated members of the B complex (apart from vitamin C), and the newly discovered fat-soluble vitamins were given alphabetic designations (eg, vitamins D, E, K). Apart from their solubility characteristics, the water-soluble vitamins have little in common from the chemical point of view.

BIOMEDICAL IMPORTANCE

Absence or relative deficiency of vitamins in the diet leads to characteristic deficiency states and diseases. Deficiency of a single vitamin of the B complex is rare, since poor diets are most often associated with **multiple deficiency states**. Nevertheless, definite syndromes are characteristic of deficiencies of specific vitamins. Among the water-soluble vitamins the following deficiency states are recognized: **beriberi** (thiamin deficiency); **cheilosis, glossitis, seborrhea,** and **photophobia** (riboflavin deficiency); **pellagra** (niacin deficiency); **peripheral neuritis** (pyridoxine deficiency); **megaloblastic anemia, methylmalonic aciduria,** and **pernicious anemia** (cobalamin deficiency); **megaloblastic anemia** (folic acid deficiency); and **scurvy** (ascorbic acid deficiency). Vitamin deficiencies are avoided by consumption of food of a wide variety in adequate amounts.

VITAMINS OF THE B COMPLEX ARE COFACTORS IN ENZYMATIC REACTIONS

The B vitamins essential for human nutrition are: (1) thiamin (vitamin B_1), (2) riboflavin (vitamin B_2), (3) niacin (nicotinic acid, nicotinamide) (vitamin B_3), (4) pantothenic acid (vitamin B_5), (5) vitamin B_6 (pyridoxine, pyridoxal, pyridoxamine), (6) biotin, (7) vitamin B_{12} (cobalamin), and (8) folic acid (pteroylglutamic acid).

Because of their water solubility, excesses of these vitamins are excreted in urine and so rarely accumulate in toxic concentrations. For the same reason, their storage is limited (apart from cobalamin) and as a consequence they **must be provided regularly.**

THIAMIN

Thiamin consists of a substituted pyrimidine joined by a methylene bridge to a substituted thiazole. (Fig 53-1.)

Active Thiamin Is Thiamin Diphosphate

An ATP-dependent thiamin diphosphotransferase present in brain and liver is responsible for the conversion of thiamin to its active form, thiamin diphosphate (pyrophosphate) (Fig 53-1).

Thiamin Diphosphate Serves as a Coenzyme in Enzymatic Reactions, Transferring an Activated Aldehyde Unit

There are two types of such reactions: (1) an **oxidative decarboxylation** of α-keto acids (eg, α-ketoglutarate, pyruvate, and the α-keto analogs of leucine, isoleucine, and valine); and (2) **transketolase reactions** (eg, in the pentose phosphate pathway). All of these reactions are **inhibited in thiamin deficiency.** In each case, the thiamin diphosphate provides a reac-

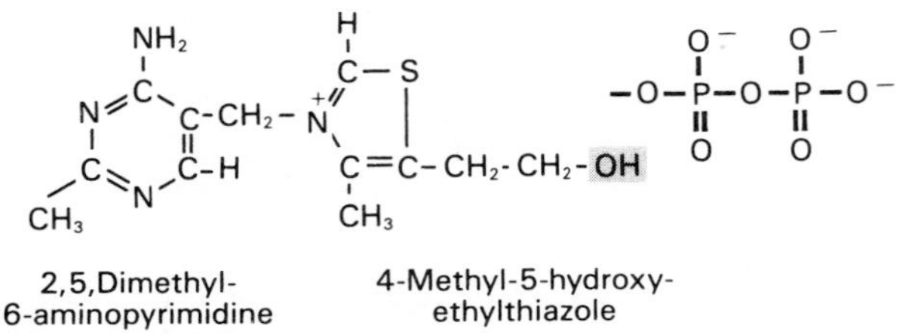

Figure 53–1. Thiamin. In thiamin diphosphate, the —OH group is replaced by pyrophosphate.

tive carbon on the thiazole that forms a carbanion, which is then free to add to the carbonyl group of, for instance, pyruvate (Fig 19–5). The addition compound then decarboxylates, eliminating CO_2. This reaction occurs in a multienzyme complex known as the pyruvate dehydrogenase complex (see Chapter 19 for further details).

The oxidative decarboxylation of α-ketoglutarate to succinyl-CoA and CO_2 (see Chapter 18) is catalyzed by an enzyme complex structurally very similar to the pyruvate dehydrogenase complex. Again, the thiamin diphosphate provides a stable carbanion to react with the α carbon of α-ketoglutarate. A similar oxidative decarboxylation of the α-ketocarboxylic acid derivatives of the branched-chain amino acids (Chapter 32) utilizes thiamin diphosphate. The role of thiamin diphosphate as a coenzyme in the transketolase reactions (Chapter 21) is very similar to that described above for the oxidative decarboxylations.

Lack of Thiamin Causes Beriberi & Related Deficiency Syndromes

In the thiamin-deficient human, thiamin diphosphate-dependent reactions are prevented or severely limited, leading to accumulation of the substrates of the reactions, eg, pyruvate, pentosugars, and the α-

ketocarboxylate derivatives of the branched-chain amino acids leucine, isoleucine, and valine.

Thiamin is present in almost all plant and animal tissues commonly used as food, but the content is usually small. Unrefined cereal grains and meat are good sources of the vitamin. **Beriberi** is caused by carbohydrate-rich/low-thiamin diets, eg, polished rice or other highly refined foods such as sugar and white flour acting as the staple food source. Early symptoms include peripheral neuropathy, exhaustion, and anorexia, which lead on to edema and cardiovascular, neurological, and muscular degeneration. **Wernicke's encephalopathy** is a condition associated with thiamin deficiency. It is found frequently in chronic alcoholics consuming little other food. Certain raw fish contain a heat-labile enzyme (thiaminase) that destroys thiamin, but this is not considered to be critical in human nutrition.

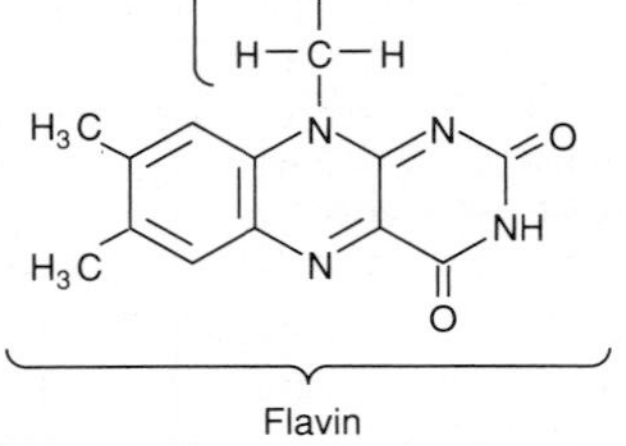

RIBOFLAVIN

Riboflavin consists of a heterocyclic isoalloxazine ring attached to the sugar alcohol, ribitol (See Fig 52–2.). It is a colored, fluorescent pigment that is relatively heat stable but decomposes in the presence of visible light.

Active Riboflavin Is Flavin Mononucleotide (FMN) & Flavin Adenine Dinucleotide (FAD)

FMN is formed by ATP-dependent phosphorylation of riboflavin (Fig 53–2), whereas FAD is synthesized by a further reaction with ATP in which the AMP moiety of ATP is transferred to FMN (Fig 53–3).

FMN and FAD Serve as Prosthetic Groups of Oxidoreductase Enzymes

These enzymes are known as **flavoproteins.** The prosthetic groups are usually tightly but not covalently bound to their apoproteins. Many flavoprotein enzymes contain one or more metals, eg, molybdenum and iron, as essential cofactors and are known as **metalloflavoproteins.**

Flavoprotein enzymes are widespread and are represented by several important oxidoreductases in mammalian metabolism, eg., **α-amino acid oxidase** in amino acid deamination (see p 273), **xanthine oxidase**

Figure 53–2. Riboflavin. In riboflavin phosphate (flavin mononucleotide, FMN), the —OH is replaced by phosphate.

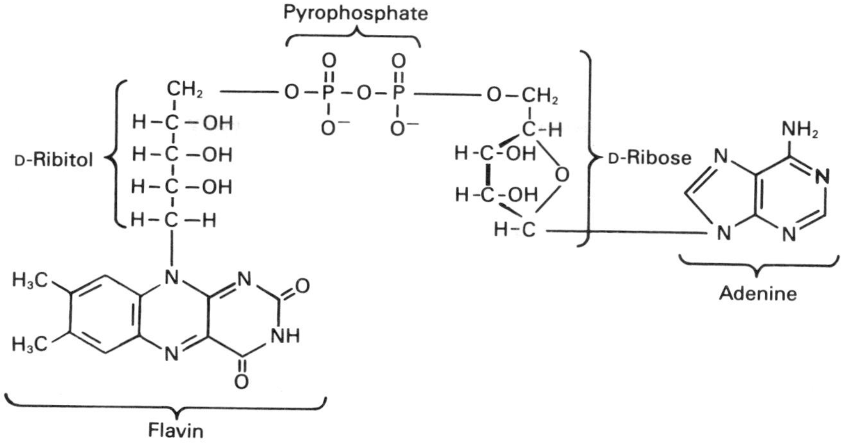

Figure 53–3. Flavin adenine dinucleotide (FAD).

in purine degradation (see p 345), **aldehyde dehydrogenase** in the degradation of aldehydes, **mitochondrial glycerol-3-phosphate dehydrogenase** in transporting reducing equivalents from the cytosol into mitochondria (see p 120), **succinate dehydrogenase** in the citric acid cycle (see p 159), **acyl-CoA dehydrogenase** and the **electron-transferring flavoprotein** in fatty acid oxidation (see p 207), and **dihydrolipoyl dehydrogenase** in the oxidative decarboxylation of pyruvate and α-ketoglutarate (see p 168); **NADH dehydrogenase** is a major component of the respiratory chain in mitochondria (see p 114). All of these enzyme systems are impaired in riboflavin deficiency.

In their role as coenzymes, flavoproteins undergo reversible reduction of the isoalloxazine ring to yield the reduced forms $FMNH_2$ and $FADH_2$ (see Fig 13–3).

Lack of Riboflavin Causes a General Nonfatal Deficiency Syndrome

In view of its widespread metabolic functions it is surprising that riboflavin deficiency does not lead to major life-threatening conditions. However, when there is deficiency, various symptoms are seen, including angular stomatitis, cheilosis, glossitis, seborrhea, and photophobia.

Riboflavin is synthesized by plants and microorganisms, but not by mammals. Yeast, liver, and kidney are good sources of the vitamin, which is absorbed in the intestine by a phosphorylation-dephosphorylation sequence in the mucosa. Hormones (eg, thyroid hormone and ACTH), drugs (eg, chlorpromazine, a competitive inhibitor), and nutritional factors affect the conversion of riboflavin to its cofactor forms. Because of its light sensitivity, riboflavin deficiency may occur in newborn infants with hyperbilirubinemia who are treated by phototherapy.

NIACIN

Niacin is the generic name for nicotinic acid and nicotinamide either of which may act as a source of the vitamin in the diet. Nicotinic acid is a monocarboxylic acid derivative of pyridine (Fig 53–4).

Active Niacin Is Nicotinamide Adenine Dinucleotide (NAD+) and Nicotinamide Adenine Dinucleotide Phosphate (NADP+)

Nicotinate is the form of niacin required for the synthesis of NAD+ and NADP+ by enzymes present in the cytosol of most cells. Therefore any dietary nicotinamide must first undergo deamidation to nicotinate (Fig 53–4). In the cytosol, nicotinate is converted to **desamido-NAD+** by reaction first with 5-phosphoribosyl 1-pyrophosphate (PRPP) and then by adenylation with ATP. The amido group of glutamine then contributes to form the coenzyme NAD+. This may be phosphorylated further to form NADP+.

NAD+ and NADP+ Are Coenzymes of Many Oxidoreductase Enzymes

The nicotinamide nucleotides play a widespread role as coenzymes to many dehydrogenase enzymes occurring both in the cytosol, eg, **lactate dehydrogenase,** and within the mitochondria, eg, **malate dehydrogenase.** They are therefore key components of many metabolic pathways affecting carbohydrate, lipid, and amino acid metabolism. Generally, **NAD-linked dehydrogenases** catalyze oxidoreduction reactions in **oxidative pathways,** eg, the citric acid cycle, whereas **NADP-linked dehydrogenases** or reductases are often found in pathways concerned with **reductive syntheses,** eg, the pentose phosphate pathway.

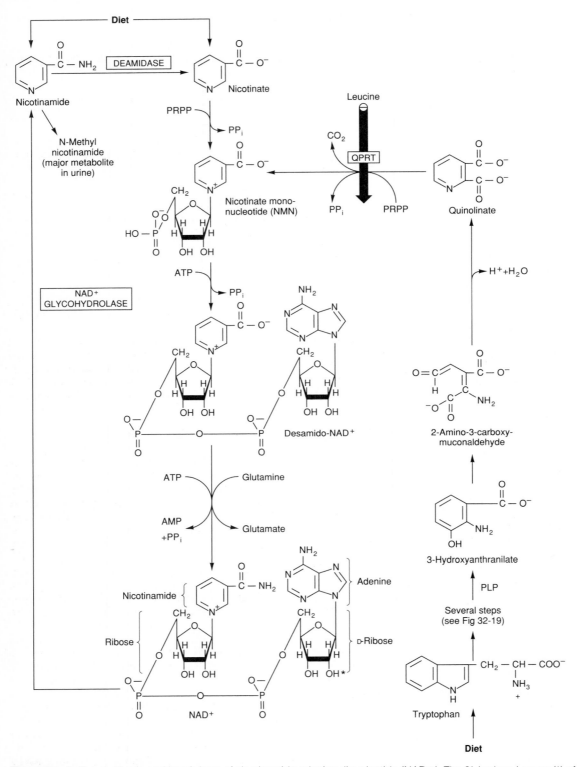

Figure 53–4. The synthesis and breakdown of nicotinamide adenine dinucleotide (NAD^+). The 2′-hydroxyl group (*) of the adenosine moiety is phosphorylated in nicotinamide dinucleotide phosphate ($NADP^+$). Humans, but not cats, can provide all of their niacin requirement from tryptophan if there is a sufficient amount in the diet. Normally, about two-thirds comes from this source. PRPP, 5-phosphoribosyl-1-pyrophosphate QPRT, quinolinate phosphoribosyl transferase. PLP, pyridoxal phosphate.

The mechanism of oxidoreduction involves a reversible addition of a hydride ion (H^-) to the pyridine ring plus the generation of a free hydrogen ion (H^+) (see Fig 13–5), eg,

$$NAD^+ + AH_2 \leftrightarrow NADH + H^+ + A$$

Lack of Niacin Causes the Deficiency Syndrome Pellagra

Symptoms include weight loss, digestive disorders, dermatitis, depression, and dementia.

Niacin is found widely in most animal and plant foods. However, assessment of niacin value of a food must take account of the fact that the essential amino acid **tryptophan** can be converted to NAD^+ (see Fig 53–4). For every 60 mg of tryptophan, 1 mg equivalent of niacin can be generated. Thus, **in order to produce niacin deficiency a diet must be poor in both available niacin and tryptophan.** Such criteria occur in populations dependent on maize (corn) as the staple food, resulting in the condition of **pellagra.** In maize, niacin is in fact present, but in a bound unavailable form, niacytin, from which niacin can be released by pretreatment with alkali. Dependence on sorghum is also pellagragenic, due not to low tryptophan but to a high **leucine** content. Apparently, excess dietary leucine can bring about niacin deficiency by inhibiting **quinolinate phosphoribosyl transferase,** a key enzyme in the conversion of tryptophan to NAD^+ (see Fig 53–4). It is also of note that **pyridoxal phosphate,** the active form of vitamin B_6, is involved as a cofactor in the pathway of synthesis of NAD^+ from tryptophan (see Fig 53–4) and therefore vitamin B_6 deficiency can potentiate a deficiency in niacin.

Other conditions leading to symptoms of pellagra include administration of some drugs such as **isoniazid, malignant carcinoid syndrome** in which tryptophan metabolism is diverted to serotonin, and **Hartnup disease,** in which tryptophan absorption is impaired.

Nicotinic acid (but not nicotinamide) has been used therapeutically for lowering plasma cholesterol. This is due to inhibition of the flux of FFA from adipose tissue, which leads to less formation of the cholesterol-bearing lipoproteins, VLDL, IDL, and LDL (See p 259).

PANTOTHENIC ACID

Pantothenic acid is formed by combination of pantoic acid and β alanine (Fig 53–5.)

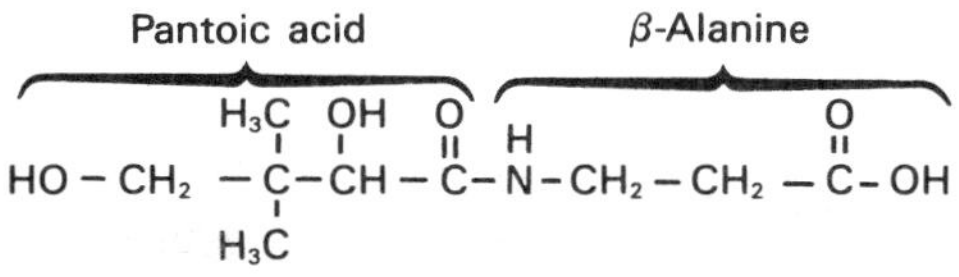

Active Pantothenic Acid Is Coenzyme A (CoA) and the Acyl Carrier Protein (ACP)

Pantothenic acid is absorbed readily in the intestines and subsequently phosphorylated by ATP to form 4'-phosphopantothenate (Fig 53–6). Addition of cysteine and removal of its carboxyl group results in the net addition of thioethanolamine, generating **4'-phospho-**

Figure 53–5. Pantothenic acid.

pantetheine, the prosthetic group of both CoA and ACP. Like the active coenzymes of so many other water-soluble vitamins, CoA contains an adenine nucleotide. Thus, 4'-phosphopantetheine is adenylylated by ATP to form **dephospho-CoA.** The final phosphorylation occurs with ATP adding phosphate to the 3'-hydroxyl group of the ribose moiety to generate **CoA** (Fig 53–6).

The Thiol Group Acts as a Carrier of Acyl Radicals in Both CoA and ACP

This occurs with CoA in reactions of the citric acid cycle (see p 155), fatty acid oxidation and synthesis (see pp 207 and 199), acetylation reactions (eg, of drugs), and cholesterol synthesis (see p 249). ACP participates in reactions concerned with fatty acid synthesis (see p 199). It is customary to abbreviate the structure of the free (ie, reduced) CoA as CoA-SH, in which the reactive SH group of the coenzyme is designated.

Deficiency of Pantothenic Acid is Rare

This is because it is widely distributed in foods, being particularly abundant in animal tissues, whole-grain cereals, and legumes. However, the **burning foot syndrome** has been ascribed to pantothenate deficiency in prisoners of war and is associated with reduced capacity for acetylation.

VITAMIN B_6

Vitamin B_6 consists of 3 closely related pyridine derivatives: **pyridoxine, pyridoxal,** and **pyridoxamine** (Fig 53–7) and their corresponding phosphates. Of these, pyridoxine, pyridoxal phosphate, and pyridoxamine phosphate are the main representatives of the vitamin in the diet. All 3 have equal vitamin activity.

Active Vitamin B_6 Is Pyridoxal Phosphate

All forms of vitamin B_6 are absorbed from the intestine, but some hydrolysis of the phosphate esters occurs during digestion. Pyridoxal phosphate is the major form transported in plasma. Most tissues contain the enzyme **pyridoxal kinase,** which is able to catalyze the phosphorylation by ATP of the unphosphorylated forms of the vitamin to their respective phosphate esters (Fig 53–8). While pyridoxal phos-

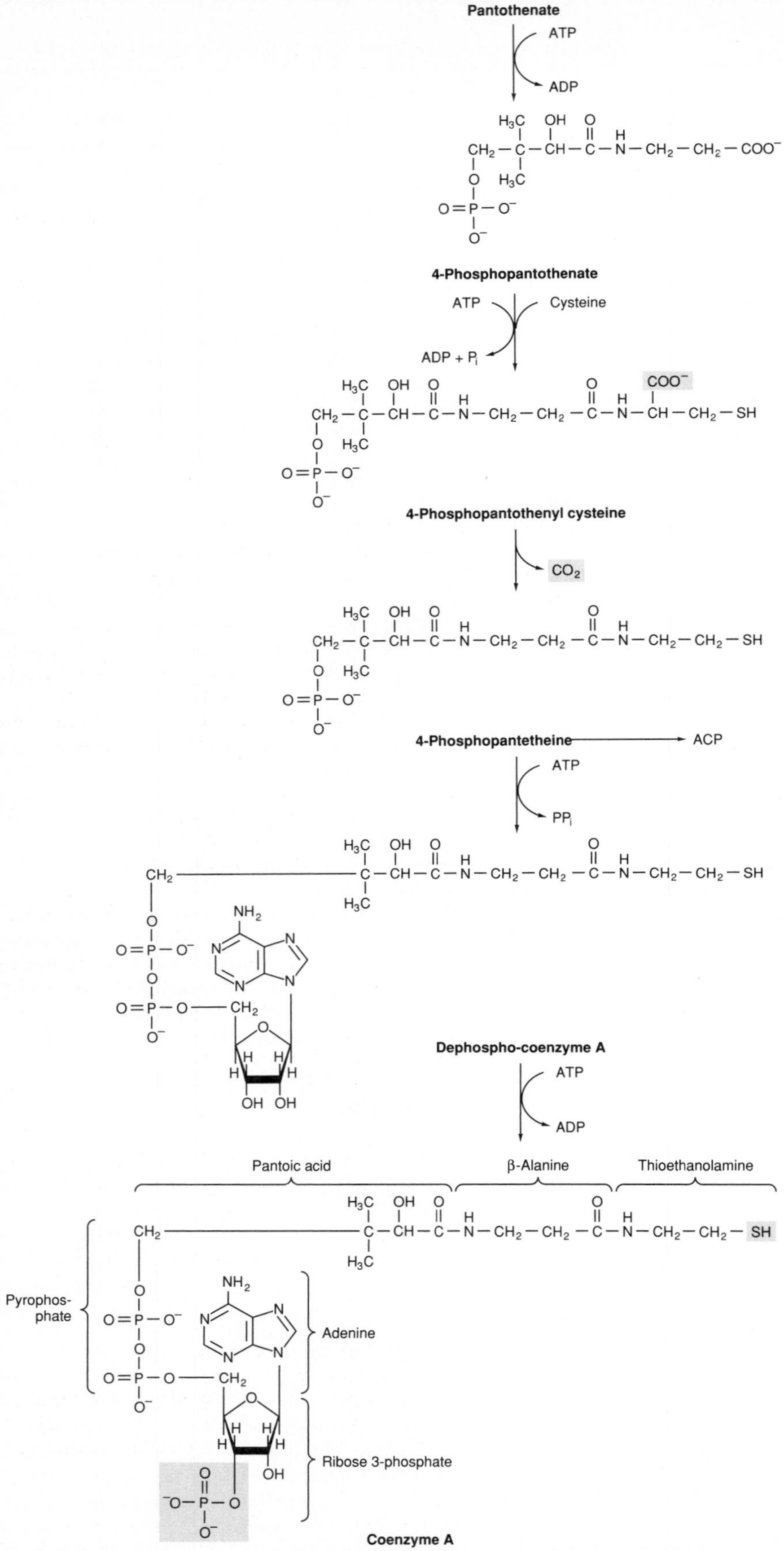

Figure 53–6. The synthesis of coenzyme A from pantothenic acid. ACP, acyl carrier protein.

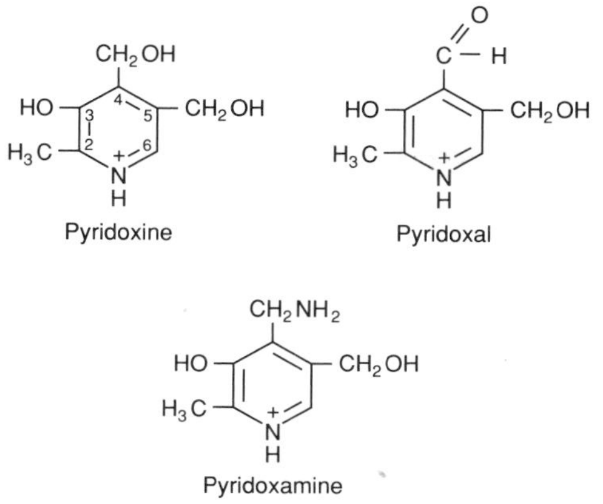

Figure 53–7. Naturally occurring forms of vitamin B_6.

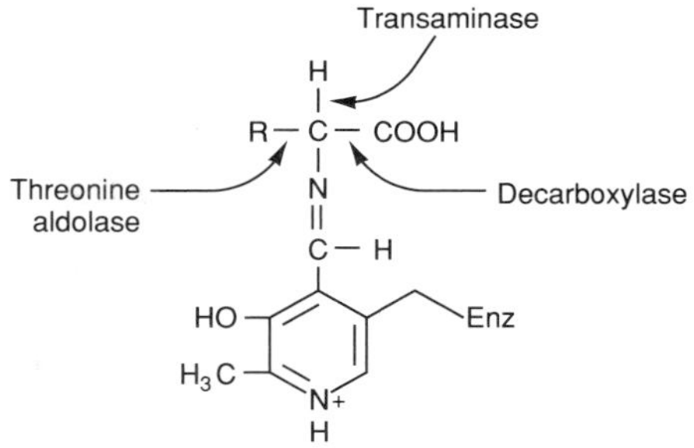

Figure 53–9. The covalent bonds of an α-amino acid that can be made reactive by its binding to various pyridoxal phosphate specific enzymes.

phate is the major coenzyme expressing B_6 activity, pyridoxamine phosphate may also act as an active coenzyme.

Pyridoxal Phosphate Is the Coenzyme of Several Enzymes of Amino Acid Metabolism

By entering into a **Schiff base** combination between its aldehyde group and the amino group of an α-amino acid (Fig 53–9), pyridoxal phosphate can facilitate changes in the 3 remaining bonds of the α-amino carbon to allow either transamination (see p 275), decarboxylation (see Fig 33–9), or aldolase activity (see Fig 32–11), respectively. The role of pyridoxal phosphate in transamination is illustrated in Fig 53–10.

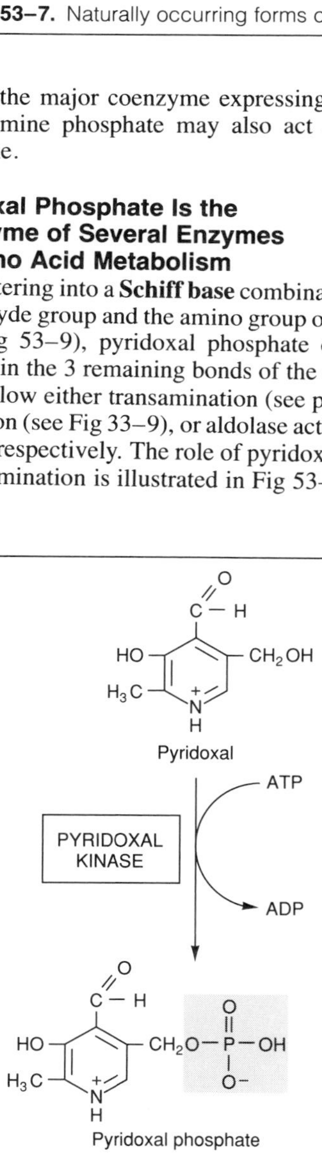

Figure 53–8. The phosphorylation of pyridoxal by pyridoxal kinase to form pyridoxal phosphate.

Pyridoxal Phosphate Is Also Involved in Glycogenolysis

The coenzyme is an integral part of the mechanism of action of phosphorylase, the enzyme mediating the breakdown of glycogen (see p 174). In this action it also forms an initial Schiff base with an ϵ-amino group of a lysine residue of the enzyme, which, however, remains intact throughout the phosphorolysis of the $1 \rightarrow 4$ glycosidic bond to form glucose 1-phosphate. Muscle phosphorylase may account for as much as 70–80% of total body vitamin B_6.

Deficiency of Vitamin B_6 May Occur During Lactation, in Alchoholics, and During Isoniazid Therapy

A deficiency due to lack of vitamin B_6 alone is rare, and any deficiency is usually part of a general deficiency of B-complex vitamins. Liver, mackerel, avocados, bananas, meat, vegetables, and eggs are good sources of the vitamin. A possibility of deficiency is recognized in nursing infants whose mothers are depleted of the vitamin due to long-term use of oral contraceptives. Alcoholics may also be deficient due to metabolism of ethanol to acetaldehyde, which stimulates hydrolysis of the phosphate of the coenzyme. A widely used antituberculosis drug, **isoniazid,** can induce vitamin B_6 deficiency by forming a hydrazone with pyridoxal (Fig 53–11).

BIOTIN

Biotin is an imidazole derivative widely distributed in natural foods (Fig 53–12.). As a large portion of the human requirement for biotin is met by **synthesis from intestinal bacteria,** biotin deficiency is caused not by simple dietary deficiency but by defects in utilization.

Biotin is a Coenzyme of Carboxylase Enzymes

Biotin functions as a component of specific multisubunit enzymes (Table 53–1) that catalyze carbox-

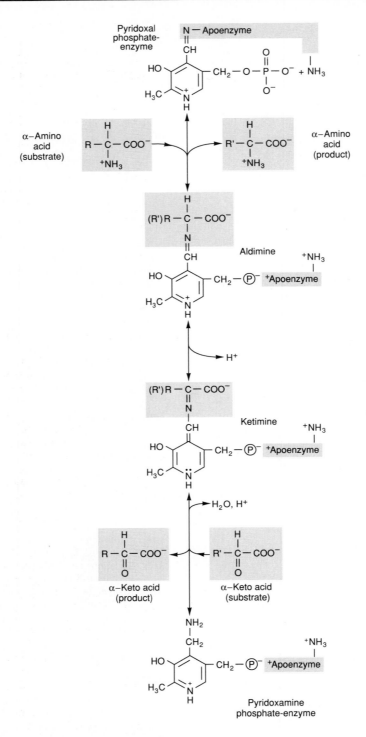

Figure 53–10. The role of pyridoxal phosphate coenzyme in the transamination of an α-amino acid. The first phase involves the production of the corresponding α-keto acid and pyridoxamine phosphate enzyme, followed by reversal of the process using a new α-keto acid as substrate. It will be noted that initially pyridoxal phosphate binds to its apoenzyme by a Schiff base link between its aldehyde group and an amino group of the enzyme (an ε-amino group of a lysine residue) and via an ionic bond between its phosphate and the enzyme. The α-amino group of a substrate amino acid displaces the ε-amino group, forming a new Schiff base.

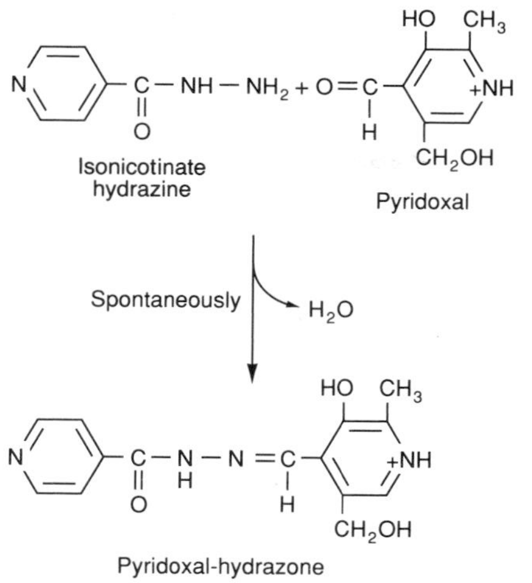

Figure 53–11. The formation of the rapidly excreted pyridoxal-hydrazone from pyridoxal and isonicotinate hydrazine (isoniazid).

Table 53–1. Biotin-dependent enzymes in animals.

Enzyme	Role
Pyruvate carboxylase (see p 179)	First reaction in pathway that converts 3-carbon precursors to glucose (gluconeogenesis)
	Replenishes oxaloacetate for citric acid cycle
Acetyl-CoA carboxylase (see p 199)	Commits acetate units to fatty acid synthesis by forming malonyl-CoA
Propionyl-CoA carboxylase (see p 181)	Converts propionate to succinate, which can then enter citric acid cycle
β-Methylcrotonyl-CoA carboxylase (see p 303)	Catabolism of leucine and certain isoprenoid compounds

ylase reactions. A carboxylate ion is attached to the N^1 of the biotin, generating an activated intermediate, **carboxybiotin-enzyme** (Fig 53–13). This step requires HCO_3^-, ATP, Mg^{2+}, and acetyl-CoA (as an allosteric effector). The activated carboxyl group is then transferred to the substrate of the reaction, eg, pyruvate.

Consumption of Raw Eggs Can Cause Biotin Deficiency

Egg white contains a heat-labile protein, **avidin,** which combines very tightly with biotin, preventing its absorption and inducing biotin deficiency. The symptoms include depression, hallucination, muscle pain, and dermatitis. Absence of the enzyme **holocarboxylase synthase,** which attaches biotin to the lysine residue of the carboxylase apoenzymes, also causes biotin deficiency symptoms, including accumulation of substrates of the biotin-dependent enzymes, which can be detected in urine. These metabolites include lactate, β-methylcrotonate, β-hydroxyisovalerate, and β-hydroxypropionate. In some cases children with this deficiency have immune deficiency diseases.

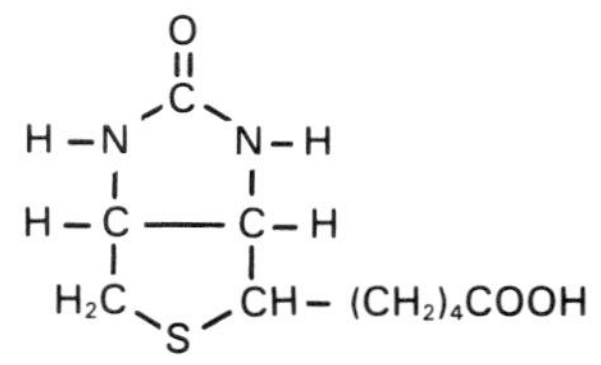

Figure 53–12. Biotin.

VITAMIN B₁₂

Vitamin B_{12} (cobalamin) has a complex ring structure (corrin ring), similar to a porphyrin ring, to which is added a **cobalt ion** at its center (Fig 53–14). The vitamin is synthesized exclusively by microorganisms. Thus, it is absent from plants—unless they are contaminated by microorganisms—but is conserved in animals in the liver, where it is found as methylcobalamin, adenosylcobalamin, and hydroxocobalamin. Liver is therefore a good source of the vitamin, as is yeast. The commercial preparation is cyanocobalamin.

Intrinsic Factor Is Necessary for Absorption of Vitamin B₁₂

The intestinal absorption of vitamin B_{12} is mediated by receptor sites in the ileum that require it to be bound by a highly specific glycoprotein, **intrinsic factor,** secreted by parietal cells of the gastric mucosa. After absorption, the vitamin is bound by a plasma protein, **transcobalamin II,** for transport to the tissues. It is stored in the liver, which is unique for a water-soluble vitamin, bound to transcobalamin I.

The Active B₁₂ Coenzymes are Methylcobalamin & Deoxyadenosylcobalamin

After transport in the blood, free cobalamin is released into the cytosol of cells as **hydroxocobalamin.** It is either converted in the cytosol to **methylcobalamin** or enters mitochondria for conversion to 5'-**deoxyadenosylcobalamin.**

Deoxyadenosylcobalamin Is the Coenzyme for the Conversion of Methylmalonyl-CoA to Succinyl-CoA (See Fig 53–15.)

This is a key reaction in the pathway of conversion of propionate to a member of the citric acid cycle and

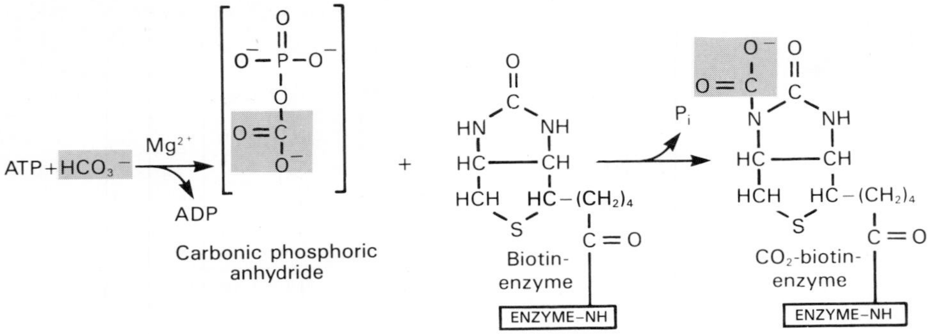

Figure 53–13. Formation of the CO_2-biotin enzyme complex. For its participation in pyruvate carboxylation see Fig 21–1.

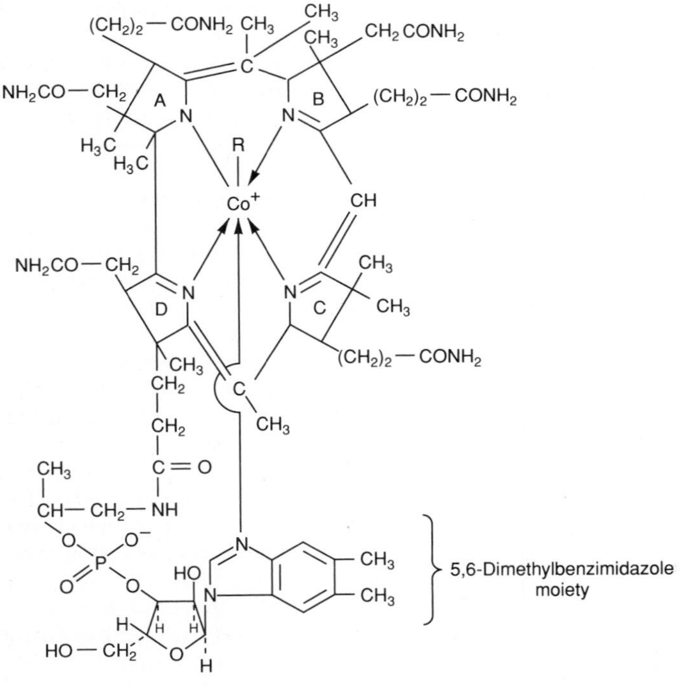

Figure 53–14. Vitamin B_{12} (cobalamin). R may be varied to give the various forms of the vitamin, eg, R = CN in cyanocobalamin; R = OH in hydroxocobalamin; R = 5′-deoxyadenosyl in 5′-deoxyadenosylcobalamin; and R = CH_3 in methylcobalamin.

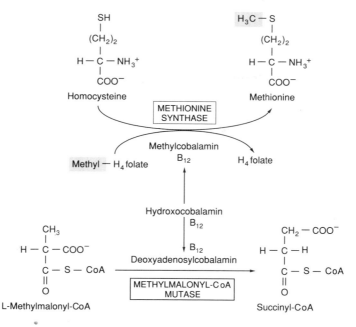

Figure 53–15. The 2 important reactions catalyzed by vitamin B_{12} coenzyme-dependent enzymes.

is, therefore, of significance in the process of **gluco-neogenesis** (see p 181). It is of particular importance in ruminants, as propionate is a major product of microbial fermentation in the rumen.

Methylcobalamin Is Coenzyme in the Combined Conversion of (1) Homocysteine to Methionine and (2) Methyl Tetrahydrofolate to Tetrahydrofolate (See Fig 53–15.)

In this reaction, the methyl group bound to cobalamin is transferred to homocysteine to form methionine and the cobalamin then removes the methyl group from N^5-**methyltetrahydrofolate** to form tetrahydrofolate. The metabolic benefits of this reaction are that stores of methionine are maintained and tetrahydrofolate is made available to participate in purine, pyrimidine, and nucleic acid syntheses.

Deficiency of Vitamin B_{12} Leads to Megaloblastic Anemia

When absorption is prevented by **lack of intrinsic factor** (or by gastrectomy) the condition is called **pernicious anemia.** Vegans are at risk of actual dietary deficiency as the vitamin is found only in foods of animal origin or from microorganisms, food contaminated with microorganisms being of benefit from this point of view. The deficiency leads to impairment of the methionine synthase reaction. Anemia results from impaired DNA synthesis affecting formation of the nucleus of new erythrocytes. This is due to impaired purine and pyrimidine synthesis resulting from tetrahydrofolate deficiency as a consequence of folate

being trapped as methyltetrahydrofolate (the "folate trap") (see Fig 53–15). Homocystinuria and methylmalonic aciduria also occur. The neurologic disorder associated with vitamin B_{12} deficiency may be secondary to a relative deficiency of methionine.

Four inherited disorders of cobalamin metabolism have been described. Two affect synthesis of deoxyadenosylcobalamin only; in the other two, patients are unable to synthesize either deoxyadenosylcobalamin or methylcobalamin.

FOLIC ACID

Folacin is the generic name covering folic acid and related substances having the biochemical activity of folic acid.

Folic acid, or folate, consists of the base **pteridine** attached to one molecule each of *P*-**aminobenzoic acid** (PABA) and **glutamic acid** (Fig 53–16). Animals are not capable of synthesizing PABA or of attaching glutamate to pteroic acid and, therefore, require folate in their diet; yeast, liver, and leafy vegetables are major sources. In plants, folic acid exists as a polyglutamate conjugate consisting of a γ-linked polypeptide chain of 7 glutamate residues. In the liver, the major folate is a pentaglutamyl conjugate.

Active Folate Is Tetrahydrofolate (H_4folate)

Folate derivatives in the diet are cleaved by specific intestinal enzymes to monoglutamyl folate for absorption. Most of this is reduced to **tetrahydrofolate** in the intestinal cell (Fig 53–17) by the enzyme **folate re-**

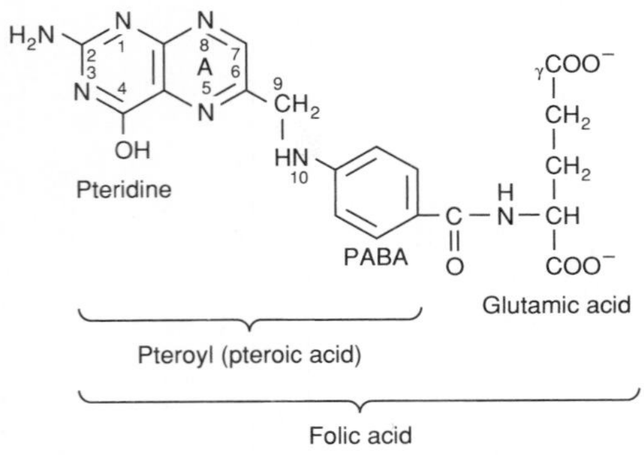

Figure 53–16. The structure and numbering of atoms of folic acid.

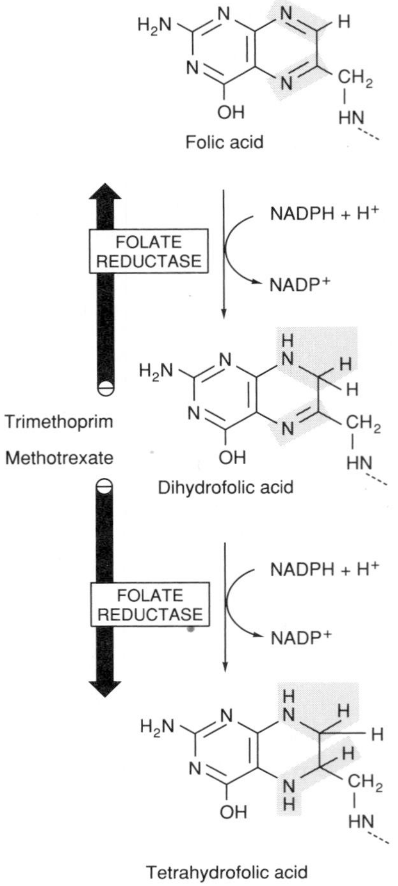

ductase, which uses NADPH as donor of reducing equivalents. Tetrahydrofolate polyglutamates are probably the functional coenzymes in tissues.

H₄folate Is the Carrier of Activated One-Carbon Units

The one-carbon units carried by H_4folate represent a series in various states of oxidation, viz, **methyl, methylene, methenyl, formyl, and formimino.** All are metabolically interconvertible (Fig 53–18).

Serine is the major source of a one-carbon unit in the form of a methylene group, which it transfers reversibly to H_4folate to form glycine and **N^5,N^{10}-methylene-H_4folate,** which plays a central role in one-carbon unit metabolism. It can be reduced to **N^5-methyl-H_4folate,** which has an important role in methylation of homocysteine to methionine involving methylcobalamin as a cofactor (see Fig 53–15). Alternatively, it can be oxidized to **N^5,N^{10}-methenyl-H_4folate,** which can then be hydrated to either **N^{10}-formyl-H_4folate** or to **N^5-formyl-H_4folate.** The latter is also known as **folinic acid,** a stable form that can be used for administration of reduced folate.

Formiminoglutamate (Figlu), a catabolite of histidine, transfers its formimino group to H_4folate to form **N^5-formimino-H_4folate.** In folate deficiency, Figlu will accumulate after oral challenge with histidine.

Folate Deficiency Causes Megaloblastic Anemia

The explanation for this is similar to that advanced above to account for the effects of deficiency of vitamin B_{12}. N^5,N^{10}-Methylene-H_4folate provides the methyl group in the formation of thymidylate, a necessary precursor of DNA synthesis and erythrocyte formation (Fig 58–19). Concomitant with the reduction of the methylene to the methyl group, there is oxidation of H_4folate to dihydrofolate, which must be reconverted to H_4folate for further use. Therefore, cells that synthesize thymidylate (for DNA) are particularly vulnerable to inhibitors of folate reductase such as **methotrexate** (Figs 53–17 and 53–19).

Figure 53–17. The reduction of folic acid to dihydrofolic acid and dihydrofolic acid to tetrahydrofolic acid by the enzyme folate reductase. Trimethoprim is a selective inhibitor of folate reductase in gram-negative bacteria and has little effect on the mammalian enzyme. It is therefore used as an antibiotic. Methotrexate (amethopterin) binds more strongly and is used as an anticancer drug.

ASCORBIC ACID (VITAMIN C)

The structure of ascorbic acid (Fig 53–20) is reminiscent of glucose, from which it is derived in the majority of mammals (see Fig 22–3). However, in primates, including man, and a number of other animals, eg, guinea pigs, some bats, birds, fishes, and invertebrates, the absence of the enzyme L-gulonolactone oxidase prevents this synthesis.

Active Vitamin C Is Ascorbic Acid Itself, Which Is a Donor of Reducing Equivalents in Certain Key Reactions

When ascorbic acid acts as a donor of reducing equivalents it is oxidized to dehydroascorbic acid, which itself can act as a source of the vitamin. Ascorbic acid is a reducing agent with a hydrogen potential

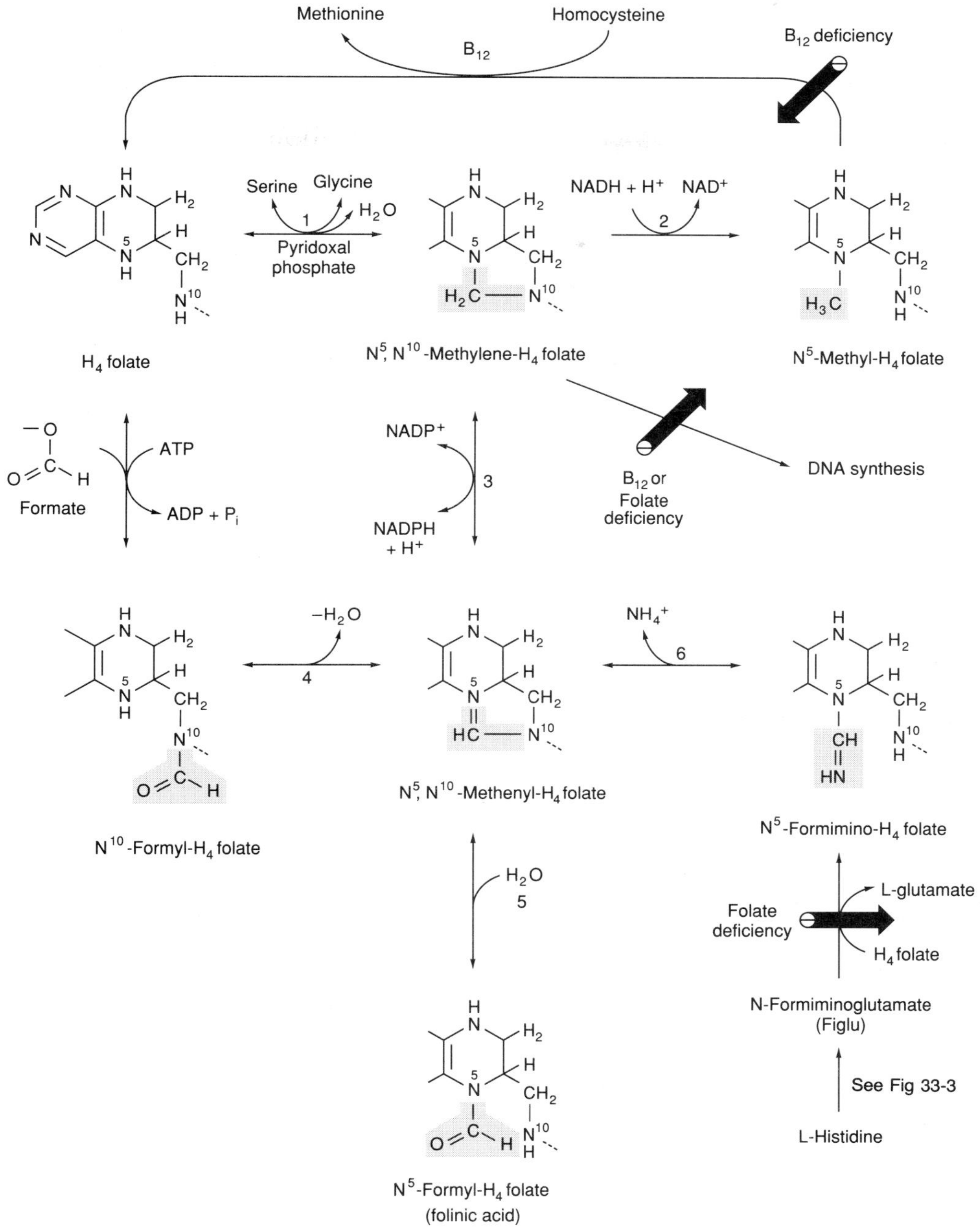

Figure 53–18. The interconversions of one-carbon units attached to tetrahydrofolate.

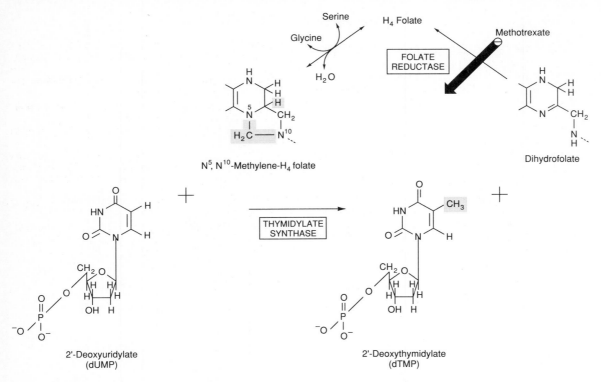

Figure 53–19. The transfer of a methyl group from N_5,N_{10}-methylene-H_4folate to deoxyuridylate to generate deoxythymidylate and dihydrofolate (H_2folate).

of +0.08 V, making it capable of reducing such compounds as molecular oxygen, nitrate, and cytochromes a and c. The mechanism of action of ascorbic acid in many of its activities is far from clear, but the following are some of the better documented processes requiring ascorbic acid.

1: Hydroxylation of proline in **collagen synthesis** (see Chapter 59 and Fig 30–11).

2: In the **degradation of tyrosine,** the oxidation of *p*-hydroxyphenylpyruvate to homogentisate requires vitamin C, which may maintain the reduced state of cop-

per necessary for maximal activity (see Fig 32–13 (old)). The subsequent step is catalyzed by homogentisate dioxygenase, which is a ferrous iron-containing enzyme that also requires ascorbic acid.

3: In the **synthesis of epinephrine from tyrosine** at the dopamine β-hydroxylase step (see p 513).

4: In **bile acid formation** at the initial 7α-hydroxylase step (see p 248).

5: The **adrenal cortex** contains large amounts of vitamin C, which is rapidly depleted when the gland is stimulated by adrenocorticotropic hormone. The rea-

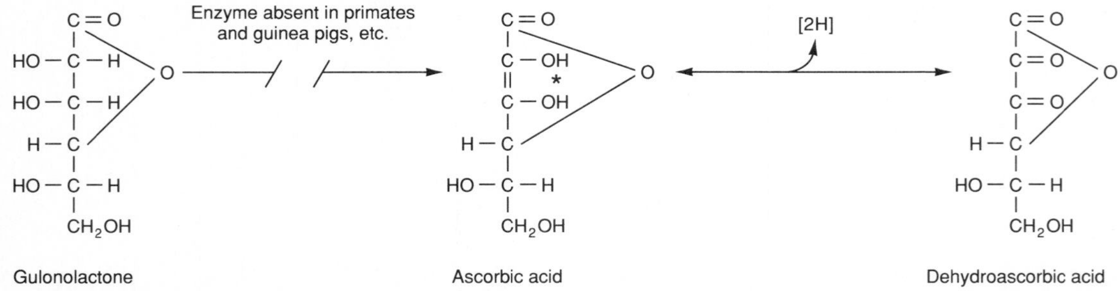

Figure 53–20. Ascorbic acid, its source in nonprimates, and its oxidation to dehydroascorbic acid *, ionizes in ascorbate.

son for this is obscure, but steroidogenesis involves several reductive syntheses.

6: The **absorption of iron** is significantly enhanced by the presence of vitamin C.

7: Ascorbic acid may act as a general water-soluble **antioxidant** (see p 142) and may inhibit the formation of nitrosamines during digestion.

Ascorbic Acid Deficiency Causes Scurvy

Scurvy is the classical syndrome of vitamin C deficiency. It is related to defective collagen synthesis which is indicated by subcutaneous and other hemorrhages, muscle weakness, soft swollen gums, and loose teeth; and is cured by consumption of fruits and fresh vegetables. The normal stores of vitamin C are sufficient to last 3–4 months before signs of scurvy appear.

REFERENCES

Benkovic SJ: On the mechanism of action of folate and biopterin-requiring enzymes. *Annu Rev Biochem* 1980;**49:**227.

Olson RE et al (editors): *Present Knowledge in Nutrition,* 5th ed. The Nutrition Foundation, Inc, 1984.

Passmore R, Eastwood, MA: *Human Nutrition and Dietetics,* Churchill Livingstone, 1986.

Rivlin RS: Hormones, drugs and riboflavin. *Nutr Rev* 1979;**37:**241.

Rosenberg L: Disorders of propionate, methylmalonate, and cobalamin metabolism. Pages 474–497 in: *The Metabolic Basis of Inherited Disease,* 5th ed. Stanbury JB et al (editors). McGraw-Hill, 1983.

Rubin RH, Swartz MN: Trimethroprim-sulfamethoxazole. *N Engl J Med* 1980;**303:**426.

Sebrell WH Jr: History of pellagra. *Fed Proc* 1981;**40:**1520.

Seetharam B, Alpers DH: Absorption and transport of cobalamin (vitamin B_{12}). *Annu Rev Nutr* 1982;**2:**343.

54

Structure and Function of the Lipid-Soluble Vitamins

Peter A. Mayes, PhD, DSc

INTRODUCTION

The lipid-soluble (fat-soluble) vitamins are **apolar hydrophobic** molecules, which are all **isoprene derivatives** (Fig 54–1). They cannot be synthesized by the body in adequate amounts and must, therefore, be supplied by the diet. They require **normal fat absorption** to be occurring in order to be absorbed efficiently. Once absorbed, they must be transported in the blood, like any other apolar lipid, in **lipoproteins** or **specific binding proteins.** The lipid-soluble vitamins have diverse functions, eg, vitamin A, vision; vitamin D, calcium and phosphate metabolism; vitamin E, antioxidant; vitamin K, blood clotting. Although once thought of as a vitamin, in the true sense **vitamin D (cholecalciferol) is a hormone.**

BIOMEDICAL IMPORTANCE

Because of their lipid nature, conditions affecting the digestion and absorption of the lipid-soluble vitamins such as steatorrhea and disorders of the biliary system can all lead to deficiencies. Dietary deficiencies affect the functions of the vitamins mentioned above; viz, vitamin A deficiency causes **night blindness** and **xerophthalmia;** vitamin D deficiency leads to **rickets** in young children and **osteomalacia** in adults; in vitamin E deficiency, which is rare, **neurologic disorders** and **anemia** of the newborn may arise; vitamin K deficiency, again very rare in adults, leads to **bleeding** and **hemorrhage** of the newborn. Because of the body's ability to store surplus lipid-soluble vitamins, **toxicity** can result from excessive intake of vitamins A and D. A role in **cancer prevention** has been ascribed to both vitamin A and β-carotene, provitamin A.

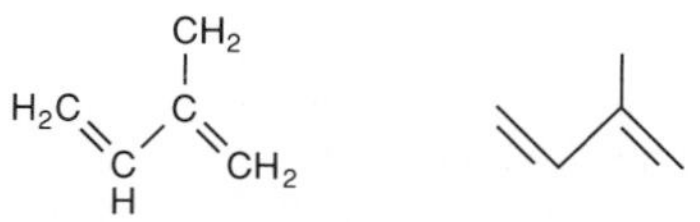

Figure 54–1. Two representations of the isoprene unit.

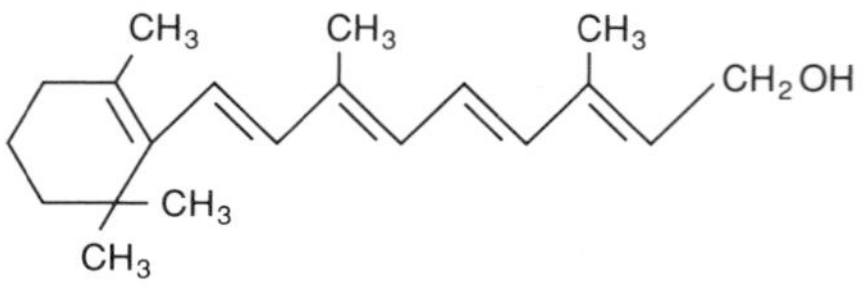

Figure 54–2. Retinol (vitamin A).

VITAMIN A

Vitamin A, or retinol, is a polyisoprenoid compound containing a cyclohexenyl ring. (Fig 54–2.). Vitamin A is a generic term referring to all compounds from animal sources that exhibit the biologic activity of vitamin A. They are **retinol, retinoic acid,** and **retinal.** Only retinol has full vitamin A activity, the others fulfilling some, but not all, vitamin A functions. The term **retinoids** has been used to describe both the natural forms and the synthetic analogs of retinol.

Vitamin A Has a Provitamin, β-Carotene

In vegetables, vitamin A exists as a **provitamin** in the form of the yellow pigment **β-carotene,** which consists of two molecules of retinal joined at the aldehyde end of their carbon chains (Fig 54–3). However, because β-carotene is not efficiently metabolized to vitamin A, weight for weight, β-carotene is only one-sixth as effective a source of vitamin A as retinol.

Digestion of Vitamin A Accompanies That of Lipids, Followed by Transformations in the Intestinal Mucosa

Retinol esters dissolved in the fat of the diet are dispersed in bile droplets and hydrolyzed in the intestinal lumen, followed by absorption directly into the intestinal epithelium. Ingested β-carotenes may be oxidatively cleaved by **β-carotene dioxygenase** (Fig 54–3). This cleavage utilizes molecular oxygen, is enhanced by the presence of bile salts, and generates 2 molecules of **retinaldehyde (retinal).** Also, in the in-

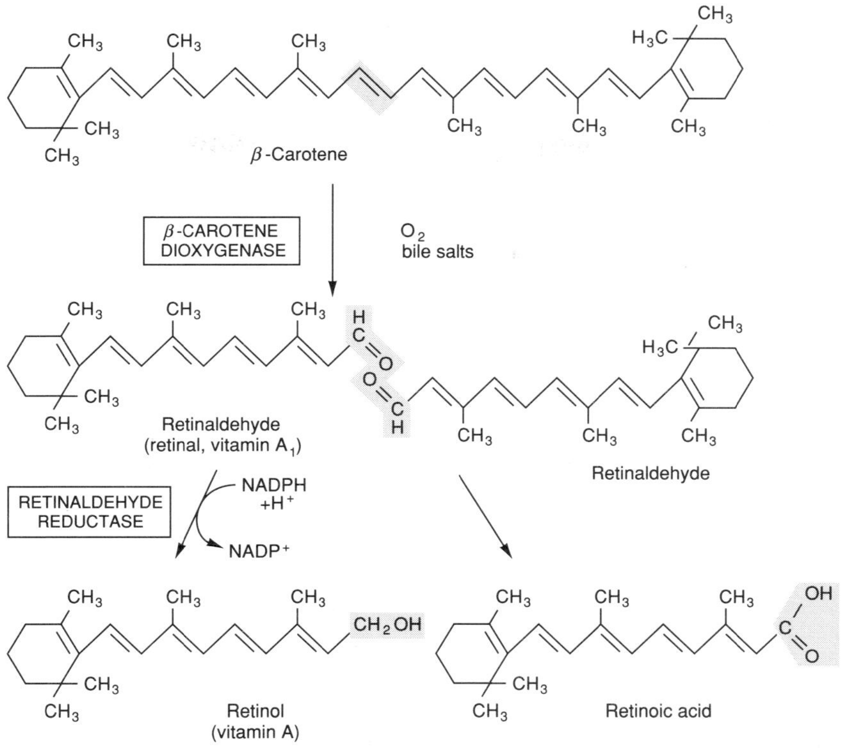

Figure 54–3. β-Carotene and its cleavage to retinaldehyde. The reduction of retinaldehyde to retinol and the oxidation of retinaldehyde to retinoic acid are also shown.

testinal mucosa, retinal is reduced to retinol by a specific **retinaldehyde reductase** utilizing NADPH. A small fraction of the retinal is oxidized to **retinoic acid.** Most of the retinol is esterified with saturated fatty acids and incorporated into lymph chylomicrons (see p 237), which enter the bloodstream. These are converted to chylomicron remnants, which are taken up by the liver together with their content of retinol. Carotenoids may escape some of these processes and pass directly into the chylomicrons.

Vitamin A Is Stored in the Liver & Released Into Blood Attached to Binding Proteins

In the liver, vitamin A is stored as an ester in the **lipocytes,** probably as a lipoglycoprotein complex. For transport to the tissues, it is hydrolyzed and the retinol bound to **apo-retinol binding protein** (RBP). The resulting **holo-RBP** is processed in the Golgi apparatus and secreted into the plasma. Retinoic acid is transported in plasma bound to albumin. Once inside extrahepatic cells, retinol is bound by a **cellular retinol binding protein** (CRBP).

Vitamin A toxicity occurs after the capacity of RBP has been exceeded and the cells are exposed to **unbound retinol.**

Retinol, Retinal, and Retinoic Acid Each Have Their Own Unique Biologic Function

Retinol and retinal are interconverted in the presence of NAD- or NADP-requiring dehydrogenases or reductases, present in many tissues. However, **once formed from retinal, retinoic acid cannot be converted back to retinal, or to retinol.** Thus, retinoic acid can support growth and differentiation **but cannot replace retinal in its role in vision or retinol in its support of the reproductive system.**

Retinol Acts Like a Steroid Hormone

When retinol is taken up into CRBP it is transported about the cell and binds to **nuclear proteins,** where it is probably involved in the **control of the expression of certain genes.** Thus, in this respect vitamin A behaves in a manner similar to that of steroid hormones. The requirement of vitamin A for normal reproduction may be ascribed to this function.

Retinal Is a Component of the Visual Pigment Rhodopsin

Rhodopsin occurs in the rod cells of the retina, which are responsible for vision in poor light. **11-*cis-***

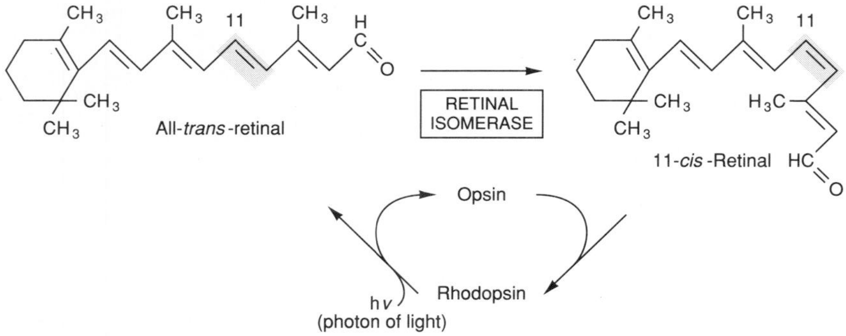

Figure 54–4. 11-*cis*-Retinal, formed from all-*trans*-retinal, combines with opsin to form rhodopsin in the rod cell of the eye. The absorption of a photon of light by rhodopsin causes it to bleach, generating opsin and all-*trans*-retinal. Retinal is required to maintain this cycle of reactions.

Retinal, an isomer of **all-*trans*-retinal,** is specifically bound to the visual protein **opsin** to form **rhodopsin** (Fig 54–4). When rhodopsin is exposed to light, it dissociates as it bleaches and forms all-*trans*-retinal and opsin. This reaction is accompanied by a conformational change that induces a **calcium ion channel** in the membrane of the rod cell. The rapid influx of calcium ions triggers a nerve impulse, allowing light to be perceived by the brain.

Retinoic Acid Participates in Glycoprotein Synthesis

This may account, in part, for the action of retinoic acid in promoting growth and differentiation of tissues. It has been proposed that retinoyl phosphate functions as a carrier of oligosaccharides across the lipid bilayer of the cell, by way of an enzymatic ***trans-cis* isomerization** analogous to that described above in the *trans-cis* isomerization of rhodopsin. The evidence that retinoic acid is involved in glycoprotein synthesis is compelling, since a deficiency of vitamin A results in the accumulation of abnormally low molecular weight oligosaccharide-lipid intermediates of glycoprotein synthesis (see Chapter 57).

Lack of Vitamin A Causes Characteristic Deficiency Symptoms

These are due to malfunction of the various cellular mechanisms in which retinoids participate. One of the first indications of vitamin A deficiency is **defective night vision,** which occurs when liver stores are nearly exhausted. Further depletion leads to **keratinization** of epithelial tissues of the eye, lungs, gastrointestinal, and genitourinary tracts, coupled with reduction in mucous secretion. Deterioration in the tissues of the eye, **xerophthalmia,** leads to blindness. Vitamin A deficiency occurs mainly on poor basic diets coupled with a lack of vegetables that would otherwise provide the provitamin, **β-carotene.**

Both Retinoids and Carotenoids Have Anticancer Activity

Many human cancers arise in epithelial tissues that depend on retinoids for normal cellular differentiation. Some epidemiologic studies have shown an inverse relationship between the vitamin A content of the diet and the risk of cancer, and experiments have shown that retinoid administration diminishes the effect of some carcinogens.

β-Carotene is an antioxidant and may play a role in trapping peroxy free radicals in tissues at low partial pressures of oxygen. The ability of β-carotene to act as an antioxidant is due to the stabilization of organic peroxide free radicals within its conjugated alkyl structure (Fig 54–5). Since β-carotene is effective at low oxygen concentrations, it complements the antioxidant properties of vitamin E, which is effective at higher oxygen concentrations (see p 141). The antioxidant properties of these two lipid-soluble vitamins may well account for their possible anticancer activity.

VITAMIN D

Vitamin D is a steroid prohormone. It is represented by a group of steroids that occur chiefly in animals but also in plants and yeast. By various metabolic changes in the body they give rise to a hormone known as **calcitriol,** which plays a central role in **calcium and phosphate metabolism** (see p 495).

D Vitamins Are Generated From the Provitamins Ergosterol & 7-Dehydrocholesterol by the Action of Sunlight

Ergosterol occurs in plants and 7-dehydrocholesterol in animals. Ergosterol differs from 7-dehydrocholesterol only in its side chain, which is unsaturated and contains an extra methyl group (Fig 54–6). Ultraviolet irradiation from sunlight cleaves the B ring of both compounds. **Ergocalciferol** (vitamin D_2) is formed in plants, whereas in animals, **cholecalciferol** (vitamin D_3) is formed in exposed skin. Both

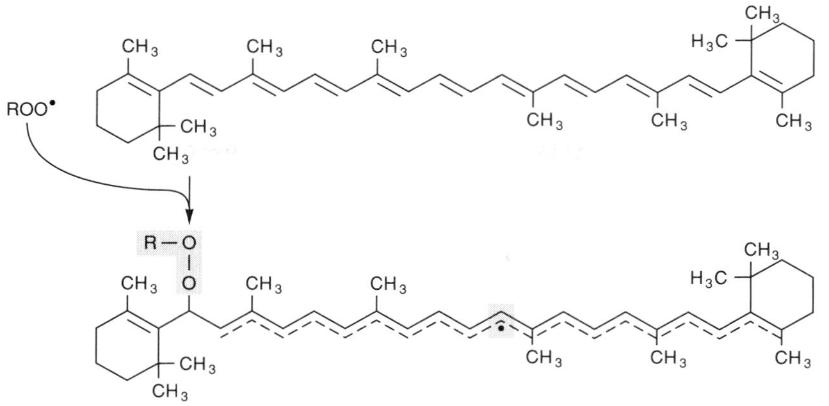

Figure 54–5. The formation of a resonance-stabilized carbon-centered radical from a peroxyl radical (ROO•) and β-carotene. (Slightly modified and reproduced, with permission, from Burton GW, Ingold KU: β-Carotene: An unusual type of lipid antioxidant. *Science* 1984;**224**:569. Copyright © 1984 by The American Association for the Advancement of Science.)

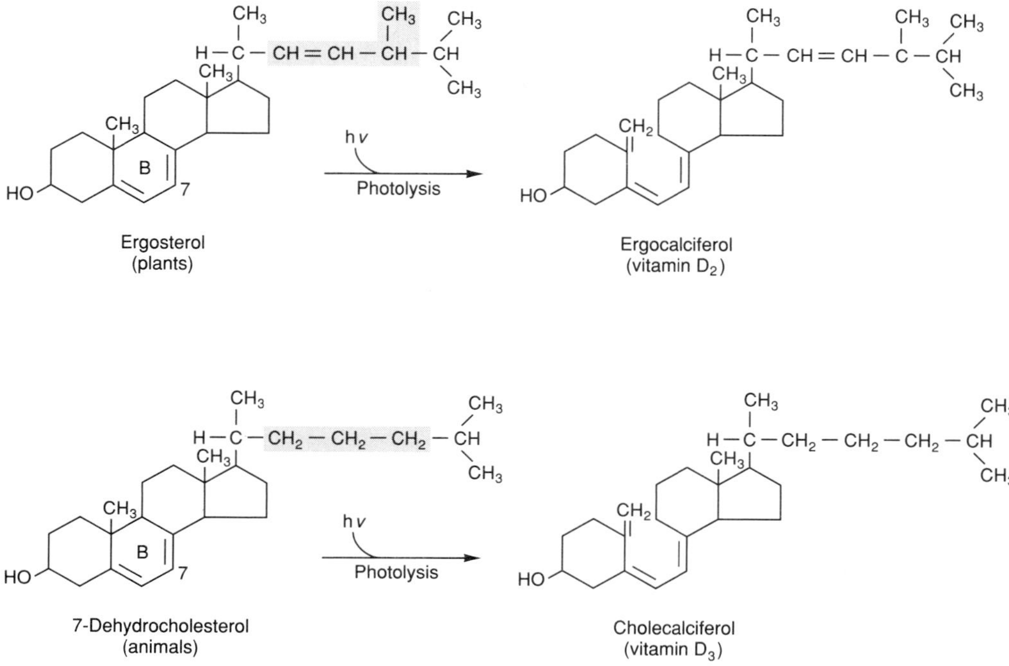

Ergosterol
(plants)

Ergocalciferol
(vitamin D$_2$)

7-Dehydrocholesterol
(animals)

Cholecalciferol
(vitamin D$_3$)

Figure 54–6. Ergosterol and 7-dehydrocholesterol and their conversion by photolysis to ergocalciferol and cholecalciferol, respectively.

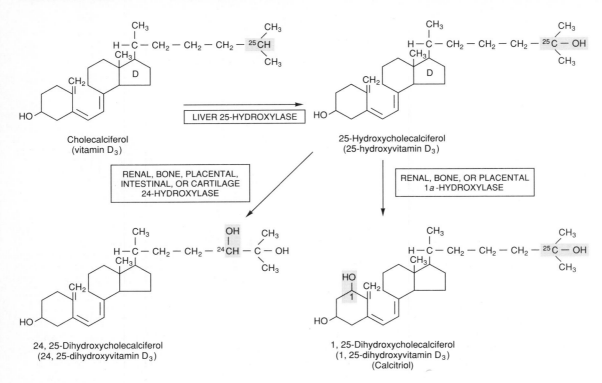

Figure 54–7. Cholecalciferol can be hydroxylated at the C_{25} position by a liver enzyme. The 25-hydroxycholecalciferol is further metabolized to 1α,25-dihydroxycholecalciferol or to 24,25-dihydroxycholecalciferol. The levels of 24,25-dihydroxy-cholecalciferol and 1,25-dihydroxycholecalciferol are regulated in a reciprocal manner.

vitamins are of equal potency, giving rise to D_2 calcitriol and D_3 calcitriol, respectively. Only the latter pathway will be described.

Both the Liver and the Kidney Are Involved in Calcitriol Synthesis

In the liver, dietary vitamin D_3 (or D_2) is taken up from the blood (bound to a specific globulin) after absorption from micelles in the intestine. **It is hydroxylated on the 25 position** by vitamin D_3-25-hydroxylase, an enzyme of the endoplasmic reticulum considered to be rate limiting (Fig 54–7). **25-hydroxyD$_3$** is the major form of vitamin D in the circulation and the major storage form in the liver. A significant fraction of 25-hydroxyD$_3$ undergoes enterohepatic circulation and disturbances of this process can lead to deficiency of vitamin D.

In the renal tubules, bone, and placenta, the 25-hydroxyD$_3$ is further hydroxylated in position 1 by 25-hydroxyD$_3$1-hydroxylase, a mitochondrial enzyme. The product is **1α,25-dihydroxyD$_3$ (calcitriol),** the **most potent vitamin D metabolite.** Its production is **regulated by its own concentration, parathyroid hormone, and serum phosphate.**

25-HydroxyD$_3$ can also be hydroxylated at the 24 position by a mitochondrial enzyme present in renal tubules, cartilage, intestine, and placenta. The level of the product 24,25-dihydroxyD$_3$ is reciprocally related to the level of 1,25-dihydroxyD$_3$ in serum.

For further details of the regulation and role of calcitriol in calcium and phosphate metabolism, see Chapter 48.

Deficiency of Vitamin D Causes Rickets and Osteomalacia

Rickets occurs in young children and osteomalacia in adults who are not exposed to sunlight or who do not receive adequate amounts of vitamin D in the diet. It is due to softening of bones resulting from lack of calcium and phosphate. Oily fish, egg yolk, and liver are good sources of the vitamin.

VITAMIN E (TOCOPHEROL)

There are several naturally occurring tocopherols. All are isoprenoid substituted 6-hydroxychromanes or tocols (Fig 54–8).

D-α-Tocopherol has the widest natural distribution and the greatest biologic activity. Other tocopherols of dietary significance are indicated in Table 54–1.

Active Fat Absorption Promotes the Absorption of Vitamin E

Impaired fat absorption leads to vitamin E deficiency because tocopherol is found dissolved in the fat of the diet and is liberated and absorbed during fat

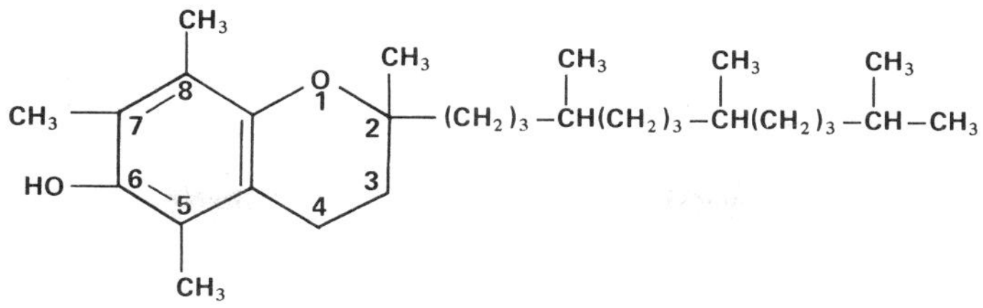

Figure 54–8. α-Tocopherol.

digestion. Furthermore, it is **transported in the blood by lipoproteins,** first, by incorporation into chylomicrons, which distribute the vitamin to the tissues containing lipoprotein lipase and to the liver in chylomicron remnants, and second, by export from the liver in very low density lipoproteins. It is stored in adipose tissue.

Vitamin E Is a Most Important Natural Antioxidant

Vitamin E appears to be the first line of defense against **peroxidation of polyunsaturated fatty acids** contained in cellular and subcellular membrane phospholipids (see p 141). The phospholipids of mitochondria, endoplasmic reticulum, and plasma membranes possess affinities for α-tocopherol and the vitamin appears to concentrate at these sites. The tocopherols act as antioxidants, breaking free-radical chain reactions as a result of their ability to transfer a phenolic hydrogen to a peroxyl free radical of a peroxidized polyunsaturated fatty acid (Fig 54–9, and see p 141). The phenoxy free radical formed then reacts with a further peroxyl free radical). Thus, α-tocopherol does not readily engage in reversible oxidation; the chromane ring and the side chain are oxidized to the non-free-radical product shown in Fig 54–10. This oxidation product is conjugated with glucuronic acid via the 2-hydroxyl group and excreted in the bile. Unlike some other vitamins, eg, niacin, B_{12}, and folate, tocopherol is not recycled after carrying out its function but must be replaced totally to continue its biologic role in the cell. The antioxidant action of tocopherol is effective at high oxygen concentrations, and thus it is not surprising that it tends to be concentrated in those lipid structures that are exposed to the highest O_2 partial pressures, eg, the erythrocyte membrane and the membranes of the respiratory tree.

Vitamin E and Selenium Act Synergistically & Reduce the Body's Requirement for Each Other

Glutathione peroxidase, of which selenium is an integral component (see p 193), provides a second line of defense against peroxides before they can propagate in chain reactions, damaging membranes and other cell components. Thus, tocopherol and selenium reduce each other's requirement or reinforce each other in their actions against lipid peroxides. In addition, selenium is required for normal pancreatic function, which is necessary for the digestion and absorption of lipids, including vitamin E. Conversely, vitamin E reduces selenium requirements by preventing loss of selenium from the body or maintaining it in an active form.

Deficiency of Vitamin E May Give Rise to Anemia of the Newborn

There is a possible need for supplementation of tocopherol in the diet of pregnant and lactating women and for newborn infants where anemia can arise as a result of insufficiency of vitamin E. Anemia may be due to decreased production of hemoglobin and a shortened erythrocyte life span.

The requirement for vitamin E is increased with greater intake of polyunsaturated fat. Intake of mineral oils, exposure to oxygen (as in oxygen tents), or diseases leading to inefficient lipid absorption may cause deficiencies of the vitamin leading to **neurologic disorder.**

Vitamin E is destroyed by commercial cooking and food processing, including deep-freezing. Wheat germ, sunflower seed and safflower seed oils, and corn and soya bean oil are all good sources of the vitamin. Although fish liver oils are rich sources of vitamins A and D, they have insignificant amounts of vitamin E.

Table 54–1. Naturally occurring tocopherols of dietary significance.

Tocopherol	Substituents
Alpha	5,7,8-Trimethyl tocol
Beta	5,8-Dimethyl tocol
Gamma	7,8-Dimethyl tocol
Delta	8-Methyl tocol

Figure 54–9. The chain-breaking antioxidant activity of tocopherols (TocOH) toward peroxyl radicals (ROO•).

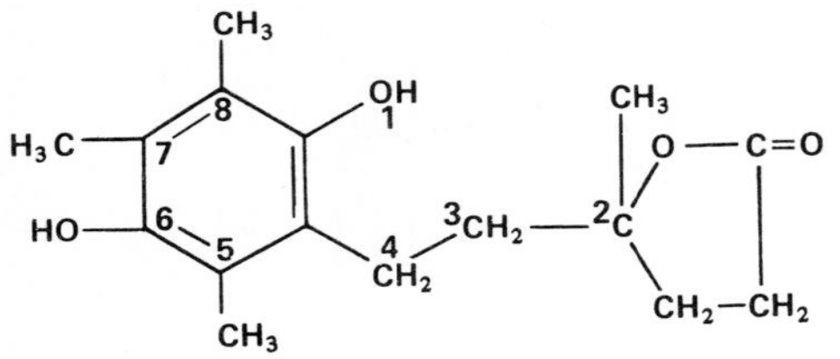

Figure 54–10. The oxidation product of α-tocopherol. The numbers allow one to relate the atoms to those in the parent compound.

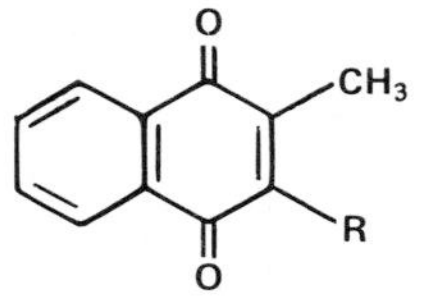

2-Methyl-1,4-naphthoquinone

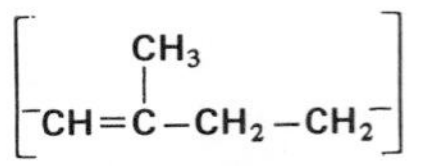

Isoprenoid unit

Figure 54–11. Substituted naphthoquinone and the isoprenoid unit. In the vitamins K, the R substitutions are polyisoprenoids.

VITAMIN K

Vitamins belonging to the K group are poly-isoprenoid-substituted napthoquinones (Fig 54–11.). Menadione (K_3), the parent compound of the vitamin K series (Fig 54–12), is not found naturally but if administered it is alkylated in vivo to one of the menaquinones (K_2). Phylloquinone (K_1) is the major form of vitamin K found in plants. Menaquinone-7 is one of the series of polyprenoid unsaturated forms of vitamin K found in animal tissues and synthesized by bacteria in the intestine.

Absorption of Vitamin K Requires Normal Fat Absorption

Fat malabsorption is the commonest cause of vitamin K deficiency. The naturally occurring K deriva-

tives are absorbed only in the presence of bile salts, like other lipids, and are distributed in the bloodstream via the lymphatics, in chylomicrons. Menadione, being water soluble, is absorbed **even in the absence of bile salts,** passing directly into the hepatic portal vein. Although vitamin K accumulates initially in the liver, its hepatic concentration declines rapidly and storage is limited.

Vitamin K Is Required for the Biosynthesis of Blood Clotting Factors

Vitamin K has been shown to be involved in the maintenance of normal levels of **blood clotting factors II, VII, IX and X,** all of which are synthesized in the liver initially as **inactive precursor proteins** (see Chapter 58).

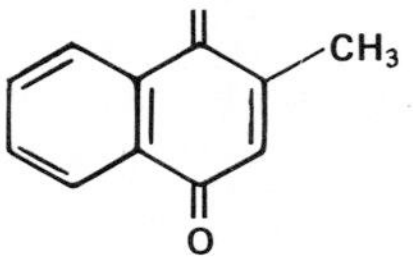

Menadione (vitamin K_3)

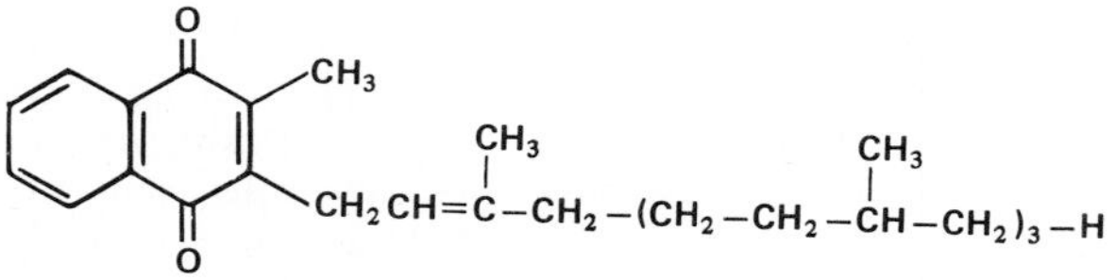

Phylloquinone (vitamin K_1, phytonadione, Mephyton)

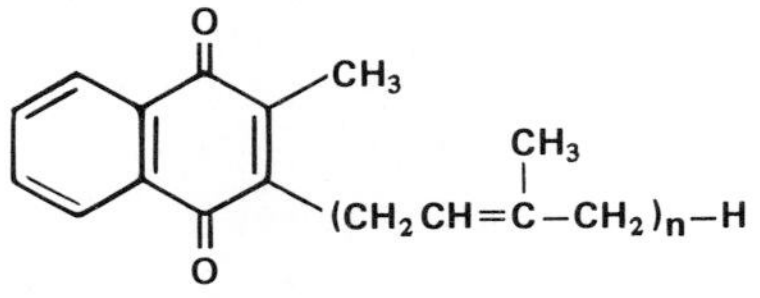

Menaquinone-n (vitamin K_2; n = 6, 7, or 9)

Figure 54–12. The naturally occurring vitamins K.

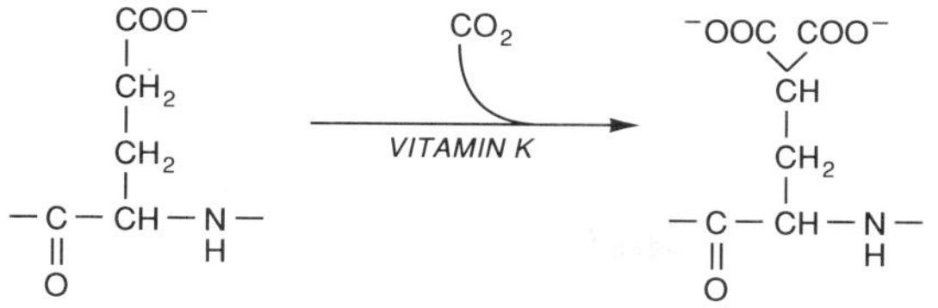

Figure 54–13. Carboxylation of a glutamate residue catalyzed by vitamin K.

Vitamin K Acts as Cofactor of the Carboxylase That Forms γ-Carboxyglutamate Residues in Precursor Proteins

Generation of the biologically active clotting factors involves the **posttranslational modification** of glutamate (Glu) residues of the precursor proteins to γ-carboxyglutamate (Gla) residues by a specific vitamin K-dependent carboxylase (Fig 54–13). Prothrombin (factor II) contains 10 of these residues, which allow **chelation of calcium** in a specific protein-calcium-phospholipid interaction, essential to their biologic role (Fig 54–14). Other proteins containing K-dependent Gla residues have now been identified in several tissues.

The Vitamin K Cycle Allows Reduced Vitamin K to be Regenerated

The vitamin K-dependent carboxylase reaction occurs in the endoplasmic reticulum of many tissues and requires molecular oxygen, carbon dioxide (not HCO_3^-), and the **hydroquinone** (reduced) form of vitamin K. In the endoplasmic reticulum of liver there exists a **vitamin K cycle** (Fig 54–15) in which the **2,3-epoxide** product of the carboxylation reaction is converted by 2,3-epoxide reductase to the quinone form of vitamin K, using an, as yet, unidentified dithiol reductant. This reaction is **sensitive to inhibition by 4-hydroxydicoumarin (dicumarol) types of anticoagulant,** such as **warfarin** (Fig 54–16). Subsequent reduction of the quinone form to the hydroquinone by

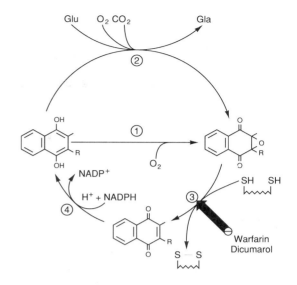

Figure 54–15. Vitamin K-related metabolic activities in liver. The locus of action of the dicumarol-type anticoagulants is indicated. The details of some of the reactions are still uncertain. 1, monooxygenase; 2, carboxylase; 3, 2,3-epoxide reductase; 4, reductase. (Modified and reproduced, with permission, from Suttie JW: The metabolic role of vitamin K. *Fed Proc* 1980;**39**:2730.)

NADH completes the vitamin K cycle for regenerating the active form of the vitamin.

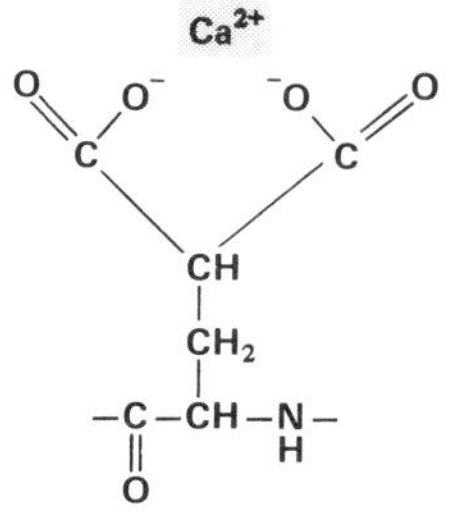

An important therapeutic use of vitamin K is as an antidote to poisoning by dicumarol-type drugs. The quinone forms of vitamin K will bypass the inhibited epoxide reductase and provide a potential source of the active hydroquinone form of vitamin K.

Hemorrhagic Disease of the Newborn Is Caused by Deficiency of Vitamin K

Vitamin K is widely distributed in plant and animal tissues used as food, and production of the vitamin by the microflora of the intestine virtually ensures that dietary deficiency does not occur in adults. However, newborn infants are vulnerable to the deficiency, because the placenta does not pass the vitamin to the fetus efficiently and the gut is sterile immediately after birth. In normal infants, the plasma concentration decreases immediately after birth but recovers after food is ab-

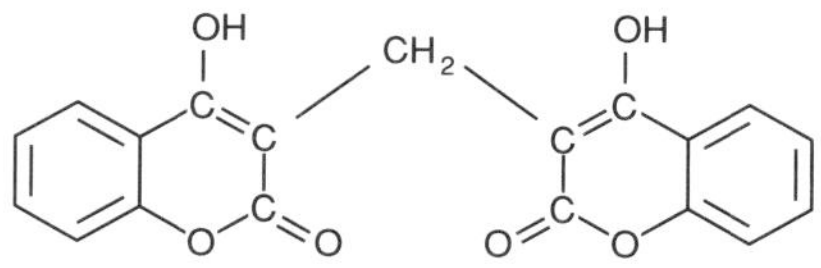

Figure 54–14. The chelation of calcium ion by the γ-carboxyglutamyl residue in clotting-factor proteins.

Figure 54–16. Dicumarol (bishydroxycoumarin; 3,3′-methylene-bishydroxycoumarin).

sorbed. If the prothrombin level drops too low, the hemorrhagic syndrome may appear.

Vitamin K deficiency can be caused by fat malabsorption, which may be associated with pancreatic dysfunction, biliary disease, atrophy of the intestinal mucosa, or any cause of steatorrhea. In addition, **sterilization of the large intestine by antibiotics** can result in deficiency when dietary intake is limited.

REFERENCES

Adams JS, et al: Vitamin-D synthesis and metabolism after ultraviolet irradiation of normal and vitamin-D-deficient subjects. *N Engl J Med* 1982;**306:**722.

Goodman DS: Vitamin A and retinoids in health and disease. *N Engl J Med* 1984;**310:**1023.

Norman AW, Roth J, Orci L: The vitamin D endocrine system. *Endocr Rev* 1982;**3:**331.

Olson RE et al (editors): *Present Knowledge in Nutrition,* 5th ed, The Nutrition Foundation, Inc, Washington, DC, 1984.

Passmore R, Eastwood MA: *Human Nutrition and Dietetics,* Churchill Livingstone, 1986.

Scott, ML: Advances in our understanding of vitamin E. *Fed Proc* 1980;**39:**2736.

Sokol RJ: Vitamin E deficiency and neurologic disease. *Annu Rev Nutr* 1988;**8:**351.

Sporn MB, Roberts AB: Role of retinoids in differentiation and carcinogenesis. *Cancer Res* 1983;**43:**3034.

Suttie JW: The metabolic role of vitamin K. *Fed Proc* 1980; **39:**2730.

Whitlon DS, et al: Mechanism of coumarin action: Significance of vitamin K epoxide reductase inhibition. *Biochemistry* 1978;**17:**1371.

Nutrition

55

Peter A. Mayes, PhD, DSc

INTRODUCTION

The science of nutrition examines the qualitative and quantitative requirements of the diet necessary to maintain good health. Virtually all components of the diet needed to maintain life are known, since it is possible to sustain humans or other animals on chemically defined diets. However, there is still considerable discussion and controversy surrounding the quantitative requirements of each component of the diet, particularly as this varies with the age, sex, and lifestyle of the individual. Metabolic biochemistry provides much of the understanding of modern concepts of nutrition and has been discussed in earlier parts of this book. In particular, the biochemical roles played by the water-soluble and lipid-soluble vitamins have been described in the preceding 2 chapters.

BIOMEDICAL IMPORTANCE

Overt nutritional deficiency is rare in more affluent populations, although some degree of nutritional deficiency may be present among the poor or the elderly and among groups with specialized nutritional requirements, eg, growing children, pregnant or lactating women, ill and convalescing patients, alcoholics, or individuals on restricted diets from necessity, eg, patients on intravenous feeding, or choice, eg, vegans. In more deprived populations, overt deficiencies are more widespread, eg, deficiency of protein (**kwashiorkor**), of vitamins (vitamin A in **xerophthalmia**), of minerals (iron, giving rise to **anemia**), and of energy (**starvation**). Malabsorption may lead to deficiency of nutrients, and cause pathologic conditions; eg, malabsorption of vitamin B_{12} and folate causes **anemia.** Although **obesity** has always been associated with dietary excess, the concept of excess intake of particular nutrients and their association with the prevalence of certain diseases in developed societies is gaining recognition, eg, atherosclerosis and coronary heart disease, diabetes, cancer of the breast and colon, cerebrovascular disease and strokes, and cirrhosis of the liver.

NUTRITIONAL REQUIREMENTS CAN NOW BE DEFINED

Table 55–1 summarizes nutritional requirements.

ENERGY IS REQUIRED TO POWER ALL BODY FUNCTIONS

The mammalian body requires nutrients sufficient to provide free energy to manufacture the daily requirement of high-energy phosphate (mainly ATP) and reducing equivalents (2H) needed to power all body functions (see Fig 17–1).

Carbohydrate and Fat are the Principal Energy Sources in the Diet

Energy-yielding nutrients are provided by dietary carbohydrate, fat, and, to a lesser extent, protein in widely varying proportions among different human populations. Consumption of alcohol can also provide a significant proportion of energy intake.

A constant body weight under conditions of unaltered energy requirement indicates that there is enough energy in the diet for immediate needs.

The amount of energy available in the major food sources is indicated in Table 55–2. The large energy content per gram of fat compared to that of protein or carbohydrate and the relatively high energy content of alcohol are 2 notable facts. The recommended energy intake per head for selected groups of people is given in Table 55–3.

Several Factors Affect Expenditure of Energy

Under conditions of **energy equilibrium** (calorie balance), energy intake must equal energy expenditure. Energy expenditure varies widely in different conditions and may be measured by placing an animal inside an insulated chamber and measuring the energy output represented by heat loss and excretory products. It is usually more convenient to measure **oxygen consumption,** since under most conditions 1 L of O_2 consumed accounts for approximately 4.83 kcal (20 kJ) of energy expended.

The energy expended by an individual depends on 4 main factors:

(1) The **basal metabolic rate** is the energy expenditure necessary to maintain basic physiologic functions under standardized conditions; the subject should be at rest, awake, and in a warm environment, and measurements should be taken at least 12 hours after the last meal. The basal metabolic rate is proportionate to lean

Table 55–1. Essential nutritional requirements.

	Humans	Selected Differences in Other Species
Amino acids	Histidine,[1] isoleucine, leucine, lysine, methionine (cysteine[3]), phenylalanine (tyrosine[3]), threonine, tryptophan, valine	Arginine[2] required by growing rats. Glycine required in chicks and taurine in cats. Most amino acids not essential in ruminants; requirement spared in other herbivores with substantial population of microorganisms in the gut.
Fatty acids	Linoleic acid (arachidonic acid[3]), α-linolenic acid[4]	Arachidonic acid is a specific requirement in cats.
Vitamins Water-soluble	Ascorbic acid (C), biotin,[5] cobalamin (B_{12}), folic acid, niacin, pantothenic acid, pyridoxine (B_6), riboflavin (B_2), thiamin (B_1)	Most mammals can synthesize ascorbic acid, but it is essential in the diet of primates, guinea pigs, and Indian fruit bats. Water-soluble vitamins are not essential in ruminants; requirements are spared in other herbivores with substantial populations of microorganisms in the gut.
Fat-soluble	Vitamin A, D[6], E, K[5]	Most species can utilize β-carotene as a source of vitamin A (retinol); must be supplied as retinol in cats.
Minerals Macrominerals	Calcium, chloride, magnesium, phosphorus, potassium, sodium	
Microminerals (trace elements)	Chromium, copper, iodine, iron, manganese, molybdenum, selenium, zinc	Silicon, vanadium, nickel, arsenic, fluoride, and tin have been shown to be essential in various species and may be required in humans. Cobalt is required for synthesis of cobalamin by ruminal microorganisms.
Fiber	Required for optimal health	
Water	The most critical component of the diet	
Energy	Utilization of carbohydrates, fats, and protein in variable proportions	

[1] Required in infants and probably in children and in adults.
[2] May be partly essential in infants.
[3] Cysteine, tyrosine, and arachidonic acid spare the requirement for methionine, phenylalanine, and linoleic acid, respectively.
[4] Workers disagree whether α-linolenic acid is essential in the human diet.
[5] Synthesized by intestinal microorganisms; therefore, dietary requirement uncertain.
[6] Exposure of the skin to sunlight reduces dietary requirement.

body weight and to surface area. It is higher in males than females, in young children, and in people with **fever** and **hyperthyroidism.** It is lower in **hypothyroidism** and in **starvation.**

(2) The **thermogenic effect** (specific dynamic action) of food is equivalent to about 5–10% of total energy expenditure and is attributed to the energy expenditure due to digestion and to any stimulation of metabolism caused by the influx of new substrate.

(3) **Physical activity** is the largest variable affecting energy expenditure; the range is over 10-fold between resting and maximum athletic activity.

(4) When **environmental temperature** is low, it causes increased energy expenditure owing to shivering and to nonshivering thermogenesis in animals having brown fat (see p 248). At temperatures above blood heat, extra energy is expended in cooling.

PROTEINS ARE REQUIRED TO SUPPLY SPECIFIC AMINO ACIDS & NITROGEN FOR THE SYNTHESIS OF KEY NITROGENOUS COMPOUNDS

Protein normally provides the body's requirement for amino acid nitrogen and amino acids themselves.

Dietary protein is digested and enters the circulation as individual amino acids. The body requires 20 amino acids to synthesize specific proteins and other nitrogen-containing compounds such as purines, pyrimidines, and heme.

Essential Amino Acids Are Required Amino Acids That Cannot Be Synthesized by the Body & Must Therefore Be Supplied by the Diet

There are 9 essential amino acids in humans: histidine, isoleucine, leucine, lysine, methionine, phenylalanine, threonine, tryptophan, and valine (Table 55–1). Two other amino acids, cysteine and tyrosine, may be formed from the essential amino acids methionine and phenylalanine, respectively. If insufficient methionine and phenylalanine are present in the diet, they spare the requirement for methionine and phenylalanine.

As long as sufficient amounts of essential amino acids are present in the diet, the remaining amino acids required for protein synthesis and other purposes can be formed through transamination and other reactions (see p 276).

Table 55–2. Heats of combustion and energy available from the major food sources.*

	Energy kcal/g (kJ/g)		
	Heat of Combustion (Bomb Calorimeter)	**Human Oxidation**	**Standard Conversion Factors†**
Protein	5.4 (22.6)	4.1 (17.2)‡	4 (17)
Fat	9.3 (38.9)	9.3 (38.9)	9 (38)
Carbohydrate	4.1 (17.2)	4.1 (17.2)	4 (17)
Ethanol	7.1 (29.7)	7.1 (29.7)	7 (29)

*Adapted from Davidson S et al: *Human Nutrition and Dietetics,* 7th ed. Churchill Livingstone, 1979.
†Conversion factors are obtained by rounding off heats of combustion and correcting for estimates of absorption efficiency.
‡Protein oxidation corrected for loss of amino groups excreted in urine.

Table 55–3. Recommended energy intake for men and women.*

Category	Age (years)	Weight (kg)	Weight (lb)	Energy Needs Mean (kcal)	Energy Needs Range (kcal)	(MJ)
Men	23–50	70	154	2900	2300–3100	12.1
Women	23–50	55	120	2200	1600–2400	9.2
Pregnant				+300		
Lactating				+500		

*Data from *Recommended Dietary Allowances,* 10th ed. Food and Nutrition Board, National Research Council—National Academy of Sciences, 1989.

Nitrogen Balance Is Maintained by Dietary Intake (See also Chapter 31.)

When its diet is such that an adult animal is in a state of metabolic equilibrium, dietary protein is required to replace the essential amino acids and amino acid nitrogen lost during metabolic turnover. Nitrogen is lost in the urine, feces, saliva, desquamated skin, hair, and nails. The daily requirements for total protein and essential amino acids in humans are set forth in Table 55–4. When these requirements are calculated on the basis of body weight, the extra growth needs of infants and children are clearly evident. Pregnancy, lactation, tissue repair after injury, recovery from illness, and increased physical activity are other conditions requiring more dietary protein. For most situations, **a diet in which 12% of the energy is supplied as protein is adequate.**

The Efficiency With Which Dietary Protein Is Used Determines the Total Quantity of Protein Required

This quantity is affected by 3 major factors: **protein quality, energy intake,** and **physical activity.**

Protein Quality: The quality of protein is measured by comparing the proportions of essential amino acids in a food with the proportions required for good nutrition. The closer the 2 numbers are, the higher the pro-

Table 55–4. Estimated protein and amino acid requirements and intakes in humans.*

	Requirement (mg/kg body weight/d)			Intake (g/d)	
	Infant (4–6 months)	**Child (10–12 years)**	**Adult**	**Adult (70 kg) Allowance***	**Estimated US Adult Intake†**
Protein	1100	1000	800	56	101
Animal	...	...	...	...	71
Vegetable	...	...	...	...	30
Essential amino acids	(3–4 months)				
Histidine	28	?	10	0.70	?
Isoleucine	70	28	10	0.70	5.3
Leucine	161	42	14	0.98	8.2
Lysine	103	44	12	0.84	6.7
Methionine (and cysteine)	58	22	13	0.91	2.1
Phenylalanine (and tyrosine)	125	22	14	0.98	4.7
Threonine	87	28	7	0.49	4.1
Tryptophan	17	3.3	3.5	0.25	1.2
Valine	93	25	10	0.70	5.7

*Data from *Recommended Dietary Allowances,* 10th ed. Food and Nutrition Board, National Research Council—National Academy of Sciences, 1989.
†Data from Munro HN, Crim M: The proteins and amino acids. In: Goodhart RS, Shils ME: *Modern Nutrition in Health and Disease,* 6th ed. Lea & Febiger, 1980.

tein quality. Egg and milk proteins are **high-quality proteins** that are efficiently used by the body and are used as reference standards against which other proteins can be compared. Meat protein is of high protein quality, whereas several proteins from plants used as major food sources are relatively deficient in certain essential amino acids, eg, tryptophan and lysine in maize (corn), lysine in wheat, and methionine in some beans. In a mixed diet, a deficiency of an amino acid in one protein is made up by its abundance in another; such proteins are described as **complementary**; eg, the protein of wheat and beans combined provides a satisfactory amino acid intake. Under such circumstances, a greater *total* amount of protein must be consumed to satisfy requirements. Amino acids that are not incorporated into new protein and are unnecessary for immediate requirements cannot be stored and are rapidly degraded, and the nitrogen is excreted as urea and other products.

Energy Intake: The energy derived from carbohydrate and fat affects protein requirements because it spares the use of protein as an energy source. To use expensive (high-quality) dietary protein efficiently and to reduce requirements for it to a minimum, it is necessary to ensure adequate provision of energy from nonprotein sources, some of which should be carbohydrate in order to spare protein from gluconeogenesis.

Physical Activity: Physical activity increases nitrogen retention from dietary protein.

Protein-Energy Malnutrition Causes Marasmus & Kwashiorkor

Protein-energy malnutrition (PEM) includes a range of disorders of starvation and malnutrition that involve other nutrients such as vitamins and minerals in addition to protein deficiency. In severe form, it occurs in growing children, usually under 5 years of age, in underdeveloped areas of Asia, Africa, and South America. Two extreme forms are recognized, marasmus and kwashiorkor (see Chapter 62). In **marasmus,** there is generalized wasting due to deficiency of **both energy and protein,** whereas in **kwashiorkor,** which is characterized by edema, there is a deficiency in both the **quantity and the quality of protein,** although energy intake may be adequate. Conditions are often encountered that are intermediate between typical marasmus and typical kwashiorkor.

GLUCOSE REQUIREMENTS CAN BE MET BY MANY CARBOHYDRATES

Glucose is specifically required by many tissues but does not have to be provided as such in the diet, since other dietary carbohydrates are readily converted to glucose, either during digestion (eg, starch) or subsequently in the liver (eg, fructose, galactose; see pp 194 and 197). Glucose is also formed from the glycerol moiety of fats and from glucogenic amino acids by gluconeogenesis (see p 179). Although a minimum daily intake of carbohydrate (50–100 g) is recommended in humans to prevent **ketosis** (see p 263) and loss of muscle protein, a balanced diet should contain more polysaccharide in order to reduce the amount of fat that would otherwise be required for energy. The major foodstuffs containing carbohydrates are described in Chapter 15.

FIBER IS REQUIRED FOR OPTIMAL HEALTH

Dietary fiber denotes all plant cell wall components that cannot be digested by an animal's own enzymes, eg, cellulose, hemicellulose, lignin, gums, pectins, and pentosans. In herbivores such as ruminants, fiber (mainly as cellulose) is a major source of energy after it is digested by microorganisms to acetate, propionate, and butyrate, which are absorbed into the portal vein. Colonic fermentation may also make some contribution to human energy requirements (2–7% on low-fiber intakes). Gases such as CH_4, CO_2, and H_2 are also produced.

In humans, a high-fiber diet exerts beneficial effects by aiding water retention during passage of food along the gut and thereby producing larger, softer feces. A high-fiber diet is associated with reduced incidence of **diverticulosis, cancer of the colon, cardiovascular disease,** and **diabetes mellitus.** The more insoluble fibers such as cellulose and lignin found in wheat bran are beneficial with regard to colonic function, whereas the more soluble fibers found in legumes and fruit, eg, gums and pectins, lower blood cholesterol, possibly by binding bile acids and dietary cholesterol. The soluble fibers also slow stomach emptying, and they delay and attenuate the postprandial rise in blood glucose, with consequent reduction in insulin secretion. This effect is beneficial to diabetics and to dieters because it reduces the rebound fall in blood glucose that stimulates appetite.

LIPIDS ARE REQUIRED AS A VEHICLE FOR LIPID-SOLUBLE VITAMINS & TO SUPPLY ESSENTIAL FATTY ACIDS

Although lipid frequently provides a significant proportion of the dietary requirement for energy, this is not an essential function.

Apart from increasing the palatability of food and producing a feeling of satiety, dietary lipid has 2 essential functions in mammalian nutrition. It acts as the dietary vehicle for the lipid-soluble vitamins, and it supplies **essential polyunsaturated fatty acids** that the body is unable to synthesize. Three polyunsaturated fatty acids have been recognized as essential in the diet of at least some animals: **linoleic acid** (ω6, 18:2), **α-linolenic acid** (ω3, 18:3), and **arachidonic** acid ($\omega6$, 20:4). These acids are found in the lipids of plant and animal food (see Table 16–2). See also Chapter 25 for a discussion of their metabolism.

In humans, arachidonic acid may be formed from linoleic acid and is not essential if sufficient linoleic acid is present in the diet. Debate continues about whether α-linolenic acid is truly essential in humans (but see p 220). **Deficiency of linoleic acid** is rare but may occur in infants on skim milk diets and in patients fed intravenously on lipid-free diets.

A principal function of essential fatty acids is to serve as precursors of **leukotrienes, prostaglandins,** and **thromboxanes** (see Fig 25–6), which function as "local hormones." A dietary intake in which 1–2% of the total energy requirement is supplied as essential fatty acid prevents clinical deficiency.

There Is an Association Between Fat Consumption & Disease

Numerous studies have shown a correlation between **coronary heart disease,** blood cholesterol, and the consumption of fat, particularly of saturated fat (see Chapter 28). High fat consumption is also associated with cancer of the breast and colon. The main source of saturated fat in the human diet is the meat of ruminants, dairy products, and hard margarine. Cholesterol is found only in foods of animal origin and particularly in egg yolk.

VITAMINS CARRY OUT A VARIETY OF BIOCHEMICAL FUNCTIONS

Vitamins are organic nutrients that are required in small quantities for many different biochemical functions and generally cannot be synthesized by the body and must, therefore, be supplied by the diet. Humans require either milligram or microgram quantities of each vitamin per day. Vitamins are classified into two main groups: the **water-soluble vitamins,** more fully described in Chapter 53, and the **lipid-soluble vitamins,** more fully described in Chapter 54.

Water-soluble vitamins include the vitamin B complex (thiamin, riboflavin, niacin, pantothenic acid, vitamin B_6, biotin, vitamin B_{12}, and folic acid) and ascorbic acid (vitamin C). Water-soluble vitamins are absorbed into the hepatic portal vein, and any surplus is excreted in the urine. There is thus little storage of the free vitamin, which in most instances needs to be continually supplied in the diet. Some storage of folic acid occurs in the liver. Depletion may take several months for ascorbic acid and several years for B_{12} (also stored in the liver). Excess intake is generally well tolerated except for side effects occurring with large doses of niacin in the form of nicotinic acid, ascorbic acid, or pyridoxine (vitamin B_6).

The **fat-soluble vitamins** (vitamins A, D, E, and K) are found in food lipids of both plant and animal origin. They are digested with fat and absorbed by the intestine and incorporated into chylomicrons. They are consequently transported mainly in chylomicron remnants, initially to the liver, which is a major store of vitamins A, D, and K. Adipose tissue is the major storage source of vitamin E. Fat-soluble vitamins are not excreted in the urine and, if taken in excess, are toxic (particularly vitamins A and D).

Some diseases concerned with cofactor metabolism that respond to treatment with specific vitamins are listed in Table 55–5.

Nonavailability of vitamins, whether due to dietary or other reasons, eg, defects in absorption, results in **characteristic deficiency syndromes.** Details of these are given in Chapters 53 and 54.

Table 55–5. Vitamin-responsive syndromes. Examples of specific defects in vitamin cofactor metabolism that can be corrected by vitamin therapy, usually requiring very large doses.*

Vitamin	Disease	Biochemical Defect
Biotin	Propionic acidemia	Propionyl-CoA carboxylase
Vitamin B_{12}	Methylmalonic aciduria	Formation of cobamide coenzyme
Folic acid	Folate malabsorption	Folic acid transport
Niacin	Hartnup disease	Tryptophan transport
Pyridoxine (vitamin B_6)	Infantile convulsions Cystathioninuria Homocystinuria	Glutamic acid decarboxylase (?) Cystathioninase Cystathionine synthase
Thiamin	Hyperalaninemia Thiamin-responsive lactic acidosis	Pyruvate dehydrogenase

*From: Herman RH, Stifel FB, Greene HL: Vitamin-deficient state and other related diseases. In: *Disorders of the Gastrointestinal Tract; Disorders of the Liver; Nutritional Disorders.* Dietschy JM (editor). Grune & Stratton, 1976.

Table 55–6. Essential macrominerals: Summary of major characteristics.

Elements	Functions	Metabolism[1]	Deficiency Disease or Symptoms	Toxicity Disease or Symptoms[2]	Sources[3]
Calcium	Constituent of bones, teeth; regulation of nerve, muscle function.	Absorption requires calcium-binding protein. Regulated by vitamin D, parathyroid hormone, calcitonin, etc.	Children: rickets. Adults: osteomalacia. May contribute to osteoporosis.	Occurs with excess absorption due to hypervitaminosis D or hypercalcemia due to hyperparathyroidism, or idiopathic hypercalcemia.	Dairy products, beans, leafy vegetables.
Phosphorus	Constituent of bones, teeth, ATP, phosphorylated metabolic intermediates. Nucleic acids.	Control of absorption unknown (vitamin D?). Serum levels regulated by kidney reabsorption	Children: rickets. Adults: osteomalacia.	Low serum Ca^{2+}:P_i ratio stimulates secondary hyperthyroidism; may lead to bone loss.	Phosphate food additives.
Sodium	Principal cation in extracellular fluid. Regulates plasma volume, acid-base balance, nerve and muscle function, Na^+/K^+-ATPase.	Regulated by aldosterone.	Unknown on normal diet; secondary to injury or illness.	Hypertension (in susceptible individuals).	Table salt; salt added to prepared food.
Potassium	Principal cation in intracellular fluid; nerve and muscle function, Na^+/K^+-ATPase.	Also regulated by aldosterone.	Occurs secondary to illness, injury, or diuretic therapy; muscular weakness, paralysis, mental confusion.	Cardiac arrest, small bowel ulcers.	Vegetables, fruit, nuts.
Chloride	Fluid and electrolyte balance; gastric fluid		Infants fed salt-free formula. Secondary to vomiting, diuretic therapy, renal disease.		Table salt.
Magnesium	Constituent of bones, teeth; enzyme cofactor (kinases, etc).		Secondary to malabsorption or diarrhea, alcoholism.	Depressed deep tendon reflexes and respiration.	Leafy green vegetables (containing chlorophyll).

[1] In general, minerals require carrier proteins for absorption. Absorption is rarely complete; it is affected by other nutrients and compounds in the diet (eg, oxalates and phytates that chelate divalent cations). Transport and storage also require special proteins. Excretion occurs in feces (unabsorbed minerals) and in urine, sweat, and bile.
[2] Excess mineral intake produces toxic symptoms. Unless otherwise specified, symptoms include nonspecific nausea, diarrhea, and irritability.
[3] Mineral requirements are met by a varied intake of adequate amounts of whole-grain cereals, legumes, leafy green vegetables, meat, and dairy products.

MINERALS ARE REQUIRED FOR BOTH PHYSIOLOGIC & BIOCHEMICAL FUNCTIONS

Minerals may be divided arbitrarily into 2 groups: (1) **macrominerals,** which are required in amounts greater than 100 mg/d, and (2) **microminerals (trace elements)**, which are required in amounts less than 100 mg/d. Table 55–6 provides a summary of the properties of the macrominerals, and Table 55–7 does the same for the microminerals.

RECOMMENDED DIETARY ALLOWANCES (RDA)

A review of the daily needs for essential nutrients has been published by the Food and Nutrition Board of the National Research Council as **Recommended Dietary Allowances** (Table 55–8). The allowances are to provide for individual variations among most normal persons as they live under their usual environmental conditions. They do not allow for extra requirements in illness or pathologic disorders. **Diets should be based on a variety of foods,** both to cover known requirements and to provide other nutrients for which human requirements have been less well defined. Table 55–8 covers protein, 10 vitamins, and 6 minerals. Too few data are available to ascribe RDAs to the remaining vitamins and minerals. However, ranges of intake of these nutrients that appear safe and adequate are given in Table 55–9.

All nutritional requirements must be met to prevent deficiency diseases and ill health. Ignorance or poor economic conditions are almost always the underlying cause of failure to satisfy nutritional requirements. On the other hand, certain common diseases are associated with *excess* intake of nutrients. Obesity generally

Table 55–7. Essential microminerals (trace elements): Summary of major characteristics.

Elements	Functions	Metabolism[1]	Deficiency Disease or Symptoms	Toxicity Disease or Symptoms[1]	Sources[2]
Chromium	Trivalent chromium, a constituent of "glucose tolerance factor."		Impaired glucose tolerance; secondary to parenteral nutrition.		
Cobalt	Required only as a constituent of vitamin B_{12}.	As for vitamin B_{12}.	Vitamin B_{12} deficiency.		Foods of animal origin.
Copper	Constituent of oxidase enzymes: cytochrome c oxidase, etc.	Transported by albumin; bound to ceruloplasmin.	Anemia (hypochromic, microcytic); secondary to malnutrition, Menke's syndrome.	Rare; secondary to Wilson's disease.	
Iodine	Constituent of thyroxine, triiodothyronine.	Stored in thyroid as thyroglobulin	Children: cretinism. Adults: goiter and hypothyroidism, myxedema.	Thyrotoxicosis, goiter.	Iodized salt, seafood
Iron	Constituent of heme enzymes (hemoglobin, cytochromes, etc).	Transported as transferrin; stored as ferritin or hemosiderin: lost in sloughed cells and by bleeding.	Anemia (hypochromic, microcytic).	Siderosis; hereditary hemochromatosis.	Iron cookware.
Manganese	Cofactor of hydrolase, decarboxylase, and transferase enzymes. Glycoprotein and proteoglycan synthesis.		Unknown in humans.	Inhalation poisoning produces psychotic symptoms and Parkinsonism.	
Molybdenum	Constituent of oxidase enzymes (xanthine oxidase).		Secondary to parenteral nutrition.		
Selenium	Constituent of glutathione peroxidase.	Synergistic antioxidant with vitamin E.	Marginal deficiency when soil content is low; secondary to parenteral nutrition, protein-energy malnutrition.	Megadose supplementation induces hair loss, dermatitis, and irritability.	
Zinc	Cofactor of many enzymes: lactate dehydrogenase, alkaline phosphatase; carbonic anhydrase, etc.		Hypogonadism, growth failure, impaired wound healing, decreased taste and smell acuity; secondary to acrodermatitis enteropathica, parenteral nutrition.	Gastrointestinal irritation, vomiting.	
Fluoride[3]	Increases hardness of bones and teeth.		Dental caries; osteoporosis(?).	Dental fluorosis.	Drinking water.

[1]Excess mineral intake produces toxic symptoms. Unless otherwise specified, symptoms include nonspecific nausea, diarrhea, and irritability.
[2]Mineral requirements are met by a varied intake of adequate amounts of whole-grain cereals, legumes, leafy green vegetables, meat, and dairy products.
[3]Fluoride is essential for rat growth. While not proved to be strictly essential for human nutrition, fluorides have a well-defined role in prevention and treatment of dental caries.

reflects excess intake of energy and is often associated with the development of **non-insulin-dependent diabetes mellitus. Atherosclerosis** and **coronary heart disease** are associated with diets high in total fat and saturated fat. **Cancer** of the breast, colon, and prostate correlates with high fat intake. High incidence of **cerebrovascular disease** and **hypertension** is associated with a high salt intake.

Many committees throughout the world have investigated the composition of human diets and made recommendations for improvement. These may be summarized as follows: (1) If overweight, total energy intake should be reduced to achieve optimum weight. (2) There should be a general shift away from fat consumption to consumption of more carbohydrate. (3) A greater proportion of carbohydrate should be con-

Table 55–8. Recommended daily dietary allowances. (Revised 1989.) Designed for the maintenance of good nutrition of the majority of healthy people in the USA.

	Age (years)	Weight (kg)	Weight (lb)	Height (cm)	Height (in)	Protein (g)	Vitamin A (µg RE)[1]	Vitamin D (µg)[2]	Vitamin E (mg α-TE)[3]	Vitamin K (µg)	Vitamin C (mg)	Thiamin (mg)	Riboflavin (mg)	Niacin (mg NE)[4]	Vitamin B6 (mg)	Folate (µg)	Vitamin B12 (µg)	Calcium (mg)	Phosphorus (mg)	Magnesium (mg)	Iron (mg)	Zinc (mg)	Iodine (µmg)	Selenium (µg)
Infants	0.0–0.5	6	13	60	24	13	375	7.5	3	5	30	0.3	0.4	5	0.3	25	0.3	400	300	40	6	5	40	10
	0.5–1.0	9	20	71	28	14	375	10	4	10	35	0.4	0.5	6	0.6	35	0.5	600	500	60	10	5	50	15
Children	1–3	13	29	90	35	16	400	10	6	15	40	0.7	0.8	9	1.0	50	0.7	800	800	80	10	10	70	20
	4–6	20	44	112	44	24	500	10	7	20	45	0.9	1.1	12	1.1	75	1.0	800	800	120	10	10	90	20
	7–10	28	62	132	52	28	700	10	7	30	45	1.0	1.2	13	1.4	100	1.4	800	800	170	10	10	120	30
Males	11–14	45	99	157	62	45	1000	10	10	45	50	1.3	1.5	17	1.7	150	2.0	1200	1200	270	12	15	150	40
	15–18	66	145	176	69	59	1000	10	10	65	60	1.5	1.8	20	2.0	200	2.0	1200	1200	400	12	15	150	50
	19–24	72	160	177	70	58	1000	10	10	70	60	1.5	1.7	19	2.0	200	2.0	1200	1200	350	10	15	150	70
	25–50	79	174	176	70	63	1000	5	10	80	60	1.5	1.7	19	2.0	200	2.0	800	800	350	10	15	150	70
	51+	77	170	173	68	63	1000	5	10	80	60	1.2	1.4	15	2.0	200	2.0	800	800	350	10	15	150	70
Females	11–14	46	101	157	62	46	800	10	8	45	50	1.1	1.3	15	1.4	150	2.0	1200	1200	280	15	12	150	45
	15–18	55	120	163	64	44	800	10	8	55	60	1.1	1.3	15	1.5	180	2.0	1200	1200	300	15	12	150	50
	19–24	58	128	164	65	46	800	10	8	60	60	1.1	1.3	15	1.6	180	2.0	1200	1200	280	15	12	150	55
	25–50	63	138	163	64	50	800	5	8	65	60	1.1	1.3	15	1.6	180	2.0	800	800	280	15	12	150	55
	51+	65	143	160	63	50	800	5	8	65	60	1.0	1.2	13	1.6	180	2.0	800	800	280	10	12	150	55
Pregnant						60	800	10	10	65	70	1.5	1.6	17	2.2	400	2.2	1200	1200	320	30	15	175	65
Lactating						65	1300	10	12	65	95	1.6	1.8	20	2.1	280	2.6	1200	1200	355	15	19	200	75

[1] Retinol equivalents. 1 retinol equivalent = 1 µg retinol or 6 µg β-carotene.
[2] As cholecalciferol. 10 µg cholecalciferol = 400 IU of vitamin D.
[3] α-Tocopherol equivalents. 1 mg α-tocopherol = 1 α-TE.
[4] 1 NE (niacin equivalent) is equal to 1 mg of niacin or 60 mg of dietary tryptophan.

Table 55–9. Estimated safe and adequate daily dietary intakes of selected vitamins and minerals.

	Age (years)	Biotin (µg)	Pantothenic Acid (mg)	Copper (mg)	Manganese (mg)	Fluoride (mg)	Chromium (mg)	Molybdenum (µg)	Sodium (mg)	Potassium (mg)	Chloride (mg)
Infants	0–0.5	10	2	0.4–0.6	0.3–0.6	0.1–0.5	0.01–0.04	15–30	115–350	350–925	275–700
	0.5–1	15	3	0.6–0.7	0.6–1.0	0.2–1.0	0.02–0.06	20–40	250–750	425–1275	400–1200
Children and adolescents	1–3	20	3	0.7–1.0	1.0–1.5	0.5–1.5	0.02–0.08	25–50	325–975	550–1650	500–1500
	4–6	25	3–4	1.0–1.5	1.5–2.0	1.0–2.5	0.03–0.12	30–75	450–1350	775–2325	700–2100
	7–10	30	4–5	1.0–2.0	2.0–3.0	1.5–2.5	0.05–0.2	50–150	600–1800	1000–3000	925–2775
	11+	30–100	4–7	1.5–2.5	2.0–5.0	1.5–2.5	0.05–0.2	75–250	900–2700	1525–4575	1400–4200
Adults		30–100	4–7	1.5–3.0	2.0–5.0	1.5–4.0	0.05–0.2	75–250	1100–3300	1875–5625	1700–5100

Both tables from: *Recommended Dietary Allowances*, 10th ed. food and Nutrition Board, National Research Council–National Academy of Science, 1989.

sumed as complex carbohydrates and less as sugars. (4) A greater proportion of dietary fat should be in the form of polyunsaturated and monounsaturated fat and less as saturated fat. (5) Consumption of cholesterol and salt should be reduced. (6) Dietary fiber should be increased.

REFERENCES

Cummings JH: Dietary Fibre. *Brit Med Bull* 1981;**37**:65.

Forbes JM: Metabolic aspects of the regulation of voluntary food intake and appetite. *Nutr Res Rev* 1988;**1**:145.

Kritchevsky D: Dietary Fiber. *Annu Rev Nutr* 1988;**8**:301.

Nestle M: *Nutrition in Clinical Practice.* Jones Medical Publications, 1985.

Nielsen FH: Nutritional significance of the ultratrace elements. *Nutr Rev* 1988;**46**:337.

Olson RE et al (editors): *Present Knowledge in Nutrition,* 5th ed. The Nutrition Foundation Inc, Washington, DC, 1984.

Passmore R, Eastwood MA: *Human Nutrition and Dietetics,* 8th ed. Churchill Livingstone, 1986.

Woo R, Daniels-Kush R, Horton, ES: Regulation of energy balance. *Annu Rev Nutr* 1985;**5**:411.

56

Digestion & Absorption

Peter A. Mayes, PhD, DSc

INTRODUCTION

Most foodstuffs are ingested in forms that are unavailable to the organism, since they cannot be absorbed from the digestive tract until they have been broken down into smaller molecules. This disintegration of the naturally occurring foodstuffs into assimilable forms constitutes the process of digestion.

The chemical changes incident to digestion are accomplished with the aid of hydrolase enzymes of the digestive tract that catalyze the hydrolysis of native proteins to amino acids, of starches to monosaccharides, and of triacylglycerols to monoacylglycerols, glycerol, and fatty acids. In the course of these digestive reactions, the minerals and vitamins of the foodstuffs are also made more assimilable.

A systematic account of the nature and functions of the gastrointestinal hormones is given in Chapter 52.

BIOMEDICAL IMPORTANCE

Some clinical conditions arise from defects in the digestive processes such as **ulceration** by gastric HCl, or diminished secretion of HCl causing **achlorhydria.** Defects of bile secretion lead to **gallstones** or possibly defective lipid digestion. Malabsorption of nutrients is due to a wide variety of defects and often leads to **nutritional deficiency,** eg, malabsorption of vitamin B_{12} and folate causes **anemia;** defects in calcium and magnesium lead to **tetany** and vitamin D malabsorption leads to **rickets** and **osteomalacia;** and a general **malabsorption syndrome** includes these and other defects. Lactase deficiency gives rise to **milk intolerance,** and defects in absorption of neutral amino acids are involved in **Hartnup disease.**

DIGESTION BEGINS IN THE ORAL CAVITY

Saliva, secreted by the salivary glands, consists of about 99.5% water. It acts as a lubricant for mastication and for swallowing. Adding water to dry food provides a medium in which food molecules can dissolve and in which hydrolases can initiate digestion.

Mastication subdivides the food, increasing its solubility and surface area for enzyme attack. The saliva is also a vehicle for the excretion of certain drugs (eg, ethanol and morphine), of inorganic ions such as K^+, Ca^{2+}, HCO_3^-, thiocyanate (SCN^-), and iodine, and of immunoglobulins (IgA).

The pH of saliva is usually about 6.8, although it may vary on either side of neutrality.

Saliva Contains an Amylase & a Lipase

Salivary amylase is capable of bringing about the hydrolysis of starch and glycogen to maltose; however, this is of little significance in the body because of the short time it can act on the food. Salivary amylase is readily inactivated at pH 4.0 or less, so that digestive action on food in the mouth will soon cease in the acid environment of the stomach. In many animals, a salivary amylase is entirely absent. A **lingual lipase** is secreted by the dorsal surface of the tongue (Ebner's glands).

THE DIGESTION OF PROTEIN BEGINS IN THE STOMACH

The gastric secretion is known as **gastric juice.** It is a clear, pale yellow fluid of 0.2–0.5% HCl, with a pH of about 1.0. The gastric juice is 97–99% water. The remainder consists of mucin and inorganic salts, the digestive enzymes (pepsin and rennin), and a lipase.

Hydrochloric Acid Denatures Protein & Kills Bacteria

The **parietal (oxyntic) cells** are the source of gastric HCl, which originates according to the reactions shown in Fig 56–1. The process is similar to that of the "chloride shift" described for the red blood cell. There is also a resemblance to the renal tubular mechanisms for secretion of H^+, wherein the source of H^+ is also the **carbonic anhydrase**-catalyzed formation of H_2CO_3 from H_2O and CO_2. An alkaline urine often follows the ingestion of a meal ("alkaline tide"), as a result of the formation of bicarbonate in the process of hydrochloric acid secretion. Secretion of H^+ into the

Figure 56–1. Production of gastric hydrochloric acid. $\ominus$, H^+,K^+-ATPase.

lumen is an active process driven by a membrane-located H^+,K^+-**ATPase,** which, unlike the Na^+/K^+-ATPase, is ouabain insensitive. Parietal cells contain numerous mitochondria needed to generate the ATP used for powering the H^+,K^+-ATPase. HCO_3^- passes into the plasma in exchange for Cl^-, which is coupled to the secretion of H^+ into the lumen.

As a result of contact with gastric HCl, proteins are denatured; ie, the tertiary protein structure is lost as a result of the destruction of hydrogen bonds. This allows the polypeptide chain to unfold, making it more accessible to the actions of proteolytic enzymes (proteases). The low pH also has the effect of destroying most microorganisms entering the gastrointestinal tract.

Pepsin Initiates Protein Digestion

This is the major digestive function of the stomach. Pepsin is produced in the chief cells as the inactive **zymogen, pepsinogen.** This is activated to pepsin by H^+, which splits off a protective polypeptide to expose active pepsin; and by pepsin, which rapidly activates further molecules of pepsinogen (**autocatalysis**). Pepsin transforms denatured protein into proteoses and then peptones, which are large polypeptide derivatives. Pepsin is an **endopeptidase,** since it hydrolyzes peptide bonds within the main polypeptide structure rather than adjacent to N- or C-terminal residues, which is characteristic of **exopeptidases.** It is specific for peptide bonds formed by aromatic amino acids (eg, tyrosine) or dicarboxylic amino acids (eg, glutamate).

Rennin (Chymosin, Rennet) Causes the Coagulation of Milk

It is important in the digestive processes of infants, because it prevents the rapid passage of milk from the stomach. In the presence of calcium, rennin changes the casein of milk irreversibly to a **paracasein** which is

then acted on by pepsin. Rennin is reported to be absent from the stomach of adults. It is used in the making of cheese.

Lipases Continue the Digestion of Triacylglycerols

The heat of the stomach is important in liquidizing the bulk of dietary lipids; emulsification takes place aided by peristaltic contractions. The stomach contains a **gastric lipase** capable of hydrolyzing triacylglycerols of short and longer chain length. However, the **lingual lipase** can continue its activity at the low pH of the stomach, where because of the retention time of 2–4 hours, about 30% of dietary triacylglycerol may be digested. Lingual lipase is more active on triacylglycerols having shorter chain fatty acids and more specific for the ester linkage in the *sn*-3 position rather than position 1. Milk fat contains short- and medium-chain fatty acids, which tend to be esterified in the *sn*-3 position. Therefore, milk fat seems to be a particularly good substrate for this enzyme. The released hydrophilic short-chain fatty acids are absorbed via the stomach wall and enter the portal vein, whereas longer-chain fatty acids dissolve in the fat droplets and pass on to the duodenum.

DIGESTION CONTINUES IN THE INTESTINE

The stomach contents, or **chyme,** are intermittently introduced during digestion into the duodenum through the pyloric valve. The **alkaline** content of pancreatic and biliary secretions neutralizes the acid of the chyme and changes the pH of this material to the alkaline side; this shift of pH is necessary for the activity of the enzymes contained in pancreatic and intestinal juice, but **it inhibits further action of pepsin.**

Table 56–1. The composition of hepatic and of gallbladder bile.

| | Hepatic Bile (as secreted) | | Bladder Bile |
	Percent of Total Bile	Percent of Total Solids	Percent of Total Bile
Water	97.00	. . .	85.92
Solids	2.52	. . .	14.08
Bile acids	1.93	36.9	9.14
Mucin and pigments	0.53	21.3	2.98
Cholesterol	0.06	2.4	0.26
Esterified and non-esterified fatty acids	0.14	5.6	0.32
Inorganic salts	0.84	33.3	0.65
Specific gravity	1.01	. . .	1.04
pH	7.1–7.3	. . .	6.9–7.7

Bile Emulsifies, Neutralizes, & Excretes Cholesterol & Bile Pigments

In addition to many functions in intermediary metabolism, the **liver,** by producing bile, plays an important role in digestion. The gallbladder stores bile produced by the liver between meals. During digestion, the gallbladder contracts and supplies bile rapidly to the duodenum by way of the common bile duct. The pancreatic secretions mix with the bile, since they empty into the common duct shortly before its entry into the duodenum.

A. Composition of Bile: The composition of hepatic bile differs from that of gallbladder bile. As shown in Table 56-1, the latter is more concentrated.

B. Properties of Bile:

1. Emulsification–The bile salts have considerable ability to lower surface tension. This enables them to emulsify fats in the intestine and to dissolve fatty acids and water-insoluble soaps. The presence of bile in the intestine is an important adjunct to accomplish the digestion and absorption of fats as well as the absorption of the fat-soluble vitamins A, D, E, and K. When fat digestion is impaired, other foodstuffs are also poorly digested, since the fat covers the food particles and prevents enzymes from attacking them. Under these conditions, the activity of the intestinal bacteria causes considerable putrefaction and production of gas.

2. Neutralization of acid–In addition to its function in emulsification, the bile, having a pH slightly above 7, neutralizes the acid chyme from the stomach and prepares it for digestion in the intestine.

3. Excretion–Bile is an important vehicle for bile acid and cholesterol excretion, but it also removes many drugs, toxins, bile pigments, and various inorganic substances such as copper, zinc, and mercury.

4. The limited solubility of cholesterol in bile is a cause of gallstone formation–Free cholesterol is, for practical purposes, insoluble in water; consequently, it is incorporated into a phospholipid-bile salt micelle (see p 145). Indeed, phosphatidylcholine, the predominant phospholipid in bile, is itself insoluble in aqueous systems but can be dissolved by bile salts in micelles. The large quantities of cholesterol present in the bile of humans are solubilized in these water-soluble mixed micelles, allowing cholesterol to be transported in bile via the biliary tract to the intestine. However, **the actual solubility of cholesterol in bile depends on the relative proportions of bile salt, phosphatidylcholine, and cholesterol.** The solubility also depends on the water content of bile. This is especially important in dilute hepatic bile.

Using triangular coordinates (Fig 56–2), Redinger and Small were able to determine the maximum solubility of cholesterol in human gallbladder bile. Reference to the figure indicates that any triangular point falling above the line ABC would represent a bile whose composition is such that cholesterol is either supersaturated or precipitated.

It is believed that at some time during the life of a patient with gallstones there is formed an abnormal

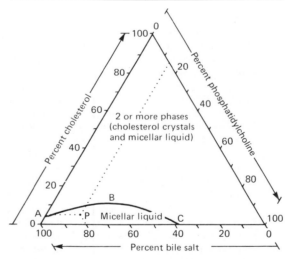

Figure 56–2. Method for presenting 3 major components of bile (bile salts, phosphatidylcholine, and cholesterol) on triangular coordinates. Each component is expressed as a percentage mole of total bile salt, phosphatidylcholine, and cholesterol. Line ABC represents maximum solubility of cholesterol in varying mixtures of bile salt and phosphatidylcholine. Point P represents normal bile composition, containing 5% cholesterol, 15% phosphatidylcholine, and 80% bile salt, and falls within the zone of a single phase of micellar liquid. Bile having a composition falling above the line would contain excess cholesterol in either supersaturated or precipitated form (crystals or liquid crystals). (Reproduced, with permission, from Redinger RN, Small DM: Bile composition, bile salt metabolism, and gallstones. *Arch Intern Med* 1972;**130**:620. Copyright © 1972. American Medical Association.)

bile that has become supersaturated with cholesterol. With time, various factors such as infection, for example, serve as seeding agents to cause the supersaturated bile to precipitate the excess cholesterol as crystals. Unless the newly formed crystals are promptly excreted into the intestine with the bile, the crystals will grow to form stones. When the activities of key enzymes in bile acid formation were measured in the livers of patients with gallstones, cholesterol synthesis was elevated but bile acid synthesis was reduced, causing liver cholesterol concentrations to increase. It seems that decreased 7α-hydroxylase activity leads to a diminished enterohepatic bile acid pool that signals the liver to produce more cholesterol. The bile then becomes overloaded with cholesterol, which is unable to dissolve completely in the mixed micelles.

The above information concerning cholesterol solubility has been used in attempts to dissolve gallstones or to prevent their further formation. **Chenodeoxycholic acid** appears to offer specific medical treatment for asymptomatic radiolucent gallstones in functioning gallbladders because of its specific inhibition of HMG-CoA reductase in the liver, with consequent reduction in cholesterol synthesis.

5. Bile pigment metabolism—The origin of the bile pigments from hemoglobin is discussed in Chapter 34.

The Pancreatic Secretion Contains Enzymes for Attacking All the Major Foodstuffs

Pancreatic secretion is a nonviscid watery fluid that is similar to saliva in its content of water and contains some protein and other organic and inorganic compounds—mainly Na^+, K^+, HCO_3^-, and Cl^-, but Ca^{2+}, Zn^{2+}, HPO_4^{2-}, and SO_4^{2-} are also present in small amounts. The pH of pancreatic secretion is distinctly alkaline, 7.5–8.0 or higher.

Many enzymes are found in pancreatic secretion; some are secreted as zymogens.

Trypsin, Chymotrypsin, & Elastase Are Endopeptidases: The proteolytic action of pancreatic secretion is due to the 3 **endopeptidases** trypsin, chymotrypsin, and elastase, which attack proteins and polypeptides released from the stomach, to produce polypeptides, peptides, or both. **Trypsin** is specific for peptide bonds of basic amino acids, and **chymotrypsin** is specific for peptide bonds containing uncharged amino acid residues, such as aromatic amino acids. **Elastase,** in spite of its name, has rather broad specificity in attacking bonds next to small amino acid residues such as glycine, alanine, and serine. All 3 enzymes are secreted as zymogens. Activation of **trypsinogen** is due to another proteolytic enzyme, **enterokinase,** secreted by the intestinal mucosa. This hydrolyzes a lysine peptide bond in the zymogen, releasing a small polypeptide that allows the molecule to unfold as active trypsin. Once trypsin is formed, it will attack not only additional molecules of trypsinogen but also the other zymogens in the pancreatic secretion, **chymotrypsinogen, proelastase,** and **procarboxy-**

peptidase, liberating chymotrypsin, elastase, and **carboxypeptidase,** respectively.

Carboxypeptidase is an Endopeptidase: The further attack on the polypeptides produced by the action of endopeptidases is carried on by the exopeptidase **carboxypeptidase,** which attacks the carboxy-terminal peptide bond, liberating single amino acids.

Amylase Attacks Starch & Glycogen: The starch-splitting action of pancreatic secretion is due to a pancreatic **α-amylase.** It is similar in action to salivary amylase, hydrolyzing starch and glycogen to maltose, maltotriose [three α-glucose residues linked by $\alpha(1 \rightarrow 4)$ bonds], and a mixture of branched $(1 \rightarrow 6)$ oligosaccharides (α-limit dextrins), nonbranched oligosaccharides, and some glucose.

Lipase Attacks the Primary Ester Link of Triacylglycerols: The pancreatic lipase acts at the oil-water interface of the finely emulsified lipid droplets formed by mechanical agitation in the gut in the presence of the products of lingual lipase activity, bile salts, **colipase** (a protein present in pancreatic secretion), **phospholipids,** and **phospholipase A$_2$** (also present in the pancreatic secretion). Phospholipase A$_2$ and colipase are secreted in pro-forms and require activation by tryptic hydrolysis of specific peptide bonds. Ca^{2+} is necessary for phospholipase A$_2$ activity. A limited hydrolysis of the ester bond in the 2 position of the phospholipid by phospholipase A$_2$ (see Fig 25–5) results in the binding of lipase to the substrate interface and a rapid rate of hydrolysis of triacylglycerol. Colipase binds to the bile salt-triacylglycerol/water interface, providing a high-affinity anchor for the lipase. The complete hydrolysis of triacylglycerols produces glycerol and fatty acids. However, the second and third fatty acids are hydrolyzed from the triacylglycerols with increasing difficulty. Pancreatic lipase is virtually specific for the hydrolysis of primary ester linkages, ie, at positions 1 and 3 of triacylglycerols. During fat digestion, the aqueous or "micellar phase" contains mixed **disklike micelles** and **liposomes** of bile salts saturated with lipolytic products (see Fig 15–34).

Because of the difficulty of hydrolysis of the secondary ester linkage in the triacylglycerol, it is probable that the digestion of triacylglycerol proceeds by removal of the terminal fatty acids to produce 2-monoacylglycerol. Since this last fatty acid is linked by a secondary ester bond, its removal requires isomerization to a primary ester linkage. This is a relatively slow process; as a result, **2-monoacylglycerols are major end products of triacylglycerol digestion,** and less than one-fourth of the ingested triacylglycerol is completely broken down to glycerol and fatty acids (Fig 56–3).

Cholesteryl Esters Are Broken Down by a Specific Hydrolase: Under the conditions within the lumen of the intestine, **cholesteryl ester hydrolase** (cholesterol esterase) catalyzes the hydrolysis of cholesteryl esters, which are thus absorbed from the intestine in a nonesterified, free form.

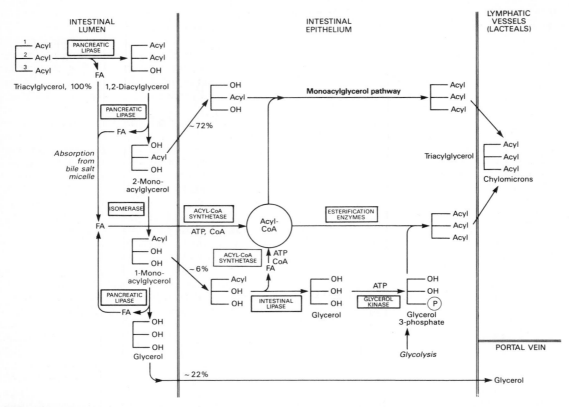

Figure 56–3. Digestion and absorption of triaclyglycerols. FA, long-chain fatty acid. (Modified from Mattson FH, Volpenheim RA: The digestion and absorption of triglycerides. *J Biol Chem* 1964;**239**:2772.)

Ribonuclease (RNase) and Deoxyribonuclease (DNase) are Responsible for the Digestion of Dietary Nucleic Acids: see Chapters 38 and 39.

Phospholipase A$_2$ Hydrolyzes the Ester Bond in the 2 Position of Glycerophospholipids of both biliary and dietary origins to form lysophospholipids.

Intestinal Secretions Complete the Digestive Process

The intestinal juice secreted by the glands of Brunner and of Lieberkühn contains digestive enzymes, including the following:

(1) **Aminopeptidase,** which is an exopeptidase attacking peptide bonds next to N-terminal amino acids of polypeptides and oligopeptides; and **dipeptidases** of various specificity, some of which may be within the intestinal epithelium. The latter complete digestion of dipeptides to free amino acids.

(2) Specific **disaccharidases** and **oligosaccharidases,** ie, **α-glucosidase (maltase)**, which removes single glucose residues from α(1 → 4) linked oligosaccharides and disaccharides, starting from the nonreducing ends; **isomaltase (α-dextrinase)**, which hydrolyzes 1 → 6 bonds in α-limit dextrins; **β-galactosidase (lactase)** for removing galactose from lactose; **sucrase**

for hydrolyzing sucrose; and **trehalase** for hydrolyzing trehalose.

(3) A **phosphatase,** which removes phosphate from certain organic phosphates such as hexose phosphates, glycerophosphate, and the nucleotides derived from the diet and the digestion of nucleic acids by nucleases.

(4) **Polynucleotidases,** which split nucleic acids into nucleotides.

(5) **Nucleosidases** (nucleoside phosphorylases) catalyze the phosphorolysis of nucleosides to give the free nitrogen base plus a pentose phosphate.

(6) The intestinal secretion is also said to contain a **phospholipase** that attacks phospholipids to produce glycerol, fatty acids, phosphoric acid, and bases such as choline.

The Major Products of Digestion Are Assimilated

The final result of the action of the digestive enzymes described is to reduce the foodstuffs of the diet to forms that can be absorbed and assimilated. These end products of digestion are, for carbohydrates, the monosaccharides (principally glucose); for proteins, the amino acids; for triacylglycerol, the fatty acids, glycerol, and monoacylglycerols; and for nucleic acids, the nucleobases, nucleosides, and pentoses.

The plant cell wall polysaccharides and lignin of the

Table 56–2. Summary of digestive processes.

Source of Secretion and Stimulus for Secretion	Enzyme	Method of Activation and Optimal Conditions for Activity	Substrate	End Products or Action
Salivary glands: Secrete saliva in reflex response to presence of food in oral cavity.	Salivary amylase	Chloride ion necessary. pH 6.6–6.8	Starch Glycogen	Maltose plus 1:6 glucosides (oligosaccharides) plus maltotriose.
Lingual glands	Lingual lipase	pH range 2.0–7.5; optimal, 4.0–4.5	Short-chain primary ester link at *sn*-3	Fatty acids plus 1,2-diacylglycerols.
Stomach glands: Chief cells and parietal cells secrete gastric juice in response to reflex stimulation and action of gastrin.	Pepsin A (fundus) Pepsin B (pylorus)	Pepsinogen converted to active pepsin by HCl. pH 1.0–2.0.	Protein	Peptides.
	Rennin	Calcium necessary for activity. pH 4.0.	Casein of milk	Coagulates milk.
Pancreas: Presence of acid chyme from the stomach activates duodenum to produce (1) secretin, which hormonally stimulates flow of pancreatic juice; (2) cholecystokinin, which stimulates the production of enzymes.	Trypsin	Trypsinogen converted to active trypsin by enterokinase of intestine at pH 5.2–6.0. Autocatalytic at pH 7.9.	Protein Peptides	Polypeptides. Dipeptides.
	Chymotrypsin	Secreted as chymotrypsinogen and converted to active form by trypsin. pH 8.0.	Protein Peptides	Same as trypsin. More coagulating power for milk.
	Elastase	Secreted as proelastase and converted to active form by trypsin.	Protein Peptides	Polypeptides. Dipeptides.
	Carboxypeptidase	Secreted as procarboxypeptidase, activated by trypsin.	Polypeptides at the free carboxyl end of the chain	Lower peptides. Free amino acids.
	Pancreatic amylase	pH 7.1.	Starch Glycogen	Maltose plus 1:6 glucosides (oligosaccharides) plus maltotriose.
	Lipase	Activated by bile salts, phospholipids, colipase. pH 8.0.	Primary ester linkages of triacylglycerol	Fatty acids, monoacylglycerols, diacylglycerols, glycerol.
	Ribonuclease		Ribonucleic acid	Nucleotides.
	Deoxyribonuclease		Deoxyribonucleic acids	Nucleotides.
	Cholesteryl ester hydrolase	Activated by bile salts.	Cholesteryl esters	Free cholesterol plus fatty acids.
	Phospholipase A_2	Secreted as proenzyme, activated by trypsin and Ca^{2+}.	Phospholipids	Fatty acids, lysophospholipids.
Liver and gallbladder: Cholecystokinin, a hormone from the intestinal mucosa—and possibly also gastrin and secretin—stimulate the gallbladder and secretion of bile by the liver.	(Bile salts and alkali)		Fats—also neutralize acid chyme	Fatty acid-bile salt conjugates and finely emulsified neutral fat-bile salt micelles and liposomes.
Small intestine: Secretions of Brunner's glands of the duodenum and glands of Lieberkühn.	Aminopeptidase		Polypeptides at the free amino end of the chain	Lower peptides. Free amino acids.
	Dipeptidases		Dipeptides	Amino acids.
	Sucrase	pH 5.0–7.0.	Sucrose	Fructose, glucose.
	Maltase	pH 5.8–6.2.	Maltose	Glucose.
	Lactase	pH 5.4–6.0.	Lactose	Glucose, galactose.

(*continued*)

Table 56–2 (cont'd). Summary of digestive processes.

Source of Secretion and Stimulus for Secretion	Enzyme	Method of Activation and Optimal Conditions for Activity	Substrate	End Products or Action
Small intestine: Secretions of Brunner's glands of the duodenum and glands of Lieberkühn.	Trehalase		Trehalose	Glucose.
	Phosphatase	pH 8.6.	Organic phosphates	Free phosphate.
	Isomaltase or 1:6 glucosidase		1:6 glucosides	Glucose.
	Polynucleotidase		Nucleic acid	Nucleotides.
	Nucleosidases (nucleoside phosphorylases)		Purine or pyrimidine nucleosides	Purine or pyrimidine bases, pentose phosphate.

diet that cannot be digested by mammalian enzymes constitute **dietary fiber** and make up the bulk of the residues from digestion. Fiber performs an important function in adding bulk to the diet and is discussed in the preceding chapter. Table 56–2 summarizes the major digestive processes.

ABSORPTION FROM THE GASTROINTESTINAL TRACT RESULTS IN PASSAGE OF NUTRIENTS INTO THE HEPATIC PORTAL VEIN OR THE LYMPHATICS

There is little absorption from the stomach, although considerable gastric absorption of ethanol is possible.

The small intestine is the main digestive and absorptive organ. About 90% of the ingested foodstuffs are absorbed in the course of passage through the small intestine, and water is absorbed at the same time. Considerably more water is absorbed after the foodstuffs pass into the large intestine, so that the contents, which were fluid in the small intestine, gradually become more solid in the colon.

There are 2 pathways for the transport of materials absorbed by the intestine: the **hepatic portal system,** which leads directly to the liver, transporting water-soluble nutrients, and the **lymphatic vessels,** which lead to the blood by way of the thoracic duct and transport lipid-soluble nutrients.

Carbohydrates Are Absorbed as Monosaccharides

The products of carbohydrate digestion are absorbed from the jejunum into the blood of the portal venous system in the form of monosaccharides, chiefly as hexose (glucose, fructose, mannose, and galactose) and as pentose sugars (ribose). The oligosaccharides (compounds derived from starches that yield 3–10 monosaccharide units upon hydrolysis) and the disaccharides are hydrolyzed by appropriate enzymes derived from the mucosal surfaces of the small intestine, which may include pancreatic amylase adsorbed onto the mucosa. There is little free disaccharidase activity in the intestinal lumen. Most of the activity is associated with small "knobs" on the brush border of the intestinal epithelial cell.

Two mechanisms are responsible for the absorption of monosaccharides: **active transport** against a concentration gradient and **simple diffusion.** However, the absorption of some sugars does not fit clearly into one or the other of these mechanisms. The molecular configurations that seem necessary for active transport, which are present in glucose and galactose, are the following: The OH on carbon 2 should have the same configuration as in glucose, a pyranose ring should be present, and a methyl or substituted methyl group should be present on carbon 5. Fructose is absorbed more slowly than glucose and galactose. Its absorption appears to proceed by diffusion with the concentration gradient. Normally, there is little fructose in blood apart from that derived from the diet.

Active Absorption of Glucose Is Powered by the Sodium Pump

The brush border of the enterocyte contains several transporter systems, some very similar to those of the renal brush border membranes, which specialize in the uptake of the different amino acids and sugars. A carrier has been postulated that binds both glucose and Na^+ at separate sites and that transports them both through the plasma membrane of the intestinal cell. It is envisaged that both glucose and Na^+ are released into the cytosol, allowing the carrier to take up more "cargo." The Na^+ is transported down its concentration gradient and at the same time causes the carrier to transport glucose against its concentration gradient. The free energy required for this active transport is obtained from the hydrolysis of ATP linked to a sodium pump that expels Na^+ from the cell in exchange for K^+ (Fig 56–4). The active transport of glucose is inhibited by ouabain (a cardiac glycoside), an inhibitor of the sodium pump, and by phlorhizin, a known inhibitor of glucose reabsorption in the kidney tubule. The ratio of Na^+:glucose transported varies but may be consistent with 2 carriers—a 1:1 and a 3:1 carrier operating in parallel. There is also an Na^+-independent carrier of glucose.

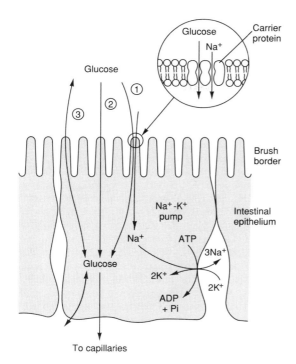

Figure 56–4. Transport of glucose across the intestinal epithelium. Active glucose transport is coupled to the Na⁺-K⁺ pump, 1, or to an Na⁺-independent system, 2. Diffusion is represented by 3.

Hydrolysis of polysaccharides, oligosaccharides, and disaccharides is rapid; therefore, the absorptive mechanisms for glucose and fructose are quickly saturated. A conspicuous exception is the hydrolysis of lactose, which proceeds at only half the rate for sucrose, accounting for the fact that digestion of lactose does not lead to saturation of the transport mechanisms for glucose and galactose.

Defects in Enzymes of Carbohydrate Digestion Cause Specific Disorders

Lactase Deficiency: Intolerance to lactose, the sugar of milk, may be attributable to a deficiency of lactase. The syndrome should not be confused with intolerance to milk resulting from a sensitivity to milk proteins, usually to the β-lactoglobulin. The signs and symptoms of lactose intolerance are the same regardless of the cause. These include abdominal cramps, diarrhea, and flatulence. They are attributed to accumulation of lactose, which is osmotically active, so that it holds water, and to the fermentative action on the sugar of the intestinal bacteria which produce gases and other products that serve as intestinal irritants.

There are 3 types of lactase deficiency:

1. Inherited lactase deficiency—In this syndrome, which is relatively rare, symptoms of intolerance develop very **soon after birth.** The feeding of a lactose-free diet results in disappearance of the symptoms. The occurrence of lactose in the urine is a prominent feature of this syndrome, which appears to be attributable to an effect of lactose on the intestine.

2. Secondary low-lactase activity—Because digestion of lactose is limited even in normal humans, intolerance to milk is not uncommon as a consequence of intestinal diseases. Examples are tropical and nontropical (celiac) sprue, kwashiorkor, colitis, and gastroenteritis. The disorder may be noted also after surgery for peptic ulcer.

3. Primary low-lactase activity—This is a relatively common syndrome, particularly among nonwhite populations. Since intolerance to lactose was not a feature of the early life of adults with this disorder, it is presumed to represent a gradual decline in activity of lactase in susceptible individuals.

Sucrase Deficiency: There is an inherited deficiency of the disaccharidases sucrase and isomaltase. These 2 deficiencies coexist, because sucrase and isomaltase occur together as a complex enzyme. Symptoms occur in early childhood and are the same as those described in lactase deficiency.

Disacchariduria: An increase in the excretion of disaccharides may be observed in some patients with disaccharidase deficiencies. As much as 300 mg or more of disaccharide may be excreted in the urine of these people and in patients with intestinal damage (eg, sprue).

Monosaccharide Malabsorption: There is a congenital condition in which glucose and galactose are absorbed only slowly, owing to a defect in the carrier mechanism. Because fructose is not absorbed via the carrier, its absorption is normal.

The Products of Lipid Digestion Are Absorbed From Bile Salt Micelles

The 2-monoacylglycerols, fatty acids, and small amounts of 1-monoacylglycerols leave the oil phase of the lipid emulsion and diffuse into the mixed micelles and liposomes consisting of bile salts, phosphatidylcholine, and cholesterol, furnished by the bile (Fig 56–2). Because the micelles are soluble, they allow the products of digestion to be transported through the aqueous environment of the intestinal lumen to the brush border of the mucosal cells, where they are absorbed into the intestinal epithelium. The bile salts pass on to the ileum, where most are absorbed into the **enterohepatic circulation** (see Chapter 27). Phospholipids of dietary and biliary origin (eg, phosphatidylcholine) are hydrolyzed by phospholipase A₂ of the pancreatic secretion to fatty acids and lysophospholipids, which are also absorbed from the micelles. Cholesteryl esters are hydrolyzed by cholesteryl ester hydrolase of the pancreatic juice, and the free cholesterol, together with most of the biliary cholesterol, is absorbed through the brush border after transportation in the micelles. Over 98% of dietary lipid is normally absorbed.

Within the intestinal wall, 1-monoacylglycerols are further hydrolyzed to produce free glycerol and fatty acids by a lipase, which is distinct from pancreatic lipase. 2-monoacylglycerols are reconverted to triacylglycerols via the **monoacylglycerol pathway** (Fig 56–3). The utilization of fatty acids for resynthesis of triacylglycerols first requires their "activation."

It is likely that the synthesis of triacylglycerols proceeds in the intestinal mucosa in a manner similar to that which takes place in other tissues, as described on p 226. The absorbed lysophospholipids, together with much of the absorbed cholesterol, are also reacylated with acyl-CoA to regenerate phospholipids and cholesteryl esters.

The free glycerol released in the intestinal lumen is not reutilized but passes directly to the portal vein. However, the glycerol released within the intestinal cells can be reutilized for triacylglycerol synthesis after activation to glycerol 3-phosphate by ATP. Thus, **all long-chain fatty acids absorbed in intestinal wall mucosal cells are utilized in the re-formation of triacylglycerols.**

Triacylglycerols, having been synthesized in the intestinal mucosa, are not transported to any extent in the portal venous blood. Instead, the great majority of absorbed lipids, including phospholipids, cholesteryl esters, cholesterol, and fat-soluble vitamins, generate chylomicrons that form a milky fluid, the **chyle,** that is collected by the lymphatic vessels of the abdominal region and passed to the systemic blood via the thoracic duct (see also Fig 26–3).

The majority of absorbed fatty acids of more than 10 carbon atoms in length, irrespective of the form in which they are absorbed, are found as esterified fatty acids in the lymph of the thoracic duct. Fatty acids with carbon chains **shorter than 10–12 carbons** are transported in the portal venous blood as unesterified (free) fatty acids.

Of the plant sterols (phytosterols), none are absorbed from the intestine except activated ergosterol (provitamin D).

Chyluria is an abnormality in which the patient excretes milky urine because of the presence of an abnormal connection between the urinary tract and the lymphatic drainage system of the intestine, a so-called chylous fistula. In a similar abnormality, **chylothorax,** there is an abnormal connection between the pleural space and the lymphatic drainage of the small intestine that results in the accumulation of milky pleural fluid. Feeding triacylglycerols in which the fatty acids are of medium chain length (<12 carbons) in place of dietary fat results in a disappearance of chyluria. In chylothorax, the use of triacylglycerol with short-chain fatty acids results in the appearance of clear pleural fluid. Both these effects are due to the fact that medium-chain fatty acids are absorbed into the hepatic portal vein rather than as chylomicrons in the thoracic duct.

The Products of Protein Digestion Are Absorbed as Individual Amino Acids

Under normal circumstances, the dietary proteins are almost completely digested to their constituent amino acids, and these end products of protein digestion are then rapidly absorbed from the intestine into the portal blood. It is possible that some hydrolysis, eg, of dipeptides, is completed in the intestinal wall. Animals may be successfully maintained when a complete amino acid mixture is fed to them.

The natural (L) isomer—but not the D isomer—of an amino acid is actively transported across the intestine from the mucosa to the serosa; vitamin B_6 (pyridoxal phosphate) may be involved in this transfer. This active transport of the L-amino acids is energy-dependent, as evidenced by the fact that 2,4-dinitrophenol, the uncoupler of oxidative phosphorylation (see p 116), inhibits transport. Amino acids are transported through the brush border by a multiplicity of carriers, many having Na^+-dependent mechanisms similar to the glucose carrier system (Fig 56–4). Of the Na^+-dependent carriers, there is a neutral amino acid carrier, a phenylalanine and methionine carrier, and a carrier specific for imino acids such as proline and hydroxyproline. Na^+-independent carriers specializing in the transport of neutral and lipophilic amino acids (eg, phenylalanine and leucine) or of cationic amino acids (eg, lysine) have been characterized.

Clinical aspects: Individuals in whom an immunologic response to ingested protein occurs must be able to absorb some unhydrolyzed protein, since digested protein is nonantigenic. This is not entirely undocumented, since the antibodies of the colostrum are known to be available to the infant.

There is increasing support for the hypothesis that the basic defect in **nontropical sprue** is located within the mucosal cells of the intestine and permits the polypeptides resulting from the peptic and tryptic digestion of gluten, the principal protein of wheat, not only to exert a local harmful effect within the intestine but also to be absorbed into the circulation and thus to elicit the production of antibodies. It has been established that circulating antibodies to wheat gluten or its fractions are frequently present in patients with nontropical sprue. The harmful entity is a polypeptide composed of 6 or 7 amino acids.

These observations on a disease entity that is undoubtedly the adult analog of **celiac disease** in children advance the possibility that protein fragments of larger molecular size than amino acids are absorbed from the intestine under certain conditions.

Table 56–3. Site of absorption of nutrients.

Site	Nutrient
Jejunum	Glucose and other monosaccharides; some disaccharides Monoacylglycerols, fatty acids, glycerol, cholesterol Amino acids, peptides Vitamins, folate Electrolytes, iron, calcium, water
Ileum	Bile acids Vitamin B_{12} Electrolytes Water

Table 56–4. Summary of disturbances due to malabsorption.

Sign or Symptom	Substance Malabsorbed
Anemia	Iron, vitamin B_{12}, folate
Edema	Products of protein digestion
Tetany	Calcium, magnesium, vitamin D
Osteoporosis	Calcium, products of protein digestion, vitamin D
Milk intolerance	Lactose
Bleeding, bruising	Vitamin K
Steatorrhea (fatty stools)	Lipids and fat-soluble vitamins
Hartnup disease (defect in intestinal neutral amino acid carrier)	Neutral amino acids

Tables 56–3 and 56–4 summarize the sites of intestinal absorption of some common nutrients and some disorders resulting from malabsorption, respectively.

BACTERIA IN THE LARGE INTESTINE CAUSE PUTREFACTION & FERMENTATION

Most ingested food is absorbed from the small intestine. The residue passes into the large intestine. Here considerable absorption of water takes place, and the semiliquid intestinal contents gradually become more solid. During this period, considerable bacterial activity occurs. By fermentation and putrefaction, the bacteria produce various gases, such as CO_2, methane, hydrogen, nitrogen, and hydrogen sulfide, as well as acetic, lactic, propionic, and butyric acids. The bacterial decomposition of phosphatidylcholine may produce choline and related toxic amines such as neurine.

Choline

Neurine

Many amino acids undergo decarboxylation as a result of the action of intestinal bacteria to produce toxic amines (ptomaines).

$$R-C(COOH)(NH_2)(H) \xrightarrow[\text{DECARBOXYLASE}]{\text{BACTERIAL}} RCH_2NH_2 + CO_2$$

A ptomaine

Such decarboxylation reactions produce cadaverine from lysine; agmatine from arginine; tyramine from tyrosine; putrescine from ornithine; and histamine from histidine. Many of these amines are powerful vasopressor substances.

The amino acid tryptophan undergoes a series of reactions to form indole and methylindole (skatole), the substances particularly responsible for the odor of feces.

Indole

Skatole

The sulfur-containing amino acid cysteine undergoes a series of transformations to form mercaptans such as ethyl and methyl mercaptan as well as H_2S.

Ethyl mercaptan

Methyl mercaptan

$$CH_3SH \xrightarrow{[2H]} CH_4 + H_2S$$

Methyl mercaptan

Methane and hydrogen sulfide

The large intestine is a source of considerable quantities of **ammonia,** a product of bacterial activity on nitrogenous substrates. This is absorbed into the portal circulation, but under normal conditions it is rapidly removed from the blood by the liver. In **liver disease,** this function of the liver may be impaired, in which case the concentration of ammonia in the peripheral blood will rise to toxic levels. It is believed that ammonia intoxication may play a role in the genesis of hepatic coma in some patients. The oral administration of **neomycin** has been shown to reduce the quantity of ammonia delivered from the intestine to the blood, owing to the antibacterial action of the drug. The feeding of **high-protein diets** to patients suffering from advanced liver disease, or the occurrence of gastrointestinal hemorrhage in such patients, may contribute to the development of ammonia intoxication. Neomycin is also beneficial under these circumstances.

Intestinal Bacteria Are Also Beneficial

The intestinal flora may comprise as much as 25% of the dry weight of the feces. In herbivora, whose diet consists largely of cellulose, the intestinal or ruminal bacteria are essential to digestion, since they decompose the polysaccharide and make it available for absorption. In addition, these symbiotic bacteria accomplish the synthesis of essential amino acids and vitamins. In humans, although the intestinal flora is not as important as in the herbivora, nevertheless some nutritional benefit is derived from bacterial activity in the synthesis of certain vitamins, particularly vitamins K and B_{12}, and possibly other members of the B complex, which are made available to the body.

REFERENCES

Börgstrom B: The micellar hypothesis of fat absorption: Must it be revisited? *Scand J Gastroenterol* 1985;**20:**389.

Gardner M: Gastrointestinal absorption of intact proteins. *Annu Rev Nutr* 1988;**8:**329.

Gargouri Y et al: Kinetic assay of human gastric lipase on short- and long-chain triacylglycerol emulsions. *Gastroenterology* 1986;**91:**919.

Foltmann B: Gastric proteinases. *Essays Biochem* 1981;**17:** 52.

Nelson GJ, Ackman RG: Absorption and transport of fat in mammals with emphasis on N-3 polyunsaturated fatty acids. *Lipids* 1988;**23:**1005.

Steven BR, Kaunitz JD, Wright EM: Intestinal transport of amino acids and sugars. *Annu Rev Physiol* 1984;**46:**417.

Tso P: Gastrointestinal digestion and absorption of lipid. *Adv Lipid Res* 1985;**21:**143.

Wolfe MM, Soll AH: The physiology of gastric acid secretion. *New Engl J Med* 1988;**319:**1707.

Glycoproteins & Proteoglycans

57

Robert K. Murray, MD, PhD

INTRODUCTION

Glycoproteins are proteins that have oligosaccharide (glycan) chains covalently attached to their polypeptide backbones. **Glycosaminoglycans** are polysaccharides built up of repeating disaccharide units, generally composed of an amino sugar (either glucosamine or galactosamine, which may or may not be sulfated) and a uronic acid (glucuronic acid or iduronic acid). Glycosaminoglycans were formerly called **mucopolysaccharides.** They usually occur covalently attached to protein; the complex of one or more glycosaminoglycans with protein is called a **proteoglycan. Glycoconjugates** and **complex carbohydrates** are equivalent terms used to denote molecules containing one or more carbohydrate chains covalently linked to protein or lipid; glycoproteins, proteoglycans, and glycolipids all fall within these categories.

BIOMEDICAL IMPORTANCE

Almost all the **plasma proteins** of humans, except albumin, are glycoproteins. Many **proteins of cellular membranes** (see Chapter 43) contain substantial amounts of carbohydrate. A number of the **blood group substances** are glycoproteins, whereas others are glycosphingolipids. Certain **hormones** (eg, chorionic gonadotropin) are glycoproteins. **Cancer** is increasingly recognized as a disorder resulting from abnormal genetic regulation (see Chapter 61). The major problem in cancer is metastasis, the phenomenon whereby cancer cells leave their tissue of origin (eg, the breast), migrate through the bloodstream to some distant site in the body (eg, the brain), and grow there in a completely unregulated manner, with catastrophic results for the affected individual. Many cancer researchers think that alterations in the structures of glycoconjugates on the surfaces of cancer cells are responsible, at least in part, for the phenomenon of metastasis. A **number of diseases** (the mucopolysaccharidoses) are due to deficiencies of the activities of specific lysosomal enzymes that degrade particular glycosaminoglycans; as a consequence, one or more of them accumulate in the tissues, resulting in various signs and symptoms. Hurler's syndrome is one such condition.

GLYCOPROTEINS OCCUR WIDELY & PERFORM NUMEROUS FUNCTIONS

Glycoproteins occur in most organisms, from bacteria to humans. Many animal viruses also contain glycoproteins, some of which have been much investigated, in part because they are suitable for biosynthetic studies.

Numerous proteins of diverse functions are glycoproteins (Table 57–1); their carbohydrate content ranges from 1% to over 85% by weight. The precise roles oligosaccharide chains play in the functions of glycoproteins are still not clear despite extensive research; some suggested roles are listed in Table 57–2.

Table 57–1. Some functions served by glycoproteins.

Structural molecules
Cell walls
Collagen, elastin
Fibrins
Bone matrix
Lubricants and protective agents
Mucins
Mucous secretions
Transport molecules for
Vitamins
Lipids
Minerals and trace elements
Immunologic molecules
Immunoglobulins
Histocompatibility antigens
Complement
Interferon
Hormones
Chorionic gonadotropin
Thyrotropin (TSH)
Enzymes
Proteases
Nucleases
Glycosidases
Hydrolases
Clotting factors
Cell attachment/recognition sites
Cell-cell
Virus-cell
Bacterium-cell
Hormone receptors
Antifreeze in antarctic fishes
Lectins

Table 57–2. Some functions of the oligosaccharide chains of glycoproteins.*

Modulate physicochemical properties, eg, solubility, viscosity, charge, and denaturation
Protect against proteolysis, from inside and outside the cell
Affect proteolytic processing of precursor proteins to smaller products
Are involved in biologic activity, eg, of hCG
Affect insertion into membranes, intracellular migration, sorting, and secretion
Affect embryonic development and differentiation
May affect sites of metastasis selected by cancer cells

*Adapted from Schachter H: Biosynthetic controls that determine the branching and heterogeneity of protein-bound oligosaccharides. *Biochem Cell Biol* 1986;**64**:163.

OLIGOSACCHARIDE CHAINS CONTAIN BIOLOGIC INFORMATION

An enormous number of glycosidic linkages can be generated between sugars. For example, 3 different hexoses may be linked to each other to form over 1000 different trisaccharides. The conformations of the sugars in oligosaccharide chains vary depending upon their linkages and proximity to other molecules with which the oligosaccharides may interact. A widely held belief is that oligosaccharide chains encode considerable **biologic information** and that this depends upon the constituent sugars, their sequences, and their conformations. There is also evidence that certain **drugs** and **toxins** interact with specific oligosaccharide chains of cell surface glycoconjugates (eg, cholera toxin interacts with that of the ganglioside G_{M1}; see Chapter 16). In addition, mannose 6-phosphate residues appear to **target** newly synthesized lysosomal enzymes to that organelle (see below).

TECHNIQUES ARE AVAILABLE FOR DETECTION, PURIFICATION, & STRUCTURAL ANALYSIS OF GLYCOPROTEINS

Detection

Glycoproteins can be resolved from complete mixtures (eg, cell membranes) by **SDS-PAGE** (see Chapter 3) and detected by staining with the **periodic acid-Schiff (PAS) reagent,** which detects aldehyde groups in sugars produced by the action of periodic acid. They can also be detected by showing, through the use of **SDS-PAGE-radioautography,** that they incorporate radioactivity when cells or tissues are incubated with suitable radioactive sugars.

Purification and Structural Analysis

As for other molecules, it is first necessary to purify glycoproteins before attempting to determine their structures. This can generally be accomplished using conventional methods of protein purification. **Affinity columns using lectins** (see below) have also proved valuable in the purification of certain glycoproteins. Their carbohydrate composition is determined following acid hydrolysis using analyses by **gas-liquid chromatography-mass spectrometry (GLC-MS)**. Various methods have been used in the past to determine the detailed structure of their oligosaccharide chains. At present, the combined use of GLC-MS and **high-resolution NMR spectrometry** often makes possible complete determination of the structures (ie, sugar composition, details of linkages and of their anomeric natures) of the glycan chains of most glycoconjugates. Details of the **linkages** between the sugars of glycoproteins (which are beyond the scope of this chapter) are of fundamental importance in determining the structures and functions of these molecules.

SEVEN SUGARS PREDOMINATE IN HUMAN GLYCOPROTEINS

Although about 200 monosaccharides are found in nature, only 7 are commonly found in the oligosaccharide chains of glycoproteins (Table 57–3). Most of these sugars were described in Chapter 14. N-Acetylneuraminic acid (NeuAc) is found at the termini of oligosaccharide chains, usually attached to subterminal galactose (Gal) or N-acetylgalactosamine (GalNAc) residues. The other sugars listed are generally found in more internal positions.

NUCLEOTIDE SUGARS ACT AS SUGAR DONORS IN MANY BIOSYNTHETIC REACTIONS

The first nucleotide sugar to be reported was uridine diphosphate glucose (UDP-Glc). The common nucleotide sugars involved in the biosynthesis of glycoproteins are listed in Table 57–3; the reasons some contain UDP and other guanosine diphosphate (GDP) or cytidine monophosphate (CMP) are obscure. Many but not all of the glycosylation reactions involved in the biosynthesis of glycoproteins utilize these compounds (see below). The anhydro nature of the linkage between the phosphate group and the sugars is of the high-energy, high-group-transfer-potential type. The sugars of these compounds are thus "**activated**" and can be transferred to suitable acceptors provided appropriate transferases are available.

The nucleotide sugars are formed in the cytosol, generally from reactions involving the corresponding nucleoside triphosphate. Formation of uridine diphos-

Table 57–3. The principal sugars found in human glycoproteins. The structures of most of the sugars listed are illustrated in Chapter 15.

Sugar	Type	Abbreviation	Nucleotide Sugar	Comments
Galactose	Hexose	Gal	UDP-Gal	Often found subterminal to NeuAc in N-linked glyco-proteins. Also found in core trisaccharide of proteogly-cans.
Glucose	Hexose	Glc	UDP-Glc	Present during the biosyn-thesis of N-linked glyco-proteins but not usually present in mature glyco-proteins.
Mannose	Hexose	Man	GDP-Man	Common sugar in N-linked glycoproteins.
N-acetylneuraminic acid	Sialic acid (9 C atoms)	NeuAc	CMP-NeuAc	Often the terminal sugar in both N- and O-linked gly-coproteins. Other types of sia-lic acid are also found, but NeuAc is the major species found in humans.
Fucose	Deoxyhexose	Fuc	GDP-Fuc	May be external in both N- and O-linked glycoproteins or linked to the GlcNAc residue attached to Asn in N-linked species.
N-Acetylgalactosamine	Aminohexose	GalNAc	UDP-GalNAc	Present in both N- and O-linked glycoproteins.
N-Acetylglucosamine	Aminohexose	GlcNAc	UDP-GlcNAc	The sugar attached to the polypeptide chain via Asn in N-linked glycoproteins; also found at other sites in the oligosaccharides of these pro-teins.

phate galactose (UDP-Gal) requires the following 2 reactions:

$$\boxed{\begin{array}{c} \text{UDP-Glc} \\ \text{PYROPHOSPHORYLASE} \end{array}}$$

UTP + Glucose 1-phosphate $\longleftrightarrow$
UDP-Glc + Pyrophosphate

$$\boxed{\begin{array}{c} \text{UDP-Glc} \\ \text{EPIMERASE} \end{array}}$$

UDP-Glc $\longleftrightarrow$ **UDP-Gal**

Because many glycosylation reactions occur within the lumen of the Golgi apparatus, carrier systems (per-meases, transporters) transport nucleotide sugars across the Golgi membrane. Systems transporting UDP-Gal, GDP-Man, and CMP-NeuAc into the cister-nae of the Golgi apparatus have been described. They are **antiport** systems; ie, the influx of one molecule of nucleotide sugar is balanced by the efflux of one mole-cule of the corresponding nucleotide (eg, UMP, GMP, or CMP) formed from the nucleotide sugars. This mechanism ensures an adequate concentration of each

nucleotide sugar inside the Golgi. UMP is formed from UDP-Gal in the above process as follows:

$$\boxed{\begin{array}{c} \text{GALACTOSYL-} \\ \text{TRANSFERASE} \end{array}}$$

UDP-Gal + Protein $\longrightarrow$ **Protein–Gal + UDP**

$$\boxed{\begin{array}{c} \text{NUCLEOSIDE} \\ \text{DIPHOSPHATE} \\ \text{PHOSPHATASE} \end{array}}$$

UDP $\longrightarrow$ **UMP + P$_i$**

EXOGLYCOSIDASES & ENDOGLYCOSIDASES ARE USEFUL IN STUDIES OF GLYCOPROTEINS

A number of **glycosidases** of defined specificity have proved useful in examining structural and func-tional aspects of glycoproteins (Table 57–4). These enzymes act at either external (exoglycosidases) or in-ternal (endoglycosidases) positions of oligosaccharide chains. Examples of exoglycosidases are **neuramini-dases** and **galactosidases;** their sequential use re-

Table 57–4. Some glycosidases used to study the structure and functions of glycoproteins. The enzymes are available from a variety of sources and are often specific for certain types of glycosidic linkages and also for their anomeric natures. The sites of action of endoglycosidases F and H are shown in Fig 57–4. F acts on both high-mannose and complex oligosaccharide chains, whereas H acts on the former.

Enzymes	Type
Neuraminidases	Exoglycosidase
Galactosidases	Exoglycosidase
Endoglycosidase F	Endoglycosidase
Endoglycosidase H	Endoglycosidase

moves terminal NeuAc and subterminal Gal residues from most glycoproteins. **Endoglycosidases F** and **H** are examples of the latter class; these enzymes cleave the oligosaccharide chains at specific GlcNAc residues close to the polypeptide backbone (ie, at internal sites) and are thus useful in producing large oligosaccharide chains for structural analyses. A glycoprotein can be treated with one or more of the above glycosidases to analyze the effects on its biologic behavior of removal of specific sugars.

Experiments performed by Ashwell and his colleagues in the early 1970s strongly suggested that exposure of Gal residues by treatment with neuraminidase led to the rapid disappearance of ceruloplasmin from the plasma. Further work demonstrated that liver cells contain a receptor which recognizes the galactosyl moiety of many desialylated (ie, free of NeuAc) plasma proteins and leads to their endocytosis. This work clearly indicated that an individual sugar could play an important role in governing at least one of the biologic properties (ie, time of residence in the circulation) of certain glycoproteins.

LECTINS CAN BE USED TO PURIFY GLYCOPROTEINS & TO PROBE THEIR FUNCTIONS

Lectins are proteins of plant origin that bind one or more specific sugars. Their precise roles in plants are still under investigation. Many lectins have been purified, and more than 40 are commercially available (Table 57–5). Lectins have found a number of uses in biochemical research, including the following: (1) **In purification and analysis of glycoproteins.** Lectins such as concanavalin A (Con A) can be attached covalently to inert supporting media such as Sepharose. The resulting Sepharose-Con A may be used for

Table 57–5. Three lectins and the sugars with which they interact. In most cases, lectins show specificity for the anomeric nature of the glycosidic linkage (α or β); this is not indicated in the table.

Lectin	Abbreviation	Sugars
Concanavalin A	Con A	Man and Glc
Soybean lectin		Gal and GalNAc
Wheat germ agglutinin	WGA	GlcNac and NeuAc

the purification of glycoproteins that contain oligosaccharide chains which interact with Con A. (2) **As tools for probing the surfaces of cells.** Since lectins recognize specific sugars, they can be used to probe, at a general level, the sugar residues exposed on the surface membranes of cells. Numerous studies have been performed using lectins to compare the surfaces of normal and cancer cells (see Chapter 61). A general finding has been that smaller amounts of certain lectins are required to cause agglutination of tumor cells than of normal cells. This suggests that the organization or structure of a number of glycoproteins on the surfaces of tumor cells may be different from that found in normal cells. (3) **To generate mutant cells lacking certain enzymes of oligosaccharide synthesis.** When mammalian cells in tissue culture are exposed to appropriate concentrations of certain lectins (eg, Con A), most of them are killed, but a few resistant cells survive. Such cells are often found to lack certain enzymes involved in oligosaccharide synthesis. The cells are resistant, presumably because they do not produce one or more surface glycoproteins that would interact with the lectin used. The use of such mutant cells has been important in elucidating a number of aspects of glycoprotein biosynthesis.

THERE ARE 4 MAJOR CLASSES OF GLYCOPROTEINS

Glycoproteins can be divided into 4 classes, based on the nature of the linkage between their polypeptide chains and their oligosaccharide chains: (1) those containing a Ser (or Thr)-GalNAc linkage (Fig 57–1); (2) proteoglycans containing a Ser-Xyl linkage; (3) col-

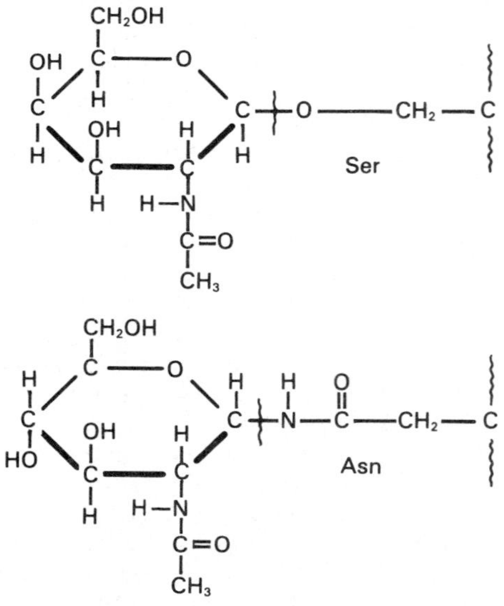

Figure 57–1. Linkage of N-acetylgalactosamine to serine and of N-acetylglucosamine to asparagine.

lagens containing a hydroxylysine (Hyl)-Gal linkage; and (4) glycoproteins containing an Asn-GlcNAc linkage (Fig 57–1).

Classes 1, 2, and 3 are joined to the corresponding amino acids via **O-glycosidic linkages** (ie, a linkage involving an OH in the side chain of an amino acid and a sugar residue). The fourth class involves an **N-glycosidic linkage** (ie, a linkage involving the N of the amide group of asparagine and a sugar residue). Because classes 2 and 3 are relatively uncommon, the term **O-linked glycoproteins** is often used, as it is here, to refer only to members of class 1. The members of class 4 are called **N-linked glycoproteins.** Other minor classes exist. The number of oligosaccharide chains attached to one protein can vary from one to 30 or more, with the sugar chains ranging from 2 or 3 residues in length to much larger structures. Some glycoproteins contain *both* N- and O-glycosidic linkages.

O-Linked Glycoproteins Contain an Asn-Y-Ser(Thr) Tripeptide Sequence

Most of the O-glycosidic links occur to the free OH groups of Ser and Thr residues of the polypeptide in a tripeptide sequence of Asn-Y-Ser(Thr), where Y is an amino acid other than aspartate. This specific tripeptide sequence is very common in proteins, but *not* every such sequence is glycosylated. The decision to glycosylate such Ser and Thr residues is based on the protein conformation surrounding that tripeptide as it emerges through the endoplasmic reticulum.

Occurrence and Structure of Oligosaccharide Chains of O-Linked Glycoproteins: Many glycoproteins of this class are found in **mucins.** However, O-glycosidic linkages are also found in some membrane and circulating glycoproteins. As indicated above, the most common sugar attached directly to the Ser or Thr residue is GalNAc (Fig 57–1). Usually a Gal or a NeuAc residue is attached to the GalNAc. The structures of 2 typical oligosaccharide chains of glycoproteins of this class are shown in Fig 57–2. Many variations are found.

Biosynthesis of O-Linked Glycoproteins: The polypeptide chains of these and other glycoproteins are encoded by mRNA species; because most glycoproteins are membrane-bound or secreted, they are generally translated on membrane-bound polyribosomes (see Chapter 40). The sugars of the oligosaccharide chains of the O-glycosidic type of glycoprotein are built up by the **stepwise donation of sugars from nucleotide sugars,** such as UDP-GalNAc, UDP-Gal, and CMP-NeuAc. The enzymes catalyzing this type of reaction are membrane-bound **glycoprotein glycosyltransferases.** The synthesis of each such enzyme is controlled by one specific gene. Generally, synthesis of one specific type of linkage requires the activity of a correspondingly specific transferase (ie, the "**one linkage, one glycosyl transferase**" hypothesis). The enzymes catalyzing addition of the inner sugar residues are located in the endoplasmic reticulum, and

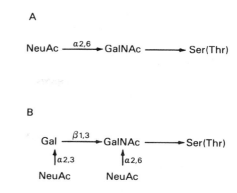

Figure 57–2. Structures of 2 O-linked oligosaccharides found in (*A*) submaxillary mucins and (*B*) fetuin and in the sialoglycoprotein of the membrane of human red blood cells. (Modified and reproduced, with permission, from Lennarz WJ: *The Biochemistry of Glycoproteins and Proteoglycans.* Plenum Press, 1980.)

addition of the first sugars occurs during translation (ie, **cotranslational** modification of the protein occurs). The enzymes adding the terminal sugars (such as NeuAc) are located in the Golgi apparatus.

N-Linked Glycoproteins Contain an Asn-GlcNAc Linkage

The class is distinguished by the presence of the Asn-GlcNAc linkage (Fig 57–1). It is **the major class** of glycoproteins and has been much studied, since the most readily accessible glycoproteins (eg, plasma proteins) mainly belong to this group. It includes both **membrane-bound** and **circulating** glycoproteins. The principal difference between this and the previous class, apart from the nature of the amino acid to which the oligosaccharide chain is attached (mainly Asn versus Ser), concerns their biosynthesis.

Classes of N-Linked Glycoproteins and Their Structures: There are 3 major classes of N-linked glycoproteins: **complex, hybrid,** and **high-mannose** (Fig 57–3). Each type shares **a common pentasaccharide** ([Man]$_3$[GlcNAc]$_2$, shown within the boxed area in Fig 57–3 and also depicted in Fig 57–4), but they differ in their outer branches. The presence of the common pentasaccharide is explained by the fact that all 3 classes share an initial common mechanism of biosynthesis. Glycoproteins of the complex type generally contain terminal NeuAc residues and underlying Gal and GlcNAc residues, the latter often constituting the disaccharide lactosamine. (The presence of repeating **lactosamine units,** [Galβ1-4GlcNAcβ1-3]$_n$, characterizes a fourth class of glycoprotein, the **polylactosamine** class; it is important because the I/i blood group substances belong to it.) The majority of complex-type oligosaccharides contain 2 (Fig 57–3), 3, or 4-outer branches, but structures containing 5 branches have also been described. The oligosaccharide branches are often referred to as antennae, so that bi-, tri-, tetra-, and penta-antennary structures may all be

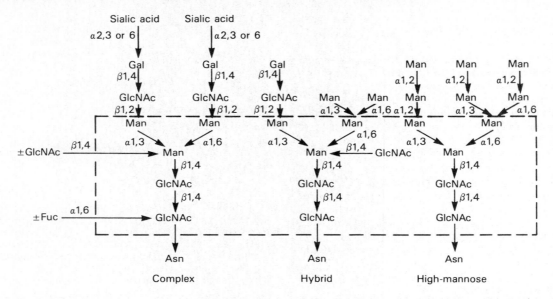

Figure 57–3. Structures of the major types of asparagine-linked oligosaccharides. The boxed area encloses the pentasaccharide core common to all N-linked glycoproteins. (Reproduced, with permission, from Kornfeld R, Kornfeld S: Assembly of asparagine-linked oligosaccharides. *Annu Rev Biochem* 1985;**54**:631.)

found. A bewildering number of chains of the complex type exist, and that indicated in Fig 57–3 is only one of many. Other complex chains may terminate in Gal or Fuc. High-mannose oligosaccharides typically have 2–6 additional Man residues linked to the pentasaccharide core. Hybrid molecules contain features of both of the 2 other classes.

Overview of the Biosynthesis of N-Linked Glycoproteins: Leloir and his colleagues described the occurrence of an **oligosaccharide-pyrophosphoryl-dolichol (oligosaccharide-P-P-Dol),** which subsequent research showed to play a key role in the biosynthesis of N-linked glycoproteins. The oligosaccharide chain of this compound generally has the structure $(Glc)_3(Man)_9(GlcNAc)_2$–R (R = P-P-Dol). The sugars of this compound are first assembled on the pyrophosphoryl-dolichol backbone, and the oligosaccharide chain is then transferred en bloc to suitable Asn

residues of acceptor apoglycoproteins during their synthesis on membrane-bound polyribosomes. To form an oligosaccharide chain of the **complex** type, the Glc residues and 6 of the Man residues are removed, forming the core pentasaccharide $(Man)_3(GlcNac)_2$– (Fig 57–4). The sugars characteristic of complex chains (GlcNAc, Gal, NeuAc) are then added by the action of individual glycosyltransferases, mostly located in the Golgi apparatus. To form **high-mannose** chains, only the Glc residues plus or minus certain of the peripheral Man residues are removed. The phenomenon whereby the glycan chains of N-linked glycoproteins are first partially degraded and then in some cases rebuilt is referred to as **oligosaccharide processing.** Rather surprisingly, the initial steps involved in the biosynthesis of the N-linked glycoproteins differ markedly from those involved in the biosynthesis of the O-linked glycoproteins. The former involves the oligosaccharide-P-P-Dol; the latter, as described earlier, does not.

The process can be broken down into 2 stages: (1) assembly and transfer of oligosaccharide-P-P-dolichol, and (2) processing of the oligosaccharide chain.

1. Assembly and transfer of oligosaccharide-P-dolichol–Polyisoprenol compounds exist in both bacteria and eukaryotic tissues. They participate in the synthesis of bacterial cell walls and in the biosynthesis of N-linked glycoproteins. The polyisoprenol used in eukaryotic tissues is **dolichol,** which is, next to rubber, the longest naturally occurring hydrocarbon made up of a single repeating unit. Dolichol is made up of 17–20 repeating isoprenoid units (Fig 57–5).

Before it participates in the biosynthesis of oligosaccharide-P-P-Dol, dolichol must first be phosphory-

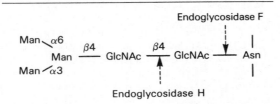

Figure 57–4. Schematic diagram of the pentasaccharide core common to all N-linked glycoproteins and to which various outer chains of oligosaccharides may be attached. The sites of action of endoglycosidases F and H are also indicated.

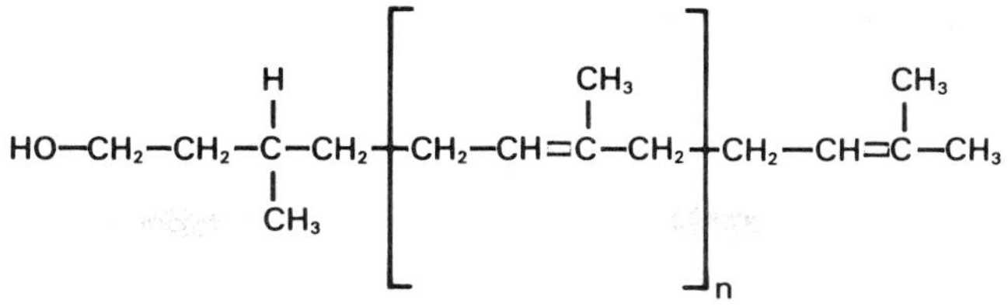

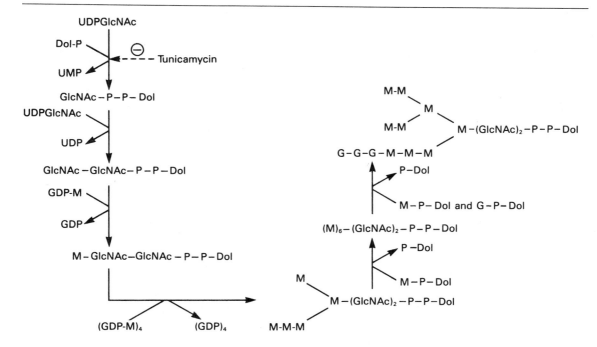

Figure 57–5. The structure of dolichol. The phosphate in dolichol phosphate is attached to the primary alcohol group at the left-hand end of the molecule. The group within the brackets is an isoprene unit (n = 17–20 isoprenoid units).

lated to form dolichol phosphate (Dol-P) in a reaction catalyzed by **dolichol kinase** and using ATP as the phosphate donor.

GlcNAc-pyrophosphoryl-dolichol (GlcNac-P-P-Dol) is the key lipid that acts as an **acceptor** for other sugars in the assembly of oligosaccharide-P-P-Dol. It is synthesized in the membranes of the endoplasmic reticulum from Dol-P and UDP-GlcNAc in the following reaction:

Dol-P + UDP-GlcNAc → GlcNAc-P-P-Dol + UMP

The above reaction—which is the first step in the assembly of oligosaccharide-P-P-Dol—and the other later reactions are summarized in Fig 57–6.

The essential features of the subsequent steps in the assembly of oligosaccharide-P-P-dolichol are as follows:

a. A second GlcNAc residue is added to the first, again using UDP-GlcNAc as the donor.

b. Five Man residues are added, using GDP-mannose as the donor.

c. Four additional Man residues are next added, using Dol-P-Man as the donor. Dol-P-Man is formed by the following reaction

Dol-P + GDP-Man → Dol-P-Man + GDP

d. Finally, the **3 peripheral glucose residues** are donated by Dol-P-Glc, which is formed in a reaction analogous to that just presented, except that Dol-P and UDP-Glc are the substrates.

It should be noted that the first 7 sugars (2 GlcNAc and 5 Man residues) are donated by nucleotide sugars, whereas the last 7 sugars (4 Man and 3 Glc residues) added are donated by dolichol sugars. The net result of

Figure 57–6. Pathway of biosynthesis of dolichol-P-P-oligosaccharide. The specific linkages formed are indicated in Fig 57–7. Note that the internal mannose residues are donated by GDP-mannose, whereas the more external mannose residues and the glucose residues are donated by dolichol-P-mannose and dolichol-P-glucose. UDP, uridine diphosphate; Dol, dolichol; P, phosphate; UMP, uridine monophosphate; GDP, guanosine diphosphate; M, mannose; G, glucose.

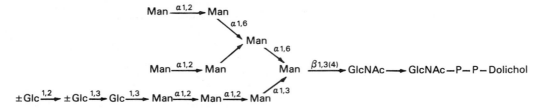

Figure 57–7. Structure of dolichol-P-P-oligosaccharide. (Reproduced, with permission, from Lennarz WJ: *The Biochemistry of Glycoproteins and Proteoglycans.* Plenum Press, 1980.)

the above is assembly of the compound illustrated in Fig 57–7 and referred to in shorthand as $(Glc)_3$ $(Man)_9$ $(GlcNAc)_2$-P-P-Dol.

The oligosaccharide linked to dolichol-P-P is transferred en bloc to form an N-glycosidic bond with one or more specific Asn residues of an acceptor protein emerging from the luminal surface of the membrane of the endoplasmic reticulum. The reaction is catalyzed by an "**oligosaccharide transferase,**" a membrane-associated enzyme. The transferase will recognize and transfer any glycolipid with the general structure R–$(GlcNAc)_2$-P-P-Dol. Glycosylation occurs at the Asn residue of an Asn-X-Ser/Thr tripeptide sequence, where X is any amino acid except possibly proline or aspartate. A tripeptide site contained within a β turn is favored. Only about one-third of the Asn residues that are potential acceptor sites are actually glycosylated. The acceptor proteins are of both the secretory and integral membrane class. Cytosolic proteins are rarely glycosylated. The transfer reaction and subsequent processes in the glycosylation of N-linked glycoproteins, along with their subcellular locations, are depicted in Fig 57–8.

The other product of the oligosaccharide transferase reaction is dolichol-P-P, which is subsequently converted to dolichol-P by a phosphatase. The dolichol-P can serve again as an acceptor for the synthesis of another molecule of oligosaccharide-P-P-dolichol.

2. Processing of the oligosaccharide chain–

a. Early phase–The various reactions involved are indicated in Fig 57–8. The oligosaccharide transferase catalyzes reaction 1 (see above). Reactions 2 and 3 involve the removal of the terminal Glc residue by glucosidase I and of the next 2 Glc residues by glucosidase II, respectively. In the case of **high-mannose** glycoproteins, the process may stop here, or up to 4 Man residues may also be removed. However, to form **complex** chains, additional steps are necessary, as follows. Four external Man residues are removed in reactions 4 and 5 by 2 different mannosidases. In reaction 6, a GlcNAc residue is added to one of the Man residues by GlcNAc transferase I. The action of this latter enzyme permits the occurrence of reaction 7, a reaction catalyzed by yet another mannosidase (Golgi α-mannosidase II) and which results in a reduction of the Man residues to the core number of 3 (Fig 57–4).

An important additional pathway is indicated in reactions I and II of Fig 57–8. This involves enzymes

destined for **lysosomes.** Such enzymes are targeted to the lysosomes by a specific chemical marker. In reaction I, a residue of GlcNAc-1-P is added to carbon 6 of one or more specific Man residues of these enzymes. The reaction is catalyzed by a GlcNAc phosphotransferase, which uses UDP-GlcNAc as the donor and generates UMP as the other product:

| **GlcNAc PHOSPHOTRANSFERASE** |

UDP-GlcNAc + **Man–Protein** →
GlcNAc-1-P-6-Man–Protein + **UMP**

In reaction II the GlcNAc is removed by the action of a phosphodiesterase, leaving the Man residues phosphorylated in the 6 position:

| **PHOSPHODIESTERASE** |

GlcNAc-1-P-6-Man–Protein →
P-6-Man–Protein + **GlcNAc**

A Man-6-P receptor, located in the Golgi apparatus, binds the Man-6-P residue of these enzymes and directs them to the lysosomes. Fibroblasts from patients with **I-cell disease** (see below) are severely deficient in the activity of the GlcNAc phosphotransferase.

b. Late phase–To assemble a typical complex oligosaccharide chain, additional sugars must be added to the structure formed in reaction 7. Hence, in reaction 8, a second GlcNAc is added to the peripheral Man residue of the other arm of the bi-antennary structure shown in Fig 57–8; the enzyme catalyzing this step is GlcNAc transferase II. Reactions 9, 10, and 11 involve the addition of Fuc, Gal, and NeuAc residues at the sites indicated, in reactions catalyzed by fucosyl, galactosyl, and sialyl transferases, respectively.

Subcellular sites. As indicated in Fig 57–8, the endoplasmic reticulum and the Golgi apparatus are the major sites involved in glycosylation processes. Addition of the oligosaccharide occurs in the rough endoplasmic reticulum during or after translation. Removal of the Glc and some of the peripheral Man residues also occurs in the endoplasmic reticulum. The Golgi apparatus is composed of *cis,* medial, and *trans* cisternae; these can be separated by appropriate centrifugation procedures. Vesicles containing glycoproteins appear to bud off in the endoplasmic reticulum and are transported to the *cis* Golgi. Various studies have shown that the enzymes involved in glycoprotein

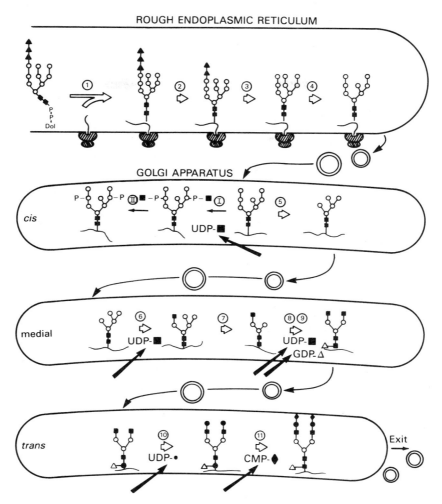

Figure 57–8. Schematic pathway of oligosaccharide processing. The reactions are catalyzed by the following enzymes: **1,** oligosaccharyltransferase; **2,** α-glucosidase I; **3,** α-glucosidase II; **4,** endoplasmic reticulum α1,2 mannosidase; I, N-acetylglucosaminylphosphotransferase; II, N-acetylglucosamine-1-phosphodiester α-N-acetylglucosaminidase; **5,** Golgi apparatus α-mannosidase I; **6,** N-acetylglucosaminyltransferase I; **7,** Golgi apparatus α-mannosidase II; **8,** N-acetylglucosaminyltransferase II; **9,** fucosyltransferase; **10,** galactosyltransferase; **11,** sialyltransferase. ■, N-acetylglucosamine; ○, mannose; ▲, glucose; △, fucose; ●, galactose; ◆, sialic acid. (Reproduced, with permission, from Kornfeld R, Kornfeld S: Assembly of asparagine-linked oligosaccharides. *Annu Rev Biochem* 1985;**54:**631.)

processing show differential locations in the cisternae of the Golgi. As indicated in Fig 57–8, Golgi α-mannosidase I (catalyzing reaction 5) is located mainly in the *cis* Golgi, whereas GlcNAc transferase I (catalyzing reaction 6) appears to be located in the medial Golgi, and the fucosyl, galactosyl, and sialyl transferases (catalyzing reactions 9, 10, and 11) are located in the *trans* Golgi.

Regulation of the Glycosylation of Glycoproteins Involves Many Enzymes & Enzyme Sequences

It is evident that glycosylation of glycoproteins is a complex process, involving a large number of enzymes. One index of its complexity is that 7 distinct

GlcNAc transferases involved in glycoprotein biosynthesis have been reported, and a number of others are theoretically possible. Multiple species of the other glycosyltransferases (eg, sialyltransferases) also exist. Controlling factors of the first stage (ie, oligosaccharide assembly and transfer) of N-linked glycoprotein biosynthesis include (1) the presence of suitable acceptor sites in proteins, (2) the tissue level of Dol-P, and (3) the activity of the oligosaccharide transferase.

Factors involved in the regulation of oligosaccharide processing include the following: (1) The activities in cells of the various hydrolases and transferases involved in processing are important determining factors of the types of oligosaccharide chains (eg, complex or high-mannose) formed. Obviously, if a particular transferase is not present in a tissue, the corresponding sugar linkage will not be synthesized. (2) Certain enzymes can act only if another enzyme has acted previously. For example, prior action of GlcNAc transferase I (Fig 57–8) is necessary for the action of Golgi α-mannosidase II. (3) The activities of various transferases can wax and wane systematically during development, explaining in part how different oligosaccharides may be formed during different stages of an organism's life cycle. (4) The fact that the various glycosyltransferases have different intracellular locations plays a role in regulation. For example, if a protein is destined for insertion into the membranes of the endoplasmic reticulum (eg, HMG-CoA reductase), it will never travel to the Golgi cisternae to encounter the various processing enzymes located there. Accordingly, it is not surprising to discover that HMG-CoA reductases is a high-mannose type of glycoprotein. (5) Another important factor is protein conformation. Closely related viral glycoproteins exhibit different types of oligosaccharide chains when grown in the same cells. The best explanation appears to be that these proteins must exhibit different conformations that affect the extent of processing. (6) There are considerable differences in the profiles of processing enzymes displayed by cells from different species. The oligosaccharides of one particular Sindbis virus vary (eg, from high-mannose to complex) depending upon the host cell in which the virus is grown. (7) An area of great current interest involves analysis of the activities of glycoprotein-processing enzymes in various types of **cancer cells.** There is some evidence that cancer cells synthesize different oligosaccharide chains (eg, they often exhibit greater branching) from those made in control cells; correlating the activity of a particular processing enzyme with the **metastatic properties** of some cancer cells (see Chapter 61) is of great interest.

Several Substances Inhibit Processes Involved in Glycosylation

A number of compounds are known to inhibit various reactions involved in glycoprotein processing. **Tunicamycin, deoxynojirimycin,** and **swainsonine** are 3 such agents. The reactions that they inhibit are indicated in Table 57–6. These agents can be used experimentally to inhibit various stages of glycoprotein biosynthesis and to study the effects of specific

alterations upon the process. For instance, if cells are grown in the presence of tunicamycin, no glycosylation of their normally N-linked glycoproteins will occur. In certain cases, this has been shown to increase the susceptibility of these proteins to proteolysis. Inhibition of glycosylation does not appear to have a consistent effect upon the secretion of glycoproteins that are normally secreted.

I-CELL DISEASE RESULTS FROM MISDIRECTION OF LYSOSOMAL ENZYMES

As explained above, Man-6-P serves as a chemical marker to **target** lysosomal enzymes to that organelle. Analysis of cultured fibroblasts derived from patients with I-cell (inclusion cell) disease played a large part in revealing the above role of Man-6-P. I-cell disease is a rare condition in which cultured cells lack almost all of the normal lysosomal enzymes; the lysosomes thus accumulate many different types of undegraded molecules. However, samples of serum from patients with the disease contain very high activities of lysosomal enzymes, suggesting that lysosomal enzymes are being synthesized but failing to reach their proper intracellular destination. Cultured cells from patients with the disease can take up exogenously added lysosomal enzymes obtained from normal subjects, indicating that the cells contain a normal receptor for uptake of lysosomal enzymes. In addition, it suggests that lysosomal enzymes from patients with I-cell disease might lack a recognition marker. Further studies revealed that lysosomal enzymes from normal individuals carried the Man-6-P recognition marker described above. Cultured cells from patients with I-cell disease were then found to be deficient in the activity of the GlcNAc phosphotransferase, explaining how the lysosomal enzymes failed to acquire the Man-6-P marker. Thus, biochemical investigations of this disease not only led to elucidation of its basis but also contributed significantly to knowledge of how newly synthesized proteins are targeted to specific organelles, in this case the lysosome.

BLOOD GROUP ANTIGENS

There are Many Blood Group Systems

At least **21 human blood group systems are recognized,** the better known of which include the ABO, Lewis, and Rhesus systems. The term **blood group applies to a defined system of red blood cell antigens** (blood group substances) **controlled by a genetic locus** having a variable number of alleles (eg, A, B, and O in the ABO system). The term **blood type refers to the antigenic phenotype,** usually recognized by the use of appropriate antibodies. For purposes of blood transfusion, it is important to know the basics of the ABO and Rh systems. However, knowledge of blood

Table 57–6. Three inhibitors of enzymes involved in the glycosylation of glycoproteins and their sites of action.

Inhibitor	Site of Action
Tunicamycin	Inhibits the enzyme catalyzing addition of GlcNAc to dolichol-P, the first step in the biosynthesis of oligosaccharide-P-P-dolichol
Deoxynojirimycin	Inhibitor of glucosidases I and II
Swainsonine	Inhibitor of mannosidase II

group systems is also of biochemical, genetic, immunologic, anthropologic, obstetric, pathologic, and forensic interest. Here, we shall only discuss, quite briefly, some key features of the ABO system. From a biochemical viewpoint, the major interests in the ABO substances have been in isolating and determining their **structures,** elucidating their pathways of **biosynthesis,** and determining the **natures of the products of the A, B, and O genes.**

The ABO System

This system was first discovered by Landsteiner in 1900, when investigating the basis of compatible and incompatible transfusions in humans. The **plasma membranes of the red blood cells of most individuals contain blood group substances of type A, type B, type AB, or type O.** Individuals of type A have anti-B antibodies in their plasma and will thus agglutinate type B or type AB blood. Individuals of type B have anti-A antibodies, and will agglutinate type A or type AB blood. Type AB blood has neither anti-A nor anti-B antibodies and has been designated the universal recipient. Type O blood has neither A nor B substances, and has been designated the universal donor. The explanation of these findings is related to the fact that the body does not usually produce antibodies to its own constituents. Thus, individuals of type A do not produce antibodies to their own blood group substance, A, but do possess antibodies to the foreign blood group substance, B, probably because similar structures to it are present in microorganisms to which the body is exposed early in life. As individuals of type O do not have either A or B substances, they possess antibodies to both of these foreign substances. The above description has been simplified considerably; for example, there are 2 subgroups of type A, A_1 and A_2.

The ABO Substances Are Glycosphingolipids & Glycoproteins and Differ in Structure by Single Sugars

The ABO substances have been shown to be **complex oligosaccharides** present in most cells of the body and also in certain secretions. On red blood cells, the ABO substances appear to be mostly **glycosphingolipids,** whereas in secretions they are present as **glycoproteins.** Their presence in secretions is determined by a gene designated **Se** (for **secretor**), which codes for a specific fucosyl (Fuc) transferase in secretory organs, such as the exocrine glands, but not in red blood cells. Individuals of SeSe or Sese genotypes secrete A or B antigens (or both), whereas individuals of the sese genotype do not secrete A or B substances, but their red blood cells can express the A and B antigens.

H Substance is the Biosynthetic Precursor of Both the A & the B Substances

The ABO substances have been isolated and their structures determined; very much simplified versions, showing only their nonreducing ends, are presented in Fig 57–9. It is important first to appreciate the structure of the **H substance,** as it is the precursor of both the A and B substances and it is **the blood group substance found in persons of type O.** H substance itself is formed by the action of a **fucosyltransferase** that catalyzes the addition of the terminal fucose in $\alpha 1,2$ linkage onto the terminal Gal residue of its precursor:

$$\text{GDP-Fuc} + \text{Gal-}\beta\text{R} \rightarrow \text{Fuc-}\alpha1,2\text{-Gal}\beta\text{-R} + \text{GDP}$$

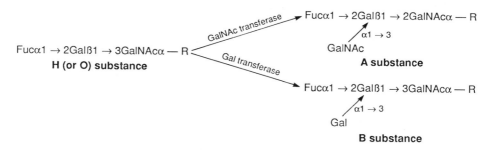

Figure 57–9. Diagrammatic representation of the structures of the H, A, and B blood group substances. R represents a long complex oligosaccharide chain, joined either to ceramide where the substances are glycosphingolipids, or to the polypeptide backbone of a protein via a serine or threonine residue where the substances are glycoproteins. It should be noted that the blood group substances are biantennary; ie, they have two arms, formed at a branch point (not indicated) between the GalNAcα—R, and only one arm of the branch is shown. Thus, the H, A, and B substances each contain two of their respective short oligosaccharide chains shown above. The AB substance contains one type A chain and one type B chain.

The H locus codes for this fucosyltransferase. The h allele of the H locus codes for an inactive fucosyltransferase; therefore, individuals of the hh genotype cannot generate H substance, the precursor of the A and B antigens. Thus, individuals of the hh genotype will be erythrocyte blood type O, even though they may contain the enzymes necessary to make the A or B substances (see below).

The A Gene Encodes a GalNAc Transferase, the B Gene a Gal Transferase

In comparison with H substance (Fig 57–9), **A substance** contains an additional GalNAc and **B substance** an additional Gal, linked as indicated. Anti-A antibodies are directed to the additional GalNAc residue found in the A substance, and anti-B antibodies are directed toward the Gal residue in B substance. In view of the structural findings, it is not surprising that A substance can be synthesized in vitro from O substance in a reaction catalyzed by a GalNAc transferase, employing UDP-GalNAc as the sugar donor. Similarly, blood group B can be synthesized from O substance by the action of a Gal transferase, employing UDP-Gal. It is crucial to appreciate that the **product of the A gene is the GalNAc transferase** that adds the terminal GalNAc to the O substance. Similarly, the **product of the B gene is the Gal transferase** adding the Gal residue to the O substance. **Individuals of type AB possess both enzymes** and thus have 2 oligosaccharide chains (cf legend to Fig 57–9), one terminated by a GalNAc and the other by a Gal. **Individuals of type O possess neither transferase,** and thus H substance is their ABO blood group substance.

PROTEOGLYCANS AND GLYCOSAMINOGLYCANS

The Glycosaminoglycans Found in Proteoglycans Are Built Up of Repeating Disaccharides

Proteoglycans are proteins that contain covalently linked glycosaminoglycans (GAGs). The proteins bound covalently to GAGs are called **core proteins;** they have proved difficult to isolate and characterize, but the use of recombinant DNA technology is beginning to yield important information about their structures. The amount of carbohydrate in a proteoglycan is usually much greater than is found in a glycoprotein and may comprise up to 95% of its weight. There are at least 7 GAGs: hyaluronic acid, chondroitin sulfate, keratan sulfates I and II, heparin, heparan sulfate, and dermatan sulfate. **A GAG is an unbranched polysaccharide made up of repeating disaccharides,** one component of which is always an **amino sugar** (hence the name GAG), either D-glucosamine or D-galactosamine. The other component of the repeating disaccharide (except in the case of keratan sulfate) is a **uronic acid,** either L-glucuronic acid (GlcUA) or its 5-epimer, L-iduronic acid (IdUA). With the exception of hyaluronic acid, all the GAGs contain **sulfate groups,** either as O-esters or as N-sulfate (in heparin and heparan sulfate). Hyaluronic acid affords another exception, in that there is no clear evidence that it is attached covalently to protein, as the definition of a proteoglycan given above specifies. Both GAGs and proteoglycans have proved difficult to work with in the past, partly because of their complexity. However, they are major components of the ground substance, they have a number of important biologic roles, and they are involved in a number of disease processes, so that interest in them is steadily increasing.

The Biosynthesis of Glycosaminoglycans Involves Attachment to Core Proteins, Chain Elongation, and Chain Termination

Attachment to Core Proteins: The linkage between GAGs and their core proteins is generally one of 3 types.

(1) An O-glycosidic bond between xylose (Xyl) and Ser, **a bond that is unique to proteoglycans.** This linkage is formed by transfer of a Xyl residue to Ser from UDP-xylose. Two residues of Gal are then added to the Xyl residue, forming a **link trisaccharide** –Gal-Gal-Xyl-Ser. Further chain growth of the GAG occurs on the terminal Gal.

(2) An O-glycosidic bond forms between GalNAc and Ser(Thr), present in keratan sulfate II. This bond is formed by donation to Ser (or Thr) of a GalNAc residue, employing UDP-GalNAc as its donor.

(3) An N-glycosylamine bond between GlcNAc and the amide nitrogen of Asn, as found in N-linked glycoproteins. Its synthesis is believed to involve dolichol-P-P-oligosaccharide.

The synthesis of the core proteins occurs in the ER, and formation of at least some of the above linkages also occurs there. Most of the later steps in the biosynthesis of GAG chains and their subsequent modifications occur in the Golgi apparatus.

Chain Elongation: Appropriate nucleotide sugars and highly specific Golgi-located glycosyltransferases are employed to synthesize the oligosaccharide chains of GAGs. The "one enzyme, one linkage" relationship appears to hold here, as in the case of certain types of linkages found in glycoproteins. The enzyme systems involved in chain elongation are capable of high-fidelity reproduction of complex GAGs.

Chain Termination: This appears to result from (1) **sulfation,** particularly at certain positions of the sugars, and (2) the **progression** of the growing GAG chain **away from the membrane site** where catalysis occurs.

Further Modifications: After formation of the GAG chain, numerous chemical modifications occur, such as the **introduction of sulfate** groups onto GalNAc moieties and the **epimerization of GlcUA to IdUA** residues. The enzymes catalyzing sulfation are designated **sulfotransferases** and use 3′-phosphoadenosine-5′-phosphosulfate (PAPS, active sulfate) as the sulfate donor. These Golgi-located enzymes are

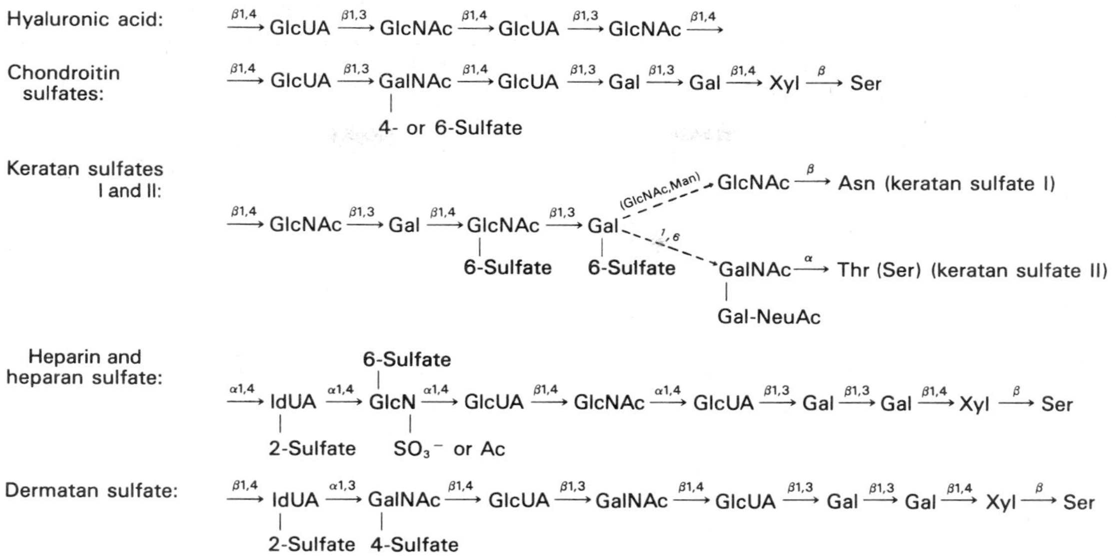

Hyaluronic acid: $\xrightarrow{\beta 1,4}$ GlcUA $\xrightarrow{\beta 1,3}$ GlcNAc $\xrightarrow{\beta 1,4}$ GlcUA $\xrightarrow{\beta 1,3}$ GlcNAc $\xrightarrow{\beta 1,4}$

Chondroitin sulfates: $\xrightarrow{\beta 1,4}$ GlcUA $\xrightarrow{\beta 1,3}$ GalNAc $\xrightarrow{\beta 1,4}$ GlcUA $\xrightarrow{\beta 1,3}$ Gal $\xrightarrow{\beta 1,3}$ Gal $\xrightarrow{\beta 1,4}$ Xyl $\xrightarrow{\beta}$ Ser

                |
                4- or 6-Sulfate

Keratan sulfates I and II:
$\xrightarrow{\beta 1,4}$ GlcNAc $\xrightarrow{\beta 1,3}$ Gal $\xrightarrow{\beta 1,4}$ GlcNAc $\xrightarrow{\beta 1,3}$ Gal $\cdots$ (GlcNAc,Man) — GlcNAc $\xrightarrow{\beta}$ Asn (keratan sulfate I)

             | | $\cdots 1,6$ GalNAc $\xrightarrow{\alpha}$ Thr (Ser) (keratan sulfate II)

           6-Sulfate   6-Sulfate |

                                                Gal-NeuAc

Heparin and heparan sulfate:
$\xrightarrow{\alpha 1,4}$ IdUA $\xrightarrow{\alpha 1,4}$ GlcN $\xrightarrow{\alpha 1,4}$ GlcUA $\xrightarrow{\beta 1,4}$ GlcNAc $\xrightarrow{\alpha 1,4}$ GlcUA $\xrightarrow{\beta 1,3}$ Gal $\xrightarrow{\beta 1,3}$ Gal $\xrightarrow{\beta 1,4}$ Xyl $\xrightarrow{\beta}$ Ser

         | | 6-Sulfate (above GlcN)

        2-Sulfate SO_3^- or Ac

Dermatan sulfate:
$\xrightarrow{\beta 1,4}$ IdUA $\xrightarrow{\alpha 1,3}$ GalNAc $\xrightarrow{\beta 1,4}$ GlcUA $\xrightarrow{\beta 1,3}$ GalNAc $\xrightarrow{\beta 1,4}$ GlcUA $\xrightarrow{\beta 1,3}$ Gal $\xrightarrow{\beta 1,3}$ Gal $\xrightarrow{\beta 1,4}$ Xyl $\xrightarrow{\beta}$ Ser

      | |

   2-Sulfate   4-Sulfate

Figure 57–10. Summary of structures of proteoglycans and glycosaminoglycans. (GlcUA, D-glucuronic acid; IdUA, L-iduronic acid; GlcN, D-glucosamine; GalN, D-galactosamine; Ac, acetyl, Gal, D-galactose; Xyl, D-xylose; Ser, L-serine; Thr, L-threonine; Asn, L-asparagine; Man, D-mannose; NeuAc, N-acetylneuraminic acid.) The summary structures are qualitative representations only and do not reflect, for example, the uronic acid composition of hybrid glycosaminoglycans such as heparin and dermatan sulfate, which contain both L-iduronic and D-glucuronic acid. Neither should it be assumed that the indicated substituents are always present, eg, whereas most iduronic acid residues in heparin carry a 2-sulfate group, a much smaller proportion of these residues are sulfated in dermatan sulfate. The presence of linkage trisaccharides (-Gal-Gal-Xyl-) in the chondroitin sulfates, heparin, and heparan and dermatan sulfates is shown. (Slightly modified and reproduced, with permission, from Lennarz WJ: *The Biochemistry of Glycoproteins and Proteoglycans.* Plenum Press, 1980.)

highly specific, and distinct enzymes catalyze sulfation at different positions (eg, carbons 2, 3, 4, and 6) on the acceptor sugars. An epimerase catalyzes conversions of glucuronyl to iduronyl residues.

The Various Glycosaminoglycans Exhibit Subtle Differences in Structure & Have Characteristic Distributions

The seven GAGs named above differ from each other in a number of the following properties: amino sugar composition, uronic acid composition, linkages between these components, the chain length of the disaccharides, the presence or absence of sulfate groups and their positions of attachment to the constituent sugars, the nature of the core proteins to which they are attached, the nature of the linkage to core protein, their tissue and subcellular distribution, and their biologic functions.

The structures (see Fig 57–10), the associations with core and other proteins to form proteoglycans, and the distributions of each of the GAGs will now be briefly discussed. The major features of these 7 GAGs are summarized in Table 57–7.

Hyaluronic Acid: Hyaluronic acid consists of an unbranched chain of repeating disaccharide units containing **GlcUA and GlcNAc.** There is no firm evidence that it is linked to protein, as are the other GAGs. Hyaluronic acid is present in bacteria and is widely distributed among various animals and tissues, includ-

ing **synovial fluid,** the vitreous body of the eye, and loose connective tissue.

Chondroitin Sulfates (Chondroitin 4-Sulfate and Chondroitin 6-Sulfate): Proteoglycans linked to chondroitin sulfate by the Xyl-Ser O-glycosidic bond are very prominent components of **cartilage.** The repeating disaccharide is similar to that found in hyaluronic acid, containing **GlcUA** but with **GalNAc** replacing GlcNAc. The GalNAc is substituted with sulfate at either its 4 or its 6 position, with approximately one sulfate being present per disaccharide unit. Each chain contains some 40 disaccharide units and thus has a MW of about 20,000. Many such chains are attached to a single protein molecule, generating massive proteoglycans of high MW (eg, in nasal cartilage approximately 2.5×10^6).

The chondroitin sulfates associate tightly with **hyaluronic acid** with the aid of 2 "**link proteins**", that bind hydrophobically to both hyaluronic acid and to the core protein, generating very large aggregates in connective tissue (see Figs 57–11 and 57–12).

Keratan Sulfates I and II: As shown in Fig 57–10, the keratan sulfates consist of repeating **Gal-GlcNAc** disaccharide units containing sulfate attached to the 6 position of GlcNAc or occasionally of Gal. Type I is abundant in **cornea** and type II is found along with chondroitin sulfate attached to hyaluronic acid in **loose connective tissue.** Types I and II have different attachments to protein (Fig 57–10).

Heparin: The repeating disaccharide contains

Table 57–7. Major properties of the glycosaminoglycans.

GAG	Sugars	Sulfate	Linkage to protein	Location
HA	GlcNAc, GlcUA	Nil	No firm evidence	Synovial fluid, vitreous humor, loose connective tissue
CS	GalNAc, GlcUA	4′ or 6′ of GalNAc	Xyl-Ser; associated with HA via link proteins	Cartilage, bone, cornea
KSI	GlcNAc, Gal	6′ of GlcNAc 6′ of Gal	GlcNAc-Asn	Cornea
KS II	GlcNAc, Gal	Same as KS I	GalNAc-Thr	Loose connective tissue
Heparin	GlcN, IdUA	N of GlcN 6′of GlcN 2′ of IdUA	Ser	Mast cells
Heparan sulfate	GlcN, GlcUA	N of GlcN 6′ of GlcN	Xyl-Ser	Skin fibroblasts, aortic wall
Dermatan sulfate	GalNAc, IdUA, (GlcUA)	4′ of GalNAc 2′ of IdUA	Xyl-Ser	Wide distribution

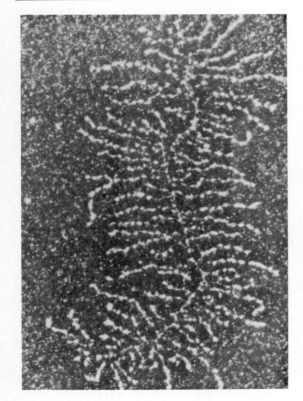

Figure 57–11. Darkfield electron micrograph of a proteoglycan aggregate of intermediate size in which the proteoglycan subunits and filamentous backbone are particularly well extended. (Reproduced, with permission, from Rosenberg L, Hellman W, Kleinschmidt AK: Electron microscopic studies of proteoglycan aggregates from bovine articular cartilage. *J Biol Chem* 1975;**250**:1877.)

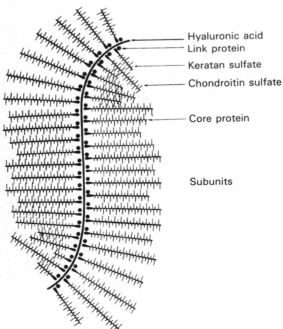

Hyaluronic acid
Link protein
Keratan sulfate
Chondroitin sulfate

Core protein

Subunits

Figure 57–12. Schematic representation of proteoglycan aggregate. (Reproduced, with permission, from Lennarz WJ: *The Biochemistry of Glycoproteins and Proteoglycans.* Plenum Press, 1980.)

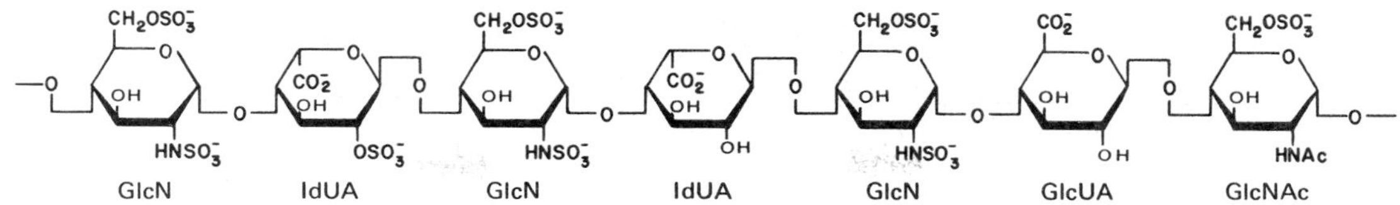

Figure 57–13. Structure of heparin. The polymer section illustrates structural features typical of heparin; however, the sequence of variously substituted repeating disaccharide units has been arbitrarily selected. In addition, non-O-sulfated or 3-O-sulfated glucosamine residues may also occur. (Modified, redrawn, and reproduced, with permission, from Lindahl U et al: Structure and biosynthesis of heparinlike polysaccharides. *Fed Proc* 1977;**36**:19.)

glucosamine (GlcN) and either of the 2 **uronic acids** (Fig 57–13). Most of the amino groups of the GlcN residues are **N-sulfated,** but a few are **acetylated.** The GlcN also carries a C_6 sulfate ester.

Approximately 90% of the uronic acid residues are **IdUA.** Initially, all of the uronic acids are GlcUA, but a **5-epimerase** converts approximately 90% of the GlcUA residues to IDUA after the polysaccharide chain is formed. The protein molecule of the heparin proteoglycan is unique, consisting exclusively of serine and glycine residues. Approximately two-thirds of the serine residues contain GAG chains, usually with a MW of 5000–15,000, but occasionally much higher. Heparin is found in the granules of **mast cells** and also in liver, lung, and skin.

Heparan Sulfate: This molecule is present on many **cell surfaces** as a proteoglycan and is extracellular. It contains **GlcNAc** with fewer N-sulfates than heparin, and, unlike heparin, its predominant uronic acid is **GlcUA.**

Dermatan Sulfate: This is widely distributed in animal tissues. Structurally, it resembles both the chondroitin sulfates and heparan sulfate. Its structure is similar to that of chondroitin sulfate, except that in place of a GlcUA in β-1,3 linkage to GalNAc, it contains an **IdUA** in an α-1,3 linkage to **GalNAc.** Formation of the IdUA occurs, as in heparin and heparan sulfate, by **5-epimerization of GlcUA.** Because this is regulated by the degree of sulfation, and sulfation is incomplete, dermatan sulfate contains both IdUA-GalNAc and GlcUA-GalNAc disaccharides.

Deficiencies of Enzymes That Degrade Glycosaminoglycans & Glycoproteins Result in Mucopolysaccharidoses & Mucolipidoses

Both Exo- & Endo- Glycosidases Degrade GAGs: Like most other biomolecules, GAGs are subject to turnover, being both synthesized and degraded. In adult tissues, GAGs generally exhibit relatively slow turnover, their half-lives being in the order of days to weeks.

Understanding of the degradative pathways for glycoproteins, proteoglycans, and GAGs has been greatly aided by discoveries of the specific enzyme deficiencies of inborn errors of metabolism. Two groups of diseases whose study has contributed greatly are the **mucopolysaccharidoses and the mucolipidoses.** The mucolipidoses were originally thought to affect mucopolysaccharides (GAGs) or glycyosphingolipids, but degradation of the oligosaccharide chains of glycoproteins can also be affected in these conditions. All of them, with one exception, share a common mechanism; **deficiency of the activity of a degradative enzyme** (usually located in the lysosomes) results in a **marked accumulation of the substrate** in various tissues. This can result in organ enlargement, disturbances of the structure of bone and skin, mental retardation, and other features, depending upon the severity of the deficiency and the tissue distribution of the substrate. The **one exception is I-cell disease, in which a mislocation of lysosomal enzymes results,** as described earlier in this chapter. Table 57–8 lists the biochemical defects in these and related disorders. Most of the mucolipidoses listed are characterized by increased excretion of various fragments of glycoproteins in the urine, which accumulate because of the metabolic block.

Degradation of the polysaccharide chains of GAGs is carried out by **endoglycosidases, exoglycosidases, and sulfatases.**

Hyaluronidase is a widely distributed endoglycosidase that cleaves hexosaminidic linkages. From hyaluronic acid, the enzyme will generate a tetrasaccharide with the structure $(GlcUA-β1,3-GlcNAc-β1,4)_2$. Hyaluronidase acts on both hyaluronic acid and chrondroitin sulfate. A deficiency of this enzyme has not apparently been described. The tetrasaccharide described above can be degraded further by a β-glucuronidase and β-N-acetylhexosaminidase.

β-Glucuronidase is an exoglycosidase that removes both GlcUA and IdUA from nonreducing termini of tetrasaccharides or larger polysaccharides. Its substrates include dermatan sulfate, heparan sulfate, chondroitin sulfate, and hyaluronic acid. In inherited deficiency of this enzyme in humans, the first 3 of these GAGs are excreted in the urine (in which they can be detected), but hyaluronic acid is not. Apparently, there are other pathways that can degrade the tetrasaccharide produced from hyaluronic acid by hyaluronidase.

β-D-Acetylhexosaminidase is an exoglycosidase

Table 57–8. Biochemical defects and diagnostic tests
in mucopolysaccharidoses, mucolipidoses, and related disorders.*

Name	Alternate Designation	Enzymatic Defect	Urinary Metabolites
Mucopolysaccharidoses			
Hurler, Scheie, Hurler/Scheie	MPS I	α-L-Iduronidase	DS, HS
Hunter	MPS II	Iduronate sulfatase	DS, HS
Sanfilippo A	MPS III A	HS N-sulfatase (sulfami-dase)	HS (±)
Sanfilippo B	MPS III B	α-N-acetylglucosamini-dase	HS
Sanfilippo C	MPS III C	Acetyltransferase	HS
Morquio	MPS IV	N-acetylgalactosamine 6-sulfatase	KS
Morquio-like	None	β-Galactosidase	KS
Maroteaux-Lamy	MPS VI	N-acetylgalactosamine 4-sulfatase (arylsulfatase B)	DS
β-Glucuronidase deficiency	MPS VII	β-Glucuronidase	DS, HS (±)
Unnamed disorder	MPS VIII	N-acetylglucosamine 6-sulfatase	HS, KS
Mucolipidoses and related disorders			
Sialidosis	ML I	Sialidase (neuraminidase)	GF
I-cell disease	ML II	UDP-N-acetylglucosamine: glycoprotein N-acetylglu-cosaminylphosphotransfer-ase (acid hydrolases thus lack phosphomannosyl residue)	GF
Pseudo-Hurler polydystrophy	ML HI	As for ML II but deficiency is incomplete	GF
Multiple sulfatase deficiency	None	Arylsulfatase A and other sulfatases	DS, HS
Mannosidosis	None	α-Mannosidase	GF
Fucosidosis	None	α-L-Fucosidase	GF

MPS = mucopolysaccharidosis; ML = mucolipidosis; DS = dermatan sulfate; KS = keratan sulfate; HS = heparan sulfate; GF = glycoprotein fragments.

*Modified, with permission, from DiNatale P, Neufeld EF: The biochemical diagnosis of mucopolysaccharidoses, mucolipidosis and related disorders. In: *Perspectives in Inherited Metabolic Diseases.* Vol 2. Barra B et al (editors). Editones Ermes (Milan), 1979. Fibroblasts, leukocytes, tissues, amniotic fluid cells, or serum can be used for the assay of many of the above enzymes.

present in many mammalian tissues. It cleaves Glc-NAc and GalNAc, when in β-linkage, from the non-reducing termini of polysaccharides. The substrates for this enzyme include gangliosides and chondroitin sulfates, hyaluronic acid, dermatan sulfate, and keratan sulfates I and II. There are 2 isozymes of β-D-acetylhexosaminidase. The **A isozyme** consists of 2 different subunits, α and β, while the **B isozyme** contains only β subunits. In **Tay-Sachs disease, the α subunit is defective,** and thus only the A isozyme is inactive. In **Sandhoff disease, the β subunit is defective,** resulting in a deficiency of both A and B isozymes. In both of these diseases, death usually results in infancy because of the severe accumulation of the ganglioside GM_2 in the brain. These diseases are classified as **sphinglipidoses** (see Table 25–1) rather than mucopolysaccharidoses, because the major biomolecule affected is a ganglioside.

β-Galactosidases (exoglycosidases) exist in several forms in animal tissues. Both chondroitin sulfate and keratan sulfate contain β-galactosides and thus are substrates for acid galactosidases. When acid β-galactosidase is deficient, both keratan sulfate and glycoprotein fragments accumulate, along with GM_1 ganglioside (see Chapter 26).

α-L-Iduronidase is a lysosomal exoglycosidase

that removes IdUA from the nonreducing terminus of polysaccharide chains. This enzyme is deficient in **Hurler's syndrome. Iduronate sulfatase** is a specific exoenzyme that cleaves the C_2 sulfate from an IdUA residue at the nonreducing end of heparin, heparan sulfate, and dermatan sulfate. An inherited deficiency of this enzyme is the cause of **Hunter's syndrome.**

The activities of most of the enzymes mentioned above, along with the others listed in Table 57–8, can usually be measured in skin fibroblasts, leukocytes, amniotic cells, and possibly serum. Analyses for the presence of increased amounts of GAGs in the **urine** are also of **diagnostic use.**

Proteoglycans Have Numerous Functions

General Aspects: As indicated above, proteoglycans are remarkably complex molecules and are found in every tissue of the body, mainly in the **extracellular matrix or "ground substance".** There they are associated with each other and also with the other major structural components of the matrix, collagen and elastin, in quite specific manners. Some proteoglycans bind to collagen and others to elastin. These interactions are important in **determining the structural organization** of the matrix. In addition, some of them **interact with certain adhesive proteins,** such as fibronectin and laminin (see Chapter 59), also found in the matrix. The **GAGs present in the proteoglycans are polyanions,** and hence **bind polycations and cations** such as Na and K. This latter ability **attracts water** by osmotic pressure into the extracellular matrix and contributes to its **turgor.** GAGs also **gel** at relatively low concentrations. Because of the long extended nature of the polysaccharide chains of GAGs and their ability to gel, the **proteoglycans can act as sieves,** restricting the passage of large macromolecules into the extracellular matrix, but allowing relatively free diffusion of small molecules. Again, **because of their extended structures and the huge macromolecular aggregates that they often form,** they occupy a very large volume of the matrix, relative to proteins.

Some Functions of Specific GAGs &/or Proteoglycans: Hyaluronic acid is especially high in concentration in embryonic tissues and is thought to play an important role in **permitting cell migration** during morphogenesis and wound repair. Its ability to attract water into the extracellular matrix and thereby "loosen it up" may be important in this regard. The high concentrations of hyaluronic acid and chondroitin sulfates present in cartilage contribute to its **compressibility.**

Chondroitin sulfates are located at sites of calcification in endochondral bone. This proteoglycan is located inside certain **neurons** and may provide an endoskeletal structure, helping to maintain their shape.

Both **keratan sulfate I** and **dermatan sulfate** are present in the cornea. They lie between collagen fibrils, and play a critical role in permitting **corneal transparency.** Changes in proteoglycan composition

are found in corneal scars, which disappear when the cornea heals. The presence of dermatan sulfate in the **sclera** may also play a role in maintaining the overall shape of the eye.

Heparin is an important **anticoagulant.** It binds with factors IX and XI, but its most important interaction is with **plasma antithrombin III.** The 1:1 binding of heparin to this plasma protein greatly accelerates the ability of the latter to inactive serine proteases, particularly **thrombin.** The binding of heparin to lysine residues in antithrombin III appears to induce a conformational change in the latter that favors its binding to the serine proteases. Heparin can also bind specifically to **lipoprotein lipase** present in capillary walls, causing a release of this enzyme into the circulation.

Certain proteoglycans (eg, heparan sulfate) are associated with the **plasma membrane of cells,** with their core proteins actually spanning that membrane. In it they may act as **receptors** and may also participate in the **mediation of cell growth** and **cell-cell communication.** The **attachment of cells** to their substratum in culture is mediated, at least in part, by heparan sulfate. This proteoglycan is also found in the basement membrane of the kidney, along with type IV collagen and laminin (see Chapter 59), where it plays a major role in determining the charge selectiveness of glomerular filtration.

Proteoglycans are also found in intracellular locations such as the **nucleus;** their function in this organelle has not been elucidated. They have also been shown to be present in some **storage or secretory granules,** such as the chromaffin granules of the adrenal medulla. It has been postulated that they play a role in the release of the contents of such granules. The various functions of GAGs are summarized in Table 57–9.

Associations With Major Diseases & With Aging: Hyaluronic acid may be important in **permitting tumor cells to migrate** through the extracellular

Table 57–9. Some functions of GAGs and/or proteoglycans.*

Structural components of the extracellular (EC) matrix
Specific interactions with collagen, elastin, fibronectin, laminin, and other proteins of the matrix
As polyanions, they bind polycations and cations
Contribute to the characteristic turgor of various tissues
Act as sieves in the EC matrix
Facilitate cell migration (HA)
Role in compressibility of cartilage in weight-bearing (HA, CS)
Role in corneal transparency (KS I and DS)
Structural role in sclera (DS)
Anticoagulant (heparin)
Components of plasma membranes, where they may act as receptors and participate in cell adhesion and cell-cell interactions (eg, HS)
Determines charge-selectivity of renal glomerulus (HS)
Components of synaptic and other vesicles (eg, HS)

*Abbreviations: HA, hyaluronic acid; CS, chondroitin sulfate; KS I, keratan sulfate I; DS, dermatan sulfate; HS, heparan sulfate.

matrix. Tumor cells can induce fibroblasts to synthesize greatly increased amounts of this GAG, thereby perhaps facilitating their own spread. Some tumor cells have less heparan sulfate at their surfaces, and this may play a role in the **lack of adhesiveness** that these cells display.

The **intima of the arterial wall** contains hyaluronic acid and chondroitin sulfate, dermatan sulfate, and heparan sulfate proteoglycans. Of these proteoglycans, **dermatan sulfate binds plasma low density lipoproteins.** In addition, dermatan sulfate appears to be the major GAG **synthesized by arterial smooth muscle cells.** Because it is these cells that proliferate in atherosclerotic lesions in arteries, dermatan sulfate may play an important role in the **development of the atherosclerotic plaque.**

In various types of **arthritis,** proteoglycans may act as **autoantigens,** thus contributing to the pathology of these conditions. The amount of **chondroitin sulfate** in cartilage **diminishes with age,** whereas the amounts of **keratan sulfate and hyaluronic acid increase.** These changes may contribute to the development of **osteoarthritis.** Changes in the amounts of certain GAGs in the **skin** are also observed with age and help to account for the characteristic changes noted in this organ in the elderly.

REFERENCES

Buckwalter JA, Rosenberg LC: Electron microscopic studies of cartilage proteoglycans. *J Biol Chem* 1982;**257;**9830.

DiNatale P, Neufeld EF: The biochemical diagnosis of mucopolysaccharidoses, mucolipidosis and related disorders. In: *Perspectives in Inherited Metabolic Diseases.* Vol 2. Barra B et al (editors). Editiones Ermes (Milan), 1979.

Höök M et al: Cell-surface glycosaminoglycans. *Annu Rev Biochem* 1984;**53:**847.

Jaques LB: Heparin: An old drug with a new paradigm. *Science* 1979;**206:**528.

Kornfeld R, Kornfeld S: Assembly of asparagine-linked oligosaccharides. *Annu Rev Biochem* 1985;**54:**631.

Kornfeld S, Sly WS: Lysosomal storage defects. *Hosp Pract* (Aug) 1985;**20:**71.

Lennarz WJ: *The Biochemistry of Glycoproteins and Proteoglycans.* Plenum Press, 1980.

Poole AR: Proteoglycans in health and disease: structures and functions. *Biochem J* 1986;**236:**1.

Poole AR et al: Proteoglycans from bovine nasal cartilage: Immunochemical studies of link protein. *J. Biol Chem* 1980;**255:**9295.

Schachter H: Biosynthetic controls that determine the branching and heterogeneity of protein-bound oligosaccharides. *Biochem Cell Biol* 1986;**64:**163.

Plasma Proteins, Immunoglobulins, & Clotting Factors

58

Elizabeth J. Harfenist, PhD, & Robert K Murray, MD, PhD

INTRODUCTION

The blood circulates in what is virtually a closed system of blood vessels. Blood consists of **solid elements, the red and white blood cells and the platelets,** suspended in a liquid medium, the **plasma.** As indicated below, blood—and plasma in particular—performs many functions that are absolutely critical for the maintenance of health.

Once the blood has clotted (coagulated), the remaining liquid phase is called **serum.** Serum lacks the clotting factors (including fibrinogen) that are normally present in plasma but have been consumed during the process of coagulation. Serum does contain some degradation products of clotting factors—products that have been generated during the coagulation process and thus are *not* normally present in plasma.

BIOMEDICAL IMPORTANCE

The fundamental role that the blood plays in the maintenance of **homeostasis** and the ease with which blood can be obtained have meant that the study of its constituents has been of central importance in the development of biochemistry and clinical biochemistry. **Hemoglobin, albumin, the immunoglobulins,** and the various **clotting factors** are among the most studied of all proteins. Changes in the amounts of various **plasma proteins** occur in many diseases and can be monitored by electrophoresis. Alterations of the activities of certain **enzymes** found in plasma are of diagnostic use in a number of pathologic conditions (see the Appendix). **Hemorrhagic and thrombotic states** can pose serious medical emergencies and **thromboses** in the coronary and cerebral arteries are major causes of death in many parts of the world. Rational management of these conditions requires a clear understanding of the bases of blood clotting and fibrinolysis.

THE BLOOD HAS MANY FUNCTIONS

The functions of blood—all except specific cellular ones such as oxygen transport and cell-mediated immunologic defense—are carried out by plasma and its constituents. They are enumerated in Table 58–1.

Plasma consists of water, electrolytes, metabolites, nutrients, proteins, and hormones. The water and electrolyte composition of plasma is practically the same as that of all extracellular fluids. Laboratory determinations of serum levels of Na^+, K^+, Ca^{2+}, Cl^-, and carbon dioxide and of blood pH (see the Appendix) are very important in assisting in the management of many patients.

Plasma Contains a Very Complex Mixture of Proteins

The **concentration of total protein** in human plasma is approximately 7–7.5 g/dL and comprises the major part of the solids of the plasma. The proteins of the plasma are actually a very complex mixture that include not only **simple proteins,** but also conjugated proteins such as **glycoproteins** and various types of **lipoproteins.** There are thousands of **antibodies** present in human plasma, although the amount of any one antibody is usually quite low under normal circumstances. The relative dimensions and molecular

---

Table 58–1. The major functions of the blood.

1. **Respiration**–transport of oxygen from the lungs to the tissues and of CO_2 from the tissues to the lungs
2. **Nutrition**–transport of absorbed food materials
3. **Excretion**–transport of metabolic waste to the kidneys, lungs, skin, and intestines for removal
4. Maintenance of normal **acid-base balance** in the body
5. Regulation of **water balance** through the effects of blood on the exchange of water between the circulating fluid and the tissue fluid
6. Regulation of **body temperature** by the distribution of body heat
7. **Defense** against infection by the white blood cells and circulating antibodies
8. Transport of **hormones** and regulation of metabolism
9. Transport of **metabolites**
10. **Coagulation**

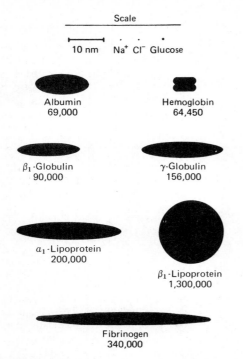

Scale

10 nm   Na$^+$ Cl$^-$ Glucose

Albumin
69,000

Hemoglobin
64,450

β_1-Globulin
90,000

γ-Globulin
156,000

a_1-Lipoprotein
200,000

β_1-Lipoprotein
1,300,000

Fibrinogen
340,000

Figure 58–1. Relative dimensions and molecular weights of protein molecules in the blood (Oncley).

weights of some of the most important plasma proteins are shown in Fig 58–1.

The **separation of individual proteins** from a complex mixture is frequently accomplished by the use of various solvents or electrolytes (or both) to remove different protein fractions in accordance with their solubility characteristics. This is the basis of the so-called **salting-out methods,** which find some usage in the determination of protein fractions in the clinical laboratory. Thus, one can separate the proteins of the plasma into 3 major groups, **fibrinogen, albumin, and globulins,** by the use of varying concentrations of sodium or ammonium sulfate.

The most common method of analyzing plasma proteins is by **electrophoresis.** There are many types of electrophoresis, each using a different supporting medium. In clinical laboratories, cellulose acetate is widely used as a supporting medium. Its use permits resolution, after staining, of plasma proteins into 5 bands, designated albumin, α_1, α_2, β, and γ, fractions, respectively (Fig 58–2). The stained strip of cellulose acetate (or other supporting medium) is referred to as an **electrophoretogram.** The amounts of these 5 bands can be conveniently quantified by use of densitometric scanning machines. Table 7 of the Appendix lists the major plasma proteins found in the α_1, α_2, β, and γ bands and the Appendix also indicates various diseases in which the amounts of albumin, globulin, and fibrinogen are altered.

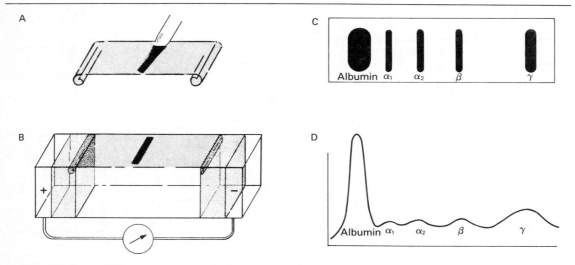

Figure 58–2. Technique of cellulose acetate zone electrophoresis. *A:* Small amount of serum or other fluid is applied to cellulose acetate strip. *B:* Electrophoresis of sample in electrolyte buffer is performed. *C:* Separated protein bands are visualized in characteristic positions after being stained. *D:* Densitometer scanning from cellulose acetate strip converts bands to characteristic peaks of albumin, α_1-globulin, α_2-globulin, β-globulin, and γ-globulin. (Reproduced, with permission, from Stites DP, Stobo JD, Wells JV [editors]: *Basic & Clinical Immunology,* 6th ed. Appleton & Lange, 1987.)

The Concentration of Protein in Plasma Is Important in Determining the Distribution of Fluid Between Blood & Tissues

Blood plasma is by definition an intravascular fluid. On the arterial side of the circulation, the intravascular **hydrostatic pressure** generated by the heart and large vessels is 20–25 mm Hg greater than the hydrostatic pressure in the tissue spaces. In order to prevent too much intravascular fluid from being forced into the extravascular tissue spaces, the hydrostatic pressure is opposed by an intravascular **colloid osmotic pressure** generated by the plasma proteins. If the concentration of plasma proteins is markedly diminished (eg, due to severe protein malnutrition), fluid is not attracted back into the intravascular compartment and accumulates in the extravascular tissue spaces, a condition known as **edema.** Edema has many causes; protein deficiency is one of them.

Plasma Proteins Have Been Studied Extensively

Because of the relative ease with which they can be obtained, plasma proteins have been studied extensively in both humans and animals. Considerable information is available about the biosynthesis, turnover, structure, and functions of the major plasma proteins. Alterations of their amounts and of their metabolism in many disease states have also been investigated. In recent years, many of the genes for plasma proteins have been cloned and their structures determined.

Many of the preparations listed in Table 2–7 have been used in the study of the plasma proteins. The preparation of **antibodies** specific for the individual plasma proteins has greatly facilitated their study, allowing the **precipitation and isolation of pure proteins** from the complex mixture present in tissues or plasma. In addition, the use of **isotopes** has made possible the determination of their pathways of biosynthesis and of their turnover rates in plasma. The following are **6 generalizations** that have emerged from studies of plasma proteins.

1. Most of the Plasma Proteins Are Synthesized in the Liver: This has been established by experiments at the whole animal level (eg, hepatectomy) and by use of the isolated perfused liver preparation, of liver slices, of liver homogenates, and of in vitro translation systems using preparations of mRNA extracted from liver. However, the γ-globulins are synthesized in plasma cells and certain plasma proteins are synthesized in other sites, such as endothelial cells.

2. Plasma Proteins Are Generally Synthesized on Membrane-Bound Polyribosomes: They then traverse the major secretory route in the cell (RER → SER → Golgi → plasma membrane) prior to entering the plasma. Thus, most plasma proteins are synthesized as **preproteins** and initially contain N-terminal signal peptides (cf Chapter 43). They are usually subjected to various posttranslational modifications (proteolysis, glycosylation, phosphorylation, etc) as they travel through the cell. Transit times through the hepatocyte

from the site of synthesis to the plasma vary from 30 min to several hours or more for individual proteins.

3. Almost All of the Plasma Proteins Are Glycoproteins: Accordingly, they contain either N- or O-linked oligosaccharide chains, or both (see Chapter 57). Albumin is the major exception; it does not contain sugar residues. The oligosaccharide chains have various functions (see Table 57–2). An important finding is that removal of terminal sialic acid residues from certain plasma proteins (eg, ceruloplasmin) by exposure to neuraminidase can markedly shorten their half-lives in plasma (see Chapter 57).

4. Many of the Plasma Proteins Exhibit Polymorphism: A polymorphism is a Mendelian or monogenic trait that exists in the population in at least two phenotypes, neither of which is rare (ie, neither of which occurs with a frequency of less than 1–2%) (Vogel & Motulsky, 1986). The ABO blood group substances (Chapter 57) are the best known examples of human polymorphisms. Human plasma proteins that exhibit polymorphism include α_1-antitrypsin, haptoglobin, transferrin, ceruloplasmin, and immunoglobulins. The polymorphic forms of these proteins were often first recognized by use of starch gel electrophoresis, in which each polymorphic form shows a characteristic migration. Analyses of these human polymorphisms have proved to be of genetic and anthropologic interest and, in certain cases (eg, α_1-antitrypsin, see below), of clinical interest.

5. Each Plasma Protein Has a Characteristic Half-Life in the Circulation: The half-life of a plasma protein can be determined by labeling the isolated pure protein with ^{131}I under mild, nondenaturing conditions. This isotope unites covalently with tyrosine residues in the protein. The labeled protein is freed of unbound ^{131}I and its specific activity (dpm per mg protein) determined. A known amount of the radioactive protein is then injected into a normal adult subject and samples of blood are taken at various time intervals for determinations of their radioactivities. The values for radioactivity are plotted against time and the half-life of the protein (the time for the radioactivity to decline from its peak value to one-half of its peak value) can be calculated from the resulting graph, discounting the times for the injected protein to equilibrate (mix) in the blood and in the extravascular spaces. The half-lives obtained for albumin and haptoglobin in normal healthy adults are approximately 20 and 5 days, respectively. In certain diseases, the half-life of a protein may be markedly altered. For instance, in some gastrointestinal diseases such as regional ileitis (Crohn's disease), very considerable amounts of plasma proteins, including albumin, may be lost into the bowel through the inflamed intestinal mucosa. Patients with this condition have a **protein-losing gastroenteropathy** and the half-life of injected iodinated albumin in these subjects may be reduced to as little as one day.

6. The Levels of Certain Proteins in Plasma Increase During Acute Inflammatory States or Secondary to Certain Types of Tissue Damage: These proteins are called **acute phase proteins (or**

reactants) and include C-reactive protein (CRP, named because it reacts with the C polysaccharide of pneumococci), α_1-antitrypsin, haptoglobin, α_1-acid glycoprotein, and fibrinogen. The elevations of the levels of these proteins vary from as little as 50% to as much as 1000-fold in the case of CRP. Their levels are also usually elevated during chronic inflammatory states and in patients with cancer. These proteins are believed to play a role in the body's response to inflammation. For example, C-reactive protein can stimulate the classical complement pathway and α_1-antitrypsin can neutralize certain proteases released during the acute inflammatory state. Recent work indicates that interleukin 1 (IL-1), a polypeptide released from mononuclear phagocytic cells, is the principal—but not the sole—stimulator of the synthesis of the majority of acute phase reactants by hepatocytes. Other molecules such as IL-6 are also involved and they as well as IL-1 appear to work at the level of gene transcription.

The following section presents basic information regarding 6 of the major human plasma proteins, including the immunoglobulins. The lipoproteins are discussed in Chapter 27.

Albumin Is the Major Protein in Human Plasma

This is the major protein of human plasma (approximately 4.5 g/dL), has a MW of approximately 69,000, and comprises some 60% of the total plasma protein. Some 40% of albumin is present in the plasma, and the other 60% is present in the extracellular space. The liver produces about 12 g of albumin per day, representing about 25% of total hepatic protein synthesis and half of all of its secreted protein. Albumin is initially synthesized as a **preproprotein.** Its signal peptide is removed as its passes into the cisternae of the rough ER, and a hexapeptide at the resulting N-terminus is subsequently cleaved off further along the secretory pathway. The synthesis of albumin is depressed in a variety of diseases, particularly those of the liver (see the Appendix under Proteins). The plasma of patients with liver disease often shows a decrease in the ratio of albumin to globulins (decreased A:G ratio). The synthesis of albumin decreases relatively early in conditions of protein malnutrition, such as kwashiorkor.

Mature human albumin consists of one polypeptide chain of 585 amino acids and contains 17 disulfide bonds. By the use of proteases, albumin can be subdivided into 3 **domains,** which have different functions. Albumin has an **ellipsoidal shape,** which means that it does not increase the viscosity of the plasma as much as an elongated molecule like fibrinogen does. Because of its relatively low MW (approximately 69,000) and high concentration, albumin is thought to be responsible for 75–80% of the **osmotic pressure** of human plasma. Electrophoretic studies have shown that the plasma of certain humans lacks albumin. These subjects are said to exhibit **analbuminemia.** One cause of this condition is a mutation that affects splicing. Subjects with analbuminemia show only

moderate edema, despite the belief that albumin is the major determinant of plasma osmotic pressure. It is thought that the amounts of the other plasma proteins increase and compensate for the lack of albumin.

Another important function of albumin is its ability to **bind various ligands.** These include free fatty acids (FFA), calcium, certain steroid hormones, bilirubin, and some of the plasma tryptophan. In addition, albumin binds approximately 10% of the plasma copper, the rest being attached to ceruloplasmin (see below). A variety of drugs, including sulfonamides, penicillin G, dicoumarol, and aspirin, are bound to albumin; this has important pharmacologic implications.

Preparations of human albumin are of use in the treatment of hemorrhagic shock and of burns.

Haptoglobin Binds Extracorpuscular Hemoglobin, Preventing Free Hemoglobin From Entering the Kidney

Haptoglobin (Hp) is a plasma glycoprotein that binds extracorpuscular hemoglobin (Hb) in a tight noncovalent complex (Hb-Hp). The amount of Hp in human plasma ranges from 40 to 180 mg of Hb-binding capacity per deciliter. Approximately 10% of the Hb that is degraded each day is released into the circulation, and is thus extracorpuscular. The other 90% is present in old, damaged red blood cells, which are degraded by cells of the histiocytic system. The MW of Hb is approximately 65,000, whereas the MW of the simplest polymorphic form of Hp (Hp 1-1) found in humans is approximately 90,000. Thus, the Hb-Hp complex has a MW of approximately 155,000. Free Hb passes through the glomerulus of the kidney, enters the tubules, and tends to precipitate therein (as can happen after a massive incompatible blood transfusion, when the capacity of Hp to bind Hb is grossly exceeded) (see Fig 58–3). However, the Hb-Hp complex is too large to pass through the glomerulus. The function of Hp thus appears to be to **prevent loss of free Hb** into the kidney. This conserves the valuable iron present in Hb, which would otherwise be lost to the body.

Human Hp exists in three polymorphic forms, known as **Hp 1-1, Hp2-1, and Hp 2-2.** Hp 1-1 migrates in starch gel electrophoresis as a single band, whereas Hp 2-1 and Hp 2-2 exhibit much more complex band patterns. Two genes, designated Hp^1 and Hp^2, direct these three phenotypes, with Hp 2-1 being the heterozygous phenotype. No significant functional differences among the polymorphic forms of Hp have been established.

The levels of Hp in human plasma vary and are of some diagnostic use. **Low levels of Hp** are found in patients with **hemolytic anemias.** This is explained by the fact that whereas the half-life of Hp is approximately 5 days, the half-life of the Hb-Hp complex is some 90 min, the complex being rapidly removed from plasma by hepatocytes. Thus, when Hp is bound to Hb, it is cleared from the plasma some 80 times faster than normally. Accordingly, the level of Hp falls

A. Hb → Kidney → Excreted in urine or precipitates in tubules;
(MW 65,000)   iron is lost to body

B. Hb   +   Hp → Hb : Hp complex ⇸ Kidney
(MW 65,000) (MW 90,000) │
│ (MW 155,000)
↓
Catabolized by liver cells;
iron is conserved and reused

Figure 58–3. Different fates of free hemoglobin and of the hemoglobin-haptoglobin complex.

rapidly in situations where Hb is constantly being released from red blood cells, such as occur in hemolytic anemias. Hp is an acute phase protein and its **plasma level is elevated in a variety of inflammatory states.**

Certain other plasma proteins bind heme, but not Hb. **Hemopexin** is a β_1-globulin that binds free heme. **Albumin** will bind some metheme (ferric heme), to form methemalbumin, which then transfers the metheme to hemopexin.

Transferrin Shuttles Iron Around the Blood Stream to Sites Where it Is Needed & Thus Plays a Central Role in Iron Metabolism

Transferrin (Tf) is a β_1-globulin of approximate MW 80,000. It is a glycoprotein and is synthesized in the liver. More than 20 polymorphic forms of Tf have been found. Tf plays a central role in the body's metabolism of iron because it **transports iron** (2 moles of Fe^{3+} per mole of Tf) in the circulation to sites where iron is required, eg, from the gut to the bone marrow and other organs. Iron is very important in the human body, because of its occurrence in many hemoproteins, such as hemoglobin, myoglobin, and the cytochromes. Iron is ingested in the diet and its absorption as Fe^{2+} is tightly regulated at the level of the intestinal mucosa. Under normal circumstances, the body guards its content of iron zealously, so that a healthy adult male loses about 1 mg per day, which is replaced by absorption. Adult females are more prone to states of iron deficiency, because some may lose excessive blood during menstruation. The amounts of iron in Tf and in various other body compartments are shown in Table 58–2.

Approximately 20 mL of of red blood cells are catabolized per day, releasing some 25 mg of iron into the body. **Free iron is toxic, but association with Tf diminishes its potential toxicity and also directs iron to where it is required in the body.** There are **receptors** on the surfaces of many cells for Tf. The protein binds to these receptors and is internalized by receptor-mediated endocytosis (cf the fate of LDL, Chapter 27). The acid pH inside the lysosome causes the iron to dissociate from the protein. However, unlike the protein component of LDL, apoTf is not degraded within the lysosome. Instead, it remains associated with its receptor, returns to the plasma membrane, dissociates from its receptor, reenters the plasma, picks up more iron, and again delivers the iron to needy cells.

Problems of Iron Metabolism: Attention to iron metabolism is **particularly important in women,** for the reason mentioned above. Also, in **pregnancy** allowances must be made for the growing fetus. **Older people** with poor dietary habits ("tea and toasters") may also develop iron deficiency. Iron deficiency anemia due to inadequate intake, inadequate utilization, or excessive loss of iron is one of the commonest conditions seen in medical practice.

The concentration of Tf in plasma is approximately 300 mg/dL. This amount of Tf can bind some 300 μg of iron per deciliter, so that this represents the **total iron-binding capacity** of plasma. However, the protein is normally only one-third saturated with iron. In **iron-deficiency anemia,** the protein is even less saturated with iron, whereas in conditions of storage of **excess iron** in the body (eg, hemochromatosis) the saturation with iron is much greater than one-third.

Ferritin: Ferritin is another protein that is important in the metabolism of iron. In normal conditions, it stores iron so that it can be called upon for use as conditions require. In conditions of excess iron (eg, hemochromatosis), body stores of iron are greatly increased and much more ferritin is present in the tissues, such as liver and spleen. Ferritin contains approximately 23% iron and apoferritin (the protein moiety free of iron) has a MW of approximately 440,000. Ferritin is comprised of 24 subunits of 18,500 MW, which surround in a micellar form some 3000–4500 ferric atoms. Normally, there is a little ferritin in human plasma. However, in patients with excess iron, the amount of ferritin in plasma is markedly elevated. The amount of ferritin in plasma can be conveniently measured by a sensitive and specific radioimmunoassay and serves as a good index of body iron stores.

Hemosiderin: Hemosiderin is a somewhat ill-de-

Table 58–2. Distribution of iron in a 70-kg adult male.

Transferrin	3–4 mg
Hb in red blood cells	2500 mg
In myoglobin and various enzymes	300 mg
In stores (ferritin and hemosiderin)	1000 mg
Absorption	1 mg/d
Losses	1 mg/d

In an adult female of similar weight, the amount in stores would generally be less (eg, 100–400 mg) and the losses would be greater (eg, 1.5–2 mg/d).

Table 58–3. Laboratory tests of value in assessing patients with disorders of iron metabolism.

Red blood cell count and estimation of hemoglobin
Determinations of plasma iron, total iron binding capacity (TIBC), and % transferrin saturation
Determination of ferritin in plasma by radioimmunassay
Prussian blue stain of tissue sections
Determination of amount of iron (μg/g) in a tissue biopsy

fined molecule; it appears to be a partly degraded form of ferritin, but still containing iron. It can be detected by histologic stains (eg, Prussian blue) for iron and its presence is checked for histologically when excessive storage of iron occurs.

Primary Hemochromatosis: Primary hemochromatosis is a genetic disorder characterized by excessive storage of iron in tissues, leading to tissue damage. The cause appears to be excessive absorption of iron from the intestinal mucosa, although little is known about the molecular nature of the abnormality. The organs most affected are liver, skin, and pancreas; patients with this disorder usually develop cirrhosis of the liver. In addition, they may acquire pigmentation of the skin and diabetes mellitus (bronzed diabetes). The disorder can be kept under control by periodic withdrawals of blood (phlebotomies).

Table 58–3 summarizes laboratory tests that are useful in the assessment of patients with abnormalities of iron metabolism. Also see the Appendix under Iron, Serum and Iron-Binding Capacity, Serum.

Ceruloplasmin Binds Copper & Is Associated With Wilson's Disease, A Condition of Copper Toxicosis

Ceruloplasmin is an α_2-globulin of approximately 160,000 MW, with a concentration in plasma of approximately 30 mg/dL. It has a blue color because of its **copper** content and carries 90% of the copper present in plasma. Each molecule of ceruloplasmin binds six atoms of copper very tightly, so that the copper is not readily exchangeable. **Albumin** carries the other 10% of the plasma copper. However, the tightness of binding of copper to albumin is less than that of the copper bound to ceruloplasmin. Consequently, albumin donates its copper to tissues more readily than ceruloplasmin and appears to be more important than

ceruloplasmin in copper transport in the human body. Ceruloplasmin exhibits an **oxidase** activity, but its significance is not clear.

The amount of ceruloplasmin in plasma is decreased in liver disease. Ceruloplasmin bears a special relationship to **Wilson's disease** (hepatolenticular degeneration), although it appears unlikely that the primary cause of this disease is an abnormality of ceruloplasmin. Wilson's disease is an inherited condition in which copper fails to be excreted in the bile, possibly because the lysosomes are unable to excrete copper derived from the breakdown of ceruloplasmin into the bile (Fig 58–4). Copper thus accumulates progressively in the body, particularly in liver, brain, kidney, and red blood cells, and the disease may be regarded as an inability to maintain a near-zero copper balance resulting in **copper toxicosis.** The increase of copper in liver cells inhibits the coupling of copper to apoceruloplasmin and leads to low levels of the protein in plasma (<20 mg/dL). As the amount of copper accumulates, patients may develop a hemolytic anemia, chronic liver disease, and a neurologic syndrome due to accumulation of copper in the basal ganglia and other brain centers. A frequent clinical finding is the **Kayser-Fleischer ring.** This is a green or golden pigment ring around the cornea due to deposition of copper in Descemet's membrane. If Wilson's disease is suspected, a liver biopsy should be performed; a value for liver copper of over 250 μg/g dry weight along with a plasma level of ceruloplasmin of under 20 mg/dL is diagnostic. Treatment consists of a diet low in copper and judicious administration of D-**penicillamine,** which chelates copper and is excreted in the urine, thus depleting the body of the excess of this mineral.

Copper is the **metal cofactor** for a number of important enzymes, such as cytochrome oxidase, tyrosinase, and copper-dependent superoxide dismutase. The body of the normal adult contains 100–150 mg of copper, located mostly in bone, liver, and muscle. The daily intake of copper is 2–4 mg and absorption occurs in the upper small intestine. Copper is carried to the liver bound to albumin, taken up by liver cells, and partly excreted in the bile. Copper also leaves the liver attached to ceruloplasmin, which is synthesized in that organ.

Copper Deficiency: Deficiency of copper is relatively rare and results in **one type of anemia. Menkes'**

A. Normal: Cu → Bile
Cu + Apoceruloplasmin → Plasma ceruloplasmin

B. Wilson's disease: ↓Cu excretion into bile
↓
Cu accumulates in liver
↓
Inhibits its own coupling to
apoceruloplasmin → Low plasma ceruloplasmin

Figure 58–4. Schematic representation of fate of copper in (a) normal liver and (b) in the liver of a patient with Wilson's disease. The excretion of copper into the bile is impaired in Wilson's disease, possibly because of a defect in excretion by lysosomes of Cu derived from the catabolism of ceruloplasmin.

A. Active elastase + α_1–AT → Inactive elastase: α_1–AT complex → No proteolysis of lung → No tissue damage

B. Active elastase + ↓ or no α_1–AT → Active elastase → Proteolysis of lung → Tissue damage

Figure 58–5. Scheme illustrating (a) normal inactivation of elastase by α_1-AT and (b) situation where the amount of α_1-AT is substantially reduced, resulting in proteolysis by elastase leading to tissue damage.

syndrome ("kinky" or "steely" hair disease) is another disorder of copper metabolism. It is X-linked, affects male infants, and is generally fatal. Affected infants exhibit brittle, kinky hair, impairment of growth, mental retardation, and widespread arterial aneurysms. Evidence has been obtained that intestinal absorption of copper is deficient, although the biochemical nature of the lesion has not been defined. Blood levels of copper and of ceruloplasmin are low in affected individuals.

Deficiency of α_1-Antiproteinase (α_1-Antitrypsin) Is Associated with Emphysema & One Type of Liver Disease

α_1-Antiproteinase was formerly called **α_1-antitrypsin (α_1-AT)**, and this name is retained here. It has a MW of approximately 45,000 and is the major component of the α_1 fraction of human plasma. It is synthesized in the liver and is the principal protease inhibitor (Pi) of human plasma. It **inhibits trypsin, elastase, and certain other proteases** by forming complexes with them. The protein is **highly polymorphic;** the multiple forms can be separated by electrophoresis. The major genotype is MM and its phenotypic product is PiM.

There are two areas of clinical interest concerning α_1-AT. A deficiency of this protein has a role in certain cases (approximately 5%) of **emphysema.** This occurs mainly in subjects with the **ZZ genotype,** who synthesize PiZ. Considerably less of this protein is secreted, as compared to PiM. When the amount of α_1-AT is deficient and polymorphonuclear white blood cells increase in the lung (eg, during pneumonia), the affected individual lacks a countercheck to proteolytic damage of the lung by proteases such as elastase (Fig 58–5). Of considerable interest is that a particular methionine (residue 358) of α_1-AT is involved in its binding to proteases. **Smoking** oxidizes this methionine to methionine sulfoxide, and thus inactivates it. As a result, affected molecules of α_1-AT no longer neutralize proteases. This is particularly devastating in patients (PiZZ phenotype) who already have low levels of α_1-AT. The further diminution in α_1-AT brought about by smoking results in increased proteolytic destruction of lung tissue, accelerating the development of emphysema. Intravenous administration of α_1-AT has been proposed as an adjunct in the treatment of patients with emphysema due to α_1-AT deficiency. Attempts are being made, using the techniques of protein engineering (cf Chapter 42), to replace methionine 358 by another residue that would not be subject to oxidation. The resulting "mutant" α_1-AT would thus afford protection against proteases for a much longer period of time than would native α_1-AT.

Deficiency of α_1-AT is also implicated in one type of **cirrhosis** (α_1-AT deficiency liver disease). In this condition, molecules of the ZZ phenotype accumulate in the cisternae of the ER of hepatocytes, perhaps because of an inability of hepatocytes to secrete this particular type of α_1-AT. By mechanisms that are not understood, this results in cirrhosis (accumulation of massive amounts of collagen, resulting in fibrosis) of the liver.

THE PLASMA IMMUNOGLOBULINS PLAY A MAJOR ROLE IN THE BODY'S DEFENSE MECHANISMS

The immune system of the body consists of 2 major components, **B lymphocytes and T lymphocytes.** The B lymphocytes are mainly derived from bone marrow cells in higher animals and from the bursa of Fabricius in birds. The T signifies lymphocytes of thymic origin. The **B cells** are responsible for the synthesis of circulating, humoral antibodies, also known as **immunoglobulins.** The **T cells** are involved in a variety of important **cell-mediated immunologic processes,** such as graft rejection, hypersensitivity reactions, and defense against malignant cells and many viruses. This section considers only the plasma immunoglobulins, which are synthesized mainly in **plasma cells,** specialized cells of B cell lineage that synthesize and secrete immunoglobulins into the plasma.

All Immunoglobulins Contain a Minimum of 2 Light & 2 Heavy Chains

All immunoglobulin molecules consist of 2 identical light (L) chains (MW 23,000) and 2 identical heavy (H) chains (MW 53,000–75,000) held together as a tetramer (L_2H_2) by disulfide bonds (Fig 58–6). Each chain can be divided conceptually into specific **domains,** or regions, that have structural and functional significance. The half of the **light (L) chain** toward the carboxyl terminus is referred to as the **constant region (C_L),** while the amino-terminal half is the **variable region** of the light chain (V_L). Approximately one-quarter of the **heavy (H) chain** at the amino terminus is referred to as its variable region (V_H), and the other three-quarters of the heavy chain are referred to as the constant regions (C_H1, C_H2, C_H3) of

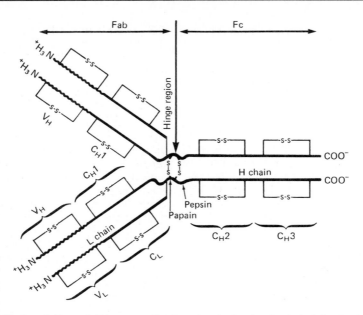

Figure 58–6. A simplified model for an IgG human antibody molecule showing the 4-chain basic structure and domains. V indicates variable region; C, the constant region; and the vertical arrow, the hinge region. Thick lines represent H and L chains; thin lines represent disulfide bonds. (Modified and reproduced, with permission, from Stites DP, Stobo JD, Wells JV [editors]: *Basic & Clinical Immunology*, 6th ed. Appleton & Lange, 1987.)

that H chain. The portion of the immunoglobulin molecule that **binds the specific antigen** is formed by the amino-terminal portions (variable regions) of both the H and L chains—ie, the V_H and V_L domains. The domains of the protein chains do not simply exist as linear sequences of amino acids but form globular regions with secondary and tertiary structure in order to effect binding of specific antigens.

As depicted in Fig 58–6, digestion of an immunoglobulin by the enzyme papain produces 2 antigen-binding fragments (**Fab**) and one crystallizable fragment (**Fc**). The area in which papain cleaves the immu-

noglobulin molecule—ie, the region between the C_H1 and C_H2 domains—is referred to as the **hinge region.**

All Light Chains Are Either of Kappa or Lambda Type

There are 2 general types of light chains, kappa (κ) and lambda (λ), which can be distinguished on the basis of structural differences in their C_L regions (Table 58–4). A given immunoglobulin molecule always contains two κ or two λ light chains, **never a mixture of κ and λ.** In humans, the κ chains are more frequent than λ chains in immunoglobulin molecules.

Table 58–4. Properties of human immunoglobulin chains.*

Designation	H Chains					L Chains		Secretory Component	J Chain
	γ	α	μ	δ	ϵ	κ	λ	SC	J
Classes in which chains occur	IgG	IgA	IgM	IgD	IgE	All classes	All classes	IgA	IgA, IgM
Subclasses or sub-types	1,2,3,4	1,2	1,2	. . .	. . .	. . .	1,2,3,4	. . .	. . .
Allotypic variants	Gm(1)–(25)	A2m(1), (2)	. . .	. . .	. . .	Km(1)–(3)†	. . .	. . .	. . .
Molecular weight (approximate)	50,000‡	55,000	70,000	62,000	70,000	23,000	23,000	70,000	15,000
V region subgroups	V_HI–V_HIV					V_κI–V_κIV	V_λI–V_λVI		
Carbohydrate (average percentage)	4	10	15	18	18	0	0	16	8
Number of oligo-saccharides	1	2 or 3	5	?	5	0	0	?	1

*Reproduced, with permission, from Stites DP, Stobo JD, Wells JV (editors): *Basic & Clinical Immunology*, 6th ed. Appleton & Lange, 1987.
†Formerly Inv(1)–(3).
‡60,000 for γ3.

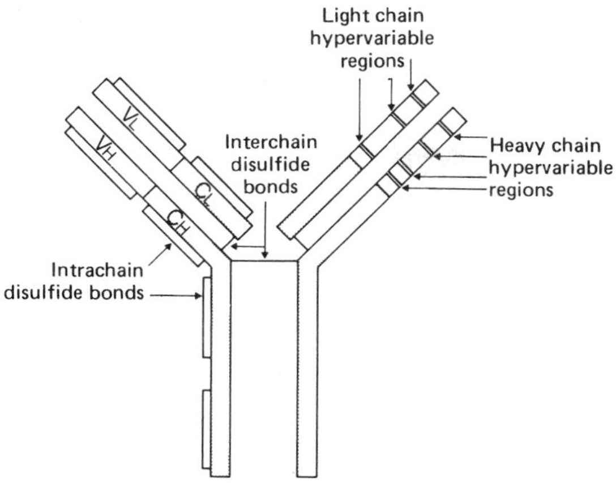

Figure 58–7. Schematic model of an IgG molecule showing approximate positions of the hypervariable regions in heavy and light chains. (Modified and reproduced, with permission, from Stites DP, Stobo JD, Wells JV [editors]: *Basic & Clinical Immunology,* 6th ed. Appleton & Lange, 1987.)

There Are 5 Types of Heavy Chains & These Determine to Which Class an Immunoglobulin Belongs

Five classes of H chains have been found in humans, and these classes can be distinguished by differences in their C_H regions (Table 58–4). The 5 classes of H chains are designated γ, α, μ, δ, and ϵ and vary in molecular weight from 50,000 to 70,000 (Table 58–4). The μ and ϵ chains each have four C_H domains rather than the usual 3. The type of H chain determines the class of immunoglobulin and thus its effector function. There are 5 immunoglobulin classes: **IgG, IgA, IgM,** **IgD,** and **IgE.** As shown in Table 58–4, many of the H chain classes can be further divided into subclasses on the basis of subtle structural differences in the C_H regions.

No Two Variable Regions Are Identical

The variable regions of immunoglobulin molecules consist of the V_L and V_H domains and are quite heterogeneous. In fact, no 2 variable regions from different humans have been found to have identical amino acid sequences. However, there are discernible patterns between the regions from different individuals, and these shared patterns have been divided into 3 main groups based on the degree of amino acid sequence homology. There is a V_κ group for kappa L chains, a V_λ group for lambda L chains, and a V_H group for the H chains. At higher resolution, there are even subgroups within each of these 3 groups.

Thus, within the variable regions there are some positions that are relatively invariable to account for the groups and subgroups. Upon comparing variable regions from different light chains of the same group or subgroup or different heavy chains from the same group or subgroup, it is apparent that there are **hypervariable regions** interspersed between the relatively invariable (subgroup-determining) positions (Fig 58–7). L chains have 3 hypervariable regions (in V_L), and H chains have 4 (in V_H).

The Constant Regions Determine Class-Specific Effector Functions

The constant regions of the immunoglobulin molecules, particularly the C_H2 and C_H3 (and C_H4 of IgM and IgE), which constitute the Fc fragment, are responsible for the class-specific effector functions of the different immunoglobulin molecules (Table 58–5, bottom part).

Table 58–5. Properties of human immunoglobulins.*

	IgG	IgA	IgM	IgD	IgE
H chain class	γ	α	μ	δ	ϵ
H chain subclass	$\gamma1, \gamma2, \gamma3, \gamma4$	$\alpha1, \alpha2$	$\mu1, \mu2$		
L chain type	κ and λ	κ and λ	κ and λ	κ and λ	κ and λ
Molecular formula	γ_2L_2	α_2L_2† or $(\alpha_2L_2)_2$SC§J‡	$(\alpha_2L_2)_5$J‡	δ_2L_2	ϵ_2L_2
Sedimentation coefficient (S)	6–7	7	19	7–8	8
Molecular weight (approximate)	150,000	160,000† 400,000**	900,000	180,000	190,000
Electrophoretic mobility (average)	γ	Fast γ to β	Fast γ to β	Fast γ	Fast γ
Complement fixation (classic)	+	0	++++	0	0
Serum concentration (approximate; mg/dL)	1000	200	120	3	0.05
Placental transfer	+	0	0	0	0
Reaginic activity	?	0	0	0	++++
Antibacterial lysis	+	+	+++	?	?
Antiviral activity	+	+++	+	?	?

*Reproduced, with permission, from Stites DP, Stobo JD, Wells JV (editors): *Basic & Clinical Immunology,* 6th ed. Appleton & Lange, 1987.
†For monomeric serum IgA.
‡J chain.
§Secretory component.
**For secretory IgA.

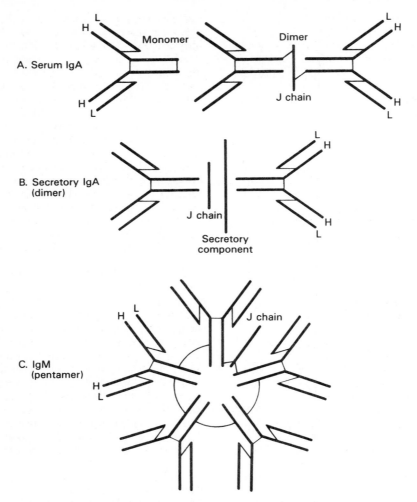

Figure 58–8. Highly schematic illustration of polymeric human immunoglobulins. Polypeptide chains are represented by thick lines; disulfide bonds linking different polypeptide chains are represented by thin lines. (Reproduced, with permission, from Stites DP, Stobo JD, Wells JV [editors]: *Basic & Clinical Immunology,* 6th ed. Appleton & Lange, 1987.)

Some immunoglobulins such as immune IgG exist only in the basic tetrameric structure, while others such as IgA and IgM can exist as higher-order polymers of 2,3 (IgA) or 5 (IgM) tetrameric units (Fig 58–8).

The L chains and H chains are synthesized as separate molecules and are subsequently assembled within the B cell or plasma cell into mature immunoglobulin molecules, all of which are **glycoproteins** (Table 58–4).

Both Light & Heavy Chains Are Products of Multiple Genes

Each immunoglobulin light chain is the product of at least 3 separate structural genes: a variable region (V_L) gene, a joining region (J) gene (bearing no relationship to the J chain of IgA or IgM), and a constant region (C_L) gene. Each heavy chain is the product of at least 4 different genes; a variable region (V_H) gene, a diversity region (D) gene, a joining region (J) gene, and a constant region (C_H) gene. Thus, the "one gene, one protein" concept is invalid. The molecular mechanisms responsible for the generation of the single immunoglobulin chains from multiple structural genes are discussed in Chapters 38 and 41.

Antibody Diversity Depends on Gene Rearrangements

Each person is capable of generating antibodies directed against perhaps 1 million different antigens. The generation of such immense **antibody diversity** appears to depend upon the **combinations of the various structural genes** contributing to the formation of each immunoglobulin chain and upon a high frequency of **somatic mutational events** in the rearranged V_H and V_L genes.

Class Switching Occurs During Immune Responses

In most humoral immune responses, antibodies with identical specificity but of different classes are generated in a specific chronologic order in response to the immunogen (immunizing antigen). A single type of immunoglobulin light chain can combine with an antigen-specific μ chain to generate a specific **IgM** molecule. Subsequently, the same antigen-specific light chain combines with a γ chain with an identical V_H region to generate an **IgG** molecule with antigen specificity identical to that of the original IgM molecule. Subsequently, the same light chain can combine with an α heavy chain, again containing the identical V_H region, to form an **IgA** molecule with identical antigen specificity. These 3 classes (IgM, IgG, and IgA) of immunoglobulin molecules against the same antigen have identical variable domains of both their light (V_L) chains and heavy (V_H) chains and are said to share an **idiotype.** The different class **isotypes** are determined by the C_H regions combined with the same antigen-specific V_H region. One aspect of the genetic regulatory mechanisms responsible for the switching of the C_H region gene is discussed in Chapter 41.

Both Over- & Underproduction of Immunoglobulins May Result in Disease States

Disorders of immunoglobulins include increased production of specific classes of immunoglobulins or even specific immunoglobulin molecules, the latter by clonal tumors of plasma cells called **myelomas. Hypogammaglobulinemia** may be restricted to a single class of immunoglobulin molecules (eg, IgA or IgG) or may involve underproduction of all classes of immunoglobulins (IgA, IgD, IgE, IgG, and IgM). The disorders of immunoglobulin levels are almost without exception due to disordered rates of immunoglobulin production or secretion, for which there can be many causes.

HEMOSTASIS INVOLVES BLOOD VESSELS, PLATELETS, & FACTORS THAT BOTH CLOT PLASMA & LYSE SUCH CLOTS

Hemostasis Has Four Phases

Hemostasis is the cessation of bleeding that follows interruption of vascular integrity. It encompasses **blood clotting (coagulation)** and involves blood vessels, platelets, and plasma proteins that cause both clotting and dissolution of clots.

There are four phases to hemostasis:

1. The first phase is **constriction** of the injured vessel to diminish blood flow distal to the injury.

2. The second phase consists of the formation of a loose and temporary **platelet plug** at the site of injury. Platelets bind to **collagen** at the site of vessel wall injury and are **activated by thrombin** (the mechanism of activation of platelets is described below), formed in the coagulation cascade at the same site, or by **ADP** released from other activated platelets. Upon activation, platelets change shape and, **in the presence of fibrinogen,** they **aggregate** to form the platelet plug.

3. The third phase of hemostasis is the formation of a **fibrin mesh** or clot that entraps the platelet plug (white thrombus) and/or red cells (red thrombus) forming a more stable thrombus.

4. The fourth phase is the partial or complete **dissolution of the clot by plasmin.** In normal hemostasis there is a dynamic steady state in which thrombi are constantly being formed and dissolved.

There Are 3 Types of Thrombi

Three types of thrombi or clots are distinguished.

1. The **white** thrombus is composed of platelets and fibrin and is relatively poor in erythrocytes. It forms at the site of an injury or abnormal vessel wall, particularly in areas where blood flow is rapid (arteries).

2. The second type of thrombus is the **red** thrombus which consists primarily of red cells and fibrin. The red thrombus morphologically resembles the clot formed in a test tube and may form in vivo in areas of retarded blood flow or stasis with or without vascular injury, or it may form at a site of injury or in an abnormal vessel in conjunction with an initiating platelet plug.

3. A third type of clot is a disseminated **fibrin deposit** in very small blood vessels or capillaries.

All 3 clots contain fibrin in variable proportions.

We shall first describe the coagulation pathway leading to the formation of fibrin. Later we shall briefly describe some aspects of the involvement of platelets and blood vessel walls in the overall process of coagulation. This separation of clotting factors and platelets is artificial, as both play intimate and often mutually interdependent roles in coagulation, but it facilitates description of the overall processes involved.

Both Intrinsic & Extrinsic Pathways Result in the Formation of Fibrin

Two pathways lead to fibrin clot formation, the **intrinsic** and the **extrinsic** pathways. The initiation of the **red thrombus** in an area of **restricted blood flow** or in response to an **abnormal vessel wall without tissue injury** is carried out by the **intrinsic pathway.** Initiation of the fibrin clot in response to **tissue injury** is carried out by the **extrinsic pathway.** These pathways converge in a **final common pathway** involving the activation of prothrombin to thrombin and the thrombin-catalyzed cleavage of fibrinogen to form the fibrin clot. The intrinsic, extrinsic, and final common pathways are complex and involve many different proteins (Fig 58–9, Table 58–6). In general, as shown in Table 58–7, these proteins can be classified into four types: (1) **zymogens** of serine-dependent proteases, which become activated during the process of coagulation; (2) **cofactors;** (3) **fibrinogen;** and (4) a **transglutaminase,** which stabilizes the fibrin clot.

The Intrinsic Pathway Leads to Activation of Factor X

The intrinsic pathway (Fig 58–9) involves factors XII, XI, IX, VIII, and X as well as prekallikrein, high-MW kininogen, Ca^{2+}, and platelet phospholipids. It

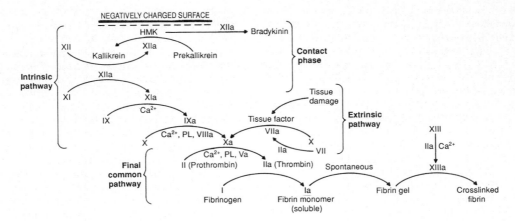

Figure 58–9. Intrinsic and extrinsic pathways of blood coagulation leading to an insoluble fibrin clot. HMK, high MW kininogen; PL, phospholipid.

results in the production of factor Xa (by convention activated clotting factors are referred to by use of the suffix a).

This pathway **commences with the "contact phase"** in which prekallikrein, high-MW kininogen, factor XII, and factor XI are exposed to a **negatively charged activating surface. Collagen** on the exposed surface of a blood vessel probably provides this site in vivo, whereas glass or kaolin can be used for in vitro tests of the intrinsic pathway. When the components of the contact phase assemble on the activating surface, **factor XII** is activated to **factor XIIa** upon proteolysis by **kallikrein.** This factor XIIa, generated by kallikrein, attacks prekallikrein to generate more kallikrein, setting up a reciprocal activation. Factor XIIa, once formed, activates **factor XI to XIa,** and also releases **bradykinin** (a nonapeptide with potent vasodilator action) from high MW kininogen.

Factor XIa in the presence of Ca^{2+} activates **factor IX** (MW 55,000, a zymogen containing the vitamin K-dependent γ-carboxyglutamate [Gla] residues), to the serine protease, **factor IXa.** This in turn cleaves an Arg-Ile bond in **factor X** (MW 56,000) to produce the 2-chain serine protease, **factor Xa.** This latter reaction requires the assembly of components, called **the tenase complex, on the surface of activated platelets: Ca^{2+} and factor VIIIa, as well as factors IXa and X.** It should be noted that in all reactions involving the **Gla-containing zymogens** (factors II, VII, IX, and X), the **Gla residues** in the amino terminal regions of the molecules serve as **high-affinity binding sites for Ca^{2+}**. For the assembly of the tenase complex the platelets must first be activated to expose the **acidic**

Table 58–6. Numerical system for nomenclature of blood clotting factors. The numbers indicate the order in which the factors have been discovered and bear no relationship to the order in which they act.

Factor	Common Name(s)
I	Fibrinogen } These factors are usually referred
II	Prothrombin } to by their common names.
III	Platelet phospholipid } These are usually not re-
IV	Ca^{2+} } ferred to as coagulation factors.
V	Proaccelerin, labile factor, accelerator (Ac-) globulin
VII	Proconvertin, serum prothrombin conversion accelerator (SPCA), cothromboplastin
VIII	Antihemophilic factor A, antihemophilic globulin (AHG)
IX	Antihemophilic factor B, Christmas factor, plasma thromboplastin component (PTC)
X	Stuart-Prower factor
XI	Plasma thromboplastin antecedent (PTA)
XII	Hageman factor
XIII	Fibrin stabilizing factor (FSF), fibrinoligase

Table 58–7. The functions of the blood coagulation factors.

1. Zymogens of serine proteases

Factor XII	Binds to exposed collagen at site of vessel wall injury; activated by high MW kininogen and kallikrein.
Factor XI	Activated by factor XIIa.
Factor IX	Activated by factor XIa in presence of Ca^{2+}.
Factor VII	Activated by thrombin in presence of Ca^{2+}.
Factor X	Activated on surface of activated platelets by tenase complex (Ca^{2+}, factors VIIIa and IXa) and by factor VIIa in presence of tissue factor and Ca^{2+}.
Factor II	Activated on surface of activated platelets by prothrombinase complex (Ca^{2+}, factors Va and Xa). [Factors II, VII, IX and X are Gla-containing zymogens.]

2. Cofactors

Factor VIII	Activated by thrombin; factor VIIIa is a cofactor in the activation of factor X by factor IXa.
Factor V	Activated by thrombin; factor Va is a cofactor in the activation of prothrombin by factor Xa.

3. Fibrinogen

Factor I	Cleaved by thrombin to form fibrin clot.

4. Thiol-dependent transglutaminase

Factor XIII	Activated by thrombin in presence of Ca^{2+}; stabilizes fibrin clot by covalent crosslinking.

(anionic) phospholipids, phosphatidyl serine and phosphatidyl inositol, that are normally on the internal side of the plasma membrane of resting, nonactivated platelets. **Factor VIII** (MW 330,000), a glycoprotein, is not a protease precursor, but is a cofactor that serves as a receptor for factors IXa and X on the platelet surface. Factor VIII is activated by minute quantities of thrombin to form **factor VIIIa,** which is in turn inactivated upon further cleavage by thrombin.

The Extrinsic Pathway Also Leads to Activation of Factor X But by a Different Mechanism

Factor Xa occurs at the site where the intrinsic and extrinsic pathways converge (see Fig 58–9) and lead into the final common pathway of blood coagulation. The extrinsic pathway involves tissue factor, factor VII, X, and Ca²⁺ and results in the production of factor Xa. It is initiated at the site of tissue injury with the release of **tissue factor** (abundant in placenta, lung, and brain) (Fig 58–9), which acts as a cofactor in the **factor VIIa-catalyzed activation of factor X.** Factor VIIa cleaves the same Arg-Ile bond in factor X that is cleaved by the tenase complex of the intrinsic pathway. **Factor VII** (MW 53,000), a circulating Gla-containing glycoprotein synthesized in the liver, is a zymogen, but it has rather a high endogenous activity; this activity is increased by conversion to the active serine protease, factor VIIa, by **thrombin or factor Xa.** Activation of factor X provides an important link between the intrinsic and extrinsic pathways.

The Final Common Pathway of Blood Clotting Involves Activation of Prothrombin to Thrombin

In the final common pathway, **factor Xa,** produced by either the intrinsic or the extrinsic pathway, activates **prothrombin** (factor II) to **thrombin** (factor IIa), which then converts fibrinogen to fibrin (Fig 58–9).

The activation of prothrombin, like that of factor X, occurs on the surface of **activated platelets,** and requires the assembly of a **prothrombinase complex,** consisting of **platelet anionic phospholipids, Ca²⁺, factor Va, factor Xa, and prothrombin.**

Factor V (MW 330,000) a glycoprotein with homology to factor VIII and ceruloplasmin, is synthe-sized in the liver, spleen, and kidney and is found in platelets as well as in plasma. It functions as a cofactor in a manner similar to that of factor VIII in the tenase complex. When activated to **factor Va** by traces of **thrombin,** it binds to specific receptors on the platelet membrane (Fig 58–10) and forms a complex with **factor Xa and prothrombin.** It is subsequently inactivated by further action of thrombin, thereby providing a means of limiting the activation of prothrombin to thrombin. **Prothrombin** (MW 72,000; Fig 58–11) is a single-chain glycoprotein synthesized in the liver. The amino-terminal region of prothrombin (1 in Fig 58–11) contains 10 Gla residues, and the serine-dependent active protease site (indicated by the arrowhead) is in the carboxy-terminal region of the molecule. Upon binding to the complex of factors Va and Xa on the platelet membrane, prothrombin is cleaved by factor Xa at two sites to generate the active, 2-chain thrombin molecule, which is then released from the platelet sur-face. The A and B chains of thrombin are held together by a disulfide bond.

Prothrombin can also be activated by **staphylocoag-ulase** as the result of a simple conformational altera-tion not involving cleavage of the molecule.

The Conversion of Fibrinogen to Fibrin Is Catalyzed by Thrombin

Fibrinogen (factor I, MW 340,000; see Figs 58–9 and 58–12 and Table 58–7) is a soluble plasma glycoprotein, 47.5 nm in length that consists of 3 non-identical pairs of polypeptide chains $(A\alpha B\beta,\gamma)_2$ covalently linked by disulfide bonds. The Bβ and γ chains contain asparagine-linked complex oligosac-charides. All 3 chains are synthesized in the liver; the 3 structural genes involved are on the same chromosome and their expression is coordinately regulated in hu-mans. The amino-terminal regions of the 6 chains are held in close proximity by a number of disulfide bonds, while the carboxy-terminal regions are spread apart, giving rise to a highly asymmetrical, elongated molecule (Fig 58–12). The A and B portions of the Aα and Bβ chains, designated **fibrinopeptides A (FPA) and B (FPB),** respectively, at the amino-terminal ends of the chains, bear excess negative charges as a result of the presence of aspartate and glutamate residues, as well as an unusual tyrosine O-sulfate in FPB. **These**

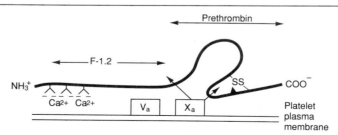

Figure 58–10. Diagrammatic representation of the binding of factors V_a, X_a, Ca²⁺, and prothrombin to the plasma membrane of the activated platelet. The sites of cleavage of prothrombin by factor X_a are indicated by 2 arrows. The part of prothrombin destined to form thrombin is labelled prethrombin.

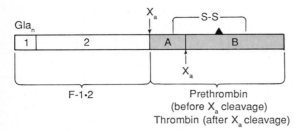

Figure 58–11. Diagrammatic representation of prothrombin. The amino-terminus is to the left; region 1 contains all 10 Gla residues. The sites of cleavage by factor X_a are shown and the products named. The site of the catalytically active serine residue is indicated by ▲. The A and B chains of active thrombin (shaded) are held together by the disulfide bridge.

negative charges contribute to the solubility of fibrinogen in plasma and also serve to prevent aggregation by causing electrostatic repulsion between fibrinogen molecules.

Thrombin (MW 34,000) is a serine protease present in plasma that hydrolyzes the 4 Arg-Gly bonds between the fibrinopeptides and the α and β portions of the Aα and Bβ chains of fibrinogen (Fig 58–13A). The release of the fibrinopeptides by thrombin generates fibrin monomer, which has the subunit structure $(\alpha,\beta,\gamma)_2$. Since FPA and FPB contain only 16 and 14 residues, respectively, the fibrin molecule retains 98% of the residues present in fibrinogen. The removal of the fibrinopeptides exposes binding sites that allow the molecules of fibrin monomers to aggregate spontaneously in a regularly staggered array forming an insoluble fibrin clot. It is the formation of this insoluble fibrin polymer that traps platelets, red cells, and other components to form the white or red thrombi. This initial fibrin clot is rather weak, held together only by the noncovalent association of fibrin monomers.

Thrombin, in addition to converting fibrinogen to fibrin, also **converts factor XIII to factor XIIIa.** This factor is a highly specific **transglutaminase** that covalently crosslinks fibrin molecules by forming peptide bonds between the γ-carboxyl groups of glutamine and ϵ-amino groups of lysine (Fig 58–13B),

yielding a more stable fibrin clot with increased resistance to proteolysis.

The Concentration of Circulating Thrombin Must Be Carefully Controlled or Disastrous Clots May Occur

In normal hemostasis the concentration of active thrombin must be carefully controlled to prevent the formation of potentially catastrophic clots. This is achieved in 2 ways. Thrombin circulates as its inactive precursor, **prothrombin,** which is activated as the result of a **cascade of enzymatic reactions,** each converting an **inactive zymogen to an active enzyme** and leading finally to the conversion of prothrombin to thrombin (Fig 58–9). At each point in the cascade, feedback mechanisms produce a delicate balance of activation and inhibition. The concentration of factor XII in plasma is approximately 30 μg/mL, while that of fibrinogen is 3 mg/mL, with intermediate clotting factors increasing in concentration as one proceeds down the cascade, showing that the clotting cascade provides **amplification.** The second means of controlling thrombin activity is the inactivation of any thrombin formed by **circulating inhibitors,** the most important of which is **antithrombin III** (see below).

The Activity of Antithrombin III, an Important Inhibitor of Thrombin, Is Increased by the Anticoagulant Heparin

Four naturally occurring thrombin inhibitors exist in normal plasma. The most important is **antithrombin III,** which contributes approximately 75% of the antithrombin activity. Antithrombin III can also inhibit the activities of factors IXa, Xa, XIa, and XIIa. α_2**-Macroglobulin** contributes most of the remainder of the antithrombin activity, with **heparin cofactor II** and α_1**-antitrypsin** (α_1-antiproteinase) acting as minor inhibitors under physiologic conditions.

The endogenous activity of antithrombin III is greatly potentiated by the presence of acidic proteoglycans such as **heparin** (see Chapter 57). These bind to a specific cationic site of antithrombin III, inducing a conformational change and promoting its binding to thrombin as well as to its other substrates. This is the

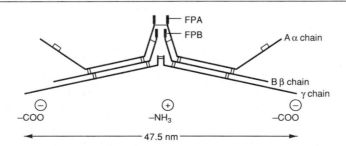

Figure 58–12. Diagrammatic representation (not to scale) of fibrinogen showing pairs of Aα, Bβ, and γ chains linked by disulfide bonds. FPA Fibrinopeptide A, FPB Fibrinopeptide B.

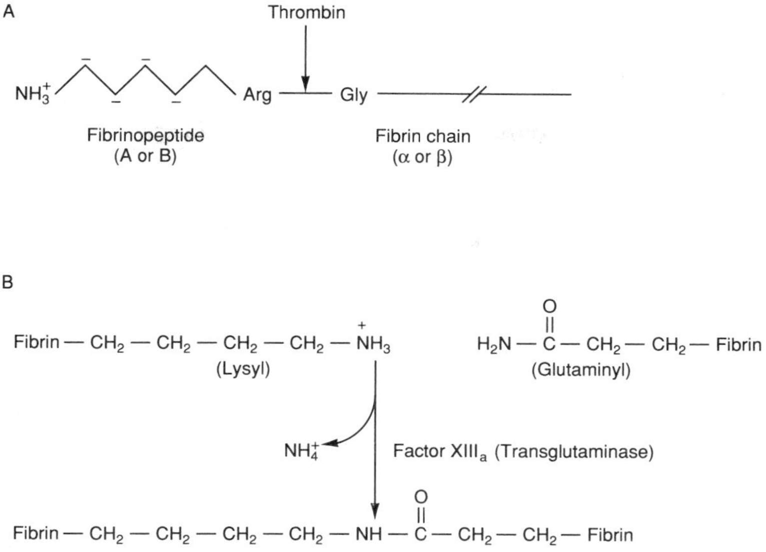

Figure 58–13. Formation of a fibrin clot. A. Thrombin-induced cleavage of Arg-Gly bonds of the Aα and Bβ chains of fibrinogen to produce fibrinopeptides (left-hand side) and the α and β chains of fibrin monomer (right-hand side). B. Crosslinking of fibrin molecules by activated factor XIII (factor XIIIa).

basis for the use of heparin in clinical medicine to inhibit clotting. In addition, heparin in low doses appears to coat the endothelial lining of blood vessels, perhaps thereby reducing activation of the intrinsic pathway. The anticoagulant effects of heparin can be antagonized by strongly cationic polypeptides such as **protamine,** which bind strongly to heparin, thus inhibiting its binding to antithrombin III. Individuals with **inherited deficiencies** of antithrombin III are prone to develop frequent and widespread clots, providing evidence that antithrombin III has a physiologic function and that the clotting system in humans is normally in a dynamic state.

The Coumarin Anticoagulants Inhibit the Vitamin K-Dependent Carboxylation of Factors II, VII, IX, & X

The coumarin drugs (eg, dicoumarol), which are used as anticoagulants, inhibit the vitamin K-dependent carboxylation of Glu to Gla residues in the amino-terminal regions of factors II, VII, IX, and X. These factors, all of which are synthesized in the liver, are dependent upon the Ca^{2+}-binding properties of the Gla residues for their normal function in the coagulation pathways. The coumarins act by inhibiting the reduction of the quinone derivatives of vitamin K to the active hydroquinone forms. Thus, the administration of vitamin K will bypass the coumarin-induced inhibition and allow maturation of the Gla-containing factors. Reversal of coumarin inhibition by vitamin K requires 12–24 hours, whereas reversal of the anticoagulant effects of heparin by protamine is almost instantaneous.

Hemophilia A Is Due to a Genetically Determined Deficiency of Factor VIII

A number of inherited deficiencies of the clotting system are found in humans. The most common deficiency is that of **factor VIII, causing hemophilia A,** an X-chromosome-linked disease, that has played a major role in the history of the royal families of Europe. **Hemophilia B is due to a deficiency of factor IX;** it has an almost identical clinical picture to that of hemophilia A, but the two conditions can be separated on the basis of specific assays that distinguish between the two factors.

The **gene** for human factor VIII has been cloned, and is one of the largest so far studied, measuring 186 kb in size and containing 26 exons. A variety of **lesions** have been detected leading to diminished activity of factor VIII; these include partial gene deletions and point mutations resulting in premature chain termination. **Prenatal diagnosis by** DNA analysis after chorionic villus sampling is now possible.

In recent years, **treatment** of patients with hemophilia A has consisted of administration of cryoprecipitates (enriched in factor VIII) prepared from individual donors or lyophilized factor VIII concentrates prepared from plasma pools of up to 5000 donors. It is hoped to produce in the near future sufficient factor VIII by **recombinant DNA technology** to treat subjects with hemophilia A. This has become important, since a number of patients with hemophilia A have developed AIDS subsequent to replacement therapy with lyophilized factor VIII concentrates prepared from pooled plasma; factor VIII prepared by recombinant DNA technology would not be associated with this problem.

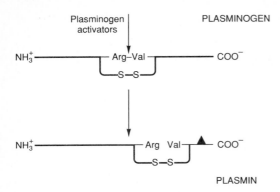

Figure 58–14. Activation of plasminogen. The same Arg-Val bond is cleaved by all plasminogen activators to give the 2-chain plasmin molecule. ▲ indicates the serine residue of the active site. The 2 chains of plasmin are held together by a disulfide bridge.

Fibrin Clots Are Dissolved by Plasmin

As stated above, the coagulation system is normally in a state of dynamic equilibrium in which fibrin clots are constantly being laid down and dissolved. **Plasmin,** the serine protease mainly responsible for degrading fibrin and fibrinogen, circulates in the form of its inactive zymogen, **plasminogen** (MW 90,000), and any small amounts of plasmin that are formed in the fluid phase under physiologic conditions are rapidly inactivated by the fast-acting plasmin inhibitor, **α_2-antiplasmin.** Plasminogen binds to both fibrinogen and fibrin and thus becomes incorporated in clots as they are produced; since plasmin that is formed when bound to fibrin is protected from α_2-antiplasmin, it remains active. Activators of plasminogen of various types are found in most body tissues and all cleave the same Arg-Val bond in plasminogen to produce the 2-chain serine protease, plasmin (Fig 58–14).

Tissue plasminogen activator (TPA) is a serine protease that is released into the circulation from vascular endothelium under conditions of injury or stress and is catalytically inactive unless bound to **fibrin.** Upon binding to fibrin, TPA cleaves plasminogen within the clot to generate plasmin, which in turn digests the fibrin to form soluble degradation products and thus dissolves the clot. Neither plasmin nor the plasminogen activator can remain bound to these degradation products, and so they are released into the fluid phase where they are inactivated by their natural inhibitors. **Prourokinase** is the precursor of a second activator of plasminogen, **urokinase,** which does not display the same high degree of selectivity for fibrin. Urokinase, which is secreted by certain epithelial cells lining excretory ducts (eg, renal tubules), is probably involved in lysing any fibrin that is deposited in such ducts.

TPA Synthesized by Recombinant DNA Technology Is Under Worldwide Investigation for the Treatment of Coronary Thrombosis

At present a different activator of plasminogen, **streptokinase,** is used therapeutically as a fibrinolytic agent. However, it is less selective than TPA, activating plasminogen in the fluid phase (where it can degrade circulating **fibrinogen**) as well as plasminogen that is bound to a fibrin clot. The amount of plasmin produced by therapeutic doses of streptokinase may exceed the capacity of the circulating α_2-antiplasmin, causing fibrinogen as well as fibrin to be degraded and resulting in the bleeding often encountered during fibrinolytic therapy. Because of its **selectivity for degrading fibrin,** there is considerable therapeutic interest in the use of TPA, produced by recombinant DNA technology, to restore the patency of coronary arteries following thrombosis. If administered early enough, before irreversible damage of heart muscle occurs, TPA could significantly reduce the mortality from myocardial damage following coronary thrombosis. So far, in various clinical trials, TPA has been found to cause some bleeding and it remains to be determined whether it will prove superior to the much cheaper streptokinase in the treatment of acute coronary thrombosis.

There are a number of disorders, including cancer and shock, in which the concentrations of **plasminogen activators** increase. In addition, the **antiplasmin activities** contributed by α_1-antitrypsin and α_2-antiplasmin may be impaired in diseases such as cirrhosis of the liver. Since certain bacterial products, such as streptokinase, are capable of activating plasminogen, they may be responsible for the **diffuse hemorrhage** sometimes observed in patients with disseminated bacterial infections.

Activation of Platelets Involves Stimulation of the Polyphosphoinositide Pathway

Platelets must undergo 3 processes in order for hemostasis to occur: (1) **adhesion** to exposed collagen in blood vessels (2) **release** of the contents of their granules, and (3) **aggregation.**

Adhesion of platelets to collagen is mediated by **von Willebrand's factor,** a glycoprotein secreted by endothelial cells into plasma. This protein acts as a secure bridge between a glycoprotein on platelet surfaces (Gplb) and collagen fibrils in the walls of blood vessels. It thus prevents platelets from being detached from vessel walls by the shearing forces developed in blood vessels.

Platelet Activation: Normal platelets are in the unstimulated state. During coagulation, involved platelets become activated. **Activation** is a complex phenomenon, which includes **changes in platelet shape, increased movement, liberation of contents of their granules,** and **aggregation. Thrombin,** formed from the coagulation cascade, initiates platelet activation in

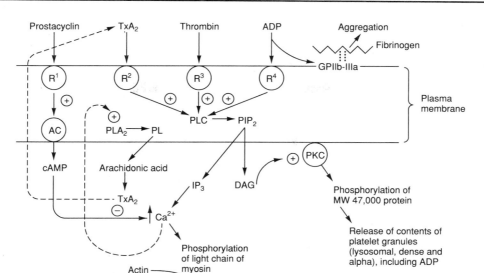

Figure 58–15. Diagrammatic representation of platelet activation. The external environment, the plasma membrane, and the inside of a platelet are depicted from top to bottom. GP, glycoprotein; R^1, R^2, R^3, R^4, various receptors; AC, adenylate cyclase; PLA_2, phospholipase A_2; PL, phospholipids; PLC, phospholipase C; PIP_2, phosphatidyl inositol 4,5-bisphosphate; cAMP, cyclic AMP; PKC, protein kinase C; TxA_2, thromboxane A_2; IP_3, inositol triphosphate; DAG, diacylglycerol.

vivo by interacting with its receptor on the plasma membrane (Fig 58–15). The further events leading to platelet activation are an example of **transmembrane signaling,** in which a chemical messenger outside the cell generates effector molecules inside the cell. In this instance, thrombin acts as the external chemical messenger (stimulus or agonist). The interaction of thrombin with its receptor stimulates the activity of a **phospholipase C** in the plasma membrane. This enzyme hydrolyzes **phosphatidyl inositol 4,5-bisphosphate** (PIP_2, a polyphosphoinositide) to form the two internal effector molecules, **diacylglycerol** (DAG) and **inositol triphosphate** (IP_3).

Hydrolysis of PIP_2 is also involved in the action of many hormones (see Chapter 45) and drugs. DAG stimulates **protein kinase C,** which phosphorylates **a platelet protein** of MW 47,000. Phosphorylation of this protein results in **release of the contents of** the various types of platelet granules (lysosomes, dense granules, and alpha granules). **ADP** released from dense granules can also activate platelets, resulting in activation of additional platelets. In addition, ADP modifies the platelet surface so that **fibrinogen** (formed in the coagulation cascade) will attach to a complex of two glycoproteins (GPIIb and GP IIIa) on the platelet surface. Molecules of fibrinogen then link adjacent platelets to each other, forming a platelet **aggregate.** IP_3 attracts Ca^{2+} into the cytosol and interacts with calmodulin and myosin light chain kinase, leading to **phosphorylation of the light chains of**

myosin. These chains then interact with **actin,** causing movement and changes of platelet shape.

Activation of a platelet **phospholipase A_2** by increased levels of Ca^{2+} results in liberation of **arachidonic acid** from platelet phospholipids, leading to the formation of **thromboxane A_2** (Chapter 25), which in turn can further activate phospholipase C, promoting platelet aggregation. Activated platelets, besides forming a platelet plug, are required, via their phospholipids, for the activation of factors X and II in the coagulation cascade (see Fig 58–9).

Blood Vessels Synthesize Prostacyclin & Other Compounds That Affect Clotting & Thrombosis

The **endothelial cells** in the walls of blood vessels make important contributions to the overall regulation of clotting and thrombosis. As described in Chapter 25, these cells **synthesize prostacyclins** (PGI_2), which are potent **inhibitors of platelet aggregation,** opposing the action of thromboxanes. Prostacyclins probably act by stimulating the activity of **adenylate cyclase** in the surface membranes of platelets. The resulting increase of intraplatelet cAMP opposes the increase in the level of intracellular Ca^{2+} produced by IP_3 and thus inhibits platelet activation (Fig 58–15). Endothelial cells play other roles in the regulation of thrombosis. For instance, these cells **metabolize ADP,** which opposes its aggregating effect on platelets. In addition, these cells appear to **synthesize**

Table 58–8. The major features of hemostasis and coagulation.

---

*Blood vessels, platelets, and clotting factors are all involved.

*The intrinsic pathway of coagulation is activated by collagen exposed on the surface of blood vessels.

*The extrinsic pathway is activated by tissue factor released by tissue damage.

*The intrinsic and extrinsic pathways both lead to production of factor Xa and then share a final common pathway resulting in the production of fibrin from fibrinogen by the action of thrombin.

*Many clotting factors exist as zymogens of serine proteases; these zymogens are activated by proteolysis.

*Genetically determined absences or alterations of clotting factors occur, causing various hemorrhagic diseases such as hemophilia A.

*Inhibitors of clotting factors (eg, antithrombin III) occur in plasma and help regulate coagulation.

*A number of the key reactions of coagulation occur on the surface of platelets; platelets are activated by the action of thrombin, collagen, and ADP, and their aggregation depends upon interaction with fibrinogen.

*The overall process of coagulation constitutes a cascade, in which considerable amplification occurs.

*For factors II, VII, IX, and X to be active, a vitamin K-dependent carboxylation of certain of their glutamate residues to form γ-carboxyglutamate (Gla) residues must occur; these Gla residues bind Ca^{2+}, an important participant in coagulation.

*Blood vessels synthesize prostacyclin, which inhibits platelet activation and aggregation; blood vessels also synthesize other compounds that play a role in the regulation of thrombosis.

*Clots are formed of fibrin, which is strengthened by crosslinking reactions catalyzed by a transglutaminase; fibrin clots are dissolved by the action of plasmin produced from plasminogen.

---

heparan sulfate, which binds some clotting factors, and they also synthesize plasminogen activators, which may help dissolve clots.

Analysis of the mechanisms of uptake of atherogenic lipoproteins, such as LDL, by endothelial, smooth muscle, and monocytic cells of arteries, along with detailed studies of how these lipoproteins damage such cells is a key area of study in elucidating the mechanisms of atherosclerosis (cf Chapter 28).

LABORATORY TESTS OF BLOOD COAGULATION ARE AVAILABLE

A variety of laboratory tests are available to measure the 4 phases of hemostasis described above. They include platelet count, bleeding time, partial thromboplastin time (PTT), prothrombin time (PT), thrombin time, concentration of fibrinogen, fibrin clot stability, and measurement of fibrin degradation products. The reader is referred to a textbook of physiology or hematology for a discussion of these tests.

The major features of hemostasis and coagulation are summarized in Table 58–8.

REFERENCES

Bloom AL, Thomas DP (editors): *Haemostasis and Thrombosis,* 2nd ed. Churchill Livingstone, 1987.

Colman RW, Hirsh J, Marder VJ, Salzman EW (editors): *Hemostasis and Thrombosis,* 2nd ed. JB Lippincott Co, 1987.

Dugaiczyk A, Law SW, Dennison OE: Nucleotide sequence and the encoded amino acids of human serum albumin mRNA. *Proc Natl Acad Sci USA* 1982;**79**:71.

Gitschier J et al: Characterization of the human factor VIII gene. *Nature (London)* 1984;**312**:326.

Handin RI. Chapter 54, Bleeding and thrombosis, p 266, in: *Harrison's Principles of Internal Medicine* 11th ed. Braunwald, E, Isselbacher, KJ, Petersdorf RG et al (editors). McGraw-Hill, 1987.

Mariani G (editor): *Pathophysiology of Plasma Protein Metabolism.* Plenum Press, 1984.

McKee PA: Hemostasis and disorders of blood coagulation, in: *The Metabolic Basis of Inherited Disease,* 5th ed. Stanbury JB et al (editors). McGraw-Hill, 1983.

Reed RG, Peters T Jr. Chapter 14, Plasma proteins, pp 435–464, in: *Clinical Biochemistry Reviews,* Vol 3. Goldberg DM (editor). John Wiley & Sons, 1982.

Ruffner DE, and Dugaiczyk A: Splicing mutation in human hereditary analbuminemia. *Proc Natl Acad Sci USA* 1988; **85**:2125.

Stamatoyannopoulos G, Nienhuis AW, Leder P, Majerus PW (editors): *The Molecular Basis of Blood Diseases.* WB Saunders, 1987.

Stites DP, Stobo JD, Wells JV: *Basic & Clinical Immunology,* 6th ed. Appleton & Lange, 1987.

Vogel F, Motulsky AG. *Human Genetics,* 2nd ed. Springer-Verlag, 1986.

Contractile & Structural Proteins

59

Victor W Rodwell, PhD, Robert K Murray, MD, PhD, & Frederick W Keeley, PhD*

INTRODUCTION

Protein molecules may serve functions other than catalysis. The regulatory, signal transmission, and recognition functions of protein molecules have been described in earlier chapters. Protein molecules also provide important **transducing** and **structural** functions. Some of these latter roles, which depend upon the fibrous nature of specific protein molecules, are reviewed in this chapter.

BIOMEDICAL IMPORTANCE

Our understanding of the molecular basis of major genetic diseases of structural and contractile proteins received significant impetus in 1986 with the successful cloning of the gene for **Duchenne-type muscular dystrophy,** an achievement that holds high promise for accurate diagnosis and possible therapy for this disease (see Case no 6, Chapter 62). While many molecular diseases of proteins arise from mutations in the structural genes that code for that protein (eg, hemoglobin), proteins subject to posttranslational modification present additional sites for genetic deficiency diseases. For example, **several human diseases result from genetic defects in processing of precursor forms of collagen** (osteogenesis imperfecta, Ehlers-Danlos syndrome and Marfan's syndrome). Defects in collagen may also arise when posttranslational modification is inhibited by the lack of a cofactor such as ascorbic acid (vitamin C), which is required for hydroxylation of peptide-bound prolyl and lysyl residues to hydroxyproline and hydroxylysine. The clinical presentation of weakened connective tissue typical of classical **scurvy** can be explained by such a deficiency.

MUSCLE TRANSDUCES CHEMICAL ENERGY INTO MECHANICAL ENERGY

Muscle is the major biochemical transducer (machine) that converts potential (chemical) energy into kinetic (mechanical) energy. Muscle, the largest single tissue in the human body, comprises somewhat less than 25% of body mass at birth, more than 40% in the young adult, and somewhat less than 30% in the aged adult.

An effective **chemical-mechanical transducer** must meet several requirements: (1) There must exist a constant supply of chemical energy. In vertebrate muscle, ATP and creatine phosphate supply chemical energy. (2) There must be a means of regulating the mechanical activity—ie, the speed, duration, and force of contraction in the case of muscle. (3) The machine must be connected to an operator, a requirement met in biologic systems by the nervous system. (4) There must be a way of returning the machine to its original state.

Muscle is a pulling, not a pushing machine. Therefore, a given muscle must be antagonized by another group of muscles or another force such as gravity or elastic recoil.

In vertebrates, the above requirements and the specific needs of the organisms are met by 3 types of muscles: skeletal muscle, cardiac muscle, and smooth muscle. Both **skeletal** and **cardiac muscle** appear **striated** upon microscopic observation; **smooth muscle** is **nonstriated.** Although skeletal muscle is under **voluntary** nervous control, the control of both cardiac and smooth muscle is **involuntary.**

The Sarcolemma of Muscle Cells Contains ATP, Phosphocreatine, & Glycolytic Enzymes

Striated muscle is composed of multinucleated muscle fiber cells surrounded by an electrically excitable membrane, the **sarcolemma.** An individual muscle fiber cell, which may extend the entire length of the muscle, contains a bundle of many **myofibrils** arranged in parallel, embedded in intracellular fluid termed **sarcoplasm.** Within this fluid is contained glycogen, the high-energy compounds ATP and phosphocreatine, and the enzymes of glycolysis.

The Sarcomere Is the Functional Unit of Muscle

The **sarcomere** is repeated along the axis of a fibril at distances of 1500–2300 nm (Fig 59–1). When the myofibril is examined by electron microscopy, alter-

*Research Institute, Hospital for Sick Children, Toronto, and Department of Biochemistry, University of Toronto

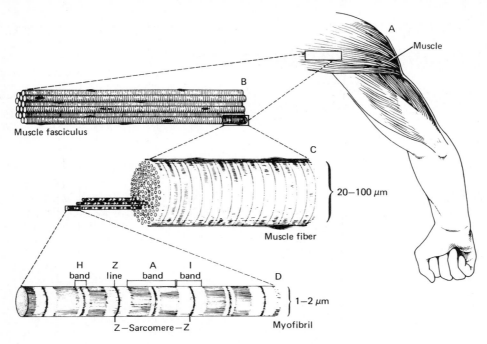

Figure 59–1. The structure of voluntary muscle. (Drawing by Sylvia Colard Keene. Reproduced, with permission, from Bloom W, Fawcett DW: *A Textbook of Histology,* 10th ed. Saunders, 1975.)

nating dark and light bands (A bands and I bands) can be observed. The central region of the A band (the H zone) appears less dense than the rest of the band. The I band is bisected by a very dense and narrow Z line (Fig 59–2).

The striated appearance of voluntary and cardiac muscles in light microscopic studies results from their high degree of organization in which most muscle fiber cells are aligned so that their sarcomeres are in parallel register (Fig 59–1).

The Protein of Thick Filaments Is Myosin

When **cross sections** of a myofibril are examined in an electron micrograph, it appears that each myofibril is constructed of 2 types of longitudinal filaments. One type (the thick filament), confined to the A band, contains chiefly the protein **myosin.** These filaments are about 16 nm in diameter and arranged in cross section as a hexagonal array (Fig 59–2).

Thin Filaments Contain Actin, Tropomyosin, & Troponin

The other filament (thin filament) lies in the I band and extends also into the A band but not into the H zone of the A band (Fig 59–2). The thin filaments contain the proteins **actin, tropomyosin,** and **troponin.** In the

A band, the thin filaments are arranged around the thick (myosin) filament as a secondary hexagonal array. Each thin filament lies symmetrically between 3 thick filaments, and each thick filament is surrounded symmetrically by 6 thin filaments (Fig 59–2).

The thick and thin filaments interact via cross-bridges that emerge at intervals of 14 nm along the thick filaments. As depicted in Fig 59–2, the cross-bridges or "arrowheads" on the thick filaments have opposite polarities at the 2 ends of the filaments. The 2 poles of the filaments are separated by a 150-nm segment (the M band) that is free of projections.

Interdigitating Filaments Slide Past One Another During Muscle Contraction

When muscle contracts, there is no change in the lengths of the thick filaments or of the thin filaments, but the **H zone and the I bands shorten.** Thus, the **arrays of interdigitating filaments must slide past one another during muscle contraction.** The **cross-bridges generate and sustain the tension.** The tension developed during muscle contraction is proportionate to the filament overlap and thereby the number of cross-bridges. Each cross-bridge head is connected to the thick filament via a flexible fibrous segment that can bend outward from the thick filament to accommodate the interfilament spacing.

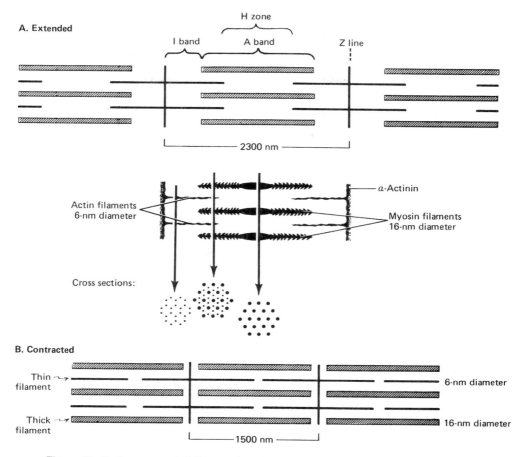

Figure 59–2. Arrangement of filaments in striated muscle. *A:* Extended. *B:* Contracted.

ACTIN & MYOSIN ARE MAJOR MUSCLE PROTEINS

The mass of a fresh muscle fibril is made up of 75% water and more than 20% protein. The 2 major muscle proteins are actin and myosin.

Monomeric (globular) actin (G-actin) is a 43,000-MW globular protein that comprises 25% of muscle protein by weight. At physiologic ionic strength and in the presence of magnesium, G-actin **polymerizes** noncovalently to form an insoluble double helical filament called F-actin (Fig 59–3). The **F-actin** fiber is 6–7 nm thick and has a pitch or repeating structure every 35.5 nm. Neither G- nor F-actin exhibits any catalytic activity.

Four Additional Proteins Perform Key Functions

In striated muscle, there are 4 other proteins that are minor in terms of their mass contribution but important in terms of their function. **Tropomyosin** is a fibrous molecule that consists of 2 chains, alpha and beta, that attach to the F-actin in the groove between the 2 polymers (Fig 59–3). **Tropomyosin is present in all mus-** cle and musclelike structures. The **troponin** system is unique to **striated muscle** and consists of 3 separate proteins. **Troponin T (TpT)** binds to tropomyosin as well as to the other 2 troponin components (Fig 59–3). **Troponin I (TpI)** inhibits the F-actin-myosin interaction and also binds to the other components of troponin. **Troponin C (TpC)** is a calcium-binding protein that has a primary and secondary structure as well as a function quite analogous to **calmodulin,** a protein widely spread in nature. Four molecules of calcium ion are bound per molecule of troponin C or calmodulin, and both protein molecules have a molecular weight of 17,000. The thin filament of striated muscle consists of F-actin, tropomyosin, and the 3 components of troponin: TpC, TpI, and TpT (Fig 59–3). The repeat distance of the tropomyosin and troponin system is 38.5 nm.

Myosin contributes 55% of muscle protein by weight and forms the thick filaments. Myosin is an asymmetric hexamer with a molecular weight of 460,000. The myosin has a **fibrous portion** consisting of 2 intertwined helices, each with a **globular head** portion attached at one end (Fig 59–4). The **hexamer** consists of one pair of heavy chains (MW 200,000) and

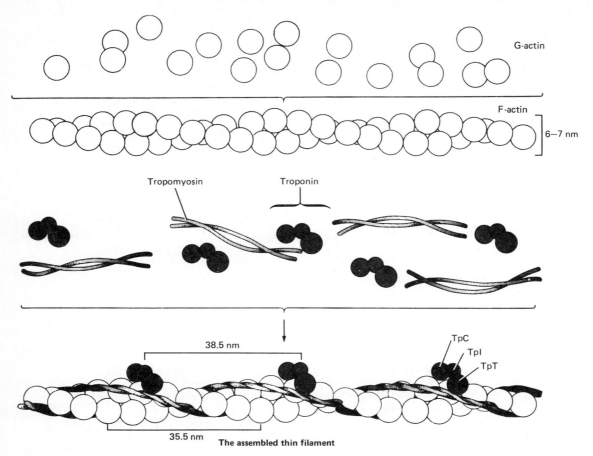

Figure 59–3. Schematic representation of the thin filament, showing the spatial configuration of the 3 major protein components—actin, tropomyosin, and troponin.

2 pairs of light chains (MW 15,000–27,000). Skeletal muscle myosin exhibits **ATP-hydrolyzing (ATPase) activity** and binds to F-actin, an insoluble molecule.

Much Has Been Learned From Partial Digestion of Myosin

When myosin is digested with trypsin, 2 myosin fragments (meromyosins) are generated. Light meromyosin (LMM) consists of aggregated, insoluble α-helical fibers (Fig 59–14). LMM exhibits no ATPase activity and will not bind to F-actin.

Heavy meromyosin (HMM) is a 340,000-MW soluble protein that has both a fibrous portion and a globular portion (Fig 59–14). HMM exhibits **ATPase activity** and **binds to F-actin.** The digestion of HMM with papain generates 2 subfragments, S-1 and S-2. The S-2 is fibrous in character, exhibits no ATPase activity, and does not bind to F-actin.

S-1 has a molecular weight of 115,000, exhibits **ATPase activity,** and in the absence of ATP will **bind to and decorate actin with "arrowheads"** (Fig 59–15). Although both S-1 and HMM exhibit ATPase activity, that **catalytic activity is accelerated 100- to**

200-fold by the addition of F-actin. As discussed below, F-actin greatly enhances the rate at which myosin ATPase releases its products, ADP and P_i. Thus, although F-actin does not affect the hydrolysis step per se, its ability to **promote release of the ATPase products** greatly accelerates the overall rate of catalysis.

α-Actinin is a protein in the Z line to which the ends of the F-actin molecules of the thin filaments attach (Fig 59–2).

ATP-Driven Dissociation of Myosin Heads From Thin Filaments Powers Contraction

How can ATP hydrolysis produce macroscopic movement? Muscle contraction consists of the **cyclic attachment and detachment of the globular head portion of myosin to the F-action filament.** The attachment is followed by a change in the actin-myosin interaction, so that the actin filaments and the myosin filaments slide past one another. The energy is supplied indirectly by ATP, which is hydrolyzed. ATP hydrolysis by the myosin ATPase is greatly accelerated by the binding of the myosin head to F-actin. The

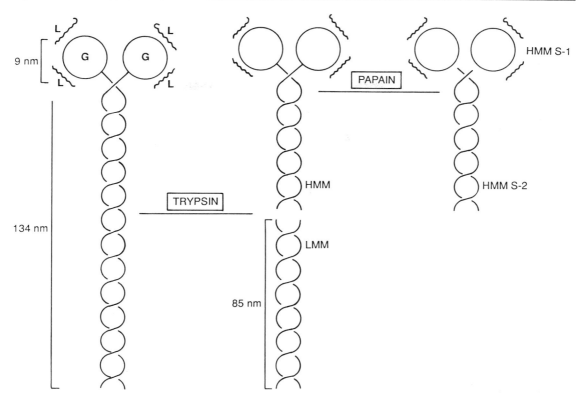

Figure 59–4. Diagram of a myosin molecule showing the 2 intertwined α helices (fibrous portion), the globular region (G), the light chains (L), and the effects of proteolytic cleavage by trypsin and papain. HMM heavy meromyosin; LMM light meromyosin; S-1 subfragment 1; S-2 subfragment 2.

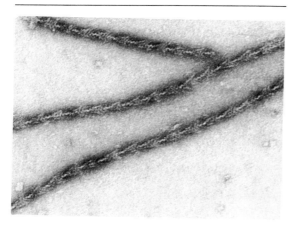

Figure 59–5. The decoration of actin filaments with the S-1 fragments of myosin to form "arrowheads." (Courtesy of Professor James Spudich, Stanford University.)

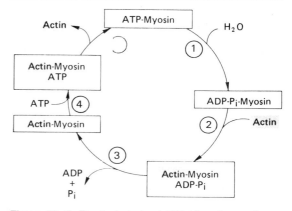

Figure 59–6. The hydrolysis of ATP drives the cyclic association and dissociation of actin and myosin in 5 reactions described in the text. (Modified from Stryer L: *Biochemistry*, 2nd ed. Freeman, 1981.)

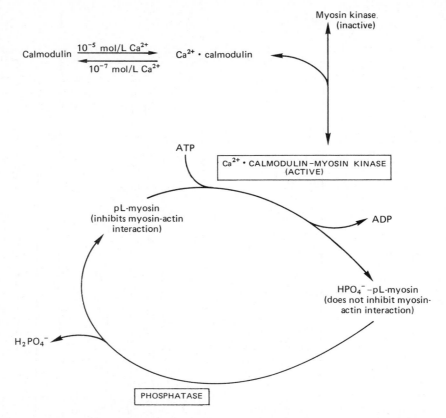

Figure 59–7. Regulation of smooth muscle contraction by Ca^{2+}. Adapted from Adelstein RS. Eisenberg R: Regulation and kinetics of actin-myosin ATP interaction. *Annu Rev Biochem* 1980:**49**:921.)

biochemical cycle of muscle contraction consists of 5 steps (Fig 59–16): (1) The myosin head alone can hydrolyze ATP to ADP + P_i, but it cannot release the products of this hydrolysis. Thus, the hydrolysis of ATP by the myosin head alone is stoichiometric rather than catalytic. (2) The myosin head containing ADP and P_i can rotate freely through large angles in order to locate and bind to F-actin, making an angle of about 90 degrees with the fiber axis. The interaction (3) promotes the release of ADP and P_i from the actin-myosin complex. Because the conformation of lowest energy for the actomyosin bond is 45 degrees, the myosin changes its angle from 90 degrees to about 45 degrees by **pulling the actin** (10–15 nm) toward the center of the sarcomere. (4) A new ATP molecule binds to the myosin-F actin complex. Myosin-ATP has a poor affinity for actin, and thus the myosin (ATP) **head is released** (5) from the F-actin. This last step is **relaxation,** a process clearly **dependent upon the binding of ATP** to the actin-myosin complex. The ATP is again hydrolyzed by the myosin head, but without releasing ADP + P_i, to continue the cycle.

It should be clear that **ATP dissociates the myosin head from the thin filament and powers the contraction.** The efficiency of this contraction is about 50%; that of the internal combustion engine is less than 20%.

Ca^{2+} Plays a Key Role in Regulation of Muscle Contraction

The contraction of muscles from all sources occurs by the general mechanism described immediately above. Muscles from different organisms and from different cells and tissues within the same organism may have different molecular mechanisms responsible for the regulation of their contraction and relaxation. In all systems, **Ca^{2+} plays a key regulatory role.** There are 2 general mechanisms of regulation of muscle contraction: actin-based and myosin-based.

Actin-Based Regulation Occurs in Striated Muscle

Actin-based regulation of muscle occurs in vertebrate skeletal and cardiac muscles, both **striated.** In the general mechanism described above, the only potentially limiting factor in the cycle of muscle contraction might be ATP. The skeletal muscle system is inhibited at rest and is deinhibited to activate contraction. The **inhibitor of striated muscle is the troponin system,** which is bound to tropomyosin and F-actin in the thin filament (Fig 59–3). In striated muscle, there is no control of contraction (or ATPase as a biochemical indicator of contraction) unless the tropomyosin-troponin systems are present along with the actin and myosin filaments. As described above, tropomyosin lies along the groove of F-actin, and the 3 components of troponin—TpT, TpI, and TpC—are bound to the F-actin-tropomyosin complex. TpI prevents binding of the myosin head to its F-actin attachment site either by altering the conformation of F-actin via the tropomyosin molecules or by simply rolling tropomyosin into a position that directly blocks the sites on F-actin to which the myosin heads attach. Either way prevents activation of the myosin ATPase that is mediated by binding of the myosin head to F-actin. Hence, the TpI system blocks the contraction cycle at step 2 of Fig 59–6. This accounts for the inhibited state of relaxed striated muscle.

Ca^{2+} Mediates Excitation of Muscle Contraction

In resting muscle sarcoplasm, the concentration of Ca^{2+} is $10^{-7} - 10^{-8}$ mol/L. Calcium is sequestered in the sarcoplasmic reticulum, a network of fine membranous sacs, by an active transport system utilizing a Ca^{2+}-binding protein called **calsequestrin**. The sarcomere is surrounded by an **excitable membrane** that has transverse (T) channels closely associated with the sarcoplasmic reticulum. When the sarcomere membrane is excited, such as by the occupation of an acetylcholine receptor by acetylcholine, **Ca^{2+} is rapidly released** into the sarcoplasm from the sarcoplasmic reticulum. The Ca^{2+} concentration in sarcoplasm rapidly rises to 10^{-5} mol/L. The Ca^{2+} binding sites on TpC in the thin filament are quickly occupied by Ca^{2+}. The TpC·4Ca^{2+} interacts with TpI and TpT to alter this interaction with tropomyosin. Accordingly, tropomyosin simply moves out of the way or alters the F-actin conformation so that the myosin head ADP-P$_i$ can interact with F-actin to start the contraction cycle.

Relaxation occurs when (1) sarcoplasm Ca^{2+} falls below 10^{-7} mol/L owing to its resequestration in the sarcoplasmic reticulum by an energy-dependent Ca^{2+} pump; (2) TpC·4Ca^{2+} loses its Ca^{2+}; (3) troponin, via its interaction with tropomyosin, inhibits further myosin head-F-actin interaction; and (4) in the presence of ATP, the myosin head detaches from the F-actin to induce relaxation. Thus, **Ca^{2+} controls muscle contraction by an allosteric mechanism** mediated in muscle by TpC, TpI, TpT, tropomyosin, and F-actin.

Extracellular Fluid Supplies Ca^{2+}

In cardiac muscle, the extracellular fluid is a major source of Ca^{2+} for excitation. In the absence of Ca^{2+} in the extracellular fluid, cardiac muscle will cease contracting (beating) within 1 minute; skeletal muscle can contract for hours without extracellular Ca^{2+}.

The loss of ATP in the sarcoplasm has 2 major effects: (1) The Ca^{2+} pump in the sarcoplasmic re-

ticulum ceases to maintain the low sarcoplasm Ca^{2+} concentration. Thus, the interaction of the myosin heads with F-actin is promoted. (2) The ATP-dependent detachment of myosin heads from F-actin cannot occur, and "rigor mortis" sets in.

Muscle contraction is a delicate dynamic balance of the attachment and detachment of myosin heads to F-actin, subject to fine regulation via the nervous system.

Ca^{2+} Also Regulates Contraction in Smooth Muscle

While all muscles contain actin, myosin, and tropomyosin, **only vertebrate striated muscles contain the troponin system.** Thus, the mechanisms which regulate contraction must differ in various contractile systems.

Smooth muscles have molecular structures similar to those in striated muscle, but the sarcomeres are not aligned so as to generate the striated appearance. Smooth muscles contain α-actinin and tropomyosin molecules, as do skeletal muscles. They do not have the troponin system, and the light chains of smooth muscle myosin molecules differ from those of striated muscle myosin. However, like striated muscle, **smooth muscle contraction is regulated by Ca^{2+}.**

Phosphorylation of Myosin p-Light Chains Initiates Smooth Muscle Contraction

When smooth muscle myosin is bound to F-actin in the absence of other muscle proteins such as tropomyosin, there is no detectable ATPase activity. This absence of ATPase is quite unlike the situation described for striated muscle myosin and F-actin, which has abundant ATPase activity. Smooth muscle myosin contains a light chain (p-light chain) that prevents the binding of the myosin head to F-actin. The p-light chain must be phosphorylated before it allows F-actin to activate myosin ATPase. The **phosphorylation of p-light chains commences the attachment-detachment contraction cycle of smooth muscle.**

Calmodulin·4Ca^{2+}-Activated Myosin Light Chain Kinase Phosphorylates the p-Light Chain

Smooth muscle sarcoplasm contains a **myosin light chain kinase that is calcium-dependent.** The Ca^{2+} activation of myosin light chain kinase requires binding of **calmodulin·4Ca^{2+}** to its 105,000-MW kinase subunit (Fig 59–7). The calmodulin·4Ca^{2+}-activated light chain kinase phosphorylates the p-light chain, which then ceases to inhibit the myosin-F-actin interaction. The contraction cycle then begins.

Smooth Muscle Relaxes When the Ca^{2+} Concentration Falls Below 10^{-7} Molar

Relaxation of smooth muscle occurs when (1) sarcoplasm Ca^{2+} falls below 10^{-7} mol/L. The Ca^{2+} dissociates from calmodulin, which in turn dissociates from the myosin light chain kinase, (2) inactivating the

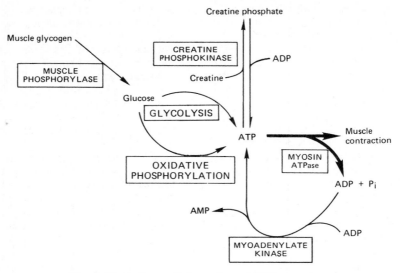

Figure 59–8. The multiple sources of ATP in muscle.

kinase. (3) No new phosphates are attached to the p-light chain, and light chain protein phosphatase, which is continually active and calcium-independent, removes the existing phosphates from the p-light chain. (4) Dephosphorylated myosin p-light chain then inhibits the binding of myosin heads to F-actin and the ATPase activity. (5) The myosin head detaches from the F-actin in the presence of ATP, but it cannot reattach because of the presence of dephosphorylated p-light chain; hence, relaxation occurs.

Table 59–1 summarizes and compares the regulation of actin-myosin interactions (activation of myosin ATPase) in striated and smooth muscles.

The myosin light chain kinase is not directly affected or activated by cAMP. However, the usual cAMP-activated protein kinase (see Chapter 45) can phosphorylate the myosin light chain kinase (*not* the p-light chain itself). The phosphorylated myosin light chain kinase exhibits a significantly lower affinity for calmodulin·Ca^{2+} and thus is less sensitive to activation. Accordingly, an increase in cAMP dampens the contraction response of smooth muscle to a given elevation of sarcoplasm Ca^{2+}. This molecular mechanism can explain the relaxing effect of β-adrenergic stimulation on smooth muscle.

Phenothiazines Bind Calmodulin & Relax Smooth Muscle

The **phenothiazines,** widely used antipsychotic drugs, bind to calmodulin and prevent its attachment to

Table 59–1. Actin-myosin interactions in striated and smooth muscle.

	Striated Muscle	Smooth Muscle (and Nonmuscle Cells)
Proteins of muscle filaments	Actin Myosin (hexamer) Tropomyosin Troponin (TpI, TpT, TpC)	Actin Myosin (hexamer)* Tropomyosin
Spontaneous interaction of F-actin and myosin *alone* (spontaneous activation of myosin ATPase by F-actin)	Yes	No
Inhibitor of F-actin-myosin interaction (inhibitor of F-actin-dependent activation of ATPase)	Troponin system (TpI)	Unphosphorylated myosin p-light chain
Contraction activated by	Ca^{2+}	Ca^{2+}
Direct effect of Ca^{2+}	$4Ca^{2+}$ bind to TpC	$4Ca^{2+}$ bind to calmodulin
Effect of protein-bound Ca^{2+}	TpC · $4Ca^{2+}$ antagonizes TpI inhibition of F-actin-myosin interaction (allows F-actin activation of ATPase).	Calmodulin · $4Ca^{2+}$ activates myosin light chain kinase that phosphorylates myosin p-light chain. The phosphorylated p-light chain no longer inhibits F-actin-myosin interaction (allows F-actin activation of ATPase).

*Light chains of myosin are different in striated and smooth muscles.

calcium-dependent enzymes. Phenothiazines also relax smooth muscle.

Striated muscle from mollusks such as the scallop exhibits a myosin-based regulation of contraction. Like myosin and F-actin from smooth muscle, that from scallops also exhibits no ATPase, an effect of the inhibitory properties of the "regulatory" light chain of scallop myosin. The inhibition of scallop actin-myosin interaction is relieved when Ca^{2+} binds directly to a specific site on the myosin molecule. This regulation does not require covalent modification of myosin or the addition of a separate protein such as calmodulin or TpC to be Ca^{2+}-dependent.

Phosphorylation Plays Key Roles in Smooth Muscular Contraction

As described above, the phosphorylation of the light chain of smooth muscle myosin alleviates its inhibitory effect on the actin-myosin interaction and thereby commences the contraction cycle. The phosphate on the myosin light chains may form a chelate with the Ca^{2+} bound to the tropomyosin-TpC-actin complex, leading to an increased rate of formation of cross-bridges between the myosin heads and actin.

Phosphorylation of myosin heavy chains probably is a prerequisite for their assembly into the thick filaments in skeletal muscle, smooth muscle, and nonmuscle cells (see below).

The TpI and a peptide component of the sarcoplasmic reticulum Ca^{2+} pump in cardiac muscle can be phosphorylated by cAMP-dependent protein kinase. There is a rough correlation between the phosphorylation of TpI and the increased contraction of cardiac muscle induced by catecholamines. This mechanism may account for the inotropic effects (increased contractility) of the β-adrenergic compounds on the heart.

Several Mechanisms Replenish Muscle ATP Stores

The ATP required as the constant energy source for the contraction-relaxation cycle of muscle can be generated by glycolysis, oxidative phosphorylation, creatine phosphate, or two ADP molecules (Fig. 59–8). The ATP stores in skeletal muscle are short-lived during contraction, providing energy probably for less than 1 second of contraction. In **slow skeletal muscle,** which has abundant O_2 stores in myoglobin, **oxidative phosphorylation** is the major source of ATP regeneration. **Fast** skeletal muscles regenerate ATP from glycolysis, mainly.

Creatine Phosphate Constitutes a Major Energy Reserve

The **phosphagen** creatine phosphate prevents the rapid depletion of ATP by providing a readily available high-energy phosphate, which is all that is necessary to re-form ATP from ADP. Creatine phosphate is formed from ATP and creatine at times when the muscle is relaxed and ATP demands are not so great. The enzyme catalyzing the phosphorylation of creatine is creatine phosphokinase (CPK), a muscle-specific en-

zyme with clinical utility in the detection of acute or chronic disorders of muscle.

Skeletal Muscle Contains Large Stores of Glycogen

Skeletal muscle sarcoplasm contains large **glycogen** stores, located in granules close to the I bands. The release of glucose from glycogen is dependent upon a specific muscle glycogen phosphorylase (see Chapter 20). To generate glucose 6-phosphate for glycolysis in skeletal muscle, glycogen phosphorylase b must be activated to phosphorylase a via phosphorylation by phosphorylase b kinase (see Chapter 20). Ca^{2+} promotes the activation of phosphorylase b kinase, also by phosphorylation. Thus, Ca^{2+} **both activates muscle contraction and activates a pathway to provide necessary energy.** Muscle glycogen phosphorylase b is missing in **McArdle's disease**, a glycogen storage disease.

Muscle Generates ATP by Oxidative Phosphorylation

ATP synthesis via oxidative phosphorylation requires oxygen supply. Muscles that have high oxygen demand as a result of sustained contraction (such as to maintain posture) store oxygen in **myoglobin** (see Chapter 7). Because of the heme moiety to which oxygen is bound in myoglobin, muscles containing myoglobin are red, Table 59–2 compares some of the properties of fast (or white) skeletal muscle with slow (or red) skeletal muscle.

Myokinase Interconverts Adenine Mono-, Di-, & Triphosphates

The enzyme myokinase catalyzes formation of one ATP molecule and one AMP from two ADP molecules. This reaction (Fig 59–8) is coupled with the hydrolysis of ATP by myosin ATPase during muscle contraction.

Deamination of Adenine by Working Muscle Releases Ammonia

The immediate source of ammonia in skeletal muscle is AMP, which is deaminated to IMP, catalyzed by adenylate deaminase. IMP may be converted back to

Table 59–2. Characteristics of fast and slow skeletal muscle.

	Fast Skeletal Muscle	Slow Skeletal Muscle
Myosin ATPase	High	Low
Energy utilization	High	Low
Color	White	Red
Myoglobin	No	Yes
Contraction rate	Fast	Slow
Duration	Short	Prolonged

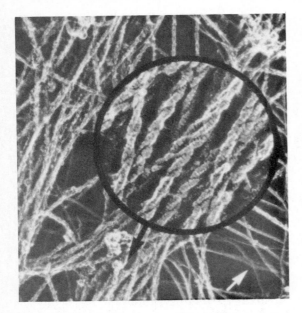

Figure 59–9. Replica of a freeze-dried cytoskeleton that was exposed to the myosin subfragment 1 (S-1) before quick-freezing. Nearly all the filaments in the lengthwise bundles, and many of the intervening filaments, have been thickened and converted into ropelike double helices (see *inset*). However, some of the filaments that travel by themselves, in between the bundles, remain totally undecorated (arrow); these are presumably intermediate filaments. × 70,000; *Inset.* × 200,000. (Reproduced, with permission, from Heuser JE, Kirschner MW: Filament organization revealed in platinum replicas of freeze-dried cytoskeletons. *J Cell Biol* 1980;**86**:212.)

AMP by reactions utilizing aspartate and catalyzed by adenylosuccinate synthetase and adenylosuccinase (see Chapter 36).

SKELETAL MUSCLE CONSTITUTES THE MAJOR PROTEIN RESERVE OF THE BODY

In humans, skeletal muscle protein is the major non-fat source of stored energy. This explains the very large losses of muscle mass, particularly in adults, resulting from prolonged caloric undernutrition.

The study of tissue protein breakdown in vivo is difficult, because amino acids released during intracellular breakdown of proteins can be extensively reutilized for protein synthesis within the cell, or the amino acids may be transported to other organs where they enter anabolic pathways. However, actin and myosin are methylated following their synthesis, forming peptidyl **3-methylhistidine.** During intracellular breakdown of actin and myosin, 3-methylhistidine is released and excreted into the urine. The urinary output of the methylated amino acid provides a reliable index of the rate of myofibrillar protein breakdown in the musculature of human subjects.

Skeletal muscle is active in the degradation of certain amino acids as well as in the synthesis of others. In mammals, muscle appears to be the primary site of catabolism of the branched-chain amino acids. Muscle oxidizes leucine to CO_2 and converts the carbon skeletons of aspartate, asparagine, glutamate, isoleucine, and valine into amphibolic intermediates of the tricarboxylic acid cycle. The capacity of muscles to degrade branched-chain amino acids increases 3- to 5-fold during fasting and in diabetes.

Muscle also synthesizes and releases large amounts of alanine and glutamine. These compounds are synthesized utilizing amino groups that are generated in the breakdown of branched-chain amino acids, and the amino nitrogen is then transferred to α-ketoglutarate and to pyruvate by transamination. Glycolysis from exogenous glucose provides almost all of the pyruvate for synthesis of alanine. These reactions constitute the so-called glucose-alanine cycle, wherein alanine from muscle is utilized in hepatic gluconeogenesis while simultaneously bringing amino groups to the liver for removal as urea.

CYTOSKELETONS PERFORM MULTIPLE CELLULAR FUNCTIONS

Non-muscle cells perform mechanical work, including self-propulsion, morphogenesis, cleavage, endocytosis, exocytosis, intracellular transport, and changing cell shape. These cellular functions are carried out by an extensive intracellular network of filamentous structures constituting the **cytoskeleton.** The cell cytoplasm is not a sac of fluid, as once thought. Essentially all eukaryotic cells contain 3 types of filamentous structures: **actin filaments** (7–9.5 nm in diameter), **microtubules** (25 nm), and **intermediate filaments** (10–12 nm). Each of these types of filaments can be distinguished biochemically and electron microscopically.

Non-muscle Cells Contain Actin

The G-actin protein isolated from non-muscle cells has a molecular weight of about 43,000 and contains N-methylhistidyl residues. In the presence of magnesium and potassium chloride, this actin will spontaneously polymerize to form the double helical **F-actin filaments** like those seen in muscle. There are at least 2 types of actin in nonmuscle cells: β-actin and γ-actin. Both types can coexist in the same cell and probably even copolymerize in the same filament. In the cellular cytoplasm, actin forms **microfilaments** of 7–9.5 nm that frequently exist as bundles of tangled-appearing meshwork. The bundles of microfilaments are prominent just underlying the plasma membrane of resting cells and are there referred to as **stress fibers.** These stress fibers will decorate with the S-1 portion of myosin to reveal their double helical character (Fig 59–9). The stress fibers disappear as cell motility in-

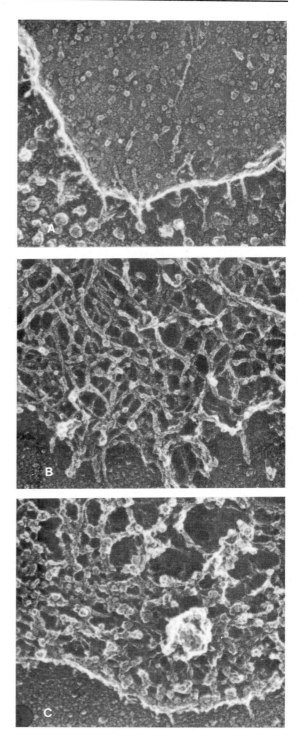

Figure 59–10 (at right). Three moderately high-powered views of ruffles or lamellipodia from fibroblasts that were fixed while whole (in *A*), were extracted with Triton before fixation (in *B*), or extracted with Triton after fixation (in *C*). In *A*, the plasma membrane is intact, and no internal structure can be seen. In *B*, the plasma membrane has been removed and an underlying web of "kinky" filaments revealed. In other experiments, these filaments decorate with S-1, but they are much more concentrated and much more extensively interdigitated than actin in other regions of the cell. In *C*, the plasma membrane has again been removed, but only after the cell was fixed with aldehyde. The delicate meshwork of underlying filaments appears coarser after the chemical fixation. *A*, × 140,000. *B* and *C*, × 115,000. (Reproduced, with permission, from Heuser JE, Kirschner MW: Filament organization revealed in platinum replicas of freeze-dried cytoskeletons. *J Cell Biol* 1980;**86**:212.)

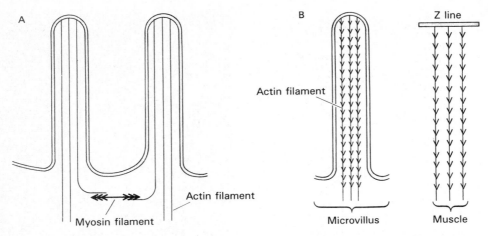

Figure 59–11. Microvilli are tiny cytoplasmic protrusions that extend out from the epithelial cells lining the small intestine, greatly increasing the surface for the absorption of nutrients. Microvilli contain both actin and myosin filaments and are known to contract much like muscle cells, and so they provide a convincing example of nonmuscle movement mediated by sliding filaments of actin and myosin. As is shown in A, bundles of actin filaments project upward inside each microvillus; the myosin filaments are localized at the base of the microvilli. In B, the orientation of the actin filaments was determined by treating the microvilli with isolated head fragments from muscle myosin, termed heavy meromyosin; these fragments retain the ability to bind to actin filaments. When the head fragments are applied to muscle cells, they form "arrowhead" complexes with the actin filaments that point in the direction of the filaments. When heavy meromyosin was added to microvilli, the head fragments formed arrowhead complexes with the actin filaments that pointed downward from the attachment sites in the tips of the microvilli. The actin filaments within the microvilli are therefore analogous to the actin filament arrays of muscle cells. (Reproduced, with permission, from Lazarides E, Revel JP: The molecular basis of cell movement. *Sci Am* [May] 1979;**240**:100.)

creases or upon the malignant transformation of the cell by chemicals or oncogenic viruses.

Cellular Microprojections Contain Actin Filaments

Microfilaments are also tightly packed in a meshwork pattern underlying the leading edge or "ruffle" of a motile cell (Fig 59–10). Actin microfilaments are found in all cellular microprojections such as filopodia and microvilli. For instance, the microvilli of intestinal mucosal cells contain 20–30 actin microfilaments arranged longitudinally within the microvilli (Fig 59–11). At the base of the microvilli, myosin filaments exist and are capable of pulling together the actin filaments projecting into the microvilli. The contraction process does not involve any change of length of actin or myosin and thus must occur, as in muscle, by the sliding filament mechanism. As in smooth muscle, activation of the actin-myosin interaction and thereby contraction, is mediated by phosphorylation of the myosin light chain.

Actin and myosin are also both found between the spindle poles and the chromosomes and along the cleavage furrow of mitotic telophase.

Actin microfilaments are associated with other muscle-like proteins in non-muscle cells. **α-Actinin** is present at the plasma membrane sites to which microfilaments attach, such as the tips of microvilli. The geodesic domes—cytoskeletal scaffolding surrounding the nuclei of eukaryotic cells—consist of actin, α-

actinin, and tropomyosin. α-Actinin is also found along actin microfilaments themselves.

Myosin is also found along the actin fibers but as filaments thinner and shorter than in muscle that seem to play a role in maintenance of the filamentous character of actin.

Tropomyosin participates in the formation of the geodomes surrounding nuclei. Tropomyosin along actin microfilaments seems to serve a structural rather than a motility function.

Specialized Proteins Regulate Nonmuscle Actin

The regulation of nonmuscle actin function depends upon several specialized proteins. **Profilin** prevents the polymerization of G-actin even in the presence of the proper concentrations of magnesium and potassium chloride. **Filamin** promotes the formation of an actin microfilament meshwork. **Tropomyosin** promotes the formation of bundles of actin stress fibers. **α-Actinin** promotes the attachment of actin microfilaments to membranes, substratum, and other cell organelles. **Cytochalasin,** a naturally occurring peptide, breaks microfilaments and prevents their polymerization. This response provides a diagnostic test for the existence or function of microfilaments.

The actual motility of cells appears to be led by the **ruffle membrane,** or lamellipodium, that contains fingerlike projections called filopodia. The ruffle attaches at its tip to the substratum via the filopodia, and

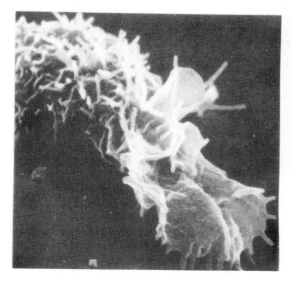

Figure 59–12. Individual cells in tissue culture are depicted. The delicate feathery structure at the bottom right is a "ruffle," or lamellipodium, which marks the leading edge of the cell. A cell is shown from an oblique angle as it moves across the substrate, extending its ruffle to form new adhesions. (Reproduced, with permission, from Lazarides E, Revel JP: The molecular basis of cell movement. *Sci Am* [May] 1979;**240**:100.)

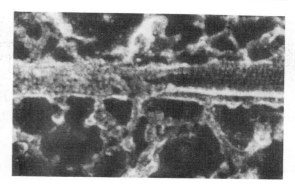

Figure 59–13. High magnification of a microtuble that was fractured and deep etched after quick-freezing. The left half of the field illustrates the outer surface of the microtubule, which displays longitudinal bands of bumps spaced 55 nm apart, which may represent the microtubule's protofilaments. To the right, the microtubule is fractured open to reveal its inner luminal walls, which display characteristic oblique striations separated by 40 nm. The reticulum surrounding the microtubule is thought to be unpolymerized tubulin and microbutule-associated proteins. (Reproduced, with permission, from Heuser JE, Kirschner MW: Filament organization revealed in platinum replicas of freeze-dried cytoskeletons. *J Cell Biol* 1980;**86**:212.)

the cell then seems to pull in its rear margins. The ruffle releases and folds back over the top of the cell as new filopodia attach to the substratum (Fig 59–12).

Microtubules Contain α- & β-Tubulin

Microtubules, an integral component of the cellular cytoskeleton, consist of cytoplasmic tubes 25 nm in diameter and of indefinite length. Microtubules are necessary for the formation and function of the **mitotic spindle** and thus are present in all eukaryotic cells. Microtubules carry out additional cellular functions. They are responsible for the intracellular movement of endocytotic and exocytotic vesicles. They form the major structural component of **cilia and flagella.** Microtubules are a major protein component of **axons and dendrites,** where they maintain the structure and participate in the axoplasmic flow of material along these neuronal processes.

Microtubules are cylinders of 13 longitudinally arranged **protofilaments,** each consisting of dimers of **α-tubulin** and **β-tubulin** (Fig 59–13). α-Tubulin (MW 53,000) and β-tubulin (MW 55,000) are closely related protein molecules. The tubulin dimers assemble into protofilaments and subsequently into sheets and then cylinders, as depicted in Fig 59–14. The assembly of tubulin into microtubules requires two **GTP** molecules per tubulin dimer. Two proteins termed high-molecular-weight (HMW) protein and Tau promote the formation of microtubules but are not re-

quired for assembly. Calmodulin and phosphorylation may both play roles in microtubule assembly.

Microtubules "grow" with a polarity from specific sites (centrioles) within cells. On each chromatid of a chromosome (see Chapter 37) there exists a kinetochore that serves as a point of origin for microtubular growth. Many abnormalities of chromosomal segregation result from abnormal structure or function of kinetochores. The centrosome, which is at the center of the mitotic poles, also nucleates microtubular formation. The movement of chromosomes during anaphase of mitosis is dependent upon microtubules, but the molecular mechanism has not been delineated.

Several Alkaloids Block Microtubule Assembly

Certain alkaloids can prevent microtubule assembly. These include **colchicine** and its derivative **demecolcine** (used for treatment of acute gouty arthritis), **vinblastine** (a *Vinca* alkaloid used for treating cancer), and **griseofulvin** (an antifungal agent).

Cell Flagella & Cilia Are Microtubules Specialized for Motility

At the base of all eukaryotic **flagella** and **cilia** is a structure called the **basal body** that is identical to the centriole and acts as a nucleation center for the formation of the 9-doublet array of microtubules in the flagella and cilia. These microtubular structures are specialized for motility. Each member of a doublet shares a common wall of 3 protofilaments with its partner, and the doublets are connected by a flexible

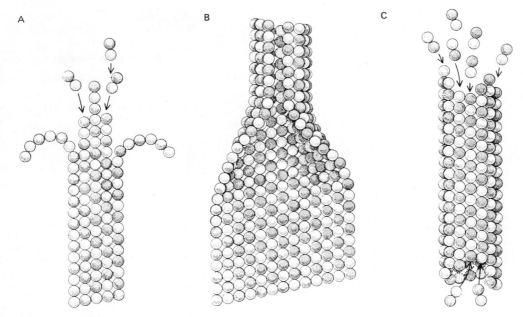

Figure 59–14. Assembly of microtubules in the laboratory begins with 2 protein molecules, α-tubulin and β-tubulin, which are globular molecules (probably more ovoid than these highly schematic spheres). The tubulins form dimers, or double molecules. If the dimers are present in a high enough concentration, they associate to form various intermediate structures, including double rings, spirals, and stacked rings; the equilibrium is biased in favor of either the isolated dimers or the intermediate structures, depending on the conditions. The next steps are not well established. It seems that the rings or spirals open up to form strands, called protofilaments, of linearly associated dimers, which assemble side by side in a sheet (*A*); sometimes the ends of protofilaments curve. When a sheet is wide enough, it forms a tube, perhaps by curling up (*B*). Once a short tube has formed (*C*), it is lengthened by the addition of dimers preferentially at one end. (Reproduced, with permission, from Dustin P: Microtubules. *Sci Am* [Aug] 1980;**243**:67.

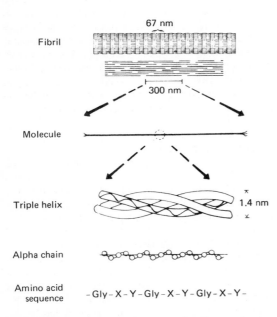

Figure 59–15. Molecular features of collagen structure from primary sequence up to the fibril. (Slightly modified and reproduced, with permission, from Eyre DR: Collagen: Molecular diversity in the body's protein scaffold. *Science* 1980;**207**:1315. Copyright © 1980 by the American Association for the Advancement of Science.)

protein, **nexin.** Movement is effected by the sliding of the doublets past one another causing distortion of the cilium in waves. Connected to one of the doublets in a cilium is a large protein, **dynein,** which possesses an ATPase necessary for the microtubular doublet sliding movement.

Intermediate Filaments Differ From Microfilaments & Microtubules

There exists an intracellular fibrous system of filaments with an axial periodicity of 21 nm and a diameter of 8–10 nm that are distinct from microfilaments (6 nm in diameter) and microtubules (23 nm in diameter). There are 6 major classes of these filaments that share an antigenic determinant and exhibit diameters **intermediate** in size between actin microfilaments and microtubules. Each intermediate filament consists of biochemically and immunologically distinct subunits. Intermediate filaments seem to form relatively **stable components** of the cytoskeleton, not undergoing rapid assembly and disassembly and not disappearing during mitosis as do actin and many microtubular filaments. Table 59–3 summarizes some properties and distributions of intermediate filaments.

There are 2 types of **keratin,** I and II, comprising 10–20 different polypeptides that are as different from one another as they are from the other 4 classes of

Table 59–3. Classes of intermediate filaments and their distributions.

Proteins	MW (Thousands)	Diameter (nm)	Distributions
Keratin type I and type II (tonofilaments)	40–65 (10–20 major proteins)	8	Epithelial cells (never cells of mesenchymal origin).
Desmin	50–55	10	Muscle (Z lines).
Vimentin	52	10	Mesenchymal and nonmesenchymal cells, eg, muscle, glial cells, epithelial cells.
Neurofilament	200 150 70	10	Neurons.
Glial filament	51	10	Glial cells.

intermediate filament proteins—desmin, vimentin, neurofilament, and glial filament. These latter 4 classes have a high degree of homology among them. A keratin filament will contain at least 2 different keratin polypeptides, whereas the other 4 classes of intermediate filaments are homopolymers. Each of the intermediate filaments consists of 4 α-helical domains separated from one another by regions of β-pleated sheets and flanked on both ends by nonhelical terminal domains. The nonhelical terminal domains are involved in end-to-end extension of protofilaments and side-to-side interprotofibrillar interactions. The ends of the microfibrillar keratins can be crosslinked through disulfide bonding to form insoluble filaments, such as those characteristic of wool.

COLLAGEN IS THE MOST ABUNDANT PROTEIN IN THE ANIMAL WORLD

Collagen, the major component of most connective tissues, constitutes approximately 25% of the protein of mammals. It provides an extracellular framework for all metazoan animals and exists in virtually every animal tissue. More than **10 distinct types** of collagen have been identified in mammalian tissues. Although several of these are present only in small proportions, they may play important roles in determining the physical properties of the tissues.

All collagen types have a **triple helical structure.** Each polypeptide subunit or alpha chain is twisted into a **left-handed helix** of 3 residues per turn (Fig 59–15). Three of these alpha chains are then wound into a **right-handed super-helix,** forming a rodlike molecule 1.4 nm in diameter and about 300 nm long. A striking characteristic of collagen is the occurrence of **glycine residues at every third position** of the triple helical portion of the alpha chain. This is necessary because glycine is the only amino acid small enough to be accommodated in the limited space available down the central core of the triple helix. This repeating structure, represented as **(Gly-X-Y)n, is an absolute requirement for the formation of the triple helix.** While X and Y can be any other amino acids, about 100 of the X positions are **proline** and about 100 of the Y positions are **hydroxyproline.** Hydroxyproline is formed by the posttranslational hydroxylation of peptide-bound proline residues catalyzed by the enzyme **prolyl hydroxylase,** whose cofactors are **ascorbic acid** (vitamin C) and **α-ketoglutarate.** Lysines in the Y position may also be posttranslationally modified to **hydroxylysine** through the action of **lysyl hydroxylase,** an enzyme with similar cofactors. Some of these hydroxylysines may be further modified by the addition of **galactose or galactosyl-glucose through an O-glycosidic linkage, a glycosylation site that is unique to collagen.**

Collagen types that form long rodlike fibers in tissues are assembled by lateral association of these triple helical units into a **"quarter-staggered"** alignment such that each is displaced longitudinally from its neighbor by slightly less than one-quarter of its length (Fig 59–15). This arrangement is responsible for the banded appearance of these fibers in connective tissues. Collagen fibers are further stabilized by the formation of **covalent crosslinks,** both within and between the triple helical units. These crosslinks form through the action of **lysyl oxidase,** a copper-dependent enzyme that oxidatively deaminates the ε-amino groups of certain lysine and hydroxylysine residues, yielding reactive aldehydes. Such aldehydes can form aldol condensation products with other lysine- or hydroxylysine-derived aldehydes, or form Schiff bases with the ε-amino groups of unoxidized lysines or hydroxylysines. These reactions, after further chemical rearrangements, result in the stable, covalent crosslinks that are important for the **tensile strength** of the fibers.

Several collagen types do not form banded fibers in tissues (Table 59–4). At the molecular level, these collagens are characterized by interruptions of the triple helix with stretches of protein lacking Gly-X-Y repeat sequences. These non-Gly-X-Y sequences result in areas of globular structure interspersed in the triple helical structure. **Type IV collagen,** the best characterized example of collagens with discontinuous triple helices, is an important component of **basement membranes** where it forms a meshlike network.

Collagen Undergoes Extensive Posttranslational Modifications

Newly synthesized collagen undergoes extensive posttranslational modification before becoming part of

Table 59–4. Some examples of the genetically distinct vertebrate collagens. At least 10 different molecules containing 12 genetically distinct α chains are present in higher animals.*

Type	Molecular Formula	Native Polymer	Tissue Distribution	Distinctive Features
I	$[\alpha 1(\text{I})]_2\alpha 2$	Fibrils	Skin, tendon, bone, dentin, fascia; most abundant.	Low content of hydroxylysine; few sites of hydroxylysine glycosylation; broad fibrils.
II	$[\alpha 1(\text{II})]_3$	Fibrils	Cartilage, nucleus pulposus, notochord, vitreous body.	High content of hydroxylysine; heavily glycosylated; usually thinner fibrils than type I.
III	$[\alpha 1(\text{III})]_3$	Fibrils	Skin (especially fetal skin), uterus, blood vessels; "reticulin" fibers generally.	High content of hydroxyproline; low content of hydroxylysine; few sites of hydroxylysine glycosylation; interchain disulfides between cysteines at the carboxyl end of the helix.
IV	$[\alpha_1(\text{IV})]_2\alpha_2(\text{IV})$ and $[\alpha_1(\text{IV})]_3$	Meshlike network	Kidney glomeruli, lens capsule; Descemet's membrane; basement laminae of all epithelial and endothelial cells?	Very high content of hydroxylysine; almost fully glycosylated; low alanine content; retains procollagen extensions.
V	$\alpha_1(\text{V}), \alpha_2(\text{V})$, and $\alpha_3(\text{V})$ chains; variable stoichiometry	Small fibrils	Widespread in small amounts; basement laminae of blood vessels and smooth muscle cells; exoskeleton of fibroblasts and other mesenchymal cells.	High content of hydroxylysine; heavily glycosylated; low alanine content.

*Modified and reproduced, with permission, from Eyre DR: Collagen: Molecular diversity in the body's protein scaffold. *Science* 1980;**207**: 1315. Copyright © 1980 by the American Association for the Advancement of Science.

a mature, extracellular collagen fiber (Table 59–5). Like most secreted proteins, collagen is synthesized on ribosomes in a precursor form, **preprocollagen,** which contains a leader or signal sequence that directs the polypeptide chain into the vesicular space of the endoplasmic reticulum. As it enters the endoplasmic reticulum this leader sequence is enzymatically removed. Hydroxylation of proline and lysine residues and glycosylation of hydroxylysines in this **procollagen** molecule also take place at this site. The procollagen molecule contains polypeptide extensions (**extension peptides**) of 20,000–35,000 MW at both its amino- and carboxy-terminal ends, neither of which is present in mature collagen. Both of these extension peptides contain cysteine residues. While the amino-terminal propeptide forms only intrachain disulfide bonds, the carboxy-terminal propeptides form both intrachain and interchain disulfide bonds. **Formation of these disulfide bonds assists in the registration of the three collagen molecules to form the triple helix,** winding from the carboxy-terminal end. After formation of the triple helix, no further hydroxylation of proline or lysine, or glycosylation can take place.

Following secretion from the cell by way of the Golgi apparatus, extracellular enzymes called **procollagen aminoproteinase** and **procollagen carboxyproteinase** remove the extension peptides at amino- and carboxy-terminal ends, respectively. Cleavage of these propeptides may occur within crypts or folds in the cell membrane. Once the propeptides are removed, the triple helical collagen molecules, containing approximately 1000 amino acids per chain, spontaneously assemble into collagen fibers. These are further stabilized by the formation of **inter- and intrachain crosslinks** through the action of lysyl oxidase, as described previously.

The same cells that secrete collagen also secrete **fibronectin,** a large glycoprotein present on cell surfaces, in the extracellular matrix, and in blood. Fibronectin binds to aggregating precollagen fibers and alters the kinetics of fiber formation in the pericellular matrix. Associated with fibronectin and procollagen in this matrix are the **proteoglycans** heparan sulfate and chondroitin sulfate (see Chapter 57). In fact, type IX collagen, a minor collagen type from cartilage, contains attached proteoglycan chains. Such interactions may serve to regulate the formation of collagen fibers and to determine their orientation in tissues.

Table 59–5. Order and location of processing of the fibrillar collagen precursor.

Intracellular
1. Cleavage of signal peptide
2. Hydroxylaton of Y-prolyl residues and some Y-lysyl residues; glycosylation of some hydroxylysyl residues
3. Formation of intrachain and interchain S-S bonds in extension peptides
4. Formation of triple helix

Extracellular
1. Cleavage of amino- and carboxy-terminal propeptides
2. Assembly of collagen fibers in quarter-staggered alignment
3. Oxidative deamination of ε-amino groups of lysyl and hydroxylysyl residues to aldehydes
4. Formation of intra- and interchain crosslinks via Schiff bases and aldol condensation products

Several Genetic Diseases Result in Defective Assembly, Stabilization, or Crosslinking of Collagen

The inherited diseases that result in abnormal collagen constitute an increasing number of variants of at

Table 59–6. General features of 4 heritable diseases that affect collagen.

Disease	Clinical Features	Biochemical Causes*
1. Osteogenesis imperfecta (multiple types)	Abnormal fragility of bones	Various abnormalities in genes coding for procollagen α1(I)
2. Ehlers-Danlos syndrome (multiple types)	Hypermobile joints, skin abnormalities	Abnormality in gene for procollagen α1(III), deficiency of lysyl hydroxylase, deficiency of procollagen N-proteinase
3. Marfan's syndrome	Skeletal, ocular and cardiovascular abnormalities	Mutation in gene for procollagen α2(I) resulting in longer procollagen chains
4. Menkes' syndrome	Kinky hair, growth retardation	Lysyl oxidase deficiency due to abnormal copper metabolism

*Only some of the most clearly elucidated causes of these diseases are listed.

least 4 different types of syndromes: **osteogenesis imperfecta, Marfan's syndrome, Ehlers-Danlos syndrome, and Menkes' (kinky-hair) syndrome.** Each of these syndromes is comprised of a number of variants. Originally, these diseases were defined on the basis of having similar phenotypes, but as the understanding of collagen structure and function has increased, it has become evident that similar molecular defects may present dissimilar clinical syndromes and vice versa. Basically, **assembly, stabilization** of the triple helix, or **crosslinking** of collagen are affected by the lesions. The defects include mutations causing **little or no synthesis** of procollagen chains, production of **shortened** procollagen chains, production of **lengthened** procollagen chains, and deficiencies of **processing enzymes** (eg, lysine hydroxylase or procollagen N-proteinase). So far, only defects in **type I** and **type III** procollagen molecules have been recognized. Menkes' syndrome affects copper metabolism (see Chapter 58), and thus secondarily affects the activity of lysine oxidase and resultant crosslinking. Table 59–6 summarizes major features of the above 4 syndromes and Fig 59–16 illustrates how 3 of them affect the structure of type I procollagen.

ELASTIN CONFERS EXTENSIBILITY & RECOIL ON LUNG, BLOOD VESSELS & LIGAMENTS

Elastin is a connective tissue protein that is responsible for properties of **extensibility** and **elastic recoil** in tissues. Although not as widespread as collagen, elastin is present in large amounts, particularly in tissues that require these physical properties, eg, **lung, large arterial blood vessels, and some elastic ligaments.** Smaller quantities of elastin are also found in skin, ear cartilage, and several other tissues. In contrast to collagen, there appears to be only **one genetic type of elastin,** although variants arise by **differential processing of the hnRNA** for elastin (see Chapter 39). Elastin is synthesized as a soluble monomer of 70,000 MW called **tropoelastin.** Some of the prolines of tropoelastin are hydroxylated to **hydroxyproline** by prolyl hydroxylase, although hydroxylysine and glycosylated hydroxylysine are not present. Unlike collagen, tropoelastin is not synthesized in a pro- form with extension peptides. Furthermore, elastin does *not* contain repeat Gly-X-Y sequences, triple helical structure, or carbohydrate moieties.

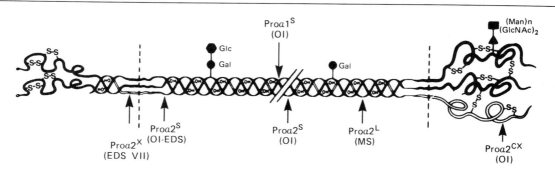

Figure 59–16. Approximate locations of mutations in the structure of type I procollagen. EDS, Ehlers-Danlos syndrome; MS, Marfan's syndrome; OI, osteogenesis imperfecta. Proα1s and proα2s, shortened proα chains; proα2X, poorly defined mutation altering the structure of proα, chains; proα2L, lengthened proα chains; proα2CX, mutation altering the structure of the C-propeptides of proα chains. (Reproduced, with permission, from Prockop DJ, Kivrikko KL: Heritable diseases of collagen. *N Engl J Med* 1984;**311**:376).

After secretion from the cell, certain lysyl residues of tropoelastin are oxidatively deaminated to aldehydes by **lysyl oxidase,** the same enzyme involved in this process in collagen. However, the major crosslinks formed in elastin are the **desmosines,** which result from the condensation of three of these lysine-derived aldehydes with an unmodified lysine to form a tetrafunctional crosslink unique to elastin. Once crosslinked in its mature, extracellular form elastin is **highly insoluble and extremely stable,** with a very low turnover rate. Elastin exhibits a variety of **random coil** conformations that permit the protein to **stretch and subsequently recoil** during the performance of its physiologic functions.

Table 59–7 summarizes the major differences between collagen and elastin.

REFERENCES

Adelstein RS, Eisenberg R: Regulation and kinetics of actin-myosin ATP interaction. *Annu Rev Biochem* 1980;**49:** 921.

Barany M, Barany K: Phosphorylation of the myofibrillar proteins. *Annu Rev Physiol* 1980;**42:**275.

Caplan A: Cartilage. *Sci Am* (Oct) 1984;**250:**84.

Eyre DR et al: Cross-linking in collagen and elastin. *Annu Rev Biochem* 1984;**53:**717.

Fuchs E, Hanukoglu I: Unraveling the structure of intermediate filaments. *Cell* 1983;**34:**332.

Heuser JE, Kirschner MW: Filament organization revealed

Table 59–7. Major differences between collagen and elastin.

Collagen	Elastin
1. Many different genetic types	One genetic type
2. Triple helix	No triple helix; random coil conformations permitting stretching
3. $(Gly-X-Y)_n$ repeating structure	No $(Gly-X-Y)_n$ repeating structure
4. Presence of hydroxylysine	No hydroxylysine
5. Contains carbohydrate	No carbohydrate
6. Intramolecular aldol crosslinks	Intramolecular desmosine crosslinks
7. Presence of extension peptides during biosynthesis	No extension peptides present during biosynthesis

in platinum replicas of freeze-dried cytoskeletons. *J Cell Biol* 1980;**86:**212.

Kleinman HK, Klebe RJ, Martin GR: Role of collagenous matrices in the adhesion and growth of cells. *J Cell Biol* 1981;**88:**473.

Lazarides E: Intermediate filaments: A chemically heterogeneous, developmentally regulated class of proteins. *Annu Rev Biochem* 1981;**51:**219.

Lazarides E, Revel JP: The molecular basis of cell movement. *Sci Am* (May) 1979;**240:**100.

Prockop DJ, Kivirikko KI: Heritable diseases of collagen. *N Engl J Med* 1984;**311:**376.

Rosenbloom J: Elastin: relation of protein and gene structure to disease. *Lab Invest* 1984;**51:**605.

Sandberg LB, Soskel NT, Leslie JG: Elastin structure, biosynthesis, and relation to disease states. *N Engl J Med* 1981;**304:**566.

Metabolism of Xenobiotics

60

Robert K. Murray, MD, PhD

INTRODUCTION

Increasingly, humans are subjected to exposure to various foreign chemicals (xenobiotics), whether they be drugs, food additives or pollutants, etc. The situation is well summarized in the following quotation: "As crude a weapon as the cave man's club, the chemical barrage has been hurled against the fabric of life." (Rachel Carson, 1907–1964). Carson was one of the first individuals to recognize the threat posed to survival of all life forms by indiscriminate chemical pollution and other abuses of the natural resources of the Earth. This is probably the most pressing matter facing mankind, and unless it is confronted squarely there will no longer be, in the not too distant future, any biochemistry on earth with which to be concerned. A knowledge of how xenobiotics are handled at the cellular level is one important aspect in learning how to cope with the chemical onslaught.

BIOMEDICAL IMPORTANCE

Knowledge of the metabolism of xenobiotics is basic to a rational understanding of pharmacology, toxicology, cancer research, and drug addiction. All of these areas involve administration of, or exposure to, xenobiotics.

HUMANS ENCOUNTER MANY XENOBIOTICS WHICH MUST BE METABOLIZED IN ORDER TO BE EXCRETED

A **xenobiotic** (Gk *xenos,* strange) is a compound that is foreign to the body. The principal classes of xenobiotics of medical relevance are **drugs, chemical carcinogens** and **various compounds** that have found their way into our **environment** by one route or another, such as polychlorinated biphenyls (PCBs) and certain insecticides.

Most of these compounds are subject to metabolism (chemical alteration) in the human body, with the **liver** being the main organ involved; occasionally a xenobiotic may be excreted unchanged. The metabolism of xenobiotics can be divided into two phases:

In **phase 1,** the major reaction involved is **hydroxylation,** catalyzed by members of a class of enzymes referred to as monooxygenases or cyto-

chrome P-450 species. Other types of reactions in phase 1 include **reduction** and **hydrolysis.**

In **phase 2,** the hydroxylated or other compounds produced in phase 1 are converted by specific enzymes to various polar metabolites by **conjugation** with glucuronic acid, sulfate, acetate, glutathione, or certain amino acids, or by **methylation.**

The overall purpose of the two phases of metabolism of xenobiotics is to increase their **water solubility** (**polarity**) and thus **facilitate their excretion** from the body. Very hydrophobic xenobiotics would persist in adipose tissue almost indefinitely if they were not converted to more polar forms. In certain cases, phase 1 metabolic reactions convert xenobiotics from inactive to biologically active compounds. In these instances, the original xenobiotics are referred to as **prodrugs or procarcinogens.** In other cases, additional phase 1 reactions (eg, further hydroxylation reactions) convert the active compounds to less active or inactive forms, prior to conjugation. In yet other cases, it is the conjugation reactions themselves that convert the active products of phase 1 reactions to less active or inactive species, which are subsequently excreted in the urine or bile. In a very few cases, conjugation may actually **increase** the biologic activity of a xenobiotic.

Detoxification is sometimes used to refer to many of the reactions involved in the metabolism of xenobiotics. However, it is not always an appropriate term, because, as mentioned above, in some cases the reactions to which xenobiotics are subject actually increase their biologic activity and toxicity.

CYTOCHROME P-450 HYDROXYLATES XENOBIOTICS IN PHASE 1 OF THEIR METABOLISM

Hydroxylation is the chief reaction involved in phase 1. The responsible enzymes are called **monooxygenases** or cytochrome P-450 species. The reaction catalyzed by a monooxygenase (or cytochrome P-450 species) is:

$$RH + O_2 + NADPH + H^+ \rightarrow R\text{–}OH + H_2O + NADP \quad (1)$$

RH above can represent a very wide variety of drugs, carcinogens, pollutants (such as a mixture of PCBs), and certain endogenous compounds, such as steroids. In many cases the endogenous substrates of cytochrome P-450 species have not been defined, but it seems unlikely that these enzymes evolved just to deal with exogenous compounds. The actual reaction mechanism is complex and will not be described here. However, it has been shown by the use of $^{18}O_2$, that one atom of oxygen enters R—OH and one atom enters water. This dual fate of the oxygen accounts for the former naming of monooxygenases as **mixed function oxidases.** The reaction catalyzed by cytochrome P-450 can also be represented as:

$$\text{Reduced cyt. P-450} \curvearrowright \text{Oxidized cyt. P-450} \qquad (2)$$

$$RH + O_2 \rightarrow R\text{—}OH + H_2O$$

The major monooxygenases in the ER are **cytochrome P-450** species. The name cytochrome P-450 is used because the enzyme was discovered when it was noted that preparations of microsomes that had been chemically reduced and then exposed to carbon monoxide exhibited a distinct peak at 450 nm. This enzyme is extremely important because it has been estimated that approximately 50% of the drugs that patients ingest are metabolized by species of cytochrome P-450. The same enzyme also acts on various carcinogens and pollutants.

Important Points Concerning Cytochrome P-450 Species

The following are important points concerning cytochrome P-450 species.

1. Like hemoglobin, they are **hemoproteins.**

2. They are present in highest amount in the **membranes of the endoplasmic reticulum** (ER) (microsomal fraction) of **liver,** in which membranes they can comprise approximately 20% of the total protein. They are also found in other tissues. In the **adrenal,** they are found in mitochondria as well as in the ER; the various hydroxylases present in that organ play an important role in steroid biosynthesis (see Chapter 49).

3. There are **at least 6 closely related species** of cytochrome P-450 present in liver ER, each with wide and somewhat overlapping substrate specificities, that act on a very wide variety of drugs, carcinogens, and other xenobiotics, in addition to endogenous compounds such as certain steroids. The genes for these species of cytochrome P-450 have been isolated and studied in detail in recent years.

4. **NADPH,** not NADH, is involved in the reaction mechanism of cytochrome P-450. The enzyme that uses NADPH to yield the reduced cytochrome P-450, shown in the left-hand side of equation (2), is called **NADPH-cytochrome P-450 reductase.** This enzyme is an important component of the cytochrome P-450 system.

5. Lipids are also components of the cytochrome P-450 system. The preferred lipid is **phosphatidyl choline,** which is the major lipid found in membranes of the ER.

6. Most species of cytochrome P-450 are **inducible.** For instance, the administration of phenobarbital (PB) or of many other drugs, causes a hypertrophy of the smooth ER and a 3- to 4-fold increase of the amount of cytochrome P-450 within 4–5 days. The mechanism of induction has been studied extensively and involves increased transcription of mRNA for cytochrome P-450.

Induction of this enzyme has important clinical implications, as it is one biochemical mechanism of **drug interaction.** A drug interaction has occurred when the effects of one drug are altered by the prior or concurrent administration of another. To illustrate this, suppose that a patient is taking the anticoagulant dicoumarol to prevent blood clotting. This drug is metabolized by the cytochrome P-450 system. Then suppose it is discovered that the patient also needs PB, to treat some other condition, such as a certain type of epilepsy. Accordingly, administration of PB is started but the dose of dicoumarol is not changed. After 5 days or so, the level of cytochrome P-450 in the patient's liver will be elevated 3- to 4-fold. This in turn means that the dicoumarol will be metabolized much more quickly than before, and its dosage will have become inadequate. Therefore, the dose must be increased if it is to be therapeutically effective. To pursue this example further, a problem could arise later on if the patient is taken off PB, but the increased dosage of dicoumarol is continued. The patient will be at risk of bleeding since the high dose of dicoumarol will be even more active than it was previously, because the level of cytochrome P-450 will decline once PB has been stopped.

7. One of the species of cytochrome P-450 has its characteristic absorption peak not at 450 nm, but at 448 nm. It is called **cytochrome P-448.** This species appears to be relatively specific for the metabolism of polycyclic aromatic hydrocarbons (PAHs) and related molecules; for this reason it is called **aromatic hydrocarbon hydroxylase** (AHH). This enzyme is very important in the metabolism of PAHs and in carcinogenesis produced by these agents. For example, in the lung it may be involved in the conversion of inactive PAHs (procarcinogens), inhaled by smoking, to active carcinogens by hydroxylation reactions. Smokers have higher levels of this enzyme in some of their cells and tissues than do nonsmokers. Some reports have suggested that the activity of this enzyme may be elevated (induced) in the **placentae** of women who smoke, thus potentially altering the quantities of metabolites of PAHs (some of which could be harmful) to which the fetus is exposed.

CONJUGATION REACTIONS PREPARE XENOBIOTICS FOR EXCRETION IN PHASE 2 OF THEIR METABOLISM

In phase 1 reactions, xenobiotics are generally converted to more polar, hydroxylated derivatives. In phase 2 reactions, these derivatives are conjugated with molecules such as glucuronic acid, sulfate, or glutathione. This renders them even more water soluble and they are eventually excreted in the urine or bile.

There Are at Least 5 Types of Phase 2 Reactions

Glucuronidation: The glucuronidation of bilirubin was discussed in Chapter 34. The reactions whereby xenobiotics are glucuronidated are essentially similar. **UDP-glucuronic acid** is the glucuronyl donor, and a variety of **glucuronyl transferases,** present in both the ER and cytosol, are the catalysts. Molecules such as 2-acetylaminofluorene (a carcinogen), aniline, benzoic acid, meprobromate, phenol, and many steroids are excreted as glucuronides. The glucuronide may be attached to oxygen, nitrogen, or sulfur groups of the substrates. Glucuronidation is probably the most frequent conjugation reaction.

Sulfation: Some alcohols, arylamines, and phenols are sulfated. The **sulfate donor** in these and other biologic sulfation reactions (eg, sulfation of steroids, glycosaminoglycans, glycolipids, and glycoproteins) is adenosine 3'-phosphate-5'-phosphosulfate (PAPS) (cf Chapter 26); this compound is called **active sulfate.**

Conjugation With Glutathione: Glutathione (γ-glutamylcysteinylglycine) is a tripeptide consisting of glutamic acid, cysteine, and glycine (see Fig 4–4). Glutathione is commonly abbreviated to GSH; the SH indicates the sulfhydryl group of its cysteine and is the business part of the molecule. A number of potentially toxic **electrophilic xenobiotics** (such as certain carcinogens) are conjugated to the nucleophilic GSH, in reactions that can be represented:

$$R + GSH \rightarrow R\text{--}S\text{--}G$$

where R = an electrophilic xenobiotic. The enzymes catalyzing these reactions are called **glutathione S-transferases** and are present in high amounts in liver cytosol and in lower amounts in other tissues. A variety of glutathione S-transferases are present in human tissue. They exhibit different substrate specificities and can be separated by electrophoretic and other techniques. If the potentially toxic xenobiotics were not conjugated to GSH, they would be free to combine covalently with DNA, RNA, or cell protein and could thus lead to serious cell damage. GSH is thus an important **defense mechanism** against certain toxic compounds, such as some drugs and carcinogens. If the levels of GSH in a tissue such as liver are lowered (as can be achieved by the administration to rats of certain compounds that react with GSH), then that tissue can be shown to be more susceptible to injury by various chemicals that would normally be conjugated to GSH.

Glutathione-conjugates are subjected to further metabolism prior to excretion. The glutamyl and glycinyl groups belonging to glutathione are removed by specific enzymes and an acetyl group (donated by acetyl-CoA) is added to the amino group of the remaining cysteinyl moiety. The resulting compound is a **mercapturic acid,** a conjugate of L-acetylcysteine, which is then excreted in the urine.

Glutathione has **other important functions** in human cells, apart from its role in xenobiotic metabolism.

(1) It participates in the **decomposition of potentially toxic hydrogen peroxide** in the reaction catalyzed by glutathione peroxidase (see Chapter 21). (2) It is an important intracellular **reductant,** helping to maintain essential SH groups of enzymes in their reduced state. When GSH acts as a reducing agent, its SH becomes oxidized and forms a disulfide link with another molecule of glutathione:

$$GSH + GSH \leftrightarrow G\text{--}S\text{--}S\text{--}G$$

The product G–S–S–G is oxidized glutathione. In turn, G–S–S–G can be reduced to GSH by the action of **glutathione reductase,** in a reaction using NADPH:

$$G\text{--}S\text{--}S\text{--}G + NADPH + H^+ \rightarrow 2\,GSH + NADP$$

This reaction is particularly important in red blood cells. If levels of NADPH are low in these cells, as occurs when the activity of G6P dehydrogenase is low (see Chapter 21), then reduced GSH is not regenerated in adequate amounts by the above reaction. The reduced levels of GSH allow peroxides to accumulate inside red blood cells, and hemolysis can occur due to their oxidative effect on the lipids of the red cell membrane. (3) A metabolic cycle involving GSH as a carrier has been implicated in the **transport of certain amino acids across membranes** in the kidney. The first reaction of the cycle is:

$$\text{Amino acid} + GSH \rightarrow \gamma\text{-Glutamyl amino acid} + \text{Cysteinylglycine}$$

This reaction helps to transfer certain amino acids across the plasma membrane, the amino acid being subsequently hydrolyzed from its complex with GSH and the GSH being resynthesized from cysteinylglycine. The enzyme catalyzing the above reaction is γ-**glutamyltransferase** (GGT). It is present in the plasma membrane of renal tubular cells and in the ER of hepatocytes. The enzyme has diagnostic value, because it is released into the blood from hepatic cells in various hepatobiliary diseases (see Appendix, under γ-Glutamyl Transpeptidase).

Other Reactions: The two most important are acetylation and methylation. **Acetylation** is represented by:

$$X + \text{Acetyl-CoA} \rightarrow \text{Acetyl-X} + \text{CoA}$$

where X represents a xenobiotic. As for other acetylation reactions, acetyl-CoA (active acetate) is the acetyl donor. These reactions are catalyzed by acetyltransferases present in the cytosol of various tissues, particularly liver. The drug **isoniazid,** used in the treatment of tuberculosis, is subject to acetylation. Polymorphic types of acetyltransferases exist, resulting in individuals who are classified as **slow and fast acety-**

lators, and influence the rate of clearance of drugs such as isoniazid from blood. Slow acetylators are more subject to certain toxic effects of isoniazid, because the drug persists longer in these individuals.

A few xenobiotics are subject to **methylation** by methyltransferases, employing S-adenosylmethionine (see Fig 31–22) as the methyl donor.

THE ACTIVITIES OF XENOBIOTIC-METABOLIZING ENZYMES ARE AFFECTED BY AGE, SEX, & OTHER FACTORS

Various factors affect the activities of the enzymes metabolizing xenobiotics. Such factors include the following:

(1) The activities of these enzymes may differ substantially among **species.** This is important as it means that results on, for example, the possible toxicity or carcinogenicity of xenobiotics cannot be extrapolated freely from one species to another.

(2) There are also significant differences in enzyme activities among **individuals,** many of which appear to be due to **genetic factors.**

(3) The activities of some of these enzymes vary according to **age and sex.**

(4) Intake of various xenobiotics such as phenobarbital, PCBs, or certain hydrocarbons can cause **enzyme induction.** It is thus very important to know whether or not an individual has been exposed to these inducing agents in evaluating biochemical responses to xenobiotics.

(5) Metabolites of certain xenobiotics can **inhibit** the activities of xenobiotic-metabolizing enzymes. The above considerations help explain the often appreciable differences in responses to xenobiotics (such as drugs, toxic chemicals, and carcinogens) noted among different species and among individuals of the same species.

RESPONSES TO XENOBIOTICS INCLUDE PHARMACOLOGIC, TOXIC, IMMUNOLOGIC, & CARCINOGENIC EFFECTS

Xenobiotics are metabolized in the body by the reactions described above. When the xenobiotic is a **drug,** phase 1 reactions may produce its active form or may diminish or terminate its action if it is pharmacologically active in the body without prior metabolism. The diverse effects produced by drugs comprise the area of study of pharmacology; here it is important to appreciate that drugs act through biochemical mechanisms.

Certain xenobiotics are very **toxic,** even at low levels (eg, cyanide). On the other hand, there are few xenobiotics, including drugs, that do not exert some toxic effects if sufficient amounts are administered. The toxic effects of xenobiotics cover an extremely wide spectrum. However, there are three general types of effects (see Fig 60–1) that will be mentioned briefly here, because of their relationship to xenobiotic metabolism.

The first of these is **cell injury,** which can be severe enough to result in cell death. There are many mecha-

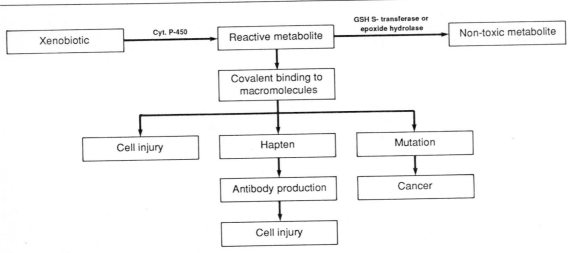

Figure 60–1. Simplified scheme showing how metabolism of a xenobiotic can result in cell injury, immunologic damage, or cancer. In this instance, the conversion of the xenobiotic to a reactive metabolite is catalyzed by a species of cytochrome P-450 and the conversion of the reactive metabolite (eg, an epoxide) to a nontoxic metabolite is catalyzed by either a GSH S-transferase or epoxide hydrolase.

nisms by which xenobiotics injure cells. The one considered here is covalent binding to cell macromolecules of reactive species of xenobiotics produced by metabolism. These macromolecular targets include DNA, RNA, and protein. If the macromolecule to which the reactive xenobiotic binds is essential for short-term cell survival, eg, a protein or enzyme involved in some critical cellular function such as oxidative phosphorylation or regulation of the permeability of the plasma membrane, then it is apparent that severe effects on cellular function could become evident quite rapidly.

Second, the reactive species of a xenobiotic may bind to a protein, modifying it and altering its antigenicity. The xenobiotic is said to act as a **hapten,** ie, a small molecule that by itself does not stimulate antibody synthesis but will combine with antibody once formed. The resulting antibodies can then damage the cell by several immunologic mechanisms that grossly perturb normal cellular biochemical processes.

Third, reactions of activated species of chemical carcinogens with DNA are thought to be of great importance in **chemical carcinogenesis** (see Chapter 61). Some chemicals (eg, benzpyrene) require activation by monooxygenases in the ER to become carcinogenic (they are thus called **indirect carcinogens**). The activities of the monooxygenases and of other xenobiotic-metabolizing enzymes present in the ER thus help to determine whether such compounds become carcinogenic or are "detoxified." Other chemicals (eg, various alkylating agents) can react directly (**direct carcinogens**) with DNA, without undergoing intracellular chemical activation.

The enzyme **epoxide hydrolase** is of interest because it can exert a protective effect against certain carcinogens. The products of the action of certain monooxygenases on some procarcinogen substrates are **epoxides.** Epoxides are highly reactive and mutagenic and/or carcinogenic. Epoxide hydrolase, like cytochrome P-450 also present in the membranes of the ER, acts on these compounds, converting them into much less reactive dihydrodiols. The reaction catalyzed by epoxide hydrolase can be represented as follows:

$$-\overset{|}{\underset{}{C}}-\overset{|}{\underset{}{C}}- \ + \ H_2O \rightarrow -\overset{|}{\underset{HO}{C}}-\overset{|}{\underset{OH}{C}}-$$

$$\text{Epoxide} \qquad\qquad \text{Dihydrodiol}$$

REFERENCES

Katzung BG (editor): *Basic & Clinical Pharmacology,* 4th ed. Appleton & Lange, 1989.

Klaassen CD, Amdur MO, Doull J (editors): *Casarett & Doull's Toxicology,* 3rd ed. Macmillan, 1986.

Nebert DW, Gonzalez FJ: P450 genes: structure, evolution and regulation. *Annu Rev Biochem* 1987;**56**:945.

Nebert DW, Gonzalez FJ: P450 genes and evolutionary genetics. *Hosp Pract* (March) 1987;**22**:63.

61

Cancer, Oncogenes, & Growth Factors

Robert K. Murray, MD, PhD

INTRODUCTION

Cancer cells are characterized by 3 properties: (1) diminished or unrestrained control of growth; (2) invasion of local tissues, and (3) spread, or metastasis, to other parts of the body. Cells of benign tumors also show diminished control of growth but do not invade local tissue or spread to other parts of the body.

In this chapter, we discuss some biochemical aspects of cancer. The key issues are to explain in biochemical terms the uncontrolled growth of cancer cells and their ability to invade and metastasize. The genes controlling growth and interactions with other normal cells are apparently abnormal in structure or regulation in cancer cells. Information on how cell growth—both normal and pathologic—is controlled is limited, and knowledge of specific genes involved in growth regulation is even more meager. Little is known as yet about the biochemical basis of metastasis, so that coverage of this topic will be brief. At least some types of cancer (eg, certain leukemias) can be regarded as examples of abnormal differentiation. Again, astonishingly little is known about the molecular basis of differentiation. However, many workers in this area think that further research on oncogenes and growth factors will provide insight into the nature of the disturbed control of growth, of differentiation (where applicable), and of cell-cell interaction exhibited by cancer cells. Thus, both of these topics will be discussed in some detail.

BIOMEDICAL IMPORTANCE

Cancer is the second most common cause of death in the USA after cardiovascular disease. Humans of all ages develop cancer, and a wide variety of organs are affected. The incidence of many cancers increases with age, so that as people live longer, more will develop the disease. Apart from individual suffering, the economic burden to society is immense.

BIOCHEMICAL LABORATORY TESTS ARE ESSENTIAL IN THE MANAGEMENT OF PATIENTS WITH CANCER

Biochemical laboratory tests help in the management of patients with cancer. Many cancers are associated with the abnormal production of enzymes, proteins, and hormones (see the discussion of the progression of tumors, below), which can be measured in plasma or serum. These molecules are known a **tumor markers.** Measurement of some tumor markers is now an integral part of management of some types of cancer (Table 61–1). Applications of tumor markers in diagnosis and management of cancer are listed in Table 61–2. Three major conclusions have emerged from the study of tumor markers (McIntire, 1984): (1) No single marker is useful for all types of cancer or for all patients with a given type of cancer. For this reason, the use of a battery of tumor markers is sometimes advantageous. (2) Markers are most often detected in advanced stages of cancer rather than early stages, when they would be more helpful. (3) Of the uses of markers listed in Table 61–2, the most successful have been the monitoring of responses to therapy and the detection of early recurrence.

Table 61–1. Clinically useful tumor markers.*

Marker	Associated Cancer
Carcinoembryonic antigen (CEA)	Colon, lung, breast, pancreas
Alpha-fetoprotein (AFP)	Liver, germ cell
Human chorionic gonadotropin (hCG)	Trophoblast, germ cell
Calcitonin (CT)	Thyroid (medullary carcinoma)
Prostatic acid phosphatase (PAP)	Prostate

*Adapted, with permission, from McIntire KR: Tumor markers: How useful are they? *Hosp Pract* (Dec) 1984;**19**:55.

Table 61–2. Applications of tumor markers.*

Detection: Screening in asymptomatic persons.
Diagnosis: Differentiating malignant from bening conditions.
Monitoring: Predicting effect of therapy and detecting recurrent cancer.
Classification: Choosing therapy and predicting tumor behavior (prognosis).
Staging: Defining extent of disease.
Localization: Nuclear scanning of injected radioactive antibodies.
Therapy: Cytotoxic agents directed to marker-containing cells.

*Adapted, with permission, from McIntire KR: Tumor markers: How useful are they? *Hosp Pract* (Dec) 1984;**19**:55.

PHYSICAL, CHEMICAL, & BIOLOGIC AGENTS CAN CAUSE CANCER

Agents causing cancer fall into 3 broad groups: radiant energy, chemical compounds, and viruses.

Radiant Energy Can Be Carcinogenic

Ultraviolet rays, x-rays, and γ-rays are mutagenic and carcinogenic. These rays damage DNA in several ways. Ultraviolet radiation may cause pyrimidine dimers to form. Apurinic or apyrimidinic sites may form by elimination of corresponding bases. Single- and double-strand breaks or crosslinking of strands may occur. Damage to DNA is presumed to be the basic mechanism of carcinogenicity with radiant energy, but the details are unclear. Repair of DNA is discussed in Chapter 38. Apart from direct effects on DNA, x-rays and γ-rays cause **free radicals** to form in tissues. The resultant ·OH, superoxide, and other radicals can interact with DNA and other macromolecules, leading to molecular damage and thereby probably contributing to carcinogenic effects of radiant energy.

Many Chemicals Are Carcinogenic

A wide variety of chemical compounds are carcinogenic (Table 61–3); the structures of 3 of the most widely studied are shown in Fig 61–1. Most of the

Table 61–3. Some chemical carcinogens.

Class	Compound
Polycyclic aromatic hydrocarbons	Benzo[a]pyrene, dimethylbenzanthracene
Aromatic amines	2-Acetylaminofluorene, N-methyl-4-aminoazobenzene (MAB)
Nitrosamines	Dimethylnitrosamine, diethylnitrosamine
Various drugs	Alkylating agents (eg, cyclophosphamide), diethylstilbestrol
Naturally occurring compounds	Dactinomycin, aflatoxin B₁
Inorganic compounds	Arsenic, asbestos, beryllium, cadmium, chromium

compounds listed in Table 61–3 have been tested by administration to rodents or other animals. However, many substances are associated with the development of cancer in humans. It is estimated that up to 80% of human cancers are caused by environmental factors, principally chemicals. Exposure to such compounds can occur because of a person's **occupation** (eg, benzene, asbestos); **diet** (eg, aflatoxin B_1, which is produced by the mold *Aspergillus flavus* and sometimes found as a contaminant of peanuts and other foodstuffs); **life-style** (eg, cigarette smoking); or in other ways (eg, certain therapeutic **drugs** can be carcinogenic). We shall present only a few important generalizations that have emerged from the study of chemical carcinogenesis.

Structure: Both organic and inorganic molecules may be carcinogenic (Table 61–3). The diversity of these compounds indicates that they do not possess one common structural feature that confers carcinogenicity.

Action: The organic carcinogens have been the most thoroughly studied. Some, such as nitrogen mustard and β-propiolactone, have been found to interact directly with target molecules (**direct carcinogens**), but others require prior metabolism to become carcinogenic (**procarcinogens**) (cf. Chapter 60). The process whereby one or more enzyme-catalyzed reactions convert procarcinogens to active carcinogens is called **metabolic activation**. Any intermediate compounds formed are **proximate carcinogens,** and the final compound that reacts with cellular components (eg, DNA) is the **ultimate carcinogen.** The sequence is thus:

Procarcinogen → Proximate carcinogen A → Proximate carcinogen B → Ultimate carcinogen

The procarcinogen itself is not a chemically reactive species, whereas the ultimate carcinogen is often highly reactive. At least 2 reactions are required to convert the procarcinogen 2-acetylaminofluorene (2-AAF) to the ultimate carcinogen, the sulfate ester of N-hydroxy-AAF. An important generalization is that ultimate carcinogens are usually **electrophiles** (ie, molecules deficient in electrons), which readily attack nucleophilic (electron-rich) groups in DNA, RNA, and proteins.

Metabolism of Chemical Carcinogens: The metabolism of procarcinogens and other xenobiotics involves monooxygenases and transferases (see Chapter 60). The enzymes responsible for metabolic activation of procarcinogens are principally species of cytochrome P-450, located in the endoplasmic reticulum. These are the **same enzymes that are involved in the metabolism of other xenobiotics,** such as drugs and environmental pollutants (eg, PCBs). The polycyclic aromatic hydrocarbons are of great interest in chemical carcinogenesis. The particular monooxygenase involved in their metabolism is named **cytochrome P-448,** or **aromatic hydrocarbon hydroxylase.** The

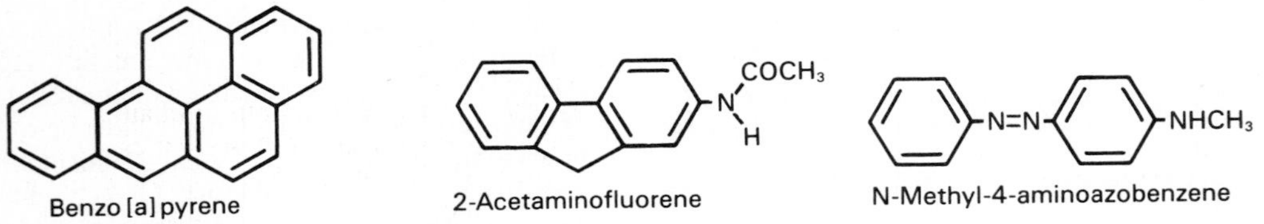

Figure 61–1. Structures of 3 important experimentally used chemical carcinogens.

activities of the enzymes metabolizing chemical carcinogens are affected by a number of factors, such as species, genetic considerations, age, or sex. The variations in activities of these enzymes help explain the often appreciable differences in the carcinogenicity of chemicals among different species and individuals of the same species. Many of the above points are discussed in more detail in Chapter 60.

Covalent Binding: When chemical carcinogens are administered to animals or placed in cultured cells, it can be shown (eg, by using radioactive carcinogens) that they or their derivatives generally bind covalently to cellular macromolecules, including DNA, RNA, and proteins. The chemical natures of the adducts formed by interaction of certain ultimate carcinogens with their target molecules have been determined. Most interest has focused on products formed with DNA. Carcinogens have been found to interact with the purine, pyrimidine, or phosphodiester groups of DNA. The most common site of attack is guanine, and the addition of various carcinogens to the N_2, N_3, N_7, O_6, and O_8 atoms of this base has been observed.

Damage to DNA: The covalent interaction of direct carcinogens or ultimate carcinogens with DNA can result in several types of damage; this damage can be repaired, as described in Chapter 38.

Despite the existence of repair systems, certain modifications of DNA by chemical carcinogens persist for relatively long periods of time. It is possible that these persistent unrepaired lesions are of special importance in generating mutations critical to carcinogenesis.

Mutagens: Most chemical carcinogens are mutagens. This has been demonstrated using the Ames assay (see below) and other tests. At a molecular level, transitions, transversions, and other types of mutation (see Chapters 38 and 40) have been shown to occur following exposure of certain bacteria to ultimate carcinogens. It has been assumed that some types of cancer are due to mutations in somatic cells that affect key regulatory processes. Direct evidence of this has now been obtained (see the discussion of oncogenes, below).

Since testing the carcinogenicity of chemicals in animals is slow and expensive, assays for screening the potential carcinogenicity of chemical compounds have been developed. Many are based on detection of the mutagenicity of chemical carcinogens. Such assays are more rapid and less expensive than detecting tumors in animals. None is ideal, since the ultimate test

of a carcinogen is to show that it causes tumors in animals. However, one assay based on detecting mutagenicity, the **Ames assay,** has proved useful in screening for potential carcinogens. This assay uses a specially constructed strain of *Salmonella typhimurium* that has a mutation (His$^-$) in a gene that codes for one of the enzymes involved in the synthesis of histidine. Thus, these particular salmonellae cannot synthesize histidine, which must be present in the medium for growth to occur. When a mutation caused by a carcinogen occurs at the site of the His$^-$ mutation, the latter mutation can restore its reading sequence, converting it to His$^+$. The progeny from bacteria containing such a reverse mutation can now synthesize histidine and thus grow in a medium lacking it. Such salmonellae can be detected as readily observable and quantifiable colonies growing on agar plates.

One problem with the use of bacteria in mutagenicity tests is that they **do not contain the spectrum of monooxygenases** found in higher animals. Thus, if a compound requires activation to become a mutagenic or carcinogenic species, this may not occur when bacteria are used. Ames circumvented this problem by incubating the agents to be tested in a postmitochondrial supernatant of rat liver (the S-9 fraction, which is the supernatant fraction after centrifuging a rat liver homogenate at 9000 g for a suitable period of time). The S-9 fraction contains most of the various monooxygenases and other enzymes required to activate potential mutagens and carcinogens.

The Ames assay identifies approximately 90% of known carcinogens. It is becoming routine to test newly synthesized chemicals by this assay, particularly if they are to be introduced commercially or widely used in industry. Compounds giving a positive reaction should undergo further testing, including assessment of carcinogenicity in animals.

Initiation & Promotion: In certain organs such as skin and liver, it has been shown that carcinogenesis can be divided into at least 2 stages. The classic example is skin. Typically, identical areas of the skin of a group of mice are painted once with benzo[a]pyrene. If no other subsequent treatment is used, no skin tumors develop (Fig 61-2). However, if the application of benzo[a]pyrene is followed by several applications of croton oil, many tumors subsequently develop. Applications of croton oil alone (ie, no pretreatment with benzo[a]pyrene) do not result in skin tumors. Many other variants of this basic protocol have been carried out, permitting the following conclusions: (1) The

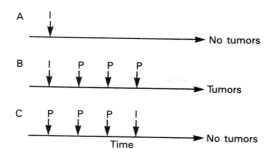

Figure 61–2. Diagrammatic representation of the stages of initiation and promotion of chemical carcinogenesis in skin. *A:* One application of the initiator (eg, benzo[a]pyrene) is made to the skin of a number of mice. *B:* The application of the initiator is followed by a number of applications of a promoter (eg, croton oil) at, for instance, weekly intervals. *C:* The promoter is applied first, and then the initiator is applied. Benign tumors of the skin (papillomas) may appear in about 100 days; malignant tumors (carcinomas) take about 1 year to appear. Many other informative variants of this protocol have been performed, all substantiating the basic concepts of initiation and promotion. I, initiator; P, promoter.

stage of carcinogenesis caused by application of benzo[a]pyrene is called **initiation;** this stage appears to be rapid and irreversible. It is presumed to involve an irreversible modification of DNA, perhaps resulting in one or more mutations. Benzo[a]pyrene is thus called an **initiating agent.** (2) The second, much slower (ie, months or years) stage of carcinogenesis, resulting from application of croton oil, is called **promotion.** Croton oil is thus a promoting agent, or **promoter.** Promoters are incapable of causing initiation. (3) Most carcinogens are capable of acting as **both** initiating and promoting agents.

A large number of compounds, including phenobarbital and saccharin, can act as promoters in various organs. The active agent of croton oil is a mixture of phorbol esters. The most active phorbol ester is 12-0-tetradecanoylphorbol-13-acetate (TPA), which has numerous effects. The most interesting finding has been that protein kinase C can act as a receptor for TPA. Stimulation of the activity of this enzyme by interaction with TPA may result in the phosphorylation of a number of membrane proteins, leading to effects on transport and other functions. This important result ties in the action of certain tumor promoters to the field of transmembrane signaling (see the discussion of growth factors, below). Many tumor promoters appear to act by causing alterations of gene expression, but the precise mechanisms by which promoters influence the initiated cell to become a tumor cell remain to be determined.

DNA Is the Critical Macromolecule in Carcinogenesis

DNA is the critical target molecule in carcinogenesis. The following facts support this conclusion. (1)

Cancer cells beget cancer cells, ie, the essential changes responsible for cancer are transmitted from mother to daughter cells. This is consistent with the behavior of DNA. (2) Both irradiation and chemical carcinogens damage DNA and are capable of causing mutations in DNA. (3) Many tumor cells exhibit abnormal chromosomes. (4) Transfection experiments (see below) indicate that purified DNA (oncogenes) from cancer cells can transform normal cells into (potential) cancer cells. However, epigenetic factors may also play a role in carcinogenesis.

Some DNA & RNA Viruses Are Carcinogenic

Oncogenic viruses contain either DNA or RNA as their genome (Table 61–4). Only a few important features of the major members of these 2 classes will be described here.

Polyomavirus and **SV40 viruses** have played an important role in the development of current ideas about viral oncogenesis. They are both small (containing a genome of about 5 kb), and their circular genomes code for only about 5–6 proteins. Under certain circumstances, infection of appropriate cells with these viruses can result in malignant transformation. Specific viral proteins are known to be involved. In the case of SV40, these proteins (often called **antigens,** because they were detected by immunologic methods) are known as T ("large T") and t ("small t"), and in the case of polyomavirus, they are known as T, mid-T, and t. (T refers to the fact that the first of these proteins was detected in a tumor.) How these proteins cause malignant transformation is still under investigation; the T antigens are known to bind tightly to DNA and cause alterations in gene expression. These proteins show cooperative effects, suggesting that alteration of more than one reaction or process is required for transformation.

Some types of adenovirus are known to cause transformation of certain animal cells. There is considerable interest in the Epstein-Barr virus, since it is associated with Burkitt's lymphoma and nasopharyngeal carcinoma in humans. Herpes simplex virus (type 2)

Table 61–4. Some important tumor viruses.

Class	Members
DNA viruses Papovavirus	Polyomavirus, SV40 virus, papillomavirus
Adenovirus	Adenoviruses 12, 18, and 31
Herpesvirus	Epstein-Barr virus, herpes simplex type 2 virus
Hepadnavirus	Hepatitis B virus
RNA viruses Retrovirus type C	Murine sarcoma and leukemia viruses, avian sarcoma and leukemia viruses, human T cell leukemia viruses I and II
Retrovirus type B	Mouse mammary tumor virus

has been associated with cancer of the cervix, and hepatitis B virus may be associated with some cases of liver cancer in humans.

Since much of the knowledge of oncogenes obtained in recent years has emerged from the study of RNA-containing tumor viruses, the subsequent discussion of oncogenes contains frequent references to these viruses.

BOTH MORPHOLOGIC & BIOCHEMICAL CHANGES OCCUR UPON MALIGNANT TRANSFORMATION

When cultured cells are infected with certain oncogenic viruses, they may undergo malignant transformation. The most important morphologic and biochemical changes occurring upon transformation are listed in Table 61–5. These changes affect cell shape, motility, adhesiveness to the culture dish, growth, and a number of biochemical processes. They are interpreted as reflecting the primary processes that cause—and the secondary changes that result from—conversion from the normal to the malignant state. The ability to equate transformation approximately with acquisition of malignant properties has been of tremendous importance in cancer research. However, acquisition by cells of the changes collectively known as transformation does not necessarily mean that such cells will display the same biologic properties as tumor cells in vivo; cells must yield tumors when injected into a suitable host animal.

ONCOGENES PLAY A CRUCIAL ROLE IN CARCINOGENESIS

Oncogenes are genes capable of causing cancer. Their discovery has had a major impact on research on

Table 61–5. Some change shown by cultured cells which suggest that malignant transformation has occurred (eg, after infection by an oncogenic virus). The crucial test of malignancy is the ability of cells to grow into tumors in vivo.

Alterations of morphology: Transformed cells often have a much rounder shape than control cells.
Increased cell density (loss of contact inhibition of growth): Transformed cells often form multilayers, while control cells usually form a monolayer.
Loss of anchorage dependence: Transformed cells can grow without attachment to the surface of the culture dish and will often grow in agar.
Loss of contact inhibition of movement: Transformed cells grow over one another, while normal cells stop moving when they come into contact with each other.
A variety of biochemical changes, including an increased rate of glycolysis, alterations of the cell surface (eg, changes in the composition of glycoproteins or glycosphingolipids), and secretion of certain proteases.
Alterations of cytoskeletal structures, such as actin filaments.
Diminished requirement for growth factors and, often, increased secretion of certain growth factors into the surrounding medium.

the fundamental mechanisms involved in carcinogenesis. Oncogenes were first recognized as unique genes of tumor-causing viruses that are responsible for the process of transformation (**viral oncogenes**).

Oncogenes of Rous Sarcoma Virus: Analyses of the oncogene of the Rous sarcoma virus and its product have been particularly revealing. The genome of this retrovirus contains 4 genes named *gag, pol, env,* and *src.* This can be shown schematically as follows:

The *gag* gene codes for group-specific antigens of the virus, *pol* for the reverse transcriptase that characterizes retroviruses, and *env* for certain glycoproteins of the viral envelope. A **protein-tyrosine kinase** was shown to be the product of *src* (ie, the sarcoma-causing gene) that is responsible for transformation. This finding was of fundamental importance. It revealed a specific biochemical mechanism (ie, abnormal phosphorylation of a number of proteins) that could explain, at least in part, how a tumor virus could cause the **pleiotropic effects** of transformation. The **critical cell proteins,** whose abnormal phosphorylation presumably leads to transformation, are still to be defined. **Vinculin,** a protein found in focal adhesion plaques (structures involved in intercellular adhesion), is one candidate. The abnormal phosphorylation of vinculin in focal adhesion plaques could help explain the rounding-up of cells and their diminished adhesion to the substratum and to one another observed during transformation (Table 61–5). Certain **glycolytic enzymes** appear to be target proteins for the *src* protein-tyrosine kinase; this is in keeping with the observation that transformed cells often show increased rates of glycolysis. The product of *src* may also catalyze phosphorylation of **phosphatidylinositol** to phosphatidylinositol mono- and bisphosphate. When phosphatidylinositol 4,5-bisphosphate is hydrolyzed by the action of phospholipase C, 2 second messengers are released: inositol triphosphate and diacylglycerol (see Chapter 45). The first compound mediates release of Ca^{2+} from intracellular sites of storage (eg, the endoplasmic reticulum). Diacylglycerol stimulates the activity of the plasma membrane-bound protein kinase C, which in turn phosphorylates a number of proteins, some of which may be components of ion pumps. Specifically, it has been proposed that mild alkalinization of the cell, brought about by activation of an Na^+/H^+ antiport system (see Chapter 42), could play a role in stimulating mitosis. Thus, the product of *src* may affect a large number of cellular processes by its ability to phosphorylate various target proteins and enzymes and by stimulating the pathway of synthesis of the polyphosphoinositides.

Protein-Tyrosine Kinases in Normal & Transformed Cells: The observation that Rous sarcoma virus contained a protein-tyrosine kinase stimulated

Table 61–6. Some oncogenes of retroviruses.*

Oncogene	Retrovirus	Origin	Oncogene Product	Subcelluar Location
abl	Abelson murine leukemia virus	Mouse	Protein-tyrosine kinase	Plasma membrane
erb-B	Avian erythroblastosis virus	Chicken	Truncated EGF receptor	Plasma membrane
fes	Feline sarcoma virus	Cat	Protein-tyrosine kinase	Plasma membrane
fos	Murine sarcoma virus	Mouse	?	Nucleus
myc	Myelocytoma virus 29	Chicken	DNA-binding protein	Nucleus
sis	Simian sarcoma virus	Monkey	Truncated PDGF (B chain)	Membranes ? Secreted
src	Rous sarcoma virus	Chicken	Protein-tyrosine kinase	Plasma membrane

Tyr, tyrosine; EGF, epidermal growth factor; PDGF, platelet-derived growth factor.
*Modified and reproduced, with permission, from Franks LM, Teich NM (editors): *Introduction to the Cellular and Molecular Biology of Cancer.* Oxford Univ Press, 1986.

much research on the phosphorylation of tyrosine. It is now known that **many normal cells** contain protein-tyrosine kinase activity. The amount of phosphotyrosine in most normal cells is low but is usually elevated in cells transformed by an oncogenic virus containing a protein-tyrosine kinase, although the amount is still relatively small (~1% of the total phosphoamino acids [mainly phosphoserine, phosphothreonine, and phosphotyrosine] in such cells). Certain receptors (eg, for epidermal growth factor, insulin, and platelet-derived growth factor) found in both normal and transformed cells have protein-tyrosine activities that are stimulated upon interaction with their ligands (see the discussion of growth factors, below). Protein-tyrosine kinase activities thus play important roles in both normal and transformed cells.

Oncogenes of Other Retroviruses: In addition to the oncogenes of Rous sarcoma virus, approximately 20 oncogenes of other retroviruses have been recognized. About half of the products of these viral oncogenes are protein kinases, mostly of the tyrosine type. Some viral oncogenes are listed in Table 61–6, along with their products. While some of those listed encode protein kinases, the remainder encode various other proteins with interesting biologic activities. The product of the *erb*-B gene of avian erythroblastosis virus is a truncated form of the receptor for epidermal growth factor, and that of the *sis* oncogene of simian sarcoma virus is a truncated B chain of platelet-derived growth factor. The product of the oncogene (*fms*) of one type of viral isolate of feline sarcoma virus is a macrophage colony-stimulating factor. On the other hand, the product of the *myc* oncogene, originally found in chicken myelocytoma viruses, is a DNA-binding protein, which may affect the control of mitosis. The product of the *ras* oncogene of murine sarcoma viruses binds GTP, has GTPase activity, and appears to be related to the proteins that regulate the activity of the important plasma membrane enzyme, adenylate cyclase (see Chapter 45).

Proto-Oncogenes: A key issue raised by the dis-

covery of viral oncogenes relates to their origin. Use of nucleic acid hybridization (see Chapter 36) revealed that normal cells contained DNA sequences similar—if not identical—to those of the viral oncogenes. Thus, the viruses apparently incorporated cellular genes into their genomes during their passages through cells. The retention of such genes in their genomes indicated that they must confer a selective advantage on the affected viruses, presumably related to the altered growth properties of transformed cells.

The cellular sequences were found to be conserved in a wide range of eukaryotic cells, suggesting that they were important components of normal cells. In addition, mRNA species and proteins derived from these normal sequences could be detected at various stages of their development or life cycles. The genes present in normal cells thus have been designated proto-oncogenes, and their products are believed to play important roles in normal differentiation and other cellular processes.

Oncogenes From Tumor Cells: Experiments using DNA extracted from tumors have also provided evidence for the existence of oncogenes. The method used for detecting such cellular oncogenes is called **gene transfer** or **DNA transfection.** It depends on the fact that certain genes present in tumors can cause transformation of "normal" cultured cells. DNA is isolated from tumor cells and added to recipient cells, often a line of mouse fibroblasts known as NIH/3T3 cells. DNA isolated from tumor cells is precipitated with calcium phosphate (to facilitate endocytosis) and added to NIH/3T3 cells in tissue culture. The cells are observed microscopically over a period of 1–2 weeks for formation of foci of transformed cells. If transformation occurs, the NIH/3T3 cells change their morphology from flat to rounded cells that grow in characteristic foci. The procedure is repeated several times using DNA extracted from the transformed cells, thus reducing the amount of DNA not involved in transformation that was transfected and facilitating identification (eg, by the Southern blotting technique, using a suitable probe [see Chapter 36]) of the specific gene

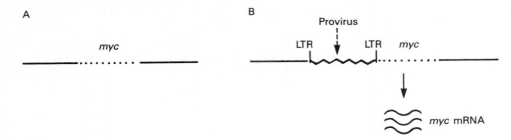

Figure 61–3. Schematic representation of how promoter insertion may activate a proto-oncogene. *A:* Normal chicken chromosome, showing an inactive *myc* gene. *B:* An avian leukemia virus has integrated in the chromosome in its proviral form, adjacent to the *myc* gene. Its right-hand long terminal repeat (LTR), containing a strong promoter, lies just upstream of the *myc* gene and activates that gene, resulting in transcription of *myc* mRNA. For simplicity, only one strand of DNA is depicted and other details have been omitted.

involved. Some 20 or so different cellular oncogenes have been recognized in this manner, with a number of them being related to *ras* oncogene of murine sarcoma viruses. These cellular oncogenes either are identical to normal genes or show very small structural differences from their normal counterparts (see below). In the former case, the regulation of their expression may be abnormal in cancer cells.

Abbreviations for Cellular & Viral Oncogenes: The abbreviation c-*onc* (cellular oncogene, eg, c-*ras*) is used to designate an oncogene present in tumor cells. The species present in normal cells—ie, its proto-oncogene—can be conveniently referred to as the corresponding c-*onc* proto-oncogene (eg, the c-*ras* proto-oncogene). Similarly, a viral oncogene is designated v-*onc* (viral oncogene, eg, v-*ras*), with its proto-oncogene being referred to as a v-*onc* proto-oncogene (eg, the v-*ras* proto-oncogene).

Proto-Oncogenes are Activated to Oncogenes by Various Mechanisms

Five mechanisms will be discussed that alter the expression or structure of proto-oncogenes and thus

participate in their becoming oncogenes. For the sake of convenience, the process whereby transcription of a gene is increased (from zero or a relatively low level) will be designated as **activation.** Familiarity with the mechanisms involved in activation is crucial for understanding contemporary thinking about carcinogenesis.

Promoter Insertion: Certain retroviruses lack oncogenes (eg, avian leukemia viruses) but may cause cancer over a longer period of time—months rather than days—than those which do contain oncogenes. As for other retroviruses, when these particular viruses infect cells, a DNA copy (cDNA) of their RNA genome is synthesized by reverse transcriptase, and the cDNA is integrated into the host genome. The integrated double-stranded cDNA is called a **provirus.** The cDNA copies of retroviruses are flanked at both ends by sequences named **long terminal repeats,** as are certain **transposons ("jumping genes")** found in bacteria and plants (see Chapter 39). The long terminal repeat sequences appear to be important in the mechanism of proviral integration, and they can act as promoters of transcription (see Chapter 40). Following infection of chicken B lymphocytes by certain avian leukemia viruses, the proviruses become integrated

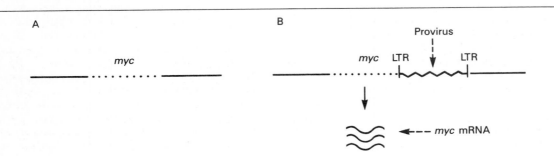

Figure 61–4. Schematic representation showing how enhancer insertion may activate a proto-oncogene. *A:* Normal chicken chromosome, showing an inactive *myc* gene. *B:* An avian leukemia virus has integrated in the chromosome in its proviral form, adjacent to the *myc* gene. However, in this instance, the site of integration is just downstream of the *myc* gene and it cannot act as a promoter (Fig 61–6). Instead, a certain proviral sequence acts as an enhancer element, leading to activation of the upstream *myc* gene and its transcription. For simplicity, only one strand of DNA is depicted and other details have been omitted.

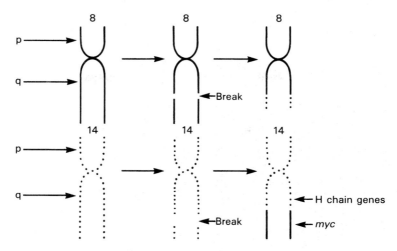

Figure 61–5. Schematic representation of the reciprocal translocation involved in Burkitt's lymphoma. The chromosomes involved are 8 and 14. A segment from the end of the q arm of chromosome 8 breaks off and moves to chromosome 14. The reverse process moves a small segment from the q arm of chromosome 14 to chromosome 8. The *myc* gene is contained in the small piece of chromosome 8 that was transferred to chromosome 14; it is thus placed next to genes transcribing the heavy chains of immunoglobulin molecules and itself becomes activated.

near the *myc* gene. The *myc* gene is activated by an upstream, adjacent viral long terminal repeat acting as a promoter, resulting in transcription of both the corresponding *myc* mRNA and translation of its product in such cells (Fig 61–3). A B cell tumor ensues, although the precise role of the products of the *myc* gene in the overall process in not clear. Similar events occur following infection of various cells with other retroviruses.

Enhancer Insertion: In some cases, the provirus is inserted downstream from the *myc* gene, or upstream from it but oriented in the reverse direction; nevertheless, the *myc* gene becomes activated (Fig 61–4). Such activation cannot be due to promoter insertion, since a promoter sequence must be upstream of the gene whose transcription it increases and the sequence must be in the correct 5' to 3' direction. Instead, enhancer sequences (see Chapters 40 and 42) present in the long terminal repeat sequences of the retroviruses under consideration appear to be involved.

The above 2 mechanisms—promoter and enhancer insertion—commonly operate in viral carcinogenesis. Proto-oncogenes other than *myc* are also probably involved.

Chromosomal Translocations: As mentioned earlier, many tumor cells exhibit chromosomal abnormalities. One type of chromosomal change seen in cancer cells is translocation. The basis of a translocation is that a piece of one chromosome is split off and then joined to another chromosome. If the second chromosome donates material to the first, the translocation is said to be **reciprocal.** Characteristic translocations are found in a number of tumor cells. One important translocation is the **Philadelphia chromosome,** involving chromosomes 9 and 22 and occurring in chronic granulocytic leukemia.

Burkitt's lymphoma is a fast-growing cancer of human B lymphocytes. In certain cases, an example of a reciprocal translocation (Fig 61–5) is found that has illuminated the mechanisms of activation of potential cellular oncogenes. Chromosomes 8 and 14 are involved. The segment of chromosome 8 that breaks off and moves to chromosome 14 contains the *myc* gene. As shown in Fig 61–6, the transposition places the previously inactive *myc* gene under the influence of enhancer sequences in the genes coding for the heavy chains of immunoglobulins. This juxtaposition results in activation of transcription of the *myc* gene. Apparently, synthesis of greatly increased amounts of the DNA-binding protein coded for by the *myc* gene acts to "drive" or "force" the cell toward becoming malignant, perhaps by an effect on the regulation of mitosis. This mechanism is similar to enhancer insertion, except that chromosomal translocation (rather than integration of provirus) is responsible for placing the proto-oncogene (ie, *myc*) under the influence of an enhancer.

Gene Amplification: Amplification of certain genes (see Chapter 39) is found in a number of tumors. One method of bringing about gene amplification in tumors is by administration of the anticancer drug methotrexate, an inhibitor of the enzyme dihydrofolate reductase. Tumor cells can become resistant to the action of this drug. The basis of this phenomenon is that the gene for dihydrofolate reductase becomes amplified, resulting in an increase of the activity of the enzyme (up to 400-fold). The amplified genes, measuring up to 1000 kb or more in length, may be detected as **homogeneously staining regions** on a specific chromosome. Alternatively, they are detected as **double-minute chromosomes,** which are minichromosomes lacking centromeres. The precise relationship of ho-

A

B

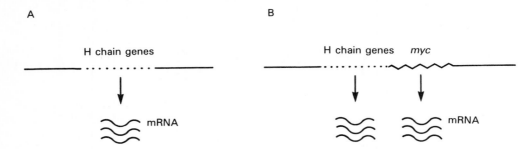

Figure 61–6. Schematic representation showing how the translocation involved in Burkitt's lymphoma may activate the *myc* proto-oncogene. *A:* A small segment of chromosome 14 prior to the translocation. The segment shown contains the genes encoding regions of heavy chains of immunoglobulins. *B:* Following the translocation, the previously inactive *myc* gene is placed under the influence of enhancer sequences in the genes encoding the heavy chains and is thus activated, resulting in transcription. For simplicity, only one strand of DNA is depicted and other details have been omitted.

mogeneously staining regions to double-minute chromosomes is under investigation. Certain cellular oncogenes can also be amplified in like manner and are thus activated. There is evidence suggesting that increased amounts of the products of certain oncogenes (such as c-*ras*) produced by gene amplification may play a role in the progression of tumor cells to a more malignant state (see the discussion of the progression of tumors, below).

Single-Point Mutation: The v-*ras* oncogene was originally detected in certain murine (ie, rat and mouse) retroviruses. Its product, a protein (p21) of MW 21,000, appears to be related to the G proteins that modulate the activity of adenylate cyclase (see above and Chapter 45) and thus plays a key role in cellular responses to many hormones and drugs. Analyses by DNA sequencing of the c-*ras* proto-oncogene from normal human cells and of the c-*ras* oncogene from a cancer of the human bladder showed that they differed solely in one base, resulting in an amino acid substitution at position 12 of p21. This intriguing result has been confirmed by analyses of c-*ras* genes from other human tumors. In each case the results were consistent; the gene isolated from the tumor exhibited only a single-point mutation, in comparison with the c-*ras* proto-oncogene from normal cells. The position of the mutation varied somewhat, so that other amino acid substitutions were observed. These mutations in p21 appear to affect its conformation and to diminish its activity as a GTPase. The lower activity of GTPase could result in chronic stimulation of the activity of adenylate cyclase, which normally is diminished when GDP is formed from GTP (see Chapter 45). The resulting stimulation of the activity of adenylate cyclase can result in a number of effects on cellular metabolism exerted by the increased amount of cAMP affecting the activities of various cAMP-dependent protein kinases. These events may assist in tipping the balance of cellular metabolism toward a state favoring transformation or its maintenance.

General Comments on Activation of Oncogenes: Of the 5 mechanisms described above, the

first 4 (promoter insertion, enhancer insertion, chromosome translocation, and gene amplification) involve an increase in amount of the product of an oncogene due to increased transcription but no alteration of the structure of the product of the oncogene. Thus, it appears that increased amounts of the product of an oncogene may be sufficient to push a cell toward becoming malignant. The fifth mechanism, single-point mutation, involves a change in the structure of the product of the oncogene but not necessarily any change in its amount. This finding implies that the presence of a structurally abnormal key regulatory protein in a cell may also be sufficient to tip the scale toward cancer.

When considering the role of oncogenes in cancer, it is important to bear in mind that oncogenes have been isolated from only about 15% of human tumors. It is likely that their activation in at least some cases may be only a secondary occurrence associated with transformation, rather than a causal event. Their involvements in chemically induced experimental tumors are just beginning to be explored. However, recent work has shown that activation of c-*ras* in rat mammary cancers induced by nitrosomethylurea was apparently due to a specific G → A transition type of mutation, demonstrating that oncogenes are probably involved in chemical carcinogenesis. In addition, because a single dose of nitrosomethylurea was used (without any promoter), the above mutation may be an important event in the initiation stage of chemical carcinogenesis. Much further research is needed to examine the possible involvement of oncogenes in the phenomena of initiation, promotion, tumor progression, and metastasis.

Mechanisms of Action of Oncogenes: The following are 3 mechanism by which the products of oncogenes may stimulate growth (Fig 61–7). (1) **They may act on key intracellular pathways involved in growth control,** uncoupling them from the need for an exogenous stimulus. Relevant examples (described above) are the product of *src* acting as a protein-tyrosine kinase, the product of *ras* acting to stimulate the activity of adenylate cyclase, and the product of

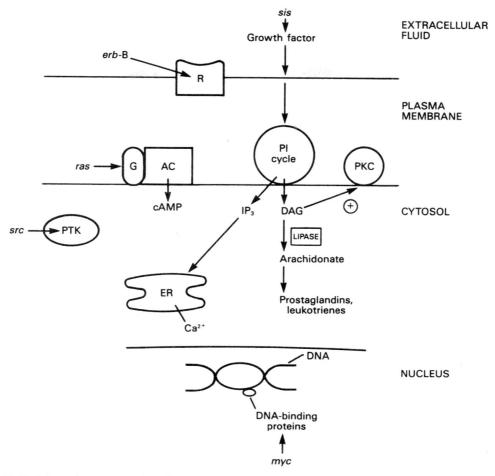

Figure 61–7. Schematic representation of mechanisms by which the products of certain oncogenes may alter cellular metabolism and thereby stimulate growth. Cyclic AMP can affect a number of cellular processes by activating cAMP-dependent protein kinases. Protein-tyrosine kinases and protein kinase C both can affect the activities of a number of target proteins. Ca^{2+} has numerous cellular effects, as do prostaglandins and leukotrienes produced from arachidonate. R, receptor; G, G protein; AC, adenylate cyclase; cAMP, cyclic AMP; PI, phosphatidylinositol; PKC, protein kinase C; PTK, protein-tyrosine kinase; IP_3, inositol triphosphate; DAG, diacylglycerol; ER, endoplasmic reticulum.

myc acting as a DNA-binding protein. Each of these could affect the control of mitosis, the first 2 by events involving phosphorylation of key regulatory proteins. A major deficiency in our knowledge of cell growth is that remarkably little is known about molecular aspects of the regulation of mitosis, even in normal cells. (2) **The products of oncogenes may also imitate the action of a polypeptide growth factor** or (3) **imitate an occupied receptor for a growth factor** (see below). Other mechanisms whereby oncogenes stimulate growth are currently being defined.

POLYPEPTIDE GROWTH FACTORS ARE MITOGENIC

The subject of growth factors is one of the most rapidly developing in contemporary biomedical sci-

ence. A variety of such factors have been isolated and partly characterized (Table 61–7). Until recently, only very small quantities of most growth factors were available for study. However, the genes for a number of growth factors have now been cloned, confirming their separate identities and making usable amounts available through recombinant DNA technology. The growth factors known to date affect many different types of cells, eg, cells from the blood, nervous system, mesenchymal tissues, and epithelial tissues. They exert a mitogenic response on their target cells, although special conditions may be necessary to demonstrate this, such as depriving cells in culture of serum so that they have become quiescent prior to exposure to a growth factor. **Platelet-derived growth factor (PDGF)**, released from the α granules of platelets, probably plays a role in normal wound healing. Various growth factors appear to play key roles in

Table 61–7. Some polypeptide growth factors.*

Growth Factor	Source	Function
Epidermal growth factor (EGF)	Mouse salivary gland	Stimulates growth of many epidermal and epithelial cells
Erythropoietin	Kidney, urine	Regulates development of early erythropoietic cells
Insulinlike growth factors I and II (IGF-I and IGF-II, also named somatomedins C and A)	Serum	Stimulate sulfate incorporation into cartilage, are mitogenic for chondrocytes, and exert insulinlike effects on many cells
Interleukin-1 (IL-1)	Conditioned media	Stimulates production of IL-2
Interleukin-2 (IL-2)	Conditioned media	Stimulates growth of T cells
Nerve growth factor (NGF)	Mouse salivary gland	Tropic effect on sympathetic and certain sensory neurons
Platelet-derived growth factor (PDGF)	Platelets	Simulates growth of mesenchymal and glial cells
Transforming growth factor (TGF-α)	Conditioned media of transformed or tumor cells	Is similar to EGF
Transforming growth factor (TGF-β)	Kidney, platelets	Exerts both stimulatory and inhibitory effects on certain cells

*Modified and reproduced, with permission, from Franks LM, Teich NM (editors): *Introduction to the Cellular and Molecular Biology of Cancer.* Oxford Univ Press, 1986.

regulating differentiation of stem cells to form various types of mature hematopoietic cells. Growth inhibitory factors also exist (eg, **transforming growth factor [TGF-β]** may exert inhibitory effects on the growth of certain cells). Thus, chronic exposure to increased amounts of a growth factor or to decreased amounts of a growth inhibitory factor could alter the balance of cellular growth.

Growth Factors Act in Endocrine, Paracrine, or Autocrine Manners

Growth factors may operate in 3 general ways (see also Chapter 44): (1) Their effects may be **endocrine;** that is to say, like hormones, they may be synthesized elsewhere in the body and pass in the circulation to their target cells. (2) They may be synthesized in certain cells and secreted from them to affect neighboring cells. However, the cells that synthesize the growth factor are not themselves affected, because they lack suitable receptors. This mode of action is called **paracrine.** (3) Certain growth factors can affect the cells that synthesize them. This third mode of action is called **autocrine.** For instance, a factor may be secreted and then attach to its cell of origin, provided that cell possesses appropriate receptors. Alternatively, if a certain amount of the factor is not secreted, its presence inside the cell may directly stimulate various processes.

Growth Factors Act on Mitosis Via Transmembrane Signal Transduction

Relatively little is known about how growth factors operate at the molecular level. Like polypeptide hormones (see Chapter 45), they must transmit a message across the plasma membrane to the interior of the cell (**transmembrane signal transduction**). In the case of growth factors, the message will ultimately affect one or more processes involved in mitosis. Most growth factors have high-affinity protein receptors on the plasma membrane of target cells. The genes for the receptors for **epidermal growth factor (EGF)** and **insulin** have been cloned and models of the structures of the receptors constructed. They have short membrane-spanning segments and external and cytoplasmic domains of varying lengths. The ligands bind to the external domains. A number of receptors (eg, for EGF, insulin, and PDGF) have been found to exhibit **protein-tyrosine kinase activities,** reminiscent of the product of the v-*src* gene (see above). This kinase activity, located in the cytoplasmic domains, causes autophosphorylation of the receptor protein and also phosphorylates other target proteins. The receptor-ligand complexes are subjected to endocytosis in coated vesicles (see LDL receptor, Chapter 27); it is not yet clear whether the receptors recirculate to the cell surface. The precise events resulting in transmembrane signaling are still under investigation and differ among the various factors. The case of PDGF will be described as an example. Phospholipase C is stimulated following exposure of cells to PDGF, resulting in hydrolysis of phosphatidylinositol 4,5-bisphosphate to form inositol triphosphate and diacylglycerol (see v-*src,* above, and Fig 44–5). These 2 second messengers can affect intracellular release of Ca^{2+} and stimulation of the activity of protein kinase C, respectively, thus affecting a large number of cellular reactions. The subsequent hydrolysis of diacylglycerol by phospholipase A_2, liberating arachidonic acid, can also result in the production of prostaglandins and leukotrienes, which themselves may exert many biologic activities (see Chapter 24). Exposure of cells to PDGF can result in rapid (minutes to 1–2 hours) activation of certain cellular proto-oncogenes (eg, c-*myc* and c-*fos*). It seems likely that gene activation, whether of normal genes or proto-oncogenes, is involved in the action of most growth factors.

Growth Factors & Oncogenes Interact in Several Ways

The products of several oncogenes are either growth factors or parts of the receptors for growth factors (Table 61–6).

The B chain of PDGF contains 109 amino acids. It seems likely that the B chain is biologically active as a homodimer, without involvement of the A chain. The discovery that v-*sis* encodes 100 of the 109 amino acids of the B chain of PDGF revealed a direct relationship between oncogenes and growth factors. It also suggested that autocrine stimulation by PDGF—supplying a chronic mitogenic stimulus—could be an important factor in the mechanism of transformation of cells by v-*sis*. Indeed, many cultured tumor cells are known to secrete growth factors into their surrounding media and also to possess receptors for these molecules.

Sequence analysis of v-*erb*-B revealed that it encoded a truncated form of the receptor for EGF, with much of the external domain of the receptor being deleted but the protein-tyrosine kinase activity being retained. It has been suggested that the abnormal form of the receptor for EGF encoded by v-*erb*-B may be continuously active when present in cells, simulating an occupied receptor. As in the case of autocrine stimulation by PDGF, this could result in a chronic mitogenic signal, "driving" cells toward the transformed state.

Transforming growth factor (TGF-β) was originally thought to be a positive growth factor, since it caused fibroblasts to behave as if they were transformed. TFG-β is now known to inhibit the growth of most cell types, except fibroblasts. It will inhibit the growth of the monkey kidney cells that synthesize it. TGF-β may activate the *sis* gene in fibroblasts. How it produces its inhibitory effects on other cells remains to be established. As we shall now see, there is other evidence that certain genes code for products that slow cell growth and also that tumor-suppressive genes exist. Thus, the control of cell growth is very complex, and both positive and negative regulatory factors are involved.

A LOSS OF GROWTH-SUPPRESSOR GENES (ANTI-ONCOGENES) IS IMPORTANT IN CERTAIN TYPES OF CANCER

Recently, genes other than oncogenes have been recognized to play a major role in the causation of at least some types of cancers. These are **growth-suppressor genes.** They operate quite differently from oncogenes, in that a **loss of these suppressor genes removes certain mechanisms of growth control.** An important model for understanding such genes has been the tumor known as **retinoblastoma.** Retinoblasts are precursor cells of cones, photoreceptor cells in the retina. Cytologic analyses of chromosomes from human retinoblastomas revealed that they often showed a deletion in the long arm of chromosome 13. This suggested that a critical gene in this area of chromosome 13 was deleted. The gene has been named *Rb* and it is believed that its product exerts a growth inhibitory effect on normal cells, which is lost in retinoblastomas. Genetic studies have indicated that **both copies of the *Rb* gene must be inactivated for cancer to develop.**

It now appears that there may be quite a number of genes whose products are negative regulators of cell growth and both copies of which must be inactivated for cancer to occur. Such growth-suppressor genes have also been called **anti-oncogenes** or **tumor-suppressing genes.** Loss of anti-oncogenes has already been shown to be important in the genesis of one type of breast cancer, of Wilm's tumor of the kidney, and of small-cell carcinoma of the lung. Recently, the normal *Rb* gene has been cloned, so that it should be possible to determine the precise nature of its gene product. Other studies have already suggested that the product of the *Rb* gene is nuclear in location and that it may act by **modulating gene expression.** The search for anti-oncogenes and the identification of their modes of action is a major effort in contemporary cancer research. Of therapeutic interest, the exciting possibility exists that introducing the normal chromosomes which specify negative growth regulatory factors into certain tumor cells may correct their altered growth rate.

THE MALIGNANCY OF TUMOR CELLS TENDS TO PROGRESS

Once a cell becomes a tumor cell, the composition and behavior of its progeny do not remain static. Instead, there is a tendency for malignancy to increase. This is manifested by increasingly abnormal karyotypes, increasing rates of growth, and increasing tendency to invade and metastasize. The important phenomenon of **progression** appears to reflect a fundamental instability of the genome of tumor cells. The activation of additional oncogenes may be involved. Cells with faster rates of growth have a selective advantage.

It is important to distinguish the biochemical profiles of cells that have **just been transformed** from those of **fast-growing,** highly malignant tumor cells. The former cells may show few differences from normal cells, apart from key changes leading to cancer (eg, activation of one or more oncogenes) and those changes generally associated with transformation (Table 61–5). Analyses of such cells can disclose the key biochemical alterations that result in transformation. The biochemical profile of highly malignant cells may be very different from that of normal cells. Many changes in enzyme profile and other biochemical parameters have occurred (Table 61–8), some of which are secondary to rapid growth rate and others probably due to chromosomal instability. These fast-growing cells tend to **maximize the anabolic processes** involved in growth (eg, DNA and RNA synthesis), cut down on catabolic functions (eg, catabolism of

Table 61–8. Biochemical changes often found in fast-growing tumor cells.

Increased activity of ribonucleotide reductase.
Increased synthesis of RNA and DNA.
Decreased catabolism of pyrimidines.
Increased rates of aerobic and anaerobic glycolysis.
Alterations of isozyme profiles, often to a fetal pattern.
Synthesis of fetal proteins (eg, carcinoembryonic antigen).
Loss of differentiated biochemical functions (eg, diminished synthesis of specialized proteins).
Inappropriate synthesis of certain growth factors and hormones.

Table 61–9. Some changes that have been detected at the surfaces of malignant cells.*

Alterations of permeability
Alterations in transport properties
Diminished adhesion
Increased agglutinability by many lectins
Alterations of the activities of a number of enzymes (eg, certain proteases)
Alterations of surface charge
Appearance of new antigens
Loss of certain antigens
Alterations of the oligosaccharide chains of glycoproteins
Changes of glycolipid constituents

*Adapted from Robbins JC, Nicolson GL: Surfaces of normal and transformed cells. In: *Cancer: A Comprehensive Treatise.* Vol 4. Becker FF (editor). Plenum Press, 1975.

pyrimidines), and dispense with the differentiated functions shown by their normal ancestors. In other words, they are concentrating almost exclusively upon growth. They also show biochemical changes that reflect **altered gene regulation,** such as the synthesis of certain fetal proteins (some of which can be used as tumor markers; see Table 61–1) and the inappropriate manufacture of growth factors or hormones. Analyses of such cells are unlikely to reveal the initial key events responsible for transformation, partly because these events are obscured by a myriad of changes secondary to progression. However, knowledge of the biochemical profile of such cells is extremely **important in choosing chemotherapy,** as it is exactly this type of cell that the oncologist is usually called upon to deal with.

METASTASIS IS THE MOST DANGEROUS PROPERTY OF TUMOR CELLS

Metastasis is the spread of cancer cells from a primary site of origin to other tissues where they grow as secondary tumors, and it is the major problem presented by the disease. Metastasis is a complex phenomenon to analyze in humans, and knowledge of its biochemical basis is quite restricted. Because it reflects a failure in cell-cell interaction, much attention has naturally focused on comparisons of the biochemistry of the surfaces of normal and malignant cells. Many changes have been documented at the surfaces of malignant cells (Table 61–9), although not all are directly relevant to the problem of metastasis.

At present, considerable research is devoted to developing suitable animal model systems for the study of metastasis. Many studies also are being done to uncover the possible roles of certain proteases (eg, type 4 collagenase) and of certain glycoproteins and glycosphingolipids of the cell surface in the phenomenon of metastasis. For instance, it is possible that changes in the oligosaccharide chains of cell glycopro-

teins (secondary to alterations of activities of specific glycoprotein glycosyltransferases) may be critical in permitting metastasis to occur. Elucidation of the biochemical mechanisms involved in metastasis could provide a basis for the rational development of more effective anticancer therapies.

REFERENCES

Barbacid M: Oncogenes and human cancer: Cause or consequence? *Carcinogenesis* 1986;**7**:1037.

Bishop JM: The molecular genetics of cancer. *Science* 1987; **235**:305.

Croce CM, Klein G: Chromosome translocations and human cancer. *Sci Am* (March) 1985;**252**:54.

Darnell J, Lodish H, Baltimore D: *Molecular Cell Biology.* Scientific American Books, 1986.

Feldman M, Eisenbach L: What makes a tumor cell metastatic? *Sci Am* (Nov) 1988;**259**:60.

Franks LM, Teich N: *Introduction to the Cellular and Molecular Biology of Cancer.* Oxford Univ Press, 1986.

Hunter T, Cooper JA: Protein-tyrosine kinases. *Annu Rev Biochem* 1985;**54**:897.

Massague J: The transforming growth factors. *Trends Biochem Sci* 1985;**10**:237.

McIntire KR: Tumor markers: How useful are they? *Hosp Pract* (Dec) 1984;**19**:55.

Nicolson GL: Cell surface molecules and tumor metastasis: Regulation of metastatic phenotypic diversity. *Exp Cell Res* 1984;**150**:3.

Pitot HC: *Fundamentals of Oncology,* 3rd ed. Marcel Dekker, 1986.

Sporn MB, Roberts AB: Autocrine growth factors and cancer. *Nature* 1985;**313**:745.

Tannock IF, Hill R (editors): *The Basic Science of Oncology.* Pergamon Press, 1987.

Weinberg, RA: Finding the anti-oncogene. *Sci Am* (Sept) 1988;**259**:44.

Biochemistry and Disease

Robert K. Murray, MD, PhD

INTRODUCTION

In Chapter 1 it was stated that a knowledge of biochemistry is important for the understanding and **maintenance of health** and for the understanding and effective **treatment of diseases.** Put another way, a knowledge of biochemistry offers a rational framework, based on the experimental approach (as opposed to a received dogma) with which to view health and disease. Numerous examples of biochemical abnormalities resulting in loss of health and development of disease have been encountered throughout this text. However, these examples have necessarily appeared in a random fashion.

This final chapter has the following objectives:

1. To reemphasize the importance of **defining the precise biochemical abnormalities** that lead to loss of health or to the development of disease because, once these are understood, rational therapy may follow.

2. To discuss briefly the **causes of disease** from a biochemical perspective and to emphasize that these **operate by perturbing** either the **structures** of certain biomolecules (eg, DNA) or the **biochemical reactions and processes** in which they are involved.

3. To present a number of important points about **biochemical aspects of disease** in general, and briefly to cover **genetic diseases** in particular.

4. To illustrate the application of biochemistry to medicine by the use of **selected case histories** that cover the major categories of disease.

5. To help **bridge the gap** between learning basic biochemistry and applying it to the benefit of your patients.

ELUCIDATION OF THE BIOCHEMICAL BASES OF HEALTH AND DISEASE GENERALLY LEADS TO RATIONAL TREATMENT

Elucidation of the biochemical bases of health and disease generally leads to rational treatment. It should be evident from reading this text that the maintenance of health is dependent on an adequate intake of water, calories, vitamins, and certain minerals, amino acids, and fatty acids. To select one of these groups of compounds, the recognition of the fundamental roles played by vitamins in metabolism and of the results of their deficiencies has made the treatment of vitamin deficiencies rational, eg, the treatments of scurvy, rickets, and beri-beri by administration of vitamin C, vitamin D, and thiamine, respectively. This illustrates the basic medical principle that **it is generally necessary to know the cause and mechanisms involved in generating a disease in order to have rational and effective treatment for it.** At the present time, the causes of many important diseases, such as Alzheimer's disease, atherosclerosis, rheumatoid arthritis and schizophrenia, are still unknown. Thus, the **treatments of such conditions are generally symptomatic and empirical,** not based on correcting the underlying abnormality and generally not very effective. **The economic and human costs of such ignorance are immense.**

However, **even if the molecular nature of a disease has been shown, it may be impossible to institute appropriate treatment** because of limitations—technological or other—in the ability to correct the basic abnormality. This is the case with **sickle cell anemia,** for which it is not yet possible to correct the underlying abnormality in DNA. The pace of research on gene therapy suggests that this statement may not apply for very long to this disease or to many other genetic disorders.

ALL DISEASES HAVE A BIOCHEMICAL BASIS

Table 1–1 lists the major causes of disease. Life as we know it on Earth depends upon biochemical reactions; if these cease, death results. Health depends upon the regulated, harmonious functioning of the thousands of biochemical reactions and processes that occur in normal cells and that operate to maintain the constancy (with regard to pH, osmolality, concentration of electrolytes, etc) of the internal environment. **Disease results from perturbations of either the structures,** eg, ultimately DNA in genetic diseases, **or the amounts of certain biomolecules** or **perturbations of important biochemical reactions and processes.** These perturbations, temporary or permanent, are induced by the causes listed in Table 1–1 and often lead to severe alterations of the internal environment,

Table 62–1. Examples of the involvements of various biomolecules in diseases.

Biomolecule	Property Affected	Disease	Fundamental Cause
DNA	Structure	HbS	Mutation
RNA	Structure	Thalassemia (certain types)	Mutation leading to faulty splicing of mRNA
Protein	Structure/function	HbS	Mutation
Lipid (GM_2)	Amount (increase)	Tay-Sachs	Mutation resulting in defective hexosaminidase A
Polysaccharide (glycogen)	Amount (increase)	Glycogen storage disease	Mutation in gene for enzyme degrading glycogen (eg, phosphorylase)
GAG (dermatan and heparan sulfates)	Amount (increase)	Hurler's syndrome	Mutation resulting in defective iduronidase
Electrolyte (Cl^-)	Amount in sweat (increase)	Cystic fibrosis	A mutation in a membrane protein affecting transport of Cl^-
Water	Amount (decrease)	Cholera	Infection of small intestine by *Vibrio cholerae* leading to massive loss of water and electrolytes

for which **compensatory mechanisms** can only operate for a finite period of time. Case histories illustrating aspects of one disease generated by each of the causes listed in Table 1–1 are presented below.

The above view of diseases is **reductionist** and simplistic. Diseases are seen as abnormalities in the structure and function of cells, organs, and systems, generated by biochemical mechanisms. **However, patients experience illnesses,** which reflect underlying disease processes but also **reflect changes in being, as well as cultural and other factors.** Physicians must treat the whole patient, taking into account social, psychological, cultural, economic, and other factors, but always relying upon a sound knowledge of biochemical, physiologic, and pathologic mechanisms.

SIX POINTS TO NOTE WHEN DISEASES ARE CONSIDERED FROM A BIOCHEMICAL STANDPOINT

In this section, a number of general points will be presented that are important when considering diseases from a biochemical viewpoint.

(1) Many Diseases are Determined Genetically: This topic is of such importance that it will receive separate treatment below.

(2) All of the Classes of Biomolecules Found in Cells Are Affected in Structure, Function, or Amount in One or Another Disease: This point is illustrated with brief examples in Table 62–1. **Biomolecules can be affected in a primary or a secondary manner;** in genetic diseases, the primary defect resides in DNA and the structures, functions, or amounts of the other biomolecules are affected secondarily.

(3) Biochemical Perturbations That Cause Disease May Occur Rapidly or Slowly: Some diseases progress rapidly. For instance, death can occur within minutes or less after a massive coronary thrombosis (cf Case no. 4 below). This reflects the fact that most tissues (brain and heart in this particular instance) are very sensitive to **lack of oxygen and fuel** (eg,

glucose for the brain). A vivid example of the reliance upon oxygen is the fact that cyanide (which inhibits cytochrome oxidase) kills within a few minutes. Massive **loss of water and electrolytes** in cholera (cf Case no. 3) can threaten life within hours of onset of the disease. In general, rapid large alterations of the amounts or distribution in the body of certain electrolytes (eg, K) become hazardous very quickly, at least in part because of the sensitivity of myocardial muscle to such changes. Severe alterations of **pH** can also only be tolerated for a short time. On the other hand, **it may take years for the buildup of a biomolecule** (eg, due to lack of a lysosomal enzyme responsible for its normal degradation) to affect organ function. An example of this is the relatively slow accumulation of sphingomyelin in liver and spleen that occurs in mild cases of Niemann-Pick disease. The above are examples of the time-honored clinical classification of diseases into **acute** and **chronic** categories.

(4) Diseases Can Be Caused by Deficiency or Excess of Certain Biomolecules: This statement is well illustrated by consideration of deficiency and excess of **vitamin A** (see Chapter 54). Deficiency of this vitamin results in night blindness. On the other hand, excessive intake of vitamin A can result in acute or chronic states of toxicity. Similarly, deficiency of **vitamin D** results in rickets, but excess results in a potentially serious hypercalcemia. In thinking about nutritional deficiencies, it is useful to **consider primary (poor diet) and secondary causes of deficiency.** General causes of secondary deficiency include (1) inadequate absorption, (2) increased requirement, (3) inadequate utilization, and (4) increased excretion. Each of these four general causes can be brought about by a number of diseases or conditions.

(5) Almost Every Cell Organelle Has Been Involved in the Genesis of Various Diseases: This statement is illustrated in Table 62–2, which lists some of the organelles implicated in various diseases.

(6) Different Biochemical Mechanisms Can Produce Similar Pathologic, Clinical, and Laboratory Findings: The body has a rather limited

Table 62–2. Involvements of the major intracellular organelles in various diseases.

Organelle	Disease(s)	Mechanism
Nucleus	Most genetic diseases	Mutations in DNA.
Mitochondrion	Several diseases including Leber's hereditary optic neuropathy & mitochondrial myopathies	Mutations in mitochondrial DNA affecting the structures of proteins (such as NADH dehydrogenase) that are encoded by the mitochondrial genome.
ER	Chemical toxicities, such as following intake of CCl_4	Enzymes in the ER such as cytochrome P-450 activate various chemicals to potentially toxic species.
Golgi	I-cell disease	Absence of GlcNAc phosphotransferase in the Golgi results in lysosomal enzymes being misdirected and secreted by affected cells.
Plasma membrane	Metastasis of cancer cells	Changes in the oligosaccharides of glycoproteins in this organelle are thought to be important in permitting metastasis.
Lysosome	Lysosomal storage diseases	Decreased activities (due to mutations) of the various hydrolases present in lysosomes result in accumulation of various biomolecules.
Peroxisome	Zellweger (cerebrohepatorenal) syndrome and others	Decreased biogenesis of peroxisomes and decreased activity of certain peroxisomal enzymes, such as dihydroxyacetone phosphate acyltransferase.

number of ways of reacting to the disease-causing perturbations listed in Table 1–1. These ways of reacting are generally called **pathologic processes;** the most important of them are listed in Table 62–3. These processes, however, can be produced by a number of different stimuli. For example, many quite different bacteria and viruses can cause acute or chronic **inflammation.** Similarly, **hepatomegaly** can occasionally arise from accumulation of glucosylceramide, but much more commonly it is due to heart failure or metastases. **Fibrosis of the liver** (cirrhosis) can result from chronic intake of ethanol, excess of copper (Wilson's disease), excess of iron (primary hemochromatosis) or a deficiency of α_1-antitrypsin. In addition, a variety of inborn errors of metabolism can cause **mental retardation,** and many conditions can result in **ketosis.** Another example of different biochemical lesions producing a similar end point is when **the local concentration of a compound exceeds its solubility point,** because of excessive formation or decreased removal. This can result in its precipitation to form a **calculus** (stone). Calcium oxalate, magnesium ammonium phosphate, uric acid, and cystine may all form **renal calculi,** but they accumulate for different biochemical reasons. The overall point is that quite distinct biochemical causes can produce the same pathologic finding (eg, cirrhosis), clinical finding (eg, mental retardation), or laboratory finding (eg, ketosis). However, it is usually possible to distinguish among diseases that share some common findings by the history, physical examination, and appropriate laboratory tests.

THE MOLECULAR BASES OF MOST GENETIC DISEASES MAY BE REVEALED WITHIN THE NEXT DECADE

More than 3000 diseases have a genetic basis. Genetic diseases have been estimated to account for approximately 10% of hospitalized children in a number of centers and many of the chronic diseases that afflict adults (eg, diabetes mellitus, atherosclerosis) have an

Table 62–3. The major pathologic processes by which the body reacts to disease-causing perturbations.

Pathologic Process	One Cause	Example of Disease	Example of Biomolecule(s) Involved
Inflammation, acute or chronic	Infections, bacterial or viral	Pneumonia	Mediators of inflammation (prostaglandins, leukotrienes)
Degenerations	Various chemicals	Fatty liver	Ethanol
Enlargement of an organ (eg, liver)	Accumulation of a compound	Gaucher's disease	Glucosylceramide
Atrophy (decrease in size)	Diminished blood supply	Atrophy of a kidney	Decrease of various nutrients supplied by the blood
Anemia	Lack of vitamin or mineral	Iron deficiency anemia	Iron
Neoplasia	Irradiation	Various leukemias	DNA damaged by irradiation
Cell death	Diminished blood supply	Myocardial infarction	Lack of oxygen
Fibrosis	Often follows cell death	Cirrhosis of the liver	Collagen accumulates
Formation of a calculus	High local concentration of a compound	Renal calculus in gout	Uric acid

important genetic component. The advent of recombinant DNA and methods for sequencing DNA have revolutionized genetics, and medical genetics in particular. It has been predicted that, through the use of these approaches, the molecular bases of a majority of the genetic diseases will have been elucidated by the year 2000.

Chromosomal, Monogenic, & Multifactorial Disorders Comprise the Major Classes of Genetic Disease

Genetic diseases can be arranged into 3 classes: (1) chromosomal disorders, (2) monogenic (classic Mendelian) disorders, and (3) multifactorial disorders in which a number of genes are involved (polygenic).

The **chromosomal disorders** will not be discussed here in any detail but include conditions in which there is an **excess or loss** of chromosomes, **deletion** of part of a chromosome, or a **translocation.** The best known condition is trisomy 21 (Down syndrome). They can be recognized by analysis of the karyotype (chromosomal pattern) of an individual. Chromosomal translocations have been shown to be important in activating oncogenes (Chapter 61).

The **monogenic disorders** involve single mutant genes. They are classified as (1) autosomal dominant, (2) autosomal recessive, and (3) X-linked. The term **dominant** is used to denote that the mutation will be clinically evident even if only one chromosome is affected (heterozygous), whereas **recessive** denotes that both chromosomes must be affected (homozygous). **X-Linked** indicates that the mutation is present on the X chromosome. As females have two X chromosomes, they may be either heterozygous or homozygous for the affected gene. Thus X-linked inheritance in females can be dominant or recessive. On the other hand, males have only one X chromosome, so that they will be affected if they inherit the mutant gene. Each of these 3 classes has its own characteristic pattern of inheritance; a textbook of medical genetics should be consulted for details.

Multifactorial disorders involve the action of a number of genes. The pattern of inheritance of these conditions does not follow classical Mendelian genetics. Less is known about this class, but it is assuming increasing importance because common adult diseases such as ischemic heart disease and hypertension are members of this group. Table 62–4 lists some important examples of each of the above 3 classes of genetic disease.

Mutations in the Mitochondrial Genome Also Cause Disease

Human mitochondrial DNA measures approximately 16.5 kilobases and codes for 13 proteins, all components of oxidative phosphorylation. Mitochondrial DNA differs from nuclear DNA in 3 ways: (1) it is transmitted by the mother, (2) it contains few introns and, (3) its genetic code is slightly different from that of nuclear DNA. It also has a high rate of mutation and exhibits polymorphism. It has become clear that mutations in mitochondrial DNA are involved in the causation of some cases of **mitochondrial myopathies** and also in **Leber's hereditary optic neuropathy** (LHON). The former are characterized by muscle weakness, biochemical abnormalities in the mitochondria, and "ragged red fibers" in muscle biopsies, whereas LHON is characterized by rapid bilateral loss of central vision due to neuroretinal degeneration. Various **deletions** of mitochondrial DNA have been found in cases of mitochondrial myopathies. A **transition** from G to A at position 11778 of the mitochondrial genome, affecting the structure of subunit 4 of NADH dehydrogenase, has been shown to be one cause of LHON.

Genetic Diseases Produce Their Pathologic Consequences by Affecting DNA, RNA, Proteins, & Cell Function

A mutation in a structural gene may affect the structure of the encoded protein, whether it be an enzyme or a noncatalytic protein. **If an enzyme is affected, an inborn error of metabolism may result.** The concept of an inborn error of metabolism was first proposed by the English clinician, Sir Archibald Garrod, in the ear-

Table 62–4. Examples of each of the 3 major classes of genetic disease.

Class	Example	Comment
Chromosomal	Trisomy 21 (Down syndrome)	Prevalence increases with maternal age
Chromosomal	Chronic myelogenous leukemia	Presence of Philadelphia chromosome
Monogenic	Familial hypercholesterolemia	Autosomal dominant; mutation in gene for LDL receptor
Monogenic	Huntington's chorea	Autosomal dominant; diagnostic probe now available
Monogenic	Cystic fibrosis	Autosomal recessive; majority of cases are due to deletion of a Phe residue in a membrane protein regulating Cl⁻ transport
Monogenic	Hb S	Autosomal recessive; mutation of Glu→Val at β^6 position of globin
Monogenic	PKU	Autosomal recessive; mutation in gene for Phe hydroxylase
Monogenic	Duchenne muscular dystrophy	X-Linked; affects synthesis of dystrophin
Monogenic	Hemophilia	X-Linked; affects synthesis of factor VIII (AHG)
Multifactorial	Ischemic heart disease	Complex genetics; study of DNA polymorphisms for lipoproteins holds promise for resolving genetic susceptibility
Multifactorial	Essential hypertension	A single-gene theory also has its proponents

ly 1900s, based on his studies of alkaptonuria, albinism, cystinuria, and pentosuria. An inborn error of metabolism is a genetic disorder in which a specific enzyme is affected, producing a metabolic block, which may have pathologic consequences as explained below. The following depicts a metabolic block:

$$
\begin{array}{ccc}
 & \textbf{Increased X,Y} & \\
\textbf{E} & \uparrow & \textbf{*E} \\
\textbf{S} \rightarrow \textbf{P} & \textbf{Increased S} \ —\!\!\parallel\!\!\rightarrow \ \textbf{Decreased P} \\
\textbf{Normal} & & \textbf{Block}
\end{array}
$$

where *E = mutant enzyme and X,Y = alternative products of the metabolism of S. As shown, a block can have 3 results: (1) decreased formation of the product P, (2) accumulation of the substrate S behind the block, and (3) increased formation of metabolites (X,Y) of the substrate S, resulting from its accumulation. Any one of these 3 results may have pathologic effects. In the case of **phenylketonuria** (**PKU**) (see Chapter 32), the mutant enzyme is phenylalanine hydroxylase, resulting in the following situation:

Increased phenylpyruvic acid

$$\uparrow$$

Increased phenylalanine — $\parallel$ → Decreased tyrosine

Thus, patients with PKU synthesize less tyrosine (they are often fair skinned, because tyrosine is used for the synthesis of melanin), have increased plasma levels of phenylalanine, and also exhibit increased amounts of phenylpyruvate and other metabolites of phenylalanine in their body fluids and urine. The precise cause of toxicity in PKU is not clear and may reflect decreased availability of tyrosine for protein and neurotransmitter synthesis in the brain and also inhibitory effects of the high level of phenylanine on the transport of other amino acids into brain. The crucial point is that **either or both decreased formation of product and accumulation of substrate** or other metabolites behind a block, as well as **alterations of feedback regulation** (eg, if an allosteric site on an enzyme is affected by mutation), **alter the flux through metabolic pathways,** leading to pathologic effects.

However, it is also important to understand that **some inborn errors of metabolism are essentially harmless.** These are usually blocks in peripheral areas of metabolism, where neither diminished formation of product nor accumulation of its precursor perturbs the cell (eg, pentosuria). On the other hand, inborn errors of the TCA cycle are virtually unknown; because of its central importance in metabolism, any block in this cycle would have disastrous consequences for a cell, and would probably result in its death at a very early stage of development.

If a structural gene for a **noncatalytic protein is affected by a mutation,** in many cases a mutant protein is synthesized. Even a change of one amino acid, as in the case of HbS (see Chapter 7), can have disastrous pathologic consequences. The mutant protein

Table 62–5. Levels at which the pathologic effects of genetic diseases are expressed. All genetic diseases are due to changes in DNA. However, it is useful to consider their pathologic consequences at the levels indicated. Some diseases are shown acting at 2 levels, and most of the defects in nonenzymatic proteins listed affect organ functions. Treatment of genetic diseases can also be aimed at the different levels shown. (Adapted from Stanbury et al: *Metabolic Basis of Inherited Disease*, 5th ed. Table 1–8. McGraw-Hill, 1983.)

DNA: Altered nuclear DNA (various types of mutations) Altered mitochondrial DNA (various types of mutations)
RNA: Altered splicing (eg certain cases of thalassemia)
Proteins: a) Altered enzymes: decreased activity, absent, increased activity (rare)
b) Altered nonenzymatic proteins:
　Transport: albumin (analbuminemia), Hb (HbS)
　Protective: γ-globulin (agammaglulinemia), fibrinogen (afibrinogenemia)
　Structural: collagen (altered in various collagen diseases)
　Hormonal: thyroglobulin (deficient in certain cases of familial goiter)
　Contractile: dystrophin (decreased or absent in Duchenne Muscular Dystrophy (DMD))
　Receptor: LDL receptor (deficient or altered in familial hypercholesterolemia)
Cell & organ consequences:
　Deficiency of product: melanin (albinism)
　Accumulation of toxic precursor: phenylalanine or various keto acids (PKU)
　Disordered feedback regulation: of porphyrin synthesis (acute intermittent porphyria)
　Altered membrane function: LDL receptor (familial hypercholesterolemia), altered renal transporter for cystine (cystinuria)
　Altered compartmentalization: decreased Glc NAc phosphotransferase leads to misdirection (secretion) of lysosomal enzyme (I-cell disease)
　Effect on cell or tissue architecture
　　Cell shape: sickle cells (HbS)
　　Altered organelle: decrease of peroxisomes (Zellweger syndrome)
　　Altered extracellular matrix: altered collagen (various collagen diseases)

may not function properly (certain mutant hemoglobins), **may aggregate** (HbS), or **may move very slowly** through the cell (eg, α_1-antitrypsin).

Table 62–5 summarizes different levels at which the pathologic consequences of genetic diseases can be exerted. The levels shown are not mutually exclusive; for example, all genetic diseases depend upon changes in the structure of DNA.

Early Diagnosis of Certain Inborn Errors is Imperative if Permanent Damage Is To Be Avoided

What should make one consider when diagnosing an inborn error of metabolism? This is an important issue as treatment must be commenced immediately in certain inborn errors, otherwise the infant will be permanently damaged (eg, PKU and galactosemia). A number of useful cues are listed in Table 62–6.

The sources of material that can be analyzed and the major tests used in investigating patients or fetuses suspected of having genetic diseases are listed in Table

Table 62–6. Cues for considering the diagnosis of an inborn error of metabolism.

Family history of a genetic disease
Positive screening test (eg, for PKU)
Sick newborn with other conditions (infections, cardiovascular, etc) excluded
History of poor physical and mental development and of failure to thrive
Physical findings such as enlarged organs (eg, hepatomegaly in Gaucher's and Niemann-Pick disease)
Unusual odor from breath or urine (eg, maple syrup urine disease)
Low blood sugar (eg, glycogen storage disease)
Low blood pH (eg, due to accumulation of organic acids in maple syrup urine disease)

Table 62–7. Major tests of use in the diagnosis of genetic diseases.

Material: Plasma, red cells, white cells, fibroblasts, urine, organ biopsy, chorionic villus sample, amniotic cells
General Tests: Blood and urine glucose, blood pH, blood ammonia, amino acids in plasma and urine, detection of organic acids in urine, various color tests for miscellaneous metabolites
Specific Tests: Measurement of low amount of product or elevated amount of precursor (eg, phenylalanine in PKU)
Measurement of activity of candidate enzyme in red cells, white cells, or tissue biopsy
Electrophoresis (eg, for HbS)
Analyses by Southern blotting of restriction fragment length polymorphisms (RFLPs) and other features of DNA structure linked to or causing specific diseases (eg HbS, Huntington's chorea, DMD, etc.)
Identification of novel metabolite in urine or plasma by GLC-MS

62–7. Chorionic villus sampling and amniocentesis apply only to investigation of the fetus.

SUCCESSFUL TREATMENT IS AVAILABLE FOR SOME GENETIC DISEASES

In general, treatments of genetic diseases employ one of the four following strategies: (1) attempts to **correct the metabolic consequences** of the disease by administration of the missing product or limiting the availability of substrate; (2) attempts to **replace the absent enzyme or protein** or to increase its activity; (3) attempts to **remove excess of a stored compound;** (4) attempts to **correct the basic genetic abnormality.** Examples of each strategy are listed in Table 62–8.

Some of these approaches are quite effective for certain disorders—eg, the dietary treatments of PKU and galactosemia, replacement therapy for hemophilia and agammaglobulinemia, and removal of iron by periodic bleedings in hemochromatosis. However, attempts at enzyme replacement therapy have so far met with mostly limited success. Problems include obtaining good sources of human enzymes (placentae have been useful in some instances), targetting enzymes in sufficient amounts to the appropriate organ, and maintaining their activities in tissues if they are rapidly degraded. This is particularly difficult in the case of brain, where administered enzymes must be made to cross the blood-brain barrier. Many attempts have been made to use liposomes to deliver enzymes (or DNA and other molecules) to target organs, but so far success has not been spectacular.

Table 62–8. Major classes of treatments available for genetic diseases. The treatments are divided into the following 4 classes: (1) attempts to correct the metabolic consequences of the disease by administering the missing product or by limiting the availability of substrate; (2) attempts to replace the missing enzyme or protein or to increase its activity; (3) attempts to remove excess of a stored compound; (4) attempts to correct the basic genetic abnormality. (Adapted from Stanbury et al: *Metabolic Basis of Inherited Disease,* 5th ed. Table 1-12. McGraw-Hill, 1983.)

Class of Treatment	Principle	Disease	Treatment or Comment
1	Replacement of missing product	Familial goiter	Administration of L-thyroxine
1	Limitation of substrate	PKU	Diet low in phenylalanine
2	Replace mutant enzyme	Gaucher's disease	Injections of β-glucosidase
2	Replace missing protein	Hemophilia	Injections of factor VIII (AHG)
3	Increase activity of mutant enzyme by supplying large amounts of cofactor	Methylmalonic aciduria	Injections of vitamin B_{12}
3	Increase activity of mutant enzyme by induction	Criggler-Najjar syndrome	Administration of phenobarbital
4	Replace diseased organ carrying defective gene by normal organ	Galactosemia*	Liver transplant
4	Introduction of enzyme into somatic or germ cells by gene therapy (eg, using retroviral vector)	Many possible candidates	Strictly very experimental at the present time

*A diet low in lactose is first-line treatment for infants with galactosemia who have been diagnosed early.

Because of the tremendous progress in the field of recombinant DNA (see Chapter 42), a great deal of thought and experimental effort is being given to strategy (4). **Gene therapy could involve (i) gene replacement, (ii) correction, or (iii) augmentation.** In (i), the mutant gene would be removed and replaced with a normal gene. In (ii), only the mutated area of the affected gene would be corrected, the remainder being left unchanged. Neither of these two approaches has been developed. Augmentation involves introduction of foreign genetic material into a cell to compensate for the defective product of the mutant gene. For instance, it is already possible to introduce a normal gene into affected cells (eg, by transfection, microinjection, or carriage by a viral vector) and to demonstrate its expression. However, it should be noted that such cells contain both the mutant and the foreign gene. Moreover, the foreign gene is introduced at random sites on chromosomes, may interrupt (by insertional mutagenesis) the expression of certain host genes and is not subject to normal regulatory mechanisms. Many ingenious attempts are being made to develop methods of **targetting** genes to specific sites. Most attention has focussed on using retroviral vectors to correct certain genetic diseases in bone marrow cells, which can be manipulated both in vitro and in vivo. The **transgenic** approach is also feasible, but again the problem of precision of insertion has to be solved, along with many other technical and ethical issues. Nevertheless, gene therapy is a key area in medical research, and rapid progress is likely.

CASE HISTORIES

Introduction

In the following section, 8 case histories are presented, each representing one of the 8 causes of disease listed in Table 1–1. The basic idea is to illustrate the value of a knowledge of biochemistry in the understanding of all diseases by presenting specific human examples. The list of causes in Table 1–1 could have been extended; for instance, justifiable additions would have been "neoplastic" and "psychological." On the other hand, the causes of neoplasms can be considered under "physical" (cancer due to irradiation), "biologic" (considering the role of oncogenic viruses in certain types of cancer), and "chemical" (approximately 80% of human tumors appear to be due to this cause). Also, with regard to psychological diseases, there has been great interest recently in the finding that certain cases of schizophrenia and manic-depression may have a genetic basis. Most of the diseases described are common, or relatively common in a global sense. However, two (Cases no. 1 and no. 7) are relatively rare. They are included because they nicely illustrate the cause of disease under which they are listed and also because they illustrate 2 biologic principles—the importance both of DNA **repair** and of antibodies as **protective** mechanisms. A schematic diagram summarizing the biochemical mechanisms in-

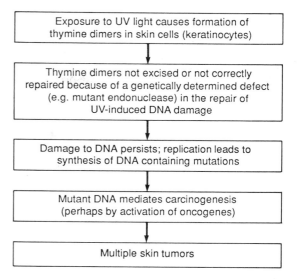

Figure 62–1. Summary of mechanisms involved in the causation of xeroderma pigmentosa.

volved in the causation of each of the 8 diseases discussed is included with each case history.

Case No. 1: Xeroderma pigmentosa, Fig 62–1

Classification: Physical, exposure to UV irradiation

History and Physical Examination: A 12-year-old boy presented at Dermatology Clinic with a skin tumor on his right cheek. He had always avoided exposure to sunlight because it made his skin blister. His skin had scattered areas of hyperpigmentation and other areas where it looked mildly atrophied. Because of the presence of a skin tumor at such a young age, the history of avoidance of sunlight and the other milder skin lesions, the dermatologist made a tentative diagnosis of xeroderma pigmentosa (pigmented dry skin).

Laboratory Investigations: Histologic examination of the excised tumor showed that it was a squamous carcinoma (a common type of skin cancer in older people, but not in a boy of this age). A sample of skin was taken for the preparation of fibroblasts. A research lab in the hospital specialized in radiobiology and was set up to measure the amount of thymine dimers formed following exposure to UV light. The patient's fibroblasts and control fibroblasts were exposed to UV light and cell samples were taken at 8-hourly intervals for a total of 32 hours postirradiation. Extracts of DNA were prepared and the numbers of dimers remaining at each time point indicated were determined. Whereas only 24% of the dimers formed persisted in DNA extracted from the normal cells at 32 hours, approximately 95% were found in the extract from the patient's cells at 32 hours. This confirmed the diagnosis of xeroderma pigmentosa.

Comments: Xeroderma pigmentosa (XP) is a relatively rare autosomal recessive condition in which the **mechanisms for repair of DNA subsequent to damage inflicted by UV irradiation are impaired.**

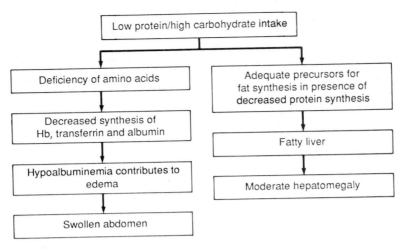

Figure 62–2. Summary of mechanisms involved in the causation of kwashiorkor.

The major damage inflicted to DNA by UV irradiation is the formation of thymine dimers, where covalent bonds are formed between carbons 5 and 5 and carbons 6 and 6 of adjacent intrachain thymine residues. These can be removed by an endonuclease. One cause of XP appears to be a defect in this enzyme. Detailed studies have shown that there are various subgroups of XP, and the precise enzymes involved in the overall repair of UV damage in human cells have still to be elucidated. However, if UV damage is not repaired, mutations in DNA will result and may cause cancer. Patients with XP often suffer from a variety of skin cancers from an early age. The parents of this boy were told that he would have to be watched throughout life for the development of new skin cancers. In addition, he was advised to continue to avoid sunlight and to use an appropriate sunscreen ointment. Although XP is a rare condition, the existence of a variety of mechanisms for repairing DNA following exposure to different types of irradiation and to chemical damage is of great protective importance. Without their existence, life on this planet would be fraught with even more danger than it already is! For instance, it has been estimated that patients with XP have a 1000-fold greater chance of developing skin cancer than do normal individuals.

Case No. 2: Kwashiorkor, Fig 62–2

Classification: Nutritional, deficiency of protein

History and Physical Exam: A 2-year-old girl in a developing country was brought by her mother to the out-patients' department of the local hospital. The mother had 4 children, the last of whom was 1 year old, and was being breastfed. The family lived in a dilapidated shack some miles out of town and the father had found it very difficult to obtain more than occasional employment. Their main subsistence food was a starchy gruel, high in carbohydrate and low in protein. The family had been able to get milk and meat only very rarely during the previous 2 years. The mother stated that the daughter had been eating poorly for

the past month, had intermittent diarrhea for some time, and had become quite irritable and apathetic.

On examination the patient was found to be both underweight for her height and small for her age. She was pale, irritable, and weak. Her midarm circumference was also well below normal, indicative of protein-energy malnutrition (PEM). Her skin was flaky and her hair dry and brittle. The abdomen was distended and the liver was moderately enlarged. Generalized edema was evident. The doctor on duty made the diagnoses of kwashiorkor and diarrhea.

Lab Results: Only a few analyses were performed because of limited facilities. Hb was 6.0 g/dL (normal 13–15 g/dL). Both the levels of total serum protein (4.4 g/dL, normal 6–8 g/dL) and of albumin (2.0 g/dL, normal 3.5–5.5 g/dL) were low. Stool culture was positive for a gram-negative anaerobe.

Treatment: In many cases it is undesirable to treat patients with mild protein-energy malnutrition (PEM) in hospital, because this only increases the chance of picking up an infection. However, in view of the weakness and severe edema, this child was admitted. Because of the diarrhea, the patient was started on a suitable broad-spectrum antibiotic and was initially given frequent oral feedings of isotonic dextrose-saline with other mineral supplements. This was switched to 4-hourly feedings of a high-energy milk-based diet within 2 days, followed later by a multivitamin preparation and mineral supplements. The child improved steadily and was discharged 10 days later. The mother was told to bring the child once a week to the hospital's small nutrition clinic, where a nutritionist gave instructions on how to improve dietary quality. The child's weight, height, and midarm circumference were recorded at each visit to monitor progress. In addition, free cartons of powered milk were given to the family at these visits.

Discussion: PEM is the commonest nutritional disorder in the world. It has been estimated that as many as 1 billion people suffer from various degrees of

Table 62–9. Differences between kwashiorkor and marasmus.

	Kwashiorkor	Marasmus
Edema	Present	Absent
Hypoalbuminemia	Present, may be severe	Mild
Fatty liver	Present	Absent
Levels of insulin	Maintained	Low
Levels of epinephrine	Normal	High
Muscle wasting	Absent or mild	May be very severe
Body fat	Diminished	Absent

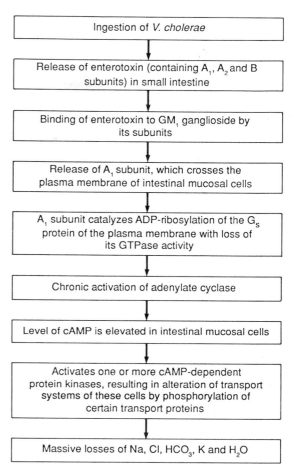

Figure 62–3. Summary of mechanisms involved in the causation of the diarrhea of cholera.

severity of PEM. **Kwashiorkor** is at one end of the spectrum of PEM, in which protein deficiency predominates. **Marasmus** is at the other end, with energy deficiency playing the major role. The distinction between kwashiorkor and marasmus is not sharp and intermediate cases of **marasmic kwashiorkor** are common. The major differences between kwashiorkor and marasmus are indicated in Table 62–9. The hallmarks of kwashiorkor are edema, hypoalbuminemia, and fatty liver.

Kwashiorkor is the word used by members of the Ga tribe in Ghana to describe "the sickness the older gets when the next child is born." It follows weaning from breast milk and exposure to a **diet low in protein and high in carbohydrate.** The edema, poor skin and hair, and fatty liver seen in kwashiorkor are mainly due to protein deficiency. **Deficiencies of vitamins and minerals** are often found in patients with kwashiorkor. **Hormones** are important in the generation of PEM. The exposure to a **high intake of carbohydrate keeps levels of insulin high and levels of epinephrine and cortisol low in kwashiorkor,** as opposed to marasmus. The combination of low insulin and high cortisol greatly favors catabolism of muscle; thus, muscle wasting is greater in marasmus than in kwashiorkor. Also, because of the lower levels of epinephrine, fat is not mobilized to the same extent in kwashiorkor. The low dietary intake of protein in kwashiorkor results in **decreased synthesis of plasma proteins,** especially albumin and transferrin, and also decreased synthesis of hemoglobin. Impaired protein synthesis in the liver along with sufficient dietary carbohydrate to assure lipid synthesis lead to the accumulation of triglycerides in the liver (**fatty liver**). The **immune system is impaired in PEM,** particularly T-cell function. Individuals with PEM are thus very susceptible to infections (eg, causing diarrhea), and infections worsen the situation further by placing a higher metabolic demand on the body, eg, through fever.

Kwashiorkor is entirely **preventable** if children are given a well-balanced diet containing adequate amounts of protein and of the essential amino acids.

Case No. 3: Cholera, Fig 62–3

Classification: Biologic, bacterial

History and Physical Exam: A 21-year-old female student working in a developing country suddenly began to pass profuse watery stools almost con-

tinuously. She soon started to vomit, her general condition declined abruptly, and she was rushed to the local village hospital. On admission she was cyanotic, skin turgor was poor, BP was 70/50 (normal 120/80), and her pulse was rapid and weak. The doctor on duty diagnosed cholera, took a stool sample for culture, and started treatment immediately.

Treatment: This consisted of the IV administration of a solution made up in the hospital, containing 5 g NaCl, 4 g Na HCO$_3$ and 1 g KCl per liter of pyrogen-free distilled water. This was initially given rapidly at a rate of 100 mL/kg, until her BP was normal and her pulse was strong. She was also started on tetracycline. On the second day she was able to take the oral rehydration solution (ORS) recommended by the World Health Organization (WHO) for the treatment of cholera, consisting of 20 g glucose, 3.5 g NaCl, 2.5 g NaHCO$_3$, and 1.5 g KCl per liter of drinking water. Solid food was reinstituted on the fourth day after admission. She continued to recover rapidly and was discharged 7 days after admission.

Discussion: Cholera is an important infectious disease that is endemic in certain Asian countries and also in other parts of the world. It is due to *Vibrio cholerae*, a bacterium which secretes a protein **enterotoxin.** The enterotoxin is made up of one A subunit (comprised of one A_1 and one A_2 peptide joined by a disulfide link) and five B subunits, and has a MW of approximately 84,000. In the small intestine, the toxin attaches by means of the B subunits to the **ganglioside GM_1** (the structure of GM_1 is shown in Fig 15–20) present in the plasma membrane of mucosal cells; the A subunit then dissociates, and the A_1 peptide passes across to the inner aspect of the plasma membrane. It catalyzes the **ADP-ribosylation** (using NAD as donor) of the G_s regulatory protein, which inhibits the GTPase activity and fixes it in its active form. Thus, **adenylate cyclase** becomes chronically activated (cf Chapter 46, p. 475). This results in an elevation of **cAMP,** which is thought to activate a protein kinase that phosphorylates one or more membrane proteins involved in active transport. The consequence of this chain of events is that **absorption of NaCl** into the intestinal cells by a neutral NaCl transport cosystem is inhibited and active **secretion of Cl** is stimulated. These events lead to massive secretion of **Na and water** into the lumen of the small intestine, producing the liquid stools characteristic of cholera. The histologic structure of the small intestine remains remarkably unaffected, despite the outpouring of not only Na and water, but also Cl, HCO_3 and K. It is the loss of these constituents that result in the marked fluid loss, low blood volume, acidosis, and K depletion that are found in serious cases of cholera and that can prove fatal unless appropriate replacement therapy (as described above) is begun immediately. The recognition and easy availability of appropriate replacement fluids, such as ORS, has led to tremendous improvement in the treatment of cholera. It should be noted that **glucose** is an essential component of ORS; whereas the cholera toxin inhibits absorption of NaCl by intestinal cells, it does not inhibit **glucose-facilitated Na transport** into these cells.

Case No. 4: Myocardial infarction, Fig 62–4

Classification: Oxygen lack

History and Physical Exam: A 46-year-old businessman was admitted to the emergency department of his local hospital complaining of severe retrosternal pain of 2 hours' duration. He had previously been admitted to hospital once for treatment of a small myocardial infarction (MI), but despite this continued to smoke heavily. His blood pressure was 150/90 (his normal = 140/80), pulse 60/min, and he was sweating quite profusely. There was no evidence of cardiac failure. Because of the admitting diagnosis of myocardial infarction, he was given morphine to relieve his pain and apprehension and immediately transferred to a cardiac care unit, where continuous monitoring by ECG was started at once.

Lab Tests: The initial **ECG** showed S-T segment elevation and other changes in certain leads, indicative of an acute anterior transmural left ventricular infarction. Blood was taken at 4 hours and subsequently at regular intervals for measurement of the **MB isozyme**

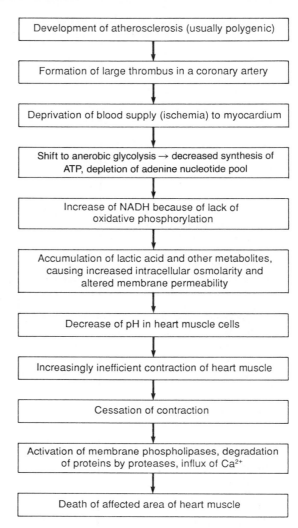

Figure 62–4. Summary of mechanisms involved in the causation of an acute myocardial infarction. The arrows below the boxes do not in all cases imply a strict causal connection.

of **CK;** at 4 hours the activity of this isozyme was slightly elevated and at 12 hours the elevation was 4-fold. The plasma cholesterol was moderately elevated (6.5 mmol/L) and triglyceride levels were normal.

Treatment: The attending cardiologist, after reviewing all aspects of the case, decided to administer **streptokinase** (SK) by cardiac catheterization, because of the diagnosis of anterior transmural MI seen within 4 hours of the onset of symptoms. Chest pain began to disappear after 12 hours and the patient felt increasingly comfortable. He was discharged from hospital 10 days later under the care of his family doctor with the advice to start a cholesterol lowering regimen (including a reduction of intake of saturated fat and administration of a drug inhibiting HMG-CoA reductase) and to stop smoking.

Discussion: A transmural MI is generally caused by an occlusive or near-occlusive **thrombus** lying in close proximity to an atherosclerotic plaque. Generally the diagnosis can be made from the clinical history, the ECG results, and serial measurement of the CK-MB isozyme (see Appendix). The overall aims of treatment are to prevent death from cardiac arrhythmias by administration of appropriate drugs and to limit the size of the MI. In this case, the decision was made to limit the size of the infarct by intracoronary administration of SK, which can dissolve the thrombus (see Chapter 58). This enzyme is considerably less expensive than tissue plasminogen activator (TPA), and several studies indicate that it may be just as effective. For long-term therapy, measures to reduce the plasma cholesterol were started by prescribing a drug that inhibits HMG-CoA reductase.

The causes of the **atherosclerotic lesion** in the coronary artery that led to the thrombus can be discussed only very briefly here; a textbook of pathology should be consulted for the details. Atherosclerosis is predisposed to by high levels of LDL, low levels of HDL, unknown polygenic factors, and a variety of risk factors such as hypertension, elevated levels of cholesterol, and smoking; this patient had an elevated level of cholesterol and was a heavy smoker. The intima of arteries is initially affected and macrophages, plasma lipoproteins, GAGs, collagen, and calcium accumulate in a lesion called a **fatty streak.** Platelets and fibrin can become deposited on the luminal aspect of the blood vessel and monoclonally derived smooth muscle cells present in the medial layer of the artery grow into the intimal lesion, attracted by growth factors released by macrophages and platelets (eg, platelet-derived growth factor). The overall lesion is now an intimal **plaque.** Hemorrhage and local inflammation can occur into the plaque, leading to rupture of its surface and exposure of its underlying constituents to the blood. Platelets will adhere to the exposed collagen and a thrombus is initiated (see Chapter 58).

If the thrombus occludes 90% of the vessel wall, blood flow through the affected vessel may cease (total ischemia) and capillary hemoglobin will be very rapidly **depleted of oxygen.** The normal metabolism of the myocardium is **aerobic,** with most of its ATP being derived from **oxidative phosphorylation.** The anoxia secondary to total ischemia results in a switch of myocardial metabolism to **anaerobic glycolysis,** which generates only about 1/10th of the ATP produced by oxidative phosphorylation. Not only does this switch in metabolism occur, but the flow of substrates into the myocardium via the blood and the removal of metabolic products from it are also greatly reduced. **This accumulation of intracellular metabolites increases the intracellular osmotic pressure,** resulting in cell swelling, affecting the permeability of the plasma membrane. Thus, the affected myocardium exhibits a **depletion of ATP, an accumulation of lactic acid, the development of severe acidosis and a marked reduction in contractile force.** Synthesis of macromolecules and of nucleotides ceases under these

metabolic conditions. The **accumulation of lactate and H+ ions inhibits glycolysis at the level of glyceraldehyde phosphate dehydrogenase and oxidized NAD becomes deficient** as it is not regenerated by oxidative phosphorylation. As the level of ATP drops, the level of ADP rises initially. However, **ADP is converted to AMP** by muscle adenylate kinase, and the AMP is further degraded by adenosine deaminase to adenosine. Adenosine itself is further converted to inosine and other products of purine catabolism. All of this **markedly depletes the adenine nucleotide pool,** which is a key component of normal cellular metabolism. It is known that the level of ATP in the canine myocardium drops to approximately 10% of control values after about 40 min of severe ischemia. The exhaustion of the adenine nucleotide pool **coincides with the development of irreversible cellular damage,** but does not necessarily cause it. At present it is not possible to state what the precise metabolic changes are that commit a cell irreversibly to dying. Various studies have implicated **depletion of ATP, activation of intracellular phospholipases** (which will produce damage to cell membranes), **activation of proteases,** and **accumulation of intracellular Ca²⁺.**

Generally speaking, the earlier attempts are made to reestablish perfusion through an area of ischemic myocardium, the better. By 6 hours, irreversible damage has probably been done and some cells are probably irreversibly damaged by as little as 1 hour of complete ischemia. Thus, administration of SK or TPA must be prompt.

The biochemical events that occur if **reperfusion** of an ischemic area of myocardium is established (eg, following administration of SK) are also of great biochemical and clinical interest. Reperfusion can itself lead to cell death, a condition known as **reperfusion injury.** One mechanism whereby this can occur is that damage to the **plasma membrane** (eg, to various of its ion pumps) may have occurred during the period of ischemia, seriously altering its permeability properties. This will affect the membrane potential and will also permit a flood of compounds such as Ca^{2+} to enter from the plasma. **High levels of intracellular Ca²⁺ can wreak havoc inside the cell by activating or inhibiting various enzymes in an unregulated manner.** There is also considerable interest in the possibility that **free radicals,** eg, highly reactive partially reduced metabolites of oxygen such as superoxide (O_2^-) and hydroxyl radicals (·OH), may play a role in reperfusion injury. ·OH radicals are particularly reactive. They can be generated by myocardial cells or by circulating blood cells, such as polymorphonuclear leukocytes (PMNLs). They damage cells by causing lipid peroxidation, breakage of DNA strands, and oxidation of SH groups in proteins. Superoxide anion is generated by the transfer of a single electron to O_2:

$$O_2 + e^- \rightarrow O_2^-$$

in reactions catalyzed by cytochrome P-450, xanthine oxidase, and the respiratory burst oxidase (NADPH

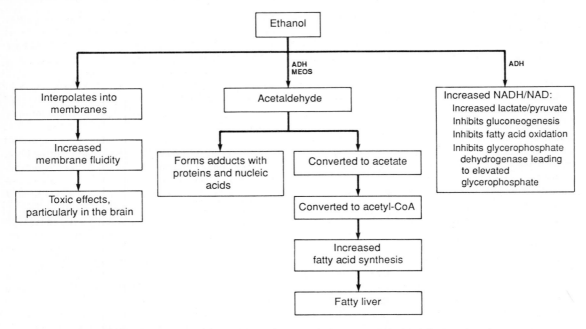

Figure 62–5. Summary of mechanisms involved in the causation of toxicity by ethanol.

oxidase) present in polymorphonuclear leukocytes. The enzyme **superoxide dismutase** (SOD) catalyzes the following reaction:

$$O_2^- + O_2^- + 2H^+ \rightarrow H_2O_2 + O_2$$

It thus scavenges superoxide anions, converting them to the less reactive hydrogen peroxide. Experiments have been performed to determine whether administration of SOD during reperfusion can protect the myocardium from reperfusion injury, but the results have been equivocal. The role, if any, of superoxide anions in reperfusion injury thus remains to be settled; however, at present there is considerable interest in the role of free radicals in many types of cell injury and disease.

Case No. 5: Acute intoxication with ethanol, Fig 62–5

Classification: Chemical

History and Physical Exam: A 52-year-old male was admitted to the emergency department in coma. Apparently, he had become increasingly depressed after the death of his wife one month previously. Prior to her death he had been a moderate drinker, but his consumption of alcohol had increased markedly over the previous few weeks. He had also been eating poorly. His married daughter had dropped around to see him on Sunday morning and had found him unconscious on the living room couch. Two empty bottles of rye whisky were found on the living room table. On examination, he could not be roused, his breathing was deep and noisy, alcohol could be smelled on his breath, and his temperature was 35.5 °C (normal 37.4). The diagnosis on admission was coma due to excessive intake of alcohol.

Lab Tests: The pertinent lab results on blood were: alcohol 500 mg/dL, glucose 2.7 mmol/L (normal 3.6–6.1), lactate 8.0 mmol/L (normal 0.5–2.2), and pH of 7.21 (normal 7.4). These results were consistent with the admitting diagnosis, accompanied by a metabolic acidosis.

Treatment: Because of the very high level of blood alcohol and the coma, it was decided to start **hemodialysis** immediately. This directly eliminates the toxic ethanol from the body but is only required in very serious cases of ethanol toxicity. In this case, the level of blood alcohol fell rapidly and the patient regained consciousness later the same day. Intravenous **glucose** (5%) was administered after dialysis was stopped to counteract the hypoglycemia that this patient exhibited. The patient made a good recovery and was referred for psychiatric counseling.

Discussion: Excessive consumption of alcohol is a major health problem in most societies. The present case deals with the acute, toxic effects of a very large intake of ethanol. Another related problem, which will not be discussed here but which has many biochemical aspects, is the development of liver cirrhosis in individuals who maintain a high intake of ethanol (eg, 80 g of absolute ethanol daily) for more than 10 years.

From a biochemical viewpoint, the major question concerning the present case is how does ethanol produce its diverse acute effects, including coma, lactic acidosis, and hypoglycemia? The clinical viewpoint is how best to treat this condition.

The metabolism of ethanol has been described previously (Chapter 27); it occurs mainly in the liver and involves 2 routes. The first and major route uses **alcohol dehydrogenase and acetaldehyde dehydrogenase,** thus converting ethanol via acetaldehyde to acetate, which is then converted to acetyl-CoA. Reduced $NADH^+ + H^+$ is produced in both of these reactions. The intracellular **NADH/NAD ratio** can thus be altered appreciably by ingestion of large amounts of ethanol. In turn, this can affect the K_{eq} of a number of important metabolic reactions that use these 2 cofactors. High levels of NADH favor **formation of lactate** from pyruvate, accounting for the lactic acidosis. This diminishes the concentration of pyruvate (required for the pyruvate carboxylase reaction) and thus **inhibits gluconeogenesis.** In severe cases, when liver glycogen is depleted and no longer available for glycogenolysis, this will result in **hypoglycemia.** The second route involves a **microsomal cytochrome P-450** (microsomal ethanol oxidizing system, MEOS), also producing acetaldehyde. **Acetaldehyde** is a highly reactive molecule and can form adducts with proteins, nucleic acids, and other molecules. It appears likely that its ability to react with various molecules is involved in the causation of the toxic effects of ethanol. Ethanol also appears to be able to **interpolate into biological membranes,** expanding them and increasing their fluidity. When the membranes affected are excitable, this results in alterations of their action potentials, impairs active transport across them, and also affects neurotransmitter release. All of these depress cerebral function and, if severe enough, can produce coma and death from respiratory paralysis.

Case No. 6: Duchenne muscular dystrophy, Fig 62–6

Classification: Genetic

History and Physical Examination: A 4-year-old boy was brought to clinic at Children's Hospital. His mother was very concerned because she had noticed that her son was walking awkwardly, fell over frequently, and had difficulty climbing stairs. There were no siblings but the mother had a brother who died at age 19 of muscular dystrophy. The examining physician noted muscle weakness in both the pelvic and shoulder girdle. Modest enlargement of the calf muscles was noted. Because of the muscle weakness and its distribution, the pediatrician made a tentative diagnosis of Duchenne muscular dystrophy (DMD).

Lab and Other Tests: The activity in serum of creatine phosphokinase was markedly elevated. Electromyographic findings were characteristic of muscular dystrophy, whereas nerve conduction studies were normal. A biopsy of calf muscle showed areas of muscle necrosis and some variation in the size of muscle fibers. Using a cDNA probe available for dystrophin, a Southern blot revealed an absence of the fragment corresponding to the gene for this protein. In addition, analysis of a small portion of the muscle biopsy by 2-dimensional electrophoresis showed an absence of the protein dystrophin.

Discussion: The family history, the typical distribution of muscle weakness, the elevation of the activity in serum of CPK, the electromyographic findings, the muscle biopsy and the lab findings revealing abnormalities in the gene for dystrophin confirmed the

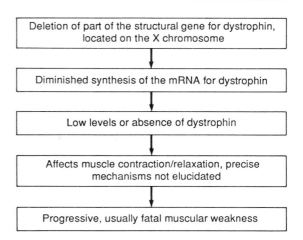

Figure 62–6. Summary of mechanisms involved in the causation of Duchenne muscular dystrophy.

pediatrician's provisional diagnosis of DMD. This is a severe **X-chromosome-linked** degenerative disease of muscle. It has an incidence of approximately 1 in 3500 live male births. It affects young boys, who first show loss of strength in their proximal muscles, leading to a waddling gait, to difficulty in standing up, and eventually to very severe weakness. Various studies led to localization of the defect to the middle of the short arm of the X chromosome and to subsequent identification of a **segment of DNA that was deleted** in patients with DMD. Using the corresponding non-deleted segment from normal individuals, a cDNA was isolated derived by reverse transcription from a transcript (mRNA) of 14 kb that was expressed in fetal and adult skeletal muscle. This was cloned and the protein product was identified as **dystrophin,** a high MW (about 400 kd) rod-shaped protein of approximately 3685 amino acids. Dystrophin was absent in electrophoretic gels of muscle from patients with DMD and from mice with an X-linked muscular dystrophy (mdx). Antibodies against dystrophin were used to study its **localization** in muscle; it appears to be located in the **sarcolemma** (plasma membrane) of normal muscle and was absent or markedly deficient in the sarcolemma of patients with DMD. A deficiency of dystrophin also appears to be involved in the etiology of **Becker's muscular dystrophy,** another milder type of muscular dystrophy, possibly an allelic variant of DMD.

Dystrophin appears to have 4 domains, 2 of which are similar to domains present in α-actinin and 1 to a domain in **spectrin.** Spectrin is a cytoskeletal membrane protein that helps red blood cells change their shape during passage through capillaries. It has been speculated that dystrophin may similarly allow muscle cells to contract/relax via membrane changes, or that it may be involved in the stabilization of the sarcolemma (plasma membrane). Further studies have shown that the gene coding for dystrophin is one of the **largest human genes** recognized to date, a fact that helps explain the observation that approximately one-third of cases of DMD are new mutations. Attempts are being made to produce dystrophin by recombinant DNA technology and perhaps eventually administer it to patients. Assay of dystrophin cannot be used in prenatal diagnosis as the protein is expressed only by muscle cells, and not by amniotic cells or samples of chorionic villus. However, the availability of cDNA probes for dystrophin facilitates **prenatal diagnosis** of DMD by chorionic villus sampling or amniocentesis. The classification of the various types of MD will also be facilitated by determining in which types dystrophin is affected, and in which types it does not appear to play a role. The demonstration of dystrophin as a cause of DMD has been one of the major accomplishments of the application of the new molecular biology to human disease.

Treatment: At present, no specific therapy for DMD exists. Treatment was thus essentially symptomatic. The boy was encouraged to exercise and regular attendance at a specialized muscular dystrophy clinic was commenced, so that complications could be treated immediately as they arose. The mother was advised to seek genetic counselling. Hopefully, the advances referred to above in elucidating the cause of this disease will lead, in the near future, to more specific therapy.

Case No. 7: X-Linked agammaglobulinemia, Fig 62–7

Classification: Immunologic, genetic

History and Physical Exam: A 3-year-old boy was referred to Children's Hospital with an elevated temperature, pain in the chest, and difficulty in breathing. His family doctor had appropriately diagnosed pneumonia. Because an older brother had previously been diagnosed as having X-linked agammaglobulinemia, the attack of pneumonia raised the suspicion in his doctor's mind that the younger brother might also lack the potential to synthesize gammaglobulins.

Lab Results: Electrophoretic analysis disclosed that the levels of albumin and α- and β-globulins were normal, but only a trace of γ-globulin was detected. B lymphocytes were markedly low in the blood (approximately 1% of normal) but the T-cell count was normal. A biopsy of a lymph node revealed an absence of plasma cells. Other components of the immune system, eg, serum complement levels, were normal. *Streptococcus pneumoniae* was demonstrated in samples of sputum.

Discussion: Disorders of the immune system can affect either the B-cell system (circulating immunoglobulins) or the T-cell system (cell-mediated immune responses). Only the former system has been discussed in this text (Chapter 58), and this case has been selected because it shows dramatically the loss of protection against infection by bacteria (here pneumococci) that occurs when levels of plasma immunoglobulins are markedly depressed. Phenotypically, this is an immunologic condition, although the underlying cause is genetic.

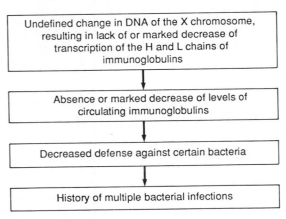

Figure 62–7. Summary of mechanisms involved in the causation of X-linked agammaglobulinemia.

The family history, the presence of only a few circulating B lymphocytes, the severe depression of circulating immunoglobulins, and the lack of plasma cells confirmed the provisional diagnosis of X-linked agammaglobulinemia made by the family doctor. X-Linked agammaglobulinemia affects young boys (like DMD, Case no. 6) and is a relatively uncommon condition, with approximately 200 cases having been recorded. The precise nature of the defect in this condition has not been established but involves the X chromosome. Although only a few circulating B cells are present, normal numbers of pre-B cells are present in bone marrow. Interestingly, when B lymphocytes from patients with this disorder are fused with mouse myeloma cells, the resulting hybrid cells secrete human immunoglobulin heavy (H) and light (L) chains. This demonstrates that the structural genes for H and L chains are present in B lymphocytes, but for some reason are not expressed.

Case No. 8: Diabetes mellitus (type I) with ketoacidosis, Fig 62–8

Classification: Endocrine deficiency, lack of insulin

History and Physical Exam: A 14-year-old girl was admitted to Children's Hospital in coma. Her mother stated that the girl had been in good health until approximately 2 weeks previously, when she developed a sore throat and moderate fever. She subsequently lost her appetite and generally did not feel well. Several days prior to admission she began to complain of undue thirst and also started to get up several times during the night to urinate. Her family doctor was out of town and the mother felt reluctant to contact another physician. However, on the day of admission the girl had started to vomit, had became drowsy and difficult to arouse, and accordingly had been brought to the emergency department. On examination she was dehydrated, her skin was cold, she was breathing in a deep sighing manner (Kussmaul's respiration), and her breath had a fruity odor. Her blood pressure was 90/60 and her pulse rate was 115/min.

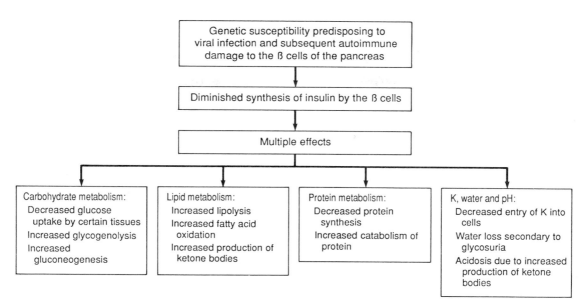

Figure 62–8. Summary of mechanisms involved in the causation of the ketoacidosis of type I diabetes mellitus.

She could not be aroused. A diagnosis of severe type I insulin-dependent diabetes mellitus with resulting ketoacidosis and coma was made by the intern on duty.

LAB RESULTS:
(A) PLASMA:
Glucose 35 mmol/L (3.6–6.1)
β-Hydroxybutyrate 13.0 mmol/L (<0.25)
Acetoacetate 2.8 mmol/L (<0.2)
Bicarbonate 5 mmol/L (24–28)
Urea 12 mmol/L (2.9–8.9)
Arterial blood H^+ ion 89 nmol/L [pH 7.05]
(44.7–45.5)[pH 7.35–7.45]
Potassium 5.8 mmol/L (3.5–5.0)
Creatinine 160 μmol/L (60–132)
(B) URINE:
Glucose + + + +
Ketone Bodies + + + +

The above lab results confirmed the admitting diagnosis.

Treatment: The most important measures in treatment of diabetic ketoacidosis (DKA) are **IV administration of insulin and saline fluids.** This patient was given IV insulin (10 u/h) added to 0.9% NaCl. Glucose was withheld until the level of plasma glucose fell below 250 mg/dL. **KCl** was also administered cautiously, monitoring plasma K levels every hour initially. Continual monitoring of K levels is extremely important in the management of DKA, because inadequate management of K balance is the major cause of death in this condition. Bicarbonate is not needed routinely in the treatment of DKA but may be required if acidosis is very severe.

Discussion: The precise cause of type I insulin-

dependent diabetes mellitus has not been elucidated. One likely chain of events is the following. Patients with this type of diabetes have a **genetic susceptibility, which may predispose to a viral infection.** The infection and consequent inflammatory reaction apparently alter the antigenicity of the surface of the pancreatic β cells and set up an **autoimmune reaction,** involving both cytotoxic antibodies and T lymphocytes. This leads to widespread **destruction of β cells,** resulting in type I diabetes mellitus. Perhaps the sore throat that this patient had several weeks before admission reflected the initiating viral infection.

The marked hyperglycemia, glucosuria, ketonemia, and ketonuria confirmed the diagnosis of DKA. The low pH indicated a severe acidosis due to the greatly increased production of acetoacetic acid and β-hydroxybutyric acid. The low levels of bicarbonate and pCO_2 confirmed the presence of a metabolic acidosis with partial respiratory compensation (by hyperventilation). The values of urea and creatinine indicated some renal impairment (due to diminished renal perfusion because of low blood volume secondary to dehydration), dehydration, and increased degradation of protein. A high plasma level of potassium is often found in DKA, due to a lowered uptake of potassium by cells in the absence of insulin.

Thus, the clinical picture in DKA reflects the abnormalities in the metabolism of **carbohydrate, of lipid, and of protein** that occur when plasma levels of insulin are sharply reduced. The causes of the disturbances of **K, water, and H ion** metabolism found in DKA are also summarized in Fig 62–8. The increased **osmolality** of plasma due to hyperglycemia also contributes to the development of coma in DKA.

It should be apparent that the rational management

of a patient with DKA depends upon a thorough familiarity with the actions of insulin.

REFERENCES

Connor JM, Ferguson-Smith MA: *Essential Medical Genetics,* 2nd ed. Blackwell Sci Publ, 1987.

Friedmann, T: Progress toward human gene therapy. *Science* 1989;**244:**1275.

Halliwell B (editor): *Oxygen Radicals and Tissue Injury.* Proceedings of an Upjohn Symposium, 1988.

Harding, AE: The mitochondrial genome—breaking the magic circle. *N Engl J Med* 1989;**320:**1341. (This issue contains 3 other articles on mitochondrial DNA and human diseases.)

Jennings RB, Reimer KA: Pathobiology of acute myocardial ischemia. *Hosp Pract* (Jan) 1989;**24:**89.

Koenig M, Monaco AP, Kunkel LM: The complete sequence of dystrophin predicts a rod-shaped cytoskeletal protein. *Cell* 1988;**53:**219.

Lieber CS: Biochemical and molecular basis of alcohol-induced injury to liver and other tissues. *New Engl J Med* 1988;**319:**1639.

Robbins SL, Kumar V: *Basic Pathology,* 4th ed. WB Saunders Co, 1987.

Rubin E, Farber, JL (editors): *Pathology,* 1st ed. JB Lippincott Co, 1988.

Scriver CR, et al (editor): In: *The Metabolic Basis of Inherited Disease.* 6th ed. McGraw Hill Book Co, 1989.

The Merck Manual, 15th ed. Merck, Sharp & Dohme, 1987.

Torun B, Viteri FE: In: *Modern Nutrition in Health and Disease.* 7th ed, Shils ME, Young VR (editors). Lea & Febiger, 1988.

Vogel F, Motulsky AG: *Human Genetics,* 2nd ed. Springer-Verlag, 1986.

Weatherall DJ: *The New Genetics and Clinical Practice,* 2nd ed. Oxford Univ Press, 1985.

Appendix*

CHEMICAL CONSTITUENTS OF BLOOD & BODY FLUIDS

Validity of Numerical Values in Reporting Laboratory Results

The value reported from a clinical laboratory after determination of the concentration or amount of a substance in a specimen represents the best value obtainable with the method, reagents, instruments, and technical personnel involved in obtaining and processing the material.

Accuracy is the degree of agreement of the determination with the "true" value (eg, the known concentration in a control sample). **Precision** denotes the reproducibility of the analysis and is expressed in terms of variation among several determinations on the same sample. **Reliability** is a measure of the congruence of accuracy and precision.

Precision is not absolute but is subject to variation inherent in the complexity of the method, the stability of reagents, the accuracy of the primary standard, the sophistication of the equipment, and the skill of the technical personnel. Each laboratory should maintain data on precision (reproducibility) that can be expressed statistically in terms of the standard deviation from the mean value obtained by repeated analyses of the same sample. For example, the precision in determination of cholesterol in serum in a good laboratory may be the mean value ± 5 mg/dL. The 95% confidence limits are ± 2 SD, or ± 10 mg/dL. Thus, any value reported is "accurate" within a range of 20 mg/dL. Thus, the reported value 200 mg/dL means that the true value lies between 190 and 210 mg/dL. For the determination of serum potassium with a variance of 1 SD of ± 0.1 mmol/L, values ± 0.2 mmol could be obtained on the same specimen. A report of 5.5 could represent at best the range 5.3–5.7 mmol/L. That is, the 2 results—5.3 and 5.7 mmol/L—might be obtained on analysis of the same sample and still be within the limits of precision of the test.

Physicians should obtain from the laboratory the values for the variation of a given determination as a basis for deciding whether one reported value represents a change from another on the same patient.

Interpretation of Laboratory Tests

Normal values are those that fall within 2 standard deviations from the mean value for the normal population. This normal range encompasses 95% of the population. Many factors may affect values and influence the normal range; by the same token, various factors may produce values that are normal under the prevailing conditions but outside the 95% limits determined under other circumstances. These factors include **age; race; sex; environment; posture; diurnal** and **other cyclic variations; fasting** or **postprandial state, foods eaten; drugs;** and **level of exercise.**

Normal or reference values vary with the method employed, the laboratory, and conditions of collection and preservation of specimens. The normal values established by individual laboratories should be clearly expressed to ensure proper interpretation.

Interpretation of laboratory results must always be related to the condition of the patient. A low value may be the result of deficit or of dilution of the substance measured, eg, low serum sodium. Deviation from normal may be associated with a specific disease or with some drug consumed by the subject—eg, elevated serum uric acid levels may occur in patients with gout or may be due to treatment with chlorothiazides or with antineoplastic agents. The reader should consult an appropriate text for lists of drugs interfering with chemical tests.

Values may be influenced by the method of collection of the specimen. Inaccurate collection of a 24-hour urine specimen, variations in concentration of the randomly collected urine specimen, hemolysis in a blood sample, addition of an inappropriate anticoagulant, and contaminated glassware or other apparatus are examples of causes of erroneous results.

Note: Whenever an unusual or abnormal result is obtained, all possible sources of error must be considered before responding with therapy based on the laboratory report. Laboratory medicine is a specialty, and experts in the field should be consulted whenever results are unusual or in doubt.

Effect of Meals & Posture on Concentration of Substances in Blood

Meals: The usual normal values for blood tests have been determined by assay of "fasting" specimens collected after 8–12 hours of abstinence from food. With few exceptions, water is usually permitted as desired.

---

*Reproduced, with permission, from Krupp MA, Schroeder SA, Tierney LM Jr: *Current Medical Diagnosis & Treatment 1987.* Appleton & Lange, 1987.

Few routine tests are altered from usual fasting values if blood is drawn 3–4 hours after breakfast. When blood is drawn 3–4 hours after lunch, values are more likely to vary from those of the true fasting state (ie, as much as +31% for AST [SGOT], −5% for lactate dehydrogenase, and lesser variations for other substances). Valid measurement of triglyceride in serum or plasma requires abstinence from food for 10–14 hours.

Posture: Plasma volume measured in a person who has been supine for several hours is 12–15% greater than in a person who has been up and about or standing for an hour or so. It follows that measurements performed on blood obtained after the subject has been lying down for an hour or more will yield lower values than when blood has been obtained after the same subject has been upright. An intermediate change apparently occurs with sitting.

Validity of Laboratory Tests*

The clinical value of a test is related to its specificity and sensitivity and the incidence of the disease in the population tested.

Sensitivity means percentage of positive results in patients with the disease. The test for phenylketonuria is highly sensitive: a positive result is obtained in all who have the disease (100% sensitivity). The carcinoembryonic antigen (CEA) test has low specificity: only 72% of those with carcinoma of the colon provide a positive result when the disease is extensive, and only 20% are positive with early disease. Lower sensitivity occurs in the early stages of many diseases—in contrast to the higher sensitivity in well-established disease.

Specificity means percentage of negative results among people who do not have the disease. The test for phenylketonuria is highly specific: 99.9% of the normal individuals give a negative result. In contrast, the CEA test for carcinoma of the colon has a variable specificity: about 3% of nonsmoking individuals give a false-positive result (97% specificity), whereas 20% of smokers give a false-positive result (80% specificity). The overlap of serum thyroxine levels between hyperthyroid patients and those taking oral contraceptives or those who are pregnant is an example of a change in specificity from that prevailing in a different set of individuals.

The **predictive value** of a positive test defines the percentage of positive results that are true positives. This is related fundamentally to the incidence of the disease. In a group of patients on a urology service, the incidence of renal disease is higher than in the general population, and the serum creatinine level will have a higher predictive value in that group than for the general population.

---

*This section is an abridged version of an article by Krieg AF, Gambino R, Galen RS: Why are clinical laboratory tests performed? When are they valid? *JAMA* 1975;**233**:76. Reprinted from the Journal of the American Medical Association. Copyright 1975, American Medical Association. See also Galen RS, Gambino SR: *Beyond Normality: The Predictive Value and Efficiency of Medical Diagnosis.* Wiley, 1975.

Formulas for definitions:

$$\text{Sensitivity} = \frac{\text{True positive}}{\text{True positive} + \text{false negative}} \times 100$$

$$\text{Specificity} = \frac{\text{True negative}}{\text{True negative} + \text{false positive}} \times 100$$

Predictive value

$$= \frac{\text{True positive}}{\text{True positive} + \text{false negative}} \times 100$$

Before ordering a test, attempt to determine whether test sensitivity, specificity, and predictive value are adequate to provide useful information. To be useful, the result should influence diagnosis, prognosis, or therapy; lead to a better understanding of the disease process; and benefit the patient.

SI Units (*Système International d'Unités*)

A "coherent" system of measurement has been developed by an international organization designated the General Conference of Weights and Measures. An adaptation has been tentatively recommended by the Commission on Quantities and Units of the Section on Clinical Chemistry, International Union of Pure and Applied Chemistry. SI units are in use in some European countries, and the conversion to SI will continue if the system proves to be helpful in understanding physiologic mechanisms.

Eight fundamental measurable properties of matter (with authorized abbreviations shown in parentheses) were selected for clinical use.

length: metre (m)
mass: kilogram (kg)
amount: of substance: mole (mol)
time: second (s)
thermodynamic temperature: kelvin (K)
electric current: ampere (A)
luminous intensity: candela (cd)
catalytic activity: katal (kat)

Derived from these are the following measurable properties:

mass concentration: kilogram/l (kg/L)
mass fraction: kilogram/kilogram (kg/kg)
volume fraction: l/l (L/L)
volume: cubic metre (m^3); for clinical use, the unit will be the (L)
substance concentration: mole/l (mol/L)
molality: mole/kilogram (mol/kg)
mole fraction: mole/mole (mol/mol)
pressure: pascal (Pa) = newton/m^2

Decimal factors are as follows:

Number	Name	Symbol
10^{12}	tera	T
10^9	giga	G
10^6	mega	M
10^3	kilo	k

Number	Name	Symbol
10^2	hecto	h
10^1	deca	da
10^{-1}	deci	d
10^{-2}	centi	c
10^{-3}	milli	m
10^{-6}	micro	μ
10^{-9}	nano	n
10^{-12}	pico	p
10^{-15}	femto	f
10^{-18}	atto	a

"Per"—eg, "per second"—is often written as the negative exponent. Per second thus becomes $\cdot s^{-1}$; per meter squared, $\cdot m^{-2}$; per kilogram, $\cdot kg^{-1}$. *Example:* $cm/s = cm \cdot s^{-1}$; $g/m^2 = g \cdot m^{-2}$; etc.

In anticipation that the SI system may be adopted in the USA in the next several years, values are reported here in the traditional units with equivalent SI units following in parentheses.

COMMON CLINICAL VALUES IN TRADITIONAL & SI MEASUREMENTS*

For information regarding precautions to be taken in collecting specimens of blood or urine for analysis (eg, use of whole blood, plasma, or serum; use of suitable anticoagulants etc) the reader should consult the biochemistry laboratory in a hospital.

Albumin, Serum, or Plasma:
See Proteins, Serum or Plasma.

Aminotransferases, Serum:
Normal (varies with method): AST (SGOT), 6–25 IU/L at 30 °C; SMA, 10–40 IU/L at 37 °C; SMAC, 0–41 IU/L at 37 °C. ALT (SGPT), 3–26 IU/L at 30 °C; SMAC, 0–45 IU/L at 37 °C.

A. Physiologic Basis: Aspartate aminotransferase (AST; SGOT), alanine aminotransferase (ALT; SGPT), and lactic dehydrogenase are intracellular enzymes involved in amino acid or carbohydrate metabolism. These enzymes are present in high concentrations in muscle, liver, and brain. Elevations of concentrations of these enzymes in the blood indicate necrosis or disease, especially of these tissues.

B. Interpretation:

1. Elevated after myocardial infarction (especially AST); acute infectious hepatitis (ALT usually elevated more than AST); cirrhosis of the liver (AST usually elevated more than ALT); and metastatic or primary liver neoplasm. Elevated in transudates associated with neoplastic involvement of serous cavities. AST is elevated in muscular dystrophy, dermatomyositis, and paroxysmal myoglobinuria.

2. Decreased with pyridoxine (vitamin B_6) deficiency (often as a result of repeated hemodialysis), renal insufficiency, and pregnancy.

Ammonia, Blood:
Normal (Conway): 10–110 μg/dL whole blood. (SI: 12–65 μmol/L.)

A. Physiologic Basis: Ammonia present in the blood is derived from 2 principal sources: (1) In the large intestine, putrefactive action of bacteria on nitrogenous materials releases significant quantities of ammonia. (2) In the process of protein metabolism, ammonia is liberated. Ammonia entering the portal vein or the systemic circulation is rapidly converted to urea in the liver. Liver insufficiency may result in an increase in blood ammonia concentration, especially if protein consumption is high or if there is bleeding into the bowel.

B. Interpretation: Blood anemia is elevated in hepatic insufficiency or with liver bypass in the form of a portacaval shunt, particularly if protein intake is high or if there is bleeding into the bowel.

Amylase, Serum:
Normal (varies with method): 80–180 Somogyi units/dL serum. (One Somogyi unit equals amount of enzyme that will produce 1 mg of reducing sugar from starch at pH 7.2.) 0.8–3.2 IU/L.

A. Physiologic Basis: Normally, small amounts of amylase (diastase), molecular weight about 50,000, originating in the pancreas and salivary glands, are present in the blood. Inflammatory disease of these glands or obstruction of their ducts results in regurgitation of large amounts of enzyme into the blood and increased excretion via the kidney.

B. Interpretation:

1. Elevated in acute pancreatitis, pseudocyst of the pancreas, obstruction of pancreatic ducts (carcinoma, stone, stricture, duct sphincter spasm after morphine), and mumps. Occasionally elevated in renal insufficiency, in diabetic acidosis, and in inflammation of the pancreas from a perforating peptic ulcer. Rarely, combination of amylase with an immunoglobulin produces elevated serum amylase activity (macroamylasemia) because the large molecular complex (molecular weight at least 160,000) is not filtered by the glomerulus.

2. Decreased in acute and chronic hepatitis, in pancreatic insufficiency, and occasionally, in toxemia of pregnancy.

Amylase may also be measured in samples of urine. It is elevated in the same situations in which the activity of serum amylase is elevated.

Bicarbonate Serum or Plasma:
Normal: 24–28 meq/L. (SI: 24–28 mmol/L.)

A. Physiologic Basis: Bicarbonate-carbonic acid buffer is one of the most important buffer systems in maintaining normal pH of body fluids. Bicarbonate and pH determinations on arterial whole blood serve as a basis for assessing "acid-base balance."

---

*The values listed in this section and in the following section have been gleaned from many sources. Values will vary with method and individual laboratory.

B. Interpretation:

1. Elevated in-

a. Metabolic alkalosis (arterial blood pH increased) due to ingestion of large quantities of sodium bicarbonate, protracted vomiting of acid gastric juice, or accompanying potassium deficit.

b. Respiratory acidosis (arterial blood pH decreased) due to inadequate elimination of CO_2 (leading to elevated P_{CO_2}) because of pulmonary emphysema, poor diffusion in alveolar membrane disease, heart failure with pulmonary congestion or edema, or ventilatory failure due to any cause, including oversedation, narcotics, or inadequate artificial respiration.

2. Decreased in-

a. Metabolic acidosis (arterial blood pH decreased) due to diabetic ketoacidosis, lactic acidosis, starvation, persistent diarrhea, renal insufficiency, ingestion of excess acidifying salts or methanol, or salicylate intoxication.

b. Respiratory alkalosis (arterial blood pH increased) due to hyperventilation (decreased P_{CO_2}).

Bilirubin, Serum:

Normal: Total, 1.1–1.2 mg/dL (SI: 3.5–19 μmol/L). Direct (glucuronide), 0.1–0.4 mg/dL. Indirect (unconjugated), 0.2–0.7 mg/dL. (SI: direct, up to 7 μmol/L; indirect, up to 12 μmol/L.)

A. Physiologic Basis: Destruction of hemoglobin yields bilirubin, which is conjugated in the liver to the diglucuronide and excreted in the bile. Bilirubin accumulates in the plasma when liver insufficiency exists, biliary obstruction is present, or the rate of hemolysis increases. Rarely, abnormalities of enzyme systems involved in bilirubin metabolism in the liver (eg, absence of glucuronyl transferase) result in abnormal bilirubin concentrations.

B. Interpretation:

1. Direct and indirect forms of serum bilirubin are elevated in acute or chronic hepatitis; biliary tract obstruction (cholangiolar, hepatic, or common ducts); toxic reactions to many drugs, chemicals, and toxins; and Dubin-Johnson and Rotor's syndromes.

2. Indirect serum bilirubin is elevated in hemolytic diseases or reactions and absence or deficiency of glucuronyl transferase, as in Gilbert's disease and Crigler-Najjar syndrome.

3. Direct and total bilirubin can be significantly elevated in normal and jaundiced subjects by fasting 24–48 hours (in some instances even 12 hours) or by prolonged caloric restriction.

Calcium, Serum:

Normal: Total, 8.5–10.3 mg/dL or 4.2–5.2 meq/L. Ionized, 4.2–5.2 mg/dL or 2.1–2.6 meq/L. (SI: total, 2.1–2.6 mmol/L; ionized, 1.05–1.3 mmol/L.)

A. Physiologic Basis: Endocrine, renal, gastrointestinal, and nutritional factors normally provide for precise regulation of calcium concentration in plasma and other body fluids. Since some calcium is bound to plasma protein, especially albumin, determination of the plasma albumin concentration is necessary before the clinical significance of abnormal serum calcium levels can be interpreted accurately.

B. Interpretation:

1. Elevated in hyperparathyroidism, secretion of parathyroidlike hormone by malignant tumors, vitamin D excess, milk-alkali syndrome, osteolytic disease such as multiple myeloma, invasion of bone by metastatic cancer, Paget's disease of bone, Boeck's sarcoid, immobilization, and familial hypocalciuria. Occasionally elevated with hyperthyroidism and with ingestion of thiazide drugs.

2. Decreased in hypoparathryoidism, vitamin D deficiency (rickets, osteomalacia), renal insufficiency, hypoproteinemia, malabsorption syndrome (sprue, ileitis, celiac disease, pancreatic insufficiency), severe pancreatitis with pancreatic necrosis, and pseudohypoparathyroidism.

Analysis of the daily excretion of calcium in the urine is also of value in certain clinical situations, such as hyperparathyroidism.

Ceruloplasmin & Copper, Serum

Normal: Ceruloplasmin, 25–43 mg/dL (SI: 1.7–2.9 μmol/L); copper, 100–200 μg/dL (SI: 16–31 μmol/L).

A. Physiologic Basis: About 5% of serum copper is loosely bound to albumin and 95% to ceruloplasmin, an oxidase enzyme that is an α_2 globulin with a blue color. In Wilson's disease, serum copper and ceruloplasmin are low and urinary copper is high.

B. Interpretation:

1. Elevated in pregnancy, hyperthyroidism, infection, aplastic anemia, acute leukemia. Hodgkin's disease, cirrhosis of the liver, and with use of oral contraceptives.

2. Decreased in Wilson's disease (accompanied by increased urinary excretion of copper), malabsorption, nephrosis, and copper deficiency that may accompany total parenteral nutrition.

Chloride, Serum or Plasma

Normal: 96–106 meq/L. (SI: 96–106 mmol/L.)

A. Physiologic Basis: Chloride is the principal inorganic anion of the extracellular fluid. It is important in maintenance of acid-base balance even though it exerts no buffer action. When chloride as HCl or NH_4Cl is lost, alkalosis follows; when chloride is retained or ingested, acidosis follows. Chloride (with sodium) plays an important role in control of osmolarity of body fluids.

B. Interpretation:

1. Elevated in renal insufficiency (when Cl intake exceeds excretion), nephrosis (occasionally), renal tubule acidosis, hyperparathyroidism (occasionally), ureterosigmoid anastomosis (reabsorption from urine in gut), dehydration (water deficit), and overtreatment with saline solution.

2. Decreased in gastrointestinal disease with loss of gastric and intestinal fluids (vomiting, diarrhea, gastrointestinal suction), renal insufficiency (with salt deprivation), overtreatment with diuretics, chronic

Table 1. Lipidemia: Ranges of population (USA) for serum concentrations of cholesterol (C), triglyceride (TG), low-density lipoprotein cholesterol (HDL-C), and high-density lipoprotein cholesterol (HDL-C).*

Age	C (mg/dL)	TG (mg/dL)	LDL-C (mg/dL) Upper Limit	HDL-C (mg/dL)	
				Male	Female
<29	120–240	10–140	170	45 ± 12	55 ± 12
30–39	140–270	10–150	190		
40–49	150–310	10–160	190		
>49	160–330	10–190	210		

*Reproduced, with permission, from Krupp MA et al: *Physician's Handbook,* 21st ed. Lange, 1985.

respiratory acidosis (emphysema), diabetic acidosis, excessive sweating, adrenal insufficiency (NaCl loss), hyperadrenocorticism (chronic K^+ loss), and metabolic alkalosis ($NaHCO_3$ ingestion; K^+ deficit).

Cholesterol, Serum or Plasma

Normal: 150–280 mg/dL. (SI: 3.9–7.2 mmol/L.) See Table 1.

A. Physiologic Basis: Cholesterol concentrations are determined by metabolic functions, which are influenced by heredity, nutrition, endocrine function, and integrity of vital organs such as the liver and kidney. Cholesterol metabolism is intimately associated with lipid metabolism.

B. Interpretation:

1. Elevated in familial hypercholesterolemia (xanthomatosis), hypothyroidism, poorly controlled diabetes mellitus, nephrotic syndrome, chronic hepatitis, biliary cirrhosis, obstructive jaundice, hypoproteinemia (idiopathic, with nephrosis or chronic hepatitis), and lipidemia (idiopathic, familial).

2. Decreased in acute hepatitis and Gaucher's disease. Occasionally decreased in hyperthyroidism, acute infections, anemia, malnutrition, and apolipoprotein deficiency.

Creatine Phosphokinase (CPK), Serum

Normal (varies with method): 10–50 IU/L at 30 °C.

A. Physiologic Basis: CPK splits creatine phosphate in the presence of ADP to yield creatine and ATP. Skeletal and heart muscle and brain are rich in the enzyme.

B. Interpretation:

1. Elevated in the presence of muscle damage such as with myocardial infarction, trauma to muscle, muscular dystrophies, polymyositis, severe muscular exertion (jogging), hypothyroidism, and cerebral infarction (necrosis). Following myocardial infarction, serum CPK concentration increases rapidly (within 3–5 hours), and it remains elevated for a shorter time after the episode (2 or 3 days) than does AST or LDH.

2. Not elevated in pulmonary infarction or parenchymal liver disease.

Creatine Phosphokinase Isoenzymes, Serum

See Table 2.

A. Physiologic Basis: CPK consists of 3 proteins separable by electrophoresis. Skeletal muscle is characterized by isoenzyme MM, myocardium by isoenzyme MB, and brain by isoenzyme BB.

B. Interpretation: CPK isoenzymes are increased in serum. CPK-MM is elevated in injury to skeletal muscle, myocardial muscle, and brain; in muscle disease (eg, dystrophies, hypothyroidism, dermatomyositis, polymyositis); in rhabdomyolysis; and after severe exercise. CPK-MB is elevated soon (within 2–4 hours) after myocardial infarction and for up to 72 hours afterward (high levels are prolonged with extension of infarct or new infarction); also elevated in extensive rhabdomyolysis or muscle injury, severe muscle disease, Reye's syndrome, or Rocky Mountain spotted fever. CPK-BB is occasionally elevated in severe shock, in some carcinomas (especially oat cell carcinoma or carcinoma of the ovary, breast, or prostate), or in biliary atresia.

Creatine, Urine (24 Hours)

Normal: See Table 3.

A. Physiologic Basis: Creatine is an important constituent of muscle, brain, and blood; in the form of creatine phosphate, it serves as a source of high-energy phosphate. Normally, small amounts of creatine are excreted in the urine, but in states of elevated catabolism and in the presence of muscular dystrophies, the rate of excretion is increased.

Table 2. Creatine kinase isoenzymes.

Isoenzyme	Normal Levels % of Total
(Fastest) Fraction 1, BB	0
Fraction 2, MB	0–3
(Slowest) Fraction 3, MM	97–100

Table 3. Urine creatine and creatinine, normal values (24 hours).*

	Creatine	Creatinine
Newborn	4.5 mg/kg	10 mg/kg
1–7 months	8.1 mg/kg	12.8 mg/kg
2–3 years	7.9 mg/kg	12.1 mg/kg
4–4½ years	4.5 mg/kg	14.6 mg/kg
9–9½ years	2.5 mg/kg	18.1 mg/kg
11–14 years	2.7 mg/kg	20.1 mg/kg
Adult male	0–50 mg	25 mg/kg
Adult female	0–100 mg	21 mg/kg

B. Interpretation:

1. Elevated in muscular dystrophies such as progressive muscular dystrophy, myotonia atrophica, and myasthenia gravis; muscle wasting, as in acute poliomyelitis, amyotrophic lateral sclerosis, and myositis manifested by muscle wasting; starvation and cachectic states; hyperthyroidism; and febrile diseases.

2. Decreased in hypothyroidism, amyotonia congenita, and renal insufficiency.

Creatinine, Serum or Plasma

Normal: 0.7–1.5 mg/dL. (SI: 60–132 μmol/L.)

A. Physiologic Basis: Endogenous creatinine is excreted by filtration through the glomerulus and by tubular secretion at a rate about 20% greater than clearance of inulin. The Jaffe reaction measures chromogens other than creatinine in the plasma. Because the chromogens are not passed into the urine, the measurement of creatinine in the urine is about 20% less than chromogen plus creatinine in plasma, providing, fortuitously, a compensation for the amount secreted. Thus, inulin and creatinine clearances for clinical purposes are comparable and creatinine clearance is an acceptable measure of glomerular filtration rate—except that with advancing renal failure, creatinine clearance exceeds inulin clearance owing to the secretion of creatinine by remaining renal tubules.

B. Interpretation: Creatinine is elevated in acute or chronic renal insufficiency, urinary tract obstruction, and impairment of renal function induced by some drugs. Materials other than creatinine may react to give falsely high results with the alkaline picrate method (Jaffe reaction): acetoacetate, acetone, β-hydroxybutyrate, α-ketoglutarate, pyruvate, glucose, bilirubin, hemoglobin, urea, and uric acid. Values below 0.7 mg/dL are of no known significance.

Creatinine, Urine

Normal: See Table 3.

Glucose, Serum or Plasma

Normal: Fasting "true" glucose, 65–110 mg/dL. (SI: 3.6–6.1 mmol/L.)

A. Physiologic Basis: The glucose concentration in extracellular fluid is normally closely regulated, with the result that a source of energy is available to tissues, and no glucose is excreted in the urine. Hyper-

glycemia and hypoglycemia are nonspecific signs of abnormal glucose metabolism.

B. Interpretation:

1. Elevated in diabetes, hyperthyroidism, adrenocortical hyperactivity (cortical excess), hyperpituitarism, and hepatic disease (occasionally).

2. Decreased in hyperinsulinism, adrenal insufficiency, hypopituitarism, hepatic insufficiency (occasionally), functional hypoglycemia, and by hypoglycemic agents.

γ-Glutamyl Transpeptidase or Transferase, Serum

Normal: Males, <30 mU/mL at 30 °C. Females, <25mU/mL at 30 °C. Adolescents, <50mU/mL at 30 °C.

A. Physiologic Basis: γ-Glutamyl transferase (GGT) is an extremely sensitive indicator of liver disease. Levels are often elevated when transaminases and alkaline phosphatase are normal, and it is considered more specific than both for identifying liver impairment due to alcoholism.

The enzyme is present in liver, kidney, and pancreas and transfers C-terminal glutamic acid from a peptide to other peptides or L-amino acids. It is induced by alcohol.

B. Interpretation: Elevated in acute infectious or toxic hepatitis, chronic and subacute hepatitis, cirrhosis of the liver, intrahepatic or extrahepatic obstruction, primary or metastatic liver neoplasms, and liver damage due to alcoholism. It is elevated occasionally in congestive heart failure and rarely in postmyocardial infarction, pancreatitis, and pancreatic carcinoma.

Iron, Serum

Normal: 50–175 μg/dL. (SI: 9–31.3 μmol/L.)

A. Physiologic Basis: Because of diurnal variation with highest values in the morning, fasting morning blood specimens are desirable. Iron concentration in the plasma is determined by several factors, including absorption from the intestine; storage in intestine, liver, spleen, and marrow; breakdown or loss of hemoglobin; and synthesis of new hemoglobin.

B. Interpretation:

1. Elevated in hemochromatosis, hemosiderosis (multiple transfusions, excess iron administration), hemolytic disease, pernicious anemia, and hypoplastic anemias. Often elevated in viral hepatitis. Spuriously elevated if patient has received parenteral iron during the 2–3 months prior to determination.

2. Decreased in iron deficiency; with infections, nephrosis, and chronic renal insufficiency; and during periods of active hematopoiesis.

Iron-Binding Capacity, Serum

Normal: Total, 250–410 μg/dL. (SI: 45–73 μmol/L.) Percent saturation, 20–55%.

A. Physiologic Basis: Iron is transported as a complex of the metal-binding globulin transferrin (siderophilin). Normally, this transport protein carries an amount of iron that represents about 30–40% of its capacity to combine with iron.

B. Interpretation of Total Iron-Binding Capacity:

1. Elevated in iron deficiency anemia, with use of oral contraceptives, in late pregnancy, and in infants. Occasionally elevated in hepatitis.

2. Decreased in association with decreased plasma proteins (nephrosis, starvation, cancer), chronic inflammation, and hemosiderosis (transfusions, thalassemia).

C. Interpretation of Saturation of Transferrin:

1. Elevated in iron excess (iron poisoning, hemolytic disease, thalassemia, hemochromatosis, pyridoxine deficiency, nephrosis, and, occasionally, hepatitis).

2. Decreased in iron deficiency, chronic infection, cancer, and late pregnancy.

Lactate Dehydrogenase (LDH), Serum, Serous Fluids, Spinal Fluid, or Urine

Normal (varies with method): Serum, 55–140 IU/L at 30 °C; SMA, 100–225 IU/L at 37 °C; SMAC, 60–200 IU/L at 37 °C. Serous fluids, lower than serum. Spinal fluid, 15–75 units (Wroblewski); 6.3–30 IU/L. Urine, less than 8300 units/8 hr (Wroblewski).

A. Physiologic Basis: LDH catalyzes the interconversion of lactate and pyruvate in the presence of NADH or $NADH_2$. It is distributed generally in body cells and fluids.

B. Interpretation: Elevated in all conditions accompanied by tissue necrosis, particularly those involving acute injury of the heart, red cells, kidney, skeletal muscle, liver, lung, and skin. Marked elevations accompany hemolytic anemias, the anemias of vitamin B_{12} and folate deficiency, and polycythemia rubra vera. The course of rise in concentration over 3–4 days followed by a slow decline during the following 5–7 days may be helpful in confirming the presence of a myocardial infarction; however, pulmonary infarction, neoplastic disease, and megaloblastic anemia must be excluded. Although elevated during the acute phase of infectious hepatitis, enzyme activity is seldom increased in chronic liver disease.

Lactate Dehydrogenase (LDH) Isoenzymes, Serum

Normal: See Table 4.

A. Physiologic Basis: LDH consists of 5 separable proteins, each made of tetramers of 2 types, or subunits, H and M. The 5 isoenzymes can be distinguished by kinetics, electrophoresis, chromatography, and immunologic characteristics. By electrophoretic separation, the mobility of the isoenzymes corresponds to serum proteins α_1, α_2, β, γ_1, and γ_2. These are usually numbered 1 (fastest moving), 2, 3, 4, and 5 (slowest moving). Isoenzyme 1 is present in high concentrations in heart muscle (tetramer H H H H) and in erythrocytes and kidney cortex; isoenzyme 5 in skeletal muscle (tetramer M M M M) and liver.

B. Interpretation: In myocardial infarction, the α isoenzymes are elevated—particularly LDH 1–to yield a ratio of LDH 1:LDH 2 of greater than 1. Similar α isoenzyme elevations occur in renal cortex infarction and with hemolytic anemias.

LDH 5 and 4 are relatively increased in the presence of acute hepatitis, acute muscle injury, dermatomyositis, and muscular dystrophies.

Lipase, Serum

Normal: 0.2–1.5 units.

A. Physiologic Basis: A low concentration of fat-splitting enzyme is present in circulating blood. In the presence of pancreatitis, pancreatic lipase is released into the circulation in higher concentrations, which persist, as a rule, for a longer period than does the elevated concentration of amylase.

B. Interpretation: Serum lipase is elevated in acute or exacerbated pancreatitis and in obstruction of pancreatic ducts by stone or neoplasm.

Magnesium Serum

Normal: 1.8–3 mg/dL or 1.5–2.5 meq/L. (SI: 0.75–1.25 mmol/L.)

A. Physiologic Basis: Magnesium is primarily an intracellular electrolyte. In extracellular fluid, it affects neuromuscular irritability and response. Magnesium deficit may exist with little or no change in extracellular fluid concentrations. Low magnesium levels in plasma have been associated with tetany, weaknesses, disorientation, and somnolence.

B. Interpretation

1. Elevated in renal insufficiency and in overtreatment with magnesium salts.

2. Decreased in chronic diarrhea, acute loss of enteric fluids, starvation, chronic alcoholism, chronic hepatitis, hepatic insufficiency, excessive renal loss (diuretics), and inadequate replacement with parenteral nutrition. May be decreased in and contribute to persistent hypocalcemia in patients with hypoparathyroidism and when large doses of vitamin D and calcium are being administered.

Phosphatase, Acid, Serum

Normal values vary with method: 0.1–0.63 Sigma units. (SI: 36–175 nmol/s/L.)

A. Physiologic Basis: Phosphatases active at pH 4.9 are present in high concentration in the prostate gland, erythrocytes, platelets, reticuloendothelial cells, liver, spleen, and kidney. A variety of isoenzymes have been found in these tissues and serum and account

Table 4. Lactate dehydrogenase isoenzymes.

	Isoenzyme	Percentage of Total (and Range)
(Fastest)	1 (α_1)	28 (15–30)
	2 (α_2)	36 (22–50)
	3 (β)	23 (15–30)
	4 (γ_1)	6 (0–15)
(Slowest)	5 (γ_2)	6 (0–15)

for different activities operating against different substrates.

B. Interpretation: In the presence of carcinoma of the prostate, the prostatic fraction of acid phosphatase may be increased in the serum, particularly if the cancer has spread beyond the capsule of the gland or has metastasized. Palpation of the prostate will produce a transient increase. Total acid phosphatase may be increased in Gaucher's disease, malignant tumors involving bone, renal disease, hepatobiliary disease, diseases of the reticuloendothelial system, and thromboembolism. Fever may cause spurious elevations.

Phosphatase, Alkaline, Serum

Normal (varies with method): Bessey-Lowry, children, 2.8–6.7 units; Bessey-Lowry, adults, 0.8–2.3 units. Adults, King-Armstrong, 5–13 units: 24–71 IU/L at 30 °C; SMA, 30–85 IU/L at 37 °C; SMAC, 30–115 IU/L at 37 °C.

A. Physiologic Basis: Alkaline phosphatase is present in high concentration in growing bone, in bile, and in the placenta. In serum, it consists of a mixture of isoenzymes not yet clearly defined. The isoenzymes may be separated by electrophoresis; liver alkaline phosphatase migrates faster than bone and placental alkaline phosphatase, which migrate together.

B. Interpretation:

1. Elevated in-

a. Children (normal growth of bone).

b. Osteoblastic bone disease—Hyperparathyroidism, rickets and osteomalacia, neoplastic bone disease (osteosarcoma, metastatic neoplasms), ossification as in myositis ossificans, Paget's disease (osteitis deformans), and Boeck's sarcoid.

c. Hepatic duct or cholangiolar obstruction due to stone, stricture, and neoplasm.

d. Hepatic disease resulting from drugs such as chlorpromazine and methyltestosterone.

e. Pregnancy.

2. Decreased in hypothyroidism and in growth retardation in children.

Phosphorus, Inorganic, Serum

Normal: Children, 4–7 mg/dL. (SI: 1.3–2.3 mmol/L.) Adults, 3–45. mg/dL. (SI: 1–1.5 mmol/L.)

A. Physiologic Basis: The concentration of inorganic phosphate in circulating plasma is influenced by parathyroid gland function, action of vitamin D, intestinal absorption, renal function, bone metabolism, and nutrition.

B. Interpretation:

1. Elevated in renal insufficiency, hypoparathyroidism, and hypervitaminosis D.

2. Decreased in hyperparathyroidism, hypovitaminosis D (rickets, osteomalacia), malabsorption syndrome (steatorrhea), ingestion of antacids that bind phosphate in the gut, starvation or cachexia, chronic alcoholism (especially with liver disease), hyperalimentation with phosphate-poor solutions, carbohydrate administration (especially intravenously), renal tubular defects, use of thiazide diuretics, acid-base

disturbances, diabetic ketoacidosis (especially during recovery), and genetic hypophosphatemia. Occasionally decreased during pregnancy and with hypothyroidism.

Potassium, Serum or Plasma

Normal: 3.5–5 meq/L. (SI: 3.5–5 mmol/L.)

A. Physiologic Basis: Potassium concentration in plasma determines neuromuscular and muscular irritability. Elevated or decreased concentrations impair the capability of muscle tissue to contract.

B. Interpretation:

1. Elevated in renal insufficiency (especially in the presence of increased rate of protein or tissue breakdown); adrenal insufficiency (especially hypoaldosteronism); hyporeninemic hypoaldosteronism; use of spironolactone; too rapid administration of potassium salts, especially intravenously; and use of triamterene or phenformin.

2. Decreased in-

a. Inadequate intake (starvation).

b. Inadequate absorption or unusual enteric losses–Vomiting, diarrhea, malabsorption syndrome, or use of sodium polystyrene sulfonate resin.

c. Unusual renal loss–Secondary to hyperadrenocorticism (especially hyperaldosteronism) and to adrenocorticosteroid therapy, metabolic alkalosis, use of diuretics such as chlorothiazide and its derivatives and the mercurials; renal tubular defects such as the de Toni-Fanconi syndrome and renal tubular acidosis; treatment with antibiotics that are excreted as anions (carbenicillin, ticarcillin); use of phenothiazines, amphotericin B, and drugs with high sodium content; and use of degraded tetracycline.

d. Abnormal redistribution between extracellular and intracellular fluids—Familial periodic paralysis or testosterone administration.

Proteins, Serum or Plasma
(Includes Fibrinogen)

Normal: See Interpretation, below.

A. Physiologic Basis: Concentration of protein determines colloidal osmotic pressure of plasma. The concentration of protein in plasma is influenced by the nutritional state, hepatic function, renal function, occurrence of disease such as multiple myeloma, and metabolic errors. Variations in the fractions of plasma proteins may signify specific disease.

B. Interpretation:

1. Total protein, serum–Normal: 6–8 g/dL. (SI:

Table 5. Protein fractions as determined by electrophoresis.

	Percentage of Total Protein
Albumin	52–68
α_1 globulin	2.4–4.4
α_2 globulin	6.1–10.1
β globulin	8.5–14.5
γ globulin	10–21

Table 6. Gamma globulins by immunoelectrophoresis.

IgA	90–450 mg/dL
IgG	700–1500 mg/dL
IgM	40–250 mg/dL
IgD	0.3–40 mg/dL
IgE	0.006–0.16 mg/dL

60–80 g/L.) See albumin and globulin fractions, below, and Table 5.

2. Albumin, serum or plasma–Normal: 3.5–5.5 g/dL. (SI: 33–55 g/L.)

a. Elevated in dehydration, shock, hemoconcentration, and administration of large quantities of concentrated albumin "solution" intravenously.

b. Decreased in malnutrition, malabsorption syndrome, acute or chronic glomerulonephritis, nephrosis, acute or chronic hepatic insufficiency, neoplastic disease, and leukemia.

3. Globulin, serum or plasma–Normal: 2–3.6 g/dL. (SI: 20–36 g/L.) (See Tables 6 and 7.)

a. Elevated in hepatic disease, infectious hepatitis, cirrhosis of the liver, biliary cirrhosis, and hemochromatosis; disseminated lupus erythematosus; plasma cell myeloma; lymphoproliferative disease; sarcoidosis; and acute or chronic infectious diseases, particularly lymphogranuloma venereum, typhus, leishmaniasis, schistosomiasis, and malaria.

b. Decreased in malnutrition, congenital agammaglobulinemia, acquired hypogammaglobulinemia, and lymphatic leukemia.

4. Fibrinogen, plasma–Normal: 0.2–0.6 g/dL. (SI: 2–6 g/L.)

a. Elevated in glomerulonephritis, nephrosis (occasionally), and infectious diseases.

b. Decreased in disseminated intravascular coagulation (accidents of pregnancy such as placental ablation, amniotic fluid embolism, and violent labor; meningococcal meningitis; metastatic carcinoma of the prostate and occasionally of other organs; and leuke-

Table 7. Some constituents of globulins.

Globulin	Representative Constituents
α_1	Thyroxine-binding globulin Transcortin Glycoprotein Lipoprotein Antitrypsin
α_2	Haptoglobin Glycoprotein Macroglobulin Ceruloplasmin
β	Transferrin Lipoprotein Glycoprotein
γ	γG γD γM γE γA

mia), acute and chronic hepatic insufficiency, and congenital fibrinogenopenia.

Sodium, Serum or Plasma

Normal: 136–145 meq/L. (SI: 136–145 mmol/L.)

A. Physiologic Basis: Sodium constitutes about 140 of the 155 meq of cation in plasma. With its associated anions it provides the bulk of osmotically active solute in the plasma, thus affecting the distribution of body water significantly. A shift of sodium into cells or a loss of sodium from the body results in a decrease of extracellular fluid volume, affecting circulation, renal function, and nervous system function.

B. Interpretation:

1. Elevated in dehydration (water deficit), central nervous system trauma or disease, and hyperadrenocorticism with hyperaldosteronism or corticosterone of corticosteroid excess.

2. Decreased in adrenal insufficiency; in renal insufficiency, especially with inadequate sodium intake; in renal tubular acidosis; as a physiologic response to trauma or burns (sodium shift into cells); in unusual losses via the gastrointestinal tract, as with acute or chronic diarrhea or with intestinal obstruction or fistula; and in unusual sweating with inadequate sodium replacement. In some patients with edema associated with cardiac or renal disease, serum sodium concentration is low even though total body sodium content is greater than normal; water retention (excess ADH) and abnormal distribution of sodium between intracellular and extracellular fluid contribute to this paradoxic situation. Hyperglycemia occasionally results in shift of intracellular water to the extracellular space, producing a dilutional hyponatremia. (Artifact: When measured by the flame photometer, serum or plasma sodium will be decreased in the presence of hyperlipidemia or hyperglobulinemia; in these disorders, the volume ordinarily occupied by water is taken up by other substances, and the serum or plasma will thus be "deficient" in water and electrolytes. In the presence of hyperglycemia, serum sodium concentration will be reduced by 1.6 meq/L per 100 mg/dL glucose above 200 mg/dL because of shifts of water into extracellular fluid.)

Thyroxine (T_4), Total (TT_4), Serum

Normal: Radioimmunoassay (RIA), 5–12 μg/dL (SI: 65–156 nmol/L); competitive binding protein (CPB) (Murphy-Pattee), 4–11 μg/dL (SI: 51–142 nmol/L).

A. Physiologic Basis: The total thyroxine level does not necessarily reflect the physiologic hormonal effect of thyroxine. Levels of thyroxine vary with the concentration of the carrier proteins (thyroxine-binding globulin and prealbumin), which are readily altered by physiologic conditions such as pregnancy and by a variety of diseases and drugs. Any interpretation of the significance of total T_4 depends upon knowing the concentration of carrier protein either from direct measurement or from the result of the erythrocyte or resin uptake of triiodothyronine (T_3) (see below). It is the concentration of free T_4 and of T_3 that determines hormonal activity.

B. Interpretation:

1. Elevated in hyperthyroidism, with elevation of thyroxine-binding proteins, and at times with active thyroiditis or acromegaly.

2. Decreased in hypothyroidism (primary or secondary) and with decreased concentrations of thyroxine-binding proteins.

Thyroxine, Free, Serum

Normal (equilibrium dialysis): 0.8–2.4 ng/dL. (SI: 0.01–0.03 nmol/L.) May be estimated from measurement of total thyroxine and resin T_3 uptake.

A. Physiologic Basis: The metabolic activity of T_4 is related to the concentration of free T_4. T_4 is apparently largely converted to T_3 in peripheral tissue. (T_3 is also secreted by the thyroid gland.) Both T_4 and T_3 seem to be active hormones.

B. Interpretation:

1. Elevated in hyperthyroidism and at times with active thyroiditis.

2. Decreased in hypothyroidism.

Thyroxine-Binding Globulin (TBG), Serum

Normal (radioimmunoassay): 2–4.8 mg/dL.

A. Physiologic Basis: TBG is the principal carrier protein for T_4 and T_3 in the plasma. Variations in concentration of TBG are accompanied by corresponding variations in concentration of T_4 with intrinsic adjustments that maintain the physiologically active free hormones at proper concentration for euthyroid function. The inherited abnormalities of TBG concentration appear to be X-linked.

B. Interpretation:

1. Elevated in pregnancy, in infectious hepatitis, and in hereditary increase in TBG concentration.

2. Decreased in major depleting illness with hypoproteinemia (globulin), nephrotic syndrome, cirrhosis of the liver, active acromegaly, estrogen deficiency, and hereditary TBG deficiency.

Triiodothyronine (T_3) Uptake, Serum; Resin (RT_3U) or Thyroxine-Binding Globulin Assessment (TBG Assessment)

Normal RT_3U, as percentage of uptake, of ^{125}I-T_3 by resin, 25–36%; RT_3U ratio (TBG assessment) expressed as ratio of binding of ^{125}I-T_3 by resin in test serum/pooled normal serum, 0.85–1.15.

A. Physiologic Basis: When serum thyroxine-binding proteins are normal, more TBG binding sites will be occupied by T_4 in T_4 hyperthyroidism, and fewer binding sites will be occupied in hypothyroidism. ^{125}I-labeled T_3 added to serum along with a secondary binder (resin, charcoal, talc, etc) is partitioned between TBG and the binder. The binder is separated from the serum, and the radioactivity of the binder is measured for the RT_3U test. Since the resin takes up the non-TBG-bound radioactive T_3, its activity varies

inversely with the numbers of available TBG sites, ie, RT_3U is increased if TBG is more nearly saturated by T_4 and decreased if TBG is less well saturated by T_4.

B. Interpretation:

1. RT_3U and RT_3U ratio are increased when available sites are decreased, as in hyperthyroidism, acromegaly, nephrotic syndrome, severe hepatic cirrhosis, and hereditary TBG deficiency.

2. RT_3U and RT_3U ratio are decreased when available TBG sites are increased, as in hypothyroidism, pregnancy, the newborn, infectious hepatitis, and hereditary increase in TBG.

Transaminases

See Aminotransferases, above.

Triglycerides, Serum

Normal: <165 mg/dL. (SI: < 1.65 g/L.) See also Table 1.

A. Physiologic Basis: Dietary fat is hydrolyzed in the small intestine, absorbed and resynthesized by the mucosal cells, and secreted into lacteals in the form of chylomicrons. Triglycerides in the chylomicrons are cleared from the blood by tissue lipoprotein lipase (mainly adipose tissue), and the split products are absorbed and stored. Free fatty acids derived mainly from adipose tissue are precursors of the endogenous triglycerides produced by the liver. Transport of endogenous triglycerides is in association with β-lipoproteins, the very low density lipoproteins. In order to ensure measurement of endogenous triglycerides, blood must be drawn in the postabsorptive state.

B. Interpretation: Concentration of triglycerides, cholesterol, and lipoprotein fractions (very low density, low-density, and high-density) is interpreted collectively. Disturbances in normal relationships of these lipid moieties may be primary or secondary in origin.

1. Elevated (hyperlipoproteinemia)-

a. Primary–Type I hyperlipoproteinemia (exogenous hyperlipidemia), type II hyperbetalipoproteinemia, type III broad beta hyperlipoproteinemia, type IV hyperlipoproteinemia (endogenous hyperlipidemia), and type V hyperlipoproteinemia (mixed hyperlipidemia).

b. Secondary–Hypothyroidism, diabetes mellitus, nephrotic syndrome, chronic alcoholism with fatty liver, ingestion of contraceptive steroids, biliary obstruction, and stress.

2. Decreased (hypolipoproteinemia)-

a. Primary–Tangier disease (α-lipoprotein deficiency), abetalipoproteinemia, and a few rare, poorly defined syndromes.

b. Secondary–Malnutrition, malabsorption, and, occasionally, with parenchymal liver disease.

Urea Nitrogen & Urea, Blood, Plasma, or Serum

Normal: Blood urea nitrogen, 8–25 mg/dL (SI: 2.9–8.9 mmol/L). Urea, 21–53 mg/dL (SI: 3.5–9 mmol/L).

A. Physiologic Basis: Urea, an end product of

protein metabolism, is excreted by the kidney. The urea concentration in the glomerular filtrate is the same as in the plasma. Tubular reabsorption of urea varies inversely with rate of urine flow. Thus, urea is a less useful measure of glomerular filtration than is creatinine, which is not reabsorbed. Blood urea nitrogen varies directly with protein intake and inversely with the rate of excretion of urea.

B. Interpretation:

1. Elevated in–

a. Renal insufficiency–Nephritis, acute and chronic; acute renal failure (tubular necrosis); and urinary tract obstruction.

b. Increased nitrogen metabolism associated with diminished renal blood flow or impaired renal function–Dehydration (from any cause) and upper gastrointestinal bleeding (combination of increased protein absorption from digestion of blood plus decreased renal blood flow).

c. Decreased renal blood flow–Shock, adrenal insufficiency, and, occasionally, congestive heart failure.

2. Decreased in hepatic failure, nephrosis not complicated by renal insufficiency, and cachexia.

Uric Acid, Serum or Plasma

Normal: Males, 3–9 mg/dL (SI: 0.18–0.53 mmol/L); females, 2.5–7.5 mg/dL (SI: 0.15–0.45 mmol/L).

A. Physiologic Basis: Uric acid, an end product of purine metabolism, is excreted by the kidney. Gout, a genetically transmitted metabolic error, is characterized by an increased plasma or serum uric acid concentration, an increase in total body uric acid, and deposition of uric acid in tissues. An increase in uric acid concentration in plasma and serum may accompany increased purine catabolism (blood dyscrasias, therapy with antileukemic drugs), use of thiazide diuretics, or decreased renal excretion.

B. Interpretation:

1. Elevated in gout, preeclampsia-eclampsia, leukemia, polycythemia, therapy with antileukemic drugs and a variety of other agents, renal insufficiency, glycogen storage disease (type I), Lesch-Nyhan syndrome (X-linked hypoxanthine-guanine phosphoribosyltransferase deficit), and Down's syndrome. The incidence of hyperuricemia is greater in Filipinos than in whites.

2. Decreased in acute hepatitis (occasionally), treatment with allopurinol, and treatment with probenecid.

Measurement of the daily urinary excretion of uric acid is also of use in certain clinical situations, such as gout.

Uric Acid, Urine

Normal: 350–600 mg/24 h on a standard purine-free diet. (SI: 2.1–3.6 mmol/24 h.) Normal urinary uric acid/creatinine ratio for adults is 0.21–0.59; maximum of 0.75 for 24-hour urine while on purine-free diet.

A. Precautions: Diet should be free of high-purine foods prior to and during 24-hour urine collection. Strenuous activity may be associated with elevated purine excretion.

B. Physiologic Basis: Elevated serum uric acid may result from overproduction or diminished excretion.

C. Interpretation:

1. Elevated renal excretion occurs in about 25–30% of cases of gout due to increased purine synthesis. Excess uric acid synthesis and excretion are associated with myeloproliferative disorders. Lesch-Nyhan syndrome (hypoxanthine-guanine phosphoribosyltransferase deficit) and some cases of glycogen storage disease are associated with uricosuria.

2. Decreased in renal insufficiency, in some cases of glycogen storage disease (type I), and in any metabolic defect producing either lactic acidemia or β-hydroxybutyric acidemia. Salicylates in doses of less than 2–3 g/d may produce renal retention of uric acid.

NORMAL LABORATORY VALUES

(Blood [B], Plasma [P], Serum [S], Urine [U])

MAJOR CHEMICAL CONSTITUENTS OF BLOOD, PLASMA, OR SERUM

(Values vary with method used.)

Acetone and acetoacetate: [S] 0.3–2 mg/dL (3–20 mg/L).

Aminotransferases:

Aspartate aminotransferase (AST; SGOT) (varies with method used): [S] 6–25 IU/L at 30 °C; SMA, 10–40 IU/L at 37 °C; SMAC, 0–41 IU/L at 37 °C.

Alanine aminotransferase (ALT: SGPT) (varies with method used): [S] 3–26 IU/L at 30 °C; SMAC, 0–45 IU/L at 37 °C.

Ammonia: [B] < 110 μg/dL (<65 μmol/L) (diffusion method).

Amylase: [S] 80–180 units/dL (Somogyi). Values vary with method used.

α_1-Antitrypsin: [S] > 180 mg/dL.

Ascorbic acid: [P] 0.4–1.5 mg/dL (23–85 μmol/L).

Base, total serum: [S] 145–160 meq/L (145–160 mmol/L).

Bicarbonate: [S] 24–28 meq/L (24–28 mmol/L).

Bilirubin: [S] Total, 0.2–1.2 mg/dL (3.5–20.5 μmol/L). Direct (conjugated), 0.1–0.4 mg/dL (< 7 μmol/L). Indirect, 0.2–0.7 mg/dL (< 12 μmol/L).

Calcium: [S] 8.5–10.3 mg/dL (2.1–2.6 mmol/L). Values vary with albumin concentration.

Calcium, ionized: [S] 4.25–5.25 mg/dL; 2.1–2.6 meq/L (1.05–1.3 mmol/L).

β-Carotene: [S, fasting] 50–300 μg/dL (0.9–5.58 μmol/L).

Ceruloplasmin: [S] 25–43 mg/dL (1.7–2.9 μmol/L).

Chloride: [S or P] 96–106 meq/L (96–106 mmol/L).

Cholesterol: [S or P] 150–265 mg/dL (3.9–6.85 mmol/L). (See Lipid fractions.) Values vary with age.

Cholesteryl esters: [S] 65–75% of total cholesterol.
CO_2 content: [S or P] 24–29 meq/L (24–29 mmol/L).
Copper: [S or P] 100–200 μg/dL (16–31 μmol/L).
Cortisol: [P] 8:00 AM, 5–25 μg/dL (138–690 nmol/L); 8:00 PM, < 10 μg/dL (275 nmol/L).
Creatine phosphokinase (CPK): [S] 10–50 IU/L at 30 °C.
Creatine phosphokinase isoenzymes: See Table 2.
Creatinine: [S or P] 0.7–1.5 mg/dL (62–132 μmol/L).
Cyanocobalamin: [S] 200 pg/mL (148 pmol/L).
Epinephrine: [P] Supine, < 100 pg/mL (< 550 pmol/L).
Ferritin: [S] Adult women, 20–120 ng/mL; men, 30–300 ng/mL. Child to 15 years, 7–140 ng/mL.
Folic acid: [S] 2–20 ng/mL (4.5–45 nmol/L). [RBC] > 140 ng/mL (> 318 nmol/L).
Glucose: [S or P] 65–110 mg/dL (3.6–6.1 mmol/L).
Haptoglobin: [S] 40–170 mg of hemoglobin-binding capacity.
Iron: [S] 50–175 μg/dL (9–31.3 μmol/L).
Iron-binding capacity: [S] Total, 250–410 μg/dL (44.7–73.4 μmol/L). Percent saturation, 20–55%.
Lactate: [B, special handling] Venous, 4–16 mg/dL (0.44–1.8 mmol/L).
Lactate dehydrogenase (LDH): (Varies with method.) [S] 55–140 IU/L at 30 °C; SMA, 100–225 IU/L at 37 °C; SMAC, 60–200 IU/L at 37 °C.
Lipase: [S] < 150 U/L.
Lipid fractions: [S or P] Desirable levels: HDL cholesterol, > 40 mg/dL; LDL cholesterol, < 180 mg/dL; VLDL cholesterol, > 40 mg/dL. (To convert to mmol/L, multiply by 0.026.)
Lipids, total: [S] 450–1000 mg/dL (4.5–10 g/L).
Magnesium: [S or P] 1.8–3 mg/dL (.075–1.25 mmol/L).
Norepinephrine: [P] Supine, < 500 pg/mL (< 3 nmol/L).
Osmolality: [S] 280–296 mosm/kg water (280–296 mmol/kg water).
Oxygen:
  Capacity: [B] 16–24 vol %. Values vary with hemoglobin concentration.
  Arterial content: [B] 15–23 vol %. Values vary with hemoglobin concentration.
  Arterial % saturation: 94–100% of capacity.
  Arterial P_{o2} (P_{ao2}): 80–100 mm Hg (10.67–13.33 kPa) (sea level). Values vary with age.
P_{aCO2}: [B, arterial] 35–45 mm Hg (4.7–6 kPa).
pH (reaction): [B, arterial] 7.35–7.45 (H^+ 44.7–45.5 nmol/L).
Phosphatase, acid: [S] 1–5 units (King-Armstrong), 0.1–0.63 units (Bessey-Lowry).
Phosphatase, alkaline: [S] Adults, 5–13 units (King-Armstrong), 0.8–2.3 (Bessey-Lowry); SMA, 30–85 IU/L at 37 °C; SMAC, 30–115 IU/L at 37 °C.
Phosphorus, inorganic: [S, fasting] 3–4.5 mg/dL (1–1.5 mmol/L).
Potassium: [S or P] 3.5–5 meq/L (3.5–5 mmol/L).
Protein:
  Total: [S] 6–8 g/dL (60–80 g/L).
  Albumin: [S] 3.5–5.5 g/dL (35–55 g/L).
  Globulin: [S] 2–3.6 g/dL (20–36 g/L).
  Fibrinogen: [P] 0.2–0.6 g/dL (2–6 g/L).
  Separation by electrophoresis: See Table 5.
Pyruvate: [B] 0.6–1 mg/dL (70–114 $\mu\mu$mol/L).
Sodium: [S or P] 136–145 meq/L (136–145 mmol/L).
Specific gravity: [B] 1.056 (varies with hemoglobin and protein concentration). [S] 1.0254–1.0288 (varies with protein concentration).
Transferrin: [S] 200–400 mg/dL (23–45 μmol/L).

Triglycerides: [S] < 165 mg/dL (1.9 mmol/L). (See Lipid fractions).
Urea nitrogen: [S or P] 8–25 mg/dL (2.9–8.9 mmol/L).
Uric acid: [S or P] Men, 3–9 mg/dL (0.18–0.54 mmol/L); women, 2.5–7.5 mg/dL (0.15–0.45 mmol/L).
Vitamin A: [S] 15–60 μg/dL (0.53–2.1 μmol/L).
Vitamin B_{12}: [S] > 200 pg/mL (> 148 pmol/L).
Vitamin D: [S] Cholecalciferol (D_3): 25-Hydroxycholecalciferol, 8–55 ng/mL (19.4–137 nmol/L); 1,25-dihydroxycholecalciferol, 26–65 pg/mL (62–155 pmol/L); 24,25-dihydroxycholecalciferol, 1–5 ng/mL (2.4–12 nmol/L).
Zinc: [S] 50–150 μg/dL (7.65–22.95 μmol/L).

HORMONES, SERUM OR PLASMA

Measurements of the following hormones are used in appropriate clinical situations. The reader should consult a hospital laboratory for the normal range of values.

Pituitary:
  Growth hormone (GH)
  Thyroid-stimulating hormone (TSH)
  Follicle-stimulating hormone (FSH)
  Luteinizing hormone (LH)
  Corticotropin (ACTH)
  Prolactin
  Somatomedin C
  Antidiuretic hormone (ADH; vasopressin)
Adrenal:
  Aldosterone: ng/dL (56–250 pmol/L); increased when upright.
  Cortisol
  Deoxycortisol
  Dopamine
  Epinephrine
  Noreprinephrine
  See also Miscellaneous Normal Values.
Thyroid:
  Thyroxine, free (FT_4)
  Thyroxine, total (TT_4)
  Thyroxine-binding globulin capacity
  Triiodothyronine (T_3)
  Reverse triiodothyronine (rT_3)
  Triiodothyronine uptake (RT_3U)
  Calcitonin
  Parathyroid: Parathyroid hormone levels vary with method and antibody. Correlate with serum calcium.
Islets:
  Insulin
  C peptide
  Glucagon
Stomach:
  Gastrin
  Pepsinogen
Kidney:
  Renin activity
Gonad:
  Testosterone, free
  Testosterone, total
  Estradiol (E_2)
  Progesterone
Placenta:
  Estriol (E_3)
  Chorionic gonadotropin

MISCELLANEOUS LABORATORY TESTS

Measurements of the following miscellaneous compounds are used in appropriate clinical situations. The reader should consult a hospital laboratory for the normal range of values.

Adrenal hormones and metabolites:
Catecholamines
11,17-Hydroxycorticoids
17-Ketosteroids
Metanephrine
Vanillylmandelic acid (VMA)
Fecal fat
Lead
Porphyrins:
Delta-aminolevulinic acid
Coproporphyrin
Uroporphyrin
Porphobilinogen
Urobilinogen
Urobilinogen, fecal

REFERENCES

Friedman RB et al: Effects of diseases on clinical laboratory tests. *Clin Chem* 1980;**26(Suppl 4):**1D.

Hansten PD, Lybecker LA: Drug effects on laboratory tests. In: *Basic & Clinical Pharmacology*, 2nd ed. Katzung BG (editor). Lange, 1984.

Lippert H, Lehmann HP: *SI Units in Medicine: An Introduction to the International System of Units With Conversion Tables and Normal Ranges.* Urban & Schwarzenberg, 1978.

Lundberg GD, Iverson C, Radulescu G (editors): Now read this: The SI units are here. (Editorial). *JAMA* 1986;**255:** 2329.

Powsner EK: SI quantities and units for American medicine. *JAMA* 1984;**252:**1737.

Scully RE et al: Normal reference laboratory values: Case records of the Massachusetts General Hospital. *N Engl J Med* 1986;**314:**39.

Sonnenwirth AC, Jarett L: *Gradwohl's Laboratory Methods and Diagnosis,* 8th ed. Vols 1 and 2. Mosby, 1980.

ABBREVIATIONS ENCOUNTERED IN BIOCHEMISTRY

A(Å) Angstrom units(s) (10^{-10} m, 0.1 nm)
AA Amino acid
α-**AA** *α*-Amino acid
ACTH Adrenocorticotropic hormone, adrenocorticotropin, corticotropin
Acyl-CoA An acyl derivative of coenzyme A (eg, butyryl-CoA)
ADH Alcohol dehydrogenase
ADH Antidiuretic hormone (vasopressin)
AHG Antihemophilic globulin
Ala Alanine
ALA Aminolevulinic acid
AMP Adenosine monophosphate
Arg Arginine

Asn Asparagine
Asp Aspartic acid
ATP Adenosine triphosphate
BAL Dimercaprol (British anti-lewisite)
cAMP 3′,5′-Cyclic adenosine monophosphate, cyclic AMP
CBG Corticosteroid-binding globulin
CBZ Carbobenzoxy
CCCP *m*-Chlorocarbonyl cyanide phenylhydrazone
CCK(PZ) Cholecystokinin (pancreozymin)
CDP Cytidine diphosphocholine
Cer Ceramide
cGMP 3′,5′-Guanosine monophosphate, cyclic GMP
CI Chain-initiating
CMP Cytidine monophosphate; 5′-phosphoribosyl cytosine
CoA·SH Free (uncombined) coenzyme A. A pantothenic acid-containing nucleotide that functions in the metabolism of fatty acids, ketone bodies, acetate, and amino acids.

$$\overset{\text{O}}{\underset{\|}{}}$$

CoA·S·C·CH₃ Acetyl-CoA, "activated acetate." The form in which acetate is "activated" by combination with coenzyme A for participation in various reactions
CPK Creatine phosphokinase
CRH (CRF) Corticotropin-releasing hormone
CRP C-reactive protein
CTP Cytidine triphosphate
Cys Cysteine
D- Dextrorotatory
D₂ (vitamin) Ergocalciferol
D₃ (vitamin) Cholecalciferol
1,25(OH)₂-D₃ 1,25-Dihydroxycholecalciferol
dA Deoxyadenosine
dC Deoxycytosine
dG Deoxyguanosine
DNA Deoxyribonucleic acid
DNP Dinitrophenol
Dopa 3,4-Dihydroxyphenylalanine
DPG Diphosphoglycerate (bisphosphoglycerate)
DPN Diphosphopyridine nucleotide (now replaced by NAD)
dT Deoxythymidine
dTMP Deoxythymidine 5′-monophosphate
dUMP Deoxyribose uridine 5′-phosphate
E Enzyme (also Enz)
E.C. Enzyme commission number (IUB) system
EDTA Ethylenediaminetetraacetic acid. A reagent used to chelate divalent metals
Enz Enzyme (also E)
Eq Equivalent
eu Enzyme unit
FAD Flavin adenine dinucleotide (oxidized)
FADH₂ Flavin adenine dinucleotide (reduced)
FDA Food & Drug Administration
FFA Free fatty acids
Figlu Formiminoglutamic acid
FMN Flavin mononucleotide
FP Flavoprotein
FSF Fibrin stabilizing factor
FSH Follicle-stimulating hormone
FSHRH (FSHRH) Follicle-stimulating hormone-releasing hormone
g Gram(s)
g Gravity
Gal Galactose

GalNAc N-Acetylgalactosamine
GDP Guanosine diphosphate
GFR Glomerular filtration rate
GH Growth hormone
GHRH (GHRF) Growth hormone-releasing hormone
GHRIH (GHRIF) Growth hormone releasing-inhibiting hormone (somatostatin)
GLC Gas-liquid chromatography
Glc Glucose
GlcNAc N-Acetylglucosamine
GlcUA Glucuronic acid
Gln Glutamine
Glu Glutamic acid
Gly Glycine
GMP Guanosine monophosphate
GnRH Gonadotropin-releasing hormone
GTP Guanosine triphosphate
Hb Hemoglobin
hCG Human chorionic gonadotropin
hCS Human chorionic somatomammotropin
HDL High-density lipoproteins
H$_2$folate Dihydrofolate
H$_4$folate Tetrahydrofolate
His Histidine
HMG-CoA β-Hydroxy-β-methylglutaryl-CoA
Hyl Hydroxylysine
Hyp 4-Hydroxyproline
ICD Isocitric dehydrogenase
IDL Intermediate-density lipoproteins
IDP Inosine diphosphate
IF Initiation factor (for protein synthesis)
Ile Isoleucine
IMP Inosine monophosphate; hypoxanthine ribonucleotide
INH Isonicotinic acid hydrazide (isoniazid)
ITP Inosine triphosphate
ITyr Monoiodotyrosine
I$_2$Tyr Diiodotyrosine
IU International unit(s)
IUB International Union of Biochemistry
α-KA α-Keto acid
kcal Kilocalorie (calorie)
α-KG α-Ketoglutarate
kJ Kilojoule
K$_m$ Substrate concentration producing half-maximal velocity (Michaelis constant)
L- Levorotatory
LCAT Lecithin:cholesterol acyltransferase
LD Lactate dehydrogenase (see also LDH)
LDH Lactic dehydrogenase
LDL Low-density lipoproteins
Leu Leucine
LH Luteinizing hormone
LHRH (LHRF) Luteinizing hormone-releasing hormone
LLF Laki-Lorand factor
LTH Luteotropic hormone
Lys Lysine
M Molar
MAO Monoamine oxidase
MCH Mean corpuscular hemoglobin
MCHC Mean corpuscular hemoglobin concentration
MCV Mean corpuscular volume
Met Methionine
mol Mole(s)
MRF Melanocyte releasing factor
MRH Melanocyte releasing hormone
MRIH Melanocyte release-inhibiting hormone

mRNA Messenger RNA
MSH Melanocyte-stimulating hormone
MW Molecular weight
NAD Nicotinamide adenine dinucleotide (oxidized)
NADH Nicotinamide adenine dinucleotide (reduced)
NADP Nicotinamide adenine dinucleotide phosphate (oxidized)
NADPH Nicotinamide adenine dinucleotide phosphate (reduced)
NDP Any nucleoside diphosphate
NeuAc N-acetylneuraminic acid
NTP Any nucleoside triphosphate
OA Oxaloacetic acid
OD Optical density
P$_i$ Inorganic phosphate (orthophosphate)
PCV Packed cell volume
Phe Phenylalanine
PIH (PIF) Prolactin release-inhibiting hormone
PL Pyridoxal
PLP Pyridoxal phosphate
PP$_i$ Inorganic pyrophosphate
PRH (PRF) Prolactin releasing hormone
PRIH (PRIF) Prolactin release-inhibiting hormone
PRL Prolactin
Pro Proline
PRPP 5-Phosphoribosyl-1-pyrophosphate
PTA Plasma thromboplastin antecedent
PTC Plasma thromboplastin component
RBC Red blood cell
RDA Recommended daily allowance
RE Retinol equivalents
RNA Ribonucleic acid
RQ Respiratory quotient
rRNA Ribosomal RNA
S (Sf) units Svedberg units of flotation
SDA Specific dynamic action
SDS Sodium dodecyl sulfate
Ser Serine
SGOT Serum glutamic oxaloacetic transaminase
SGPT Serum glutamic pyruvic transaminase
SH Sulfhydryl
SLR *Streptococcus lactis* R
SPCA Serum prothrombin conversion accelerator
SRIH (SRIF) Somatostatin (growth hormone release-inhibiting hormone)
sRNA Soluble RNA (same as tRNA, which term is preferred)
STP Standard temperature and pressure (273° absolute, 760 mm Hg)
T$_3$ Triiodothyronine
T$_4$ Tetraiodothyronine; thyroxine
TEBG Testosterone-estrogen-binding globulin
TG Triacylglycerols (formerly called triglycerides)
Thr Threonine
TLC Thin layer chromatography
Tm$_{Ca}$ Tubular maximum for calcium
Tm$_G$ Tubular maximum for glucose
TPN Triphosphopyridine nucleotide (now replaced by NADP)
TRH (TRF) Thyrotropin-releasing hormone
Tris Tris(hydroxymethyl)aminomethane, a buffer
tRNA Transfer RNA (see also sRNA)
Trp Tryptophan
TSH Thyroid-stimulating hormone; thyrotropin
Tyr Tyrosine
UDP Uridine diphosphate
UDPGal Uridine diphosphate galactose

UDPGlc Uridine diphosphate glucose
UDPGlcUA Uridine diphosphoglucuronic acid
UDPGluc Uridine diphosphoglucuronic acid
UMP Uridine monophosphate; uridine-5′-phosphate; uridylic acid
UTP Uridine triphosphate

V_{max} Maximal velocity
Val Valine
VHDL Very high density lipoproteins
VLDL Very low density lipoproteins
VMA Vanillylmandelic acid
vol % Volumes percent

Index